Advances in Sol-Gel Derived Materials and Technologies

Series Editors

Michel A. Aegerter
Michel Prassas

For further volumes:
http://www.springer.com/series/8776

The International Sol-Gel Society (ISGS)

The goals of the **ISGS** are to **educate**, **federate**, and **disseminate**.

- **Educate** scientists and technologists from around the world, and from both academia and industry, through conferences such as the biannual International Sol-Gel Conference, topical workshops organised in conjunction with various conferences or at the demand of specific interest groups, and summer schools, which provides comprehensive teaching for both students and experienced researchers in a relaxed atmosphere. The ISGS also provides travel scholarship to students to attend these conferences.
- **Federate** scientists from all around the planet by organising events (conferences, schools, etc..) where they can meet and discuss everything sol-gel. The ISGS also acts as a point of contact for a wide range of people, from sculptors interested in shaping aerogels, to a Chinese industrial consortium interested in learning about hard coatings, or researchers from emerging economies who are looking for specific technical information. In collaboration with our partner, the Sol-Gel Gateway (www.solgel.com), we also aim to provide a virtual meeting place for everyone interested in sol-gel. The recent revamping of the ISGS website (www.isgs.org) provides us with a more interactive structure where members can participate in the life of the ISGS. Our quarterly newsletter also encourages people to contribute stories and opinions and help us build a stronger sense of community in sol-gel science.
- **Disseminate** information about sol-gel science, technology and products through the society website, the quarterly newsletter, our partnership with the Sol-Gel Gateway, and the publication of the *Journal of Sol-Gel Science and Technology*, the official journal of the ISGS. The journal's 2009 Impact Factor of 1.39 ranks 5th of 25 in the category Materials Science, Ceramics.

This new series of books on sol-gel topics, with volumes on aerogels, sol-gel characterisation, hybrids, nanocomposites, sol-gel coatings for energy applications, and membranes currently in development, will be edited and written by top scientists in the field, and **fits perfectly with the education and dissemination mission of the ISGS**. The books will provide both novices and experts with new ideas and concepts on various aspects of sol-gel science and technology. They will also provide the sol-gel community with a solid foundation for exploring new, unchartered scientific territories in the 21st century.

Christophe Barbé
President of the International Sol-Gel Society

Editor-in-Chief
Michel A. Aegerter
Editors
Nicholas Leventis
Matthias M. Koebel

Aerogels Handbook

Springer

Editors
Michel A. Aegerter
Résidence Vert-Pré
Ch. des Placettes 6
CH-1041 Bottens
Switzerland
michel.aegerter@bluewin.ch

Nicholas Leventis
Missouri University of Science & Technology
Department of Chemistry
Rolla, MO 65409, USA
leventis@mst.edu

Matthias M. Koebel
Empa
Building Technologies Laboratory
CH-8600 Dübendorf
Switzerland
matthias.koebel@empa.ch

ISBN 978-1-4419-7477-8 e-ISBN 978-1-4419-7589-8
DOI 10.1007/978-1-4419-7589-8
Springer New York Dordrecht Heidelberg London

Library of Congress Control Number: 2011921715

Printed on acid-free paper

Springer is part of Springer Science+Business Media (www.springer.com)

Preface

Aerogels are rather old materials that were invented back in 1931 by Steven Kistler at the College of the Pacific in Stockton, California. The first aerogels were synthesized from silica gels by replacing the liquid component with a gas. They appeared as quite revolutionary solid-state materials because of their extremely low density and their outstanding physical properties, especially for thermal and acoustical insulation, and were marketed by Monsanto in the form of granular material a few years later. However, their development was slowed almost to a standstill for about three decades, mainly because of the time-consuming and labor-intensive solvent-exchange steps. That changed with a major technological breakthrough by Teichner at the University of Claude Bernard in Lyon, who replaced Kistler's sodium silicate by alkoxysilanes and thus eliminated simultaneously the inorganic salt byproducts from the gels and the need for a water-to-alcohol exchange step. New developments followed rapidly as many scientists joined the field. However, the field remained mostly of scientific interest till the mid-eighties when the alcohol in the gels was replaced by liquid carbon dioxide before supercritical drying, a major advance in safety eliminating the explosion hazards due to the alcohol during the preparation process. For mostly historical reasons, the recent development of aerogels followed a somewhat distinct route from the fast growing sol–gel field, as specific International Symposia were organized regularly from 1985 to 2003. More recently, a few 1–2 day, international meetings were held in which companies could divulge a limited amount of technical information. Therefore, when in November, 2008, Dr. David Packer, Executive Editor at Springer, asked the Editor-in-Chief of the Journal of Sol–Gel Science and Technology, Prof. Dr. M. A. Aegerter, about his interest to edit a Handbook on Aerogels, his answer was of course quite positive since no such book summarizing the development and the state-of-the-art of these outstanding materials had been published before.

Realizing that such task would be impossible to realize alone, he asked for the collaboration of two renowned scientists in the field, Prof. Nicholas Leventis from the Missouri University of Science and Technology (MST), Rolla (USA) and Dr. Matthias M. Koebel, head of the Building Material Group at the Swiss Federal Laboratories for Materials Testing and Research (EMPA), Duebendorf (Switzerland), who readily accepted the offer. Working together, the editorial structure of the Handbook was rapidly built up keeping in mind that it should present for *the first time in a single book* the state-of-the-art in the development, processing, and properties of inorganic, organic and composite aerogels, the most important techniques of characterization as well as a multidisciplinary description of the use, recent applications and the main products commercialized today by companies. Out of 52 prospective leading authors in the field contacted by the editors in early 2009, 41 accepted to write specific overviews with exhaustive lists of references presenting not only their own research and development but also those realized by other international colleagues in each selected area. This Handbook serves consequently as an authoritative source for a broad audience of individuals involved in research, products development and use of aerogels as well as for advanced undergraduate and graduate students in many fields. It presents a rather exhaustive coverage of the processing and properties of most types of aerogels developed till now from the original silica-based systems, to nonsilicate inorganics, natural, and synthetic organic compounds such as carbon aerogels, the most recent composite systems involving cross-linked, aerogel-polymer, interpenetrating hybrid networks which exhibit remarkable mechanical properties of strength and flexibility as well as more exotic aerogels based on clays, chalcogenides, phosphides, doped quantum dots, and chitin. Many scientific and industrial applications are also reported in the field of electronics, chemistry, mechanics, engineering, energy production and storage, sensors, medicine, biology, pharmaceutics, nanotechnology, military and aerospace, nuclear waste, C-sequestration, oil and gas recovery, thermal insulation, as well as household uses for which a conservative estimated market annual growth rate of around 70% is expected over the next 5 years.

The **Aerogels Handbook** presents 42 contributions arranged in 16 parts. Part I is dedicated to the **History of aerogels** with one contribution numbered Chap. 1. Parts II–VI, entitled **Materials and processing** summarize the development of the different types of aerogels: *Inorganic-silica based aerogels* with 4 contributions (Chaps. 2–5), *Inorganic-non silicate aerogels* with 3 contributions (Chaps. 6–8), *Organic-natural and synthetic aerogels* with 4 contributions (Chaps. 9–12), *Composite aerogels* with 4 contributions (Chaps. 13–16) and finally *Exotic aerogels* containing 4 contributions (Chaps. 17–20). The basic properties of aerogels (structural, mechanical, thermal and modeling) are then described in Part VII entitled **Properties** which comprises 4 contributions (Chaps. 21–24). Parts VIII to XIV describe rather exhaustively many recent **Applications** in fields such as *Energy* (2 contributions; Chaps. 25 and 26), *Chemistry and Physics* (3 contributions; Chaps. 27–29), *Biomedical and pharmaceutical* (2 contributions; Chaps. 30 and 31), *Space and airborne* (2 contributions; Chaps. 32 and 33), *Metal industry* (1 contribution; Chap. 34), *Art* (1 contribution; Chap. 35) and *Other* (1 contribution; Chap. 36). Finally many **Commercial products** are described in Part XV where 5 contributions written by US, Chinese, and German companies are presented (Chaps. 37–41). All the contributions were evaluated by two reviewers.

The Handbook also contains:

A *List of contributors* arranged alphabetically with, for each author, their affiliation and Email addresses.

A brief section entitled *Concluding remarks and outlook* written by the editors (Part XVI, Chap. 42).

The *Subject index* has been worked out by the editors and is partly based on the author's suggestions; it should assist the readers in finding references to a particular topic. Effort has been made to realize an index as comprehensive and useful as possible. Cross-references are also given to terms of related interest, and are found after the entry for the first-level term to which they apply. The numbers in parenthesis correspond to a chapter; when there is a substantial discussion in a chapter, the number appears in bold.

Finally a section entitled *Glossary, acronyms, and abbreviations* has also been included. The words with their definition also have been in great part suggested by the authors. All these words and acronyms are written in italic in the text of the different contributions.

The editors express their sincere thanks to all authors of the contributions included in this Handbook for their precious time in writing excellent overviews. We also extend our gratitude to Dr. David Packer, Executive Editor at Springer, who suggested the edition of this Handbook and accepted the publication of more than half of the illustrations in color.

Michel A. Aegerter
Nicholas Leventis
Matthias M. Koebel

Contents

Part XVI Conclusion

List of Contributors

Patrick Achard, Prof. Dr., Center for Energy and Processes, MINES ParisTech, 1 Rue Claude Daunesse, B.P. 207, 06 904 Sophia Antipolis Cedex, France
Email: patrick.achard@mines-paristech.fr

Michel A. Aegerter, Prof. Dr., Ch. des Placettes, 6, 1041 Bottens, Switzerland
Email: michel.aegerter@bluewin.ch

Ann M. Anderson, Prof. Dr., Department of Mechanical Engineering, Union College, 807 Union Street, Schenectady, NY 12308, USA
Email: andersoa@union.edu

Uzma K. H. Bangi, Air Glass Laboratory, Department of Physics, Shivaji University, 416004 Kolhapur, Maharashtra, India
Email: uzmayesh@yahoo.co.in

Theodore F. Baumann, Dr., Lawrence Livermore National Laboratory-LLNL, Livermore, CA 94551, USA
Email: baumann2@llnl.gov

Lassaad Ben Hammouda, Dr., Département de Chimie, Faculté des Sciences de Tunis, Campus Universitaire, 2092 ElManar, Tunisia
Email: bhlassaad@gmail.com

Massimo Bertino, Prof. Dr., Department of Physics, Virginia Commonwealth University, Richmond, VA 23284-2000, USA
Email: mfbertino@vcu.edu

Stephanie L. Brock, Prof. Dr., Department of Chemistry, Wayne State University, 5101 Cass Avenue, Detroit, MI 48202, USA
Email: sbrock@chem.wayne.edu

Mary K. Carroll, Prof. Dr., Department of Chemistry, Union College, 807 Union Street, Schenectady, NY 12308, USA
Email: carrollm@union.edu

Maria Francesca Casula, Dr., Dipartimento di Scienze Chimiche and INSTM, Università di Cagliari, Complesso Universitario di Monserrato, S.S. 554 Bivio per Sestu, 09042 Monserrato, CA, Italy
Email: casulaf@unica.it

Anna Corrias, Prof. Dr., Dipartimento di Scienze Chimiche, Università di Cagliari, Complesso Universitario di Monserrato, S.S. 554 Bivio per Sestu, 09042 Monserrato, CA, Italy
Email: corrias@unica.it

Nicolas de la Rosa-Fox, Prof. Dr., Facultad de Ciencias, Departamento de Fisica de la Materia Condensada, Universidad de Cádiz, 11510 Cádiz, Spain
Email: nicolas.rosafox@uca.es

Hans-Peter Ebert, Dr., Functional Materials for Energy Technology, Bavarian Center for Applied Energy Research, ZAE Bayern Am Hubland, 97074 Würzburg, Germany
Email: ebert@zae.uni-wuerzburg.de

Luis Esquivias, Prof. Dr., Departamento de Física de la Materia Condensada, Universidad de Sevilla, 41012 Sevilla, Spain
Email: luisesquivias@us.es

Alexander E. Gash, Dr., Lawrence Livermore National Laboratory-LLNL, Livermore, CA 94551, USA
Email: gash2@llnl.gov

Lev D. Gelb, Prof. Dr., Department of Materials Science and Engineering, University of Texas at Dallas, Richardson, TX 75080, USA
Email: lev.gelb@utdallas.edu

Abdelhamid Ghorbel, Prof. Dr., Département de Chimie, Faculté des Sciences de Tunis, Campus Universitaire, 2092 ElManar, Tunisia
Email: Abdelhamid.Ghorbel@fst.rnu.tn

Dayong Y. Guan, Dr., Shanghai Key Laboratory of Special Artificial Microstructure Materials and Technology, Department of Physics, Tongji University, 1239 Siping Road, 20092 Shanghai, P.R. China
Email: dayong415@163.com

Masahiko Hashimoto, Advanced Technology Research Laboratory, Panasonic Corporation, 3-4 Hikaridai, Seika-cho, Soraku-gun, Kyoto 619-0237, Japan
Email: hashimoto.masahiko@jp.panasonic.com

Hiroshi Hirashima, Prof. Dr., Faculty of Science and Technology, Keio University, 3-14-1, Hiyoshi, Kohoku-ku, Yokohama, Japan
Email: hirasima@applc.keio.ac.jp
Email: hirasima@e03.itscom.net

Chengli Jin, R & D Center, Nano Hi-tech Co.Ltd, Shanxi Road 488, Shaoxing, Zhejiang, China
Email: jin_cl@163.com

Steven M. Jones, Dr., Jet Propulsion Laboratory, California Institute of Technology, 4800 Oak Grove Drive, MS 125-109, Pasadena, CA 91109, USA
Email: steven.m.jones@jpl.nasa.gov

Matthias M. Koebel, Dr., Laboratory for Building Technologies, Empa – Swiss Federal Laboratories for Materials Science and Technology, 8600 Duebendorf, Switzerland
Email: matthias.koebel@empa.ch

Nicholas Leventis, Prof. Dr., Department of Chemistry, Missouri University of Science and Technology, Rolla, MO 65409, USA
Email: leventis@mst.edu

Xipeng Liu, Dr., Northboro Research and Development Center, Saint Gobain, SA, 9 Goddard Road, Northboro, MA 01532, USA
Email: xipeng.liu@saint-gobain.com

Hongbing Lu, Prof. Dr., Department of Mechanical Engineering, The University of Texas at Dallas, 800 W. Campbell Road, EC-38, Richardson, TX 75080, USA
Email: hongbing.lu@utdallas.edu

Huiyang Luo, Dr., Department of Mechanical Engineering, The University of Texas at Dallas, Richardson, TX 75080, USA
Email: huiyang.luo@utdallas.edu
Bruce McCormick, CEO, Savsu Technologies, 1 High Country Rd, Santa Fe, NM 87508, USA
Email: b.mccormick@savsu.com
Mary Ann B. Meador, Dr., NASA Glenn Research Center, Mailstop 49-3, Cleveland, OH 44135, USA
Email: maryann.meador@nasa.gov
Imene Mejri, Dr., Département de Chimie, Faculté des Sciences de Tunis, Campus Universitaire, 2092 ElManar, Tunisia
Email: mejriimene@gmail.com
Robert Mendenhall, CTO, American Aerogel Corporation, 460 Buffalo Road, Rochester, NY 14611, USA
Email: rmendenhall@americanaerogel.com
Ioannis Michaloudis, Dr., Visual Artist, 13, Gortynias Street, 136 75 Athens, Greece
Email: info@michalous.com
Email: michalou@alum.mit.edu
Tom Miller, Cabot Aerogel, 157 Concord Road, Billerica, MA 01821, USA
Email: tom_miller@cabot-corp.com
Barbara Milow, Dr., Institute of Materials Physics in Space DLR, German Aerospace Center, 51147 Cologne, Germany
Email: barbara.milow@dlr.de
V. Morales-Flórez, Dr., Instituto de Ciencias de Materiales de Sevilla, CSIC-US, 41092 Sevilla, Spain
Email: victor.morales@icmse.csic.es
Sudhir Mulik, Dr., Georgia-Pacific Chemicals, 2883 Miller Road, Decatur, GA 30038, USA
Present address: The Dow Chemical Company, 727 Norristown Road, P.O. Box 904, Spring House, PA 19477, USA
Email: sudhir.mulik@gmail.com
Digambar Y. Nadargi, Dr., Laboratory for Building Technologies, Empa, Swiss Federal Laboratories for Materials Science and Technology, CH-8600, Dübendorf, Switzerland
Email: digambar_nadargi@yahoo.co.in
Hidetomo Nagahara, Advanced Technology Research Laboratory, Panasonic Corporation, 3-4 Hikaridai, Seika-cho, Soraku-gun, 619-0237, Kyoto, Japan
Email: nagahara.hidetomo@jp.panasonic.com
G. M. Pajonk, Prof. Dr., Laboratoire des Matériaux et Procédés Catalytiques, Université Claude Bernard Lyon 1, 43 Boulevard du 11 novembre 1918, 69622 Villeurbanne, France
Email: pajonk1939@laposte.net
Jean Phalippou, Prof. Dr., CNRS-Université Montpellier II, Place E Bataillon, 34095, Montpellier Cedex 5, France
Email: jean.phalippou698@orange.fr
Alain C. Pierre, Prof. Dr., Université Lyon 1, CNRS, UMR 5256, IRCELYON, Institut de recherches sur la catalyse et l'environement de Lyon, 2 Avenue Albert Einstein, 69626 Villeurbanne Cedex, France
Email: Alain.Pierre@ircelyon.univ-lyon1.fr

M. Piñero, Prof. Dr., Departamento de Fisica Aplicada, CASEM, Universidad de Cádiz, 11510 Cádiz, Spain
Email: manolo.piniero@uca.es

A. Venkateswara Rao, Prof. Dr., Air Glass Laboratory, Department of Physics, Shivaji University, 416004 Kolhapur, Maharashtra, India
Email: avrao2012@gmail.com

A. Parvathy Rao, Dr., Air Glass Laboratory, Department of Physics, Shivaji University, 416004 Kolhapur, Maharashtra, India
Email: parvathy_shivaj@yahoo.co.in

Lorenz Ratke, Prof. Dr. h.c, Institute of Materials Physics in Space DLR, German Aerospace Center, 51147 Cologne, Germany
Email: lorenz.ratke@dlr.de

Gudrun Reichenauer, Dr., Functional Materials for Energy Technology, Bavarian Center for Applied Energy Research, ZAE Bayern, Am Hubland, 97074 Würzburg, Germany
Email: reichenauer@zae.uni-wuerzburg.de

Jerome Reynes, Dr., CNRS-Université Montpellier II, Place E Bataillon, 34 095 Montpellier Cedex 5, France
Email: reynes.jerome@neuf.fr

Arnaud Rigacci, Dr., MINES ParisTech, CEP, Centre Energétique et Procédés, 1 rue Claude Daunesse, BP 207, 06904 Sophia Antipolis, France
Email: arnaud.rigacci@mines-paristech.fr

William M. Risen, Jr., Prof. Dr., Department of Chemistry, Brown University, Providence, RI 02912, USA
Email: William_Risen_Jr@brown.edu

David A. Rubenstein, Prof. Dr., School of Mechanical and Aerospace Engineering, Oklahoma State University, Stillwater, OK 74078, USA
Email: david.rubenstein@okstate.edu

Jeffrey Sakamoto, Prof. Dr., Chemical Engineering and Materials Science Department, Michigan State University, 2527 Engineering Building, East Lansing, MI 48824, USA
Email: jsakamot@egr.msu.edu

Joe H. Satcher Jr., Dr., Lawrence Livermore National Laboratory-LLNL, Livermore, CA 94551, USA
Email: satcher1@llnl.gov

Frank Schneider, Dr.-Ing., OKALUX GmbH, Am Jöspershecklein 1, 97828 Marktheidenfeld, Germany
Email: fschneider@okalux.de

Jun Shen, Prof. Dr., Department of Physics, Tongji University, 1239 Siping Road, 20092 Shanghai, P.R. China
Email: shenjun67@tongji.edu.cn

Randall L. Simpson, Dr., Lawrence Livermore National Laboratory-LLNL, P.O. Box 808 L-160, Livermore, CA 94551, USA
Email: simpson5@llnl.gov

Irina Smirnova, Prof. Dr.-Ing., Technische Universität Hamburg-Harburg, Institut für Thermische Verfahrenstechnik-V8, Eißendorferstr 38, 21073 Hamburg, Germany
Email: irina.smirnova@tu-harburg.de

Chariklia Sotiriou-Leventis, Prof. Dr., Department of Chemistry, Missouri University of Science and Technology, Rolla, MO 65409, USA
Email: cslevent@mst.edu

Hilary Thorne-Banda, Cabot Corporation, 157 Concord Road, Billerica, MA 01821, USA
Email: hilary_banda@cabot-corp.com

Thierry Woignier, Dr., UMR CNRS 6116 -UMR IRD 193- IMEP- IRD-PRAM, 97200 Le Lamentin, Martinique, France; CNRS-Université de Montpellier II, Place E. Bataillon, 34095 Montpellier Cedex 5, France
Email: woignier@univ-montp2.fr

Chunhua Jennifer Yao, Dr., Research and Development Department, Stepan Company, 22 W Frontage Road, Northfield, IL 60093, USA
Email: jyao@stepan.com

Wei Yin, Prof. Dr., School of Mechanical and Aerospace Engineering, Oklahoma State University, 218 Engineering North, Stillwater, OK 74078, USA
Email: wei.yin@okstate.edu

Hiroshi Yokogawa, Dr., Advanced Materials Development Department, New Product Technologies Development Department, Panasonic Electric Works Co., Ltd, 1048, Kadoma, Osaka 571-8686, Japan
Email: yokogawa@panasonic-denko.co.jp

Mohamed Kadri Younes, Dr., Département de Chimie, Faculté des Sciences de Tunis, Campus Universitaire, 2092 ElManar, Tunisia
Email: younes.kadri@laposte.net

Hongtao Yu, Dr., Department of Chemistry, Wayne State University, 5101 Cass Avenue, Detroit, MI 48202, USA
Email: hongtao.yu@asu.edu

Part I

History of Aerogels

1

History of Aerogels

Alain C. Pierre

Abstract This chapter presents a historical account of the progressive development of the solid materials known as aerogels. The founding work of Kistler is first summarized. His precursory work attracted the attention of scientists who focused on the physics and chemistry of aerogels, reviewed in the second section. The latter studies allowed for the understanding of how a very high open porosity solid could be maintained when drying a gel and how some specific properties such as transparency or a hydrophobic character could be granted to these materials. In turn, a better knowledge of the scientific basis behind aerogels led to determining their technical characteristics and developing their applications, as addressed in the third section. The most recent developments summarized in the last section aim at a progressive transfer of these applications toward the industry.

1.1. The Founding Studies by Kistler

The term *aerogel* was first introduced by Kistler in 1932 to designate gels in which the liquid was replaced with a gas, without collapsing the gel solid network [1]. While wet gels were previously dried by evaporation, Kistler applied a new *supercritical drying* technique, according to which the liquid that impregnated the gels was evacuated after being transformed to a *supercritical fluid*. In practice, *supercritical drying* consisted in heating a gel in an autoclave, until the pressure and temperature exceeded the critical temperature T_c and pressure P_c of the liquid entrapped in the gel pores.

This procedure prevented the formation of liquid–vapor meniscuses at the exit of the gel pores, responsible for a mechanical tension in the liquid and a pressure on the pore walls, which induced gel shrinkage. Besides, a *supercritical fluid* can be evacuated as gas, which in the end lets the "dry solid skeleton" of the initial wet material. The dry samples that were obtained had a very open porous texture, similar to the one they had in their wet stage.

Overall, aerogels designate dry gels with a very high relative or specific pore volume, although the value of these characteristics depends on the nature of the solid and no official convention really exists (Chap. 21). Typically, the relative pore volume is of the order of 90% in the most frequently studied silica aerogels [2] (Chap. 2), but it may be

A. C. Pierre • Institut de recherches sur la catalyse et l'environement de Lyon, Université Lyon 1, CNRS, UMR 5256, IRCELYON, 2 Avenue Albert Einstein, 69626 Villeurbanne Cedex, France
e-mail: Alain.Pierre@ircelyon.univ-lyon1.fr

M.A. Aegerter et al. (eds.), *Aerogels Handbook*, Advances in Sol-Gel Derived Materials and Technologies, DOI 10.1007/978-1-4419-7589-8_1,

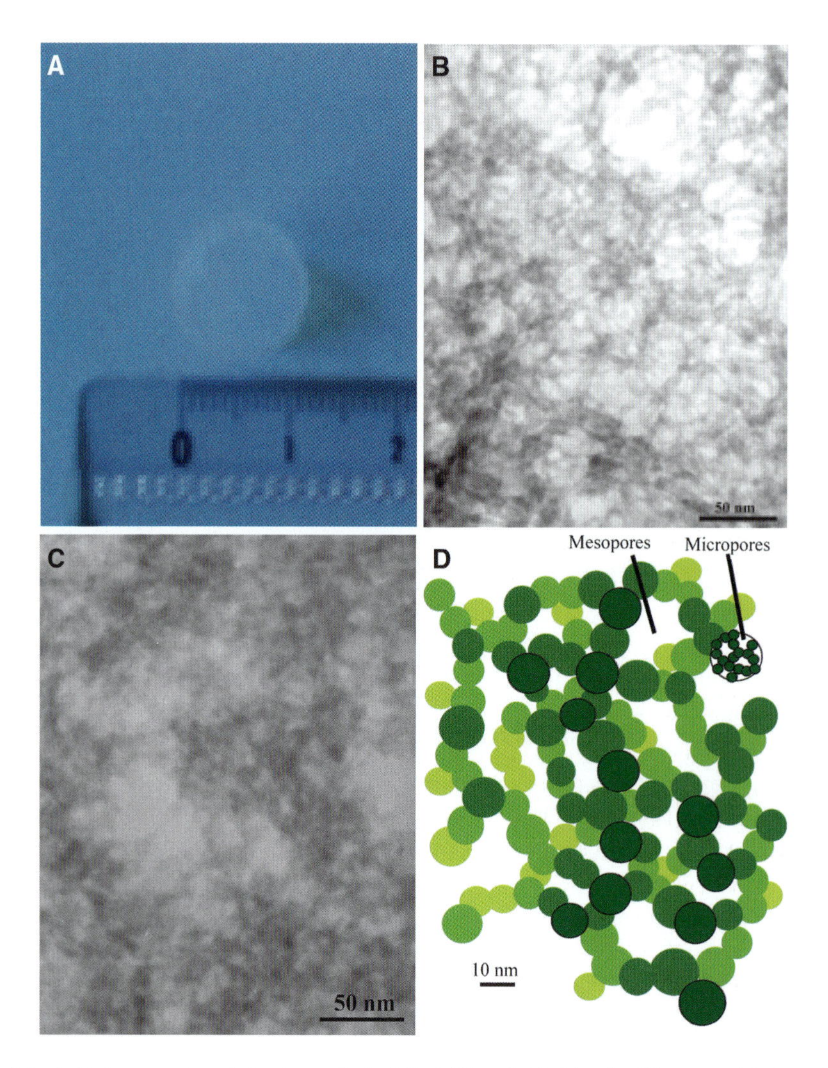

Figure 1.1. **A.** Y_2O_3 aerogel made by the method of Gash [3] and dried by the CO_2 supercritical method [4]; **B.** *TEM* micrograph of the aerogel in **A.** (Pierre A, unpublished photograph, see also [4]); **C.** *TEM* micrograph of a SiO_2 aerogel made from 80% TMOS and 40% methyltrimethoxysilane, dried by the CO_2 supercritical method (Pierre A, unpublished photograph, see also [5]); **D.** Technical modeling of a local domain in an oxide aerogel, inspired from local zones in the *TEM* micrographs **B.** and **C.**

significantly lower in other types of aerogels. This porous texture is well illustrated in Figure 1.1, which shows a Y_2O_3 aerogel, transmission electron micrographs (*TEM*) of this aerogel and of a SiO_2 aerogel, plus a technical art illustration of a local domain in an aerogel solid network, inspired from these two micrographs.

By considering recent developments in the synthesis of aerogels, it is indeed more realistic to define these materials with reference to the initial idea of Kistler, simply as gels in which the liquid has been replaced by air, with very moderate shrinkage of the solid network. This definition enlightens the main difference between aerogels and *xerogels*. The term *xerogel* is defined by IUPAC as an "Open network formed by the removal of all swelling agents from a gel" [6], but it was first introduced by Freundlich to designate shrinking (or swelling) gels [7]. And indeed, the capillary stresses previously mentioned may contract a wet gel down to 30% or less of its initial volume [2a] and hence to a much lower value than in aerogels [2b].

In his founding work, Kistler synthesized a series of aerogels of very different nature [1]. Besides silica aerogels, which could be relatively easily handled without any fracture damage (Chap. 2), he succeeded in making alumina aerogels, which proved to be mechanically very weak. He also synthesized more exotic aerogels of tungstic, ferric, or stannic oxide and nickel tartrate, a list to which organic aerogels of cellulose (Chap. 9), nitrocellulose, gelatine, agar, or egg albumin must be added. Besides, the fluids in which supercritical drying was performed were also quite diverse. The silica gels were made from sodium silicate (water glass, Na_2SiO_3) (Chap. 5), and water was found to disperse the gel to a powder near the aqueous medium critical point. To maintain a monolithic character, the wet gels were actually dialyzed to replace water by ethanol, so that Kistler initiated *supercritical drying* in this alcohol. But Kistler also applied the supercritical method to other fluids: ethyl ether, propane, and even liquid CO_2 in an unsuccessful attempt to make rubber aerogels. At last, Kistler rapidly perceived the potential industrial applications of aerogels, for instance, as catalysts, thickening agents, insulating materials, or water repellents when they were hydrophobic (Chap. 3). He deposited several patents: one of them was assigned to the Monsanto Chemical Company, which started the industrial production of silica aerogels (Chap. 2) commercialized under the name of Santocel® [8]. He also patented the first hydrophobic silica aerogels made by sylilation with trichloromethyl silane, for use as water repellents [9] (Chap. 3). Hence, Kistler initiated most of the directions in which aerogels were later developed.

After Kistler, the chemical composition of the materials that were made as aerogels has progressively diversified. The list of aerogel silicates in which silica was a major component steadily increased [10]. A large range of simple oxides or binary oxides were investigated by Teichner et al. [11, 12]. Borate aerogels were prepared at the Sandia National Laboratories [13]. Pekala, at the Lawrence Livermore National Laboratory in Berkeley, developed new organic aerogels made by polycondensation of resorcinol–formaldehyde (Chap. 11) for a laser project [14]. These aerogels had a low thermal conductivity of 0.012 $W\ m^{-1}\ K^{-1}$ (Chap. 23). Their pyrolysis in an inert atmosphere produced a carbon aerogel that conducted electricity [15] (Chaps. 24 and 36). Examples of an *RF* aerogel, the carbon aerogel derived from it, and the fractal structure of this carbon aerogel observed under a *TEM* are illustrated in Figure 1.2.

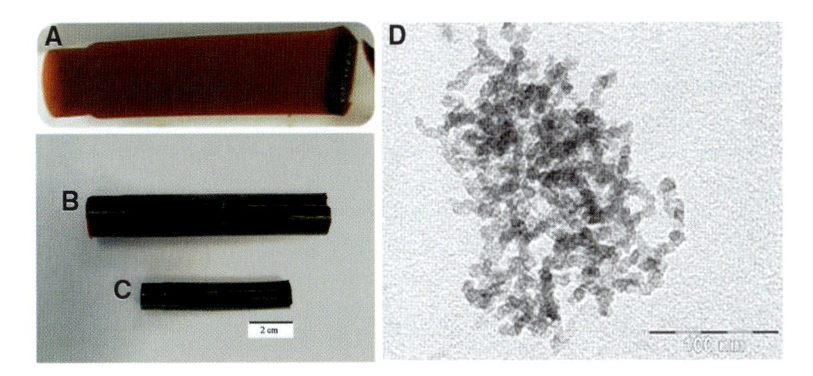

Figure 1.2. Resorcinol Formaldehyde (*RF*) and carbon aerogels. Courtesy Sandrine Berthon-Fabry Mines ParisTech/Centre Energétique et Procédés. Rue Claude Daunesse, BP 207 F-06904 Sophia-Antipolis, France. **A.** *RF* gel before drying; **B.** *RF* aerogel after direct CO_2 *supercritical drying*; **C.** carbon aerogel derived from the *RF* aerogel in **B.** by pyrolysis at 1,050°C under nitrogen gas flow; **D.** transmission electron micrograph of the carbon aerogel in **C.** See also [16].

The evolution in the nature of aerogel materials also followed the progress in sol–gel chemistry, with the synthesis of hybrid organosilica aerogels [17] and some non-oxide aerogels (e.g., sulfides) [18] (Chap. 17).

1.2. Further Studies on the Synthesis Chemistry of Aerogels

Initially, the chemical compounds termed *precursors*, which contained the cations M from which an oxide gel was made, were essentially metallic salts. The sodium metasilicate Na_2SiO_3, initially used by Kistler [1], was cheap. Hence, an industrial process based on this precursor was developed for some time by BASF [19]. Simple metallic salts also remained interesting when a more elaborate precursor was not easily available. More recently, the use of metallic salts as sol–gel *precursors* has seen a renewed interest when hydrolyzed in solution in an organic solvent, in which a slow "proton scavenger" such as an epoxide was added [3] (Chap. 8). Nice aerogel monoliths were obtained in this way with Cr, Fe, Al, Zr, and other cations.

However, elimination of the sodium chloride formed in the silica gels derived from Na_2SiO_3, followed by exchange of water for a fluid such as ethanol, required lengthy dialysis steps. Hence, the *precursors* at present mostly favored are the *alkoxide*s $M(OR)_n$, a type of chemicals first used by Ebelmen to synthesize silica *xerogel*s [20]. In these compounds, R designates an alkyl group and OR an *alkoxide* group. Alkoxides are often available in solution in their parent alcohol, making it possible to directly dry these gels by the supercritical method in such solvents. Peri [21] was the first to open this route with tetraethyl orthosilicate $Si(OEt)_4$ (or *TEOS*) as the SiO_2 aerogel precursor. More extensive works were focused in this direction by the Teichner group, on SiO_2 (Chap. 2) as well as on a larger range of previously mentioned compositions [11, 12].

More importantly, water was now a reactant that could be added in controlled proportion in the sol–gel process, altogether with various gelation catalysts. Consequently, the physical chemistry of hydrolysis and condensation of these *precursors* could be easily studied, which in turn permitted to progressively develop reproducible silica aerogel monoliths (Chap. 2). Among the most outstanding results, Woignier studied the structure change during high-temperature supercritical extraction in alcohol [22]. Vacher in Montpellier [23] and Schaefer et al. at the Sandia Laboratories in New Mexico investigated the fractal nature of the silica aerogel network by small-angle X-ray scattering (SAXS) and small-angle neutron scattering (*SANS*) [24].

A kinetic growth model making easier to control the texture of silica aerogels was also developed at the Sandia National Laboratories [2c] and a two-step acid–base catalysis process permitted to design very low-density silica aerogel monoliths [25]. Later on, Tillotson and Hrubesh modified this two-step process, by replacing the alcohol with an aprotic solvent, by distillation, in order to prevent reverse hydrolysis reactions [26]. The latter method produced the silica aerogel monoliths with the lowest density ($3\ kg\ m^{-3}$) and was 99.8% porous (by volume) [26]. Tiles of these aerogels are now used by NASA in space shuttle flights [27] (Chap. 32). When a sol–gel precursor such as *TEOS*, the solvent, and water have a limited solubility in each other, Zarzycki and Esquivias et al. showed that the mixture could advantageously be submitted to ultrasonic vibrations. This operation favored the contacts between all reactants, and gels with a different texture, termed "sonogels," were then obtained [28, 29] (Chap. 20).

Presently, high-temperature *supercritical drying* (or *HOT*) in alcohol is distinguished from low-temperature *supercritical drying* (*COLD*) in CO_2. The critical conditions are indeed very different depending on the fluid that impregnates the wet gel as indicated in Table 1.1. Hence, the properties of the aerogels, in particular their hydrophobicity, can be very different depending on the *supercritical drying* fluid used (Chap. 3). Various detailed procedures regarding the drying protocol, such as the heating-pressurization path, the atmosphere above the samples, and the duration of each step, were investigated. These parameters, summarized by Pajonk [31], are important to control during the drying process, in order to minimize differential stresses.

Table 1.1. Critical point parameters of a few common fluids [30]

Fluid	Formula	T_c(°C)	P_c(MPa)
Water	H_2O	374.1	22.04
Ethanol	C_2H_5OH	243.0	6.38
Methanol	CH_3OH	239.4	8.09
Carbon dioxide	CO_2	31.0	7.37
Freon® 116	$(CF_3)_2$	19.7	2.97

Besides *alkoxide*s, many more *precursors* are now available, including oxy*alkoxides* $O_xM(OR)_y$ and a long list of $XSi(OR)_3$ precursor molecules, in which the ligand X is an organic group. The latter ligand may itself bring interesting characteristics, such as hydrophobic properties or the possibility to undertake its own polymerization (Chap. 13). Consequently, new hydrophobic aerogels (Chap. 3) and hybrid organic–inorganic gels could be designed (Chap. 13). The solvent characteristics were also investigated and it was shown that they induced important textural effects in the final dry aerogel. Regarding cations other than Si, a partial charge model was proposed by Livage et al. [32], according to which the positive partial charge δ^+ on cations such as Ti, Zr, or Al was significantly higher than on Si. Hence, the hydrolysis reactions with water, which replaces OR ligands by OH ones, were much faster with *alkoxide*s of these cations than with the Si *alkoxide*s. In these conditions, condensation of the hydrolyzed *precursors* generally led to dense oxides or hydroxides. To obtain gels, and subsequently dry them to aerogels, it was necessary to replace some OR groups by ligands that could not be hydrolyzed, for instance, because of chelation with the cation M, or new slower hydrolysis techniques had to be developed as done by Gash et al. [3].

In turn, the evolution in the sol–gel chemistry permitted to progress the synthesis of "ambigel" type aerogels, by ambient pressure drying (Chaps. 5 and 7) without any use of a *supercritical drying* autoclave, as described by Land et al. [33]. In the simplest approach, organic compounds could be added in the liquid, in which gelation was performed, to decrease the capillary drying stresses. Some of these additives, in particular glycerol, formamide, dimethyl formamide, oxalic acid, and tetramethylammonium hydroxide, are known as "Drying Control Chemical Additives" (*DCCA*) [34]. Depending on their polar or apolar and protic or aprotic characteristics, *DCCA* permitted to obtain uncracked dry monoliths with a relative pore volume as high as 97.4% [35]. Einarsrud at the Norwegian Institute of Technology showed that wet gels were also strengthened by the deposition of some silica in the necks between the nanoparticles constituting the solid network, when these wet gels were aged in *alkoxide*/alcohol solutions [36]. Wet gels can also be made hydrophobic by silylation of their pore surface, as done first by Kistler [9] and later investigated by Schwertfeger et al. [37] (Chap. 3). In this operation, –M–OR or –M–OH functionalities are replaced by hydrophobic –M–O–Si–X ones. The silylation operation can be achieved by chemical treatment of the wet gel after gelation or by using a precursor that brings the ≡Si–X functionalities. A consequence of such hydrophobic surface groups was that a gel was no longer submitted to strong capillary contraction stresses during solvent evaporation. Moreover, the ≡Si–X functionalities made impossible the formation of new siloxane ≡Si–O–Si≡ bonds by condensation. In turn, this explained that, in a study by Smith et al. at the Sandia National Laboratory, a shrinking gel during evaporation springed back to 96.9% of its wet volume when completely dry (Chap. 22). In comparison, an equivalent unsilylated dry *xerogel* occupied 30.1% of the initial wet gel volume [38].

Cryogels are obtained by freezing gels or solutions of their *precursors*. The main problem with this technique is that the nucleation and growth of solvent crystals may eventually destroy or at least distort the gel network, producing very large pores [39]. This problem was attenuated by using solvents with a low expansion coefficient and a high sublimation pressure and also by applying rapid freezing in liquid nitrogen, also known as flash freezing, at cooling rates over 10 Ks^{-1}. Many small crystals are then nucleated, and separate crystallization of the various components is prevented [40]. SiO_2 and Al_2O_3 powders with a texture close to those of aerogels were obtained using this technique, by Pajonk et al. [41].

1.3. Technical Characterization of Aerogels and Development of Their Applications

Besides the applications proposed by Kistler, a range of new applications of aerogels was rapidly pointed out concerning Cerenkov radiators (Chap. 28), immobilization medium of rocket propellants, insect killers, and catalysts or catalyst supports with a high specific surface area [11, 12]. Regarding the latter type of applications, NiO–Al_2O_3 and CuO–Al_2O_3 aerogels could be reduced to produce Ni–Al_2O_3 or Cu–Al_2O_3 aerogel composites with dispersed fine metal particles (Chap. 16).

To better develop these applications, aerogels had therefore become sufficiently important to discuss their characteristics and properties in symposia [42–47]. Reviews were also regularly published on the applications of aerogels [48–53]. The thermal and sound conduction properties of aerogels [54] (Chap. 23) were extensively studied at the University of Würzburg in Germany, and low-density silica aerogels were rapidly found to be excellent thermal and sound insulators [54, 55]. Hence, a major type of applications concerned the development of efficient thermal insulators (Chap. 26). For windows, a substantial amount of effort was directed toward the fabrication of transparent monolithic aerogel panels in the European JOULE II and III (HILIT project) and the two French ADEME programs (PACTE projects) [56]. Transparent glass panels, e.g., with dimension 55 cm × 55 cm × 2 cm, were made by the Airglass Ltd. company in Sweden, as part of the European EUROSOL program [56]. Examples of transparent SiO_2 aerogel cylinders and a tile, as well as monolithic TiO_2 gels and their porous structure, are illustrated in Figure 1.3.

Hydrophobic silica aerogel insulators with an acceptable transparency were also developed [57] (Chap. 3). Aerogel granules are much less costly than flat panels to produce and, although they are less transparent, they can also make very efficient translucent roofs or walls insulation. Aerogel insulators have also been studied for cooling or heating systems, including piping [53], and as new furnace crucibles to permit a better control of the

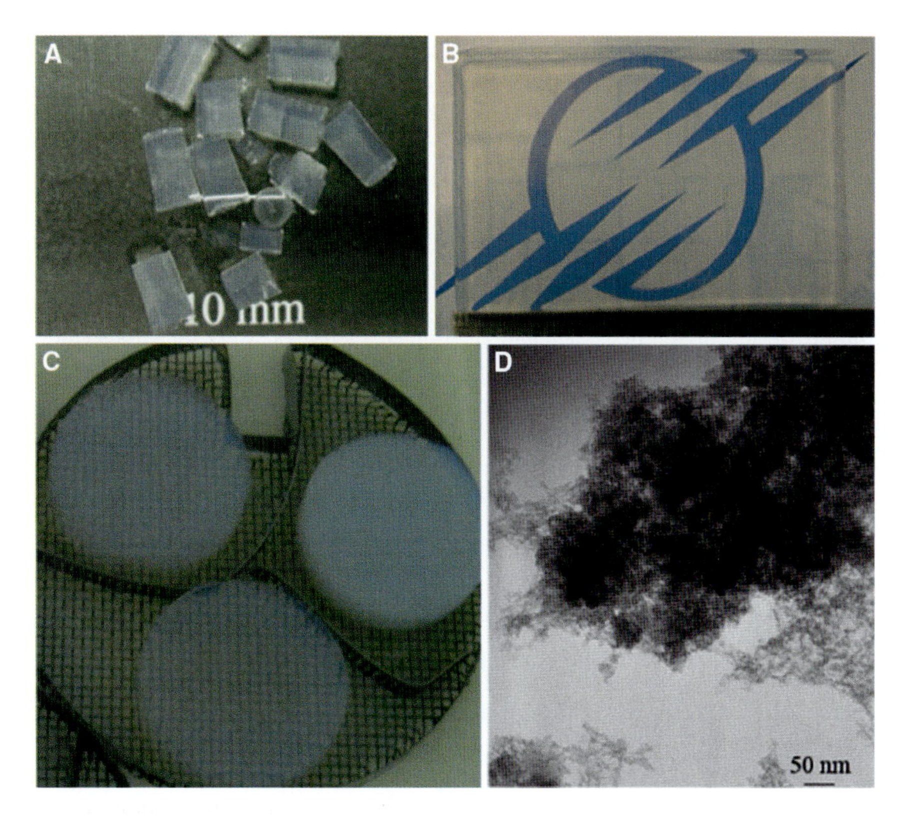

Figure 1.3. Oxide gels and aerogels. Courtesy Rigacci A. Mines ParisTech/Centre Energétique et Procédés. Rue Claude Daunesse, BP 207 F-06904 Sophia-Antipolis, France. **A.** SiO_2 aerogel small cylinders obtained by supercritical CO_2 drying of gels synthesized from *TEOS* through a two-step process, first catalysis with H_2SO_4 and then followed by aqueous NH_3, as part of the PhD thesis of Bisson A and the French ADEME Project "ISOGEL". **B.** SiO_2 aerogel tiles (1 cm thick) obtained by supercritical CO_2 drying of gels synthesized through a two-step process, first catalysis with H_2SO_4 and then followed by HF, as part of the PhD thesis of Masmoudi Y and the European project "HILIT+". **C.** TiO_2 gels obtained by substoichiometric hydrolysis of titanium tert-butoxide through HNO_3 catalysis, before drying. **D.** Transmission electron micrograph of the TiO_2 aerogel obtained by CO_2 *supercritical drying* of the gels in **C.**; the micrographs were taken by Berger M-H, Mines ParisTech/Centre des Matériaux Pierre-Marie Fourt. RN.446 F-91000 Evry, France. Both materials in **C.** and **D.** were elaborated as part of the PhD of D'Elia D and a French Carnot M.I.N.E.S project.

temperature gradient, required to grow monocrystalline metals or semiconductors [58] (Chap. 34). Borate, organic, carbon, and silica aerogels were used for some time as microspheres to contain mixtures of liquid deuterium and tritium in inertial confinement fusion experiments (*ICF*) [59]. Another group of applications, on which the attention was focused, concerned acoustic insulation (Chap. 13) and the development of ultrasonic transducers (Chap. 33), range finders, speakers [60], and anechoic chambers [61].

Although silica aerogels were sufficiently strong to be handled, they remained extremely brittle materials (Chap. 2). Their mechanical properties were therefore investigated in detail to improve this drawback (Chap. 22). Woignier and Phalippou in Montpellier [62] and LeMay et al. in Berkeley [63] established a scaling relationship between the Elastic Young's modulus E and the density ρ depending on the synthesis chemistry. An interesting consequence of this study was that aerogels of low density could easily be compressed and used to design parts able to absorb the energy involved in shock compression [64] (Chap. 13). The progressive absorption of shock energy by silica aerogel monoliths was used by the Jet Propulsion Laboratory to collect comet dust in the space [27]. Other applications of this type are being considered, for instance, to collect aerosol particles [65], to protect space mirrors, or to design tank baffles [51] (Chap. 32). Some techniques were also studied to improve the mechanical resistance of silica aerogels (Chaps. 2 and 22). One method consists in aging the aerogels in liquid media, particularly in *alkoxides* [66]. Inorganic powders or organic binders can also be added, although they make the silica aerogels more opaque [67] (Chaps. 13 and 15). Aerogel plates able to support the ceramic tiles of a floor, while offering an excellent sound attenuation characteristic, were for instance designed [52]. Aerogel particles can be used as filler of paints, varnishes, and tire elastomers to provide them with some hardness, resistance to wear, or some thickening characteristics [52]. Partially sintered aerogels, which resist mechanically better to capillary stresses, are convenient to store or transport liquids such as rocket fuel [68]. They can also be used to adsorb or extract some chemical components, for instance, to purify waste water, to confine radioactive waste [51] (Chap. 29), to filter gases [69], or to design heat storage equipments by adsorption of water [70]. Organic polymers, which comprise strong –(C–C)– covalent bonds, can also be directed to form relatively strong monolithic aerogels [71] (Chap. 13). The latter monoliths can have a thermal conductivity of the same order of magnitude as silica aerogels at room temperature and they are less brittle, although they are usually not transparent [59].

The optical transparency of silica aerogels was studied for use in optical devices and the so-called two-step procedure previously mentioned, acid-catalyzed hydrolysis, followed by base-catalyzed condensation, was found to be very efficient [72] (Chap. 2). The refraction index could be tuned to controlled values in a range from 1.01 to 1.03, very convenient for applications in *Cerenkov counters*, as previously shown by the Teichner group [73]. These counters permit to identify electrically charged particle in high-energy physics experiments, in cosmology, and to distinguish antimatter from matter particles in the cosmic rays (Chap. 28). Hydrophobic silica aerogel tiles (Chap. 3) of reproducible and controlled dimension and density were developed for this task by D. Poelz for the Deutches Elektronen Synchrotron (DESY) in Hamburg [74] and by Henning and Hardel in Sweden for the European Organization for Nuclear Research (CERN) in Geneva [75]. A company, Airgals AB, was created to produce 60 cm $\times$ 60 cm tiles. Similar tiles were designed for Ring Imaging Cerenkov (RICH) counters in the HERMES experiment at the DESY-HERA German facility [76] and for Cerenkov detectors of the threshold type, such as the BELLE detector at the KEK B-Factory in Japan [77] or the KEDR detector in Russia [78]. Another application of silica aerogels, due to their refractive index close to that of air, is as cladding films, with a thickness of the order of 20 μm, on optical fibers. Such films increase the light trapping fraction by a multiplication factor of 4 [79]. Aerogel coatings also attenuate the Fresnel scattering losses on solar cells and improve their efficiency [80]. For other applications, aerogels can be used to entrap or encapsulate a very large span of molecules, for instance, photoluminescent dopants [81], lanthanides for lasers [82], or a large amount of small ions such as Li^+ for high-temperature electrical batteries of high capacity [53, 83] or high-energy positive electrodes [84]. Carbon aerogels are themselves electrical conductors; hence, they were investigated for applications as electrodes in supercapacitors, capacitive deionization units, and fuel cells [85] (Chap. 36). Because of their high specific surface area, they can store more electrical energy than conventional capacitors. A proportion of Ru $>$ 50 wt% could also be deposited, which increased the specific capacitance of the carbon aerogels to values greater than 250 F/g [86]. This energy could, moreover, be released very rapidly to provide a high instant power [87]. Given their high specific pore volume, carbon aerogels can also be used to capture a large amount of ionic compounds, such as NaCl or other aqueous pollutants [49, 85], by electrolytic methods. In a similar domain, silica aerogels were proposed for the storage of long-life actinide wastes because they are chemically very stable with time and possess a very large specific porous volume. They can also be easily converted to dense vitreous silica at a relatively low temperature (~1,000°C) after a short heat treatment [88] (Chap. 29).

Besides, silica or organic aerogels are dielectric materials, with a dielectric constant that can be tuned to a value close to 1. This property makes them interesting for ultra-large-scale integrated circuits for high-speed computers [89] or as electret materials [90]. For ultrasonic applications in gases, porous piezoelectric transducers with a low acoustic impedance are required, and "PZT" aerogel monoliths, for instance, with a composition $PbZr_{0.53}Ti_{0.47}O_3$ and porosity up to 90 vol%, were developed [91] (Chap. 33).

Because of their high specific surface area, aerogels are a priori good materials for applications in catalysis and they have been studied for oxidation [92], nitroxidation [93], selective reduction [94], polymerization [95], and selective hydrogenation reactions [96]. In the latter case, catalysis is mostly due to the metal atoms, and aerogels are efficient as supports [97]. However, active metal particles supported on oxide aerogels can also be directly prepared from a multicomponent aerogel, as previously mentioned [98] (Chap. 16). Another domain of interest in catalysis is the use of aerogels as encapsulation medium of enzymes [99, 100]. Such solid biocatalysts can be used for biochemical synthesis or as the active component in biosensors [101] (Chap. 27). Aerogels also offered an opportunity to perform new fundamental studies, for instance, to study interactions in fluids such as ^3He–^4He mixtures near their critical temperature, because their high specific surface area induced strong perturbations in the fluids [102].

1.4. Recent Aerogel Developments

During the last decade, the synthesis techniques previously used were refined. For instance, a new rapid supercritical extraction (RSCE) method was experimented, in which the sol–gel *precursors* were themselves brought to a supercritical state inside a pressurized mold [103]. The mechanism of structuration by acoustic cavitation in sonogel [104] (Chap. 20), as well as condensation in the pores by light scattering [105], was studied. The ambient pressure drying method was extended to aerogels obtained from sodium silicate [106] (Chap. 5). The technique of Gash, based on slow deprotonation, was further extended to a range of materials comprising SnO_2 [107], Y_2O_3-stabilized ZrO_2 aerogels [108] (Chap. 6), aerogel *thermites* with dispersed Al nanoparticles [109] (Chap. 25), SiO_2 aerogels with dispersed metal nanoparticles [110] (Chap. 16) or with CdS nanoparticles grown by lithography inside the aerogel [111] (Chap. 19), and chitosan–SiO_2 hybrid aerogels with dispersed gold particles [112] (Chap. 18).

The types of aerogels developed were themselves in continuation of the materials previously studied. A nonlimitative list includes superhydrophobic aerogels (Chap. 4) with a water contact angle higher than 150° and self-cleaning properties [113], sulfated zirconia aerogels [114] (Chap. 6), TiO_2 aerogels [115] (Chap. 7), aerogel composites [115], and more exotic chalcogenide aerogels [116] (Chap. 17). To these, one must add organic aerogels based on cellulose [117] (Chap. 9), resorcinol–formaldehyde (Chap. 11), the carbon aerogels derived from them [118] (Chap. 36), allophanes (amorphous clay) [119] (Chap. 12), and polyurethane [120] (Chap. 10). Hybrid organic–inorganic aerogels were developed based on isocyanate crosslinked vanadia aerogels [121], CuO resorcinol–formaldehyde aerogels [122] (Chap. 14), and epoxy or polystyrene-linked SiO_2 aerogels with improved mechanical properties [123, 124] (Chaps. 13, 15, and 22).

A significant effort was directed at a better knowledge of the aerogel properties. This effort concerned their mechanical properties [125] (Chap. 22) including their simulation [126] (Chap. 24), for instance, the mechanical properties of hybrid organic–inorganic aerogels as a function of their organic content [127] (Chaps. 13 and 15). The luminescent, conducting, or magnetic properties of aerogels also deserved much interest, in particular for energy storage applications [128, 129]. The thermal properties were themselves further studied (Chap. 23), in particular regarding carbon aerogels [130] (Chap. 36).

Overall, aerogels have now reached a stage where the focus is on their applications. These concern the containment of nuclear waste [131] (Chap. 29), CO_2 trapping [132], water-repellent coatings [133], chemical sensors [134] (Chap. 27), heterogeneous catalysts [135, 136], metal casting molds [137] (Chap. 34), acoustic transducers [138] (Chap. 33), energy storage devices [139], *thermites* [109], pharmaceutical drug carriers [140] (Chaps. 30 and 31), nonflammable cryogenic insulator [141], and confinement media to study the interactions in superfluids [142].

Aerogels are now commercialized by several companies, among them the Cabot Corp. as aerogel particulate insulators for windows [143] (Chap. 38), Nano HiTech (Chap. 40) and Okagel also (Chap. 41) as insulating products, the American Aerogel Corp. as open-cell foam materials [144] (Chap. 39), the Birdair Company as membranes with aerogel inserts [145], and the Aspen Aerogels Company as flexible insulation products [146] (Chap. 4). The future of aerogels certainly rests on their efficient commercialization and much success is wished to these companies (Chap. 37).

Acknowledgments

The author is very grateful to researchers in the Mines ParisTech centers, France, and in particular to Berthon-Fabry Sandrine, Rigacci A, and Berger M-H for providing photographs of aerogels included in this chapter.

References

1. Kistler SS (1932) Coherent expanded aerogels. J Phys Chem 36:52–64
2. Brinker CJ, Scherer GW (1990) Sol-gel science. The physics and chemistry of sol-gel processing, Academic Press, New-York: (a) p461; (b) p523; (c) p97
3. Gash AE, Tillotson TM, Satcher Jr. JH, Hrubesh LW, Simpson RL (2001) New sol-gel synthetic route to transition and main-group metal oxide aerogels using inorganic salt *precursors*. J Non Cryst Solids 285:22–28
4. Eid J, Pierre AC, Baret G (2005) Preparation and characterization of transparent Eu doped Y_2O_3 aerogel monoliths, for application in luminescence. J Non Cryst Solids 351:218–227
5. Buisson P, Pierre AC (2006) Immobilization in quartz fiber felt reinforced silica aerogel improves the activity of *Candida rugosa* lipase in organic solvents. J Mol Catal B Enzym 39:77–82
6. Alemánl J, Chadwick AV, He J, Hess M, Horie K, Jones RG, Kratochvíl P, Meisel I, Mita I, Moad G, Penczek S, Stepto RFT (2007) Definitions of terms relating to the Structure and processing of sols, gels,

Networks, and inorganic–organic hybrid materials (IUPAC recommendations 2007). Pure Appl Chem 79:1801–1829
7. Freundlich H (1923) Colloid and capillary chemistry, Duttom Ed., New-York
8. Kistler SS (1941) Aerogels. Patent US 2249767 assigned to Monsanto Chemical Co
9. Kistler SS (1952) Water Repellent Aerogels. Patent US 2589705
10. Woignier T, Phalippou J, Zarzycki J (1984) Monolithic aerogels in the systems SiO_2-B_2O_3, SiO_2-P_2O_5, SiO_2-B_2O_3-P_2O_5. J Non Cryst Solids 63:117–130
11. Nicolaon GA, Teichner SJ (1968) Preparation of silica aerogels from methyl orthosilicate in alcoholic medium, and their properties. Bull Soc Chim Fr: 1906–1911
12. Teichner SJ (1986) Aerogels of inorganic oxides, Springer Proc Phys 6:22–30
13. Brinker CJ, Ward KJ, Keefer KD, Holupka E, Bray PJ, Pearson RK (1986) Synthesis and structure of borate based aerogel. In Aerogels. Springer Proc Phys 6:57–67
14. Pekala RW (1989) Organic aerogels from the polycondensation of resorcinol with formaldehyde. J Mater Sci 24:3221–3227
15. Pekala RW, Mayer ST, Kaschmitter JL, Kong FM (1994) Carbon aerogels: an update on structure, properties, and applications In Attia YA (ed) Sol-gel Process Appls, Plenum, New-York 369–377
16. Marie J, Berthon-Fabry S, Chatenet M, Chainet E, Pirard R, Cornet N, Achard P (2007) Platinum supported on resorcinol–formaldehyde based carbon aerogels for PEMFC electrodes: Influence of the carbon support on electrocatalytic properties. J Appl Electrochem 37:147–153
17. Sanchez C, Ribot F (1994) Design of hybrid organic–inorganic materials synthesized via sol-gel chemistry. New J Chem 18:1007–1047
18. Stanic V, Etsell TH, Pierre AC, R.J. Mikula RJ (1997) Metal Sulfide Preparation from a Sol-Gel Product and Sulfur. J Mater Chem 7:105–107
19. Broecker FJ, Heckmann W, Fischer F, Mielke M, Schroeder J, Stange A (1986) Structural analysis of granular silica aerogel. Springer Proc Phys 6:160–166
20. Ebelmen M (1846) Recherches sur les combinaisons des acides borique et silicique avec les éthers. Ann Chim Phys 16:129–166; (1847) Sur l'hyalite artificielle et l'hydrophane. C R Acad Sci Paris 25:854–856
21. Peri JB (1966) Infrared study of OH and NH2 groups on the surface of a dry silica aerogel. J Phys Chem 70:2937–2945
22. Woignier T, Phalippou J, Quinson JF, Pauthe M, Laveissiere F (1992) Physicochemical transformation of silica gels during hypercritical drying. J Non-Cryst Solids 145:25–32
23. Vacher R, Phalippou J, Pelous J, Woignier (eds) (1989) On the fractal structure of silica aerogels. In Vacher R, Phalippou J, Pelous J, Woignier T (eds) Proceedings of the Second International Symposium on Aerogels (ISA2), Rev. Phys. Appl. Colloq, 24-C4:127–131
24. Schaefer DW, Wilcoxon JP, Keefer KD, Bunker BC, Pearson RK, Thomas IM, Miller DE, (1987) Origin of porosity in synthetic materials. In Banavar JR, Koplik J, Winkler KW (eds) Phys chem porous media 2, American Institute of Physics 154: 63–80
25. Brinker CJ, Keefer KD, Schaefer DW, Ashley CS (1982) Sol-gel transition in simple silicates. J Non-Cryst Solids 48:47–64
26. Tillotson TM, Hrubesh LW (1992) Transparent ultralow-density silica aerogels prepared by a two-step sol-gel process J Non-Cryst Solids 145:44–50
27. Tsou P (1995) Silica aerogel captures cosmic dust intact. J Non-Cryst Solids 186:415–427
28. De la Rosa-Fox N, Esquivias L, Craievich AF, Zarzycki J (1990) Structural study of silica sonogels. J Non-Cryst Solids 121: 211–15
29. Zarzycki J (1994) Sonogels. Heterog Chem Rev 1:243–253
30. Matson DW, Smith RD (1989) Supercritical fluid technologies for ceramic-processing applications. J Am Ceram Soc 72:871–881
31. Pajonk GM (1994) A short history of the preparation of aerogels and carbogels. In: Attia YJ (ed) Sol-Gel Processing and Applications, Plenum Press, New-York, 201–209
32. Livage J, Henry M, Sanchez C (1988) Sol-Gel Chemistry of Transition Metal Oxides. Prog Solid State Chem 18:259–341
33. Land VD, Harris TM, Teeters DC (2001) Processing of low-density silica gel by critical point drying or ambient pressure drying. J Non-Cryst Solids 283:11–17
34. Zarzycki J, Prassas M, Phalippou J (1982) Synthesis of glasses from gels: the problem of monolithic gels. J Mater Sci 17:3371–3379
35. Hench, LL (1986) Use of dying control chemical additives (*DCCAs*). In controlling sol-gel processing. In: Hench LL, Ulrich DR (eds) Science of Ceramic Chemical Processing, Wiley, New-York, 52–64

36. Hæreid S, Einarsrud MA, Scherer GW (1994) Mechanical strengthening of TMOS-based alcogels by aging in silane solutions. J Sol-Gel Sci Tech 3:199–204
37. Schwertfeger F, Glaubitt W, Schubert U (1992) Hydrophobic aerogels from tetramethoxysilane/methyltrimethoxysilane mixtures J Non-Cryst Solids 145:85–89
38 Smith DM, Stein D, Anderson JM, Ackermann W (1995) Preparation of low-density *xerogels* at ambient pressure. J Non-Cryst Solids 186:104–112
39. Kocklenberg R, Mathieu B, Blacher S, Pirard R, Pirard JP, Sobry R, VandenBossche G (1998) Texture control of freeze-dried resorcinol-formaldehyde gels. J Non-Cryst Solids 225:8–13
40. Tretyakov YD, Shlyakhtin OA (1999) Recent progress in cryochemical synthesis of oxide materials. J Mater Chem 9:19–24
41. Pajonk GM, Repellin-Lacroix M, Abouarnadasse S, Chaouki J, Klvana D (1990) From sol-gel to aerogels and cryogels. J. Non-Cryst. Solids. 121: 66–67
42. Fricke J (ed) (1986) Aerogels – Proceedings of the First International Symposium, Wurzburg, FRG, Sept. 23–25, 1985, Springer-Verlag, Berlin
43. Vacher R, Phalippou J, Pelous J, Woignier T (eds) (1989) Proceedings of the Second International Symposium on Aerogels (ISA2), Rev. Phys. Appl. Colloq, 24–C4
44. Fricke J (ed) (1992) Proceedings of the Third International Symposium on Aerogels (ISA 3), J Non-Cryst Solids 145
45. Pekala RW, Hrubesh LW (eds) (1995) Proceedings of the Fourth International Symposium on Aerogels (ISA 4), J Non-Cryst Solids 186
46. Phalippou J, Vacher R (eds) (1998) Proceedings of the Fifth International Symposium on Aerogels (ISA 5), J Non-Cryst Solids 225
47. Ashley CS, Brinker CJ, Smith DM (eds) (2001) Aerogels 6. Proceedings of the Sixth International Symposium on Aerogels (ISA6), Albuquerque, NM, USA; 8–11 October 2000. J Non-Cryst Solids 285
48. Fricke J, Emmerling A (1992) Aerogels – preparation, properties, applications. Struct Bonding (Berlin) 77:37–87
49. Fricke J, Emmerling A (1998) Aerogels – Recent progress in production techniques and novel applications. J Sol-Gel Sci Technol 13:299–303
50. Burger T, Fricke J (1998) Aerogels: Production, modification and applications. Berichte der Bunsen Gesellschaft Phys Chemi Chem Phys 102:1523–1528
51. Hrubesh LW (1998) Aerogel applications. J Non-Cryst Solids 225:335–342
52. Schmidt M, Schwertfeger F (1998) Applications for silica aerogel products. J Non-Cryst Solids 225:364–368
53. Husing N, Schubert U (1998) Aerogels – Airy materials: Chemistry, structure, and properties. Angew Chem Int Ed 37:23–45
54. Caps R, Doell G, Fricke J, Heinemann E, Hetfleisch J (1989) Thermal transport in monolithic silica aeroge. In Vacher R, Phalippou J, Pelous J, Woignier T (eds) Proceedings of the Second International Symposium on Aerogels (ISA2), Rev. Phys. Appl. Colloq, 24-C4:113–118
55. Gronauer M, Kadur A, Fricke J (1986) Mechanical and acoustic properties of silica aeroge. In ISA1 Springer Proc Phys 6:167–173
56. Pajonk GM (2003) Some applications of silica aerogels. Colloid Polym Sci 38:4407–4413
57. Venkateswara Rao A, Pajonk GM (2001) Effect of methyltrimethoxysilane as a co-precursor on the optical properties of silica aerogels. J Non-Cryst Solids 285:202–209
58. Tscheuschner D, Ratke L (1999) Crystallization of InSb in aerogel crucibles. Cryst Res Technol 34:167–174
59. Kim KK, Jang KY (1991) Hollow silica spheres of controlled size and porosity by sol-gel processing. J Am Ceram Soc 74:1987–1992
60. Gerlach R, Kraus O, Fricke J, Eccardt PC, Kroemer N, Magori V (1992) Modified silica aerogels as acoustic impedance matching layers in ultrasonic devices. J Non-Cryst.Solids 145:227–232
61. Forest L, Gibiat V, Woignier T (1998) Biot's theory of acoustic propagation in porous media applied to aerogels and alcogels. J Non-Cryst Solids 225:287–292
62. Woignier T, Phalippou J (1989) Scaling law variation of the mechanical properties of silica aerogels. In Vacher R, Phalippou J, Pelous J, Woignier T (eds) Proceedings of the Second International Symposium on Aerogels (ISA2), Rev. Phys. Appl. Colloq, 24-C4:179–184
63. LeMay JD, Tillotson TM, Hrubesh LW, Pekala RW (1990) Microstructural dependence of aerogel mechanical properties Mat Res Soc Sym Proc 180:321–324
64. Holmes NC, Radousky HB, Moss MJ, Nellis WJ, Henning S (1984) Silica at ultrahigh temperature and expanded volume. Appl Phys Lett 45:626–628
65. Guise MT, Hosticka B, Earp BC, Norris PM (1995) An experimental investigation of aerosol collection utilizing packed beds of silica aerogel microspheres. J Non-Cryst Solids 285:317–322

66. Einasrud MA, Dahle M, Lima S, Hæreid S (1995) Preparation and properties of monolithic silica *xerogels* from *TEOS*-based alcogels aged in silane solutions. J Non-Cryst Solids 186:96–103
67. Lu X, Wang P, Arduini-Schuster MC, Kuhn J, Büttner D, Nilsson O, Heinemann U, Fricke J (1992) Thermal transport in organic and opacified silica monolithic aerogels. J Non-Cryst Solids 145:207–210
68. Gesser HD, Goswani PC (1989) Aerogels and related porous materials. Chem Rev 89:765–788
69. Emmerling A, Gross J, Gerlach R, Goswin R, Reichenauer G, Fricke J, Haubold HG (1990) Isothermal sintering of silica aerogels. J Non-Cryst Solids 125:230–243
70. Aristov YI, Restuccia G, Tokarev MM, Cacciola G (2000) Selective Water Sorbents for Multiple Applications, 10. Energy Storage Ability. React Kineti Catal Lett 69:345–353
71. Barral K (1998) Low-density organic aerogels by double-catalysed synthesis. J Non-Cryst Solids 225:46–50
72. Venkateswara Rao A, Haranath D, Pajonk GM, Wagh PB (1998) Optimisation of *supercritical drying* parameters for transparent silica aerogel window applications. Mater Sci Technol 14: 1194–1199
73. Cantin M, Casse M, Koch L, Jouan R, Mestran P, Roussel D, Bonnin F, Moutel J, Teichner SJ (1974) Silica aerogels used as Cherenkov radiators. Nucl Instrum Methods 118:177–182
74. Poelz G, Riethmueller R (1982) Preparation of silica aerogel for Cherenkov counters. Nuc Instr Meth 195:491–503
75. Henning S, Svensson L (1981) Production of silica aerogel. Phys Scr 23:697–702
76. Nappi E (1998) Aerogel and its applications to RICH detectors. Nucl Phys B Proc Suppl 61B:270–276
77. Sumiyoshi T, Adachi I, Enomoto R, Iijima T, Suda R, Yokoyama M, Yokogawa H (1998) Silica aerogels in high energy physics. J Non-Cryst Solids 225:369–374
78. Barnyakov MY, Bobrovnikov VS, Buzykaev AR, Danilyuk AF, Ganzhur SF, Goldberg II, Kolachev GM, Kononov SA, Kravchenko EA, Minakov GD, Onuchin AP, Savinov GA, Tayursky VA (2000) Aerogel Cherenkov counters for the KEDR detector. Nucl Instrum Methods Phys Res Sect A 453:326–330
79. Sprehn GA, Hrubesh LW, Poco JF, Sandler PH (1997) Aerogel-clad optical fiber. US Patent, US5684 907, Chem Abstr 127, P339050n
80. Hrubesh LW, Poco JF (1995) Thin aerogel films for optical, thermal, acoustic and electronic applications. J Non-Cryst Solids 188:46–53
81. Leventis N, Elder IA, Rolison DR, Anderson ML, Merzbacher CI (1999) Durable Modification of Silica Aerogel Monoliths with Fluorescent 2,7-Diazapyrenium Moieties. Sensing Oxygen near the Speed of Open-Air Diffusion. Chem Mater 11:2837–2845
82. Tillotson TM, Sunderland WE, Thomas IM, Hrubesh LW (1994) Synthesis of lanthanide and lanthanide-silicate aerogels. J Sol-Gel Sci Technol 1:241–249
83. Passerini S, Coustier F, Giorgetti M, Smyrl WH (1999) Li-Mn-O aerogels. Electrochem Solid-State Lett 2:483–485
84. Owens BB, Passerini S, Smyrl WH (1999) Lithium ion insertion in porous metal oxides. Electrochim Acta 45:215–224
85. Pekala RW, Farmer JC, Alviso CT, Tran TD, Mayer ST, Miller JM, Dunn B (1998) Carbon aerogels for electrochemical applications. J Non-Cryst Solids 225:74–80
86. Miller JM, Dunn B (1999) Morphology and Electrochemistry of Ruthenium/Carbon Aerogel Nanostructures. Langmuir 15:799–806
87. Gouerec P, Miousse D, Tran-Van F, Dao LH (1999) Characterization of pyrolyzed polyacrylonitrile aerogel thin films used in double-layer supercapacitors. J New Mater Electrochem Syst 2:221–226
88. Woignier T, Reynes J, Phalippou J, Dussossoy JL, Jacquet-Francillon N (1998) Sintered silica aerogel: a host matrix for long life nuclear wastes. J Non-Cryst Solids 225:353–357
89. Kawakami N, Fukumoto Y, Kinoshita T, Suzuki K, Inoue K (2000) Preparation of highly porous silica aerogel thin film by supercritical drying. Jpn J Appl Phys Part 2 39:L182–L184
90. CaoY, Xia ZF, Li Q, Shen J, Chen LY, Zhou B (1998) Study of porous dielectrics as electret materials. IEEE Trans Dielectr Electr Insul 5:58–62
91. Sinko K, Cser L, Mezei R, Avdeev M, Peterlik H, Trimmel G, Husing N, Fratzl P (2000) Structure investigation of intelligent aerogels. Physica B 276:392–393
92. Matis G, Juillet F, Teichner SJ (1976) Catalytic oxidation of paraffins on nickel oxide-based catalysts. I. Selectivity in the partial oxidation of isobutane and propane. Bull Soc Chim Fr 1633–1636
93. Pajonk GM (1991) Aerogel catalysts. Appl Catal 72:217–276
94. Willey RJ, Lai H, Peri JB (1991) Investigation of iron oxide-chromia-alumina aerogels for the selective catalytic reduction of nitric oxide by ammonia. J Catal 130:319–331
95. Fanelli AJ, Burlew JV, Marsh GB (1989) The polymerization of ethylene over titanium tetrachloride supported on alumina aerogels: low-pressure results. J Catal 116:318–324

96. Blanchard F, Pommier B, Reymond JP, Teichner SJ (1983) New Fischer-Tropsch catalysts of the aerogel type. In: Poncelet G, Grange P, Jacobs PA (eds) Studies in Surface Science and Catalysis, vol. 16 Preparation of Catalysts III, Elsevier, Amsterdam, 395–407
97. Klvana D, Chaouki J, Kusohorski D, Chavarie C, Pajonk GM (1988) Catalytic storage of hydrogen: hydrogenation of toluene over a nickel/silica aerogel catalysts in integral flow conditions. Appl Catal 42:121–130
98. Lacroix M, Pajonk G, Teichner SJ (1981) Activation for catalytic reactions of the silica gel by hydrogen spillover. In: Seiyama T, Tanabe K (eds) Studies in Surface Science and Catalysis, vol. 7 New Horizons in Catalysis, Elsevier, Amsterdam, 279–290
99. Antczak T, Mrowiec-Bialon J, Bielecki S, Jarzebski AB, Malinowski JJ, Lachowski AI, Galas E (1997) Thermostability and esterification activity in silica aerogel matrix and in organic solvents. Biotechnol Techn 11:9–11
100. Pierre M, Buisson P, Fache F, Pierre AC (2000) Influence of the drying technique of silica gels on the enzymatic activity of encapsulated lipase. Biocatal Biotransform 18:237–251
101. Power M, Hosticka B, Black E, Daitch C, Norris P (2001) Aerogels as biosensors: viral particle detection by bacteria immobilized on large pore aerogel. J Non-Cryst Solids 285:303–308
102. Ma J, Kim SB, Hrubesh LW (1993) Phase separation of helium-3-helium-4 mixture in aerogel. J Low Temp Phys 93:945–955
103. Roth TB, Anderson AM, Carroll MK (2008) Analysis of a rapid supercritical extraction aerogel fabrication process: Prediction of thermodynamic conditions during processing J. Non-Cryst Solids 354:3685–3693
104. de la Rosa-Fox N, Morales-Florez V, Pinero M, Esquivias L (2009) Nanostructured sonogels. Key Eng Mater 391:45–78
105. Reichenauer G, Manara J, Weinlaeder H (2007) Strong light scattering upon capillary condensation in silica aerogels. Stud Surf Sci Catal 160:25–32
106. Bangi Uzma KH, Parvathy Rao A, Hirashima H, Venkateswara RaO A (2009) Physico-chemical properties of ambiently dried sodium silicate based aerogels catalyzed with various acids J Sol-Gel Sci Technol 50:87–97
107. Baumann TF, Kucheyev SO, Gash AE, Satcher JH Jr (2005) Facile synthesis of a crystalline, high-surface-area SnO2 aerogel. Advanced Mater 17:1546–1548
108. Chervin CN, Clapsaddle BJ, Chiu HW, Gash AE, Satcher JH Jr, Kauzlarich SM (2005) Aerogel Synthesis of Yttria-Stabilized Zirconia by a Non-Alkoxide Sol-Gel Route Chem Mater 17:3345–3351
109. Gash AE, Pantoya M, Satcher JH, Simpson RL (2008) Nanostructured energetic materials: aerogel thermite composites. Polymer Preprints (American Chemical Society, Division of Polymer Chemistry) 49:558–559
110. Hund JF, Bertino MF, Zhang G, Sotiriou-Leventis C, Leventis N (2004) Synthesis of homogeneous alloy metal nanoparticles in silica aerogels. J Non-Cryst Solids 350:9–13
111. Bertino MF, Gadipalli RR, Story JG, Williams CG, Zhang G, Sotiriou-Leventis C, Tokuhiro AT, Guha S, Leventis N (2004) Laser writing of semiconductor nanoparticles and quantum dots. Appl Phys Letters 85:6007–6009
112. Kuthirummal N, Dean A, Yao C, Risen W (2008) Photo-formation of gold nanoparticles: Photoacoustic studies on solid monoliths of Au(III)-chitosan-silica aerogels Spectrochim Acta A: Mol Biomol Spectroscopy 70A:700–703
113. Venkateswara Rao A, Latthe Sanjay S, Nadargi Digambar Y, Hirashima H, Ganesan V (2009) Preparation of MTMS based transparent superhydrophobic silica films by sol-gel method J Colloid Interf Sci 332:484–490
114. Mejri I, Younes MK, Ghorbel A (2006) Comparative study of the textural and structural properties of the aerogel and *xerogel* sulphated zirconia J Sol-Gel Sci Technol 40:3–8
115. Suzuki Y, Berger M-H, D'Elia D, Ilbizian P, Beauger C, Rigacci A, Hochepied J-F, Achard P (2008) Synthesis and microstructure of a novel TiO2 aerogel-TiO2 nanowire composite NANO 3:373–379
116. Yao Q, Arachchige IU, Brock SL(2009) Expanding the Repertoire of Chalcogenide Nanocrystal Networks: Ag2Se Gels and Aerogels by Cation Exchange Reactions J Amer Chem Soc 131:2800–2801
117. Fischer F, Rigacci A, Pirard R, Berthon-Fabry S, Achard P(2006) Cellulose-based aerogels. Polymer 47:7636–7645
118. Mulik S, Sotiriou-Leventis C, Leventis N (2008) Macroporous Electrically Conducting Carbon Networks by Pyrolysis of Isocyanate-Cross-Linked Resorcinol-Formaldehyde Aerogels Chem Mater 20:6985–6997
119. Chevallier T, Woignier, T, Toucet J, Blanchart E, Dieudonne P (2008) Fractal structure in natural gels: effect on carbon sequestration in volcanic soils J Sol-Gel Sci Techno 48:231–238
120. Rigacci A, Marechal JC, Repoux M, Moreno M, Achard P (2004) Preparation of polyurethane-based aerogels and *xerogels* for thermal superinsulation. J Non-Cryst Solids 350:372–378

121. Luo H, Churu G, Fabrizio EF, Schnobrich J, Hobbs A, Dass A, Mulik S, Zhang Y, Grady BP, Capecelatro A, et al (2008) Synthesis and characterization of the physical, chemical and mechanical properties of isocyanate-crosslinked vanadia aerogels. J Sol-Gel Sci Technol 48:113–134
122. Leventis N, Chandrasekaran N, Sadekar A, Sotiriou-Leventis C, Lu H (2009) One-Pot Synthesis of Interpenetrating Inorganic/Organic Networks of CuO/Resorcinol-Formaldehyde Aerogels: Nanostructured Energetic Materials J Amer Chem Soc 131:4576–4577
123. Meador MAB, Weber AS, Hindi A, Naumenko M, McCorkle L, Quade D, Vivod SL, Gould GL, White S, Deshpande K (2009) Structure-Property Relationships in Porous 3D Nanostructures: Epoxy-Cross-Linked Silica Aerogels Produced Using Ethanol as the Solvent Appl Mater Interfaces 1:894–906
124. Nguyen BN, Meador MAB, Tousley ME, Shonkwiler B, McCorkle L, Scheiman DA, Palczer A (2009) Tailoring Elastic Properties of Silica Aerogels Cross-Linked with Polystyrene. Appl Mater Interfaces 1:621–630
125. Lu H, Luo H, Mulik S, Sotiriou-Leventis C, Leventis N (2008) Compressive behavior of crosslinked mesoporous silica aerogels at high strain rates Polymer Preprints (Amer Chem Soc) 49:515–516
126. Gelb LD (2007) Simulating Silica Aerogels with a Coarse-Grained Flexible Model and Langevin Dynamics J Phys Chem C 111:15792–1580
127. Morales-Florez V, Toledo-Fernandez JA, Rosa-Fox N, Pinero M, Esquivias L (2009) Percolation of the organic phase in hybrid organic-inorganic aerogels J Sol-Gel Sci Technol 50:170–175
128. Baudrin E, Sudant G, Larcher D, Dunn B, Tarascon J-M (2006) Preparation of Nanotextured VO2[B] from Vanadium Oxide Aerogels. Chem Mater 18:4369–4374
129. Carpenter EE, Long JW, Rolison DR, Logan MS, Pettigrew K, Stroud RM, Kuhn LT, Hansen BR, Moerup S (2006) Magnetic and Mossbauer spectroscopy studies of nanocrystalline iron oxide aerogels. J Appl Phys 99:08N711/1-08N711/3
130. Drach V, Wiener M, Reichenauer G, Ebert H.-P, Fricke J (2007) Determination of the Anisotropic Thermal Conductivity of a Carbon Aerogel-Fiber Composite by a Non-contact Thermographic Technique. Internat J Thermophys 28:1542–1562
131. Woignier T, Primera J, Lamy M, Fehr C, Anglaret E, Sempere R, Phalippou J (2004) The use of silica aerogels as host matrices for chemical species. J Non-Cryst Solids 350:299–307
132. Santos A, Ajbary M, Toledo-Fernandez JA, Morales-Florez V, Kherbeche A; Esquivias L (2008) Reactivity of CO2 traps in aerogel-wollastonite composites J Sol-Gel Sci Technol 48:224–230
133. Latthe Sanjay S, Nadargi Digambar Y, Venkateswara Rao A (2009) TMOS based water repellent silica thin films by co-precursor method using TMES as a hydrophobic agent. Appl Surf Sci 255:3600–3604
134. Plata DL, Briones YJ, Wolfe RL, Carroll MK, Bakrania SD, Mandel SG, Anderson AM (2004) Aerogel-platform optical sensors for oxygen gas. J Non-Cryst Solids 350:326–335
135. Bali S, Huggins FE, Huffman GP, Ernst RD, Pugmire RJ, Eyring EM (2009) Iron Aerogel and Xerogel Catalysts for Fischer-Tropsch Synthesis of Diesel Fuel Energy & Fuels 23:14–18
136. Gasser-Ramirez JL, Dunn BC, Ramirez DW, Fillerup EP, Turpin GC, Shi Y, Ernst RD, Pugmire RJ, Eyring EM, Pettigrew KA, et al. (2008) A simple synthesis of catalytically active, high surface area ceria aerogels. J Non-Cryst Solids 354:5509–5514
137. Steinbach S, Ratke L (2007) The microstructure response to fluid flow fields in Al-cast alloys Trans Indian Inst Metals 60:167–171
138. Nagahara H, Suginouchi T, Hashimoto M (2006) Acoustic properties of nanofoam and its applied air-borne ultrasonic transducers. Proc IEEE Ultrason Symp 3:1541–1544
139. Long JW, Fischer AE, McEvoy TM, Bourg ME, Lytle JC, Rolison DR (2008) Self-limiting electropolymerization en route to ultrathin, conformal polymer coatings for energy storage applications. PMSE Preprints (2008), 99, 772–773
140. Guenther U, Smirnova I, Neubert RH H (2008) Hydrophilic silica aerogels as dermal drug delivery systems – Dithranol as a model drug. Eur Jo Pharmac Biopharmac 69:935–942
141. Begag R, Fesmire JE, Sonn JH (2008) Nonflammable, hydrophobic aerogel composites for cryogenic applications. Thermal Cond 29:323–333
142. Nakagawa H, Kado R, Obara K, Yano H, Ishikawa O, Hata T, Yokogawa H, Yokoyama M (2007) Equal-spin-pairing superfluid phase of 3He in an aerogel acting as an impurity Phys Rev B: Condensed Matter Mater Phys 76:172504/1-172504/4
143. Bauer U, Darsillo MS, Field RJ, Floess JK, Frundt J, Rouanet S, Doshi DA (2008) Aerogel particles and methods of making same PCT Pat. Int. Appl., WO 2008115812 A2 20080925
144. Albert DF, Andrews GR, Bruno JW (2002) Organic, open cell foam materials, their carbonized derivatives and methods for production. U.S. Pat. Appl. Publ. US 2002064642

145. Augustyniak MJ, Hamilton KP, Kalkstein HC (2008) Manufacture of architectural membrane structures based on aerogel-containing 3-ply composites U.S. Pat. Appl. Publ. US 2008229704 A1 20080925
146. Lee JK, Gould GL, Rhine W (2009) Polyurea based aerogel for a high performance thermal insulation material J Sol-Gel Sci Technol 49:209–220

Part II

Materials and Processing: Inorganic – Silica Based Aerogels

2

SiO_2 Aerogels

Alain C. Pierre and Arnaud Rigacci

Abstract This chapter focuses on one of the most studied aerogel materials, silica aerogels. It aims at presenting a brief overview of the elaboration steps (sol–gel synthesis, aging, and drying), the textural and chemical characteristics (aggregation features, porosity, and surface chemistry), the main physical properties (from thermal, mechanical, acoustical, and optical, to biological, medical, etc.), and a rather broad panel of related potential applications of these fascinating nanostructured materials. It cannot be considered as an exhaustive synopsis but must be used as a simple tool to initiate further bibliographic studies on silica aerogels.

2.1. Elaboration

2.1.1. Sol–Gel Synthesis

The physics and chemistry involved in the synthesis of silica gels were detailed in books [1, 2], and many reviews on aerogels with particular focus on silica aerogels have already been published [3–10]. Schematically, a nanostructured solid network is formed in a liquid reaction medium as a result of a *polymerization* process, which creates *siloxane bridges* (≡Si–O–Si≡) between Si atoms delivered by *precursor* molecules. Such transformations are the equivalent of a *polymerization* process in organic chemistry, where direct bonds between the carbon atoms of organic precursors are established leading to linear chains or branched (crosslinked) structures, depending on the type of reactive monomers and *crosslinkers* used. Dispersed solid colloidal silica particles (i.e., nanoparticles with a size well below 1 μm) or "more or less" linear *oligomers* are formed in the early stage of the sol–gel process. In the second stage, these elementary objects can link with each other while still in the solvent, such as to make up a three-dimensional (3D) open network structure termed a gel, only limited by the container. The continuous transformation of a sol to a gel constitutes the *gelation* process. The brutal change from the liquid to the solid stage is termed the sol–gel transition. The gels that are obtained are termed either colloidal or polymeric depending on the nature of the building blocks of which the network is composed

A. C. Pierre (✉) • Université Lyon 1, CNRS, UMR 5256, IRCELYON, Institut de recherches sur la catalyse et l'environnement de Lyon, 2 avenue Albert Einstein, 69626 Villeurbanne, France
e-mail: Alain.Pierre@ircelyon.univ-lyon1.fr
A. Rigacci • MINES ParisTech, CEP, Centre Energétique et Procédés, 1 rue Claude Daunesse, BP 207, 06904 Sophia Antipolis, France
e-mail: arnaud.rigacci@mines-paristech.fr

M.A. Aegerter et al. (eds.), *Aerogels Handbook*, Advances in Sol-Gel Derived Materials and Technologies, DOI 10.1007/978-1-4419-7589-8_2,

of and whether they are nanoparticulate or more linear (polymer-like). For both stages, the driving reactions are *hydrolysis* and water and/or alcohol *condensation*.

For silica gels, a first important precursor is sodium metasilicate Na_2SiO_3, also termed waterglass [11, 12], which was previously used by Kistler to produce the first silica aerogels reported in the literature [13]. This precursor reacts with an acid such as HCl according to the reactions of the type shown in (2.1) below. A salt is produced, which must be eliminated by tedious dialysis or by exchange for H^+ through an acidic ion exchange column [14].

$$Na_2SiO_3 + 2HCl + (x - 1)H_2O \rightarrow SiO_2 \cdot xH_2O + 2NaCl \quad (2.1)$$

At present, some of the brand new works conducted on the silica system in aqueous solution concern the use of agricultural wastes, such as rice hull ash, as an inexpensive silica source [15, 16].

However, the Si precursors most frequently used nowadays are *alkoxide*s of the $Si(OR)_4$ type, in which R and OR designate alkyl and *alkoxide* groups, respectively. Often, R is a methyl group CH_3 (or Me). An often used precursor is hence termed TetraMethOxySilane (or *TMOS*) [17]. Another common material has four ethyl rests C_2H_5 (or Et) as R groups, in which case the precursor is termed TetraEthOxySilane (or *TEOS*) [18]. The first inorganic gels synthesized were indeed silica gels made by Ebelmen from such precursors in 1846, except that they were not dried by a supercritical method to produce aerogels [19]. A much larger list of *alkoxide*-derived precursors and mixtures of them are used today, comprising, for instance, polyethoxydisiloxane (PEDS) [20, 21], methyltrimethoxysilane (*MTMS*) [22, 23], methyltriethoxysilane (*MTES*) [24], 3-(2-aminoethylamino) propyltrimethoxysilane (EDAS) [25], *N*-octyltriethoxysilane [26], dimethyldiethoxysilane [27], and perfluoroalkysilane (PFAS) [28]. Some precursors (such as EDAS) include built-in chemical functionality, which can then be used to modify the resulting gel materials with appropriate chemical synthetic strategies.

All these precursors are characterized by the existence of Si–O polar covalent bonds. This bond can in a first approximation be described as being $\approx$ 50% covalent, according to a description introduced by Pauling [29]. Such a characteristic explains the differential ability to build a random – (M–O) – network, between Si atoms and other cations M. The covalent character of the Si–O bond is sufficient to permit a wide distribution of the $\equiv$Si–O–Si$\equiv$ angle values [30], leading to a "random 3D network" similar to the network known to prevail in silica glass.

In real life, Si *alkoxide*s are often available as complexes in solution in their parent alcohol and are typically polymerized to a smaller or larger extent. Even if some monolithic but quite dense aerogel-like materials can be synthesized through ultrasonic-assisted solventless sol–gel routes (Chap. 20) [31, 32], their *polymerization* is mostly carried out in an organic solvent through simultaneous *hydrolysis* (2.2) and poly*condensation* of water (2.3) and alcohol (2.4) so that water becomes a reactant added in controlled proportion to drive the *hydrolysis* reaction (2.2).

$$\equiv Si - OR + H_2O \rightarrow \equiv Si - OH + R - OH \quad (2.2)$$

$$\equiv Si - OH + HO - Si \equiv \rightarrow \equiv Si - O - Si \equiv + H_2O \quad (2.3)$$

$$\equiv Si - OR + HO - Si \equiv \rightarrow \equiv Si - O - Si \equiv + R - OH \quad (2.4)$$

The *hydrolysis* mechanisms involve first a nucleophilic attack of oxygen lone pairs of the H_2O molecule on the Si atoms [33]. Because of the polarized Si–O bonds, the silicon atoms hold a partial positive electronic charge δ^+, which in turn determines the kinetics of the nucleophilic attack and hence of the overall *hydrolysis* reaction. In *alkoxide*s, the Si atoms carry a relatively moderate partial positive charge (e.g., $\delta^+ \approx 0.32$ in $Si(OEt)_4$ by comparison with ≈ 0.65 and 0.63, respectively, in $Zr(OEt)_4$ and $Ti(OEt)_4$ [33]). As a result, the global *gelation* kinetics of $Si(OR)_4$ *alkoxide*s is very slow, unless the *hydrolysis* and *condensation* steps of Si are catalyzed either by bases that carry strong negative charges (e.g., OH^-, but also strong *Lewis base*s such as F^- ions) or by acids (e.g., H^+) in which case the reaction mechanism changes completely. In practice, the relative magnitudes of the *hydrolysis* and *condensation* rates are sufficiently slow to permit a relatively independent control. Overall, silica gels with a *texture* closer to that of polymeric gels derived from organic chemistry are obtained if the *hydrolysis* rate is faster than the *condensation* one. This is usually the case under acidic catalysis. In the literature, a large range of acids have proven to be useful catalysts, including HCl [27, 34], HF [3, 20, 28], or carboxylic acids [22, 34, 35].

On the other hand, proton acceptors, i.e., bases, accelerate the *condensation* reactions more than *hydrolysis*, which then favors the formation of denser colloidal silica particles and *colloidal gels*. Aqueous NH_3 is the basic catalyst most frequently used [22, 36, 37], but a *Lewis base* such as NaF or NH_4F also has its advantages, in particular when used with precursors such as *MTMS* [23, 38].

Additional parameters that affect the properties of the final aerogel are the nature of the solvent, which most typically is an alcohol [21, 27, 28, 34, 37] but can be acetone [35] or ethyl acetoacetate [20]. In addition, there is also the molar ratio "Si-precursor to water" as well as the concentration of Si precursor in the solvent and the catalyst nature and concentration, which are important parameters too. In short, these parameters rule the *nanostructuration* of the resulting gel (primary particles size, pores size distribution, fractality, *tortuosity*, density, etc.). To better control this *nanostructuration*, researchers have developed *two-step processes* so that they can successively favor one or another type of catalysis [39], a trick which is used, for example, to elaborate ultralight silica aerogels [40] or to achieve a good compromise between low *thermal conductivity* and high optical transparency [41]. The latter aerogels were synthesized first with H_2SO_4 catalysis of *TEOS* in ethanol under a substoichiometric molar ratio $H_2O/TEOS = 1.8$, followed by a second catalysis step with HF in ethylacetoacetate. Subsequent ^{29}*Si NMR* studies have shown that the intermediate precondensed species were rather polymeric, while the final gels were more colloidal (Figure 2.1). More recently, new developments are focused on the use of new classes of nonaqueous polar solvents such as

Figure 2.1. High-resolution scanning electron micrographs (HR-SEM) of silica aerogels synthesized by a two-step process (first catalysis with H_2SO_4 followed by HF catalysis) (*courtesy of Grillon F, MINES ParisTech, Evry, France and Rigacci A*).

ionic liquids [42–45]. Such solvents are expected to permit the synthesis of new porous *textures* and species of inorganic/organic hybrid nanomaterials.

2.1.2. Ageing

Before drying, silica gels are often aged via different processes. The aim of this step generally is to mechanically reinforce the tenuous solid skeleton generated during the sol–gel process. The majority of studies dedicated to these strengthening treatments were performed by Prof. Mari-Ann Einarsrud and coworkers at NTNU University, Trondheim, Norway [20, 46]. *Aging* schematically consists in taking advantage of *syneresis* and/or *Ostwald ripening* mechanisms by modifying the composition of the liquid phase contained in pores of the silica gel. Adding water [47] and/or monomeric alkoxysilanes such as *TEOS* [11, 20, 38, 46, 48–50] can significantly enhance surface reactions and primarily those involving the residual hydroxy/alcoxy groups. Consequently, supplementary *condensation* reactions and *dissolution/reprecipitation* of silica can occur. The associated kinetics depends on the pH and the nature of the solvent. Generally, the particles "neck" area, the average pore size, and the apparent density of the gel increase through aging treatments. If properly controlled, these morphological changes can significantly improve the mechanical properties (*E*, *K*, *MOR*...) and the liquid *permeability* (D). In one case, for example, an augmentation of the shear modulus by up to 23 times was reported [51].

Recently, it has been shown that other types of successful *aging* treatments exist to simultaneously increase the *permeability* and mechanical properties. These include, for example, the addition of larger precursor molecules, e.g., polyethoxydisiloxanes [20] or simply adding a dilute HF solution without additional silica precursor [50] (Figure 2.2). It has also been demonstrated that monitoring the temperature and simply performing a thermal *aging* of the wet gel in water can be a key factor to decrease the gel microporosity before drying [52].

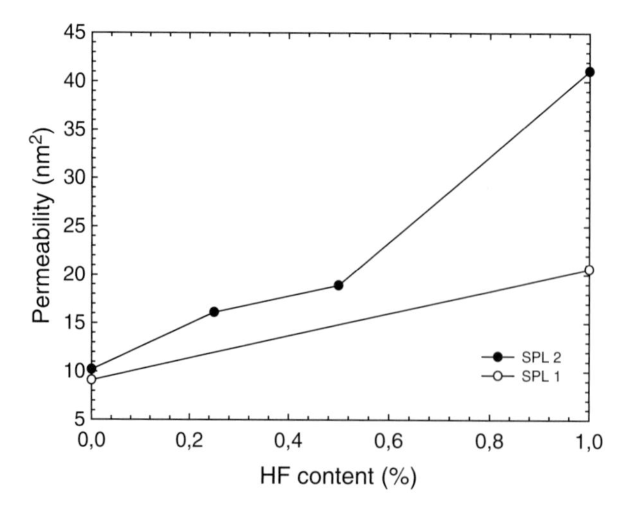

Figure 2.2. Permeability of silica wet gels (i.e., before drying) as a function of HF concentration in the aging bath composed of ethanol (SPL1, *white marks*) or ethylacetoacetate (SPL2, *solid marks*) containing 3 vol% water. After [50].

2.1.3. Drying

Capillary stresses inevitably occur whenever gas–liquid menisci appear at the pore boundaries, during evaporation of the pores' liquid. Even if the siloxane bonds have approximately 50% covalent character, which makes the silica gels much more capable of resisting to evaporative effects than other oxide gels, particular attention must be paid to this really tricky step. As illustrated in a number of review articles [53, 54], three main routes are commonly used for drying:

1. *Freeze-drying* (which necessitates to bypass the triple point)
2. Evaporation (which implies crossing the liquid–gas equilibrium curve)
3. Supercritical drying (which necessitates to bypass the critical point)

In general, *freeze-drying* and evaporation applied to finely nanostructured silica gels have not yet proved to produce monolithic aerogels. *Freeze-drying* leads to cracked pieces or even powder-like products [55]. Evaporation without specific surface (e.g., *sylilation* [56]) and/or *aging* [46] treatments results in "dense" (e.g., $\rho > 0.25$ g cm^{-3} [57]) and even cracked materials, so-called *xerogels*, as discussed in the preceding Chap. 1 (Figure 2.3). As shown by Phalippou et al. [58], the densification during evaporation comes from the *condensation* of remaining reactive silica species [see (2.3) and (2.4)]. When submitted to *capillary stresses*, initially far distant *hydroxyl* and/or alkoxy groups can come close enough to one another to react and generate new siloxane bonds, thus leading to irreversible shrinkage (Figure 2.4), because of the inherent flexibility of the silica chains.

Supercritical drying, on the other hand, permits to eliminate *capillary stresses*. Hence, this process produces monolithic silica aerogels of rather large dimensions (Figure 2.5), if required by a targeted application [59] (Figure 2.6).

Figure 2.3. Appearance of cracks during evaporative drying under ambient conditions (here observed with mesoporous silica wet gels impregnated with ethylacetoacetate at times $t_1 = 5$ min and $t_2 = 20$ min, respectively, after the beginning of evaporation t_0). *Courtesy of Rigacci A.*

Supercritical drying can be performed (1) in organic solvents in their supercritical state (generally alcohol as the pore liquid and consequently above 260°C if ethanol is used) according to a so-called *HOT* process [61] or (2) in supercritical CO_2 at a temperature slightly above the critical temperature of CO_2 ~31°C according to a so-called *COLD* process. Application of the *COLD* process to silica gels was investigated by Tewari et al. [62]. For this purpose, the liquid that impregnated the wet gels had to be exchanged with CO_2, either in the normal liquid state [18] or directly in the supercritical state [63]. Indeed,

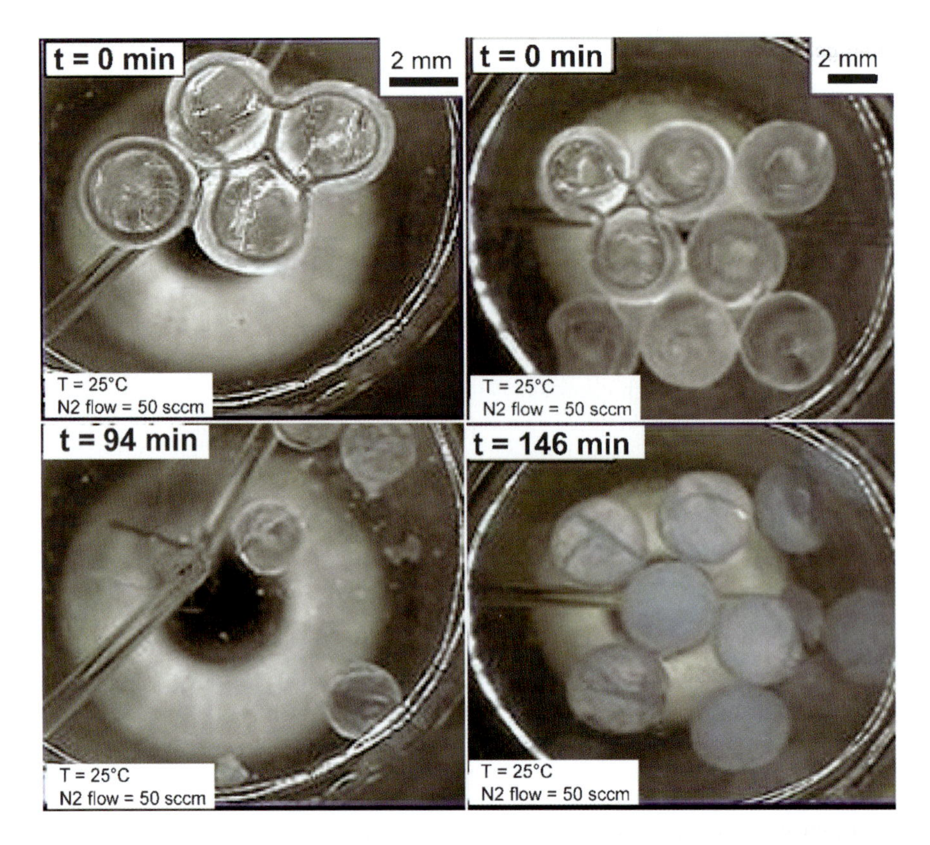

Figure 2.4. Comparison of the shrinkage behavior, from the wet (**upper photographs**) to the dry (**lower photographs**) states, occurring during evaporative drying of native, e.g., untreated (**left-hand side**) and sylilated (**right-hand side**) silica gels. *Courtesy of Rigacci A.*

Figure 2.5. Monolithic silica aerogels obtained after supercritical CO_2 extraction. *Courtesy of Rigacci A.*

Figure 2.6. Large monolithic silica aerogel monoliths integrated in demonstration glazing (**left side**) and window (**right side**) prototypes [60]. *Courtesy of K.I. Jensen and J.M. Schultz, DTU, Lyngby, Copenhagen.*

the interdiffusion of CO_2 with methanol or ethanol, and with it the exchange, is slow and is significantly accelerated when CO_2 is in the supercritical state [64, 65].

However, although perfect monolithic silica aerogels can be elaborated by both *HOT* and *COLD supercritical drying* routes, the supercritical way could remain too time-consuming to be widely exploited on an industrial scale to produce this type of samples. Indeed, the low gel *permeability* results in rather slow CO_2 washing and vessel depressurization steps [66, 67], in particular for thick gel plates. To speed up the CO_2 washing, simple molecular diffusion must be assisted by forced convection, for example, by integrating compression–decompression cycles into the process [68]. However, if an accelerated depressurization is required, gels must be significantly strengthened prior to drying. If not, they will experience cracks even at low depressurization rates (Figure 2.7).

Figure 2.7. Typical depressurization crack (perpendicular to the largest surface) experimented by the silica gel during supercritical drying (illustrated here on a 1 cm thick wet silica tile having a liquid *permeability* between 5 and 10 nm^2, dried with supercritical CO_2 at 313 K and 90 bar, and submitted to an autoclave depressurization of 0.15 bar min^{-1}). *Courtesy of Rigacci A.*

To accelerate the supercritical drying process, a "rapid supercritical extraction method," in which the silica sol or the precursors were directly gelled inside the container under *HOT* supercritical conditions, was investigated in the mid-1990s by Poco et al. [69] and subsequently by Gross et al. [70], Scherer et al. [71], and Gauthier et al. [72]. Even though successful in the case of small samples, this technique does yet not permit to elaborate large crack-free, low-density monolithic silica aerogels. Currently, to try solving the fluid exchange difficulties associated with the standard supercritical CO_2 routes, one of the major challenges concerns the direct synthesis of the silica gel in supercritical CO_2 by a water-free process [73–75].

Because the supercritical drying methods have remained considerably costly, some specific subcritical [76] and *ambient pressure drying* methods were developed to synthesize silica aerogels. In this case, the *capillary stresses* depend on surface tension and viscosity of the pore liquid, drying rate, and wet gel *permeability*. In order to reduce their negative impact on the drying, Drying Control Chemical Additives (*DCCA*) such as polyethylene glycol (PEG) [17, 77], polyvinyl alcohol (PVA) [78, 79], glycerol [80], or surfactants [81] have been used. They interfere with the *hydrolysis* products of the respective Si precursors and permit to control the pore size and pore volume as well as their distribution. In any case, one of the key points relies on the introduction of incondensable species in the system via the *sylilation* of the silica gel [56] or the use of specific Si precursor such as *MTMS* [22] in order to promote a so-called spring-back effect when the solvent front retreats and *capillary stresses* are released [54]. To conclude, after some trial and error *ambient pressure drying* was applied with great success to the synthesis of silica aerogels from *alkoxide*s [82–84], as well as from waterglass [11, 56], and is today the most promising manufacturing technique for SiO_2 aerogels. Densities below 0.1 g cm^{-3}, for a total specific pore volume sometimes larger than that of CO_2-dried samples, could be obtained, something which would have been unimaginable 20 years ago.

2.1.4. Synthesis Flexibility

Besides these synthesis and processing methods, it must be emphasized that the flexibility of sol–gel processes permits to enlarge the selection of silica aerogel-based materials, which is currently accessible. Bulk architecture can be tailored by templating techniques [85]. The gel chemistry can be modified by *grafting*, either during [86] (Figure 2.8) or after *gelation* [87]. Composites and nanocomposites can be elaborated by impregnation of *foam*s (Figure 2.9) or fibrous networks (Figure 2.10), by dispersion of particles [88], powders [89], or polymers [90], or by synthesis of mixed silica-based oxides [91, 92]. Organic silica hybrids [93] can also be made either by many techniques such as *cogelation* and crosslinking [94] or by reaction with functionalized particles [95].

After the drying stage, a wide panel of posttreatments can also be applied to increase the huge application potential of silica aerogel-based materials. For example, chemical modifications by *grafting* in solution after re-impregnation [96] or in a gaseous a*tmo*-*s*phere [97], skeleton coating by chemical vapor infiltration [98], impregnation of the bulk porosity with reactive species (Figure 2.11), embedding in polymers (Figure 2.12), mechanical engineering by milling, cutting, laser micromachining [100], and thermal processing such as sintering [101] can be performed to target specific applications.

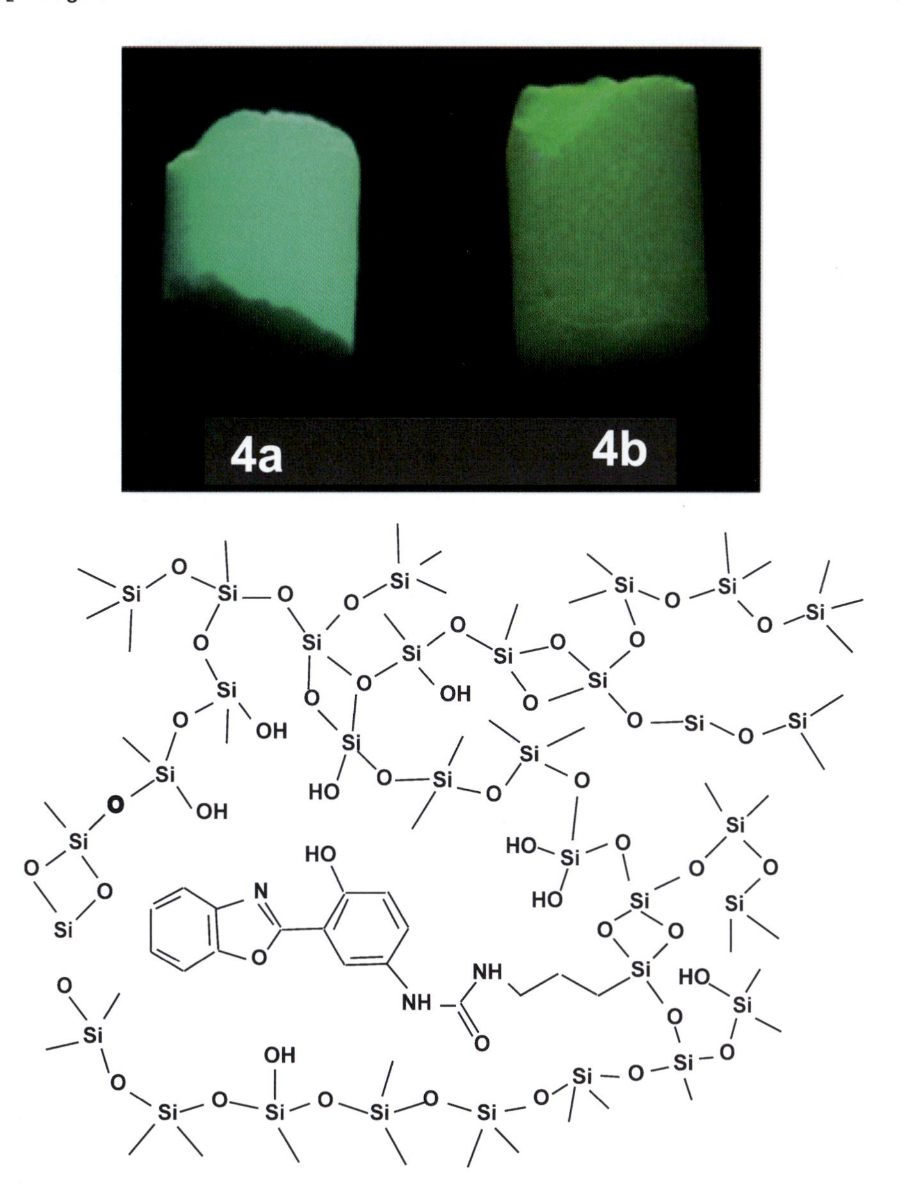

Figure 2.8. Monolithic fluorescent silica aerogels obtained by reaction of silyl-functionalized benzazoles dyes with polyethoxydisiloxane in isopropanol under HF catalysis [86]. *Courtesy of Stefani W (UFRGS, Porto Alegre, Brazil) and Rigacci A.*

2.2. Main Properties and Applications of Silica Aerogels

2.2.1. Texture

Silica aerogels are *amorphous materials*. They have a skeletal density, as measured by *Helium pycnometry* [102] ~2 g cm^{-3}, close to that of amorphous silica (2.2 g cm^{-3}). They typically have a pore volume above 90% of their whole monolith volume. Some ultraporous

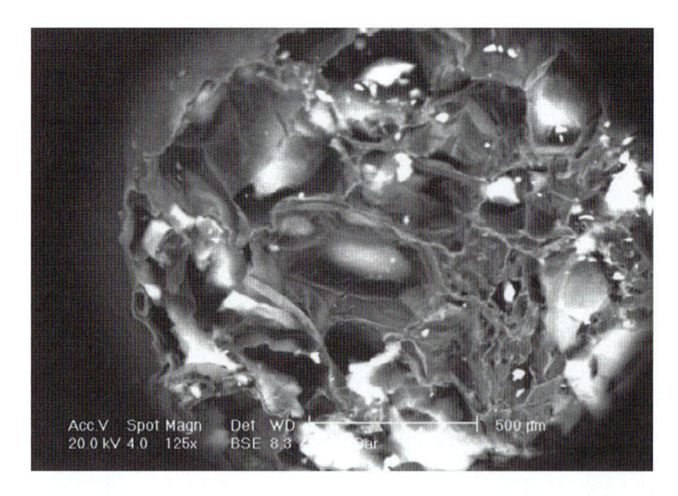

Figure 2.9. Scanning electron microscopy of an organic–inorganic composite obtained by impregnation of a cellular open-cell organic *foam* with a silica sol. *Courtesy of RepouxM., MINES ParisTech, CEMEF, Sophia Antipolis, France and Rigacci A.*

Figure 2.10. Blanket-type composite obtained by casting of a silica sol on an unwoven mineral fiber network. *Courtesy of Repoux M, MINES ParisTech, CEMEF, Sophia Antipolis, France and Rigacci A.*

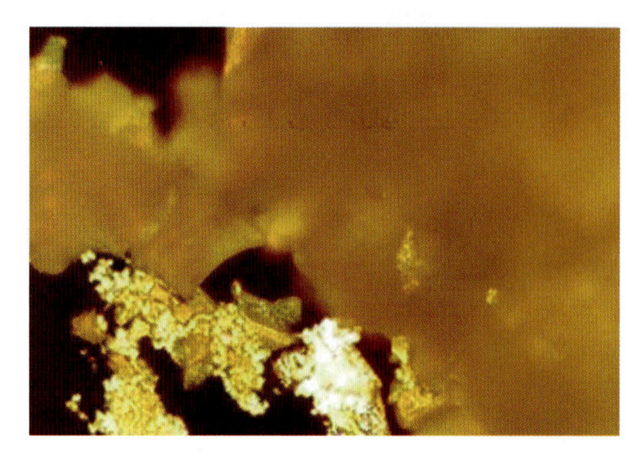

Figure 2.11. Polarized light microscopy of polyacrylate–silica aerogel composites obtained by photo*polymerization* of octylcyanobiphenyl liquid crystal infiltrated in the porosity of the aerogel (in the *cliché*, liquid crystal and silica aerogel are, respectively, the brilliant and opaque phases). *Courtesy of Pesce da Silveira N (UFRGS, Porto Alegre, Brazil) and Rigacci A.* [99].

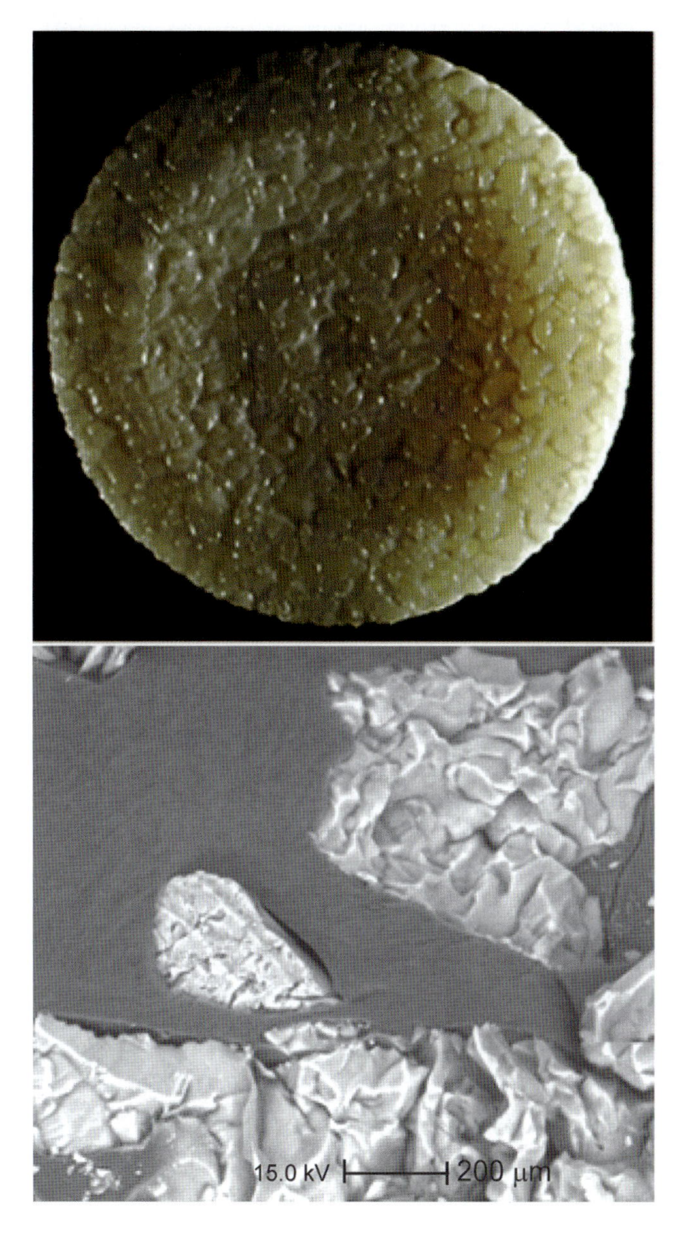

Figure 2.12. Photography (**top**) and scanning electron microscopy (**bottom**) of sylilated silica aerogel beads embedded by a polymer. *Courtesy of Rigacci A.*

and ultralight silica aerogels can be synthesized by *two-step process*, and a density as low as 0.003 g cm^{-3} has been reported [40]: these are the lightest silica aerogels that can be found in the literature.

Silica aerogels are usually largely mesoporous, with interconnected pore sizes typically ranging from 5 to 100 nm and an average pore diameter between 20 and 40 nm. Micropores (i.e., pore sizes < 2 nm) become significant in aerogels synthesized under acid catalysis conditions or having undergone particular treatments [52]. The associated *specific surface area* is rather high, typically from 250 to 800 m^2 g^{-1} and can exceed 1,000 m^2 g^{-1} [28].

As already underlined in the first part of the text, the mesoporosity can be controlled through the sol–gel process conditions, for example, when a two-step catalysis is applied to standard *alkoxide*s such as *TMOS* [36, 82, 103]. If these simple tetra-*alkoxide*s are mixed with a more exotic functionalized Si precursor, differences in the *hydrolysis* and *condensation* rates of the two precursors may drastically influence the final *texture* of the material. For example, when the functionalized precursor carries basic moieties such as an amine in 3-(2-aminoethylamino)propyltrimethoxysilane (EDAS), 3-aminopropyltriethoxysilane (AES), or 3-aminopropyltrimethoxysilane (AMS), these functionalized precursors can act as *nucleation center*s for *condensation* and can lead to generation of large *macropores* [104, 105].

The architecture of silica gel networks is often described as that of a fractal geometry [106] and it is possible to distinguish mass from surface fractalities. In the former case, the mass M of a gel inside a sphere of radius R, centered about a random point in the gel network, is a statistical function of R of the type (2.5).

$$M \approx R^f, \tag{2.5}$$

where f is termed the *fractal dimension* because this is not an integer.

For a surface fractal object, the surface area A follows the law (2.6):

$$A \approx R^{fS} \tag{2.6}$$

The *fractal dimension* can be experimentally determined by *adsorption* of molecules of different cross-sectional area, by small or ultra-small-angle X-ray *scattering* (*SAXS*/*USAXS*) or by small-angle neutron *scattering* (*SANS*) spectroscopy [35, 107]. These experimental results support various theoretical fractal models summarized by Brinker and Scherer [1, 108] (Chap. 24). In theory, true fractal structures can only exist near the *gel point*. In real silica gels, a fractal description only applies over a limited length scale range from one to hundreds of nm, which in detail depends on the exact structure of the aerogel. Silica aerogels made by Einarsrud et al., for instance, had a fractal network in the microporous range with an average mass fractal exponent $f \approx 1.9$ consistent with a "Diffusion Limited Cluster Aggregation" (*DLCA*) model [20]. *SAXS*/*USAXS* and *SANS* are also widely used to characterize elementary particles and/or distributions of clusters that constitute the silica skeleton as well as other structural features such as the *specific surface area*. These results generally give larger values than their *BET* analogues (Chap. 21).

Even if solid and porous networks of silica aerogels are essentially tailored during the sol–gel step, posterior treatments also influence the overall structure of the materials. As already mentioned in Sect. 2.1.3, some shrinkage occurs during evaporative drying due to the presence of intense *capillary stresses*, but this also sometimes holds for supercritical drying. Indeed, some irreversible shrinkage also occurs due to the restructuration of the gel network by dissolution–reprecipitation of SiO_2 in supercritical media and also because of the stresses during solvent exchange processes [35, 40]. All these kinds of shrinkages affect the porous network, but when applied to a wet gel, supercritical drying mostly affects the larger pores that involve network dimensions beyond the fractal scale, while drying by evaporation to obtain a xerogel in addition leads to a drastic shrinkage of the *mesopores*. Furthermore, concerning supercritical drying, the silica aerogels obtained by the *HOT* process tend to possess a lower specific microporosity but a similar specific mesoporosity to those obtained by the *COLD* process [35, 109, 110].

The porous network characteristics (specific pore volume, mean pore size, pore size distribution, etc.) reported in the literature are usually obtained either by *adsorption* of nitrogen [111] or by non-intrusive mercury *porosimetry* [112] and are discussed in detail in Chap. 21. An example of pore size distribution obtained by non-intrusive mercury *porosimetry* is presented in Figure 2.13.

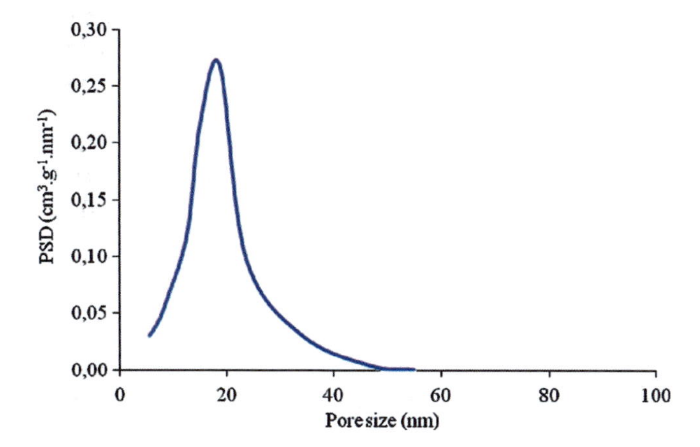

Figure 2.13. Pore size distribution of a 0.15 g cm^{-3} silica aerogel obtained by non-intrusive mercury *porosimetry* on samples synthesized in alcohol via a *two-step process* with *TetraEthOxySilane* (using a buckling constant k_f of 28 nm $MPa^{1/4}$). *Courtesy of Rigacci A.*

2.2.2. Chemical Characteristics

After drying, untreated (e.g., not sylilated) silica aerogels contain residual reactive groups (e.g., alkoxy and *hydroxyl* groups). Fourier-transformed infrared spectra (*FTIR*) reveals a *silanol* surface coverage (≡Si–OH) of typically 4–6 groups per nm^2 [113], which gives them a certain hydrophilic behavior. This parameter is influenced by the specific sol–gel conditions used as well as by the drying route. Concerning drying, *silanol* groups condense during evaporation while *HOT* supercritical processing induce re-*esterification* of the *silanol* functionalities. This makes both *xerogels* and aerogels more hydrophobic after drying is complete.

Indeed, for many applications, it is necessary that aerogels be hydrophobic to not absorb (even adsorb) water vapor. The methods to prepare hydrophobic silica aerogels were intensively studied by many authors (Schwertfeger et al. [56], Schuth et al. [114], Venkateswara Rao et al. [115–119], Lee et al. [14], Kim and Hyun [120], Shi et al. [121], and Hwang et al. [12]). Good reviews of the methods to synthesize such materials and of the characteristics achieved are presented in Chaps. 3 and 4. Briefly, to elaborate hydrophobic silica aerogels, the pore surface must be covered with nonpolar side functions such as $\equiv Si–CH_3$, which can be introduced by several methods. For instance, a silicon precursor containing at least one nonpolar chemical group, of the type $RSi(OR')_3$ where R and R' are alkyl groups, can be cogelled in various proportions with *TMOS* or *TEOS*. For example, when some hybrid gels are synthesized with *TMOS* and *MTMS*, the *silanol* concentration can be decreased. *MTMS*, for instance, is more difficult to condense than *TMOS*. Hence, the gel network is then mostly made by *TMOS*, while in the second stage $\equiv Si–CH_3$ end groups cover the pore surface [122]. Of course, hydrophobization can also be realized after supercritical drying, for instance, by treating the aerogel with hexamethyldisilazane (*HMDS*) [123], but it is now far less studied than *sylilation* in solution. As already mentioned in Sect. 2.1.3 dedicated to drying, these *sylilation* methods permit to promote spring-back effect and to obtain ambient-dried aerogel while giving them specific surface properties.

2.2.3. Physical Properties and Some Related Applications

Thermal Conductivity

One of the major characteristics of silica aerogels is their very low *thermal conductivity*, typically of the order of 0.015 W m^{-1} K^{-1} at ambient temperature, pressure, and relative humidity. These values are significantly lower than the conductivity of air under the same conditions, e.g., 0.025 W m^{-1} K^{-1}. Thus, silica aerogels are among the best-known thermal insulating materials [124]. Besides, silica is nonflammable and silica aerogels are amorphous.

Moreover, silica aerogels can be made optically transparent, although they are also very brittle. Consequently, they present an amazing application potential for opaque or transparent insulating components [125, 126] as well as daylighting devices, if they can be mechanically reinforced. The physics governing their thermal properties, as well as their use in thermal insulation (Chap. 26) and the first related commercial products (Part XV), is described later in the handbook (Chap. 23). As an example for a high-tech product, it shall be underlined that silica aerogels find ever increasing use as thermal insulator in the aeronautical and aerospace domains [127]. They were, for instance, used in the recent PATHFINDER MARS mission to insulate the Sojourner Mars Rover. During the mission, the nocturnal temperature dropped down to −67°C, while a stable inside temperature of 21°C was maintained. This permitted to protect the Rover's very sensitive electronics from damage by the cold. For a similar program termed European Retrieval Carrier (EURECA) satellite, the use of aerogels has been investigated [128].

Of course, it must be underlined here that applications in space are not at all limited to thermal insulation. Indeed, silica aerogels can be applied to collect *aerosol* particles [129], to protect space mirrors, or to design *tank baffles* [130, 131]. The applications in space were reviewed by Jones [132] (Part XI). The most recent project, Stardust, successfully returned to earth in January 2006. This mission provided samples of a recently deflected comet named Wild-2, which are being examined in various laboratories all over the world.

Optical Properties

The optical transmission and *scattering* properties of silica aerogels constitute another group of important characteristics, sometimes in conjunction with their thermal properties when a transparent thermal insulation is targeted, such as in windows. A first review of this subject was published by Pajonk [133]. More recent works comprise those of Buzykaev et al. [134], Danilyuk et al. [135], Venkastewara Rao et al. [115], Jensen et al. [136], Schultz et al. [59], and Adachi et al. [137]. The transparency and visible light transmittance of silica aerogels can be high, although they all tend to scatter the transmitted light to some extent, which reduces their optical quality [138]. The *Rayleigh scattering* due to the solid gel network heterogeneities in the nanometer range is responsible for a yellowish coloration of silica aerogel observed in transmission and a bluish coloration when observed in reflection mode against a dark background. The *scattering* due to heterogeneities in the micrometer range is responsible for a blurred deformation of optical images [139]. A two-step *gelation* catalysis procedure, previously described, was found to give satisfactory transparency results. Typically, aerogels made from *TMOS* in methanol can be obtained with an optical transmittance ratio up to 93% (for ~1 cm thick aerogels) at a wavelength of 900 nm [140, 141]. Silicon precursors prehydrolyzed with an acid catalyst and a substoichiometric water molar ratio $w = n_{H_2O}/n_{Si} < 2$ are available on the market [124, 142] and make it possible to control a more uniform porous *texture*, as is needed for optical applications. Transparency ratio up to 90% together with specific extinction coefficient of the order of 15 m^{-1} [143] can be obtained with 1 cm thick aerogels synthesized with such prepolymerized precursors (Figure 2.14).

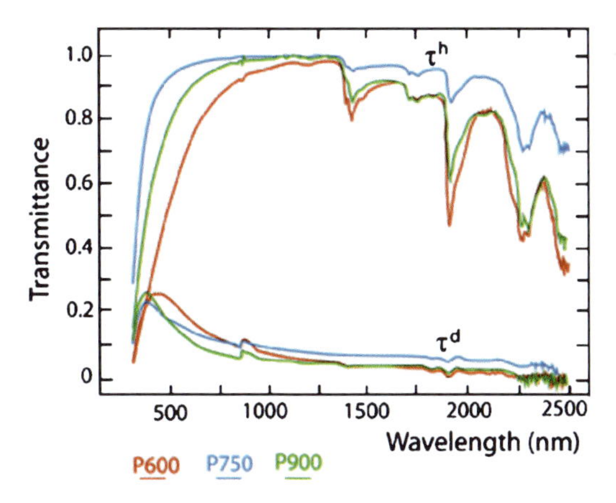

Figure 2.14. Hemispherial (τ^h) and diffuse (τ^d) transmittances of 1 cm thick plates of silica aerogels synthesized by a *two-step process*, from prepolymerized silica precursors made by PCAS, Longjumeau, France, with a precursor volume fraction in solution of 0.5, dried by a supercritical CO_2 method. After Rigacci et al. [143].

The *refractive index* of silica aerogels increases with their density ρ (kg m^{-3}) according to (2.7) [144]:

$$n - 1 = 2.1 \times 10^{-4}\rho. \tag{2.7}$$

Hence, n is very close to 1 [145], which makes aerogels excellent materials to apply in the radiator of *Cerenkov counters* [146]. In a Cerenkov counter, the radiator is a low-density medium such as an aerogel slab, in which electrically charged elementary particles travel with a velocity "v" higher than the velocity of light "c" and from where they radiate (emit) photons. An analysis of these photons can therefore be used to derive the velocity of the particles and hence their nature. This was one of the well-known historical uses of silica aerogels. Kharzheev published a recent review on the use of silica aerogels, for this type of application, with a description of their optical and physical characteristics [147]. Moreover, a summary of the operational experience from 1998 to 2007, with silica aerogel tiles used in *RICH*-type *Cerenkov counters*, for the HERMES experiment at the DESY – HERA facility, was published by de Leo [148]. The average *refractive index* value (n) of these tiles at a wavelength of 633 nm was equal to 1.0304 ± 0.0004. The aerogel optical quality was characterized by its light *attenuation length* Λ at 400 nm. This characteristic is defined as the distance into an aerogel, where the probability that a photon having this wavelength has not been absorbed has dropped by $1/e$ (where $e = 2.71828$). Clearly, this property depends on the aerogel density and it increased from $\Lambda \approx 1$ cm in the 1980s, to ≈ 2 cm in the early 1990s, to reach ~4.5 cm in recent larger aerogel tiles ($20 \times 20 \times 4$ cm^3).

Acoustic Properties

The acoustic properties of silica aerogels are closely related to their thermal insulation properties. The acoustic propagation in aerogels depends on the interstitial gas nature and pressure, the aerogel density, and more generally the *texture* [149]. Silica aerogels are indeed excellent acoustic insulators. The propagation of an acoustic wave is attenuated both in amplitude and velocity because the wave energy is progressively transferred from the gas to the aerogel solid network, over the entire aerogel workpiece thickness [150]. The longitudinal *acoustic velocity* is typically of the order of 100 m s^{-1} [151, 152], which makes silica aerogels suitable for applications in acoustic devices.

Mechanical Properties

A fourth group of important characteristics has to deal with their mechanical properties (Chap. 22). The *compressive strength*, *tensile strength*, and *elastic modulus* of silica aerogels are very low and largely depend on the *network connectivity* [153] and aerogel density [154]. Indeed, silica aerogels can easily be elastically compressed when the porous *texture* is studied, for example, by the *capillary stresses* that they are subjected to during nitrogen *adsorption* or desorption [155] or by compression during mercury *porometry* and *thermoporometry* [83, 156]. The magnitude of the contraction can reach ~50% by length [157].

As previously mentioned, many synthetic routes are investigated with the goal to strengthen silica aerogels, the most popular approaches of which are *aging* [51] and hybridization. For example, hybrid silica aerogels can be made for strengthening purposes by mixing a silicon *alkoxide* with hybrid precursors such as polydimethyl siloxane (*PDMS*). Gels obtained in this way are termed *ORMOSIL* hybrids. They have a more rubber-like flexibility. With 20% (by mass) *PDMS*, they can be elastically compressed to 30% (by volume) with no damage [158]. Some flexible silica aerogels with superhydrophobic properties were also made from *MTMS* by A. Venkateswara Rao et al. [116]. At last, polymeric crosslinking of the silica skeleton can also permit to drastically increase the initial mechanical characteristics [159].

In addition, the mechanical properties are sensitive to the synthesis chemistry of the gel, the environment, and the storage history [160]. With age, the *compressive strength* and elastic moduli increase while the strain at fracture decreases. The environment is responsible for *subcritical crack growth*, which proceeds by a stress corrosion mechanism analogous to the phenomenon observed in dense silica glass subjected to alkaline aqueous solutions. Crack growth can be considerably reduced in *hydrophobic aerogels* [161], although an apparently fully hydrophobic character does not totally impede water molecules from reaching the crack tip [162].

In spite of their low density and brittle characteristics, the mechanical properties of silica aerogels are quite promising for some rare but specific applications. As an illustration, their good compressibility can be used advantageously for absorbing the kinetic energy involved in a compressive shock [134, 163]. Silica aerogels are useful low *shock impedance* materials to confine few eV plasmas, to capture frozen states of minerals generated under high shock pressures, or as medium to study the mixing of fluids requiring X-ray-induced shocks [134, 163, 164].

Dielectric Properties

The *relative dielectric constant* of silica aerogels can be as low as 1.1 [165]. Hence, thin film silica aerogels could be and are being considered as super-low dielectric constant material for integrated circuits in computers. It is also possible to modify the surface of silica aerogel to obtain good *electret* materials (i.e., materials that produce a permanent external electric field) [166].

Entrapment, Release, Sorption, and Storage

The combination of a high specific pore volume, with – in some specific cases – a relatively resistant solid SiO_2 network, can also be advantageously used to entrap a large variety of molecules or nanoparticles. The entrapped species can be luminescent molecules or particles [167–172].

Hollow silica aerogel droplets were also proposed by Kim et al. [173], Jang et al. [174], and Kim and Jang [175] for the inertial confinement of fluids (*ICF*), namely, mixtures of liquid deuterium and tritium, as the target in fusion experiments under a powerful laser. The droplets were generated by a double nozzle reactor; they were stable to radiation and could be wetted by deuterium and tritium. In aerogels, the gel network acts as an impurity for the fluids that strongly interact with the solid surface. Hence, large fluid NMR signals with long polarization lifetimes can be recorded, revealing a very weak specific wall relaxation [176]. This effect can be used to study either the gel itself or the trapped fluid. The gel network was, for instance, studied by impregnation with liquid ^{131}Xe near its critical temperature (T_c of approximately 289 K). Magnetic resonance images were obtained, which made it possible to differentiate between aerogels of different densities and hydration levels [177]. As far as studies on fluid flow are concerned, the interaction between *superfluid* ^{3}He and the silica aerogel network has received significant attention and was reviewed by Halperin et al. [178]. The solid network introduced disorders in these interactions, thus allowing us to suppress the *superfluid* transition temperature.

Similarly, when an aerogel is impregnated with a *nematic liquid crystal*, the gel network randomness destroys the crystal long-range order and transforms the liquid crystal into a *glassy state* [179].

Biomaterials can also be successfully immobilized inside aerogels. Besides an early study by Antczak et al. [180], Pierre et al. [181, 182] described the in situ encapsulation of the *Pseudomonas cepacia* lipase into a hydrophobic silica aerogel on which they tested various enzyme catalyzed *esterification* and *transesterification* reactions. The authors observed that, contrary to the same silica matrix but in the form of an ambient-dried aerogel, the material showed a much higher activity. Because of the larger pore size, each lipase molecule was indeed able to operate as a free isolated molecular catalyst, while agglomeration of the catalyst was made impossible by dispersion in the aerogel network. Bacteria can also be trapped inside aerogels and still remain alive. Power et al. [183] described a *biosensor* under the form of an *aerosol* composed of aerogel dust particles, containing

Escherichia coli and the Green Fluorescent Protein (GFP) obtained from the jellyfish Aequorea Victoria. When a virus, like the bacteriophage T7 polymerase promoter also under the form of an *aerosol*, contacted the bacteria, a green fluorescent light was emitted.

The controllable pore size and high specific pore volume of silica aerogels make them also ideal candidates for releasing medical drugs or agriculture chemicals (fungicides, herbicides, and pesticides) [184] in a controlled fashion. Hydrophilic silica aerogels can be loaded with chemicals during the sol–gel synthesis process or by posttreatment of dried aerogels [185].

Inversely, aerogels can be used to adsorb or extract some chemical compounds, for instance, to treat waste water, to confine radioactive waste [134], or to filter gases [186]. Silica aerogels were proposed by Reynes et al. [187] for the storage of long life actinide wastes because they are chemically very stable with time on stream and they possess a very large relative pore volume. Besides, aerogels could easily be converted to vitreous silica after a short heat treatment at $\approx$ 1,000°C [188, 189]. The experiments also showed that these aerogels were able to store the waste much longer than the conventional borosilicate glass.

Silica aerogels impregnated with $CaCl_2$, LiBr, and $MgCl_2$ salts have been tested too as water sorbents for heat storage at low temperature. Their energy storage ability E measured by differential scanning calorimetry (*DSC*) can reach 4.0 kJ g^{-1}, which is much higher than for common sorbents such as *zeolites* and unimpregnated silica gels [190]. Partially sintered aerogels, which are mechanically stronger to resist *capillary stresses*, are convenient to store or transport liquids such as rocket fuel [191]. Hazardous liquids such as red fuming nitric acid and unsymmetric 1,1-dimethyl hydrazine (UDMH), both used as rocket fuels, have been stored with silica aerogels [192].

Aerogel particles can also be used as the dispersed phase in composite materials, such as *elastomers* for tires or paints. The incorporation of aerogels will provide them with additional hardness, resistance to wear, and also exert a thickening effect on the mixture [135]. As colloidal abrasives, they even show some insecticide properties because they remove the protecting lipid layer on insects [193].

2.3. Conclusion

Clearly, silica aerogels occupy a dominant place within aerogels. They are used in many important applications and several chapters of this handbook are dedicated to some of them (Chaps. 26, 28, 30–34) or to a detailed review of important properties that sustain these applications (Chaps. 21–23). These fields are still rapidly growing, primarily because of their potential impact in sustainable technology developments. Moreover, new possible applications are appearing such as CO_2 sequestration, as recent studies have shown that enhanced carbonation reactions could be observed in SiO_2-based aerogels under ambient conditions [89]. The amazing class of silica aerogel materials has for a long time been considered as the most fascinating examples of nanostructured porous materials. It appears now that a massive commercialization may be imminent while fundamental studies still continue.

Acknowledgments

One of the authors would like to warmly acknowledge the European Commission, the French Agency for Environment and Energy Management (ADEME), the French National Research Agency (ANR), and ARMINES (The Contract Research Association of MINES

Schools) for their financial support since the early 1990s through different projects (such as HILIT, HILIT+, PACTE Aerogels, and ISOCOMP), the historical partners among which PCAS (Longjumeau, France), CSTB (Grenoble, France), LACE (Lyon, France), NTNU (Trondheim, Norway), ULG (Liège University, Belgium), as well as Yasmine Masmoudi and Antoine Bisson plus Jihan Jabbour for their respective PhD and Post-Doc researches on silica aerogels, and, last but not least, Patrick Achard, Head of the EM&P team in which A. Rigacci has been studying aerogels for energy applications since 1994.

References

1. Brinker CJ, Scherer GW (1990) Sol-Gel Science. The Physics and Chemistry of Sol-Gel Processing. Academic Press, New-York
2. Pierre AC (1998) Introduction to Sol-Gel Processing Kluwer, Boston
3. Soleimani DA, Abbasi MH (2008) Silica aerogel; synthesis, properties and characterization. J Mater Proc Technol 199:10–26
4. Pierre AC, Pajonk GM (2002) Aerogels and their applications. Chem Rev 102:4243–4265
5. Ambekar AP, Bagade P (2006) A review on: 'aerogel - world's lightest solid'. Popular Plastics & Packaging, 51:96–102
6. Kocon L, Phalippou J (2005) Aerogels. Material aspect Techniques de l'Ingenieur, Sciences Fondamentales, AF196:AF3610/1-AF3610/21
7. Carraher CE Jr (2005) General topics: silica aerogels-properties and uses. Polymer News, 30(12), 386–388
8. Carraher CE Jr (2005) Silica aerogels - synthesis and history. Polymer News, 30:62–64
9. Pajonk GM (2003) Some applications of silica aerogels. Colloid and Polymer Science, 281:637–651
10. Akimov YK (2003) Fields of Application of Aerogels (Review) Instruments and Experimental Techniques (Translation of Pribory i Tekhnika Eksperimenta), 46:287–299
11. Venkateswara Rao A, Parvathy Rao A, Kulkarni MM (2004) Influence of gel aging and Na_2SiO_3/H_2O molar ratio on monolithicity and physical properties of water–glass-based aerogels dried at a*tmos*pheric pressure. J Non-Cryst Solids 350:224–229
12. Hwang S-W, Jung H-H, Hyun S-H, Ahn Y-S (2007) Effective preparation of crack-free silica aerogels via ambient drying. J Sol Gel Sci Technol 41, 139–146
13. Kistler SS (1932) Coherent expanded aerogels. J Phys Chem 36:52–64
14. Lee CJ, Kim GS, Hyun SH (2002) Synthesis of silica aerogels from waterglass via new modified ambient drying. J Mater Sci 37:2237–2241
15. Tang Q, Wang T (2005) Preparation of silica aerogel from rice hull ash by supercritical carbon dioxide drying. J Supercrit Fluids 35:91–94
16. Li T, Wang T (2008) Preparation of silica aerogel from rice hull ash by drying at a*tmos*pheric pressure. Materials Chemistry and Physics 112:398–401
17. Nakanishi K, Minakuchi H, Soga N, Tanaka N (1998) Structure Design of Double-Pore Silica and Its Application to HPLC. J Sol Gel Sci Technol 13:163–169
18. Wagh PB, Begag R, Pajonk GM, Venkasteswara Rao A., Haranath D (1999) Comparison of some physical properties of silica aerogel monoliths synthesized by different precursors. Mater Chem Phys 57:214–218
19. Ebelmen M (1846) Recherches sur les combinaisons des acides borique et silicique avec les éthers. Ann Chim Phys 16:129–166; (1847) Sur l'hyalite artificielle et l'hydrphane. C R Acad Sci Paris 25:854–856
20. Einarsrud MA, Nilsen E, Rigacci A, Pajonk GM, Buathier S, Valette D, Durant M, Chevalier P, Nitz P, Ehrburger-Dolle F (2001) Strengthening of silica gels and aerogels by washing and aging processes J. Non-Cryst. Solids 285:1–7
21. Deng Z, Wang J, Wei J, Shen J, Zhou B, Chen L (2000) Physical Properties of Silica Aerogels Prepared with Polyethoxydisiloxane. J Sol Gel Sci Technol 19:677–680
22. Venkastewara Rao A, Bhagat SD, Hirashima H, Pajonk GM (2006) Synthesis of flexible silica aerogels using methyltrimethoxysilane (MTMS) precursor. J Colloid Interface Sci 300:279–285
23. El Rassy H, Buisson P, Bouali B, Perrard A, Pierre AC (2003) Surface Characterization of Silica aerogels with Different Proportions of Hydrophobic Groups, dried by the CO_2 Supercritical Method; Langmuir, 19:358–363
24. Harreld JH, Ebina T, Tsubo N, Stucky G (2002) Manipulation of pore size distributions in silica and ormosil gels dried under ambient pressure conditions. J Non-Cryst Solids 298:241–251

25. Allié C, Pirard R, Lecloux AJ, Pirard JP (1999) Preparation of low-density *xerogels* through additives to TEOS-based alcogels. J Non-Cryst Solids 246:216–228
26. Rodriguez SA, Colon LA (1999) Investigations of a sol–gel derived stationary phase for open tubular capillary electrochromatography. Anal Chim Acta 397:207–215
27. Venkateswara Rao A, Haranath D (1999) Effect of methyltrimethoxysilane as a synthesis component on the hydrophobicity and some physical properties of silica aerogels. Microporous Mesoporous Mater 30:267–273
28. Zhou B, Shen J, Yuehua W, Wu G, Ni X (2007) Hydrophobic silica aerogels derived from polyethoxydisiloxane and perfluoroalkylsilane. Mater Sci Eng C 27:1291–1294
29. Pauling L (1960) The Nature of Chemical Bonds, 3rd ed. Cornell University Press, USA
30. Mozzi RL, Warren BE (1969) Structure of vitreous silica. J Appl Cryst 2:164–172
31. de la Rosa-Fox N, Esquivias L, Craievich AF, Zarzycki J (1990) Structural study of silica sonogels. J Non-Cryst Solids 121:211–215
32. Vollet DR, Nunes LM, Donatti DA, Ibanez Ruiz A, Maceti H (2008) Structural characteristics of silica sonogels prepared with different proportions of TEOS and *TMOS*. J Non-Cryst Solids 354:1467–1474
33. Livage J, Henry M, Sanchez C (1988) Sol-Gel Chemistry of Transition Metal Oxides. Prog. Solid State Chem.18:259–341
34. Nicolaon GA, Teichner SJ (1968) Preparation of silica aerogels from methyl orthosilicate in alcoholic medium, and their properties. Bull Soc Chim Fr 1906–1911
35. Moner-Girona M, Roig A, Molins E (2003) Sol-Gel Route to Direct Formation of Silica Aerogel Microparticles Using Supercritical Solvents. J Sol Gel Sci Technol 26:645–649
36. Mezza P, Phalippou J, Sempere R (1999) Sol–gel derived porous silica films. J Non-Cryst Solids 243:75–79
37. Dieudonne P, Hafidi Alaoui A, Delord P, Phalippou J (2000) Transformation of nanostructure of silica gels during drying. J Non-Cryst Solids 262:155–161
38. Suh DJ, Park TJ, Sonn JH, Lim JC (1999) Effect of aging on the porous texture of silica aerogels prepared by NH4OH and NH4F catalyzed sol-gel process. J Mater Sci Lett 18:1473–1475
39. Boonstra AH, Bernards TNM (1988) The dependence of the gelation time on the hydrolysis time in a two-step SiO_2 sol-gel process. J Non-Cryst Solids 105:207–213
40. Kocon L, Despetis F, Phalippou J (1998) Ultralow density silica aerogels by alcohol supercritical drying. J Non-Cryst Solids 225:96–100
41. Begag R, Pajonk GM, Elaloui E, Chevalier B (1999) Synthesis and properties of some monolithic silica carbogels produced from polyethoxydisiloxanes dissolved in ethylacetoacetate and acid catalysis. Materials Chemistry and Physics 58:256–263
42. Dai S, Ju YH, Gao HJ, Lin JS, Pennycook SJ, Barnes CE (2000) Preparation of silica aerogel using ionic liquids as solvents. Chem Commun 243–244
43. Karout A, Pierre AC (2009) Influence of ionic liquids on the texture of silica aerogels. J Sol Gel Sci Technol 49:364–372
44. Karout A, Pierre AC (2007), Silica *xerogels* and aerogels synthesized with ionic liquids, J. Non-Cryst Solids 353:2900–2909
45. M.V. Migliorini, R.K. Donato, M.A. Benvegnu, R.S. Gonçalves, H.S. Schrekker (2008) J. Sol-Gel Sci Technol 48:272–276
46. Einarsrud M-A, Kirkedelen MB, Nilsen E, Mortensen K, Samseth J (1998) Structural development of silica gels aged in TEOS. J of Non-Cryst Solids 231:10–16
47. S. Haereid, E. Nilsen, V. Ranum, M.-A. Einarsrud (1997) Thermal and temporal aging of two steps acid-base catalyzed silica gels in water/ethanol solutions, J. Sol-Gel Sci and Tech 8:153–157
48. Smitha S, Shajesh P, Aravind PR, Rajesh Kumar S, Krishna Pillai P, Warrier KGK (2006) Effect of aging time and concentration of aging solution on the porosity characteristics of subcritically dried silica aerogels. Microporous Mesoporous Mater 91:286–292
49. Estella J, Echeverria JC, Laguna M, Garrido JJ (2007) Effects of aging and drying conditions on the structural and textural properties of silica gels. Microporous Mesoporous Mater 102:274–282
50. Strøm RA, Masmoudi Y, Rigacci A, Petermann G, Gullberg L, Chevalier B, Einarsrud M-A (2007) Strengthening and aging of wet silica gels for up-scaling of aerogel preparation. J Sol Gel Sci Technol 41:291–298
51. Einasrud MA, Dahle M, Lima S, Hæreid S (1995) Preparation and properties of monolithic silica *xerogels* from TEOS-based alcogels aged in silane solutions. J Non-Cryst Solids 186:96–103
52. Reichenauer G (2004), Thermal aging of silica gels in water. J Non-Cryst Solids 350:189–195
53. Pajonk GM (1989) Drying methods preserving the textural properties of gels. Revue de Physique Appliquée 24(C4):13–22

54. Bisson A, Rigacci A, Lecomte D, Rodier E, Achard P (2003) Drying of silica gels to obtain aerogels : phenomenology and basic techniques, Progrcss in Drying Technologies, Vol. 4, - special issue of Drying Technology, 21, Number 4: 593–628
55. Egeber ED, Engel J (1989) Freeze-drying of silica gels prepared from siliciumethoxide. Revue de Physique Appliquée 24(C4):23–28
56. Schwertfeger F, Frank D, Schmidt M (1998) Hydrophobic waterglass based aerogels without solvent exchange or supercritical drying. J Non-Cryst Solids 225:24–29
57. Allié C, Tcherkassova N, Ferrauche F, Lambert S, Heinrich B, Pirard R, Pirard JP (2006), Multigram scale synthesis and characterization of low-density silica *xerogels*. J Non-Cryst Solids 352:2763–2771
58. Duffours L, Woignier T, Phalippou J (1995) Plastic behavior of aerogels under isostatic pressure. J Non-Cryst Solids 186:321–327
59. Schultz JM, Jensen KI, Kristiansen FH (2005) Super insulating aerogel glazing. Solar Mater Solar Cells 89:275–285
60. Jensen KI, Kristiansen FH, Schultz JM (2005) Highly superinsulating and light transmitting aerogel glazing for superinsulating windows, Publi Final Report HILIT+ (Eu contract ENK-CT-2002-00648)
61. Henning S (1985) Large-scale production of Airglass. In: Fricke J (ed) Aerogels, Springer-Verlag, Berlin, Heidelberg, New-York, Tokyo
62. Tewari PH, Hunt AJ, Lofftus KD (1985) Ambient-temperature supercritical drying of transparent silica aerogels. Mater Lett 3:363–367
63. van Bommel MJ, de Haan AB (1995) Drying of silica aerogel with supercritical carbon dioxide. J Non-Cryst Solids 186:78–82
64. Novak Z, Knez Z, Hadolin M (1999) Drying of silica aerogels with liquid and supercritical CO2. Recents Progres en Genie des Procedes 13:115–122
65. Wawrzyniak P, Rogacki G, Pruba J, Bartczak Z (2001) Effective diffusion coefficient in the low temperature process of silica aerogel production. J Non-Cryst Solids 285: 50–56
66. Scherer GW (1994) Stress in aerogel during depressurization of autoclave : I. Theory. J of Sol-Gel Science and Technology 3:127–139
67. Woignier T, Scherer GW (19994) Stress in aerogel during depressurization of autoclave : II. Silica gels. J Sol Gel Sci Technol 3:141–150
68. Lee K, Begag R (2001) Rapid aerogel production process, WO 01/28675 A1
69. Poco JF, Coronado PR, Pekala RW, Hrubesh LW (1996) A rapid supercritical extraction process for the production of silica aerogels. Mater Res Soc Symp Proc 431:297–302
70. Gross J, Coronado PR, Hrubesh LW (1998) Elastic properties of silica aerogels from a new rapid supercritical extraction process. J Non-Cryst Solids 225:282–286
71. Scherer GW, Gross J, Hrubesh LW, Coronado PR (2002) Optimization of the rapid supercritical extraction process for aerogels. J Non-Cryst Solids 311:259–272
72. Gauthier BM, Bakrania SD, Anderson AM, Carroll MK (2004) A fast supercritical extraction technique for aerogel fabrication. J Non-Cryst Solids 350:238–243
73. Loy DA, Russick EM, Yamanaka SA, Baugher BM, Shea KJ (1997) Direct Formation of Aerogels by Sol-Gel Polymerizations of Alkoxysilanes in Supercritical Carbon Dioxide. Chem Mat 9:2264–2268
74. Sharp KG(1994) A two-component, non-aqueous route to silica gel. J Sol Gel Sci Technol 2:35–41
75. Moner-Girona M, Roig A, Molins E, Llibre J (2003) Sol-Gel Route to Direct Formation of Silica Aerogel Microparticles Using Supercritical Solvents. J Sol Gel Sci Technol 26:645–649
76. Kirkbir F, Murata H, Meyers D, Ray Chaudhuri S (1998) Drying of aerogels in different solvents between a*tmos*pheric and supercritical pressures. J Non-Cryst Solids 225:14–18
77. Martin J, Hosticka B, Lattimer C, Norris PM (2001) Mechanical and acoustical properties as a function of PEG concentration in macroporous silica gels. J Non-Cryst Solids 285:222–229
78. Reetz MT, Zonta A, Simpelkamp J(1996) Efficient immobilization of lipases by entrapment in hydrophobic sol-gel materials. Biotechnol Bioeng 49:527–534
79. Pierre M, Buisson P, Fache F, Pierre A (2000) Influence of the drying technique of silica gels on the enzymatic activity of encapsulated lipase. Biocatal Biotransformation 18:237–251
80. Venkastewara Rao A, Kulkarni MM (2003) Effect of glycerol additive on physical properties of hydrophobic silica aerogels. Mater Chem Phys 77:819–825
81. Anderson MT, Sawyer PS, Rieker T (1998) Surfactant-templated silica aerogels. Microporous Mesoporous Mater 20:53–65
82. Parvathy Rao A, Pajonk GM, Venkastewara Rao A (2005) Effect of preparation conditions on the physical and hydrophobic properties of two step processed ambient pressure dried silica aerogels. J Mater Sci 40:3481–3489

83. Deshpande R, Hua, DW, Smith DM, Brinker CJ (1992) Pore structure evolution in silica gel during aging/ drying. III. Effects of surface tension. J Non-Cryst Solids 144:32–34
84. Land VD, Harris TM, Teeters DC (2001) Processing of low-density silica gel by critical point drying or ambient pressure drying. J Non-Cryst Solids 283:11–17
85. Hüsing N, Schubert U (2002) Aerogels. In: Ullmann's Encyclopedia of Industrial Chemistry, 6th edn. Wiley
86. Rodembusch FS, Campo LF, Stefani V, Rigacci A (2005) The first silica aerogel fluorescent by excited state intramolecular proton transfer mechanism (ESIPT). J Mater Chem 15:1537–1541
87. Smith DM, Deshpande R, Brinker CJ (1992) Preparation of low-density aerogels at ambient pressure. Mat Res Soc Symp Proc Vol. 271 567–572
88. Kuhn J, Gleissner T, Aruini-Schuster MC, Korder S, Fricke J (1995) Integration of mineral powders into SiO_2 aerogels. J Non-Cryst Solids 186:291–295
89. Santos A, Ajbary M, Toldeo-Fernandez JA, Morales-Florez V, Kherbeche A, Esquivias L (2008) Reactivity of CO_2 traps in aerogel-wollastonite composite. J Sol Gel Sci Technol 48:224–230
90. Kulkarni MM, Bandyopadhyaya R, Bhattacharya B, Sharma A (2006) Microstructural and mechanical properties of silica-PEPEG polymer composite *xerogels*. Acta Materialia 54:5231–5240
91. Vicarini MA, Nicolaon GA, Teichner SJ (1970) Propriétés texturales d'aérogels minéraux mixte préparés par hydrolyse simultanée de deux alcoolates métalliques en solution dans un milieu organique. Bulletin de la Société Chimique de France 10:3384–3387
92. Cao S, Yao N, Yeung KL (2008) Synthesis of freestanding silica and titania-silica aerogels with ordered and disordered mesopores. J Sol-Gel Sci and Tech 46:323–333
93. Mosquera MJ, de los Santos DM, Valdez-Castro L, Esquivias L (2008) New route for producing crack-free *xerogels*: obtaining uniform pore size. J Non-Cryst Solids 354:645–650
94. Meador MA, Fabrizio EF, Ilhan F, Dass A, Zhang G, Vassilarias P, Johnston JC, Leventis N (2005) Cross-linking amine-modified silica aerogels with epoxies : mechanically strong lightwight prorous materials. Chem Mater 17:10851098
95. Patwardhan SV, Mukherjee N, Durstock MF, Chiang LY, Clarson SJ (2002) Synthesis of C-60 fuellerene-silica hybrid nanostructures. Journal of Inorganic and Organometallic Polymers 12:49–55
96. Silveira F, Pires GP, Petry CF, Pozebon D, Stedile FC, dos Santos JHZ, Rigacci A (2007) Effect of silica texture on grafting metallocene catalysts. Journal of Molecular Catalysis A: Chemicals 265:167–176
97. Marzouk S, Rachdi F, Fourati M, Bouaziz J (2004) Synthesis and grafting of silica aerogels. Colloids and Surfaces A: Physicochem. Eng. Aspects 234:109–116
98. Lee D, Stevens PC, Zeng SQ, Hunt AJ (1995) Thermal characterization of carbon-opacified silica aerogels. J of Non-Cryst Solids 186:285–290
99. Silveira N, Ehrburger-Dolle F, Rochas C, Rigacci A, Vargas Pereira F, Westfahl H (2005) Smectic ordering in polymer liquid crystal-silica aerogels nanocomposites, J Thermal analysis and Calorimetry 79:579–585
100. Bednarczyk S, Bechir R, Baclet P (1999) Laser micromachining of small objects for high energy laser experiments. Applied Physics A (A69):495–500
101. Prassas M, Phalippou J, Zarzycki J (1986) Sintering of monolithic silica aerogels. In: Hench LL and Ulrich DR (eds) Science of Ceramic Chemical Processing, Wiley, New York
102. Ayral A, Phalippou A, Woignier T (1992) Skeletal density of silica aerogels determined by helium pycnometry. J Materials Science 27:1166–1170
103. Hafidi Alaoui A, Woignier T, Phalippou J, Scherer GW (1998) Room Temperature Densification of Aerogel by Isostatic Compression. J Sol Gel Sci Technol 13:365–369
104. Venkastewara Rao A, Kulkarni MM (2002) Hydrophobic properties of *TMOS*/TMES-based silica aerogels. Mater Res Bull 37:1667–1677
105. Allié C, Pirard R, Pirard J-P (2002) The role of the main silica precursor and the additive in the preparation of low-density *xerogels*. J Non-Cryst Solids 311:304–313
106. Mandelbrot BB (1977) Fractals: Form, Chances and Dimensions. Freeman, San Francisco
107. Platzer WJ, Bergkvist M (1993) Bulk and surface light scattering from transparent silica aerogel. Solar Energy Mater Solar Cells 31:243–251
108. Kolb M, Botet R, Jullien R (1983) Scaling of kinetically growing clusters, Phys Rev Lett 51:1123–1127
109. Ehrburger-Dolle F, Dallamano J, Elaloui E, Pajonk G (1995) Relations between the texture of silica aerogels and their preparation. J Non-Cryst Solids 186:9–17
110. Yoda S, Ohshima S (1999) Supercritical drying media modification for silica aerogel preparation. J Non-Cryst Solids 248:224–234
111. Reichenauer G, Scherer GW (2001) Nitrogen sorption in aerogels, J. Non-Cryst Solids 285:167–174

112. Pirard R, Blacher s, Brouers f, Pirard JP (1995) Interpretation of mercury porosimetry applied to aerogels; J Mater Res 10:2114–2119
113. Calas S (1997) Surface et porosité dans les aérogels de silice : étude structurale et texturale. Thèse de doctorat, Université de Montpellier
114. Schuth F, Sing KSW, Weitkamp J. (Eds.) (2002) Handbook of Porous Solids. Wiley-VCH Verlag, Weinheim, Germany. 3:2014
115. Venkastewara Rao A, Nilsen E, Einarsrud MA (2001) Effect of precursors, methylation agents and solvents on the physicochemical properties of silica aerogels prepared by a*tmos*pheric pressure drying method. J Non-Cryst Solids 296:165–171
116. Venkastewara Rao A, Hegde ND, Hirashima H (2007) Absorption and desorption of organic liquids in elastic superhydrophobic silica aerogels. J Colloid Interface Sci 305:124–132
117. Venkastewara Rao A, Pajonk GM, Bhagat SD, Barboux P (2004) Comparative studies on the surface chemical modification of silica aerogels based on various organosilane compounds of the type R_nSiX_{4-n}. J Non-Cryst Solids 350:216–223
118. Parvathy Rao A, Venkateswara Rao A, Pajonk GM (2005) Hydrophobic and Physical Properties of the Two Step Processed Ambient Pressure Dried Silica Aerogels with Various Exchanging Solvents. J Sol Gel Sci Technol 36:285–292
119. Parvathy Rao A, Venkastewara Rao A, Pajonk GM (2007) Hydrophobic and physical properties of the ambient pressure dried silica aerogels with sodium silicate precursor using various surface modification agents. Appl Surf Sci 253:6032–6040
120. Kim GS, Hyun SH (2003) Effect of mixing on thermal and mechanical properties of aerogel-PVB composites. J Mater Sci 38:1961–1966
121. Shi F, Wang L, Liu J (2006) Synthesis and characterization of silica aerogels by a novel fast ambient pressure drying process. Mater Lett 60:3718–3722
122. Hüsing N Schubert U, Misof K, Fratzi P (1998) Formation and Structure of Porous Gel Networks from $Si(OMe)_4$ in the Presence of $A(CH_2)nSi(OR)_3$ (A = Functional Group). Chem Mater 10:3024–3032
123. Venkateswara Rao A, Kulkarni MM, Pajonk GM, Amalnerkar DP, Seth T (2003) Synthesis and Characterization of Hydrophobic Silica Aerogels Using Trimethylethoxysilane as a Co-Precursor J Sol Gel Sci Technol 27:103–109
124. Yoldas BE, Annen MJ, Bostaph J (2000) Chemical engineering of aerogel morphology formed under nonsupercritical conditions for thermal insulation. Chem Mater 12:2475–2484
125. Wolff B, Seybold G, Krueckau FE (1989) Thermal insulators having density 0.1 to 0.4 g/cm^3, and their manufacture BASF-G, Eur Pat Appl EP 0340707
126. Quenard D, Chevalier B, Sallee H, Olive F, Giraud D (1998) Heat transfer by conduction and radiation in building materials: review and new developments Rev Metall Cahier Inf Tech 95: 1149–1158
127. Fesmire JE, Sass JP (2008) Aerogel insulation applications for liquid hydrogen launch vehicle tanks. Cryogenics 48:223–231
128. Tsou P (1995) Silica aerogel captures cosmic dust intact. J Non-Cryst Solids 186:415–427
129. Guise MT, Hosticka B, Earp BC, Norris PM (1995) An experimental investigation of aerosol collection utilizing packed beds of silica aerogel microspheres. J Non-Cryst Solids 285:317–322
130. Hrubesh LW (1998) Aerogel applications. J Non-Cryst Solids 225:335–342
131. Schmidt M, Schwertfeger F (1998) Applications for silica aerogel products. J Non-Cryst Solids 225:364–368
132. Jones SM (2006) Aerogel: Space exploration applications. J Sol-Gel Sci Technol 40:351–357
133. Pajonk GM (1998) Transparent silica aerogels. J Non-Cryst Solids 225:307–314
134. Buzykaev AR, Danilyuk AF, Ganzhur SF, Kravchenko EA, Onuchin AP (1999) Measurement of optical parameters of aerogel. Nucl Instr Meth Phys Res A 433:396–400
135. Danilyuk AF, Kravchenko EA, Okunev AG, Onuchin AP, Shaurman SA (1999) Synthesis of aerogel tiles with high light scattering length. Nucl Instr Meth Phys Res A 433:406–407
136. Jensen KI, Schultz JM, Kristiansen FH (2004) Development of windows based on highly insulating aerogel glazings. J Non-Cryst Solids 350:351–357
137. Adachi I, Fratina S, Fukushim T, Gorisek A, Iijima T, Kawai H, Konishi M, Korpar S, Kozakai Y, Krizan P, Matsumoto T, Mazuka Y, Nishida S, Ogawa S, Ohtake S, Pestotnik R, Saitoh S, Seki T, Sumiyoshi T, Tabata M, Uchida Y, Unno Y, Yamamoto S (2005) Study of highly transparent silica aerogel as a *RICH* radiator. Nucl Instr Meth Phys Res A 553:146–151
138. Duer K, Svendsen S (1998) Monolithic silica aerogel in superinsulating glazings Sol Energy 63:259–267
139. Husing N, Schubert U (1998) Aerogels - Airy materials: Chemistry, structure, and properties. Angew Chem Int Ed 37:23–45

140. Tajiri K, Igarashi K (1998) The effect of the preparation conditions on the optical properties of transparent silica aerogels. Sol Energy Mater Sol Cells 54:189–195
141. Venkateswara Rao A, Haranath D, Pajonk GM, Wagh PB (1998) Optimisation of supercritical drying parameters for transparent silica aerogel window applications. Mater Sci Technol 14: 1194–1199
142. Pajonk GM, Elaloui E, Achard P, Chevalier B, Chevalier JL, Durant M (1995) Physical properties of silica gels and aerogels prepared with new polymeric precursors, J Non-Cryst Solids 186:1–8
143. Rigacci A, Ehrburger-Dolle F, Geissler E, Chevalier B, Sallée H, Achard P, Barbieri O, Berthon S, Bley F, Livet F, Pajonk GM, Pinto N, Rochas C (2001), Investigation of the multi-scale structure of silica aerogels by *SAXS*. J. Non-Cryst. Solids 285:187–193
144. Sumiyoshi T, Adachi I, Enomoto R, Iijima T, Suda R, Yokoyama M, Yokogawa H (1998) Silica aerogels in high energy physics. J Non-Cryst Solids 225:369–374
145. Poelz G, Riethmueller R (1982) Preparation of silica aerogel for *Cherenkov counters*. Nuc Instr Meth 195:491–503
146. Yokogawa H, Yokoyama M (1995) Hydrophobic silica aerogels. J Non-Cryst Solids 186:23–29
147. Kharzheev YN (2008) Use of silica aerogels in *Cherenkov counters*. Physics of Particles and Nuclei, 39:107–135
148. De Leo R (2008) Long-term operational experience with the HERMES aerogel *RICH* detector. Nuclear Instruments & Methods Phys Res A: Accelerators, Spectrometers, Detectors, and Associated Equipment 595:19–22
149. Forest L, Gibiat V, Woignier T (1998) Biot's theory of acoustic propagation in porous media applied to aerogels and alcogels. J Non-Cryst Solids 225:287–292
150. Conroy JFT, Hosticka B, Davis SC, Smith AN, Norris PM (1999) Evaluation of the acoustic properties of silica aerogels MD (Am Soc Mech Eng) 82:25–33
151. Burger T, Fricke J (1998) Aerogels: Production, modification and applications. Berichte der Bunsen Gesellschaft Phys Chemi Chem Phys 102:1523–1528
152. Gross J, Fricke J (1992) Ultrasonic velocity measurements in silica, carbon and organic aerogels J Non-Cryst Solids 145:217–222
153. Ma H-S, Roberts AP, Prévost J-H, Jullien R, Scherer GW (2000) Mechanical structure-property relationship of aerogels. J Non-Cryst Solids 277:127–141
154. Woignier T, Phalippou J (1989) Scaling law variation of the mechanical properties of silica aerogels. Revue de Physique Appliquée 24(C4):179–184
155. Scherer GW, Smith DM, Qiu X, Anderson JM (1995) Compression of aerogels. J Non-Cryst Solids 186: 316–320
156. Bisson A, Rodier E, Rigacci A, Lecomte D, Achard P (2004) Study of evaporative drying of treated silica gels. J Non-Cryst Solids 350:230–237
157. Fricke J, Emmerling A (1998) Aerogels - Recent progress in production techniques and novel applications. J Sol-Gel Sci Technol 13:299–303
158. Kramer SJ, Rubio-Alonso F, Mackenzie JD (1996) Organically modified silicate aerogels, "Aeromosils". Mater Res Soc Symp Proc 435:295–300
159. Capadona LA, Meador MA, Alunni A, Fabrizio EF, Vassilaras P, Leventis N (2006) Flexible low-density polymer crosslinked silica aerogels. Polymer 47:5754–5761
160. Parmenter KE, Milstein F (1998) Mechanical properties of silica aerogels. J Non-Cryst Solids 223:179–189
161. Despetis F, Etienne P, Phalippou J (2000) Crack speed in ultraporous brittle amorphous material Phys Chem Glasses 41:104–106
162. Etienne P, Despetis F, Phalippou J (1998) Subcritical crack velocity in silica aerogels J Non-Cryst Solids 225: 266–271
163. Holmes NC, Radousky HB, Moss MJ, Nellis WJ, Henning S (1984) Silica at ultrahigh temperature and expanded volume. Appl Phys Lett 45:626–628
164. Amendt P, Glendinning SG, Hammel BA, Landen OL, Murphy TJ, Suter LJ, Hatchett S, Rosen MD, Lafittte S, Desenne D, Jadaud JP (1997) New methods for diagnosing and controlling hohlraum drive asymmetry on Nova. Phys Plasmas 4:1862–1871
165. Kawakami N, Fukumoto Y, Kinoshita T, Suzuki K, Inoue K (2000) Preparation of highly porous silica aerogel thin film by supercritical drying. Jpn J Appl Phys Part 2 39:L182–L184
166. CaoY, Xia ZF, Li Q, Shen J, Chen LY, Zhou B (1998) Study of porous dielectrics as *electret* materials. IEEE Trans Dielectr Electr Insul 5:58–62
167. Charlton A, McKinnie IT, Meneses-Nava MA, King TA (1992) A tunable visible solid state laser. J Mod Opt 39:1517–1523

168. Zhou B, Wang J, Zhao L, Shen J, Deng ZS, Li YF (2000) Preparation of C60-doped silica aerogels and the study of photoluminescence properties. J Vac Sci Technol B 18:2001–2004
169. Shen J, Wang J, Zhou B, Deng ZS, Weng ZN, Zhu L, Zhao L, Li YF (1998) Photoluminescence of fullerenes doped in silica aerogels. J Non-Cryst Solids 225:315–318
170. Leventis N, Elder IA, Rolison DR, Anderson ML, Merzbacher CI (1999) Durable Modification of Silica Aerogel Monoliths with Fluorescent 2,7-Diazapyrenium Moieties. Sensing Oxygen near the Speed of Open-Air Diffusion. Chem Mater 11:2837–2845
171. Bockhorst M, Heinloth K, Pajonk GM, Begag R, Elaloui E (1995) Fluorescent dye doped aerogels for the enhancement of Cherenkov light detection. J Non-Cryst Solids 186:388–394
172. Barnik MI, Vasilchenko VG, Golovkin SV, Medvedkov AM, Solovev AS, Yudin SG (2000) Scintillation properties of materials based on liquid crystals in static and dynamic states. Instrum Exp Techn 43:602–611
173. Kim NK, Kim K, Payne DA, Upadhye RS (1988) Fabrication of hollow silica aerogel spheres by a droplet generation method and sol-gel processing. J Vac Sci Technol A 7:1181–1184
174. Jang KY, Kim K, Upadhye RS (1990) Hollow silica spheres of controlled size and porosity by sol-gel processing. J Vac Sci Technol A 8:1732–1735
175. Kim KK, Jang KY (1991) Hollow silica spheres of controlled size and porosity by sol-gel processing. J Am Ceram Soc 74:1987–1992
176. Tastevin G, Nacher PJ, Guillot G (2000) NMR of hyperpolarised 3He gas in aerogel Physica B 284–288 Part 1:291–292
177. Pavlovskaya G, Blue AK, Gibbs SJ, Haake M, Cros F, Malier L, Meersmann T (1999) Xenon-131 Surface Sensitive Imaging of Aerogels in Liquid Xenon near the Critical Point J Magn Reson 137:258–264
178. Halperin WP, Gervais G, Yawata K, Mulders N. (2003) Impurity phases of *superfluid* 3He in aerogel. Physica B: Condensed Matter (Amsterdam, Netherlands), 329–333(Pt. 1):288–291
179. Feldman DE (2000) Quasi-Long-Range Order in Nematics Confined in Random Porous Media Phys Rev Lett 84:4886–4889
180. Antczak T, Mrowiec-Bialon J, Bielecki S, Jarzebski AB, Malinowski JJ, Lachowski AI, Galas E (1997) Thermostability and *esterification* activity in silica aerogel matrix and in organic solvents. Biotechnol Techn 11:9–11
181. Pierre M, Buisson P, Fache F, Pierre AC (2000) Influence of the drying technique of silica gels on the enzymatic activity of encapsulated lipase. Biocatal Biotransform 18:237–251
182. Buisson P, Hernandez C, Pierre M, Pierre AC (2001) Encapsulation of lipases in aerogels. J Non-Cryst Solids 285:295–302
183. Power M, Hosticka B, Black E, Daitch C, Norris P (2001) Aerogels as *biosensors*: viral particle detection by bacteria immobilized on large pore aerogel. J Non-Cryst Solids 285:303–308
184. Bernik DL (2007) Silicon based materials for drug delivery devices and implants. Recent Patents on Nanotechnology, 1:186–192
185. Smirnova I, Arlt W (2004) Synthesis of silica aerogels and their application as drug delivery system. In Brunner, G (ed) Supercritical Fluids as Solvents and Reaction Media: 381–427
186. Emmerling A, Gross J, Gerlach R, Goswin R, Reichenauer G, Fricke J, Haubold HG (1990) Isothermal sintering of silica aerogels. J Non-Cryst Solids 125:230–243
187. Reynes J, Woignier T, Phalippou J, Dussossoy JL (1999) Host material for nuclear waste storage. Adv Sci Technol 24:547–550
188. Woignier T, Reynes J, Phalippou J, Dussossoy JL, Jacquet-Francillon N (1998) Sintered silica aerogel: a host matrix for long life nuclear wastes. J Non-Cryst Solids 225:353–357
189. Reynes J, Woignier T, Phalippou J (2001) Permeability measurement in composite aerogels: application to nuclear waste storage J Non-Cryst Solids 285:323–327
190. Aristov YI, Restuccia G, Tokarev MM, Cacciola G (2000) Selective Water Sorbents for Multiple Applications, 10. Energy Storage Ability. React Kinet Catal Lett 69:345–353
191. Gesser HD, Goswani PC (1989) Infrared study of OH and NH2 groups on the surface of a dry silica aerogel. Chem Rev 89:765–788
192. Pajonk GM, Venkateswara Rao A (2001) From sol-gel chemistry to the applications of some inorganic and/or organic aerogels. Recent Res Develop Non-Crystalline Solids 1:1–20
193. Loschiavo SR (1988) Availability of food as a factor in effectiveness of a silica aerogel against the merchant grain beetle (Coleoptera: Cucujidae). J Econom Entomoly 81:1237–1240

3

Hydrophobic Silica Aerogels: Review of Synthesis, Properties and Applications

Ann M. Anderson and Mary K. Carroll

Abstract There are many applications for which a material must be water-resistant. Silica aerogels can have unusual properties, including high surface area, low density, low thermal conductivity, and good optical translucency. This combination of properties makes hydrophobic silica aerogels attractive materials for use in applications ranging from transparent insulation systems to drug delivery platforms. These aerogel materials have been prepared using a wide variety of techniques, including incorporation of silica precursors with non-polar substituents into the sol–gel matrix and surface modification of the matrix following gelation. In this chapter we describe the different aerogel synthesis methods, present a discussion of techniques for measuring hydrophobicity and review the extensive literature on hydrophobic silica aerogels, including information on their physical properties and applications.

3.1. Introduction

Aerogels are sol–gel materials that are dried in such a way as to avoid pore collapse, leaving an intact solid nanostructure in a material that is 90–99% air by volume. The high porosity of the resulting material gives rise to its unusual physical properties. Silica aerogels can be made with low thermal conductivity (<0.02 W/mK), high surface area (>1,000 m^2/g), low densities (<0.05 g/cm^3) and they can be made translucent. This unique combination of thermal, optical, and structural properties makes it an attractive material for a variety of applications ranging from platforms for chemical sensors (Plata et al. [1]) to thermal insulation (Fesmire [2]) to comet dust collectors (Gougas et al. [3]). Unmodified aerogels prepared from two common precursors, tetramethoxysilane (*TMOS*) and tetraethoxysilane (*TEOS*) are somewhat *hydrophilic*, due to surface hydrolysis. As a result, if they are exposed to ambient air, particularly in humid environments, they can deteriorate over time. Structural instability limits the applications of hydrophilic silica aerogels, which is one reason that there has been substantial research into the synthesis of *hydrophobic* silica

A. M. Anderson • Department of Mechanical Engineering, Union College, 807 Union Street, Schenectady, NY 12308, USA
e-mail: andersoa@union.edu
M. K. Carroll • Department of Chemistry, Union College, 807 Union Street, Schenectady, NY 12308, USA
e-mail: carrollm@union.edu

M.A. Aegerter et al. (eds.), *Aerogels Handbook*, Advances in Sol-Gel Derived Materials and Technologies, DOI 10.1007/978-1-4419-7589-8_3,

aerogels. Furthermore, hydrophobic aerogels can be used in a number of emerging application areas such as low drag hydrophobic surfaces [4] and drug delivery systems [5].

In this chapter, we present a review of the synthesis techniques for, properties of and applications of hydrophobic silica aerogels. In the following section, we review the techniques used to make silica aerogels. This is followed by sections devoted to the fabrication of hydrophobic aerogels, techniques for measuring hydrophobicity, a review of the literature on hydrophobic silica aerogels, and a summary of application areas for hydrophobic aerogels.

3.2. Aerogel Fabrication Techniques

There are a number of different processes used to prepare silica aerogels (Chap. 2). All of these first involve polymerization of one or more silicon-containing precursors to form a solid silica sol–gel matrix in which the pores are filled with solvent byproducts of the reaction. To obtain an aerogel from the wet sol gel, it is then necessary to extract the solvent without effecting pore collapse, leaving the silica nanostructure intact and dry. This can be done using freeze drying, surface modification with ambient drying, or supercritical extraction. This section reviews the techniques used to fabricate silica aerogels. We first discuss the sol–gel processing step. This is followed by a discussion of the different techniques used to dry the wet gel.

3.2.1. Forming the Wet Sol Gel

Silica sol gels are generally formed via hydrolysis and polycondensation reactions of silica precursors in the presence of an acid and/or base catalyst (Figure 3.1). The resulting silica matrix is highly porous, and the pores of the sol gel are filled with the solvent byproducts of the hydrolysis and polymerization reactions. If the solvent mixture can be removed from the wet sol gel without substantial structural collapse, an aerogel is formed.

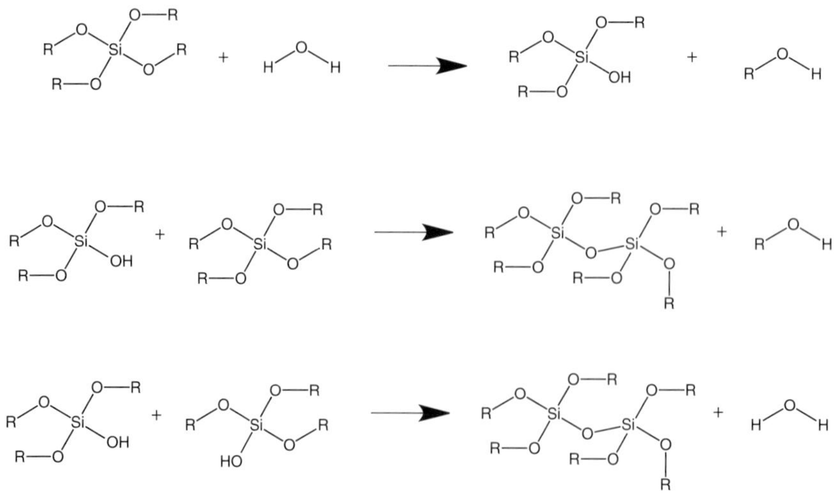

Figure 3.1. Schematic representation of reactions taking place during formation of silica sol gels.

Table 3.1. Precursor chemicals used in the production of silica aerogels

Precursor	Line structure	Abbreviation
Tetramethoxysilane $Si(OCH_3)_4$	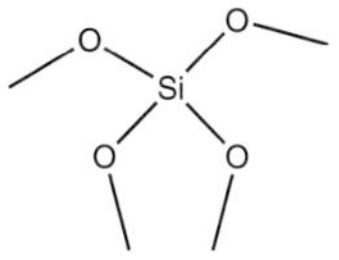 O, Si, O, O	*TMOS*
Tetraethoxysilane $Si(OC_2H_5)_4$	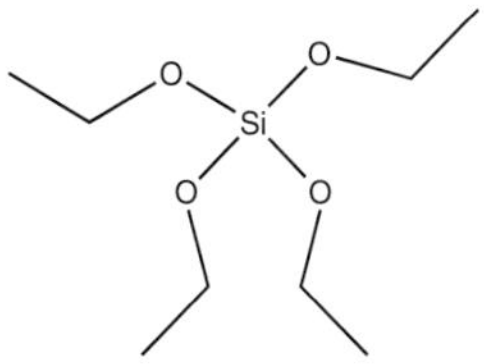 O, O, Si, O, O	*TEOS*
Triethoxysilane $Si(OC_2H_5)_3H$	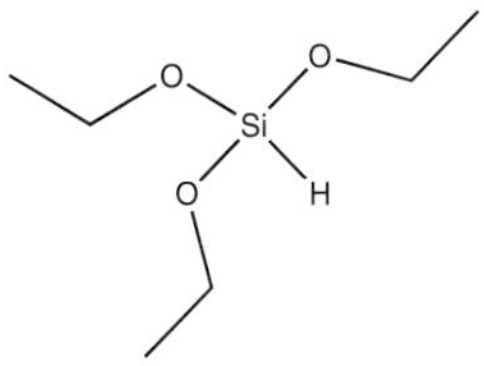 O, O, Si, O, H	TriEOS
Methyltrimethoxysilane $Si(OCH_3)_3CH_3$	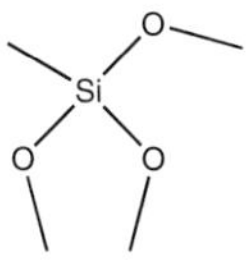 O, Si, O, O	*MTMS*
Sodium Metasilicate Na_2SiO_3 and related salts	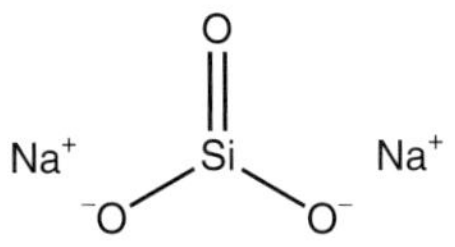 O, Na^+, Si, Na^+, ^-O, O^-	SS
Methyl Silicate 51 $CH_3OSi(OCH_3)_{24}OCH_3$		MS51
Polyethoxydisiloxane (E-40)		PEDS
Silbond H-5 (Ethyl Polysilicate) $C_2H_5O–Si(OC_2H_5)_2O_2–xC_2H_5$		SH5

The sol–gel recipe includes one or more silicon-containing precursor moieties. The most commonly used silica precursors, *TMOS* and *TEOS*, have four identical alkoxide groups in a tetrahedral arrangement around the central silicon atom (see Table 3.1). Following hydrolysis, the polymerization reaction can occur in all four directions, leading to Si–O–Si linkages, and a bulk material that is composed of silica (SiO_2). If one of the alkoxide "branches" off the silicon atom is replaced by a different functional group that is incapable of undergoing a condensation reaction, then that functional group will remain covalently bound to a silicon atom within the sol–gel matrix (see Figure 3.2). Consequently, the surface properties of the resulting aerogel will differ from an aerogel prepared solely from *TMOS* or *TEOS*. Aerogels have been prepared using the derivatized *TMOS* and *TEOS* precursors *MTMS* and TriEOS (Table 3.1). It is also common to utilize recipes with more than one precursor, or with subsequent surface modification, as is discussed later in this chapter.

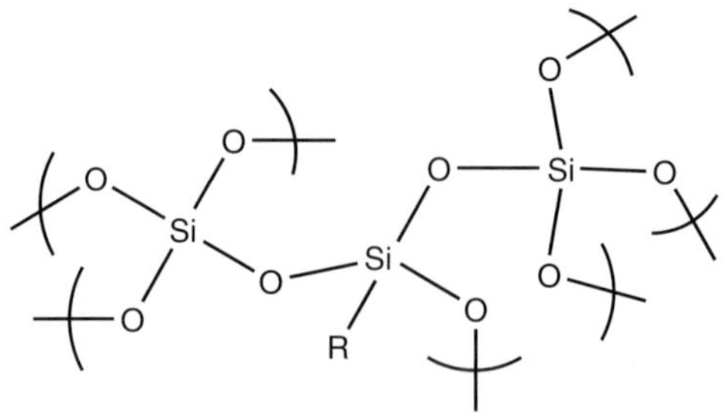

Figure 3.2. Schematic representing a section of silica matrix for which one-third of the silicon-containing precursors had an unreactive side chain, R.

It is also possible to employ "pre-polymerized" species (see Table 3.1) in the precursor recipe. Three examples are the *TMOS*-based Methyl Silicate 51 (Fuso Chemical Co., Ltd), *TEOS*-based polyethoxydisiloxane (E-40), and Silbond H-5 (Silbond Corp.). Silbond H-5 is a viscous mixture that is prepared via incomplete reaction of *TEOS* in ethanol with water in the presence of catalyst. Water is the limiting reactant here, so complete gelation does not occur. Instead, a mixture of soluble silicon alkoxy-oxide species is formed. These can then be reacted with additional water and catalyst to yield sol gels that can subsequently be processed to produce aerogels (Yokogawa and Yokoyama [6], Miner et al. [7], Zhang et al. [8], or Zhou et al. [9]).

Some researchers prepare silica aerogels using relatively inexpensive silicate salts, rather than alkoxysilanes. The salt sodium metasilicate (Na_2SiO_3), more commonly known as "sodium silicate" or "water glass," is soluble in water. The metasilicate (SiO_3^{2-}) ion is basic, but can be converted to an acid form via separation on an ion-exchange column (see, for example, Schwertfeger et al. [10] or Lee et al. [11]), or direct reaction with acid in solution (Bangi et al. [12]). The resulting Si–OH groups subsequently polymerize to form a hydrophilic sol–gel. Since the precursor salt does not itself contain any non-polar groups, to prepare hydrophobic silica aerogels from silicate salts requires surface modification following gelation.

All of the above methods result in a mixture that is initially in the liquid phase and subsequently gels to form a solid matrix. The precursor mixture is typically poured into a mold or glass container in which it is allowed to gel. Depending on the selection of drying process, the wet sol gel may undergo a series of solvent exchanges. The next step is to dry the wet gel to form an aerogel.

3.2.2. Drying the Wet Gel

To make an aerogel from a wet gel, it is necessary to remove the liquid solvent from the solid sol–gel matrix. This presents a challenge because the delicate nanostructure of an unmodified wet gel cannot withstand the large capillary forces that exist during drying due to surface tension at the liquid–vapor interface. As the wet gel dries, the meniscus retreats and capillary forces cause fracture to the nano-structure. Unmodified sol gels dried under ambient conditions shrink to about 1/8 of their initial volume due to this pore collapse; the resulting materials are termed "xerogels" (dry gels). If this drying process occurs slowly, intact xerogel monoliths can be produced. But to make an aerogel, one must avoid crossing the vapor–liquid phase boundary. The current methods used to avoid pore collapse in aerogel fabrication can be categorized into three general techniques. Each is designed to minimize or eliminate the capillary forces due to surface tension effects. They are (a) ambient pressure drying following surface modification, (b) freeze drying, and (c) supercritical drying techniques. A detailed overview of these methods can be found in Bisson et al. [13]. A brief summary of each is included below.

Ambient Pressure Drying Processes

These drying techniques are designed to dry the wet gel at ambient pressure. They require chemical processing with lengthy solvent exchange to either reduce the capillary forces acting on the nanostructure, or to increase the ability of the nanostructure to withstand those forces (either by strengthening the structure or by making it more flexible). Prakash et al. [14], [15] manipulated the surface chemistry of a *TEOS*-based wet gel to promote reversible shrinkage or a "spring-back" effect using a solvent exchange with hexane, followed by a surface modification with a *silylation* process with *TMCS* (see Table 3.2 for information about the silylating agents). This method has been adapted by a number of other researchers who have studied the use of different silylating agents. Schwertferger et al. [10] demonstrated the fabrication of a sodium silicate-based, *TMCS* treated, ambient dried aerogel. Another technique used by Haerid and Einarsrud [16] ages the gels in alkoxide/alcohol solutions to stiffen the microstructure to avoid pore collapse. Bhagat et al. [17] show that it is possible to make flexible aerogel monoliths that "spring back" after low-temperature drying using methyltrimethoxysilane (*MTMS*) as a precursor material. They show that the sol–gel nanostructure of *MTMS*-formulated aerogels is flexible, which allows it to withstand the capillary forces during ambient pressure drying without damage. Another approach is to synthesize polymer cross-linked silica aerogels that can withstand the capillary forces during drying under ambient pressure conditions (Leventis et al. [18]).

Table 3.2. Co-Precursors (CP), silylating chemicals (S), and organic modifiers (OM) used in the production of hydrophobic silica aerogels

Co-precursor, silylating agent or organic modifier	Line structure	Abbreviation	Method[a]
TMOS derivatives			
Methyltrimethoxysilane $Si(OCH_3)_3CH_3$	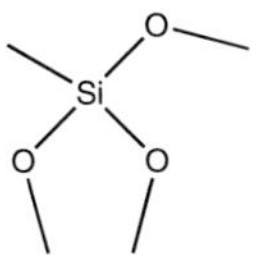	*MTMS*	CP, S
Ethyltrimethoxysilane $Si(OCH_3)_3C_2H_5$	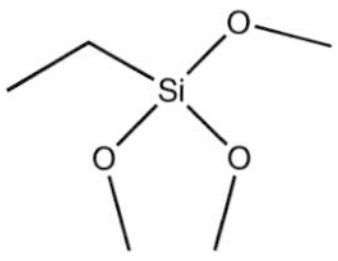	ETMS	CP
Propyltrimethoxysilane $Si(OCH_3)_3C_3H_7$	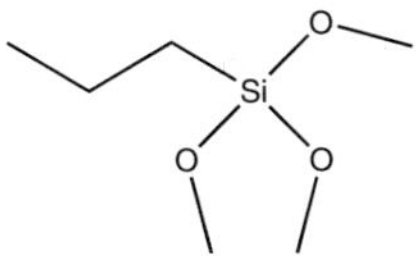	PTMS	CP

(continued)

Table 3.2 (continued)

Co-precursor, silylating agent or organic modifier	Line structure	Abbreviation	Method[a]
3,3,3-Trifluoropropyl-trimethoxysilane $Si(OCH_3)_3(CH_2)_2CF_3$	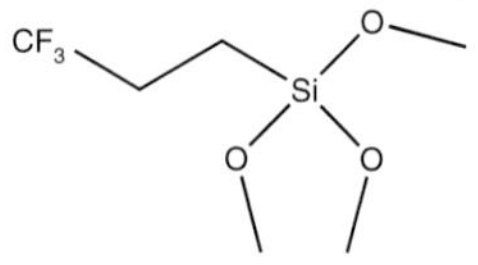 CF_3, O, Si, O, O	FPTMS	CP
n-Butyltrimethoxysilane $Si(OCH_3)_3C_4H_9$	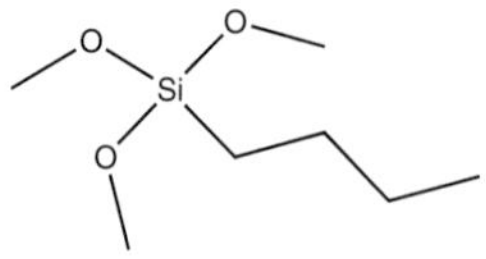 O, O, Si, O	BTMS	CP
Hexadecyltrimethoxysilane $H_3C(CH_2)_{15}Si(OCH_3)_3$	 O, Si, O, O	HDTMS	CP
Vinyltrimethoxysilane $H_2C{=}CHSi(OCH_3)_3$	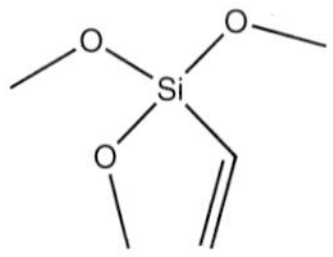 O, O, Si, O	VTMS	CP, S
Phenyltrimethoxysilane $Si(OCH_3)_3C_6H_5$	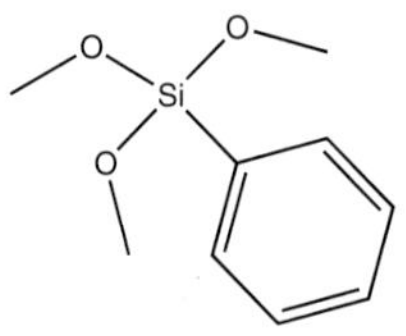 O, O, Si, O	PhTMS	CP, S
Trimethoxysilylpropylmethacrylate $H_2C{=}C(CH_3)CO_2(CH_2)_3Si(OCH_3)_3$	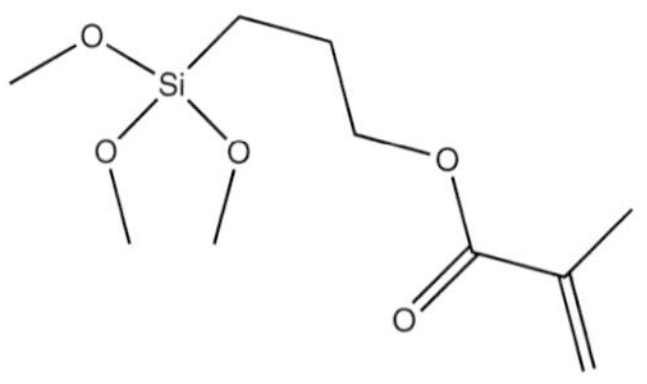 O, Si, O, O, O, O	TMSPMA	CP
Dimethyldimethoxysilane $Si(OCH_3)_2(CH_3)_2$	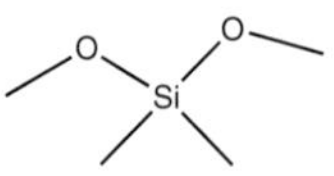 O, O, Si	DMDMS	S
Trimethylmethoxysilane $Si(OCH_3)(CH_3)_3$	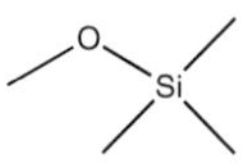 O, Si	TMMS	CP, S
TEOS derivatives			
Methyltriethoxysilane $Si(OC_2H_5)_3CH_3$	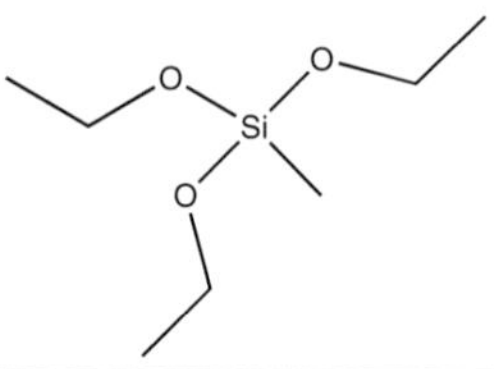 O, O, Si, O	MTES	S

(continued)

Table 3.2 (continued)

Co-precursor, silylating agent or organic modifier	Line structure	Abbreviation	Method[a]
Ethyltriethoxysilane $Si(OC_2H_5)_3C_2H_5$	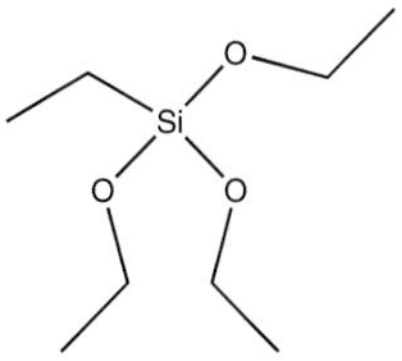	ETES	CP
Phenyltriethoxysilane $Si(OC_2H_5)_3C_6H_5$	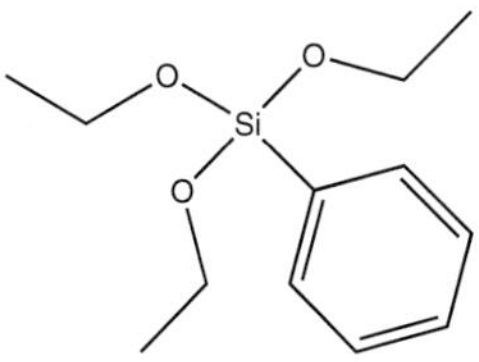	PhTES	CP, S
Vinyltriethoxysilane $H_2C{=}CHSi(OC_2H_5)_3$	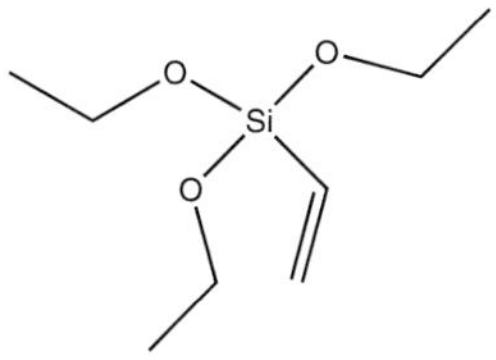	VTES	CP
Aminopropyltriethoxysilane $H_2N(CH_2)_3Si(OC_2H_5)_3$	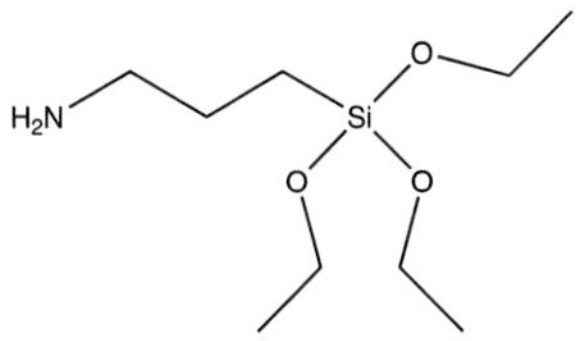	APTES	CP
Diethylphosphatoethyltriethoxysilane $(CH_3CH_2)_2PO(CH_2)_2Si(OC_2H_5)_3$	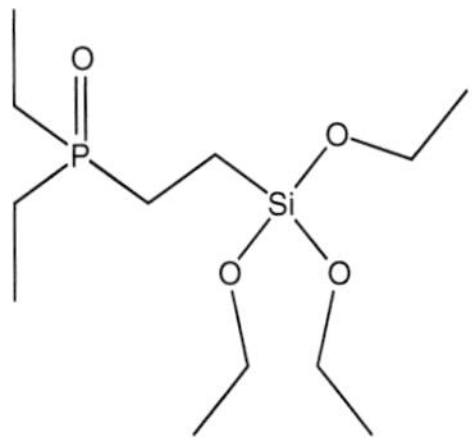	PPTES	CP
Dimethyldiethoxysilane $Si(OC_2H_5)_2(CH_3)_2$	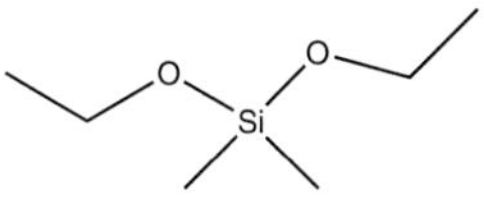	DMDS	CP
Trimethylethoxysilane $Si(CH_3)_3(OCH_2CH_3)$	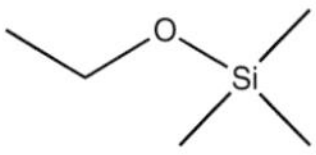	TMES	CP
Other silanes			
Dimethylchlorosilane $Si(CH_3)_2HCl$	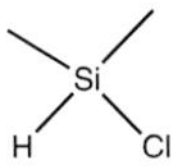	DMCS	CP

(continued)

Table 3.2 (continued)

Co-precursor, silylating agent or organic modifier	Line structure	Abbreviation	Method[a]
Dimethyldichlorosilane $Si(CH_3)_2Cl_2$	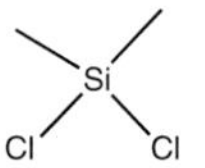	DMDC	CP
Trimethylchlorosilane $Si(CH_3)_3Cl$	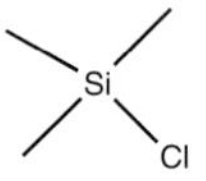	*TMCS*	CP, S
Trimethylbromosilane $Si(CH_3)_3Br$	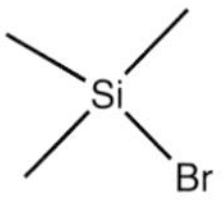	TMBS	CP
Trimethylsilyl Chloroacetate $Si(CH_3)_3OCOCH_2Cl$	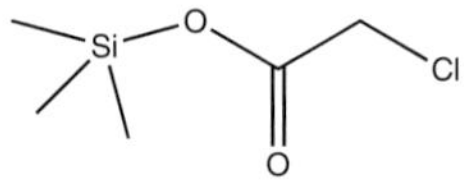	TMSCA	CP
1H,1H,2H,2H-Perfluorooctyl-dimethylchlorosilane $CF_3(CF_2)_5(CH_2)_2\ Si(CH_3)_2Cl$	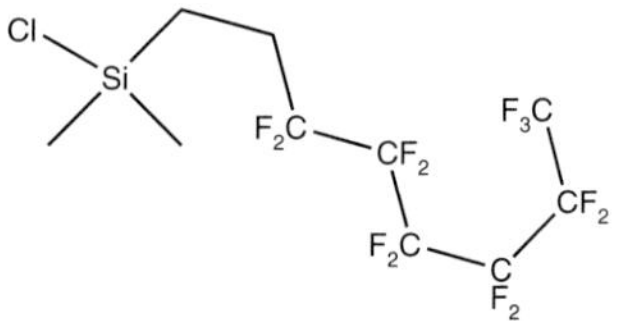	PFAS	CP, S
Hexamethyldisilazane $HN(Si(CH_3)_3)_2$	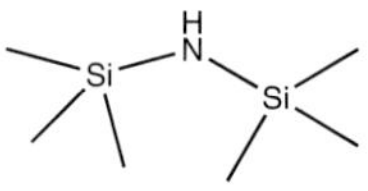	HMDZ	CP, S
Hexamethyldisiloxane $O(Si(CH_3)_3)_2$	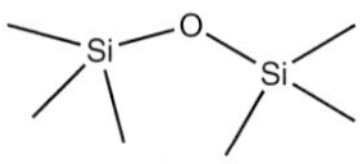	HMDSO	S
Bis(*trimethylsilyl)acetamide* $CH_3C{=}NSi(CH_3)_3OSi(CH_3)_3$	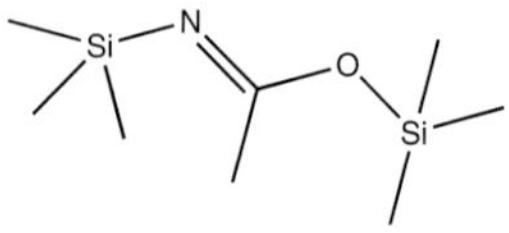	BTSA	S
Organic modifiers			
Tris(hydroxylmethyl)Aminomethane $NH_2C(CH_2OH)_3$	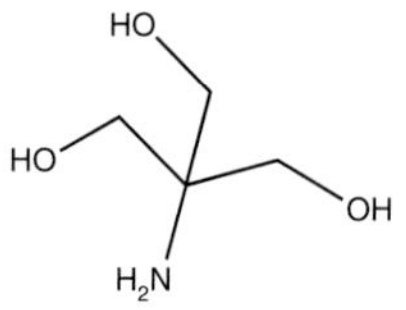	TAM	CP
Styrene $C_6H_5CHCH_2$	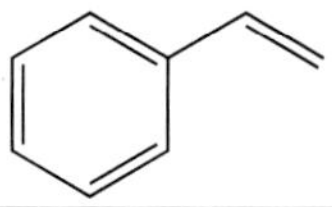	PS	OM

(continued)

Table 3.2 (continued)

Co-precursor, silylating agent or organic modifier	Line structure	Abbreviation	Method[a]
p-Chloromethylstyrene $CH_2ClC_6H_4CHCH_2$	Cl	CMS	OM
2,3,4,5,6-pentafluorostyrene $C_6F_5CHCH_2$	F F F F F	PFS	OM

[a]*CP* co-precursor, *OM* organic modifier, *S* silylation.

Freeze-Drying

Freeze-drying a wet gel results in a cryogel. Pajonk et al. [19] and Kalinen et al. [20] describe freeze-drying processes which result in an opaque aerogel powder. This technique removes excess solvent through sublimation. The wet gel is "frozen" and then the solvent is allowed to sublimate at lower pressures. Kalinen et al. [20] show that the ice microcrystals that form during the freeze-drying process lead to more macroporous aerogels (as compared to those dried through supercritical extraction). A thorough description of the theory related to freeze-drying can be found in Scherer [21].

Supercritical Extraction Methods

The supercritical extraction methods avoid the liquid/vapor boundary line by bringing the solvent above its supercritical point and then removing it from the sol–gel matrix as a supercritical fluid. In this state there is no liquid–vapor interface and therefore no capillary stresses due to the receding meniscus. The supercritical extraction technique was first developed by Kistler [22] in the 1930s. There are several types of supercritical extraction methods in use. They include high temperature, low temperature, and rapid supercritical techniques, each of which is described in more detail below.

High-temperature alcohol supercritical extraction techniques (*ASCE*) bring the wet gel to the supercritical state of the solvent (usually methanol or ethanol) in an autoclave or other pressure vessel. This involves high pressures (above 8 MPa) and temperatures (above 260°C). A number of studies have been performed to examine the effects of solvent fill volume, pre-pressure, and other processing parameters (see for example: Phalippou et al. [23], Danilyuk et al. [24], Pajonk et al. [25]). The low-temperature extraction techniques (*CSCE*) are based on supercritical extraction of CO_2, which has a lower critical-point temperature than the alcohol mixture that remains in the sol–gel pores after polymerization. The *CSCE* methods require a series of solvent exchanges, first with non-polar solvents, then with liquid CO_2 prior to supercritical extraction, which can then take place at the critical point of CO_2 (see for example Tewari et al. [26] or Van Bommel and de Haan [27]). The advantages of this technique are a lower critical temperature and a more stable solvent; however, additional steps are added to the process, thereby lengthening the time required to prepare aerogels. Because the critical pressure requirement is not changed significantly compared to the *ASCE* methods (CO_2 has a similar critical pressure to methanol and ethanol), this process requires the use of pressure vessels. In addition, the solvent exchange process becomes a size deterrent, as the diffusion kinetics of the solvent exchange are dependent upon the size of the gel.

Rapid supercritical extraction (*RSCE*) techniques use a *confined mold* in either a pressure vessel (Poco et al. [28], Scherer et al. [29]) or a *hydraulic hot press* (Gauthier et al. [30]). These techniques are one-step precursor-to-aerogel processes and yield aerogels in as little as 3 h. In the Poco et al. method, the liquid precursor chemicals and catalyst are poured into a two-piece mold which is then heated rapidly. The pressure is initially set by fastening the two mold parts together with properly tensioned bolts, or by applying an external hydrostatic pressure inside of a larger pressure vessel, or by a combination of these two. Once the supercritical point of the alcohol is reached, the supercritical fluid is allowed to escape. The Gauthier et al. RSCE fabrication procedure uses a hydraulic hot press to seal and heat a mold containing the aerogel precursor mixture. The liquid mixture of aerogel precursors is poured into a metal mold which is then placed in the hot press. During a typical run, the hot press is closed to seal the mixture inside the mold and the press provides a compressive restraining force. The mold and mixture are then brought above the supercritical temperature and pressure of methanol. The aerogel precursors react to form a wet gel with a porous nanostructure during this heating process. Once a supercritical state is reached the press force is decreased and the supercritical fluids are released, leaving behind an aerogel. Roth et al. [31] present a model for the Gauthier et al. RSCE process that relates the process variables to internal mold conditions. Anderson et al. [32] present a study on the effects of processing parameters on aerogel properties.

Comparison of Methods

Each of the aerogel fabrication methods described above has strengths and limitations. Direct comparison of the different drying techniques is complicated by the use of different precursor recipes, different gelation conditions, and aging times, as well as different extraction methods. For example, for *CSCE* it is necessary to allow sufficient aging that there is some shrinkage of the wet gels, so that the gels can be removed from their initial container for solvent exchange and extraction. In RSCE processes, there is typically little "aging" time; however, the high temperature employed in these processes has a significant impact on the kinetics of the condensation reactions.

The main advantage of the ambient pressure methods is that they do not require expensive and potentially dangerous high-pressure equipment; however, they require multiple processing steps with solvent exchange. To date there has been limited research into the use of freeze-drying methods. These techniques require special equipment to reach the low temperatures required for sublimation of the solvent and result in aerogel powders, rather than monoliths. The main limitations of the *ASCE* techniques are the difficulties associated with obtaining the high temperatures necessary to reach the critical point of the alcohol solvent, as well as the safety concerns with operating the pressure vessel at those conditions.

CSCE has been used extensively in the fabrication of small to very large aerogel monoliths, however, it can take days to weeks to make them, and the required multiple solvent-exchange steps make the process complicated and generate considerable solvent and CO_2 waste. The RSCE methods are simpler and faster. The entire process is done in one step, and can be accomplished in hours, as opposed to multiple steps and time scales on the order of days to weeks for other methods, and these methods generate little waste. A disadvantage of RSCE methods is that they require both high temperature and high pressure.

3.3. Hydrophobic Aerogels

3.3.1. What Makes an Aerogel Hydrophobic?

A hydrophobic surface is one that repels water, whereas a hydrophilic surface attracts water. But there is a wide range of surface behavior; most surfaces are neither fully hydrophilic nor absolutely hydrophobic. One of the main determinants of the overall hydrophobicity of a sample is the extent to which the functional groups on the surface of the material interact with water. These interactions include intermolecular forces (dipole–dipole interactions and hydrogen bonds), as well as acid/base chemistry and other types of surface reactions involving water. Water will readily wet a hydrophilic surface (see Figure 3.3A), but will bead up on a hydrophobic surface (Figure 3.3B).

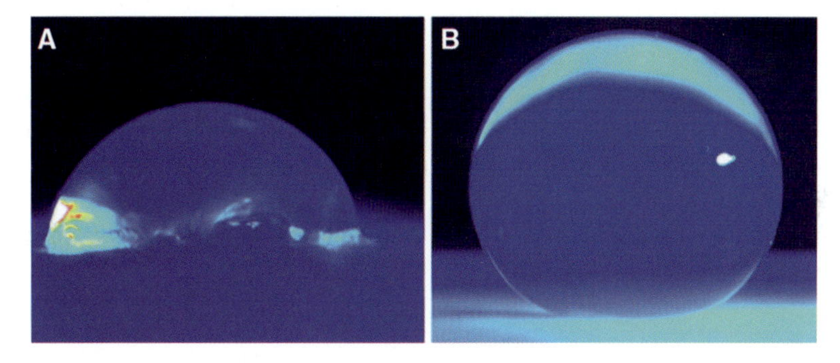

Figure 3.3. Water droplets on aerogel surfaces: **A.** a *TMOS*-based aerogel fabricated in an RSCE process and **B.** an aerogel made from a 25% by volume *TMOS* and 75% by volume *MTMS* mixture in an RSCE process. The water droplet on the *TMOS* aerogel spreads out and wets the surface, indicating a lack of hydrophobicity, whereas the droplet on the aerogel made with an *MTMS* co-precursor beads up, indicating that the surface is hydrophobic. We note that in some cases a water droplet on an unmodified *TMOS* aerogel is immediately taken up by the material. For the RSCE aerogel in **A.** the droplet remained on the surface of the aerogel.

Silica sol–gel matrices form via a series of condensation reactions of individual alkoxide molecules. The majority of the resulting material is silica (SiO_2). However, some of the alkoxide side chains remain unreacted (Si–OR) and some are left partially reacted (Si–OH) at the end of the polymerization reaction. If the silica precursors contained groups that could not react via condensation (Si–R), those groups will also be present in the aerogel matrix. When these groups (Si–OR, Si–OH, Si–R) are found at the surface of the aerogel, they can come into contact with water from the environment. And, as aerogels have unusually high surface areas compared to many other materials, there are many possible points of contact with water from the environment.

Silica aerogels prepared from *TMOS* or *TEOS* have Si–OR and Si–OH groups. The Si–OH groups render the aerogels hydrophilic. The Si–OH groups can undergo strong hydrogen bonding with water. We have observed that when *TMOS*-based aerogels are placed in water, they undergo structural collapse, rendering them unsuitable for chemical sensor applications in aqueous samples (Plata et al. [1]). Indeed, some silica aerogels are so hydrophilic as to be *hygroscopic*. Miner et al. [7] noted that significant uptake of water by silica aerogels in humid environments results in the aerogel materials becoming cloudier with time, and even fragmenting; this complicates the use of aerogel materials for thermal insulation. Moreover, the Si–OH group is a weak acid. Consequently, the average charge of these surface groups depends on the pH of the environment in which the gel is found. This property of silica is exploited in chromatography and electrophoresis applications.

The presence of a significant number of non-polar side groups (Si–R) on the surface of a silica aerogel renders it hydrophobic. Water molecules undergo only weak intermolecular forces with hydrocarbons and fluorohydrocarbons, but very strong intermolecular forces with other water molecules. This is why water beads up on PTFE surfaces and on hydrophobic aerogels, and why researchers have observed hydrophobic aerogels floating on top of water for months at a time (Rao et al. [33]) and even coating the surface of a water droplet (Hegde et al. [34]).

Measuring surface hydrophobicity is insufficient to demonstrate that the materials are impervious to water. Aerogels have highly porous nanostructures, and gas-phase species can readily pass through the aerogel matrix. Indeed, we and other groups have exploited that property of aerogels in preparing rapid-response gas-phase sensors. (See, for example, Leventis et al. [35], [36] and Plata et al. [1].) In order to demonstrate the utility of hydrophobic silica aerogels for applications in a variety of areas such as insulation (for example see Fesmire [2]), it is critically important to assess both the hydrophobicity of the interior of the aerogels and the extent to which water vapor can permeate aerogels that have hydrophobic surfaces.

There are three primary techniques used to make hydrophobic silica aerogels, all of which seek to replace the hydrophilic hydroxyl groups using surface modification. These involve either vapor phase after treatment, *co-precursor* techniques or derivatization methods.

Methoxylation or vapor-phase treatment (see Lee et al. [37] or Smirnova et al. [38]) involves heating the initially hydrophilic aerogel in the presence of methanol vapor, to convert the Si–OH groups to Si–OCH_3 groups. Figure 3.4 shows a schematic representation of the process. Aerogels are typically fabricated from *TEOS* or *TMOS* precursors using the CSCE method and then placed in a reactor which is heated to 220–240°C. Methanol vapor is then passed through the reactor for 10–40 h. Lee et al. [37] show through *FTIR* spectra that the –OH and Si–OH peaks, which exist in the initially hydrophilic samples, disappear after 10 h of surface treatment. In addition, they found that there was no significant decrease in transparency as a result of the vapor treatment. Methoxylation might occur during RSCE processing of *TMOS* aerogels, because the methanol present in the pores is brought to sufficiently high temperature. Anderson et al. [39] did not observe peaks characteristic of –OH or Si–OH stretches in the IR spectra of RSCE *TMOS* aerogels.

The co-precursor methods involve replacing some of the *TMOS*, *TEOS*, or other precursor with some quantity of *organosilane* material followed by *CSCE*, *ASCE*, or RSCE processing. Figure 3.5 shows a schematic representation of the process and Table 3.2 gives information about the different co-precursor chemicals that have been used to make hydrophobic silica aerogels. These co-precursors have at least one non-polar group bonded to the

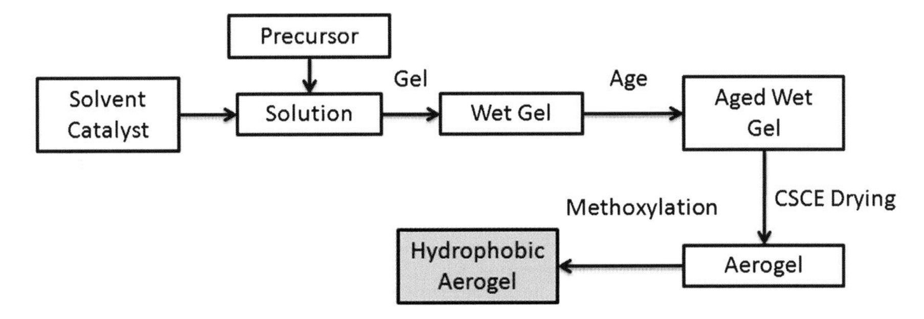

Figure 3.4. Schematic illustration of the methoxylation method of preparing hydrophobic silica aerogels. The precursor chemical, solvent, catalyst (and water) are mixed in solution and allowed to gel and age. The aged wet gel is then dried (typically through *CSCE*) and the resulting aerogel undergoes a methoxylation process in a reactor at high temperature to form a hydrophobic aerogel.

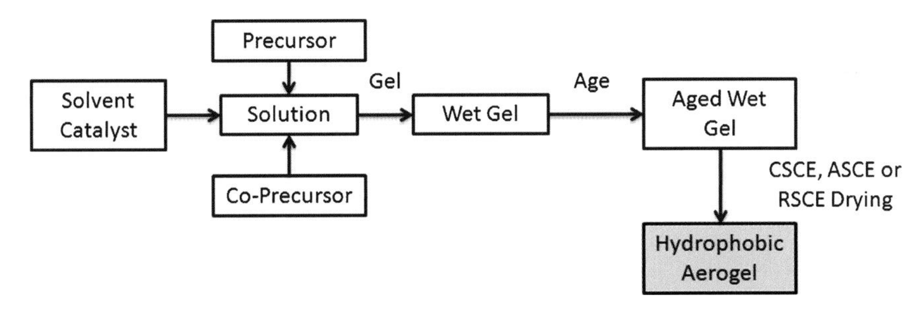

Figure 3.5. Schematic illustration of the co-precursor method of preparing hydrophobic silica aerogels. The hydrophobizing co-precursor is added to the liquid solution, the solution is allowed to gel and then the wet gel is dried by a *CSCE*, *ASCE*, or RSCE technique which results in a hydrophobic aerogel.

central silicon atom and when used will result in less polar aerogels. It is important to note that in addition to embedding particular side chains within the resulting aerogel, the use of a precursor containing a side chain that cannot participate in the condensation reactions also has an effect on the kinetics of the sol–gel formation. Because fewer branches are able to participate in the polymerization reaction, there will be fewer effective collisions in a given period of time which will result in longer gel times. Unreactive side chains can be bulky and can result in steric hindrance of the condensation reactions. In general, results show that increasing the amount of co-precursor leads to lower density, less transparent, more hydrophobic aerogels.

The third technique uses silylation to change the surface chemistry of the wet gels before drying. In this technique, wet gels are prepared using standard sol–gel chemistry. Then the wet gels undergo a number of solvent exchanges and soaking with a silylating agent which acts to modify the surface of the wet gel. Figure 3.6 shows a schematic representation of the process and Table 3.2 lists the different silylating agents used in the literature. As a result of the silylation, the nanostructure is more "elastic" and the wet gels can be dried at ambient temperature.

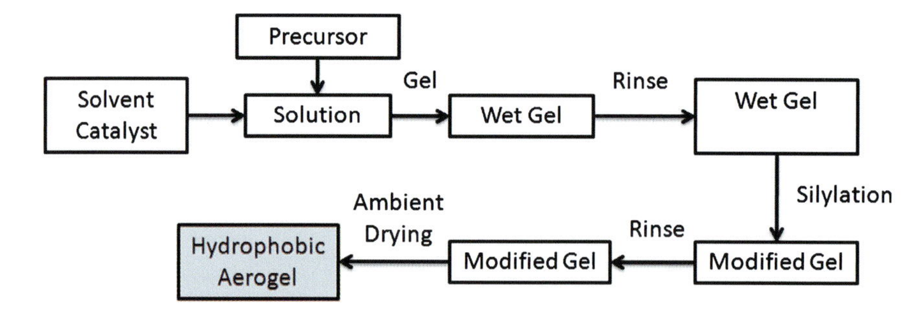

Figure 3.6. Schematic illustration of the silylation method of preparing hydrophobic silica aerogels. The precursor, solvent, catalyst (and water) are mixed and allowed to gel. The wet gel is then rinsed, soaked in a silylating agent, rinsed again and then dried at ambient pressure and elevated temperature.

3.3.2. How Do We Measure Hydrophobicity?

There are a number of techniques that can be used to measure the degree of hydrophobicity of a surface. Direct techniques such as water uptake measurements and sessile drop *contact angles* are often supplemented with spectroscopic studies using *FTIR* and *NMR* to provide evidence of structural groups present in aerogel samples and to quantitate the relative number of those groups. These methods are described below.

Water Uptake Measurement

One method used to assess the degree of hydrophobicity is to expose an aerogel to a humid environment and monitor the increase in the mass of the aerogel with time. If an aerogel is hydrophilic, its mass will increase significantly due to adsorbed water. As the humid air diffuses into the aerogel, water molecules interact strongly with the unreacted Si–OH groups. On the other hand, the mass of a truly hydrophobic aerogel should remain unchanged by exposure to humid air. These studies can be performed by placing the aerogels in a controlled humidity environment (typically >96% humidity) or by allowing the aerogels to float directly on water for a period of time (see Figure 3.7). Water uptake studies are typically carried out over a 1 to 6 month period. Schwertferger et al. [40], [41] and Husing et al. [42] monitored the behavior of hydrophilic and hydrophobic aerogels in a 20°C,

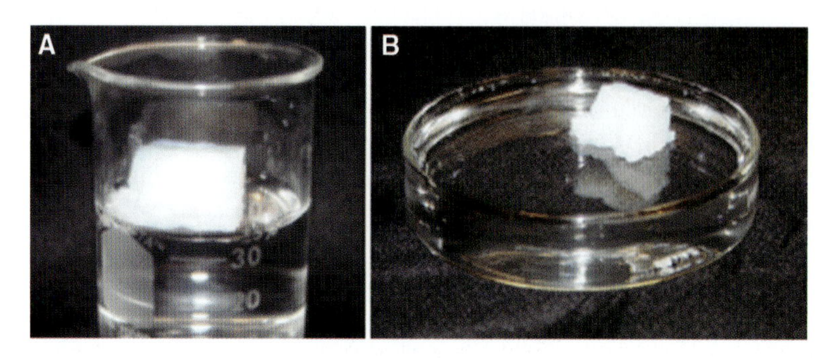

Figure 3.7. Hydrophobic aerogels floating on water.

96% humidity environment for 40 days. The hydrophilic aerogels exhibited mass increases of 10–60% within the first five days. Rao et al. [43] and Rao and Haranath [44] performed water uptake studies by floating aerogels directly on water for 3–6 months and considered an increase of less than 2% in mass as an indication that the aerogel was hydrophobic.

Contact Angle Measurement

Another technique used to assess the degree of hydrophobicity is to measure the contact angle that a sessile drop (sitting drop) makes with a surface. A *goniometer* can be used to measure the contact angle. A goniometer typically consists of an imaging system (high resolution camera) and image processing software. Commercial contact angle measurement instrumentation is available, but many research groups employ set-ups constructed in house. A small drop of water is placed on the surface of interest using a small-gauge hypodermic needle and an image is taken of the drop. The contact angle, θ, is the angle that the droplet makes with the surface as shown in Figure 3.8. Hydrophilic surfaces attract the water and the contact angle approaches 0° for highly hydrophilic surfaces. A surface is considered hydrophobic if it has a contact angle greater than 90°. *Superhydrophobic* surfaces are those surfaces on which the droplet appears to sit without any significant surface wetting and exhibits contact angles in excess of 150°. Figure 3.9 shows a 2.5-mm diameter drop of water on a hydrophobic silica aerogel prepared using the co-precursor method with a mixture of *TMOS* and *MTMS* and RSCE drying.

The contact angle can be measured using direct inspection of the image as shown in Figure 3.10A. However, it can be difficult to determine the exact contact point. There are a number of advanced image processing techniques used by commercial manufacturers that perform drop shape analysis to determine the contact angle. A simple method as recommended

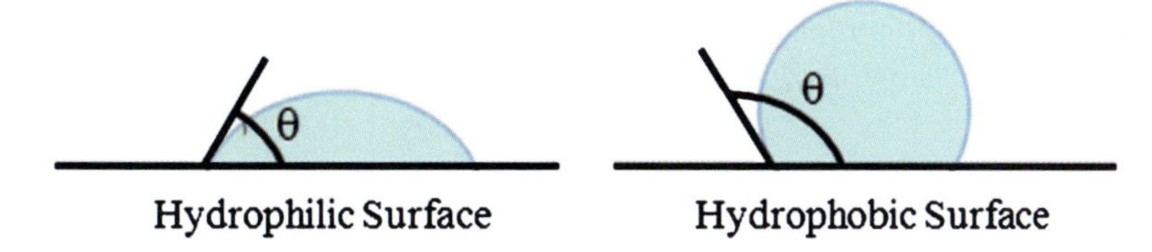

Figure 3.8. Representation of contact angles on hydrophilic and hydrophobic surfaces.

Figure 3.9. Image of a 2.5-mm diameter water drop on a hydrophobic silica aerogel monolith prepared using the co-precursor method with a mixture of *TMOS* and *MTMS* and RSCE drying.

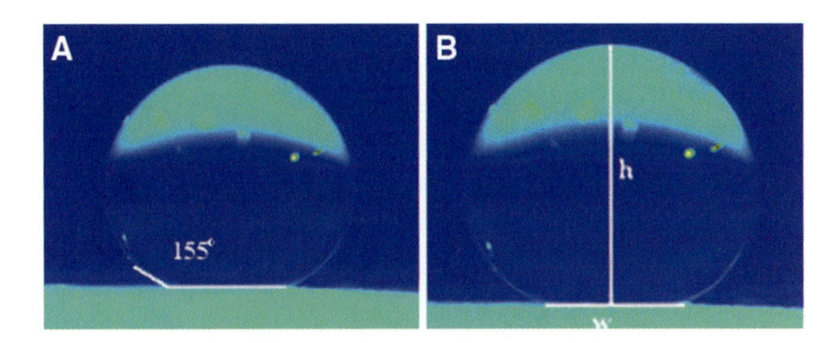

Figure 3.10. Illustration of **A.** the direct measurement technique and **B.** the height width measurement technique on a *TMOS*-based aerogel made with 50% (by volume) *MTMS* in an *RSCE* process. The direct measurement technique yields a contact angle of 155° and the height/width method yields a contact angle of 153° (for $w = 197\text{p}, h = 413\text{p}$).

by Rao and coworkers [43], [45], [46] can be used to estimate the contact angle from measurements of the drop height, h, and contact width, w (see Figure 3.10B). The contact angle is then estimated as

$$\theta = 2 * \tan^{-1}(2h/w)$$

Rao and Kulkarni [45] show that the direct measurement and height/width technique yield similar results.

Spectroscopic Measurements

Fourier Transform Infrared (*FTIR*) and solid-state Nuclear Magnetic Resonance (*NMR*) spectroscopy have been used to complement water uptake or contact angle studies. Both methods give direct structural information about the chemical bonds present in an aerogel sample. The differences in structure of hydrophobic and hydrophilic aerogels result in observed differences in their *FTIR* and *NMR* spectra.

Fourier Transform Infrared

Molecular vibrations occur in the infrared region of the electromagnetic spectrum. The frequency at which a vibration is observed depends on the masses of the atoms involved in the bond(s) and the strength of the bond(s) giving rise to the vibration. The intensity of the observed transition, which can be expressed in transmittance, reflectance, or absorbance units, is related to both the probability of the transition and the concentration of absorbing species in the sample. Signals characteristic of Si–O–H (ca. 3,200 cm^{-1}), Si–O–C–H (ca. 2,940 cm^{-1}), and Si–C–H (ca. 2,975 cm^{-1}) stretches are readily distinguished, as well as peaks due to other bond motions (rocking, bending). Water absorbs strongly in the IR, and if an aerogel sample contains a considerable amount of adsorbed water, it will appear as a strong, broad band near 3,500 cm^{-1}. It is more common to use *FTIR* qualitatively, but if appropriate standards are prepared, it is possible to compare the relative amounts of different bonds in a sample (see for example, Yokogawa and Yokoyama [6], Rao et al. [47] or Martín et al. [48]).

Nuclear Magnetic Resonance

When a sample containing nuclei that have spin >0 is placed in a magnetic field, formerly degenerate spin states for those atoms are split into slightly different energy levels.

Transitions between these nuclear spin states can then be probed using radiofrequency electromagnetic radiation of the appropriate energy. The frequency of the *NMR* signal from a nucleus depends on the extent to which that nucleus is shielded from the magnetic field by the electron cloud surrounding it. And the electron distribution around an atom depends on the bonds in which the atom is participating and the other nearby atoms. Coupling between *NMR*-active nuclei results in additional splitting of the energy levels. As a result, atoms that are of the same element but are present in different parts of the structure experience slightly different magnetic fields, and this gives rise to changes (shifts) in the observed resonance frequency. Consequently, information about the structure of a sample can be deduced through analysis of its *NMR* spectrum. There are a wide variety of *NMR* experiments and techniques, including multidimensional *NMR* spectroscopy, which can be employed to obtain detailed structural information about samples.

NMR spectroscopy of solid-state samples, such as aerogels, is more complicated than for non-viscous solutions. The rapid tumbling of molecules in solution and spinning of sample tubes averages out the anisotropic effects. But this tumbling does not occur in solid samples, so the observed shifts contain an anisotropic component. Grinding the sample to a small, uniform particle size and spinning it at the magic angle reduces these effects somewhat.

Silica aerogels can be interrogated using the spin ½ nuclei ^{1}H, ^{13}C and ^{29}Si. For example, El Rassy and Pierre [49] found that they could observe four distinct ^{1}H shifts, those due to bonds involving the hydrogen atoms in $Si–O–CH_3$ groups, Si–O–H groups, $Si–CH_3$ groups, and adsorbed water. They were able to integrate the ^{1}H spectra in order to make a quantitative assessment of the relative amounts of each group within each sample. And they observed differences in both ^{13}C and ^{29}Si *NMR* shifts due to the $Si–CH_3$ and $Si–O–CH_3$ groups.

3.4. A Review of the Literature

Table 3.3 presents a representative summary of the literature related to the synthesis of hydrophobic aerogels. The referenced papers are listed by first author (1st column) in chronological order (publication year in 2nd column). The main chemical used to fabricate the hydrophobic aerogel is listed in the precursor (3rd) column while the co-precursor or silylating agent is listed in the 5th column (see Tables 3.1 and 3.2 for information on the precursor and co-precursor chemicals). The technique used to hydrophobize the aerogel is found in the 4th column, with CP used to indicate a co-precursor method, M used to indicate methoxylation, S to indicate a silylation technique and OM used to represent organic modification. The drying method is indicated in column 6, with AMB used to represent ambient drying and *CSCE*, *ASCE* and RSCE used to indicate the type of supercritical drying, as defined above. Representative properties are included to give the reader a sense of the density, surface area, contact angle, and optical transparency. In some cases we present a range, in others a single value for what the author described as the "best" aerogel. Some authors performed thermal stability tests on the aerogels to determine the maximum temperature at which the aerogel remains hydrophobic. These temperatures are indicated in parentheses below the contact angle information. To indicate the level of transparency we include the percentage of light transmitted through the (ca. 1 cm) samples at a value near 800 nm (when reported, the actual thickness and wavelength values are indicated in the table below the transmission percentage).

Table 3.3. Summary of hydrophobic silica aerogel literature

First author and reference number	Year	Precursor chemical	Method used to make hydrophobic	Additional reagent employed	Drying method	Density (g/mL)	Surface area (m^2/g)	Hydrophobicity (contact angle or water uptake)	Optical transmission
Schwertfeger [40]	1992	*TMOS*	CP	*MTMS*	*ASCE*	0.06–0.29	505–679	Less than 1% water uptake	Decreases w/addition of *MTMS*
Pauthe [61]	1993	TriEOS	CP	–	*CSCE*	–	–	IR spect. (300°C)	–
Schwertfeger [41]	1994	*TMOS*	CP	Various[a]	*ASCE*	0.19–0.29	432–628	Less than 1% water uptake, IR spect.	–
Husing [42]	1995	*TMOS*	CP	*MTMS*+	*CSCE* and *ASCE*	0.24–0.38	505–686	Less than 1% water uptake	–
Lee [37]	1995	*TMOS* *TEOS*	M	Methanol Vapor	*CSCE*	0.03–0.38	750–870	–	70–80% (800 nm, 2.5 cm)
Yokogawa [6]	1995	MS51	CP, M	HMDZ	*CSCE*	0.04–0.06	–	–	90% (550 nm)
Schwertfeger [10]	1998	SS	S	HMDSO, *TMCS*	AMB	0.1–0.16	700	(250°C)	Translucent
A.V. Rao [33]	1998	*TMOS*	CP	TAM	*ASCE*	0.11	–	(223°C)	90% (900 nm, 1 cm)
A.V. Rao [44]	1999	*TMOS*	CP	*MTMS*	*ASCE*	–	875–1,000	Less than 2% water uptake (280°C)	10–90% (900 nm, 1 cm)
Jeong [67]	2000	*TEOS*	S	*TMCS*	AMB	0.8	950–980	157–159	–
Hrubesh [58]	2001	*TMOS*	CP	FPTMS	*ASCE*	0.2	790	150	–
A.V. Rao [50]	2001	*MTMS*	CP	*MTMS*	*ASCE*	0.052–0.08	720–1,140	50–140 (250°C)	30–92% (750 nm, 1 cm)
A.V. Rao [51]	2001	*TMOS*	CP	*MTMS*	ASCE	0.042–0.045	–	45–140	60–90% (800 nm, 1 cm)

A.V. Rao [45]	2002	*TMOS*	CP	*TMES*	*ASCE*	0.06–0.08	–	100–140	5–93% (750 nm)
A.V. Rao [53]	2002	*TMOS*	CP	*MTMS*, Glycerol	*ASCE*	0.04–0.15	–	Less than 6% uptake	45–95% (750 nm, 1 cm)
Wagh [52]	2002	*TMOS*	CP	*MTMS*	*ASCE*	0.08	–	Less than 2% water uptake (200°C)	80% (900 nm, 1 cm)
Lee [11]	2002	SS	S	*TMCS*	AMB	0.12–0.15	440–630	(350°C)	–
A.V. Rao [54]	2003	*TEOS*	CP	Various[b]	*ASCE*	0.12–0.25	–	128–136	5–65% (750 nm)
A.V. Rao [55]	2003	*TEOS*	CP	*TMES*	*ASCE*	0.11–0.15	–	110–136	5–40% (750 nm, 1 cm)
A.V. Rao [56]	2003	*TEOS*	CP	PhTES	*ASCE*	0.23–0.33	–	120–130 (520°C)	5–40%
A.V. Rao [43]	2003	*TMOS*	CP	Halides and Alkylalkoxy silanes[c]	*ASCE*	0.18–0.23 0.10–0.14	–	100–120 95–135	<8% >80% (700 nm, 1 cm)
A.V. Rao [46]	2003	*TMOS*	CP	*TMES*	*ASCE*	–	–	105–130 (300°C)	5–90% (700 nm, 1 cm)
El Rassy [91]	2003	*TMOS*	CP	*MTMS*	*CSCE*	–	740–840	25–80	–
Tillotson [82]	2004	*TMOS*	CP	FPTMS	*CSCE*	0.04–0.07	600–1,000	Less than 2.5% uptake	Opaque at high molar ratios
A.V. Rao [63]	2004	*TEOS*	S	Various[d]	*ASCE*	–	285	102–120 (520°C)	85%
			CP		*ASCE*		520	127–136 (280°C)	5%
Roig [71]	2004	*TEOS*	S	PFAS	*CSCE*	0.31	457	150	–
A.P. Rao [69]	2005	*TEOS*	S	HMDZ	AMB	0.075	–	160	90% (750 nm, 1 cm)
A.P. Rao [70]	2005	*TEOS*	S	HMDZ	AMB	0.05–0.2	400–750	130–170	55–90% (700 nm)

(continued)

Table 3.3 (continued)

First author and reference number	Year	Precursor chemical	Method used to make hydrophobic	Additional reagent employed	Drying method	Density (g/mL)	Surface area (m^2/g)	Hydrophobicity (contact angle or water uptake)	Optical transmission
Bhagat [64]	2006	*TEOS*	CP	Various[e]	*ASCE*	0.04–0.16	–	134–148	2–80% (750 nm, 1 cm)
Hegde [62]	2006	*TEOS*	CP	HDTMS	*ASCE*	0.04–0.10	–	98–152 (240°C)	0–83% (700 nm)
Ilhan [65]	2006	*TMOS*	OM	APTES and CMS with PS or FPS	*CSCE*	0.41–0.77	213–393	120 (PS) 151 (FPS)	Opaque
Shi [73]	2006	SS	S	*TMCS*	AMB	0.13–0.14	559–619	–	–
Zhang [8]	2006	PEDS	S	*TMCS*	AMB	–	700	135	–
Hegde [34]	2007	*TEOS*	CP	*TMES*	*ASCE*	0.07–0.11	–	135–155	0–65% (700 nm, 1.2 cm)
Bhagat [17]	2007	*MTMS*	CP	*MTMS*	AMB	0.06–0.30	400–520	152	–
Zhou [9]	2007	PEDS	CP	PFAS	*ASCE*	0.05–0.40	753–1,100	145	–
A.P. Rao [76]	2007	SS	S	HMDZ	AMB	0.06–0.030	–	65–165 (325°C)	60–85%
A.V. Rao [47]	2007	*MTMS*, *TEOS*	CP	Various[f]	*ASCE*	0.05–0.15	–	151–160 (520°C)	–
Štandeker [57]	2007	*TMOS*	CP	*MTMS*, *TMES*	*CSCE*	–	112–872, 732	42–173, 100–180	–
A.P. Rao [77]	2007	SS	S	Various[g]	AMB	0.16–0.06	–	60–165 (325°C)	30–88% (750 nm)
Wei [68]	2007	*TEOS*	S	TMCS	AMB	0.07–0.62	750–891	114–143	–
Martín [48]	2008	*TMOS*	CP	*MTMS*, TMSPMA	*ASCE* *CSCE*	0.11–0.16	399–880	125–160	Transparent and translucent
Bangi [12]	2008	SS	S	*TMCS*	AMB	0.08–0.2	–	130–146 (435°C)	20–60% (in visible range)
A.P. Rao [75]	2008	SS	S	*TMCS*	AMB	0.05–0.20	–	74–159	–
A.P. Rao [78]	2008	SS	S	*TMCS*	AMB	0.07–0.19	–	161–172	90–65% (700 nm)

A.P. Rao [79]	2008	SS	S	*TMCS*	AMB	0.05–0.08	476–812	145–159	85–50% (700 nm)
Shewale [81]	2008	SS	S	*TMCS* HMDZ	AMB	0.04–0.10 0.09–0.12	–	144–150 (450°C) 136–144 (430°C)	10–15% 55–75%
Gurav [72]	2009	*TEOS*	S	HMDZ	AMB	0.09–0.27	–	120–150	–
Shewale [80]	2009	SS	S	*TMCS*	AMB	0.08–0.14	–	138–146 (470°C)	70–25% (visible range)
Anderson [39]	2010	*TMOS*	CP	*MTMS*	RSCE	0.05–0.07	610–770	145–155	50–83% (800 nm, 1.2 cm)

CP co-precursor, *OM* organic modifier, *S Silylation, M* methoxylation, abbreviations used in the "Precursor" column are defined in Table 3.1; abbreviations used in the "Additional Reagent(s) Employed" column and in footnotes, below, are defined in Table 3.2; AMB ambient pressure extraction, *ASCE* alcohol supercritical extraction, *CSCE* CO_2 supercritical extraction, *RSCE* rapid supercritical extraction. The temperature indicated in the Hydrophobicity column indicates the maximum temperature at which the aerogels were hydrophobic.

[a] *MTMS*, PTMS, PhTMS, BTMS plus others.

[b] *MTMS*, MTES, DMCS, TMES, ETES, PhTES.

[c] *MTMS*, DMCS, DMDC, TMBS, *TMCS*, TMSCA, TMMS, TMES, HMDZ.

[d] *MTMS*, MTES, ETES, PhTES, DMCS, *TMCS*, TMES, HMDZ.

[e] *MTMS*, DMDS, TMES, ETES, PTMS, PhTES, PhTMS, VTMS, VTES.

[f] *MTMS*, ETES, PhTES.

[g] *MTMS*, MTES, VTMS, PhTMS, PhTES, DMDMS, TMMS, *TMCS*, BTSA, HMDZ.

3.4.1. Review of Co-precursor Methods

The majority of the work reported in the literature has focused on the use of organosilanes (or fluorinated organosilanes) as a co-precursor. The most widely studied co-precursor material is methytrimethoxysilane (*MTMS*) which differs from *TMOS* by the substitution of a methyl group for one of the methoxy groups. Some co-precursors are prohibitively expensive; however, *MTMS* is commercially available at a lower price than *TMOS*. Schwertfeger et al. [40] first demonstrated that the use of *MTMS* as a co-precursor would yield hydrophobic aerogels. Additional early studies were performed by Rao and coworkers [44], [50], [51] and Wagh and Ingale [52]. Rao and Pajonk [50] made monolithic aerogels with contact angle up to 140°, but low optical transmittance. They show that an *MTMS*/*TMOS* molar ratio of 0.7 yields good optical transmittance (80% at 750 nm for a 1-cm thick sample) with good hydrophobicity (contact angles of 110°). In an attempt to yield more uniform pore distributions in the *TMOS*/*MTMS* aerogels, Rao and Kulkarni [53] added glycerin as a drying control chemical additive (DCCA) to reduce drying stress. They found that low levels of glycerin improved the quality (lower density and higher transparency) of the resulting aerogels. Bhagat et al. [17] were able to make aerogels from 100% *MTMS* using an ambient pressure drying technique. Each *MTMS* moiety can undergo three condensation reactions, resulting in a solid silica network that contains Si–CH_3 groups. The resulting aerogels are elastic with a contact angle of 152°. Martín et al. [48] studied the mechanical strength characteristics of *TMOS*/*MTMS* aerogels and showed that the aerogels become more elastic as the amount of *MTMS* is increased but that the hardness remains unaffected. Using an *MTMS*/*TMOS* molar ratio of 1.5 they made aerogels with a contact angle of 160°. Anderson et al. [39] demonstrated that superhydrophobic aerogels (contact angles in excess of 150°) can be manufactured in as few as 15 h using a one-step, precursor-to-aerogel, RSCE method.

Another chemical commonly used as a hydrophobizing co-precursor is trimethyl-ethoxysilane (TMES). TMES is an organosilane with three methyl groups and one methoxy group. It can undergo only one condensation reaction to connect into the Si–O–Si backbone of the sol–gel matrix, leaving the $Si(CH_3)_3$ moeity unreacted in the matrix. Rao and coworkers [45], [46] investigated the use of *TMOS* with *TMES*. They found that as they increased the amount of co-precursor the aerogels became more hydrophobic and less transparent (similar to the *MTMS* results). Rao and coworkers [54], [55] used a *TEOS*-based recipe with TMES. The mixtures required more time to gel but were more hydrophobic than those made with *TMOS*. Rao et al. [56] studied *TEOS* with PhTES as a co-precursor and made hydrophobic aerogels. Hegde et al. [34] used a two-step acid/base catalysis process with *TEOS* and TMES that decreased the gelation time and resulted in higher hydrophobicity (contact angles up to 155°). Štandeker et al. [57] studied the effects of *MTMS* and TMES on hydrophobicity using *CSCE* for toxic organic compound clean-up applications and achieved contact angles of 42–173° for *MTMS*/*TMOS* molar ratios of 0.5–5 and contact angles of 100–180° for *TMES*/*TMOS* molar ratios of 0.5–5.

Several investigators have used fluorinated organosilane co-precursors. Hrubesh, Coronado and Satcher [58] and Reynolds, Coronado and Hrubesh [59], [60] incorporated FPTMS into silica aerogels using the co-precursor method and demonstrated that this type of aerogel can be used as an adsorbent for organic compounds in water. They measured contact angles of 150°. Zhou et al. [9] used a perfluoroalkylsilane (PFAS) co-precursor. The addition of the PFAS increased hydrophobicity (contact angle of 145°), density, and gelation time.

Other precursors/co-precursors that have been reported in the literature include TriEOS (Pauthe [61]) and TAM (Rao and Wagh [33]). Hegde and Rao [62] used a two-step process with *TEOS*/HDTMS, which decreased the gelation time and resulted in contact angles as high as 152°, although they found that the use of HDTMS decreased optical transmission.

Given the number of different variables (processing method, precursor, catalyst type, solvent, and water molar ratios) it can be difficult to compare the effects of co-precursor type on aerogel hydrophobicity. However, a number of papers in the literature provide direct comparison studies on the performance of different co-precursors. In general, it can be shown that, up to a limit, co-precursors with more alkyl/aryl groups will result in more hydrophobic aerogels. Rao et al. [43] studied a range of co-precursors with 0–3 functional groups in combination with *TMOS*. They saw an increase in contact angle from 95° for aerogels made with *MTMS* (a monoalkyl organosilane) to 135° for those made with hexamethyldisilazane (both Si atoms are trialkyl substituted). Rao and Kalesh [63] looked at *TEOS* with a variety of different co-precursors including TMES, *MTMS*, PhTES, MTES, ETES, and DMCS. They found that the use of *MTMS* and MTES co-precursors resulted in monolithic transparent aerogels. PhTES, TMES, and ETES aerogels were monolithic but opaque, whereas the DMCS aerogels were cracked and opaque. The contact angle increased with number of akyl or aryl groups covering the aerogel surface. Bhagat and Rao [64] looked at *TEOS* plus a range of co-precursors and found that as the chain length of alkyl groups present in the co-precursor increased from methyl to ethyl to propyl the contact angle increased. However, Anderson et al. [39] synthesized aerogels with *MTMS*, PTMS, and ETMS using a rapid supercritical extraction technique and saw no appreciable difference in contact angle for the different co-precursors at similar *TMOS* volume ratios. These RSCE fabricated aerogels also showed higher levels of hydrophobicity (compared to those in the literature) for low levels of co-precursor material.

A different approach is to make aerogel monoliths that combine a silica core with a cross-linked polystyrene shell. Ilhan et al. [65] used 3-aminopropyltriethoxysilane (APTES) as a co-precursor with *TMOS* in order to prepare an amine-derivatized silica sol gel, then reacted the amine groups with a chlorinated styrene derivative, followed by addition (free-radical) polymerization with either styrene or a 2,3,4,5,6-pentafluorostyrene (PFS). The fluorinated aerogels had higher hydrophobicity, with contact angles of 151°. To create stronger *TMOS*-based silica aerogels, Leventis et al. [66] cross-linked the wet gels with poly (hexamethylene diisocyanate) prior to CSCE processing. The resulting materials exhibited mass increases of only 4–5% when exposed to water vapor for three days, and withstood direct contact with liquids without collapse.

3.4.2. Review of Silylation Methods

Silylation has been performed using a variety of precursor materials including *TEOS*, sodium silicate, and PEDS. Due to the resulting surface passivation these techniques are generally able to use ambient drying. *TMCS* is probably the most widely used silylation agent but studies have been performed to look at the effects of other agents (see Table 3.3). Several studies have been performed using *TEOS*-based precursors with *TMCS*-based silylation. Jeong et al. [67] provide a direct comparison of *TEOS*-based, ambient-dried aerogel powders with and without silylation and show an increase in contact angle from near zero to 158° when silylation was used. Wei et al. [68] present a technique that utilizes multiple treatment steps with *TMCS* and results in low-density (0.069 g/cm^3), hydrophobic (contact angle 143°) monolithic aerogels. Other researchers have used *TEOS* with HMDZ as a silylating agent. Rao et al. [69] report on the use of a *TEOS* precursor with a two-step acid then base catalysis process (to speed up gelation) and HMDZ (in hexane) as a silylating agent with ambient drying. They were able to make superhydrophobic surfaces with high transparency and report the best quality aerogel had a contact angle of 160° and optical transmission of 90% at 700 nm. However, the process involves a number of solvent exchanges and takes about a week to complete. The work was extended in Rao et al. [70] to examine the effect of different solvents (hexane, cyclohexane, heptane, benzene, toluene, and xylene) in the solvent exchange steps. They obtained highly hydrophobic aerogels (contact angle = 172°) when xylene was used, however, the level of hydrophobicity decreased over time (down to a contact angle of ~140° 120 days later). Roig et al. [71] used PFAS in an after-treatment process through solvent exchange with both *CSCE* and ambient drying process and produced aerogels with contact angles of 150°. Gurav et al. [72] looked at reducing the number of steps required to make ambient-dried TEOS-based aerogels and produced aerogels in a four-day process with contact angles as high as 150°. Rao and Kalesh [63] compared *TEOS*-based hydrophobic aerogels made using the co-precursor and silylation methods and concluded that aerogels made using co-precursors were more hydrophobic but less transparent than those made using the silylation techniques.

Recent efforts to make aerogels from a lower-cost precursor, sodium metasilicate (frequently called "sodium silicate" or "water glass"), have resulted in hydrophobic aerogels (Chap. 5; their hydrophobic properties are reviewed here). Schwertfeger et al. [10] synthesized aerogels at ambient pressure using sodium silicate with water-soluble silylation agents (HMDSO and *TMCS*). Use of these compounds not only provides the necessary surface modification but also fills the pores with solvents more appropriate for ambient drying. The method was designed to avoid the lengthy solvent exchange step. Lee et al. [11] modified Schwertfeger's technique using solvent exchange and modification with an IPA/*TMCS*/n-Hexane solution. The IPA reacts with *TMCS* to yield isopropoxytrimethylsilane as a silylating agent. Other work with sodium silicate precursors was aimed at decreasing processing time (Shi et al. [73], Hwang et al. [74], Rao et al. [75]), studying the effects of preparation method (Rao et al. [76]), studying the effects of silylating agents (Rao et al. [77]), and studying the effects of solvents (Rao and coworkers [78], [79]). Bangi et al. [12] developed a sodium silicate processing route using tartaric acid instead of an ion-exchange process. They found that increasing the number of washings with water improved the transparency of the gels. Their best aerogel had low density (0.084 g/mL), high optical transmission (50%) and high contact angle (146°). Shewale et al. [80], [81] studied the effects of silylating agent type (HMDZ and *TMCS*) on the physical properties of silicate-based aerogels. Acetic acid was added to a sodium silicate solution to form and age the wet gel, which was then subjected to a vapor passing process to remove the sodium ions. The process involved several lengthy solvent-exchange steps and results showed that *TMCS* yielded lower density, more hydrophobic but less transparent aerogels than HMDZ.

3.4.3. Effect of Drying Method on Hydrophobicity

A study by Husing et al. [42] presents a comparison of aerogels dried using *CSCE* to those dried using *ASCE*. The aerogels were synthesized using *TMOS* with an *MTMS* co-precursor. They found that the concentration of Si–OH on the surface of the *ASCE* aerogel was negligible (making them hydrophobic) whereas the *CSCE* aerogels had non-negligible concentrations. However, Tillotson et al. [82] performed a similar comparison for aerogels made using *TMOS* with an FPTMS co-precursor and found that the water absorbing properties of the *CSCE* dried aerogels were similar to those of *ASCE* dried aerogels.

3.5. Applications

Chapter 4 reviews a number of important application areas for aerogels (see also Pierre and Pajonk [83] and Akimov [84]). Hydrophobic aerogels are beneficial in traditional application areas such as thermal insulation (Fesmire [2]), transparent window systems (Lee et al. [37]), and *Cherenkov* detectors (Gougas et al. [3]) (Chap. 28). The resistance to water vapor that the hydrophobicity offers aids in the long term stability of the aerogel material and allows it to remain both transparent (if needed) and thermally insulating. More recently, a number of new application areas have emerged that seek to specifically exploit the hydrophobic surface property in conjunction with other aerogel properties such as high surface area, high porosity, and nanoscale-sized features. This section presents a brief review of some of the ongoing work using hydrophobic aerogels in environmental, biological and superhydrophobic surface applications.

3.5.1. Environmental Clean-up and Protection

Investigators have studied the use of aerogels in environmental clean-up or protection applications. These applications take advantage of the porous, high surface area nature of the aerogel material that can be used to absorb contaminants. Adebajo et al. [85] review a range of porous materials and their absorbing properties. They describe the hydrophobicity, the *oleophilicity*, the ability to absorb the organic material at high volume and rate, and the ability for the material to be reused as important features of such materials and review the use of aerogels in this area. Figure 3.11 illustrates the oleophilic and hydrophobic nature of *TMOS/MTMS* aerogels.

Reynolds et al. [59], [60] investigated the use of hydrophobic aerogels for oil spill clean-up. They synthesized hydrophobic aerogels using the co-precursor method of combining *TMOS* with the fluorinated co-precursor FPTMS and dried the wet gels using high-temperature

Figure 3.11. Demonstration of the oleophilic and hydrophobic nature of *TMOS/MTMS* aerogels. The aerogel monolith was placed in a beaker containing oil and water. Note that the aerogel has adsorbed the yellow oil, and is floating at the top of the water. The top part of the silica aerogel retains its optical properties, indicating that this aerogel had capacity to absorb additional oil. No oil layer is visible on the water surface.

supercritical extraction. They show that the hydrophobic aerogels can absorb as much as 14 times their weight (or 237 times the weight of the CF_3 component) and are able to demonstrate the use of separation and extraction so that the aerogels could be re-used. Rao et al. [86] used superhydrophobic elastic aerogels synthesized with 100% *MTMS* to study the absorption and desorption of organic liquids (alkanes, aromatics, alcohols, and oils). They were able to demonstrate high uptake levels (10–21 times the aerogel mass) and showed that the mass of the organic liquid absorbed increases nearly linearly with an increase in the surface tension of the liquid. They were able to reuse the aerogel material at least three times.

Hrubesh et al. [58] performed a study to determine if FPTMS hydrophobic aerogel material could be used to clean organic solvents from wastewater. They performed adsorption studies with a number of solvents in water and compared the results to those achieved by granular activated carbon (GAC). They found that the hydrophobic aerogel material outperformed the GAC on a gram per gram comparison and was 30–130 times more effective. Štandeker et al. [57] synthesized hydrophobic aerogels with *TMOS* and *MTMS* or TMES co-precursors for a study on the use of these materials for the clean-up of toxic organic compounds in water. They showed that the hydrophobic aerogels have 15–400 times the capacity of GAC and that this capability persists for more than 20 adsorption/desorption cycles. Coleman et al. [87] reported on the development of an aerogel-GAC composite that can be used to clean groundwater by removing uranium. They fabricated composite materials by mixing *TMOS* with the hydrophobic sol–gel precursor FTPMS and with specified modifiers, such as H_3PO_4, $Ca(NO_3)_2$, and PPTES, gelation catalysts, and GAC in a supercritical reactor system and showed that this composite material exhibited superior performance compared to the use of GAC alone.

Rao et al. [47] studied the use of hydrophobic aerogels for surface protection (i.e. sculptures or statues) from corrosion due to interaction with the environment. Such applications would require transparent hydrophobic materials that are impervious to the polar organic solvents methanol, ethanol, and acetone and strong acids. They synthesized hydrophobic aerogels from a combination of *MTMS* and *TEOS* with ETES and PhTES co-precursors and measured contact angles with water and other solvents. They found that the ETES and PTES aerogels were impervious to aqueous solutions containing up to 20% of the solvents and 80% of the acids, whereas the *MTMS* aerogels were impervious to solutions containing up to 60% solvent or concentrated strong acid.

3.5.2. Biological Applications

Hydrophobic aerogels have also been used in the development of drug delivery systems and for enzyme encapsulation (Chaps. 30 and 31). These applications take advantages of the nano-sized structure and high surface area as well as hydrophobic nature. Simirnova et al. [5], [38] and Gorle et al. [88] have studied the feasibility of silica aerogels as platforms for drug delivery systems. They synthesized aerogels from *TMOS* using *CSCE* with a methoxylation process to hydrophobize the gels and demonstrated that the adsorption and release of the drugs studied can be controlled by the density and hydrophobicity of the aerogel materials. Aerogels have also been used to encapsulate enzymes (see for example Buisson et al. [89] or Orçaire et al. [90]). El Rassy and co-workers [49], [91], [92] reported that the hydrophobicity of the enzyme support system affects the success of such systems.

3.5.3. Superhydrophobic Surfaces

The development of superhydrophobic surfaces is another area of application for hydrophobic aerogels. Li et al. [93] provide a review of superhydrophobic surfaces including a description of how they are made and where they are used. They describe the properties of superhydrophobic surfaces (self-cleaning and anti-sticking) as having important applications for anti-fouling and anti-sticking materials, self-cleaning windshields, stain-resistant textiles, and even the manufacture of water-proof and fire-retardant clothing. Several investigators have looked at employing aerogel materials to make superhydrophobic surfaces. In this case the aerogels provide both a nanoscale surface roughness and a chemical hydrophobicity. Doshi et al. [4] demonstrated that they can make superhydrophobic surfaces from a *TMOS*-based recipe using FPTMS as a co-precursor to make an organo-silica aerogel film. Using UV/ozone treatments they are able to control the surface hydrophobicity to yield contact angles from 10 to 160°. Rao et al. [94] and Hegde et al. [34] demonstrated the transport of liquids on superhydrophobic aerogel surfaces. They prepared aerogels using *MTMS* or *TEOS/TMES* through *ASCE* drying. Using aerogel powdered surfaces they were able to make marbles of aerogel coated liquid droplets (~14 μL) which could be easily rolled on any surface, facilitating the transport of small quantities of liquid with potential applications in micro- and nanofluidic devices. Truesdell et al. [95] used aerogel material to fabricate a superhydrophobic surface. Their surface was constructed of a polydimethylsilane-patterned surface, coated with a layer of gold and then dip-coated in an aerogel material to form a superhydrophobic surface. Through measurements of the force and velocity field near the surface they showed that the drag on the surface can be reduced (Chap. 4).

3.6. Conclusion

As we have illustrated in this chapter, there are many methods for making hydrophobic silica aerogels. The sol–gel matrix can be rendered hydrophobic through inclusion in the initial sol–gel precursor mixture of co-precursors that contain non-polar side chains, or through a variety of post-gelation reactions (organic modification, silylation). Processing of the wet sol gels via ambient techniques, freeze-drying or supercritical extraction of the solvent yields hydrophobic aerogels. Dry aerogels can be rendered hydrophobic through methoxylation. The co-precursor method of making hydrophobic aerogels is relatively easy to employ. Studies have shown that silica aerogels become more hydrophobic with the addition of increasing amounts of co-precursor; however, this leads to longer gel times and reduced transparency. The use of silylation yields hydrophobic aerogels that can be dried under ambient pressure conditions. These aerogels are generally more transparent but the processing can be more complex. The combination of hydrophobicity with other properties of silica aerogels (high surface area, low density, low thermal conductivity, optical translucency, and so forth), makes these materials attractive for a variety of application areas of current interest, including chemical spill clean-up, drug delivery, and transparent insulation.

Acknowledgments

The authors thank the following Union College students for photographs used in this chapter: Emily Green, Jason Melville, and Caleb Wattley. Our own work with aerogels has been funded by grants from the National Science Foundation (NSF MRI CTS-0216153, NSF RUI CHE-0514527, NSF MRI CMMI-0722842, and NSF RUI CHE-0847901) and the American Chemical Society's Petroleum Research Fund (ACS PRF 39796-B10).

Any opinions, findings, and conclusions or recommendations expressed in this material are those of the authors and do not necessarily reflect the views of the National Science Foundation.

References

1. Plata, D L, Briones, Y J, Wolfe, R L, Carroll, M K, Bakrania, S D, Mandel, S G, & Anderson, A M (2004) Aerogel-platform optical sensors for oxygen gas. J Non-Cryst Solids 350: 326–335.
2. Fesmire, J (2006) Aerogel insulation systems for space launch applications. Cryogenics 46(2–3): 111–117.
3. Gougas, A, Ilie, D, Ilie, S, & Pojidaev, V (1999) Behavior of hydrophobic aerogel used as a cherenkov medium. Nucl Instrum Methods Phys Res, Sect A 421(1–2): 249–255.
4. Doshi, D A, Shah, P B, Singh, S, Branson, E D, Malanoski, A P, Watkins, E B (2005) Investigating the interface of superhydrophobic surfaces in contact with water. Langmuir 21(17): 7805–7811.
5. Smirnova, I, Mamic, J, & Arlt, W (2003) Adsorption of drugs on silica aerogels. Langmuir, 19(20): 8521–8525.
6. Yokogawa, H, & Yokoyama, M (1995) Hydrophobic silica aerogel. J Non-Cryst Solids 186: 23–29.
7. Miner, M R, Hosticka, B, & Norris, P M (2004) The effects of ambient humidity on the mechanical properties and surface chemistry of hygroscopic silica aerogel. J Non-Cryst Solids 350(7): 285–289.
8. Zhang, Z, Shen, J, Ni, X, Wu, G, Zhou, B, Yang, M, Gu, X, Qian, M, & Wu, Y, (2006) Hydrophobic silica aerogels strengthened with nonwoven fibers. J Macromol Sci Part A Pure Appl Chem 43(11): 1663–1670.
9. Zhou, B, Shen, J, Wu, Y, Wu, G, & Ni, X (2007) Hydrophobic silica aerogels derived from polyethoxydisiloxane and perfluoroalkylsilane. Mater Sci Eng, C 27(5–8): 1291–1294.
10. Schwertfeger, F, Frank, D, & Schmidt, M (1998) Hydrophobic waterglass based aerogels without solvent exchange or supercritical drying. J Non-Cryst Solids 225(1): 24–29.
11. Lee, C, Kim, G, & Hyun, S (2002) Synthesis of silica aerogels from waterglass via new modified ambient drying. J Mater Sci, 37(11): 2237–2241.
12. Bangi, U K H, Rao, A V, & Rao, A P (2008) A new route for preparation of sodium-silicate-based hydrophobic silica aerogels via ambient-pressure drying. Sci Technol Adv Mater 9(3): 035006.
13. Bisson, A, Rigacci, A, Lecomte, D, Rodier, E, & Achard, P (2003) Drying of silica gels to obtain aerogels: Phenomenology and basic techniques. Drying Technol 21(4): 593–628.
14. Prakash, S S, Brinker, C J, & Hurd, A J (1995) Silica aerogel films at ambient pressure. J Non-Cryst Solids 190(3): 264–275.
15. Prakash, S S, Brinker, C J, Hurd, A J, & Rao, S M (1995) Silica aerogel films prepared at ambient pressure by using surface derivatization to induce reversible drying shrinkage. Nature, 374(6521): 439–443.
16. Haereid, S, & Einarsrud, A (1994) Mechanical strengthening of *TMOS*-based alcogels by aging in silane solutions. J Sol-Gel Sci Technol 3: 1992–204.
17. Bhagat, S D, Oh, C S, Kim, Y H, Ahn, Y S, & Yeo, J G (2007) Methyltrimethoxysilane based monolithic silica aerogels via ambient pressure drying. Microporous Mesoporous Mater 100(1–3): 350–355.
18. Leventis, N, Palczer A, McCorkle L, Zhang G, & Sotiriou-Leventis C (2005) Nanoengineered silica-polymer composite aerogels with no need for supercritical fluid drying. J Sol-Gel Sci Tech 35: 99–105.
19. Pajonk, G M, Repellin-Lacroix, M, Abouarnadasse, S, Chaouki, J, & Klavana, D (1990) From sol-gel to aerogels and cryogels. J Non Cryst Solids 121: 66–67.
20. Kalinin, S, Kheifets, L, Mamchik, A, Knot'ko, A, & Vertigel, A (1999) Influence of the drying technique on the structure of silica gels. J Sol-Gel Sci Technol 15(1): 31–35.
21. Scherer, G W (1993) Freezing gels. J Non-Cryst Solids 155(1): 1–25.
22. Kistler, S S (1932) Coherent expanded aerogels. J Phys Chem 13: 52–64.
23. Phalippou, J, Woignier, T, & Prassas, M (1990) Glasses from aerogels. J Mater Sci 25(7): 3111–3117.
24. Danilyuk, A F, Gorodetskaya, T A, Barannik, G B, & Lyakhova, V F (1998) Supercritical extraction as a method for modifying the structure of supports and catalysts. React Kinet Catal Lett 63(1): 193–199.
25. Pajonk, G M, Rao, A V, Sawant, B M, & Parvathy, N N (1997) Dependence of monolithicity and physical properties of TMOS silica aerogels on gel aging and drying conditions. J Non-Cryst Solids 209(1): 40–50.

26. Tewari, P H, Hunt, A J, & Lofftus, K (1985) Ambient-temperature supercritical drying of transparent silica aerogels. Mater Lett 3(9): 363–367.
27. Van Bommel, M J, & de Haan, A B (1995) Drying of silica aerogel with supercritical carbon dioxide. J Non-Cryst Solids 186: 78–82.
28. Poco, J F, Coronado, P R, Pekala, R W, & Hrubesh, L W (1996) A rapid supercritical extraction process for the production of silica aerogels. Mat Res Soc Symp 431: 297–302.
29. Scherer, G W, Gross, J, Hrubesh, L W, & Coronado, P R (2002) Optimization of the rapid supercritical extraction process for aerogels. J Non-Cryst Solids 311(3): 259–272.
30. Gauthier, B M, Bakrania, S D, Anderson, A M, & Carroll, M K (2004) A fast supercritical extraction technique for aerogel fabrication. J Non-Cryst Solids 350: 238–243.
31. Roth, T B, Anderson, A M, & Carroll, M K (2008) Analysis of a rapid supercritical extraction aerogel fabrication process: Prediction of thermodynamic conditions during processing. J Non-Cryst Solids 354(31): 3685–3693.
32. Anderson, A M, Wattley, C W, & Carroll, M K (2009) Silica aerogels prepared via rapid supercritical extraction: Effect of process variables on aerogel properties. J Non-Cryst Solids, 355(2): 101–108.
33. Rao, A V, & Wagh, P (1998) Preparation and characterization of hydrophobic silica aerogels. Mater Chem Phys 53(1): 13–18.
34. Hegde, N D, Hirashima, H, & Rao, A V (2007) Two step sol-gel processing of TEOS based hydrophobic silica aerogels using trimethylethoxysilane as a co-precursor. J Porous Mater 14(2): 165–171.
35. Leventis, N, Elder, I A, Rolison, D R, Anderson, M L, & Merzbacher, C I (1999) Durable modification of silica aerogel monoliths with fluorescent 2,7-diazapyrenium moieties sensing oxygen near the speed of open-air diffusion. Chem Mater 11: 2837–2845.
36. Leventis, N, Rawashdeh, A-M M, Elder, I A, Yang, J, Dass, A, & Sotiriou-Leventis, C (2004) Synthesis and Characterization of Ru(II) Tris(1,10-phenanthroline)-Electron Acceptor Dyads Incorporating the 4-benzoyl-N-methylpyridinium cation or N-Benzyl-N′-methyl-viologen Improving the Dynamic Range, Sensitivity and Response Time of Sol-Gel Based Optical Oxygen Sensors. Chem Mater 16: 1493–1506.
37. Lee, K H, Kim, S Y, & Yoo, K I P (1995) Low-density, hydrophobic aerogels. J Non-Cryst Solids 186: 18–22.
38. Smirnova, I, Suttiruengwong, S, & Arlt, W (2004) Feasibility study of hydrophilic and hydrophobic silica aerogels as drug delivery systems. J Non-Cryst Solids 350: 54–60.
39. Anderson, A M, Carroll, M K, Green, E C, Melville, J M, & Bono, M S (2010) Hydrophobic silica aerogels prepared via rapid supercritical extraction. J Sol-Gel Sci Technol 53(2): 199–207.
40. Schwertfeger, F, Glaubitt, W, & Schubert, U (1992) Hydrophobic aerogels from Si (OMe)/MeSi (OMe) mixtures. J Non-Cryst Solids 145: 85–89.
41. Schwertfeger, F, Hüsing, N, & Schubert, U (1994) Influence of the nature of organic groups on the properties of organically modified silica aerogels. J Sol-Gel Sci Technol 2(1): 103–108.
42. Hüsing, N, Schwertfeger, F, Tappert, W, & Schubert, U (1995) Influence of supercritical drying fluid on structure and properties of organically modified silica aerogels. J Non-Cryst Solids 186: 37–43.
43. Rao, A V, Kulkarni, M M, Amalnerkar, D, & Seth, T (2003) Surface chemical modification of silica aerogels using various alkyl-alkoxy/chloro silanes. Appl Surf Sci 206(1–4): 262–270.
44. Rao, A V, & Haranath, D (1999) Effect of methyltrimethoxysilane as a synthesis component on the hydrophobicity and some physical properties of silica aerogels. Microporous Mesoporous Mater 30(2–3): 267–273.
45. Rao, A V, & Kulkarni, M M (2002) Hydrophobic properties of *TMOS*/TMES-based silica aerogels. Mater Res Bull 37: 1667–1677.
46. Rao, A V, Kulkarni, M, Pajonk, G, Amalnerkar, D, & Seth, T (2003) Synthesis and characterization of hydrophobic silica aerogels using trimethylethoxysilane as a co-precursor. J Sol-Gel Sci Technol 27(2): 103–109.
47. Rao, A V, Hegde, N D, & Shewale, P M (2007) Imperviousness of the hydrophobic silica aerogels against various solvents and acids. Appl Surf Sci 253(9): 4137–4141.
48. Martín, L, Ossó, J O, Ricart, S, Roig, A, García, O, & Sastre, R (2008) Organo-modified silica aerogels and implications for material hydrophobicity and mechanical properties. J Mater Chem 18(2): 207–213.
49. El Rassy, H, & Pierre, A (2005) *NMR* and IR spectroscopy of silica aerogels with different hydrophobic characteristics. J Non-Cryst Solids 351(19–20): 1603–1610.
50. Rao, A V, & Pajonk, G M (2001) Effect of methyltrimethoxysilane as a co-precursor on the optical properties of silica aerogels. J Non-Cryst Solids 285(1–3): 202–209.
51. Rao, A V, Pajonk, G, & Haranath, D (2001) Synthesis of hydrophobic aerogels for transparent window insulation applications. Mater Sci Technol 17(3): 343–348.

52. Wagh, P B, & Ingale, S V (2002) Comparison of some physico-chemical properties of hydrophilic and hydrophobic silica aerogels. Ceram Int 28(10): 43–50.
53. Rao, A V, & Kulkarni, M M (2002) Effect of glycerol additive on physical properties of hydrophobic silica aerogels. Mater Chem Phys 77(3): 819–825.
54. Rao, A V, & Kalesh, R R (2003) Comparative studies of the physical and hydrophobic properties of TEOS based silica aerogels using different co-precursors. Sci Technol Adv Mater 4(6): 509–515.
55. Rao, A V, Kalesh, R R, Amalnerkar, D P, & Seth, T (2003) Synthesis and characterization of hydrophobic *TMES/TEOS* based silica aerogels. J Porous Mater 10(1): 23–29.
56. Rao, A V, Kalesh, R, & Pajonk, G (2003) Hydrophobicity and physical properties of TEOS based silica aerogels using phenyltriethoxysilane as a synthesis component. J Mater Sci, 38(21): 4407–4413.
57. Štandeker, S, Novak, Z, & Knez, Ž (2007) Adsorption of toxic organic compounds from water with hydrophobic silica aerogels. J Colloid Interface Sci 310(2): 362–368.
58. Hrubesh, L W, Coronado, P R, & Satcher, J H (2001) Solvent removal from water with hydrophobic aerogels. J Non-Cryst Solids 285(1–3): 328–332.
59. Reynolds, J G, Coronado, P R, & Hrubesh, L W (2001) Hydrophobic aerogels for oil-spill cleanup-intrinsic absorbing properties. Energy Sources 23: 831–834.
60. Reynolds, J G, Coronado, P R, & Hrubesh, L W (2001) Hydrophobic aerogels for oil-spill cleanup-synthesis and characterization. J Non-Cryst Solids 292: 127–137.
61. Pauthe, M (1993) Hydrophobic silica CO_2 aerogels. J Non-Cryst Solids 155: 110–114.
62. Hegde, N D, & Rao, A V (2006) Organic modification of TEOS based silica aerogels using hexadecyltrimethoxysilane as a hydrophobic reagent. Appl Surf Sci 253(3): 1566–1572.
63. Rao, AV, & Kalesh, R R (2004) Organic surface modification of TEOS based silica aerogels synthesized by co-precursor and derivatization methods. J Sol-Gel Sci Technol 30(3): 141–147.
64. Bhagat, S D, & Rao, A V (2006) Surface chemical modification of TEOS based silica aerogels synthesized by two step (acid–base) sol–gel process. Appl Surf Sci 252(12): 4289–4297.
65. Ilhan, U F, Fabrizio, E F, McCorkle, L, Scheiman, D A, Dass, A, Palczer, A, Meador, M B, Johnston, J C, & Leventis, N (2006) Hydrophobic monolithic aerogels by nanocasting polystyrene on amine-modified silica. J Mater Chem 16(29): 3046–3054.
66. Leventis, N, Sotiriou-Leventis, C, Zhang, G, & Rawashdeh, A-M M (2002) Nano-engineering strong silica aerogels. Nano Lett 2: 957–960.
67. Jeong, A Y, Koo, S M, & Kim, D P (2000) Characterization of hydrophobic SiO_2 powders prepared by surface modification on wet gel. J Sol-Gel Sci Technol 19(1): 483–487.
68. Wei, T, Chang, T, Lu, S, & Chang, Y (2007) Preparation of monolithic silica aerogel of low thermal conductivity by ambient pressure drying. J Am Ceram Soc 90(7): 2003–2007.
69. Rao, A P, Pajonk, G, & Rao, A V (2005) Effect of preparation conditions on the physical and hydrophobic properties of two step processed ambient pressure dried silica aerogels. J Mater Sci 40(13): 3481–3489.
70. Rao, A P, Rao, A V, & Pajonk, G (2005) Hydrophobic and physical properties of the two step processed ambient pressure dried silica aerogels with various exchanging solvents. J Sol-Gel Sci Technol 36(3): 285–292.
71. Roig, A, Molins, E, Rodríguez, E, Martínez, S, Moreno-Mañas, M, & Vallribera, A (2004) Superhydrophobic silica aerogels by fluorination at the gel stage. Chem Commun 2004(20): 2316–2317.
72. Gurav, J L, Rao, A V, & Bangi, U K H (2009) Hydrophobic and low density silica aerogels dried at ambient pressure using TEOS precursor. J Alloys Compd 471(1–2): 296–302.
73. Shi, F, Wang, L, & Liu, J (2006) Synthesis and characterization of silica aerogels by a novel fast ambient pressure drying process. Mater Lett 60(29–30): 3718–3722.
74. Hwang, S, Jung, H, Hyun, S, & Ahn, Y (2007) Effective preparation of crack-free silica aerogels via ambient drying. J Sol-Gel Sci Technol 41(2): 139–146.
75. Rao, A P, Rao, A V, & Bangi, U K H (2008) Low thermalconductive, transparent and hydrophobic ambient pressure dried silica aerogels with various preparation conditions using sodium silicate solutions. J Sol-Gel Sci Technol 47(1): 85–94.
76. Rao, A P, Rao, A V, Pajonk, G, & Shewale, P M (2007) Effect of solvent exchanging process on the preparation of the hydrophobic silica aerogels by ambient pressure drying method using sodium silicate precursor. J Mater Sci 42(20): 8418–8425.
77. Rao, A P, Rao, A V, & Pajonk, G (2007) Hydrophobic and physical properties of the ambient pressure dried silica aerogels with sodium silicate precursor using various surface modification agents. Appl Surf Sci 253(14): 6032–6040.

78. Rao, A P, Rao, A V, & Gurav, J L (2008) Effect of protic solvents on the physical properties of the ambient pressure dried hydrophobic silica aerogels using sodium silicate precursor. J Porous Mater 15(5): 507–512.
79. Rao, A P, & Rao, A V (2008) Microstructural and physical properties of the ambient pressure dried hydrophobic silica aerogels with various solvent mixtures. J Non-Cryst Solids 354(1): 10–18.
80. Shewale, P M, Rao, A V, Gurav, J L, & Rao, A P, (2009) Synthesis and characterization of low density and hydrophobic silica aerogels dried at ambient pressure using sodium silicate precursor. J Porous Mater 16(1): 101–108.
81. Shewale, P M, Rao, A V, & Rao, A P (2008) Effect of different trimethyl silylating agents on the hydrophobic and physical properties of silica aerogels. Appl Surf Sci 254(21): 6902–6907.
82. Tillotson, T M, Foster, K G, & Reynolds, J G (2004) Fluorine-induced hydrophobicity in silica aerogels. J Non-Cryst Solids 350: 202–208.
83. Pierre, A C, & Pajonk, G M (2002) Chemistry of aerogels and their applications. Chem Rev 102: 4243–4265.
84. Akimov, Y K (2003) Fields of application of aerogels (review). Instrum Exp Tech 46(3): 287–299.
85. Adebajo, M, Frost, R, Kloprogge, J, Carmody, O, & Kokot, S (2003) Porous materials for oil spill cleanup: A review of synthesis and absorbing properties. J Porous Mater 10(3): 159–170.
86. Rao, A V, Hegde, N D, & Hirashima, H (2007) Absorption and desorption of organic liquids in elastic superhydrophobic silica aerogels. J Colloid Interface Sci 305(1): 124–132.
87. Coleman, S J, Coronado, P R, Maxwell, R S, & Reynolds, J G (2003) Granulated activated carbon modified with hydrophobic silica aerogel-potential composite materials for the removal of uranium from aqueous solutions. Environ Sci Technol 37(10): 2286–2290.
88. Gorle, B S K, Smirnova, I, & McHugh, M A (2009) Adsorption and thermal release of highly volatile compounds in silica aerogels. J Supercrit Fluids 48(1): 85–92.
89. Buisson, P, Hernandez, C, Pierre, M, & Pierre, A C (2001) Encapsulation of lipases in aerogels. J Non-Cryst Solids 285(1): 295–302.
90. Orçaire, O, Buisson, P, & Pierre, A C (2006) Application of silica aerogel encapsulated lipases in the synthesis of biodiesel by transesterification reactions. J Mol Catal B: Enzym 42(3–4): 106–113.
91. El Rassy, H, Buisson, P, Bouali, B, Perrard, A, & Pierre, A C (2003) Surface characterization of silica aerogels with different proportions of hydrophobic groups, dried by the CO_2 supercritical method. Langmuir 19(2): 358–363.
92. El Rassy, H, Maury, S, Buisson, P, & Pierre, A (2004) Hydrophobic silica aerogel–lipase biocatalysts possible interactions between the enzyme and the gel. J Non-Cryst Solids 350: 23–30.
93. Li, X M, Reinhoudt, D, & Crego-Calama, M (2009) What do we need for a superhydrophobic surface? A review on the recent progress in the preparation of superhydrophobic surfaces. Chem Soc Rev 36: 1350–1368.
94. Rao, A V, Kulkarni, M M, & Bhagat, S D (2005) Transport of liquids using hydrophobic aerogels. J Colloid Interface Sci 285: 413–418.
95. Truesdell, R, Mammoli, A, Vorobieff, P, Van Swol, F, & Brinker, C J (2006) Drag reduction on a patterned superhydrophobic surface. Phys Rev Lett 97(4): 044504.

4

Superhydrophobic and Flexible Aerogels

A. Venkateswara Rao, G. M. Pajonk, Digambar Y. Nadargi, and Matthias M. Koebel

Abstract Aerogels with reduced fragility and increased hydrophobicity have significant potential to expand their use as lightweight structural, insulating or shock absorbing materials especially in aeronautics, microelectronics, and sensing applications. In addition, there is a potential for extremely hydrophobic aerogels in oil-spill clean-up applications. This chapter describes synthesis, physico-chemical properties, and applications of flexible superhydrophobic silica aerogels that is to say silica aerogels with typical water contact angles $>150°$ and high mechanical flexibility. Such materials are accessible via a two-step sol–gel process from methyl-trialkoxysilane precursors. Extreme hydrophobicity has been obtained with measured water contact angles as high as 175°. The criticality of the water droplet size on a superhydrophobic aerogel was determined to be 2.7 mm. The velocity of the water droplet on such a superhydrophobic surface has been observed to be 1.44 m/s for 55° inclination, which is close to the free fall velocity (~1.5 m/s). Elastic and rheological properties of as-prepared aerogels are also described in this chapter. Young's modulus of the aerogels is determined by uniaxial compression test measurements. Apart from synthesis and characterization, emphasis is placed on their potential use as shock absorbing materials and efficient absorbents of oil and organic compounds in general.

4.1. Introduction

The main technological interest in aerogels originated from a need of low-density materials for high-energy physics applications with good optical transmission in the visible range and a low index of refraction [1–4] (Chap. 28). However, few attempts were made so far to use silica aerogels as shock absorbing media as they are typically very fragile and brittle. Additionally, they can be sensitive to moisture which causes them to lose their structural network integrity, with time or upon applying a mechanical load. Therefore, with

A. V. Rao • Air Glass Laboratory, Department of Physics, Shivaji University, 416004 Kolhapur, Maharashtra, India
e-mail: avrao2012@gmail.com
G. M. Pajonk • Laboratoire des Matériaux et Procédés Catalytiques, Université Claude Bernard Lyon 1, 43 Boulevard du 11 novembre 1918, 69622 Villeurbanne, France
e-mail: pajonk1939@laposte.net
D. Y. Nadargi and M. M. Koebel • Laboratory for Building Technologies, Empa, Swiss Federal Laboratories for Materials Science and Technology, 8600 Dübendorf, Switzerland
e-mail: digambar_nadargi@yahoo.co.in; matthias.koebel@empa.ch

M.A. Aegerter et al. (eds.), *Aerogels Handbook*, Advances in Sol-Gel Derived Materials and Technologies, DOI 10.1007/978-1-4419-7589-8_4,

the purpose of broadening the application range of aerogels while fully capitalizing on their outstanding properties, mechanically more robust aerogels are needed. Various methods have been tried to improve the mechanics of silica aerogels such as structural reinforcement by polymer cross-linking, carbon nanotubes, etc. [5–8]. In principle, to improve flexibility or elastic recovery in silica aerogels, it is required either to include organic linking groups in the underlying silica structure or to cross-link the skeletal gel framework through surface silanol groups by reacting them with monomers/polymers. In the cross-linking approach, elastic properties can be induced as demonstrated by Leventis et al. [9] (Chap. 13). In their study, they have used di-isocyanate to react with silanols on the surface of the silica, creating a conformal polymer coating over the entire silica skeleton. This cross-linking increases the strength of the reinforced silica aerogels over native silica aerogels by as much as three orders of magnitude while only doubling the density. However, it has been shown that elastic properties can be enhanced in noncross-linked aerogels by alternating the chemical nature of the silica backbone. Kramer et al., for example, demonstrated that an addition of up to 20% (W/W) poly(dimethylsiloxane) in tetraethoxyorthosilicate (*TEOS*)-based aerogels resulted in rubbery behavior with up to 30% recoverable compressive strain [10]. Kanamori et al., using a surfactant to control the pore size, have shown that *MTMS*-derived gels can have reversible deformation upon compression [11]. Shea and Loy have developed hybrid organic–inorganic aerogels from bridged polysilsesquioxanes, using building blocks comprised of organic bridging groups attached to two or more trialkoxysilyl groups via nonhydrolyzable carbon–silicon bonds [12–14]. All of these methods have shown possibilities and in some sense also the limitations of tailoring the mechanical properties of aerogels to one's needs.

Besides, there is a need for materials for spill clean-up of organic matter such as oils, hydrocarbons, toxic waste after accidental release. Reynolds et al. [15] demonstrated the use of hydrophobic aerogels for oil-spill clean-up. They showed that aerogels exhibit a good performance in oil-spill cleaning: when treating oil/salt–water mixtures with perfluoroalkyl (CH_2 and CF_3-groups) containing aerogels, oil amounts of up to 3.5 times that of the aerogel by volume could be absorbed completely. Hrubesh et al. [16] demonstrated the removal of organic solvents (miscible and immiscible in water) using hydrophobic aerogels. More recently, Suzana et al. [17] determined the adsorption capacity of hydrophobic aerogels for various toxic organic compounds from an aqueous phase. These aerogels remained stable even after 20 adsorption/desorption cycles and exhibited 15–400 times higher uptake capacity in comparison to granulated active carbon (GAC).

In an altogether different approach to improve the mechanics and introduce superhydrophobic properties, in this chapter, we are presenting work on the synthesis of flexible and superhydrophobic silica aerogels derived from methyltrimethoxysilane (*MTMS*) and methyltriethoxysilane (*MTES*) precursors and their usability as efficient sorbents for oil and hydrocarbons [18, 19]. The main objective of this chapter is to provide an overview of synthesis, characterization, and most promising areas of applications.

4.2. Synthesis and Characterization

Sol–gel processes in general are known to involve both hydrolysis and polycondensation reactions leading up to the formation of network structures in liquid solution with varying degrees of cross-linking. This network connectivity depends on a number of reaction parameters, which are easy to describe phenomenologically but rather cumbersome to understand in detail. Perhaps the central parameter is the type of *precursor* used for the synthesis, as it defines the chemistry at the molecular level of each building block. Commonly, *precursor* molecules with four Si–O linkages such as tetraalkoxysilanes or inorganic silicates (waterglass) are used to synthesize silica aerogels leading up to the maximum coordination and the highest possible connectivity within the silica microstructure. By reducing the number of Si–O bonding centers on each Si atom, the connectivity is reduced. The simplest and most straightforward way to do so is to substitute Si–O moieties by Si–R, where R is a nonhydrolysable alkyl group such as methyl ($-CH_3$) or ethyl (CH_2CH_3). Table 4.1 summarizes the range of alkyl–alkoxysilane compounds bearing 0–3 alkyl functional groups.

Table 4.1. Structure, silicon/oxygen ratio, R/Si ratio of the various alkyl–alkoxysilane compounds

Units	Monofunctional	Difunctional	Trifunctional	Quadrifunctional
Structure	$R_3Si{-}OR'$ (R–Si(R)(R)–OR')	$R'O{-}SiR_2{-}OR'$ (R'O–Si(R)(R)–OR')	$R'O{-}SiR(OR'){-}OR'$ (R'O–Si(R)(OR')–OR')	$R'O{-}Si(OR')_2{-}OR'$ (R'O–Si(OR')(OR')–OR')
Silicon/oxygen ratio	0.5	1	1.5	2
R/Si ratio	3	2	1	0
Examples	TMES	DMMS	*MTMS*, *MTES*	*TEOS*, TMOS

The role of replacing alkoxy by alkyl groups is twofold: first, there is the connectivity argument which was mentioned before. With increasing R-substitution, the number of possible covalent Si–O–Si linkages responsible for the silica network formation is reduced. Second, alkyl groups are carriers of hydrophobicity. During the sol-forming stage, such groups will have a tendency to orient toward one another which results in hydrophobic domains or pockets within the microstructure of the silica. This, however, does not mean that hydrophobicity of a gel network can be increased generally by using siloxanes with more alkyl functional groups: with increasing R/Si ratio, the number of bonding centers available for building up a coherent silica network decreases. Trifunctional alkoxysilane compounds such as *MTMS* and *MTES* offer a decent compromise between hydrophobic property and a stable network structure. In combination with the reduced silica network connectivity, the superhydrophobic nature leads to mechanical properties which are completely different from those of native silica aerogels: Much in anology to polymers, reducing the extent of cross-linking results in a substantial increase in the materials flexibility. Alkylalkoxysilane-derived aerogels exhibit outstanding flexibility, a fact which will be demonstrated in the section dealing with mechanical properties later on in the text.

4.2.1. Sol–Gel Synthesis and Supercritical Drying

Much like pure alkoxysilane aerogels, superhydrophobic and flexible *MTMS* and or *MTES*-derived silica aerogels are made in two steps: (1) the preparation of an alcogel by single as well as two-step sol–gel processes and (2) the supercritical drying of the wet gels to remove the interstitial solvent. The complete hydrolysis and condensation reactions of *MTMS* are shown below in Figure 4.1.

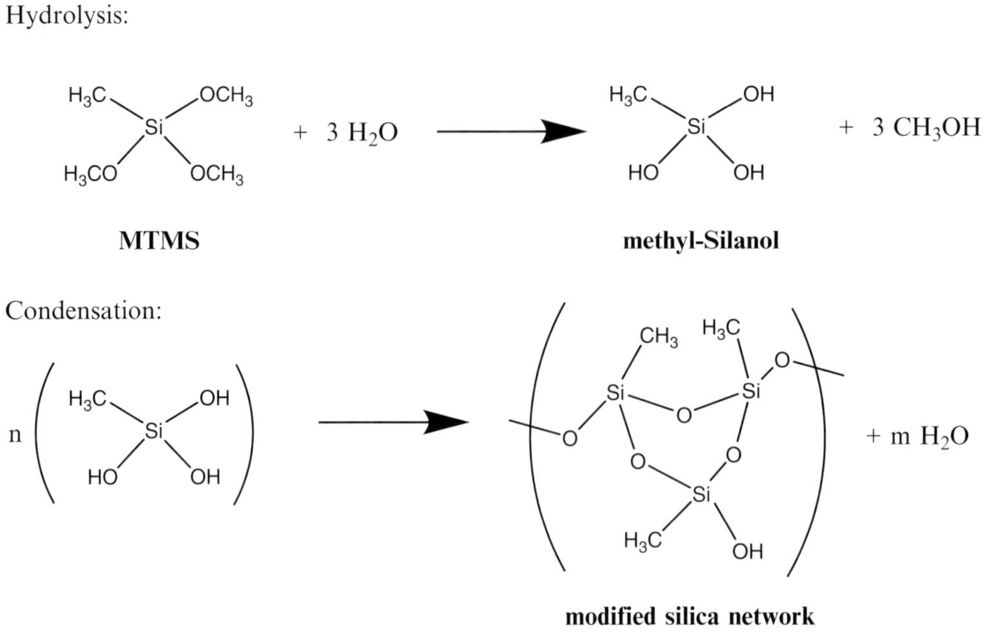

Hydrolysis:

$$\mathrm{H_3C{-}Si(OCH_3)_3} + 3\,\mathrm{H_2O} \longrightarrow \mathrm{H_3C{-}Si(OH)_3} + 3\,\mathrm{CH_3OH}$$

MTMS **methyl-Silanol**

Condensation:

$$n\,\left(\mathrm{H_3C{-}Si(OH)_3}\right) \longrightarrow \text{modified silica network} + m\,\mathrm{H_2O}$$

modified silica network

Figure 4.1. Sol–gel chemistry of methyltrimethoxysilane (*MTMS*).

Typically, the required amount of *precursor* (*MTMS* or *MTES*) is diluted in a solvent such as a water/methanol (H_2O/MeOH) mixture and partially hydrolyzed under addition of catalytic amounts of oxalic acid (0.001 M). The mixture is then set aside for 1 day. After that, the condensation of colloidal particles is initiated by addition of a 10-M ammonia solution, yielding alcogels. These are then aged for 2 days in methanol and directly dried under supercritical conditions from MeOH in an autoclave at a temperature of 265°C and a pressure of 10 MPa. After venting the vessel at these conditions, it is allowed to cool to room temperature and the aerogels are retrieved. Figure 4.2 shows a photographic image of a typical flexible silica aerogel sample.

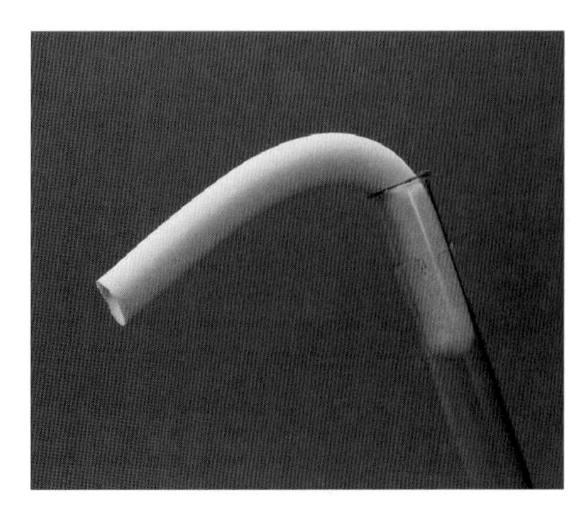

Figure 4.2. *MTMS*-based flexible silica aerogel.

4.2.2. Materials Characterization

The as-prepared aerogels were characterized by measurement of the bulk density, volume shrinkage, and porosity. Table 4.2 summarizes those physical properties for *MTMS*- and *MTES*-derived aerogels.

Table 4.2. Physical properties for *MTMS*- and *MTES*-derived aerogels

Sr. No.	Precursor amount used (mmol)	Bulk density (g/cm^3)	Volume shrinkage (%)	Porosity (%)
(a) *MTMS*-derived aerogels				
1	14	0.093	28	95
2	21	0.074	19	96
3	28	0.051	19	97
4	35	0.040	7	98
(b) *MTES*-derived aerogels				
5	6	0.090	25	94
6	9	0.079	22	95
7	12	0.072	21	96
8	16	0.061	18	97
9	19	0.050	8	97

Both, *MTMS*- and *MTES*-derived materials exhibit very high porosities >94% and bulk densities below 0.1 g/cm^3 or less than 5% of the density of bulk SiO_2. With increasing *precursor* amount used in the synthesis, the extent of volume shrinkage experienced during drying goes down. The microstructure of aerogels was visualized by means of transmission electron microscopy (*TEM*). A typical micrograph of a *MTMS* aerogel can be seen in Figure 4.3.

Figure 4.3. *TEM* image of a *MTMS*-derived superhydrophobic silica aerogel. The *scale bar* length is 200 nm.

The *TEM* image shows a dendritic-type structure of the aerogel, similar in nature to the one commonly observed in conventional silica aerogels. The average size of individual particles appears to be on the order of 20 nm. The hydrophobicity of the aerogels is quantified in terms of their water contact angle (θ) value. As-synthesized *MTMS*-derived aerogels have a contact angle of 175° for a water droplet which measures 2.4 mm in diameter. This underlines the extreme hydrophobic surface properties of this class of materials. A photograph of a water droplet on the aerogel surface is shown in Figure 4.4.

Figure 4.4. Photograph showing a water droplet placed on the surface of a superhydrophobic silica aerogel.

Infrared spectroscopy (*FTIR*) was used to identify the chemical bonding states within the material. The *FTIR* spectrum shown in Figure 4.5 indicates the presence of all chemical bonds which are expected: The Si–O–Si linkages of the silica network are represented by a strong and broad Si–O vibrational mode at 1,100 cm^{-1}. The presence of Si–CH_3

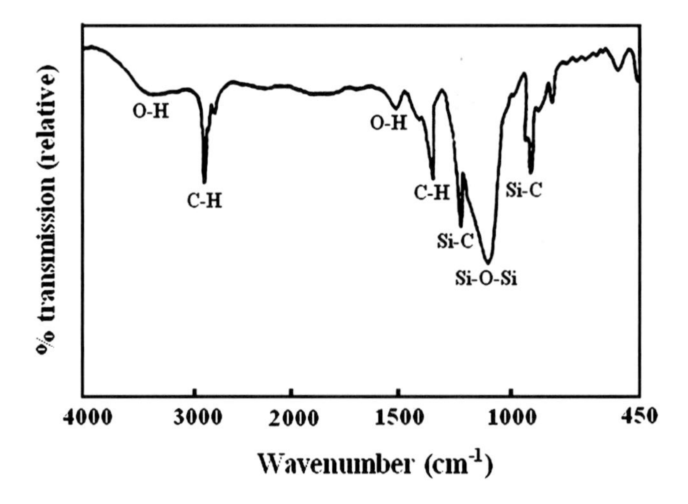

Figure 4.5. FTIR spectrum of an *MTMS*-based aerogel sample.

functionalities is confirmed by Si–C modes at 840 cm^{-1} and 1,310 cm^{-1} as well as C–H stretching and bending vibrations at 2,900 cm^{-1} and 1,400 cm^{-1} corresponding to Si–C and C–H bonds, respectively [20, 21]. Very small O–H vibrations at 3,500 cm^{-1} and 1,650 cm^{-1} indicate the presence of residual free OH groups (or adsorbed water) within the aerogel.

These aerogels offer unprecedented hydrophobic as well as mechanical character unlike native silica aerogels. Let us discuss in detail their interactions with water and hydrocarbons, and also the response under mechanical loading.

4.3. Water–Surface Interactions

Surfaces are known to interact with their environment in a variety of ways. One central concept governing such interactions is the minimization of the interfacial surface-free energy. This describes the wetting behavior of a surface by a liquid phase following Young's law [22] given below in (4.1):

$$\gamma_{sg} = \gamma_{sl} + \gamma_{lg}\cos\theta. \tag{4.1}$$

γ_{sg}, γ_{sl}, and γ_{lg} are the corresponding interfacial energies between the solid–gas, solid–liquid, and liquid–gas phases, respectively. θ denotes the contact angle defined by the tangent on a liquid droplet with respect to the neighboring solid surface.

Now, let us discuss water–surface interactions or more specifically, the interaction of superhydrophobic aerogels with water droplets. First, we shall consider the velocity of water droplets running on a smooth, hydrophobic aerogel surface at a well-defined inclination as an excellent example of minimized water–surface interaction. Second, the spontaneous coating of water with aerogel powder will be discussed as it forms a continuous and stable surface coating around a water droplet resulting in a "liquid marble". In this study, superhydrophobic *MTMS* aerogels were used as the interacting surface. The molar compositions of *MTMS*, MeOH, water, and ammonia were varied systematically over the six samples M1–M6, spanning a range of "extreme values" in the composition/chemical synthesis parameters as shown in Table 4.3.

Note that the contact angles of these materials are much higher than those of standard silica aerogels which have been hydrophobized in a subsequent step. For this reason we expect the surface–water interactions to be significantly weaker and hence the runoff velocities higher.

Measurement of runoff velocity. The runoff velocity of liquid droplets on an inclined surface coated by the superhydrophobic silica aerogel powder was quantified by recording

Table 4.3. Molar compositions of *MTMS*, MeOH, water, and ammonia used in the preparation of superhydrophobic aerogels

Sr. No.	*MTMS*:MeOH:H_2O:NH_4OH molar ratio	Sample
1	1:3.5:2:0.17	M1
2	1:3.5:8:0.17	M2
3	1:1.7:4:0.17	M3
4	1:14:4:0.17	M4
5	1:3.5:4:0.17	M5
6	1:3.5:2:0.35	M6

the transit time of a droplet rolling down a slanted surface by means of two photosensor barriers [23]. A schematic illustrating this setup is given in Figure 4.6 below.

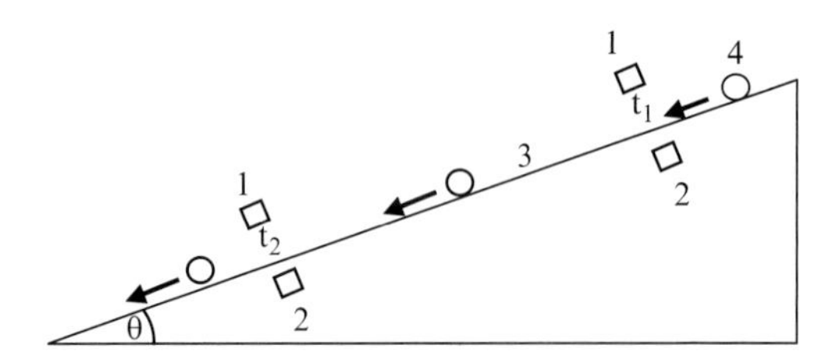

Figure 4.6. Schematic setup for water droplet velocity measurements. 1: Light emitting diodes (*LED*); 2: Photoconductive detectors; 3: Inclined plane platform; 4: Water droplet [size: ~2.7 mm (±0.2 mm)].

The measurement event is triggered by the droplet passing the upper light emitting diodes (*LED*) – photodiode pair and thus records a timestamp t1 via a computer interface. On its way down, the droplet is passing the second sensor, recording a second timestamp t2. From the time difference and the detector separation, the average velocity is evaluated.

The size of the droplet is chosen ~2.7 mm (±0.2 mm) in diameter since it is comparable to the water *capillary length* κ^{-1}:

$$\kappa^{-1} = \left(\frac{\gamma_{\mathrm{lg}}}{\rho \cdot g}\right)^{\frac{1}{2}}. \tag{4.2}$$

For water, the values for surface tension and density are $\gamma_{\mathrm{lg}} = 72.99$ N/m and $\rho = 1{,}000$ kg/m^3, respectively; g is the gravitational constant (9.81 m/s^2). Using (4.2), a critical droplet size κ^{-1} of 2.7 mm can be obtained. A droplet smaller than κ^{-1} generally remains stuck when placed on a solid surface because of the contact angle hysteresis. On the other hand, gravity would flatten a droplet of radius $r > \kappa^{-1}$.

Besides the capillary length, another key figure of merit is the work of adhesion (W_{a}). It describes the amount of work necessary to lift a droplet off the surface, hence it is associated with overcoming the surface–liquid interactions. W_{a} values can be calculated using Dupre's equation [24]:

$$W_{\mathrm{a}} = \gamma_{\mathrm{lg}}(1 + \cos\theta). \tag{4.3}$$

A list of values for various aerogel/water pairs is given in Table 4.4. As the contact angle (θ) increases (approaches to 180°), the respective works of adhesion diminish ($W_{\mathrm{a}} = 0.54$ N/m, for 173°) because the solid–liquid interactions (contacted areas) become smaller and smaller which becomes apparent upon closer inspection of (4.3). For complete nonwetting ($\theta = 180°$), the W_{a} equals to zero. This means that the drop is levitating on the surface.

4.3.1. Water Droplet Sliding

The results of the water droplet velocity measurements are summarized in Table 4.4 along with contact-angle values for the samples M1 through M6. As expected, the measured

Table 4.4. Contact angle (θ) and velocity (v) of a water droplet on a superhydrophobic aerogel coated surface for various angles of inclination

Sample $\longrightarrow$	M_1	M_2	M_3	M_4	M_5	M_6
Contact angle (θ) $\longrightarrow$	162°	160°	160°	159°	173°	162°
Work of adhesion (W_a) mN/m	3.57	4.40	4.40	4.84	0.54	3.57
Drop center deviation (δ) cm	0.03	0.033	0.035	0.05	0.024	0.03
Inclination (θ) $\neg$	v (cm/s)	v (cm/s)	v (cm/s)	v (cm/s)	v (cm/s)	v (cm/s)
5°	28.16	29.14	19.25	40	29.6	40
15°	40	42.16	34	52	37.25	52.4
22°	42	67.75	43	70	60	77
35°	75.7	84.40	64.6	83.2	83.8	97.83
52°	96.75	105	92.6	97.83	97.83	144

velocity (v) of a given water droplet increased with larger angles of inclination (φ) from 5° to 52°, irrespective of the aerogel composition. The minimum velocity of 19 cm/s is observed for $\varphi = 5°$ whereas a maximum value of 92 cm/s was reached at $\varphi = 52°$ in the case of aerogel sample M3. Similarly, minimum and maximum velocities of 40 cm/s at $\varphi = 5°$ and 144 cm/s at $\varphi = 52°$, respectively, were observed for the M6 sample.

These results are only partially consistent with the contact-angle measurements, where the highest contact angle value of 173° was obtained for the sample M5. Generally, it is expected that higher drop velocities should be obtained for more hydrophobic surfaces. However, in the present set of experiments, this was not the case because:

(a) As the drop radius r used for the contact-angle measurements ($r = 1$ mm) is much smaller than the capillary length κ^{-1} for water, the drop maintains its spherical shape to a good extent and hence leads to an overestimation of the contact angle.
(b) The roughness of the surface, or more specifically local variations in the hydrophobic surface properties, cannot be neglected as they affect the velocity of the liquid droplet.

Furthermore, the velocity of a freely falling water droplet of 2.8 mm (±0.2 mm) size in the gravitation field is calculated to ~152 cm/s. The difference in the freely falling drop (~152 cm/s) and a rolling drop (highest v ~144 cm/s) indicates that a contact zone, however small, must be forming between the drop and the solid surface, which opposes the liquid flow. It has been shown that, even for a small droplet of a liquid having contact angle 180° with a particular solid, a contact zone of radius l forms between the solid and the liquid due to the gravitational potential [25]. The mass center of the droplet is lowered by a quantity δ due its own weight. If r is the radius of the drop, then the δ and l are connected by the expression given below in (4.4):

$$\delta \approx \frac{l^2}{r}. \tag{4.4}$$

The deviation (δ) values predicted for a water droplet of 2.8 mm (±0.2 mm) diameter placed on various aerogels are given in the same table above. As the contact angle value increased from 159° to 173°, the δ value decreased from 0.05 to 0.024. To conclude, the runoff velocity of liquid droplets on an inclined surface coated by the superhydrophobic silica aerogel powder mainly depends on radius of the liquid droplet and the hydrophobic character of the surface.

4.3.2. Liquid Marbles: Superhydrophobic Aerogel-Coated Water Droplets

Next, let us look at the formation of aerogel powder-coated water droplets. These are referred to as liquid marbles from hereon. Superhydrophobic aerogels were ground to a fine powder (typical particle size ~5 μm) by crushing them with mortar and pestle. By rolling a water drop on the aerogel powder, liquid marbles were obtained. Figure 4.7 shows a typical photograph of such a marble [2.7 mm (±0.2 mm)] floating on a water surface.

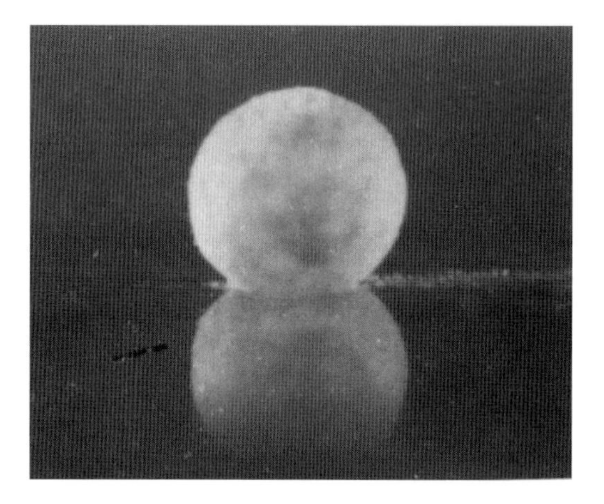

Figure 4.7. Liquid marble (2.7 mm (±0.2 mm) size) floating on a water surface.

Analogous to the water droplet sliding experiments, liquid marbles were placed on a standard float glass surface and the velocities measured at various inclinations. In the case of the liquid marbles, a similar trend is observed. The marbles prepared using M6 showed a higher velocity of 123.8 cm/s at $\varphi = 52°$ inclination, as can be seen in Table 4.5. The marbles prepared using M1 moved comparably slowly ($v = 82.7$ cm/s for $\varphi = 52°$). The drop center deviations (δ values) observed for the liquid marbles are again a central figure of merit. Interestingly, the δ values are lower as compared to those observed for the water

Table 4.5. The velocity (v) of the liquid marbles on nonadhesive side of the tape (i.e., uncoated surface) for various angles of inclination

Sample $\longrightarrow$	M_1	M_2	M_3	M_4	M_5	M_6
Drop center deviation (δ) cm	0.026	0.028	0.030	0.027	0.024	0.024
Inclination (ϕ) $\neg$	v (cm/s)	v (cm/s)	v (cm/s)	v (cm/s)	v (cm/s)	v (cm/s)
5°	31.75	30	31	31.75	29.5	46.25
15°	35.25	38	37.3	39	38.4	60.5
22°	43.87	54	49.5	52	52	76.33
35°	57.66	82	70	66	73.75	105
52°	82.75	105	89.6	105	95	123.8

droplets. It is believed that as the powder covers the surface, the water inside the marble is not able to spread and hence has to support its own weight. The velocity of the marbles is lower than that of the water droplets because the marble surface is considerably rougher than that of a water droplet. The rougher the surface of the powder forming the liquid marble, the slower will be its sliding velocity.

4.4. Mechanical and Elastic Properties

As previously mentioned, *MTMS*-derived silica aerogels exhibit unprecedented flexibility when compared with standard silica analogs. Let us discuss the properties of these outstanding materials in some detail and underline them by measurements:

Solids can be deformed by tension, torsion, shear, bending, and compression. The cause of deformation is an external load or stress which induces strain inside the body. According to Hooke's law, the stress is directly proportional to the strain [26]. In tension mode (4.5) holds,

$$\frac{F}{A} = Y\left(\frac{L - L_0}{L_0}\right) \tag{4.5}$$

where the proportionally constant Y is termed Young's modulus or the modulus of elasticity. It is a measure of the hardness, stiffness, or rigidity (softness, flexibility, or pliability) of a given solid. Small numbers indicate flexible whereas large numbers reflect stiff behavior, respectively.

Here we present the result of a uniaxial compression test study carried out using a setup depicted in Figure 4.8. In this test, a sample under investigation was kept in a wide-bore ($r_\mathrm{i} = 11.23$ mm) glass tube fixed with a rigid support in such a way, that the bottom of tube came to rest on a support plate. This arrangement allows for a stable setup and minimizes measurement artifacts. Using a cantilever, various weights were placed onto the cylindrical aerogel, which acted as a compressible piston. The aerogel sample undergoes compression, which is characterized by measuring the corresponding change in length (ΔL) with a traveling microscope. Finally, Young's modulus of the aerogel samples can be calculated according to (4.6):

$$Y = \left(\frac{Lg}{\pi r^2}\right) \cdot \left(\frac{m}{\Delta L}\right)$$

or

$$Y = \left(\frac{Lg}{\pi r^2}\right) \cdot \frac{1}{\text{slope}} \tag{4.6}$$

where L is the original length of the aerogel before deformation and r is the radius of the aerogel body (inner tube diameter).

Generally, any deformation is directly proportional to applied force, provided the forces do not exceed a certain limit called "elastic limit." Within this limit, a specimen returns to its original shape and size, due to elastic (nonplastic) deformations. In other words,

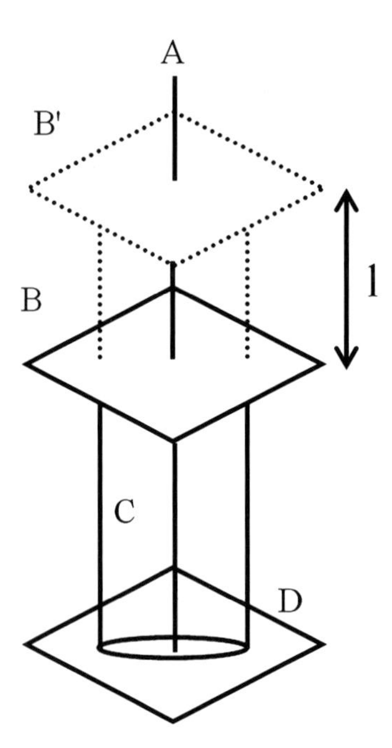

Figure 4.8. Schematic diagram of Uniaxial compression test for Young's modulus measurements. A: Vertical axis; B: platform for the application of the stress; C: silica aerogel cylinder; D: flat bottom surface; l: change in length after the application of the stress.

there is little to no hysteresis of the compression/relaxation loop. If the plastic limit is exceeded, the specimen plastically deforms, meaning that there is a measurable hysteretic behavior of the material.

In the case of *MTES* flexible aerogels, an increase in the applied load from 0.1 to 0.8 N, the material was compressed from 0.9 to 4.1 cm [27]. A further increase of the applied load by only 0.02 N (total load = 0.82 N) led to cracking of the aerogel or in other words caused irreversible damage. Figure 4.9 shows a steady increase of the compression with increasing load applied in an almost linear fashion from point O to point A. Beyond point A, the reversibility is lost and the material damaged. Hence, point A is called an elastic limit point with the corresponding compression called the elastic limit compressibility. Within the area under the curve, the inner aerogel network structure counteracts the deforming forces and recovers its original length when the load ceases to exist. When increasing the load beyond point A, the reversibility is lost and the material irreversibly damaged. This results in partial or complete loss of structural integrity, the main mode of material failure for most aerogels.

4.4.1. Effect of Synthesis Parameters on Material Elasticity

One would generally expect that the way in which the gel structure is formed will influence the mechanics of the system. The effect of the *precursor* concentrations S_1 and S_2 [MeOH/*MTMS* molar ratio (S_1) and MeOH/*MTES* molar ratio (S_2)] on the elastic properties of the aerogels is described in this section. For simplicity, the molar ratio of H_2O/*precursor* was kept constant at 8:1 and oxalic acid $(COOH)_2$ and ammonium hydroxide (NH_4OH)

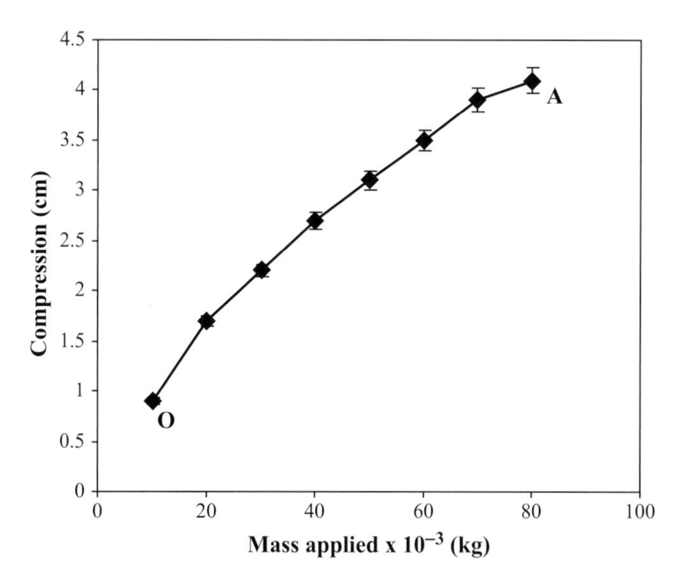

Figure 4.9. Compression of an aerogel sample due to an applied load.

catalyst concentrations of 0.001 M and 10 M were used, respectively. Generally, the density of the aerogels decreased with increasing S_1 and S_2 ratios. Young's moduli (Y) of aerogels scaled with the bulk density in much the same manner. For S_1 ratios within the range from 14 to 35, the volume shrinkage and the density of the corresponding aerogels decreased from 28 to 7% and from 100 to 40 kg/m^3, respectively. However, when using an *MTES precursor*, the volume shrinkage and the density of aerogels decreased from 25 to 8% and from 90 to 50 kg/m^3, respectively, as the S_2 ratio was varied from 6 to 19. Figure 4.10 shows the change in length of a 10-cm long aerogel cylinder (*MTMS* based) with various loads applied for three different S_1 values 21, 28, and 35. The slope of the curve is used to determine Young's modulus Y. Figure 4.11 shows the same data for *MTES*-based aerogels. For the two extreme S_1 values 21 to 35, Y was determined to be 1.4×10^5 and 3.4×10^4 N/m^2. For analogous S_2 extreme values of 6.45 and 19.35, *MTES*-derived aerogels yielded Y values of 1.5×10^5 to 3.9×10^4 N/m^2. As expected, with higher S ratio (lower density), the flexibility of the aerogels increases.

The above argument is well in agreement with the photographs of *MTMS* aerogel samples given in Figure 4.12A, B. The left-hand side image (A) shows a material with a lower S_1 ratio of 28 and the sample on the right (B) is of a material with $S_1 = 35$. Both pictures show the maximum bending deformation of the material before rupture. Again, the higher S value translates into a lower density material which is more flexible [28]. Similarly, for *MTES* aerogels, Figure 4.13A–C shows the maximum possible bending of the aerogels prepared at $S_2 = 6.4$, 12.9, and 19.3, respectively.

4.4.2. Potential Applications in Mechanical Damping

Aerogels can be fractured easily but surprisingly they have strong compressive strength. Young's modulus (measure of elasticity) of superhydrophobic flexible aerogels was compared with the Y values of metals (Iron and Indium), polymer (rubber) and native silica aerogel. Figure 4.14 shows the overall comparison of Young's moduli of these materials.

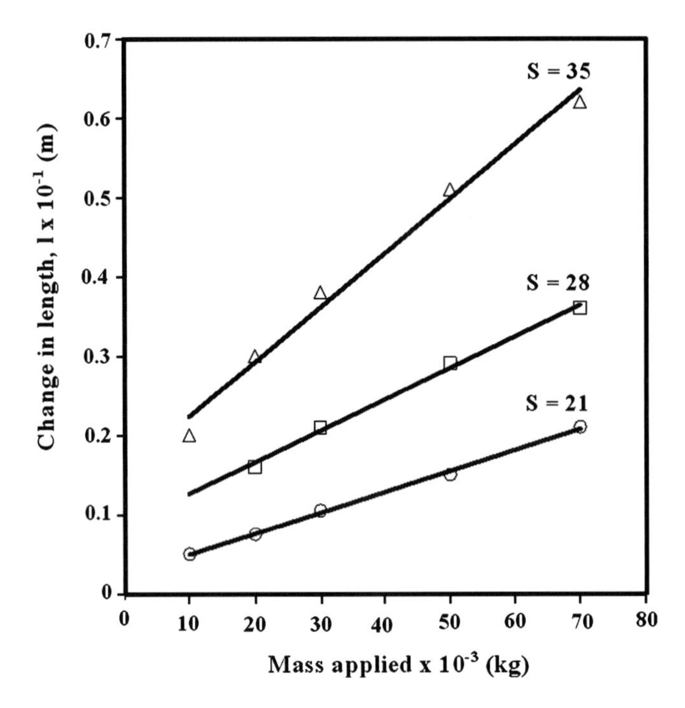

Figure 4.10. Plots of change in length against mass applied for the silica aerogels prepared with various MeOH/*MTMS* molar ratios.

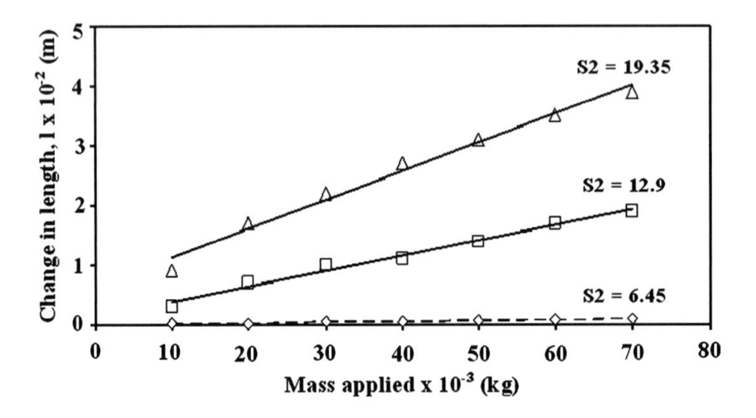

Figure 4.11. The variation in the change in length (ΔL) against the mass (m) applied on the aerogels samples prepared at $S = 6.45$, $S = 12.9$, and $S = 19.35$.

The Y value of the as-prepared aerogels was found to be 9.6×10^6 and 2.6×10^5 times smaller than those of Iron and Indium, respectively. The Y values of these aerogels are closer to those of rubber when compared with the metals mentioned above. It is only 2.6×10^2 times smaller than the Y value of rubber which is quite rarely observed in case of silica aerogels. This improvement in elastic modulus of superhydrophobic flexible aerogels with those of native silica aerogels is roughly two orders of magnitude.

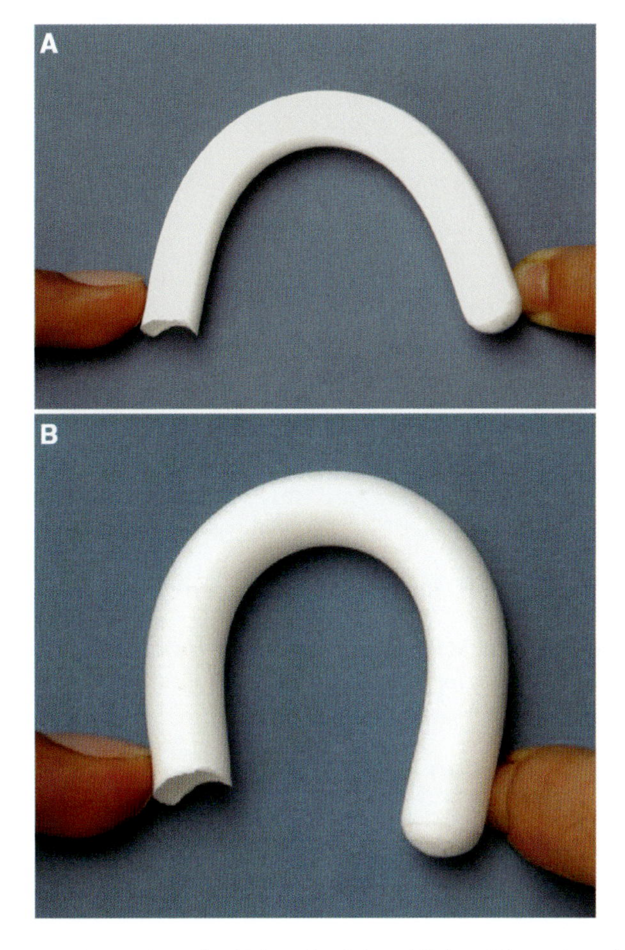

Figure 4.12. Flexible *MTMS* aerogel sample which can bend about 90° **A.** and the maximum bending of sample **B.**

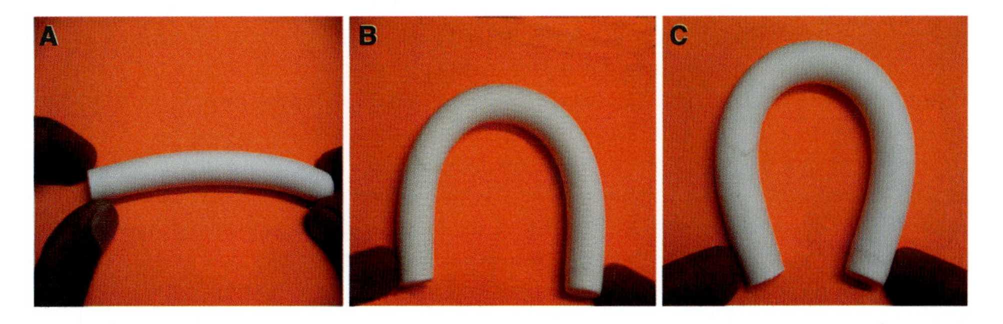

Figure 4.13. Photographs showing the maximum possible bending of the aerogels prepared at **A.** $S = 6.45$, **B.** $S = 12.96$, and **C.** $S = 19.35$.

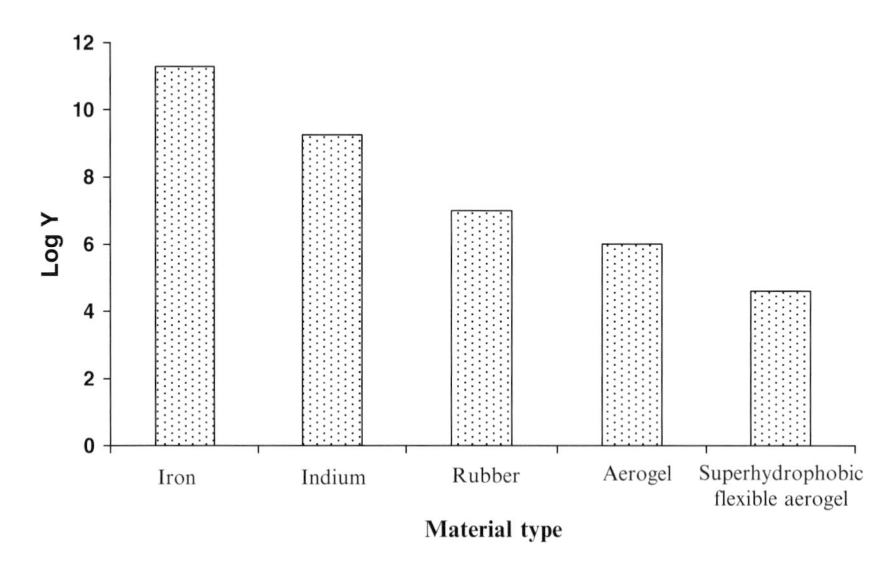

Figure 4.14. Comparison of *Y* value of superhydrophobic flexible aerogels with various materials.

Therefore, MTMS and MTES aerogels can find use in minimal load and shock absorption or damping such as for example micromechanics applications.

4.5. Hydrocarbon Sorption Behavior

With oil reserves dwindling and the advent of "peak oil," new oil fields are less and less accessible. This forces companies to expand their exploration strategy more and more toward more risky off-shore drilling. This is reflected in the tripling of offshore oil platforms worldwide since the beginning of the Millennium. Recently the risks of such installations have been demonstrated in the most macabre way with the explosion and sinking of the "Deepwater Horizon" in the spring of 2010 [29]. During clean-up a wide range of materials such as dispersants, solidifiers, absorbents, and skimmers [30–35] have been employed. Still, the oil companies have comparably little useful technology available to eliminate oil or solvent spills, particularly at sea. Hence, there is a need for new materials to adsorb, remove, and recycle oil and/or other organic liquids in the case of their accidental release. While hydrophobicity and oleophilicity are the primary determinants of physical sorption of these materials, another key element is the porosity of the adsorbent which determines to a large extent the maximum loading capacity. Additionally, retention and recovery of oil or organic solvents from the sorbents, and the reusability and recyclability of the sorbent are important factors for practical use in spill clean-up [33].

Aerogels are highly porous sol–gel-based materials. Their astonishing properties are directly related to their nanostructure as well as micro- and meso-porosity. Because of the large pore volume of aerogels, a small mass of such a solid can confine a large volume (and also a large mass) of liquid. It is because of these properties that superhydrophobic aerogels are excellent candidates for oil-spill clean-up. Their light weight also offers the possibility of controlled release from the air and thus to cover large area oil slicks within short time. In the following, one will discuss the usability of elastic superhydrophobic aerogels as efficient absorbents for oil and organic liquids.

The assessment of the practical usefulness of such systems for oil-spill clean-up is based on the following four main criteria:

- Sorption capacity of the material for various hydrocarbons
- Desorption rate and its temperature dependence
- Process reversibility and reuse of aerogels
- Economic factors

Let us start by presenting selected results of absorption and desorption studies [18] with various organic solvents.

4.5.1. Uptake Capacity

An as-prepared aerogel sample was placed in an organic solvent until it was completely soaked with the liquid. The uptake capacity of the aerogel sample was then quantified in terms of the mass of the organic liquid absorbed per unit mass (1 g) of the aerogel sample. Table 4.6 contains a list of sorption capacities for various common hydrocarbons.

Table 4.6. Mass of various organic liquids absorbed by unit mass (1 g) of the aerogel sample

Organic liquid/oil	Mass of the organic liquid/oil absorbed per unit mass (1 g) of the aerogel (g)	Moles of the organic liquid absorbed per unit mass (1 g) of the aerogel
Pentane	9.83	0.1365
Hexane	10.95	0.1215
Heptane	13.38	0.1338
Octane	15.62	0.1210
Benzene	19.92	0.2553
Toluene	20.64	0.2189
Xylene	20.37	0.1921
Methanol	14.03	0.4410
Ethanol	14.54	0.3231
Propanol	19.15	0.3191
Butanol	18.93	0.2558
Gasoline	13.82	–
Kerosene	16.45	–
Diesel	18.55	–

From the table, one concludes that from the class of common crude oil distillates, standard gasoline (95 octane) is least absorbed (13.82 g) and diesel fuel the most (18.55 g). When looking at common organic solvents, the aerogel absorbed nearly the same amount (~20 g) of the completely nonpolar, aromatic compounds benzene, toluene, and xylene. Among the alkanes, the mass of pentane absorbed was smallest (9.83 g) while that of octane absorbed was highest (15.62 g). For the alcohols, the uptake capacity was shown to increase with increasing chain length from methanol (~14 g) to butanol (~19 g). Table 4.6 also gives the number of moles of solvent absorbed per unit mass of the aerogel. The oils such as gasoline, diesel, and kerosene being the extracts from crude oil (mixtures), the molecular weights are not well defined and hence the mole amount cannot be specified.

4.5.2. Desorption Rate

The rate of desorption was determined experimentally by measuring the mass of the aerogels at regular time intervals until the liquid had evaporated completely and the original mass of the aerogel was restored. Figure 4.15 shows the photographs showing various stages of absorption and desorption of hexane from the aerogel at 30°C.

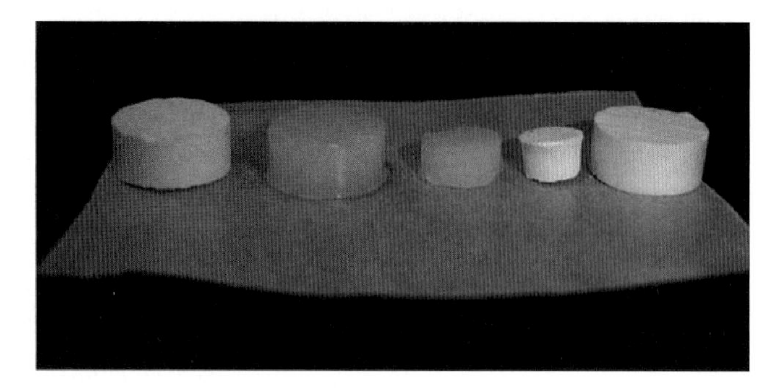

Figure 4.15. Photograph showing various stages of absorption and desorption of hexane from the same type of aerogel.

The behavior observed in the picture reminds one of an ambient drying process. As more and more solvent evaporates, the material undergoes considerable shrinkage, until the final stage where it recovers its original shape completely because of the spring-back effect. The desorption time graph given in Figure 4.16 shows the dependence of desorption time on the hydrocarbon loading (in terms of number of moles of solvent per unit mass of the aerogel). One can clearly see a linear dependence for all the three classes of organic solvents alike – alkanes, aromatics, and alcohols.

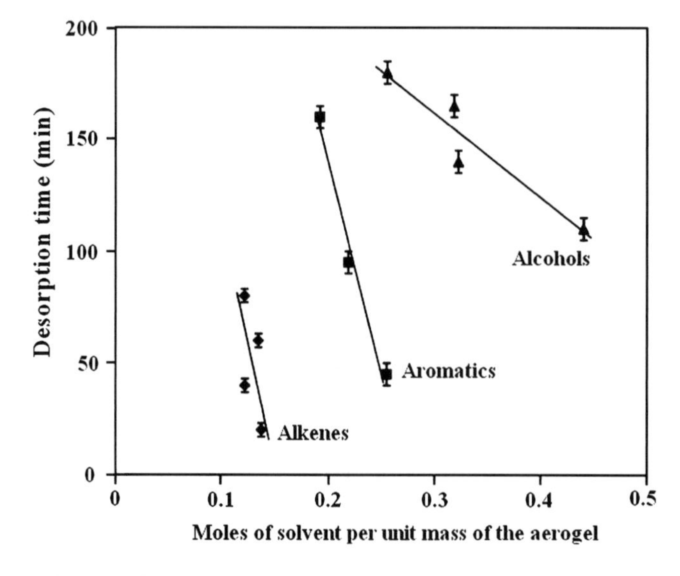

Figure 4.16. Desorption time of solvents as a function of moles of solvent absorbed per unit mass of the aerogel.

So far, we were discussing only the desorption time which is linked with the complete removal of the organic phase from the aerogel. When looking at the release kinetics for any given desorption experiment in more detail, one observes the same sort of transient behavior which is typical for the drying of aerogels from hydrocarbons (reference ambient dried). The rate of desorption is high at the beginning and slows down with time. It is further observed that the desorption rate decreases generally when going from shorter chain (pentane) to the longer chain (octane) alkanes. The same trend is also valid for the alcohols. This is due to the fact that the process of evaporation takes place in two stages. During the first stage, solvent molecules are transported from the bulk interior to the surface. The smaller the surface tension, the easier it is for the molecules to come to the surface. During the second stage, they vaporize or run off from the outside surfaces. The second process is therefore dependent on the vapor pressure of the liquid. The solvent vapor inside the bulk material also actively pushes the liquid front to the surface much like a piston.

Next, let us look at the temperature dependence of the desorption rate. Figure 4.17 shows the effect of temperature on the desorption time of standard gasoline. It was observed that with an increase in temperature, the desorption time showed a steep decrease. The desorption time for butanol, for example, was reduced from 180 min at 30°C to 14 min at 100°C. Similarly, for octane, it was reduced from 80 min at 30°C to 8 min at 100°C.

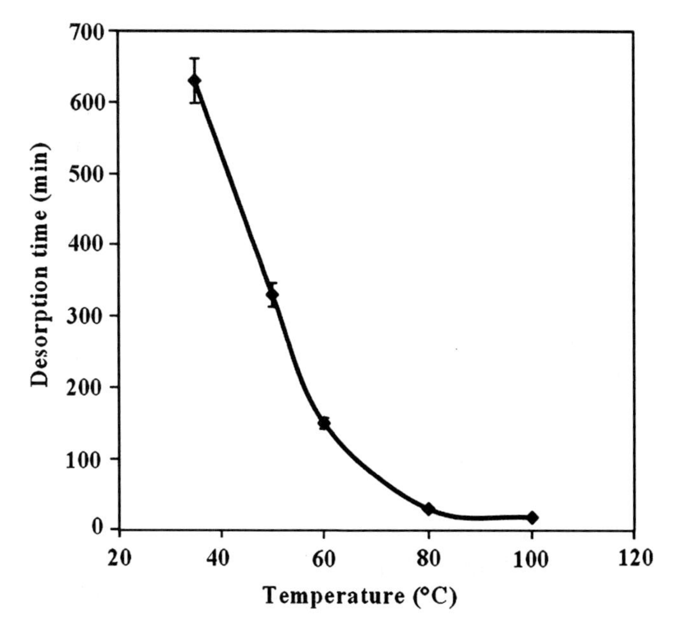

Figure 4.17. Desorption time of gasoline (95 octane rating) as a function of temperature.

With an increase in temperature from ambient to 100°C, the vapor pressure of typical organic solvents increases considerably. For example, the vapor pressure of methanol at 30°C is 280 mbar, while it is 555 mbar at 50°C. The surface tension of all those liquids decreases linearly with increasing temperature. For small temperature differences the temperature dependence [36] of the surface tension $\gamma(T)$ can be approximated as

$$\gamma(T) = \gamma_0(1 - \alpha T) \tag{4.7}$$

where γ_0 is the surface tension at 0°C, α is the linear temperature co-efficient and T the temperature in units of °C.

From a surface tension and vapor pressure point of view, it becomes easier for the liquid to effuse from the aerogel material with increasing temperature as mentioned previously. In addition, there is a non-"steady state" contribution during the heating stage, when temperature gradients can speed up this process even more.

To conclude, superhydrophobic aerogels are excellent sponges for organic solvents as proven by measurements of the uptake capacity but even more, they can be almost quantitatively retrieved by a suitable thermal treatment. From a practical point of view, the solvent should be retrieved at the lowest possible temperature in order to prevent decomposition of the solvent and degradation of the aerogel itself.

4.5.3. Process Reversibility and Reuse of Aerogels

A key factor for the practical usefulness of aerogel-based hydrocarbon sorption systems is the reusability. Because aerogels are extremely expensive to produce (Chap. 26), some of the added cost must be compensated by exactly those types of practical aspects. As part of our preliminary studies on the reversibility of the sorption and desorption of superhydrophobic aerogels with various organics, the following trends were observed:

- Pure solvents, particularly of low molecular weight (hydrocarbons, alcohols, aromatics) are absorbed and desorbed with quite good reversibility. The superhydrophobic aerogel can undergo up to 20 absorption/desorption cycles without much decrease in uptake capacity.
- Electron microscopy indicated little to no structural change in the *MTMS*-derived aerogel after a single methanol absorption/desorption cycle as shown in Figure 4.18.
- When using complex mixtures of oils with a large content of higher molecular weight compounds such as heavier fractions of petroleum distillates and/or crude oil, the reversibility goes down quite substantially. When pump oil was used as a sorbent, the uptake capacity showed a clear drop after only four absorption/desorption cycles.

These preliminary results indicate the potential usefulness of these systems for spill clean-up of typical organic solvents. Even though the performance of the aerogel sponge

Figure 4.18. Transmission electron micrographs of the aerogels before absorption **A**. and after desorption **B**. of methanol. The *scale bar* measures 200 nm.

suffers with exposure to heavier oils, not all hope is lost but more work is necessary. For example, it seems also plausible that larger fractions of long-chain organics and/or particulate matter or inclusions remain trapped inside the porous aerogel structure. Perhaps, a solvent extraction step could then recuperate the full performance of the aerogel as a sponge. For a complete evaluation, the mechanism of this performance degradation must be further investigated.

4.5.4. Economic Factors

Superhydrophobic *MTMS*- or *MTES*-derived aerogels are certainly a class of remarkable materials, but also, they are not ready for large-scale fabrication. As far as potential applications are concerned they can only compete in certain well-defined niche markets. Among the most likely applications are the removals of oil spills from water surfaces, where the oil exists in the form of a thin film, or the clean-up of organic compounds which need to be retrieved no matter the cost. Some examples could involve spills of extremely precious compounds in the chemical industry, such as active drug ingredients and fine chemicals, for example, pure enantiomer forms of selected organic compounds. Perhaps, even the removal of extremely toxic or hazardous metalloorganic compounds (e.g., alkyl derivates of heavy metals) can be absorbed by superhydrophobic aerogels and in some cases must be retrieved quantitatively. Hence, an application of this type of aerogel compound only makes sense, where its performance exceeds state of the art by a considerable amount and if the added cost of the material is compensated at least in part by its superior performance. Some of the main advantages of these materials are their tremendous uptake capacity on a per weight basis and their low density (weight). However, conventional silica-based superhydrophobic aerogels offer similar density and uptake capacities. Here, the flexible nature of alkyltrialkoxysilane-based aerogels offers an additional advantage but it also comes with a higher price tag, at least when dried supercritically. Because of their low density and Young's modulus, these materials also are promising dampers for ultrasensitive measurement equipment such as scanning microscopes (AFM, *STM*) or for microoptics and MEMS applications. This, however, requires small parts to be machined with micrometer precision.

Time will tell if these materials can find their way from the laboratory into certain niche applications. Before a complete evaluation of economic factors and possible fields of applications makes sense, the reusability and solvent extraction questions must be answered and a complete service life balance established. Only then will it be possible to weigh cost against performance.

4.6. Summary

Commonly, aerogels are brittle and fragile materials. In the case of aerogels which are made from precursors carrying internal hydrophobic groups such as *MTMS* and *MTES*, one observes a strong deviation from these classical properties of silica aerogels. The behavior of the resulting aerogels is closer to that of soft rubber or polymer-like materials. Within this chapter, we have elucidated synthesis and applications of flexible superhydrophobic silica aerogels in a single- or two-step sol–gel process without the need for additional chemical modification of the gels. This new and fascinating class of aerogels offers similar network structure but vastly different mechanical properties when compared with standard silica aerogels. Also, they regain their original shape and size after loading and offer extremely low values of Young's modulus *Y*.

These aerogels also absorbs up to 15 times their own mass of organic solvents or oil and a moderate heating of the soaked material releases the solvent quantitatively. This in combination with their superhydrophobic and flexible nature makes them promising candidates for specialty spill clean-up applications. To conclude, a facile and versatile route to this custom-tailorable class of aerogels with a unique combination of properties has been demonstrated. A generally useful strategy to tune the desired properties is to vary the *precursor* concentration in the sol–gel synthesis, as it is perhaps generally valid in sol–gel chemistry. Little work has been done so far to produce these materials by ambient drying but ambient dried materials of this class would certainly raise their attractiveness for commercialization.

References

1. Buzykaev A. et al (1996) Project of aerogel cherenkov counters for KEDR, J. Nucl. Instr. and Meth. Phys. Res. A 379: 453–456
2. Carlson P.J., Johansson K.E., Norrloy J.K., Pingot O., Tacernier S., Bogert F., Luncker L. (1979) Increased photoelectron collection efficiency of a photomultiplier in an aerogel cherenkov counter, J. Nucl. Instr. and Meth. Phys. Res. A 160: 407–410
3. Cunha P., Neves F., Lopes M. (2000) On the reconstruction of Cherenkov rings from aerogel radiators, J Nucl. Instr. and Meth. Phys. Res. A 452: 401–421
4. Jang K. Y., Kim K., Uphadye R. S.: Hollow silica spheres of controlled size and porosity by sol-gel processing, J. Am. Ceram. Soc. 74, 1987–1992 (1991)
5. Nguyen B. N., Meador M. A. B., Tousley M. E., Shonkwiler B., McCorkle L., Scheiman D. A., Palczer A.: Tailoring elastic properties of silica aerogels cross-linked with polystyrene, ACS Appl. Mater. Interfaces 1, 621–630 (2009)
6. Mulik S., Sotiriou-Leventis C., Churu G., Lu H, Leventis N.: Cross-linking 3D assemblies of nanoparticles into mechanically strong aerogels by surface-initiated free-radical polymerization, Chem. Mater. 20, 5035–5046 (2008)
7. Meador M. A. B., Weber A. S., Hindi A., Naumenko M., McCorkle L., Quade D., Vivod S. L., Gould G. L., White S., Deshpande K.: Structure-property relationships in porous 3D nanostructures: epoxy-cross-linked silica aerogels produced using ethanol as the solvent, ACS Appl. Mater. Interfaces 1, 894–906 (2009)
8. Meador M. A., Vivod S. L., McCorkle L., Quade D., Sullivan R. M., L. Ghson J., Clark N., Capaclona L. A.: Reinforcing polymer cross-linked aerogels with carbon nanofibers, J. Mater. Chem 18, 1843–1852 (2008)
9. Capadona, L. A.; Meador M. A. B., Alunni A., Fabrizio E. F., Vassilaras P., Leventis N.: Flexible, low-density polymer crosslinked silica aerogels, Polymer 47, 5754–5761 (2006)
10. Kramer S. J., Rubio-Alonso F., Mackenzie J. D.: Organically modified silicate aerogels, aeromosils, Mater. Res. Soc. Symp. Proc. 435, 295–299 (1996)
11. Kanamori K., Aizawa M., Nakanishi K, Hanada T.: New transparent methylsilsesquioxane aerogels and xerogels with improved mechanical properties, Adv. Mater. 19, 1589–1593 (2007)
12. Loy D. A., K. J. Shea: Bridged polysilsesquioxanes: highly porous hybrid organic-inorganic materials, Chem. Rev. 95, 1431–1442 (1995)
13. Loy D. A., Jamison G. M., Baugher B. M., S Myers. A., Assink R. A., Shea K. J.: Sol-gel synthesis of hybrid organic-inorganic materials. hexylene- and phenylene-bridged polysiloxanes, Chem. Mater. 8, 656–663 (1996)
14. Shea K. J., Loy D. A.: Bridged polysilsesquioxanes: molecular-engineered hybrid organic-inorganic materials, Chem. Mater. 13, 306–331 (2001)
15. Reynolds J. G., Coronado P. R., Hrubesh L. W.: Hydrophobic aerogels for oil-spill clean up - synthesis and characterization, J. Non-Cryst. Solids 292, 127–137 (2001)
16. Hrubesh L. W., Coronado P. R., Satcher Jr. J. H.: Solvent removal from water with hydrophobic aerogels, J. Non-Cryst. Solids 285, 328–332 (2001)

17. Suzana S., Zoran N., Zeljko Kenz: Adsorption of toxic organic compounds from water with hydrophobic silica aerogels, J. Colloid and Interface Science 310, 362–368 (2007)
18. A. Venkateswara Rao, Hegde N. D. Hirashima H.: Absorption and desorption of organic liquids in elastic superhydrophobic silica aerogels, J. Colloid and Interface Science 305, 124–132 (2007)
19. Gurav J. L., A. Venkateswara Rao, Nadargi D. Y., Park H. H.: Ambient pressure dried *TEOS*-based silica aerogels: good absorbents of organic liquids, J. Mater Sci 45, 503–510 (2010)
20. Hering N., Schriber K., Riedel R., Lichtenberger O., Woltersodorf J.: Synthesis of polymeric precursors for the formation of nanocrystalline Ti-C-N/amorphous Si-C-N composites, J. Appl. Organometal. Chem. 15, 879–886 (2001)
21. Laczka M., Cholwa-Kowalska K., Kogut M.: Organic-inorganic hybrid glasses of selective optical transmission, J.Non-Cryst. Solids 287, 10–14 (2001)
22. Arthur B.: The Mainstream of Physics. Wesley Publishing Company Inc., Second Edition (1962)
23. A. Venkateswara Rao, Kulkarni M. M., Bhagat S. D.: Transport of liquids using superhydrophobic aerogels, J. Colloid and Interface Science, 285, 413–418 (2005)
24. Isenberg, C.: The science of soap films and soap bubbles. In: general introduction, pp. 1–26 General publishing company, Canada (1992)
25. Fridrikhsberg D. A. (1986) A Course in Colloid Chemistry. Mir Publishers, Moscow
26. Resnick R., Halliday D., Walker J.: Fundamentals of Physics, 6th Edition, John Wiley & Sons (2001)
27. Nadargi D. Y., Latthe S. S., Hirashima H., A. Venkateswara Rao,: Studies on rheological properties of methyltriethoxysilane (*MTES*) based flexible superhydrophobic silica aerogels, J. Microporous and Mesoporous Mat. 117, 617–626 (2009)
28. A. Venkateswara Rao, Bhagat S. D., Hiroshima H., Pajonk G. M.: Synthesis of flexible silica aerogels using methyltrimethoxysilane (MTMS) precursor, J. Colloid and Interface Sci., 300 279–285 (2006)
29. Wald, M. L., Clarifying Questions of Liability, Cleanup and Consequences, NY Times, 2010, May 6th issue
30. Fingus, M. Oil spills and their cleanup. Chemistry Industry 1005–1008 (1995)
31. Arthur B (1962) The mainstream of physics. Addison Wesley Publishing Company Inc., Second Edition
32. Delaune R.D., Lindau C.W., Jugsujinda A.: Effectiveness of "Nochar" solidifier polymer in removing oil from open water in coastal wetlands, Spill Science & Technology Bulletin 5, 357–359 (1999)
33. Teas Ch., Kalligeros S, Zanikos F., Stournas S., Lois E., Anastopoulos G.: Investigation of the effectiveness of absorbent materials in oil spills clean up, Desalination 140, 259–264 (2001)
34. Doerffer J.W. (1992) Oil spill response in the marine environment. Pergamon Press, Oxford
35. Lessard RR, Demarco G: The significance of oil spill dispersants, Spill Sci Technol Bull 6, 59–68 (2000)
36. Newman F.H., Searle V.H.L.: The general properties of matter. Orient Longmans, Fifth Edition (1957)

5

Sodium Silicate Based Aerogels via Ambient Pressure Drying

A. Venkateswara Rao, G. M. Pajonk, Uzma K. H. Bangi, A. Parvathy Rao, and Matthias M. Koebel

Abstract The first step in the preparation of silica aerogels is a sol–gel process producing a gel. This is followed by drying of the gel by either supercritical drying (*SCD*) or ambient pressure drying (*APD*). Traditionally, silica aerogels are prepared by the more energy-intensive and -expensive *SCD* method using *alkoxide* precursors such as tetraethoxysilane (*TEOS*) or tetramethoxysilane (*TMOS*). This choice partly restricts the commercialization of aerogels. Recent developments have shown great potential of the *APD* as an alternative method employing sodium silicate (Na_2SiO_3) as a purely inorganic precursor. The properties of such aerogels are very similar to those obtained by more conventional methods. This chapter focuses on the preparation of sodium silicate based aerogels via *APD* and the effect of various parameters on their physicochemical properties. The process chemistry is further contrasted with factors relevant for large-scale production.

5.1. Introduction

5.1.1. Silica Aerogels

Aerogels, the lightest transparent solids known, are a class of low-density solid-state materials obtained from a gel by replacing the pore liquids with air while maintaining the network structure as it is in the gel state. They are also known as frozen smoke or air-glass and are comprised of particles with typical dimensions below 10 nm and pore sizes < 50 nm in diameters. Aerogels possess a wide variety of exceptional properties such as low *thermal conductivity* (~0.01 W/m.K), high porosity (~99%), high optical transmission (90%) in the visible region, high *specific surface area* (1,000 m^2/g), low dielectric constant (~1.0–2.0), low refractive index (~1.05), and low sound velocity (100 m/s) [1–4]. Owing to these properties, aerogels find applications in a number of fields such as thermal insulation, space technology, catalysis, acoustics, filtration, particle detectors, and electronics (Parts 8, 9, 10, 11).

A. V. Rao, U. K. H. Bangi, and A. P. Rao • Air Glass Laboratory, Department of Physics, Shivaji University, 41600 Kolhapur, Maharashtra, India
e-mail: avrao2012@gmail.com, uzmayesh@yahoo.co.in, parvathy_shivaj@yahoo.co.in
G. M. Pajonk • Laboratoire des Matériaux et Procédés Catalytiques, Université Claude Bernard Lyon 1, 43 Boulevard du 11 novembre 1918, 69622 Villeurbanne, France
e-mail: pajonk1939@laposte.net
M. M. Koebel • Building Materials Group, Department of Building Science and Technology, Empa – Swiss Federal Laboratories for Materials Science and Technology, Dübendorf, Switzerland
e-mail: matthias.koebel@empa.ch

M.A. Aegerter et al. (eds.), *Aerogels Handbook*, Advances in Sol-Gel Derived Materials and Technologies, DOI 10.1007/978-1-4419-7589-8_5,

Silica aerogels were first prepared by S. S. Kistler at the College of the Pacific, in 1931, using sodium silicate as a precursor and supercritical drying [5]. However, because of the very elaborate and time-consuming procedures of the supercritical drying, there was only a little follow-up work done in the field. Later on, in 1968, a team of researchers headed by Prof. S. J. Teichner at the University of Lyon, France, succeeded in producing silica aerogels using silicon *alkoxides* namely tetraethoxysilane (*TEOS*) and tetramethoxysilane (*TMOS*) within one day [6]. However, these *alkoxide* precursors are more costly, an important factor when considering large-scale commercialization. Therefore, for an industrial production, there is a need to fabricate aerogels using a low-cost precursor in combination with a rapid and cheap drying method. A combination of sodium silicate and ambient pressure drying (APD) is perhaps the most promising route to low-cost silica aerogels. In 1995, Prakash et al. reported the preparation of ambient pressure-dried sodium silicate based films by surface chemical modification of wet silica films prior to drying [7, 8]. In continuation to this, Deshpande [9] and Schwertfeger [10] have produced aerogels from a *waterglass* precursor with Na^+ ion exchange followed by surface chemical modification and an APD step. Further, Kang and Choi [11], Jeong et al. [12], Wei et al. [13], and Kim and Hyun [14] have prepared similar aerogels employing the same method. However, the processing steps during the preparation of aerogels include many washing and exchange steps. This results in a lengthy process and limits the fast production of aerogels. Further, in this method of preparation, trapped sodium ions, which reduce the optical transmission and lead to poor material properties of the aerogels, were removed by ion exchange only. However, there are other alternatives to remove sodium ions, which are less expensive and just as easy to carry out as washing with water or water/vapor passing. Thus, attempts have been made to improve material properties by ion-exchange and water washing. Recently, silica aerogels have been synthesized using sodium silicate, employing an inexpensive and easy method involving (1) preparation and aging of a gel, (2) ion-exchange/water washing, (3) solvent exchange/surface modification of the gel, and (4) subsequent *APD* [15–22] (see also Chap.1).

The overall goal of this chapter is to describe ways to prepare low-density silica aerogels, particularly with emphasis on the elaboration of less time and cost-intensive synthesis methods, and to point out factors that are relevant in the large-scale production of silica aerogels. Using a factorial design set of experiments, various process parameters have been varied over a well-defined range to tune and tailor the desired properties of the aerogels. The selection of parameters discussed here are the concentration of sodium silicate in solution (the H_2O/Na_2SiO_3 molar ratio), the pH of the sol, the gel aging period, the duration of the water washing period, the exchange solvent used and the washing periods, the concentration of the silylating agent, the *silylation* period, and the drying temperature. All these parameters tremendously influence the structural and physicochemical properties of the resulting aerogels.

5.1.2. Why Use Sodium Silicate?

Sodium silicate (or sodium metasilicate), Na_2SiO_3, also known as *waterglass* or liquid glass is an inorganic compound that readily dissolves in water. A saturated solution is viscous with a density around 1.4 g/cm^3 and has a pH around 12.5. *Waterglass* is synthesized commercially by reacting quartz sand with sodium hydroxide and/or sodium carbonate at elevated temperature and pressure. Given the wide abundance and inexpensive nature of these reactants, *waterglass* is perhaps the least expensive industrial silica source. The polar nature of the molecule (Si–O$^-$ and Na$^+$ ion pairs) on one hand makes it dissolve easily in water and on the other hand prevents the spontaneous formation of larger silica polycondensates or *gelation* due to electrostatic effects. Besides, it is easy to handle and poses no flammability hazard like silicon *alkoxides* such as *TEOS* or *TMOS* do. It is also chemically long-term stable under standard conditions of use and storage. Hence, this type of precursor combines most of the key advantages for the preparation of aerogels at an industrial scale. With aerogel still being significantly more expensive than other forms of silica, cost reduction for upscaling is the major challenge for industrial manufacturers. A large fraction of the cost of the final aerogel product manufacturing cost is due to the silica precursor itself, up to 50% for *alkoxide*-based synthesis methods. Hydrophobization agents typically come second, assuming that the washing and exchange solvents are completely reused and that the system operates more or less loss-free. Currently, the Cabot Nanogel product (Chap. 38) is manufactured from *waterglass* precursors in a close to ideal manufacturing process.

5.1.3. The Need for Ambient Pressure Drying

Supercritical drying is the standard method of synthesizing aerogels. For most aerogels, this particular method is the only way to obtain true aerogels. Even though significant research effort is put into fabricating non-silica aerogels via ambient drying, (Chap. 7) this proves a difficult task. In the case of silica, true aerogels with virtually identical properties of those of supercritically dried analogs can be obtained. The ambient drying from an organic solvent proves more economic for mass production for two main reasons:

1. Supercritical drying requires moderate to high pressures and temperatures whereas ambient drying is done at ambient pressure and typical temperatures from room temperature to 200°C. Nowadays, the supercritical hydrocarbon drying has mostly been replaced by supercritical CO_2 drying; primarily because of the low temperature (31°C) and lower pressure (73 bar) requirements and reduced flammability risk. Aspen's aerogel blanket products are dried by the latter method. Still, energy use as well as instrumentation setup and maintenance cost favors ambient drying.
2. Even the more moderate supercritical CO_2 method is a batch type process and represents a bottleneck in a modern, continuous production process. In other words, for drying, individual batches of the material must be transferred to an autoclave, exchanged with liquid CO_2 solvent, dried under supercritical conditions, and retrieved from the reactor to make space for the next batch. Continuous, supercritical drying installations are not yet used in industrial aerogel production.

To summarize, from an industrial manufacturing point of view, *APD* is clearly the more efficient technology. This fact seems to have been recognized also by the academic community: if one peruses the literature on *waterglass*-based silica aerogels published since 2005, a majority of these articles deal with issues related to "ambient" or "non-supercritical" drying, indicating that people have realized the importance of this field of research.

5.1.4. Necessity of Surface Chemical Modification

In analogy to *alkoxide*-derived silica aerogels, ambient drying of their waterglass-based sister compounds is possible only if the internal surfaces are *hydrophobic*. Following a surface modification step, the repulsion of the nonpolar alkyl (typically methyl) groups, and their strong interaction with organic solvents typically used during ambient drying results in a significant reduction of the capillary pressure exerted on the silica network and hence the original gel structure of the final aerogel product suffers little to no collapse as the drying proceeds. Secondly, the structure of aerogels can deteriorate with time due to absorption of water from humid air or direct liquid exposure. If the pores of an aerogel are filled with water and the water evaporates again, this represents a second drying cycle and the structure will collapse at least partially. Hence, water uptake imposes certain limits on long-term applications of these materials. For this reason, hydrophobization of the inner surfaces of aerogels is also a countermeasure to prevent aging effects.

Typically, the hydrophobization treatment employs silylating agents, which graft themselves onto surface OH groups and introduce hydrophobic chemical functionalities. Organosilane compounds of the type R_nSiX_{4-n} (R: alkyl group, X: halide or alkoxy group, $n = 1$–3) such as trimethylchlorosilane (*TMCS*) or hexamethyldisilazane (*HMDS*) are commonly used for surface modification of the wet gels. Note that these surface modification agents are mostly immiscible with water as they contain nonpolar alkyl groups. In addition, they tend to react with protic solvents, particularly with water, thus becoming mostly inactive in carrying out their main mission: functionalizing the silica surface.

5.2. Preparation of Sodium Silicate Based Aerogels via Ambient Pressure Drying

Waterglass-based silica aerogels are commonly fabricated by a typical sol–gel process exemplified in Figure 5.1.

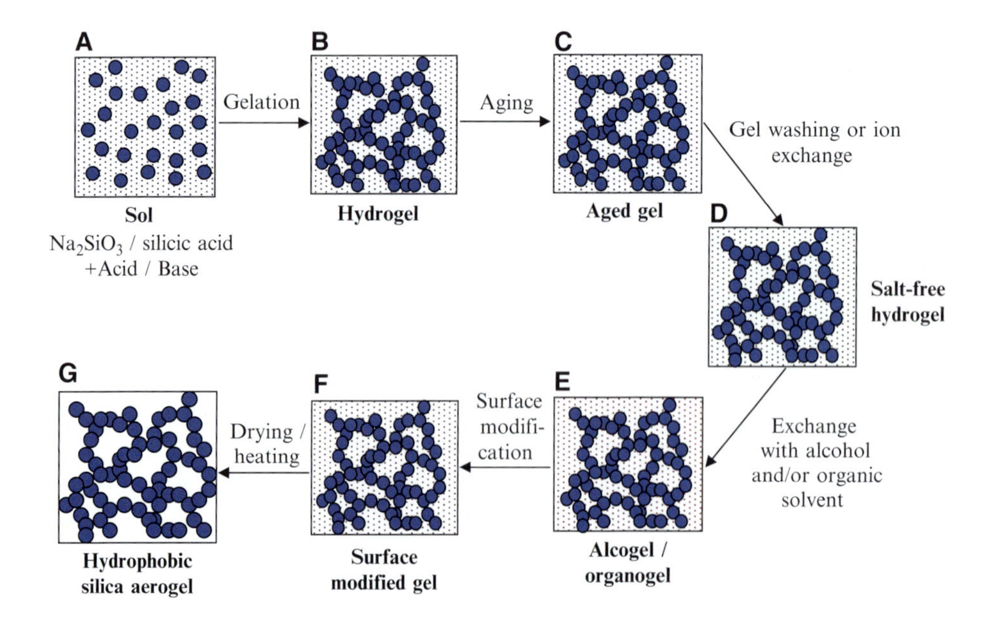

Figure 5.1. Schematic illustration of the synthetic steps used to manufacture silica aerogels from *waterglass*.

A number of different strategies are known today, which can produce ambient dried silica aerogels from *waterglass*. Generally, one can distinguish:

1. "Conventional" *gelation* followed by solvent exchange and hydrophobization and
2. Co-precursor methods. In the latter, the surface modification agent is added together with the silica precursor (*waterglass*) during *gelation*

Looking back at the times of the discovery of aerogels by Kistler (Chap. 1), there has been a substantial development in the process efficiency: all recently developed techniques are by far faster and more economic. However, most of the time required for preparation is still spent during washing, solvent exchange, surface modification, and drying steps, i.e., during the stages (c) through (g). The drying step at ambient pressure requires a more or less fixed amount of time and hence offers less potential for speeding up the synthesis. This leaves room for optimization for gel-washing, solvent exchange, and hydrophobization. It is in exactly those steps where most synthetic methods digress. Let us briefly peruse the most recent developments that were able to greatly reduce the overall processing time:

- The use of a combined solvent exchange and hydrophobization (*CSH*): Often, the hydrogel is immersed into a mixture of an alcohol (isopropyl alcohol (*IPA*)/alkane (hexane or heptane)/*TMCS* solution. In this way, steps (e) and (f) are combined [23, 24].
- In some instances, aging of the gels is done in a water alcohol or acetone mixture in order to facilitate and/or speed up the subsequent *CSH* treatment [25, 26].
- Hwang and coworkers have optimized the *CSH* method to obtain crack-free monolithic silica aerogel blocks [27].
- Bhagat and coworkers [28] have employed a *waterglass/HMDS* coprecursor method to synthesize superhydrophobic silica aerogel powders in a total preparation time of less than 5 h.

Note that the synthesis time is contingent upon the complete exchange of the pore fluid or the penetration of the entire gel volume, which relies on the diffusion of solvents and reagents [29]. The co-precursor method offers the advantage of an even faster surface modification and improved internal hydrophobicity. *Gelation* followed by surface derivatization usually takes more time. However, the derivatization method is more often employed in the synthesis of sodium silicate based aerogels via APD because only few silylating agents are miscible with sodium silicate or silicic acid solutions. In the following lines, we discuss the chemistry of *waterglass*-based silica gels and their synthesis.

5.2.1. Gel Preparation by the Sol–Gel Route

Silica gels are synthesized from molecular silicon-containing precursors. Two common methods are used to initiate the *gelation* of a *waterglass* solution:

1. Acidifcation or partial neutralization of a sodium silicate solution by addition of a *Brønsted acid*.
2. Replacement of sodium ions Na^+ by protons H^+ by means of an ion exchange resin in its acid form, thus forming a silicic acid solution and initiating *gelation* by addition of a *Lewis base* (F^-) or Brønsted base (OH^-).

Method (1) is a so-called single step process. Adjusting the pH to a value between 5 and 9 is equivalent with a partial neutralization of the sodium silicate. Commonly, the term acid catalysis is used to describe this process. Strictly speaking, this is only partially correct, because the addition of acid serves the primary purpose of partially neutralizing the alkaline sodium silicate solution and to lower the pH value. Method (2) is a classical two-step process. In the language of tetraalkoxysilane (*TEOS*, *TMOS*) chemistry, this is often referred to as an acid-base catalyzed synthesis. For the *waterglass* system, the same language is commonly encountered in the literature to describe the various stages of sol and gel formation. In the following, let us discuss the silica gel formation from *waterglass*. The two key steps in this process are neutralization and *condensation*. Figure 5.2 shows the neutralization of silicate to silicic acid H_2SiO_3. In a second step, the formation of dimer species with one equivalent of silicic acid (A) or sodium silicate (B) respectively is shown.

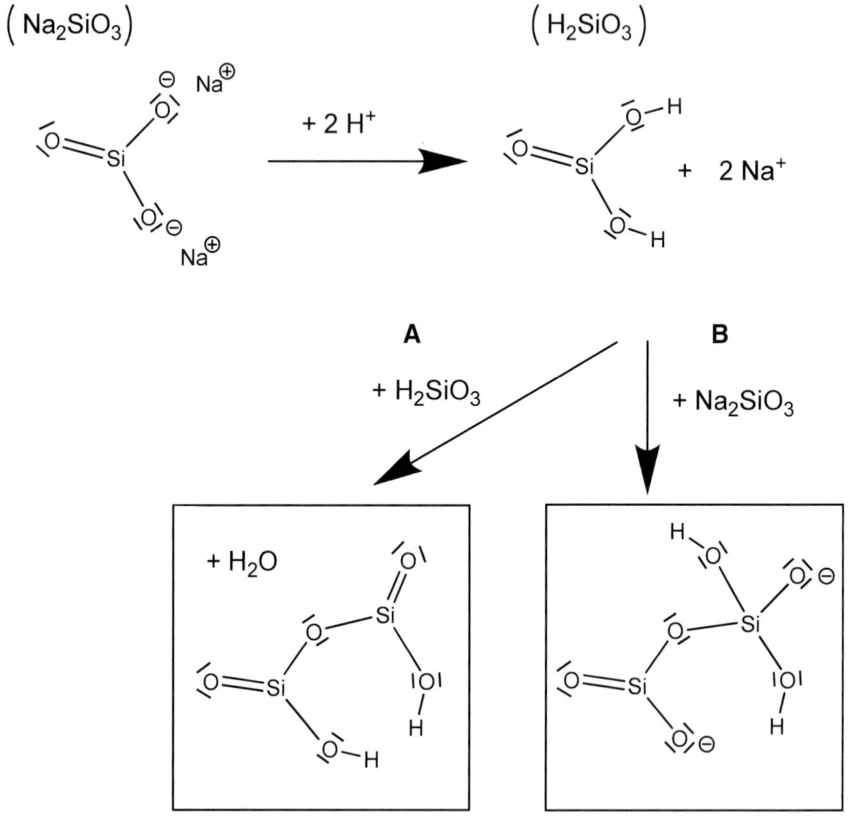

Figure 5.2. Acidification of a sodium silicate molecule to produce silicic acid and reaction with another molecule of **A.** silicic acid or **B.** sodium silicate.

Neutralization

In alkoxysilane chemistry (Chap. 2), the *gelation* process is initiated by the *hydrolysis*, which leads to the formation of hydroxylated silica (silicic acid) species. By definition, *hydrolysis* is a chemical reaction in which chemical bonds in a molecule are broken and a water molecule enters to become a part of the final product, commonly resulting in OH functional groups. In the *waterglass* case, no actual *hydrolysis* takes place, but the activation of molecular silicates or silicic acid precursor species is controlled by simple acid base chemistry, that is, through partial protonation of $Si–O^-$ centers. This is achieved by addition of H^+ from an acid source. Alternatively, complete ion exchange (Na^+ by H^+) can be done before the reaction that converts the sodium silicate into a silicic acid solution. The importance of the sodium ion removal will be discussed later on. The free silicic acid is then treated with a small equivalent of base to increase the pH value. This is again a partial neutralization reaction but this time starting with the acid form. In both cases, the partial neutralization initiates condensation reactions, which lead to *gelation*. The formation of free Si–OH and the overcoming of electrostatic repulsion interactions between silica species are prerequisites for forming larger silica polycondensates or colloidal silicate particles, which in turn form the three-dimensional gel network.

Condensation

The chemistry of aqueous silicate and silicic acid systems is quite complex: repulsion and ion pairing are known to govern intermolecular interactions in alkali silicate solutions [30]. More than 20 different oligomer species of *waterglass* containing 2–8 silicon atoms have been identified by means of ^{29}Si nuclear magnetic resonance (*NMR*) spectroscopy [31]. Alkaline silicates tend to form ring and cage-like structures. The condensation kinetics depend strongly on the pH value since they are governed by electrostatic interactions of charged molecular species and clusters: At pH > 10, i.e., where sodium silicate is the prevalent species, condensation reactions are slow, because the negative charges $Si–O^-$ prevent the coming together of the solvated silicate ions and because of an increased competition with the backward reaction (dissolution). In other words, *waterglass* solutions do not gel unless the pH value is lowered. In addition, the condensation of Na_2SiO_3 species with themselves or with a free silicic acid molecule, as shown in Figure 5.2B, is disfavored from an entropic point of view because two molecules come together and form one. If two molecules of metasilicic acid dimerize, as shown in Figure 5.2A, there is a substantially smaller kinetic barrier for this process. Because H_2SiO_3 is electroneutral (the isoelectric point of bulk silica ranges from 2 to 4) compared to metasilicate, there is significantly less electrostatic repulsion, and the reaction is also favored from an entropy point of view (formation of H_2O as a reaction product). This is certainly a simplified view and the complete picture is far more complex, but the example of the dimer formation should serve an illustrative purpose here.

At pH < 4, *waterglass* exists mainly in its protonated form that is as silicic acid oligomers. The simplest forms are meta-silicic acid H_2SiO_3 and its hydrated form orthosilicic acid H_4SiO_4. Even at low pH, silicic acid tends to dimerize ($H_6Si_2O_7$ from orthosilicic acid and $H_2Si_2O_5$ from metasilicic acid). Trimers such as $H_8Si_3O_{10}$ are formed by addition of another molecule of silicic acid [32]. The stepwise condensation of small silicic acid oligomers is believed to produce mostly ring-like structures with three to six silicon atoms and a largest possible number of Si–O–Si linkages. Numerical calculations [33] predict a constant increase in the enthalpy of formation of only about 17 kJ/mol for each additional molecule of H_4SiO_4 added. For this reason, the nucleation of colloidal silica particles is believed to be occurring around a critical cluster size containing three to six silicon atoms.

Now let us take a look what happens at more neutral pH values and why condensation of smaller colloids happens primarily in that acidity range. So far, one has established that silicon oxides exist in the form of small molecular clusters at high and low pH. In alkaline solution, electrostatic repulsion and the lack of free hydroxyl leaving groups prevent the formation of larger polycondensates and network structures. In the case of silicic acids, it is believed that polycondensation of colloidal particles in solution does indeed occur, but that the formation (forward reaction) is outweighed by the dissolution of larger silica species because of the tremendous competition for the protonation oxide groups. A second factor that helps explain slower polycondensation rates altogether is the lack of $Si–O^-$ groups at low pH, which are better nucleophiles than free silanol groups. In conclusion, it seems indeed plausible that condensation reactions and hence also the *gelation* of silica occur preferably in the pH range from 5 to 9. Looking back at our simplified model in Figure 5.2, the neutral case can be illustrated by the reaction (b) of one silicic acid and silicate monomer. The intermediate pH regime is accessible from either side. In other words, partial neutralization of *waterglass* (1) produces hydrogels in much the same way as (2) the addition of moderate amounts of base to silicic acid solutions. The time required for gel formation, as soon as either acid or base is added to the precursor, is called gelation time. Because the kinetics of the network formation also governs its final structure, the *gelation* pH and the type and concentration of ions present in solution are known to affect the final aerogel properties. This will be further explained in Sect. 5.3, which describes the influence of various reaction parameters.

It seems important to mention once more that the chemistry of silicates in aqueous solution is rather complex and that the language adapted from alkoxysilane sol–gel chemistry used to describe formation of hydrogels from *waterglass* fails to paint an accurate picture of what is happening at the molecular level. Even though there are similarities between those two systems, the limitation of such analogies made must always be kept in mind (Chap. 2).

5.2.2. Washing/Solvent Exchange/Surface Modification

Ambient pressure-dried silica aerogels are prepared from hydrocarbon exchanged gels and require the use of hydrophobization agents to minimize forces exerted onto the gels during drying and to prevent structural collapse (reference chapter). The washing and drying of *waterglass* silica gels are again very similar to that of silicon *alkoxide*-derived aerogels. Even more, in this case, the processes are truly quite the same because they start from virtually identical silica hydrogels or alcogels, respectively. The main difference is that *waterglass*-based silica are typically synthesized in purely aqueous solutions and alkoxysilane gels are typically synthesized in alcoholic media. The choice of pore liquid affects not only the aging process (higher water content changes kinetics of dissolution and reprecipitation of silica from and onto the gel network) but also the solvent exchange steps. As described earlier, the washing, solvent exchange, and surface modification steps (Figure 5.1C–F) can be done separately, which is a more time-consuming process. Most recent preparative methods [23–28] combine those 2–3 steps in a single one, typically by a treatment with a solution containing an alcohol, a saturated linear hydrocarbon, and a hydrophobization agent, for example, *IPA*/heptane/*TMCS*.

A schematic of the hydrophobization of a silica surface is shown below in Figure 5.3: Silanol groups on the gel surface are replaced by nonpolar alkyl groups through reaction with an alkyl-derivatized silane compound such as *TMCS*.

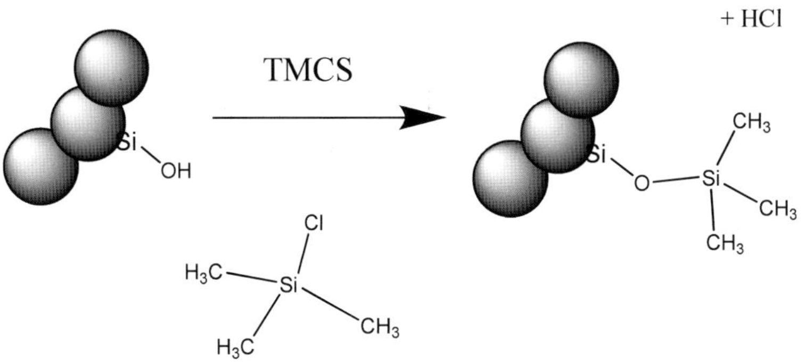

Figure 5.3. Hydrophobization of a surface silanol group in a silica gel (represented by the three colloidal particles) with trimethylcholorosilane *TMCS*.

Combined Washing, Solvent Exchange, and Hydrophobization (*CSH*)

The combined approach has its advantages and disadvantages. Let us recall that the equilibration of pore liquids inside a gel is controlled by diffusion of solvent and/or reagent. Thus, the treatment time necessary to ensure a complete reaction increases with the square of the smallest dimension of a given gel body. For this reason, it is generally more economical to produce smaller particles or granular samples rather than larger monoliths. Also, the time savings become apparent if we consider a single step process in comparison with three consecutive ones. On the other hand, there is the question of the hydrophobization agent consumption in a combined approach. Common hydrophobization agents such as *HMDS* or *TMCS* are known to react with free hydroxyl groups. If these reagents are added to aqueous or alcoholic media, they are known to react very rapidly to form the silanol or at moderate rates to produce alkylether compounds, respectively. The reaction of *TMCS* with water and alcohols R–OH, R=H, alkyl) is shown below in (5.1) and (5.2).

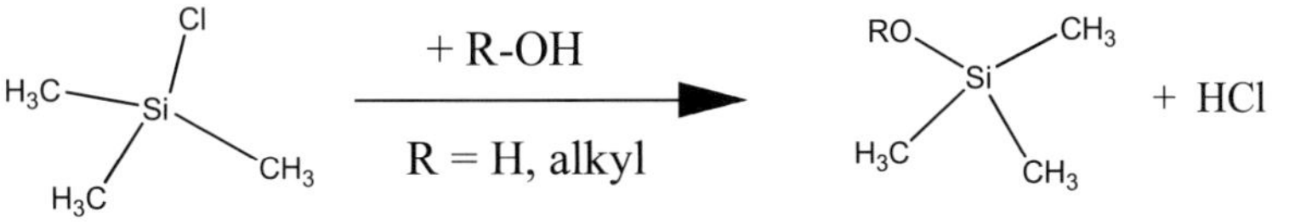

$$(H_3C)_3Si\text{-}Cl \xrightarrow[R\,=\,H,\ alkyl]{+\,R\text{-}OH} (H_3C)_3Si\text{-}OR + HCl \qquad (5.1)$$

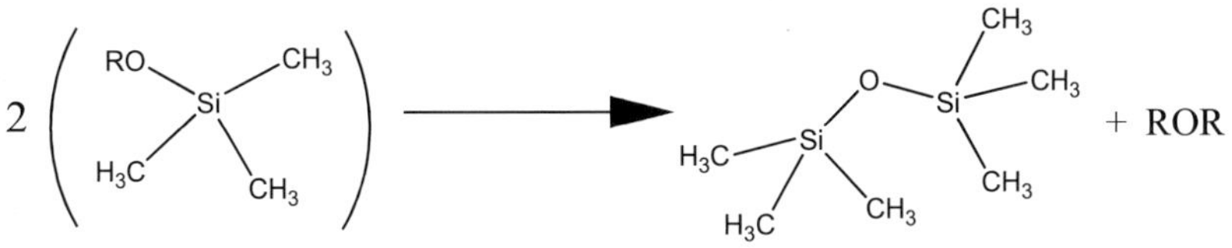

$$2\left((H_3C)_3Si\text{-}OR\right) \longrightarrow \underset{\text{HMDSO}}{(H_3C)_3Si\text{-}O\text{-}Si(CH_3)_3} + ROR \qquad (5.2)$$

During the exchange of the pore water, the water and alcohol solvent both consume *TMCS*. If large amounts of pore water are present, the formation of silanols is the preferred reaction [(5.1), R=H). These silanols and the alcohol are ambifunctional molecules, which means that they promote the phase transfer and hence promote the solvent exchange from the aqueous to organic medium. A large fraction of the hydrophobization agent used in this process is therefore lost because it settles at the bottom of the vessel (outside the gel) in the aqueous/alcohol mixed phase. For a silica aerogel with an average surface OH coverage of 4.5 per nm^2, a specific surface area of 650 m^2/g, and a density of 0.1–0.15 g/cm^3, one would require 2–3 mmol of *TMCS* for complete surface hydrophobization per cm^3 of wet gel. Typical preparation conditions for a single step exchange/hydrophobization with *IPA*/*TMCS*/hexane require 0.25–0.4 mol equivalents per pore water. This translates to a five- to tenfold excess of the hydrophobization agent if we assume quantitative conversion of all free surface OH groups as shown in Figure 5.3. This means that the combined method uses the surface modification agents in a very inefficient manner. For large-scale industrial processes, where material expenses are the main contributor to the total product cost, this is indeed a substantial disadvantage. From this point of view, using an inexpensive *waterglass* precursor but wasting a majority of the most expensive component in the manufacture, the hydrophobization agent, appears like a shot in the dark. At the industrial level, the direct CSH approach is only realistic, if the loss of surface modification agents can be minimized, for example, via continuous recycling.

At the laboratory scale, the cost of the chemicals used is normally of secondary importance, but speedy processing can be quite important. Consequently, CSH is best suited for the fabrication of smaller specimens, especially at the laboratory scale.

Consecutive Washing, Solvent Exchange, and Hydrophobization Steps

The traditional method of washing, solvent exchange, and hydrophobization relies on a number of steps with the simple goal to exchange the pore fluid from aqueous to alcohol or acetone and then to an organic hydrocarbon. As mentioned previously, the overall procedure is time consuming, but is efficient from the point of view of minimized consumption of the hydrophobization agent. [25] describes the synthesis of crack-free silica aerogel monoliths of approximately 4 cm diameter and > 1 cm thickness using several consecutive washing steps. Most importantly, before the surface modification treatment, the gel liquid is pure heptane. In other words, most of the pore water and alcohol have been removed from the system, and it is only under these conditions that the hydrophobization agent, in this case *HMDS*, reacts only with surface silanol groups (Figure 5.3) and is not consumed by undesired side reactions (5.1) and (5.2). Because of the large contribution of surface modification agents to the total cost, this route seems to be the method of choice for the synthesis of aerogels at an industrial scale.

5.2.3. Drying of Modified Gels

Drying of the surface-modified gels from an organic solvent phase is typically done in two steps. The first is a low-temperature drying step at close to ambient temperatures (25°C–50°C) typically lasting one to two days. The second step is done at high temperatures between 150°C and 200°C. The last step is the most energy intensive of the *APD* process. A list of typical physical properties of ambient dried *waterglass*-based silica aerogels is given below in Table 5.1.

Today, the *APD* process yields aerogels that offer identical properties to those dried supercritically. However, the elaboration of every synthetic method starts with arduous trials and optimization studies until this point is reached. From that point of view, the supercritical drying method is certainly less labor-intensive because there are fewer steps to the final product and the drying is not as sensitive to process parameter changes as in the *APD* case.

Table 5.1. Collection of typical properties of silica aerogels derived from *waterglass* by *APD*

Physical property	
Gelation time (minutes)	1–150
Density (g/cm^3)	0.07–0.15
Porosity (%)	92–97
BET surface area (m^2/g)	550–750
Cumulative pore volume (cm^3/g)	2.0–2.3
Average pore diameter (nm)	10–15

In the following lines, we try to explain the influence of a number of selected reaction parameters, demonstrating their effect on essential physical properties such as density, *thermal conductivity*, optical transmission, *specific surface area*, and *hydrophobicity*. Figure 5.4 below shows a typical transmission electron micrograph of an ambient dried silica aerogel made from *waterglass*. It clearly reveals the nanoscopic building blocks and the high porosity of the material.

Figure 5.4. *TEM* Image of a *waterglass*-based silica aerogel revealing the mesoporous pore structure.

5.3. Effects of Various Process Parameters on the Physicochemical Properties of the Aerogels

5.3.1. Effect of the Sodium Silicate Concentration in the Sol

The concentration of sodium silicate is the most central reaction parameter. It determines the amount of silica per added sol volume and hence also the "packing density" of silica units in the gel network. In an experimental study in our laboratory [19],

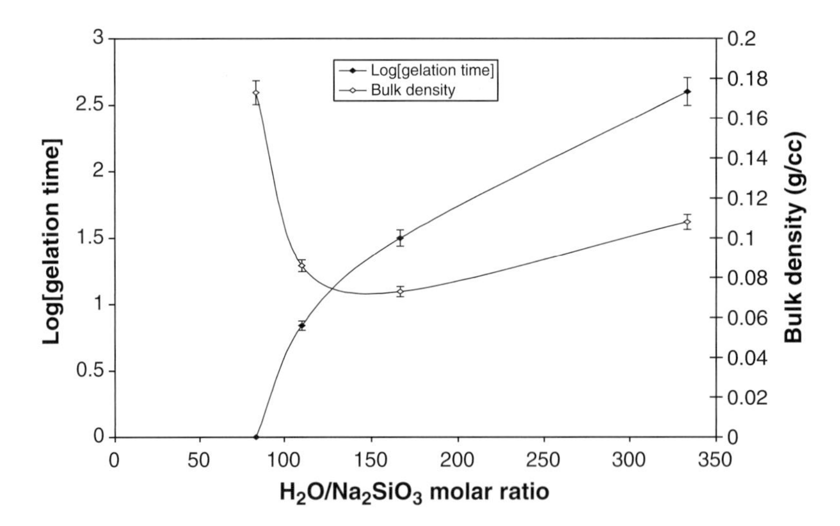

Figure 5.5. Effect of H_2O/Na_2SiO_3 molar ratio (A) on *gelation* and density of aerogels.

the H_2O/Na_2SiO_3 molar ratio was varied from 83.3 to 333.3 as shown in Figure 5.5. It is observed that with increasing silicate concentration in the sol, the gelation time t_g increases from 1 min to 6 h. This is due to the fact that with increasing dilution (H_2O/Na_2SiO_3 ratio), the mean separation between condensating silica species becomes larger and larger, and hence the collision frequency is lower. That also implies that the random three-dimensional gel structure takes longer to form. The bulk density of the final aerogel first decreases and then increases with increasing dilution of the *waterglass* sol. The right hand slope of the density curve is attributed to the fact that too little silica is present in dilute systems. Hence the network created in this case is thin and fragile and collapses during drying. At a H_2O/Na_2SiO_3 molar ratio of 160, the structure seems strong enough to survive the consequent exchange, hydrophobization, and drying procedure and marks the ideal ratio to obtain the highest porosities. By further increasing the *waterglass* solution concentration (H_2O/Na_2SiO_3 molar ratio < 150), the network not only becomes more robust but also contains larger colloidal silica particles and higher silica volume fractions in the fresh gel. An aerogel that is made from large colloids therefore shows a smaller pore size. Hence, smaller pore size, reduced *microporosity*, and pronounced weak points in the network due to different size particles give rise to higher density aerogels.

5.3.2. Effect of Sol pH

As explained in some detail in the discussion of the sol–gel chemistry of *waterglass*, condensation reactions and hence the formation of gels occur fastest at moderate pH values. The addition of acid to a sodium silicate or base to a silicic acid solution considerably raises the kinetics of condensation and network formation. The influence of the pH of the sol on the optical transmission, gelation time, porosity, and density of the final product is given below in Figures 5.6 and 5.7. The data are taken from an earlier study [34]. A sodium silicate solution was diluted to produce a SiO_2/H_2O ratio of 370 and ion exchanged to form a silicic acid solution with a pH value around 2. The pH of the sol was then adjusted by addition of a 1 M NaOH solution and the gelation time recorded. The gels were aged and processed by a

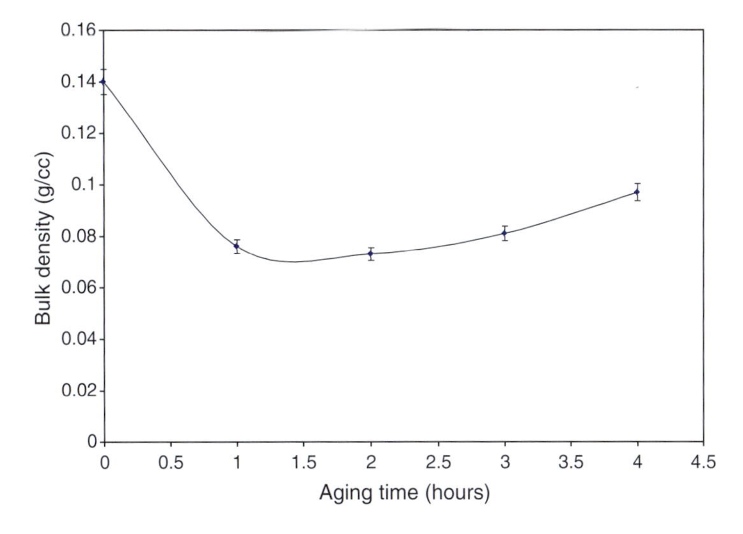

Figure 5.6. Effect of the aging period on the bulk density of the final aerogel.

Figure 5.7. Photographic image of hexane exchanged gels at the beginning (**left**) and at the end (**middle**) of the hydrophobization treatment. The gels clearly float. Contact angles above 140° of aerogels obtained in this way indicate superhydrophobic properties (**right**).

consecutive washing, exchange, and hydrophobization process and dried at ambient pressure. Table 5.2 shows the effect of the sol pH on the above-mentioned properties.

Both the gelation time t_g and the optical transmission decrease at higher pH. At pH < 4, colloidal silica particles are believed to carry a positive surface charge [35, 36], resulting in columbic interparticle repulsion. Hence, the gelation time increases and the probability of the fusing together of the silica particles is smaller. Two other factors that disfavor or slow down the growth of a gel network at low pH are, first, the strong competing effect of H^+, which shifts the equilibrium of the condensation reaction toward the side of free silanol groups, and second, the absence of $Si–O^-$, which is a better nucleophile.

When assembled at low pH (<4), aerogels exhibit a comparably low porosity $<90\%$, whereas at pH values from 5 to 8, the porosity is above 95%. In this study, the gelation times

Table 5.2. Effect of the gelation pH on the gelation time and aerogel properties

	t_g [min]	Density [g/cm^3]	Porosity [%]	Visual appearance
3	32	0.25	82	Transparent
4	12	0.15	92	Transparent
5	5	0.1	95	Less transparent
6	2.5	0.05	97.5	Semitransparent
7	1.6	0.06	97	Opaque
8	1.2	0.065	97	Opaque

at pH > 6 were extremely fast, this means that there is little time for the system to condensate into a perfect, well-distributed structure. At such short gelation times, the formation of larger pores becomes more likely. Those act as scattering centers for light in the visible range of the spectrum and promote opacity of the final aerogel.

5.3.3. Effect of Aging (t_a) and Washing (t_w) Periods

In this section, the aging and water washing steps are discussed. For this purpose, gels were synthesized using partially neutralized sodium silicate solutions with a silicate to water ratio of 1:167. Fresh gels were aged for various periods of time from 0.5 to 2 h in 0.5 h intervals. Aging of the gel before drying helps to strengthen the network integrity and thereby reduces the risk of fracture during the subsequent processing steps [37]. The effect of gel aging period on the density of silica aerogels is shown in Figure 5.6.

It is observed that with an increase in the aging period, the density of aerogel decreases first and then increases again after a critical aging time of around 1 h. This is due to the fact that the polycondensation (Si–O–Si bond formation) of silica species still continues even after the *gelation* has occurred. For short t_g, primarily Q_2 species exist at the point of *gelation*, where Q_n denotes a silicon atom with n bridging Si–O–Si bonds. Owing to this, a fresh gel network is still extremely fragile and cannot withstand external forces or pressure exerted without suffering damage. From *NMR* studies, it is known that the fraction of Q_3 and Q_4 species increases with time. A higher connectivity of the silica network increases the network strength, in analogy to a crosslinking process in polymer chemistry. Aging also causes coarsening of the gel network leading to the growth of thicker interparticle necks due to dissolution and reprecipitation of silicate oligomers and particles [38]. Thus, continued polymerization and coarsening of the gel stiffen and strengthen the gel network, therefore contributing to minimum shrinkage during drying, yielding low density silica aerogels. The observed increase in the bulk density for $t_a > 3$ h is mostly due to increased condensation. With increasing aging period, the average Q_n value increases, or in other words, more bridging bonds are formed. Those take up less space than the two Si–OH groups from which they originated. This means that with increasing time, the network contracts again as it maximizes the connectivity [39]. This explains the higher shrinkage and hence higher aerogel density, which are accompanied by longer aging [19, 20].

Next, let us look at the effect of the gel washing treatment: water washing (Figure 5.1C) is an economic alternative to the use of an ion exchange resin. The goal of both processes is to remove sodium ions Na^+ because they affect the structural and optical transmission properties of *waterglass*-derived aerogels. For ion-exchange, the sodium silicate solution (or gel with excess water) is passed through an ion exchange resin to replace Na^+ ions in the solution by protons H^+. To regenerate the resin's capacity when depleted, a regeneration step in the form of mineral acid immersion is needed to replace Na^+ ions again by H^+. In an industrial process, the extra time spent for water washing needs to be weighed against the purchase, use, and regeneration of the ion exchange resin.

For any washing or solvent exchange step, there are two distinctly different modes of operation:

1. Diffusion-limited exchange (the standard case)
2. Exchange forced by fluid flow

In the diffusion-limited case, the exchange time is limited by the diffusion coefficient of the solvent(s) and the pore size and structure of the gel. Diffusive processes are strongly temperature dependent. This is the reason why washing and exchange steps during *APD* processes are often carried out at moderate temperatures in the 40–70°C range.

Forced exchange by a fluid flow can result in a substantial speeding up of washing and exchange steps. On the other hand, its applicability is quite limited in terms of the geometry of the gel body. Forced flow requires the buildup of pressure on one side of the gel in order to press out the pore liquid on another side. This is best visualized in the case of a forced exchange with water or water vapor in a tube that contains a cylindrical gel body. That one has to fit snug into the tube and be held in place so that the buildup of a pressure is possible. The fluid flow rate is given by the applied pressure as the driving force for the exchange (which must be quite low given the fragility of the gels) and the permeability of the gel. According to the *Carman–Kozeny* equation, the *permeability* (*D*) of a gel is given by [38]:

$$D \propto (1 - \rho_b) r, \tag{5.3}$$

where ρ_b is the bulk density and r is the pore radius. For a fixed ρ_b, larger r values come with an increase in permeability making solvent exchange and *silylation* easier. Recently, we have reported on the use of water vapor passing [20] as one method of forced hydrogel washing. In principle, washing of cylindrical gels in a tube can also be done with plain water; however, in that case, it is more difficult to precisely control the pressure. In an industrial process, forced water or vapor exchange is quite often impractical. In such cases, diffusion-limited "standard" washing cycles are the method of choice.

Today, the diffusion limited exchange is still the method of choice for reasons of practicality. Table 5.3 shows the effect of the gel washing time of an aged hydrogel at a volume ratio of water to gel of 1:1. It clearly shows that the sodium content in the gel goes down with increasing washing time. With less Na^+ present in solution, also the density of the final aerogel will be lowered. On the other hand, the optical transmission of gels with reduced sodium content is improved. If the sodium concentration in the gels were reduced below 1%, the gel transparency could be still improved significantly.

For the gels studied here, a washing time of 18 h is needed. After that, no significant change in the properties is observed. This is because the gel and water phases reach equilibrium. If the gel is washed once with the same volume of pure water, the equilibrium composition will come to lay around half of the initial Na^+ concentration. Hence, for a complete removal, multiple washings are required [40]. In addition, the use of larger volume equivalents of the washing fluid or a continuous process with a countercurrent flow can help

Table 5.3. The effect of the hydrogel washing time with water on selected properties of the aerogel

Water washing time [h]	Density [g/cm^3]	Na^+ content [%]	Optical transmission [%]
0	0.11	10	50
6	0.105	7	56
12	0.1	6.5	58
18	0.098	5.5	60
24	0.098	5.5	60
30	0.098	5.5	60
36	0.097	5.5	60

to make washing processes more efficient. Also, the time required for diffusion-limited water washing depends on the gel permeability and the size of the gel body or particles. From this point of view, granular systems require significantly shorter washing times.

5.3.4. Effect of the Type of Exchange Solvent Used

The pore fluid of a salt-free hydrogel needs to be exchanged with an organic solvent, because hydrophobization agents are insoluble in water and are consumed by side reactions with water and alcohols as shown in (5.1) for *TMCS*. When switching from a protic to an aprotic (hydrocarbon) medium, alcohols are ideal intermediary solvents, as their bifunctional nature (polar /nonpolar) promotes miscibility of water and organic phase. What seems at first surprising is the fact that the choice of alcohol has a tremendous effect on the pore structure and therefore also on the final materials properties. Here, we summarize the results of a recent study, which was carried out using Na^+-free (silicic acid derived) gels [41]. Aged hydrogels were first exchanged with water, then with the intermediary alcohol solvent, and finally modified with a mixture of alcohol, *TMCS*, and hexane. The volume ratio of silicic acid:NH_3(1N):alcohol:*TMCS*:hexane was kept constant at 1:0.004:0.4:0.4:0.4. The results are summarized in Table 5.4.

The Table 5.4 data suggest that the density and volume shrinkage of the aerogels increase and the porosity decreases in the following order of alcohols used: isopropanol < methanol < ethanol < propanol < isobutanol < butanol < hexanol. Isopropanol and methanol definitely produce the best aerogels. The chain length of the hydrocarbon end

Table 5.4. Effect of the type of intermediary solvent used to exchange the pore fluid from aqueous to organic medium

Solvent/formula	Volume shrinkage [%]	Density [g/cm^3]	Porosity [%]	Mean pore diameter [nm]/FWHM [nm]
Methanol/CH_3OH	37	0.10	94.9	11.5/4.5
Ethanol/C_2H_5OH	55	0.155	92.1	10/7.0
n-Propanol/C_3H_7OH	57.5	0.16	91.8	12/8.0
Isopropanol/$(CH_3)_2CHOH$	30.7	0.07	96.6	14/5.0
n-Butanol/C_4H_9OH	62.5	0.17	91.3	7.5/5.0
Isobutanol/$(CH_3)_2CH_2CH_2OH$	61	0.165	91.5	15/7.0
n-Hexanol/$C_6H_{13}OH$	65	0.19	90.0	–

influences both vapor pressure and the surface tension and wetting behavior of the solvent/gel interface. Generally, low vapor pressure and low surface tension solvents decrease the surface tension in the gel and produce less shrinkage. With increasing chain length of the alcohol, the hydrophobic part becomes dominant over the hydrophilic one. This results in poor coordination with the water and thus also in a reduced exchange efficiency. The reason why isopropanol and methanol are performing so well is that they carry an extremely compact hydrophobic site in the form of $-CH(CH_3)_2$ and $-CH_3$ group, respectively, which are covalently attached to the polar OH group. It is this compactness that allows the perfect ambipolar nature to take its effect as a mediator between phases. Less compact alkyl chains do not succeed in the same way, resulting in more pronounced structural collapse. This is also seen in the pore size distributions. Isopropanol and methanol both offer optimal pore sizes between 10 and 15 nm for a lightweight and high porosity aerogel but even more importantly, they both show extremely narrow size distributions with a full width at half maximum FWHM value of 5 nm or less.

5.3.5. Effect of the Amount of Silylating Agent Used and the Duration of the Silylation Treatment

As previously mentioned, it makes little sense to use an inexpensive silica precursor to manufacture silica aerogels and then to overcompensate the cost savings by using large amounts of pricy hydrophobization agents. In the first place, losses by unwanted *hydrolysis* with water should be minimized having implications on the washing and exchange steps as described before. Hence, from a practical realization point of view, it is important to know the minimum amount of silylating agent needed per equivalent of SiO_2 for each preparation method. This section describes *sylilation* process parameters.

To study the effect of the concentration and exposure time of the silylating agent on the physical properties of aerogels, gels were prepared by adding 0.024 mol equivalents of acetic acid to a 0.2 M sodium silicate solution. The gels were washed thrice with water to remove Na^+, exchanged with methanol and washed three times with hexane. They were then sylilated in a solution containing the surface modification agent *HMDS* in hexane. This procedure is the complete multistep process (Figure 5.1) executed as a series of consecutive steps (as opposed to *CSH*). During the hydrophobization treatment, the gels typically begin to float in the hexane solution. This can be seen in the photographs below in Figure 5.7.

When prepared and treated identically a variation of the *HMDS*/Na_2SiO_3 molar ratio in the range from 1.8 to 6.9, a distinct difference of the aerogel properties becomes apparent (Figure 5.8). At a ratio below 4, the density of the gel continuously drops from approximately 0.117 down to 0.095 g/cm^3 and the optical transparency from 70% to 60%. This is attributed to the incomplete surface modification of the silica surface by hydrophobic trimethylsilyl groups, which leads to greater volume shrinkage during drying. As the amount of *HMDS* is raised above five equivalents, the aerogel properties rapidly deteriorate. When using a sevenfold molar excess, the density shoots up to >0.2 g/cm^3 and the transparency drops below 20%. This trend can be explained by an increase in steric crowding of the nonhydrolyzable methyl groups and the formation of micelle-like *HMDS*-rich subphases, leading to a decrease in the surface modification efficiency [21]. This again leads to structural collapse, which results in higher density and opacity.

This study demonstrates that the consecutive washing method uses significantly less hydrophobization agent than the combined *CSH* method: decent aerogel materials can be

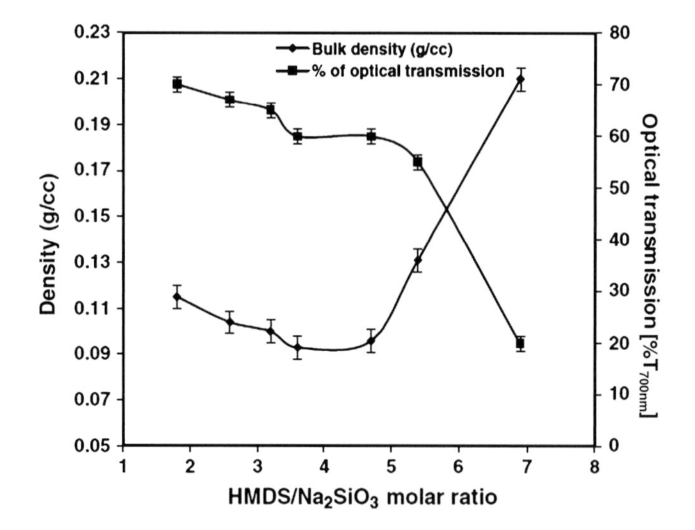

Figure 5.8. Effect of HMDS/Na_2SiO_3 molar ratio on bulk density and % of optical transmission of aerogels.

obtained at a molar ratio of 2.5. The *CSH* method uses typically a tenfold excess of hydrophobization agent, which translates into a significant cost penalty for a commercial fabrication process.

In addition to the amount, the *silylation* period also plays a significant role in the surface modification treatment. A study was carried out with *silylation* by *HMDS* with 3.6 equivalents. The treatment time ranged from 6 to 36 h in 6 h steps while keeping all other synthesis parameters the same as previously. In the plot of bulk density versus *silylation* period as shown in Figure 5.9, one can see that the bulk density of the aerogels decreases with treatment times to reach a minimum value of about 0.085 g/cm^3 after 30 h. This means

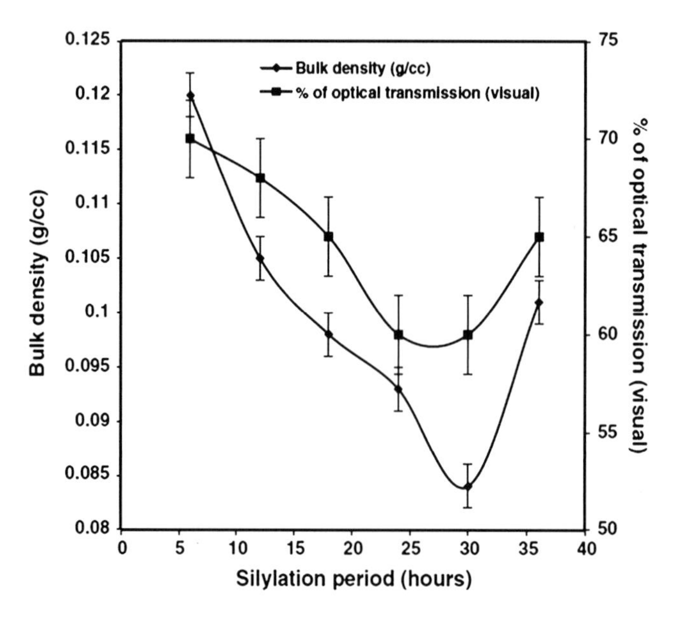

Figure 5.9. Effect of the silylation period on density and optical transmission (%) of the aerogel.

that the diffusion of *HMDS* into the most inner pores of the gel body and the kinetics of the surface modification reaction are quite slow. Other sylilation agents such as *TMCS* are more reactive and are expected to react more quickly. Aside from indirect methods such as the characterization of aerogel properties, the extent of the hydrophobization can be rapidly quantified by contact angle measurements and IR spectroscopic techniques. The latter is less quantitative but provides direct evidence for the derivatization by alkyl groups through the presence of C–H and Si–C vibrational signatures.

Naturally, also the type of silylation agent used plays a crucial role in determining the final properties of aerogels obtained by APD. *HMDS* and TMCS are the ones most commonly used, and that for a reason. They yield the lowest density and highest porosity materials and also are easily accessible through standard commercial retail channels. For a more comprehensive comparison of physicochemical properties of identically prepared sodium silicate aerogels silylated with various surface modification agents, the reader is referred to [16].

5.3.6. Effect of Drying Temperature

Drying involves the evaporation of the pore solvent, typically hydrocarbons and alcohols, from the pores of gel network. Drying at high temperature induces a so-called springback effect of silylated gels. This leads to a lowering of the final density of the dried silica aerogels. Springback implies that the gels contract and then expand again upon drying [39]. Figure 5.10 shows this very effect in the form of the large variation in the gel volume as a function of the drying temperature. A maximum contraction of almost 90% of the total volume at a temperature around 100°C is followed again by expansion, which above 150°C brings back the gel to > 90% of its original volume [17, 42]. After drying, a translucent aerogel product is obtained.

Figure 5.10. Volume change observed during *APD* process with a maximum contraction occurring around 100°C (**left**) and a photographic image of the dried aerogel material (**right**).

5.3.7. General Comments About Parameter Optimizations

So far, we have analyzed a number of parameters of the whole fabrication process of aerogels from silicates without going into too much detail. The preparation of sodium silicate aerogels by *APD* contains a total of 4–7 main steps (depending on whether consecutive washing, solvent exchange, and hydrophobization or *CSH* are used), each of which comprises at least one main process variable. Hence, it is extremely challenging to optimize all steps involved in the preparation, or in other words, the devil is in the details. Another fact often neglected is that the ideal parameter combination for one particular step, such as for example the hydrophobization, may not be generally valid and depends on the previous history, that is, on the choice of preparation steps, methods and the set of parameters used before. Just like in human psychology, everything is connected to everything else. For these reasons, the use of high-throughput methods such as synthesis robots could be advantageous. This technique has great potential to identify optimal parameter sets and drastically reduces experimental error.

When done properly, systematic studies of sol–gel synthesis and processing can facilitate the search for new materials, particularly silica aerogels, with desired properties such as low thermal conductivity, high optical transparency, improved mechanical properties, or reduced dust release behavior.

5.3.8. Silica Aerogels as Thermal Insulating Materials

Aerogels are known for their outstanding thermal insulation properties combined with light weight, chemical resistivity, and nonflammability. This makes them ideal candidates for both ambient and high temperature insulation applications. For a detailed discussion on the use of superinsulating aerogels, the reader is referred to Chap. 23 and Part VIII. In general, heat is transferred by radiation, conduction, and convection. Due to the small pore sizes of aerogels, the convective and gas conduction pathways are virtually eliminated. What remains are primarily radiative heat transfer and conduction through the silica network structure. Silica aerogels are the best-known aerogel insulation systems and so far also the only ones that are commercially exploited. Products by Aspen Aerogels and Cabot Corporation offer thermal conductivity values as low as 0.013 W/(mK). Aerogel insulation products are successful in niche markets; however, with increasing popularity and production volumes, there will be room for growth of respective market shares, and new products will become affordable for applications that are not yet covered today. In this respect, *APD waterglass*-based aerogels have a tremendous potential for upscaling and cost reduction for the two main reasons widely discussed in this text: the process and cost advantage of ambient pressure drying and the low cost of *waterglass* as a precursor for the gel synthesis. Therefore, it seems reasonable to assume that there will be an ongoing interest in *waterglass*-based aerogels in the near future, both academically and from a manufacturing point of view. The emerging of several new production facilities for silica aerogels worldwide and the fact that most new ones are using *waterglass* precursors and *APD* speak for themselves.

5.4. Conclusions

Waterglass-derived, *APD*-dried aerogels offer virtually the same properties as their alkoxysilane and/or supercritically dried analogs. From an economical and large-scale production point of view, there is still a significant potential for cost reduction. Especially the washing/solvent exchange and hydrophobization steps influence the final cost of the aerogels. The sol–gel parameters discussed in the text demonstrate the importance for step-by-step optimization of all relevant processes and parameters. This strongly reflects on the physical properties of silica aerogels fabricated under optimized conditions. The first step of the gel formation is a partial neutralization of the *waterglass* or silicic acid precursor. When the pH-value of the sol is kept in the range between 4 and 9, *gelation* occurs within minutes to hours, leading to well-developed silica networks. Aging for 1–3 h followed by washing the gel with water for 24 h produces low density and semitransparent silica aerogels. The solvent exchange from aqueous to organic phase is best done with isopropanol or methanol as intermediary alcohol as they work best to remove water from the pores and are accompanied by low capillary pressures. The concentration of the silylating agent and the *silylation* period both have striking effects on the surface modification efficiency. This is reflected in the final properties of the aerogels after drying. Ambient dried silica aerogels made from *waterglass* have great potential for a rapid expansion in the worlds growing aerogel insulation and other specialty niche markets.

References

1. Hüsing, N.; Schubert, U.; Aerogels - Airy Materials: Chemistry, Structure, and Properties, Angew. Chem. Int. Ed. 1998, 37 (1/2), 22–45
2. Fricke, J.; Aerogels - highly tenuous solids with fascinating properties, J. Non-Cryst. Solids 1988, 100, 169–173
3. Pajonk, G. M., Transparent silica aerogels, J. Non-Cryst. Solids 1998, 225, 307–314
4. Kocon, L. ; Despetis, F. ; Phalippou, J., J., Ultralow density silica aerogels by alcohol supercritical drying, J. Non-Cryst. Solids 1998, 225, 96–100
5. Kistler, S. S., Coherent expanded aerogels and jellies, Nature 1931, 127, 741
6. Nicolaon, G. A., Teichner, S. J., On a new process of preparation of silica xerogels and aerogels and their textural properties, Bull. Soc. Chem. France 1968, 5, 1900
7. Smith, D. M., Deshpande, R., Brinker, C. J., Preparation of low-density aerogels at ambient pressure, Mater. Res. Soc. Sympo. Proceedings 1992, 271, 567
8. Prakash, S. S., Brinker, C. J., Hurd, A. J., Rao, S. M., Silica aerogel films prepared at ambient pressure by using surface derivatization to induce reversible drying shrinkage, Nature 1995, 374, 439–443
9. Deshpande, R.; Smith, D.; Brinker, C. J., Preparation of high porosity xerogels by chemical surface modification, US Pat. No. 5,565,142, 1996
10. Schwertfeger, F., Frank, D., Schmidt, M., (1998) Hydrophobic waterglass based aerogels without solvent exchange or supercritical drying, J. Non-Cryst Solids 1998, 225, 24–29
11. Kang, S. K.; Choi, S. Y., Synthesis of low-density silica gel at ambient pressure: Effect of heat treatment, J. Mater. Sci. 2000, 35(19), 4971–4976
12. Jeong, A. Y.; Goo, S. M.; Kim, D. P., Characterization of hydrophobic SiO_2 powders prepared by surface modification on wet gel, J. Sol-Gel. Technol. 2000, 19, 483–487
13. Wei, T.-Y.; Chang, T.-F.; Lu, S.-Y.; Chang, Y. C., Preparation of monolithic silica aerogel of low thermal conductivity by ambient pressure drying, J. Am. Ceram. Soc. 2007, 90(7), 2003–2007
14. Kim, G. S.; Hyun, S. H., Effect of mixing on thermal and mechanical properties of aerogel-PVB composites, J. Mater. Sci. 2003, 38(9), 1961–1966
15. Rao, A. P.; Pajonk, G. M.; Rao, A. V., Effect of preparation conditions on the physical and hydrophobic properties of two step processed ambient pressure dried silica aerogels, J. Mater. Sci. 2005, 40(13), 3481–3489
16. Rao, A. P.; Rao, A. V.; Pajonk, G. M., Hydrophobic and physical properties of the ambient pressure dried silica aerogels with sodium silicate precursor using various surface modification agents, J. Applied Surface Sci. 2007, 253, 6032–6040
17. Bangi, U. K. H.; Rao, A. V.; and Rao, A. P., A new route for preparation of sodium silicate based hydrophobic silica aerogels via ambient-pressure drying, Sci. Technol. Adv. Mater 2008, 9, 035006 (10pp)

18. Bangi, U. K.H.; Rao, A. P.; Hirashima, H.; Rao, A. V., Physico-chemical properties of ambiently dried sodium silicate based aerogels catalyzed with various acids, J. Sol-Gel Sci. Technol. 2009, 50, 87–97
19. Shewale, P. M.; Rao, A. V.; Gurav, J. L.; Rao, A. P., Synthesis and characterization of low density and hydrophobic silica aerogels dried at ambient pressure using sodium silicate precursor, J. Porous Mater. 2009, 16 (1), 101–108
20. Gurav, J. L.; Rao, A. V.; Rao, A. P.; Nadargi, D. Y.; Bhagat, S. D., Physical properties of sodium silicate based silica aerogels prepared by single step sol-gel process dried at ambient pressure, J. Alloys and Compounds 2009, 476, 397–402
21. Shewale, P. M.; Rao, A. V.; Rao, A. P.; Bhagat, S. D., Synthesis of transparent silica aerogels with low density and better hydrophobicity by controlled sol–gel route and subsequent atmospheric pressure drying, J. Sol-Gel Sci. Technol. 2009, 49, 285–292
22. Schwertfeger, F.; Emmerling, A.; Gross, J.; Schubert, U.; Fricke, J.; Y. A. Attia (Editor.), Sol-Gel Processing and Applications, Plenum press New York 1994, pp. 43
23. Lee, C. J.; Kim, G. S.; Hyun, S. H., Synthesis of silica aerogels from Waterglass via new modified ambient drying, J. Mater. Sci. 2002, 37, 2237–2241
24. Hwang, S.-W.; Kim, T.-Y.; Hyun, S.-H., Optimization of instantaneous solvent exchange/surface modification process for ambient synthesis of monolithic silica aerogels, J. Colloid Interface Sci. 2008, 322, 224–230
25. Hwang, S.-W.; Kim, T.-Y.; Hyun, S.-H., Effect of surface modification conditions on the synthesis of mesoporous crack-free silica aerogel monoliths from waterglass via ambient drying, Micropor. Mesopor. Mater. 2010, 130, 295–302
26. Shi, F.; Wang, L.; Liu, J., Synthesis and characterization of silica aerogels by a novel fast ambient drying process, Mater. Lett. 2006, 60, 3718–3722
27. Hwang, S.-W.; Jung, H.-H.; Hyun, S.-H., Ahn, Y.-S., Effective preparation of crack-free silica aerogels via ambient drying, J. Sol-Gel Sci. Technol. 2007, 41, 139–146
28. Bhagat, S. D.; Kim, Y.-H.; Suh, K.-H.; Ahn, Y.-S., Yeo, J.-G.; Han, J.-H., Superhydrophobic silica aerogel powders with simultaneous surface modification, solvent exchange and sodium ion removal from hydrogels, Micropor. Mesopor. Mater., 2008, 112, 504–509
29. Yokagawa, H.; Yokoyama, M., Hydrophobic silica aerogels, J. Non-Cryst. Solids 2005, 186, 23–29
30. Provis, J. L.; Duxson, P.; Lukey, G. C.; Separovic, F.; Kriven, W. M.; van Deventer, S. J, Modelling speciation in highly concentrated alkaline silicate solutions, Ind. Eng. Chem. Res., 2005, 44, 8899–8908
31. Swaddle, T. W. ; Salerno, J. Tregloan, P.A., Aqueous aluminates, silicates and aluminosilicates, Chem. Soc. Rev. 1994, 23, 319–325
32. Icopini, G. A.; Brantley, S. L.; Heaney, P. J; Kinetics of silica oligomerization and nanocolloid formation as a function of pH and ionic strength, Geochim. Cosmochim. Acta 2005, 69(2), 293–303
33. West J. K.; Hench L. L., Molecular-orbital models of silica rings and their vibrational spectra., J. Am. Ceramic Soc. 1995, 78 (4), 1093–1096
34. Rao, A. P.; Rao, A.V.; Bangi, U. K. H., Low thermalconductive, transparent and hydrophobic ambient pressure dried silica aerogels with various preparation conditions using sodium silicate solutions, J. Sol-gel Sci. Technol. 2008, 47, 85–94
35. Gerber, T., Himmel, B., Hubert, C., WAXS and SAXS investigation of structure formation of gels from sodium water glass, J. Non-Cryst Solids 1994, 175, 160–168
36. Knoblich, B., Gerber, T., Aggregation in SiO_2 sols from sodium silicate solutions, J. Non-Cryst Solids 2001, 283, 109–113
37. Sarawade, P. B.; Kim, J.-K.; Park, J.-K.; Kim, H.-K., Influence of solvent exchange on the physical properties of sodium silicate based aerogel prepared at ambient pressure, Aerosol Air Qual. Res. 2006, 6(1), 93–105
38. Brinker, C. J.; Scherer, G. W., Sol-Gel Science, Academic Press, San Diego 1990, pp. 358
39. Brinker, C. J.; Scherer, G. W., The Physics and Chemistry of Sol-Gel Processing, Academic Press, New York 1990, pp. 373
40. Vogel A.I., Quantitative Inorganic Analysis, ELBS and Longman, U.K. 1939, pp. 891
41. Rao, A. P.; Rao , A.V.; Gurav, J. L., Effect of protic solvents on the physical properties of the ambient pressure dried hydrophobic silica aerogels using sodium silicate precursor, J. Porous Materials 2008, 15, 507–512
42. Ameen, K. B.; Rajasekar, T.; Rajasekharan, T.; Rajasekharan, M. V., The effect of heat-treatment on the physico-chemical properties of silica aerogel prepared by sub-critical drying technique, J. Sol-gel Sci. Technol. 2008, 45(1), 9–15

Part III

Materials and Processing: Inorganic – Non-silicate Aerogels

6

ZrO_2 Aerogels

Lassaad Ben Hammouda, Imene Mejri, Mohamed Kadri Younes, and Abdelhamid Ghorbel

Abstract Mesoporous zirconia aerogels are very interesting materials with important properties used for several applications such as ceramics and catalysis. The textural and structural properties of zirconia aerogel depend on the preparation parameters such as the hydrolysis ratio, the acid and zirconium precursor concentrations, and the gel aging. The extraction of the solvent under its supercritical conditions (high-temperature supercritical conditions) or under the supercritical conditions of CO_2 (low-temperature supercritical condition) leads to aerogel zirconia with different properties. In fact, aerogel zirconia dried under high-temperature supercritical conditions is well crystallized and exhibits a large surface area. However, aerogels obtained with low-temperature supercritical conditions are amorphous and present a low surface area. Zirconia aerogels doped by metals, such as platinum, iron, and copper, or ions, in particular sulfate, phosphate, and tungstate anions, exhibit very important catalytic properties in many reactions such as *n*-alkane isomerization and Fischer–Tropsch synthesis. Doping zirconia with other oxides leads to materials with very interesting properties, allowing its use in fuel cells, thermal barrier coatings, oxygen sensors and many other high-temperature applications.

6.1. Introduction

Mesoporous zirconia is a particulate oxide with properties for several applications such as catalyst or catalyst support [1, 2] and nanocomposite materials (Chap. 25) [3, 4]. Moreover, zirconia nanoparticles are subject of extensive studies dealing with the preparation of piezoelectric, electro-optic, dielectric materials with wide use, and hybrids for solid oxide fuel cells (*SOFCs*) [3, 5–8]. Zirconia is also a material in ceramic industry [9, 10].

In order to improve the performance (physicochemical characteristics in general as well as reactivity for catalysis or mechanical properties for ceramics) of prepared materials containing zirconium oxide, the most important characteristics requested are high surface area and phase stability of zirconia nanoparticles.

For many years, several studies were devoted to the synthesis of zirconia materials, which exhibit interesting textural properties (surface area, porosity, particle size, etc.).

L. B. Hammouda, I. Mejri, M. K. Younes, and A. Ghorbel • Dèpartement de Chimie, Faculté des Sciences de Tunis, Campus Universitaire, 2092 ElManar, Tunisia
e-mail: bhlassaad@gmail.com; mejriimene@gmail.com;
younes.kadri@laposte.net; Abdelhamid.Ghorbel@fst.rnu.tn

M.A. Aegerter et al. (eds.), *Aerogels Handbook*, Advances in Sol-Gel Derived Materials and Technologies, DOI 10.1007/978-1-4419-7589-8_6,

For this purpose, different preparation methods (precipitation, electrolysis, sol–gel process, etc.) and zirconium precursors (chloride, oxychloride, nitrate, and alkoxides) have been used. However, it is widely known that supercritical drying (*SCD*) of zirconia gels is one of the more promising processes allowing the enhancement of textural and structural properties (Chap. 21) of zirconium oxide phases.

In recent years, it has been reported that the preparation of zirconium oxide using the sol–gel method allowed a better control of texture and structural features. In fact, many authors [11, 12] indicated that zirconia *aerogels* obtained by a high-temperature *SCD* exhibit increased surface areas, a large pore volume, and a temperature-stable tetragonal phase (Chap. 23). In addition, the *SCD* procedure has led to metal oxide nanoparticles with distinct morphological features (polyhedral shapes at the nanoscale), giving rise to the high surface reactivity [13].

Therefore, samples possess higher surface areas and pore volumes and smaller crystallite domains (Chap. 21) than samples prepared by precipitation even after calcination at relatively high temperature of 500°C. Note that zirconia *aerogels* are typically stabilized in their tetragonal crystal phase, so that they maintain this crystallographic structure, even after calcination at temperatures as high as 700°C (Chap. 23). Conventional mesoporous samples are converted from tetragonal to monoclinic under similar calcination treatment.

Furthermore, the hydrolysis and condensation behavior of zirconium alkoxides differs fundamentally from those of silicon (Chap. 2), aluminum, and titanium alkoxides (Chap.7) [14–16]. The sol–gel process of zirconium alkoxides is complicated and difficult to control because of their increased sensitivity to various sol–gel synthetic parameters. In particular, the gelation rate is sharply dependent on temperature in contrast to other sol–gel systems. Hence, the gelation times and sol–gel product properties are not consistent unless the sol–gel temperature can be controlled very carefully. Another factor to be considered is how to reduce the greater reactivity of zirconia alkoxides with water. When a stoichiometric amount of water with acid and alcohol is directly added to an alkoxide solution, localized condensation often occurs, resulting in translucent or sometimes opaque gels. Hence, alkoxide solution must be acidified before the hydrolysis step to prepare transparent homogeneous gels rather than precipitates. The formation of transparent polymeric *alcogels* was found to be critical for obtaining the high surface areas and narrow pore size distributions typical for *aerogels* [17].

6.2. Preparation of Zirconia Aerogels

Zirconia *aerogels* were among the first to be prepared by controlled hydrolysis of zirconium alkoxides in nonaqueous media followed by SCD [18] (Chap. 1). Later this method became a standard and was extended to many other metal oxides [19, 20].

In fact, zirconia *aerogels* with very high surface areas exceeding 500 m^2/g were prepared by Bedilo et al. [21], using a sol–gel method with high-temperature SCD.

Zhao et al. [22] characterized and compared the microstructure and properties of zirconia *aerogels* prepared by electrolyzing zirconium oxychloride solutions at room temperature followed by supercritical CO_2(l) extraction or *freeze-drying* (Chap. 21). Their results showed that the high surface area (640 $m^2\ g^{-1}$) *aerogel* produced by the supercritical CO_2(l) drying process (*S-aerogel*) was a mesoporous and transparent monolithic material with an average pore size of 9.7 nm, whereas samples prepared by the *freeze-drying* process (*F-aerogel*) were opaque white powders with a microporous structure (pore size, 0.59 nm).

Zhang et al. [23] reported the preparation of nanoporous ZrO_2 *aerogels* using zirconyl nitrate ($ZrO(NO_3)_2 \cdot 5H_2O$) as a precursor by a hydrothermal method followed by SCD technique. Obtained *aerogels* are nanoporous with an average pore diameter in the range between 5 and 60 nm. This preparation method leads to developed specific surface areas as high as 916.5 m^2/g and homogeneous pore size distributions.

The synthesis of zirconia nanomaterials by a one-step sol–gel route under supercritical CO_2(l) drying using zirconium alkoxides and *acetic acid* was reported by Sui et al. [24]. Their materials exhibited high surface areas (up to 399 m^2/g) and porosity, while the calcined materials demonstrated tetragonal and/or monoclinic nanocrystallites. Either a translucent or opaque monolith was obtained. Investigation by electron microscopy showed that the translucent monolithic ZrO_2 exhibited a well-defined mesoporous structure, while the opaque monolith, formed using added alcohol as a *co-solvent*, was composed of loosely compacted nanospherical particles with a diameter of approximately 20 nm.

Stöcker et al. [25] reported the preparation of zirconia *aerogels* by a sol–gel process using *zirconium n-butoxide* under acid catalysis. The nitric acid-to-alkoxide ratio and the alcoholic *solvent* were varied. The *solvent* was removed by means of one of two different methods: high-temperature *SCD* and low-temperature extraction with supercritical CO_2(l). Their results show that depending on the kind of alcoholic *solvent*, the nitric acid-to-alkoxide ratio, and the drying method, specific surface areas of the mesoporous *aerogels* varied from 55 to 205 ($m^2\ g^{-1}$) after calcination in air at 500°C. The high-temperature supercritically dried *aerogels* showed larger pores (17–65 nm) and *BET* surface areas (143–205 $m^2\ g^{-1}$) compared to the *aerogels* dried by low-temperature extraction (20 nm and 55–112 $m^2\ g^{-1}$, respectively). The width of the size distribution of all high-temperature supercritically dried gels became smaller with increasing amounts of nitric acid. All *aerogels* dried by low-temperature extraction with supercritical CO_2(l) were amorphous as confirmed by *XRD* measurements (Chap. 21). However, upon calcination in air at 500°C, these amorphous *aerogels* crystallized to form tetragonal zirconia. The high-temperature supercritically dried *aerogels* contained predominantly tetragonal zirconia. The fraction of the residual monoclinic ZrO_2 phase depends on different parameters. It increases with an increase in the nitric acid-to-alkoxide ratio, the calcination temperature, and the use of long-chain branched alcohol *solvent*, such as *t-butanol*. On the other hand, the same research group [25] reported that the use of *ethanol* as *solvent* resulted in the highest specific surface areas and pore volumes (up to 1.5 $cm^3 g^{-1}$), whereas the use of *t-amylalcohol* caused a shift of the maxima of the broad pore size distributions from 30 to 70 nm. For the corresponding *xerogels*, prepared via the same procedure but dried at ambient temperature, the use of butanol resulted in a maximum pore size of 3 nm, whereas *t-amylalcohol* led to a bimodal pore size distribution with maxima at 3 and 15 nm. The variation of the acid-to-alkoxide ratio in the range between 0.08 and 0.12 at a molar stoichiometry (H_2O/Zr) of 4 did not significantly influence the structural properties of the resulting *aerogels*. All uncalcined *aerogels* contained crystalline ZrO_2, whereas the corresponding uncalcined *xerogels* were X-ray amorphous and crystallized only after calcination up to 300°C.

6.3. Impact of Preparation Parameters on the Textural and Structural Properties of Zirconia Aerogels

The effects of different synthesis parameters on the properties of zirconia *aerogels* have been investigated by many research groups. Obtained results show that the gelation occurs more slowly either by increasing the acid concentration or by decreasing the hydrolysis ratio or the *Zr(OR)*$_4$ precursor concentration [26]. Furthermore, the increase in the aging time of the gel resulted in surface area development, but had little effect on pore structure. It is also possible to control the pore size distributions by varying the concentrations of acid and zirconium precursors.

6.3.1. Influence of Acid Concentration

The hydrolysis of zirconium alkoxides is very fast. The rate of the sequential condensation can be effectively influenced by the concentration of an acid catalyst. Suh et al. [17] indicated that four main types of products can be obtained. If no acid or not enough acid is added, immediate precipitation occurs. The precipitates usually do not possess desirable porous structure and surface areas [27]. In a certain concentration range, quick formation of a rigid polymeric gel containing some precipitate particulates can also be observed. Further increase of the acid amount progressively results in the increase of the gelation time and the formation of optically transparent rigid and finally clear soft *alcogels*. If the acid concentration is sufficiently high, condensation can be completely avoided.

6.3.2. Influence of Hydrolysis Ratio (H_2O/Zr)

The amount of water added can also affect the final properties of the zirconia *aerogels* [28]. In theory, four molecules of water should be required to completely hydrolyze each *Zr(OR)*$_4$ molecule. However, complete hydrolysis is not necessary to form polymeric gels. In fact, 2–4 water molecules per zirconium atom are used to form hydrous oxides. For example, Ko et al. [27] reported that hydrolysis molar ratio ($H_2O/Zr = 2$) appears to yield the highest surface areas after low-temperature *SCD* followed by calcination at 500°C. However, with the higher water ratios, larger concentrations of acid were required to prevent precipitation.

6.3.3. Influence of Zirconium Precursor Concentration

It has also been reported [21] that the concentration of the zirconium precursor is an important synthetic parameter that exerts a great effect on the properties of the obtained *aerogels*. It is apparent that higher precursor concentrations favor olation reactions and more dense *aerogels* with smaller pores are formed. A concentration between 0.25 and 0.5 M is ideal, thus allowing a relatively rapid formation of a rigid gel at relatively low acid concentrations and lead to high porosities. Increasing precursor concentrations require large amounts of acid to obtain clear and rigid gels, leading to *aerogels* with important surface areas that drastically decrease after calcination. In addition, the higher the precursor concentration, the lower the average pore radius and the smaller the pore volume.

6.3.4. Influence of the Supercritical Drying Temperature

Concerning the effect of the *SCD* temperature, which must be 20°C higher than the critical temperature of the *solvent* [28], many research groups claimed that higher *SCD* temperatures appear to be beneficial. In fact, Bedilo et al. [21] reported, for example, that surface areas of calcined zirconia samples increase with *SCD* temperature. The highest *SCD* temperature (295°C), used by this research group, allowed the synthesis of zirconia *aerogel* with the broadest pore radius and a symmetrical pore size distribution. These results indicate that higher *SCD* temperatures can cause restructuration of the gels and somewhat lowered surface areas initially, but increased thermal stability (Chap. 23).

6.3.5. Zirconia Aerogels Obtained by High- or Low-Temperature SCD

Zirconia *aerogels* are obtained by *SCD* of prepared gels at high or low temperature. Furthermore, during a high-temperature *SCD* procedure, *solvent*, which is usually alcohol, is converted into the supercritical state by heating in an autoclave. In the low-temperature *SCD*, the *solvent* is exchanged with supercritical CO_2. The properties of *aerogels* obtained by the two methods may be very different [28].

In fact, Baiker et al. [25] compared zirconia *aerogels* prepared by high-temperature *SCD* and low-temperature *SCD*. Their results show that the *aerogels* supercritically dried at high temperature exhibited larger pores (17–65) and *BET* surface areas (143–205 $m^2\ g^{-1}$) compared to the *aerogels* dried by low-temperature extraction (20 nm and 55–112 $m^2\ g^{-1}$, respectively). Furthermore, all *aerogels* dried by low-temperature extraction with supercritical CO_2 were X-ray amorphous. However, the high-temperature supercritically dried *aerogels* developed tetragonal zirconia.

It is also important to note that results obtained by different researchers indicated that surface areas of calcined zirconia *aerogels* were not much higher than those of precipitated samples when supercritical CO_2 was used for drying. In fact, after calcination at 500°C surface areas were mostly below 100 $m^2\ g^{-1}$ [17, 25, 27, 29].

On the other hand, when compared to CO_2 SCD, alcohol SCD is believed to be effective in the production of high-surface-area zirconia *aerogels* with narrow pore size distributions due to accelerated aging at the high drying temperature [25, 27]. The pore size distributions of such high-surface-area *aerogels* did not vary much upon calcination up to 500°C. This effect was correlated to the sintering at higher temperatures, which may cause the rearrangement of pore volume toward large pores, possibly by pore coalescence [30].

6.3.6. Advantages of Zirconia Aerogels Compared to Xerogels

Compared to *xerogels*, zirconia *aerogel* samples showed higher surface areas and pore volumes and exhibited smaller crystallites. In fact, the collapse of the gel framework due to the surface tension in the liquid occurs during rapid drying of the gels. However, SCD avoids this collapse and preserves the porous network.

Many researchers noted that independent of the zirconium precursor and the preparation method (precipitation, sol–gel, etc.), obtained zirconia *xerogels*, dried at moderate temperature, were X-ray amorphous and crystallized only after calcination, at least, at 300°C, while the high-temperature SCD led to samples exhibiting tetragonal nanocrystallites.

Baiker et al. [25] also reported that zirconia *xerogels* and *aerogels*, obtained by sol–gel process using *ethanol* as *solvent*, exhibited Brönsted and Lewis acid sites. However, with *xerogels*, the density of acid sites on the surface was significantly lower. This behavior was attributed to the higher amounts of organic residues that persisted in and on the *xerogels* up to 500°C and thus blocked the acid sites partially.

Tyagi et al. [31] synthesized nanocrystalline zirconia *aerogel* by sol–gel technique and SCD using *n*-propanol as *solvent* at and above supercritical temperature (235–280°C) and pressure (48–52 bar). Zirconia *xerogel* samples have also been prepared by conventional thermal drying method to be compared with the supercritically dried samples. SCD of zirconia gel was observed to give thermally stable, nanocrystalline, tetragonal zirconia *aerogels* having high specific surface area and porosity with narrow and uniform pore size distribution as compared to thermally dried zirconia. With SCD, zirconia samples show the formation of only mesopores, whereas in thermally dried samples, substantial amounts of micropores are observed along with mesopores. The samples prepared using SCD yield nanocrystalline zirconia with smaller crystallite size (4–6 nm) as compared to higher crystallite size (13–20 nm) observed with thermally dried zirconia.

6.3.7. Influence of the Gel Aging

Alcogel aging is known to be an important parameter for sol–gel synthesis process [28]. For zirconia *aerogels*, long aging time has been recommended [17]. It was found that prolonging the aging time can enhance the final surface area from about 95 up to 111 $m^2\ g^{-1}$, without having any noticeable effect on pore structure [20].

6.4. Applications of Zirconia Aerogels

6.4.1. Zirconia Aerogels and Catalysis

Zirconia *aerogels* are known to possess several properties required for applications as a catalyst or catalyst support. Its surface is known for its acidic and basic sites as well as for its oxidizing and reducing chemical properties [32]. In addition, zirconia *aerogels* exhibit adequate structural and textural characteristics. These properties confer a good dispersion of a catalytically active phase, which enhances catalytic activity, thermal stability, and resistance to poisoning. The most reported catalytic applications of zirconia *aerogels* are related to the modification of their surfaces by metals or ions, in particular sulfate, phosphate, and tungstate anions.

Zirconia *aerogels* doped with metals have been investigated for different catalytic applications. Pajonk et al. [33] showed that zirconia *aerogels* are a promising support for active metallic centers and are also by themselves interesting catalysts for reactions involving hydrogen, toward isomerization and/or hydrogenation of alkenes at low temperature. In fact, it was reported that catalyst activation at 430°C in vacuum leads to the best results in *cis*–trans isomerization in the temperature range of 80–200°C. It was also shown that, at temperatures below or equal to 150°C, the maximum of selectivity is reached. Selective poisoning experiments by NH_3 or CO_2 were carried out in order to identify the catalytic sites needed for the isomerization of *n*-butene.

Kalies et al. [34] studied the hydrogenation of formate species on a Pt–zirconia *aerogel* catalyst. Zirconia was synthesized by the hydrolysis of a solution of zirconium isopropylate in isopropanol. The *solvent* was evacuated by SCD. Zirconia was further impregnated with a solution of hexachloroplatinic acid in isopropanol to obtain a 0.5 wt% loading of Pt. The Pt-impregnated zirconia was then dried under supercritical isopropanol conditions. All solids were X-ray amorphous. The reaction was monitored by *FTIR* and showed that the formate species produced after the adsorption of CO/He was hydrogenated by H_2 into methoxy moieties only when Pt was present on zirconia. These species were further converted into methane through a reverse spillover mechanism [34]. On pure zirconia, the formate species were not hydrogenated because of the inability of zirconia to dissociate molecular hydrogen.

Zirconia *aerogel* has also been used as support for metal such as copper. For example, Cu/ZrO_2 *aerogels* were prepared using sol–gel methods by controlled precipitation at 295 K from zirconyl nitrate and cupric nitrate precursors and then tested in the hydrogenation reaction of CO toward methanol by Sun et al. [35]. These *aerogels* exhibited high surface areas (up to 250 $m^2\ g^{-1}$) even after reduction in H_2 at 573 K. X-ray diffraction and transmission electron microscopy showed that the fresh *aerogels* were amorphous and composed of clusters of particles less than 5 nm in size. A fraction of these primary particles grew to about 10 nm after 24 h of the catalyzed reaction. The fresh *aerogels* were very active and stable in this methanol synthesis reaction. A relatively strong interaction of the support with the Cu species is evidenced by *TPR* and this appears to correlate well with the obtained high activity. Hence, this activity could not only be simply attributed to the high surface area of such *aerogel* samples but also to the appropriate interaction between Cu and zirconia. Doped ZrO_2 *aerogels* with rhodium or yttrium are also very active in methane oxidation [36].

Nanoscale iron-doped zirconia solid solution *aerogels* were prepared by Chen et al. via a simple *ethanol* thermal route using zirconyl nitrate and iron nitrate as starting materials, followed by an SCD process [37]. The results showed that the obtained iron-doped solid solutions crystallized in the zirconia metastable tetragonal phase, which exhibits excellent dispersion and high solubility with iron oxide. Furthermore, when the Fe/(Fe + Zr) ratio, noted x, is lower than 0.10, all of the Fe^{3+} ions can be incorporated into ZrO_2, by substituting Zr^{4+}, to form $Zr_{1-x}Fe_xO_y$ solid solutions. Moreover, for the first time, an additional hydroxyl group band that is not present in pure ZrO_2 is observed by *DRIFT* for the $Zr(Fe)O_2$ solid solution. These $Zr_{1-x}Fe_xO_y$ solid solutions are excellent catalysts for the *solvent*-free *aerobic oxidation* of n-hexadecane using air as oxidant under ambient conditions. The $Zr_{0.8}Fe_{0.2}O_y$ solid solution catalyst demonstrated the best catalytic properties in this reaction. In fact, the conversion of n-hexadecane reached 36.2%, with a selectivity of 48 and 24% toward ketones and alcohols, respectively. It is also important to note that this catalyst can be recycled five times without significant loss of activity.

Fischer–Tropsch synthesis reaction, over cobalt supported on ZrO_2–SiO_2 *aerogels*, was investigated by Wang et al. [38]. The results showed that, under favorable conditions allowing the formation of long-chain hydrocarbons, cobalt-catalyzed Fischer–Tropsch reaction appears to be structure sensitive. Moreover, it was reported that support influences significantly the catalytic behavior. In fact, when cobalt was supported on zirconia-coated silica *aerogel*, heavy products were obtained from syngas. In this case, C_5^+ yield could reach 150 g Nm^{-3} (CO+H_2) under the optimal conditions (T = 293 K, P = 2.0 MPa, GHSV = 500 h^{-1}). However, if cobalt catalyst was supported on ZrO_2–SiO_2 mixed *aerogel*, it was shown to produce middle distillate and the yield of C_5–C_{20} products for this catalyst was about 120 g Nm^{-3} (CO+H_2).

Interesting acidic properties of zirconia *aerogels* incite researchers to use them in many reactions. Surface *acidity* of these materials was enhanced by doping with anions such as sulfate, phosphate, and tungstate. The sulfate groups were the most used dopants of zirconia *aerogels* in order to reach very high acid catalysts. These groups were introduced in the zirconia *aerogel* via different methods. It was found that the addition of sulfate groups retards the crystallization of zirconia and stabilizes tetragonal phase when sulfated zirconia *aerogels* were treated at high temperature. Furthermore, sulfate groups migrate to *aerogel* surface and the strong Zr–O–S bond combined with hydroxyl groups Zr–OH create strong *Brönsted acidity* [39].

Phosphate ions were also used to increase *acidity* on the surface of zirconia *aerogel*. Boyse et al. [40] prepared zirconia–phosphate *aerogels* by two methods: a one-step sol–gel synthesis followed by *SCD* and an incipient wetness impregnation synthesis of a calcined zirconia *aerogel*. They reported that zirconia–phosphate *aerogels* possess Brönsted acid sites, and the phosphate species were claimed to be responsible of their generation.

The nature of surface *acidity* of zirconia *aerogels* and sulfated zirconia *aerogels* was studied by many research groups. In fact, sulfated zirconia *aerogels*, with definite atomic ratio S/Zr and hydrolysis ratio H_2O/Zr, showed that the *Kelvin probe* value of pure zirconia *aerogels* is around 200 mV. However, this value increases up to 1200 mV for sulfate-doped catalysts. The modification of the work function is probably due to the charge transfer from the zirconium to the oxygen species, responsible of the increase of *Lewis acidity* [41]. Correlation between *XPS*, surface potential measurements, and isopropanol dehydration reaction results showed that when zirconia was doped by sulfate groups, its surface becomes more acidic in terms of *Lewis acidity* and that the oxygen species in the sample exist in many types, one of which is related to solid *acidity* [42]. This type of oxygen species, probably of the hydroxyl groups, is different from the oxygen species of zirconia network and the oxygen of sulfate groups. Consequently, *acidity* of sulfated zirconia is mainly due to the strong Lewis acid sites on the surface, which can convert to *Brönsted acidity* by chemisorption of water or reactant.

Sulfated zirconia *aerogel* was essentially obtained by sol–gel method. However, sulfate groups were introduced from different precursors. In their work, Ward et al. [27] added sulfuric acid to zirconium *n*-propoxide precursor before hydrolyzing with water. Obtained gels were then dried under supercritical conditions. In the drying step, the *solvent* is exchanged with supercritical CO_2 or evacuated at high temperature, where the *solvent* is converted into the supercritical state by heating in an autoclave. Bedilo et al. [43] synthesized sulfated zirconia *aerogel* by a sol–gel method followed by high-temperature SCD. Sulfur was introduced during the gel formation step. Resulting samples exhibited high surface areas and pore volumes. Impregnation of these zirconia *aerogels* with $(NH_4)_2SO_4$ led to high catalytic activity in the *n-butane isomerization* reaction, compared to conventional precipitated zirconia. Impregnated *aerogels* retain more surface sulfates after calcination. The most active samples contain a close monolayer surface coverage with sulfate groups. Over monolayer surface, sulfates hindered the isomerization activity, whereas bulk sulfates exerted little effect.

Studies focusing on the use of sulfated zirconia have shown that reaction taking place at the strong acid sites also needs the metallic phase such as platinum for burning-off *coke* produced over sulfated zirconia, stabilizing the reaction intermediates of the isomerization.

Ghorbel et al. [11] reported that *aerogel* and *xerogel* sulfated zirconia exhibit different structural, textural, and catalytic properties at various calcination temperatures. In fact, they prepared different sulfated zirconia samples as follows: zirconium (IV) propoxide, dissolved in propanol, was sulfated with concentrated H_2SO_4 to obtain a molar ratio of S/Zr = 0.5. Water was then slowly added drop wise to obtain a gel with a hydrolysis ratio (H_2O/Zr) of 3. For the next step, the wet gel was dried either by simple evaporation in an oven to give a *xerogel* called XZS0.5H3 or under supercritical conditions of the *solvent* (P = 51 bar and temperature = 263.6°C) to give an *aerogel* named *AZS*0.5H3. The resulting solids were then calcined under oxygen at different temperatures in the range 300–700°C with a heating rate of 3°C min^{-1} during 3 h. Results obtained (Figure 6.1) showed that after calcination at 560°C the *aerogel* develops only the ZrO_2 tetragonal phase, whereas the *xerogel* XZS0.5H3 contains both the monoclinic and the tetragonal phases. Heating at higher temperature causes the transition of the tetragonal phase into the monoclinic one for all the samples by loss of sulfur, but the tetragonal phase remains significantly more stable in the *aerogel*.

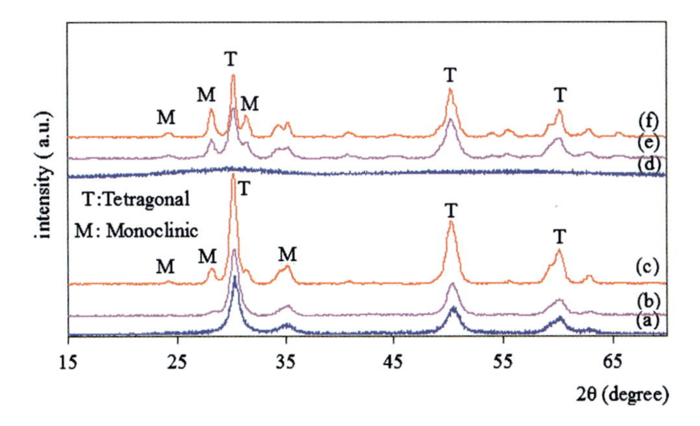

Figure 6.1. *XRD* patterns of the *aerogel* and the *xerogel* calcined at different temperatures: **A.** *AZS*0.5H3; **B.** *AZS*0.5H3–560; **C.** *AZS*0.5H3–700; **D.** XZS0.5H3 **E.** XZS0.5H3–560; **F.** XZS0.5H3–700 [11].

Characterization by *XPS* indicates that the loss of sulfur at higher temperature is easier for the *xerogel*. The ability of the *aerogel AZS*0.5H3 to retain sulfur at higher temperature explains its better stability and confers it good catalytic performances in the *n*-hexane isomerization reaction (Table 6.1).

Table 6.1. Activity (10^{-8} mol $g^{-1}s^{-1}$) of *AZS*0.5H3 and XZS0.5H3 in *n*-hexane isomerization reaction [11]

Calcination temperature (°C)	*AZS*0.5H3			XZS0.5H3		
	Reaction temperature (°C)			Reaction temperature (°C)		
	170	200	220	170	200	220
560	106	265	340	0.7	18	40
600	11	82	213	–	1.6	6
650	7	65.2	151	–	1.7	4.5
700	2.1	15.6	77.7	–	0.5	0.4

The mixed oxide including sulfated zirconia and silica or alumina was also studied. Mesoporous silica-supported nanocrystalline sulfated zirconia catalysts were prepared via the sol–gel process using an in situ sulfation and dried in an autoclave under supercritical conditions of the *solvent* ($T = 265$°C, $P = 51$ bar) by Akkari et al. [44]. The influence of the S/Zr molar ratio on the properties of the prepared catalysts has been studied. The synthesized solids were characterized using *XRD*, N_2 physisorption, *TG-DTA/MS*, sulfur chemical analysis, and adsorption–desorption of pyridine and tested in the gas-phase acid-catalyzed isomerization reaction of *n*-hexane. It has been noted that the gelation process is highly affected by the sulfate loading. Two gelation mechanisms were evidenced depending on the

S/Zr molar ratio. The first one was observed when $0.15 \leq S/Zr \leq 0.5$ and was characterized by a relatively high gelation rate. This mechanism favors the formation of two types of mesopores and a low percentage of retained sulfur. The second gelation mechanism occurs for higher S/Zr ratios: $0.5 \leq S/Zr \leq 1.2$. In that case, slower gelation rates are observed leading to materials with reduced *BET* surface area, but higher amount of retained sulfur (Figure 6.2).

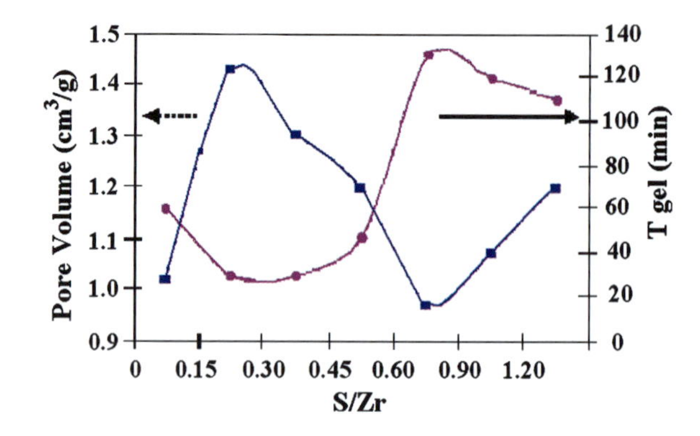

Figure 6.2. Effect of S/Zr molar ratio on the pore volume and gelation time [44].

Appreciable catalytic properties were observed for the sample prepared with the highest S/Zr ratio (Figure 6.3), which presents the smallest size of sulfated zirconia crystallites and shows both Brönsted and Lewis acid sites on its surface.

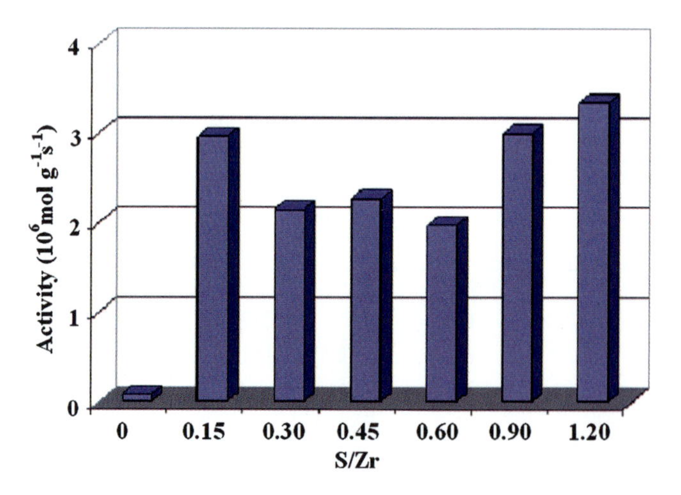

Figure 6.3. The evolution of *n*-hexane activity as a function of the S/Zr ratio for the samples calcined at 600°C [44].

Mesoporous *aerogels* based on sulfated zirconia doped with chromium have been synthesized by sol–gel method and dried in supercritical conditions of the *solvent* by Ghorbel et al. [45]. The characterization results revealed that ZrO_2 tetragonal phase is stabilized by both sulfate groups and metal at high temperature. The physisorption of N_2 showed that the sol–gel procedure coupled to the drying under supercritical conditions of the *solvent* provides to the zirconia *aerogels* a developed porosity. The important role of chromium metal phase relies on the improvement of *n*-hexane isomerization reaction by burning-off *coke* produced over sulfated zirconia (Figure 6.4).

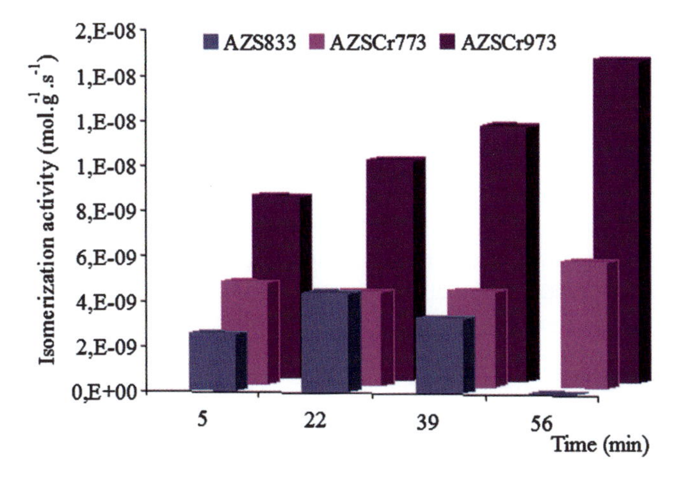

Figure 6.4. Activity of AZS (*aerogel*-sulfated zirconia) and AZSCr (*aerogel*-sulfated zirconia doped with chromium) in *n*-hexane isomerization reaction [45].

The calcination temperature plays an important role in the stabilization of a particular chromium oxidation state. This state seems to be responsible of the improvement of the catalytic performances of sulfated zirconia *aerogels* doped with chromium (Figure 6.5) [45].

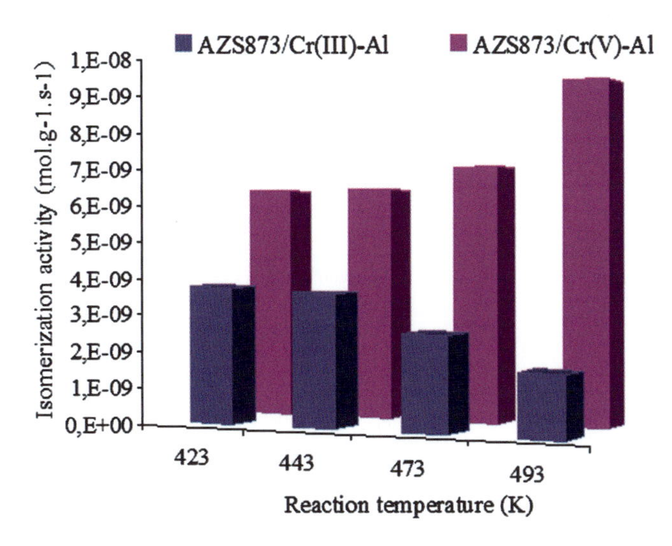

Figure 6.5. Influence of chromium oxidation state on the catalytic activity of sulfated zirconia *aerogels* doped with chromium; **A.** AZ873/Cr(III)–Al; **B.** AZ873/Cr(V)–Al [45].

In order to enhance stability and *acidity*, zirconia *aerogels* were also doped with tungsten. Boyse et al. [46] reported that zirconia–tungstate *aerogels* prepared by one-step synthesis required a more elevated activation temperature in order to expel tungstate from the bulk of zirconia before dispersion on the surface. The variation of the activation temperature allowed the study of the transition between a catalytically inactive and active material and consequently identification of the active species. Activity in *n-butane isomerization* coincided with the presence of a larger and stronger population of Brönsted acid sites on the surface. The preparation method affected the activation behavior of zirconia–tungstate materials but not the active species in *n-butane isomerization* reaction.

The same researcher [47] found that the presence of silica, both dispersed in zirconia and segregated on the surface of zirconia, retarded sintering caused by heat treatment, thereby increasing the surface area of the *aerogels*. On poorly mixed zirconia–silica, tungstate effectively dispersed onto the surface regardless of the existence of silica-rich patches. The ensuing reduction of activity during *n-butane isomerization* reaction, compared to zirconia–tungstate, established that tungstate groups on the patches did not enhance active sites. On well-mixed zirconia–silica, the presence of siloxane species in the support further reduced catalytic activity by extending the influence of silica in zirconia–tungstate. The lower catalytic activities of these materials correlated with the presence of weaker Brönsted acid sites than that on the surface of zirconia–tungstate. This work demonstrates that silica as a dopant is effective in increasing the surface area of zirconia–tungstate but not in enhancing *n-butane isomerization* activity.

Recently, sulfated zirconia *aerogels* were also doped with cerium. In order to optimize the catalytic performances of these materials, prepared by the sol–gel method, Mejri et al. [48] studied the impact of the *solvent* evacuation mode on their properties. The *xerogel* solids (*XZC*) obtained by ordinary gel drying and calcined at different temperatures exhibit a very low surface areas. On the contrary, the *aerogels* (AZC) obtained by *solvent* evacuation under supercritical conditions show a more developed surface areas (Table 6.2).

Table 6.2. Textural properties of *aerogel* (AZC)- and *xerogel* (XZC)-sulfated zirconia doped with cerium [48]

Samples	Calcination temperature (°C)	S_{BET} ($m^2\ g^{-1}$)	D_{moy} (Å)
AZC-560	560	119	213
AZC-650	650	87	202
XZC-560	560	<5	0
XZC-650	650	23	69

Both *aerogels* and *xerogels* exhibit the tetragonal phase of zirconia and/or the zirconium–cerium solid solution phase. *Aerogels* presented more developed superficial loading of Ce^{4+} and higher *acidity*, which explain their good activity in *n*-hexane isomerization reaction in the whole temperature range investigated (Table 6.3).

Derived zirconia *aerogels* catalysts were tested in many other reactions. In fact, Ferino et al. [49] studied the catalytic performances of prepared zirconia *xerogels* and *aerogels*, using zirconium *n*-propoxide as precursor, in the dehydration of methylpentan-2-ol reaction. They found that *xerogel* gives tetragonal zirconia upon calcination, during which a mesoporous system is formed. The crystal phase depends on the presence of oxygen during the cooling step in the case of the *aerogel*, whose texture is partially retained upon calcination. Both kinds of catalysts have well-balanced concentrations of acid and base sites, but the

Table 6.3. Activity and selectivity of *aerogel*- and *xerogel*-sulfated zirconia doped with cerium in *n*-hexane isomerization reaction [48]

Sample	Activity (10^{-8} mol g^{-1} s^{-1})			Selectivity (%)		
	Reaction temperature (°C)			Reaction temperature (°C)		
	170	200	220	170	200	220
AZC-560	7	50	113	88	96	98
AZC-650	3	45	142	96	99	99
XZC-560	0	0	0	0	0	0
XZC-650	0	1	9	0	90	91

acid sites are weaker in comparison with the basic ones. At 603 K the initial conversion of 4-methylpentan-2-ol over the calcined *xerogel* and *aerogel* was 45% and 63%, respectively. However, the selectivity to 4-methylpent-1-ene is 77% for both samples. Furthermore, these authors noted that the *aerogel* catalyst was quite stable during operation, whereas changes in activity and selectivity were observed for the *xerogel* sample.

6.4.2. Zirconia Aerogels and Ceramics

Zirconia is a very interesting material in ceramic industry, with exceptional electrical, thermal (Chap. 23), and mechanical properties allowing its use in fuel cells, thermal barrier coatings, oxygen sensors, and many other high-temperature applications. In fact, zirconia is the most used material in thermal barrier coatings because of its important toughness, high-temperature stability in oxidizing and reducing environments, and low thermal conductivity. So, the operational temperature of the engine can be increased, resulting in a much higher yield [50].

However, in all these applications, this material is doped with other oxides to stabilize the tetragonal and cubic phases at low temperatures, but pure zirconia is rarely used.

Doping zirconia with lower dopant concentrations allows the stabilization of the tetragonal phase, which enhances mechanical toughness. But if high dopant concentrations were used, the cubic phase could be stabilized at room temperature, leading to a very high ionic conductivity [51].

It has been reported [52] that the hardness for monoclinic zirconia is approximately 9.2 GPa, which increased slightly to values approaching 11 GPa when the tetragonal phase was stabilized by addition of 1.5 mol% yttria [52]. However, if an important amount of yttria was used, the hardness values increased to 15 GPa [53].

On the other hand, when binary zirconia–silica *aerogel* is considered, the resulting materials exhibit superior chemical resistance to alkaline attack [54], fracture toughness [55], and optical properties.

Furthermore, doping zirconia with Y^{3+}, Ce^{3+} increases the fracture toughness of the obtained material due to the stabilization of the very small tetragonal particles. Moreover, when these latter particles were added to a matrix of another material, a very high toughness for the composite was obtained. This toughening mechanism was explained by the very close density of monoclinic and tetragonal structures of zirconia avoiding the important volume increase during transformation, which becomes more sluggish when the amount of dopant is increased because the stability of the tetragonal phase is higher.

It has been also noted that the stability of the tetragonal structure can be controlled by two other factors: the particles sizes [56, 57] and the preparation conditions [17, 21]. In fact, the synthesis of zirconia *aerogels* by high-temperature SCD allows an increase in the stability of the very small tetragonal particles needed to enhance mechanical properties and point defects. Furthermore, these characteristics are also needed for the use of zirconia aerogels in the oxygen sensors and in the solid oxide fuel cells (SOFCs). In this case the cubic phase of zirconia must be first stabilized using Y_2O_3 or CaO as dopants.

6.4.3. Zirconia Aerogels and Solid Oxide Fuel Cells

Yttria-stabilized zirconia (*YSZ*) is an interesting high-temperature oxide conductor extensively used as SOFCs. In fact, it has been noted that the incorporation of 8–10% of Y_2O_3 in the matrix of the zirconium oxide leads to the stabilization of the cubic phase of zirconia at low temperature, while the charge compensation is balanced by the formation of oxygen vacancies that are necessary for thermally activated conduction processes. However, in addition to the composition and the structure, particles sizes and homogeneity of precursor powders are important factors to consider when preparing electrolytes with *YSZ*. In fact, the synthesis of homogenous nanoparticles of zirconia, which expose a high surface area, is desirable to enhance the conductivity in *YSZ*, and in this case, it has been reported that SCD is an interesting process allowing the preparation of zirconia *aerogels* with high surface areas.

Recently, Chervin et al. [58] reported that homogeneous, nanocrystalline powders of *YSZ* were prepared using a nonalkoxide sol–gel method (Chap. 8). In fact, monolithic gels, free of precipitation, were obtained by addition of propylene oxide to aqueous solutions of Zr^{4+} and Y^{3+} chlorides at room temperature. The gels were dried with supercritical CO_2(l), resulting in amorphous *aerogels* that crystallized into stabilized tetragonal or cubic ZrO_2 phase following calcination at 500°C. *Aerogel* prepared by this method showed a porous network structure with a high surface area (409 $m^2\ g^{-1}$). The crystallized *YSZ* maintained high surface area (159 $m^2\ g^{-1}$) upon formation of homogeneous nanoparticles (10 nm). Ionic conductivity at 1,000°C of sintered *YSZ* (1500°C, 3 h) was $0.13 \pm 0.02\ \Omega^{-1}\ cm^{-1}$. Activation energies for the conduction processes from 1000 to 550°C and 550–400°C were 0.95 ± 0.09 and 1.12 ± 0.05 eV, respectively.

6.5. Conclusion

Zirconia *aerogels* were prepared by different methods, in particular by the sol–gel process, using different zirconium precursors, alcoholic *solvents*, and synthesis conditions. The *solvents* were removed by means of one of two different methods: *SCD* at high-temperature or low-temperature extraction with supercritical CO_2.

The effects of different synthesis parameters on the properties of zirconia *aerogels* prepared by sol–gel process have been investigated by many research groups. In fact, taking into account that the hydrolysis of zirconium alkoxides is very fast, it has been noted that the rate of the condensation process can be influenced by the concentration of an acid catalyst leading to zirconia *aerogels* with different textural properties. The amount of added water can also affect the final properties of the zirconia *aerogels*. It has been also reported that the concentration of the zirconium precursor is an important synthetic parameter which exerts a great effect on the properties of the obtained *aerogels*.

The textural properties (surface area and porosity) of obtained samples depend considerably on the preparation conditions. However, in all cases the high-temperature supercritically dried zirconia *aerogels* showed larger porosity and higher surface areas compared to the *aerogels* dried by low-temperature extraction with supercritical CO_2.

Furthermore, zirconia *aerogels* dried by low-temperature extraction with supercritical CO_2(l) were amorphous. However, the high-temperature supercritically dried samples contained predominantly tetragonal zirconia.

Compared to *xerogels*, zirconia *aerogel* samples showed higher surface areas and pore volumes and exhibited smaller crystallites.

Zirconia *aerogels* are known to possess several properties needed for applications as a catalyst or catalyst support. The most reported catalytic applications of zirconia *aerogels* are related to the modification of their surfaces by metals or ions, in particular sulfate, phosphate, and tungstate anions.

Zirconia is a very interesting material in ceramic industry, with exceptional electrical, thermal (Chap. 23), and mechanical properties (Chap. 22 Mechanical characterization of aerogels) allowing its use in fuel cells, thermal barrier coatings, oxygen sensors, and many other high-temperature applications. In all these applications, this material is doped with other oxides to stabilize the tetragonal or cubic phase at low temperatures.

References

1. Morterra C, Cerrato G, DiCiero S, Signoretto M, Pinna F, Strukul G (1997) Platinum-promoted and unpromoted sulfated zirconia catalysts prepared by a one-step aerogel procedure: 1. Physico-chemical and morphological characterization. J Catal 165: 172–83
2. Signoretto M, Oliva L, Pinna F, Strukul G (2001) Synthesis of sulfated-zirconia aerogel: Effect of the chemical modification of precursor on catalyst porosity. J Non-Cryst Solids 290: 145–152
3. Noma T, Yoshimura M, Somiya S, Kato M, Shibataand M, Seto H (1988) Advances in ceramics, Vol. 24. Science and technology of zirconia III. American Ceramic Society, Westerville
4. Kong Y M, Bae C J, Lee S H, Kim H W, Kim H E (2005) Improvement in biocompatibility of ZrO_2-Al_2O_3 nano-composite by addition of HA. Biomaterials 26: 509–519
5. Ji Y, Kilner J A, Carolan M F (2005) Electrical properties and oxygen diffusion in yttria-stabilised zirconia (YSZ)-$La_{0.8}Sr_{0.2}MnO_3\pm\delta$ (LSM) composites. Sol Stat Ionic 176: 937–943
6. Yamahara K, Sholklapper T Z, Jacobson C P, Visco S J, de Jonghe L C (2005) Ionic conductivity of stabilized zirconia networks in composite SOFC electrodes. Sol Stat Ionic 176: 1359–1364
7. Diamant Y, Chappel S, Chen S. G, Melamed O, Zaban A (2004) Core-shell nanoporous electrode for dye sensitized solar cells: The effect of shell characteristics on the electronic properties of the electrode. Coord Chem rev 248: 1271–1276
8. Subbarao E C, Maiti H S (1984) Solid electrolytes with oxygen ion conduction. Sol Stat Ionic 11: 317–338
9. Morita Y, Nakata K, Kim Y H, Sekino T, Niihara K, Ikeuchi K (2004) Wear properties of alumina/zirconia composite ceramics for joint prostheses measured with an end-face apparatus. Bio-Med Mater Eng 14: 263–270
10. Minh N Q (1993) Ceramic fuel cells. J Am Ceram Soc 76: 563–588
11. Mejri I, Younes M K, Ghorbel A, Eloy P, Gaigneaux E M (2006) Comparative study of the sulfur loss in the xerogel and aerogel sulfated zirconia calcined at different temperatures: effect on n-hexane isomerization. Stud Surf Sci Catal 162: 953–960
12. Ward D A, Ko E I (1994) One-Step Synthesis and Characterization of Zirconia-Sulfate Aerogels as Solid Superacids. J Catal 150: 18–33

13. Stark J V, Park D G, Lagadic I, Klabunde K J (1996) Nanoscale metal oxide particles/clusters as chemical reagents. Unique surface chemistry on magnesium oxide as shown by enhanced adsorption of acid gases (sulfur dioxide and carbon dioxide) and pressure dependence. Chem Mater 8: 1904–1912
14. Bradley D C, Carter D G (1961) Metal oxide alkoxide polymers: part I. the hydrolysis of some primary alkoxides of zirconium. Can J Chem 39: 1434–1443
15. Bradley D C, Carter D G (1962) Metal oxide alkoxide polymers: part III the hydrolysis of secondary and tertiary alkoxides of zirconium. Can J Chem 40: 15–21
16. Yoldas B E (1982) Effect of variations in polymerized oxides on sintering and crystalline transformations. J Am Ceram Soc 65: 387–393
17. Suh D J, Park T J (1996) Sol-gel strategies for pore size control of high-surface-area transition-metal oxide aerogels. Chem Mater 8: 509–513
18. Vicarini M A, Nicolaon G A, Teichner S J, (1970) Propriétés texturales et structurales des aerogels d'oxydes de titane et de zirconium prepares en milieu homogène ou hétérogène. Bull Soc Chem France 2:1651–1664
19. Schwarz J A, Coutescu C, Coutescu A (1995) Methods for preparation of catalytic materials. Chem Rev 95: 477–510
20. Suh D J, Park T J (2002) Synthesis of high-surface-area zirconia aerogels with a well-developed mesoporous texture using CO_2 supercritical drying. Chem Mater 14: 1452–1954
21. Bedilo A F, Klabunde K J (1997) Synthesis of high surface area zirconia aerogels using high temperature supercritical drying. J Nano Mater 8: 119–135
22. Zhao Z, Chen D, Jiao X (2007) Zirconia aerogels with high surface area derived from sols prepared by electrolyzing zirconium oxychloride solution: Comparison of aerogels prepared by freeze-drying and supercritical CO_2(l) extraction. J Phys Chem C 111: 18738–18743
23. Zhang H X, He X D, Li Y, Hong C Q (2006) Preparation of nano-porous zirconia aerogel. J Aero Mater 26: 337–338
24. Sui R, Rizkalla A S, Charpentier P A (2006) Direct synthesis of zirconia aerogel nanoarchitecture in supercritical CO_2. Langmuir 22: 4390–4396
25. Stöcker C, Baiker A (1998) Zirconia aerogels: Effect of acid-to-alkoxide ratio, alcoholic solvent and supercritical drying method on structural properties. J Non Crist Sol 223: 165–178
26. Ben Hamouda L, Ghorbel A (2006) New process to control hydrolysis step during sol-gel preparation of sulfated zirconia catalysts. J Solgel Sci Techn 39: 123–130
27. Ward D A, Ko E I (1993) Synthesis and structural transformation of zirconia aerogels; Chem Mater 5: 956–969
28. Schneider M, Baiker A (1995) Aerogels in catalysis. Catal Rev Scien Eng 37: 515–556
29. Mrowiec J, Pajak L, Jarzebski A B, Lachowski A I, Malinowski J J (1998) Preparation effects on zirconia aerogel morphology. J Non Cryst Sol 225: 115–119
30. Lorenzano-Porras C F, Reeder D H, Annen M J, Carr P W, McCormick A V (1995) Unusual sintering behaviour of porous chromatographic zirconia produced by polymerization-induced colloid aggregation. Ind Eng Chem Res 34: 2719–2727
31. Tyagi B, Sidhpuria K, Shaik B, Jasra R V (2006) Synthesis and characterization of nanocrystalline mesoporous zirconia using supercritical drying. J Nano Nanotech 6: 1584–1593
32. Ferino I, Casula M F, Corrias A, Cutrufello M G, Monaci R, Paschina G (2000) 4-Methylpentan-2-ol dehydration over zirconia catalysts prepared by sol- gel. Phys Chem Chem Phys 2: 1847–1854
33. Pajonk G M , El Tanany A (1992) Isomerization and hydrogenation of butene-1 on a zirconia aerogel catalyst. React kine catal letters 47: 167–175
34. Kalies H, Pinto N, Pajonk G M, Bianchi D (2000) Hydrogenation of formate species formed by CO chemisorption on a zirconia aerogel in the presence of platinum. Appl Catal A 202: 197–205
35. Sun Y, sermon P A (1994) Cu-doped ZrO_2 aerogel: A novel catalyst for CO hydrogenation to CH_3OH. Topic catal 1: 145–151
36. Sermon P A, Self V A, Sun Y (1997) Doped-ZrO_2 Aerogels: Catalysts of Controlled Structure and Properties. J Sol Gel Sci Techn 8: 851–856
37. Chen L, Hu J, Ryan M R, (2008) Catalytic Properties of Nanoscale Iron-Doped Zirconia Solid-Solution Aerogels. Chem Phys Chem 9: 1069–1078
38. Zhang Y, Xiang H, Zhong B, Wang Q (1999) ZrO_2-SiO_2 aerogel supported cobalt catalysts for the synthesis of long-chain hydrocarbon from syngas. Petro Sci Techn 17: 981–998
39. Orlovic A M, Janackovic D T, Skala D U, (2005) Aerogels in Catalysis New Developments in Catalysis Research. Nova Science Publishers, Hauppauge NY
40. Boyse R A (1996) Preparation and characterization of zirconia-phosphate aerogels. Catal letters 38: 225–230

41. Younes M K, Ghorbel A, Rives A, Hubaut R (2000) Study of acidity of aerogels ZrO_2-SO_4^{2} by isopropanol dehydration reaction, surface potential and X-ray photoelectron spectroscopy. J Sol Gel Sci Techn 19: 817–819
42. Younes M K, Ghorbel A, Rives A, Hubaut R (2004) Acidity of sulphated zirconia aerogels: Correlation between XPS studies, surface potential measurements and catalytic activity in isopropanol dehydration reaction. J Sol Gel Sci Techn 32: 349–352
43. Bedilo A F, Klabunde K J (1998) Synthesis of catalytically active sulfated zirconia aerogels. J catal 176: 448–458
44. Akkari R, Ghorbel A, Essayem N, Figueras F, (2007) Synthesis and characterization of mesoporous silica-supported nano-crystalline sulfated zirconia catalysts prepared by a sol-gel process: Effect of the S/Zr molar ratio. Appl Catal A 328: 43–51
45. Raissi S, Younes M K, Ghorbel A (2009) Synthesis and characterization of aerogel sulphated zirconia doped with chromium: n-hexane isomerization. J Por Mater DOI 10.1007/s10934-009-9289-0
46. Boyse R A (1997) Crystallization behaviour of tungstate on zirconia and its relationship to acidic properties: I. Effect of preparation parameters. J Catal 171: 191–207
47. Boyse R A, Ko E I (1998) Crystallization behaviour of tungstate on zirconia and its relationship to acidic properties: II. Effect of silica. J Catal 179: 100–110
48. Mejri I, Younes M K, Ghorbel A, Eloy P, Gaigneaux E M (2008) Effect of the evacuation mode of solvent on the textural, structural and catalytic properties of sulfated zirconia doped with cerium. Stud Surf Sci Catal 174: 493–496
49. Ferino I, Casula M F, Corrias A, Cutrufello M G, Monaci R, Paschina G (2000) 4-Methylpentan-2-ol dehydration over zirconia catalysts prepared by sol- gel. Phys Chem Chem Phys 2: 1847–1854
50. Richerson D W, (2006) Modern Ceramic Engineering Properties Processing and Use in Design 3/e Taylor and Francis Group Boca Raton
51. Shackelford J F, Doremus R H, (2008) Ceramic and Glass Materials Springer
52. Bravo-Leon A, Morikawa Y, Kawahara M, Mayo M J (2002) Fracture toughness of nanocrystalline tetragonal zirconia with low yttria content Acta Mater 50: 4555–4562
53. Sakuma T, Yoshizawa Y I, Suto H, (1985) The microstructure and mechanical properties of yttria-stabilized zirconia prepared by arc-melting. J Mater Sci 20: 2399–2407
54. Paul A, (1982) Chemistry of Glasses Chapman and Hall. London
55. Nogami M, Tomozawa M (1986) ZrO_2-transformation-toughned glass-ceramics prepared by the sol gel process from metal alkoxides. J Am Ceram Soc 69: 99–102
56. Stefanic G, Music S (2002) Factors influencing the stability of low temperature tetragonal ZrO_2. Croa. Chem Acta 75: 727–767
57. Maiti H S, Gokhale K V G K, Subbarao E C (1972) Kinetics and burst phenomenon in ZrO_2 transformation. J Am Ceram Soc 55: 317–322
58. Chervin C N, Clapsaddle B J, Chiu H W, Gash A E, Satcher Jr J H, Kauzlarich S M (2006) Role of cyclic ether and solvent in a non-alkoxide sol-gel synthesis of yttria-stabilized zirconia nanoparticles. Chem Mater 18: 4865–4874

7

Preparation of TiO_2 Aerogels-Like Materials Under Ambient Pressure

Hiroshi Hirashima

Abstract Highly porous titania is attractive because of various applications such as photocatalysts. Mesoporous titania gels can be prepared by the sol–gel method with templating. The incorporated surfactant micelles or polymer aggregates in wet gels prevent the shrinkage while drying under ambient pressure. The *specific surface area*, *porosity*, and *pore size*, depending on the preparation conditions, for example, species of templating materials, are much larger than those of xerogels but not larger than those of aerogels. The performance of porous titania, for example, the photocatalytic activity, can be improved with suitable pore structure for specific applications.

7.1. Introduction

In recent years, a great interest in highly porous ceramics has been growing. Among them, highly porous silica, silica aerogel, and mesoporous silica are well known [1, 2]. Besides silica, other functional oxides with a large surface area, such as transition metal oxides for catalysts, have also attracted a great interest [2, 3]. Especially, titania have created considerable interest for their potential application such as photocatalysts [4, 5], electrodes of wet solar cells [6], gas sensors [7], and hydrophilic coatings for antifogging mirrors [8], etc. The performance of these devices depends on the surface properties of titania. The controlled porosity and microstructure of porous materials are important for specific application. Microstructure of the catalyst also affects the catalytic activity [9, 10]. It is, therefore, important to control the surface structure of the titania ceramics to enhance the efficiency of the devices and to increase the catalytic activity of the photocatalyst.

Two kinds of titania aerogels have been discussed: pure titania and mixed- or composite oxides. The better photocatalytic activity of titania aerogel was reported already in early 1970s [3]. In order to improve its catalytic activity, addition of other transition metal oxides to titania aerogels, such as vanadia, was discussed [11, 12]. Also, composite aerogels with highly adsorbable oxides, silica, and alumina were reported [13–18].

H. Hirashima • Faculty of Science and Technology, Keio University, 3-14-1, Hiyoshi, Kohoku-ku, Yokohama, Japan
e-mail: hirasima@applc.keio.ac.jp; hirasima@e03.itscom.net

M.A. Aegerter et al. (eds.), *Aerogels Handbook*, Advances in Sol-Gel Derived Materials and Technologies, DOI 10.1007/978-1-4419-7589-8_7,

Various methods have been proposed for the preparation of porous titania: sol–gel method and direct deposition from aqueous solutions [19, 20]. The sol–gel methods have some disadvantages such as high cost of raw materials, large shrinkage during processing, impurity content such as residual organics and carbon, and long processing time. However, the sol–gel method is one of the most appropriate technologies to prepare highly porous oxides. Usually, highly porous materials can be obtained by the supercritical drying or freeze-drying of wet gels. In addition, alternative methods to prepare highly porous xerogels, such as aerogels-like materials under ambient pressures, have been developed using *surface modification* of wet gels and *templating*. Several authors have proposed the sol–gel method using organics in order to modify the surface of the wet gels [21–25] to minimize the surface energy resulting in the high porosity of dried gels. These methods are usually applied to prepare highly porous silica. But also, these can be applied to nonsilica ceramics. In addition, *templating* with polymers and surfactants has been proposed as a new ambient-pressure method [26–33]. For example, a new method, immersion of wet gel into the surfactant solutions before drying to prepare mesoporous titania has been proposed. The pore size and grain shape have been shown to depend on the surfactant species [27–32]. A method for fabricating porous titania has also been reported using hydrophilic polymer such as polyethylene glycol as templates [27], as well as hydrophobic polymers such as polystyrene [33]. The surface area, pore volume, and the pore size increased as a result of surfactant- or polymer templating.

Although these methods have disadvantages, that is, high cost and long processing time, they are still attractive because of the possibility to obtain the "tailored materials" with desired pore structures and a large surface area. In this section, some examples of the ambient pressure preparation of highly porous aerogel-like titania are described.

7.2. Principles

Shrinkage of gels during drying and resulting crack-formation is the most important problem of the sol–gel processing as wet gels shrink during drying due to the capillary force arising at the hydrophilic surface of gels and solvent, the latter being usually hydrophilic.

To minimize the capillary force, *freeze-drying* and *supercritical drying* are well-known methods, as well as the modification of sols with organic species to make hydrophobic surface in order to prevent the shrinkage [21–24]. In the early stage of drying, the gel shrinks with decreasing the volume of pore-liquid. Then, after the pore-liquid is removed, the capillary forces disappear and the volume of gels comes back almost to the same volume as the wet gels (spring back effect), resulting in highly porous xerogels (*ambient pressure aerogels*).

Recently, highly porous materials prepared under ambient-pressure conditions, such as *mesoporous* silica by surfactant-templating, which has an ordered structure of pores, have been developed. Surfactant micelles or polymer aggregates incorporated in wet gels prevent the shrinkage of gels during drying. The size and shape of the pores depend on the size and shape of the micelles or aggregates which have been added. The *template* materials, surfactants or polymers, can be removed easily by heating or solvent extraction.

The same ambient-pressure methods can be applied to prepare highly porous titania. Especially, the templating method is attractive for its possibility of formation of controlled pore structures which are appropriate to specific applications.

7.3. Templating with Polymer and Surfactant: Methods

The templated highly porous titania, bulk and film, were prepared by two methods: Mixing and Immersion. *Templating materials*, surfactants, hydrophilic- and hydrophobic polymers were incorporated into wet gels before drying by these methods (Figure 7.1).

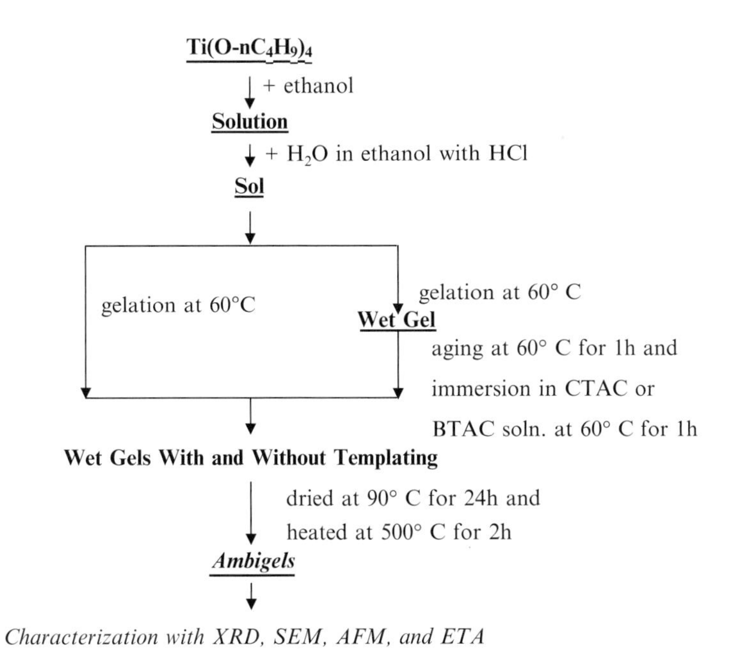

Figure 7.1. Preparation of mesoporous titania gels by surfactant templating (an example).

Titania wet gels were prepared by hydrolysis of $Ti(O\text{-}nC_4H_9)_4$ (*TNB*, 99.9%, Soekawa Rikagaku, Tokyo) in ethanol (99.5%, Junsei Chemical, Japan), using HCl as catalyst. Ethylacetoacetate (*EACAC*) was used as the chemical modifier to control the hydrolysis late.

7.3.1. Templating by the Mixing Method [27, 28]

Surfactant, cetyltrimethylammonium chloride (*CTAC*, Kanto Chemical Inc. Japan) or benzyltrimethylammonium chloride (*BTAC*, Junsei Chemical Co. Japan), or polymer, polyethyleneglycol (*PEG*, M.W. = 4,000, Sigma Chemical Co.), was added to the starting solution for the gels. Also, Pluronic P123, M.W. 5,800, $HO(EO)_{20}(PPO)_{70}(EO)_{20}H$, or Pluronic F127, M.W. 8,400, $HO(EO)_{108}(PPO)_{70}(EO)_{108}H$, was used as templates. Here, EO is CH_2CH_2O and PPO is $CH_2CH(CH_3)O$. The wet gels were aged at 60°C for 0.5 h, and dried at 90°C for 24 h (surfactant-modified gels).

The wet gels prepared without addition of surfactant or polymer were also aged and dried to obtain the xerogels. The dried gels were annealed at 500°C for 2 h.

7.3.2. Templating by the Immersion Method [28–32]

Bulk titania wet gels, prepared by hydrolysis of *TNB* in ethanol solution with addition of *EACAC* using acid catalyst, were aged at 90°C for 1 h, and the wet gels were immersed into an ethanol solution of surfactants, *CTAC* or *BTAC*, at 60°C for 1 h, and dried at 90°C for 24 h (surfactant-modified gels) [28, 29]. The dried gels were annealed at 500°C for 2 h.

The templated titania gel films were prepared by the following method. From the practical point of view, highly porous titania film is important. Titania sols were prepared by hydrolysis of *TNB* in ethanol solution with addition of EACAC using acid catalyst. The gel films were prepared by spin coating at 2,500 rpm for 30 s on silica glass substrates. Spin coating was repeated up to five times. The wet gel films were immersed into an ethanol solution of surfactants, *CTAC* or *BTAC*, at 60°C for 1 h, and dried at 90°C for 24 h (surfactant-modified gel films) [30–32]. The dried gel films were annealed at 500°C for 2 h. The preparation methods are summarized in Figure 7.2.

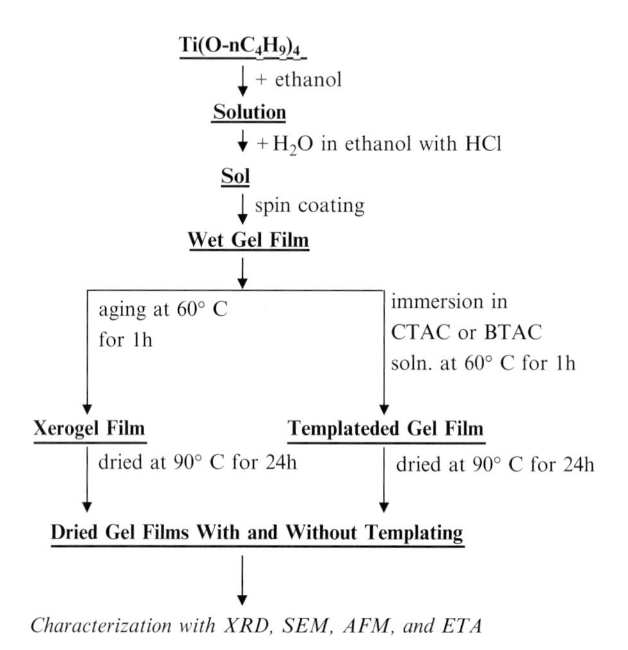

Figure 7.2. Preparation of mesoporous titania gel films by surfactant templating (an example).

Polymer-templated titania gels were also prepared. Monolithic titania wet gels were immersed in a solution of hydrophilic polymers, triblock-copolymer, poly(ethylene oxide)-poly(propylene oxide)-poly (ethylene oxide) [*Pluronic*, M.W. 12,600, Sigma Chemical Co.] or a *PEG* (M.W. = 4,000 and 20,000, Sigma Chemical Co.), under an atmospheric pressure before drying [29]. The dried gel films were annealed at 500°C for 2 h.

The bulk gels and gel films templated with hydrophobic polymer were prepared by the following Immersion methods [33].

(a) *Bulk gels*: Titania wet gels were prepared from TNB by hydrolysis in tetrahydrofuran (*THF*)/*EACAC* solutions with acid catalyst. The molar ratio of [*TNB*]:[H_2O]:[*THF*]:[*EACAC*] was 1:4:15:1, and H_2O was added as 1N HCl solution.

Monolithic gels were obtained in 20 min at room temperature. After aging at 50°C for 1 h, the wet gels were immersed into polystyrene solution. Immersion conditions are the following: the concentration of polystyrene (*PS*, M.W. 230,000, Aldrich Chemical Co.) was 0.1–11 g in 30 ml of *THF*, and immersion time was 1–100 h. The wet gels were also aged and dried without immersion to obtain the xerogels.

(b) *Gel films*: The coating sols were prepared by hydrolysis of *TNB* in *THF* using HNO_3 as a catalyst. The molar ratio of [*TNB*]:[H_2O]:[*THF*]:[*EACAC*] was 1:2:80:1. Then, the additive, the hydrophobic polymer *PS* (M.W. = 2,30,000), was added to the sols (the mixing method). Separately, amounts of *PS* were added to the solution at such amount that the concentration of *PS* in *THF* becomes 0.5 wt%, 2.07 wt%, 7.2 wt%, respectively. Gel films were coated on soda-lime-silicate glass plates (Matsunami Glass, 75 mm × 25 mm × 1 mm) by spinning at 2,500 rpm for 30 s. Spinning was repeated up to five times at room temperature. The gel films without additives were also prepared. After drying at 80°C for 24 h, the gel films were annealed at 500°C for 2 h.

7.3.3. Preparation of Aerogels-Like Materials [34]

Titania wet gels were prepared by hydrolysis of *TNB* in a methanol solution with acid catalyst. The molar ratio of *TNB*:H_2O:methanol:HNO_3 used for the synthesis was 1:13.4:127:0.06. After aging at room temperature for 1 day, the wet gels were dried at 90°C for 24 h under an atmospheric pressure (the xerogel), dried after immersion in a surfactant solution (the modified gel), or supercritically extracted with CO_2 at 60°C and 24.1 Mpa for 4 h (in this paper called "the aerogel"). After drying, the gels were annealed at temperatures up to 700°C.

7.3.4. Characterization of Dried- and Annealed Gels

The porous structure of the gels has been characterized by N_2-adsorption, scanning electron microscopy (*SEM*), atomic force microscopy (*AFM*), and emanation thermal analysis (*ETA*) [16, 20]. The porosity of films was evaluated by *refractive index* measurements.

In order to evaluate the effect of the pore structure, measurements of the photocatalytic activity of anatase films were made for some samples using a flow-type reactor. A black light was used as source of light with an intensity about 10 MW (wave length 365 nm) at the film surface. The apparent surface area of films was 50 cm^2, and humid air with 50% relative humidity, containing 1 ppm NO, was passed through the reactor at a flow rate of 3.0 l min^{-1}. The NO_x concentration was monitored with a chemiluminescent NO_x analyzer [35].

7.4. Templating with Polymer and Surfactant: Results

Residual organics and template materials, surfactants and polymers, can be removed simply by heating at temperatures up to 500°C. The heated gels at temperatures up to 600°C consist of anatase nanoparticles. Anatase titania is known to have higher photocatalytic activity than other crystalline phases of titania. At temperatures higher than 600°C, the phase transformation, from anatase to rutile, and grain growth, resulting in the decreases of porosity and surface area, are observed.

Some examples of the pore properties for the dried and annealed gels are summarized in Table 7.1.

Table 7.1. Effects of surfactant- and polymer templating on porous properties of sol–gel-derived titania (bulk), annealed at 500°C for 2 h

Sample	*BET* surface area ($m^2\ g^{-1}$)	Pore volume ($cm^3\ g^{-1}$)	Average pore diameter (nm)
Xerogel	39.9	0.0704	4.4
CTAC	56.9	0.163	8.4
BTAC	53.3	0.188	11.0
Pluronic	59.4	0.178	4.6
PEG(4,000)	45.3	0.086	5.0
PEG(20,000)	21.8	0.054	4.6
Polystyrene	85	0.27	12.0
CO_2-Aerogel	88.4	0.474	16.5

The *specific surface area* of the surfactant- and polymer-templated gels, that is, the ambient pressure aerogels, is nearly the same as that of the aerogels prepared by the supercritical extraction of CO_2. Also, the pore size and the pore volume, that is, the *porosity*, are larger (twice or more) than those of the xerogels, but smaller than those of the supercritically extracted aerogels, estimated to be ca. 55%. The pore size and pore volume depended on the properties of surfactants and polymers, molecular size, and micelle- or aggregate shape. Also, the concentration of polymers in the immersion solutions, as well as the kind of template materials, affected the pore size and pore volume (Table 7.2).

Table 7.2. Preparation condition and properties of TiO_2 xerogels with and without polymer templating, annealed at 500°C for 2 h

Sample no.	Polymer[a] (wt%)	t^b (h)	Immersion temp. (°C)	Specific surface area ($m^2\ g^{-1}$)	Pore volume ($cm^3\ g^{-1}$)	R^c (nm)	$R_{max}{}^d$ (nm)
Xerogel	–	–	–	29.6	0.038	4.5	3.6
PEG-1	0.4	1	25	36.4	0.069	3.6	3.4
PEG-2	4	1	25	45.3	0.086	5.0	4.9
Plu-1	2	1	25	33.5	0.053	5.2	4.0
Plu-2	10	1	25	32.4	0.044	5.3	7.0
Plu-3	10	5	25	32.1	0.044	4.5	7.1
Plu-4	10	24	25	43.1	0.065	5.6	3.8
Plu-5	10	1	75	33.5	0.048	4.8	6.5
Plu-6	30	1	25	56.6	0.081	4.8	7.5
Plu-7	45	1	25	58.8	0.088	5.0	7.6

[a] Amount of polymer, PEG(4000) or Pluronic P123, in the unit of the wt% calculated from the average molecular mass.

[b] Immersion time in polymer solution.

[c] Average pore diameter.

[d] Pore diameter at the peak of pore size distribution curves.

Surfactants, such as *CTAC*, can form large micelles and are consequently more effective than hydrophilic polymers to increase pore size. Hydrophobic polymers are also effective in comparison with hydrophilic polymers. The amphiphilic block copolymer, such as Pluronic P123, is also effective (Table 7.2, Figure 7.3). The repulsive force between the hydrophobic polymer and the gel surface (oxide surface) results in the high porosity and large pore size. Thus, large surface area and large pore volume are obtained for gels templated with hydrophobic polymer such as polystyrene. To obtain the large pores, as well as large surface area and large pore volume of the gel, hydrophobic polymers are useful as well as surfactant micelles.

Although the *porosity* and the pore size are smaller than those of the aerogels, the catalytic activity is improved. Some examples are shown in Table 7.3.

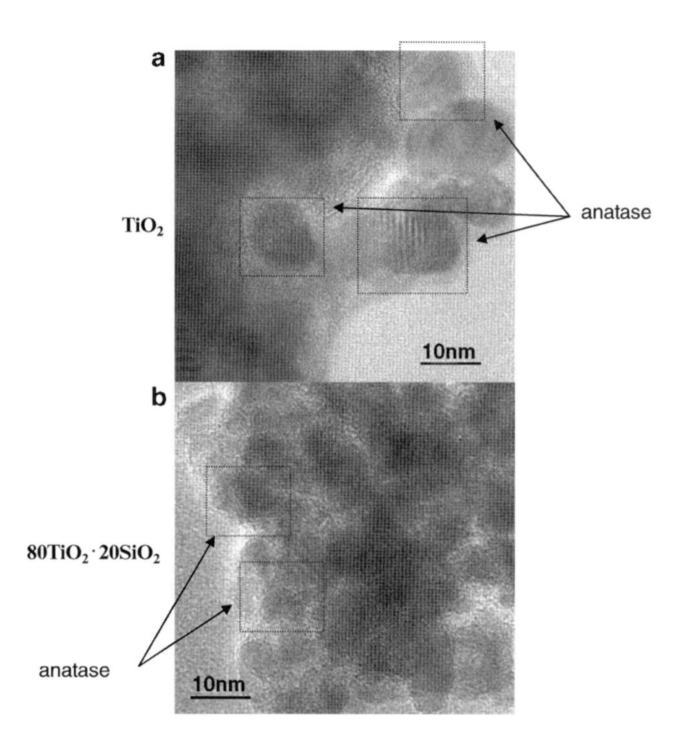

Figure 7.3. TEM images for the surfactant-templated xerogels after calcination at 400°C. Pluronic P123 was used as template.

Table 7.3. Photocatalytic activities of anatase TiO_2 films with and without surfactant templating, annealed at 500°C for 2 h [19]

Sample	Surfactant (mol l^{-1})	Porosity (%)	Removal of NO (%)	Generation of NO_2 (%)
Xerogel	–	34.2	4	3
BTAC-0.1	0.1	49.7	5.5	5
CTAC-0.0004	0.0004	47.6	3.5	3
CTAC-0.1	0.1	63.1	14	9.5
CTAC-0.5	0.5	63.1	15	10

7.5. Conclusions

It is possible to obtain highly porous titania under ambient pressure through a simple method, surfactant- or polymer templating, without special equipments. The template materials can be incorporated easily using an immersion method. The *porosity* and pore size depend on the species of the templating materials and on the amounts of incorporated polymers and surfactants. Surfactants, which form large- and defined micelles, and hydrophobic polymers, are effective templating materials to obtain highly porous titania under ambient pressure. With the appropriate conditions, porous titania suitable for specific applications can be prepared by templating under ambient pressure. However, some disadvantages such as the long- and complex processes and high cost of templating materials are still remaining. It will be necessary to find excellent applications corresponding to their high cost.

References

1. Fricke J (1986) Aerogels – a fascinating class of high-performance porous solids. in "Aerogels" Ed. Fricke J, Springer, Berlin: 2–19
2. Huesing N, Shubert U (1998) Aerogels – airy materials. Angew Chem Int Ed 37:22–45
3. Teichner S J (1986) Aerogels of inorganic oxides. in in "Aerogels" Ed. Fricke J, Springer, Berlin: 22–30
4. Matthews R W (1987) Photooxidation of organic impurities in water using thin films of titanium dioxide. J Phys Chem 91: 3328–3333
5. Matthews L R, Avnir D, Modestov A D, Sampath S, Lev O (1997) The incorporation of titanium into modified silicates for solar photodegradation of aqueous species. J Sol-Gel Sci Tech 8:619–623
6. O'Regan B, Graetzel M (1991) A low-cost, high-efficiency solar cell base on dye-sensitized colloidal TiO_2 films. Nature 353:737–740
7. Traversa E, Di Vona M L, Licoccia S, Sacerdoti M (2000) Sol-gel nanosized semiconducting titania-based powders for thick-film gas sensors. J.Sol-Gel Sci Tech 19:193–196
8. Wang R, Hashimoto k, Fijishima A, Chikuni M, Kojima E, Kitamura A, Shimohigishi M, Watanabe T (1997) Light-induced amphiphilic surfaces. Nature 388:431–432
9. Hirashima H, Kojima C, Kohama K, Imai H (1997) Application of alumina aerogels as catalysts. J Sol-Gel Sci Techn 8:843–846
10. Hirashima H, Imai H, Balek V (1998) Characterization of alumina gel catalysts by emanation thermal analysis (ETA). J Sol-Gel Sci Techn 19:399–402
11. Willey R J, Wang C-T, Peri J B (1995) Vanadium-titanium oxide aerogel catalysts. J Non-Cryst Solids 186:408–414
12. Hong-Van C, Zegaoui O, Pichat P (1998) Vanadia-titania aerogel deNOx catalysts. J Non-Cryst Solids 225:157–162
13. Beghi M, Chiurlo P, Costa L, Plladino M, Pirini M F (1992) Structural investigation of the silica-titania gel/glass transition. J Non-Cryst Solids 145:175–179
14. Yoda S, Tasaka Y, Uchida K, Kawai A, Oshima S, Ikazaki F (1998) TiO2-impregnated SiO2 aerogels by alcohol supercritical drying with zeolite. J Non-Cryst Solids 225:105–110
15. Yoda S, Otake K, Takabayashi Y, Sugeta T, Sato T (2001) Effects of supercritical impregnation conditions on the properties of silica-titania aerogels. J Non-Cryst Solids 285:8–12
16. Pietron J J, Rolison D R (2001) Electrochemically induced surface modification of titanols in a 'nanoglued' titania aerogel composite film. J Non-Cryst Solids 285:13–21
17. Pietron J J, Rolison D R (2004) Improving the efficiency of titania aerogel-based photovoltaic electrodes by electrochemically grafting isopropyl moieties on the titania surface. J Non-Cryst Solids 350:107–112
18. Tursiloadi S, Imai H, Hirashima H (2004) Preparation and characterization of mesoporous titania-alumina ceramic by modified sol-gel method. J Non-Cryst Solids 350:271–276
19. Shimizu K, Imai H, Hirashima H, Tsukuma K (1999) Low-temperature synthesis of anatase thin films on glass and organic substrates by direct deposition from aqueous solutions. Thin Solid Films 351:220–224

20. Imai H, Takei Y, Shimizu K, Matsuda M, Hirashima H (1999) Direct preparation of anatase TiO_2 nanotubes in porous alumina membranes. J Mater Chem. 9:2971–2972
21. Smith D.M, Stein D, Anderson J M, Ackerman W (1995) Preparation of low-density xerogels at ambient pressure. J Non-Cryst Solids 186:104–112
22. Prakash S S, Brinker C J, Hurd A.J (1995) Silica aerogel films at ambient pressure. J Non-Cryst Solids 190:264-275
23. Land V, Harris T.M, Teeters D C (2001) Processing of low-density silica gel by critical point drying or ambient pressure drying. J Non-Cryst Solids 283:11–17
24. Kajihara K, Nakanishi K, Tanaka K, Hirano K, Soga N (1998) Preparation of macroporous titania films by a sol-gel dip-coating method from the system containing poly(ethylene glycol). J Am Ceram Soc 81:2670–2676
25. Negishi N, Takeuchi K, Ibusuki T (1997) The surface structure of titanium dioxide thin film photocatalyst. Appl Surf Sci 121/122:417–420
26. Takahashi R, Nakanishi K, Soga N, (1995) Effects of aging and solvent exchange on pore structure of silica gels with interconnected macropores. J Non-Cryst Solids 189:66–76
27. Yusuf M M, Imai H, Hirashima H (2003) Preparation of mesoporous titania by templating with polymer and surfactant and its characterization. J Sol-Gel Sci Technol 28:97–104
28. Hirashima H, Imai H, Balek V (2001) Preparation of mesoporous TiO_2 gels and their characterization. J Non-Cryst Solids 285:96–100
29. Yusuf M M, Chimoto Y, Imai H, Hirashima H (2003) Preparation and characterization of porous titania by modified sol-gel method. J Sol-Gel Sci Technol 26:635–640
30. Yusuf M M, Imai H, Hirashima H (2001) Preparation of mesoporous TiO_2 thin films by surfactant templating. J Non-Cryst Solids 285:90–95
31. Yusuf M M, Imai H, Hirashima H (2002) Preparation of porous titania film by modified sol-gel method and its application to photocatalyst. J Sol-Gel Sci Technol 25:65–74
32. Hirashima H, Imai H, Miah M Y, Bountseva I M, Beckman I N, Balek V (2004) Preparation of mesoporous titania gel films and their characterization. J Non-Cryst Solids (2004) 350: 266–270
33. Miah M Y (2002) Preparation, characterization and application of porous titania by sol-gel method. PhD Thesis, Keio University, Japan pp.79–101
34. Tursiloadi S, Yamanaka Y, Hirashima H (2006) Thermal evolution of mesoporous titania prepared by CO_2 supercritical extraction. J Sol-Gel Sci Technol 38:5–12
35. Negishi N, Takeuchi, K, Ibusuki T (1997) The surface structure of titanium dioxide thin film photocatalyst. Appl Surface Sci 121/122:417–420

8

A Robust Approach to Inorganic Aerogels: The Use of Epoxides in Sol–Gel Synthesis

Theodore F. Baumann, Alexander E. Gash, and Joe H. Satcher Jr.

Abstract Over the last decade, the diversity of metal oxide materials prepared using sol–gel techniques has increased significantly. This transformation can be attributed, in part, to the development of the technique known as epoxide-initiated gelation. The process utilizes organic epoxides as initiators for the sol–gel polymerization of simple inorganic metal salts in aqueous or alcoholic media. In this approach, the epoxide acts as an acid scavenger in the sol–gel reaction, driving the hydrolysis and condensation of hydrated metal species. This process is general and applicable to the synthesis of a wide range of metal oxide aerogels, xerogels, and nanocomposites. In addition, modification of synthetic parameters allows for control over the structure and properties of the sol–gel product. This method is particularly amenable to the synthesis of multi-component or composite sol–gel systems with intimately mixed nanostructures. This chapter describes both the reaction mechanisms associated with epoxide-initiated gelation as well as the variety of materials that have been prepared using this technique.

8.1. Introduction

Aerogels [1, 2] are a special class of open-cell foams that exhibit many interesting properties, such as low mass densities, continuous porosities, and high surface areas. These unique properties are derived from the aerogel microstructure, which typically consists of three-dimensional networks of interconnected nanometer-sized primary particles. Because of their unusual chemical and textural properties, aerogels have been investigated for a wide variety of applications, including catalysis, sorption, insulation, energy storage, and even for cosmic dust collection (Parts VIII–XV). To further expand the utility of these materials, recent efforts have focussed the development of new synthetic processes that can be used to tailor both the composition and structure of aerogels for different applications. Aerogels are typically prepared using sol–gel chemistry, a process that involves the transformation of molecular precursors into highly cross-linked inorganic or organic gels that can then be dried using special techniques to preserve the tenuous solid network. For organic and carbon

T. F. Baumann, A. E. Gash (✉), and J. H. Satcher Jr. • Lawrence Livermore National Laboratory-LLNL, Livermore, CA 94551, USA
e-mail: baumann2@llnl.gov; gash2@llnl.gov; satcher1@llnl.gov

M.A. Aegerter et al. (eds.), *Aerogels Handbook*, Advances in Sol-Gel Derived Materials and Technologies, DOI 10.1007/978-1-4419-7589-8_8,

aerogels, this transformation involves the polymerization of multifunctional organic species into three-dimensional polymer networks [3] (Chap. 11). For inorganic aerogels, especially metal oxide aerogels, the sol–gel process begins with the hydrolysis and condensation of reactive metal-based precursors. For many years, the most commonly used precursors in the synthesis of metal oxide aerogels were metal alkoxides (M(OR)x), due mainly to their reactivity under a variety of conditions [4]. Metal oxide aerogels derived from alkoxides are synthesized through the hydrolysis of monomeric alkoxide precursors in an alcohol, typically catalyzed by a mineral acid (e.g., HCl) or base (e.g., NH_3). This process has been instrumental in the synthesis of various main group and transition metal oxide aerogels, such as silicon dioxide and titanium dioxide (Chap. 2). For the majority of elements, however, alkoxide complexes can be expensive, difficult to obtain, or very unstable, precluding their use in the preparation of oxidic aerogels.

More recently, a new approach for the fabrication of metal oxide aerogels was introduced that greatly expanded the compositional range accessible in these unique materials. The process, commonly termed epoxide-initiated gelation, utilizes organic epoxides as initiators for the sol–gel polymerization of simple inorganic metal salts in aqueous or alcoholic media [5]. In this approach, the epoxide acts as an acid scavenger in the sol–gel reaction, driving the hydrolysis and condensation of hydrated metal species. The slow and uniform increase in the pH of the sol–gel solution leads to the formation of the extended metal oxide network structure, while the protonated epoxide is consumed through an irreversible ring-opening reaction. This approach offers many advantages in the preparation of metal oxide aerogels. First, this technique utilizes simple metal salts (e.g., metal nitrates or halides) as precursors in the sol–gel reaction, eliminating the need for organometallic precursors such as metal alkoxides. As a result, this approach allows for the preparation of many main group, transition metal, and rare earth metal oxide aerogels that were impracticable using traditional sol–gel chemistry. In addition, the process is flexible and allows for control over the microstructure of the gel network through modification of the synthetic parameters. Because of the mild reaction conditions employed with this technique, the epoxide approach is amenable for the fabrication of mixed metal oxides or aerogel composites. The objective of this chapter is to provide an overview of epoxide-initiated gelation: the mechanisms associated with gel formation, the synthetic parameters that influence these mechanisms, and the different aerogel compositions that have been prepared with this approach.

8.2. Mechanisms of Epoxide-Initiated Gelation

In general, the sol–gel preparation of metal oxide aerogels involves three basic steps (1) formation of a stable colloid, (2) gelation, and (3) drying, as illustrated in Figure 8.1. Wet gels of metal oxides are prepared through the hydrolysis and condensation of inorganic precursors, leading to the formation of the three-dimensional inorganic network. To preserve the original gel structure, the solvent that resides within the pores of the solid network is removed under supercritical conditions to generate the metal oxide aerogel. Although the techniques employed for drying an aerogel can have a significant impact on the properties of the material, these details are outside of the scope of this report and will not be covered here. Rather, this section will focus mainly on the underlying mechanisms of epoxide-initiated gelation that drive the formation of the porous metal oxide network structure.

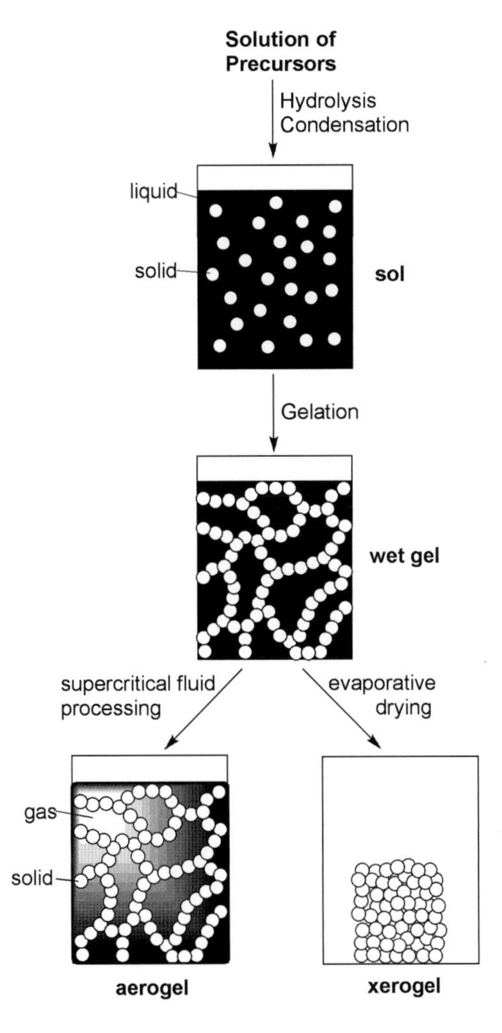

Figure 8.1. Preparation of aerogels using sol–gel chemistry.

8.2.1. Sol Formation and Gelation

A colloid is defined as a dispersion of finely divided particles in a homogeneous medium [6]. By convention, colloidal particles are smaller than 500 nm in size and consist of 10^3–10^9 atoms. Because of their small size, these particles are small enough that they remain suspended indefinitely due to Brownian motion, a random walk resulting from momentum imparted by collision with molecules of the suspending medium. A dispersion of colloidal particles in a liquid medium is termed a sol. Sols can be prepared through condensation, a process that involves the nucleation and growth of particles in solution. The size and properties of these particles depend on the relative rates of these two processes. Sol formation is favored when the rate of nucleation is high and the rate of crystal (particle) growth is low. The generation of sols also requires controlled conditions such that the resulting sol is stable toward agglomeration and precipitation. Several factors, such as polarity of the solvent, ionic strength of the reaction medium, and temperature, can be used to manipulate both the formation and stability of the sol. Gelation is the process whereby a free-flowing sol is converted into a three-dimensional solid network enclosing the solvent medium. The point of gelation is typically identified by an abrupt rise in viscosity and an elastic response to stress. This conversion can be initiated in several ways, such as a change in the ionic strength of the solution or through the removal of the reaction solvent. For preparation of aerogels, however, gelation is most conveniently induced through a change in the pH of the reaction solution. Under controlled conditions, the pH change reduces the electrostatic barrier to agglomeration and promotes inter-cluster cross-linking, leading to the formation of the three-dimensional gel network.

8.2.2. Hydrolysis and Condensation of Metal Ions

The sol–gel chemistry of metal salts is more complex than that of metal alkoxides because of the numerous molecular species that can be formed depending on the oxidation state of the metal, the pH of the reaction solution and the concentration of the reactants. Since the sol–gel polymerization of inorganic salts varies widely among the different metal ions, this section will only present a general summary of the topic. For a detailed description of the mechanism of condensation and gelation, the reader is referred to the following review on the sol–gel chemistry of transition metal oxides [7].

The aqueous chemistry of metal cations is quite complex owing to the occurrence of hydrolysis reactions that convert ions to new ionic species or to precipitates. When dissolved in pure water, a cation M^{z+} becomes solvated by the surrounding water molecules. For transition metal ions, charge transfer occurs from the filled $3a_1$ bonding orbital of the water molecule to the empty d orbital of the transition metal. This interaction causes the positive partial charge on the hydrogen to increase, making the coordinated water molecule more acidic. Depending on the magnitude of the charge transfer, the following hydrolysis equilibria are established (Scheme 8.1):

$$[M(OH_2)]^{z+} \rightleftharpoons [M(OH)]^{(z-1)+} + H^+ \rightleftharpoons [M{=}O]^{(z-2)+} + 2H^+$$

Scheme 8.1. Hydrolysis equilibria for hydrated metal ions.

The above equilibria show three types of ligands present in noncomplexing aqueous media: aquo (OH_2), hydroxo (OH), and oxo (=O) ligands. The degree of hydrolysis depends on several factors, including charge density and electronegativity of the metal, the coordination number of the aquo-complex, and the pH of the reaction solution. For example, under similar conditions, low-valent cations ($z < 4$) tend to yield aquo, hydroxo, or aquo-hydroxo complexes, whereas high-valent cations ($z > 5$) form oxo or oxo-hydroxo complexes. Tetravalent metals can form any of the possible complexes depending on the pH of the solution.

Condensation of these solvated metal ions can proceed through two possible mechanisms. Olation is the process in which one or more hydroxy bridges are formed between two metal centers. For coordinatively saturated hydroxo-aquo precursors, olation occurs through a nucleophilic substitution mechanism, where the hydroxo group is the nucleophile and water is the leaving group. Oxolation is the process in which an oxo bridge is formed between two metal centers. In the case, where the metal is coordinatively unsaturated, oxolation occurs rapidly via nucleophilic addition leading to edge- or surface-shared polyhedra. For coordinatively saturated metals, oxolation proceeds by a two-step substitution reaction between oxyhydroxy precursors involving nucleophilic addition followed by water elimination to form an M–O–M bond. These charged precursors, however, cannot condense indefinitely to form a solid phase of metal oxide. As electron-donating water molecules are eliminated from the metal centers during the substitution reactions, the hydroxo ligands become less nucleophilic and condensation stops. Depending on the nature of the metal and reaction conditions, condensation is typically limited to the formation of dimers and tetramers. In order to obtain a condensed species (i.e., sols and gels), a change in the reaction conditions is required, such as a change in temperature or solution pH. These variables control the growth and aggregation of the metal oxide species throughout the transition from the sol state to the gel state.

8.2.3. Epoxide-Initiated Gelation

In the epoxide-initiated gelation process, organic epoxides are used as initiators for the hydrolysis and condensation processes described above that lead to the formation of the three-dimensional gel network. Essentially, the epoxides serve as proton scavengers in the sol–gel reaction, analogous to their use in organic synthesis [8]. This process involves the protonation of the epoxide oxygen by an acid followed by the opening of the epoxide ring through nucleophilic attack by the conjugate base, as shown in Figure 8.2. For example, introduction of an epoxide to a solution containing the aquo ion $[Fe(H_2O)_6]^{3+}$ ($pK_a \approx 3$) leads to the protonation of the epoxide ring and formation of the aquo-hydroxo Fe^{3+} complex. The iron complex can now undergo substitution and addition reactions to form condensed iron oxide species. The protons generated in these reactions are consumed by more epoxide, and, in the presence of a suitable nucleophile (e.g., the counterion of the metal salt), the protonated epoxide undergoes irreversible ring-opening reactions. The net effect of this process is an overall increase in the pH of the reaction solution, as evidenced by the plot of measured pH versus time following addition of epoxide to an aqueous solution of ferric chloride (Figure 8.3). The addition of bases (e.g., OH^-, CO_3^{2-}, or NH_3) to aqueous solutions of metal ions has been a popular route to form condensed metal oxide species [7]. The problem with using such an approach in the synthesis of aerogels is that the condensed metal oxide species are almost instantly precipitated from the solution as a result of the rapid reaction of the base with the solvated metal ions. With the epoxide method, the solvated metal complex and the epoxide mix to produce a homogeneous solution before a significant

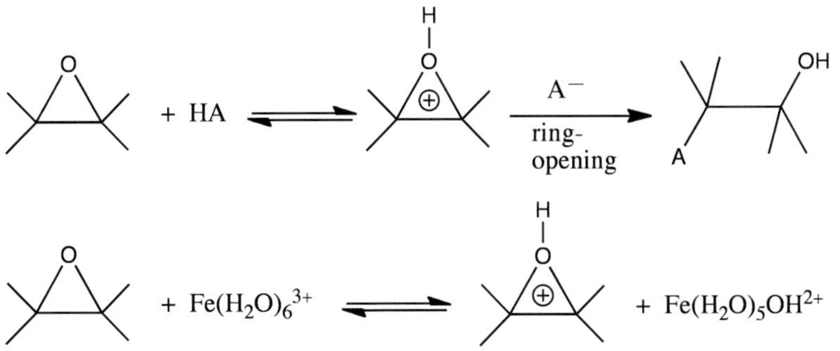

Figure 8.2. Protonation and ring opening of an epoxide (where Rn = -H, -alkyl, -aryl) in the presence of an acid, HA.

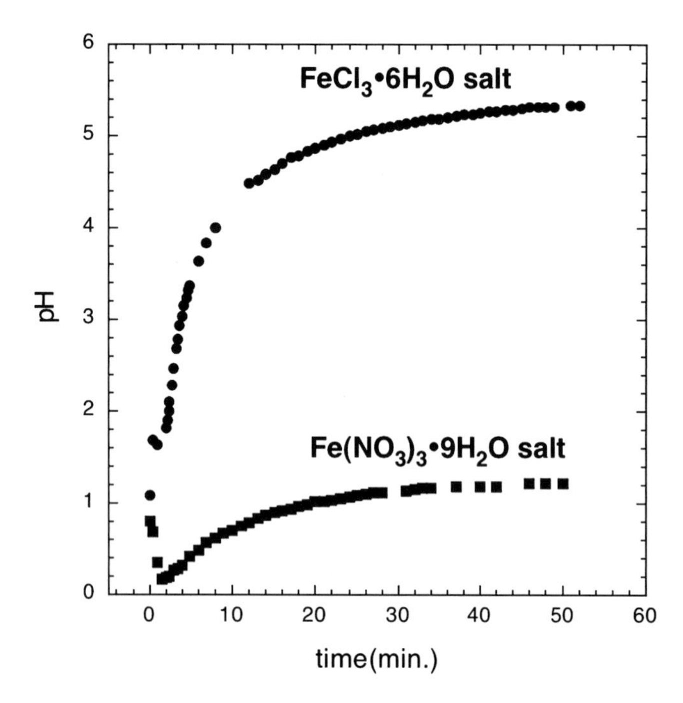

Figure 8.3. Plot of pH as a function of time for aqueous solutions of Fe(III) salts following addition of propylene oxide.

increase in pH occurs. The relatively slow and uniform increase in the reaction pH allows for controlled olation and oxolation reactions to occur, leading to the formation of a stable metal oxide sol and eventually the metal oxide gel network.

With the epoxide-initiated gelation approach, synthetic variables such as the choice of epoxide, the anion of the metal salt, and solvent used in the reaction can all have a profound impact on network formation and, thus, the properties of the resultant aerogel. For example, the size of the epoxide ring (three- or four-membered cyclic ether rings) as well as ring substituents will affect its reactivity with the hydrated metal ions in the sol–gel reaction. The rate at which the pH of the sol–gel reaction changes will, in turn, influence the nucleation of the condensed phase and growth of the network structure. These effects can be illustrated through examination of Fe(III) oxide aerogels prepared using epoxide-initiated gelation [9]. Iron oxide aerogels can be synthesized through the addition of an epoxide to an ethanolic or aqueous solution of an Fe^{3+} salt, such as $FeCl_3 \cdot 6H_2O$ or $Fe(NO_3)_3 \cdot 9H_2O$. The addition of 1,2-epoxides, such as propylene oxide, to Fe^{3+} sol–gel solutions leads to the formation of gels consisting of poorly crystallized or amorphous iron oxyhydroxide phases, such as ferrihydrite (Figure 8.4A). By comparison, the use of 1,3-epoxides, such as trimethylene oxide or 2,2-dimethyloxetane, under the same reaction conditions leads to the formation of the akaganeite phase (β-FeOOH) (Figure 8.4B). The structural differences observed in these systems can be explained by the different reactivities of the two epoxide types, whose molecular structures are shown in Scheme 8.2.

Using geometric arguments, the larger cyclic ether ring of the 1,3-epoxide is less strained and, therefore, less reactive than that of the 1,2-epoxide. As a result, reactions associated with ring opening of 1,3-epoxides are likely to occur at slower rates than those of 1,2-epoxides. This trend can be seen in the gelation times observed for Fe(III) oxide systems

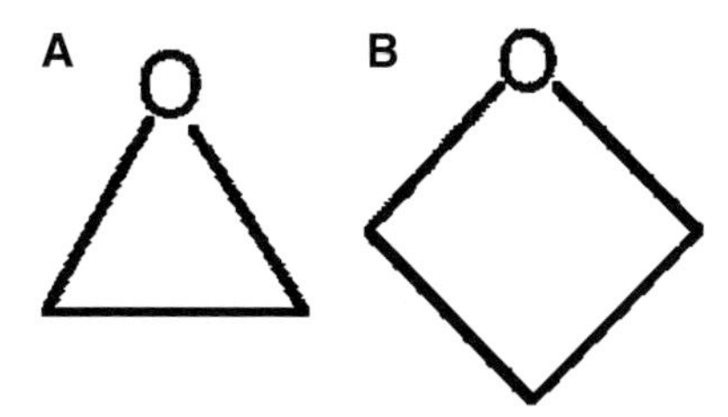

Scheme 8.2. The molecular structures of **A.** 1,2-epoxides (shown: ethylene oxide) and **B.** 1,3 epoxides (shown: oxetane or 1,3-oxacyclopropane).

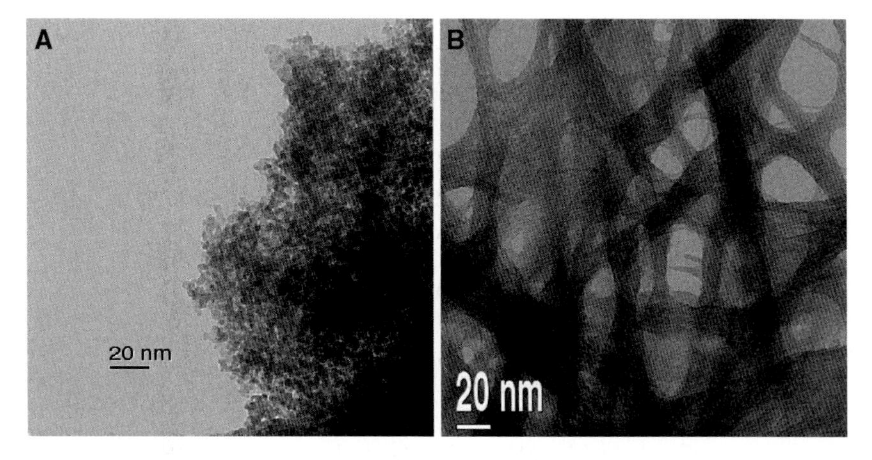

Figure 8.4. Transmission electron micrographs of iron oxide aerogels prepared from $FeCl_3$ using **A.** propylene oxide and **B.** trimethylene oxide as the gelation initiator.

prepared with both three- and four-membered ring epoxides (Table 8.1). The gelation times for the 1,3-epoxides are markedly longer than those for the 1,2-epoxide derivatives, reflecting the disparate reactivity of these reagents. In the case discussed above, the slower rise in the pH with the 1,3-epoxides allows for the growth of the highly reticulated akaganeite network. Interestingly, the elastic moduli (i.e., stiffness) of the akaganeite aerogels are over an order of magnitude greater than those of the iron oxide aerogels prepared using the 1,2-epoxides. The reactivity of epoxides also depends on the number and type of substituents on the epoxide ring. The presence of functional groups on the α-carbon atoms can strongly influence the ring-opening process through both steric and electronic effects. As a result, gelation times can vary significantly for different 1,2-epoxide derivatives, depending on the ring substituent (Table 8.1).

The anion of the metal salt can also have a profound impact on the structure and properties of aerogels prepared by epoxide-initiated gelation. Previous work on the sol–gel chemistry of transition metal oxides has shown that anions play both a chemical and physical role in the homogeneous precipitation of metal oxides [7]. Some anions can be strongly coordinated to the metal cations, thus changing the reactivity of the solvated metal species toward hydrolysis and condensation. In addition, once the colloidal particles have formed, the anions can also modify the aggregation process through changes to the double-layer composition and the ionic strength of the solution. In general, the interaction of the anion with the metal center is determined by the electronegativity of the anion relative to the

Table 8.1. Summary of gel times (*t*gel) for iron (III) oxide gels prepared with different epoxides[a]

Type	Epoxide	*t*gel (minutes)
1,2-Epoxides	Butadiene monoxide	0.33
	Cyclohexene oxide	0.45
	cis-2,3-epoxybutane	0.72
	Propylene oxide	1.5
	1,2-Epoxybutane	2.5
	1,2-Epoxypentane	4.8
	2,3-Epoxy(propyl)benzene	27
	Glycidol	62
	Epifluorohydrin	82
	Epichlorohydrin	85
	Epibromohydrin	109
1,3-Epoxides	Trimethylene oxide	480
	3,3-Dimethyloxetane	1,400

[a]Reaction conditions: $Fe(NO_3)_3 \cdot 9H_2O$ [0.37 M], ethanol, molar ratio of epoxide/Fe = 11.

ligated water molecules. The influence of anions on network formation can be seen in the case of aluminum oxide aerogels prepared by epoxide-initiated gelation [10]. The synthesis of these materials involves the addition of propylene oxide to ethanolic solutions of hydrated aluminum salts, either $AlCl_3 \cdot 6H_2O$ or $Al(NO_3)_3 \cdot 9H_2O$. Characterization of the different materials showed that aerogels prepared from hydrated $AlCl_3$ possessed microstructures containing highly reticulated networks of pseudoboehmite (AlOOH) leaflets or sheets, 2–5 nm wide and of varying lengths, whereas the aerogels prepared from hydrated $Al(NO_3)_3$ were amorphous with microstructures comprised of interconnected spherical particles with diameters in the 5–15 nm range (Figure 8.5). The difference in microstructure results in distinct physical and mechanical properties for each type of aerogel. In particular, the alumina aerogels with the nanoleaflet microstructure are optically transparent and

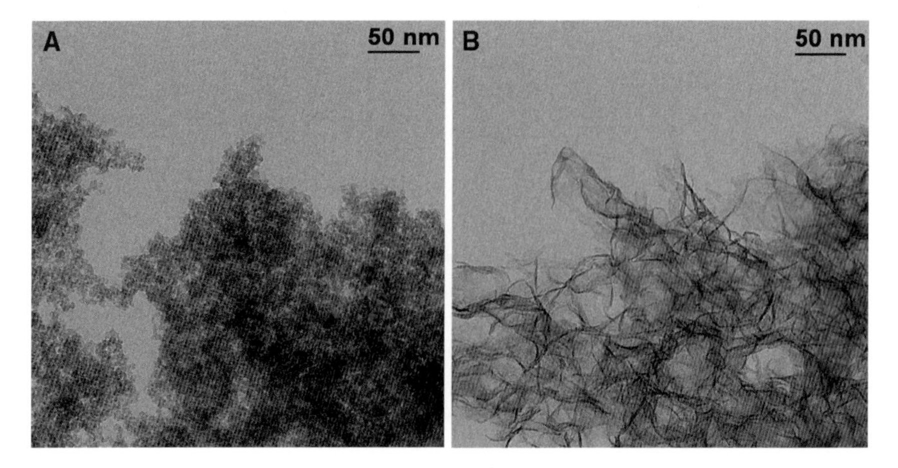

Figure 8.5. Transmission electron micrographs of alumina aerogels prepared using propylene oxide and different Al(III) precursors, **A.** $Al(NO_3)_3$ and **B.** $AlCl_3$.

significantly stiffer than those with the colloidal network [11]. On the basis of these observations, the chloride and nitrate anions appear to have different effects on the formation of the alumina network. In sol–gel chemistry, the chloride anion is typically considered "noncomplexing" and, therefore, one would expect that Cl^- would not be involved in the formation of the alumina particles. The nitrate anion, on the other hand, can exhibit "weak-complexing" ability and may be involved in the hydrolysis and condensation reactions.

In addition to interactions with the solvated metal ions, the anion of the metal salt can also affect the formation of condensed phases through its role in the epoxide ring-opening reaction. The ring-opening process involves two steps: (1) protonation of the epoxide oxygen by the acidic aquo ion and (2) subsequent attack of the ring carbons by a suitable nucleophile (Figure 8.2). Therefore, the rate at which protons are irreversibly consumed in this reaction is determined by the relative nucleophilicity of the species present. In many cases, the anion of the metal salt can serve as the nucleophile for the ring-opening reaction, but its efficacy in this process depends on a number of factors, including the chemical nature of the anion as well as the reaction solvent. For example, iron oxide gels can be prepared through the addition of propylene oxide to an aqueous solution of $FeCl_3 \cdot 6H_2O$ while gelation does not occur under the same conditions when $Fe(NO_3)_3 \cdot 9H_2O$ is used as the precursor [5]. This result can be understood in terms of the nucleophilic character of the anions in each reaction mixture. When $Fe(NO_3)_3 \cdot 9H_2O$ is used as the precursor, the potential nucleophiles for the ring-opening reaction are the nitrate ion and water. Under these conditions, water is a better nucleophile than the nitrate ion. Therefore, water preferentially attacks the ring carbons of the protonated epoxide and ring opening produces 1,2-propanediol. This reaction, however, regenerates a proton into solution, and thus the pH of the solution does not change appreciably (Figure 8.6A). By contrast, when $FeCl_3 \cdot 6 H_2O$ is utilized as the precursor, the potential nucleophiles in the reaction mixture are the chloride ion and water. In this case, chloride is the better nucleophile and the protonated epoxide is ring-opened by the chloride ion to yield 1-chloro-2-propanol (Figure 8.6B). No protons are regenerated in this reaction, leading to an increase in the pH of the reaction mixture and the formation of the iron oxide network. These processes have been verified both by characterization of the ring-opened products using nuclear magnetic resonance (NMR) spectroscopy as well as by measurement of the solution pH for each of these reactions. As shown in Figure 8.3, only a slight rise in pH occurs over time upon addition of epoxide to an aqueous solution of $Fe(NO_3)_3$. Gelation can be initiated in this system through addition of

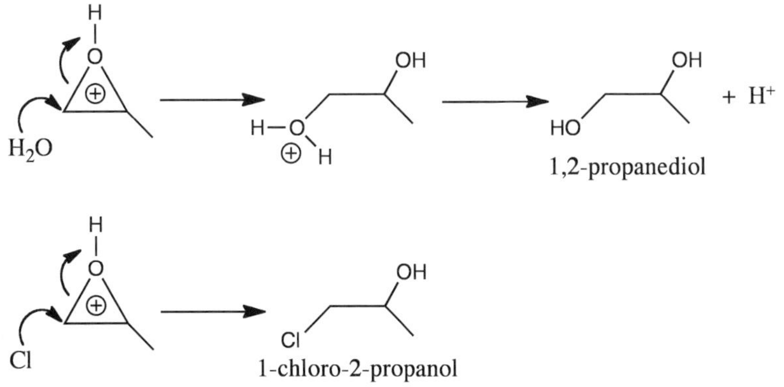

Figure 8.6. Different epoxide ring opening mechanisms.

more nucleophilic anions, such as chloride or bromide, to the aqueous iron nitrate solution. When ethanol is used as the reaction solvent, however, iron oxide gels can be prepared directly from $Fe(NO_3)_3$ as the nucleophilicity of the nitrate ion increases in nonaqueous media. In this case, the attack of the protonated epoxide ring by the nitrate ion leads to the formation of the nitrate-ester of the alcohol (i.e., 1- and 2-nitrooxy-2-propanol) and a net consumption of protons from the reaction solution. With regard to reaction solvent, it is also important to note the importance of water in the epoxide-initiated gelation method. The formation of condensed metal oxide species from metal salts necessitates the presence of an oxygen source (i.e., water) in the reaction medium. In our experience, successful gel synthesis using the epoxide method requires the addition of water, either as the reaction solvent (or cosolvent) or through the use of hydrated metal salts. The aforementioned examples underscore the importance of considering each of these reaction variables when applying this process to the preparation of metal oxide aerogels. Therefore, a basic understanding of how each of these variables influences structure formation can be a powerful tool for tailoring the structure and properties of the aerogel.

8.3. Aerogel Materials by Epoxide-Initiated Gelation

The epoxide-initiated approach has been used to prepare a wide variety of metal oxide, mixed metal oxide, and composite aerogel materials, as illustrated in Figure 8.7. This section is intended to provide a general overview of the different aerogel materials prepared using

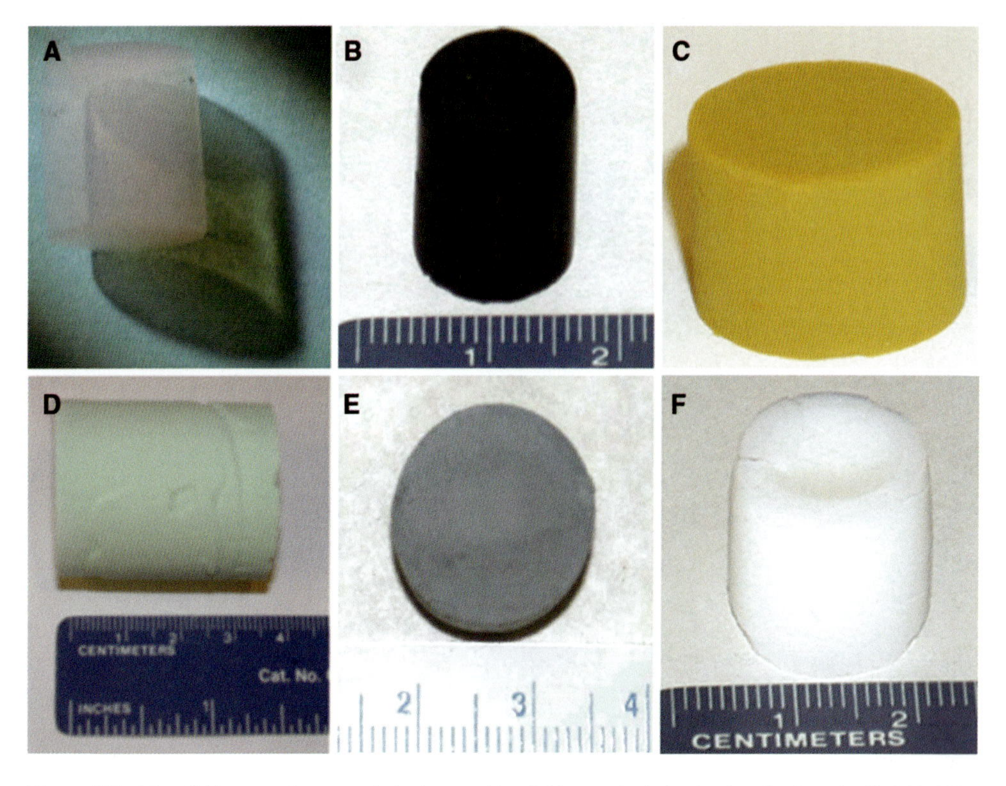

Figure 8.7. Monolithic aerogels prepared via the epoxide addition method: **A.** alumina, **B.** chromia, **C.** β-FeOOH (akaganeite), **D.** Ni(II)-based aerogel, **E.** tungsten oxide, and **F.** tin oxide.

epoxide-initiated gelation that have been reported in the literature. For more detailed information on the specific properties and applications of these materials, the reader is referred to the original published report.

8.3.1. Metal Oxide Aerogels

A number of oxidic aerogel materials derived from main group elements have been prepared using the epoxide approach. As described in the previous section, the addition of propylene oxide to an ethanolic or aqueous solutions of hydrated Al^{3+} salts yields alumina gels that can then be supercritically dried using CO_2 [10, 12–16]. Because of their thermal stability and high surface areas, in alumina aerogels of this type, the major interest has been as high-temperature catalyst supports. Epoxide-initiated gelation has also been applied to the synthesis of tin oxide aerogels [17]. These materials were prepared through treatment of an aqueous solution of $SnCl_4 \cdot 5H_2O$ with propylene oxide. Structural characterization showed that the material consists of interconnected networks of crystalline (rutile) SnO_2 nanoparticles. These materials, because of their high surface areas, are of particular interest for sensor design. Gallium and indium oxide aerogels have also been prepared using the epoxide approach [18]. For example, the treatment of an ethanolic solution containing $GaCl_3 \cdot 6H_2O$ or $InCl_3 \cdot 6H_2O$ with propylene oxide produces opaque white alcogels that can then be dried through supercritical extraction.

For transition metal oxide aerogels, the most extensively studied composition prepared by the epoxide method is iron (III) oxide. As described in the previous section, the structure and properties of the iron oxide network can be controlled through modification of synthetic parameters. The versatility of this approach has allowed for the synthesis of iron oxide materials for a variety of applications, including energetic materials [19, 20], magnetic nanostructures [21–24], and catalysis [25, 26]. The epoxide method has also been applied for the preparation of other transition metal oxide aerogels. Amorphous chromia aerogels can be prepared through the addition of epoxides to ethanolic solutions of Cr^{3+} salts [18]. High surface area ruthenium oxide aerogels have been synthesized through the reaction of propylene oxide or 1,2-epoxybutane with hydrated ruthenium chloride and are of interest for energy storage applications [27]. Titania aerogels prepared by the epoxide method [28] have been investigated as high surface area photocatalyst [29]. The syntheses of metal oxide aerogels from other transition elements, including zirconium, hafnium, niobium, tantalum, and tungsten, have also been reported [18]. The chloride salts of these elements can be reacted with propylene oxide in ethanol to yield the corresponding alcogel. Although the epoxide approach works quite well for the preparation of oxidic aerogels from higher-valent ions of the early- and mid-row transition metals, the process has proven more problematic for the synthesis of monolithic aerogels derived from divalent ions of the late 3d transition metals, such as Ni^{2+}, Cu^{2+}, and Zn^{2+}. This issue is likely related to the low relative acidities of these hydrated cations. The acidity of aquo-ions depends on charge, with M^{4+} cations typically more acidic than M^{3+} cation and M^{2+} cations being weakly acidic. For example, the $Ni(H_2O)_6^{2+}$ species (pK_a ~6.2–10.2) is a much weaker acid than the corresponding Fe^{3+} species (pK_a ~3.0). Due to the lower relative acidities of these aquo-ions, the processes that drive epoxide-initiated gelation are significantly slowed or are not favored for the formation of a stable gel network. This effect is qualitatively observed in the relatively faster rates of gel formation for the higher-valent metal ions as compared to those of the lower-valent species, under comparable experimental conditions. Nevertheless, recent work has shown that, through modification of the reaction conditions, the epoxide approach can indeed be used for the synthesis of these metal oxide aerogels. The use of 2-propanol, instead of water or ethanol, in the epoxide reaction allowed for the synthesis of Cu^{2+}- and Zn^{2+}-based aerogels [30, 31]. The as-prepared aerogels were then converted to their respective crystalline metal oxide phases through calcination at elevated temperatures. Similarly, monolithic Ni^{2+}- and Co^{2+}-based aerogels have been prepared through the reaction of the respective hydrated chloride salt with propylene oxide [32, 33].

The epoxide-initiated gelation approach has also been applied to the synthesis of lanthanide and actinide oxide aerogels. For example, nanocrystalline cerium oxide aerogels have been prepared by reacting Ce(III) salts with propylene oxide [34]. Because of their high surface areas and electronic conductivities, these materials are of interest as catalysts in solid oxide fuel cells. Aerogel materials have also been prepared from the chloride salts of other lanthanide elements [35, 36]. Oxidic aerogels derived from actinide elements, such as thorium and uranium, have also been reported [37, 38]. Materials of this type have received attention for their use in the production of nuclear fuel materials. As an example, low-density uranium oxide (UO_3) aerogels can be prepared through the addition of propylene oxide to an ethanolic solution of uranyl nitrate $[UO_2(NO_3)_2 \cdot 6H_2O]$.

8.3.2. Mixed Metal Oxide and Composite Aerogels

One of the advantages of epoxide-initiated gelation is that this approach provides a versatile and relatively straightforward route to the preparation of binary or ternary oxides. Similar to the single-component systems, mixed metal oxide gels can be readily prepared through the addition of epoxides to solutions containing two or more metal salts. When synthesizing these mixed-metal oxide aerogels, hydrolysis and condensation reactions can yield a variety of different network architectures within the aerogel framework. For example, the condensed phase can be comprised of separate interpenetrating networks of the two metal oxides, –M1–O–M1– and –M2–O–M2–, or mixed phases of the two materials, –M1–O–M2–. Alternatively, one of the metal oxides can exist as discrete entities (i.e., nanoparticles) supported by the primary oxide structure. In general, the composition and bonding motif of the gel structure is primarily a function of the reaction stoichiometry of the inorganic precursors and the relative rates of hydrolysis of the metal ions. Numerous examples of mixed metal oxide aerogels prepared by epoxide-initiated gelation have been reported in the literature. Solid state electrolyte materials, such as yttria-stabilized zirconia (YSZ), have been prepared using this approach [39, 40]. Strontium-doped lanthanum manganite (LSM) materials for use as cathodes in solid oxide fuel cells have also been fabricated using the epoxide method [41]. Mixed metal oxides with the spinel structure (AB_2O_4), such as $ZnFe_2O_4$ and $CoAl_2O_4$, have been synthesized using propylene oxide as gelation agents [42, 43]. With a similar approach, novel alumina catalysts containing copper and zinc oxide have been synthesized and investigated as catalysts for methanol reforming [44]. Examples of interpenetrating aerogel networks have also been reported. For example, the epoxide method has been utilized in the design of interpenetrating iron oxide and silicon dioxide frameworks for use as energetic materials [45–47]. Examination of these materials using element-specific TEM clearly illustrates the intimate mixing of the distinct chemical components on the nanometer scale that can be achieved using this approach (Figure 8.8). A similar strategy was utilized in the synthesis of interpenetrating metal oxide–polymer aerogel networks that can then be converted to porous metal or metal oxide structures through calcination [48, 49] (Chap. 14).

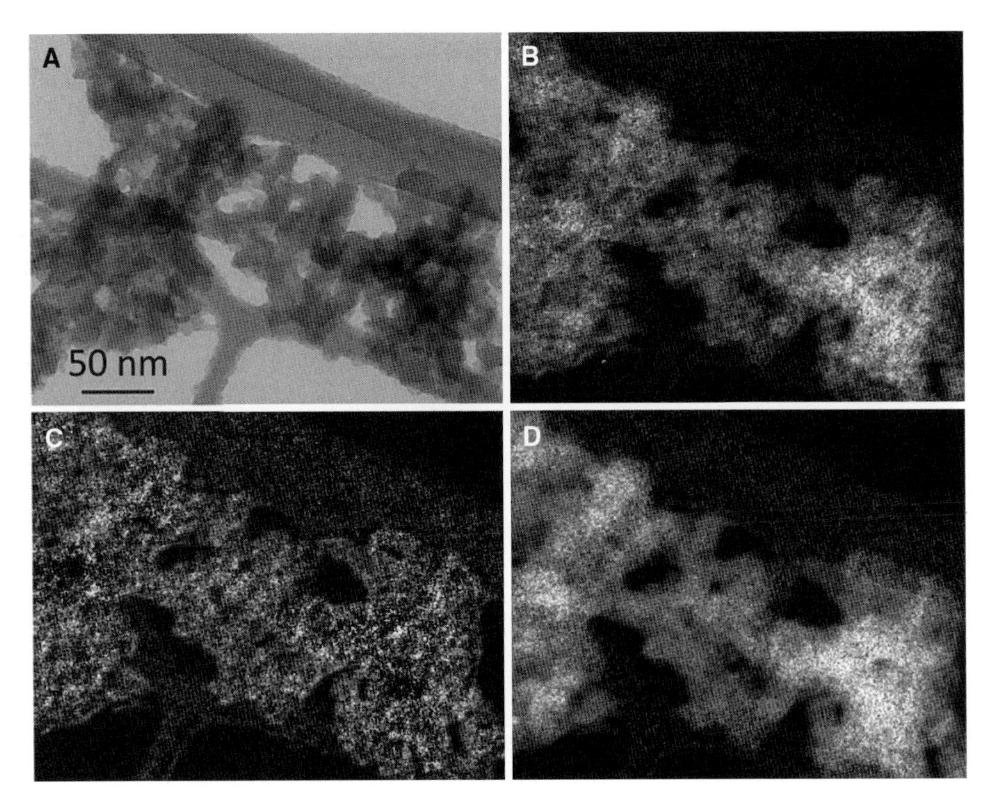

Figure 8.8. Bright field TEM image **A.** and Fe **B.**, Si **C.**, and O **D.** element-specific TEM maps of an Fe/Si = 2 aerogel prepared using TMOS as the silica precursor and TMO as the gelation promoter.

The epoxide-initiated gelation process can also play an important role in the fabrication of aerogel composites. Composites are formed by the combination of two or more distinct phases into a new material that exhibits enhanced physical, chemical, or mechanical properties relative to those of the individual components. Sol–gel chemistry allows for the assembly of the different phases on the nanometer scale, and therefore, provides the opportunity to engineer nanocomposite aerogel architectures for a variety of applications. Much like in the synthesis of the mixed metal oxide systems, aerogel composites can be readily prepared through the addition of the dispersed phase(s) to the sol–gel reaction mixture prior to gelation [50]. Because of the mild reaction conditions and the flexibility of the process, epoxide-initiated gelation allows for the incorporation of a wide variety of dispersed phases into the aerogel network. One area of research where this approach has been successfully demonstrated has been the design of energetic nanocomposites [51, 52] (Chap. 25). Using epoxide-initiated gelation, energetic formulations have been prepared where solid fuel particles, such as aluminum metal, are dispersed within an oxidizing framework, such as iron oxide [19]. Systems such as these are synthesized through the gelation of ethanolic iron (III) salt solution containing a suspension of nanometric aluminum powder. As the fibrous iron (III) oxide network forms, the insoluble aluminum phase is encapsulated and immobilized, leading to uniform dispersion of the aluminum particles within the oxide framework. The intimate mixing of the iron oxide and aluminum metal phases within the nanocomposite is shown in Figure 8.9. Mixtures of this type are known as thermites and undergo solid-state

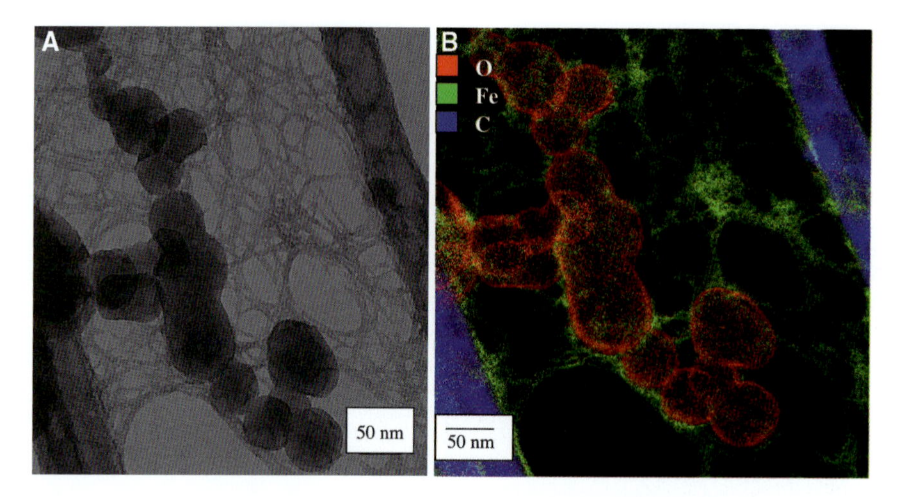

Figure 8.9. **A.** Bright field TEM image and **B.** element-specific TEM image of an iron(III) oxide aerogel/ nanometric aluminum composite prepared using the epoxide-initiated gelation method.

oxidation/reduction reactions that release a significant amount of energy at rates much higher than those observed for systems prepared through the mixing of micrometer-sized component phases [46, 47]. The approach used for the fabrication of these energetic composites should be applicable to the incorporation of other nanometric dispersed phases, such as carbon nanotubes or catalytic metal particles, into the aerogel framework.

8.4. Summary

Epoxide-initiated gelation is a general and straightforward technique for the synthesis of a variety of metal oxide aerogels, xerogels, and nanocomposites. The process utilizes organic epoxides as initiators for the hydrolysis and condensation of hydrated metal species. The epoxide approach serves as a complementary process to traditional sol–gel chemistry and greatly expands the compositional range accessible in these unique materials. Through judicious selection of epoxide, metal salt, and reaction solvent, this synthetic methodology can be used to tailor the important structural characteristics such as chemical composition, network morphology, and crystallinity that strongly influence the bulk physical properties of the resulting aerogel. In addition, this epoxide-initiated method is amenable to the synthesis of multicomponent and composite sol–gel systems with intimately mixed nano- and/or microstructures.

Acknowledgments

Work was performed under the auspices of the U.S. Department of Energy by Lawrence Livermore National Laboratory under Contract DE-AC52-07NA27344.

References

1. Hüsing N, Schubert U (1998) Aerogels-airy materials: Chemistry, structure, and properties. Angew Chem Int Ed 37:22–47.
2. Pierre AC, Pajonk GM (2002) Chemistry of aerogels and their applications. Chem Rev 102:4243–4265.

3. Pekala RW (1989) Organic aerogels from the polycondensation of resorcinol with formaldehyde. J Mater Sci 24:3221–3227.
4. Brinker CJ, Scherer GW (1990) Sol-gel science: The physics and chemistry of sol-gel processing. Academic Press, USA.
5. Gash AE, Tillotson TM, Satcher JH, Poco JF, Hrubesh LW, Simpson RL (2001) Use of epoxides in the sol-gel synthesis of porous iron oxide monoliths from Fe(III) salts. Chem Mater 13:999–1007.
6. Jirgensons B, Straumanis ME (1962) A short textbook of colloid chemistry. The MacMillan Company, USA.
7. Livage J, Henry M, Sanchez C (1988) Sol-gel chemistry of transition metal oxides. Prog Solid St Chem 18:259.
8. Dobinson B, Hoffman W, Stark BP (1969) The determination of epoxide groups. Permagon Press, Oxford.
9. Gash AE, Satcher JH, Simpson RL (2003) Strong akaganeite aerogel monolith using epoxides: Synthesis and characterization. Chem Mater 15:3268–3275.
10. Baumann TF, Gash AE, Chinn SC, Sawvel AM, Maxwell RS, Satcher JH (2005) Synthesis of high-surface-area alumina aerogels without the use of alkoxide precursors. Chem Mater 17:395–401.
11. Kucheyev SO, Baumann TF, Cox CA, Wang YM, Satcher JH, Hamza AV, Bradby JE (2006) Nanoengineering mechanically robust aerogels via control of foam morphology. Appl Phys Lett 89:041911.
12. Itoh H, Tabata T, Kokitsu M, Okazaki N, Imizu Y, Tada A (1993) Preparation of SiO2-Al2O3 gels from tetraethoxysilane and aluminum chloride: A new sol-gel method using propylene oxide as a gelation promoter. J Ceramic Soc Jap 101:1081.
13. Gan L, Xu Z, Feng Y, Chen L (2005) Synthesis of alumina aerogels by ambient drying method and control of their structures. J Porous Mater 12:317–321.
14. Tokudome Y, Fujita K, Nakanishi K, Miura K, Hirao K (2007) Synthesis of monolithic Al2O3 with well-defined macropores and mesostructured skeletons via the sol-gel process accompanied by phase separation. Chem Mater 19:3393–3398.
15. Hund JF, McElfresh J, Frederick CA, Nikroo A, Greenwood AL, Luo W (2007) Fabrication and characterization of aluminum oxide aerogel backlighter targets. Fusion Sci Tech 51:701–704.
16. Tokudome Y, Nakanishi K, Hanada T (2009) Effect of La addition on thermal microstructural evolution of macroporous alumina monolith prepared from ionic precursors. J Ceramic Soc Japan 117:351–355.
17. Baumann TF, Kucheyev SO, Gash AE, Satcher JH (2005) Facile synthesis of a crystalline, high-surface-area SnO_2 aerogel. Adv Mater 17:1546–1548.
18. Gash AE, Tillotson TM, Satcher JH, Hrubesh LW, Simpson RL (2001) New sol-gel synthetic route to transition and main-group metal oxide aerogels using inorganic salt precursors. J Non-Cryst Solids 285:22–28.
19. Tillotson TM, Gash AE, Simpson RL, Hrubesh LW, Satcher JH, Poco JF (2001) Nanostructured energetic materials using sol-gel methodologies. J Non-Cryst Solids 285:338–345.
20. Prakash A, McCormick AV, Zachariah MR (2004) Aero-sol-gel synthesis of nanoporous iron oxide particles: A potential oxidizer for nanoenergetic materials. Chem Mater 16:1466–1471.
21. Long JW, Logan MS, Rhodes CP, Carpenter EE, Stroud RM, Rolison DR (2004) Nanocrystalline iron oxide aerogels as mesoporous magnetic architectures. J Amer Chem Soc 126:16879–16889.
22. Park C, Magana D, Stiegman AE (2007) High-quality Fe and γ-Fe2O3 magnetic thin films from an epoxide-catalyzed sol-gel process. Chem Mater 19:677–683.
23. Carpenter EE, Long JW, Rolison DR, Logan MS, Pettigrew K, Stroud RM, Kuhn LT, Hansen BR, MØrup S (2006) Magnetic and Mössbauer spectroscopy studies of nanocrystalline iron oxide aerogels. J Appl Phys 99:08N711.
24. Cui, H, Ren W (2008) Low temperature and size controlled synthesis of monodispersed γ-Fe2O3 nanoparticles by an apoxide assisted sol-gel route. J Sol-Gel Sci Technol 47:81–84.
25. Bali S, Huggins FE, Huffman GP, Ernst RD, Pugmire RJ, Eyring EM (2009) Iron aerogel and xerogel catalysts for Fischer-Tropsch synthesis of diesel fuel. Energy & Fuels 23:14–18.
26. Bali S, Turpin GC, Ernst RD, Pugmire RJ, Singh V, Seehra MS, Eyring EM (2008) Water gas shift catalysis using iron aerogels doped with palladium by the gas-phase incorporation method. Energy & Fuels 22:1439–1443.
27. Suh DJ, Park T, Kim W, Hong I (2003) Synthesis of high-surface-area ruthenium oxide aerogels by non-alkoxide sol-gel route. J Power Sources 117:1–6.
28. Kucheyev SO, van Buuren T, Baumann TF, Satcher JH, Willey TM, Meulenberg RW, Felter TE, Poco JF, Gammon SA, Terminello LJ (2004) Electronic structure of titania aerogels from soft x-ray absorption spectroscopy. Phys Rev B 69:245102.
29. Chen L, Zhu J, Liu Y, Cao Y, Li H, He H, Dai W, Fan K (2006) Photocatalytic activity of epoxide sol-gel derived titania transformed into nanocrystalline aerogel powders by supercritical drying. J Mol Catal A 255:260–268.

30. Sisk CN, Hope-Weeks LJ (2008) Copper(II) aerogels via 1,2-epoxide gelation. J Mater Chem 18:2607–2610.
31. Gao YP, Sisk CN, Hope-Weeks LJ (2007) A sol-gel route to synthesize monolithic zinc oxide aerogels. Chem Mater 19:6007–6011.
32. Gash AE, Satcher JH, Simpson RL (2004) Monolithic nickel(II)-based aerogels using an organic epoxide: The importance of the counterion. J Non-Cryst Solids 350:145–151.
33. Wei T, Chen C, Chang K, Lu S, Hu C (2009) Cobalt oxide aerogels of ideal supercapacitive properties prepared with an epoxide synthetic route. Chem Mater 21:3228–3233.
34. Laberty-Robert C, Long JW, Lucas EM, Pettigrew KA, Stroud RM, Doescher MS, Rolison DR (2006) Sol-gel derived ceria nanoarchitectures: Synthesis, characterization and electrical properties. Chem Mater 18:50–58.
35. Tillotson TM, Sunderland WE, Thomas IM, Hrubesh LW (1994) Synthesis of lanthanide and lanthanide-silicate aerogels. J Sol-Gel Sci Tech 1:241.
36. Zhang HD, Li B, Zheng QX, Jiang MH, Tao XT (2008) Synthesis and characterization of monolithic Gd2O3 aerogels. J Non-Cryst Solids 354:4089–4093.
37. Reibold RA, Poco JF, Baumann TF, Simpson RL, Satcher JH (2003) Synthesis and characterization of a low-density urania (UO3) aerogel. J Non-Cryst Solids 319:241–246.
38. Reibold RA, Poco JF, Baumann TF, Simpson RL, Satcher JH (2004) Synthesis and characterization of a nanocrystalline thoria aerogel. J Non-Cryst Solids 341:35–39.
39. Chervin CN, Clapsaddle BJ, Chiu HW, Gash AE, Satcher JH, Kauzlarich SM (2005) Aerogel synthesis of yttria-stabilized zirconia by a non-alkoxide sol-gel route. Chem Mater 17:3345–3351.
40. Chervin CN, Clapsaddle BJ, Chiu HW, Gash AE, Satcher JH, Kauzlarich SM (2006) Role of cyclic ether and solvent in a non-alkoxide sol-gel synthesis of yttria-stabilized zirconia nanoparticles. Chem Mater 18:4865–4874.
41. Chervin CN, Clapsaddle BJ, Chiu HW, Gash AE, Satcher JH, Kauzlarich SM (2006) A non-alkoxide sol-gel method for the preparation of homogeneous nanocrystalline powders of La0.85Sr0.15MnO3. Chem Mater 18:1928–1937.
42. Brown P, Hope-Weeks LJ (2009) The synthesis and characterization of zinc ferrite aerogels prepared by epoxide addition. J Sol-Gel Sci Technol 51:238–243.
43. Cui H, Zayat M, Levy D (2005) Sol-gel synthesis of nanoscaled spinels using propylene oxide as a gelation agent. J Sol-Gel Sci Technol 35:175–181.
44. Guo Y, Meyer-Zaika W, Muhler M, Vukojevic S, Epple M (2006) Cu/Zn/Al xerogels and aerogels prepared by a sol-gel reaction as catalysts for methanol synthesis. Eur. J Inorg Chem 23:4774–4781.
45. Clapsaddle BJ, Gash AE, Satcher JH, Simpson RL (2003) Silicon oxide in an iron(III) oxide matrix: The sol-gel synthesis and characterization of Fe-Si mixed oxide nanocomposites that contain iron oxide as a major phase. J Non-Cryst Solids 331:190–201.
46. Clapsaddle BJ, Sprehn DW, Gash AE, Satcher JH, Simpson RL (2004) A versatile sol-gel synthesis route to metal-silicon mixed oxide nanocomposites that contain metal oxides as the major phase. J Non-Cryst Solids 350:173–181.
47. Zhao L, Clapsaddle BJ, Satcher JH, Schaefer DW, Shea KJ (2005) Integrated chemical systems: The simultaneous formation of hybrid nanocomposites of iron oxide and organo silsesquioxane. Chem Mater 17:1358–1366.
48. Leventis N, Chandrasekaran N, Sotiriou-Leventis C., Mumtaz A (2009) Smelting in the age of nano: Iron aerogels. J Mater Chem 19:63–65.
49. Du A, Zhou B, Shen J, Xiao S, Zhang Z, Liu C, Zhang M (2009) Monolithic copper oxide aerogel via dispersed inorganic sol-gel method. J Non-Cryst Solids 355:175–181.
50. Morris CA, Anderson ML, Stroud RM, Merzbacher CI, Rolison DR (1999) Silica Sol as a nanoglue: Flexible synthesis of composite aerogels. Science 284:622.
51. Plantier KB, Pantoya ML, Gash AE (2005) Combustion wave speeds of nanocomposite Al/Fe_2O_3: The effects of Fe_2O_3 particle synthesis techniques. Combustion and Flame 140:299.
52. Prentice D, Pantoya ML, Gash AE (2006) Combustion wave speeds of sol-gel-synthesized tungsten trioxide and nano-aluminum: The effect of impurities on flame propagation. Energy & Fuels 20:2370.

Part IV

Materials and Processing: Organic – Natural and Synthetic Aerogels

9

Monoliths and Fibrous Cellulose Aerogels

Lorenz Ratke

Abstract Cellulose aerogels can be produced by using several methods, yielding materials with extremely low densities. Their structure can be described as a type of nanofelt, which means that the elementary fibrils of cellulose are arranged in a random three-dimensional (3D) network. Aerogels made either by freeze or supercritical drying of dissolved cellulose nanofibrils can be transformed into filaments, establishing the first open porous filament for possible textile applications. They can also be converted to carbon aerogel monoliths and filaments opening up new fields of application. The review describes the different methods developed by several research groups worldwide to produce low density cellulose monoliths and filaments. It presents the microstructures obtained with various methods and the properties of cellulose aerogels.

9.1. Introduction

Cellulose in its various modifications is a natural linear macromolecule produced mainly by plants in huge amounts per year, approximately a billion tons. Chemically it is a chain of 1–4-linked β-D-glucopyranose, which means that the β-D-glucopyranose ring in its chair conformation is connected via oxygen bridges to a linear chain and hydrogen bonds stiffen the chain. Crystalline cellulose can exhibit essentially four crystallographically distinct polymorphic modifications [1]. The walls of plant cells produce cellulose from glucose units which are a result of photosynthesis. Cellulose fibers in the cell walls together with hemi-cellulose and lignin are responsible for the extremely good mechanical properties of this three-phase composite allowing especially trees to grow to huge sizes. For commercial applications, cellulose is obtained from cotton, bast fibers, flax, hemp, sisal, and jute or wood. The production of cellulose fibers generally needs a separation of the cellulose from *lignin* and *hemicellulose* and other constituents of plants cells. This can be done chemically and processes were then developed to produce a viscous liquid that can be spun. *Regeneration* leads to fibers consisting of pure cellulose (rayon and *viscose process* [2, 3]). Many derivatives of cellulose in the form of esters and ethers were developed having a broad variety of applications (e.g., cellulose-acetates and nitrocellulose). Cellulose is the

L. Ratke • Institute of Materials Physics in Space DLR, German Aerospace Center,
51147 Cologne, Germany
e-mail: lorenz.ratke@dlr.de

M.A. Aegerter et al. (eds.), *Aerogels Handbook*, Advances in Sol-Gel Derived Materials and Technologies, DOI 10.1007/978-1-4419-7589-8_9,

basis for papermaking. Its fibers have high strength and durability. As Zugenmaier [1] writes, they are readily wetted by water, exhibit considerable swelling when saturated, and are *hygroscopic*, i.e., they absorb appreciable amounts of water, when exposed to the atmosphere. Even in the wet state, natural cellulose fibers show almost no loss in strength, which is important for textile applications. Cellulose of different natural sources exhibits different degrees of polymerization. Cellulose from wood *pulp* has between 6,000 and 10,000 monomer units, and cotton between 10,000 and 15,000. Chemical modifications during processing lead often to a reduction of the degree of polymerization giving products below 1,000 monomer units [1].

All native cellulose is organized into fringes and *fibrils* [4], which means that there are areas of crystalline order being intermixed with amorphous such as ones as shown schematically in Figure 9.1A. Figure 9.1B shows a schematic breakdown of a cellulose fiber as it can be found in plant cells down to its polymeric unit.

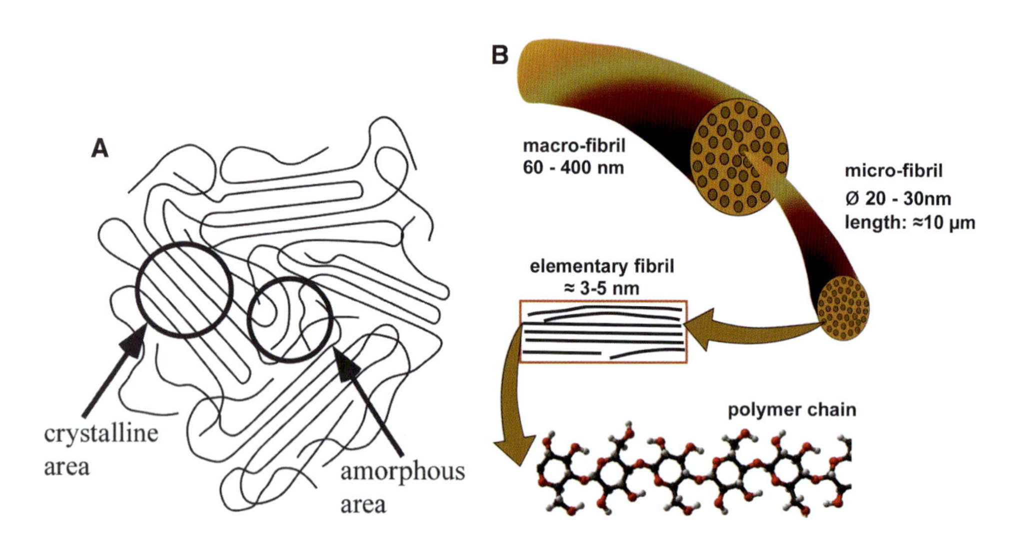

Figure 9.1. **A.** Fringe-fibril model of cellulose after Hearle [4]; see also Zugenmaier [1]. The **right** figure **B.** shows a schematic of a macro-fibril as existing in plant cells begin a composite of micro-fibrils. These consist of elementary fibrils which are made of 30–40 polymeric linear cellulose chains (picture based on the botany visual resource library [5]). The picture in figure **A.** is observed in crystalline cellulose, grown either artificially as for instance in textile fibers [1] or can be thought to mimic the structure of elementary fibrils.

Although cellulose is *hydrophilic* and considerable swelling occurs in water, it is insoluble in water and most organic solvents, but is biodegradable and as such cellulose aerogels would also have the taste of a green material. The cellulose polymer can be broken down chemically into its glucose units with concentrated acids at high temperature.

Since cellulose is chemically a very stable material, the production of aerogels from cellulose needs a technology or processing route to disintegrate the cellulose into the elementary fibrils or even down to the polymeric chain without degradation or derivatization and then to rebuild them into a suitable low density, open porous gel that can be dried to obtain a 3D structure of aerogels (Chap. 21).

9.2. Cellulose Aerogel Monoliths

Several methods are described in the literature to prepare cellulose aerogels and still many new methods are developed, partly depending on the raw material used, since *hemicellulose* needs a different process than raw cellulose or *lignocellulosic* polymer mixtures.

First attempts to preserve the swollen structure of cellulose pulp go back to the work of Weatherwax and co-workers [6] and Alince [7]. They used different pulps from paper, cotton, and rayon. The materials were swollen by destilled water, ethylenediamine, and aqueous sodium hydroxide (a standard solvent agent to produce a prematerial for the spinning of filaments in the viscose process [2, 3]). The swollen and wet materials were dried mainly by solvent exchange (water against alcohols or acetone) and then slowly evaporated. In all cases reported by them they were able to produce porous materials, but could not avoid considerable shrinkage. The materials were only characterized by adsorption isotherms and no other structural or physical property measurements were made. Weatherwax and Caulfield [6] used after solvent exchange with alcohol carbon dioxide supercritical drying to obtain cellulose aerogels with a specific surface area of around 200 m^2/g.

The first cellulose aerogels which became well-known and popular were prepared by Tan and co-workers [8]. They used cellulose acetate as a starting material and de-esterified it. The cellulose ester was *cross-linked* in an acetone solution with toluene-2,4-di-isocyanate. Tan and co-workers observed that they could form gels if the cellulose concentration was larger than five and less than 30 wt%. The supercritically dried aerogels had specific surface areas of less than 400 m^2/g and densities in the range of 100–350 kg/m^3. The amount of cellulose to volume of acetone had an effect on the density and the shrinkage as well as the ratio of cellulose to toluene cross-linker. The smaller the cellulose ester concentration and the larger the cross-linker amount the larger the shrinkage. This means that the cellulose backbone is rather rigid and does compensate shrinkage stress gradients. Their work became popular and was mentioned even in newspapers, since they measured the *impact strength* of the cellulose aerogel sheets (5 mm thick) and could show that although their material had a high porosity, its strength exceeds that of resorcinol-formaldehyde (RF) aerogels (Chap. 11). The technique to measure the strength might be questionable from an engineering point of view, since an impact test is not really appropriate for aerogels. Bending or tensile tests would be better. The test must be designed to the material [9]. The comparison with *RF-aerogels* is not indicative for a high strength material. It just shows that the RF aerogels produced by Tan are as brittle as other aerogels (Chap. 22).

Jin and co-workers [10] developed another technique to produce high-quality cellulose aerogels. Their technique avoids the utilization of toxic isocyanates and allows in contrast to the method of Tan [8] to use lower amounts of cellulose. Their technique is based on semi-crystalline raw cellulose whose morphology can be well described by a mixture of regions with highly crystalline order and unordered intermediate areas connecting the ordered ones as described above.

For the production of cellulose gels it is necessary to change the cellulose morphology as described in the introduction and the structure of the cellulose fibers themselves. This can be realized in using a suitable solvent [11–13], which must be able to break the extensive system of hydrogen bonds along the cellulose polymer chain without degrading or inducing derivate reactions of the polymer chains.

Jin et al. [10] used a so-called *salt-hydrate melt* as a dissolving agent, being a mixture of water and $Ca(SCN)_2$ at a composition close to the coordination number of the salt cation

($Ca(SCN)_2 \cdot 4H_2O$). The dissolving capabilities of salt-hydrate melts were first described and successfully realized by Phillip et al. [14] using the donor–acceptor concept, which describes the interactions between a polar solvent and the hydrogen bonds. Because of its *polarity* and *acidity* and its low melting point, being far below the crystalline amorphous transition point of cellulose/water mixtures (320°C), a salt-hydrate melt dissolves cellulose to a colloidal system [11, 12]. The hot solution of calcium thiocyanate and water $Ca(SCN)_2$: H_2O with 1:4 mol/mol has a melting temperature around 110–120°C and undergoes a reversible *sol–gel transition* at approximate 80°C [10–12]. A schematic of the process developed by Jin and co-workers [10] is shown in Figure 9.2.

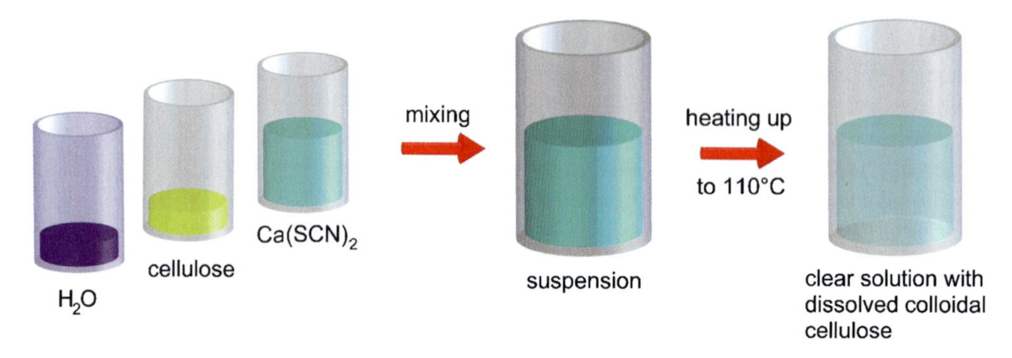

Figure 9.2. Schematic of the process developed by Jin and co-workers [10] to produce cellulose aerogels.

Jin and co-workers prepared the gels by quickly spreading the hot solution on a glass plate to form a layer of 1.0 mm thickness. After the solution solidified, the gel plate was immersed in a methanol bath, which extracted the salt and regenerated cellulose as a gel. The gel was washed with deionized water until the electrical conductivity became negligible. Residual ions left in the pore space yield beautiful colors such as pink, green, blue, and others, but they cause shrinkage in the long run [15].

The films prepared by Jin and co-workers were dried by *freeze drying* using slightly different procedures. The samples were either freeze dried regularly after immersion into liquid nitrogen or after a series of *solvent exchanges* with various alcohols prior to liquid nitrogen cooling or they were contacted with a copper plate kept at liquid nitrogen temperature and then transferred to the freeze drying unit.

Jin and co-workers did not use *supercritical drying*, which is reflected somewhat in their results. The specific surface area was measured by nitrogen adsorption and the microstructure looked at in a scanning electron microscopic. The tensile strength was measured on thin but large samples.

Essential to their process was the use of low cellulose concentration (0.5–3 wt%), and therefore they obtained extremely low-density aerogels. Their density increased with cellulose content. Their data suggest a parabolic increase with concentration, which in view of newer results suggest that the low cellulose content materials exhibited larger shrinkage (see below). The same is true for the *specific surface area*, which increases with increasing cellulose content from around 105–200 m^2/g from 0.5 wt% cellulose to 3 wt%. Freeze drying always leads to materials with higher density and lower surface area.

The variation of *tensile strength* with cellulose content is shown in Figure 9.3. In contrast to the statement of Tan [8], the strength of these aerogels is rather low. A value of

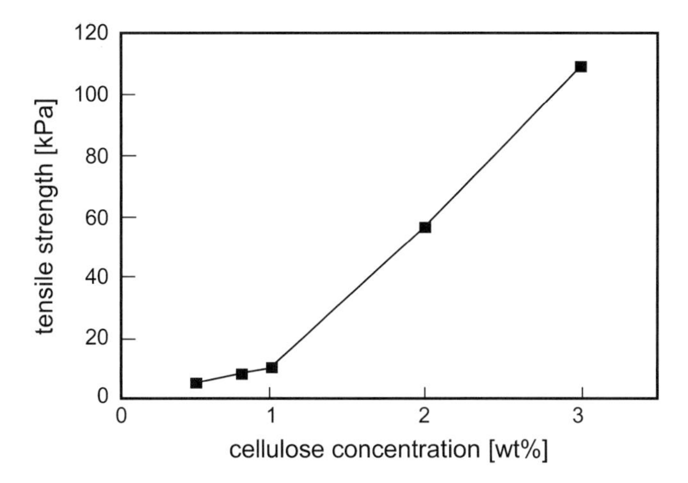

Figure 9.3. Strength of cellulose aerogels as prepared by Jin and co-workers [10] using a salt-hydrate melt to dissolve cellulose (calcium thiocyanate/water) and regeneration of the cellulose in alcohols (redrawn from [10] different units used here).

120 kPa is easily achieved with RF aerogels (Chaps. 11 and 22) if dried subcritically [16]. The low strength might also be a result of the process, since the fibrils within the aerogel were not cross-linked compared to those produced by Tan [8], and therefore a felt-like material was tested, in which the strength might be limited due only to a mechanical interlinkage of the fibrils (cellulose filaments can reach strengths of around 1–2 MPa in contrast, see below). As an example of the extensive microstructural analysis, Figure 9.4 shows a comparison of the microstructures as revealed by *SEM* for a material with 2 wt% cellulose. One is freeze dried (Figure 9.4 left panel), the other solvent exchange dried (Figure 9.4 right panel).

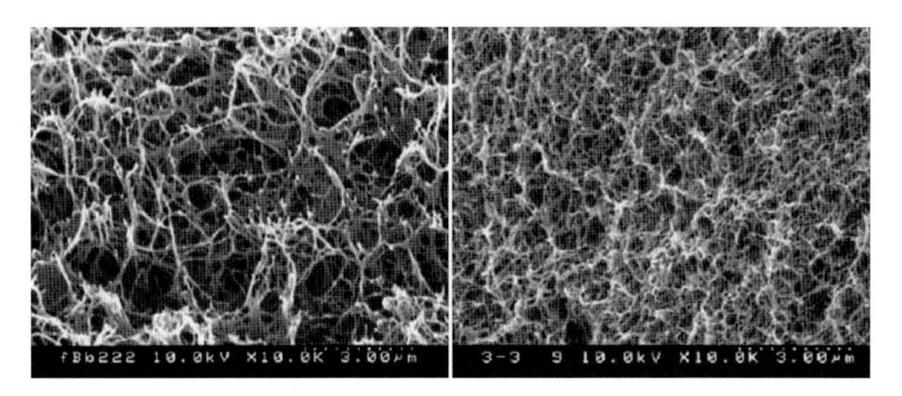

Figure 9.4. Microstructure of a cellulose aerogel as prepared by Jin and co-workers [10] (with permission of Elsevier). The samples shown have a cellulose concentration of 2 wt%. The figure in the **left panel** shows a regularly freeze dried sample, in the **right** a sample rapidly freeze dried as described above with gel casting onto a copper plate kept at liquid nitrogen temperature.

In order to prepare carbon aerogels from cellulose precursors Ishida and co-workers [17] used classical methods to dissolve microcrystalline cellulose in sulphuric acid or sodium hydroxide water solutions. The resultant aqueous suspension of cellulose with

0.2–1 wt% solid content was rapidly solidified by spraying onto a copper plate, which was partially submerged in liquid nitrogen. A very fine powder material was obtained after either freeze or supercritically drying (after solvent exchange with ethanol and carbon dioxide). The solid powder had a specific surface area ranging from 32 to 178 m^2/g. It was shown that rapid *solidification* and freeze drying could preserve the wet gel structure yielding powder particles with the highest surface area. SEM analysis performed revealed a fibrous open, sponge-like network of relatively thick cellulose fibers (with thickness around 5 μm). Pyrolysis leads to finer network structures with specific areas above 500 m^2/g and consists of carbon ribbons, being around 100–400 nm wide (by SEM).

In 2006 two papers appeared using different methods to produce cellulose aerogels. Fischer and co-workers [18] prepared cellulose aerogels from acetates and cross-linked these by urethane bonds using a polyphenylisocyanate. Their complex procedure yields gels in around 7 days at room temperature. The supercritically dried aerogels have specific surface areas in the range of 150–250 m^2/g and densities between 250 and 850 kg/m^3. These aerogels are rather dense and a more specific discussion can be found in Chap. 10.

Innerlohinger [19] produced aerogels from cellulose which was dissolved in *N*-methyl-morpholine-*N*-oxide (NMMO). NMMO is an ionic liquid that can dissolve cellulose directly and is the basis of the technical process that produces Lyocell or Tencel fibers. The region of the water/NMMO/cellulose phase diagram, where complete dissolution is possible, was intensively studied by Cibik and others [20, 21]. Cellulose was obtained from 13 different pulps by Lenzing AG having different *degrees of polymerization* ranging from 180 to 4,600. The typical process of Innerlohinger and co-workers consisted of a solution cellulose and NMMO, shaping, regeneration and washing in alcohol and water, solvent exchange and supercritical drying with carbon dioxide. They used cellulose solutions with 0.5–13 wt% and produced three different shapes of aerogels: large cylindrical monoliths with 26 mm in diameter, spherical beads, 2–4 mm in diameter, and thin films by casting the hot aerogel solution onto glass plates.

Their materials were analyzed with *nitrogen adsorption*, SEM, and density measurements. From their extensive studies, a few results are reproduced here. Figure 9.5 shows the density of aerogels as a function of initial cellulose content. The density increases in

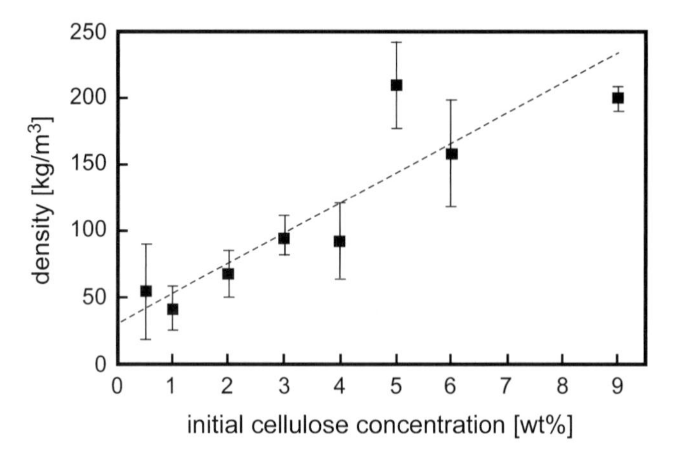

Figure 9.5. Density of cellulose aerogels prepared by Innerlohinger et al. [19] using NMMO as a solvent and supercritical drying with carbon dioxide (redrawn on the basis of their data using different units).

an almost linear fashion with the cellulose content. The lowest density is achieved with 50 kg/m^3, which still is rather high compared to polyurethanc foams.

Innerlohinger et al. [19] reported a rather large shrinkage of their wet gels around 40%. Such values are extremely large compared to supercritical drying of other aerogels (Part II) and show that their gels are rather soft and plastically deformable. The reported internal surface areas range between 100 and 400 m^2/g. The SEM figures shown in the paper reveal a more flake-like appearance of the cellulose fibrils being strongly interconnected. Therefore, it seems that their technique to dissolve cellulose can be improved especially with respect to the solvent exchange and regeneration procedures (see below), which have a marked influence on the aerogel structure and properties.

Liebner and co-workers [22, 23] report in a research letter on their approach to also use NMMO and *N*-benzyl-morpholine-*N*-oxide (NBnMO) as solvents. From NMMO solutions with cellulose contents between 1 and 12 wt%, dimensionally stable cellulose gels were produced. The cellulose solutions were produced at temperatures varying from 100 to 120°C increasing with the cellulose content. The cellulose contents were 3, 5, 10, and 12 wt% and thus rather large compared to the salt-hydrate melt route. They performed the regeneration with ethanol or ethanol-dimethylsulfoxide (DMSO) mixtures in different concentrations. The solution, regeneration, and solvent exchange routes could take up to several days. Their aerogels exhibit densities in a range from 50 to 260 kg/m^3 and specific surface areas from 172 to 284 m^2/g. All their regenerated cellulose gels were dried supercritically. The SEM microstructures presented look, compared to other cellulose aerogel structures, rather dense and not as open as aerogels prepared by other authors.

Hoepfner et al. [24] used the process of Jin [10] to produce monolithic cellulose aerogels. They also used $Ca(SCN)_2$ as described above with a water concentration near the coordination number of the cation and dissolved microcrystalline cellulose in that solvent by increasing the temperature of the salt-hydrate melt to 110–120°C until the solution became clear. The hot cellulose solution was poured into polyacrylic multi-wall sheets with a cross section of 15 × 15 mm yielding monoliths with a length of around 100 mm (Figure 9.6). The cellulose is then regenerated by insertion in ethanol. Freeze drying was used after

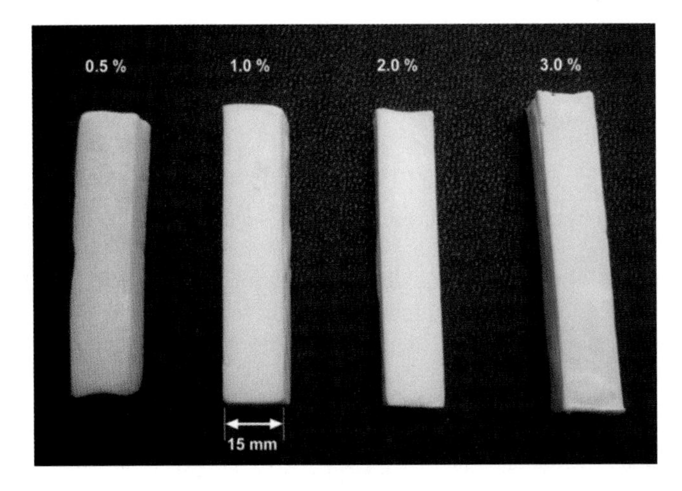

Figure 9.6. Cellulose aerogels with different concentrations of cellulose as prepared by Hoepfner and co-workers [24] using the procedure of Jin [10]. All cellulose aerogels exhibit a *white* appearance. They are soft and easily compressible by gentle pressure.

rapid solidification of the wet gel in liquid nitrogen and supercritical drying after solvent exchange by CO_2.

The monolithic aerogels shown in Figure 9.6 are white, soft, and easily deformable by doing the fingernail test. Figure 9.7 shows the average density of aerogels aged in ethanol. The density varies between 10 and 60 kg/m^3 with the cellulose concentrations varying between 0.5 and 3 wt%. The linear relation shows that only the cellulose concentration in the salt-hydrate melt determines the density. The temperature of the melt bath, the annealing time before cooling to the gel point and not the aging time in ethanol play a detectable role. Note that the density is only eight times higher than the density of air and is in the range of styrofoams [25].

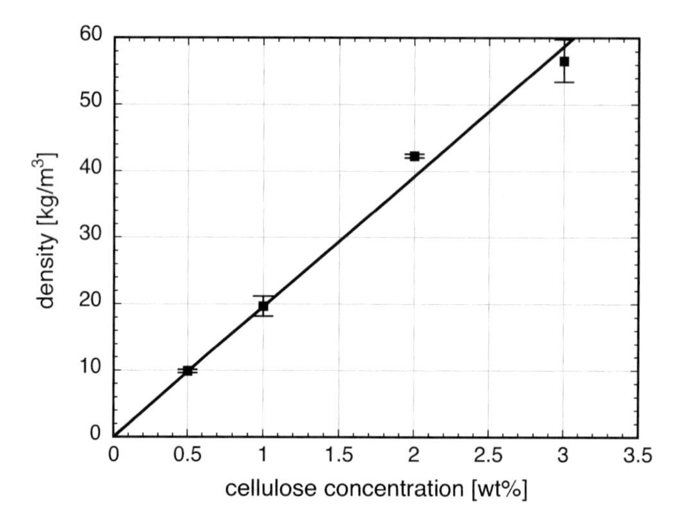

Figure 9.7. Density of cellulose aerogels prepared by Hoepfner et al. [24] using the route of Jin [10].

A linear variation of the aerogel density of cellulose concentration should be observed on simple theoretical grounds. If ρ_A denotes the density of the aerogel and w_c is the salt hydrate melt weight fraction having a density ρ_L, then for small concentration, $w_c << 1$, the aerogel density should obey the relation,

$$\rho_A \cong \rho_L w_c \tag{9.1}$$

which has also been observed by Innerlohinger [19], whose data, however, do not cross the coordinate origin at zero concentration.

They also observe a rather large shrinkage during supercritical drying, being not as large as those of Innerlohinger, but around 15% and varying linearly with cellulose concentration. A large shrinkage at the lower concentration shows that the gel network is not very stable and can be compacted by a large amount. With higher cellulose content the shrinkage is smaller showing that the network is more stable.

It should be emphasized here that Jin and co-workers [10] did not observe a linear relation between density of the aerogels and the cellulose concentration. This might be an indication that their process was not optimized, for instance, with respect to an aging treatment, solvent exchange and stress gradients induced by freeze drying might have affected the aerogel structure, especially at low cellulose concentrations.

Using nitrogen physisorption, Hoepfner and co-workers measured specific surface areas of 210 m^2/g, which were independent of the cellulose concentration. A simple calculation shows that the independence is theoretically expected, if the drying procedure does not harm the gels. To understand this they assumed that all cellulose nanofibrils can be described as cylinders with average radius $<R>$ and length $<L>$. The number density of fibrils in the aerogel shall be n_F. Then the specific surface area per mass is

$$S_m = \frac{2\pi \langle R \rangle \langle L \rangle n_F}{\rho_A} \tag{9.2}$$

neglecting the faces of the cylinders, which is allowed if the fibril length is much larger than the radius. Using the relation $\Phi_c = n_F <v_c>$, with $<v_C>$ the average volume of a cellulose fibril $<v_C> = \pi \langle R \rangle^2 \langle L \rangle$ they obtain the result

$$S_m = \frac{2}{\langle R \rangle \rho_c} \tag{9.3}$$

Assuming that the fibril radius in the aerogel network is essentially independent of the cellulose concentration, the relation clearly states that the specific surface area is a constant independent of the cellulose concentration. They could show that the network structure becomes finer and the pores become smaller, the larger the cellulose concentration is.

Figure 9.8 shows the SEM picture of SCD aerogels prepared with 2 wt% cellulose. The picture clearly shows that the nano-felt indeed consists of cellulose fibrils having thicknesses in the range of 20–50 nm.

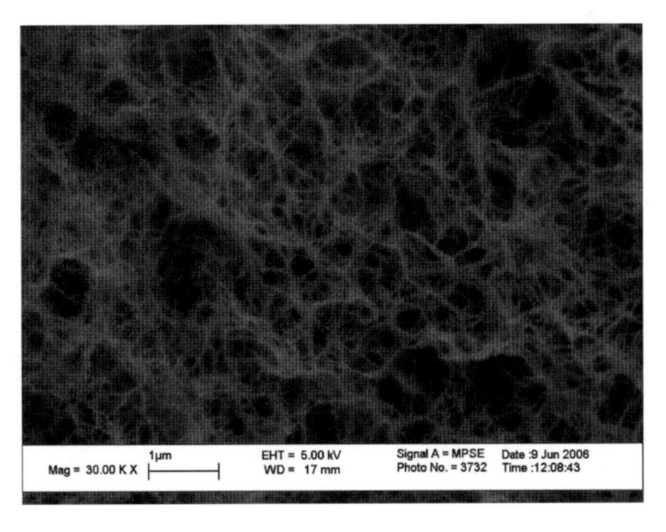

Figure 9.8. SEM picture of a supercritically dried aerogel with 2 wt% cellulose as prepared by Hoepfner et al. [24] having a density of 42 kg/m^3.

Gavillon and Budtova reported on the preparation and the morphology of an aerogel-like pure cellulose material [26]. The material was called by them Aerocellulose, and it was obtained from cellulose dissolved in nonpolluting solvents, such as aqueous NaOH or

N-methyl-morpholine-*N*-oxide (NMMO) monohydrate, and dried supercritically with CO_2. The studies were focused on the preparation of aerogelic cellulose from cellulose/NaOH/water solutions and on the influence of the preparation conditions on their morphology. As a starting material they used several microcrystalline celluloses with degree of polymerization ranging from 180 to 950 (Solucell cellulose from Lenzing R&D labs, Austria). The cellulose was dissolved and swollen in mixtures of water and sodium hydroxide at low temperatures. The cellulose concentrations were varied between 3 and 8 wt%. The solutions were cast into cylindrical molds of 8–10 mm diameter and 30–40 mm length. After gelation the samples were regenerated in water. During regeneration the samples shrunk by about 10% in volume. The gelation time varied exponentially with temperature. Although at 10°C the gelation takes around 3 days, at 40°C it takes only 6 min. The regenerated and swollen cellulose gels were subjected to a solvent exchange using acetone, which was then replaced with CO_2 and supercritically dried.

The resultant aerogels show porosities above 90%, specific surface areas above 200 m^2/g, and densities in the range of 120–140 kg/m^3. A special *surfactant* (Simusol, alkylpolyglycoside, APG) was also used in various concentrations, and it could be shown that it changes the morphology from a more open porous network of thin fibers being similar to those reported by Hoepfner et al. [24] and Innerlohinger [19] to a more clamped compact one, often observed in freeze dried samples [10, 24]. The surfactant reduced the density slightly and increased the average pore size by an order of magnitude. They generally observe that aerogels from cellulose gels regenerated in water at ambient temperature show a fibrillar structure while those regenerated at 70°C show a "cloudy" structure irrespectively of the cellulose solvent. This was attributed to a higher diffusivity of all components in the gelling system being accelerated at higher temperatures. Probably because of this, cellulose chains have less time to "pack" and rapid precipitation causes a more disorganized structure and a cloudy appearance.

Generally an NaOH/water swollen cellulose liquid is the starting solution to spin fibers in the viscose process. Therein cellulose is treated with a water/sodium hydroxide solution and then is mixed with carbon disulfide to form *cellulose xanthate*. The resulting viscose is extruded into an acid bath through a spinneret to make rayon. The acid converts the viscose back into cellulose [2, 3]. Therefore Gavillon and Budtova tested if the acidity of the regenerating bath (adjusted as in the industrial process with sulfuric acid) changes the aerogels. The higher the regenerating bath acidity, the lower are both the average pore diameter and the total porosity and thus the density.

Pinnow et al. [27] prepared cellulose aerogels from *cellulose carbamate* gelled in sodium hydroxide or dilute sulfuric acid at higher temperatures. The aerogels were characterized with scanning electron microscopy, small-angle X-ray scattering, mercury intrusion, and nitrogen adsorption. They were able to produce aerogels with a density as low as 60 kg/m^3 from cellulose carbamate having a broad pore size distribution ranging from 0.5 nm to 1 mm. The morphology of macro- and meso-pores down to a diameter of 10 nm was determined by SEM. Quantitative characterization of the pore structure was realized by combining the results of *mercury intrusion*, nitrogen sorption, and SAXS experiments (Chap. 21). The methods are complementary and, by the combination of all these methods, parameters of the whole pore structure of the aerogel samples from macro- to micropore sizes are obtained. The effects of different preparation and processing parameters including drying and pyrolysis were studied by these methods. They showed that the preparation of the gel state from cellulose carbamate is an interesting alternative to the thiocyanate route to produce highly porous cellulose materials with a low density and a specific surface of about 430 m^2/g.

Cai et al. [28] were able to prepare transparent cellulose aerogels using an aqueous alkali hydroxide/urea solution as the dissolving and gelling agent. Figure 9.9 shows one of their impressive results, a transparent cellulose aerogel.

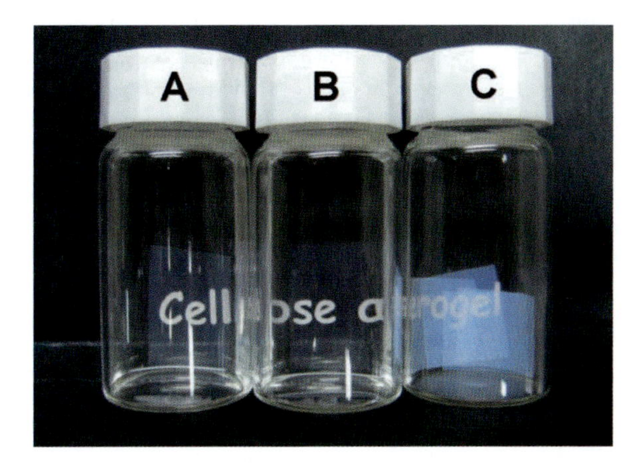

Figure 9.9. Photographs of cellulose gels according to reference [28]: A – hydrogel in water; B – alcogel in EtOH; and C – aerogel obtained by supercritical carbon dioxide drying (with permission of Wiley-VCH Verlag GmbH &Co).

They used cellulose of different sources and NaOH/urea/water or LiOH/urea/water solutions to dissolve the cellulose at low temperature by vigorously stirring and ultracentrifugation to remove air bubbles. The amount of cellulose was varied between 0.5 and 7 wt%. The clear solution was cast onto a glass plate yielding sheets of 0.5 mm thickness. The material was regenerated in various solutions such as alcohols and dilute sulfuric acid. Regenerated cellulose was washed with water, a solvent exchange performed with ethanol and subsequently with carbon dioxide taken out supercritically. Nitrogen adsorption was used to measure the specific surface area, SEM to inspect the microstructure, TEM to look into the microscopic details, X-ray diffraction to investigate the degree of crystallinity, and Young's modulus to determie also the density.

Table 9.1 shows an overview of their results as a short extract to several properties and process parameters. This excerpt already shows that the optical transmission can be better than 80% which comes into the range of inorganic aerogels (Chap. 2).

Nitrogen adsorption has shown a distinct hysteresis especially close to the saturation point. As pointed out by Reichenauer and Scherer [29] for silica aerogels, this is as a result of

Table 9.1. Excerpt of the results achieved by Cai and co-workers [28]

Solvent	Cellulose concentration [wt%]	Regeneration bath	Optical transmission [%]	Density [kg/m^3]	S_{BET} [m^2/g]
Aq. NaOH/urea	6	5 wt% H_2SO_4	49.1	260	364
	4	EtOH	19.6	140	260
Aq. LiOH/urea	6	H_2O	43.9	160	406
	6	EtOH	84.1	190	410
	6	H_2SO_4	60	260	381
	4	EtOH	78.6	120	406
	4	BuOH	53	400	304

the deformation on loading of the aerogel leading to an elastic/plastic behavior whose origins are capillary stresses in the mesopores (Chap. 21). The results of Table 9.1 also show that in contrast to NaOH, use of aqeous LiOH as a cellulose solvent allowed the formation of more transparent cellulose hydrogels and aerogels on regeneration by organic solvents (alcohols and acetone). Gels prepared with LiOH and alcohols as regenerators also showed a more homogeneous fine fibrillar and highly porous network structure. They confirm in a certain sense the results obtained by Jin [10], Hoepfner [24] and Innerlohinger [19] using a different route but also preparing monoliths instead of films.

The variability of the properties with the composition of the regeneration bath is demonstrated by a plot extracted from their tabulated data and shown in Figure 9.10. The strong variation of the density and the specific surface area with ethanol concentration in the bath clearly demonstrate how important it is to control precisely the regeneration conditions.

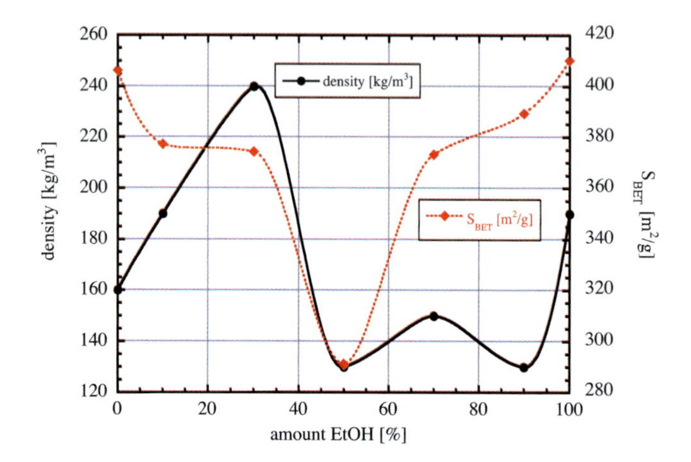

Figure 9.10. Density and specific surface area of cellulose aerogels prepared with aq. LiOH/urea and regenerated at 20°C. The cellulose concentration was 6 wt% after [30].

Instead of hydroxide or hydrate melts, Deng et al. [30] used an ionic liquid to dissolve cellulose and prepared aerogels by freeze drying. Their procedure starts with microcrystalline cellulose of a low degree of polymerization (DP 186) dissolved at 130°C in 1-butyl-3-methyl-imidazolium chloride (BmimCl) until a clear solution was obtained. The solution was cast onto a glass plate to obtain films of 1 mm thickness and then regenerated in water until no chloride was detected further. The hydrogels were dried according to different routes, like rapid freezing with liquid nitrogen, known to be best to preserve the hydrogel network structure, conventional freeze drying, typically leading to large ice crystals destroying the soft network, and furnace drying at 50°C, which would be expected to yield no aerogel at all. The results mirror the expectations and knowledge about the preparation of cellulose aerogels described above. A new aspect was the completely different dissolving agent, which, however, is well known in the context of textile fiber production [2, 3, 31]. SEM and nitrogen adsorption reveal that rapid freezing and freeze drying produce a foam having a 3D open fibrillar network structure with a specific surface area of 186 m^2/g and a porosity of 99%.

Aaltonen and Jauhiainen [32] prepared cellulose aerogels from microcrystalline cellulose, spruce wood, and from mixtures of cellulose, lignin, and xylan using an ionic liquid as a solvent. The work is interesting because of the use of a mixture of cellulose, lignin, and *xylan*, as well as direct use of wood. They mixed cellulose from 1 to 4 wt% with respect to the amount of 1-butyl-3-methylimidazolium chloride (C4mimCl which is identical to BmimCl used by Deng and co-workers [30]), lignin in the range of 1.4–3 wt% and xylan of 2 wt%. The wood spruce was used in two concentrations of 3 and 4 wt%. The dissolution was performed at 130°C and took between 2 and 27 h. The gelation of the lignocellulosic polymers was performed in water/ethanol bathes of different compositions. Gels could not be obtained in all cases, but some mixtures of cellulose and lignin, cellulose, lignin, and xylan, and also pure wood were successful. The suspensions were cast in Petri dishes giving sheets of 10 mm thickness. After gelation, pieces of 5×5 mm^2 were cut and supercritically dried after solvent exchange with ethanol and carbon dioxide. Samples were characterized by SEM, nitrogen adsorption, and density measurement.

The bulk densities of the biopolymer aerogels ranged from 25 to 114 kg/m^3 and the internal surface areas (BET) from 108 to 539 m^2/g. The aerogels exhibit the typical 3D fibrillar network structure, which varied with the processing conditions. The materials were generally mechanically soft, as noted by Hoepfner et al. [23], and did not recover their shape after slight compression with fingers. In contrast to the material described by Hoepfner [24], Aaltonen and Jauhiainen noted that their aerogels could easily be disintegrated to fibrous or powder-like material and rubbed between fingers. This is similar to some inorganic aerogels. They also observed that the aerogels obtained from wood were comparably hard and exhibited much higher mechanical strength, without, however, quantifying this.

Liebner and co-workers [33] recently extending their research on cellulose aerogels described the above [22, 23]. They used again a preparation method with hot solutions of cellulose/NMMO/water at 110–120°C and casting the clear solutions into molds. After gelation the samples were regenerated in an ethanol bath, removing the NMMO/water solution. Supercritical drying was performed with carbon dioxide. The densities were in the range of 46–69 kg/m^3 for 3 wt% of cellulose, the pore size around 10 nm, and the specific surface area around 190–310 m^2/g. The low density shows that their process preserves the wet gel structure excellently. All other methods produce 3 wt% cellulose at larger densities with the exception of the results obtained by Hoepfner et al. [24]. This pinpoints to the importance of shrinkage during regeneration and solvent exchange. Liebner et al. applied a compression test and they found that their aerogels can withstand a yield strain around 5% before fracture, which is rather large compared with inorganic aerogels; on the other hand, Young's modulus with 5–10 MPa is small. Doubling the amount of cellulose changes the elastic modulus by a factor of 10.

9.3. Cellulose Filaments for Textile Applications

Cellulose is one of the oldest materials used to produce fibers, filaments, and yarns, from which fabrics of all kinds are manufactured. Today's filament and yarn production is a field for both natural and synthetic polymers [31]. All fibers used today have a compact microstructure and a very large aspect ratio. In contrast to these, Schmenk et al. [34] and Hacker et al. [35] produced an open porous, nanostructured filament of cellulose aerogel for the first time using a sol–gel routine as described above for monoliths and different spinning techniques.

Fibers were formed using a *wet spinning* method. The hot cellulose solution was prepared as described above by the method of Jin [10] and modified by Hoepfner [24].

The hot solution was filled into a thermally insulated injection device with a volume of 200 mL and pressed with 3 MPa through a spinneret with an aperture of 250 μm diameter directly into an ethanol bath. In doing so the filaments were cooled down. Below 80°C the viscous solution gels and gets a ceraceous consistency at room temperature. Cellulose was regenerated in an ethanol bath. The regenerated fibers were dried supercritically.

The fibers were characterized with respect to their density, specific surface area, pore size distribution, microstructure (SEM), and tensile strength (using a special device to ensure fracture of the filaments and not fracture at the spanning clamps fixing the filaments). One example of supercritically dried filaments is shown in Figure 9.11. The filaments are white, soft and ductile with a diameter of 250 μm and in this respect their optical appearance is not different from the monoliths. It would be interesting to use the method of Cai et al. [28] described above in order to investigate whether transparent filaments can be produced.

Figure 9.11. Light microscopic picture of a filament of cellulose aerogel after supercritical drying having a dry density of 124 kg/m^3.

The SEM pictures from Figure 9.12A–D show the nano- to mesoscale structure of the aerogel filaments. In Figure 9.12A, a part of a fiber's surface is flaked off. The fibers exhibit the typical mantle-core structure and are well known from cellulose fiber production [31].

The mantle is usually attributed to rapid crystallization of micro-crystals of cellulose, which would be a result of the rapid cooling from around 110°C in the spinneret to room temperature in the ethanol bath below the spin hole. The structure beneath the surface looks like an open porous sponge (Figure 9.12B). The pores are irregularly arranged like channels (close to the surface they are perpendicular to) leaving the impression that the walls bounding the large pores are torn off. The average diameter of these pores is in the micrometer range, having a large volume fraction of around two-thirds of the whole filament. As observed also by Deng et al. [30], the walls are not compact, but have a fine fibrillar structure as shown in Figure 9.12C, D. The cell walls have an average thickness of a micron with fiber distances inside the walls around 80 nm. Figure 9.12D displays a small section of such a cell wall showing single, even dangling, fibrils with a thickness estimated to be around 10 nm.

The fibers are rather dense with a density of about 124 kg/m^3 compared to monoliths made of 3 wt% cellulose. This fact is astonishing with respect to the sponge-like meso-pores

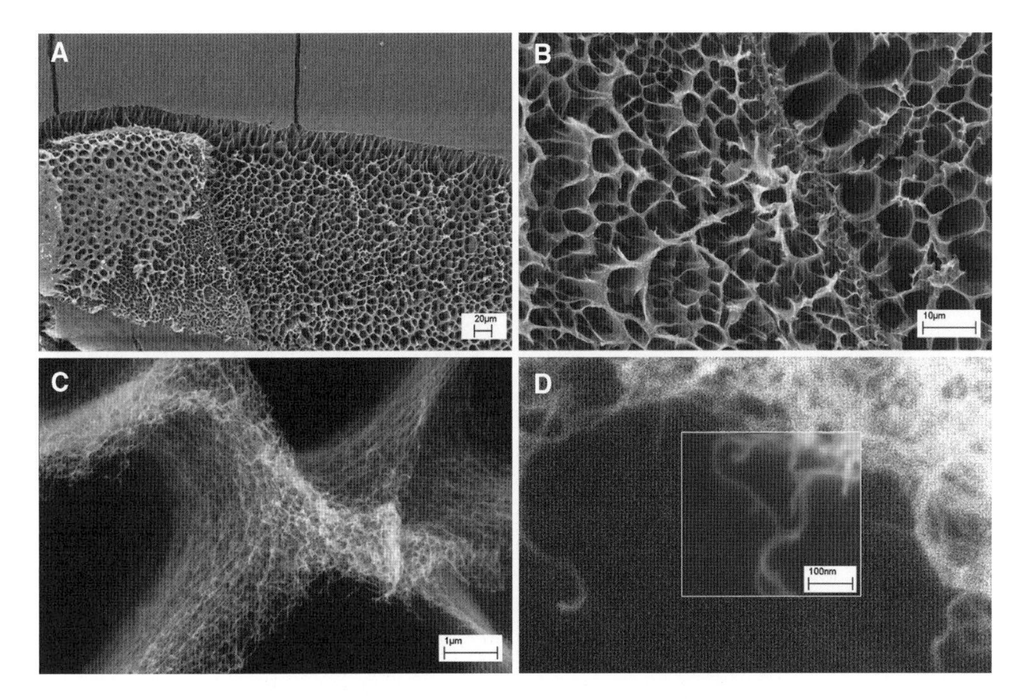

Figure 9.12. SEM picture of an aerogel filament with 3 wt% cellulose at different magnifications after [34].

and points to the fact that the aerogel filament spinning can be improved further. The specific surface area was measured to be around 180 m^2/g. The tensile strength is up to 1.4 MPa and thus higher than the strength reported for monoliths, which had been in the range of 100 kPa. The fracture strain is about 10%. The fibers may equally be characterized using the conventional textile measures. Using the standard definition they have a density of 59 *dtex* and a strength of 1.1 *cN/tex*. This value is very low compared to conventional textile fibers [31], having a strength by at least a factor of 30 higher. There is considerable space for improvement. Table 9.2 gives an overview on the results achieved for 250-μm-thick fibers.

In their recent paper Hacker et al. [35] describe the utilization of a different spinneret design adopted from conventional polymer filament fabrication. The piston-spinning plant has a standard spinneret and allows the incorporation of a godet stand, thereby facilitating the drawing of the spun fibers. This way, a much better alignment of the fibrils during spinning was observed, although the low viscosity of the hot thiocyanate/cellulose melt did not allow further stretching of the filaments in the regeneration bath. The effects of these

Table 9.2. Properties of cellulose aerogel fibers

Density	120	g/L
Specific surface area	180	m^2/g
Average pore diameter	120	Nm
Fracture strength	1.4 ± 0.2	MPa
Strength	1.1 ± 0.15	cN/tex
Elongation at fracture	9.3 ± 6.9	%
Tension length	1.2 ± 0.2	km

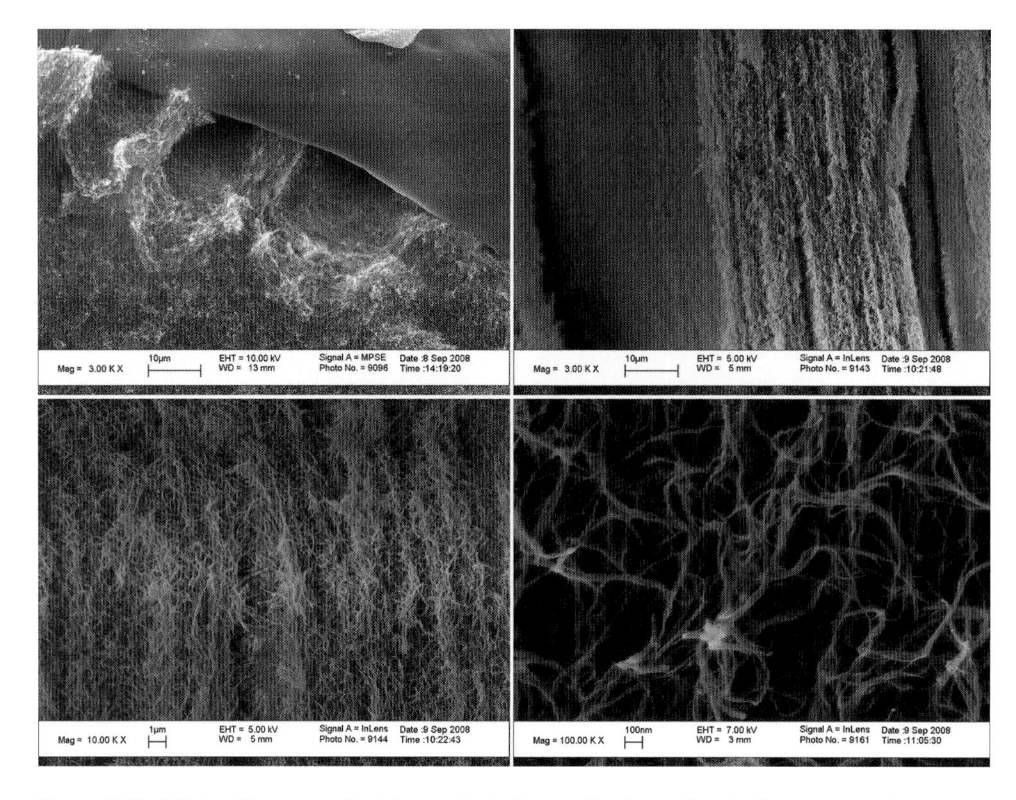

Figure 9.13. Cellulose filaments made with a standard spinneret allowing to align the filaments much better than a simple hole. At the surface there are still sponge-like channels mostly aligned parallel to the surface. The interior shows an aligned felt, which on closer inspection turns out to be irregular, especially on the nanoscale [35].

advanced experiments on the microstructure of the fibers are shown in Figure 9.13A–D. The microstructure inside the fibers varies, probably as a result of the mechanical stress.

The filaments show again a thinly dense surface layer of cellulose (Figure 9.13A), and beneath the pores they are elongated parallel to the spinning direction. The sponge-like filament interior presented above can be replaced by a structure with a preferred orientation in the direction of spinning (Figure 9.13B), although a closer inspection reveals an irregular nano-felt. This indicates that the microscopic structure can be manipulated, e.g., by drawing during the spinning process. Interestingly, the felt-like structure on the nanoscale, which is crucial for aerogels, has been preserved. Further magnification shows that the fibrils of the felt are arranged in an irregular fashion (Figure 9.13D), possibly due to the mechanism of gel-formation, and the fibrils have a diameter of around 10 nm.

9.4. Conclusions

Cellulose aerogels are at the beginning of intensive research and owing to their fascinating structure and properties they promise to have many applications. Various routes have been tested to produce aerogels. There is ample space for improvement and refinement as well as fundamental research on the kinetics of gelation in a solution of the flexible elementary fibrils, the cross-linking to stabilize and strengthen the dry aerogel structure, the fiber and filament drawing, yarn production, stability of dry cellulose aerogels against humidity, development and optimization of the processing routes with reference to the variety of possible cellulose pulp sources and characterization with respect to chemical and physical properties in relation to the nano-felt microstructure.

References

1. Zugenmaier P (2008) Crystalline Cellulose and Derivatives, Springer Series in Wood Sciences, Springer-Verlag, Berlin.
2. Marsh J T, Wood F C (1939) An Introduction to the Chemistry of Cellulose, Van Nostrand, New York.
3. Klemm D, Philipp B, Heinze T, Heinze U, Wagenknecht W (1998) Comprehensive cellulose chemistry, vols 1 and 2. Wiley-VCH, Weinheim.
4. Hearle J W S (1958) A Fringed Fibril Theory of Structure in Crystalline Polymers. J Polymer Sci 28: 432–435.
5. Botany Visual Resources Library (2001) McGraw-Hill, NY and also Uno G, Storey R, Moore R, Principles of Botany, McGraw Hill, NY.
6. Weatherwax R C, Caulfield D F (1971) Cellulose aerogels: an improved method preparing a highly expanded form of dry cellulose. Tappi 54: 985–986.
7. Alince B (1975) Porosity of swollen solvent-exchanged cellulose and its collapse during final removal. Colloid & Polymer Sci 253: 720–729.
8. Tan C, Fung M, Newman J K, Vu C (2001) Organic aerogels with very high impact strength. Adv Mater 13: 644–646.
9. Dieter G E (1986) Mechanical Metallurgy, McGraw-Hill, New York, 3rd Edition.
10. Jin H, Nishiyama, Wada Y M, Kuga S (2004), Nanofibrillar cellulose aerogels. Colloids and Surfaces A: Pysicochem. Eng. Aspects, 240: 63–67.
11. Fischer S (2003) Anorganische Salzschmelzen – ein unkonventionelles Löse- und Reaktionsmedium für Cellulose, Habilitation thesis, TU Bergakademie Freiberg, Germany.
12. Fischer, S, Leipner H, Thümmler K, Brendler E, Peters J (2003) Inorganic molten salts as solvents for cellulose. Cellulose 10: 227–236.
13. Frey M W, Theil M H (2004) Calculated phase diagrams for cellulose/ammonia/ammonium thiocyanate solutions in comparison to experimental results. Cellulose 11: 53–63.
14. Phillip B, Schleicher H, Wagenknecht W (1977) Non-aqeous solvents of cellulose. Chemtech 7: 702–709.
15. Hoepfner S, Ratke L, unpublished research.
16. Voss D, Brück S, Ratke L (2003) Aeromats – Ultraleichte Konstruktionswerkstoffe auf Aerogelbasis, in: Verbundwerkstoffe, Degischer H P, Ed., pp 505 – 509.
17. Ishida O, Kim D-Y, Kuga S, Nishiyama Y, Malcol Brown R (2004) Microfibrillar carbon from native cellulose. Cellulose 11: 475–480.
18. Fischer F, Rigacci, Pirard R, Berthon-Fabry S, Achard P (2006) Cellulose-based aerogels. Polymer 47: 7636–7645.
19. Innerlohinger J, Weber H K, Kraft G (2006) Aerocellulose: Aerogels and Aerogel-like Materials made from Cellulose. Macromol Symp 244: 126–135.
20. Cibik T (2003) Untersuchungen am System NMMO/H_2O/Cellulose, PhD thesis, TU Berlin, Germany.
21. Walker M, Zimmermann R L, Whitcombe G P, Humbert H H (1997) N-Methylmorpholinoxid (NMMO) – Die Entwicklung eines Lösemittels zur industriellen Produktion von Zellulosefasern, Lenzinger Berichte pp 76–80.
22. Liebner F, Potthast A, Rosenau T, Haimer E, Wendland M (2007) Ultralight-Weight Cellulose Aerogels from NBnMO-Stabilized Lyocell Dopes. Res Lett Mater Sci Volume 2007, Article ID 73724.
23. Liebner F, Potthast A, Rosenau T, Haimer E, Wendland M (2008) Cellulose aerogels: Highly porous, ultra-lightweight materials. Holzforschung 62: 129–135.
24. Hoepfner S, Ratke L, Milow B (2008) Synthesis and characterization of nanofibrillar cellulose aerogels. Cellulose 15: 121–129.
25. IHV (2007) Industrieverband Hartschaum eV, Heidelberg, Germany, http://www.styropor.de
26. Gavillon R, Budtova T (2008) Aerocellulose: New Highly Porous Cellulose Prepared from Cellulose-NaOH Aqeous Solutions. Biomacromolecules 9: 269–277.
27. Pinnow M, Fink H-P, Fanter C, Kunze J (2008) Characterization of Highly Porous Materials from Cellulose Carbamate. Macromol Symp 262: 129–139.

28. Cai J, Kimura S, Wada M, Kuga S, Zhang L (2008) Cellulose Aerogels from Aqueous Alkali Hydroxide–Urea Solution, ChemSusChem 1: 149 – 154.
29. Reichenauer G, Scherer G W (2001) Extracting the pore size distribution of compliant materials from nitrogen adsorption, Colloids and Surfaces A, 187–188: 41–50.
30. Deng M, Zhou Q, Du A, Van Kasteren J, Wang Y (2009) Preparation of nanoporous cellulose foams from cellulose-ionic liquid solutions. Materials Lett 63: 1851–1854.
31. Rogowin Z A (1982) Chemiefasern, Thieme Verlag, Stuttgart, Germany.
32. Aaltonen O, Jauhiainen O (2009) The preparation of lignocellulosic aerogels from ionic liquid solutions. Carbohydrate Polymers 75: 125–129.
33. Liebner F, Haimer E, Loidl D, Tschegg S, Neouze M-A, Rosenau T, Wendland M (2009) Cellulosic aerogels as ultra-lightweight materials. Part 2: synthesis and properties. Holzforschung 63: 3–11.
34. Schmenk B, Ratke L, Gries T (2008) Solution spinning process for porous cellulose aerogel filaments, In: Dörfel A (Ed.): Proceedings of the 2nd Aachen-Dresden International Textile Conference, Dresden, December 04-05, 2008. Dresden: Institute of Textile and Clothing Technology, TU Dresden, 2008.
35. Hacker H, Gries T, Popescu C, Ratke L (2009) Solution spinning process for highly porous, nanostructured cellulose fibers. Chemical Fibers Int 59: 85 – 87.

10

Cellulosic and Polyurethane Aerogels

Arnaud Rigacci and Patrick Achard

Abstract This chapter focuses on isocyanurate and cellulose-based aerogels. First, it presents the global sol–gel synthetic path by *polycondensation*. Then, it summarizes all the main results on these two families of organic aerogels. Finally, some of the recent advancements concerning their use for hybridization of silica aerogels are shortly presented. Through a brief description of the basics, together with a short overview of the main properties, this article highlights the huge potential of those two classes of urethane-based aerogels.

10.1. Introduction

First of all, by a rapid inspection of the table of contents of this Handbook, one may wonder why the cellulosic part of this chapter has not been integrated in the preceding article dedicated exclusively to cellulose aerogels (Chap. 9). The main reason is that cellulose aerogels described in detail by Prof. L. Ratke concern materials coming from drying physical gels (i.e., gels obtained by regeneration of cellulose) and are consequently rather different from the ones presented in this contribution, which are derived by drying of *chemical gels*, also referred to as covalent gels in the literature, synthesized from cellulose derivatives.

Furthermore, the reader may also wonder why biomass-based polymers, like cellulosic aerogels, are gathered together in a common contribution with materials directly derived from petrochemistry, such as polyurethane aerogels. In fact, those two aerogel families present intimate sol–gel similarities: they are both based on *polycondensation* of polyols with polyisocyanates leading to the formation of urethane groups –O–CO–NH– (Figure 10.1) that in turn produce a tridimensional polyurethane network. In fact, it can be claimed that those two systems are chemically so close to one another that some cellulosic macromolecules have already been used for the synthesis of new polyurethane foams with increased biodegradability of the final material [1].

A. Rigacci (✉) **and P. Achard** • Center for Energy and Processes, MINES ParisTech, 1 Rue Claude Daunesse, B.P. 207, 06 904 Sophia Antipolis Cedex, France
e-mail: arnaud.rigacci@mines-paristech.fr; patrick.achard@mines-paristech.fr

M.A. Aegerter et al. (eds.), *Aerogels Handbook*, Advances in Sol-Gel Derived Materials and Technologies, DOI 10.1007/978-1-4419-7589-8_10,

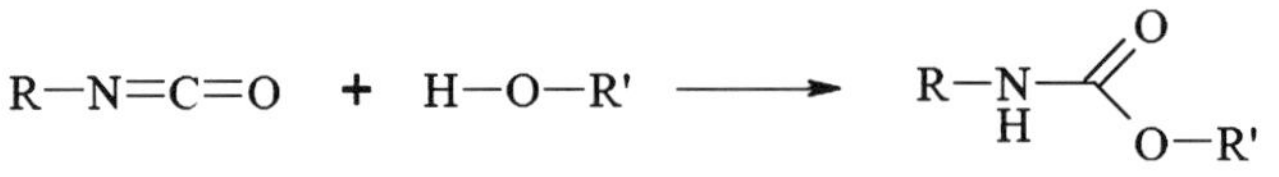

$$R{-}N{=}C{=}O \; + \; H{-}O{-}R' \longrightarrow R{-}\underset{H}{N}{-}C({=}O){-}O{-}R'$$

Figure 10.1. Scheme of urethane bonding by condensation of an alcohol with an isocyanate.

Even though polyurethane aerogels are better known than cellulose-derived aerogels, they have been developed more recently. Indeed, the first cellulosic aerogels have been synthesized by Kistler in the 1930s with nitrocellulose [2] while the first isocyanate and polyurethane-based aerogels were reported in the 1990s by Tabor [3] and by Biesman, Perrut and their co-workers [4]. Since then very few additional papers have been published on these organic aerogels, but the ever-increasing interest in biomass-based aerogels (for example, aerogels synthesized from alginate [5], starch [6, 7], peptide [8], chitin [9], *chitosan* [10], etc.) has led to an increasing interest in cellulosic aerogels as well [11–14].

Polyurethane and, to a lesser extent, cellulose-based aerogels, were initially developed for applications in thermal insulation [15, 16]. However, they have been studied for other applications as well [17]. For example, they have been considered as precursors of nanostructured carbons [18, 19] but, contrary to cellulose-based carbon aerogels (Figure 10.2) that appear promising [20, 21], no clear evidence of superior performance exists relative to resorcinol–formaldehyde-based carbon aerogels, mainly because of foaming during

Figure 10.2. Carbon aerogels obtained from pyrolysis of cellulose aerogels. **Top:** optical photography (courtesy of T. Budtova, MINES ParisTech/CEMEF, Sophia Antipolis, France). **Bottom:** *SEM* (courtesy of M. Pinnow, Fraunhofer-IAP, Natural polymer division, Golm, Germany).

Figure 10.3. Carbon aerogels obtained from pyrolysis of cellulose acetate-based aerogels. **Top:** optical photography. **Bottom:** optical microscopy [22].

pyrolysis (Figure 10.3). This is the main reason why the large majority of research on those nanostructured materials is currently focused on the organic aerogels rather than the carbons produced after pyrolysis. Based on this, this chapter is focused exclusively on pristine organic aerogels (that is before any pyrolytic treatment or conversion to carbon).

10.2. Polyurethane Aerogels

10.2.1. Synthesis

In analogy to their *resorcinol–formaldehyde* homologues, polyurethane wet gels are synthesized by *polycondensation*. Subsequently, aerogels are commonly obtained by *supercritical fluid* drying of the organic wet gels [23].

Chemical reactions involved in the sol–gel process are similar to those that lead to standard polyurethanes [24, 25]. Precursors are polyols (sometimes natural polyols like pentaerythritol or saccharose) and cross-linking is induced by using polyfunctional isocyanates (Figure 10.4). Condensation kinetics depends dramatically on the alcohol. Primary polyols react approximately ten times faster than polyols with secondary hydroxyl groups [24, 25].

Reactions take place in organic solvents (typically in dimethyl sulfoxide, *DMSO*) and are catalyzed with standard catalysts used for the synthesis of polyurethanes: metal salts and tertiary amines [26]. Polymerization mechanisms differ as a function of the catalyst and are briefly discussed below.

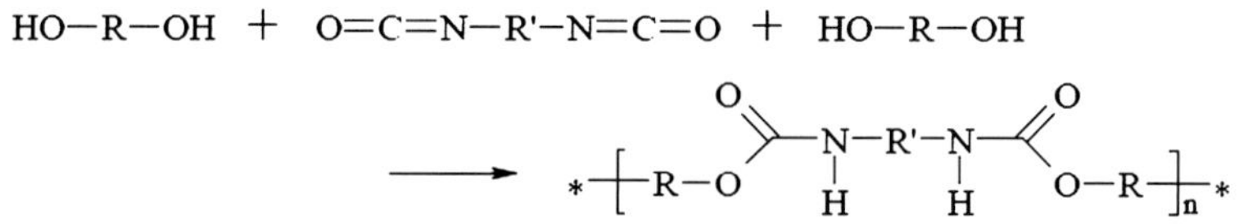

$$HO{-}R{-}OH + O{=}C{=}N{-}R'{-}N{=}C{=}O + HO{-}R{-}OH \longrightarrow *{-}\left[R{-}O{-}C(=O){-}N(H){-}R'{-}N(H){-}C(=O){-}O{-}R\right]_n{-}*$$

Figure 10.4. Urethane formation by linear (poly)condensation of diols with di-isocyanates.

Metal salts are *Lewis acids*, and induce formation of a ternary complex among the catalyst, the hydroxyl group of the polyol and the isocyanate. The presence of these complexes has been evidenced rather recently by *NMR* in the dibutyltin dilaurate catalyzed system [27] (dibutyltin dilaurate, $Sn(C_4H_9)_2(OCOC_{11}H_{23})_2$, is a classical polyurethane catalyst), but it had been suggested already from the 1960s after identification of catalyst/ hydroxyl and catalyst/isocyanate binary complexes [28–30]. The resulting condensation, which is likely to occur in the vicinity of the ternary complex, is consequently very sensitive to the steric hindrance between the three ligands.

Tertiary amines are *Lewis bases,* and are known to induce two different mechanistic pathways [31, 32]. The first route consists of the formation of a complex between the amine and the isocyanate, followed by attack by the alcohol [33]. The second mechanistic pathway is based on the formation of a complex between the amine and the hydroxyl group of the alcohol, followed by reaction with the isocyanate [34]. Whatever the mechanism might be, steric hindrance in the neighborhood of the nitrogen atom as well as the solvent basicity are the limiting factors [35]. In fact, the higher the pKa value and/or the accessibility of the lone pair of the amine nitrogen, the better the catalytic efficiency is. The latter explains the very good catalytic activity of diazobicyclo[2,2,2]octane (*DABCO*®, a classical polyurethane catalyst).

Finally, it is emphasized here that isocyanates are very reactive chemical species, which do not react only with alcohols. For example, under certain conditions, they can react with themselves at room temperature to generate dimers (even without catalyst) and trimers (Figure 10.5). Because of the electrophilic nature of the –N=C=O carbon, isocyanates react with most protic compounds too (Figure 10.6). Of course, in that regard, they can also react

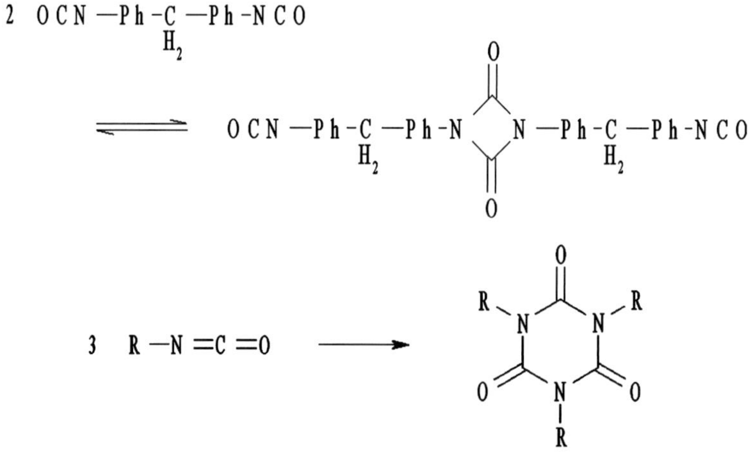

$$2\ OCN{-}Ph{-}CH_2{-}Ph{-}NCO \rightleftharpoons OCN{-}Ph{-}CH_2{-}Ph{-}N\underset{C(=O)}{\overset{C(=O)}{\diamond}}N{-}Ph{-}CH_2{-}Ph{-}NCO$$

$$3\ R{-}N{=}C{=}O \longrightarrow (RNCO)_3 \text{ (isocyanurate ring)}$$

Figure 10.5. Oligomerization of isocyanates: examples of dimerization of an aromatic di-isocyanate (**top**) and trimerization of an alliphatic isocyanate leading to an isocyanurate (**bottom**).

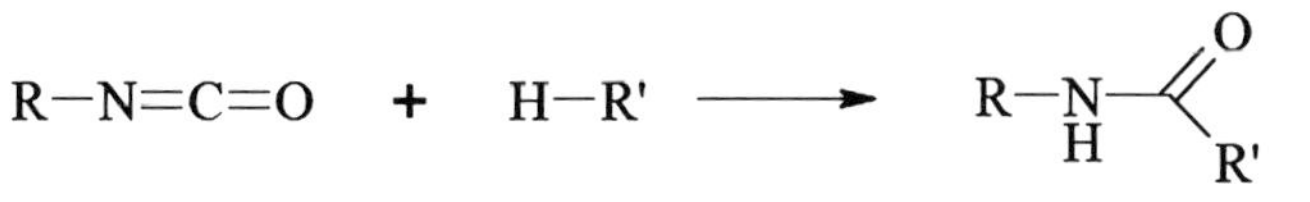

$$R{-}N{=}C{=}O \quad + \quad H{-}R' \longrightarrow R{-}NH{-}C(=O){-}R'$$

Figure 10.6. Reaction of an isocyanate with a generic protic molecule.

with water to give an instable species (carbamic acid), followed by loss of CO_2 and formation of a di-substituted urea, also called urein (Figure 10.7), which can subsequently react with another isocyanate to form a biuret (Figure 10.8). To complete this quick overview, it should be also noted that isocyanates can react with urethanes to form allophanates (Figure 10.9). All those reactions could be considered in competition with synthesis of urethanes, but they can also be controlled to our benefit, optimizing the final properties of the dry material.

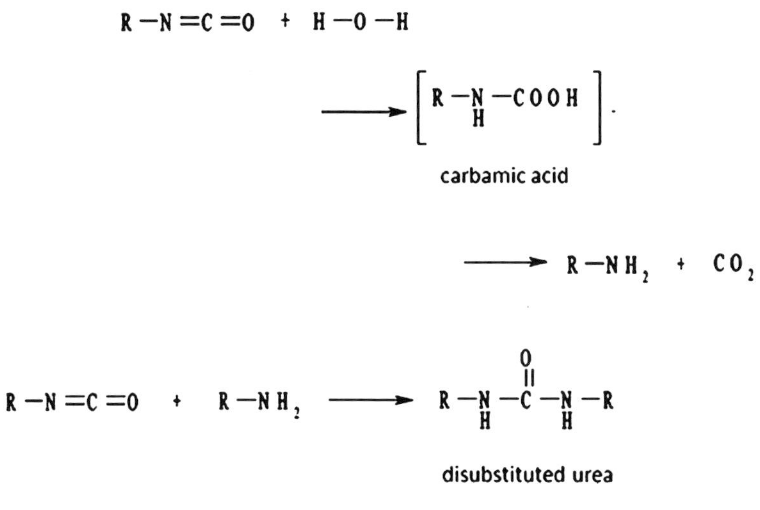

$$R{-}N{=}C{=}O \; + \; H{-}O{-}H \longrightarrow [R{-}NH{-}COOH] \longrightarrow R{-}NH_2 \; + \; CO_2$$

carbamic acid

$$R{-}N{=}C{=}O \; + \; R{-}NH_2 \longrightarrow R{-}NH{-}C(=O){-}NH{-}R$$

disubstituted urea

Figure 10.7. Reaction of an isocyanate with water.

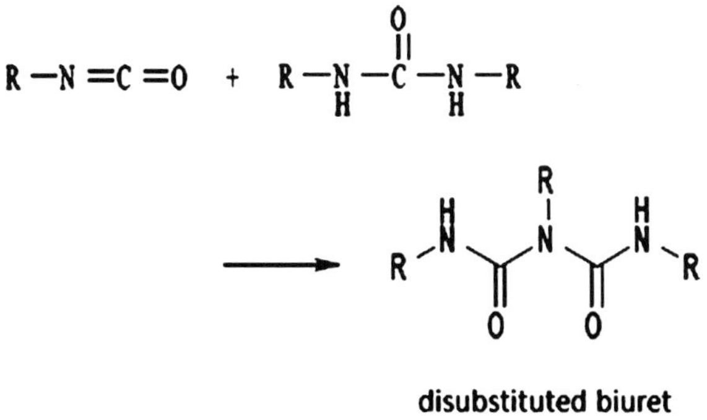

$$R{-}N{=}C{=}O \; + \; R{-}NH{-}C(=O){-}NH{-}R \longrightarrow R{-}NH{-}C(=O){-}N(R){-}C(=O){-}NH{-}R$$

disubstituted biuret

Figure 10.8. Reaction of an isocyanate with an urein leading to the formation of di-substituted biuret.

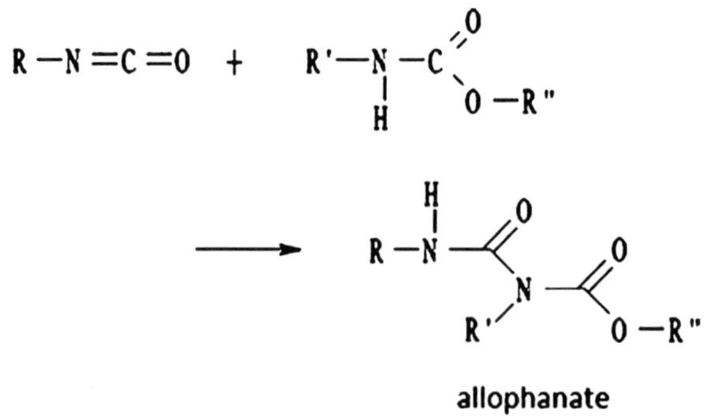

allophanate

Figure 10.9. Reaction of an isocyanate with an urethane function leading to an allophanate.

10.2.2. Process and Materials

One of the first efforts on sol–gel isocyanate-based materials was reported in the mid 1990s by Dow Chemical Co and focused on polyurea xerogels [3], as core materials for superinsulating Vacuum Insulation Panels (*VIP*) (Chap. 26). Those gels were synthesized in acetone with different polyisocyanates (polymeric diphenylmethane diisocyanates, crude toluene diisocyanate, etc.), active-hydrogen compounds (aliphatic or aromatic polyamines such as diethyl toluene diamine, diethanol amine, etc.) and a polymerization catalyst. For some of the materials, water was also included as hardener. Wet gels were simply dried by evaporation under ambient conditions. The densities of such ambient-dried aerogels and xerogels varied over a rather wide range, between 0.15 and 0.50 g/cm^3. Thermal conductivities were low under primary vacuum (0.007 W/m K at 0.2 mbar), but rather high close to the atmospheric pressure (0.033 W/m K at 0.8 bar). It should be noted that no specific information about the nanostructure of those ambient-dried materials was included in that study.

At the end of the 1990s, Biesmans, Perrut and co-workers reported on the preparation of gels synthesized in dichloromethane with an aromatic polymeric isocyanate (Suprasec DNR, a trademark of ICI Polyurethanes) and *DABCO*® as catalyst [18]. In subsequent papers [4, 15], those authors studied the effect of the gelation temperature (from room temperature to 120°C), cure time (from minutes to several days), concentration of solids in the sol and catalyst ratio on the aerogel properties (Figure 10.10). Cross-linking of the polyisocyanate through isocyanurate ring formation (Figure 10.5) or allophanate coupling at higher temperatures (Figure 10.9) was suggested to occur. All gels were dried supercritically with CO_2, according to the process of Tewari et al. [36]: they were first washed with liquid CO_2, then CO_2 was heated above its critical point and it was vented off isothermally. The aerogels obtained were lightweight and nanostructured. The density range was narrower than that reported by Tabor [3], with values typically in the range of 0.10 and 0.25 g/cm^3. The specific area and the average pore size, coming from *BET* and *BJH* analyses of N_2 sorption isotherms, respectively, were centered on 570 $\pm$ 30 m^2/g and 14 $\pm$ 3 nm. It should be noted that, for the very first time, low thermal conductivities were obtained under ambient conditions. Values as low as 0.016 $\pm$ 0.001 W/m K were measured with a heat flux meter-based method (Figure 10.11).

Patents on similar materials were published by ICI [37, 38]. The spirit of the related sol–gel synthesis was slightly different from the previous method by Perrut and

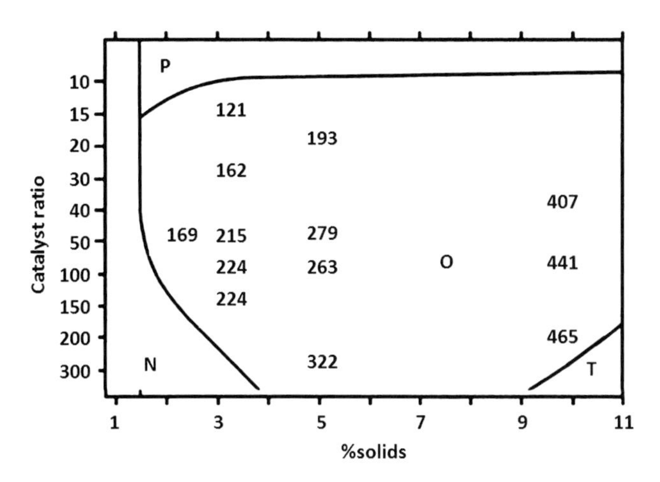

Figure 10.10. Influence of catalyst ratio (defined as the ratio between weights of isocyanate and catalyst) and concentration of the sol (% solids) on polyurethane aerogels appearance (P: powder, O: opaque, T: transparent and N: no gelation) and density (in g/cm^3) [4].

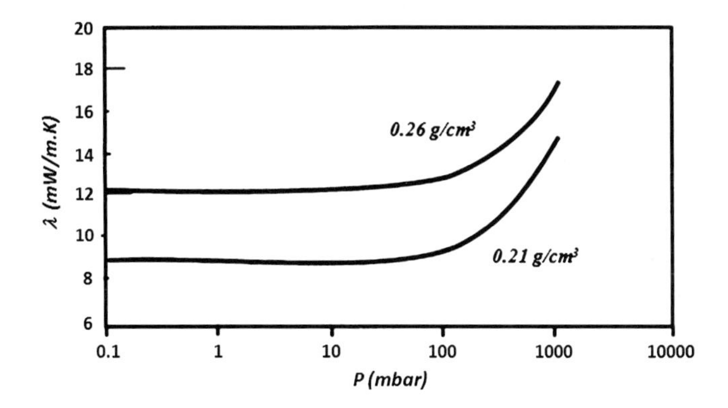

Figure 10.11. Thermal conductivity of two monolithic polyurethane aerogel plates as a function of partial vacuum of air at room temperature [4].

collaborators at SEPAREX and ICI in that the sol incorporated a copolymer containing at least one OH-based isocyanate-reactive group. The copolymers tested were taken from the families of phenolic and polyaldehyde-ketone resins, polyketones, novolaks and resols. Supercritical drying yielded aerogels with densities between 0.05 and 0.15 g/cm^3 and a minimum thermal conductivity of 0.018 W/m K at room conditions [37]. The ambient pressure-dried homologues had dramatically lower *BET* surface areas and larger thermal conductivities (a maximum of 50 m^2/g and a minimum of 29 mW/m K, respectively) but remained rather light (from 0.1 to 0.3 g/cm^3) [38].

The developments by Biesmans, Perrut and their co-workers became the starting point for further studies conducted at MINES ParisTech and CSTB. Monomeric polyols (pentaerythritol or saccharose) were cross-linked with 4,4-diphenylmethane diisocyanate in a mixture of *DMSO* and ethyl acetate (*etac*), still using *DABCO*® as catalyst [39]. Gels were washed with etac and dried according to a direct supercritical CO_2 process (i.e., washed directly with supercritical CO_2) [40]. The resulting monolithic aerogels (Figure 10.12) were

Figure 10.12. Photograph of a polyurethane aerogel obtained by supercritical CO_2 drying of a gel synthesized by cross-linking of pentaerythritol with 4,4-diphenylmethane diisocyanate under *DABCO*® catalysis.

not as insulating as the ones from ICI and SEPAREX since their thermal conductivity, measured by the hot-wire technique in ambient conditions, was equal to 0.022 W/m K. Nevertheless, those results suggested further structural characterization. *SEM* studies showed that those aerogels consisted of spherical particles (Figure 10.13) with densities, measured by *helium pycnometry*, ranging from 1.2 to 1.4 g/cm^3 [41]. Furthermore, both particle sizes (Figure 10.14) and pore diameters (Figure 10.15) seem to depend strongly on the solvent system used for gelation [39]. Some complementary ongoing work (not published yet) shows that by paying more particular attention to the reaction media (e.g., *DMSO/* acetone solution instead of *DMSO/etac*) one could improve further the thermal conductivity of those resulting aerogels further (0.020 W/m K at ambient conditions as calculated from the product of density, thermal diffusivity and specific heat). Of course, when those finely mesostructured polyurethane gels are dried through simple evaporation of the entrapped solvent (for example, acetone), the tenuous mesostructure created by the sol–gel process is lost, because of collapse of the porous network through a very pronounced drying densification (Figure 10.16).

Since that time, very few recent reports on this type of isocyanate-based aerogels can be found in the literature. The most recent ones have been published by ASPEN in 2009 [42], and concern both polyurethane and polyurea aerogels. The sol–gel synthesis is performed directly in acetone with standard isocyanates and triethylamine (*TEA*) as catalyst used at a relatively high weight percent in solution (i.e., between 5 and 10 wt%). Gels are typically dried with supercritical CO_2. The article is mainly focused on the impact of the ratio between the amine hydrogen equivalent weight and the isocyanate equivalent weight (*EW ratio*) [42] on various properties of polyurea aerogels like gelation time, drying shrinkage, final density, porosity and thermal conductivity. A *SEM* image of a representative sample is shown in

Figure 10.13. Field emission gun scanning electron microscopy (*SEM-FEG*) of sample presented at Figure 10.12 (microscopy conducted by F. Charlot at CMTC INPG, Grenoble, France).

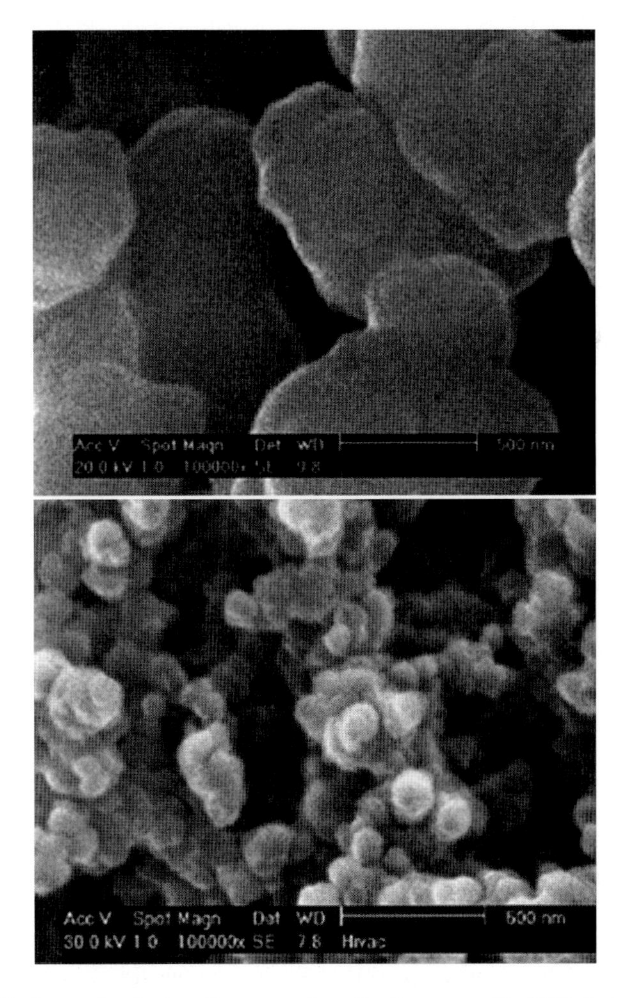

Figure 10.14. *SEM* micrographs of polyurethane aerogels coming from supercritical CO_2 drying of gels synthesized with saccharose in different solutions of *DMSO* and *etac*; with volume fraction of *DMSO* <0.3 (**top**) and >0.3 (**bottom**) (microscopy conducted by M. Repoux at MINES ParisTech/CEMEF, Sophia Antipolis, France) [39].

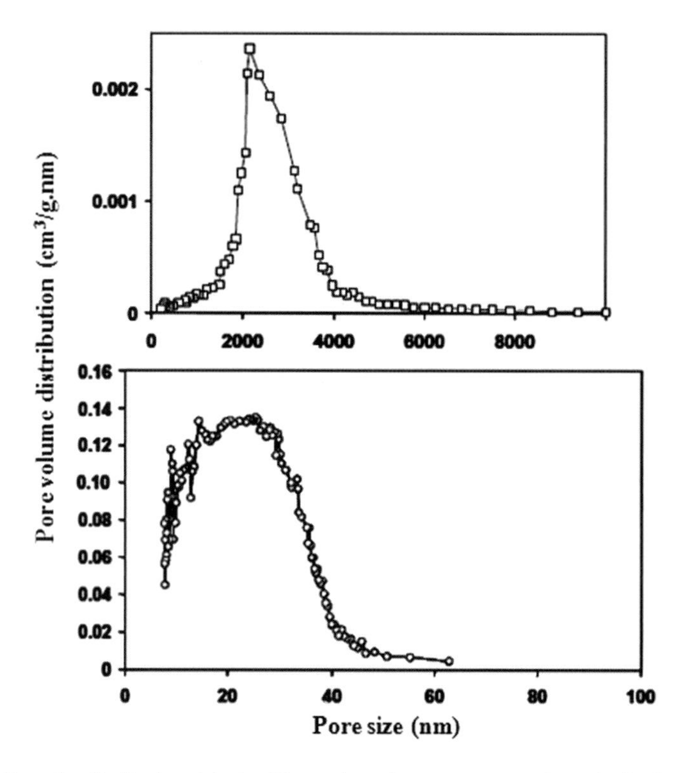

Figure 10.15. Pore size distributions (obtained by nonintrusive mercury porosimetry) of polyurethane aerogels coming from supercritical CO_2 drying of gels synthesized with pentaerythritol in different solutions of *DMSO* and *etac*; volume fraction of *DMSO* <0.3 (**top**) and >0.3 (**bottom**) (*mercury porosimetry* conducted by R. Pirard, LGC, Univervity of Liège, Belgium) [39].

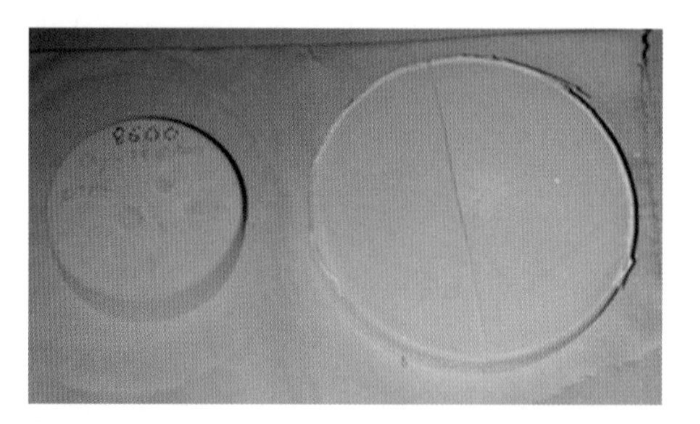

Figure 10.16. Illustration of the densification occurring during evaporative drying of polyurethane gels: xerogel (**left**) versus aerogel (**right**). Both samples come from gels synthesized with pentaerythritol in solution of *DMSO* and *etac* with a volume fraction of *DMSO* >0.3. Diameters of the xerogel and the aerogel are 3 and 5 cm, respectively.

Figure 10.17. Remarkably, low values of thermal conductivity have been obtained with these materials (e.g., as low as 0.014 W/m K as calculated from the product of density, thermal diffusivity and specific heat), suggesting them as very promising candidates for thermal superinsulation at atmospheric pressure.

Figure 10.17. *SEM* microscopy of a polyurea-based aerogel prepared with a target density of 0.1 [42].

10.2.3. Hybrids and Composites

Part of these developments on polyurethane aerogels has already been incorporated in silica-based composites and hybrids. Those studies were initiated in order to improve the mechanical properties of silica aerogels while keeping at a reasonable level some of their other desirable properties, as for example their very low thermal conductivity. Most of the related studies have been performed by the group of N. Leventis at the Missouri University of Science and Technology, USA. Initially, they used the hydroxyl functionality of the surface ≡Si–OH groups of silica to react with the NCO groups of the isocyanate. This cross-linking process is enhanced by gelation under remaining water adsorbed on silica, yielding polyurea interparticle tethers. The process results to significantly higher mechanical properties but contributes also to the density increase of the final product compared to the pristine (native) silica aerogel [43, 44]. Using organically modified silica precursors, for example with 3-aminopropyltriethoxysilane (*APTES*), this method has been extended to prefunctionalized silica gels (for example with amine-modified pore surface) [45]. Here again, the strength increase is remarkable: some cross-linked silica aerogels exhibit a 300-fold increase in strength as compared to the native silica aerogels for a nominal increase in density by a factor of 3. Some of those results are described in this Handbook in detail later by Leventis and Lu (Chap. 13).

Several other reports have been published on alternate implementation of polyurethane/silica hybrid aerogels. They are based on (a) "one-pot" gelation (i.e., not by cross-linking of preformed silica gels as in the work by Leventis, but by directly mixing silica precursors such as *TMOS* with isocyanates in adequate cogelation conditions [46]) and (b) use of "already dried" nanostructured silicas (e.g., by dispersing pyrogenic silicas and functionalized silanes as coupling agents, in isocyanate solutions [47]).

In a similar spirit, but without any consideration about covalent cross-linking, dry nanostructured silicas have been also used for the synthesis of some silica/polyurethane composites [48]. Silica nanopowders (e.g., finely divided silicas with characteristic dimensions between 10 nm and 100 μm) have been dispersed in various polyurethane sols. After gelation in the presence of these inorganic fillers, the polyurethane-based composites can be considered as superinsulating materials with thermal conductivities lower than 0.020 W/m K.

10.3. Cellulose Derivatives Aerogels

10.3.1. Synthesis

The large majority of studies dedicated to cellulosic aerogels are concentrated on pure cellulose materials (Figure 10.18). Compared to cellulose derivatives-based aerogels, they have the advantage of coming directly from vegetal biomass without need for any synthetic modification. Jin et al. were the first to use nonderivatized cellulose in 2004 to produce very porous film-like materials by dissolution and regeneration of cellulose in aqueous solution of calcium thiocyanate followed by regeneration and carefully controlled *freeze-drying* procedures [11]. Then, various routes were developed, among others, with cellulose-*NMMO* (*N*-methylmorpholine *N*-oxide solutions) [12], cellulose-NaOH/urea solutions using epichlorohydrine as cross-linker [49], aqueous cellulose-NaOH solutions (Figure 10.19) [14, 50], aqueous cellulose carbamate-NaOH solutions [51] or with cellulose dissolved in hydrated calcium thiocyanate melts [52] and even with aquagels of cellulose nanofibers [53]. Very recently, a brand new class of cellulosic aerogels based on the all-composite concept [54] has been prepared by partial dissolution of microcrystalline cellulose in LiCl/*DMAc* (*N,N*-dimethylacetamide) solutions followed by precipitation in water before *freeze-drying* [55]. The final cellulosic materials are made of thin crystals coexisting together with cellulose fibrils and exhibit very good mechanical properties (particularly, specific flexural strength and stiffness). The overview of the various existing aerogels coming from drying of physical cellulose gels can be found in the dedicated contribution by L. Ratke (Chap. 9) but it is underlined here that all those very light biomass-based aerogels are macroporous (Figure 10.19).

Figure 10.18. Photographs of aerocelluloses prepared from caustic soda solutions of cellulose (cellulose 7% w/w in aqueous NaOH 8% w/w) regenerated in water at 25°C, washed with acetone then dried with supercritical CO_2. Heights of the samples are, respectively, 3 and 9 cm (courtesy of T. Budtova, MINES ParisTech/CEMEF, Sophia Antipolis, France).

For some specific applications (for example, thermal insulation), making mesoporous cellulosic aerogels is of high interest. One of the approaches is cross-linking of cellulosic species. Because cross-linking of cellulose is necessary to be performed in heterogeneous media (mainly with formaldehyde [56], carboxylic acid [57] or epichlorohydrin [58]), one

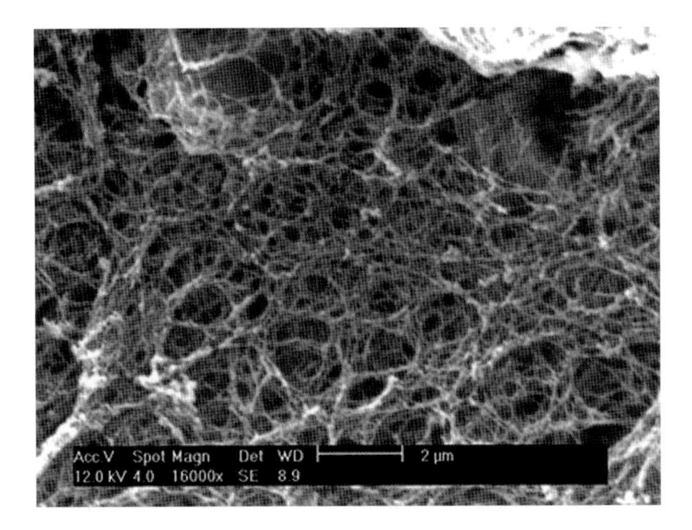

Figure 10.19. *SEM* microscopy of aerocellulose prepared from caustic soda solution of cellulose (cellulose 7% w/w in aqueous NaOH 8% w/w) regenerated in water at 25°C, washed with acetone then dried with supercritical CO_2 (courtesy of T. Budtova, MINES ParisTech/CEMEF, Sophia Antipolis, France).

must employ cellulose derivatives for this purpose. Those macromolecules are indeed soluble in water and/or organic solvents and consequently permit cross-linking in homogeneous environments. Even if some attempts of cross-linking of cellulose ethers (like carboxymethylcellulose and hydroxyethylcellulose) with divinylsulphone in sodium hydroxide solution have already been studied [59], studies until now have been exclusively focused on cross-linking of cellulose derivatives with isocyanates.

Most of the work in this area (i.e., on how to synthesize aerogels by drying cellulosic chemical gels prepared by reticulation of cellulose derivatives) was published by Tan and co-workers in 2001 [16]. The corresponding cellulosic aerogels have been prepared by cross-linking of two cellulose esters (cellulose acetate and cellulose acetate butyrate) with tolylene-2,4-diisocyanate (*TDI*) in acetone (Figure 10.20), washing the gels with acetone then with liquid CO_2 and drying under supercritical conditions.

Here, the crucial issue is identification and use of a solvent system compatible with the reactants, the gel and the extraction media (e.g., CO_2). Substituting the hydroxyl groups of cellulose with some other specific chemical groups is known to increase solubility in organic

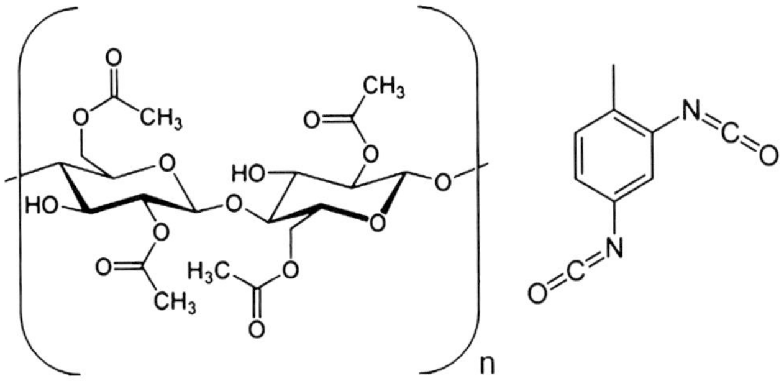

Figure 10.20. Chemical formula of cellulose acetate (**left**) and *TDI* (**right**). (Cellulose acetate butyrate formula is derived from the one of cellulose acetate by replacing the CH_3CO– groups with $CH_3CH_2CH_2CO$–.

solvents. For example, replacing hydroxyl by acetyl groups increases solubility in acetone, which is also a suitable solvent for isocyanates and polyurethanes while being miscible with CO_2. Furthermore, because of partial hydrolysis of the cellulose chains during acetylation, the number-average degree of polymerization (DP_n) of cellulose acetate is smaller than the initial one of cellulose, which is also a factor susceptible to enhance dissolution [60]. Nevertheless, acetone is a solvent of cellulose acetates only when the average acetyl degree of substitution (i.e., the average number of acetyl groups per elementary cellulosic unit) ranges between 2.2 and 2.7. Moreover, the remaining free hydroxyl groups lead to a rather poor affinity with acetone. Some aggregated zones (e.g., micelles) coexist with solvated chains (i.e., chains well-separated one from another). This behavior induces heterogeneities that are more pronounced as the concentration of cellulose acetate in the solution increases [61]. At this stage, one could expect that this solubility aspect would affect the homogeneity of the final material (i.e., the structure of the cellulosic aerogel).

Reactions involved in this sol–gel process are the same to those previously described for polyurethanes, except that a macromolecular polyol has replaced the monomeric one (Figure 10.4). The hydroxyl groups of the cellulose acetate react with the NCO groups of the isocyanates to form urethane linkages (Figure 10.21). Those reactions, which actually have been studied for a long time [62], yield to chemical species referred to as cellulose carbamates. Because of the rather large proportion of secondary alcohols in

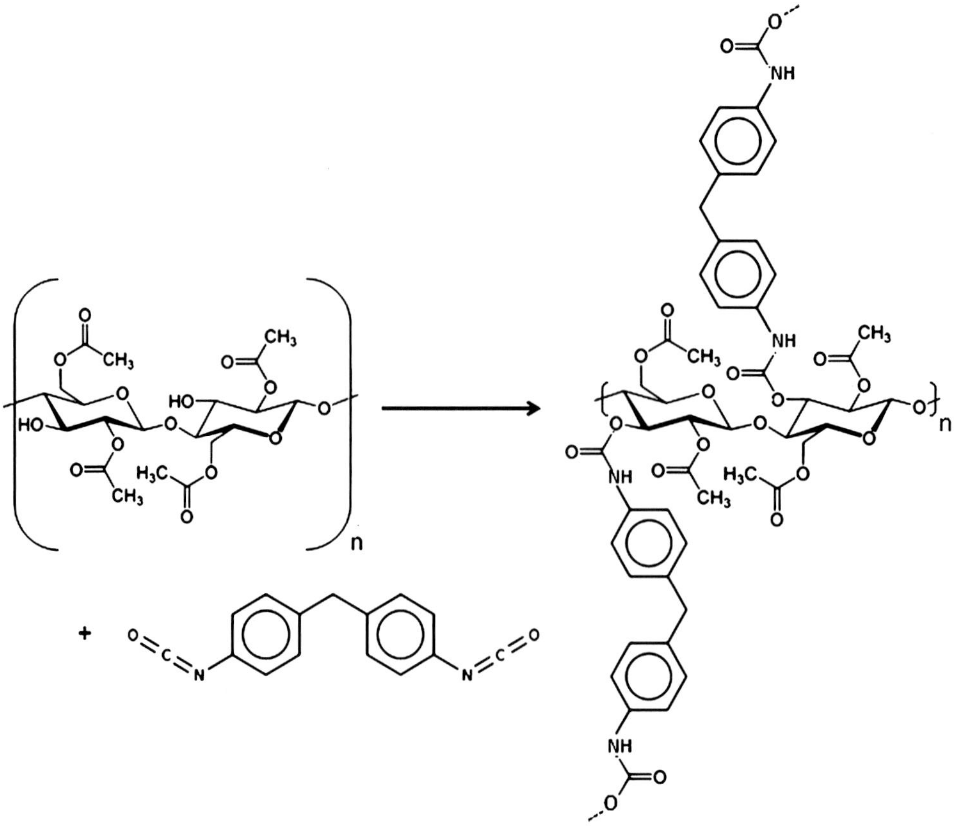

Figure 10.21. Urethane bonding by (poly)condensation of cellulose acetate with di-isocyanates [22].

cellulose diacetates (around 50% mol/mol [60]), the use of a catalyst here is absolutely necessary. In the specific case of reactions between isocyanates and cellulosic polyols, dibutyltin dilaurate is generally used for its higher catalytic activity [63].

10.3.2. Process and Materials

The sol–gel synthesis of Tan et al. [16] was performed in acetone at room temperature in the absence of water (reactants were dried in a vacuum oven and molecular sieves were used) and it was catalyzed with pyridine. Some basic parameters were varied: (a) the reactants concentration in the sol (ranging from 0.06 to 0.3 g/cm^3); (b) the cross-linking density (expressed as the molar ratio between the cellulosic repeat units and *TDI*, labeled as *C/T* and varying from 1 to 15, wherein $C/T = 2$ is the stoichiometric ratio); and (c) the drying method (evaporation versus supercritical CO_2 extraction). However, the catalysis ratio was kept constant (pyridine concentration maintained at 1% of the weight of reactants i.e., between 0.6 and 3 g/l). Kinetics was slow. Depending on the formulation of the sol, gelation time varied between 1 and 15 days. Gels dried by evaporation shrank dramatically (average shrinkage around 50% v/v), but recovered part of their initial volume by soaking in acetone. Gels dried by indirect supercritical CO_2 drying exhibited really lower but still significant shrinkage (global shrinkage between 15 and 90% in volume). Apparent densities of aerogels were between 0.10 and 0.40 g/cm^3. Even if no electron microscopy was presented, all of these materials seemed finely mesostructured as characterized by nitrogen sorption (*BET* surface areas ranged between 250 and 400 m^2/g and average *BJH* pore diameters between 12 and 18 nm), but no significant difference coming from the nature of the cellulosic precursor (acetate versus acetate butyrate) was observed. Furthermore, even if no specific structural correlation was noticeable, it appeared really clear that the apparent density increased both with the concentration of the sol and the degree of reticulation. What the authors stressed the most at that time was the huge mechanical improvement associated to these brand new aerogels compared to their silica and resorcinol–formaldehyde homologues.

Those results were considered promising and further studies were performed at MINES ParisTech [22]. First of all, the original synthetic path of Tan et al. was slightly modified: (a) because of its toxicity, *TDI* was replaced by polymethylene polyphenylene polyisocyanates (*P-MDI*) (Figure 10.22) and (b) for catalytic reasons, pyridine was substituted by dibutyltin dilaurate (however concentrations were kept rather close to the original

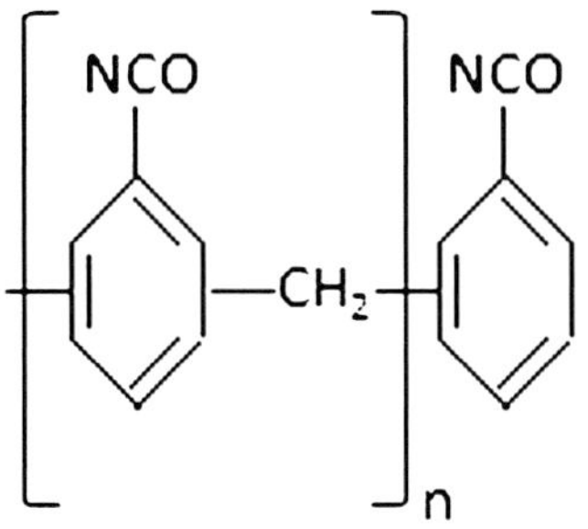

NCO NCO —CH$_2$— n

Figure 10.22. *P-MDI* [24] ($n = 1.8$ for the product used in these studies e.g., Lupranat® M20S, a polyfunctional isocyanate from BASF).

ones e.g., 6.5 g/l ± 1.5). The concentration of reactants in the solution was kept rather similar too (between 0.04 and 0.15 g/cm^3) as well as the cross-linking density (*C*/*T* varying from 1 to 10 with *C*/*T* ~2.8 being the stoichiometric ratio).

The first published results by Fischer and co-workers [13] confirmed general results obtained by Tan et al.: (a) the gels totally collapsed when dried by simple evaporation (Figure 10.23); (b) they also shrink significantly during supercritical CO_2 drying (Figure 10.24), sometimes up to 90% in volume; but, (c) the drying shrinkage is reversible as observed by soaking the aerogels in acetone (Figure 10.25). A screening routine on the solubility of the gel coupled with a simulation of the CO_2–acetone equilibrium phase at 40°C revealed that interactions between gels and their surrounding media [64] (even in supercritical form) are of the upmost importance: those gels shrink so much, because of their very poor affinity with CO_2.

Figure 10.23. Illustration of the shrinkage of cellulose acetate-based gels occurring during drying by evaporation. A wet gel (**left**) and the corresponding xerogel (**right**). The initial diameter of the wet gel is 3 cm [22].

Figure 10.24. Illustration of the shrinkage of cellulose acetate-based gels occurring during supercritical drying. A wet gel (**left**) and the corresponding aerogel (**right**). The *white bar* is 2 cm-long [22].

That paper also showed that the kinetics were accelerated significantly by using the tin-based organometallic catalyst (gelation times stand for some minutes to some hours instead of some days) and, above all, permitted to know more about the internal structure of this type of aerogels. They appeared to consist of spherical dense particles with densities, measured by *helium pycnometry,* close to 1.45 g/cm^3 forming (a) a fibrillar solid network,

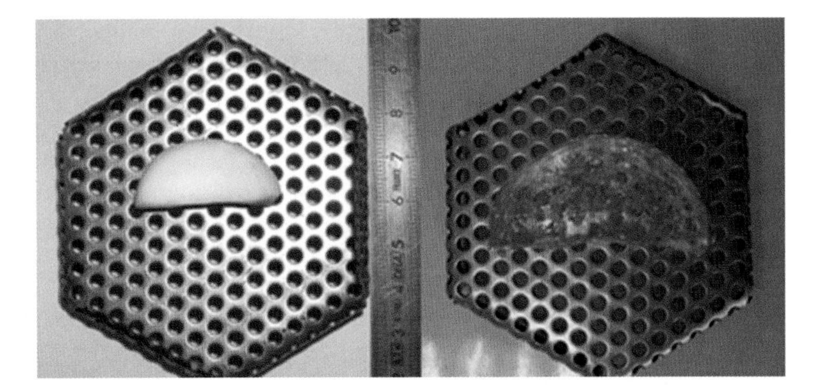

Figure 10.25. Illustration of the reversibility of the supercritical drying shrinkage of cellulose acetate-based aerogels by comparison of an aerogel as-extracted from the autoclave (**left**) and after re-impregnation with acetone (**right**). The diameters before and after re-impregnation are 3.5 and 5.5 cm, respectively [22].

comparable to a pearl-necklace structure (Figure 10.26) and (b) a mesoporous network, as suggested by the shape of the N_2 sorption isotherms (Figure 10.27) and confirmed by nonintrusive *mercury porosimetry* [65] (Figure 10.28). Fischer also properly confirmed that the cross-linking density had a drastic influence on the structure of the materials [22]. Even if it has no significant influence on the specific surface (Table 10.1), decreasing C/T from 10 to 1 induces a dramatic increase of the density and shifts the pore size distribution to smaller mesopores (Figure 10.28).

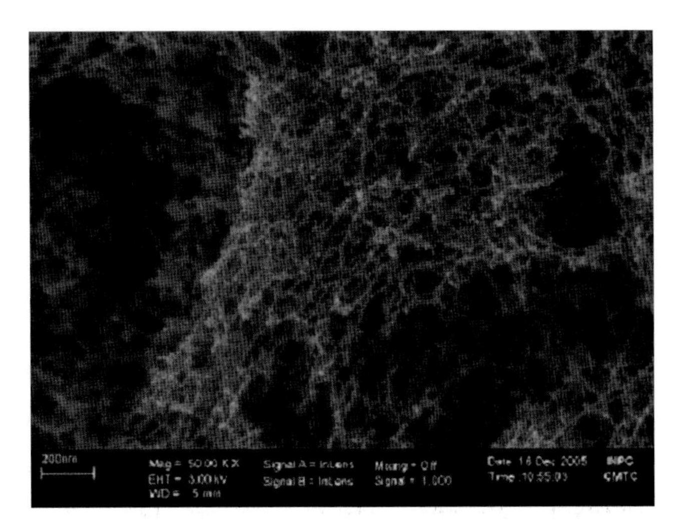

Figure 10.26. *SEM-FEG* microscopy of cellulose acetate-based aerogels (micrograph realized by F. Charlot, CMTC INPG, Grenoble, France) [66].

Particular attention has also been paid to aerogels synthesized with a high cross-linking ratio (i.e., a T/C ratio greater than the theoretical stoichiometric value or, in other words, $C/T < 2.8$). Contrary to other samples, those aerogels present rather "constant" textural properties whatever the sol–gel parameters may be (cellulose acetate concentration, DP_n, etc.). They are dense (>0.7 g/cm^3) and possess a narrow (meso)pore size distribution

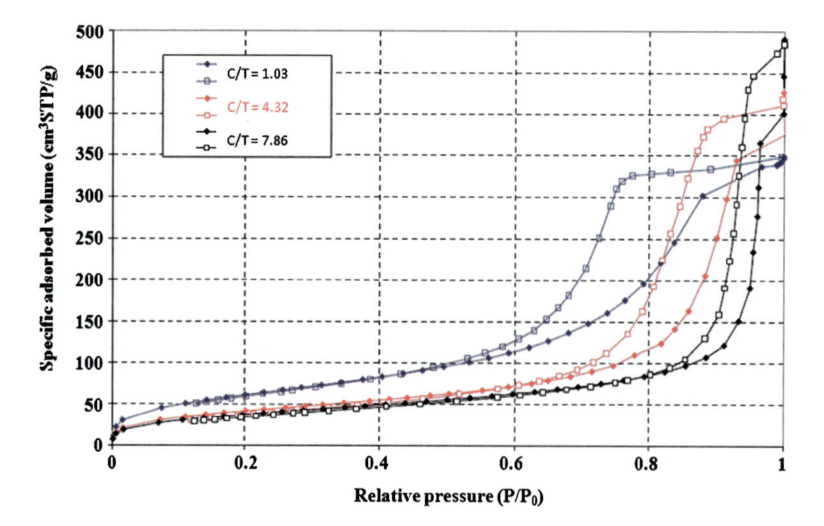

Figure 10.27. N_2 sorption isotherms of cellulose acetate-based aerogels coming from supercritical CO_2 drying of gels synthesized in extra-dry acetone by cross-linking of cellulose acetate (with a DP_n of 190) with *P-MDI* under dibutyltin dilaurate catalysis (at 8 g/l). Cellulose acetate concentration in solution is maintained constant (~80 g/l) while the cross-linking ratio (*C/T*) is varied [22].

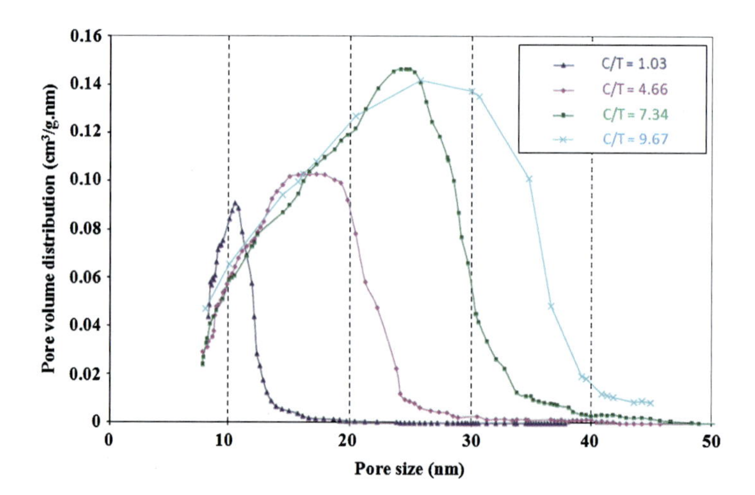

Figure 10.28. Influence of the cross-linking ratio *C/T* on pore size distribution of cellulose acetate-based aerogels determined by nonintrusive *mercury porosimetry* (using a bucking constant k_f equal to 28 nm MPa$^{1/4}$ [41]). Characteristics of the aerogels are summarized in Table 10.1 (*mercury porosimetry* conducted by R. Pirard, LGC, Univervity of Liège, Belgium) [22].

(Figure 10.29). UV–VIS spectroscopy of the acetone wash baths before drying plus elemental analysis and FT-IR characterization of the corresponding aerogels have shown that a secondary cross-linking network – based on allophanate bridging – was created when isocyanate is used in large excess.

From a more physical point of view, it was also shown that: (a) those materials were less hydrophilic than cellulose aerogels (so-called aerocellulose) and pristine cellulose

Table 10.1. Influence of the cross-linking ratio *C/T* on internal structure of cellulose acetate-based aerogels coming from supercritical CO_2 drying of gels synthesized by cross-linking of cellulose acetate with *P-MDI* under dibutyltin dilaurate catalysis [22][a]

C/T	ρ_b (g/cm^3)	ε [b]	V_{sp} [b] (cm^3/g)	V_{Hg} [c] (cm^3/g)	% V_{Hg} [d]	S_{BET} (m^2/g)	d[e] (nm)
1.03	0.85	0.40	0.46	0.39	85%	212	20
4.66	0.48	0.65	1.38	1.31	95%	252	17
7.34	0.28	0.80	2.86	2.47	86%	284	15
9.67	0.24	0.83	3.37	3.39	~100%	246	17

[a] Gels were synthesized in extra-dry acetone, with cellulose acetate with a DP_n of 115. Catalyst concentration was 5 g/l while the cellulose acetate concentration in solution was maintained constant (~40 g/l).

[b] Porosity (ε) and specific pore volume (V_{sp}) were calculated with ρ_b and ρ_s (taken equal to 1.45 g/cm^3).

[c] V_{Hg} is the partial specific pore volume measured by nonintrusive *mercury porosimetry*.

[d] The fraction of porosity characterized by nonintrusive *mercury porosimetry* (%V_{Hg}) was calculated like the ratio between V_{sp} and V_{Hg}.

[e] The mean particle size (*d*) was calculated from S_{BET} and ρ_s assuming a spherical shape and a monomodal particles distribution.

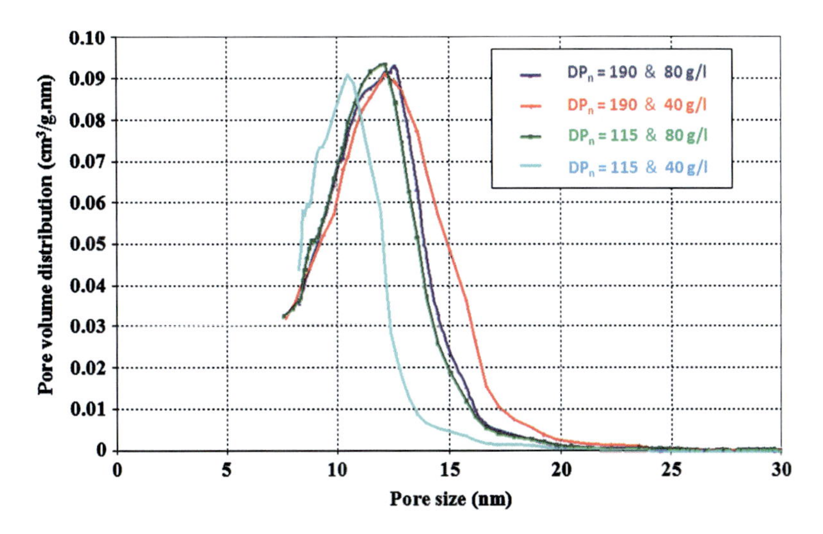

Figure 10.29. Pore size distribution of cellulose acetate-based aerogels determined by nonintrusive *mercury porosimetry* (using a bucking constant k_f equal to 28 nm MPa$^{1/4}$ [41]). Cellulose acetate-based aerogels are coming from supercritical CO_2 drying of gels synthesized in extra-dry acetone by cross-linking of cellulose acetate with *P-MDI* under dibutyltin dilaurate catalysis. Cross-linking ratio is maintained constant (*C/T* = 1.03) while cellulose acetate concentration (40 or 80 g/l), DP_n (115 or 190) and catalyst concentration (5 or 8 g/l) are varying (*mercury porosimetry* conducted by R. Pirard, LGC, Univervity of Liège, Belgium) [22].

acetate (Figure 10.30) and (b) they have a rather low thermal conductivity in ambient conditions (around 0.040 W/m K at their monolithic stage and slightly lower than 0.030 W/m K as granular bed). The recent extension to other polyurethane catalysts (*DABCO*®) has permitted to decrease the thermal conductivity of the monoliths below 0.030 W/m K.

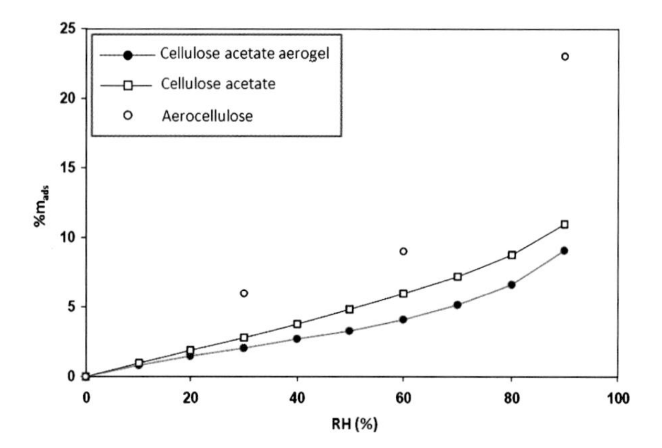

Figure 10.30. Water adsorption isotherms (one cycle) of a cellulose acetate-based aerogel, cellulose acetate and a new hyperporous cellulose so-called aerocellulose [14] at 25°C (with courtesy of Manfred Pinnow, Fraunhofer-IAP, Natural Polymer Division, Golm, Germany) [22].

10.3.3. Hybrids and Composites

The similarity between cellulose acetate and polyurethane-based aerogels lets reasonably someone to think that sol–gel coupling with silica is conceivable. For the moment, no literature on this matter can be found with cellulose derivatives. Indeed, the only studies that have been initiated in this direction have been conducted with cellulose and lignin. They are not, properly speaking, dealing with aerogels but are demonstrating the feasibility of hybridization between the relevant chemical species. Very recently, some cellulose/silica hybrids have been prepared by a sol–gel method using pulp fibers as the cellulose source and tetraethoxysilane (*TEOS*) as the silica precursor, in the presence of heteropoly acids [67]. The resulting hybrid materials exhibit covalent silica/cellulose bounded together with physical interactions [68]. A similar cross-linking approach with lignin can also be found in the literature [69]. Based on previous studies performed with *TEOS* showing that a lignin matrix can be cross-linked with silica alkoxides in organic solvents (nanodomains, where 1–2 siloxane clusters are bonded with the lignin phenyl-propane structural units) [70], lignin/silica-based hybrids with rather high specific areas (~200 m^2/g compared to ~15 m^2/g for the initial lignin) have been synthesized in aqueous media.

As already mentioned in a previous contribution (Chap. 2), a potential method to produce cellulosic/silica hybrid aerogels could be via use of ionic liquids. Recent studies have shown that cellulose [71], as well as lignin [72] aerogels, can be synthesized in 1-butyl-3-methylimidazolium chloride. Taking into account that some silica aerogels have also been prepared successfully by a sol–gel method in imidazoliums [73–76], it seems potentially really interesting to study coupling of ligno-cellulosic and silica chemistries in such new reaction media.

10.4. Conclusions

Despite their chemical similarity, polyurethane and cellulose derivative-based aerogels have not arrived at the same level of "maturity". Some of the latest studies on aerogels of polyisocyanurates, in general, and polyurethanes, in particular, have allowed synthesis of lightweight and finely mesostructured materials with promising applications in thermal insulation [42]. At present, one of the remaining challenges in this field concerns their preparation through subcritical drying routes.

The situation of cellulose derivative-based aerogels is quite different. They are only at the beginning of their evolution and more or less about everything is yet to be done to improve their properties. Among others, efforts must be focused on better control of the high shrinkage occurring during supercritical CO_2 drying. The heterogeneity of their texture (Figure 10.31) must also be carefully studied and, last but not least, their internal structure is still to be significantly refined, in order, for example, to increase their specific surface area while maintaining low bulk density (e.g., below 0.2 g/cm^3).

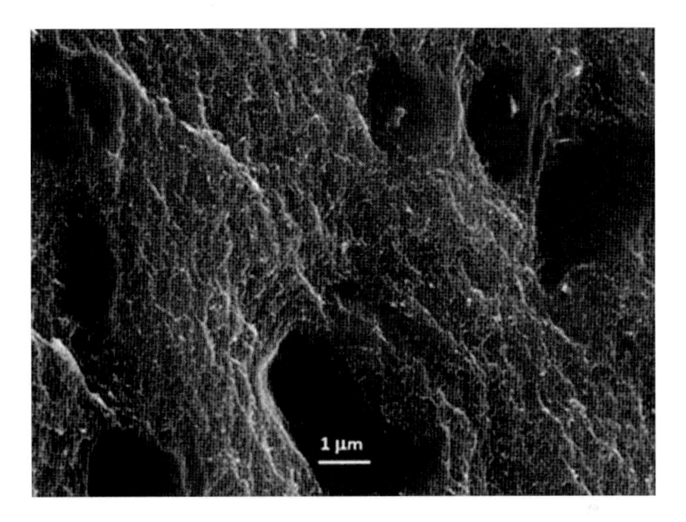

Figure 10.31. *SEM* microscopy of a cellulose acetate-based aerogel (courtesy of Manfred Pinnow, Fraunhofer-IAP, Natural Polymer Division, Golm, Germany) [22].

Despite those differences, it appears clear that those two aerogel families present a common huge potential for hybridization with their nanostructured silica homologues, which could lead to, among others, brand new reinforced superinsulating materials.

Acknowledgments

Jean-Charles Maréchal (CSTB, Grenoble, France) and Florent Fischer (formerly at MINES ParisTech/CEP, Sophia Antipolis, France and at present with SAFT R&D, Bordeaux, France) are strongly acknowledged for very close collaboration on polyurethane and cellulose acetate aerogels, respectively. René Pirard (LGC, Liège University, Belgium) is also warmly acknowledged for his high-level expertise in structural characterizations of aerogel materials as well as our daily collaborators, Claudia Hildenbrand and Pierre Ilbizian (MINES ParisTech/CEP, Sophia Antipolis, France), for their deep intervention in sol–gel studies and supercritical dryings. We also want to acknowledge Prof. Tatiana Budtova (MINES ParisTech/CEMEF, Sophia Antipolis, France) and Prof. Michel Perrut (SEPAREX, Champigneulles, France) for very fruitful discussions on cellulose and polyurethane, respectively. Finally, the authors are grateful for financial support from the French Energy and

Environment Agency (ADEME), the French Research Agency (ANR) and the European Commission for their respective financial supports on related topics.

References

1. Rivera-Armenta JL, Heinze Th, Mendoza-Martinez AM (2004) New polyurethane foams modified with cellulose derivatives. European Polymer Journal 41: 2803–2812.
2. Kistler SS (1932) Coherent expanded aerogels. J Phys Chem 63: 52–64.
3. Tabor R (1995) Microporous isocyanate-based polymer compositions and method of preparation. US Patent 5; 478–867.
4. Biesmans G, Randall D, Francais E, Perrut M (1998) Polyurethane-based organic aerogels' thermal performance. J Non-Cryst Solids 225: 36–40.
5. Escudero RR, Robitzer M, Di Renzo, F, Quignard F (2009) Alginate aerogels as adsorbents of polar molecules from liquid hydrocarbons: hexanol as probe molecule. Carbohydrate Polymers 75: 52–57.
6. Miao Z, Ding K, Wu T, Liu Z, Han B, An G, Miao S, Yang G (2008) Fabrication of 3D-networks of native starch and their application to produce porous inorganic oxide networks through a supercritical route. Microporous and Mesoporous Materials 111: 104–109.
7. Mehling T, Smirnova Guenter U, Neubert RH (2009) Polysaccharide-based [2]aerogels as drug carriers. J Non-Cryst Solids 355: 2472–2479.
8. Scanlon S, Aggeli A, Boden N, Koopmans RJ, Brydson R, Rayner CM (2007) Peptide aerogels comprising self-assembling nanofibrils, Micro & Nano Letters 2: 24–29.
9. Tsioptsias C, Michailof C, Stauropoulos G, Panayiotou C (2009) Chitin and carbon aerogels from chitin alcogels. Carbohydrate Polymers 76: 535–540.
10. Singh J, Dutta PK, Dutta J, Hunt, Macquarrie DJ, Clark JH (2009) Preparation and properties of highly soluble chitosan-L-glutamic acid aerogel derivative. Carbohydrate Polymers 76: 188–195.
11. Jin H, Nishiyama Y, Wada M, Kuga S (2004) Nanofibrillar cellulose aerogels. Colloïds and Surfaces A: Physicochem. Eng. Aspects 240: 63–67.
12. Innerlohinger J, Weber H, Kraft G (2006) Aerocellulose : aerogels and aerogel-like materials made from cellulose. Macromol. Symp. 244: 126–135.
13. Fischer F, Rigacci A, Pirard R, Berthon-Fabry S, Achard P (2006) Cellulose-based aerogels. Polymer 47: 7636–7645.
14. Gavillon R, Budtova T (2008) Aerocellulose: new highly porous cellulose prepared from cellulose-NaOH aqueous solution. Biomacromolecules 9: 269–277.
15. Biesmans G, Randall D, Francais E, Perrut M (1998) The development of polyurethane based organic aerogels and their thermal properties. In: Perrut M & Subra P (eds) Proceedings of the 5th meeting on supercritical fluids (Nice, France, 23–25 March 1998) tome 1: 5–12.
16. Tan C, Fung B, Newman JK, Vu C (2001) Organic aerogels with very high impact strength. Advanced Materials 13: 644–646.
17. Luong ND, Lee Y, Nam J-D (2008) Highly-loaded silver nanoparticles in ultrafine cellulose acetate nanofibrillar aerogel. European Polymer Journal 44: 3116–3121.
18. Biesmans G, Mertens A, Duffours L, Woignier T, Phalippou J (1998) Polyurethane based organic aerogels and their transformation into carbon aerogels. J Non-Cryst Solids 225: 64–68.
19. Guilminot E, Fischer F, Chatenet M, Rigacci A, Berthon-Fabry S, Achard P, Chainet E (2007) Use of cellulose-based carbon aerogels as catalyst support for PEM fuel cell electrodes: electrochemical characterization. Journal of Power Sources 166: 104–111.
20. Guilminot E, Gavillon R, Chatenet M, Berthon-Fabry S, Rigacci A, Budtova T (2007) New nanostructured carbons based on porous cellulose: elaboration, pyrolysis and use as platinum nanoparticles substrate for oxygen reduction electrocatalysis. Journal of Power Sources 185: 717–726.
21. Budtova T (2010) MINES ParisTech/CEMEF France, Private Communication.
22. Fischer F (2006) Synthèse et étude de matériaux nanostructurés à base d'acétate de cellulose pour applications énergétiques. PhD dissertation MINES ParisTech.
23. Perrut M, Français E (1999) Process and equipment for drying polymeric aerogel in the presence of a supercritical fluid. US patent 5; 962–539.
24. Woods G (1990) The ICI Polyurethanes book. Wiley.

25. Dieterich D, Uhlig K (2000) Polyurethanes. Ullmann's Encyclopedia of Industrial Chemistry. Wiley
26. Marotel Y (2000) Polyuréthannes. Techniques de l'ingénieur AM3425, Paris.
27. Luo SG, Tan HM, Zhang JG, Wu YJ, Pei FK, Meng XH (1997) Catalytic mechanisms of triphenyl bismuth, dibutyltin dilaurate and their combination in polyurethane-forming reaction. Journal of Applied Polymer Science 65: 1217–1225.
28. Frisch KC, Reegen SL, Floutz WV, Olivier JP (1967) Complex formation between catalysts, alcohols, and isocyanates in the preparation of urethanes. Journal of Applied Polymer Science Part A: Polymer Chemistry 5: 35–42.
29. Abbate FW, Ulrich H (1969) Urethane: organometallic catalysis of the reaction of alcohols with isocyanates. Journal of Applied Polymer Science 13: 1929–1936.
30. Reegen SL, Frisch KC (1970) Isocyanate-catalyst and hydroxyl-catalyst complex formation. Journal of Applied Polymer Science Part A: Polymer Chemistry 8: 2883–2891.
31. Baker WB, Gaunt J (1949) The mechanism of the aryl isocyanates with alcohols and amines. Part III. The base-catalysed reaction of phenyl isocyanate with alcohols. Journal of the Chemical Society 9–18.
32. Baker WB, Davies MM, Gaunt J (1949) The mechanism of the aryl isocyanates with alcohols and amines. Part IV. The evidence of infra red absorption spectra regarding alcohol-amine association in the base-catalysed reaction of phenyl isocyanate with alcohols. Journal of the Chemical Society 24–27.
33. Flynn KG, Nenortas DR (1963) Kinetics and mechanism of the reaction between phenyl isocyanate and alcohols. Strong base catalysis and deuterium effects. Journal of Organic Chemistry 28: 3527–3530.
34. Farkas A, Strohm PF (1965) Mechanism of Amine-Catalyzed Reaction of Isocyanates with Hydroxyl Compounds. Ind. Eng. Chem. Fundam. 4: 32–38.
35. Borsus JM, Merckaert P, Jérôme R, Teyssier Ph (1982) Catalysis of the reaction between isocyanates and protonic substrates. II Kinetic study of the polyurea foaming process catalysed by a series of amino compounds. Journal of Applied Polymer Science 27: 4029–4042.
36. Tewari PH, Hunt AJ, Lofftus KD (1985) Ambient-temperature supercritical drying of transparent aerogels. Materials Letters 3: 363–367.
37. Biesmans G (1999) Polyisocyanate based aerogel, US Patent 5990184.
38. Biesmans G (1999) Polyisocyanate based xerogel, US Patent 6063826.
39. Rigacci A, Maréchal JC, Repoux M, Moreno M, Achard P (2004) Elaboration of aerogels and xerogels of polyurethane for thermal insulation. J Non-Cryst Solids 35: 372–378.
40. Masmoudi Y, Rigacci A, Ilbizian P, Cauneau F, Achard P (2006) Diffusion during the supercritical drying of silica gels. Drying Technology 24: 1–6.
41. Pirard R, Rigacci A, Maréchal JC, Achard P, Quenard D, Pirard JP (2003) Characterization of porous texture of hyperporous polyurethane based xerogels and aerogels by mercury porosimetry using densification equation. Polymer 44: 4881–4887.
42. Lee JK, Gould GL, Rhine W (2009) Polyurea based aerogel for a high performance thermal insulation material. J Sol-Gel Sci Technol 49: 209–22.
43. Zhang G et al (2004) Isocyanate-crosslinked silica aerogel monoliths: preparation and characterization. Journal of Non-Crystalline Solids 350: 152–164.
44. Leventis N et al (2008) Polymer nano-encapsulation of templated mesoporous silica monoliths with improved mechanical properties. Journal of Non-Crystalline Solids 354:632–644.
45. Capadona LA, Meador MA, Alunni A, Fabrizio EF, Vassilaras P, Leventis N (2006) Flexible, low-density polymer crosslinked silica aerogels. Polymer 47: 5754–5761.
46. Yim T-J, Kim SY, Yoo K-P (2002) Fabrication and thermophysical characterization of nanoporous silica-polyurethane hybrid aerogel by sol-gel processing and supercritical solvent drying. Korean J. Chem. Eng. 19(1): 159–166.
47. Jeon HT, Jang MK, Kim BK, Kim KH (2007) Synthesis and characterization of waterborne polyurethane-silica hybrids using sol-gel process. Colloids and Surface A: Physicochem. Eng. Aspects 302: 559–567.
48. Lee K (2007) Organic aerogels reinforced with inorganic fillers. US Patent 2007/0259979.
49. Zhou J, Chang C, Zhang R, Zhang L (2007) Hydrogels prepared from unsubstituted cellulose in NaOH urea aqueous solution. Macromol. Biosci. 7: 804–809.
50. Sescousse R, Budtova T (2009) Influence of processing parameters on regeneration kinetics and morphology of porous cellulose from cellulose-NaOH-water solutions. Cellulose 16: 417–426.
51. Pinnow M, Fink H-P, Fanter C, Kunze J (2008) Characterization of Highly Porous Materials from Cellulose Carbamate. Macromol. Symp. 262: 129–139.
52. Hoepfner S, Ratke L, Milow B (2008) Synthesis and characterisation of nanofibrillar cellulose. Cellulose 15: 121–129.

53. Pääkkö M, Vapaavuori J, Silvennoinen R, Kosonen H, Ankerfors M, Lindström T, Berglund LA, Ikkala O (2008) Long and entangled native cellulose I nanofibers allow flexible aerogels and hierarchically porous templates for functionalities. Soft Matter 4: 2492–2499.
54. Nishino T, Matsuda L, Hirao K (2004) All-cellulose composite. Macromolecules 37: 7683–7687.
55. Duchemin BJC, Staiger MP, Tucker N, Newman RH (2010) Aerocellulose based on all-cellulose composites. Journal of Applied Polymer Science 115: 216–221.
56. Weatherwax RC., Caufield DF (1978) The pore structure of papers wet stiffened by formaldehyde crosslinking. Journal of colloid and interface science 67: 498–505.
57. Yang CQ (1993) Infrared spectroscopic studies of the effects of the catalyst on the ester crosslinking of cellulose by polycarboxylic acids. Journal of applied polymer science 50: 2047–2053.
58. Bai YX, Li Y-F (2006) Preparation and characterization of crosslinked porous cellulose beads. Carbohydrate polymers 64: 402–407.
59. Espositio F, Del Nobile MA, Mensitieri G, Nicolais L (1996) Water sorption in cellulose-based hydrogels. Journal of applied polymer science 60: 2403–2407.
60. Rustmeyer P (2004) Cellulose acetates: properties and applications. Macromolecular symposia 208 Issue 1. Wiley.
61. Goebel KD, Berry GC, Tanner DW (1978) Properties of cellulose acetate. III. Light scattering from concentrated solutions and films. Tensile creep and desalination studies on films. Journal of polymer science: polymer physics edition 17(6): 917–937.
62. Malm CJ, Nadeau GF (1935) Cellulose acetate carbamate, US Patent 1991107.
63. Mormann W, Michel U (2002) Improved synthesis of cellulose carbamate without by-products. Carbohydrate polymers 50: 201–208.
64. Tanaka T (1981) Gels. Sci. Am. 244: 124–138.
65. Pirard R, Blacher S, Brouers F, Pirard JP (1195) Interpretation of mercury porosimetry applied to aerogels. J. Mater. Res. 10: 2114–2119.
66. Rigacci A, Achard P (2008) Aerogels a new family of superinsulating materials for various applications fields. Trends and future of on-board energy in space systems, 24–25 november 2008, Avignon, France (CD-rom).
67. Sequeira S, Evtuguin D, Portugal I, Esculcas A (2007) Synthesis and characterization of cellulose/silica hybrids obtained by heteropoly acid catalysed sol-gel process. Materials Science and Engineering C27: 172–179.
68. Sequeira S, Evtuguin D, Portugal I (2009) Preparation and properties of cellulose / silica hybrid composites. Polymer composites 30: 1275–1282.
69. Telysheva G, Dishbite T, Evtuguin D, Mironova-Ulmane N, Lebedeva G, Andersone A, Bikovens O, Chirkova J, Belkova L (2009) Design of siliceous lignins. Novel organic/inorganic hybrid sorbent materials. Scripta Materialia 60: 687–690.
70. Telysheva G (1992) in Lignocellulosics – Science Technology, Development and Use, Kennedy J, Phillips G, Williams P (Eds). Ellis Harwood, London 643–655.
71. Deng M, Zhou Q, Du A, van Kasteren J, Wang Y (2009) Preparation of nanoporous cellulose foams from cellulose-ionic liquid solution. Materials Letters 63: 1851–1854.
72. Aaltonen O, Jauhiainen O (2009) The preparation of lignocellulosic aerogels from ionic liquid solutions. Carbohydrate Polymers 75: 125–129.
73. Dai S, Ju YH, Gao HJ, Lin JS, Pennycook SJ, Barnes CE (2000) Preparation of silica aerogel using ionic liquids as solvents. Chem Commun 243–244.
74. Karout A, Pierre AC (2007), Silica xerogels and aerogels synthesized with ionic liquids, J. Non-Cryst Solids 353: 2900–2909.
75. M.V. Migliorini, R.K. Donato, M.A. Benvegnu, R.S. Gonçalves, H.S. Schrekker (2008) Imidazolium ionic liquids as bifunctional materials (morphology controller and pre-catalyst) for the preparation of xerogel silica's. J. Sol-Gel Sci Technol 48: 272–276.
76. Karout A, Pierre AC (2009) Influence of ionic liquids on the texture of silica aerogels. J Sol Gel Sci Technol 49: 364–372.

11

Resorcinol–Formaldehyde Aerogels

Sudhir Mulik and Chariklia Sotiriou-Leventis

Abstract Resorcinol–formaldehyde (*RF*) aerogels comprise an important class of organic aerogels, and they are studied intensely for their potential uses in thermal insulation, catalysis, and as precursors of electrically conducting carbon aerogels with applications in filtration, energy storage, and the green energy initiative. This broad overview focuses on how the chemical, microscopic, as well as macroscopic characteristics of *RF* and thereby carbon aerogels can be tailored to desired application-specific structure–property relationships by varying processing conditions such as the monomer concentration, the pH, and the catalyst-to-monomer ratio.

11.1. Introduction

Aerogels comprise a special class of low-density open-cell foams with large internal void space (up to 98%, v/v), which is responsible for useful material properties such as high surface area, low thermal conductivity (2–3 orders of magnitude less than silica glass), and high acoustic impedance [1, 2]. These open-cell foams are derived from wet gels, which in turn are usually prepared by sol–gel processes and are dried using a supercritical fluid (*SCF*) such as CO_2 (Chap. 3) [3, 4]. Currently, the most promising commercial application of those materials seems to be in thermal insulation. Other potential applications of aerogels include catalysts and catalyst supports, gas filters and gas storage materials, conducting and dielectric materials, and acoustic insulation [5].

The most common type of aerogel uses inorganic metal or semimetal oxide frameworks. Purely organic aerogels were first described by Pekala and coworkers, and they were based on a framework consisting of a resorcinol–formaldehyde (*RF*) resin [6]. For quite sometime, Pekala's *RF* aerogels became synonymous with organic aerogels, but subsequently other types of materials in this class became available, mostly based on not only formaldehyde-type resins such as phenol-formaldehyde, melamine-formaldehyde, cresol-

S. Mulik • Georgia-Pacific Chemicals LLC, 2883 Miller Road, Decatur, GA 30035, USA
Present address: The Dow Chemical Company, 727 Norristown Road, P.O. Box 904, Spring House, PA 19477, USA
e-mail: sudhir.mulik@gmail.com
C. Sotiriou-Leventis • Department of Chemistry, Missouri University of Science and Technology, Rolla, MO 65409, USA
e-mail: cslevent@mst.edu

M.A. Aegerter et al. (eds.), *Aerogels Handbook*, Advances in Sol-Gel Derived Materials and Technologies, DOI 10.1007/978-1-4419-7589-8_11,

formaldehyde, phenol-furfural, but also polyimides, polyacrylamides, polyacrylonitriles, polyacrylates, polystyrenes, polyurethanes, etc. [7].

Resorcinol–formaldehyde (*RF*) aerogels were first synthesized by base-catalyzed polycondensation of resorcinol with formaldehyde in an aqueous environment using Na_2CO_3 (soda ash) as catalyst. Wet gels are dried with *SCF* CO_2. Resorcinol was considered as an ideal starting material, because even though it has three reactive sites (the 2, 4, and 6 positions of the aromatic ring) just like phenol, it is about 10–15 times more reactive than the latter and therefore it can react with formaldehyde at lower temperatures. Although the mechanism of *R*/*F* polycondensation is very different from the reaction leading to inorganic gels, the physicochemical processes leading to *RF* wet gels are quite analogous to those taking place with typical inorganic gels such as silica and titania [4]. Following Pekala's initial report on Na_2CO_3-catalyzed *RF* aerogel synthesis, several more recent literature reports discuss the synthesis of *RF* aerogels with both alkaline and acid catalysts [6, 8].

RF wet gels have been produced with different variations of essentially the same procedure. Initially, resorcinol (*R*) and formaldehyde (*F*) are mixed at the appropriate molar ratio in the presence of a catalyst (usually a basic or in very few cases an acidic). Then, the solution is heated in a closed container to a predetermined temperature for a sufficient time period to form a stable (crosslinked) wet gel. Wet gels are subsequently solvent-exchanged (washed) with a suitable organic solvent to remove the gelation solvent from the pores (in most cases water) and are dried either with *SCF* CO_2 or subcritically under ambient conditions to yield organic aerogels or xerogels, respectively. Studies show that varying the processing conditions such as the amount of catalyst (through the recorsinol-to-catalyst, *R*/*C*, ratio) results in *RF* aerogels with different nano-morphologies. *RF* aerogels prepared with higher catalyst concentrations (e.g., $R/C = 50$) led to the formation of a mesoporous skeletal framework, which is very similar to that of typical base-catalyzed silica aerogels, consisting of a pearl-necklace-like structure of secondary particles (40–70 nm), which in turn consist of primary particles (10–12 nm in diameter) (Chap. 2) [9–11]. Overall, *RF* aerogels demonstrate typical aerogel-like properties such as high mesoporosity and low thermal conductivity. In addition, however, *RF* aerogels are precursors of electrically conducting carbon aerogels, and perhaps that attribute has fueled the most intense research interest on those materials (Chap. 36).

11.2. The Resorcinol–Formaldehyde Chemistry

The resorcinol–formaldehyde polymer is classified as a phenolic resin. In fact, phenol–formaldehyde condensates are the first synthetic polymers introduced commercially (Bakelite) in the beginning of the twentieth century. Thus, the phenolic resin chemistry, on which there is a vast amount of information, can be used to understand the chemistry of the *RF* aerogel synthesis. In the first step, resorcinol reacts with formaldehyde to form hydroxymethylated resorcinol; in the second step, the hydroxymethyl groups condense with each other to form nanometer-sized clusters, which then crosslink by the same chemistry to produce a gel. The formation of clusters is influenced by typical sol–gel parameters such as the temperature, pH, and concentration of reactants. Though the first *RF* aerogel is synthesized by the aqueous polycondensation of resorcinol with formaldehyde with the use of sodium carbonate as catalyst [6, 12, 13], now several literature reports exist for aerogels made from phenol and formaldehyde [14–16].

Phenolic resins made from phenol and formaldehyde are classified as "resole" or "novolac" based on the presence of methylene ether ($–CH_2OCH_2–$) or methylene ($–CH_2–$)

linkages between the aromatic moieties, respectively. The existence of either methylene ether or methylene linkage is based on the ratio of the reactants, the pH, and the catalyst. Resorcinol undergoes most of the typical reactions of phenol reacting with formaldehyde under alkaline and/or acid conditions to produce a mixture of addition and condensation products. However, the reaction of resorcinol and formaldehyde leads predominantly to formation of methylene bridges linking two resorcinol molecules. The reason lies with the higher reactivity of resorcinol and is discussed further below.

Resorcinol is a trifunctional monomer that can add up to three equivalents of formaldehyde. Those substituted resorcinol derivatives then condense with formation of nanoclusters, which then crosslink through the surface $-CH_2OH$ groups. Generally, in case of base catalysis, the reaction is carried out in aqueous media in the presence of sodium carbonate or a similar base catalyst, and gelation requires heating at elevated temperatures for prolonged time periods. With acid catalysts, the monomers are mixed either in nonaqueous or in aqueous media with the catalyst (hydrochloric acid, acetic acid, and perchloric acid), and gelation can take place even at room temperature and is generally faster.

11.2.1. Base-Catalyzed Gelation [17]

Under basic conditions, resorcinol is deprotonated to the resorcinol anion. As shown in Scheme 11.1, due to resonance, the electron density increases at the 4 (or 6) position of resorcinol (1). Electron donation from those positions to the partially positively charged carbonyl carbon of formaldehyde leads to hydroxymethylation (addition of the $-CH_2OH$ group). In turn, hydroxymethylation activates other positions and an excess of formaldehyde leads to dihydroxymethylation (2) [18]. The base catalyst causes deprotonation of hydroxymethylated resorcinol, leading to a very reactive and unstable *o*-quinone methide intermediate (3). The *o*-quinone methide reacts further with another resorcinol molecule to form stable methylene linkage (4). The formation of *o*-quinone methides and the presence of high electron densities at the 2-, 4-, and 6-positions of resorcinol ring are the reasons for the enhanced reactivity of resorcinol compared with phenol. As a matter of fact, there is a greater tendency for continuous condensation as long as there are active sites on the resorcinol molecules or the *RF* clusters. Therefore, *RF* resins produced by the base-catalyzed reaction predominantly contain methylene-bridged novolac structures [19]. Equation (5) summarizes the overall gelation mechanism.

11.2.2. Acid-Catalyzed Gelation [9, 20]

In contrast to the base-catalyzed route, which is based on activation of the aromatic ring toward electrophilic aromatic substitution by increasing the electron donating ability of the substituents (from $-OH$ to O^-), the acid-catalyzed route is based on acceleration of the reaction by increasing the electrophilicity of formaldehyde.

As shown in Scheme 11.2, protonation of formaldehyde (6) is followed by nucleophilic attack by the π-system of resorcinol leading to hydroxymethylation of resorcinol (7). In an acidic environment, protonation of hydroxymethyl groups leads to $-OH_2^+$ group formation (8), which, as good leaving groups, cleaves further unimolecularly to form *o*-quinone methide-type intermediates (9). *o*-Quinone methides can lead to condensation via (4), or alternatively the π-system of another resorcinol can attack the protonated hydroxymethylated resorcinol moieties to form methylene linkages (10). Those

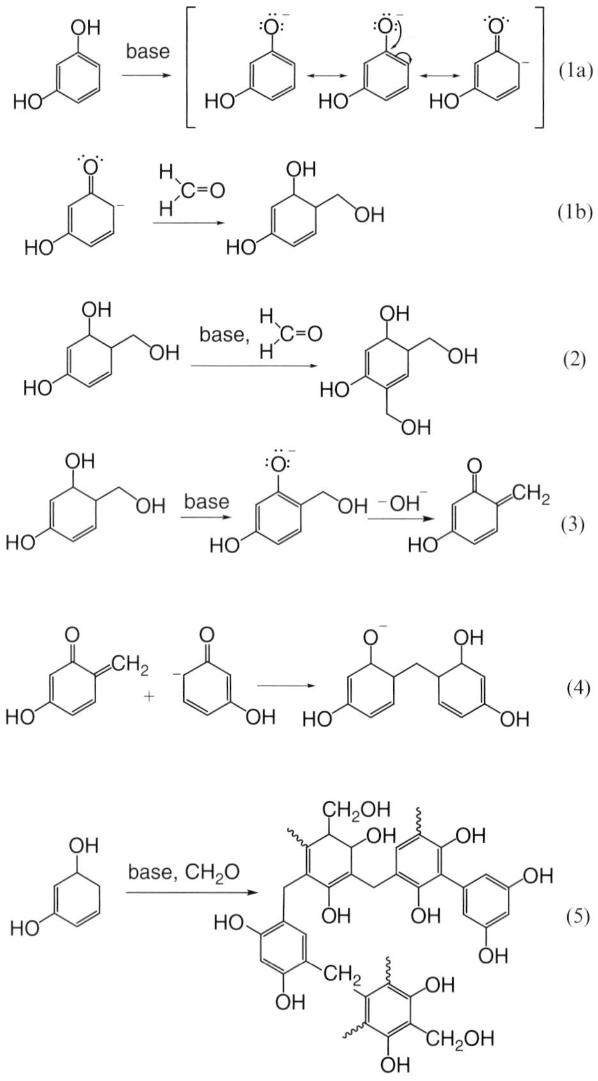

Scheme 11.1. Mechanism of the base-catalyzed *RF* aerogel synthesis.

simultaneous addition and condensation reactions result in the crosslinked structure of the resorcinol–formaldehyde aerogels (11).

11.3. *RF* Aerogels Prepared by the Base-Catalyzed Route

As discussed above, the *RF* gel formation has been associated with two major reactions:

1. Formation of hydroxymethyl derivatives of resorcinol.

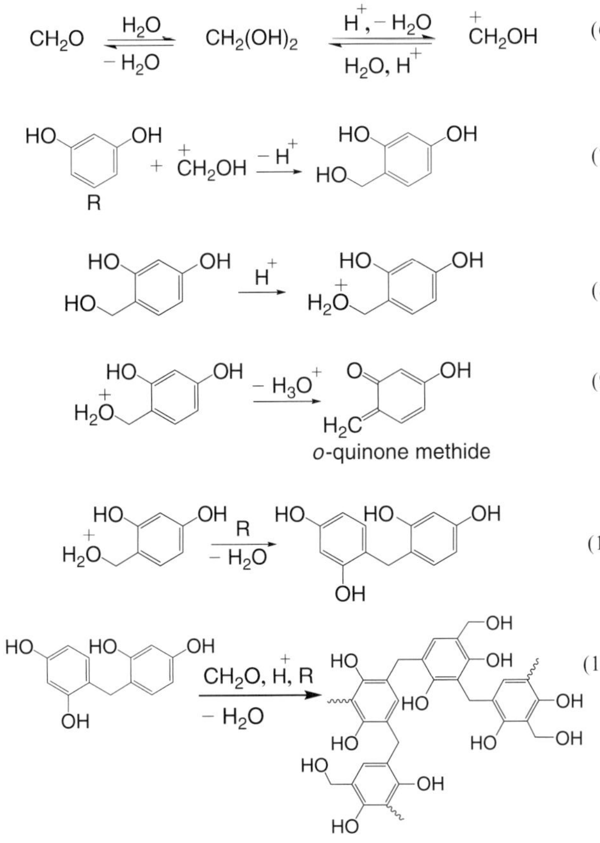

$$CH_2O \underset{-H_2O}{\overset{H_2O}{\rightleftharpoons}} CH_2(OH)_2 \underset{H_2O,\ H^+}{\overset{H^+,\ -H_2O}{\rightleftharpoons}} \overset{+}{C}H_2OH \qquad (6)$$

$$\text{R} + \overset{+}{C}H_2OH \xrightarrow{-H^+} \cdots \qquad (7)$$

$$\xrightarrow{H^+} \qquad (8)$$

$$\xrightarrow{-H_3O^+} \text{o-quinone methide} \qquad (9)$$

$$\xrightarrow[-H_2O]{R} \qquad (10)$$

$$\xrightarrow[-H_2O]{CH_2O,\ H^+,\ R} \qquad (11)$$

Scheme 11.2. Mechanism of the acid-catalyzed *RF* aerogel synthesis.

2. Condensation of those derivatives to methylene ($-CH_2-$) bridges.

Aerogels prepared by Pekala's base-catalyzed route have been studied extensively for the effect of process parameters such as the concentration of monomers and the catalyst as well as the pH of the solution. Both structure and properties such as density, surface area, particle size, and pore size distribution are influenced by those variables. Low-density *RF* aerogels (~0.03 g cm^{-3}) prepared by that method exhibit high porosities (>80%), high surface areas (400–900 $m^2\ g^{-1}$), and ultrafine pore size (<500 Å).

In particular, the effect of the resorcinol-to-catalyst (*R*/*C*) ratio on the final aerogel structure has been studied extensively. In aqueous media, this ratio has been typically varied from 50 to 300. Formation of particles connected with large necks was reported at low *R*/*C* ratios (~50). Crosslinked microspheres were observed at very high *R*/*C* ratios (~1,500). The final pore structure and the gelation time depend strongly on the sol pH; at low pHs, precipitation rather than gelation has been reported.

11.3.1. Process of Making *RF* Aerogel via Base-Catalyzed Route

Typically, based on the desirable final physical properties of the *RF* aerogels, two ratios need to be selected carefully: the resorcinol-to-catalyst ratio (*R*/*C*) and the resorcinol-to-water ratio (*R*/*W*).

Step 1: Gelation

In a typical procedure, the desirable amount of resorcinol is dissolved in water together with sodium carbonate or a similar base catalyst. Formaldehyde is added and the mixture is heated in a closed container (mold) at elevated temperatures (generally in the 80–90°C range) for prolonged period (usually a few days to several weeks) for gelation.

Step 2: Aging

After gelation, the gel is taken out from the mold and is washed/treated with dilute acid (usually, 0.1% trifluoroacetic acid) dissolved in an organic solvent such as methanol/acetone in order to increase the crosslinking density.

Step 3: Solvent Exchange

Subsequently, wet gels are washed several times with fresh methanol/acetone to remove excess water and catalyst from the pores and facilitate drying with *SCF* CO_2.

Step 4: *SCF* CO_2 Drying

Wet gels with their pores filled with the organic solvent from Step 3 are placed in an autoclave with excess of solvent. The pore-filling solvent is replaced with liquid carbon dioxide by several washing steps carried out at lower temperature (~15°C) and pressure (~900 psi). Once the organic solvent filling the pores is completely replaced with liquid CO_2, the temperature and the pressure of the autoclave are increased above the critical point of CO_2 (31°C, 1,100 psi) and the *SCF* CO_2 is vented off slowly (at 35–40°C) leaving behind the *RF* aerogels.

11.3.2. Factors Affecting the Structure and the Properties of *RF* Aerogel Prepared Through the Base-Catalyzed Route

Concentration of Monomers and Catalyst

In base catalysis, most of the literature procedures are referred to by the [resorcinol]/[catalyst] or *R*/*C* ratio [10]. The *R*/*C* ratio is the dominant factor that affects the density, surface area, and mechanical properties of *RF* aerogels. Similarly, the [resorcinol]/[water] or *R*/*W* ratio is another factor that plays crucial role in pore size distribution and porosity [21].

The *RF* aerogels prepared with high catalyst conditions (i.e., $R/C = 50$) exhibit particles 3–5 nm in diameter joined together with large necks, giving a fibrous appearance also referred to as "polymeric" *RF* aerogels. Aerogels prepared with low catalyst conditions (i.e., $R/C = 200$) exhibit particles 11–14 nm in diameter connected as in a "string of pearls" and referred to as "colloidal" *RF* aerogels [13, 22]. "Polymeric" *RF* aerogels incur substantial shrinkage during supercritical drying, and they are characterized by high surface areas and a high compressive modulus. In contrast, "colloidal" *RF* aerogels exhibit little shrinkage upon supercritical drying, and they are characterized by lower surface areas and weaker mechanical properties. *RF* aerogels prepared with the same amounts of *R*/*F* but with different *R*/*C* ratios (50, 100, 150, 200, and 300) had different densities due to shrinkage: aerogels prepared

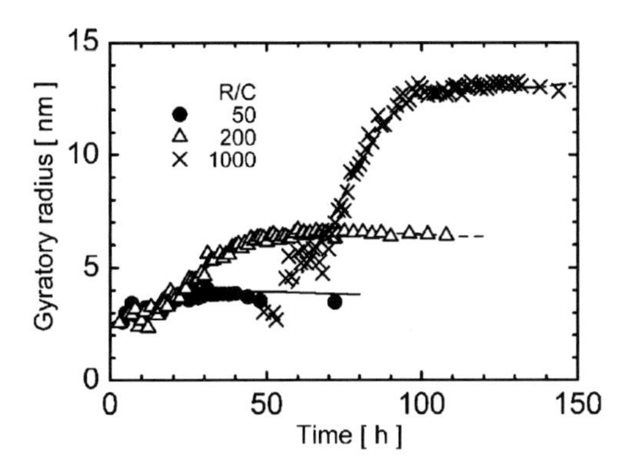

Figure 11.1. Variation in the radius of gyration versus lapsed time for the *RF* solution with K_2CO_3 catalyst at different catalyst ratios [24].

with $R/C = 50$ had the highest densities [23]. However, at equivalent densities, "polymeric" *RF* aerogels prepared at $R/C = 50$ were ~3× stiffer than "colloidal" *RF* aerogels prepared at $R/C = 200$, reflecting the size of the interparticle necks on the mechanical properties.

The pore characteristic of *RF* aerogels can be varied by the catalytic species and the catalyst ratios. Horikawa et al. used Small-Angle X-Ray Scattering (*SAXS*) to study the effect of different catalysts at different *R/C* ratios on the radius of gyration during polymerization (Figure 11.1) [24]. According to Horikawa's work, at $R/C = 50$, the RF primary particles were spherical and their size (radius of gyration) remained unchanged with time, but it changed with the chemical identity of the catalyst (K_2CO_3, $KHCO_3$, Na_2CO_3, and $NaHCO_3$). The pore size of the *RF* aerogels was found in the range of 5–20 nm and varied as a function of the *R/C* ratio [25]. The radius of gyration of the primary particles increased with increasing catalyst ratio. The *RF* primary particles were spherical for the catalyst ratios of $R/C = 50$ and 200, and they were disk shaped at the catalyst ratio of $R/C = 1{,}000$.

Tamon et al. have shown that during the polycondensation process, resorcinol and formaldehyde are consumed to form highly crosslinked particles. The aggregated particles become interconnected after gelation, showing a structure composed of rings of particles with 20–50 particles per ring [22]. Therefore, the *R/C* ratio is the principal parameter that controls the size of interconnected particles and thus the scale and size of pores in the final gel structure. By using low catalyst concentrations (i.e., high *R/C* ratios), the particle growth of *RF* gels falls into the 100 nm size range. Saliger et al. reported that with higher dilutions and higher catalyst concentrations (i.e., low *R/C* ratios), particle growth can be tailored to the micron size regime by keeping the reaction temperature low (i.e., prolonged gelation times) [26]. The addition of different solvents in the resorcinol/formaldehyde sol causes changes in the gelation time and that eventually affects the inter- and intraparticle structure (microporosity) [27].

In more recent reports, Mirzaeian et al. showed that the *R/C* ratio is also the dominant factor that controls the surface area, total pore volume, and mechanical properties of *RF* aerogels [28]. As shown in Figure 11.2, the volume of N_2 absorbed increases with the *R/C* ratio. Meanwhile, the size and number of *RF* clusters formed during the gelation process depends on the *R/C* ratio. High catalyst conditions (i.e., low *R/C* ratios) result in the

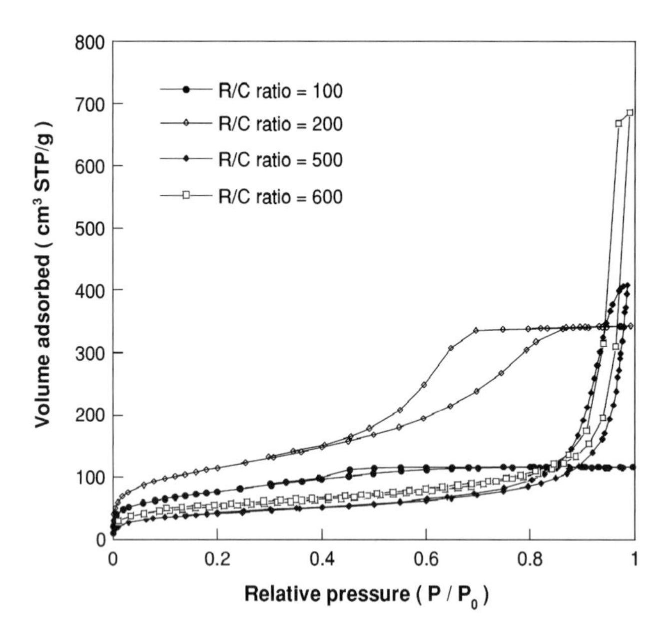

Figure 11.2. N_2 adsorption/desorption isotherms at 77°K for gels with various *R*/*C* ratios [28].

formation of small particles leading to a microporous structure; low catalyst conditions (i.e., high *R*/*C* ratios) result in larger particles leading to the development of mesoporosity.

Yamamoto et al. have showed that for fixed *R*/*W* ratios, the peak value of the pore size distribution increases with increasing *R*/*C* ratios [21]. Further, the peak value of the pore size distribution increases with decreasing *R*/*W* ratios, while the *R*/*C* ratio is kept fixed. Surface areas do not depend upon *R*/*W* ratio [29].

The Initial pH of the Sol

Many reports are available in the literature where efforts were made to evaluate the effect of the sol pH on the final physical properties of the *RF* aerogels. The general understanding is that with increasing pH, the surface area increases [12, 24, 30, 31]. Lin et al. pointed out that in the case of polymers synthesized in the presence of sodium carbonate, larger polymer particles were obtained with lower amounts of catalyst [32]. This is in agreement with the *SAXS* studies by Pekala and Schaefer at different *R*/*C* ratios [12].

As higher pH values increase the concentration of deprotonated resorcinol, they also enhance formation of hydroxymethyl derivatives of resorcinol and they produce highly branched clusters. Highly crosslinked and branched structures are less stable toward spinodal decomposition, leading to smaller and more interconnected polymer particles. On the other hand, at lower pHs, the concentration of resorcinol anions decreases, which diminishes the formation of hydroxymethyl derivatives. The less branched system would persist longer in the nucleation regime, leading to larger particles.

RF aerogels essentially display two types of pores: voids between the polymer particles corresponding to meso- and macropores and voids inside the particles corresponding to micropores. The mesopore size strongly depends on the pH: as pH decreases, both the particle size and the voids between the particles increase. In other words, mesopores become wider and tend to reach the macroporous domain (diameter larger than 50 nm). Crosslinking between particles, which is favored at high pHs, plays a role in the pore size distribution: small particles synthesized at higher pHs should be more interconnected than bigger particles formed at lower pHs. Also, higher interconnection should reduce the mesopore size. Job et al. observed that *RF* aerogels prepared in the pH range from 5.45 to 7.35 exhibited micro- and mesopores confirmed by Type I and Type IV nitrogen adsorption isotherms [30]. Those isotherms also show that the texture of the dried gel can be adjusted within a reasonable range through initial pH control.

11.4. *RF* Aerogel Prepared by Acid Catalysis

The basic problem of the base-catalyzed *RF* aerogel synthesis is the length of the gelation process that may range from a few days to weeks at a time. However, the economics of the process have been rarely questioned due to the high value applications of *RF* aerogels: pyrolysis under an inert atmosphere gives high surface area carbon aerogels for supercapacitors, fuel cells, and batteries.

Thus, the vast amount of literature on base-catalyzed *RF* aerogels overshadows the few reports on acid-catalyzed processes. The acid-catalyzed process yields *RF* aerogels similar in chemical composition to the base-catalyzed materials, but different in morphology [10, 11, 33–38]. Although the base-catalyzed route has been lengthy, the differences in morphology as well as a not so clear understanding of the mechanistic intricacies of the acid-catalyzed route caused a slow transition from base catalysis to acid catalysis. Recently, many literature studies involving *RF* aerogels have used the acid-catalyzed route for the time-efficient synthesis of *RF* aerogels.

Barbieri et al. reported *RF* aerogel synthesis by using $HClO_4$ as catalyst in acetone as the solvent at 45°C followed by 3 days of aging [10]. Merzbacher et al. used HNO_3 as catalyst in aqueous media at 80°C. This process required 2 days for gelation and 7 days of aging [34]. Along the same lines, Brandt et al. showed *RF* wet gel synthesis with acetic acid in water [11, 37]. In all these reports, gelation and aging ranged between 1 day and more than a week.

Baumann et al. reported an efficient resorcinol–propanal gel synthesis in two steps using an organic base such as triethylamine (Et_3N) in the first step and then a mineral acid such as hydrochloric acid in the second step [39]. Using ^{13}C *NMR*, Mulik et al. further investigated the necessity of the base-catalyzed step (deprotonation) and reported a time-efficient synthesis of *RF* aerogel with hydrochloric acid as catalyst in acetonitrile [20]. As shown in Figure 11.3, when no catalyst is used, only peaks corresponding to the starting materials are observed; in the presence of Et_3N, the spectra exhibit very low intensity peaks corresponding to the electrophilic aromatic substitution adduct of *R* and *F*; however, in the presence of HCl at room temperature, the spectrum is dominated by the peaks corresponding to the condensation product [40].

Figure 11.4 shows the color transition of the *R*/*F* sol in CH_3CN in the presence of HCl as catalyst: the characteristic red color of the *o*-quinone methide ($\lambda_{max} = 543$ nm) builds up quickly in short time periods, in contrast to the base-catalyzed process where it takes several hours to days.

Concurrently with the development of the *o*-quinone methide color, the viscosity of the acid-catalyzed/CH_3CN sol builds up quickly (Figure 11.5). Gelation takes place in

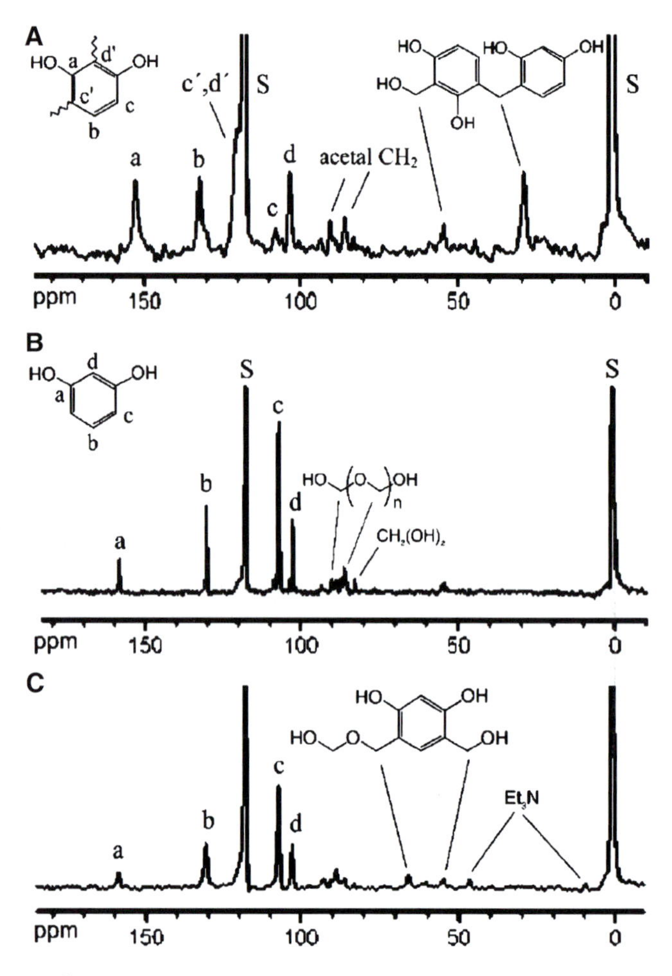

Figure 11.3. Solution ^{13}C *NMR* of resorcinol/formaldehyde mixtures in 1:2 mol ratio in CD_3CN. **A.** 15 min after mixing, using HCl as catalyst, at room temperature; **B.** 15 min after mixing, without any catalyst, at room temperature; and **C.** 75 min after mixing, with Et_3N as catalyst, at 80°C. (*S* solvent peak) [9].

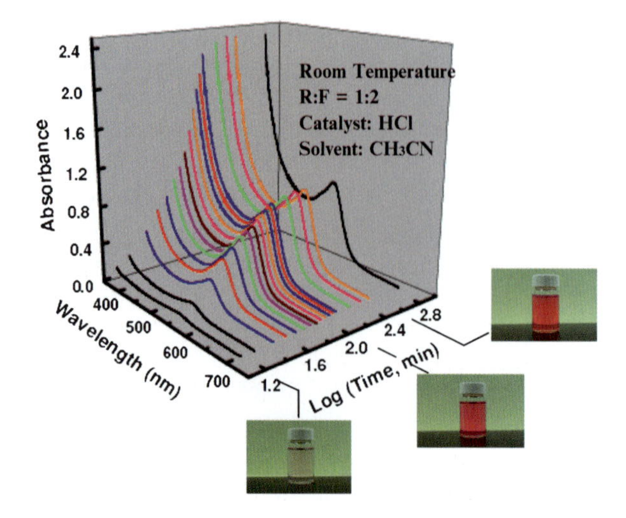

Figure 11.4. Evolution of the absorbance of an HCl-catalyzed resorcinol/formaldehyde mixture at 1:2 mol ratios in CH_3CN at room temperature. Pictures show snapshots of the red color at the specified points in time [9].

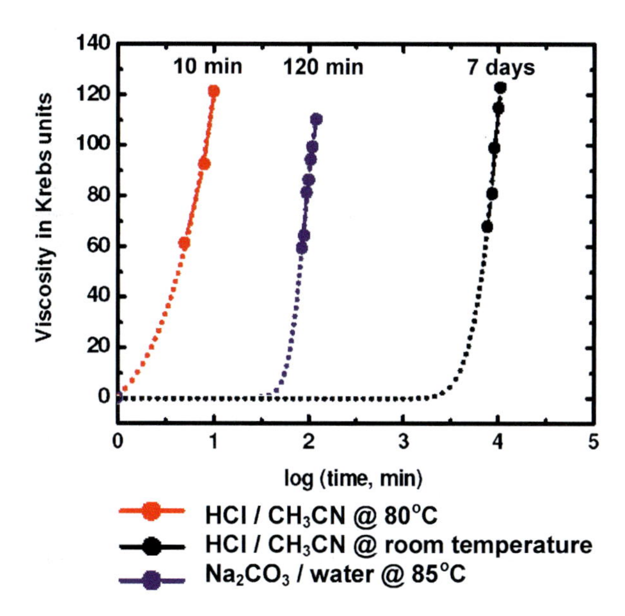

Figure 11.5. Viscosity evolution of three solutions as indicated, according to ASTM D 562-01. The Stormer Type Viscometer employed was able to register data only above 59 Krebs units. Data collection stopped at the gel point [9].

10 min at 80°C and 2 h at room temperature. An aqueous base-catalyzed sol of the same concentration gels in 7 days.

Overall, the addition of formaldehyde to resorcinol is an electrophilic aromatic substitution reaction, and as such it is accelerated by acid catalysis. Chloromethylation (i.e., addition of $–CH_2Cl$) and hydroxymethylation are the classic reactions used in organic chemistry where hydroxymethylated resorcinol is the intermediate step of the reaction [40]. Acid catalysis also promotes, in one pot, a subsequent condensation step by protonation of the newly formed $–CH_2OH$ groups, leading to the formation of *RF* polymer chains.

11.5. Property Comparison of Base- Versus Acid-Catalyzed *RF* Aerogels

11.5.1. Chemical Composition

Hydroxymethylated resorcinol in either base- or acid-catalyzed sol does not accumulate because of its instability. *p*-Hydroxymethyl groups undergo rapid conversion to carbocations or *o*-quinone methides. This explains how identical condensation products are obtained under acid or alkaline reaction conditions [19]. This fact is evident in the ^{13}C *NMR* spectra of *RF* aerogels prepared under acid- or base-catalyzed conditions [33–38].

Moudrakovski et al. reported the solid-state ^{13}C *NMR* spectra for the resorcinol–formaldehyde aerogels prepared with Pekala's base-catalyzed method at different *R*/*C* ratios [41]. As stated above, resorcinol is capable of adding formaldehyde in the 2-, 4-, and 6-ring positions. These substituted resorcinol rings condense with each other to form nanometer-sized clusters in solution. Eventually, the clusters crosslink through their

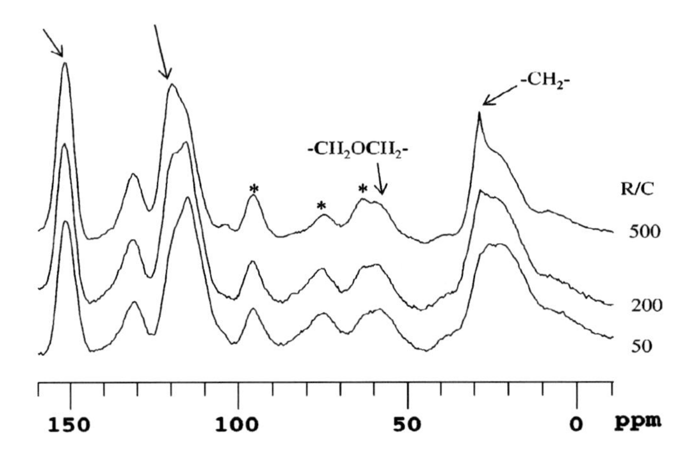

Figure 11.6. Series of solid-state ^{13}C *NMR* spectra of *RF* aerogels prepared with *R/C* ratios at 50, 200, and 500 (peaks marked with asterisks are spin side bands) [41].

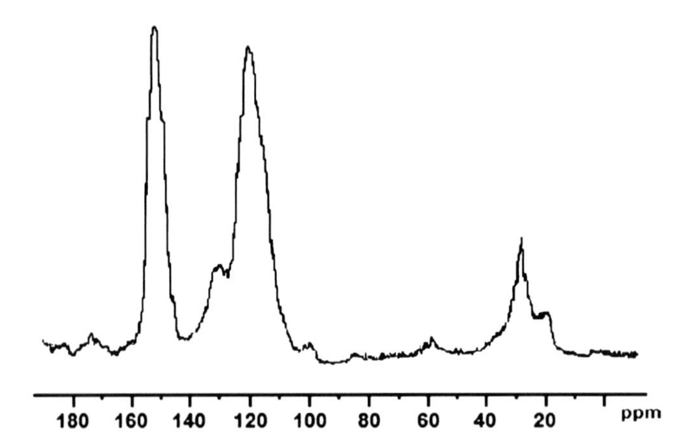

Figure 11.7. ^{13}C CPMAS *NMR* of *RF* aerogels prepared with an *R:F* ratio of 1:2 [9].

surface groups (e.g., $-CH_2OH$) to form a gel. Therefore, the final aerogels contain a mixture of the resorcinol rings having one, two, or three linkages to the adjacent rings. As shown in Figure 11.6, the resonances centered at 150, 117, 60, and 24 ppm correspond to aromatic carbons directly attached to the –OHs, aromatic carbons not directly attached to an –OH, methylene ether carbons adjacent to oxygen, and the methylene carbons, respectively.

Mulik et al. reported the solid-state ^{13}C *NMR* spectra for the resorcinol–formaldehyde aerogels prepared with the HCl-catalyzed process using *RF*/1:2 mol/mol (Figure 11.7) [20]. The resonance at 152 ppm is assigned to phenolic carbons of substituted resorcinol. The peak at 121 ppm with a shoulder at 114 ppm is assigned to aromatic carbons of mono- and disubstituted resorcinols bearing $-CH_2-$ groups in all ortho-positions relative to the phenolic –OHs. The smaller peak at 59 ppm is assigned to a small amount of CH_2-O-CH_2 bridges and apparently decreases as the relative amount of formaldehyde is increased. The broad peak at 30 ppm with shoulder at 24 ppm is assigned to the different types of $-CH_2-$ bridges.

On comparing the solid ^{13}C *NMR* spectra of base-catalyzed *RF* aerogels with those of the HCl-catalyzed process, it is evident that the latter, being completed in minutes rather than in days, yields materials that are chemically indistinguishable from those obtained by the usual base-catalyzed process in a week.

11.5.2. Morphology

SAXS studies show that acid-catalyzed *RF* aerogels have different aggregate structure from the traditional base-catalyzed samples. It is evident that even in acid catalysis, the type of solvent, the concentration of monomers, and the concentration of the catalyst play significant roles in deciding the particle size and morphology. Barbieri et al. have shown that acid-catalyzed *RF* aerogels exhibit a fractal dimension close to 2.5, which is not observed in base-catalyzed *RF* aerogels [10]. Also, the surface area in base-catalyzed aerogels is relatively high (~800 m^2/g) compared to acid-catalyzed *RF* aerogels (~300 m^2/g). Berthon et al. attributed that difference to the particle size, which in turn may be due to two different gelation mechanisms. Although the average pore diameter is of the same magnitude (10–20 nm), in acid-catalyzed *RF* aerogels the particle diameter is significantly different [20, 33]. Under acid catalysis (low water content), percolation dominates because of slow formation of the hydroxymethyl derivative of resorcinol. As soon as hydroxymethylated resorcinol is formed, it reacts with another molecule of resorcinol to form a cluster, which eventually combines with another cluster and the process continues. In base catalysis, where water is used as solvent, the hydroxymethyl derivative of resorcinol is formed fast. Highly reactive clusters grow by draining matter from their vicinity in the sol in a nucleation-like process [36]. Because of those differences in the growth mechanism, in the case of acid-catalyzed processes polymerization leads to formation of particles, followed by phase separation and eventually gelation. Along with the concentration of monomers, aging also plays a critical role in determining shrinkage [9].

Berthon-Fabry et al. have studied multiscale structure formation as a function of the concentration of monomers and of acid. The % mass ratio (percent of monomers to the total mass of the sol) was used as a controlling parameter for the effect of monomer concentration. Formulations with % mass ratio ≥ 35 and relatively low acid catalyst concentration ($R/C > 50$) gave homogenous monoliths. With a higher monomer % mass ratio (>55) and acid catalyst concentrations ($R/C < 1$), a very viscous resin-like material was obtained. Varying the concentration of monomers and acid is an effective method to produce carbon aerogels with different morphologies [11, 37].

11.6. Alternative Approaches for *RF* Aerogels

As stated above, *RF* aerogels are precursors for carbon aerogels. Carbon (C-) aerogels are made by pyrolysis (also referred to as carbonization) of *RF* aerogels at 600–2,100°C under inert atmosphere. Since C-aerogels combine electrical conductivity with typical aerogel properties such as low density, open mesoporosity (pore sizes less than 50 nm), and high surface area, they are considered for a wide variety of applications ranging from media for separations (e.g., high pressure liquid chromatography, *HPLC*), to nonreflective panels, to materials for hydrogen storage, to fuel cell applications, and to anodes in lithium ion batteries or electrodes for supercapacitors [42–46]. This large variety of potential and existing commercial applications is largely due to the tunable properties of *RF* carbon gels. The tunable properties of *RF* carbon gels such as surface area, pore volume, and pore size distribution are easily related to the synthesis and processing conditions, which produce a wide spectrum of nanostructured materials with unique properties.

Efforts to achieve and tune desirable physical properties have led to the new area of organic aerogels prepared with similar raw materials and analogous characteristics as resorcinol and formaldehyde. The major requirements for an organic aerogel include the presence of multifunctional monomers and the ability to achieve a high crosslinking density. Many literature reports have explored phenol–furfural organic aerogels as a cheaper alternative to *RF* aerogels [16, 47].

The dark red color of *RF* aerogels limits their use in certain optical applications where the material needs to transmit light or to be colorless. This specific requirement has led to the development of melamine–formaldehyde aerogels. Similar to *RF*, melamine–formaldehyde (*MF*) aerogels are lightweight and highly porous. Hexafunctional melamine reacts with formaldehyde under alkaline conditions to form hydroxymethyl (–CH_2OH) groups (12). The role of the base is to remove and shuttle the proton from positive N to negative O in –NH_2^+–CH_2O^-. The kinetics of the reaction of formaldehyde with melamine indicates that the presence of one hydroxymethyl group on N deactivates it for a second reaction by the factor of 0.6 [48]. The reaction of 6 mol of formaldehyde with 1 mol of melamine yields a mixture of products including all levels of hydroxymethylation. Following hydroxymethylation, the base catalyst is neutralized and condensation through etherification of the hydroxymethyl groups is facilitated with acid catalysis. In some cases, depending upon the final properties required, methyl or butyl alcohol is used along with the acid catalyst. There is a substantial disagreement in the literature about the mechanism of the etherification. Both specific and general acid catalyses have been suggested to be significant [48, 49].

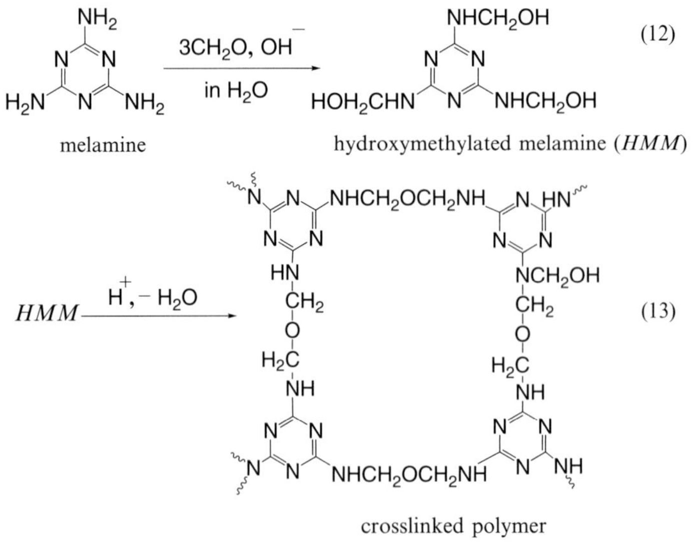

$$\text{melamine} \xrightarrow[\text{in } H_2O]{3CH_2O,\ OH^-} \text{hydroxymethylated melamine } (HMM) \quad (12)$$

$$HMM \xrightarrow{H^+,\ -H_2O} \text{crosslinked polymer} \quad (13)$$

Scheme 11.3. Gelation mechanism of the melamine–formaldehyde system.

Overall, the principal crosslinking reactions include the formation of diaminomethylene (–$NHCH_2NH$–) and diaminomethylene ether (–$NHCH_2OCH_2$–NH–) bridges [50, 51]. As an example, Scheme 11.3 shows the gelation mechanism for a reaction of 1 mol of melamine with 3 mols of formaldehyde.

All melamine–formaldehyde solutions develop a blue haze as they are cured (2 days at 50°C followed by 5 days at 95°C) [52]. This phenomenon is associated with Rayleigh scattering from *MF* clusters generated in solution. Those clusters contain surface functional groups (e.g., –CH_2OH) that eventually crosslink to form a gel. The aggregation and crosslinking processes show strong pH dependence. Finally, melamine–formaldehyde wet gels are convened to aerogels by drying with *SCF* CO_2.

Later, Wu et al. reported the self-condensation of furfural under acidic conditions; because the rate of polymerization decreases with increasing polymer chain length, they came up with phenol–formaldehyde aerogels as an alternative [15]. In all those studies it was observed that when higher quantities of formaldehyde were used, the remaining formaldehyde induced collapse of the mesoporous structure of the gels, decreasing the pore volume of the final aerogels. Whereas on the other hand, the lack of formaldehyde as a crosslinking reagent leads to weak crosslinking between the aromatic molecules, and thus, no proper gels are formed either [21]. Those observations led researchers to use other alternatives to resorcinol such as cresol [53] and 5-methyl resorcinol [54, 55]. In addition to changing the morphology and pore structure, Mulik et al. reported crosslinking of *RF* aerogels with isocyanates. The hydroxyl groups of resorcinol as well as surface hydroxymethyl groups reacted with isocyanate to form urethane and polyurea interparticle linkages. Polyurea linkages formed between RF nanoparticles hold particles from collapsing during supercritical drying resulting in lower shrinkage and hence density. Native (i.e., noncrosslinked *RF* aerogels) shrunk significantly (up to 39%), while isocyanate-crosslinked *RF* aerogels shrunk in the range of 13–22% compared with their mold dimensions [56].

In addition to varying the starting materials as a means of control over the properties and structure of *RF*-type aerogels, some researchers have reported the introduction of mesoporosity by using templating agents based on silica or polystyrene beads during gelation; those templating agents are removed from the wet gels by dissolving in HF acid or toluene [45, 57, 58].

In order to expand their potential applications, recent efforts have been focused on structural modifications by doping. One of the areas attracting significant interest is the incorporation of metals, metal oxides, and metal carbides into the *RF* framework, with the aim of modifying the structure, conductivity, and catalytic activity of the resulting carbon materials [59–63]. Job et al. incorporated metal salts of Pt, Fe, or Ni in *RF* precursor solutions and observed that the pore texture can be controlled with different salts. This finding is traced to changes in the pH of the precursor solution with different metal salts before gelation [64, 65]. It has been noted that there are often interactions present between the dopants and the resulting carbon. For example, metals can catalyze partial graphitization, and the carbon matrix in turn can act as a reducing agent for the metal precursors during heat treatment. Leventis et al. demonstrated smelting of iron oxide to form metallic iron nanoparticles [66]. The inert nature of carbon nanoparticles or carbon aerogel derived from *RF* aerogels is widely accepted as potential candidate for adsorbent materials, electrodes for capacitive deionization of aqueous solutions, electrochemical double-layer capacitors and supercapacitors, gas diffusion electrodes in proton exchange membrane (*PEM*) fuel cells, and anodes in rechargeable lithium ion batteries and ozonolysis [8, 44, 66, 67].

Incorporation of metal precursors into carbon aerogel is traced back to incorporation of metal salts into *RF* solution before the sol–gel process. Simple physical adsorption of metal salts on wet gel matrices caused leaching of the metal ions during washings required by the sol–gel process. This resorted researchers to use derivatives of resorcinol (e.g., potassium 2,4-dihydrobenzoate), which can complex with the metal ions (e.g., Fe^{3+}, Co^{3+}, and Ni^{2+}) and retain them throughout processing [62, 68, 69]. Thus, there is great interest in seeking novel dopants and the investigation of the interaction between the dopants and carbon.

11.7. Commercial Applications of *RF* Aerogels

Organic aerogels are interesting materials for different applications due to their thermal, optical, and mechanical properties. *RF* aerogels were found to be better insulators than commercial fiber glass with ~6 times more thermal resistance. When compared with silica aerogels, *RF* aerogels exhibited a very low thermal conductivity of about 0.012 W/m K under ambient conditions, whereas for silica aerogels the reported minimum value is about 0.016 W/m K [70–72]. *RF* aerogels are stiffer and stronger than silica aerogels, and this also plays a role in the manufacture of better radiative heat transfer inhibitors [73, 74].

Efficient separation of gases or liquids with nanoporous membrane is an emerging technology [75]. Aerogels may provide a convenient platform for those applications. However, organic aerogels such as *RF* give freedom not only to tune the desired porosity required as per application, but also to tune the physical shape of the material (monoliths and films) in order to make the separation process more efficient.

RF aerogels are mainly pursued as precursors for nanoparticulate carbon and carbon aerogels. Thus, *RF* aerogels provide an ability to tune properties such as surface area, pore volume, pore size distribution, dielectric constants, and electrical conductivity in the final carbon particles by tuning the initial processing conditions.

Carbon nanoparticles or carbon aerogels are receiving attention as materials for energy storage. The carbon material originated from RF aerogel has changed the perspective about uses of ultracapacitor and business aspects. Many commercial applications such as adsorbent materials, electrodes for capacitive deionization of aqueous solutions, electrochemical double-layer capacitors/supercapacitors, gas diffusion electrodes in *PEM* fuel cells, and anodes in rechargeable lithium ion batteries have all been demonstrated in the literature [8, 43, 62, 67–69, 76]. Some of those applications have already been commercialized [77].

11.8. Summary

On the basis of the processing conditions, *RF* aerogels can be nano-engineered to desired properties. Tunable properties such as the surface area, pore volume, and pore size distribution are easily related to the synthesis and processing conditions, which produce a wide spectrum of nanostructured materials with unique properties. Though *RF* aerogels are attractive as precursors for carbon aerogels, they are equally important in their own right.

The most critical factors that determine the final characteristics of the *RF* aerogels are the type and catalyst concentration (with respect to the reactants), the pH of sol, and the concentration of monomers in the sol. Increasing the *R*/*C* ratio increases the particle size with a significant effect on the surface area. Increasing the *R*/*C* ratio also results in an equivalent effect of increasing the concentration of solids in the sol. By increasing the pH of the sol, the surface area, pore volume, and electrochemical double-layer capacitance of the final carbon aerogels all increase. The proper control of the gelation and curing conditions is essential for completing the polymerization reactions and associated crosslinking of the polymerized particles. The impact of these factors is most apparent on the resulting mechanical properties of the aerogels. In addition to serving as precursors for carbon aerogels, RF aerogels receive considerable attention for use as catalyst supports, gas filters, gas storage materials, as well as acoustic and thermal insulators.

References

1. Fricke J (1988) Aerogels – highly tenuous solids with fascinating properties. J Non-Cryst Solids 100: 169–173.
2. Fricke J (1992) Aerogels and their applications. J Non-Cryst Solids 147–148: 356–362.
3. Hench L, West J (1990) The sol-gel process. Chem Rev 90: 33–72.
4. Brinker C, Scherer G (1990) Sol-Gel Science: The Physics and Chemistry of Sol-Gel Processing. Academic Press Inc.
5. Carlson G, Lewis D, McKinley K, Richardson J, Tillotson T (1995) Aerogel commercialization: technology, markets and costs. J Non-Cryst Solids 186: 372–379.
6. Pekala R (1989) Organic aerogels from the polycondensation of resorcinol with formaldehyde. J Mater Sci 24: 3221–3227.
7. Pierre A, Pajonk G (2002) Chemistry of Aerogels and Their Applications. Chem Rev 102: 4243–4265.
8. Al-Muhtaseb S, Ritter J (2003) Preparation and properties of resorcinol-formaldehyde organic and carbon gels. Adv Mater 15: 101–114.
9. Mulik S, Sotiriou-Leventis C, Leventis N (2007) Time-Efficient Acid-Catalyzed Synthesis of Resorcinol-Formaldehyde Aerogels. Chem Mater 19: 6138–6144.
10. Barbieri O, Ehrburger-Dolle F, Rieker T, Pajonk G, Pinto N, Venkateswara Rao, A (2001) Small-angle X-ray scattering of a new series of organic aerogels. J Non-Cryst Solids 285: 109–115.
11. Brandt R, Fricke J (2004) Acetic-acid-catalyzed and subcritically dried carbon aerogels with a nanometer-sized structure and a wide density range. J Non-Cryst Solids 350: 131–135.
12. Pekala R, Schaefer D (1993) Structure of organic aerogels 1. Morphology and scaling. Macromoleculres 26: 5487–5493.
13. Gebert M, Pekala R (1994) Fluorescence and light-scattering studies of sol-gel reactions. Chem Mater 6: 220–226.
14. Jirglova H, Perez-Cadenas A, Maldonado-Hodar F (2009) Synthesis and Properties of Phloroglucinol- Phenol – Formaldehyde Carbon Aerogels and Xerogels. Langmuir 25: 2461–2466.
15. Wu D, Fu R, Sun Z, Yu Z (2005) Low-density organic and carbon aerogels from the sol-gel polymerization of phenol with formaldehyde. J Non-Cryst Solids 351: 915–921.
16. Mendenhall R, Andrews G, Bruno J, Albert D (2000) Phenolic aerogels by high-temperature direct solvent extraction. U S Pat Ser No 221520.
17. Durairaj R (2005) Resorcinol: Chemistry, Technology and Applications. Springer, Germany 186–187.
18. Sprung M (1941) Reactivity of phenols toward paraformaldehyde. J Am Chem Soc 63: 334–343.
19. Pizzi A, Mittal K (2003) Resorcinol Adhesive, Handbook of Adhesive Technology: Second Ed. Marcel Dekker, Inc. New York.
20. Mulik S, Sotiriou-Levetis C, Leventis N (2006) Acid-catalyzed time-efficient synthesis of resorcinol-formaldehyde aerogels and crosslinking with isocyanates. Polym Preprints 47: 364–365.
21. Yamamoto T, Nishimura T, Suzuki T, Tamon H (2001) Control of mesoporosity of carbon gels prepared by sol-gel polycondensation and freeze drying. J Non-Cryst Solids 288: 46–55.
22. Tamon H, Ishizaka H, Mikami M, Okazaki M (1997) Porous structure of organic and carbon aerogels synthesized by sol-gel polycondensation of resorcinol with formaldehyde. Carbon 35: 791–796.

23. Fung, A W P, Reynolds G A M, Wang Z, Dresselhaus M, Dresselhaus G, Pekala R (1995) Relationship between particle size and magnetoresistance in carbon aerogels prepared under different catalyst conditions. J Non-Cryst Solids 186: 200–208.
24. (a) Horikawa T, Hayashi J, Muroyama K (2004) Controllability of pore characteristics of resorcinol-formaldehyde carbon aerogel. Carbon 42: 1625–1633. (b) Horikawa T, Hayashi J, Muroyama K (2004) Size control and characterization of spherical carbon aerogel particles from resorcinol-formaldehyde resin. Carbon 42: 169–175.
25. Pahl R, Bonse U, Pekala R, Kinney J (1991) SAXS investigations on organic aerogels. J Appl Crystallogr 24: 771–776.
26. Saliger R, Bock V, Petricevic R, Tillotson T, Geis S, Fricke J (1997) Carbon aerogels from dilute catalysis of resorcinol with formaldehyde. J Non-Cryst Solids 221: 144–150.
27. Fairen-Jimenez D, Carrasco-Marin F, Moreno-Castilla C (2008) Inter- and Intra-Primary-Particle Structure of Monolithic Carbon Aerogels Obtained with Varying Solvents. Langmuir 24:2820–2825.
28. Mirzaeian M, Hall P (2009) The control of porosity at nano scale in resorcinol formaldehyde carbon aerogels. J Mater Sci 44: 2705–2713.
29. Tamon H, Ishizaka H (1998) Porous characterization of carbon aerogels. Carbon 36:1397–1399.
30. Job N, Pirard R, Marien J, Pirard J (2004) Porous carbon xerogels with texture tailored by pH control during sol-gel process. Carbon 42: 619–628.
31. Feng Y, Miao L, Tanemura M, Tanemura S, Suzuki K (2008) Effects of further adding of catalysts on nanostructures of carbon aerogels. Mater Sci Eng B: Solid-State Materials for Advanced Technology 148: 273–276.
32. Lin C, Ritter J (1997) Effect of synthesis pH on the structure of carbon xerogels. Carbon 35: 1271–1278.
33. Conceicao F, Carrott P J M, Ribeiro Carrott M. M. L (2009) New carbon materials with high porosity in the 1–7 nm range obtained by chemical activation with phosphoric acid of resorcinol-formaldehyde aerogels. Carbon 47: 1874–1877.
34. Merzbacher C, Meier S, Pierce J, Korwin M (2001) Carbon aerogels as broadband non-reflective materials. J Non-Cryst Solids 285: 210–215.
35. Fairen-Jimenez D, Carrasco-Marin F, Moreno-Castilla C (2006) Porosity and surface area of monolithic carbon aerogels prepared using alkaline carbonates and organic acids as polymerization catalysts. Carbon 44: 2301–2307.
36. Berthon S, Barbieri O, Ehrburger-Dolle F, Geissler E, Achard P, Bley F, Hecht A.-M, Livet F, Pajonk G, Pinto N, Rigacci A, Rochas C (2001) DLS and SAXS investigations of organic gels and aerogels. J Non-Cryst Solids 285: 154–161.
37. Brandt R, Petricevic R, Proebstle H, Fricke J (2003) Acetic acid catalyzed carbon aerogels. J Porous Mater 10: 171–178.
38. Reuss M, Ratke L (2008) Subcritically dried RF-aerogels catalyzed by hydrochloric acid. J Sol-Gel Sci Technol 47: 74–80.
39. Baumann T, Satcher J, Gash A (2002) Preparation of hydrophobic organic aerogels. US Pat Appl US 2002173554 A1 20021121.
40. March J (1992) Advanced Organic Chemistry, Reactions Mechanisms and Structure Fourth Edition, Wiley: New York 548–550.
41. (a) Moudrakovski I, Ratcliffe C, Ripmeester J, Wang L, Exarhos G, Baumann T, Satcher J (2005) Nuclear Magnetic Resonance Studies of Resorcinol-Formaldehyde Aerogels. J Phys Chem B 109: 11215–11222. (b) Werstler D (1986) Quantitative carbon-13 *NMR* characterization of aqueous formaldehyde resins: 2 Resorcinol-formaldehyde resins. Polymer 27: 757–64.
42. Berthon-Fabry S, Langohr D, Achard P, Charrier D, Djurado D, Ehrburger-Dolle F (2004) Anisotropic high–surface-area carbon aerogels. J Non-Cryst Solids 350: 136–144.
43. Farmer J, Fix D, Mack G, Pekala R, Poco J (1996) Capacitive deionization of NH4ClO4 solutions with carbon aerogel electrodes. J Appl Electrochem 26: 1007–1018.
44. Petricevic R, Glora M, Fricke J (2001) Planar fiber reinforced carbon aerogels for application in PEM fuel cells. Carbon 39:857–867.
45. (a) Gierszal K, Jaroniec M (2006) Carbons with Extremely Large Volume of Uniform Mesopores Synthesized by Carbonization of Phenolic Resin Film Formed on Colloidal Silica Template. J Am Chem Soc 128: 10026-10027. (b) Tao Y, Endo M, Kaneko K (2009) Hydrophilicity-Controlled Carbon Aerogels with High Mesoporosity. J Am Chem Soc 131: 904–905.

46. Marie J, Berthon-Fabry S, Achard P, Chatenet M, Pradourat A, Chainet E (2004) Highly dispersed platinum on carbon aerogels as supported catalysts for PEM fuel cell-electrodes: comparison of two different synthesis paths. J Non-Cryst Solids 350: 88–96.
47. Pekala R (1995) Organic aerogels from the sol-gel polymerization of phenolic-furfural mixtures. US 5476878 A 19951219.
48. Wicks Z, Jones F, Pappas S (1994) Organic Coatings: Science and Technology, Vol. 1: Applications, Properties, and Performance. Wiley New York 84–87.
49. Raetzsch M, Bucka H, Ivanchev S, Pavlyuchenko V, Leitner P, Primachenko O (2004) The reaction mechanism of the transetherification and crosslinking of melamine resins. Macromol Symp 217: 431–443.
50. Pekala R (1992) Melamine-formaldehyde copolymer aerogels US 5081163 A 19920114.
51. Nguyen M, Dao L (1998) Effects of processing variable on melamine formaldehyde aerogel formation. J Non-Cryst Solids 225: 51–57.
52. Alviso C, Pekala R (1991) Melamine formaldehyde aerogels. Polym Preprints 32: 242–243.
53. Li W, Reichenauer G, Fricke J (2002) Carbon aerogels derived from cresol-resorcinol-formaldehyde for supercapacitors. Carbon 40: 2955–2959.
54. Perez-Caballero F, Peikolainen A.-L, Uibu M, Kuusik R, Volobujeva O, Koel M (2008) Preparation of carbon aerogels from 5-methylresorcinol-formaldehyde gels. Micropor Mesopor Mat 108: 230–236.
55. Peikolainen A.-L, Perez-Caballero F, Koel M (2008) Low-density organic aerogels from oil shale by-product 5-methylresorcinol. Oil Shale 25: 348–358.
56. Mulik S, Sotiriou-Leventis C, Leventis N (2008) Macroporous Electrically Conducting Carbon Networks by Pyrolysis of Isocyanate-Cross-Linked Resorcinol-Formaldehyde Aerogels. Chem Mater 20: 6985–6997.
57. Tanaka S, Katayama Y, Tate M, Hillhouse H, Miyake Y (2007) Fabrication of continuous mesoporous carbon films with face-centered orthorhombic symmetry through a soft templating pathway. J Mater Chem 17: 3639-3645.
58. Baumann T, Satcher J (2004) Template-directed synthesis of periodic macroporous organic and carbon aerogels. J Non-Cryst Solids 350: 120–125.
59. Bekyarova E, Kaneko K (2000) Structure and physical properties of tailor-made Ce, Zr-doped carbon aerogels. Adv Mater 12: 1625–1628.
60. Baumann T, Fox G, Satcher J, Yoshizawa N, Fu R, Dresselhaus M (2002) Synthesis and Characterization of Copper-Doped Carbon Aerogels. Langmuir 18: 7073–7076.
61. Baumann T, Worsley M, Han T, Satcher J (2008) High surface area carbon aerogel monoliths with hierarchical porosity. J Non-Cryst Solids 354: 3513–3515.
62. Baumann T, Satcher J (2003) Homogeneous Incorporation of Metal Nanoparticles into Ordered Macroporous Carbons. Chem Mater 15: 3745–3747.
63. Maldonado-Hodar F, Perez-Cadenas A, Moreno-Castilla C (2003) Morphology of heat – treated tungsten doped monolithic carbon aerogels. Carbon 41: 1291–1299.
64. Job N, Pirard R, Marien J, Pirard J (2004) Synthesis of transition metal-doped carbon xerogels by solubilization of metal salts in resorcinol-formaldehyde aqueous solution. Carbon 42: 3217–3227.
65. Maldonado-Hodar F, Ferro-Garcia M, Rivera-Utrilla J, Moreno-Castilla C (1999) Synthesis and textural characteristics of organic aerogels, transition metal-containing organic aerogels, and their carbonized derivatives. Carbon 37: 1199–1205.
66. Leventis N, Chandrasekaran N, Sotiriou-Leventis C, Mumtaz A (2009) Smelting in the age of nano: iron aerogels. J Mater Chem 19: 63–65.
67. Moreno-Castilla C, Maldonado-Hodar F (2005) Carbon aerogels for catalysis applications: An overview. Carbon 43: 455–465.
68. Job N, Pirard R, Vertruyen B, Colomer J, Marien J, Pirard J (2007) Synthesis of transition metal – doped carbon xerogels by cogelation. J Non-Cryst Solids 353: 2333–2345.
69. Fu R, Baumann T, Cronin S, Dresselhaus G, Dresselhaus M, Satcher J (2005) Formation of graphitic structures in cobalt - and nickel – doped carbon aerogels. Langmuir 21: 2647–2651.
70. Lu X, Arduini-Schuster M, Kuhn J, Nilsson O, Fricke J, Pekala R (1992) Thermal conductivity of monolithic organic aerogels. Science 255: 971–972.
71. Yoldas B, Annen M, Bostaph J (2000) Chemical engineering of aerogel morphology formed under nonsupercritical conditions for thermal insulation. Chem Mater 12: 2475–2484.
72. Alviso C, Pekala R, Gross J, Lu X, Caps R, Fricke J (1996) Resorcinol-formaldehyde and carbon aerogel microspheres. Mater Res Soc Sym Proc 521–525.
73. Hrubesh L, Pekala R (1994) Thermal properties of organic and inorganic aerogels. J Mater Res 9:731–738.

74. Rettelbach T, Ebert H, Caps R, Fricke J, Alviso C, Pekala R (1996) Thermal conductivity of resorcinol-formaldehyde aerogels. Therm Cond 23: 407–418.
75. Homonoff E (2000) New filtration materials for the new millennium. Book of Papers – International Nonwovens Technical Conference, Dallas, TX, United States, Sept. 26–28, 2000, 8.1–8.6.
76. Sanchez-Polo M, Rivera-Utrilla J, Mendez-Diaz J, Lopez-Penalver J (2008) Metal – doped carbon aerogels new materials for water treatments. Ind Eng Chem Res 47: 6001–6005.
77. (a) Paguio R, Takagi M, Thi M, Hund F, Nikroo A, Paguio S, Luo R, Greenwood L, Acenas O, Chowdhury S (2007) Improving the wall uniformity of resorcinol formaldehyde foam shells by modifying emulsion components. Fusion Sci Technol, 51: 682–687. (b) http://www.mkt-intl.com/aerogels/index.html (c) http://www.schafercorp.com/Company/sl/rf_aerogel.htm

12

Natural Aerogels with Interesting Environmental Features: C-Sequestration and Pesticides Trapping

Thierry Woignier

Abstract Volcanic (allophanic) soils contain amorphous clays (*allophanes*), which present completely different structures and physical properties compared to usual clays. Allophane aggregates have peculiar physical features very close to that of synthetic gels: large pore volume and pore size distribution, a high specific surface area and very large water content. These volcanic soils have exceptional carbon (C) sequestration properties and are considered as sink for green house gases (C and N). Moreover, these peculiar clays have a large ability to trap pesticides found in soils. One proposes that these interesting environmental properties can be due to the peculiar structure of the *allophane* aggregates. Because of a large irreversible shrinkage during drying, the supercritical drying technique was used to preserve the porous structure and the solid structure of allophanic soils, and the fractal structure of these natural aerogels was determined at the nanoscale. It was found that the spatial extent of the fractal aggregates depends on the allophane content in soils. One also proposes that this fractal structure, analogous to the silica gel network, could explain the high carbon, nitrogen, and pesticides content in the allophanic soils. Because of high specific surface area and low transport properties, the tortuous structure of the allophane aggregates plays the role of a labyrinth which traps C, N, and pesticides in the porosity of *allophane* aggregates.

12.1. Introduction

Soils are estimated to be the largest terrestrial pool of Carbon (C), if compared with biomass and atmosphere and the interest of the ability of soil to sequester C increases because of the threat of climate change. Carbon and nitrogen are major constituents of green house gases. C-sequestration in soils is the direct or indirect removal of carbon from the atmosphere and the storage in the soil [1, 2]. Carbon enters soils through organic matter but a part returns to the atmosphere through respiration. Respiration is the CO_2 release resulting from the decomposition of soil organic matter. It is thus important to know more about the

T. Woignier (✉) • UMR CNRS 6116-UMR IRD 193- IMEP- IRD-PRAM,
97200 Le Lamentin, Martinique, France
and
CNRS-Université Montpellier II, Place E. Bataillon, 34095 Montpellier Cedex 5, France
e-mail: woignier@univ-montp2.fr

M.A. Aegerter et al. (eds.), *Aerogels Handbook*, Advances in Sol-Gel Derived Materials and Technologies, DOI 10.1007/978-1-4419-7589-8_12,

processes which could limit soil carbon respiration. Volcanic (allophanic) soils exhibit high organic carbon concentrations [3–7] compared to other kinds of soils and a more complete understanding of the mechanisms and factors of organic carbon dynamics in allophanic soils is necessary to identify natural sinks for C-sequestration.

Chlordecone is a pesticide very resistant to degradation, that is, with a long lifetime [8]; it was used mainly for the protection of banana plantations. Now, *chlordecone* is becoming a new source of contamination for cultivated roots as it is still polluting agricultural soils. In the literature [9], it is shown that allophanic soils retain and trap more pesticides than other kinds of soils. Therefore, allophanic soils could be highly polluted but less contaminated for crops and water, and the clay microstructure should be an important physico-chemical characteristic governing the fate of the pesticide in the environment.

So, allophanic soils present interesting environmental properties in terms of C-sequestration and pesticides retention. The surface layer of these volcanic soils comprises weathering products, *allophane*, originating from ash and volcanic glasses [4–7]. *Allophanes* are amorphous alumino silicates, in which the unit cell appears as spheres with diameter between 3 and 5 nm [4]. These alumino silicates have a low bulk density (~0.5 g/cm^3) [7] and develop large specific surface areas (~700 m^2/g) [10]. It is likely that the propensity for a soil toward C-sequestration and pesticides retention could be related to the soil structure and pore features. Therefore, the correlation between retention properties and structural features, such as pores features and fractal morphology of the solid structure is essential to gain a deeper understanding. However, the experimental techniques able to give porous and structural information generally require dried solids samples. One important feature of allophanic soils is the large water content which can be as high as 300 wt% and during a classical drying, they exhibit a large irreversible shrinkage (such as silica gels) which affects their structural properties [11]. For this reason, supercritical drying has been successfully applied to dry silica gels [12–14] and other porous materials [15, 16] with only a minimum amount of shrinkage. In analogy to those systems, supercritical drying should be useful to preserve the organization of the solid phase of allophanic soils [11, 17]. Here, we discuss the preparation of natural aerogels from allophanic soils and the characterization of pores and fractal features. Finally, the permeability of these natural aerogels is calculated and the importance of the fractal structure for chemical species trapping is discussed.

12.2. Experimental

12.2.1. Samples Preparation

Soils samples, in the vicinity of the “Montagne Pelée” volcano in La Martinique (French West Indies), were used for this study. A set of samples were conserved in closed containers to avoid evaporation and were dried by supercritical drying. The apparatus used is a Balzers Critical Point Drier; the whole procedure is similar to that previously published [11, 17]. In the following, the samples are labeled by the allophanic weight percent.

The carbon and nitrogen contents were measured with a CHN (Thermofinnigan) analyzer.

The allophanic content was measured by the method of Mizota and van Reewijk [18], using Al and Si contents extracted by oxalate and pyrophosphate.

The specific surface area (S) was measured by N_2 adsorption techniques (BET analysis) with a micromeritics ASAP 2010. The BET equation was applied to the data in the p/p_0 range of 0.05–0.30. Prior to N_2 adsorption, the out gassing conditions were 24 h at 50°C, with a vacuum of 2–4 μm Hg. A cross-sectional area of 16.2 $Å^2$ for the nitrogen molecule was assumed.

The shrinkage curve during drying was measured with a previously published procedure [19]. The water saturated soil samples were placed on a balance to measure loss of water during drying. At the same time, the linear shrinkage of the height and diameter of each sample is measured by three laser spots. The data were recorded and converted to specific volume and water content [19].

The nanoscale structure was characterized by small angle X-ray scattering (SAXS) in transmission mode (Chap. 21).The experiments were carried out on solid powders placed in 1 mm diameter glass capillaries. A copper rotating anode X-ray source (operating at 4 kW) with a multilayer focusing "Osmic" monochromator giving high flux (10^8 photons/s) and punctual collimation was employed. An "Image plate" 2D detector was used. X-ray diagrams were obtained giving the scattered intensity as a function of the wave vector q.

The transmission electron micrographs were obtained with a TEM JEOL Type 1200 EX (100 KV).

12.3. Results

12.3.1. Analogy Between Allophane Aggregates and Synthetic Gels

Table 12.1 summarizes and compares the features of silica gels and allophane aggregates: it is clear that there exists an analogy between them.

Table 12.1. Analogy between synthetic and allophane aggregate features

Allophane aggregates	Synthetic silica gels
Amorphous silica alumina	Amorphous silica
Particles 3–5 nm	Particles 1–2 nm
Porosity: ≥70%	Pore volume: 70–99%
Surface area: several hundred m^2/g	Surface area: several hundred m^2/g
Fractal structure (D not known)	Fractal structure ($D = 1.8$–2.4)
High water content	High liquid content (H_2O, alcohol)
Irreversible shrinkage during drying	Irreversible shrinkage during drying

A synthetic gel of silica is a two phase solid–liquid material, amorphous and extremely porous (70–99% of porosity). The bulk density of the solid phase is typically in the range of 0.5–0.1 g/cm^3 and the specific surface area is several hundred of m^2/g [13, 14]. The gel network can be described as an assembly of clusters (~50 nm). The clusters can be fractal (Df~1.8–2.4), each being composed by aggregates of small particles (~1–2 nm) [20–22] (Chap. 2).

Allophane aggregates are amorphous silicates and have physical features very close to that of synthetic silica gels: allophanes aggregates should have a fractal geometry [23, 24] and can be described as clusters formed by hierarchical aggregation of small particles (3–4 nm) (Figures 12.1 and 12.2). These "natural gels" have large pore volume and specific surface area [7, 10].

The literature explains that allophane aggregates come from the leaching of amorphous volcanic materials (glass and ashes) [25], the weathering leading to oversaturated

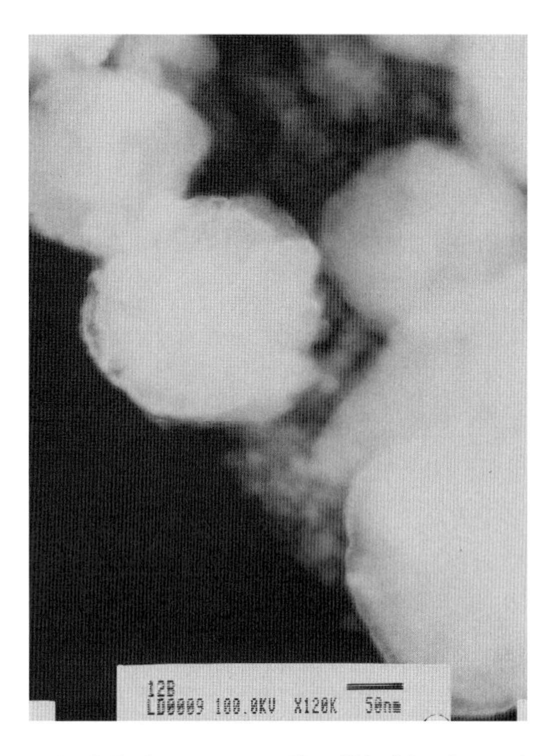

Figure 12.1. TEM micrographs of allophane aggregates. The width of the micrograph is 450 nm.

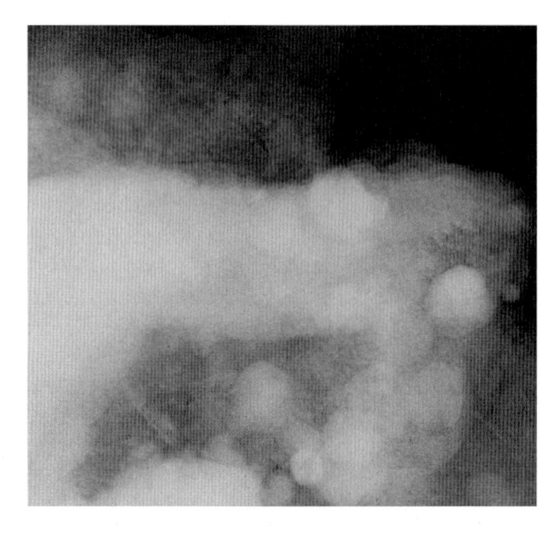

Figure 12.2. TEM micrographs of allophane aggregates. The width of the micrograph is 150 nm.

solution which forms silica aluminate sols [26]. The aggregation of colloidal particles leads to the allophane aggregates described as a "gel-like structure" [27, 28].

The formation of allophane clays and silica gels seems however to be different: a leaching mechanism for the allophane and the polymerization of silicic acid for the silica gels. However, in both cases, the network is obtained by the aggregation of colloidal particles which form loose and compliant network. It is noteworthy that the only method found in the literature to prepare synthetic allophane aggregates is the sol–gel process [29, 30]. Furthermore, the NMR and IR signatures of such sol–gel allophane aggregates are close to the ones of natural allophane aggregates [30]. For all these reasons one can consider that the allophane aggregates present in allophanic soils are "natural gels." Last but not least, during a classical drying, allophanic soils such as silica gels exhibit a large irreversible shrinkage. This point will be discussed in the following part.

12.3.2. Supercritical Drying

The experimental techniques which are able to give structural information and soil pore features generally require dried solid samples. Figure 12.3 compares the shrinkage curves during drying for allophanic soils with different allophane contents. These curves show a specific volume and water content loss 4–5 times larger for the most allophanic soils compared to the classical clay soils (allophane content 0%). In the case of allophanic soils, the shrinkage is irreversible conversely to normal clay soils (which swell and recover their volume when they are rewetted). The irreversible volume loss after drying and rewetting was measured [31] and the results show that the irreversible volume shrinkage is related to the allophane content in the soil and can be higher than 50%. Because of this large amount of liquid and the associated significant shrinkage during a classical drying, the pore structure will be strongly affected [32]. Supercritical drying is thus clearly necessary to preserve the allophane structure and correctly measure the soils pore features (Chap. 2).

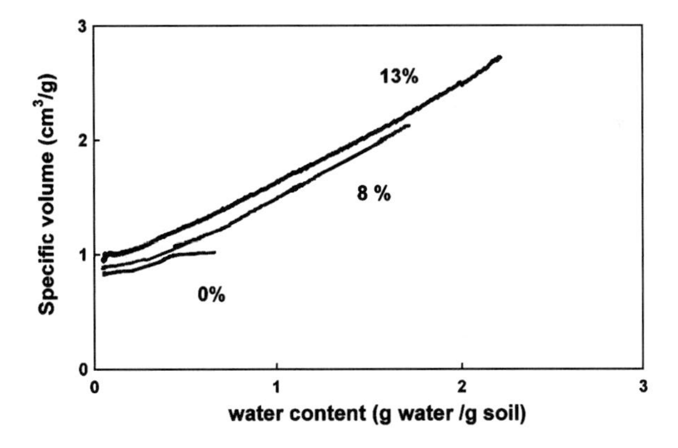

Figure 12.3. Drying curves of volcanic soils with different allophane contents.

12.3.3. Pore Properties and Fractal Structure

To understand how the allophanic soils retain more carbon and pesticides than the other clays, it is necessary to characterize the peculiar structure of the allophane aggregates. Figure 12.4 demonstrates the strong influence of the allophane content on the pore volume and specific surface area. Large specific surface area is generally the signature of an important micro-porosity contribution and thanks to Figures 12.1 and 12.2 we can describe the allophane aggregate structure as fractal in the range of a few nm up to 100 nm.

Transmission electron microscopy shows that allophane particles are nearly spherical with diameters in the range of 3.5–5 nm (by the way of comparison of the platelet-like

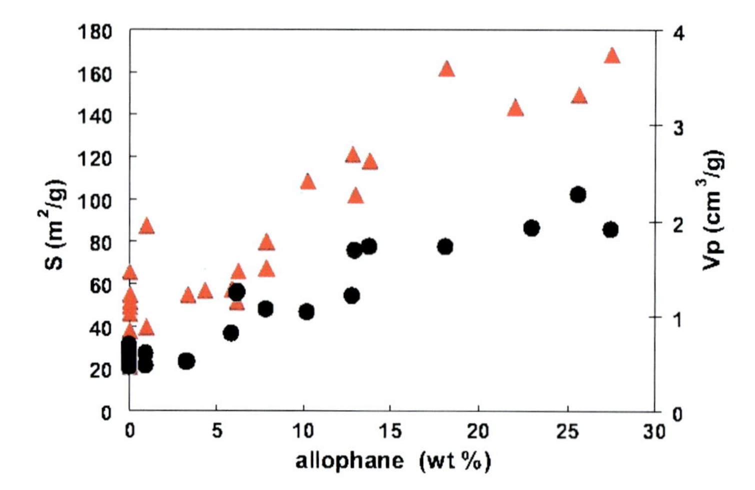

Figure 12.4. Pore volume (*black symbols*) and specific surface area (*red symbols*) versus the allophane content.

particles of kaolinite clay which are 300–3,000 nm in size) and that the allophanic particles aggregate and form clusters [4, 30]. However, TEM micrographs are not able to give quantitative information on the fractal features. To tune in on the fine structure of the soil, various samples containing different allophane contents were also characterized by SAXS.

Figure 12.5 shows the typical evolution of the scattered intensity $I(q)$ versus the wave vector (q) for the samples containing allophane aggregates on a log–log scale. The curve can be divided into different domains [22, 33, 34]. On the side of low wave vector, a little rise in

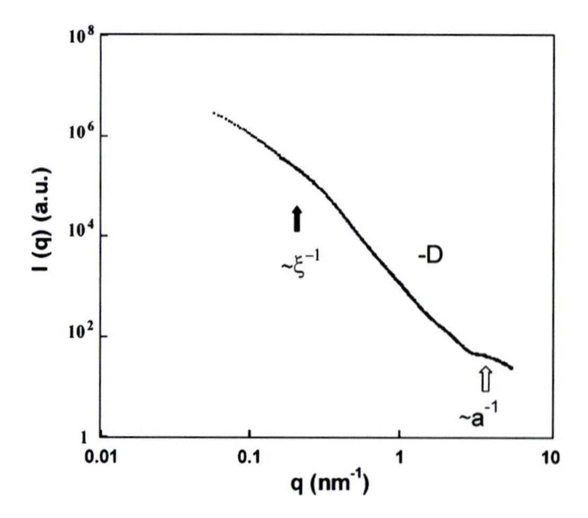

Figure 12.5. Typical SAXS curve for soil containing allophane aggregates.

the scattered intensity is observed. This rise indicates the presence of density variation in the micron range. Minerals and large pores are responsible for such a feature. In the q intermediate range, the fractal regime is observed. It spans between q values corresponding to $2\pi\xi^{-1}$ and $2\pi a^{-1}$, where ξ is the average correlation length above which the material is no more fractal (ξ may be associated to the average size of the allophanic aggregates) and a being the average allophanic particle size which built the allophane aggregates.

The power law part appearing as linear in a log–log scale has a slope related to $-D$, D the fractal dimension, which expresses the cluster compactness. The position of the two intersects (black and white arrows) allows to calculate the cluster size ξ and the particle size a. For q higher than 2 nm^{-1}, the curve shows a hump which can be attributed to the particle size (3–4 nm, white arrow). This description is in agreement with the transmission electronic microscopy results [4, 17, 23, 24]. Table 12.2 lists the evolution of ξ, D and ξ/a (the fractal extent) versus allophane content drawn from the experimental curves. The fractal extent can be considered as the size of a tortuous nano "labyrinth." Our results show that the size of the fractal labyrinth increases with the allophane content which suggests that the accessibility inside the soil microstructure will decrease.

Table 12.2. Evolution of D, ξ and ξ/a versus allophane content

Allophane weight %	ξ (nm)	D	ξ/a
3	12	2.7	4
5	23	2.6	5.7
8	32	2.7	10.6
13	35	2.5	11.7
18	45	2.6	15
26	62	2.7	15.5

12.3.4. Carbon Nitrogen and Pesticides Content in Allophanic Soils

The literature [3, 4, 6, 35] reports that allophanic soils exhibit higher organic carbon concentrations (by a factor 4) than the other clay soils (kaolinite or halloysite). However, the evolution of the carbon content as a function of the allophane content is not known. Figures 12.6 and 12.7 demonstrate that the carbon and nitrogen concentration is related to the allophane weight percent. Allophanic soils exhibit high C concentrations [6, 36] and the large C accumulation in volcanic soils is mainly attributed to a high protection of C against mineralization because of organo–Al complexes [3, 4, 37].

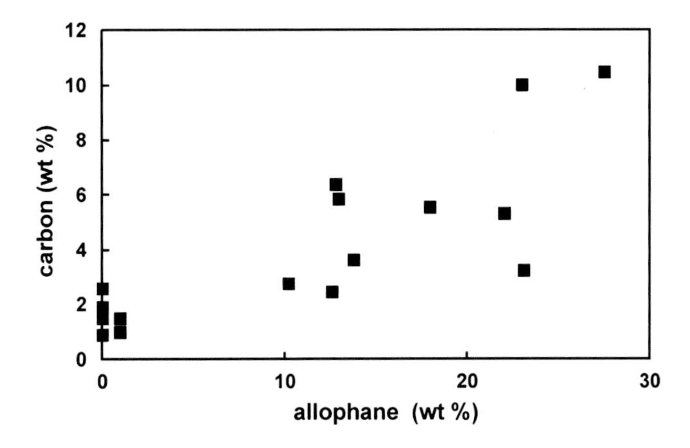

Figure 12.6. Carbon content versus allophane content.

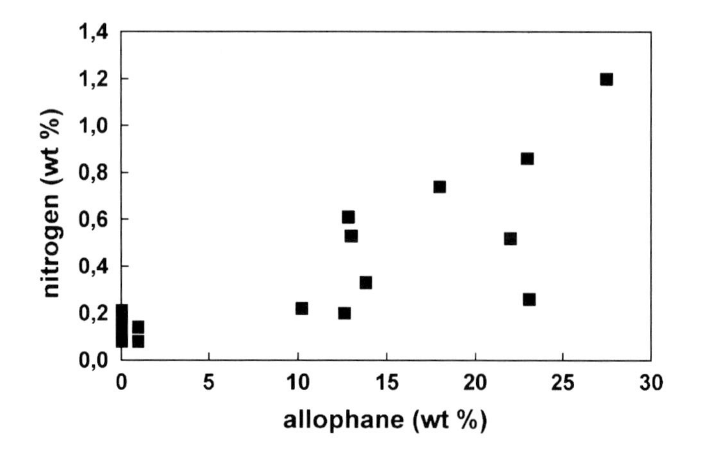

Figure 12.7. Nitrogen content versus allophane content.

However, recent results did not support the chemical protection of C bound to iron and aluminium complexes [38]. Some studies have suggested that it is the surface area of the mineral particles that can explain the slow C mineralization [39–42]. In previous studies one has shown that the carbon and nitrogen content is also related to the specific surface area [11, 17]. High specific surface areas will increase the possible adsorption sites for chemical species and certainly participate in the sequestration mechanism but the peculiar (fractal) structure of the allophane aggregate should play a role in the sequestration process. The "mesopore protection" hypothesis proposes that carbon is trapped in the tortuous porosity of the soils because the organic matter is protected from enzymatic degradation by sequestration within the mineral mesopores (2–50 nm diameter) [40, 42–44].

The stock of organic matter in soils results from the balance between roots and leaf detritus and by the efflux of carbon dioxide. So, to compare the respiration process of the different soils (C availability of the soils) the ratio g of "carbon transformed into CO_2/g per total carbon in soil" should serve as a main parameter. The experimental data [31] show that this ratio is negatively correlated with the allophane content, the presence of allophane in the soils hindering the transformation into CO_2. The slow C turnover (170,000 years) is 1,000 times larger than the values measured for several other types of soils [2, 5, 40] so, the poor ability of the allophanic soils to transform soil C in CO_2 could be the consequence of the tortuous allophane structure. The peculiar structure of the allophane aggregate features a large specific surface area favoring the fixation but hindering the chemical reaction leading to the transformation of C in CO_2.

In the case of pesticides retention, preliminary results [45] have shown that the *chlordecone* content increases with the allophane content but the *chlordecone* lability, the percent of *chlordecone* released by lixiviation experiments, decreases. So, allophanes favor the pesticides accumulation in volcanic soils but hinder the chlordecone release by weathering.

12.4. Discussions

This investigation of the pore structure details allows one to propose a possible mechanism for the C and N sequestration and pesticides retention in allophanic soils. The large specific surface area is the signature of small pore sizes. Associated to the fractal (and consequently tortuous) structure, the permeability at the scale of the allophane is very low. For chemical species located in or near the allophane aggregates, possible exchange or chemical reactions with other chemical species are difficult because of this low permeability.

The fractal dimensions found for allophone aggregates (allophane soils) are higher than the fractal dimension corresponding to the diffusion limited cluster aggregation model (DLCA, $D = 1.8$) and/or the reaction limited cluster aggregation model (RLCA, $D = 2.2$) [46] which are generally obtained for synthetic gels. However, in the case of silica aerogel, fractal dimensions close to 2.4–2.5 have been found [22]. Moreover, restructuring of the DLCA and RLCA model increases the fractal dimension [46]. The high fractal dimension that has been found in this system can be the result of a very long restructuring and structural evolution with time in the volcanic soils, corresponding to extremely long aging times in "conventional" sol–gel systems.

The fractal geometry can help to derive porous and structural properties of the allophane aggregates. In the case of a fractal aggregate it is known that the evolution of the local bulk density follows:

$$\rho(l) \propto l^{D-3} \tag{12.1}$$

When one applies the scaling relation between ξ and a (the limits of the fractal aggregates), the following relationship is obtained:

$$\rho(l) = \rho(\mathrm{a})[\mathrm{l}/\mathrm{a}]^{\mathrm{D}-3} \tag{12.2}$$

From the data in Table 12.2, the bulk density values $\rho(\xi)$ and pore volumes of the fractal aggregates can be calculated for the different samples studied. From the bulk density, one obtains the pore volume defined as

$$V_{\mathrm{p}} = \frac{1}{\rho(\xi)} - \frac{1}{\rho(a)}, \tag{12.3}$$

with $\rho(a) \approx 2.5$ g/cm^3 [33]. These data allow for an estimation of the pores associated with the allophane phase. In Figure 12.8, the pore volumes calculated based on the fractal feature data given in Table 12.2 are plotted. The red points are calculated with ξ values and $D = 2.5$ and the green ones with $D = 2.6$. Calculated pore volumes are lower but in good agreement with the data measured on the soil samples. Figure 12.8 shows that it is the peculiar fractal structure of this natural gel which is responsible for the most part for the high pore volume and likely also the high specific surface area of the allophanic soils. It gives confidence to calculate other physical properties such as permeability at the scale of the fractal aggregates.

In the literature, there exists a relationship which relates the permeability K with the relative density ρ_{r} and the mean pore size d of a porous object [13]. This relation is called the Carman–Kozeni equation

$$K \propto (1 - \rho_{\mathrm{r}})d^2 \tag{12.4}$$

This equation describes how K increases with porosity and with the size of the pores, allowing the fluid flow. This equation has been successfully applied to many types of

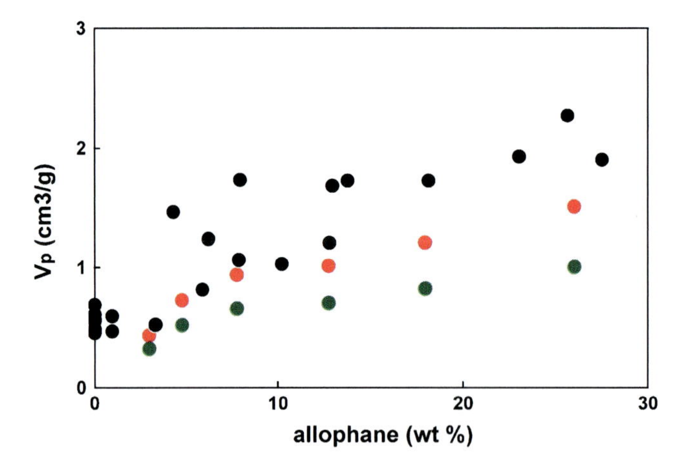

Figure 12.8. Pore volume versus allophane content: *black points* are experimental data. *Red* and *green points* are calculated from (12.3), respectively, with $D = 2.5$ and $D = 2.6$ and ξ data associated to the allophane content in Table 12.2.

granular materials [13]. In the case of a fractal aggregates, we know that the evolution of the local density goes as

$$\rho_{\mathrm{r}}(l) = \left[\frac{l}{a}\right]^{Df-3} \tag{12.5}$$

The mean pore size decreases when the length scale decreases and we find that the permeability inside the fractal aggregate varies as

$$K(l) \propto \left(1 - \left(\frac{l}{a}\right)^{Df-3}\right) l^2 \tag{12.6}$$

with $K(a) = 0$. This calculation shows that the permeability strongly decreases (3–4 orders of magnitude) when the length scale decreases between 50 and 4 nm (the limits of the allophane fractal aggregates). This means that the water and chemical species will have more difficulty to flow inside the porous and tortuous porosity of the fractal aggregates. Chemical species (C, N, pesticides) will be confined and trapped in the porosity and the chemical exchange reactions inside the fractal aggregates will be poor. This reduced permeability can be a part of the explanation concerning the retention properties of allophanic soils.

12.5. Conclusions

The porosity in volcanic soils can span a wide range of length scales including the micro-, meso- and macroporous regimes. Fluids containing inorganic and organic solutes and/or gaseous species can occupy the pores and a number of factors (size, shape, distribution and connectivity of the pore geometries) dictate how fluids migrate into and through the porosity and ultimately adsorb and react with the solid surfaces.

Allophanic soils exhibit higher carbon and nitrogen contents (among the most important components of green house gases) than those measured in other clay soils. They are thus interesting in terms of environmental properties especially because of their potential use as sinks for "greenhouse gases." Last but not least, these soils also retain pesticides in larger amounts than usual soils.

The understanding of the C-sequestration and pesticides trapping in soils necessitates to characterize the porous structure at the nanoscale range. For that, we have shown that, from a structural point of view, allophanic aggregates are very close to synthetic gels. The use of the supercritical drying to preserve the organization of the solid phase of allophanic soils has allowed to prepare natural aerogels and to get a more precise description of the porous features of the volcanic soil structure.

The low permeability calculated for fractal allophane aggregates has allows to explain the carbon sequestration and the high pesticides content of these materials. Because of the resulting low permeability, the fluid exchange is slow and the part of organic matter usually decomposed during soil respiration is lowered as the soil allophane content increases. This leads to the accumulation of carbon which under normal conditions would not be chemically transformed. These fascinating environmental properties (large water content, C-sequestration and pesticides retention) are likely due to the same factor: the fractal and tortuous structure of this natural gel.

The understanding of the retention and trapping properties in natural gels allows to propose new ideas to control the CO_2 emission and the fate of pesticides in the environment. Recently, Esquivias and co-workers [47] have proposed the synthesis of a composite silica gels containing wollastonite and this nano composite gel is able to transform CO_2 in calcite. The control of the porous properties of the composite silica gels would improve the rate of CO_2 transformation. In the case of pesticides trapping, one has shown [45] that the addition of organic matter in contaminated soils will affect the fractal structure (closing the micropores) and increase the pesticides trapping in the allophane microstructure.

References

1. Batjes, N.H. (1996) Total carbon and nitrogen in the soils of the world. European Journal of Soil Science 47(2): 151–163
2. Davidson E A, Janssens I A (2006) Temperature sensitivity of soil carbon decomposition and feedbacks to climate change. Nature 440 (9): 165–173
3. Feller C and Beare M.H. (1997) Physical control of soil organic matter dynamics in the tropics. Geoderma 79 (1-4): 69–116
4. Boudot J P, Bel Hadj, B.A. and Choné, T (1986) Carbon mineralization in Andosols and aluminium-rich highlands soils. Soil Biology & Biochemistry 18(4): 457–461
5. Wada K.J (1985). The distinctive properties of Andosol. Advances in Soil Sciences 2: 173–229
6. Basile-Doelsch I, Amundson R, Stone W E E, Masiello CA, Bottero JY, Colin F, Masin F, Borschneck D and Meunier JD (2005) Mineralogical control of organic carbon dynamics in a volcanic ash soil on La Réunion. European Journal of Soil Science 56(6): 689–703
7. Feller C, Albrecht A, Blanchart E, Cabidoche, Y M, Chevallier T, Hartmann C, Eschenbrenner V, Larre-Larrouy M C and Ndandou JF (2001) Soil organic carbon sequestration in tropical areas. General considerations and analysis of some edaphic determinants for Lesser Antilles soils. Nutrient Cycling In Agroecosystems 61(1–2): 19–31
8. Dawson G W, Weimer WC, Shupe S J (1979) Kepone-A case study of a persistent material, The American Institute of Chemical Engineers (AIChE) Symposium Series 75, 190: 366–372

9. Cabidoche Y M, Achard R, Cattan P, Clermont-Dauphin C, Massat F, Sansoulet J, (2009) Long-term pollution by chlordecone of tropical volcanic soils in the French West Indies: A simple leaching model accounts for current residue. Environmental Pollution 157: 1697–1707
10. Quentin P, Balesdent J, Bouleau A, Delaune M , Feller C (1991) Premiers stades d'altération de ponces volcaniques en climat tropical humide (Montagne Pelée , Martinique. Geoderma 79: 125–148
11. Woignier, T., Braudeau, E., Doumenc, H. and Rangon, L (2005) Supercritical drying applied to natural "gels": Allophanic soils. Journal of Sol–Gel Science and Technology. 36: 61–68
12. Kistler, S (1932) Coherent expanded aerogels. Journal of Physical Chemistry 36: 52–64
13. Brinker J F and Scherer G W (1990) Sol–Gel Science. Academic Press, N.Y.
14. Prassas M, Woignier T, Phalippou J (1990) Glasses from aerogels part 1: The synthesis of monolithic silica aerogel. J. Mater. Sci. 24: 3111–3117
15. Fillet S, Phalippou, J, Zarzycki J and Nogues JL (1986) Texture of gels produced by corrosion of radioactive waste disposal glass. Journal of Non-Crystalline Solids 82: 232–238
16. Macquet C, Thomassin J H, Woignier T (1994) Archaeological glass drying during hypercritical solvent evacuation method. Journal of Sol–Gel Science and Technology, 2: 285–291
17. Woignier T, Pochet G, Doumenc H, Dieudonne P and Duffours L (2007) Allophane: a natural gel in volcanic soils with interesting environmental properties. Journal of Sol–Gel Science and Technology 41(1): 25–30
18. Mizota C and Van Reewijk LP (1989) Clay mineralogy and chemistry of soils formed in volcanic material in diverse climatic regions Soil Monograph n°2. International Soil Reference and Information Center, Wageningen, 185 pp.
19. Braudeau E, Costantini J M , Bellier G , Colleuille H (1999) New devices and method for soil shrinkage curve measurements and characterization. Soil Sci. Soc. Amer. J., 63: 525–535
20. Lours T, Zarzycki J, Craiewich A, Dos Santos D, Aegerter M, (1987) SAXS and BET studies of aging and densification of silica aerogels. J. Non-Cryst. Solids 95&96 : 1151–1158
21. Bourret A (1988) Low density silica aerogels observed by high resolution electron microscopy. Europhysics lett. 6(8):731–737
22. Woignier T, Phalippou J, Pelous J, Courtens E (1990) Different kinds of fractal structures in Si02 aerogels. J. Non-Cryst. Solids 121:198–204
23. Adashi Y, Karube K (1999) Application of a scaling law to analysis of allophane aggregates. Colloids and surfaces A :Physicochem. Eng. Aspects 43:151–155
24. Gustafsson J P, Bhattacharya PJ, Blain DC, Fraser AR. and MacHardy W J (1995) Podzolization mechanisms and the synthesis of imogolite in northern Scandinavia. Geoderma 66: 167–174
25. Dubroeucq D, Geissert D and Quantin P (1998) Weathering and soil forming process under semi-arid conditions in two Mexican volcanic ash soils. Geoderma 86: 99–122
26. Anderson HA , Berow M L, Framer V C, Hepburn A (1982) A reassessment of podzol formation process. J. Soil Sci. 33:125–131
27. Farmer V C Lumdson D G (2001) Interaction of fulvic acid with aluminium and a proto imogolite sol. Eur. J. Soil Sci 52:177–185
28. Fieldes M, Fuckert R J (1966) The nature of allophane in soils: the significance of structural randomness in pedogenesis New Zealand J. of Science , 9(3): 608–622
29. Wada S, Eto A, Wada K (1979) Synthetic allophane and imogolite. J Soil Sci. 30: 347–355
30. Denaix L, Lamy I., Bottero JY (1999) Structure and affinity of synthetic colloidal amorphous alimino silicates and their precursors. Colloids and surfaces A :Physicochem. Eng. Aspect 158: 315–325
31. Chevallier T, Woignier T, Toucet J, Blanchart E and Dieudonné P (2008) Fractal structure in natural gels: effect on carbon sequestration in volcanic soils. Journal of Sol Gel Science and Technology 48(1–2): 231–238
32. Dorel M, Roger-Estrade J, Manichon H, Delvaux B (2000) Porosity and soil water properties of caribean volcanic ash soils. Soils use and Management 16:133–140
33. Teixeira, J(1988) Small-angle scattering by fractal systems. Journal of Applied Crystallography 21: 781–785
34. Freltof T, Kjems KJ and Sinha SK (1986) Power-law correlations and finite-size effects in silica particle aggregates studied by small-angle neutron scattering. Physical Review B 33(1): 269–275
35. Torn M, Trumbore S, Chadwick O, Vitousek P and Hendricks D (1997) Mineral control of soil organic carbon storage and turnover. Nature 389: 170–173
36. Parfitt RL, Parshotam A and Salt G J(2002) Carbon turnover in two soils with contrasting mineralogy under long-term maize and pasture. Australian Journal of Soil Research 40(1): 127–136
37. Percival H J, Parfitt R L and Scott N (2000) Factors controlling soil carbon levels in New Zealand grasslands: Is clay content important?. Soil Science Society of America Journal 64: 1623–1630

38. Buurman P, Peterse F and Almendros Martin G (2007) Soil organic matter chemistry in allophanic soils: a pyrolysis-GC/MS study of a Costa Rican Andosol catena. European Journal of Soil Science 58: 1330–1347
39. Kaiser K and Guggenberger G, 2003. Mineral surfaces and soil organic matter. European Journal of Soil Science 54(2): 219–236
40. Mayer L M (1994) Relationships between mineral surfaces and organic carbon concentrations in soils and sediments. Chemical Geology 114: 347–363
41. Mayer L M (1994). Surface area control of organic carbon accumulation in continental shelf sediments. Geochimica et Cosmochimica Acta 58: 1271–1284
42. Saggar S, Parshotam A, Sparling G P, Feltham C W and Hart P B S (1996) 14C-labelled ryegrass turnover and residence times in soils varying in clay content and mineralogy. Soil Biology & Biochemistry 28: 1677–1686
43. Mayer L M, Schick LL , Hardy KR, Wagai R and McCarthy J (2004) Organic matter in small mesopores in sediments and soils.Geochimica et Cosmochimica Acta 68(19): 3863–3872
44. Zimmerman A R, Goyne K W, Chorover J , Komarneni S. and Brantley S L (2004) Mineral mesopore effects on nitrogenous organic matter adsorption. Organic Geochemistry 35(3): 355–375
45. Jannoyer Lesueur M, Woignier T , Achard R , Calba H , (2009) Pesticide transfer from soils to plants in tropical soils: influence of clay microstructure SETAC, New Orleans 19–23 Nov 2009
46. Jullien R and Botet R (1987) Aggregation and Fractal Aggregates. World Scientific, Singapore
47. Sanchez A S, Ajabary M, Toledo Fernandez J A , Moralez Flores V, Kherbeche A, Esquivias Fedriani L M, (2008) Reactivity of C02 traps in aerogel-wollastonite composites, Journal of Sol–Gel Science and Technology, 48(1–2): 224–230

Part V

Materials and Processing: Composite Aerogels

13

Polymer-Crosslinked Aerogels

Nicholas Leventis and Hongbing Lu

Abstract Polymer-crosslinked aerogels bear a conformal polymer coating that connects covalently the skeletal nanoparticles of otherwise typical aerogels. The bulk density remains low, but the specific compressive strength of the resulting materials is higher than those of mild steel and aluminum, while the ability to store energy may surpass that of armor-grade ceramics. This chapter places polymer-crosslinked aerogels in the broader perspective of polymer/sol–gel composites, while the effects of the crosslinking chemistry and network morphology are reviewed from a mechanical property point of view.

13.1. Introduction

Aerogels were invented by Kistler in 1931 in his effort to study the microscopic structure of silica aquogels [1, 2]. Kistler preserved the network structure of the silica aquogel in the final dry object by taking the pore-filling solvent above its critical point and venting it off as a gas. This process is referred to as *supercritical fluid* (*SCF*) drying (Chap. 1). Since no gas–liquid interface residing through the gel is ever formed, no surface tension forces are ever exerted on the skeletal silica network, which keeps its shape and size, and the final low-density solid object has practically the same dimensions as the original wet gel. On the contrary, during *ambient pressure drying* of the same wet gels, surface tension forces on the skeletal framework cause excessive shrinkage and the resulting higher density materials are referred to as *xerogels*. Kistler himself in a series of publications and patents spanning from 1931 to 1943 recognized the tremendous potential of this new material in thermal insulation and in catalysis [2–8] (Chaps. 23 and 26). Kistler focused mainly on silica aerogels, and perhaps owing to his tedious preparation protocol, after Kistler, aerogels were "forgotten" for almost 25 years. They were revisited in the mid-1960s by J.B. Peri of American Oil Company and by S. Teichner's group at the University of Lyon in France starting the modern era of aerogels [9, 10]. The aerogel properties, and particularly those of silica aerogels, have been studied exhaustively [11–14], and it has been suggested that in addition to thermal insulation and catalysis, aerogels should also be appropriate for a wide

N. Leventis (✉) • Department of Chemistry, Missouri University of Science and Technology, Rolla, MO 65409, USA
e-mail: leventis@mst.edu
H. Lu • Department of Mechanical Engineering, The University of Texas at Dallas, 800 W. Campbell Road, EC-38, Richardson, TX 75080, USA
e-mail: hongbing.lu@utdallas.edu

M.A. Aegerter et al. (eds.), *Aerogels Handbook*, Advances in Sol-Gel Derived Materials and Technologies, DOI 10.1007/978-1-4419-7589-8_13,

variety of applications including acoustic insulation, dielectrics, ultrafiltration membranes, controlled drug release, oil spill cleanup kits, Cerenkov radiation detectors, and collectors of hypervelocity particles in space [15–20] (Part VIII).

Nevertheless, despite that great potential, the lack of large-scale commercial aerogel applications is also immediately apparent. The wide disparity between the real and the realizable applications is reconciled in terms of poor mechanical properties (silica aerogels are fragile materials), *hydrophilicity* (leading to environmental instability), and the need for drying from a SCF (posing cost and safety issues as *SCF* drying is a high-pressure process). Significant and remarkably successful efforts have been made to address the hydrophilicity issue and drying under ambient pressure. In fact, those two issues have been correlated and addressed together, and currently it is generally accepted that silica turned hydrophobic by surface modification with alkyl groups can also be dried under ambient pressure. This is because wet gels surface-modified with alkyl groups lack surface –OH groups. Upon drying, surface –OH groups come closer to one another, react toward new Si–O–Si bridges, and render drying shrinkage irreversible. On the contrary, although alkyl group surface-modified silica wet gels do shrink during drying, nevertheless, when all of the pore-filling solvent has evaporated, the alkyl groups repel one another and dry gels spring back to the original size of the parent wet gel. The surface modification-induced spring-back effect was pioneered by Brinker [21] and it has been adopted in several related systems. For example, Rao has reported that surface modification of sodium silicate-derived silica with trialkyl silylating agents (e.g., hexamethyldisilazane) results in only 2% volume shrinkage by ambient pressure drying [22, 23] (Chaps. 4 and 5). Rao and Kanamori reported independently flexible superhydrophobic aerogels (water contact angles 160°–164°) prepared exclusively from surface-modifying agents such as methyltrimethoxysilane (*MTMS*) and methyltriethoxysilane (*MTES*) [24–27]. Further, avoiding premature phase separation with carefully selected gelation conditions, Kanamori has shown that *MTMS*-derived transparent monolithic "xerogels" have identical properties with the corresponding aerogels obtained by *SCF* drying of identical wet gels [27].

Despite progress toward environmental stability and drying under ambient pressure, historically, the slow commercialization of aerogels is ultimately traced to their fragility. In other words, as shown by experience, aerogels can be used successfully in controlled environments as in space (Chap. 32) or in nuclear reactors as Cerenkov radiation detectors (Chap. 28) [28, 29]. Also, the one-time cost of *SCF* drying can be built into the cost structure of the product. However, the fragility issue imposes severe limitations on transportation, handling, application, and operational lifetime of the materials.

13.2. Addressing the Aerogel Fragility by Compounding with Polymers

Silica aerogels are cellular solids consisting of a pearl-necklace-like skeletal framework (Chap. 2). As it is well known in the field of ceramics, the weak points of such structures are the interparticle necks [30–33]. Along those lines, it is well established that aging of wet gels increases the strength of the final silica aerogels: Ostwald ripening with dissolution and reprecipitation of silica at surfaces with negative curvature (i.e., at the interparticle necks) results in mechanically stronger frameworks [34, 35]. However, that process is self-limiting as it strengthens the necks at the expense of the skeletal nanoparticles. A related approach involves postgelation treatment with a hydrolyzable *alkoxide* such as tetraethylorthosilicate (*TEOS*) [36]. Clearly, this approach is an improvement over simple aging, but at the end silica-reinforced silica is still silica and remains a brittle material with low tensile strength.

Aerogels could be made much more robust by increasing their tensile strength through compounding with polymers. Incorporation of polymers into aerogels has been accomplished mainly by borrowing, modifying, and adopting concepts developed for xerogels. Thus, to appreciate the issues and possibilities that exist for polymer-doped aerogels, it is necessary to place those materials within the broader perspective of polymer/sol–gel composites.

Polymers have been incorporated into silica sol–gel materials to resolve shrinkage and cracking issues during drying: as it was mentioned above, wet gels shrink a lot during ambient pressure drying; shrinking creates stresses that in turn create cracks that compromise the quality of the final *xerogels*. As polymer/sol–gel composites started being investigated from the perspective of rectifying the drying shrinkage and cracking issues, it became also apparent that those materials are usually optically transparent. This property opened possibilities for applications in optics and fueled more research on polymer/sol–gel composites.

13.3. Classification of Polymer/Sol–Gel Composites

Depending on the chemical relationship between the polymer and the surrounding silica network, polymer/sol–gel composites are placed into two categories [37]: in Class I materials, the polymer and the inorganic framework are completely independent of one another. There is no covalent bonding between them and the only interactions are through van der Waals forces, electrostatic forces, or hydrogen bonding. The two independent networks are referred to as interpenetrating (Chap. 14). In Class II materials, there is covalent bonding between the polymeric and the inorganic component. Mackenzie has further classified Class II materials into three possible models [38] (Figure 13.1): in Model 1, silica

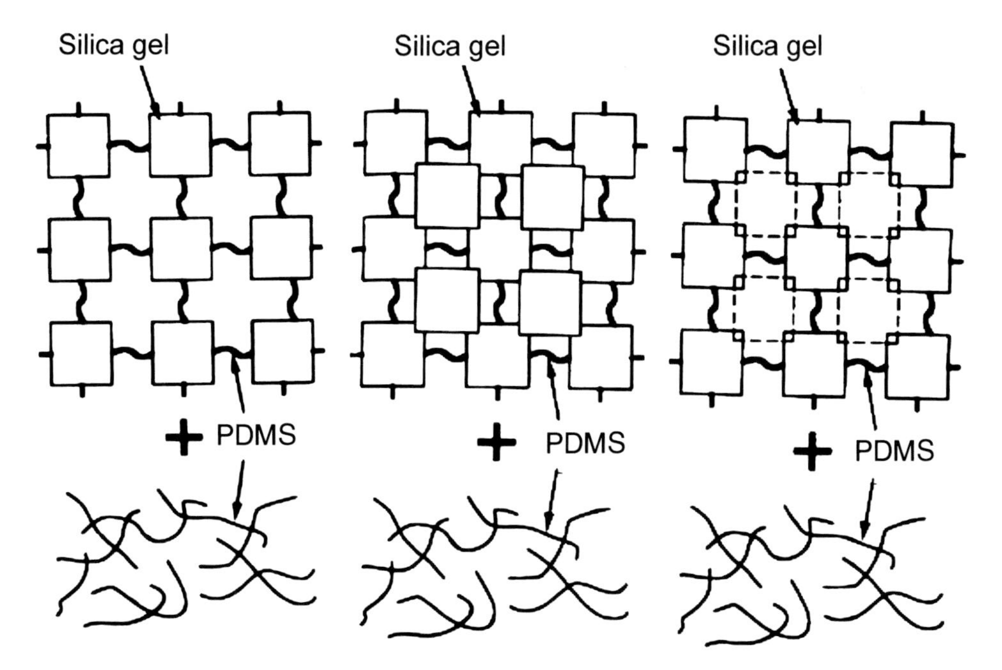

Figure 13.1. Mackenzie's three models for Class II polymer/sol–gel composites. From **left** to **right**: Model 1, Model 2, and Model 3.

particles are separated from one another, but they are connected through polymeric tethers; in Model 2, all silica particles are connected both to one another forming a continuous network and via interparticle polymeric tethers; Model 3 comprises an intermediate situation between Models 1 and 2, whereas some silica particles are connected to one another forming clusters, while covalent tethers exist between the silica particles both within the clusters and between clusters.

Interpenetrating networks (Class I materials) work sufficiently well for the purposes of xerogels. The materials are straightforward to make, mainly by including a polymer in the sol. The polymer can also be formed postgelation by polymerization of suitable monomers. In turn, monomers can be introduced in the pores by submerging a xerogel in the monomer, or they can be included in the sol and polymerization can be triggered after *gelation* is complete. Of course, in the latter case, the gelation and polymerization processes must be orthogonal, namely, the corresponding reagents should not react with one another. Since the aerogel preparation process usually requires postgelation washes (i.e., solvent exchange steps), the main issue with aerogels consisting of interpenetrating networks is the leaching of the polymer out of the pores during workup. For example, poly[methylmethacrylate] (*PMMA*) leaches out completely during processing of wet gels into aerogels [39]; on the other hand, poly[2-vinylpyridine] (*P2VP*), presumably owing to hydrogen bonding, "shows little if any leaching" [39]. Interestingly, this statement has been subject to different interpretation [40], underscoring the uncertainty associated with including polymers into a sol. Controversy put aside, a certain way to avoid leaching seems to be either via development of bonding between the polymer and the silica framework or by forming in situ a highly crosslinked (insoluble) polymer in the pores [39]. For example, while *PMMA* leaches out completely, poly[methylmethacrylate-*co*-(3-trimethoxysilyl)propylmethacrylate] (*PMMA-TMSPM*) copolymer does not: *TMSPM* develops covalent bonding with the surface of silica, and, of course, the resulting composite can no longer be classified as a Class I but rather as a Class II material. In another instance where no leaching was reported, both *N,N*-dimethylacrylamide and 1–5 mol% *N,N′*-methylenebisacrylamide were included in a tetramethylorthosilicate (*TMOS*) sol together with a free radical initiator (ammonium persulfate), leading to an interpenetrating poly[acrylamide]–silica network [39]. Mechanical characterization data show the distinct signature of the polymer [39]: for example, the stress–strain curves of *P2VP*–silica interpenetrating aerogels containing 16% (w/w) polymer show ductile yielding with a short elastic range, followed by plastic deformation and the onset of inelastic hardening. The composites show a decrease in the elastic modulus but a substantial increase (>3×) in the ultimate compressive strength. Crosslinking of *P2VP* by coordination to Cu^{2+} improves the mechanical properties further. The mechanical properties of these materials seem to be quite adequate for practical applications, but the overall impression seems to be a need for bonding between the polymer and the framework, which is a need for Class II composites.

From an aerogels perspective, Class II materials are not limited to silica. Other oxide–polymer composites are known (e.g., zirconia and tin oxide) [41] and they are very attractive, because they eliminate any possibility of leaching during postgelation washes (i.e., solvent exchange steps). One important study by Mackenzie in 1989 on silica xerogels filled completely with *PMMA* provides a direct comparison of Class I and Class II materials in terms of mechanical strength and underlines the importance of covalent bonding between the two components [42]: for Class I materials, *TEOS*-derived and meticulously dried silica xerogels were filled with methylmethacrylate/initiator (*MMA*/2% benzoyl peroxide) and polymerization was induced thermally (at 60°C for 1 week followed by 2 days at 90°C); for Class II materials, one more step was added just before the pores were filled with *MMA*: the xerogel was first treated with methacryloxypropyltrimethoxysilane, thus ensuring that all silica surfaces would be decorated with methacrylate leading to chemical bonding between the polymer filling the pores (*PMMA*) and the silica surface. The resulting Class II material clearly comprises an implementation of Model 2 in Mackenzie's classification. The compressive strength of the Class II-Model 2 version of the silica xerogel/*PMMA* composite was 22–43% higher than that of the Class I analogue [42]. Meanwhile, from a practical point of view chemical bonding between *PMMA* and a hydroxyl-rich silica/titania matrix had already been used by Schmidt since 1984 for wettable and oxygen-permeable materials for contact lenses [43]. Nevertheless, no further work was conducted on aerogels or xerogels either from a fundamental or a practical perspective.

Several in-depth studies of structure–property relationships have been conducted with hydroxyl-terminated poly[dimethylsiloxane] (*PDMS*)-filled silica aerogels. For example, Wilkes has reported that under acid-catalyzed gelation of *TEOS, PDMS* oligomers included in the sol self-condense leading to larger polymers that phase separate [44]. Silica with *PDMS* pockets have to be classified as Class I materials (even though the long *PDMS* chains are eventually connected covalently to silica framework at their ends) and they are brittle, because pockets of *PDMS* become stress concentrators [44]. If, on the other hand, the concentration of the acid catalyst is relatively high, the time period in which *PDMS* is free to self-condense is short; the gelation/polymerization process is dominated by co-condensation of silica with *PDMS*, as well as self-condensation of silica. Small-angle x-ray scattering (*SAXS*) and dynamic mechanical analysis (*DMA*) data support that shorter *PDMS* chains are better dispersed with more points of contact with the silica network. The stress–strain curve changes from a linear (brittle) to a yield-like behavior [44]. The most compliant *PDMS*-doped xerogels reported by Wilkes can be classified under Mackenzie's Model 3, although admittedly the one-pot preparation procedure and the fact that *PDMS* both self- and co-condenses with silica did not allow for much control toward any specific structure classifiable either as Model 1 or 2. *PDMS*-compounded silica gels were dried to aerogels by Mackenzie who reported rubber-like elasticity with a 30% recoverable compressive strain by incorporating 20% (w/w) *PDMS* in acid-catalyzed (HF) *TEOS*-derived aerogels [45]. Elongations at break point under tensile stress of acid-catalyzed 10% (w/w) *PDMS* aerogels (0.1 $g\,cm^{-3}$) were between 2 and 4%, which is about four times better than those of the corresponding native silica aerogels. Although Mackenzie did not report any *PDMS* leaching during workup, it seems that those materials are sensitive to the preparation protocol and leaching of PDMS was reported later by other investigators [40].

Overall, it seems that before special attention was focused to ensure connectivity of the inorganic nanoparticle framework [46], Class II aerogel composites could be classified only under either Mackenzie's Model 1 or 3. A well-known immediate predecessor of the first Class II-Model 2 aerogels consists of inorganic particles connected with tethers derived from di-isocyanates included in the sol. The isocyanate groups (–N=C=O) react in situ with the surface –OH groups of the newly formed inorganic nanoparticles, and the immediate observation is a drastic change in the gelation time [47, 48]. Thus, alumina sols normally gel in 3 days, but, in the presence of hexamethylene di-isocyanate (*HDI*), gelled within 5 min [47], probably owing to the catalytic effect of the metal ions. The bulk density of those polymer/alumina aerogel composites was significantly smaller (0.045 $g\,cm^{-3}$) than that of the corresponding native alumina aerogels (0.179 $g\,cm^{-3}$). Also, the pore size distribution of the composites was broader and the specific surface area smaller than the corresponding properties of the native alumina aerogels (209 $m^2\ g^{-1}$ vs. 594 $m^2\ g^{-1}$, respectively). The latter has been attributed to crystallization (boehmite) and coagulation of primary alumina particles [47]. On the contrary, in the presence of HDI oligomers, the gelation of either one-step base-catalyzed silica sols or two-step acid–base-catalyzed silica sols is retarded from a few minutes to about 7 days [46, 48]. This was taken as direct evidence for the reaction of surface hydroxyls with the isocyanate groups. That reaction interferes with the direct reaction between surface Si–OH that would lead to interparticle Si–O–Si bridge formation and fast gelation. To our experience, qualitatively those materials do not demonstrate any drastically improved mechanical properties over native silica [46].

13.4. Ensuring Formation of Class II-Model 2 Aerogels by Polymer Crosslinking of Preformed 3D Networks of Nanoparticles

It appears that before we visited the subject in 2002 [46], no special attention had been paid from an aerogel perspective to what Hüsing called "dual networks" [13] or to what Mackenzie described as Model 2 polymer/sol–gel composites (see Figure 13.1) [38]. Since in those materials the polymer connects points along the skeletal framework of 3D assemblies of nanoparticles that by themselves would comprise well-defined native aerogels, the resulting composites are referred to as *polymer-crosslinked aerogels*. Clearly, there are three degrees of freedom in the design of those materials: the surface chemical functionality of the skeletal nanoparticles, the chemical identity of the polymer, and the chemical identity of the skeletal framework [49].

To ensure that the polymer connects to an already fully established 3D network of nanoparticles, it was reasoned that the polymer should be applied after the silica framework is formed, pretty much in an analogous fashion to Mackenzie's *PMMA*-filled silica xerogels [42]. To our experience though, best results in terms of mechanical strength are attained when the polymer is formed in situ from monomers in the pores. This observation is in accord with Wilkes hypothesis on phase separation in long-chain *PDMS*/xerogel composites and formation of stress concentrators [44]. Now, in order for the newly formed polymer to connect particles, polymerization should either start out from or engage the surface of the skeletal nanoparticles. There should also be a mechanism by which growing polymer chains combine to form closed interparticle loops (as opposed to forming simple polymer brushes). This strategy can be implemented by introducing suitable monomers either in the *mesopores* after gelation, or in the sol with the constraint that they should not interfere with the gelation process; crosslinking should be triggerable after gelation and aging. In the last case, we refer to the gelation and crosslinking processes as orthogonal. This logic follows closely Novak's interpenetrating silica–poly[acrylamide] aerogels [38], but with additional provisions along Mackenzie's *PMMA*/xerogel composites [42] to introduce covalent bonding between the polymer and the inorganic framework. In order to avoid synthesis of monomers with trialkoxysilyl groups for every monomer considered, we opted to attach covalently a bidentate free radical initiator on the mesoporous surfaces of the inorganic framework, thus creating a generic surface that can be used with any free radical polymerizable monomer at will.

13.4.1. Crosslinking Through Postgelation Introduced Monomers

Crosslinking of Hydroxyl-Rich Surfaces with Isocyanate-Derived Polymers

In direct extension of Class II-Model 1 aerogels implemented with di-isocyanates [47, 48], the first examples of polymer-crosslinked aerogels employed the same chemistry [46]. Isocyanates used were Desmodur N3200 (a di-isocyanate) and Desmodur N3300A

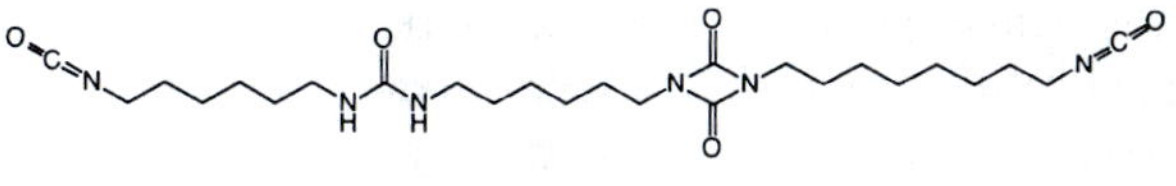

Desmodur N3200 main component

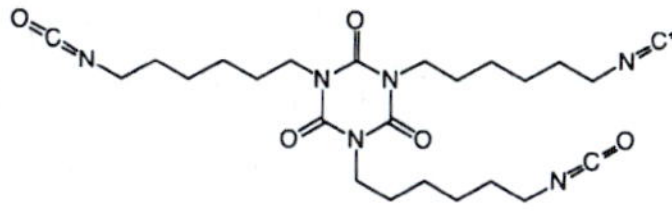

Desmodur N3300A

(a tri-isocyanate) supplied by Bayer. The isocyanate was introduced in the mesopores after gelation by washing wet gels with solutions of the monomer. The crosslinking mechanism involves both reaction of the isocyanate with the surface hydroxyl groups of silica (forming urethane linkages) and reaction of dangling isocyanates with gelation water remaining adsorbed on silica yielding amines (Scheme 13.1). Subsequently, dangling amines react with more isocyanate from the solution filling the mesopores forming interparticle polyurea tethers. Thus, the isocyanate-derived crosslinking polymer is referred to as polyurea [49, 50].

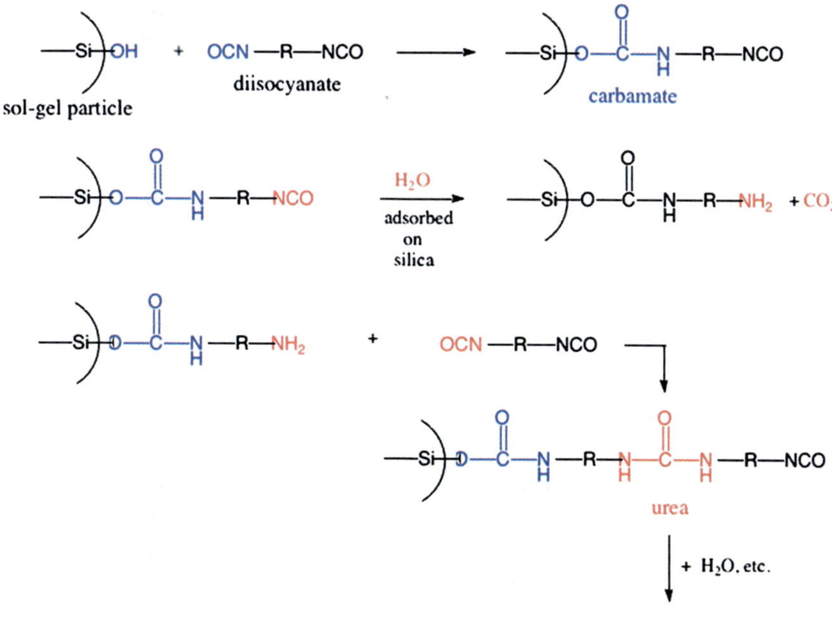

Scheme 13.1. Mechanism of crosslinking silica aerogels with a di-isocyanate [49, 50].

According to *SEM* (Figure 13.2), details on the skeletal framework remain visible, even after a bulk density increase by up to a factor of 3. The conclusion is that the skeletal silica nanoparticles have been coated conformally with polymer. This process is obviously limited by the amount of the gelation water remaining adsorbed on the skeletal framework, which in turn can be controlled to a certain extent by the number of precrosslinking washes

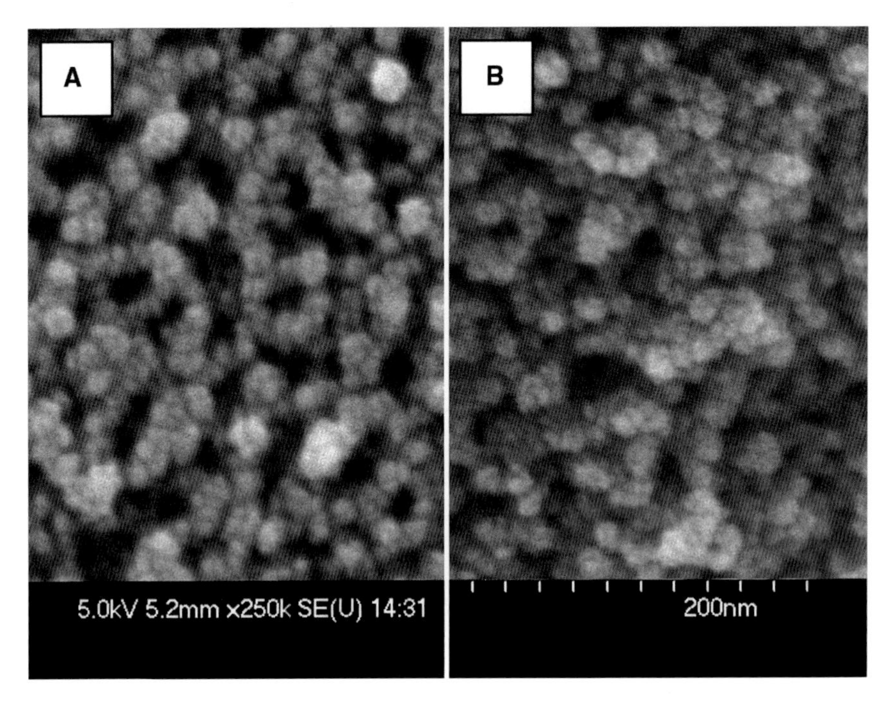

Figure 13.2. Scanning electron micrographs (*SEM*) at random spots in the interior of a fractured native silica aerogel monolith (**A.**, $\rho_b = 0.169$ g cm^{-3}) and after the skeletal network was coated (crosslinked) with Desmodur N3200-derived polyurea (**B.**, $\rho_b = 0.380$ g cm^{-3}).

with fresh solvent. Crosslinked aerogels are much less hydrophilic than native silica: for example, monolithic samples with bulk density (ρ_b) of 0.45 g cm^{-3} uptake only ~8% (w/w) water by exposure for 3 days to a moisture saturated environment, while the underlying native silica framework that yields aerogels with $\rho_b \approx 0.2$ g cm^{-3} uptake about 65% (w/w) of water. The mechanical properties are improved dramatically as well [50]. Figure 13.3A shows typical load–deformation curves in a three-point flexural bending test configuration in the spirit of ASTM D790, Procedure A (Flexural Properties of Unreinforced and Reinforced Plastics and Electronic Insulating Materials). By increasing the density, that is, by increasing the amount of the conformal polymer coating on the skeletal nanoparticles, the modulus, the ultimate stress at break point, as well as the energy absorption capability of the material (i.e., the area underneath the stress–strain curves) all increase simultaneously, meaning that at the same time the material becomes stiffer, stronger, and tougher. In terms of strength at a target density, Figure 13.3B shows that it is more efficient to start with a lower density silica skeletal framework and apply a conformal polyurea coating, rather than arrive at the target density by adding silica (i.e., by using a higher concentration of *TMOS* in the sol). Polyurea-crosslinked silica aerogels are strong enough to withstand capillary forces of evaporating liquids, making it possible to dry them under ambient pressure using low surface tension solvents (e.g., pentane and hexane) [51]. This method has been used extensively in drying the polyurea-crosslinked surfactant-templated silica aerogels described in Sect. 13.4.1.3. The absence of structural collapse during drying from low surface tension solvents has been interpreted in terms of the chemical stabilization energy gained from crosslinking, which must be higher than the work done by the surface tension forces and therefore the energy released during collapse. Since the chemical stabilization energy depends on the amount of

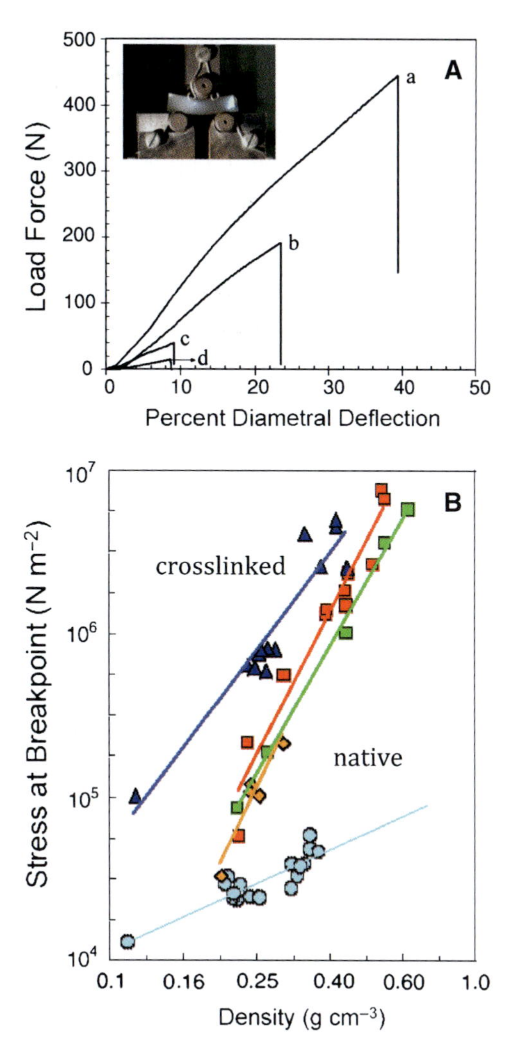

Figure 13.3. **A.** Mechanical characterization by a short beam 3-point bending (see *inset*) of polyurea-crosslinked silica aerogel monoliths and their noncrosslinked (native) counterparts; a, 0.63 $g\,cm^{-3}$; b, 0.44 $g\,cm^{-3}$; c, 0.38 $g\,cm^{-3}$, and d, 0.28 $g\,cm^{-3}$. Native samples do not register in the load–force scale shown. **B.** Cumulative data. *Dark blue triangles* and the *dark blue line* concern two-step aerogels made by acid-catalyzed hydrolysis and base-catalyzed gelation. All other samples use one-step base-catalyzed silica with different isocyanates. Multiple lines for crosslinked samples correspond to different di- and tri-isocyanate crosslinkers.

crosslinker, crosslinked silica aerogel monoliths with densities below a certain threshold (~0.3 $g\,cm^{-3}$) break into large chunks during drying from pentane, while monoliths with densities above ca. 0.4 $g\,cm^{-3}$ are dimensionally stable. The BET surface area and the average pore diameter of native samples with $\rho_b = 0.17\ g\,cm^{-3}$ are about 1,000 $m^2\,g^{-1}$ and 13.4 nm, respectively, while the corresponding values for crosslinked samples with $\rho_b = 0.39\ g\,cm^{-3}$ are 277 $m^2\,g^{-1}$ and 21.9 nm, respectively. The decrease in surface area and the increase in the average pore diameter are consistent with the crosslinking polymer blocking access to the smaller crevices of the skeletal framework (i.e., spaces within secondary particles). The BET *C* parameter (an indicator of surface polarity [52]) decreases from

80–90 to 46–48 upon crosslinking. Nevertheless, crosslinked aerogels retain a large amount of accessible porosity: based on NMR imaging (*MRI*) of crosslinked gels filled with acetone, it is calculated that samples with $\rho_b = 0.55$ g cm^{-3} consist of 80–86% (v/v) empty space vs. 89–92% in the corresponding native aerogels with $\rho_b = 0.18$ g cm^{-3}. High porosities are translated into low relative dielectric constants, falling in the range between 1.4 and 2.2 for crosslinked samples with $\rho_b = 0.48$–0.56 g cm^{-3} [50].

Crosslinking of Amine-Rich Surfaces with Isocyanate-, Epoxide-, and Styrene-Derived Polymers

Since the interparticle tethers of isocyanate-derived crosslinked aerogel consist of polyurea, and since polyureas demonstrate better mechanical properties than polyurethanes, it was reasoned to induce polyurea formation starting from the surface of the nanoparticles. That requires an amine-rich skeletal framework, which in turn is accomplished by cogelation of *TMOS* with aminopropyl triethoxysilane (*APTES*). *APTES* hydrolyzes in the same timescale as *TEOS* [53], which in turn hydrolyzes much slower than TMOS. In fact, owing to the catalytic effect of the $-NH_2$ group of *APTES*, *APTES/TMOS* sols gel extremely fast, while APTES itself does not gel. Considering those facts together leads to the conclusion that TMOS/APTES gels consist of a 3D skeletal framework of silica nanoparticles surface-modified with amines. Since the amine is a very versatile functional group, amine-modified silica was crosslinked successfully not only with polyfunctional isocyanates but also with epoxides (imitating the epoxy glue chemistry), and even with polystyrene by first attaching styrene on the surface of silica via reaction of dangling amines with chloromethyl styrene (Scheme 13.2). In all three cases, the polymer coats conformally the skeletal nanoparticles preserving the mesoporosity (Figure 13.4).

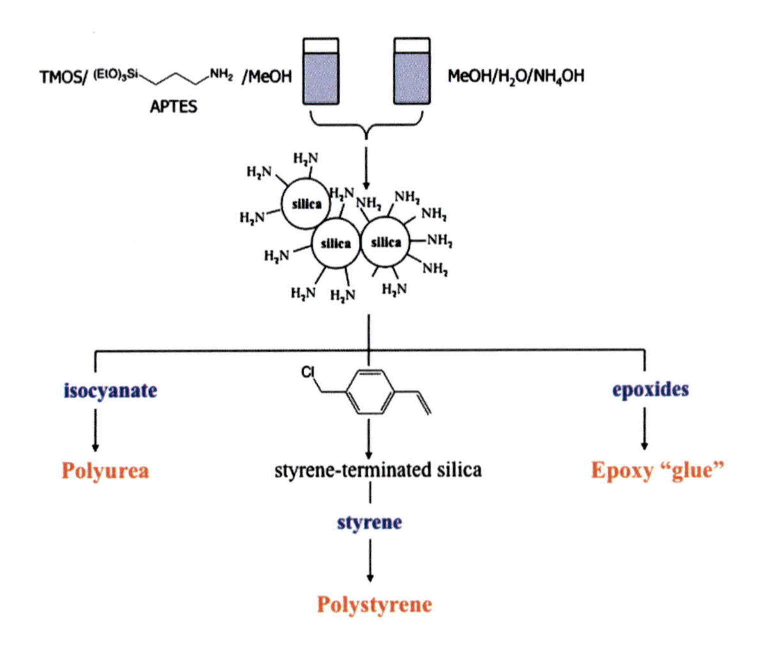

Scheme 13.2. Synthesis and crosslinking possibilities for amine-modified silica.

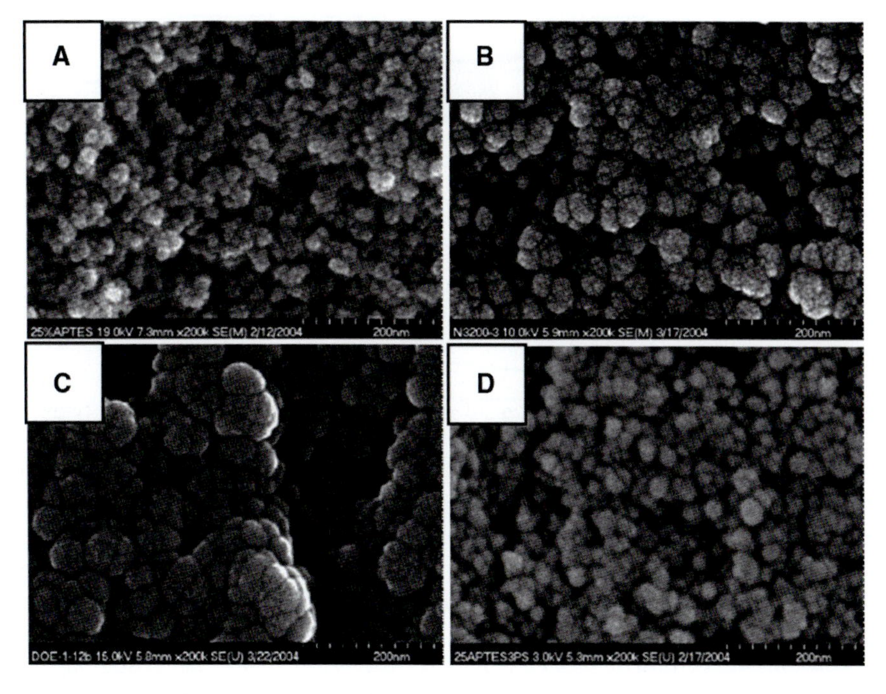

Figure 13.4. Scanning electron micrographs (*SEM*) of: **A.**, a native (noncrosslinked) amine-modified (*TMOS:APTES*, 3:1, v/v) silica aerogel ($\rho_b = 0.19$ g cm^{-3}); **B.** a silica network as in **A.**, crosslinked with polyurea ($\rho_b = 0.48$ g cm^{-3}, 261 m^2 g^{-1}); **C.** a silica network as in **A.**, crosslinked with a tri-epoxide ($\rho_b = 0.59$ g cm^{-3}, 290 m^2 g^{-1}); **D.** a silica network as in **A.**, crosslinked with polystyrene ($\rho_b = 0.48$ g cm^{-3}, 368 m^2 g^{-1}).

Polyurea-crosslinked *TMOS/APTES* aerogels by reaction with Desmodur N3200 di-isocyanate with bulk density $\rho_b = 0.48$ g cm^{-3} are translucent (as opposed to opaque, Figure 13.5) and have a thermal conductivity of about 0.041 W m^{-1} K^{-1}, which is comparable to that of glass wool [54]. Under compression, the samples do not swell or buckle showing a brief linearly elastic range (at $<4\%$ strain) followed by ductile behavior with plastic deformation (until ~40% compressive strain) and inelastic hardening

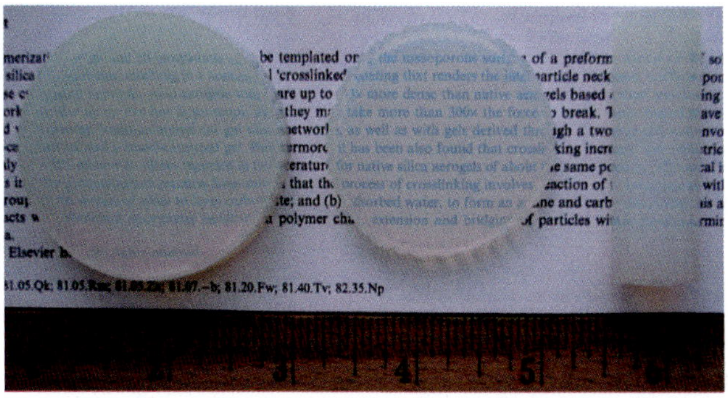

Figure 13.5. Typical amine-modified silica aerogel monoliths crosslinked with Desmodur N3200-derived polyurea ($\rho_b = 0.48$ g cm^{-3}). The disk on the **left** is about 0.5″ thick, the one in the **middle** about 0.24″ thick, and the cylinder in the **right** is similar to those used for compressive testing (Figure 13.6).

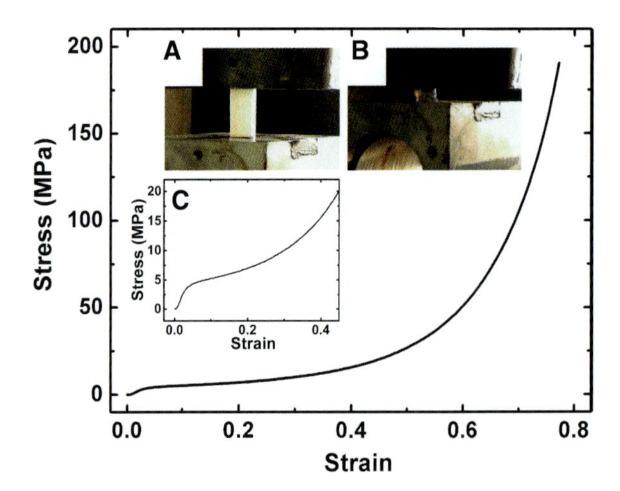

Figure 13.6. Quasi-static (strain rate = 0.0035 s^{-1}) compression testing of a polyurea-crosslinked *APTES*-modified silica aerogel cylinder ~0.5″ diameter, ~1″ long ($\rho_b = 0.48$ g cm^{-3}). **A.** The set up before testing; **B.** the set up at the point of collapse; **C.** expanded low strain range (the sample was illuminated from the *front*, and the translucency of Figure 13.5 is not apparent).

(Figure 13.6). The samples undergo lateral tensile failure at about 77% compressive strain under about 30,000 psi with an apparent Poisson ratio of 0.18 in both the linear elastic and the compaction regimes, indicating that they are absorbed within their own porosity. This is a general feature for polymer-crosslinked aerogels. The average specific compressive strength at ultimate failure is 3.89×10^5 N m kg^{-1}, which is 2.8 and 3.2 times higher than that of 4031 steel and 2024 T3 aluminum, respectively [55]. Clearly, the unique combination of translucency, high mechanical strength, and low thermal conductivity places those composite aerogels in a rather special category of multifunctional materials suitable for thermally insulating structural components. Statistical design-of-experiments (*DoE*) methods were used extensively in order to reduce the processing time of those materials by reducing or eliminating postgelation washes [56]. This could be accomplished by reaching a balance in the total silane (*TMOS* plus *APTES* in 3:1, v/v, ratio):isocyanate: water ratios. Therefore, the independent variables in the *DoE* study were chosen as the total silane concentration (s), the di-isocyanate concentration in the crosslinking bath (d), the gelation water concentration (h), and the number of postgelation washes (w). The properties monitored for the samples prepared for the *DoE* study included the bulk density, the chemical composition of the crosslinking polymer (number of monomer repeat units), the porosity, the BET surface area, the average pore diameter, the compressive modulus, the 0.2% offset yield strength, the maximum stress at break point, and the thermal conductivity [56]. Consistently with the role of the gelation water in the crosslinking mechanism (Scheme 13.1), CP MAS ^{13}C NMR data show that as the number of postgelation washes is decreased from four to none, the length of the polymer increases from three hexamethylene repeat units to more than 600. Selected material properties were plotted as a function of the four independent variables: s, d, h, w. For example, Figure 13.7 shows the variation of the density modulus and ultimate strength of polyurea-crosslinked aerogels as a function of the preparation conditions. Nonlinear least-square fitting according to a quadratic model (13.1) allowed determination of all coefficients A–O for each property monitored [56].

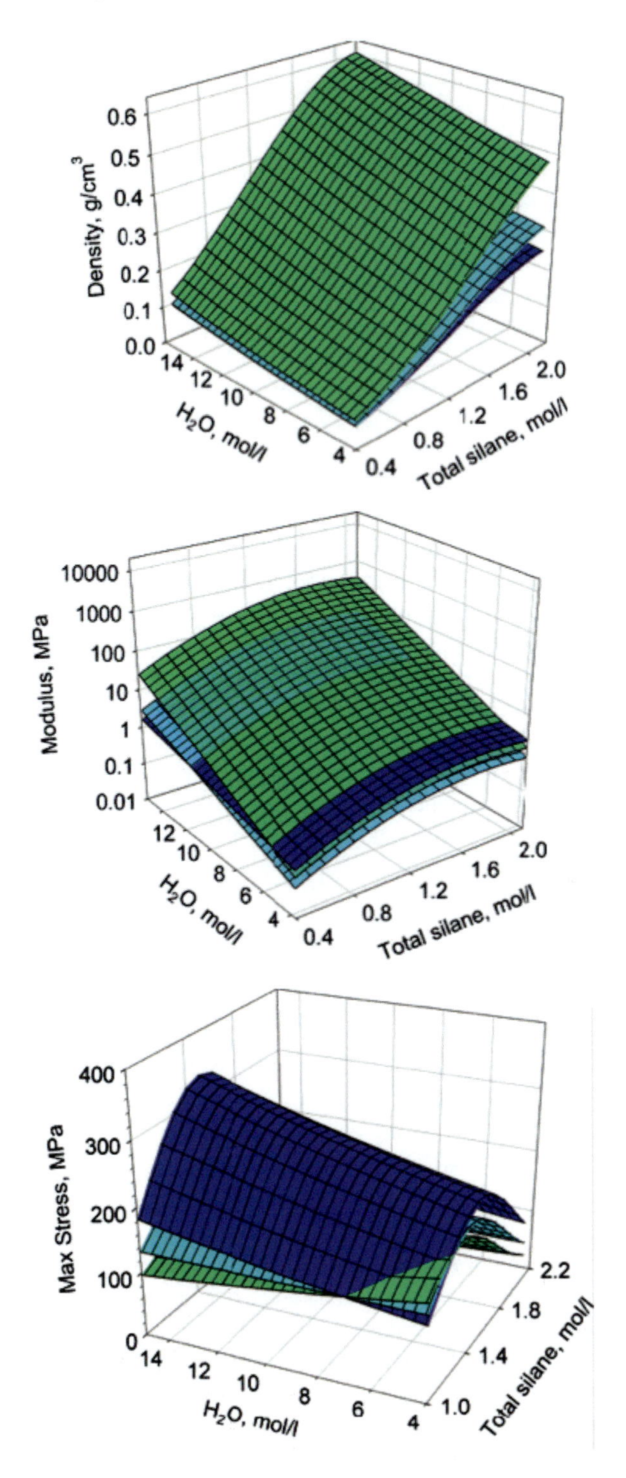

Figure 13.7. Response surface models for density, modulus, and stress at failure (by compression according to ASTM D695) of polyurea-crosslinked silica aerogels vs. total silane (*s*) and water concentration (*h*). The Desmodur N3200 concentration (*d*) was kept constant at 33.9% (w/w). *Green*: no washes; *cyan*: two washes; *blue*: four washes.

$$\begin{aligned}\text{Property} = {} & A + Bs + Cd + Dh + Ew + Fs^2 + Gd^2 + Hh^2 + Iw^2 + Jsd + Ksh \\ & + Lsw + Mdh + Ndw + Ohw\end{aligned} \tag{13.1}$$

Since the *DoE* study used four independent variables, by choosing and setting values for four desirable properties for the polyurea-crosslinked *TMOS/APTES*-derived silica aerogels (e.g., density, modulus, ultimate strength, and thermal conductivity), one can generate a system of four equations with four unknowns (s, d, h, w). Solving that system numerically yields the preparation conditions (s, d, h, w) that will produce aerogels with the four prescribed properties at the set values.

More recently, isocyanate-crosslinked *TMOS/APTES* gels has been further modified in two additional ways. First, by including up to 66 mol% of silicon coming from 1,6-bis (trimethoxysilyl)hexane (*BTMSH*), Meador created a native Class II-Model 1 network that was crosslinked with di-isocyanates and exhibited high flexibility and an ability to bend without breaking [57] in analogy to Kanamori's *MTMS* aerogels [27], but more robust (Chap. 15). Second, incorporation of up to 5% (w/w) carbon nanofibers in *TMOS/APTES* gels improves not only the mechanical properties of low-density silica aerogels ($<0.1\ \mathrm{g\,cm^{-3}}$) beyond what is achieved by polyurea crosslinking alone [58, 59], but also the processability of wet gels: reportedly, while nearly half of the lower density gels break down upon demolding, carbon fibers increase the success rate of that operation to 100%. The compressive modulus of the final fiber-reinforced crosslinked aerogels is increased threefold and the *tensile stress* at break fivefold without a significant increase in density. In higher density aerogels, particularly when the content of the crosslinking polymer is also high, the effect of the carbon fiber is smaller because the polymer alone is enough to reinforce the structure masking improvements brought about by the fibers.

In analogy to isocyanate crosslinking, which mimics polyurethane/polyurea chemistry, epoxy glue chemistry has also been adopted for crosslinking, whereas amine-modified silica reacts with multifunctional epoxies [61]. Epoxide-crosslinked aerogels based on the same *TMOS/APTES* 3:1 (v/v) formulation as above may have a density of up to $0.49\ \mathrm{g\,cm^{-3}}$ with a BET surface area of $314\ \mathrm{m^2\,g^{-1}}$ and an average pore diameter of 14.7 nm. Incorporation of *BTMSH* renders epoxide-crosslinked aerogels flexible and able to recover completely from up to 50% compressive strain [60]. CP MAS ^{13}C NMR excludes ring-opening homopolymerization of epoxides, supporting the crosslinking mechanism of Scheme 13.3.

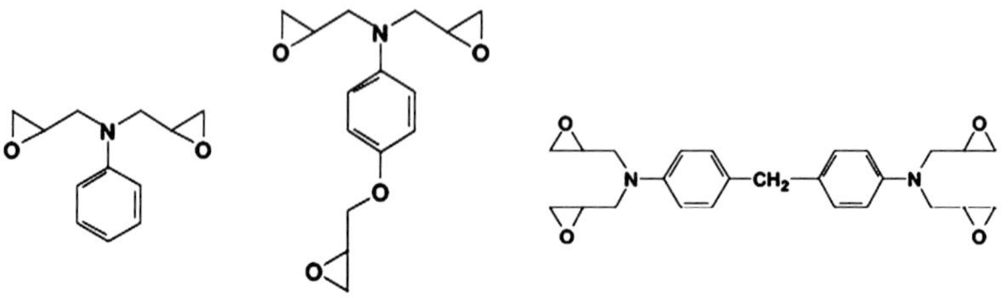

Figure 13.8 summarizes the results from a *DoE* study for density, ultimate strength at break, and modulus (by a short beam three-point bending text as shown in Figure 13.3) as a function of the preparation conditions. Using *TMOS/APTES*-derived silica, triepoxide-crosslinked samples show consistently a higher density, strength, and stiffness for almost all preparation conditions. This may be related either to the number and the spatial

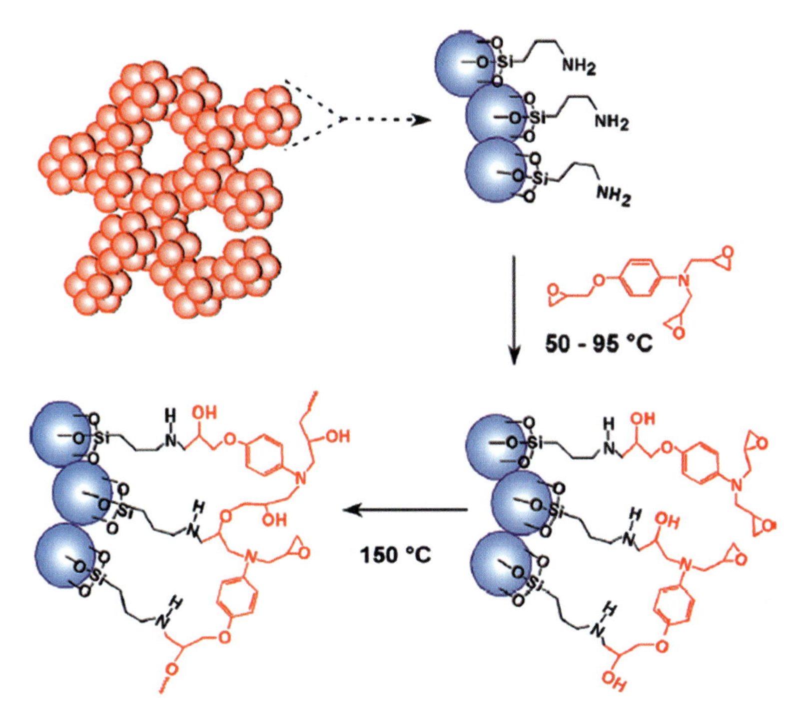

Scheme 13.3. Crosslinking mechanism of *APTES*-modified silica with a tri-epoxide [61].

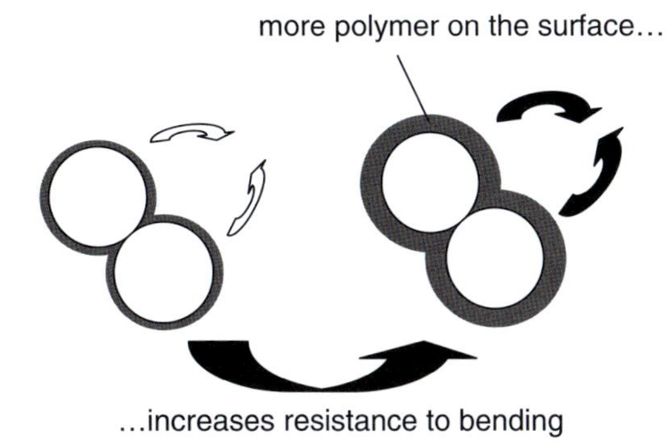

Scheme 13.4. The effect of a conformal polymer coating on the stiffness of a two-particle system.

relationship of the epoxy groups or even to the different reactivity of the specific triepoxide (note the ether linkage to the phenyl rings) [61]. In analogy to isocyanate-crosslinked *TMOS/APTES* gels (Figure 13.7), both density and modulus vary in a similar fashion as functions of the preparation parameters, meaning that stiffness varies monotonically with density for any formulation or crosslinking polymer. This is because, irrespective of the chemical identity of the polymer, increasing the particle diameter of any two-particle system by accumulation of a conformal polymer coating is expected to increase the resistance of bending (Scheme 13.4). On the other hand, the clear maxima in the stress-at-failure curves of both epoxide- and isocyanate-crosslinked samples (Figures 13.8 and 13.7, respectively) are taken as direct

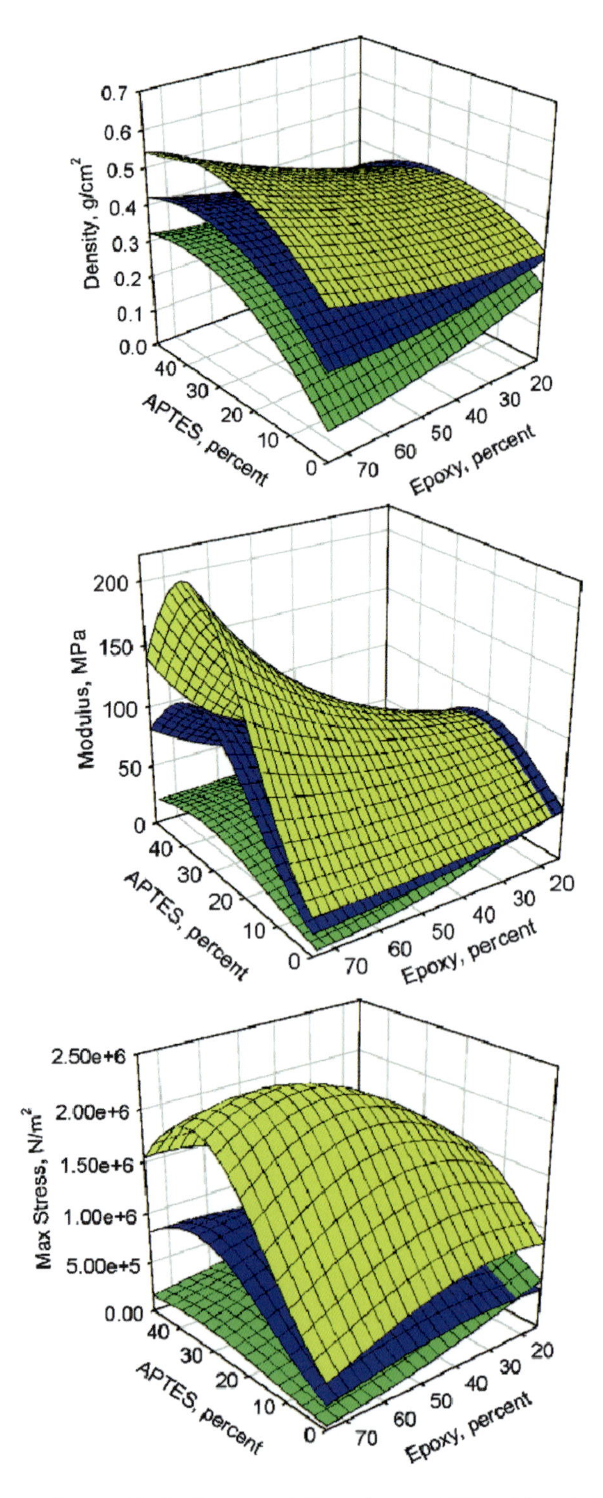

Figure 13.8. Response surface models for density, modulus, and stress at failure (via a short beam bending test method, ASTM D790) of epoxy-crosslinked silica aerogels vs. APTES and epoxide concentration with crosslinking time and bath temperature held constant at the predicted optimum value for stress at failure. *Green*: crosslinking with a di-epoxide; *blue*: with a tetra-epoxide; *yellow*: with a tri-epoxide.

evidence for the importance of building interparticle crosslinking tethers in terms of increasing the ultimate strength of the material.

Although isocyanate- and epoxide-crosslinked aerogels are much less hydrophilic than native silica, they cannot be classified as hydrophobic either. Water develops hydrogen bonding with the NH, ether, C=O, and OH groups of the interparticle tethers and ships inside the pores quickly. To impart hydrophobicity, *TMOS/APTES* gels have been crosslinked with polystyrene and pentafluorostyrene according to Scheme 13.5 [62]. First, surface amines were put to react with *p*-chloromethylstyrene; subsequently, styrene or pentafluorostyrene, together with azobisisobutyronitrile (*AIBN*) as a free radical initiator, was introduced in the *mesopores* and the samples were heated. The free radical process initiated in the mesopores engages surface-bound styrene and when radicals at the tips of polymer chains growing out of the surface meet, the radical process is terminated yielding interparticle tethers. Polystyrene-crosslinked aerogels are mechanically strong materials in analogy to their isocyanate and epoxide counterparts. Typical monoliths with a bulk density of 0.47 $g\,cm^{-3}$ show relatively high BET surface areas (370 $m^2\,g^{-1}$), small pore diameters, and thermal conductivities at 0.041 $W\,m^{-1}\,K^{-1}$. The average surface-bound polymeric tether consists of 15 monomer units (determined by both ^{13}C NMR and density change considerations) with a polydispersity of 2.05 (by gel permeation chromatography), which is higher than that of the free polymer formed in the *mesopores* (1.50), reflecting the bimodal radical termination process, whereas surface–surface and surface–solution radical combination events form shorter and longer polymer chains, respectively. Water droplets on polystyrene and poly[pentafluorostyrene]-crosslinked disks form contact angles of $121.4 \pm 0.9°$ and $151.3 \pm 1.8°$, respectively (Figure 13.9). Those contact angles do not change with time, and polystyrene-crosslinked aerogel monoliths float on water for months.

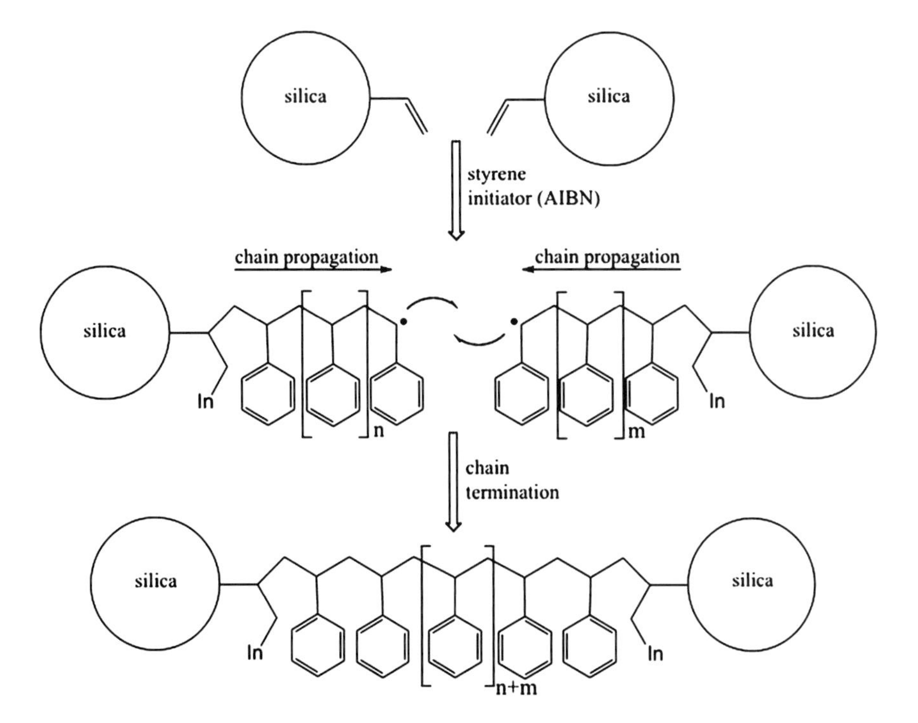

Scheme 13.5. Crosslinking mechanism of skeletal amine-modified silica nanoparticles with polystyrene.

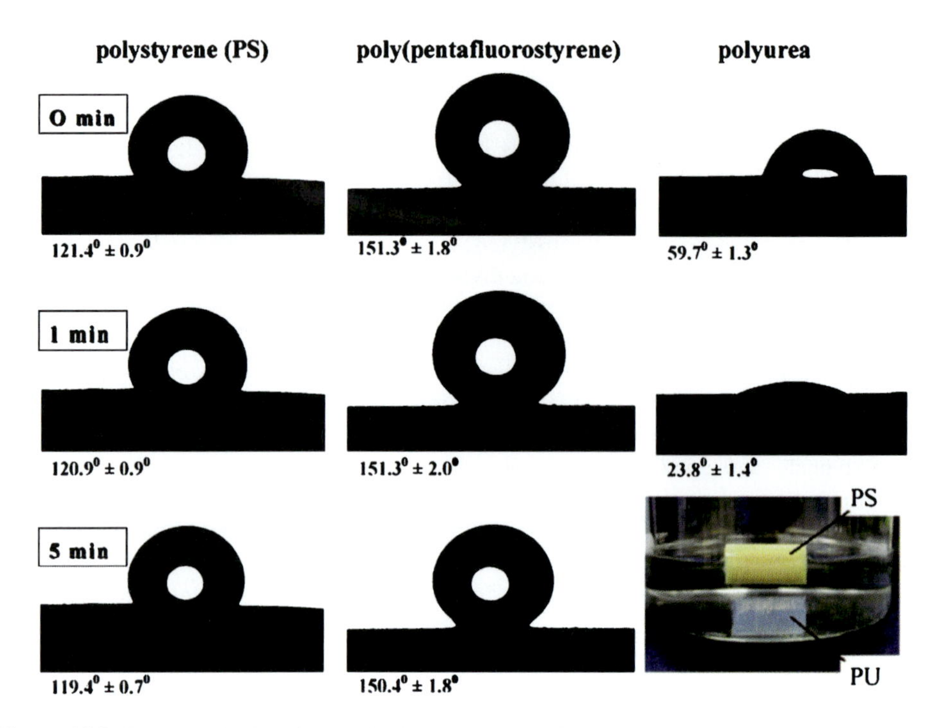

Figure 13.9. Contact angle data of water droplets on disks cut of crosslinked aerogel cylinders (all made with *TMOS/APTES* 3:1, v/v) using the crosslinkers shown on **top**. Photographs were taken immediately after the droplets were applied, 1 and 5 min later. In the case of polyurea-crosslinked samples ($\rho_b \sim 0.48$ g cm^{-3}), the water droplet was absorbed completely within 5 min. **Bottom right:** A photograph showing a polystyrene-crosslinked aerogel monolith floating on water after 4 months, while a polyurea cylinder sunk within 5 min.

By comparison (Figure 13.9), water droplets on polyurea-crosslinked aerogel disks form a contact angle of $59.7 \pm 1.3°$ and they are absorbed inside the pores within 5 min.

Derivatization of the mesoporous surface walls with styrene by reaction of *APTES* surface amines with *p*-chloromethylstyrene requires introduction of reagents in tandem by sequential washes; therefore it is time-consuming and problematic from a large-scale production perspective. Olefins have been introduced more efficiently by cogelation of *TMOS* with vinyltrimethoxysilane (*VTMS*) or trimethoxysilylpropyl methacrylate (*TMSPM*). Thus, with an eye on multifunctional sol–gel materials with potential applications in microfluidics, optics, acoustics, and rapid prototyping, Bertino used photochemically induced free radical polymerization with suitable light sources and illumination conditions and demonstrated not only bulk but also surface- and 3D-crosslinked pattern generation inside *TMOS/VTMS*-derived gels (Chap. 19) [63]. By including 9-vinylanthracene in the mesopores, the crosslinking polymer uptakes luminescent moieties and patterns become visible under black light (Figure 13.10).

With an eye on thermal insulation for space applications, Meador formulated several polystyrene-crosslinked flexible aerogels derived from a *TMOS* (or *TEOS*)/*VTMS/BTMSH* skeletal framework [64]. The best samples were identified through a statistical *DoE* study, and they were prepared with high total silane concentration (1.82 M) using *TEOS* and a high percentage of *VTMS/BTMSH*. Those samples shrunk only by 8% upon drying; their density was relatively low (0.244 g cm^{-3}), their porosity high (82%), as well as their BET surface area (221 m^2 g^{-1}). The modulus was relatively low (1.8 MPa), but samples recovered more than 98.5% of their original length after they were subjected to a 25% compressive strain twice.

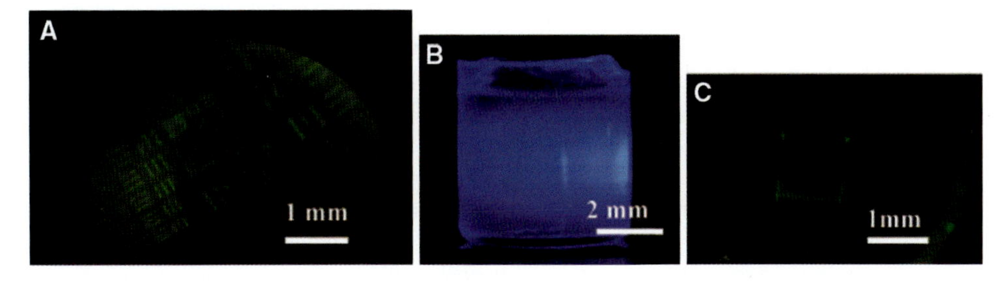

Figure 13.10. Photographs of silica monoliths modified with *VTMS* and photo-crosslinked with 9-vinylanthracene. **A.** Surface pattern produced by masking 266 nm light with a standard USAF 1951 resolution calibration mask. **B.** Pattern produced by focusing the third harmonic (355 nm) of a pulsed Nd:YAG laser inside the monolith using a 100-mm focal length lens. **C.** Pattern produced by focusing 355 nm light inside the monolith with a microscope objective lens. Samples were washed postcrosslinking with acetonitrile **B.** or methanol **A.** and **C.** Photoluminescence was exited postwashing by the 325 nm light of a He–Cd laser.

The overall polystyrene crosslinking process, whether it is implemented with *APTES*/*p*-chloromethylstyrene, *VTMS*-, or *TMSPM*-modified silica, is not different from Mackenzie's and Schmidt's copolymerization of free and surface-immobilized methylmethacrylate for filling xerogels as described above [42, 43]. The additional constraint imposed by the aerogel process is that unbound polymer formed unavoidably in the mesopores needs to be removed by postcrosslinking washes. Thus, although the polystyrene crosslinked process is fundamentally interesting in terms of increasing the strength of a 3D assembly of nanoparticles, it is also time-consuming. On the other hand, the advantage of the free radical crosslinking process is that it is orthogonal with the ionic gelation process, and therefore the two chemistries can coexist leading in principle to one-pot synthesis of crosslinked aerogels. In this regard, in order to minimize or avoid formation of free polymer in the mesopores, we resorted to surface-initiated free radical polymerization as described in Sect. 13.4.2.1.

Crosslinking of Nonsilica Aerogels and the Effect of the Inorganic Framework on Material Properties

Owing to their open mesoporous structure, aerogels have been considered as hosts for a variety of functional guests with desirable optical, electrical, magnetic, and catalytic properties for use in sensors, actuators, dielectrics, separations, and catalysis [65]. Changing the chemical identity of the inorganic framework would in effect bring about similar properties without need for external doping. Furthermore, oxide networks with morphologies different from silica could dissipate load forces differently, resulting in yet stronger materials after crosslinking. Thus, numerous metal oxide nanoparticle networks have been prepared either from the alkoxides (e.g., vanadia and titatina) or, more conveniently, from hydrated metal salts (chlorides and nitrates) via irreversible proton transfer to an epoxide, followed by ring opening according to a gelation protocol first described by Ziese in 1933 [66], then adopted by Kistler in preparing alumina, chromia, titania, and iron oxide aerogels [7], and more recently developed by Gash, Tillotson, and others in the preparation of numerous aerogels based on various transition elements in addition to the above (Ni and Cu) (Chap. 8) [67, 68]. It turns out that all nonsilica oxide aerogels are also surface-terminated with OH groups and retain gelation water; therefore they can be crosslinked with di- and tri-isocyanates in analogy to silica (Scheme 13.1). Rare earth

(*RE*) aerogels were studied in detail because of their intrinsic dielectric, magnetic, and optical (photoluminescent) properties [69]. All *RE* aerogels are nonstoichiometric amorphous materials with moderately high surface areas (368 ± 14 m^2 g^{-1} before and 156 ± 19 m^2 g^{-1} after crosslinking). The percent metal content ($58.0 \pm 2.3\%$, w/w) is significantly lower than that of pure oxides (85.4–87.9%, w/w). Even native RE aerogels contain significant amounts of carbon, mostly as carbonate (up to $9.0 \pm 2.4\%$, w/w), presumably formed during *SCF* drying (aerogels contain up to 2.5 times more carbonate than the corresponding xerogels). The gram magnetic susceptibility, χ, is a sensitive probe of the skeletal contribution to the materials properties. Aerogels have χ values lower than those of the corresponding xerogels, because of the higher carbonate content. Crosslinked aerogels exhibit χ values significantly lower than those of both xerogels and aerogels, reflecting the spin dilution by the polymer (Figure 13.11A). Despite the stoichiometric complexity, the magnetic susceptibility of *RE* aerogels varies linearly with the susceptibility of the pure oxides (Figure 13.11B), as if *RE* aerogels were also pure compounds. This is potentially a very useful feature from an applications design perspective.

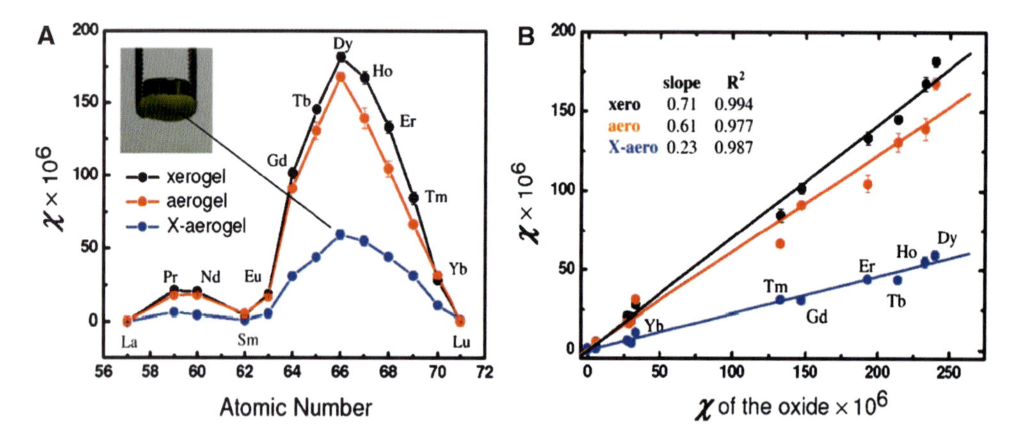

Figure 13.11. Room temperature gram magnetic susceptibilities, χ of rare earth sol–gel materials as a function of the atomic number showing the expected trend for pure compounds **A.** χs correlate linearly with the susceptibilities of the pure oxides **B.** reflecting the stoichiometric similarity within each series of materials. Slopes are proportional to the relative abundance of the metal ions. *Inset* in **A**: Polyurea-crosslinked dysprosium aerogel powder picked up by a 1-cm diameter Nd–Fe–B magnet.

Microscopically, with the exception of vanadia, all nonsilica aerogels look very similar to one another, consisting of nanoparticles, pretty much like the base-catalyzed silica of Figure 13.2, and their mechanical properties track closely those of crosslinked base-catalyzed silica. On the other hand, vanadia aerogels consist of entangled worm-like nanosized objects assembled into a bird nest-like structure (Figure 13.12) [70]. Crosslinking vanadia aerogels with Desmodur N3200 yields a conformal polymer coating around the entire skeletal framework attenuating the fine definition of the primary particles on the surface of the nanoworms (Figure 13.12B). The mechanical properties of crosslinked vanadia aerogels are impressive. Figure 13.13 compares the stress–strain curves under compression of polyurea-crosslinked nanoworm-like vanadia vs. nanoparticulate amine-modified silica (Sect. 13.4.2.2). While at room temperature (23°C), both materials show a short elastic region under small compressive strain (<4%), followed by plastic deformation with slight hardening until about 60% compressive strain, and a densification regime

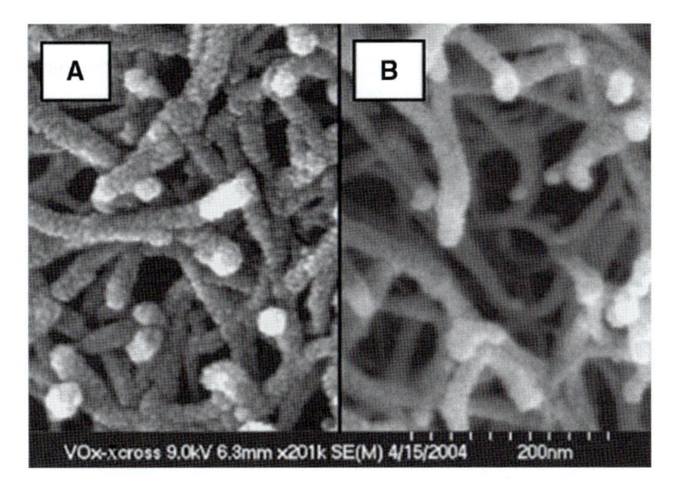

Figure 13.12. Scanning electron micrographs of native (**A.**, $\rho_b = 0.078\ g\,cm^{-3}$) and of polyurea-crosslinked (**B.**, $\rho_b = 0.420\ g\,cm^{-3}$) vanadia aerogels.

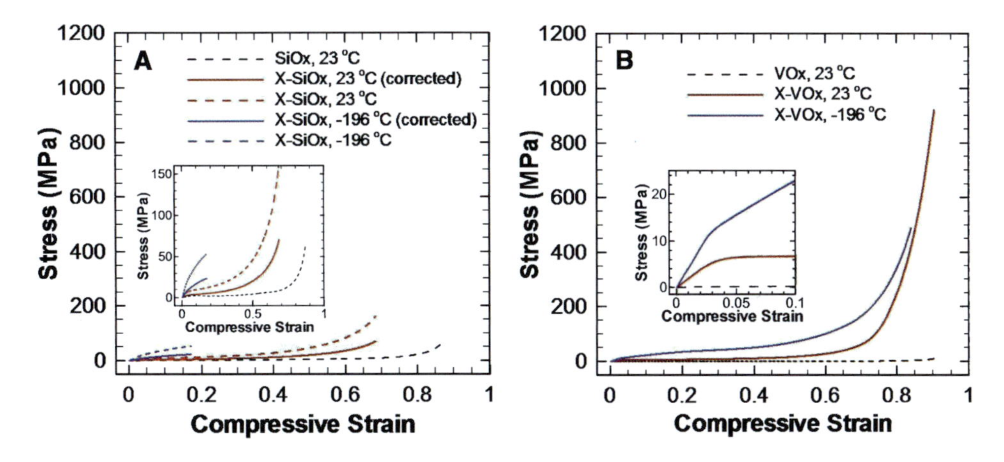

Figure 13.13. Compressive stress–strain curves of native silica (*SiO*$_x$ and *X-SiO*$_x$) and polyurea-crosslinked vanadia (*VO*$_x$ and *X-VO*$_x$) aerogels. Bulk densities; *SiO*$_x$: 0.213 $g\,cm^{-3}$; *X-SiO*$_x$: 0.548 $g\,cm^{-3}$; *VO*$_x$: 0.121 $g\,cm^{-3}$; and *X-VO*$_x$: 0.430 $g\,cm^{-3}$. **A.** Quasi-static compression of *SiO*$_x$ at a strain rate of $9 \times 10^{-4}\ s^{-1}$ at 23°C; *X-SiO*$_x$ at $5 \times 10^{-3}\ s^{-1}$ at 23°C; and *X-SiO*$_x$ at $5 \times 10^{-2}\ s^{-1}$ at −196°C; *inset*: same curves plotted using a different scale. **B.** Quasi-static compression of *VO*$_x$ at a strain rate of $9 \times 10^{-4}\ s^{-1}$ at 23°C; *X-VO*$_x$ at $5 \times 10^{-3}\ s^{-1}$ at 23°C; and *X-VO*$_x$ at $5 \times 10^{-2}\ s^{-1}$ at −196°C; *inset*: up to 10% compressive strain.

thereafter, nevertheless, unlike silica, crosslinked vanadia never fails under compression and is still capable of carrying high loads even at over 90% compressive strain. At the same bulk density ($\rho_b = 0.44\ g\,cm^{-3}$) and strain rate ($0.005\ s^{-1}$), both the Young's modulus (206 MPa) and the 0.2% offset yield strength (5.6 MPa) of polyurea-crosslinked vanadia are about double than those of polyurea-crosslinked silica (121 and 3.1 MPa, respectively). The specific ultimate strength of crosslinked vanadia (2,183 $J\,g^{-1}$) is 10× higher than that of crosslinked silica (204 MPa), and its specific energy absorption at 91% strain (192 $J\,g^{-1}$) is 6× higher than that of silica (32 $J\,g^{-1}$). In fact, the specific energy absorption of polyurea-crosslinked vanadia surpasses even that of the spider dragline silk (165 $J\,g^{-1}$, albeit under tension), which is considered the toughest natural material [71].

At cryogenic temperatures (−196°C), the stress–strain curves of crosslinked vanadia aerogels tend to be higher than those at room temperature (Figure 13.13), and the ultimate compressive strain reaches 82%, signifying that crosslinked vanadia remains ductile, a feature unusual for most bulk materials such as metals and polymers, which normally become brittle at low temperatures [72]. In sharp contrast to polyurea-crosslinked vanadia, at −196°C crosslinked amine-modified silica shatters to small pieces at 16% strain, showing a typical brittle behavior. Overall, at −196°C, the specific ultimate strength of crosslinked vanadia is 13× higher, and its specific energy absorption is 24× higher than that of polyurea-crosslinked amine-modified crosslinked silica. According to SEM (Figure 13.14), cross-linked vanadia and silica collapse very differently under compression: crosslinked vanadia appears as if the polymer has melted, flowing and fusing the worm-like objects together; on the contrary, crosslinked silica looks like as if the polymer-coated secondary particles have

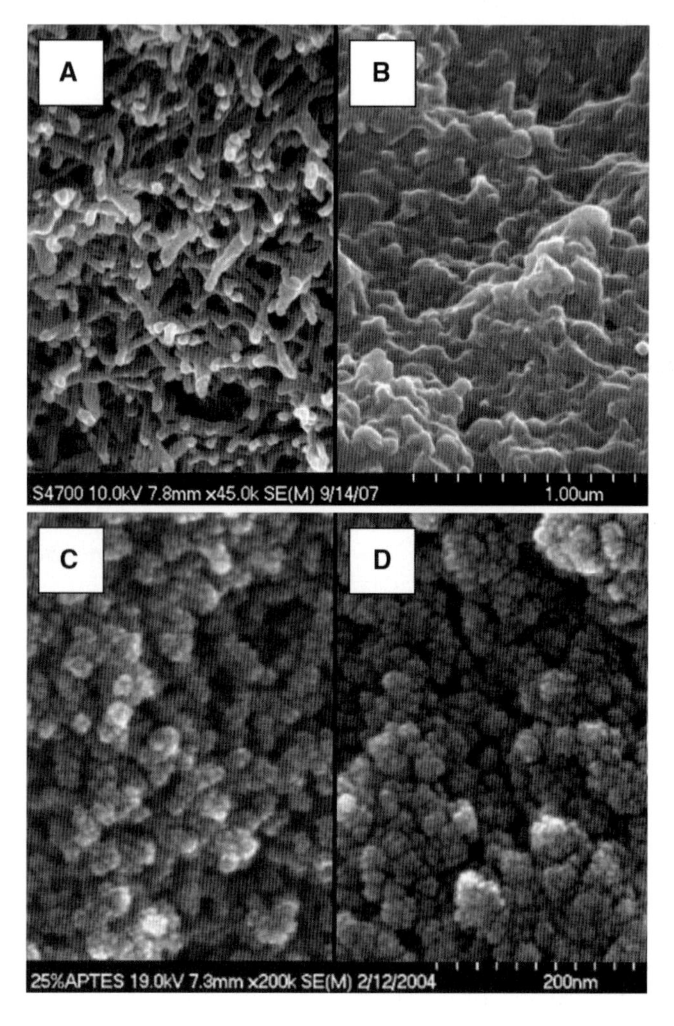

Figure 13.14. Scanning electron micrographs of polyurea-crosslinked vanadia (*X-VO$_x$*, $\rho_b = 0.444$ g cm^{-3}) and polyurea-crosslinked silica (*X-SiO$_x$*, $\rho_b = 0.480$ g cm^{-3}) before and after compression testing at room temperature (23°C). **A.** Undeformed *X-VO$_x$*. **B.** Deformed *X-VO$_x$* after compression to 91% strain at strain rate 5×10^{-3} s^{-1}. Some nanoworms are fused after compression. The specimen after compression did not break into smaller pieces, remaining as one piece. **C.** Undeformed *X-SiO$_x$*. **D.** The core of a *X-SiO$_x$* specimen after failure at 77% strain (see Figure 13.6).

been simply squeezed closer together. It has been proposed that the high stress applied at the contact points of the vanadia nanoworms depresses the melting point of polyurea, causing sintering even near cryogenic temperatures. Further, the synergistic effect of the entanglement of the skeletal nanoworms keeps the framework together for the sintering phenomena to take effect, producing a more ductile behavior than that observed with silica.

Crosslinked vanadia aerogels underline the importance of the network morphology in the mechanical behavior of crosslinked aerogels. But vanadia is expensive. Bicontinuous macro/mesoporous materials consisting of entangled worm-like nanostructures perforated by tubular mesopores (Figure 13.15A, B) have been created with acid-catalyzed silica by modification of Nakanishi's procedure [73], which in turn comprises a modification of

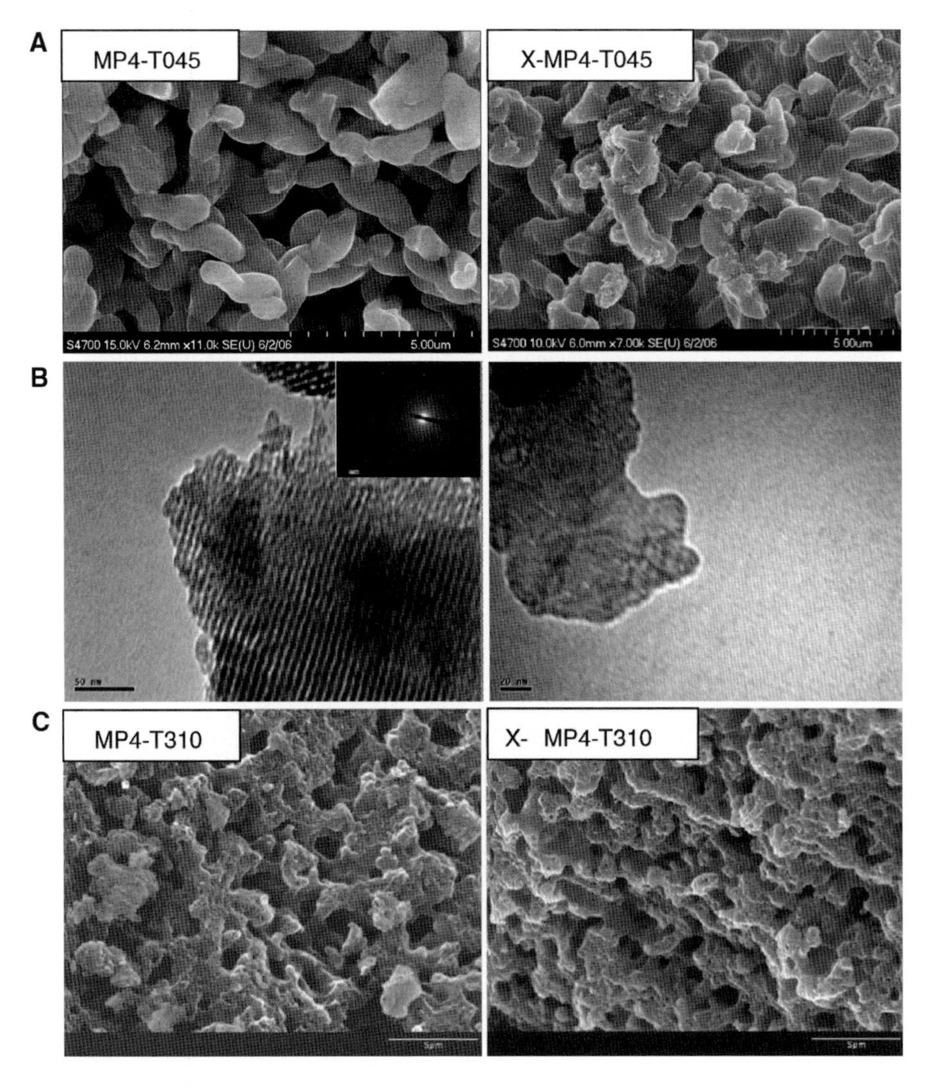

Figure 13.15. Microscopy of bicontinuous macro/mesoporous acid-catalyzed, surfactant-templated silicas. **Left column** shows native samples; **right column** shows the corresponding polyurea-crosslinked (X-) samples. (**A** and **C**) Low-resolution *SEMs* (*scale bars* at 5 μm) of samples with ordered **A.** and random (*MCF*) **C.** mesoporosity. Despite a massive polymer uptake, X-samples (contain 70–80%, w/w, polyurea) seem very similar to their native counterparts. **B.** Transmission electron micrographs (*TEM*) of the MP4-T045 and X-MP4-T045 samples shown in **A.** After polymer uptake (**right column**) the organized mesoporosity seen in the native sample at **left** is barely visible.

Stucky's use of poly[ethyleneoxide]-*block*-poly[propyleneoxide]-*block*-poly[ethyleneoxide] (*Pluronic P123*) as templating agent (surfactant) and trimethylbenzene (*TMB*) as micelle swelling agent [74–76], along the lines of M41S type of mesoporous materials [77]. The relative amounts of *Pluronic P123* and *TMB* change the nano/micromorphology of these materials predictably. For example, a relatively large amount of micelle swelling agent (*TMB*) takes all the surfactant (*Pluronic P123*) to stabilize the swelling agent as an emulsion, leading to the destruction of the ordered mesoporous structure (shown in Figure 13.15B) and to the appearance of large interconnected voids (*macropores*) in a new type of material referred to as a mesoporous cellular foam (*MCF* – Figure 13.15C) [78]. In the ordered mesoporous version of the material (Figure 13.15A, B), the concave surfaces of the 4–6 nm diameter nanotubes bring the internal –OH groups close together, which upon drying either develop hydrogen bonding with one another or condense irreversibly to form new Si–O–Si bridges, pulling the tube walls closer together causing higher shrinkage (syneresis) in the nano- rather than the micron-size regime. Those phenomena generate stresses at the interface of the two size regimes, rendering those materials inherently structurally unstable [79]. Isocyanate cross-linking stabilizes the structure against uneven shrinkage by filling the tubular *mesopores* with polyurea [79] and yields materials with vanadia-like mechanical properties [80].

Table 13.1 compares the mechanical properties under compression of crosslinked vanadia (*X-VO$_x$*) and crosslinked surfactant-templated silica (X-MP4-T045) with other strong materials. For application in armor, the figure of merit is the specific energy absorption, where both crosslinked vanadia and templated silica far surpass even silicon carbide ceramics.

Table 13.1. Comparison of selected mechanical properties under quasi-static compression of polyurea-crosslinked vanadia (*X-VO$_x$*) and acid-catalyzed surfactant-templated silica aerogels (X-MP4-T310-1) with other strong materials [81–83]

Material	Density (g cm^{-3})	Ultimate compressive strength (MPa)	Specific ultimate compressive strength (N m g^{-1})	Specific energy absorption (J g^{-1})
ABS	1.18	65	55	5
Acrylic	1.04	110	106	8
Kevlar-49 epoxy[a]	1.3	235	181	9
2024T3 Aluminum	2.87	345	120	10
4130 Steel	7.84	1,100	140	15
Alumina, AD995	3.9	2,600[b]	666	17
SiC-N Ceramics	3.02	4,500[b]	1,406	20
X-VO$_x$	0.44	96	2,183	191
X-MP4-T045	0.670	780	1,138	134

[a] Fiber volume fraction = 0.6.

[b] Values obtained under confinement at a high strain rate 1,955 s^{-1}; all other data obtained or estimated under unconfined conditions.

The data of Table 13.1 in combination with the low cost of polyurea-crosslinked acid-catalyzed surfactant-templated silica aerogels render those materials ideal candidates for impulse mitigation and energy absorption applications. Figure 13.16 compares results from high strain rate impact testing (using a split Hopkinson pressure bar (*SHPB*) at 30 kg m s^{-1}, which is 3× the muzzle momentum of a high-power 30–06 round) with *PMMA* (Plexiglas) and crosslinked templated silica: *PMMA* is shuttered to pieces, while the porous composite

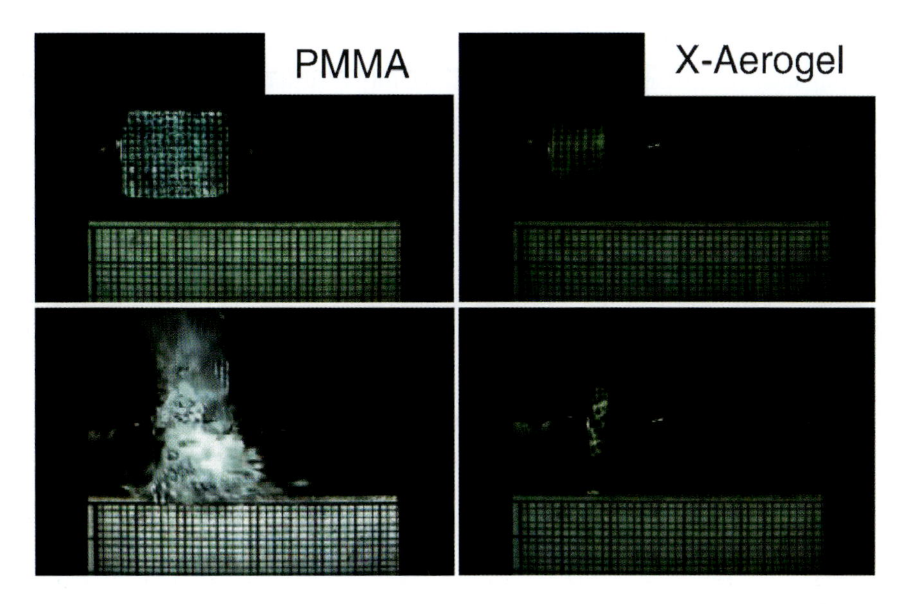

Figure 13.16. High strain rate impact testing using a split Hopkinson pressure bar (*SHPB*). **Top frames:** before impact. **Bottom frames:** after impact. Armor-grade *PMMA* ($\rho_b = 1.2$ g cm^{-3}) shutters to pieces (**bottom left**), while polyurea-crosslinked, surfactant-templated silica aerogel (X-MP4-T045, $\rho_b = 0.55$ g cm^{-3}, **bottom right**) remains in one piece.

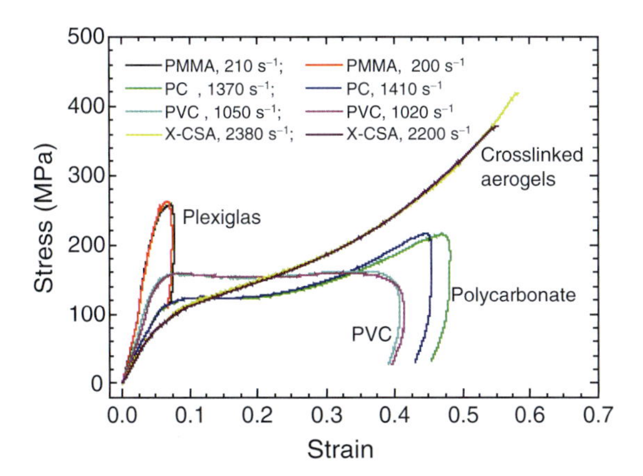

Figure 13.17. Compressive stress–strain curves of *PMMA*, polycarbonate, poly[vinylchloride] (*PVC*), and polyurea-crosslinked surfactant-templated silica aerogel (X-MP4-T310; MCF type material) at high strain rates.

remains intact. Figure 13.17 shows the stress–strain curves for crosslinked templated silica aerogels and several engineering polymers including Plexiglas and Lexan (polycarbonate) at high strain rates and confirms that although the specific crosslinked templated silica aerogels have lower density ($\rho_b = 0.985$ g cm^{-3}) than the other materials, they have a higher specific energy absorption capability (110 J g^{-1}). The Poisson's ratio remains low (~0.16) up to 60–70% compressive strain, giving low lateral tensile strains so that those materials are not prone to developing cracks under compression. It is noted further that even at 70% compressive strain, the crosslinked templated silica aerogel could still carry loads, but the experiment could not proceed further due to limitations in the loading duration time of the *SHPB* (1.5 ms).

Very important from a blast protection perspective, Figure 13.18 shows that crosslinked templated silica aerogels have superb capability to absorb acoustic (sound) waves: a panel with density 0.55 g cm^{-3} attenuates sound better than a typical polyurethane foam in the entire range from 400 to 5,000 Hz. Polyurethane foam cannot be used for practical impulse attenuation as it is soft and requires unrealistically high thickness to mitigate blast waves. On the other hand, crosslinked templated silica aerogels are stiff and strong, with extremely high specific energy absorption capabilities.

Figure 13.19A shows a photograph of a 6″ × 6″ × 1.4″ plate wrapped in Kevlar cloth with an areal density of 8.2 lb/ft^2 incorporating ~65% (by volume) polyurea-crosslinked

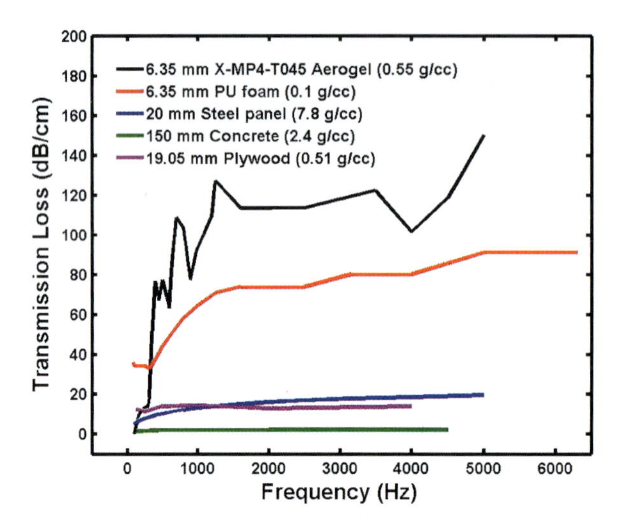

Figure 13.18. Comparison of acoustic transmission loss of polyurea-crosslinked surfactant-templated silica and several other materials.

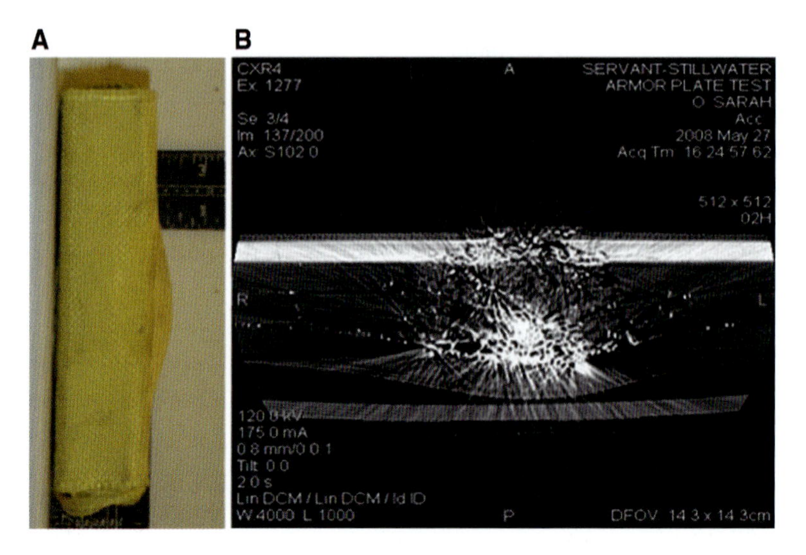

Figure 13.19. **A.** A 6″ × 6″ × 1.4″ armor plate incorporating 65% (v/v) polyurea-crosslinked surfactant-templated silica aerogel (X-MP4-T045) as the energy-absorbing layer, SiC as a front plate and Kevlar cloth for wrapping. The plate has stopped a 0.308 Winchester round (*NIJ* Level III threat) fired with a Remington 700 police sniper rifle from 50 ft (bullet entry point from the **left**). **B.** Microcomputerized tomography showing the bullet fragments captured within the plate (bullet entry point from the **top**). The backside deformation of the target plate is 0.5″ (**middle-right surface**).

Pluronic P123-templated silica aerogel in the central region after a National Institute of Justice (*NIJ*) Level III shooting test. The test bullet was a 0.308 Winchester, shot from a distance of 50 ft with a Remington 700 Police Sniper Rifle. The plate successfully stopped the bullet, with all fragments trapped inside the plate (Figure 13.19B). The backside deformation of the plate was 0.5″, much less than 1.73″ specified as the acceptable deflection in *NIJ* standard for Level III tests. It is anticipated that armor incorporating polymer-crosslinked aerogels can be optimized to reduce the areal density of the armor plate while still providing effective protection.

13.4.2. Improving the Processability of Polymer-Crosslinked Aerogels by Crosslinking in One Pot and Crosslinking in the Gas Phase

Crosslinked aerogels classified as Class II-Model 2 sol–gel materials show dramatic increase in mechanical strength with a minimal penalty in density. However, although postgelation introduction of the *crosslinker* is feasible, from a large-scale manufacturing perspective, the process includes time-consuming solvent exchange steps. The cost could be absorbed for high-end specialized applications (e.g., armor), but for mass production it would be more efficient if all postgelation washes could be avoided. There are two clearly visible possibilities to address this issue. In the first, gelation and crosslinking reagents would be included together in the sol from the beginning; the constrain is that those reagents should not react with one another (otherwise, one would end up with Class II-Model 1 materials – just like in the case of including isocyanates in alumina [47] or silica [48] sols as described in Sect. 13.3). In the second approach, crosslinking would be carried out at the final aerogel stage by introducing the monomer in the gas phase so that its diffusion in the *mesopores* would be much faster than in the liquid phase via postgelation washes (Sect. 13.4.1). Both approaches have been implemented experimentally as summarized in the following paragraphs.

Crosslinking in One Pot by Orthogonalization of the Gelation and Crosslinked Chemistries

As it was outlined in the introductory paragraph of Sect. 13.4, to deconvolute gelation from crosslinking toward a one-pot synthesis of polymer-crosslinked aerogels, the two processes have to become orthogonal. Gelation of silica (as well as of other metal oxide networks) is an acid- or base-catalyzed ionic process; therefore, for orthogonalization, the crosslinking reactions should also be either ionic with higher activation barriers than gelation or free radical processes triggered postgelation by heat or light. We have opted for the free radical route and among the two choices, namely, to decorate the surfaces of silica either with olefins ("grafting to" method) or with free radical initiators ("grafting from" method), we decided to follow the second one, because it introduces greater simplification: through careful adjustment of the amount of the initiator on the skeletal framework and of the *crosslinker* in the sol, all monomers will be consumed and all polymers will be surface-confined, thus eliminating not only washes needed to bring the monomer in the pores, but also postcrosslinking washes that otherwise are needed in order to remove unbound polymer from the mesopores. (Postcrosslinking washes are unavoidable with the "grafting to" method where the mesoporous surfaces are modified with olefins.) To implement the "grafting from" strategy, the mesoporous surfaces of silica have been modified with *AIBN* via cogelation of *TMOS* and Si–AIBN, which is a bidentate derivative of *AIBN*, in 18:1 mol ratio. Si–AIBN was synthesized in one step from azobiscyanovaleric acid and *APTES* [84].

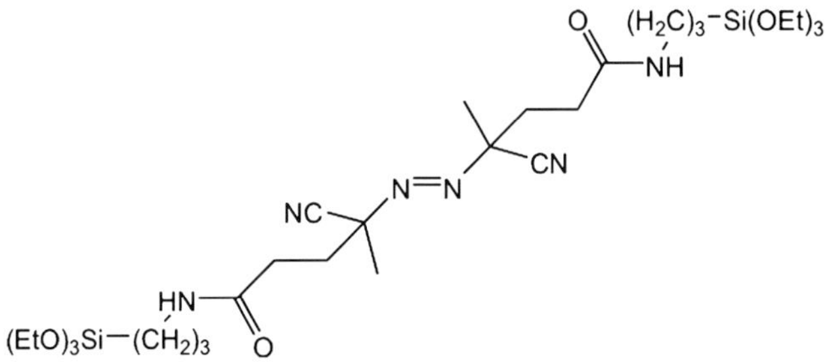

O (H$_2$C)$_3$–Si(OEt)$_3$
NH
CN
N=N
NC
HN
(EtO)$_3$Si–(CH$_2$)$_3$ O

Owing to the slower hydrolysis expected from trialkoxysilanes relative to *TMOS* [13], Si–AIBN should mostly decorate the surface of the skeletal silica nanoparticles. The activity of the surface-confined *AIBN* was demonstrated by crosslinking *TMOS*/Si–AIBN-derived wet gels with *MMA*, styrene, and divinylbenzene introduced in the nanopores by standard postgelation washes with solutions of the monomers. The density of crosslinked aerogels increases as a function of the monomer concentration, while the open aerogel structure is preserved as expected from a skeletal framework coated conformally with polymer [84]. Interestingly, this surface-initiated free radical "grafting from" method yielded on average much longer interparticle polystyrene chains (>60 monomer units) than the "grafting to" method (~15 monomer units) [62]. The mechanical properties under compression are comparable to those of polyurea-crosslinked vanadia and surfactant-templated silica aerogels. Table 13.2 summarizes and compares relevant data.

Table 13.2. Comparison of selected mechanical properties under quasi-static compression of polystyrene (*PS*) and *PMMA* "grafted from" silica aerogels via a free radical surface-initiated approach using Si–AIBN

Material	ρ_b (g cm^{-3})	Modulus (MPa)	0.2% Yield strength (MPa)	UCS (MPa) (% strain)	Specific UCS (Nm g^{-1})	Specific energy absorption (J g^{-1})
"Grafted from" *PS*-silica	0.46	148	7.2	263 (86)	55	95
"Grafted from" *PMMA*-silica	0.66	908	36.1	730 (96)	106	194
Polyurea X-linked silica	0.48	129	4.3	186 (77)	380	32
Polyurea X-linked vanadia	0.44	206	5.6	96 (91)	2,183	191
Polyurea X-linked MP4-T045	0.670	274	15.2	804 (85)	1,138	134

UCS ultimate compressive strength

A clearly visible application for the "grafted from" materials is also in ballistic protection. With a relatively low modulus (e.g., 148 MPa for polystyrene-crosslinked samples – Table 13.2) and bulk density (0.46 g cm^{-3}), it is calculated that the speed of sound (equal to [modulus/bulk density]$^{1/2}$) is 635 m s^{-1}. This value is very low for a solid material and highly desirable because it will extend the duration of an impact reducing the effective force on the material. The relatively low yield stress (7.2 MPa) will allow early activation of the energy absorption mechanism, while the long stress plateau will allow for large energy absorption.

Si–AIBN-modified gels have been used successfully in the one-pot mode with acrylonitrile. Poly[acrylonitrile] (*PAN*) is the precursor of the carbon fiber that is used in the manufacture of carbon fiber-reinforced composites [85], and it was chosen specifically for the synthesis of carbon-coated nanostructured silica, whereas the intimate contact between the two materials would lead to shape–memory–synthesis of monolithic silicon carbide aerogels [86]. Figure 13.20A shows the mesoporous morphology of the *PAN*-crosslinked aerogels whereas the monomer (acrylonitrile) was included in the sol from the beginning and its polymerization was induced by heating the Si–AIBN-modified wet gels at 60°C. The polymer coats the nanoparticles conformally so that the porosity and the BET surface area of the *PAN*-crosslinked aerogels remain high (66%, 228 m^2 g^{-1}, respectively at 0.47 g cm^{-3}). Successive pyrolyses at 225°C in air, 2,000°C in argon, and 600°C in air yield porous silicon carbide as designed (Figure 13.20B) [87].

Figure 13.20. **A.** Poly[acrylonitrile]-crosslinked silica decorated with Si–AIBN ($\rho_b = 0.47$ g cm^{-3}, 60%, w/w, polymer, 228 m^2 g^{-1}, 60%, v/v, empty space). **B.** Pure silicon carbide aerogel by pyrolysis of the poly[acrylonitrile]-crosslinked silica shown in **A.** ($\rho_b = 0.97$ g cm^{-3}, skeletal density $\rho_s = 3.12$ g cm^{-3}, 39 m^2 g^{-1}, 69%, v/v, empty space). The larger particle size is attributed to a complex mechanism that involves molten silica nanodroplets with a conformal carbon coating.

Crosslinking in the Gas Phase

Gas-phase crosslinking by a chemical vapor deposition (CVD) method of SCF-dried aerogels has been demonstrated successfully by D. Loy [88, 89]. This approach circumvents (a) all time-consuming wash steps associated with liquid-phase diffusion of the crosslinker in the mesopores and (b) all postcrosslinking washes to remove the unreacted monomer. The crosslinking monomer (methylcyanoacrylate) was introduced by gas-phase diffusion in the pores of dry regular native aerogels, as well as *APTES*-modified silica aerogels produced by cogelation of *TMOS* and *APTES* (0.25–5.0 mol% *APTES*, giving one surface amine per 11 silanols at 5.0 mol% *APTES*) [89]. Cyanoacrylate undergoes an anionic polymerization process to poly[methylcyanoacrylate] (*PMCNA*): gelation water remaining adsorbed on dry silica aerogels plays the role of the base catalyst in the case of regular *TMOS*-derived silica aerogels, while surface amines are the catalyst in the *TMOS*/*APTES*-derived silica networks. Both approaches yield a polymer shell around the silica framework. However, in the first case the polymer is not covalently bonded to the skeletal nanoparticles, while in the second one the polymer is bound to silica through the surface amines. The noncovalently bonded core-shell structure comprises a topologically unique case of a Class I interpenetrating network and brings about a 16× increase in flexural bending strength relative to the native silica samples for a 3× increase in density by polymer uptake (from 0.077 to 0.235 g cm^{-3}) [88]. When the core-shell structure is covalently bonded to the skeletal framework through the surface amines, the strength increases 31× for a comparable increase in density [89]. The resulting aerogels were strong enough to be machined into shape. From a scale-up perspective, the advantages of the gas-phase approach are obvious; conceivably, however, the downside may be that it requires handling of mechanically weak native aerogels. This issue might not be relevant with aerogel pellets and it may work exceptionally well for applications in thermal insulation.

13.5. Conclusions

Dramatic improvements in the mechanical properties of aerogels have been realized by applying conformal polymer coatings covalently bound to nanostructured inorganic skeletal frameworks. The process of applying the conformal polymer coating is referred to as crosslinking and the materials as polymer-crosslinked aerogels. New applications, not usually associated with aerogels (e.g., armor), are becoming clearly visible. Figure 13.21 uses a periodic table format to show all metal and semimetal aerogels that have been crosslinked successfully.

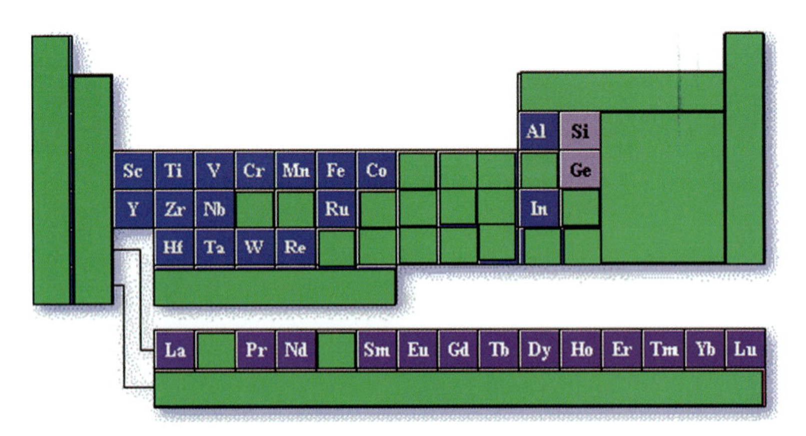

Figure 13.21. Elements whose aerogels have been crosslinked with polyurea (*PUA*).

Table 13.3 summarizes the polymeric systems that have been applied to silica. In effect, the underlying inorganic framework plays the role of a structure-directing agent (template). The mechanical property improvement is attributed to reinforcement of the interparticle necks, which are the weak points of the aerogel skeletal framework. In turn, the stabilization provided by the crosslinking polymer is attributed to the extra chemical bonds created by the interparticle polymeric tethers.

Since the mechanical properties of polymer-crosslinked aerogels are dominated by a polymer-like response (short elastic range, followed by plastic deformation and ultimately densification and inelastic hardening), the question is whether the inorganic framework is needed at all. Referring back to Kistler's early work, we note that his most durable aerogels were made of nitrocellulose [2]. Kistler did not pursue that route further, focusing rather on silica and other inorganic frameworks, perhaps because of his main interests in the structure of

Table 13.3. Summary of polymeric systems that have been used for crosslinking silica aerogels

Type of silica	X-linking method	X-linking polymer					
				Polyolefins			
		PUA[a]	Epoxies	*PS*[b]	*PMMA*[c]	*PAN*[d]	*PMCNA*[e]
Base catalyzed	Postgelation washes	×					
	Postgelation one pot						
	Gas phase[f]						×
Templated	Postgelation washes	×					
	Postgelation one pot						
	Gas phase						
$-NH_2$ modified	Postgelation washes	×	×	×[h]			
	Postgelation one pot						
	Gas phase[f]						×
–*AIBN* modified	Postgelation washes[g]			×	×		
	Postgelation one pot[g]			×	×	×	
	Gas phase						
–*VTMS* and *MTMS* modified	Postgelation washes[h]			×	×		
	Postgelation one pot[h]			×	×		
	Gas phase						

[a]Polyurea.
[b]Poly(styrene).
[c]Poly(methylmethacrylate).
[d]Poly(acrylonitrile).
[e]Poly(cyanoacrylate).
[f]Using final dry aerogels.
[g]Via surface-initiated polymerization (*SIP*) using Si–AIBN.
[h]Via free radical polymerization of the monomer with *AIBN* included in the mesopores.

silica aquogels and in catalysis [1–8]. Kistler's work probably created precedence, which, in combination with concurrent advances in xerogels, led to further developments focused on silica and other inorganic aerogels. Data presented in this chapter, however, suggest that if purely organic (i.e., polymeric) aerogels can be made with the nanostructure of silica or vanadia aerogels and with the interparticle bonding of polymer-crosslinked aerogels, then perhaps one may realize inexpensive multifunctional materials with exceptional properties. At that point aerogels will become commonplace materials from aerospace to automotive to construction.

Acknowledgments

This project has been supported by the NASA Glenn Research Center Science Advisory Board, the University of Missouri Research Board, the US National Science Foundation under CMMI-0653919/0653970 and CHE-0809562, and the Army Research office under W911NF-10-1-0476. We also wish to thank our many collaborators at the Missouri University of Science and Technology, NASA GRC, Oklahoma State University, and the University of North Texas who have made crosslinked aerogels (X-Aerogels) possible: Antonella Alunni, Prof. Massimo Bertino, Dr. Lynn Capadona, Alex Capecelatro, Naveen Chandrasekaran,

Chakkaravarthy Chidambareswarapattar, Gitogo Churu, Joe Counsil, Dr. Paul Curto, Amala Dass, Dr. Eve Fabrizio, Abigail Hobbs, Dr. U. Faysal Ilhan, Dr. J. Chris Johnston, Atul Kati, Dr. James Kinder, Dr. Huiyang Luo, Shruti Mahadik, Linda McCorkle, Dr. Mary Ann Meador, Dhairyashil Mohite, Dr. Sudhir Mulik, Anna Palczer, Vishal Patil, Prof. Abdel-Monem Rawashdeh, Prof. Samit Roy, Anand Sadekar, Dan Scheiman, Jennifer Schnobrich, Nilesh Shimpi, Prof. Chariklia Sotiriou-Leventis, Jeff Thomas, Plousia Vassilaras, Xiaojiang Wang, and Dr. Guohui Zhang.

References

1. Kistler S S (1931) Coherent expanded aerogels and jellies. Nature 127: 3211
2. Kistler S S (1932) Coherent expanded-aerogels. J Phys Chem 36: 52–63
3. Swann S Jr, Appel E G, Kistler S S (1934) Thoria aërogel catalyst: aliphatic esters to ketones. Ind Eng Chem 26: 1014–1014
4. Kistler S S, Caldwell A G (1934) Thermal conductivity of silica aërogel. Ind Eng Chem 26: 658–662
5. Kistler S S, Swann S Jr. Appel E G (1934) Aërogel catalysts – thoria: preparation of catalyst and conversions of organic acids to ketones. Ind Eng Chem 26: 388–391
6. Kistler S S (1935) The relationship between heat conductivity and structure in silica aerogel. J Phys Chem 39: 79–86
7. Kearby K, Kistler S S, Swann S Jr. (1938) Aerogel catalyst: conversion of alcohols to amines. Ind Eng Chem 30: 1082–1086
8. Kistler S S, Fisher E A, Freeman I R (1943) Sorption and surface area in silica aerogel. J Am Chem Soc 65: 1909–1919
9. Peri J B (1966) Infrared study of OH and NH, groups on the surface of a dry silica aerogel. J Phys Chem 70: 2937–2945
10. Teichner S J (1972) Method of preparing inorganic aerogels. US Patent: 3 672 833
11. Gesser H D, Goswami P C (1989) Aerogels and related porous materials. Chem Rev 89: 765–788
12. Hench L L, West J K (1990) The sol-gel process. Chem Rev 90: 33–72
13. Hüsing N, Schubert U (1998) Aerogels-airy materials: chemistry, structure and properties. Angew Chem Int Ed 37:22–45
14. Pierre A C, Pajong G M (2002) Chemistry of aerogels and their applications. Chem Rev 102: 4243–4265
15. Hrubesh L W, Poco J F (1995) Thin aerogel films for optical, thermal, acoustic and electronic applications. J Non-Cryst Solids 188: 46–53
16. Schmidt M, Schwertfeger F (1998) Applications for silica aerogel products. J Non-Cryst Solids 225: 364–368
17. Fricke J, Emmerling A (1998) Aerogels-recent progress in production techniques and novel applications. J Sol-Gel Sci Tech 13: 299–303
18. Akimov Y K (2002) Fields of application of aerogels (review). Instr Exp Techniques 46: 287–299
19. Pajonk G M (2003) Some applications of silica aerogels. Colloid Polym Sci 281: 637–651
20. Smirnova I, Suttiruengwong S, Arlt W (2004) Feasibility study of hydrophilic and hydrophobic silica aerogels as drug delivery systems. J Non-Cryst. Solids 350: 54–60
21. Prakash S S, Brinker C J, Hurd A J, Rao S M (1995) Silica aerogel films prepared at ambient pressure by using surface derivatization to induce reversible drying shrinkage. Nature 374: 439–443
22. Rao A P, Rao A V, Pajonk G M (2007) Hydrophobic and physical properties of the ambient pressure dried silica aerogels with sodium silicate precursor using various surface modification agents. Applied Surface Science 253: 6032–6040
23. Rao A P, Rao A V (2008) Microstructural and physical properties of the ambient pressure dried hydrophobic silica aerogels with various solvent mixtures. J Non-Cryst Solids 354: 10–18
24. Rao A V, Bhagat S D, Hirashima H, Pajonk G M (2006) Synthesis of flexible silica aerogels using methyltrimethoxysilane (MTMS) precursor. J Colloid Interf Sci 300: 279–285
25. Hegde N D, Rao A V (2007) Physical properties of methyltrimethoxysilane based elastic silica aerogels prepared by the two-stage sol-gel process. J Mater Sci 42: 6965–6971

26. Nadargi D Y, Latthe S S, Hirashima H, Rao A V (2009) Studies on rheological properties of methyltriethoxysilane (MTES) based flexible superhydrophobic silica aerogels. Microporous and Mesoporous Materials 117: 617–626
27. Kanamori K, Aizawa M, Nakanishi K, Hanada T (2007) New transparent methylsilsesquioxane aerogels and xerogels with improved mechanical properties. Adv Mater 19: 1589–1593
28. Jones S M (2006) Aerogel: space exploration applications. J Sol-Gel Sci Techn 40: 351–357
29. Jones S M (2007) A method for producing gradient density aerogel. J Sol-Gel Sci Techn 44: 255–258
30. She J H, Ohji T (2002) Porous mullite ceramics with high strength. J Mater Sci Lett 21: 1833–1834
31. Oh S T, Tajima K, Ando M, Ohji T (2000) Strengthening of porous alumina by pulse electric current sintering and nanocomposite processing. J Am Ceram Soc 83: 1314–1316
32. Ma H-S, Roberts A P, Prévost J-H, Jullien R, Scherer G W (2000) Mechanical structure-property relationship of aerogels. J Non-Cryst Solids 277: 127–141
33. Woignier T, Phalippou J (1988) Mechanical strength of silica aerogels. J Non-Cryst Solids 100: 404–408
34. Hæreid S, Anderson J, Einarsrud M A, Hua D W, Smith D M (1995) Thermal and temporal aging of TMOS-based aerogel precursors in water. J Non-Cryst Solids 185: 221–226
35. Lucas E M, Doescher M S, Ebenstein D M, Wald K J, Rolison D R (2004) Silica aerogels with enhanced durability, 30-nm mean pore-size, and improved immersibility in liquids. J Non-Cryst Solids 350: 244–252
36. Einarsrud M A, Kirkedelen M B, Nilsen E, Mortensen K, Samseth J (1998) Structural development of silica gels aged in TEOS. J Non-Cryst Solids 231: 10–16
37. Sanchez C, Robot F (1994) Design of hybrid organic-inorganic materials synthesized via sol-gel chemistry. New J Chem 18: 1007–1047
38. Hu Y, Mackenzie J D (1992) Rubber-like elasticity of organically modified silicates. J Mater Sci 27: 4415–4420
39. Novak B M, Auerbach D, Verrier C (1994) Low-density, mutually interpenetrating organic-inorganic composite materials via supercritically drying techniques. Chem Mater 6: 282–286
40. Gould G, Ou D, Begag R, Rhine W E (2008) Highly-transparent polymer modified silica aerogels. Polymer Preprints 49: 534–535
41. Sanchez C, Ribot F, Lebeau B (1999) Molecular design of hybrid organic-inorganic nanocomposites synthesized via sol-gel chemistry. J Mater Chem 9: 35–44
42. Pope E J A, Asami M, Mackenzie J D (1989) Transparent silica gel-PMMA composites. J Mater Res 4: 1018–1026
43. Philipp G, Schmidt H (1984) New materials for contact lenses prepared from Si- and Ti-alkoxides by the sol-gel process. J Non-Cryst Solids 63: 283–292
44. Huang H H, Orler B, Wilkes G L (1987) Structure-property behavior of new hybrid materials incorporating oligomeric species into sol-gel glasses. 3. Effect of acid content, tetraethoxysilane content, and molecular weight of poly(dimethylsiloxane). Macromolecules 20: 1322–1330
45. Kramer S J, Rubio-Alonso F, Mackenzie J D (1998) Organically modified silicate aerogels, "aeromosils." Mat Res Soc Symp Proc 435: 295–300
46. Leventis N, Sotiriou-Leventis C, Zhang G, Rawashdeh A-M M, (2002) Nanoengineering strong silica aerogels. Nano Lett 2: 957–960
47. Mizushima Y, Hori M (1994) Preparation and properties of alumina-organic compound aerogels. J Non-Cryst Solids 170: 215–222
48. Yim T-J, Kim S Y, Yoo K-P (2002) Fabrication and thermophysical characterization of nano-porous silica-polyurethane hybrid aerogels by sol-gel processing and supercritical solvent drying technique. Korean J Chem Eng 19: 159–166
49. Leventis, N (2007) Three Dimensional Core-Shell Superstructures: Mechanically Strong Aerogels. Acc Chem Res 40:874–884
50. Zhang G, Dass A, Rawashdeh A-M M, Thomas J, Counsil J A, Sotiriou-Leventis C, Fabrizio E F, Ilhan F, Vassilaras P, Scheiman D A, McCorkle L, Palczer A, Johnston J C, Meador M A B, Leventis N (2004) Isocyanate-crosslinked silica aerogel monoliths: preparation and characterization. J Non-Cryst Solids 350: 152–164
51. Leventis N, Palczer A, McCorkle L, Zhang G, Sotiriou-Leventis C (2005) Nanoengineering silica-polymer composite aerogels with no need for supercritical fluid drying. J Sol-Gel Sci Techn 35: 99–105
52. Lowen W K, Broge E C (1961) Effects of dehydration and chemisorbed materials on the surface properties of amorphous silica. J Phys Chem 65: 16–19
53. Hüsing N, Schubert U, Mezei R, Fratzl P, Riegel B, Kiefer W, Kohler D, Mader W (1999) Formation and structure of gel networks from $Si(OEt)_4/(MeO)_3Si(CH_2)_3NR'_2$ mixtures ($NR'_2{=}NH_2$ or $NHCH_2CH_2NH_2$). Chem Mater 11: 451–457

54. Katti A, Shimpi N, Roy S, Lu H, Fabrizio E F, Dass A, Capadona L A, Leventis N (2006) Chemical, physical and mechanical characterization of isocyanate-crosslinked amine-modified silica aerogels. Chem Mater 18: 285–296
55. Shigley J E, Mischke C R M (2001) Mechanical Engineering Design 6th ed, New York, p 1206
56. Meador M A B, Capadona L A, MacCorkle L, Papadopoulos D S, Leventis N (2007) Structure-property relationships in porous 3D nanostructures as a function of preparation conditions: isocyanate cross-linked silica aerogels. Chem Mater 19: 2247–2260
57. Vivod S L, Meador M A B, Nguyen B N, Quade D, Randall J (2008) Di-isocyanate crosslinked aerogels with 1,6-bis(trimethoxysilyl)hexane incorporated in silica backbone. Polym Preprints 49: 521–522
58. Vivod S L, Meador M A B, Capadona L A, Sullivan R M, Ghosn L J, Clark N, McCorkle L, Quade D J (2008) Carbon nanofiber incorporated silica based aerogels with di-isocyanate cross-linking. Polym Preprints 49: 306–307
59. Meador M A B, Vivod S L, McCorkle L, Quade D, Sullivan R M, Ghosn L J, Clark N, Capadona L A (2008) Reinforcing polymer cross-linked aerogels with carbon nanofibers. J Mater Chem 18: 1843–1852
60. Meador M A B, Weber A S, Hindi A, Naumenko M, McCorkle L, Quade D, Vivod S L, Gould G L, White S, Deshpande K (2009) Structure-property relationships in porous 3D nanostructures: epoxy-cross-linked silica aerogels produced using ethanol as the solvent. ACS Appl Mater Interfaces 1: 894–906
61. Meador M A B, Fabrizio E F, Ilhan F, Dass A, Zhang G, Vassilaras P, Johnston J C, Leventis N (2005) Crosslinking amine-modified silica aerogels with epoxies: mechanically strong lightweight porous materials. Chem Mater 17: 1085–1098
62. Ilhan U F, Fabrizio E F, McCorkle L, Scheiman D, Dass A, Palzer A, Meador M A B, Leventis N (2006) Hydrophobic monolithic aerogels by nanocasting polystyrene on amine-modified silica. J Mater Chem 16: 3046–3054
63. Wingfield C, Baski A, Betrino M F, Leventis N, Mohite D P, Lu H (2009) Fabrication of sol-gel materials with anisotropic physical properties by photo-cross-linking. Chem Mater 21: 2108–2114
64. Nguyen B N, Meador M A B, Tousley M E, Shonkwiler B, McCorkle L, Scheiman D A, Palczer A (2009) Tailoring elastic properties of silica aerogels cross-linked with polystyrene. ACS Appl Mater Interfaces 1: 621–630
65. Morris C A, Anderson M L, Stroud R M, Merzbacher C I, Rolison D R (1999) Silica sol as a nanoglue: flexible synthesis of composite aerogels. Science 284: 622–624
66. Ziese W (1933) Die Reaktion von Äthylenoxyd mit Lösungen von Erd- und Schwermetallhalogeniden. Ein neur Weg zur Gewinnung von Solen und reversiblen Gelen von Metalloxydhydraten. Ber 66: 1965–1972
67. Gash A E, Tillotson T M, Satcher J H, Hrubesh L W, Simpson R L (2001) New sol-gel synthetic route to transition and main-group metal oxide aerogels using inorganic salt precursors. J Non-Cryst Solids 285: 22–28
68. Sisk C N, Hope-Weeks J (2008) Copper(II) aerogels via 1,2-epoxide gelation. J Mater Chem 18: 2607–2610
69. Leventis N, Vassilaras P, Fabrizio E F, Dass A (2007) Polymer nanoencapsulated rare earth aerogels: chemically complex but stoichiometrically similar core-shell superstructures with skeletal properties of pure compounds. J Mater Chem 17:1502–1508
70. Leventis N, Sotiriou-Leventis C, Mulik S, Dass A, Schnobrich J, Hobbs A, Fabrizio E F, Juo H, Churu G, Zhang Y, Lu H (2008) Polymer nanoencapsulated mesoporous vanadia with unusual ductility at Cryogenic temperatures. J Mater Chem 18: 2475–2482
71. Vollrath F, Knight D P (2001) Liquid crystalline spinning of spider silk. Nature 410: 541–548
72. Wigley D A (1971) Mechanical properties of materials at low temperatures. Plenum Press, New York, NY pp 138–163
73. Amatani T, Nakanishi K, Hirao K, Kodaira T (2005) Monolithic periodic mesoporous silica with well defined macropores Chem Mater 17: 2114–2119
74. Zhao D, Feng J, Huo Q, Melosh N, Fredrickson G H, Chmelka B, Stucky G D (1998) Triblock copolymer syntheses of mesoporous silica with periodic 50–300 angstrom pores Science 279: 548–552
75. Zhao D, Huo Q, Feng J, Chmelka B F, Stucky G D (1998) Nonionic Triblock and star diblock copolymer and oligomeric surfactant syntheses of highly ordered, hydrothermally stable, mesoporous silica structures J Am Chem Soc 120: 6024–6036
76. Yang C M, Zibrowius B, Schmidt W, Schüth F (2004) Stepwise removal of the copolymer template from Mesopores and Micropores in SBA-15 Chem Mater 16: 2918–2925
77. Kresge C T, Leonowicz M E, Roth W J, Vartuli J C, Beck J S (1992) Ordered mesoporous molecular sieves synthesized by a liquid –crystal template mechanism. Nature 359: 710–712

78. Schmidt-Winkel P, Lukens W W, Zhao J D, Yang P, Chmelka B F, Stucky G D (1999) Mesocellular siliceous foams with uniformly sized cells and windows. J Am Chem Soc 121: 254–255
79. Leventis N, Mulik S, Wang X, Dass A, Sotiriou-Leventis C, Lu H (2007) Stresses at the interface of micro with nano. J Am Chem Soc 129: 10660–10661
80. Leventis N, Mulik S, Wang X, Dass A, Patil V U, Sotiriou-Leventis C, Lu H, Churu G, Capecelatro A (2008) Polymer nano-encapsulation of templated mesoporous silica monoliths with improved mechanical properties. J Non-Cryst Solids 354; 632–644
81. American Society for Metals, ASM Engineering Materials Handbook, Composites, Volume 1: ASM International: Materials Park, OH, p 178, Table 2, 1998
82. Luo H, Chen W. (2004) Dynamic compressive response of intact and damaged AD995 alumina. Intern J Appl Ceram Techn 1: 254–260
83. Luo H, Chen W, Rajendran A M (2006) Dynamic compressive response of damaged and interlocked SiC-N ceramics. J Am Ceram Soc 89: 266–273
84. Mulik S, Sotiriou-Leventis C, Churu G, Lu Hongbing, Leventis N (2008) Cross-linking 3D assemblies of nanoparticles into mechanically strong aerogels by surface-initiated free-radical polymerization. Chem Mater 20: 5035–5046
85. Morgan P (2005) Carbon fibers and their composites. CRC Press (Taylor and Francis Group), New York, NY
86. Keller N, Reiff O, Keller V, Ledoux M J (2005) High surface area submicrometer-sized β-SiC particles grown by shape memory method. Diamond & Related Materials 14: 1353–1360
87. Leventis N, Sadekar A, Chandrasekaran N, Sotiriou-Leventis C (2010) Click synthesis of monolithic silicon carbide aerogels from poly acrylonitrile-coated 3D silica networks. Chem Mater 22: 2790–2803
88. Boday D J, DeFriend K A, Wilson K V Jr, Coder D, Loy D A (2008) Formation of polycyanoacrylate-silica nanocomposites by chemical vapor deposition of cyanoacrylates on aerogels. Chem Mater 20: 2845–2847
89. Boday D J, Stover R J, Muriithi B, Keller M W, Wertz J T, DeFriend Obrey K A, Loy D A (2009) Strong, low density nanocomposites by chemical vapor deposition and polymerization of cyanoacrylates on aminated silica aerogels. ACS Appl Mater Interfaces 1: 1364–1369

14

Interpenetrating Organic/Inorganic Networks of Resorcinol-Formaldehyde/ Metal Oxide Aerogels

Nicholas Leventis

Abstract Interpenetrating native and polymer-crosslinked resorcinol-formaldehyde (*RF*) and metal oxide (*MOx*) networks have been synthesized in one pot through the catalytic effect of gelling solutions of hydrated metal ions on the gelation of *RF*. Pyrolysis under argon (Ar) induces carbothermal processes that, depending on the chemical identity of *MOx*, yield either metals in the element form or the corresponding carbides. *RF–MOx* networks in the xerogel and polymer-crosslinked aerogel (*X-aerogel*) forms undergo those processes at temperatures up to 400°C lower than in the native aerogel form. In addition to the significance of the *RF–MOx* interpenetrating networks in the design of new materials (mesoporous and macroporous monolithic metals and carbides), the effect of the compactness of the nanostructure on the activation of the carbothermal processes has important implications for process design engineering.

14.1. Introduction

In the field of material science, interpenetrating networks were developed originally for polymers and are pursued for the same basic reasons as simple polymer blends (i.e., macroscopic mixtures of polymers) or more complex derivatives such as block, graft, and AB-graft copolymers, namely, the possible synergism of the components for better thermal, mechanical, and chemical properties [1]. In most cases, the components of interpenetrating polymer networks are not actually interpenetrating at the molecular level, but instead they consist of domains of a few nanometers to a few tens of nanometers in size, reminding thus of the building blocks of typical aerogels.

Interpenetrating inorganic sol–gel networks with polymers have been pursued mainly for preventing the shrinkage and cracking problems encountered upon drying of wet inorganic gels (e.g., silica) into xerogels and with proper particle size and refractive index matching for improving the optical properties of sol–gel glasses [2–5]. Other possible applications of those systems may extend to abrasive-resistant coatings as well as to dielectrics, nonlinear optical materials, tunable lasers, and chemical/biochemical sensors [6].

N. Leventis • Department of Chemistry, Missouri University of Science and Technology,
Rolla, MO 65409, USA
e-mail: leventis@mst.edu

M.A. Aegerter et al. (eds.), *Aerogels Handbook*, Advances in Sol-Gel Derived Materials and Technologies, DOI 10.1007/978-1-4419-7589-8_14,

As reviewed early by Wilkes [6], the methodology for the synthesis of such networks is conceptually similar to what has been used in the synthesis of interpenetrating polymer networks; for example, the two networks can be formed simultaneously by including polymerizable monomers in the inorganic sol or by starting with preformed polymers. Thus, in a variation of the last method, inorganic sol–gel precursors (e.g., alkoxides) may be included in the swelling solution of crosslinked, ionomeric, or crystalline polymers followed by triggering the sol–gel reaction. If any of the synthetic procedures are carried out in relatively dilute solutions and conditions that promote gelation rather than precipitation, and if the solvent is subsequently removed without a significant volume shrinkage [e.g., by means of what is commonly referred to as supercritical fluid (*SCF*) drying: Chap. 3], the resulting low-density materials are classified as aerogels. This approach was explored early for imparting ductility, or even rubbery behavior, into brittle silica aerogels (Chaps. 4 and 15) [7–9].

Along those lines, it is well established that nanoparticles demonstrate increased reactivity relative to bulk materials [10]. In extreme cases of exothermic processes, fast reactions lead to large amounts of energy released in very short time periods with clear implications for energetic materials such as explosives, propellants, and pyrotechnics (Chap. 25) [11, 12]. These phenomena are attributed to the high surface-to-volume ratio characterizing nanoparticulate matter, thereby to the increased contact area between the reactants. This contact area is maximized in core-shell nanoparticles, but the fabrication of such structures may not be easy or feasible in certain cases [13]. Alternatively, possible memory-shape (pseudomorphic) reactions between the components could also yield new porous materials, which otherwise would be difficult to produce [14]. From that perspective, this chapter discusses interpenetrating resorcinol-formaldehyde (*RF*) and metal oxide (*MOx*) aerogels and describes a particularly convenient method for the synthesis of pure metal (e.g., Fe, Co, Ni, Sn, and Cu) and carbide aerogels (e.g., Cr_3C_2/Cr_7C_3, TiC, and HfC).

Typically, porous metals can be synthesized via a reductive pathway using polymer or surfactant templating (e.g., cases of Ag and Pt, respectively [15, 16]), by chemical vapor deposition (on carbon templates, cases of Ir and Re [17]), or by alloying and selective removal (dealloying) of one component (case of nanoporous Au by alloying with Ag and removal of the latter electrochemically [18]). More recently, low-density (~0.01 g cm^{-3}), high surface area (>250 $m^2\ g^{-1}$) monolithic metal nanofoams (Fe, Co, Ni, Cu, Ag, Au, Pd, Pt, and Ti) have been described through self-propagating combustion synthesis of selected transition metal complexes with high-nitrogen-containing ligands [19].

Another domain of interest is porous carbide materials (which are widely known in the form of ceramics). However, only few well-characterized carbide aerogels are known today. Monolithic NbC aerogels (surface area up to 74 $m^2\ g^{-1}$) have been obtained through a pseudomorphic solid-state reaction by passing a mixture of CH_4–H_2 over an Nb_2O_5 aerogel (surface area 344 $m^2\ g^{-1}$) at high temperatures (1,173–1,373 K) [14, 20]. Mo_2C and WC nanoparticles were observed in carbon aerogels after pyrolysis at 1,000°C under flowing H_2/Ar or pure N_2, respectively, of molybdenum(VI) or tungsten(VI)-doped RF aerogels [21, 22]. In the case of Mo_2C, it has been proposed that Mo(VI) is reduced to Mo(IV) (i.e., MoO_2) by the carbon matrix, and Mo(IV) is further reduced by CH_4 or CH_x species that are formed from the gasification of the carbon matrix at high temperature in the presence of H_2. In the case of tungsten carbide (WC), it appears to be formed by direct reduction of W(VI) by the carbon matrix, as expected. In the same realm of using aerogels as precursors of new materials, SiC whiskers and TiC nanoparticles have been produced by pyrolysis (above 1,300°C under Ar) of silica or titania aerogels made by mixing ethanolic silica or titania sols with carbonaceous sols (ethanol) prepared from petroleum green coke [23, 24]. More recently, an Aspen

Aerogels patent reports a multistep process to selected carbide-doped carbon aerogels from interpenetrating networks of polyimide and a variety of metal oxide aerogels [25]. The polyimide network was produced from a dianhydride with a diamine whose slow reaction in *N*-methylpyrrolidone (*NMP*) first forms a polyamic acid solution that is subsequently dehydrated to a polyimide gel by adding acetic anhydride/pyridine (according to Scheme 14.1). The metal oxide network was formed by adding an epoxide and the appropriate metal salt in the polyamic acid solution, followed by the addition of acetic anhydride/pyridine for imidization. According to that patent, pyrolysis of the resulting interpenetrating networks of the polyimide and the metal oxides under Ar yielded metal carbide aerogels, metal carbide–carbon aerogels, or carbon aerogels with dispersed metal particles, all depending on the relative stoichiometry and the chemical identity of the metal salt, respectively.

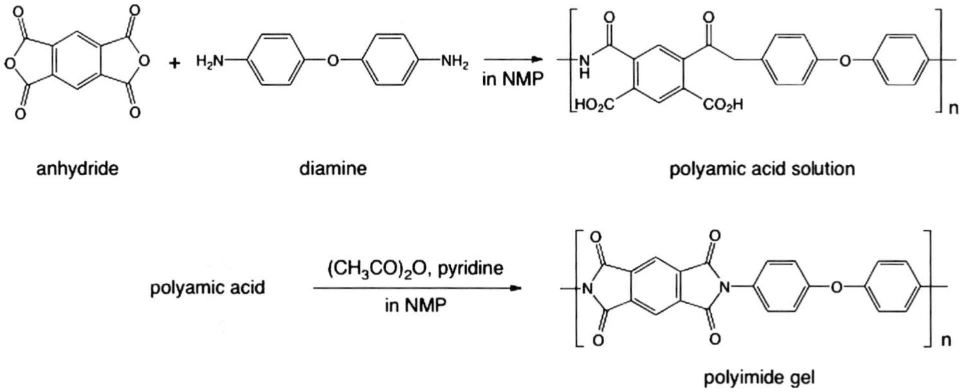

anhydride + diamine $\xrightarrow{\text{in NMP}}$ polyamic acid solution

polyamic acid $\xrightarrow[\text{in NMP}]{(CH_3CO)_2O,\ \text{pyridine}}$ polyimide gel

Scheme 14.1. Polymide synthesis according to [25].

On the other hand, the synthesis of interpenetrating *RF–MOx* networks is a versatile one-pot process based on two facts combined: (a) hydrated metal salts gelling through an epoxide route in organic media toward metal–oxide wet gels are fairly strong Brønsted acids (Chap. 8); and (b) the gelation of resorcinol (*R*) with formaldehyde (*F*) in organic media can also be (effectively) catalyzed by strong acids (Chap. 11). Pyrolysis of *RF* aerogels under inert atmosphere yields electrically conducting carbon aerogels which, owing to the chemical inertness of carbon, are pursued not only as high surface area electrodes for super-capacitors and fuel cells [26], but also as supports for catalysts (Chap. 36) [27, 28]. For the later application, the catalyst precursor is usually included as a metal salt in the *RF* sol. Since the *RF* network is produced typically by an aqueous, base-catalyzed process, a common problem is the leaching of the metal ions during the solvent exchange steps required by the aerogel fabrication process. Another issue, again related to the high solubility of the metal ions in the sol, is the segregation of large crystallites of the dopant (>15 nm in diameter) at the surface of the *RF* nodules. A successful method that halts the leaching of metal ions and yields a uniform distribution of the dopant involves the substitution of chelating potassium 2,4-dihydrobenzoate for resorcinol [29, 30]. Better dispersion of the dopant has been reported by anchoring metal ions to the *RF* network by employing complexing agents able to react and incorporate themselves within *RF* in a process referred to as "cogelation" [31]. In all cases, the level of metal doping is low and it has not been reported above 10% (w/w) [32]. Now, although most of those studies have been conducted from a catalysis perspective, one persistent observation from a material's design perspective is that in selected cases (Fe, Co, Ni, and Cu) the dopant of the resulting carbon aerogel is not a salt or the oxide but the metal itself [32–36]. It is inferred, therefore, that in those cases the metal ions have been reduced by carbon in analogy to the smelting process that has been used in extractive metallurgy for millennia.

This chapter is organized as follows: first, it reviews the synthesis of interpenetrating *RF–MOx* networks and their crosslinking with polyurea into *X-RF–MOx* systems (Chap. 13), and second, it discusses the properties of those materials from a nanostructure perspective and finishes with important findings on the activation of reactions among nanoparticles by comparing the pyrolytic behavior of native RF–MO*x* aerogels, *X-RF–MOx* aerogels, as well as of the corresponding xerogels.

14.2. Cogelation of *RF* and Metal Oxide Networks: Native, Crosslinked (X-) *RF–MOx* Aerogels, and Xerogels

RF–MOx networks are formed by adding solutions of hydrated metal chlorides, $[M(H_2O)_6]^{n+}$, with an epoxide (e.g., epichlorohydrin) in a solution of resorcinol and formaldehyde in a suitable solvent (in general ethanol/acetonitrile or N,N'-dimethylformamide: *DMF*). Solvents and conditions are shown in Table 14.1. Metals for the *RF–MOx* systems were chosen among those expected either to be reduced to the elements via reaction with carbon (e.g., Fe, Co, Ni, Sn, and Cu) or to be converted to the corresponding carbides (e.g., refractories Cr, Ti, and Hf) [37]. Scheme 14.2 shows all processes, from the synthesis of *RF–MOx* to the pyrolytic conversion to metals and carbides. It should be noted that the term "metal oxide" (*MOx*) is used loosely: detailed analysis conducted with rare earth aerogels, xerogels, and *X-aerogels* have shown that even though oxides are the dominant components, the materials also contain hydroxides (from incomplete condensation), carbonates (by reaction with CO_2), chlorides (from the original salts), and acetates (from oxidation of the solvent and ethanol) [38].

As it is well documented since the 1930s, gelation of a few hydrated metal ions (usually in the +3 or +4 oxidation states, with few exceptions like Ni^{2+} and Cu^{2+}) can be carried out through proton transfer to an epoxide that in turn undergoes ring opening to an alcohol, thereby rendering the proton transfer irreversible (Scheme 14.3) (Chap. 8) [39–44]. This process has been explored commercially primarily by the ceramics community for preparing fine nanosized powders as ceramic precursors (by peptization of the gels in water followed by precipitation) rather than by the sol–gel community [45].

Titration (Figure 14.1) of the hydrated ions, $[M(H_2O)_6]^{n+}$, with NaOH in the gelation solvents with no epichlorohydrin added confirms their acidity (14.1). Two acid dissociation constants have been identified in all cases (Figure 14.1). Epichlorohydrin in a tenfold molar excess induces gelation except in $[Co(H_2O)_6]^{2+}$ and $[Cu(H_2O)_6]^{2+}$ that just become viscous. Drying of films cast from those viscous solutions either under ambient conditions or with *SCF* CO_2 has shown the presence of nanoparticles [46]. Consistently with Scheme 14.3, the consumption of protons causes a pH increase during gelation (e.g., see Figure 14.2A). The terminal asymptotic pH value corresponds presumably to the intrinsic acidity of the skeletal *MOx* nanoparticle framework.

$$[M(H_2O)_6]^{n+} + OH^- \rightleftharpoons [M(H_2O)_5(OH)]^{(n-1)+} + H_2O \qquad (14.1)$$

Table 14.1. Gelation conditions for (a) selected metal oxides (*MOx*); (b) *RF*, catalyzed by the corresponding hydrated metal ions; and (c) *RF–MOx* interpenetrating networks

System	Oxide precursor (+ mols of H_2O)	Gelation solvent ($pK_{a,1}$; pK_{a-2})[a]	Approximate gelation time (min)[b]
Pure FeO_x network[c]	$FeCl_3 \cdot 6H_2O$	EtOH (1.19; 2.49)	10
RF with $[Fe(H_2O)_6]^{3+}$		EtOH/AN, 1:1 (v/v)[d]	50
RF–FeO_x		EtOH/AN, 1:1 (v/v)[e]	10–15
Pure CoO_x network[c]	$CoCl_2 \cdot 6H_2O$	*DMF* (5.20; 6.00)	[f]
RF with $[Co(H_2O)_6]^{2+}$		*DMF*[d]	55–60
RF–CoO_x		*DMF*[e]	40–45
Pure NiO_x network[c]	$NiCl_2 \cdot 6H_2O$	*DMF* (5.10)[g]	10
RF with $[Ni(H_2O)_6]^{2+}$		*DMF*[d]	60
RF–NiO_x		*DMF*[e]	15–20
Pure SnO_x network[c]	$SnCl_4 \cdot 5H_2O$ (1)	EtOH (1.30; 1.90)	10–15
RF with $[Sn(H_2O)_6]^{4+}$		EtOH/AN (1:1, v/v)[d]	60
RF–SnO_x		EtOH/AN (1:1, v/v)[e]	30–35
Pure CuO_x network[c]	$CuCl_2 \cdot 2H_2O$ (4)	*DMF* (3.13; 4.07)	[f]
RF with $[Cu(H_2O)_6]^{2+}$		*DMF*[d]	600
RF–CuO_x		*DMF*[e]	240
Pure CrO_x network[c]	$CrCl_3 \cdot 6H_2O$	EtOH (3.50; 4.55)	10
RF with $[Cr(H_2O)_6]^{3+}$		EtOH/AN, 1:1 (v/v)[d]	45
RF–CrO_x		EtOH/AN, 1:1 (v/v)[e]	15–20
Pure TiO_x network[c]	$TiCl_4$ (6)	EtOH (1.90; 2.55)	5
RF with $[Ti(H_2O)_6]^{4+}$		EtOH/AN, 1:1 (v/v)[d]	35
RF–TiO_x		EtOH/AN, 1:1 (v/v)[e]	10–15
Pure HfO_x network[c]	$HfCl_4$ (6)	EtOH (2.25; 3.30)	4–5
RF with $[Hf(H_2O)_6]^{4+}$		EtOH/AN, 1:1 (v/v)[d]	25–30
RF–HfO_x		EtOH/AN, 1:1 (v/v)[e]	8–10
Pure YO_x network[c]	$YCl_3 \cdot 6H_2O$	EtOH (4.70; 5.30)	8–10
RF with $[Y(H_2O)_6]^{3+}$		EtOH/AN, 1:1 (v/v)[d]	50
RF–YO_x		EtOH/AN, 1:1 (v/v)[e]	25
Pure DyO_x network[c]	$DyCl_3 \cdot 6H_2O$	EtOH (3.60; 4.35)	8–10
RF with $[Dy(H_2O)_6]^{3+}$		EtOH/AN, 1:1 (v/v)[e]	50–55
RF–DyO_x		EtOH/AN, 1:1 (v/v)[e]	20

[a]First and second dissociation constants.
[b]At 80°C.
[c]Gelled with epichlorohydrin (×10 mol excess).
[d]EtOH: ethanol; AN: acetonitrile; no epichlorohydrin added; gelled with the corresponding hydrated metal ions acting as catalysts for the *R*/*F* system.
[e]Complete system (including *R*, *F*, the hydrated metal salt, and epichlorohydrin).
[f]Did not gel; produced a thick sol.
[g]Only one dissociation constant was discernible for $[Ni(H_2O)_6]^{2+}$ in *DMF*.

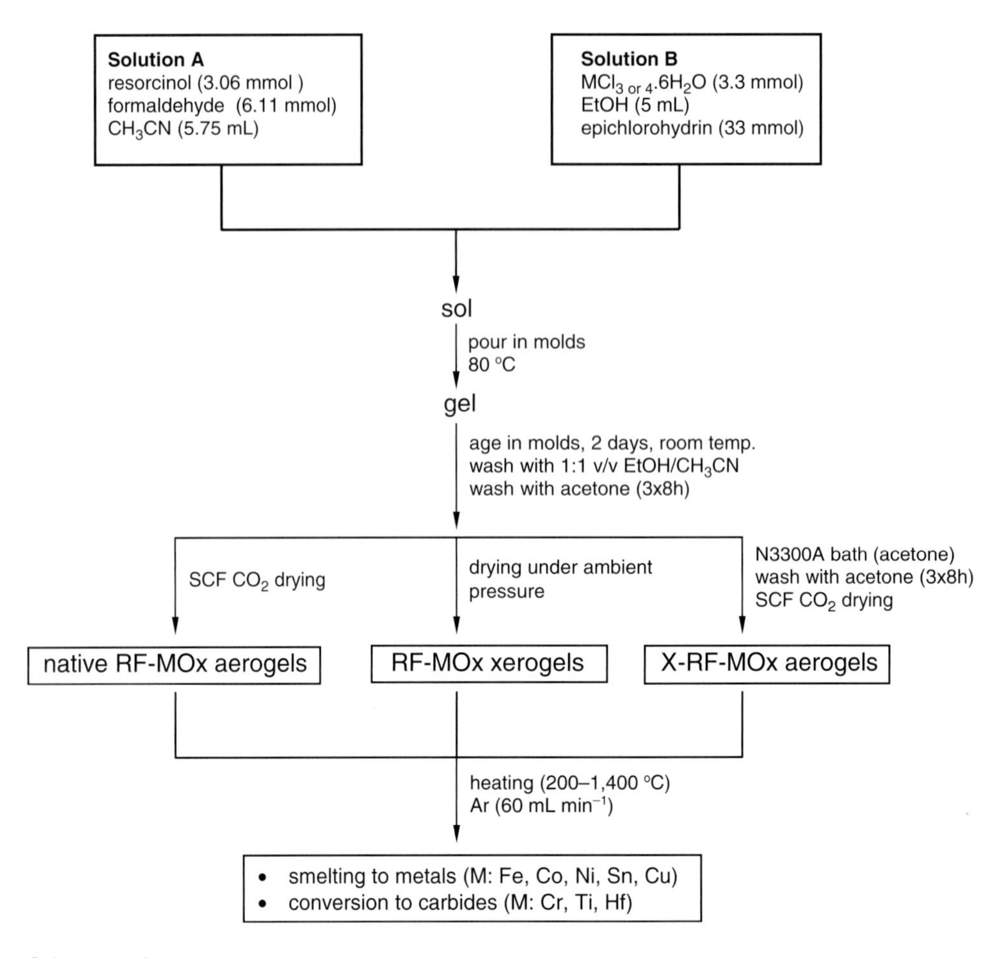

Scheme 14.2. Synthesis and pyrolysis of *RF–MOx* systems.

Uncatalyzed resorcinol-formaldehyde solutions do not gel. Typically, *RF* gel formation is induced in aqueous solutions with Na_2CO_3 (base) and takes several days [47, 48]. Base-catalyzed gelation in nonaqueous environments (alcohols) has also been described and still takes a few days [49]. On the contrary, HCl-catalyzed gelation of *R*/*F* in acetonitrile ([HCl] $\approx$ 0.363 mM, pH $\approx$ 0) occurs in about 2 h at room temperature and in about 10 min at 80°C, developing the characteristic red color of *o*-quinone methides [50, 51]; mesoporous RF aerogels obtained by either the base- or the acid-catalyzed process are chemically identical [50]. If hydrated metal ions are used as catalysts instead of HCl (i.e., mixing the R/F and $[M(H_2O)_6]^{n+}$ solutions with no epichlorohydrin added), the *R*/*F* system gels at 80°C in about 30 min to 1 h (with the exception of the copper salt; see Table 14.1), and the pH does not change drastically, consistent with the catalytic role of the acidic metal aqua complexes (Figure 14.2A). The longer gelation times of the $[M(H_2O)_6]^{n+}$-catalyzed sols relative to the HCl-catalyzed ones have been attributed to the weaker acidity of the hydrated metal ions relative to HCl, as well as the use of ethanol as solvent that is expected to reduce the concentration of formaldehyde by forming hemiacetals and acetals [49].

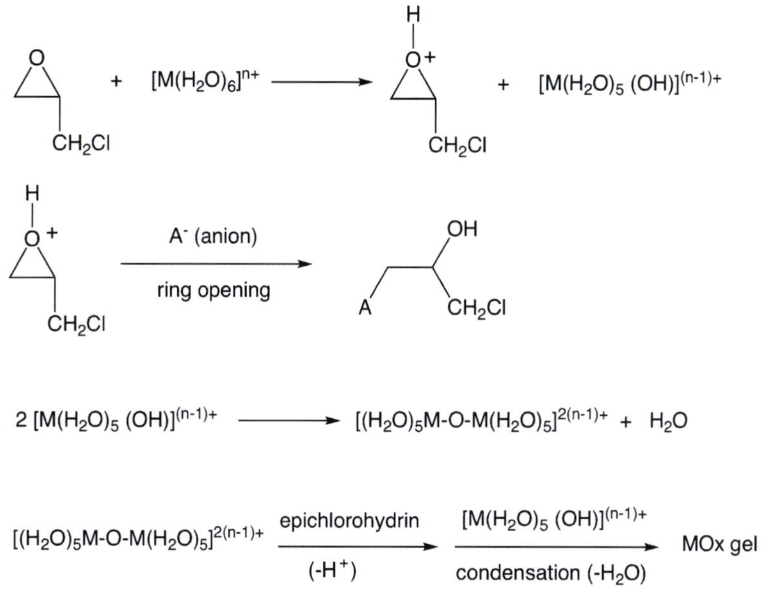

$$\text{epichlorohydrin} + [M(H_2O)_6]^{n+} \longrightarrow \text{O-protonated epichlorohydrin } (O^+\text{–H}) + [M(H_2O)_5(OH)]^{(n-1)+}$$

$$\text{O-protonated epichlorohydrin} \xrightarrow[\text{ring opening}]{A^- \text{ (anion)}} A\text{–}CH_2\text{–}CH(OH)\text{–}CH_2Cl$$

$$2\,[M(H_2O)_5(OH)]^{(n-1)+} \longrightarrow [(H_2O)_5M\text{-}O\text{-}M(H_2O)_5]^{2(n-1)+} + H_2O$$

$$[(H_2O)_5M\text{-}O\text{-}M(H_2O)_5]^{2(n-1)+} \xrightarrow[(-H^+)]{\text{epichlorohydrin}} \xrightarrow[\text{condensation } (-H_2O)]{[M(H_2O)_5(OH)]^{(n-1)+}} \text{MOx gel}$$

Scheme 14.3. Gelation mechanism of hydrated metal ions with epoxides [39–44].

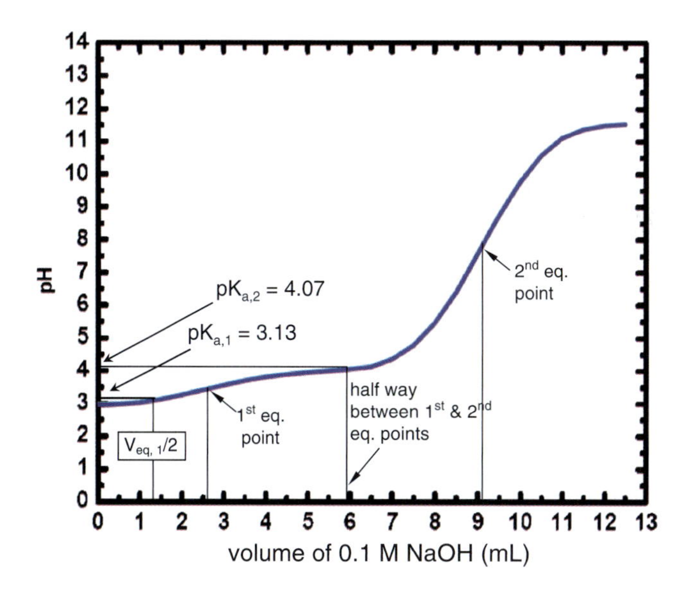

Figure 14.1. A typical titration curve for hydrated metal chlorides in the gelation solvents (case shown: $CuCl_2 \cdot 6\, H_2O$; 0.66 M in *DMF*).

Mixtures of *R*/*F* solutions with hydrated metal ions *and* epichlorohydrin gel in about the same time as the corresponding solutions of the hydrated metal ions by themselves (Table 14.1). Figure 14.3 shows typical chromatic changes taking place during the gelation process: in cases in which the metal ion is colorless (e.g., cases of Hf and Sn) or its color is quite distinct from the red color associated with the *RF* network formation (e.g., cases of Ni

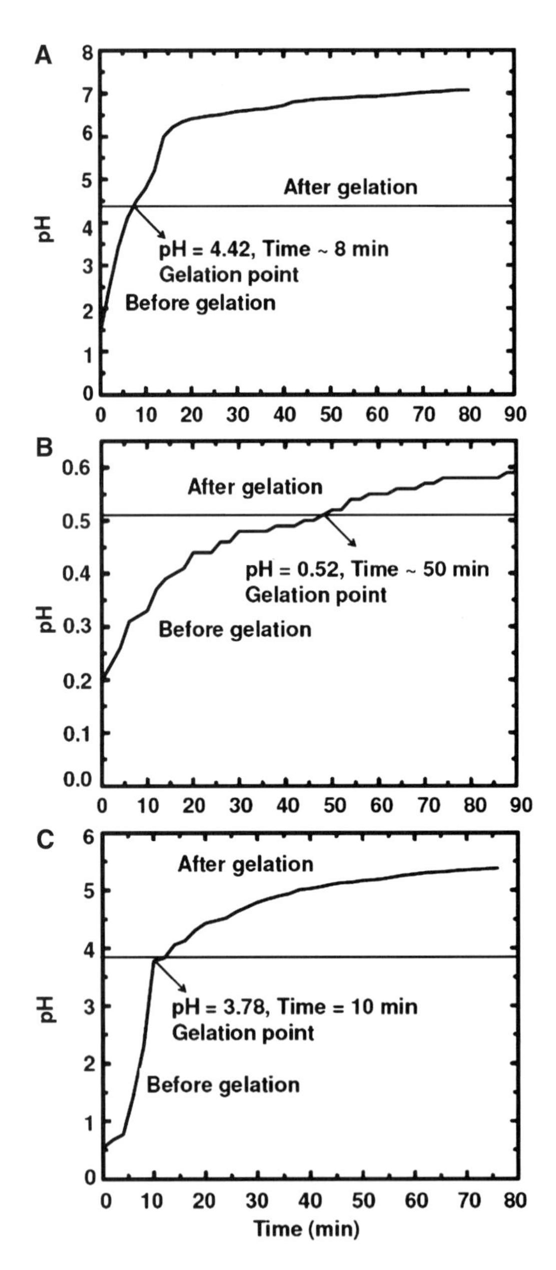

Figure 14.2. Typical pH changes during the gelation of **A.** a hydrated metal ion with epichlorohydrin (case of $[Fe(H_2O)_6]^{3+}$, 0.66 M in ethanol). **B.** *RF* in ethanol:acetonitrile using $[Fe(H_2O)_6]^{3+}$ as catalyst (0.29 M; no epichlorohydrin added). **C.** The complete *RF*/$[Fe(H_2O)_6]^{3+}$/epichlorohydrin system in ethanol/acetonitrile.

and Co), the latter appears after gelation, but within the timescale required for the acid-catalyzed gelation of *RF* by itself (Table 14.1): therefore, the formation of the two networks seems to occur independently of one another. Owing again to the irreversible reaction of protons with epichlorohydrin, the pH rises significantly during the cogelation of the organic

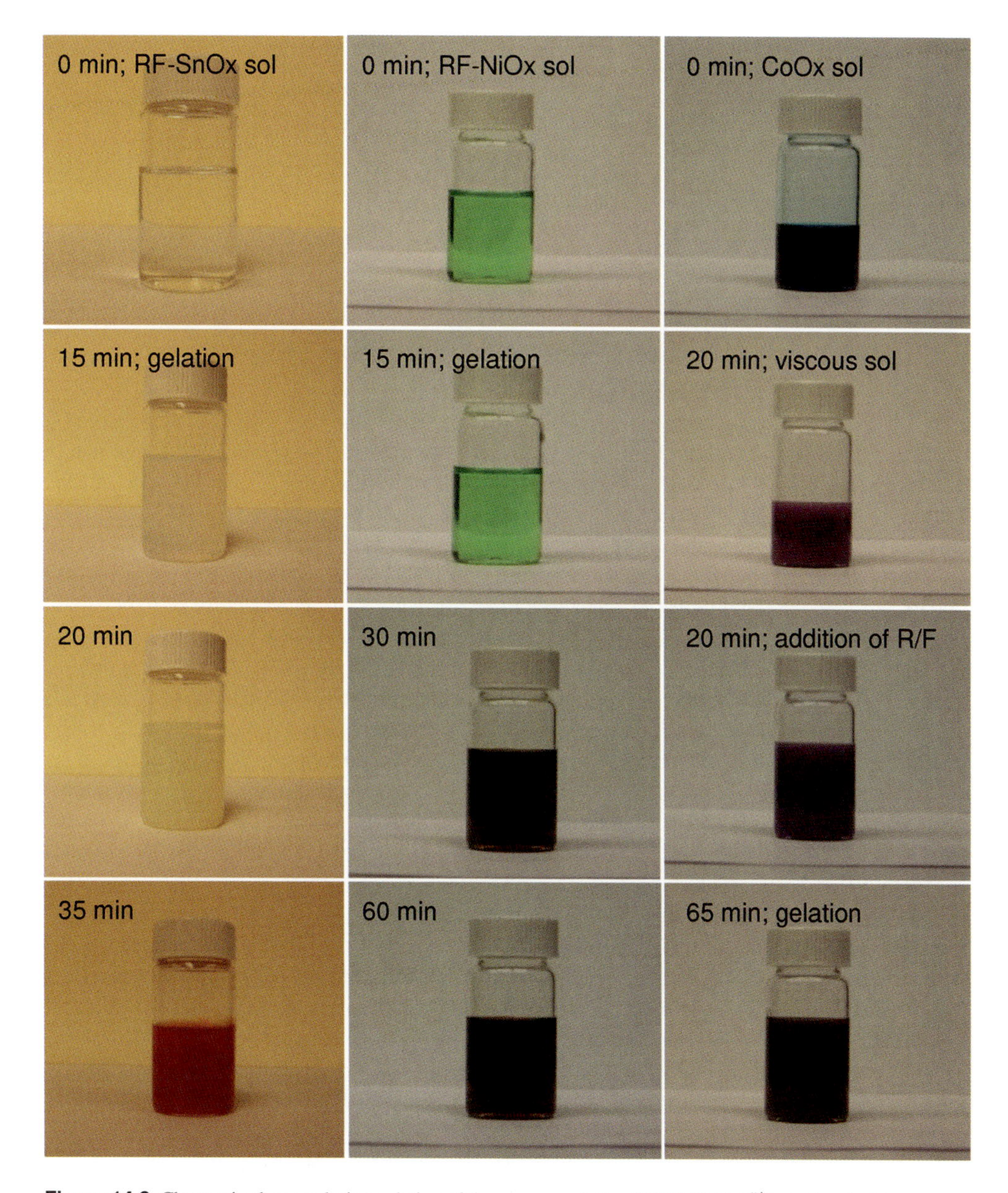

Figure 14.3. Chromatic changes during gelation of three representative *RF*/$[M(H_2O)_6]^{n+}$/epichlorohydrin systems in the solvents indicated in Table 14.1.

and inorganic networks (Figure 14.2C), and relatively long aging times (48 h) ensure complete *RF* network formation [52].

RF–MOx wet gels have been dried either in an autoclave with *SCF* CO_2 to yield native *RF–MOx* aerogels or under ambient pressure and temperature to produce *RF–MOx* xerogels. Alternatively, the *RF–MOx* skeletal framework was coated conformally (crosslinked) with a polyurea (*PUA*) derived by polymerization of a triisocyanate (Desmodur N3300A from Bayer) in an analogous fashion to what has been described for the *RF* and the *MOx* networks independently [38, 53]: wet gels are dried with *SCF* CO_2 yielding polyurea-crosslinked *X-RF–MOx* aerogels [54–57].

Desmodur N3300A: $O{=}C{=}N(H_2C)_6$, $(CH_2)_6N{=}C{=}O$, $(CH_2)_6N{=}C{=}O$ (triisocyanurate ring: N, N, N, O, O, O)

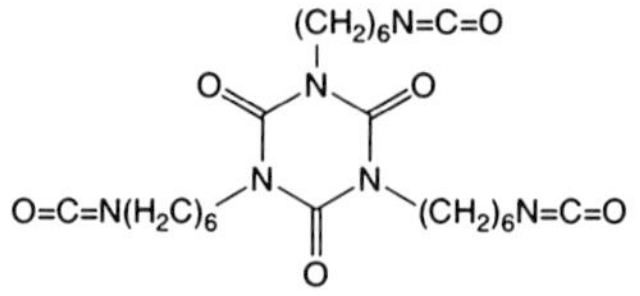

14.3. Materials Properties of Native *RF–MOx* Aerogels, Xerogels, and *X-RF–MOx* Aerogels

All three forms of the *RF–MOx* systems are monolithic. The presence of RF in native and polyurea (*PUA*) in *X-RF–MOx* aerogels was confirmed by ^{13}C solid-state NMR. Material characterization data are shown in Table 14.2. As expected [54–57], xerogels shrink 50–60% relative to the molds. Native aerogels shrink 10–30% and X-networks shrink the least (~6%). Shrinkage is reflected on the relative bulk densities (ρ_b) of xerogels and native aerogels. Skeletal densities (ρ_s) were used for evaluating the structure of the skeletal frameworks: ρ_s values of the xerogels should be equal to those of the aerogels. But they are lower (typically by 25%), a fact attributed to closed (meso)pores (see N_2 sorption data in Figure 14.5). On the other hand, experimental skeletal densities of the native and *X-aerogels* are within 5% of the weighted average values calculated from the organic (*RF*, $\rho_s = 1.45$g cm^{-3} [50]; and polyurea (*PUA*), $\rho_s = 1.201$ g cm^{-3} [53]) and inorganic (*MOx*) components (see Table 14.2),[1] signifying the absence of closed pores. Bulk and skeletal densities have been used together for calculating the porosities, Π, of the samples (Table 14.2): native *RF–MOx* aerogels consist of up to 98% (v/v) of empty space, followed by *X-aerogels* (~70%, v/v), and xerogels (34–54%, v/v).

During thermogravimetric analysis (*TGA*) in air (e.g., see Figure 14.4), native networks lose ~25% (w/w) of their mass (*RF*). Under N_2, the mass of interpenetrating networks with smeltable metals (Fe, Co, Ni, Sn, and Cu) decreases gradually and it does not level off up to 700°C. In *PUA*-crosslinked *X-RF–MOx*, the major component is the crosslinking polymer (Table 14.2). Those samples do reach constant weight (e.g., at ~500°C in the case of X-RF–FeO_x, Figure 14.4) under either air or N_2. Considered together, *TGA* data in air and under N_2 suggest a reaction between *RF* and *MOx* that proceeds slower in the native samples and is facilitated by crosslinking with *PUA*.

By scanning electron microscopy (*SEM*), all native *RF–MOx* networks appear as 3D agglomerates of nanoparticles. Figure 14.5 shows RF–CoO_x as a typical example. The microstructure is identical to what is expected of the two networks independently. In our experience, transmission electron microscopy (*TEM*) has been generally difficult with polymer/silica aerogels, because of small Z-attenuation differences (i.e., differences owing to the atomic number) and melting/decomposition of the organic component. Here, however, owing to the large Z-attenuation differences expected between *MOx* and *RF*, even in the extreme case of native RF–CoO_x aerogels in which the inorganic network does not gel by

[1] $\rho_{s,FeOx} = 3.03 \pm 0.07$ g cm^{-3}; $\rho_{s,NiOx} = 3.7 \pm 0.4$ g cm^{-3}; $\rho_{s,SnOx} = 3.0 \pm 0.1$ g cm^{-3}; $\rho_{s,CrOx} = 3.69 \pm 0.03$ g cm^{-3}; $\rho_{s,TiOx} = 3.77 \pm 0.07$ g cm^{-3}; $\rho_{s,HfOx}=3.8 \pm 0.2$ g cm^{-3}; $\rho_{s,YOx}=2.39 \pm 0.03$ g cm^{-3}; $\rho_{s,DyOx}=3.02 \pm 0.05$ g cm^{-3}.

Table 14.2. Selected properties of *RF-MOx* systems [46, 59]

Sample	Diameter (cm)[a]	Shrinkage (%)[a,b]	Bulk density ρ_b (g cm^{-3})[a]	Skeletal density ρ_s (g cm^{-3})[c]	Porosity Π (% v/v)	MO*x*:RF:PUA (w/w/w)[d]	*BET* surface area, σ (m^2 g^{-1})	Average pore diameter (nm) [half width (nm)][e]	Particle radius, *r* (nm)[f]
RF-FeOx									
Aerogel	0.946 ± 0.01	13.2 ± 1.31	0.047 ± 0.001	2.86 ± 0.12 [2.45]	98	78:22:00	298	33.8 [26.2]	3.5
Xerogel	0.38 ± 0.02	63.4 ± 0.49	1.00 ± 0.04	1.81 ± 0.01	45		184	4.8 [3.2]	9.0
X-aerogel	1.00 ± 0.03	6.16 ± 0.58	0.42 ± 0.01	1.42 ± 0.01 [1.37]	71	49:14:37	80	28.8 [33.1]	26.4
RF-CoOx									
Aerogel	0.783 ± 0.024	21.6 ± 2.51	0.082 ± 0.001	2.34 ± 0.19 [g]	97	77:23:00	143	36.3 [45.3]	9.0
Xerogel	0.43 ± 0.04	58.6 ± 0.53	0.81 ± 0.12	1.93 ± 0.04	35		123	4.8 [3.5]	12.6
X-aerogel	0.91 ± 0.02	6.16 ± 0.58	0.37 ± 0.03	1.42 ± 0.05 [g]	71	23:7:70	52	22.3 [26.2]	40.9
RF-NiOx									
Aerogel	0.69 ± 0.015	29.7 ± 1.25	0.059 ± 0.003	2.59 ± 0.13 [2.62]	98	74:26:00	309	22.9 [24.3]	3.7
Xerogel	0.55 ± 0.02	47.1 ± 0.32	1.24 ± 0.07	1.87 ± 0.00$_4$	34		h	h	
X-aerogel	0.93 ± 0.04	8.6 ± 1.1	0.397 ± 0.015	1.38 ± 0.01 [1.41]	69	20:7:73	150	17.7 [15.0]	14.5
RF-SnOx									
Aerogel	0.756 ± 0.005	24.3 ± 0.57	0.135 ± 0.031	2.58 ± 0.13 [2.34]	95	74:26:00	147	37.1 [41.4]	7.9
Xerogel	0.37 ± 0.03	64.4 ± 0.58	1.15 ± 0.08	1.85 ± 0.02	39		h	h	
X-aerogel	0.86 ± 0.01	13.8 ± 0.42	0.31 ± 0.01	1.27 ± 0.00$_4$ [1.29]	77	11:4:85	60	31.6 [34.7]	39.3
RF-CuOx									
Aerogel	0.903 ± 0.021	16.3 ± 1.92	0.083 ± 0.009	2.65 ± 0.14 [g]	98	78:22:00	189	43.6 [49.6]	6.0
Xerogel	0.496 ± 0.055	52.3 ± 1.2	1.33 ± 0.009	2.01 ± 0.29	34		159	4.7 [3.2]	9.4
X-aerogel	0.97 ± 0.02	10.5 ± 1.96	0.42 ± 0.05	1.37 ± 0.00$_3$ [g]	71	8:3.5:88.5	21	38.9 [45.2]	30.8
RF-CrOx									
Aerogel	0.89 ± 0.02	11.3 ± 1.52	0.108 ± 0.005	2.42 ± 0.01 [2.57]	96	72:28:00	472	23.9 [17.1]	2.6
Xerogel	0.51 ± 0.01	50.9 ± 1.05	1.16 ± 0.072	1.78 ± 0.04	35		117	4.8 [1.89]	14.4
X-aerogel	0.94 ± 0.02	6.6 ± 1.4	0.42 ± 0.04	1.29 ± 0.04 [1.28]	68	8.5:3.0:88.5	71	29.5 [17.9]	32.2

(continued)

Table 14.2. (continued)

Sample	Diameter (cm)[a]	Shrinkage (%)[a,b]	Bulk density ρ_b (g cm^{-3})[a]	Skeletal density ρ_s (g cm^{-3})[c]	Porosity Π (% v/v)	MO*x*:RF:PUA (w/w/w)[d]	*BET* surface area, σ (m^2 g^{-1})	Average pore diameter (nm) [half width (nm)][e]	Particle radius, *r* (nm)[f]
RF-TiOx									
Aerogel	0.86 ± 0.01	12.6 ± 0.57	0.091 ± 0.001	2.64 ± 0.11 [2.72]	96	76:24:00	230	23.9 [30.8]	4.9
Xerogel	0.43 ± 0.017	58.8 ± 0.48	0.934 ± 0.035	2.04 ± 0.21	54		145	7.2 [4.83]	10.1
X-aerogel	0.91 ± 0.01	9.0 ± 1.2	0.414 ± 0.009	1.34 ± 0.82 [1.25]	69	5.0:1.6:93.4	69	30.1 [15.1]	32.4
RF-HfOx									
Aerogel	0.853 ± 0.015	14.7 ± 1.52	0.128 ± 0.004	2.59 ± 0.01 [2.71]	95	75:25:00	282	22.9 [37.9]	4.1
Xerogel	0.45 ± 0.026	57.3 ± 0.25	1.08 ± 0.057	1.92 ± 0.18	44		159	3.1 [1.03]	9.8
X-aerogel	0.926 ± 0.025	7.3 ± 1.5	0.442 ± 0.006	1.42 ± 0.08 [1.35]	69	15:5:80	57	34.6 [32.7]	37.1
RF-YOx									
Aerogel	0.88 ± 0.01	12.6 ± 0.47	0.124 ± 0.005	2.07 ± 0.05 [2.12]	94	80.1:19.9:00	302	36.3 [32.1]	4.8
Xerogel	0.38 ± 0.019	63.2 ± 0.78	1.12 ± 0.11	1.98 ± 0.11	43		122	6.0 [1.83]	12.4
X-aerogel	0.94 ± 0.05	6.0 ± 0.18	0.330 ± 0.003	1.25 ± 0.01 [1.27]	74	10.1:2.4:87.5	99	22.3 [24.1]	24.2
RF-DyOx									
Aerogel	0.86 ± 0.05	17.0 ± 1.73	0.089 ± .008	2.55 ± 0.01 [2.41]	97	76.5:23.5:00	290	52.4 [32.8]	4.1
Xerogel	0.51 ± 0.025	50.6 ± 0.82	1.17 ± 0.031	2.18 ± 0.24	46		98	7.6 [5.3]	14.0
X-aerogel	0.93 ± 0.01	6.80 ± 1.7	0.34 ± 0.03	1.40 ± 0.04 [1.29]	68	11:3.4:85.6	132	45.7 [37.2]	15.2

[a] Average of three samples (mold diameter: 1.04 cm).

[b] Shrinkage = 100 × (mold diameter − sample diameter)/(mold diameter).

[c] Single sample, average of 50 measurements; in brackets: calculated skeletal densities based on the thermogravimetric analysis data of the next column and the skeletal densities of native *MOx*, *RF* and polyurea (PUA) aerogels prepared independently (see text).

[d] Calculated using thermogravimetric analysis (*TGA*) data in air up to 700°C (e.g., see Figure 14.4); *PUA*: Desmodur N3300A triisocyanate-derived polyurea (crosslinking polymer).

[e] By the *BJH*-desorption method; in brackets: width at half maximum.

[f] Calculated via $r = 3/\rho_s\sigma$ using experimental skeletal density data and surface areas.

[g] Could not be calculated because native CoO*x* and CuO*x* did not gel, and therefore their skeletal densities could not be determined.

[h] Unable to obtain identifiable isotherms.

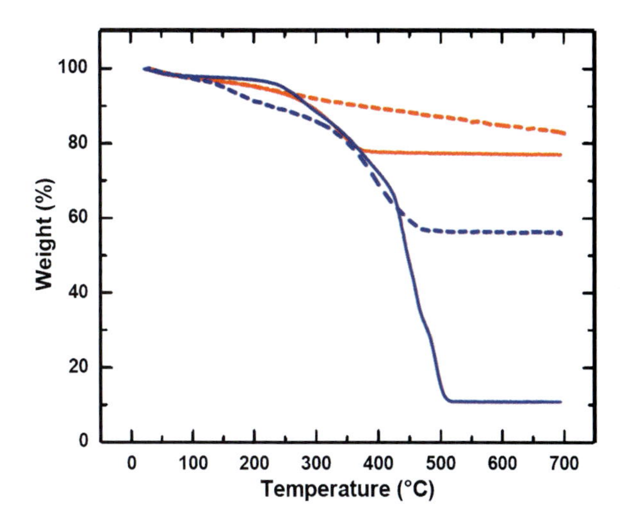

Figure 14.4. Typical thermogravimetric analysis data (*TGA*, 10°C min^{-1}) for smeltable *RF–MOx* (M: Fe, Co, Ni, Sn, and Cu). *Red*: native RF–FeO*x* ($\rho_b = 0.047 \pm 0.001$ g cm^{-3}); *blue*: X-RF–FeO_x ($\rho_b = 0.42 \pm 0.01$ g cm^{-3}). *Solid lines*: in air; *dashed lines*: in N_2.

itself, bright field *TEM* reveals an interconnected dark (inorganic) particle network surrounded by and embedded in a matrix of more "transparent" (*RF*) particles (Figure 14.6). *SEM* and *TEM* data considered together support a random distribution of interpenetrating strings of *RF* and *MOx* nanoparticles. (Further support of this model was found in the reactivities of those systems; see Sect. 14.4.)

Judging by the SEM microscopy images, xerogels also consist of nanoparticles; however, the network appears collapsed and the nanoparticles are in closer contact with one another (Figure 14.5).

With regard to the *X-RF–MOx* systems, in some cases (e.g., X-RF–HfO_x, Figure 14.11), the *X-aerogel* framework looks very similar to that of the corresponding native aerogel as expected from a conformal *PUA* coating on the skeletal framework of the native material. In some other cases (e.g., X-RF–CoO*x*, Figure 14.5), the framework is more fibrous, presumably owing to *PUA* formation catalyzed by the metal ions on the surface of the *MOx* nanoparticles. Sol–gel-based *PUA* produced by the reaction of isocyanates with water is of a fibrous morphology [53]. In the case of X-RF–CO_x (Figure 14.5), the fibers are less pronounced; however, in some other ones (e.g., X-RF–CuO_x; see Figure 14.10), they comprise a dominant feature.

The pore structure and, by extrapolation, the compactness (degree of contact between nanoparticles) are further evaluated by N_2 sorption. Representative isotherms are included in Figure 14.5 and the relevant data are summarized in Table 14.2. All three types of materials, native aerogels, xerogels, and *X-aerogels*, give typical Type IV isotherms reaching saturation. Native and *X-aerogels* show H1 hysteresis loops frequently associated with cylindrical pores. Xerogels adsorb significantly less N_2 than native and *X-aerogels*, and they show H2 desorpion loops, characteristic for ink-bottle pores, presumably formed by structural collapse during drying. Further, some ink-bottle necks may become quite narrow or clog altogether, thus explaining the lower skeletal densities of the xerogels relative to the corresponding aerogels. Average mesopore diameters were evaluated by the Barrett–Joyner–Halenda (*BJH*) method applied on the

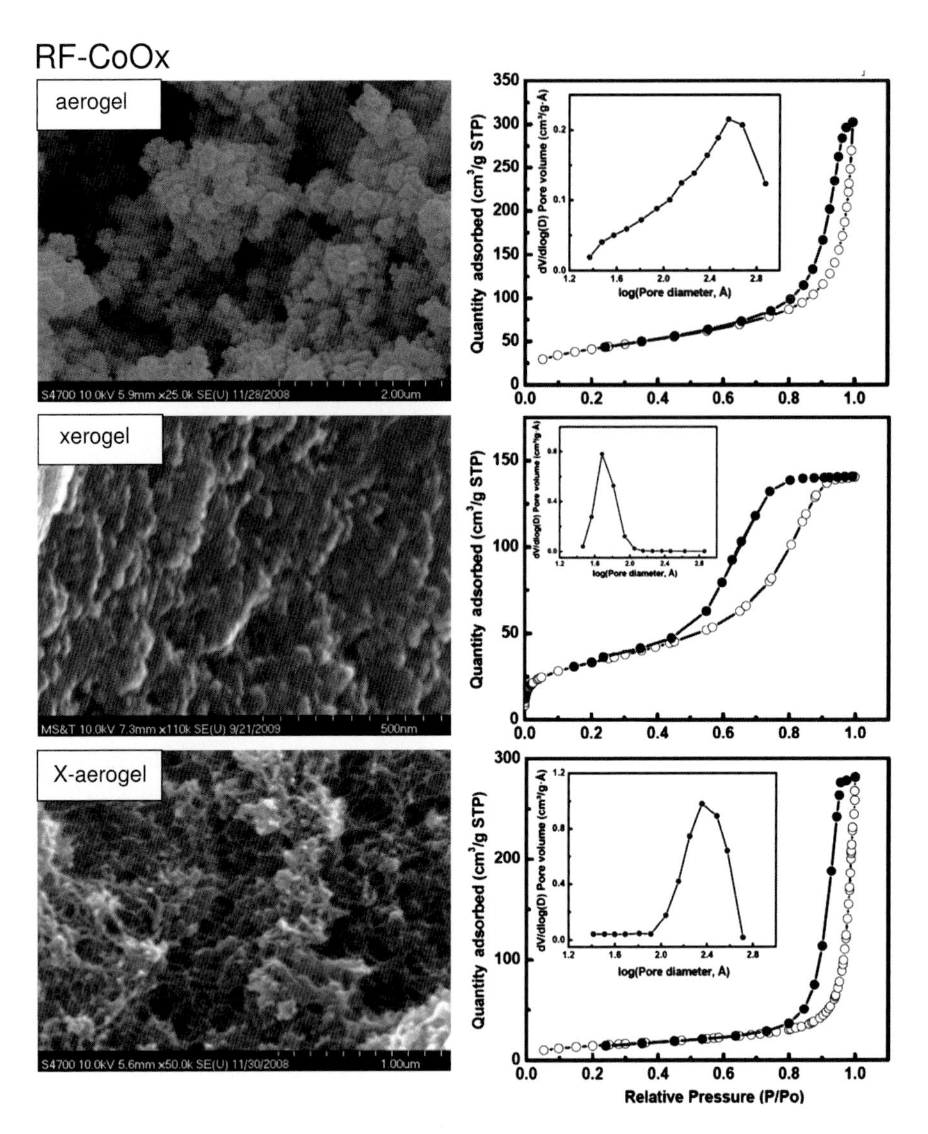

Figure 14.5. Typical *SEM* and N_2 sorption data for the *RF–MOx* systems in the native aerogel, xerogel, and *X-aerogel* forms. Case shown: the RF–CoO_x system. *Open circles*, adsorption; *dark circles*, desorption. *Insets*: *BJH* curves from the desorption branches of the isotherms.

desorption branch of the N_2 sorption isotherms (Table 14.2). Although the *BJH* method tends to underestimate the average mesopore diameters by 10–20% [58], the data appear internally consistent showing that native and *X-aerogels* have significantly larger pore diameters (18–46 nm) than the corresponding xerogels (3–8 nm), as expected. On the other hand, native samples seem to have larger pores than the corresponding *X-aerogels*, but this is a weak generalization as the pore size distributions (Table 14.2) are quite broad.

Despite the small pore diameters of the xerogels, their Brunauer–Emmett–Teller (*BET*) surface areas are still quite high, at ~50% of those of the corresponding native samples. On the other hand, despite their minimal shrinkage relative to the molds,

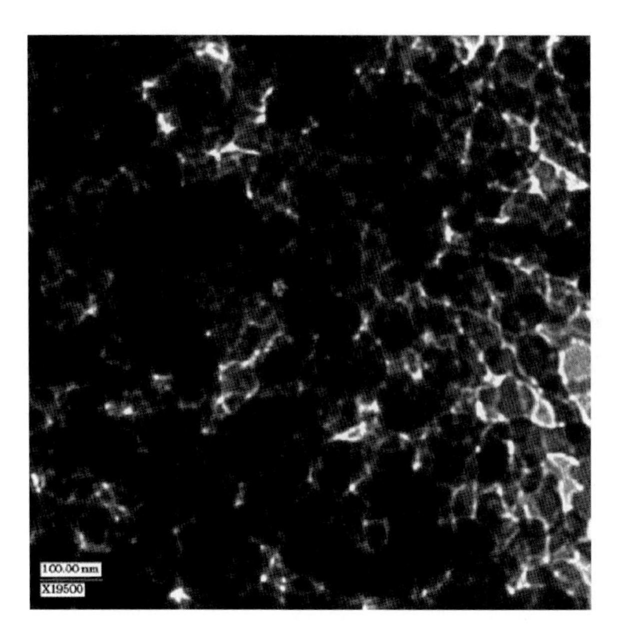

Figure 14.6. Bright field *TEM* of a native RF–CoO_x aerogel ($\rho_b = 0.082 \pm 0.001$ g cm^{-3}).

X-aerogels show greatly reduced surface areas, which is consistent with a conformal polyurea coating filling the smaller crevices of the native framework and thus blocking access of the probe (N_2) to those spaces. Within this line of reasoning, the apparent particle size of the *X-aerogels* should be significantly larger than the particle size of the native materials, a fact supported by particle radii calculations (Table 14.2). Further, the trend for the *BJH* pore diameters of the X-samples is to remain similar to those of the corresponding native samples, or even decrease (Table 14.2), and has been attributed to that the polymer does not just fill crevices on the native skeletal framework, but it keeps accumulating on the surface of the new skeletal *RF–MOx* polymer agglomerates.

Overall, based on *SEM*, *TEM*, and N_2 sorption porosimetry, the prevalent model for interpenetrating *RF–MOx* networks is one consisting of random distributions of two independent networks of nanoparticles, which in the native and the X-aerogel forms are looser, with less points of contact, while xerogels are more compact, as expected.

14.4. Reactions Between *RF* and *MOx* Nanoparticles

14.4.1. Chemical Transformations

Smeltable *RF–MOx* Systems (M: Fe, Co, Ni, Sn, and Cu)

Typical energy-dispersive X-ray spectroscopy (*EDS*) and X-ray diffraction (*XRD*) results for samples pyrolyzed in the 200–1,000°C range are exemplified by the RF–SnO_x system (Figure. 14.7). According to *EDS*, as the pyrolysis temperature increases, the percentage of the metal by weight steadily increases and the amount of both carbon and,

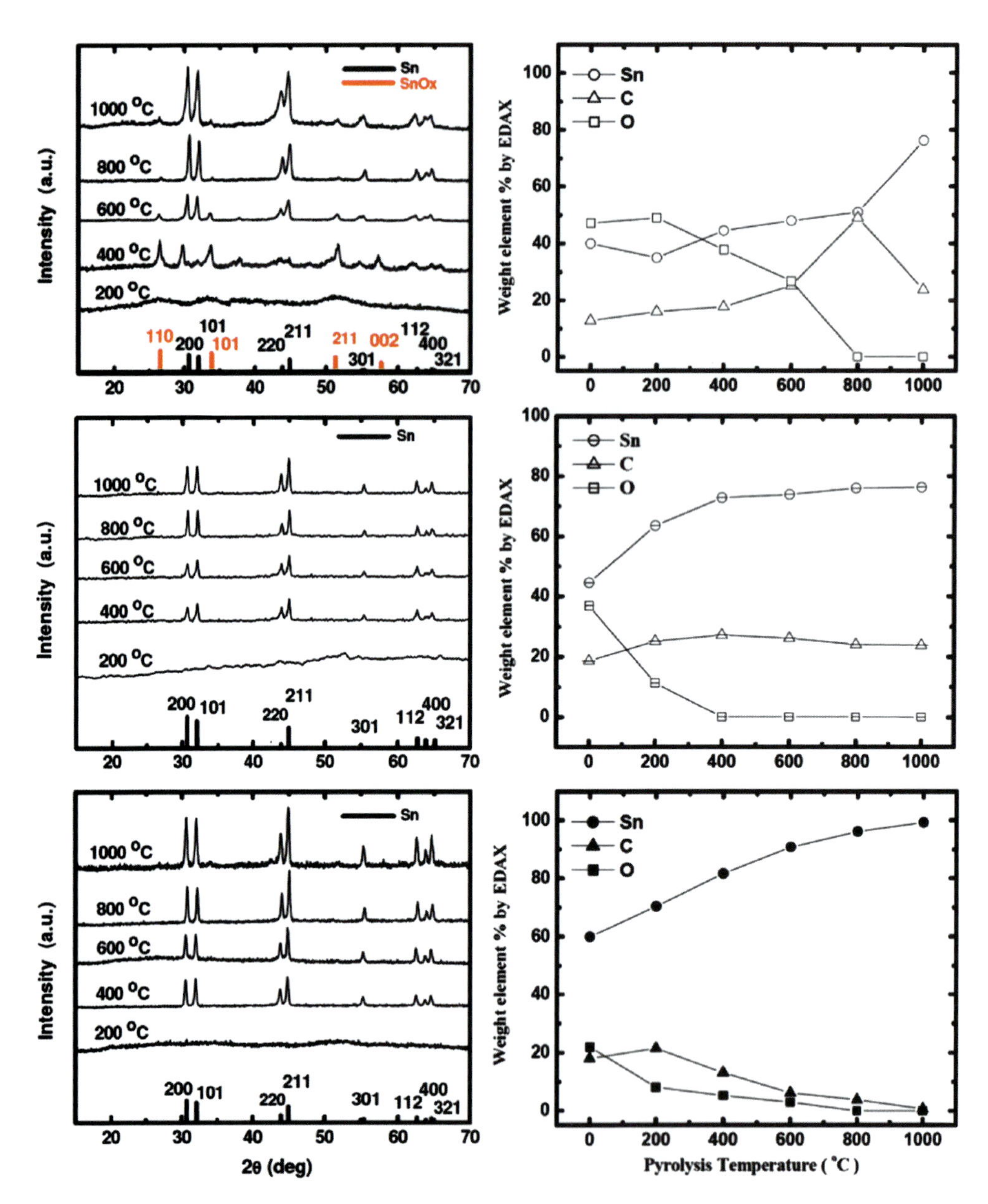

Figure 14.7. Typical *XRD* and *EDAX* data as a function of pyrolysis temperature (under Ar) for the smeltable *RF–MOx* systems in the native aerogel (**top**), xerogel (**middle**), and *X-aerogel* (**bottom**) forms. Case shown: RF–SnO_{x0}.

most importantly, oxygen decreases. In xerogels and *X-aerogels*, the oxygen is undetectable by EDS. According to quantitative *XRD* analysis for the crystalline phases developed upon pyrolysis (Figure 14.7 and Table 14.3), all systems in all three forms (native aerogels, *X-aerogels*, and xerogels) yield metals in their element forms. The RF–FeO_x system stands out as an exception, which also yields iron carbide [59]. Clearly, metal ions are reduced by the organic matter to the respective elements. Further, the smelting reaction proceeds faster

Table 14.3. Quantitative phase analysis (% w/w by *XRD*) of *RF-MOx* networks with smeltable metal oxides as a function of temperature (in parentheses: crystallite size, nm)

	200°C	400°C	600°C	800°C	1,000°C
Native RF-FeOx					
γ-Fe_2O_3	100 (4.1)	100 (16)	100 (28)	4.3 (19)	
Fe_3C				48.8 (33)	
Fe				10.9 (36)	100 (35)
$Fe_{15.1}C$				18.2	
C				17.8 (7.8)	
xero-RF-FeOx					
Fe_3C		11.4 (30)	8.8 (34)	4.2 (37)	1.6 (44)
Fe		88.6 (13)	91.2 (13)	95.8 (14)	98.4 (15)
X-RF-FeOx					
γ-Fe_2O_3	100 (4.6)	49.5 (16)	11.3 (23)	4.8 (19)	
FeO		6.1 (35)			
Fe_3C		44.5 (30)	74.5 (36)	43.7 (33)	75.0 (30)
Fe			14.2	13.3 (9.6)	25.0 (25)
$Fe_{15.1}C$				9.9	
C				28.3 (3.2)	
Native RF-CoOx					
α-Co		59 (9.5)	72.6 (11)	84.6 (11)	100 (16)
β-Co		27.2 (8.8)	20.2 (12)	8.5 (13)	
CoO		14.3 (13)	7.2 (14)	6.9 (15)	
xero-RF-CoOx					
α-Co		85.9 (13)	92.5 (13)	100 (15)	100 (15)
β-Co		14.1 (8.9)	7.5 (9.7)		
X-RF-CoOx					
α-Co		82.5 (12)	87.8 (12)	100 (14)	100 (15)
β-Co		17.5 (10)	12.2 (11)		
Native RF-NiOx					
Ni		100 (8.9)	100 (10)	100 (20)	100 (20)
xero-RF-NiOx					
Ni		100 (6.5)	100 (6.8)	100 (13)	100 (13)
X-RF-NiOx					
Ni		73.5 (26)	82.9 (28)	87.2 (29)	89.2 (32)
C		26.5 (6.1)	17.1 (6.4)	12.8 (6.5)	11.8 (6.6)
Native RF-SnOx					
Sn		44.8 (18)	73.8 (22)	85.2 (29)	91.8 (29)
SnO_2		55.2 (20)	26.2 (26)	14.8 (29)	8.2 (29)
xero-RF-SnOx					
Sn		100 (19)	100 (24)	100 (33)	100 (34)
X-RF-SnOx					
Sn		100 (34)	100 (43)	100 (48)	100 (53)
Native RF-CuOx					
Cu		58.2 (7.7)	81.5 (7.0)	87.8 (15)	87.9 (18)
CuO		41.8 (3.3)	18.5 (9.1)	12.2 (14)	12.1 (13)
xero-RF-CuOx					
Cu		100 (17)	100 (17)	100 (20)	100 (24)
X-RF-CuOx					
Cu	– (7.0)	93.8 (23)	100 (22)	100 (19)	100 (27)
CuO	– (3.3)	6.2 (3.8)			

in xerogels and *X-aerogels*, which in fact behave quite similarly: while in several native *RF–MOx* systems the metallic phase is still mixed with the oxide even after a treatment at 800–1,000°C (e.g., Fe, Co Sn, and Cu), in several xerogels and *X-aerogels* (e.g., those with M: Co, Sn, Cu), the only detectable crystalline phase above 400°C is the metallic one. Xero- and X-RF–FeO*x* also yield Fe_3C at temperatures as low as 400°C, while RF–NiO*x* comprises an exception as all three forms of the material yield Ni^0 as the only crystalline phase when heated above 400°C. The crystallite size of all crystalline phases increases with temperature, most probably due to sintering.

In native aerogels and xerogels, the only possible reaction is between solid *MOx* and the *RF*, or gases produced during carbonization of *RF*, or carbon produced terminally from *RF*. A reaction between *MOx* and gases from the decomposition of *RF* should be ruled out, because in that case there should be no substantial difference in the behavior of native aerogels and xerogels. Therefore, all reductions should be solid–solid reactions, and the significantly lower reaction temperatures of the interpenetrating xerogel networks have been attributed to their more dense structure. The different reactivity of the native aerogels and xerogels has been considered as further supporting evidence for the morphology of the *RF–MOx* networks: core-shell structures were excluded on grounds of relative reactivity considerations, because in that case the contact area between *RF* and *MOx* would have been maximized already from the wet gel stage on, and therefore, it would be independent of the subsequent drying protocol. Consequently, carbothermal reactions induced afterward in the two materials would always take place at similar temperatures.

In most native *RF–MOx* systems (with the exception of RF–NiO*x*), the quantitative reduction starts taking place at or about the onset of carbonization of *RF* (600°C); therefore, it was deemed reasonable to assume that it occurs mainly because of reaction between *MOx* and carbon. By the same token, in some native samples (RF–NiO*x*, RF–SnO_x, and RF–CuO_x) and in all xerogels, formation of the metallic face was observed below the carbonization temperature range of *RF*; hence reaction could also take place directly between the *MOx* and *RF* nanoparticles or between *MOx* and early solid decomposition products of *RF*.

Polyurea-crosslinked aerogels include a second organic component that conceivably could react directly with *MOx* yielding the metallic phase. In fact, polyurea, being a conformal coating on the *MOx* particles, could be more prone to react with the latter than *RF*, explaining conceivably the facile reaction (lower onset temperatures) of the *X-aerogels*. This possibility was investigated by the pyrolysis of X-FeO_x, X-NiO_x, and X-SnO_x aerogels crosslinked with Desmodur N3300A at 400 and 600°C. *XRD* (Figure. 14.8) shows no metallic or carbide phases for X-FeO_x even at 600°C. On the other hand, there is the formation of roughly equal amounts of Ni^0 both at 400 and 600°C (56–58%, w/w), while a major reduction of SnO_x to Sn^0 is observed already at 400°C and a complete reduction is obtained at 600°C. Thus, polyurea may indeed react directly with some of the inorganic networks. However, since RF–FeO_x xerogels show the formation of Fe^0 at as low as 400°C, while RF–NiO_x and RF–SnO_x xerogels (no polyurea present) were converted completely to Ni^0 and Sn^0 at that temperature, the safe conclusion is that the dominant reaction in the *X-RF–MOx* systems is between the *RF* and the *MOx* nanoparticles, as opposed to between *MOx* and their conformal polyurea coating. Therefore, the similar thermal behavior of *X-RF–MOx* and the corresponding xerogels has been attributed to morphological similarities.

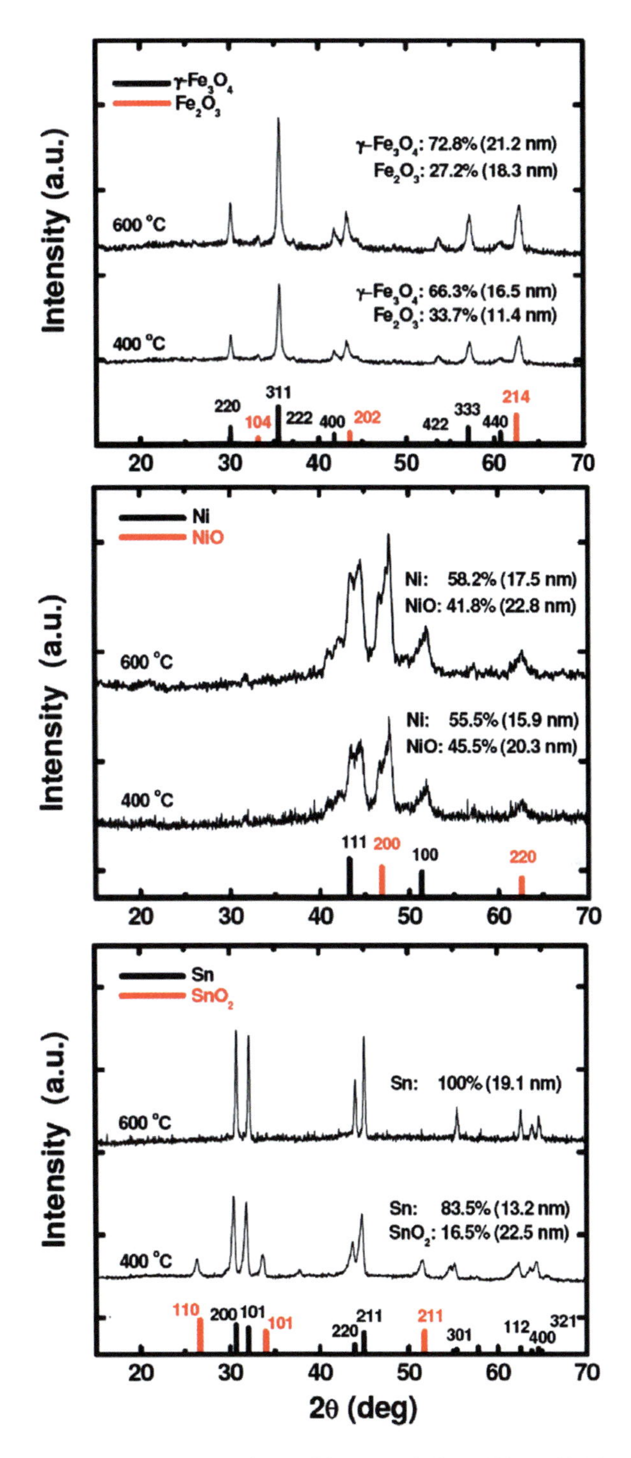

Figure 14.8. *XRD* after pyrolysis (Ar) of X-FeO_x, X-NiO_x, and X-SnO_x at 400 and 600°C.

RF–MOx Systems Yielding Carbides

Interpenetrating *RF–MOx* networks with refractories Cr, Ti, and Hf did not yield any reaction up to the 800–1,000°C range. Upon further pyrolysis up to 1,400°C, all three forms of the material yielded carbides (Figure 14.9, Table 14.4). In the same fashion as the smeltable metals, xerogels and *X-aerogels* behave similarly, yielding carbides at up to 200°C lower processing temperatures than the corresponding native aerogels. Again, the more facile reaction in xerogels relative to aerogels has been attributed to their compactness. Since the possibility of reaction of the crosslinking polymer with *MOx* is clearly excluded (at 800°C all metals still remain in their oxide forms, and all organic matter has either decomposed or transformed to carbon), the similar behavior of xerogels and *X-aerogels* was attributed to the nanomorphology of the samples.

Overall, for both types of *RF–MOx* systems, the smeltable and those that are transformed to carbides, the analogous behavior of xerogels and *X-aerogels* has been attributed to morphological similarities. But since crosslinking "freezes" the dimensions of the *X-samples* closer to the dimensions of the wet gels (refer to shrinkage data in Table 14.2), the topological relationship of the *RF* and *MOx* nanoparticles in the *X-RF–MOx* systems must be similar to or even more "loose" than that in the native samples. Thus, more facile reaction between *RF* and *MOx* in the *X-RF–MOx* systems has been attributed to the morphological changes taking place during pyrolysis.

14.4.2. Morphological Changes During Pyrolysis of the RF–MO*x* Systems

Upon pyrolysis, *X-RF* aerogels undergo morphological changes, which are attributed to physicochemical transformations of the crosslinker. Specifically, at around 200–250°C, the carbamate linkages of the polyurea tethers to the *RF* nanoparticle surfaces break and the free polymer finds itself above its melting point (~123°C) causing a partial collapse of the nanoparticulate *RF* network [53]. This process creates macropores. More importantly, *RF* nanoparticles inside the macroporous walls assume a xerogel-like (collapsed) morphology as might have been expected [53].

Similarly, Figure. 14.10 follows the structural evolution of a representative smeltable system (native and X-RF–CuO_x aerogels), while Figure. 14.11 follows the structural evolution of RF–HfO_x. All *X-aerogels* develop macropores in the same temperature range as typical *X-RF* aerogels (below 400°C). Occasional fractures during sample preparation (Figure 14.10, inset at 400°C) allows a view inside the macroporous walls revealing a typical xerogel-like morphology, just as in the case of *X-RF* aerogels [53].

14.5. Conclusions

Synthesis of *RF–MOx* interpenetrating networks is straightforward and highly versatile. Importantly, these materials can be used as precursors for the synthesis of other aerogels whose synthesis either has not been reported before or would be quite elaborate by other methods. The method can be modified easily to make films, thus comprising an economic alternative to chemical or atomic layer deposition methods (*CVD* and *ALD*, respectively), which require expensive equipment.

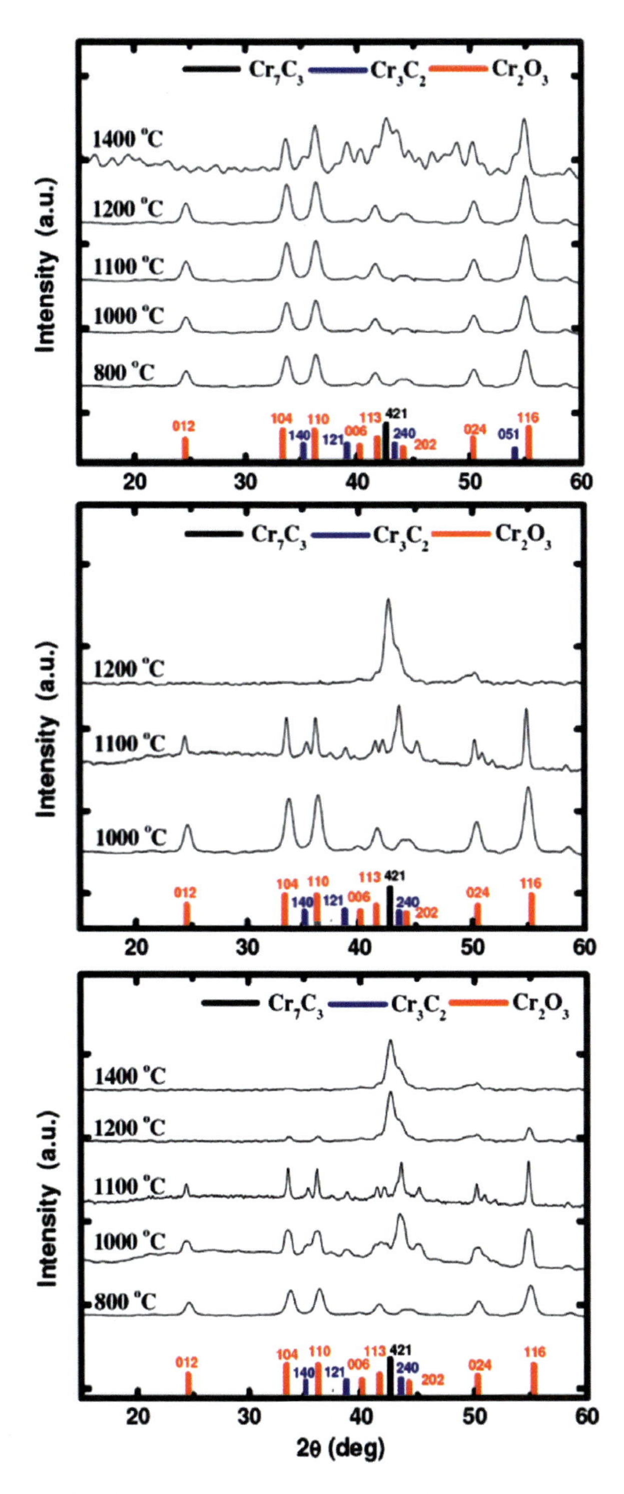

Figure 14.9. *XRD* data of RF–CrO_x as a function of the pyrolysis (Ar) temperature. **Top:** native aerogel; **middle:** xerogel; **bottom:** *X-aerogel*.

Table 14.4. Quantitative phase analysis (% w/w by *XRD*) of *RF-MOx* systems yielding carbides as a function of temperature (in parenthesis: crystallite size, nm)

	800°C	1,000°C	1,100°C	1,200°C	1,400°C
Native RF-CrOx					
Cr_2O_3	100 (11)	100 (11)	100 (13)	100 (17)	29.1 (21)
Cr_7C_3					51.2 (19)
Cr_3C_2					19.7 (11)
xero-RF-CrOx					
Cr_2O_3		100 (13)	38.9 (15)	22.9 (14)	
Cr_7C_3			19.3 (10)		
Cr_3C_2			41.8 (15)	77.1 (18)	
X-RF-CrOx					
Cr_2O_3	100 (14)	66.8 (15)	48.2 (15)	6.9 (23)	
Cr_7C_3		9.1 (6.6)	16.9 (8.7)	70.1 (11)	86.5 (12)
Cr_3C_2		24.1 (11)	34.9 (12)	23.0 (15)	13.5 (17)
Native RF-TiOx					
TiO_2	100 (9.1)	100 (9.6)	100 (10)	100 (14)	
TiC					100 (20)
xero-RF-TiOx					
TiO_2		100 (10)	100 (14)		
TiC				100 (29)	
X-RF-TiOx					
TiO_2	100 (11)	100 (12)	100 (14)		
TiC				100 (19)	100 (23)
Native RF-HfOx					
HfO_2	100 (14)	100 (14)	100 (15)	100 (15)	
HfC					100 (29)
xero-RF-HfOx					
HfO_2		100 (13)	100 (15)		
HfC				100 (29)	
X-RF-HfOx					
HfO_2	100 (10)	100 (11)	100 (11)		
HfC				100 (19)	100 (27)

Bulk mixtures of micron-sized oxides and carbon black react slowly. This has been confirmed by pyrolysis under the same conditions (3 h, under Ar) of CuO and 6-molar excess of carbon black at different temperatures. Cu^0 does not show up before 800°C and it comprises only ~11% (w/w) of the crystalline phases at that temperature; at 1,000°C, Cu^0 still comprises only 7.5% (w/w) of the crystalline phases. By comparison (Table 14.3), nanoparticulated RF–CuO_x aerogels already consist of a ~60/40 (w/w) mixture of Cu^0/CuO at 400°C, and conversion is complete by 800°C, as might have been expected. What is extraordinary, however, is that an aerogel-like, large surface-to-volume ratio is a necessary but not sufficient precondition for efficient reaction among RF (C) and MO*x* nanoparticles. That is, irrespective of the specific chemical transformations happening at the nanoparticle interfaces, a high surface-to-volume ratio must be combined with high compactness. For the RF–CuO system, for example, xerogels are completely converted to Cu^0 at 400°C and X-aerogels already consist of a 94/6 (w/w) ratio of Cu^0/CuO at that temperature.

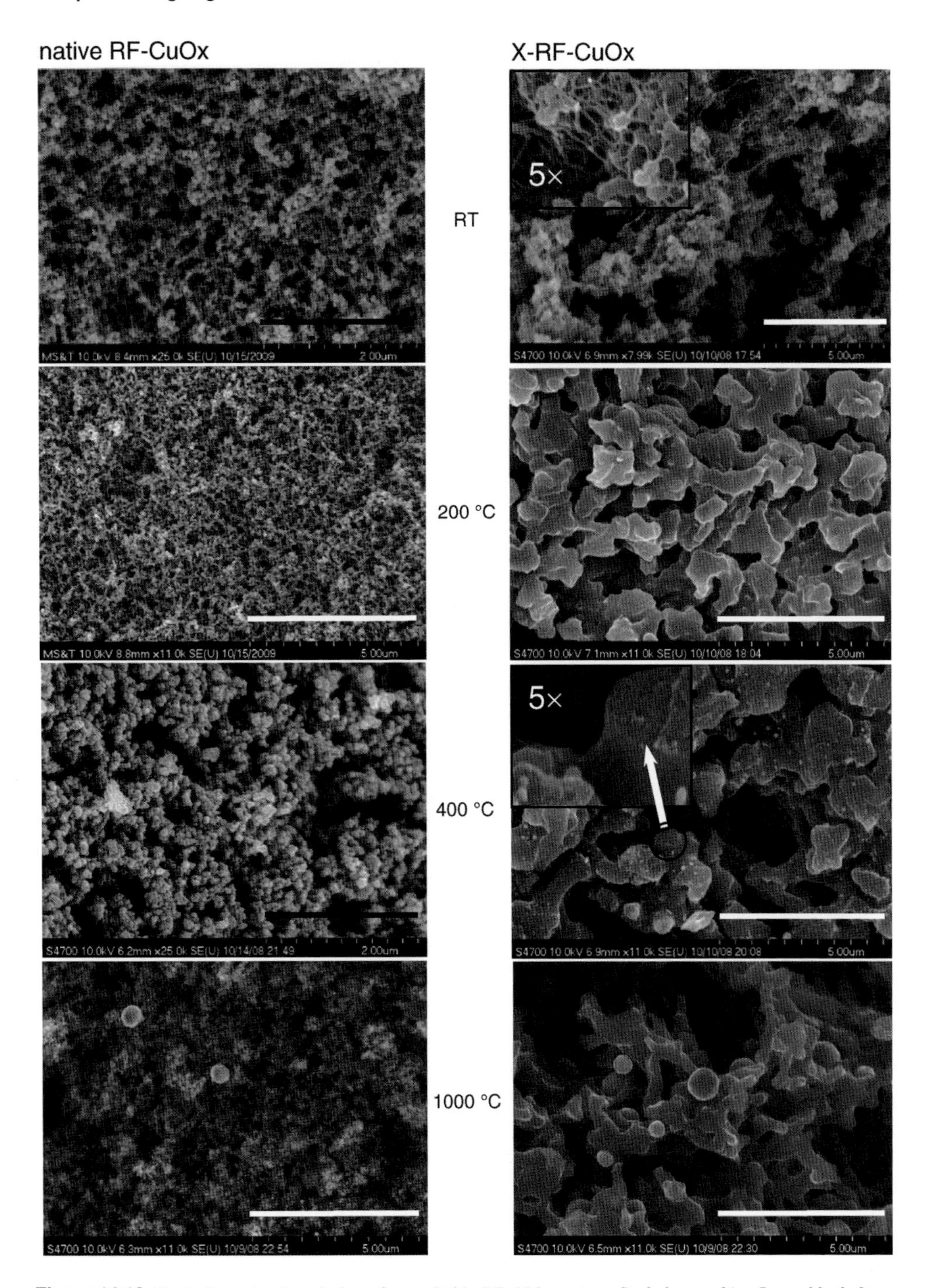

Figuer 14.10. Typical structural evolution of a smeltable *RF–MOx* system. *Scale bars*: *white*, 5 μm; *black*, 2 μm. *Arrow* points to the xerogel-like structure inside the macroporous walls [53].

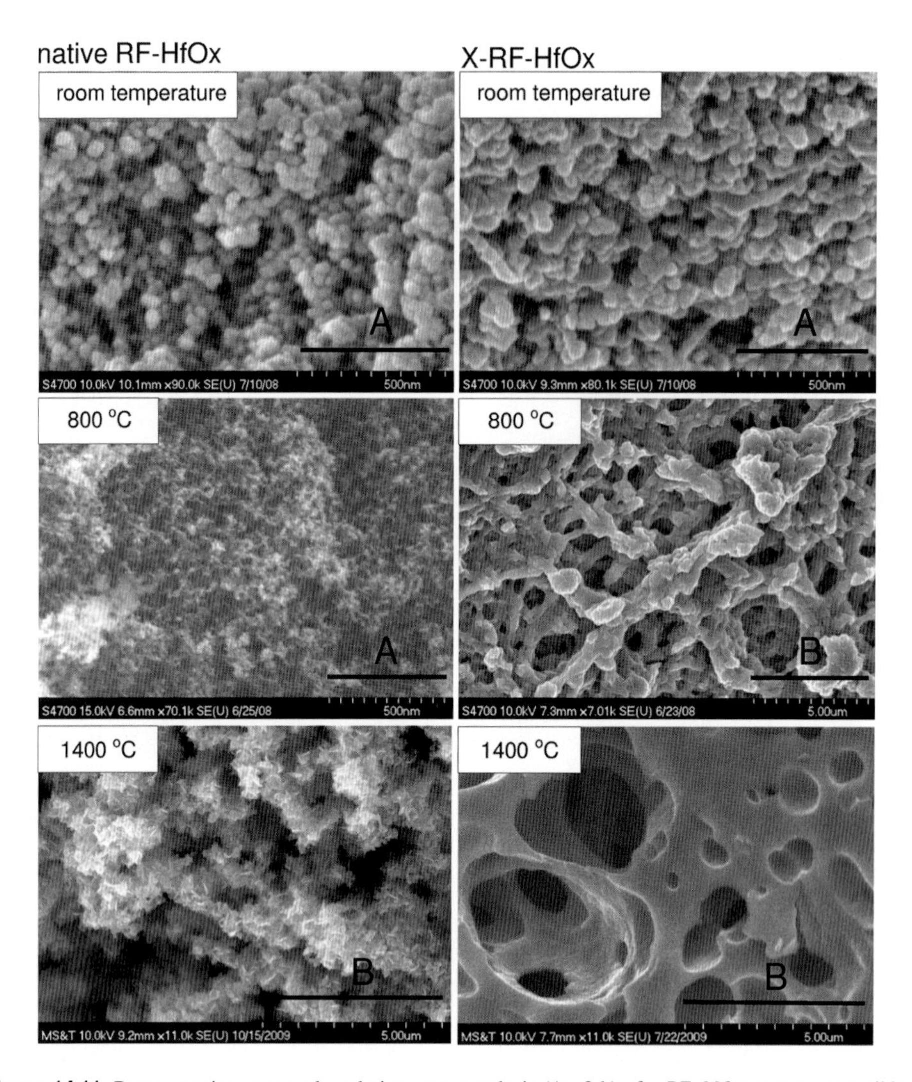

Figure 14.11. Representative structural evolution upon pyrolysis (Ar, 3 h) of a *RF–MOx* system convertible to carbide. *Scale bars*: **A.** at 500 nm;**B.** at 5 μm.

Nevertheless, it is predicted that the *degree* of compactness needed should be completely application-specific. For instance, a xerogel-like structure in Fe_2O_3/Al *thermites* should be more effective than an aerogel. However, in *thermites* that require circulation of gases (e.g., O_2 in RF–CuO_x [46, 59, 60] or the *RF*–Fe^0 system [11]), a more open aerogel- or ambigel-like [61–63] structure is expected to burn more quickly.

Acknowledgments

The author would like to thank the Army Research Office (W911NF-10-1-0476) and the National Science Foundation (CHE-0809562 and CMMI-0653919) for financial support, the Materials Research Center of Missouri S&T for support in sample characterization (*SEM* and *XRD*), co-investigator Professor Chariklia Sotiriou-Leventis, and the graduate students who have contributed to this effort: Naveen Chandrasekaran, Sudhir Mulik, and Anand G. Sadekar.

References

1. Interpenetrating polymer networks. Klempner D, Sperling L H, Utracki L A Eds, American Chemical Society, Washington DC (1994)
2. Pope E J A, Asami M, Mackenzie J D (1989) Transparent silica gel-PMMA composites. J Mater Res 4: 1018–1026
3. Philipp G, Schmidt H (1984) New materials for contact lenses prepared from Si- and Ti-alkoxides by the sol-gel process. J Non-Cryst Solids 63: 283–292
4. Huang H H, Orler B, Wilkes G L (1987) Structure-property behavior of new hybrid materials incorporating oligomeric species into sol-gel glasses. 3. Effect of acid content, tetraethoxysilane content, and molecular weight of poly(dimethylsiloxane). Macromolecules 20: 1322–1330
5. Sanchez C, Robot F (1994) Design of hybrid organic-inorganic materials synthesized via sol-gel chemistry. New J Chem 18: 1007–1047
6. Wen J, Wilkes G L (1996) Organic/inorganic hybrid network materials by the sol-gel approach. Chem Mater 8: 1667–1681
7. Hu Y, Mackenzie J D (1992) Rubber-like elasticity of organically modified silicates. J Mater Sci 27: 4415–4420
8. Novak B M, Auerbach D, Verrier C (1994) Low-density, mutually interpenetrating organic-inorganic composite materials via supercritically drying techniques. Chem Mater 6: 282–286
9. Kramer S J, Rubio-Alonso F, Mackenzie J D (1998) Organically modified silicate aerogels, "aeromosils." Mat Res Soc Symp Proc 435: 295–300
10. Henz B J, Hawa T, Zachariah M (2009) Molecular dynamics simulation of the energetic reaction between Ni and Al nanoparticles. J Appl Phys 105: 124310
11. Merzbacher C, Bernstein R, Homrighaus Z, Rolison D (2000) Thermally emitting iron aerogel composites. NRL Report: NRL/MR/1004–00-8490
12. Esmaeili B, Chaouki J, Dubois C (2009) Encapsulation of nanoparticles by polymerization compounding in a gas/solid fluidized bed reactor. AIChE Journal 55: 2271–2278
13. Zhao L, Clapsaddle B J, Satcher J H, Schaefer D W, Shea K J (2005) Integrated chemical systems: the simultaneous formation of hybrid nanocomposites of iron oxide and organo silsesquioxanes. Chem Mater 17: 1358–1366
14. Teixeira da Silva V L S, Ko E I, Schmal M, Oyama S T (1995) Synthesis of niobium carbide from niobium oxide aerogels. Chem Mater 7: 179–184
15. Walsh D, Arcelli L, Toshiyuki I, Tanaka J, Mann S (2003) Dextran templating for the synthesis of metallic and metal oxide sponges. Nat Mater 2: 386–390
16. Song Y, Yang Y, Medforth C J, Pereira E, Singh A K, Xu H, Jiang Y, Brinker C J, van Swol F, Shelnutt J (2004) Controlled synthesis of 2-D and 3-D dendritic platinum nanostructures. J Am Chem Soc 126: 635–645
17. Sherman A J, Williams B E, Delarosa M J, Laferla R. (1991) Characterization of porous cellular materials fabricated by CVD. In Mechanical Properties of Porous and Cellular Materials, Sieradzki K, Green D, Gibson L J, Eds, Materials Research Society Symposium Proceedings 207, Materials Research Society: Pittsburgh, PA, pp 141–149
18. Biener J, Hodge A M, Hamza A V, Hsiung L L, Satcher J H (2005) Nanoporous Au: a high yield strength material. J Appl Phys 97: 1–4
19. Tappan B C, Huynh M H, Hiskey M A, Chavez D E, Luther E P, Mang J T, Son S F (2006) Ultralow-Density Nanostructured Metal Foams: Combustion Synthesis, Morphology, and Composition. J Am Chem Soc 128: 6589–6594
20. Kim H S, Bugli G, Djéga-Mariadassou G (1999) Preparation and characterization of niobium carbide and carbonitride. J Solid-State Chem 142: 100–107
21. Pérez-Cadenas A F, Maldonado-Hódar F J, Moreno-Castilla C (2005) Molybdenum carbide formation in molybdenum-doped organic and carbon aerogels. Langmuir 21: 10850–10855
22. Maldonado-Hódar, Pérez-Cadenas A F, Moreno-Castilla C (2003) Morphology of heat-treated tungsten doped monolithic carbon aerogels. Carbon 41: 1291–1299
23. Li X K, Liu L, Zhang Y X, Shen Sh D, Ge Sh, Ling L Ch (2001) Synthesis of nanometre silicon carbide whiskers from binary carbonaceous silica aerogels. Carbon 39: 159–165
24. Li X K, Liu L, Ge Sh, Shen Sh D, Song J R, Zhang Y X, Li P H (2001) The preparation of Ti(C,N,O) nanoparticles using binary carbonaceous titania aerogel. Carbon 39: 827–833
25. Rhine W, Wang J, Begag R (2006) Polyimide aerogels, carbon aerogels, and metal carbide aerogels and methods of making same. US Pat No: 7,074,880

26. Al-Muhtaseb S A, Ritter J A (2003) Preparation and properties of resorcinol-formaldehyde organic and carbon gels. Adv Mater 15: 101–114
27. Moreno-Castilla C, Maldonado-Hódar F J (2005) Carbon aerogels for catalysis applications: an overview. Carbon 43: 455–465
28. Sánchez-Polo M, Rivera-Utrilla J, von Gunten U (2006) Metal-doped carbon aerogels as catalysts during ozonation processes in aqueous solutions. Water Res 40: 3375–3384
29. Baumann T F, Fox G A, Satcher J H, Yoshizawa N, Fu R, Dresselhaus M S (2002) Synthesis and Characterization of copper doped carbon aerogel. Langmuir 18: 7073–7076
30. Baumann T F, Satcher J H (2003) Homogeneous incorporation of metal nanoparticles into ordered macroporous carbons. Chem Mater 15: 3745–3747
31. Job N, Picard R, Vertruyen B, Colomer J-F, Marien J, Picard J-P (2007) Synthesis of transition metal-doped carbon xerogels by cogelation. J Non-Cryst Solids 353: 2333–2345
32. Fu R, Baumann T F, Cronin S, Dressekhaus G, Dresselhaus M S, Satcher J H (2005) Formation of graphitic structures in cobalt- and nickel-doped carbon aerogels. Langmuir 21; 2647–2651
33. Maldonado-Hódar F J, Moreno-Castilla C, Rivera-Utrilla J, Hanzawa Y, Yamada Y (2000) Catalytic graphitization of carbon aerogels by transition metals. Langmuir 16: 4367–4373
34. Yoshizawa N, Hatori H, Soneda Y, Hanzawa Y, Kaneko K, Dresselhaus M S (2003) Structure and electrochemical properties of carbon aerogels polymerized in the presence of Cu^{2+}. J Non-Cryst Solids 330: 99–105
35. Maldonado-Hódar F J, Moreno-Castilla C, Pérez-Cadenas A F (2004) Surface morphology, metal dispersion, and pore texture of transition metal-doped monolithic carbon aerogels and steam-activated derivatives. Micropor Mesopor Mater 69: 119–125
36. Steiner S A, Baumann T F, Kong J, Satcher J H, Dresselhaus M S (2007) Iron-doped carbon aerogels: novel porous substrates for direct growth of carbon nanotubes. Langmuir 23: 5161–5166
37. Cotton, F. A.; Wilkinson, G.; Murillo, C. A.; Bochmann, M. "Advanced Inorganic Chemistry Sixth Edition," Wiley: New York, N.Y. (1999)
38. Leventis N, Vassilaras P, Fabrizio E F, Dass A (2007) Polymer nanoencapsulated rare earth aerogels: chemically complex but stoichiometrically similar core-shell superstructures with skeletal properties of pure compounds. J Mater Chem 17: 1502–1508
39. Ziese W (1933) Die Reaktion von Äthylenoxyd mit Lösungen von Erd- und Schwermetallhalogeniden. Ein neur Weg zur Gewinnung von Solen und reversiblen Gelen von Metalloxydhydraten. Ber 66: 1965–1972
40. Kearby K, Kistler S S, Swann S Jr. (1938) Aerogel catalyst: conversion of alcohols to amines. Ind Eng Chem 30: 1082–1086
41. Sydney L E (1983) Preparation of sols and gels. British Patent No.: 2,111,966
42. Gash A E, Tillotson T M, Satcher J H, Hrubesh L W, Simpson R L (2001) New sol-gel synthetic route to transition and main-group metal oxide aerogels using inorganic salt precursors. J Non-Cryst Solids 285: 22–28
43. Gash A E, Tillotson T M, Satcher J H, Poco J F, Hrubesh L W, Simpson R L (2001) Use of epoxides in the sol-gel synthesis of porous Iron(III) oxide monoliths from Fe(III) salts. Chem Mater: 13, 999–1007
44. Sisk C N, Hope-Weeks J (2008) Copper(II) aerogels via 1,2-epoxide gelation. J Mater Chem 18: 2607–2610
45. Jones R W (1989) Fundamental principles of sol-gel technology. The Institute of Metals, London: UK
46. Leventis N, Chandrasekaran N, Sadekar A G, Sotiriou-Leventis C, Lu H (2009) One-pot synthesis of interpenetrating inorganic/organic networks of CuO/resorcinol-formaldehyde aerogels: nanostructured energetic materials. J Am Chem Soc 131: 4576–4577
47. Pekala R W (1989) Organic aerogels from the polycondensation of resorcinol with formaldehyde. J Mater Sci 24: 3221–3227
48. Pekala R W, Alviso C T, Kong FM, Hulsey S S (1992) Aerogels derived from multifunctional organic monomers. J Non-Cryst Solids 145: 90–98
49. Qin G, Guo S (2001) Preparation of RF organic aerogels and carbon aerogels by alcoholic sol-gel process. Carbon 39: 1935–1937
50. Mulik S, Sotiriou-Leventis C, Leventis N (2007) Time-efficient acid-catalyzed synthesis of resorcinol-formaldehyde aerogels. Chem Mater 19: 6138–6144
51. Reuß M, Ratke L (2008) Subcritically dried RF-aerogels catalysed by hydrochloric acid. J Sol-Gel Sci Technol 47: 74–80
52. Bekyarova E, Kaneko K (2000) Structure and physical properties of tailor-made Ce, Zr-doped carbon aerogels. Adv Mater 12: 1625–1628
53. Mulik S, Sotiriou-Leventis C, Leventis N (2008) Macroporous electrically conducting carbon networks by pyrolysis of isocyanate-cross-linked resorcinol-formaldehyde aerogels. Chem Mater 20: 6985–6997

54. Leventis N, Sotiriou-Leventis C, Zhang G, Rawashdeh A-M M (2002) Nano-engineering strong silica aerogels. Nano Lett 2: 957–960
55. Meador M A B, Capadona L A, MacCorkle L, Papadopoulos D S, Leventis N (2007) Structure-property relationships in porous 3D nanostructures as a function of preparation conditions: isocyanate cross-linked silica aerogels. Chem Mater 19: 2247–2260
56. Leventis N, Mulik S, Wang X, Dass A, Sotiriou-Leventis C, Lu H (2007) Stresses at the interface of micro with nano. J Am Chem Soc 129: 10660–10661
57. Leventis N (2007) Three-dimensional core-shell superstructures: mechanically strong aerogels. Acc Chem Res 40: 874–884
58. Lukens W W Jr, Schmidt-Winkel P, Zhao D, Feng J, Stucky G D (1999) Evaluating pore sizes in mesoporous materials: a simplified standard adsorption method and a simplified Broekhoff-de Boer method. Langmuir 15: 5403–5409
59. Leventis N, Chandrasekaran N, Sotiriou-Leventis C, Mumtaz A (2009) Smelting in the age of nano: iron aerogels. J Mater Chem 19: 63–65
60. Leventis N, Chandrasekaran N, Sadekar A G, Mulik S, Sotiriou-Leventis C (2010) The effect of compactness on the carbothermal conversion of interpenetrating metal oxide/resorcinol-formaldehyde nanoparticle networks to porous metals and carbides. J Mater Chem 20: 7456–7471
61. Harreld J H, Dong W, Dunn B (1998) Ambient pressure synthesis of serogel-like vanadium oxide and molybdenum oxide. Mater Res Bull 33: 561–567
62. Long J W, Swider-Lyons K E, Stroud R M, Rolison D R (2000) Design of pore and matter architectures in manganese oxide charge-storage materials. Electrochem Solid-State Lett 3: 453–456
63. Doescher M S, Pietron J J, Dening B M, Long J W, Rhodes C P, Edmondson C A, Rolison D R (2005) Using an oxide nanoarchitecture to make or break a proton wire. Anal Chem 77: 7924–7932

15

Improving Elastic Properties of Polymer-Reinforced Aerogels

Mary Ann B. Meador

Abstract Monolithic aerogels provide superior thermal insulation compared to other forms of aerogel (composites, particulate, etc.). It has also been demonstrated that monolithic aerogels can be made mechanically stronger and more durable by incorporating a conformal polymer coating on the skeletal nanostructure. However, for many applications it would be most desirable to have monolithic aerogels in a more flexible form, for example, as insulation in deployable and inflatable structures or space suits, or to wrap around a structure needing insulation. To this end, it has been found that by incorporating organic linking groups or alkyl trialkoxysilanes into the silica backbone, elastic recovery and/or flexibility is improved, while strength is maintained by the use of polymer reinforcement.

15.1. Introduction

Due to their combination of low density, high porosity, high surface area, and nanoscale pore sizes, silica aerogels are of interest for many applications including thermal and acoustic insulation, optics, catalysis, and chromatographic systems [1–3] (Chap. 2). However, potential applications of aerogel monoliths in aerospace, industry, and daily life have been restricted due to their poor mechanical properties and their extreme fragility [4]. Hence, in aerospace, especially (Chaps. 32 and 33), aerogel monoliths have been limited to a few exotic applications such as collecting hypervelocity particles from the tail of the comet Wild 2 in the Stardust Program [5] and as thermal insulation on the Mars Rover [6].

It has been shown that reinforcing silica aerogels by reacting polymer with the silanol surface to create a conformal coating over the silica skeleton is an effective way to increase mechanical strength by as much as two orders of magnitude while only doubling the density over those of native or nonreinforced aerogels [7–9] (Chap. 13). In addition, the mesoporosity of these polymer-reinforced aerogels, and hence, their superior insulation properties among other things, is maintained. Incorporating an amine onto the surface of the silica gel particles by coreacting the tetraalkoxysilane with 3-aminopropyltriethoxysilane (*APTES*), as shown in Scheme 15.1, allows for reinforcement with epoxy-terminated oligomers [10, 11] or cyanoacrylates [12] and also improves the reactivity of the silica surface toward

---

M. A. B. Meador • NASA Glenn Research Center, Mailstop 49-3, Cleveland, OH 44135, USA
e-mail: maryann.meador@nasa.gov

M.A. Aegerter et al. (eds.), *Aerogels Handbook*, Advances in Sol-Gel Derived Materials and Technologies, DOI 10.1007/978-1-4419-7589-8_15,

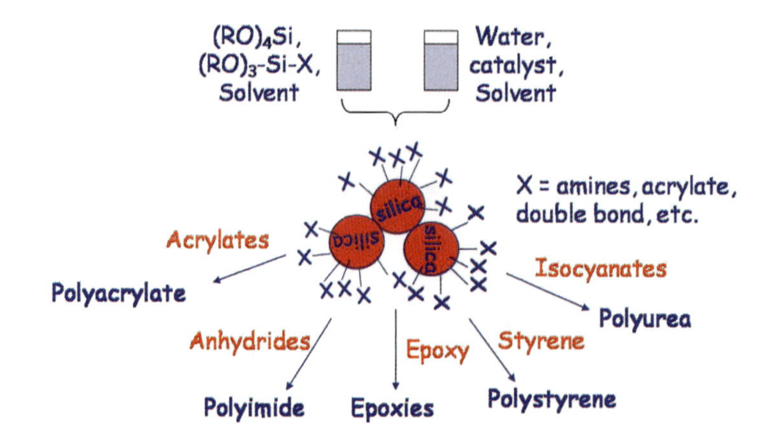

Scheme 15.1. Concept of polymer reinforcement using reactive groups on the silica surface.

isocyanates [13–15]. Expanding the silica surface chemistry to include styrene groups [16], free radical initiator [17] or vinyl [18] permits the use of polystyrene as a cross-linker. Other approaches to strengthening the silica aerogel structure by incorporation of a polymer include dispersing functionalized polymer nanoparticles in a silica network [19] (Chap. 21) and copolymerization of silica precursors with poly(methylmethacrylate) or poly(dimethylacryl-amide) [20] or poly(vinylpyrrolidone) [21].

Improvements to mechanical properties (Chap. 22) seen by reinforcing aerogels with polymer enable a whole host of weight-sensitive aerospace applications, including thermal and acoustic insulation for habitats, extravehicular activity (*EVA*) suits, launch vehicles, cryotanks, and inflatable decelerators for planetary reentry, as well as lightweight, multi-functional structures (including insulation, sound dampening, and structural support) for aircraft or rotorcraft. In addition, if manufacturing costs of polymer-cross-linked monolithic aerogel can be decreased, down to earth applications, including insulation for refrigeration, housing construction, and industrial pipelines, can be realized.

In particular, manned Mars surface applications shown in Figure 15.1 absolutely require a new insulation system. Multilayer insulation (*MLI*), usually consisting of many thin layers of polymer such as Kapton or Mylar metalized on one side with aluminum or silver, requires a high vacuum to be effective. In evacuated conditions, contact points between the separate layers in *MLI* are reduced. Hence, conduction and convection are minimized. The multiple layers reduce the last form of heat transfer, radiation, by radiating and absorbing heat from each other, effectively trapping most of the thermal energy (Chap. 23). The high vacuum required for *MLI* to be an effective insulation is abundantly available in the lunar environment and earth orbit. Mars, on the other hand, has an atmosphere consisting of mostly CO_2 with an average surface pressure of 6 torr [22]. The target thermal conductivity of materials for Mars *EVA* suits is 5 mW/m-K. Of a variety of composite materials considered in a recent study, only aerogel fiber composites come close to meeting that goal [23]. A monolithic polymer-reinforced aerogel may meet this goal, but insulation for *EVA* suits should also be durable and flexible to accommodate as much freedom of movement for the astronaut as possible. Flexible fiber-reinforced silica aerogel composite blankets were recently evaluated using thermal and mechanical cycling tests for possible use in advanced *EVA* suits [24]. During testing, thermal conductivity did not change much, but the composite aerogels were found to shed silica dust particles at unacceptable levels. Polymer reinforcement may also be a means to reduce shedding.

Figure 15.1. NASA vision of manned mission to Mars surface activities. (Reprinted with permission from SAE paper 2006-01-2235 © 2006 SAE International.)

A flexible form of polymer-reinforced aerogel would also be desirable for wrapping around structures that need to be insulated, such as cryotanks or cryogenic transfer lines. Currently, expendable cryotanks, such as the Space Shuttle's external liquid oxygen/hydrogen tank, use spray-on polyurethane foam to insulate the tank on the launch pad. Cryogenic tanks employed in space generally utilize the same *MLI* as space suits, but require additional insulation while on the launch pad since the *MLI* is not an effective insulator under ambient pressure as discussed previously. An aerogel insulation system can function reliably in both ambient pressure and high vacuum, making foam insulation unnecessary [25]. However, molding a net shape aerogel to fit around a structure can be difficult since complex molds need to be devised and the gels before polymer cross-linking can be quite fragile.

Another use for flexible durable aerogels could be as part of an inflatable decelerator used to slow spacecraft for planetary entry, descent, and landing (*EDL*) as shown in Figure 15.2 [26]. *EDL* systems used to successfully land six robotic missions on Mars from 1976 to 2008 employed a hard aeroshell heat shield and parachutes of 12–16 m in diameter. Future robotic and manned missions are much heavier and will require more drag for landing. Hence, new designs with much larger diameters (30–60 m) will be required [27]. Inflatable decelerators would stow in a small space and deploy into a large area lightweight heat shield to survive reentry [28]. Minimizing weight and thickness of the system as well as providing suitable insulation are important considerations.

Polymer-reinforced aerogels are somewhat flexible at densities below 0.05 g/cm^3, but mechanical strength is reduced at these densities [14]. However, much more flexibility can be obtained in aerogels by altering the silica backbone in some significant way. For example, Kramer et al. [29] demonstrated that including up to 20 w/w% polydimethylsiloxane (*PDMS*) in tetraethyl orthosilicate (*TEOS*)-based aerogels resulted in rubbery behavior with up to 30% recoverable compressive strain. More recently, Rao et al. [30] have demonstrated that utilizing methyltrimethoxysilane (*MTMS*) as the silica precursor and a two-step synthesis imparts extraordinary flexibility to the aerogels. The *MTMS*-derived aerogels are more flexible largely because of the resulting lower cross-link density of the silica (three alkoxy groups that can react vs. four in rigid *TMOS*- or *TEOS*-derived aerogels).

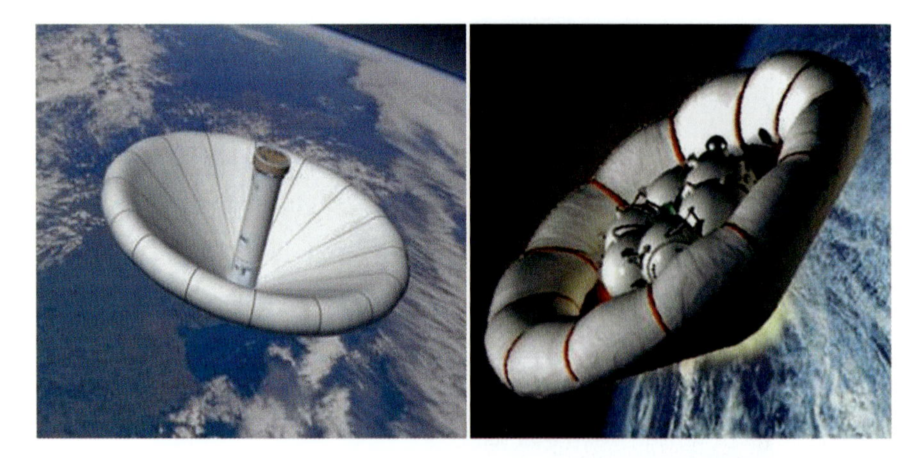

Figure 15.2. Inflatable decelerator concepts. Courtesy of the Inflatable Re-entry Vehicle Experiment (IRVE) Project.

Kanamori et al. [31], using a surfactant to control pore size and a slightly different process, have shown that *MTMS*-derived gels demonstrate reversible deformation on compression. In fact, some formulations were able to be dried ambiently. Initially, the gels shrink about 65% but spring back to nearly their original size, resulting in nearly the same density and pore structure as those dried supercritically.

Though the *MTMS*-derived aerogels are very flexible and elastic, it does not take much force to compress them. For example, Rao [28] reports a Young's modulus of only 0.03–0.06 MPa for the flexible *MTMS*-derived aerogels ranging in density from 0.04 to 0.1 g/cm^3. Kanamori [29] does not report Young's modulus, but stress–strain curves indicate that stresses of less than 1 MPa are sufficient to compress samples with bulk densities around 0.2 g/cm^3 to 25% strain.

Shea and Loy have employed bridged bis(trialkoxysilyl) monomers as precursors for silsesquioxane-derived aerogels and xerogels [32, 33]. Typically, this allowed for control of pore size directly related to the size of the bridge, with the best results obtained using a stiffer structure such as an arylene chain. More flexible bridges such as alkyl chains resulted in more compliant aerogels but tended to shrink more, reducing porosity.

More recently, we have been combining the notion of altering the underlying silica structure to add flexibility and elastic behavior and the notion of polymer reinforcement to add strength and durability. This is accomplished by substituting some of the *TEOS* or *TMOS* with organic bridged bis(trialkoxysilanes) or with *MTMS*. This work, which has been shown to be a versatile method of imparting more elastic behavior and flexibility to the aerogels, is summarized herein.

15.2. Hexyl-Linked Polymer-Reinforced Silica Aerogels

15.2.1. Di-isocyanate-Reinforced Aerogels

Silica aerogels reinforced with di-isocyanate through aminopropyl groups on the silica surface are made as shown in Scheme 15.2, where APTES and *TMOS* or *TEOS* are coreacted in acetonitrile to form a wet gel [13]. After being washed several times to remove water and alcoholic by-products of gelation, the wet gel is subsequently soaked in a solution of

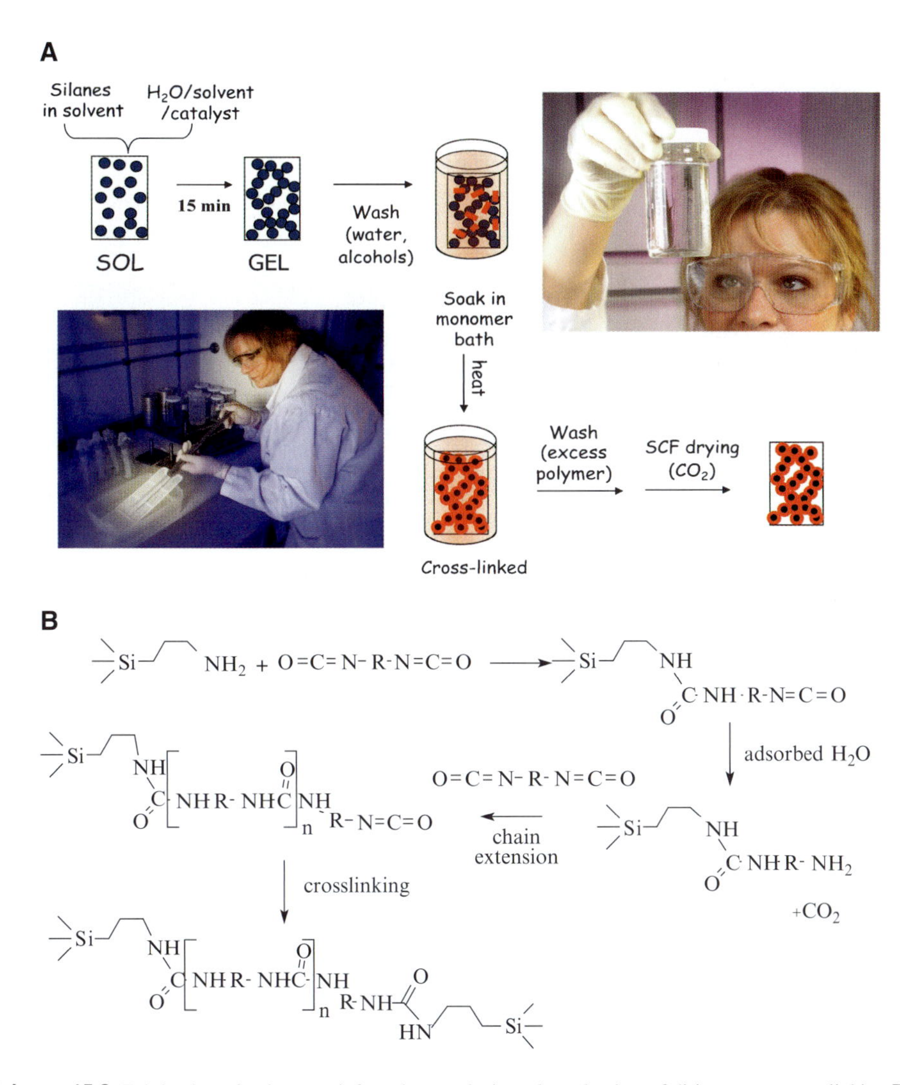

Scheme 15.2. Fabrication of polymer-reinforced aerogels **A.** and mechanism of di-isocyanate crosslinking **B.**

Desmodur N3200 [34], an oligomeric form of hexamethylene di-isocyanate (*HDI*). Heating causes reaction of the isocyanates with surface amines as shown in Scheme 15.2B. The length of the polyurea cross-links depends on both the di-isocyanate concentration in the soak solution and the amount of residual water from the hydrolysis left in the gels. This is because isocyanate reacts with water to generate an amine that can react with other isocyanates to cause chain extension.

Hexyl linkages can be incorporated into the wet gels by replacing some of the *TMOS* or *TEOS* with 1,6-bis(trimethoxysilyl)hexane (*BTMSH*), noting that each *BTMSH* molecule contributes two atoms of Si [35]. Scheme 15.3 shows a comparison of the silica gel structure (a) with 20 mol% APTES-derived Si and no hexyl linkages incorporated and (b) with about 40 mol% Si derived from *BTMSH*. A comparison of the silica gel structure with and without hexyl links is shown in Scheme 15.3. Note that the hexyl linkages serve to open up the structure and effectively reduce in size the regions of rigid silica.

A

B

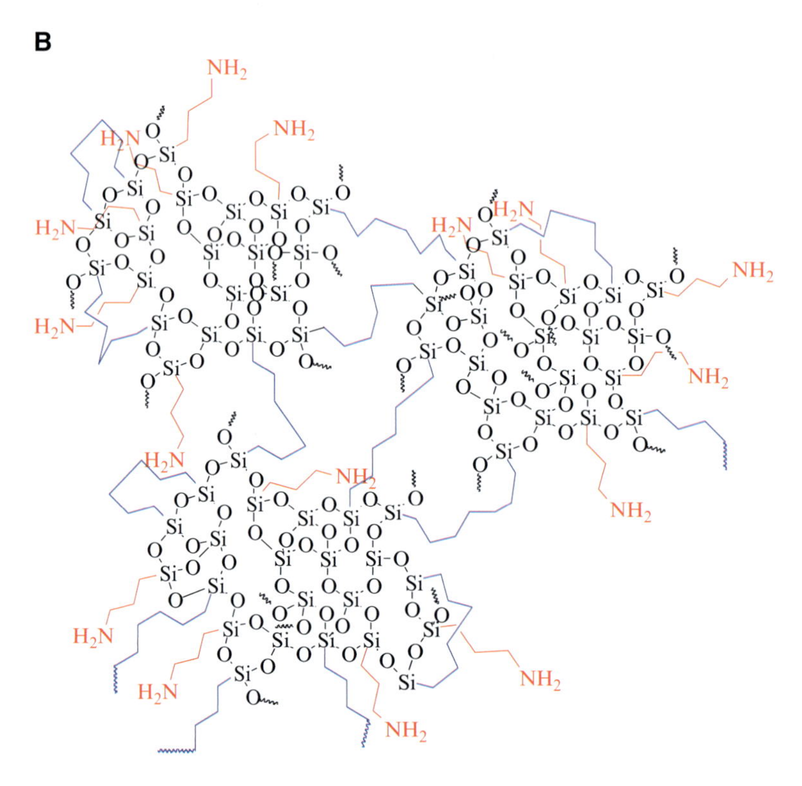

Scheme 15.3. Underlying silica backbone structure made using **A.** 20 mol% APTES and **B.** 20 mol% APTES with 40 mol% BTMSH-derived Si. Reprinted from [11], Copyright 2009 American Chemical Society.

Wet gels prepared with *BTMSH* are more resilient and easier to handle than those containing no *BTMSH* even at low silane concentrations, bending as they come out of a mold instead of breaking. Hence, the use of *BTMSH* can possibly improve the manufacturability of the aerogels. After reaction with di-isocyanate, the wet gel is resilient enough to be bent and manipulated to a great extent without breaking as shown in Figure 15.3. The resulting aerogels after drying with supercritical CO_2 can also be bent without breaking, but not to the same extent as the wet gels. This is due to a reduced plasticization of the silica/polymer network by the removal of the solvent.

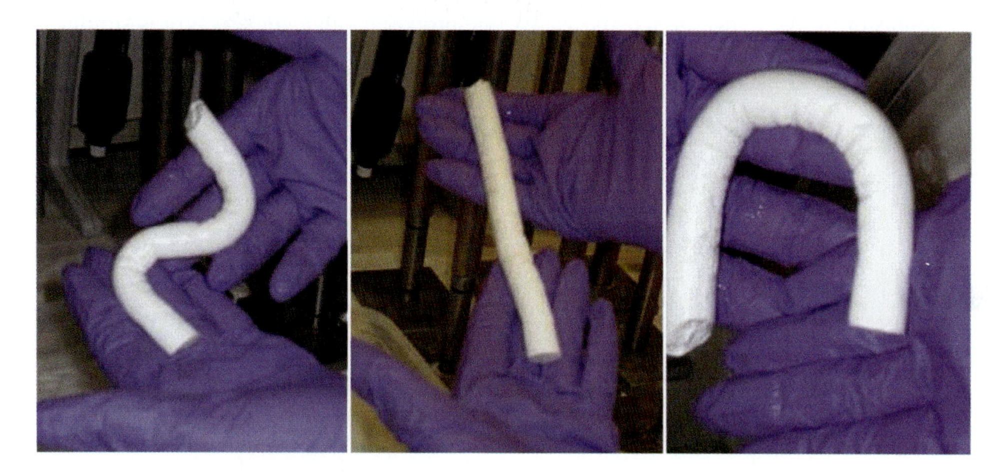

Figure 15.3. A di-isocyanate-reinforced wet gel with 35 mol% BTMSH-derived Si and 30 mol% Si from APTES is shown to be flexible and resilient before drying.

Inclusion of hexyl links does have an effect on the pore structure. As shown in Figure 15.4, the average pore diameter as measured by nitrogen sorption using the Brannauer–Emmet–Teller (BET) method is around 18 nm with a sharp pore size distribution for higher density di-isocyanate-reinforced aerogels containing no *BTMSH*. The average pore size increases and the pore size distribution broadens as silicon concentration is decreased.

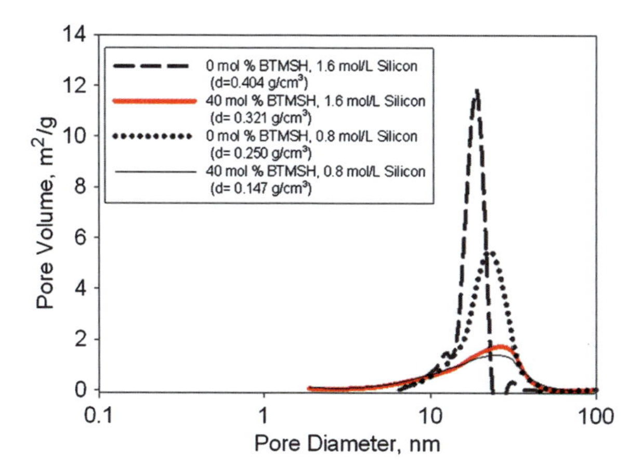

Figure 15.4. Pore volume vs. pore diameter plot for di-isocyanate-reinforced aerogels made using varying amounts of total silane concentration and mol% BTMSH-derived Si.

15.2.2. Styrene-Reinforced Aerogels

Several approaches to reinforcing silica aerogel with styrene have been studied, including incorporation of styrene [16], a free radical initiator [17], or a vinyl [18] on the silica surface. All of these approaches result in greater hydrophobicity (Chap. 3), and improvements in mechanical strength over non-cross-linked silica aerogels. A series of vinyl decorated silica gels have been fabricated by coreacting vinyltrimethoxysilane (*VTMS*), *BTMSH*, and *TMOS* in alcohol solution. Up to 50% of the *TMOS*-derived Si atoms are replaced with *BTMSH*-derived Si, noting again that each *BTMSH* contributes two atoms of Si. As shown in Scheme 15.4A, the proposed silica gel structure is similar to that previously shown for the APTES-derived structure except with a vinyl decorated surface.

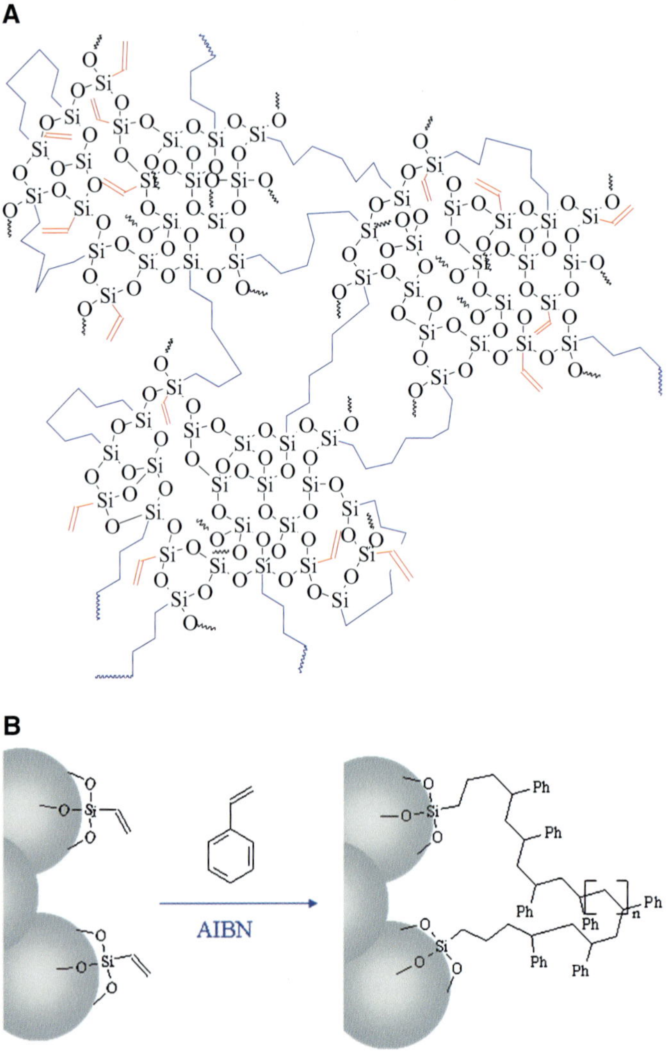

A

B

AIBN

Scheme 15.4. Silica structure incorporating hexyl links and vinyl groups for reinforcement with polystyrene. Reprinted from [18], Copyright 2009 American Chemical Society.

Cross-linking with styrene is carried out as shown in Scheme 15.4B after exchanging solvent from methanol or ethanol to chlorobenzene and soaking the wet gels in a solution of styrene monomer and radical initiator.

As a way of characterizing elastic properties (Chap. 22), styrene-reinforced aerogel monoliths are taken through two successive compression cycles to 25% strain. Stress–strain curves of successive compression tests of polymer-reinforced aerogels without hexyl-linking groups in the underlying silica typically look like that shown in Figure 15.5A. The first compression is taken to 25% strain and released and followed immediately by a second compression to 25% strain. Between the first and second compression, about 13% of the length of the sample is not recovered. In contrast, Figure 15.5B shows repeat compression tests of a styrene-reinforced aerogel made with 49 mol% of Si derived from *BTMSH*, 29 mol % derived from *VTMS*, and 22 mol% from *TMOS*. In this case, the first and second compression curves almost overlap. Nearly, all the length (99.5%) is recovered within 30 min after the second compression as shown in Figure 15.5C. Note that the compressive modulus of this hexyl-linked aerogel is over 3 MPa.

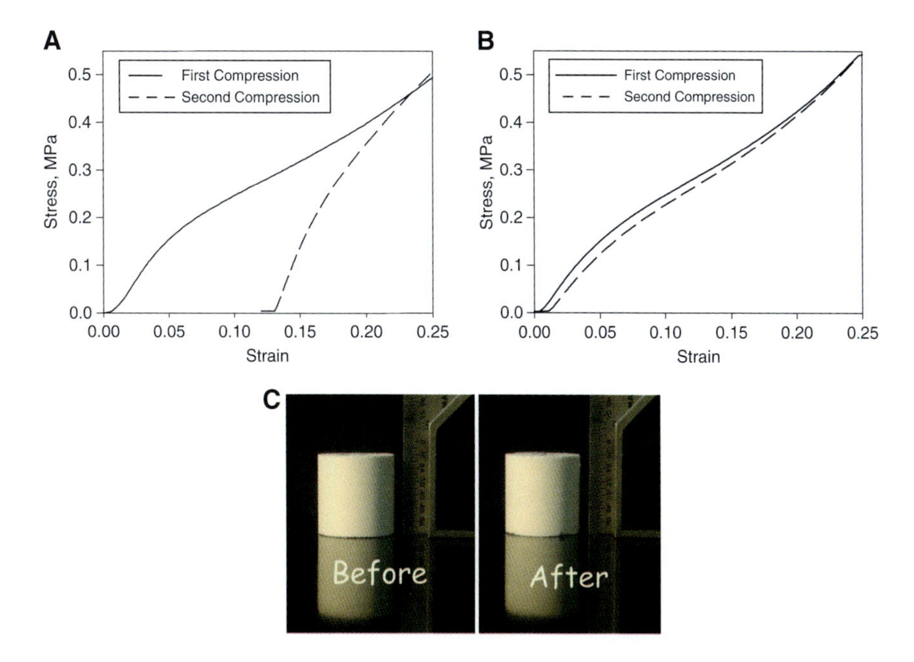

Figure 15.5. Repeat compression cycles of **A.** a styrene-reinforced aerogel monolith without flexible linking groups and (density = 0.122g/cm^3, surface area = 366 m^2/g); **B.** a styrene-reinforced monolith with 49 mol% Si derived from hexyl-linked BTMSH (density = 0.232 g/cm^3, surface area = 158 m^2/g; and **C.** the monolith from **B.** before and after two compressions. Reprinted from [18], Copyright 2009 American Chemical Society.

Interestingly, hexyl-linking groups also improve the hydrophobicity of the polystyrene-reinforced aerogels. Water droplet contact angles of styrene-reinforced aerogels have been reported to be in the range of 112–120° for samples containing no *BTMSH*. Samples with at least 29 mol% *BTMSH*-derived Si had water droplet contact angles ranging from 127 to 138°. This indicates that the hexyl group from *BTMSH* is also present on the silica surface and has a significant effect on the hydrophobic nature of the aerogels above and beyond the simple polystyrene cross-linking.

15.2.3. Epoxy-Reinforced Aerogels from Ethanol Solvent

Previously, the type of polymer reinforcement has dictated the solvent choice based on the solubility of the monomers and cure temperature of the cross-linking chemistry. In order to scale up manufacturing processes for cross-linking aerogels, it is desirable to adopt a more industrially friendly solvent. It has been demonstrated that less toxic *TEOS* and ethanol can be used to synthesize polymer-reinforced silica aerogels by cross-linking ethanol-soluble epoxies with APTES-derived amine groups (Scheme 15.5). In this manner, similar increases in mechanical properties were achieved in comparison to previously studied isocyanate cross-linked aerogels [11]. The use of ethanol as solvent in particular improves the viability of large-scale manufacturing of the polymer-cross-linked aerogels since large amounts of solvent (from initial gelation through diffusion of monomer into the alcogels and additional rinsing steps) are used in the production.

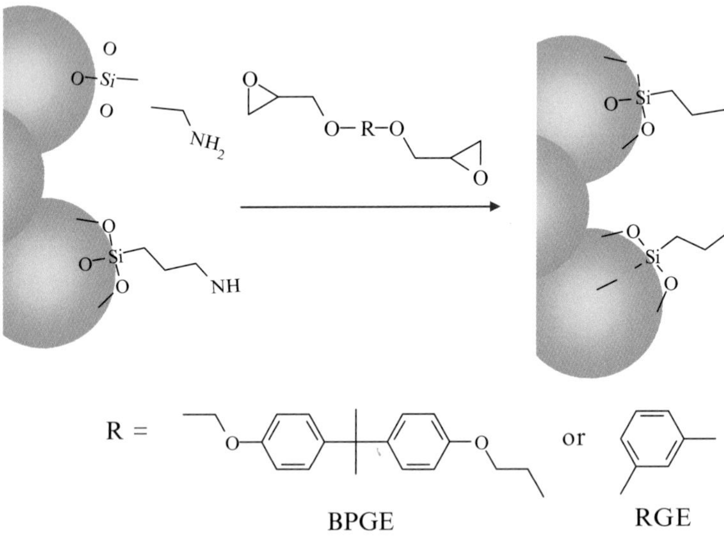

Scheme 15.5. Typical reaction scheme for crosslinking silica gels with epoxy through surface amine groups (reprinted from [11], Copyright 2009 American Chemical Society).

Varying total silane concentration and the mole fraction of APTES and *BTMSH* leads to aerogels with different pore structures as seen with both the styrene- and di-isocyanate-reinforced aerogels. Figure 15.6 shows scanning electron micrographs of four aerogel monoliths made using 15 mol% APTES. The top samples (Figure 15.6A, B) are made using 1.6 mol/l total silicon. The sample shown on the left (Figure 15.6A) is made using no *BTMSH*, while the right-hand sample (Figure 15.6B) is made using 40 mol% of Si derived from *BTMSH*. Both samples show a fine distribution of similar size particles, but in the sample containing hexyl links (Figure 15.6B), pores appear to be larger. The density of the hexyl-linked samples is also smaller, though the two samples are made with the same total moles of silicon, the same amount of amine, and the same concentration of epoxy.

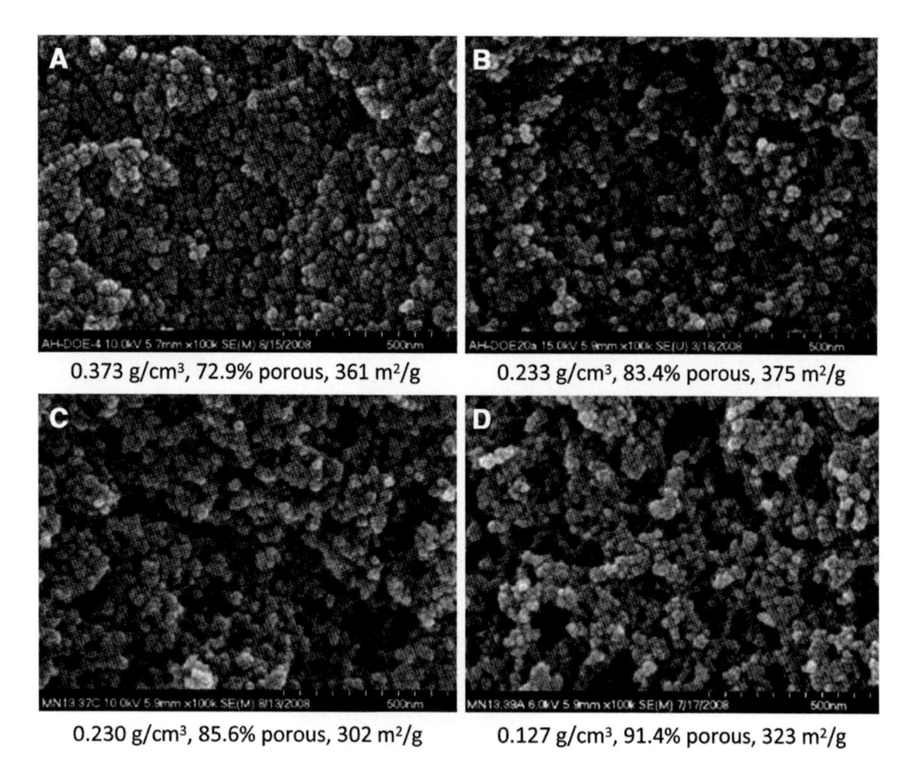

Figure 15.6. Side-by-side comparisons of micrographs of samples with (*left*) no BTMSH and (*right*) 40 mol% BTMSH-derived Si, including samples prepared with 1.6 mol/l total Si (15 mol% APTES) **A.** and **B.**; and with 0.8 mol/l total Si (15 mol% APTES) (**C** and **D**). Reprinted from [11], Copyright 2009 American Chemical Society.

The density difference is due to the fact that hexyl-linked samples tend to shrink less during processing. A similar but more pronounced effect is seen for samples made using 0.8 mol/l total silicon (Figure 15.6C, D). Again, the particle sizes are similar between the two samples, but larger pores are evident in the hexyl-linked sample shown in Figure 15.6D and the density is lower. However, BET surface areas are fairly constant for all four samples – only the distribution of pore sizes changes with *BTMSH* concentration.

Empirical models were generated for predicting properties of the epoxy-cross-linked aerogels over a wide range of densities as well as to identify and understand significant relationships between the processing parameters and final properties. Figure 15.7 exhibits empirical models for (a) density, (b) Young's modulus from compression tests, and (c) amount of recovery after compression to 25% strain. As seen in Figure 15.7A, highest density is seen when using 1.6 mol/l total Si and APTES-derived Si is 45 mol%. At the same time, Figure 15.7B shows that highest modulus (95 MPa), when no *BTMSH* is used, is achieved with APTES-derived Si at about 30 mol%.

Using *BTMSH* does appear to lower modulus with optimum modulus in this system (~40 MPa) being produced in combination with 45 mol% APTES. However, this change in modulus is at least in part due to the decrease in shrinkage (and concomitant decrease in density) using higher *BTMSH* concentrations.

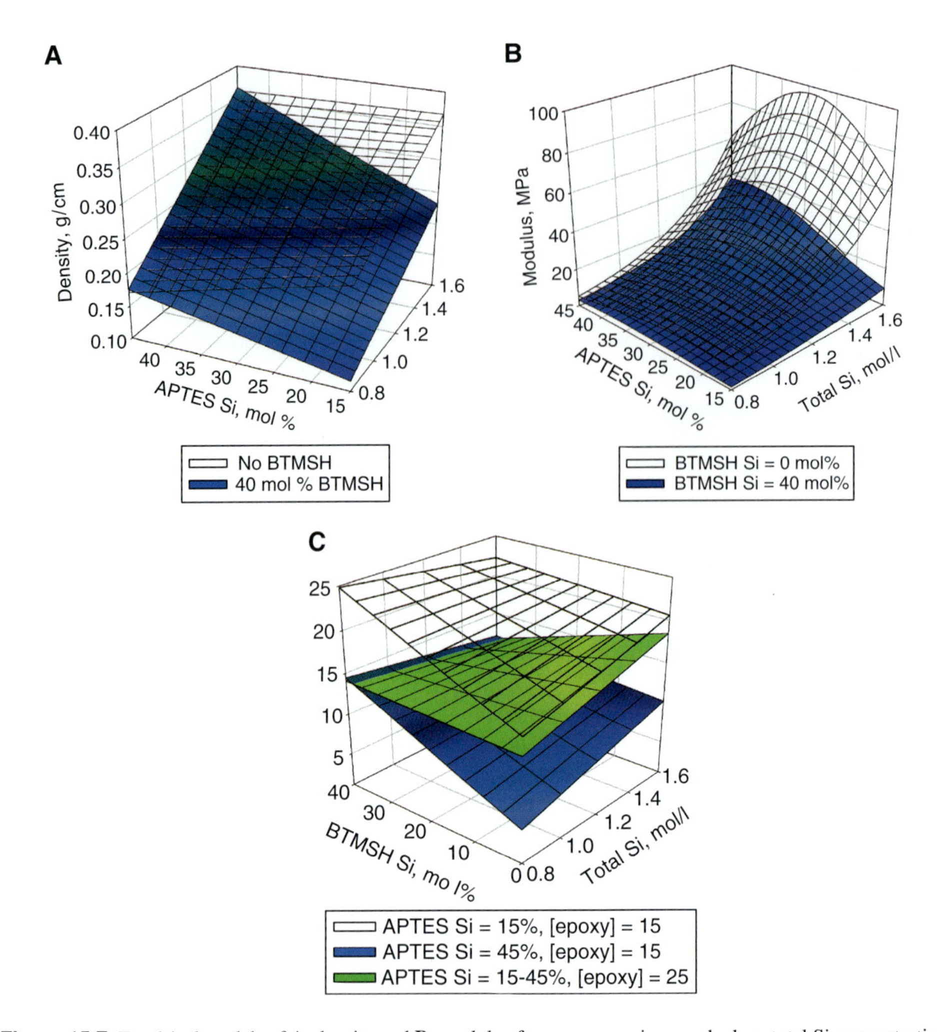

Figure 15.7. Empirical models of **A.** density and **B.** modulus from compression graphed vs. total Si concentration and fraction of APTES-derived Si, and **C.** recovered strain graphed vs. total Si concentration and BTMSH fraction. Reprinted from [11], Copyright 2009 American Chemical Society.

It is possible to achieve good, i.e., nearly complete, recovery after compression with epoxy-reinforced aerogels using 15 mol% APTES, 40 mol% Si from *BTMSH*, and 15% (w/w) epoxy in the soaking solution; this is illustrated in Figure 15.7C. In fact, across the whole range of total Si concentration (densities ranging from 0.1 to 0.23 g/cm^3 and modulus as high as 10 MPa), the monoliths prepared using these conditions recover nearly their full length after deformation to 25% strain. Similar modulus and recovery after compression is obtained in epoxy reinforced aerogels using ethyl-linked or octyl-linked bis-silanes in place of BTMSH [36].

If more flexible monoliths are desired, vastly improved elastic recovery from compression at up to as much as 50% strain can be obtained using even lower total Si concentration (0.4 mol/l total Si, 45 mol% derived from APTES, and 40 mol% from BTMSH). As shown in Figure 15.8, a 2.3 cm monolith made using this formulation after

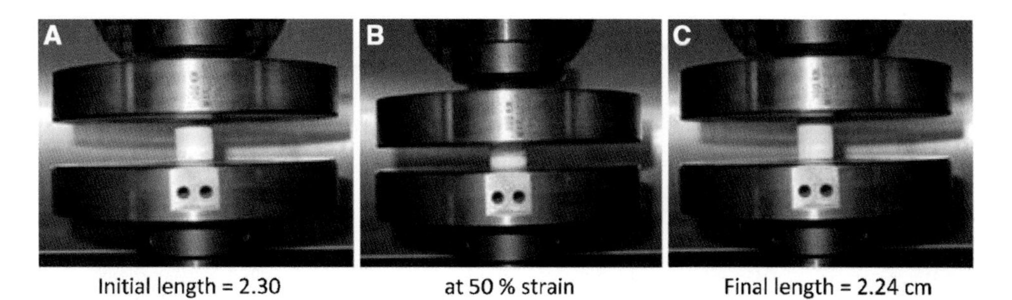

Figure 15.8. Repeat compression test to 50% strain of epoxy-reinforced monolith made using 0.4 mol/l total Si (40 mol% from BTMSH and 45 mol% from APTES). Reprinted from [11], Copyright 2009 American Chemical Society.

two compression cycles to 50% strain loses only 2.6% of initial length. At 25% strain, the sample demonstrates complete recovery after two compression cycles. However, with a density of 0.047 g/cm^3 and porosity greater than 96%, the modulus of this monolith decreases to 0.05 MPa and surface area is also low (54 m^2/g).

15.3. Alkyl Trialkoxysilane-Based Reinforced Aerogels

As pointed out in Sect. 15.1 of this chapter, *MTMS*-based aerogels have been synthesized with a surprising amount of flexibility [28, 29]. It is of interest then to see if strength and durability of the *MTMS* aerogels can be improved by introducing the notion of polymer reinforcement. In preliminary studies, formulations with *MTMS* coreacted with APTES to provide a reactive site for polymer cross-linking tend to gel in only a narrow range of total Si concentration and they shrink much more than their *TEOS*- or *TMOS*-based counterparts [37]. In contrast, *MTMS* coreacted with bis(trimethoxysilyl-propyl) amine (*BTMSPA*) readily gels across a wide range of total Si concentration [38]. The secondary amine of *BTMSPA* provides a ready reactive site for Desmodur N3300A [34], a tri-isocyanate cross-linker, as shown in Scheme 15.6.

Since only a small excess of water (ratio of water to Si ranging from 2 to 5) was used to make the aerogels in this study, chain extension of the polyurea was minimized. NMR analysis indicated that the polymer-reinforced structure is as shown with approximately a one-to-three ratio between each amine and each tri-isocyanate for samples where total Si ranged from 1.2 to 1.65 mol/l and 40 to 80 mol% of the Si from *BTMSPA*. For gels made in acetonitrile with lower total Si and 40 mol% *BTMSPA*, it was found that very little isocyanate cross-linking occurred. Monoliths made using these formulations also exhibited very low BET surface areas (<10 m^2/g), most likely due to solvent interactions between acetonitrile and polar surface groups. With methyl groups in greater abundance on the surface of the developing gel, this leads to collapse of the gel structure. Such low surface area would lead to very little secondary amine available to react with isocyanate. Using a less polar solvent such as acetone for gelation reverses this trend, leading to higher surface areas and better cross-linking.

Figure 15.9 shows a side-by-side comparison of unreinforced monoliths (left) and polymer-reinforced monoliths (right) produced using the same initial gelation conditions. All four monoliths pictured were prepared using 1.65 mol/l total Si. Figure 15.9A, B is of aerogel monoliths produced using 80 mol% Si derived from BTMSPA. As illustrated,

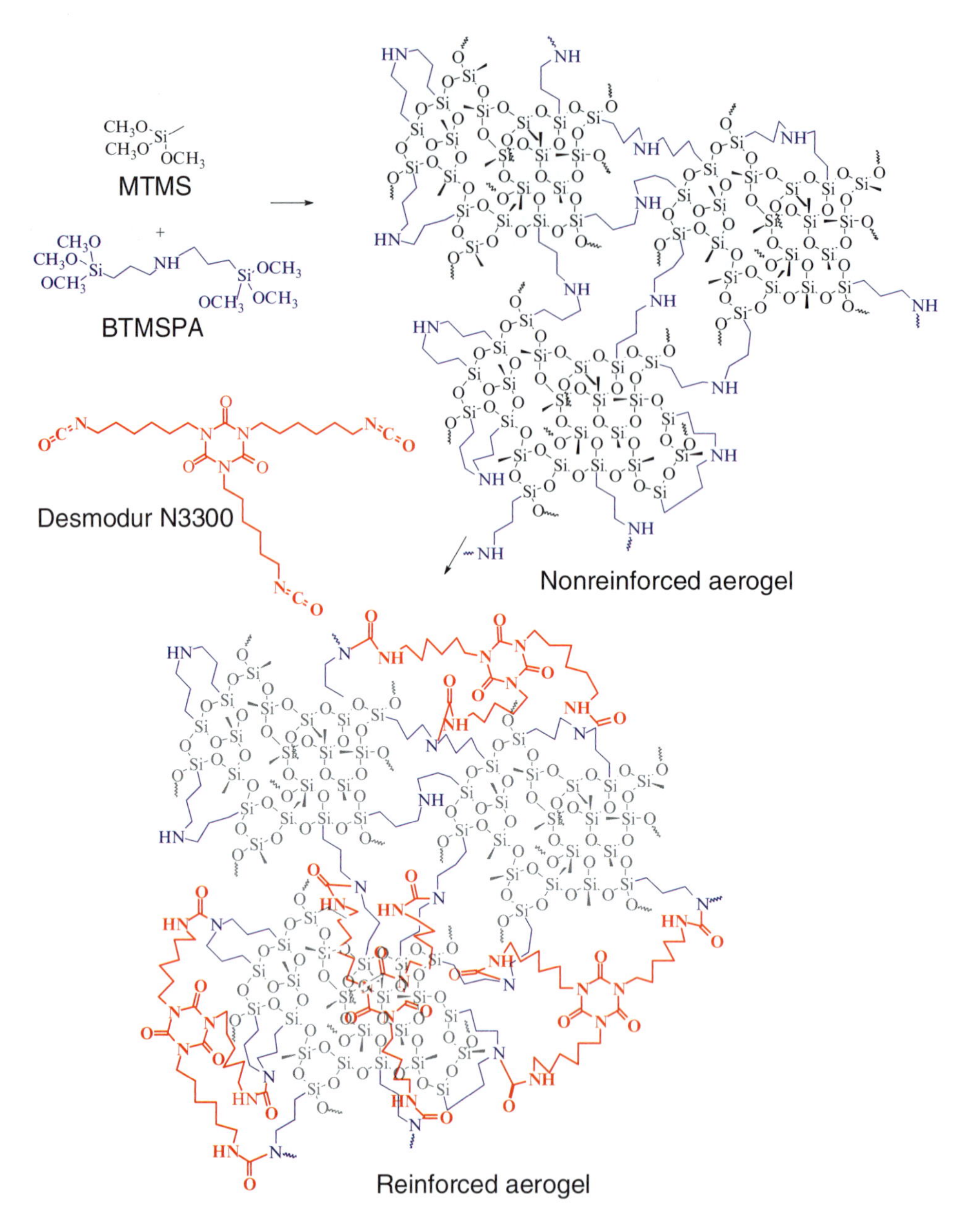

Scheme 15.6. Synthesis of MTMS-based aerogels with BTMSPA used as both a flexible linking group and site for tri-isocyanate crosslinking. Reprinted from [38], Copyright 2010 American Chemical Society.

particle sizes are very similar in appearance; however, the pores appear slightly reduced in size in the reinforced structure (Figure 15.9B). This accounts in part for the relatively large differences in density, porosity, and surface areas between the two samples. When BTMSPA mole fraction is reduced to 40 mol% (Figure 15.9C, D), both uncross-linked and polymer-reinforced aerogels exhibit a larger distribution of pore sizes than seen in Figure 15.9A, B, probably due to the different spacing formed from the connecting group in BTMSPA vs. the

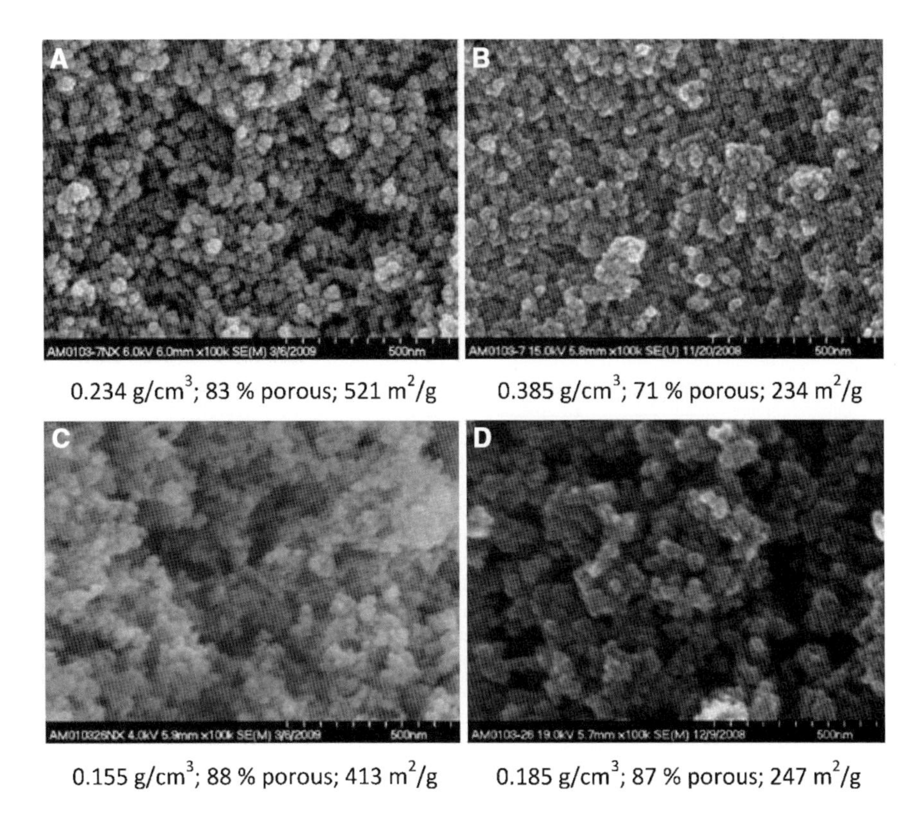

Figure 15.9. SEM images of aerogel monoliths made using 1.65 mol/l total Si: **A.** uncross-linked and **B.** polymer-reinforced aerogels prepared using 80 mol% Si from BTMSPA; and **C.** uncrosslinked and **D.** polymer-reinforced aerogels prepared using 40 mol% Si from BTMSPA. Reprinted from [38], Copyright 2010 American Chemical Society.

nonreactive methyl groups contributed by *MTMS*. Notably, the uncross-linked monolith in Figure 15.9C has a much finer particle structure and larger surface area than compared to the reinforced aerogel shown in Figure 15.9D (smoother, larger particles).

Compression testing of the monoliths made from MTMS and BTMSPA was carried out as previously described on both the unreinforced and polymer-reinforced aerogels. Though there is usually a trade-off between modulus and elastic recovery as seen with the hexyl-linked aerogels, in this study, the trade-off is very small, with even the highest modulus (84 MPa) of cross-linked aerogels exhibiting only about 3% unrecovered strain. To illustrate, stress–strain curves for repeat compression tests to 25% strain of four different polymer-reinforced aerogels are compared in Figure 15.10 (both graphs show the same four monoliths at different scales). The pairs of lines represent two subsequent stress–strain curves for each of the monoliths. The green curve labeled Test 1 is the first compression and Test 2 is the second compression for a formulation from 1.65 mol/l total Si (80 mol% from BTMSPA), with about 3% strain not recovered. In contrast, the red lines show repeated compression cycles from a formulation, made using 1.2 mol/l total Si but 80 mol% BTMSPA-derived Si. In this case, unrecovered strain is $<2\%$ (the sample recovers almost completely). Hence, a high degree of elastic recovery is present using a combination of MTMS-derived Si and the organic linking groups from BTMSPA.

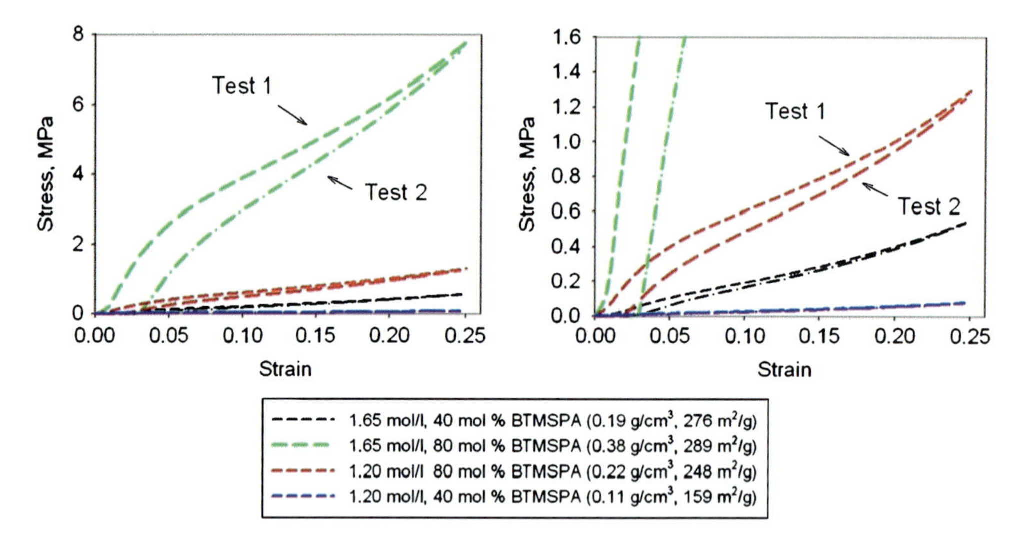

Figure 15.10. Typical stress–strain curves for a repeat compression tests on MTMS aerogels reinforced with Desmodur N3300A tri-isocyanate at different total silicon concentration and mol fraction of BTMSPA. The graphs are the same curves shown at different *y*-scales. Densities and surface areas of the monoliths made in acetonitrile are shown in *parentheses*. Reprinted from [38], Copyright 2010 American Chemical Society.

Unreinforced aerogels from the same study have about the same degree of recovery as the cross-linked aerogels shown.

Typical compressive stress–strain curves taken to failure are shown in Figure 15.11A. For unreinforced aerogels, the Young's modulus taken from the initial slope of the stress–strain curve is typically about half of that for the reinforced aerogels. In addition, the unreinforced aerogels break at about half the value of strain and the maximum stress at break is an order of magnitude lower. Hence, as seen in the graph of the response surface model shown in Figure 15.11B, the toughness, calculated from the area under the stress strain

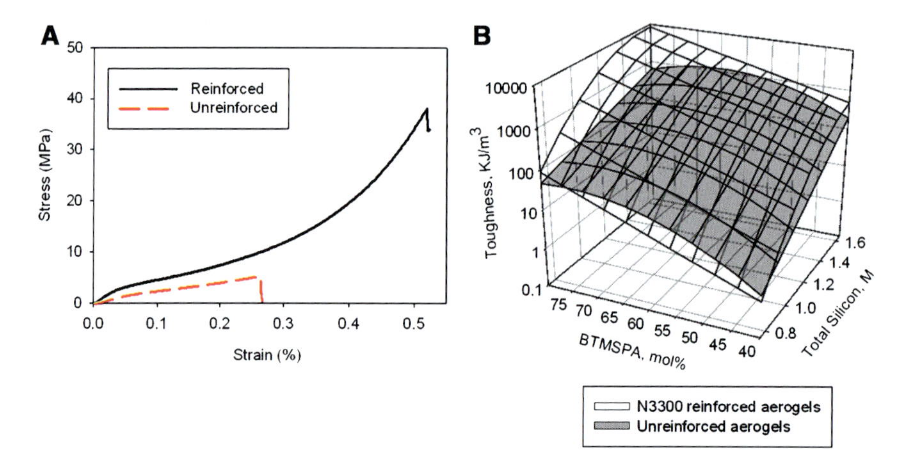

Figure 15.11. **A**. Typical stress–strain curve for compression to break and **B**. the response surface model of toughness calculated from the stress–strain curves for both crosslinked and uncrosslinked aerogel monoliths. Reprinted from [38], Copyright 2010 American Chemical Society.

curve, is as much as an order of magnitude higher for polyurea-reinforced aerogels made from *MTMS* and BTMSPA. Thus, while elastic recovery is about the same for reinforced and unreinforced aerogels using *MTMS* and BTMSPA, the polymer-reinforced aerogels are much more robust.

15.4. Future Directions

Incorporating flexible links in the silica backbone or using *MTMS* in place of *TEOS* in polymer-reinforced aerogels has been shown to be an effective way to introduce flexibility and good elastic recovery. In some cases, increasing the amount of polymer cross-linking does reduce flexibility. Thus, it stands to reason that in these cases the use of a more flexible polymer in combination with alterations made to the silica structure should increase flexibility even more. Taking this notion one step further, use of a shape memory polymer as reinforcement may give rise to a shape memory aerogel [39]. A shape memory polymer, such as a polyurethane block copolymer, is capable of changing its shape in response to a set of external stimuli, for example, pH, electric current, magnetic induction, and heating. Usually, one or more polymer blocks respond to the external stimuli, while others contribute to the desired properties. The synthesis of flexible aerogels reinforced with shape memory polyurethane cross-linkers should give rise to a shape memory effect as shown in Figure 15.12, offering the advantage of storage of the aerogel in a deformed state for transport aboard a spacecraft where storage is at a premium. Alternatively, the shape memory aerogel could be inserted into a cavity needing insulation in its deformed shape and made to recover to fill the cavity.

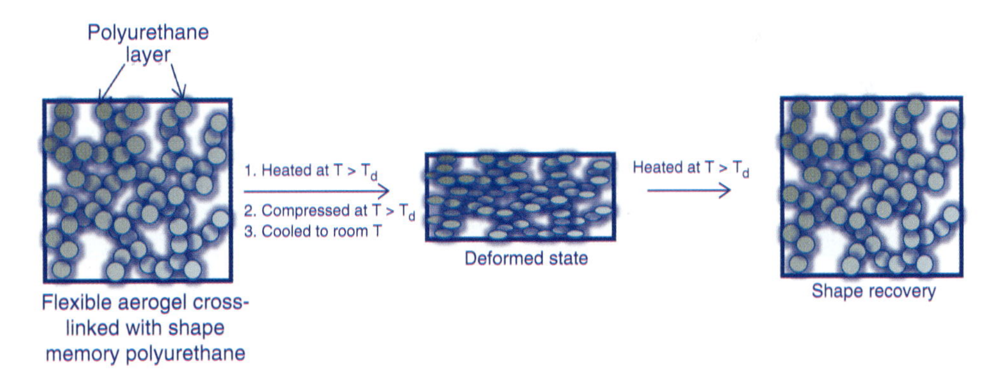

Figure 15.12. Proposed scheme to deploy a flexible aerogel reinforced with a shape memory polyurethane.

Flexible monolithic aerogels could have wide applications in various thermal insulation systems for space as previously discussed. For structures that need to be folded in a compact space for transport before deploying, such as the inflatable decelerators or habitats for the Moon or Mars, the thinner and more flexible the insulation system, the better. In such a thin form, even the polymer-reinforced aerogels with flexible linking groups as described in this chapter would probably be too fragile to stand on their own. Use of a continuous nonwoven fiber batting to further reinforce the flexible aerogels, such as in the aerogel blankets produced by Aspen Aerogels is one approach, but may still not result in a thin enough insulation for such deployable structures.

Carbon nanofiber-reinforced silica aerogel composites can be made by incorporating up to 5% (w/w) of the nanofibers into the initial sol [40]. Tensile strength of the composite aerogels using di-isocyanate as a polymer cross-link was improved by as much as a factor of 5 over di-isocyanate-cross-linked aerogels with no carbon nanofibers. In addition, the nanofiber containing gels were also easier to handle before cross-linking. Incorporating carbon nanofibers into a flexible reinforced aerogel matrix may allow the formation of thinner, more flexible aerogel sheets with improved durability (Chap. 36).

Another way to form a thin aerogel composite is to incorporate electrospun polyurethane nanofibers into a cast sol film prior to gelation of the silica-based sol [41]. By precisely controlling the gelation kinetics with the amount of base present in the formulation, nanofibers are electrospun into the sol before the onset of the gelation process and uniformly embedded in a silica network made flexible by incorporating low molecular weight oligomers of silanol-terminated *PDMS*. The final composite films, which have been fabricated to be 0.5–2 mm in thickness, are pliable and flexible as seen in Figure 15.13A. After much bending and pulling on the films by hand, small cracks develop, but as shown in Figure 15.13B, the film holds together by the nanofibers knitted together and bridging the crack. The room temperature thermal conductivity of the composite aerogels range from 13 mW/m-K for samples with bulk density of 0.17 g/cm^3 to 50 mW/m-K for samples with bulk density of 0.08 g/cm^3. Future work will involve optimization of the aerogel matrix and nanofiber chemistry, as well as nanofiber thickness to better reinforce the aerogel structure.

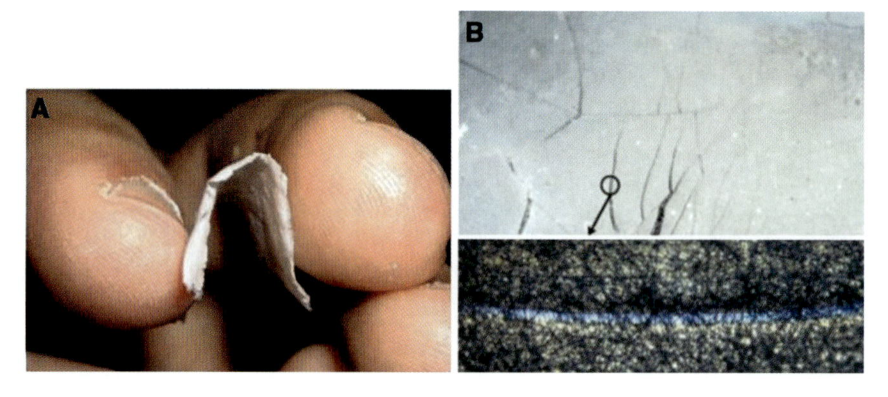

Figure 15.13. **A.** A 0.46 mm thick aerogel–nanofiber hybrid film bent almost 180°; **B.** microscope image of 2 mm thick aerogel sheet after excessive repetitive bending and pulling apart by hand. Reprinted from [41], Copyright 2009 American Chemical Society.

15.5. Conclusions

Incorporating short, flexible, organic linking groups into the silica backbone of polymer-reinforced aerogels has been shown to be a versatile way to improve elastic properties of the aerogels. The aerogels can recover from compression up to as much as 50% strain and in some cases are even flexible. The flexible linking groups may also result in greater hydrophobicity especially when combined with hydrophobic polymer cross-linkers such as polystyrene (Chap. 3) and also provide a means to tailor pore structure.

While there is a trade-off between modulus and elastic recovery in polymer-reinforced aerogels using flexible linking groups, the combination of *MTMS* and BTMSPA used in the silica backbone provides enhanced elastic properties almost independent of modulus. These aerogels recover nearly all of their length after compression to 25% strain across the whole modulus range. Reinforcing these aerogels with tri-isocyanate oligomers reacted through the secondary amine of BTMSPA results in up to an order of magnitude increase in compressive strength and toughness of the aerogel monoliths, while the overall elastic properties arising from the underlying silica structure is maintained. Use of flexible polymer-cross-linked aerogels, especially in combination with nanofiber reinforcement is a promising route to making robust aerogel thin films and sheets, enabling a multitude of aerospace applications.

References

1. Pierre A C, Pajonk G M (2002) Chemistry of aerogels and their applications. Chem Rev 102: 4243–4265.
2. Fricke J (1998) Aerogels – Highly tenuous solids with fascinating properties. J Non-Cryst Solids 100: 169–173.
3. Husing N, Schubert U (1998) Aerogels – Airy materials: Chemistry, Structure and Properties. Angew Chem Int Ed 37: 22–45.
4. Parmenter K E, Milstein F (1998) Mechanical properties of silica aerogels. J. Non-Cryst Solids 223: 179–189.
5. Tsou P, Brownlee D E, Sandford S A, Hörz F, Zolensky M E (2003) Wild 2 and interstellar sample collection and Earth return. J Geophys Res 108(E10): 8113.
6. Jones S M (2006) Aerogel: Space exploration applications. J Sol-Gel Sci Tech 40: 351–357.
7. Zhang G, Dass A, Rawashdeh A-M M, Thomas J, Counsil J A, Sotiriou-Leventis C, Fabrizio E F, Ilhan F, Vassilaras P, Scheiman D A, McCorkle L, Palczer A, Johnston J C, Meador M A B, Leventis N (2004) Isocyanate-crosslinked silica aerogel monoliths: preparation and characterization. J. Non-Cryst Solids 350:152–164.
8. Leventis N, Sotiriou-Leventis C, Zhang G, Rawashdeh A-M M (2002) Nanoengineering Strong Silica Aerogels. Nano Lett 2: 957–960.
9. Boday D J, Stover R J, Muriithi B, Keller M W, Wertz J T, Obrey K A D, Loy D A (2008) Formation of Polycyanoacrylate–Silica Nanocomposites by Chemical Vapor Deposition of Cyanoacrylates on Aerogels. Chem Mater 20: 2845–2847.
10. Meador M A B, Fabrizzio E F, Ilhan F, Dass A, Zhang G, Vassilaras P, Johnston J C, Leventis N (2005) Cross-linking amine modified silica aerogels with epoxies: Mechanically strong lightweight porous materials. Chem Mater 17: 1085–1098.
11. Meador M A B, Weber A S, Hindi A, Naumenko M, McCorkle L, Quade D, Vivod S L, Gould G L, White S, Deshpande K (2009) Structure property relationships in porous 3D nanostructures: Epoxy cross-linked silica aerogels produced using ethanol as the solvent. ACS Appl Mater Interfaces 1: 894–906.
12. Boday D J, Stover R J, Muriithi B, Keller M W, Wertz J T, Obrey K A D, Loy D A (2009) Strong, Low-Density Nanocomposites by Chemical Vapor Deposition and Polymerization of Cyanoacrylates on Aminated Silica Aerogels. ACS Appl Mater Interfaces 1: 1364–1369.
13. Meador M A B, Capadona L A, Papadopoulos D S, Leventis N (2007) Structure property relationships in porous 3D nanostructures as a function of preparation conditions: Isocyanate cross-linked silica aerogels. Chem Mater 19: 2247–2260.
14. Capadona L A, Meador M A B, Alumni A, Fabrizzio E F, Vassilaras P, Leventis N (2006) Flexible, low-density polymer crosslinked silica aerogels. Polymer 47: 5754–5761.
15. Katti A, Shimpi N, Roy S, Lu H, Fabrizio E F, Dass A, Capadona L A, Leventis N (2006) Chemical, physical and mechanical characterization of isocyanate cross-linked amine modified silica aerogels. Chem Mater 18: 285–296.
16. Ilhan U F, Fabrizio E F, McCorkle L, Scheiman D A, Dass A, Palczer A, Meador M A B, Johnston J C, Leventis N (2006) Hydrophobic monolithic aerogels by nanocasting polystyrene on amine-modified silica. J Mater Chem 16: 3046–3054.
17. Mulik S, Sotiriou-Leventis C, Churu G, Lu H, Leventis N (2008) Cross-linking 3D assemblies of nanoparticles into mechanically strong aerogels by surface-initiated free-radical polymerization. Chem Mater 20: 5035–5046.
18. Nguyen B N, Meador M A B, Tousley M E, Shonkwiler B, McCorkle L, Scheiman D A, Palczer A (2009) Tailoring elastic properties of silica aerogels cross-linked with polystyrene. ACS Appl Mater Interfaces 1: 621–630.

19. Fidalgo A, Farinha J P S, Martinho J M G, Rosa M E, Ilharco L M (2007) Hybrid silica/polymer aerogels dried at ambient pressure. Chem Mater 19: 2603–2609.
20. Novak B M, Auerbach D, Verrier C (1994) Low-density mutually interpenetrating organic–inorganic composite materials via supercritical drying techniques. Chem Mater 6: 282–286.
21. Wei T Y, Lu S Y, Chang Y C (2008) Transparent, hydrophobic composite aerogels with high mechanical strength and low high-temperature thermal conductivities. J Phys Chem B 112: 11881–11886.
22. Essex Corporation, Extravehicular Activity in Mars Surface Exploration, Final Report on Advanced Extravehicular Activity Systems Requirements Definition Study (1989) NAS9–17779.
23. Paul H L, Diller K R (2003) Comparison Thermal Insulation Performance of Fibrous Materials for the Advanced Space Suit. J Biomechanical Engineering 125: 639–647.
24. Tang H H, Orndoff E S, Trevino L A (2006) Thermal performance of space suit elements with aerogel insulation for Moon and Mars exploration. 36th International Conference on Environment Systems, July 17–20, 2006, Norfolk, Virginia AIAA 2006–01–2235.
25. Fesmire J E (2006) Aerogel insulation systems for space launch applications. Cryogenics 46: 111–117.
26. Braun R D, Manning R M (2007) Mars Exploration Entry, Descent and Landing Challenges. J Spacecraft and Rockets 44: 310–323.
27. Brown G J, Lingard J S, Darley G D, Underwood J C (2007) Inflatable aerocapture decelerators for Mars Orbiters. 19th AIAA Aerodynamic Decelerator Systems Technology Conference and Seminar 21–24 May 2007, Williamsburg, VA, AIAA 2007–2543.
28. Reza S, Hund R, Kustas F, Willcockson W, Songer J (2007) Aerocapture Inflatable decelerator (AID) for planetary entry. 19th AIAA Aerodynamic Decelerator Systems Technology Conference and Seminar 21–24 May 2007, Williamsburg, VA, AIAA 2007–2516.
29. Kramer S J, Rubio-Alonso F, Mackenzie J D (1996) Organically Modified Silicate Aerogel: Aeromosil. Mat Res Soc Symp Proc 435: 295–299.
30. Rao A V, Bhagat S D, Hirashima H, Pajonk G M (2006) Synthesis of flexible silica aerogels using methyltrimethoxysilane (MTMS) precursor. J Colloid and Interface Sci 300: 279–285.
31. Kanamori K, Aizawa M, Nakanishi K, Hanada T (2006) New transparent methylsilsesquioxane aerogels and xerogels with improved mechanical properties. Adv Mater 19: 1589–1593.
32. Shea K J, Loy D A (2001) Bridged polysilsesquioxanes. Molecular-engineered hybrid organic–inorganic materials. Chem Mater 13: 3306–3319.
33. Loy D A, Shea K J (1995) Bridged Polysilsesquioxanes. Highly porous hybrid organic–inorganic materials. Chem Rev 95: 1431–1442.
34. Supplied by Bayer Corporation.
35. Vivod S L, Meador M A B, Nguyen B N, Perry R (2009) Flexible di-isocyanate cross-linked silica aerogels with 1,6-bis(trimethoxysilyl)hexane incorporated in the underlying silica backbone. Polym Preprints 50: 119–120.
36. Randall J P, Meador M A B, Jana S C (2011) Tailoring mechanical properties of aerogels for aerospace. ACS Appl Mater and Interfaces, 3: dx.doi.org/10.1021/am200007n.
37. Meador M A B, Nguyen B N, Scherzer C S unpublished results.
38. Nguyen B N, Meador M A B, Medoro A, Arendt V, Randall J, McCorkle L, Shonkwiler B (2010) Elastic behavior of methyltrimethoxysilane based aerogels reinforced with tri-isocyanate. ACS Applied Materials and Interfaces 2: 1430–1443.
39. Jana S C, Meador M A B, Randall J P (2008) Process for forming shape-memory polymer aerogel composites. US Patent Application, 2006–854838P.
40. Meador M A B, Vivod S L, McCorkle L, Quade D, Sullivan R M, Ghosn L J, Clark N, Capadona L A (2008) Reinforcing polymer cross-linked aerogels with carbon nanofibers. J Mater Chem 18: 1843–1852.
41. Li L, Yalcin B, Nguyen B N, Meador M A B, Cakmak M (2009) Nanofiber reinforced aerogel (xerogel): synthesis, manufacture and characterization. ACS Appl Mater and Interfaces, 1: 2491–2501.

16

Aerogels Containing Metal, Alloy, and Oxide Nanoparticles Embedded into Dielectric Matrices

Anna Corrias and Maria Francesca Casula

Abstract Aerogels are regarded as ideal candidates for the design of functional nanocomposites based on supported metal or metal oxide nanoparticles. The large specific surface area together with the open pore structure enables aerogels to effectively host finely dispersed nanoparticles up to the desired loading and to provide nanoparticle accessibility as required to supply their specific functionalities.

The incorporation of nanoparticles as a way to increase the possibility of the use of aerogels as innovative functional materials and the challenges in the controlled preparation of nanocomposite aerogels is reviewed in this chapter.

16.1. Introduction

Aerogels are innovative solids that have many unique optical, insulating, and catalytic properties [1]. Tailored features of aerogels can be obtained by finely tuning the composition, the surface, and the microstructure (i.e., porosity and texture, crystallinity, and grain size) of the resulting material, which are controlled by the sol–gel parameters [2].

A way to expand the potential of aerogels to provide innovative functional materials is through the design of nanocomposites, where metal or metal oxide nanoparticles are incorporated within the aerogel matrix. In addition to broadening the range and viability of aerogels, the preparation of nanocomposites provides a means of preventing nanocrystal aggregation and growth through particle confinement, fulfilling a key requirement for any practical application. Aerogels are ideal candidates for the design of functional nanocomposites due to their large specific surface area, which can be used effectively to host finely dispersed nanoparticles up to high loadings, combined with the open pore structure, which ensures nanoparticle accessibility, which in turn is required in order to retain their specific functionalities.

Despite considerable effort in the field, major challenges remain in the control of the homogeneity, loading, size, and distribution of the nanoparticles within the host inorganic network, which in turn determine directly the electronic, optical, magnetic, and catalytic properties of nanocomposite materials.

A. Corrias and M. F. Casula • Dipartimento di Scienze Chimiche and INSTM, Università di Cagliari, Complesso Universitario di Monserrato, S.S. 554 Bivio per Sestu, 09042 Monserrato, CA, Italy
e-mail: corrias@unica.it; casulaf@unica.it

M.A. Aegerter et al. (eds.), *Aerogels Handbook*, Advances in Sol-Gel Derived Materials and Technologies, DOI 10.1007/978-1-4419-7589-8_16,

A major issue is related to the difference in chemical properties of the precursors for the nanophase (usually metal salts) and for the inorganic matrix (commonly alkoxides), as many sol–gel process parameters influence the hydrolytic polycondensation of each precursor differently [3].

Although this is a general problem in the design of any sol–gel nanocomposite, it is more critical in the case of aerogels, which require the original solvent present in the gel (usually ethanol or methanol in alcogels and water in aquogels), to be replaced by solvent exchange and then removed by supercritical solvent extraction. The supercritical drying step will be hereafter named *sc-HT* or *sc-CO_2 drying* depending on whether alcohols or carbon dioxide are supercritically evacuated (requiring typical temperatures of ~350 and 40°C respectively). This step introduces additional issues concerning solubility of the precursors and thermal stability under the supercritical drying conditions.

The different strategies that have been adopted to synthesize nanocomposite aerogels follow two general approaches, depending on whether the nanophase (or its precursor) is added during or after the sol–gel process.

The first approach includes cohydrolysis and cogelation of the nanoparticle and of the matrix precursors and cogelation of the matrix precursor together with preformed nanoparticles. This approach offers the advantage of producing materials with a controllable loading of nanoparticles throughout. On the other hand, several disadvantages should be taken into account: an accurate choice of the synthetic conditions has to be made in order to obtain a homogeneous multicomponent gel, and the nanoparticle precursors as well as the capping agents needed to stabilize preformed colloidal nanoparticles may affect the sol–gel synthesis of the matrix.

The second approach includes methods based on the addition of the nanophase after the sol–gel process and they should preserve the porous structure and morphology of the matrix. These methods include the deposition of the nanophase by impregnation, deposition–precipitation, and chemical vapor infiltration procedures. Deposition–precipitation synthesis, based on the deposition of metal hydroxides on the support by tuning the pH of the solution, as well as impregnation of a preformed aerogel by a solution of metal salts, are very straightforward and inexpensive routes. These approaches, however, suffer from two major drawbacks: (1) the poor compositional homogeneity of the resulting nanocomposites and (2) the potential damage of the support in liquid media, aerogels being brittle and fragile. Tethering the metal to a gel matrix modified by coordinating groups and soaking the alcogel or aquogel into the metal solution prior to supercritical drying have been proposed as a way to overcome respectively drawbacks described in (1) and (2).

Deposition of nanoparticles from vapor phase, as opposed to wet impregnation methods, does not alter the porous matrix and ensures that the guest phase will be distributed throughout the matrix, thanks to its open porous texture. The general applicability of this approach is limited by the availability of precursors with a sufficiently high vapor pressure to give rise to volatile by-products.

Figure 16.1 is a schematic summary of the different preparation approaches; it should be taken into account, however, that a postsynthesis treatment, most commonly a thermal treatment under controlled atmosphere, is required to induce phase separation, to promote the formation of the desired oxide or metallic nanophase, and to obtain the desired crystallinity, as will be discussed later.

In the following, some recent examples of nanocomposite aerogels will be reviewed with the aim of pointing out some significant advances in the preparation of functional

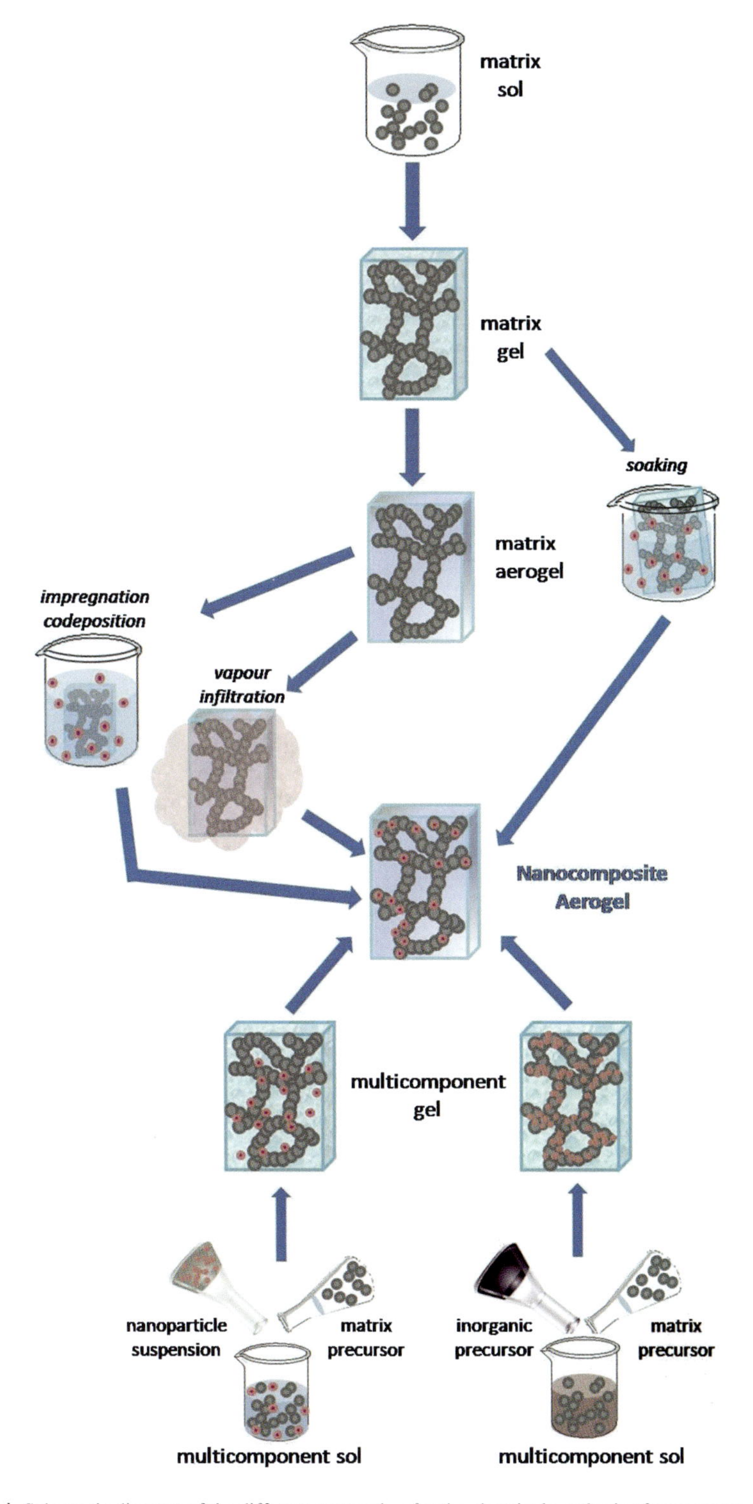

Figure 16.1. Schematic diagram of the different approaches for the chemical synthesis of nanocomposite aerogels.

nanocomposite aerogels made out of supported metal, alloy, or metal oxide nanoparticles dispersed in a dielectric matrix. Although we are aware that this chapter is not a comprehensive report of all the literature on nanocomposite aerogels, the aim of our overview is to highlight how the incorporation of nanoparticles may increase the possibility of using aerogels as innovative materials and how to select the most promising preparation route for obtaining nanocomposite aerogels for a given application. From the compositional point of view, the unique properties of silica aerogel matrices are still the most extensively exploited (Chap. 2), but alternative inorganic (Part III) and polymeric (Parts IV and V) matrices are now being envisaged, opening further routes for expanding the range of preparation protocols and functionalities of aerogels. Among the aerogels that are not covered by this chapter, nanocomposites based on carbon (Chap. 36) are particularly promising for electrochemical applications such as the design of high performance electrodes [4, 5].

16.2. Aerogel Containing Oxide Nanoparticles

Several examples are available in the literature on the synthesis of aerogels containing oxide nanoparticles embedded into dielectric matrices to be used for their magnetic, catalytic, and optical properties. While almost all of the examples refer to SiO_2-based nanocomposites, a large variety of oxide nanoparticles have been synthesized as a dispersed nanophase.

Different approaches have been tried for the synthesis of aerogels containing magnetic iron oxide nanoparticles dispersed into SiO_2 with particular emphasis on maghemite, γ-Fe_2O_3, nanoparticles, which possess interesting magnetic properties. However, even if this metastable phase is frequently kinetically stable at the nanometer scale, it proved to be difficult to obtain.

Casas et al. [6] prepared several magnetic samples (see Figure 16.2) by varying different sol–gel parameters. In particular, they used different precursors for the silica (*TEOS* or *TMOS*) in alcoholic solution (ethanol or methanol) in the presence of different amounts of water and using acid or base catalyst. They also used two different iron oxide precursors: either $Fe(NO_3)_3 \cdot 9H_2O$ or a $(FeNa(EDTA) \cdot 2H_2O)$ metallic complex in order to avoid strong bonding to the matrix and to increase pore diameters. The samples obtained with these procedures have surface areas ranging between 200 and 600 $m^2\ g^{-1}$, while *XRD* does not permit a clear identification of peaks apart from one single case in which the presence of six-line ferrihydrite, a poorly crystalline iron oxide hydroxide, became apparent. Mössbauer spectroscopy allowed identification of the iron oxide phase as ferrihydrite in some samples and magnetite in others and also indicated that the samples present *superparamagnetic relaxation* effects. *TEM* shows either spherical or acicular nanoparticles.

Fe_2O_3–SiO_2 nanocomposite aerogels were prepared by Casula et al. [7] using *TEOS* and $Fe(NO_3)_3 \cdot 9H_2O$ as precursors for the silica matrix and the iron oxide nanoparticles, respectively, while ethanol was used as solvent and nitric acid as catalyst. The obtained alcogels were submitted to *sc-HT drying* adopting two different conditions that lead to *mesoporous* and *microporous* aerogels, respectively. After supercritical drying, samples were calcined at elevated temperatures. *XRD* of the aerogels after supercritical drying showed faint peaks due to six-line ferrihydrite. After calcination, new peaks showed up and kept on growing. At 900°C, there was a mixture of different Fe(III) oxides present.

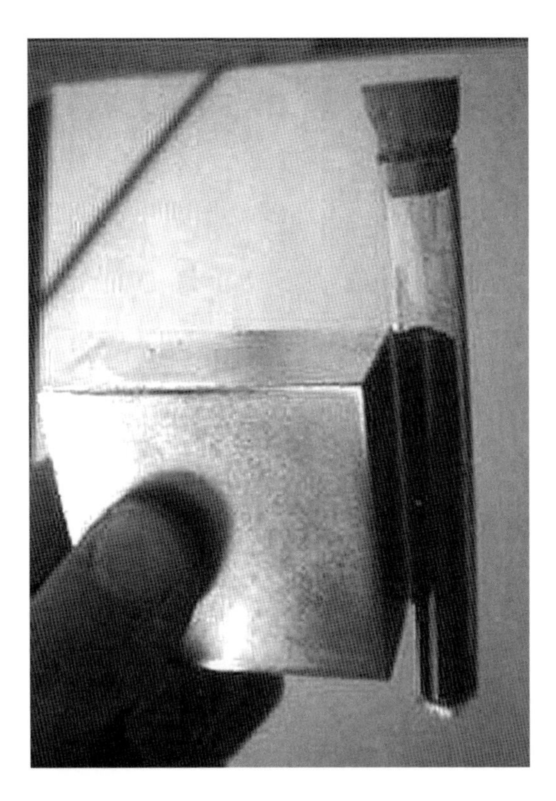

Figure 16.2. Magnetic aerogel attracted by a permanent magnet (reproduced from [6] by permission of Elsevier).

Surface areas were quite high even after calcination at 900°C (500 $m^2 g^{-1}$) and *TEM* images showed a homogeneous distribution of the nanoparticles.

Following a slightly different sol–gel method [8], in fact similar to that reported by Del Monte et al. for the preparation of xerogels [9], aerogels consisting of pure maghemite nanoparticles dispersed into silica were successfully obtained by the same group. The same precursors as in [7] were used, but gelation was performed in a closed container at 50°C. *Sc-HT drying* was performed using the conditions that led to *microporous* aerogels. After supercritical drying, calcination was directly performed at 400°C. *TEM* micrographs show nanoparticles with a narrow distribution of size (around 4 nm) located in large and irregular aggregates within the amorphous matrix, as shown in Figure 16.3. The *XRD* patterns of the calcined aerogels present broad peaks due to maghemite on top of the amorphous silica background. Mössbauer spectroscopy and magnetic measurements show the typical features of *superparamagnetic materials.*

Monoliths of iron oxide nanoparticles hosted in silica aerogels were prepared by Casas et al. using two different sol–gel routes [10]. In the first one, a preformed gel was soaked in an $Fe(NO_3)_3$ solution, while in the second case, all the precursors were mixed and the nanocomposite was obtained by cogelation. *Sc-CO_2 drying* was performed and samples were then annealed in air. *XRD* patterns, shown in Figure 16.4, show the formation of maghemite nanoparticles for samples obtained by soaking the gel, while samples obtained by cogelation show the presence of six-line ferrihydrite. Investigation of the magnetic dynamic behavior of the aerogels obtained by soaking [11] points to a *superparamagnetic*

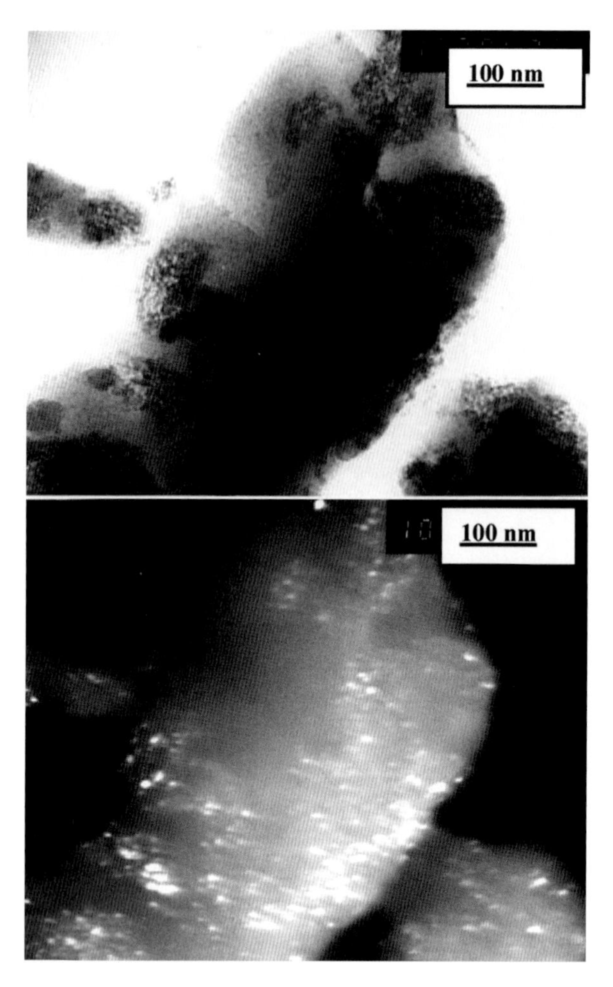

Figure 16.3. *BF* (**top**) and *DF* (**bottom**) TEM micrographs of maghemite–SiO_2 aerogel (reproduced from [8] by permission of The Royal Society of Chemistry).

behavior without interparticle interactions, indicating that the synthetic path is effective at preventing particle aggregation.

In order to avoid the introduction of water, iron oxide–silica aerogel nanocomposites were prepared by the same group by impregnation of wet gels [12], obtained under acidic catalysis, with either anhydrous ferrous acetate or anhydrous ferrous acetylacetonate. *XRD* shows peaks of a cubic oxide spinel phase with nanoparticle diameters around 5–6 nm, which is also confirmed by *TEM* (7–8 nm). Mössbauer confirmed that the nanoparticles consist of maghemite and the amount of that phase was estimated at 11–13% (w/w). In this case, the maghemite nanophase was formed without the need of postannealing treatments due to the absence of water. Surface areas of the samples were around 900 $m^2\ g^{-1}$. Mössbauer spectroscopy on the aerogels obtained by impregnation of wet gels with ferrous acetylacetonate in neutral medium [13] confirms the presence of maghemite nanoparticles (6 nm) in a nanocomposite with low apparent density (0.2 $g\,cm^{-3}$) without postannealing.

Maghemite–silica nanocomposite aerogels were prepared by van Raap et al. using different precursors (*TEOS* and *TMOS*) and different pressure–temperature drying

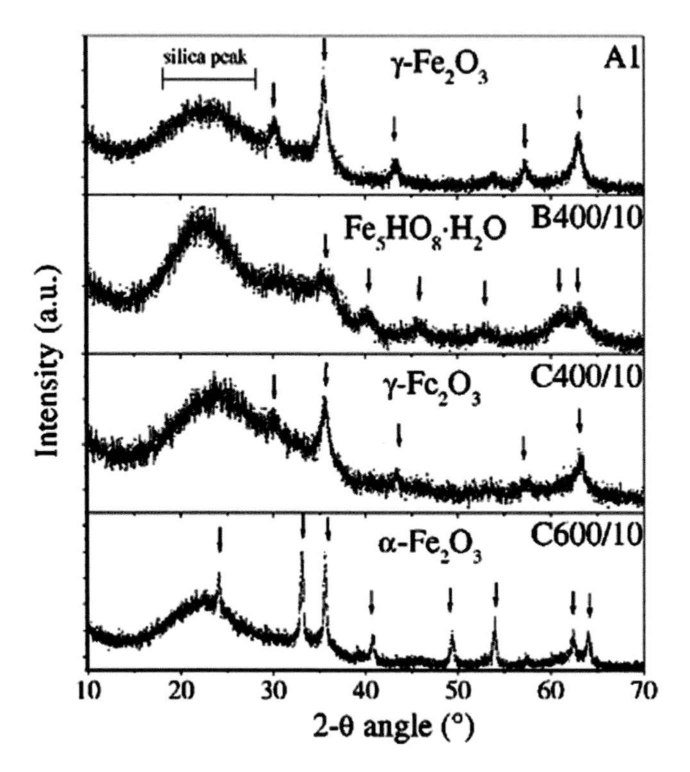

Figure 16.4. XRD patterns of iron oxide–silica aerogels obtained by soaking (A1) and by cogelation (B400) (reproduced from [10] by permission of Springer).

conditions [14]. The *XRD* patterns of the aerogels prepared with both *TEOS* and *TMOS* display weak reflection lines due to maghemite or magnetite. Mössbauer indicates that aerogels prepared using *TMOS* contain some Fe^{2+}. Those samples had also larger pores, by *TEM*.

The synthesis of high surface area, *mesoporous* iron–silicon mixed oxide aerogel was studied by Fabrizioli et al. [15] by varying different sol–gel parameters, such as gelation agents (*N,N*-diethylaniline, trihexylamine, NH_4CO_3, or NH_3 dibase), concentration of iron precursor ($Fe(NO_3)_3 \cdot 9H_2O$), type of silicon precursor (*TMOS* or *TEOS*), and postannealing treatment. In most cases, transparent gels are obtained, whereas NH_4CO_3 gives rise to hazy gels. *Sc-CO₂ drying* is followed by annealing first under flowing N_2 up to 200°C and then in air first at 200°C and then at 600°C. After calcination at 600°C, all aerogels were amorphous by *XRD* and *TEM*. Samples further calcined at 900°C showed a broad reflection, which was attributed to either maghemite or magnetite nanocrystals. EPR shows different iron species: isolated tetrahedrally coordinated iron, iron oxide clusters, and superparamagnetic iron oxide clusters of increasing size with the iron loading.

Mixed oxide nanocomposites in which iron(III) oxide is the major component and silica is the minor component were prepared by Clapsaddle et al. [16] using $FeCl_3 \cdot 6H_2O$ with either *TMOS* or *TEOS* in the presence of an organic epoxide as gelling agent (propylene oxide, PO, or trimethylene oxide, TMO) with Fe/Si molar ratios varying between 1 and 5. Drying was performed using *sc-CO_2*. This is an extension of the use of organic epoxides (Chap. 8) for the preparation of metal oxides [17], giving rise to the formation of aerogel monoliths, as shown in Figure 16.5. In order to obtain homogeneous sols, the iron salt was

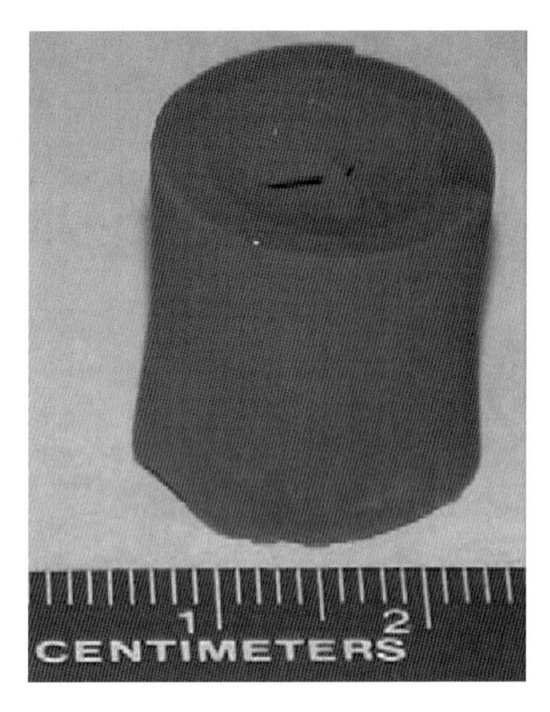

Figure 16.5. Photograph of a Fe/Si = 1 aerogel monolith (reproduced from [16] by permission of Elsevier).

dissolved in ethanol and partially reacted with the epoxide promoter before addition of *TEOS* or *TMOS*. The composites contain an iron(III) phase, which was not clearly identified, and silica. The *FTIR* results suggest the iron oxide phase to be an iron oxide hydroxide due to the presence of water and hydroxyl groups. The most pronounced effect on gelation times comes from the epoxide used; a minor influence was also observed from the Fe/Si ratio and the silica precursor. All the materials had surface areas between 350 and 450 $m^2 g^{-1}$ and were *mesoporous*. *TEM* showed that the *mesoporous* structure was due to a collection of nanoparticles clustered together, particle size depending on the epoxide used (bigger particles with TMO). Elemental maps indicate that Fe and Si are uniformly dispersed.

Several other silica-based aerogels containing transition metal oxides have also been prepared. Casula et al. [18] prepared $NiO–SiO_2$ nanocomposite aerogels using *TEOS* and Ni $(NO_3)_2 \cdot 6H_2O$ as precursors for the silica matrix and for the NiO nanoparticles, respectively, ethanol as solvent and nitric acid as catalyst. The obtained alcogels were submitted to *sc-HT drying* adopting four different conditions that differ in the heating ramp, the solvent used to fill the autoclave, and the initial overpressure in the autoclave. After supercritical drying, the samples were calcined at 773 K. The surface areas were quite high after supercritical drying and even after calcination (500–900 $m^2 g^{-1}$), while the porosity of the samples varied depending on the supercritical drying conditions. In particular, *mesoporous* or *microporous* aerogels can be obtained. The *XRD* of the samples before calcination showed peaks due to $Ni(NO_3)_2Ni(OH)_2$ and $Ni(OH)_2$. After calcination peaks due to NiO nanoparticles were observed. *TEM* showed that crystal size varied between 2 and 10 nm depending on the porosity of the aerogel.

Nanoparticles of ZnO were introduced into the pores of a preformed SiO_2 aerogel by immersion in a dilute aqueous solution of $ZnSO_4 \cdot 7H_2O$ [19], followed after a few days by addition of NH_3, which caused precipitation of $Zn(OH)_2$. Finally, thermal treatment was performed at temperatures varying between 200 and 1,150°C. *XRD* indicated the presence of ZnO with particle size increasing with the temperature of the thermal treatment. The photoluminescence intensities of the nanocomposites at about 500 nm were 10–50 times higher than that of bulk nanostructured ZnO, depending on the conditions of the thermal treatment.

Amlouk et al. [20] incorporated TiO_2 nanopowder in a silica aerogel. The nanopowder was prepared by the sol–gel process using titanium (IV) isopropoxide in a mixture of methanol and acetic acid. The nanopowder was mixed with *TEOS* and water, while ethanol and HF were used as the solvent and catalyst, giving rise to instantaneous gel formation. The gel was submitted to *sc-HT drying* and the aerogel was annealed at 1,200°C. *XRD* showed small peaks due to rutile on top of the amorphous silica background. The crystallite size was determined to be 20 nm for sample annealed at 523 K and 300 nm for sample annealed at 1,473 K. Those nanocomposites exhibited a strong near-infrared luminescence band centered at about 850 nm.

The same group prepared in a very similar way samples of silica aerogels incorporating an Al_2O_3 nanopowder [21]. *XRD* of samples annealed at 1,150°C showed the presence of peaks due to α-Al_2O_3 and δ-Al_2O_3. The crystallite size of samples annealed at 700°C was estimated at 30 nm. Photoluminescence spectra showed strong luminescent bands at about 500 and 770 nm. Other nanopowders such as ZnO and SnO_2 have been also incorporated, producing strong luminescence bands in the same regions [22].

Tin oxide nanocrystals were incorporated in the *mesoporous* network of a silica aerogel also by Wei et al. [23]. In this case, silica aerogel was obtained via ambient pressure drying after multiple surface modifications with trimethylchlorosilane. Before surface modification ethanol was replaced with *n*-hexane, which is a suitable solvent for trimethylchlorosilane and for ambient pressure drying because of its low surface tension. The deposition of tin oxide nanocrystals was achieved by immersing the aerogel first in an aqueous solution of $SnCl_4$ and then in an aqueous solution of NH_3. Annealing was required to form the oxide nanoparticles. The aerogel nanocomposite had a much lower surface area than the original silica aerogel, as expected. Moreover, after the incorporation of the nanoparticles pores lost their well-defined size and shape. *HREM* images of samples treated at 400°C showed several tin oxide nanocrystals with interlayer spacings in three different directions and some evidence of dislocation and vacancy clustering. The composite aerogels exhibit a rich photoluminescence and promising photocatalytic ability toward photodegradation of methylene blue, as shown in Figure 16.6 where conversion before and after UV exposure for the nanocomposite aerogel treated at 500°C is compared with that of a pure silica aerogel.

A 2-nm thick ZnO layer was deposited on the inner surface of ultralow density (about 0.5% of full density) nanoporous silica aerogel monoliths [24]. The deposition was performed in a viscous flow reactor using a diethyl zinc/H_2O mixture. Prior to that, a nucleation layer of amorphous Al_2O_3 was deposited in order to prepare a densely hydroxylated surface to minimize any nucleation delays. The coating produced a decrease in surface area and porosity and a corresponding increase in the apparent density. However, the coated aerogel still retained a high surface area and ultralow density. TOF-SIMS images illustrated that the coating was uniform for both the ZnO layer and the Al_2O_3 nucleation layer. The orientation and structure of the ZnO layer was illustrated by the *HREM* images, which showed a layer of crystalline ZnO after deposition at relatively low temperatures (177°C).

The preparation of free-standing *mesoporous* TiO_2–SiO_2 aerogels with hierarchical pore structure, which are expected to have applications in photocatalytic degradation

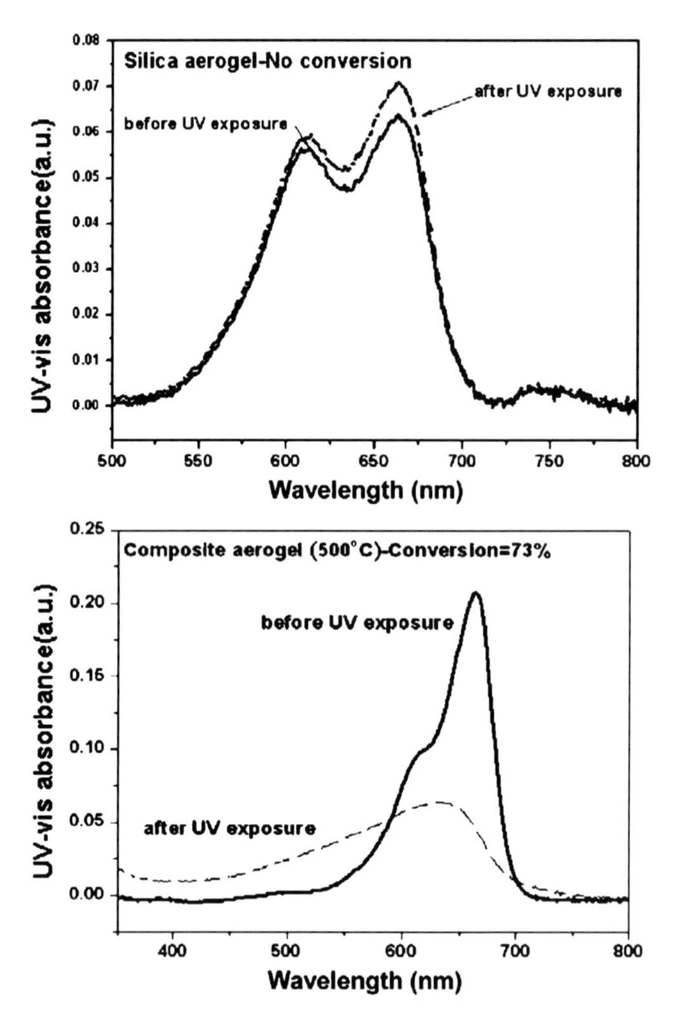

Figure 16.6. Photocatalytic characterization of pure silica and SiO_2–SnO_2 composite aerogels thermally treated at 500°C (reproduced from [23] by permission of Elsevier).

of air pollutants, was reported in [25]. The synthesis involved the use of a solution of titanium isopropoxide in ethanol as TiO_2 precursor, with nitric acid as catalyst. The silica precursor solution was prepared in two steps: the first involving the preparation of a clear polymer solution using Pluronic 123, ethanol, water, and nitric acid and the second involving the addition of *TEOS* to the polymer solution. The two precursor solutions were then mixed with varying Ti/Si ratios, and after gelation *sc-CO_2 drying* was performed. One of the samples was also submitted to *sc-HT drying*; finally, aerogels were calcined at 450°C. The preparation method bears similarities with evaporation-induced self-assembly [26], but without carrying out complete solvent evaporation. Instead the solvent was removed from the gel under supercritical conditions. The aerogels before calcination were opaque, in contrast to silica aerogel prepared without the use of P123. Small-angle *XRD* showed the presence of ordered *mesopores*, which were maintained after supercritical drying and calcination. However, after calcination, *XRD* peak intensity decreases and peaks shift toward higher angles, effects that the authors attribute to possible disorder caused by pore shrinkage. The density of the materials was between 0.2 and 0.56 g cm^{-3} depending on the Ti/Si ratio and the supercritical drying conditions, while surface areas varied between 440 and 840 m^2 g^{-1}. Finally, *TEM* indicated that ordered mesostructure was retained in most cases, but samples with higher Ti content showed TiO_2-rich domains.

All examples reported so far refer to nanocomposite aerogels where the dispersed nanoparticles were made of a simple oxide. Recently, Casula et al. reported a method to produce very highly porous $CoFe_2O_4$–SiO_2 nanocomposite aerogels with interesting magnetic properties [27, 28]. The cogelation procedure adopted to prepare nanocomposite aerogels in most cases is performed under acidic conditions in order to avoid precipitation of metal hydroxides, which leads to the formation of turbid and inhomogeneous gels. These acidic conditions generally give rise to long gelation times if gelation is performed in a closed container or to a significant solvent evaporation before gelation if it is carried out in an open container. A two-step acid–base-catalyzed procedure that uses urea (carbamide, $CO(NH_2)_2$) as the basic catalyst has been developed, allowing one to obtain low-density silica-based nanocomposite aerogels.

The precursors were *TEOS*, $Fe(NO_3)_3 \cdot 9H_2O$, and $Co(NO_3)_2 \cdot 6H_2O$, which were prehydrolyzed in acidic conditions before addition of a solution of urea. Gelation was performed in a closed container at 40°C, and *sc-HT drying* was performed. After supercritical drying, calcination treatment at temperatures between 450 and 900°C was performed. Only a minor volume decrease was observed starting from the sol to the gel and to the aerogel, as it can be seen in Figure 16.7. At the same time, the weight decreased significantly after supercritical drying so that the obtained aerogels possessed very low density (0.07 g cm^{-3}) and were *mesoporous* with surface areas around 350 m^2 g^{-1}, as assessed by N_2 physisorption measurements. *TEM* micrographs showed *macropores* that could not be assessed by physisorption and indicated that nanoparticles were homogeneously dispersed

Figure 16.7. Photographs of sol, alcogel, and aerogel for the synthesis of highly porous $CoFe_2O_4$–SiO_2 nanocomposite.

into the highly porous matrix. The surface area was still high until calcination at 900°C, while full densification was only achieved at 1,200°C. *XRD* of aerogel before calcination and after calcination at 450°C showed only faint peaks not easy to ascribe to specific phases. At 750°C, peaks due to cobalt ferrite became apparent and kept growing up to 900°C. The method was successful in preparing aerogels with different loadings of cobalt ferrite, even if the evolution toward the formation of the final nanocomposite was slightly different, being slower at lower loading. The nanoparticles displayed *superparamagnetic behavior* with blocking temperatures increasing both with loading and with calcination temperature, as reported in Figure 16.8 for the aerogel with 5% (w/w) loading of cobalt ferrite [28].

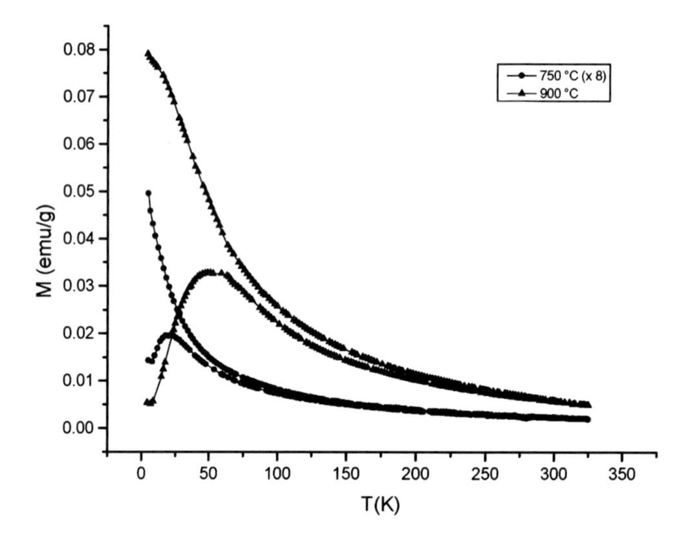

Figure 16.8. ZFC–FC magnetization curves of the $CoFe_2O_4$–SiO_2 nanocomposite aerogel with 5 wt% loading of cobalt ferrite (reproduced from [28] by permission of American Chemical Society).

A more detailed study of the evolution of these aerogels with thermal treatment was carried out with *EXAFS* and *XANES* [29, 30] spectroscopies at both the Fe and Co K-edge. The evolution was different depending on the loading of $CoFe_2O_4$. At the highest loading (10%, w/w), two separate phases were present after supercritical drying and after calcination at 450°C: Fe was present as six-line ferrihydrite, while cobalt was present in a separate phase, which was later identified as $Co_3Si_2O_5(OH)_4$ [31]. Low loading aerogels (5%, w/w) were calcined at the same temperature and contain Fe and Co in a very disordered environment. After calcination at 750°C, the 5% sample still had two separate phases, while in the 10% sample a significant amount of $CoFe_2O_4$ was already formed. At 900°C, all Co and Fe were present in the form of $CoFe_2O_4$ nanoparticles in both samples with different loadings.

This sol–gel procedure involving the use of urea as gelation agent was shown to be very versatile. By using $Ni(NO_3)_2 \cdot 6H_2O$ or $Mn(NO_3)_2 \cdot 6H_2O$ instead of $Co(NO_3)_2 \cdot 6H_2O$, nanocomposite aerogels with loading of 10 wt% of either $NiFe_2O_4$ or $MnFe_2O_4$ nanoparticles dispersed in the highly porous silica matrix were obtained [32]. The evolution of the sample was found to be similar to that observed in the $CoFe_2O_4$–SiO_2 nanocomposites, as it can be inferred from Figure 16.9 reporting the *XRD* patterns at different calcination temperatures. As also confirmed by a detailed *EXAFS* and *XANES* study on the formation of $NiFe_2O_4$–SiO_2 [33] and $MnFe_2O_4$–SiO_2 [34] highly porous nanocomposites, after

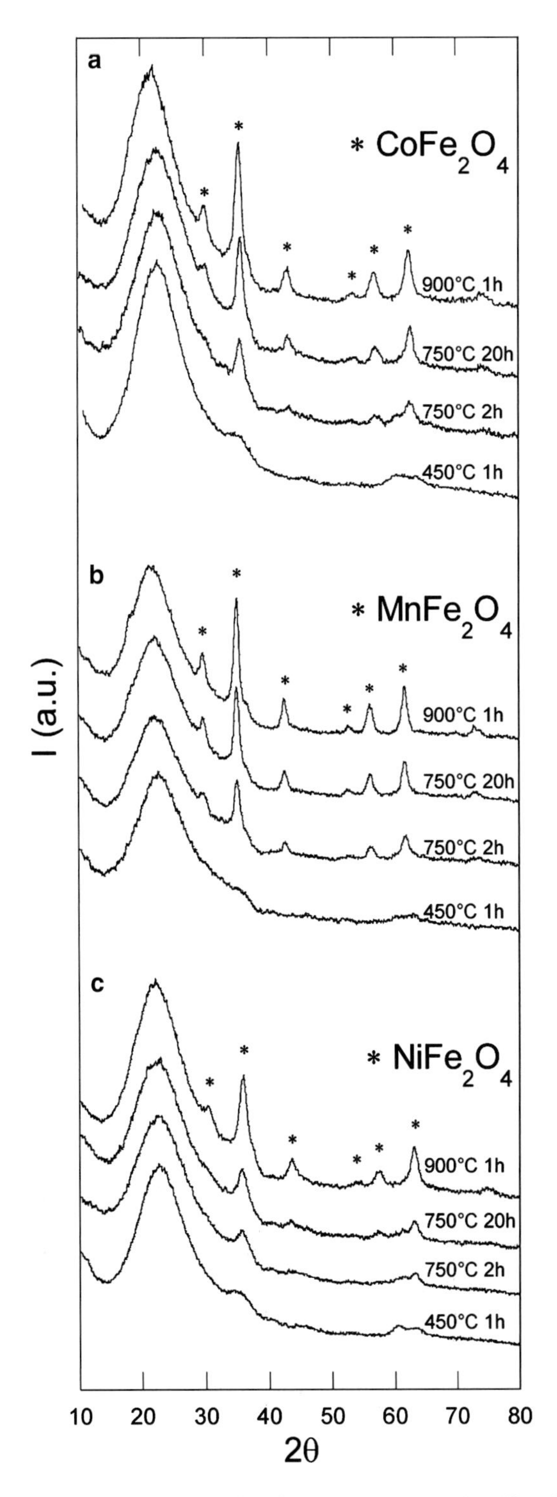

Figure 16.9. XRD patterns of $CoFe_2O_4$–SiO_2, $NiFe_2O_4$–SiO_2, and $MnFe_2O_4$–SiO_2 after different calcination treatments (reproduced from [32] by permission of American Scientific Publishers).

supercritical drying and calcination at 450°C two separate phases are present, one containing Fe and one containing either Ni or Mn. At 750°C, the ferrite phase began to form and the formation was completed either by prolonging the treatment at 750°C or by a shorter calcination at 900°C.

EXAFS and *XANES* also proved to be extremely powerful in order to study the degree of inversion in the MFe_2O_4 (M = Mn, Co, Ni) spinel nanoparticles [35]. In particular, the degree of inversion in nanosized ferrite was found to be quite close to the values generally observed in the corresponding bulk ferrites, with $MnFe_2O_4$ nanoparticles presenting the lowest degree of inversion (0.20), followed by $CoFe_2O_4$ nanoparticles (0.68), and $NiFe_2O_4$ (1.00).

Samples calcined at 450°C were at an intermediate stage toward the formation of the final $CoFe_2O_4$–SiO_2 and $NiFe_2O_4$–SiO_2 nanocomposites and were studied further [31]. To this end, additional aerogels were prepared without using the Fe precursor, i.e. containing only a single Co/Ni intermediate phase. The presence of either Co or Ni silicate hydroxide was clearly evidenced by *EXAFS* and *XANES*, while low-resolution *BF TEM* and high-resolution *BF STEM* images showed evidence that those phases had a layered structure with anisotropic nanostructures. Magnetic measurements further confirmed those findings and evidenced the anisotropic shape effect. It should be pointed out that the observed magnetic behavior was quite similar to the one reported in Co–SiO_2 aerogels [36]. However, in that case the *ZFC–FC* magnetic results were attributed to Co and CoO. The results point out that a detailed structural characterization using selective techniques, such as X-ray absorption spectroscopy, and the comparison with bulk phases is needed.

While almost all papers on nanocomposite aerogels containing metal oxide nanoparticles dispersed in an insulating matrix refer to silica-based systems, a few examples are also available for other matrices.

CuO/*RF* samples were prepared using a $CuCl_2 \cdot xH_2O$/epichlorohydrine sol to catalyze cogelation of the system (Chap. 14), using *DMF* as common solvent [37]. Drying was performed by *sc*-CO_2. The copper oxide remained trapped in the *mesoporous* voids of the *RF* network. Samples were *mesoporous* and did not present any crystallinity. After pyrolysis at 400°C, the dominant crystalline phase was cubic metallic Cu, probably by direct oxidation of RF by CuO. Similar smelting reactions were observed in analogous Fe_2O_3/*RF* composites at much higher temperatures [38], at which the *RF* network is first converted to a porous carbon network [39].

16.3. Aerogels Containing Metal and Alloy Nanoparticles

The preparation of nanocomposite aerogels made out of zero-valent metal or alloy nanoparticles in an insulating aerogel matrix has attracted a great deal of effort in view of the preparation of materials for catalysis, biomedicine, optoelectronics, and sensors. As it is the case for all sol–gel nanocomposites, silica is by far the most investigated matrix for supporting metal/alloy nanoparticles, although alumina, titania, chromia, and polymer supports have also been proposed.

On the other hand, a variety of metal nanoparticles have been considered either on the basis of the ease of reducibility, with the aim of developing optimized synthetic protocols, or in order to provide the desired functionality to the nanocomposite. Noble metal nanoparticles are the most investigated as they meet both requisites: they can be easily obtained and stored in the zero-valent state; they are catalytically active toward a wide range of chemical processes [40–44]; and they can have biomedical applications such as antimicrobial activity [45]. Moreover, metal nanoparticles exhibit characteristic surface plasmon bands and when hosted in dielectric matrices, due to the refractive index mismatch, can give rise to enhanced optical effects, namely, surface enhanced Raman scattering [46].

Transition metals and their alloys have also attracted much attention due to their magnetic properties and due to their potential in catalysis thanks to the occurrence of different oxidation states.

In the following, we will summarize how the issue of the control over the nanocomposite formation has been recently addressed.

An ideal approach for introducing a supported functional nanophase on an aerogel matrix relies on the deposition of metal nanoparticles, as the addition of the dispersed phase can be performed without interfering with the sol–gel process during the formation of the matrix, and non-aggregated metal nanoparticles are intrinsically produced. This approach was first demonstrated by Hunt et al. [47] and by Kucheyev [24] for other nanocomposite materials and has recently been used for the preparation of Ru–SiO_2 nanocomposite aerogels by atomic layer deposition of Ru on a low-density silica aerogel using bis-cyclopentadienyl ruthenium [48]. However, at present this technique can be promising only for metal deposition on very thin layers, as large-scale inhomogeneity was observed. In particular, in correspondence with the cracks in the silica substrate a high concentration of Ru was found. The authors suggest that new metal precursors with higher vapor pressures or new reactors should be designed in order to promote effective diffusion of the metal precursor in matrices with porosities on the nanometer scale and therefore extend the applicability of this approach.

A more general procedure for the preparation of metal–silica aerogel nanocomposites is based on the supercritical drying of a silica alcogel previously soaked in a metal solution.

Aerogels made out of gold supported on silica–chitosan (90:10 ratio) monoliths have been prepared by dipping the alcogels into an ethanolic solution of $HAuCl_4$ and performing *sc-CO_2 drying* [49]. In this way, a monolithic aerogel where Au(III) bridged to the biopolymer–silica support (thanks to the amino and hydroxyl groups present in the chitosan) was obtained, which was then reduced by *UV* irradiation (Chap. 18). The oxidation state of the metal and the size and size distribution of the gold nanoparticles were studied by photoacoustic spectroscopy: Compared to traditional optical absorption, this method is more sensitive and can be used on both transparent and opaque materials without the need of sample preparation, and therefore it is very suitable for monolithic aerogels. The characterization results suggest that the Au nanoparticles were mainly located on the surface of the matrix; in this respect, the role of the biopolymer-modified silica (compared to pure silica) is not clear as it does not ensure a homogeneous metal distribution by coordinating the Au(III) precursor. As structural and morphological characterization (in particular *TEM*) is missing, it is hard to provide additional evaluation on the validity of this approach, which is in any case limited to the formation of metals obtainable by photolysis-induced reduction (thermal reduction would induce pyrolysis of the polymer-modified matrix).

In an alternative approach, Morley et al. [50, 51] describe a similar procedure, which is effective for the preparation of metal–SiO_2 nanocomposites based on the use of supercritical carbon dioxide for the impregnation of a preformed silica alcogel with a suitable metal complex and for the removal of the free ligand after the reduction step. Pd– and Ag–SiO_2 aerogels, which can be used as active catalysts, were effectively obtained by this procedure where supercritical carbon dioxide was used as a recyclable solvent, which does not alter the porous structure of the matrix. However, in addition to the many steps required, a major disadvantage of this procedure is that it requires solubility of metal complexes in *sc-CO_2*.

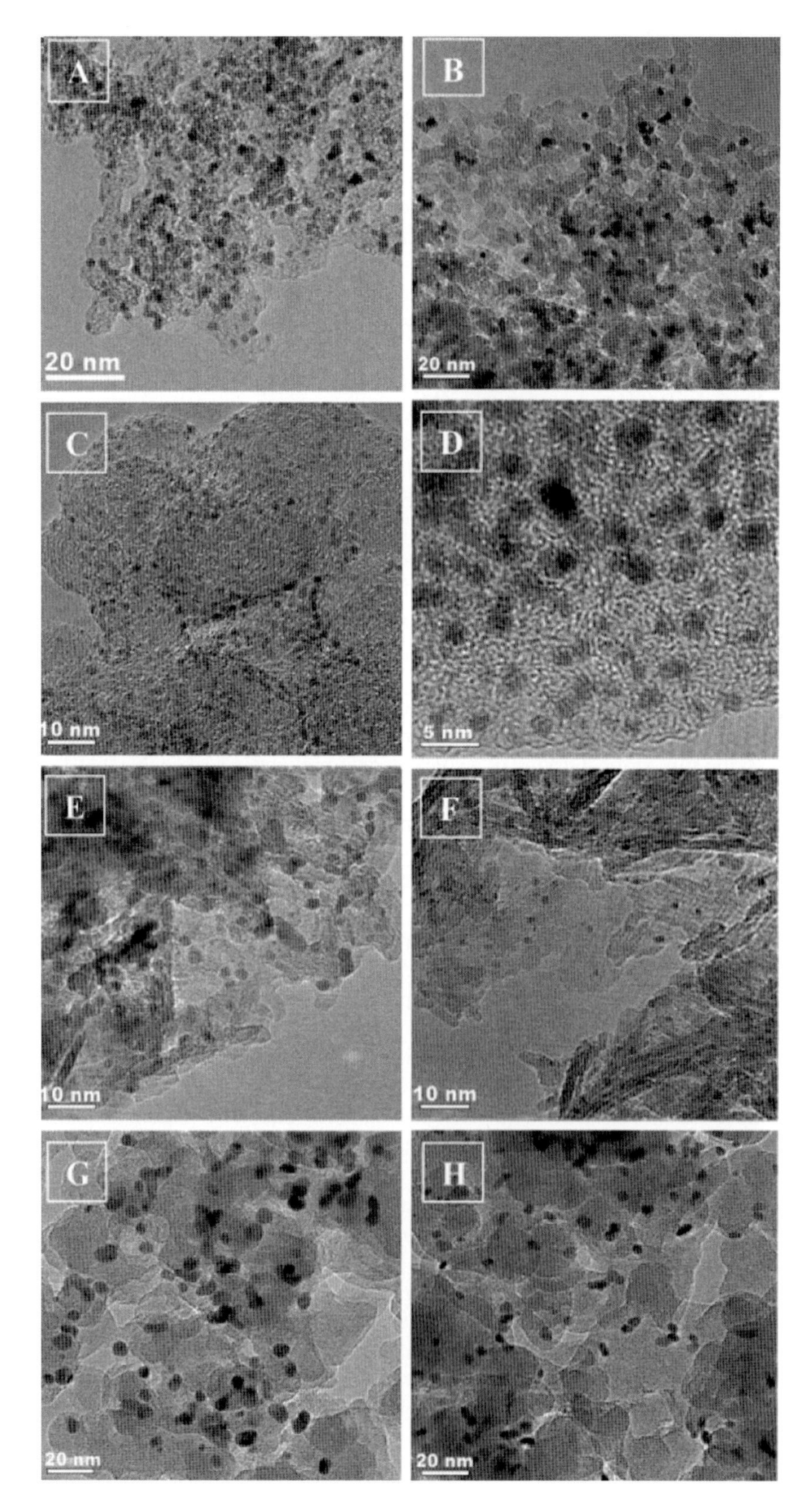

Figure 16.10. (continued)

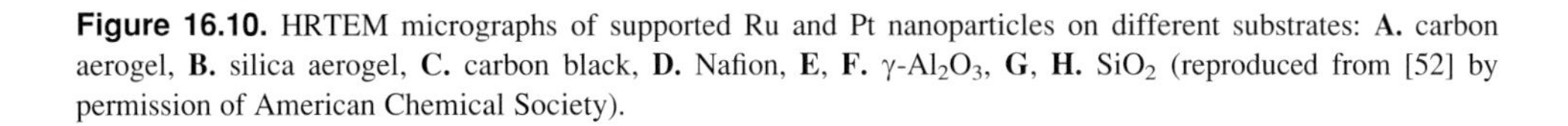

Figure 16.10. HRTEM micrographs of supported Ru and Pt nanoparticles on different substrates: **A.** carbon aerogel, **B.** silica aerogel, **C.** carbon black, **D.** Nafion, **E, F.** γ-Al_2O_3, **G, H.** SiO_2 (reproduced from [52] by permission of American Chemical Society).

The properties of *sc-CO_2* in the development of novel processes for the synthesis of nanostructured materials were also exploited in the case of Pt–SiO_2 aerogels obtained by impregnation of preformed commercial silica aerogels [52]. By making use of *sc-CO_2* as the solvent for the deposition of the metal precursor (dimethyl cyclooctadiene platinum-II), Pt–SiO_2 nanocomposites with a narrower size distribution compared to the results obtained by Morley et al. [50, 51] were obtained. However, metal nanoparticles seem to be mainly deposited on the outer surface of the silica aerogel (see Figure 16.10). By applying the same deposition procedure on different substrates (carbon black, alumina, silica, and silica aerogel), the authors were able to show that particle formation and growth was mainly governed by the strength of the interaction between the organometallic precursors and the support.

Impregnation of silica alcogels obtained by a acid/base two-step catalysis by saturated ethanolic solution of Ni(II) and Pd(II) acetylacetonates, followed by *sc-HT drying*, has been used for the preparation of Ni–SiO_2 and Pd–SiO_2 nanocomposites [53]. Metal nanoparticles of about 14 nm were obtained after supercritical drying without the need of further reduction treatment, which was tentatively ascribed to a reducing effect of supercritical ethanol. The synthetic procedure itself is quite straightforward, although time-consuming (sample preparation overall requires over 1 month) and very high surface areas (higher than 800 $m^2\ g^{-1}$) were obtained. A major drawback is in the inhomogeneous distribution of the nanoparticles within the matrix, which the authors found to be responsible for the moderate conversions obtained by using those aerogels as catalysts for *Mizoroki–Heck* C–C coupling reactions. Poor reusability of the catalyst was also observed (due to metal leaching) probably due to the weak interaction or distribution of the metal nanoparticles on the silica surface.

In order to improve the design of Ni–SiO_2 and Pd–SiO_2 aerogel catalysts, a cogelation route followed by *sc-CO_2 drying* and thermal treatment was used [43]. In that approach, complexation of Pd or Ni(II) acetates in ethanol by organofunctional siloxanes of the type $(CH_3O)_3$–Si–X (where X is a group capable of binding the metal, such as ethylenediamine or diethylenetriamine) was first carried out and then *TEOS* was added and cogelled. The corresponding aerogel obtained by *sc-CO_2 drying* was then either reduced at 500°C or first calcined and then reduced. Direct reduction of the aerogel may lead to carbon contamination, which in this case can be advantageous since metals on carbon catalysts are known to have enhanced catalytic performance in C–C coupling reactions. Good catalytic results, with conversions as high as 99%, were obtained in particular by the use of the Pd–SiO_2 aerogels, which have 1–2 nm metal nanoparticles well distributed within a *macroporous* silica matrix, with a surface area in the 500–700 $m^2\ g^{-1}$ range. That method, thanks to the use of the complexing organosilane in the sol–gel synthesis, which was first developed for xerogels [54], has been very suitable for the preparation of homogeneous nanocomposites with non-aggregated nanoparticles. A useful further implementation would be the achievement of particle size control in order to tailor the nanoparticle size for selected applications.

The design of Pd–SiO_2 active catalysts for organic reactions, such as hydrogenation of α,β-unsaturated aldehydes and C–C formation by *Heck coupling*, has been also reported [42]. By adapting a method previously proposed for pure silica aerogels [55], Pd–SiO_2 aerogels were obtained by a sol–gel procedure that makes use of an ionic liquid as the solvent, which can be removed by conventional solvent extraction techniques. In particular, a colloidal suspension of palladium nanoparticles having 1 nm size was first obtained by heating at 80°C a dispersion of palladium(II) acetate and triphenylphosphine in an ionic liquid. Once *TEOS* and formic acid were added, gelation took place in 15 min. Aging over a few weeks was required in order to strengthen the gel network, and the aerogel was finally obtained by removing the ionic liquid from the monolith by refluxing acetonitrile overnight. Although it was not possible to discuss the quality of the nanocomposites as no extensive structural investigation was presented, the advantage of that approach was related to the use of a green solvent, which enables the manufacture of aerogels by more viable and cleaner synthetic routes.

The preparation of nanocomposites by cogelation of preformed colloidal nanoparticles and of *TEOS* has also been adopted for the preparation of Ag–SiO_2 [56] and Au–SiO_2 [57] nanocomposites.

Smith et al. have prepared Ag–SiO_2 aerogels by adding a silver colloidal suspension obtained by the Creighton procedure [58] to prehydrolyzed *TEOS* and then submitting the gel to *sc-CO_2 drying*. Characterization of the aerogels was performed by optical absorption spectroscopy and the data were modeled in order to determine whether nanoparticles were surrounded by air or by the silica matrix. The calculation was based on the shift in the absorption profile, which could be ascribed to the difference in the dielectric contrast of the two environments. It was found that the fraction of Ag surface area in contact with silica was 46%. The proposed method is a promising way to determine the metal accessible area for catalytic application in nanocomposites that do not present aggregation. Unfortunately, those data were not supported by any structural and morphological characterization results.

Balkis Ameen and coworkers have reported a cost-effective preparation of Ag–SiO_2 by cogelation of *TEOS* and $AgNO_3$ under acid catalysis followed by subcritical drying of the corresponding alcogel [46]. Subcritical drying includes solvent exchange to remove water from the alcogel, aging in *TEOS* solution to improve the network stiffness and strength, again solvent exchange to absolute ethanol, and finally drying at 120°C. The resulting nanocomposite aerogels contained metallic silver, which was probably formed by free radicals of the alcohol generated within the sol system, with particle sizes in the 6–8 nm range. The particle size increased with metal loading, which was varied from 1, 5, and 25% (w/w). Those nanocomposites exhibited a large surface area for moderate loadings (surface area values drop from 850 to 300 $m^2 g^{-1}$ from pure silica to a 25%, w/w, nanocomposite) and a relatively dense texture. As the metal loading affects the average particle size, the catalytic behavior of these nanocomposites for the selective oxidation of benzene to phenol as a function of metal particle size was also investigated [40]. The highest selectivity was obtained in catalysts containing smaller nanoparticles, whereas an increase of conversion and a decrease of selectivity were observed with increasing Ag particle size.

Au–SiO_2 aerogels with a metal loading from 0.2 to 25% (w/w) were successfully obtained starting from preformed colloidal gold nanoparticles [57]. A silica alcogel obtained by a basic catalysis of *TMOS* was solvent-exchanged (from methanol to toluene) and immersed in a toluene suspension of colloidal Au nanoparticles obtained by the Brust method [59], as shown in Figure 16.11. In order to obtain aerogels, *sc-CO_2 drying* was then

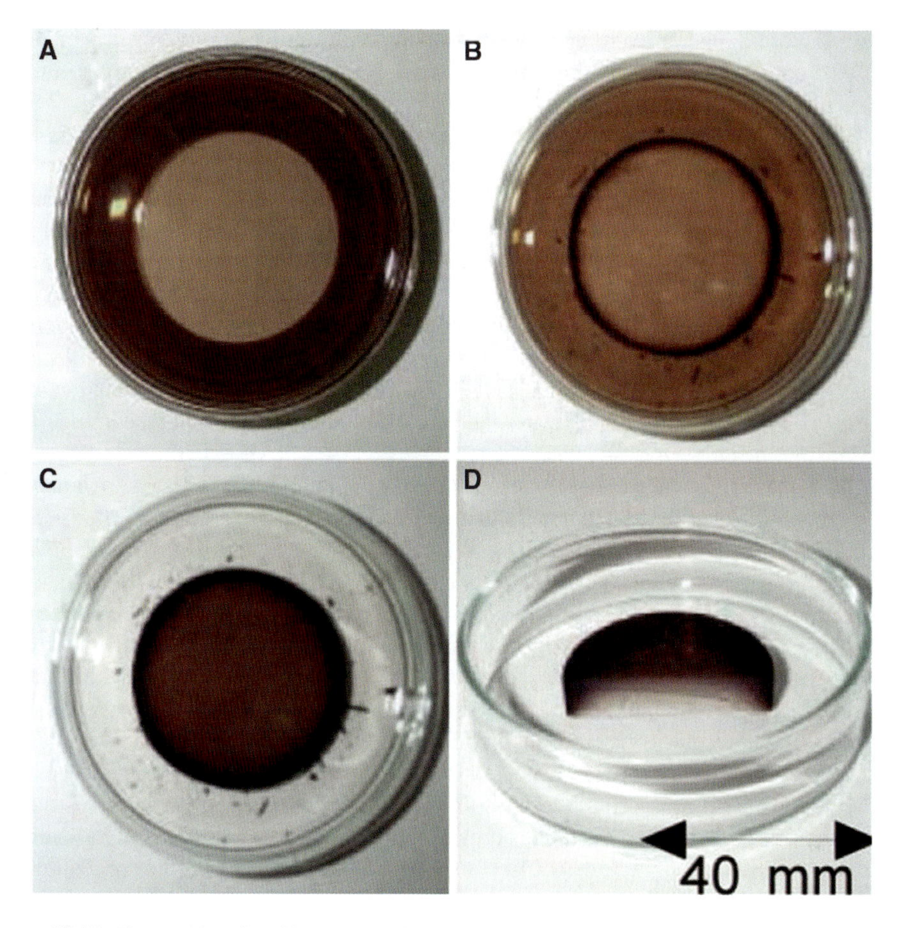

Figure 16.11. Photographs of a silica wet gel immersed in a colloidal suspension of gold nanoparticles at increasing times **A–C.** and of the Au–SiO_2 composite aerogel **D.** (reproduced from [57] by permission of Springer).

performed so that the mild drying conditions did not affect the adsorption of thiol molecules from the Au surface. The advantage of this procedure relies on the assembly of nanocomposites with both known textural features of the matrix and size of the nanoparticles. In particular, by adopting this procedure for colloidal nanoparticles with different sizes, the authors concluded that dipole–dipole-induced interactions were responsible for the adsorption of the highly polarizable Au core on silanols. In fact, larger metal nanoparticles were adsorbed more readily, whereas nanoparticle uptake was slowed down when the silica surface was modified by the addition of hydrophobic [–$Si(CH_3)_3$ groups introduced by hexamethyldisilazane].

The same authors optimized the above synthesis for the design of aerogel catalysts by preparing nanocomposites made out of Au nanoparticles supported on titania-coated silica [60]. For that purpose, once the silica alcogel was obtained and solvent exchanged, it was immersed into titanium isopropoxide and then supercritically dried with *sc*-CO_2. The powdered titania-coated aerogel was immersed in a toluene suspension of colloidal Au nanoparticles prepared by the Brust method, the nanoparticle uptake by the aerogel being monitored by the decoloration of the suspension, and finally the impregnated powder was dried by calcination at 400°C. In that case, as opposed to previous results [57], a significant decrease in pore diameter and pore volume was observed upon metal uptake by the porous matrix as a consequence of the impregnation of the aerogel into the liquid: larger pores are more fragile and collapse more easily.

Those materials are promising for the catalytic oxidation of CO, as gold has been reported to be very active as catalyst and TiO_2 is expected to increase the resistance of gold nanoparticles to sintering. The catalytic assays, reported in detail in [41], indicate that the Au nanoparticles actually grow more slowly with respect to what is observed for other catalysts: after calcination at 700°C, the nanoparticles grow from the original size of 2.6 to about 5 nm. The authors showed that the textural modification induced by impregnation of the aerogel in toluene did not affect the CO conversion and that oxidation activity increased monotonously with metal content. This latter result should indicate that all the Au nanoparticles have the same activity and is therefore ascribed to the good dispersion of the samples. At higher loadings, however, the particle size increased and the reaction rate decreased significantly.

Gold–silica aerogels were also prepared by adding commercial colloidal Au nanoparticles to about-to-gel silica sols prepared under either acid or base catalysis. After gelation, solvent exchange to acetone was performed and finally the aerogels were obtained by *sc-*CO_2 [61]. This procedure turned out to be very general in the preparation of silica-based composite aerogels and is based on the use of the about-to-gel silica sol, which acted as a "nano-glue" in the incorporation of nano- and microparticles into the silica network, and provided therefore to the dispersed phase a high stability toward further processing (exposure to solvents, temperature, and pressure) [62]. By that route, the authors were able to include into the silica network preformed colloidal nanoparticles (Au, Pt) as well as carbon black, titania, and zeolite powders. In particular, by selecting the appropriate colloidal precursor, Rolison et al. were able to immobilize polycrystalline Au nanoparticles ranging in size from 5 to 100 nm into the silica aerogel network [61]. Although the gold loading in the nanocomposite was quite low (0.1%, v/v), two main advantages of this procedure were demonstrated. First of all, the addition of the colloidal nanoparticles to the about-to-gel silica colloidal sol allowed effective gluing of the gold nanoparticles to the silica substrate, and no leaching or damage was observed even under repeated washing or during the *sc-*CO_2 step. In addition, the potential of those gold–silica composites for sensing or catalytic applications was proven by demonstrating the accessibility of the Au surface by means of direct adsorption of dye species onto the gold surface.

The same authors were able to obtain a biocomposite gel by mixing cytochrome *c* with colloidal gold prior to the addition to the about-to-gel colloidal silica [63, 64]. The heme protein self-assembled on the colloidal Au surface (with a ratio of 10,000 protein molecules per gold nanoparticle) forming assemblies of about 100 nm, which were not altered by insertion into the silica matrix or by *sc-*CO_2 *drying*. The activity of cytochrome *c* activity was largely preserved, as shown by the visible absorption band due to the metal–ligand coordination of the folded protein known as Soret band, which had an intensity $\geq$80% of that of the solution protein. It is worth noting that when cytochrome *c*–silica aerogels were prepared, both the protein superstructure and the metal–ligand coordination were damaged, as indicated by the complete disappearance of the Soret band. Indeed, the stabilization of the cytochrome *c* in the aerogel is due to the presence of gold, as specific adsorption of the heme edge onto the gold surface occurs; it is followed by protein self-assembly, which is made possible by the appropriate radius of curvature. Therefore, the authors demonstrated that the design of the appropriate metal–silica nanocomposite aerogel enabled to obtain biocomposite aerogels stable over 6 weeks. The authors applied those biocomposites to gas-phase recognition, as the protein retained its ability to bind nitrogen monoxide. Although the stability of those biocomposites was greatly improved compared to the corresponding xerogels, the response of the biocomposite varied over the ~6 weeks of performance as a consequence of further reactivity of the synthesis by-products. Such limit could not be overcome, as purification steps cannot be performed because they would lead to disruption of the protein self-assembly [63].

Magnetic composites were also prepared by adding microparticulate commercial iron to about-to-gel silica colloid; gelation took place in 5–10 min and after solvent exchange to ethanol and finally to acetone; composite aerogels were obtained by *sc-CO_2 drying* [65]. By performing gelation in a superconducting magnet, aerogels with needle-like field-imposed organization of the microparticles were obtained. As expected, the Fe microparticles could not be aligned by applying the magnetic field after gelation. Moreover, when Fe–SiO_2 composite *xerogels* were prepared by gelation in the presence of an external magnetic field followed by slow drying at ambient conditions over 2 weeks, the resulting composites did show aligned magnetic particles but were extremely fragile and brittle. By removing the needle-like arrangement of oriented Fe particles from aged iron–silica aerogels by dissolution of iron, the authors were able to generate a directional macroporosity in the micrometer regime. Selective removal of a magnetically aligned guest phase provides therefore a means to obtain sol–gel materials with extended directional porosity with a range of perspective uses such as scaffolds for biomolecules. The magnetic properties of the Fe–SiO_2 aerogels were not fully interpreted, as the aerogels unexpectedly did not show any residual macroscopic magnetization once the external field was removed. This result was attributed to the random orientation of the magnetic domains in the composite aerogel.

Magnetic nanocomposites have also been prepared by introducing preformed Fe nanoparticles obtained by fast evaporation of iron pentacarbonyl in a *TMOS* methanol-based sol [66]. The Fe–SiO_2 aerogel was obtained by *sc-HT drying* of the alcogel, obtaining a low-density aerogel with pore diameters in the 20–100 nm range. Structural characterization of the original nanoparticles indicated the presence of 15 nm particles with a Fe core and a ~2 nm layer made out of iron oxide (likely magnetite) clusters. Although no *TEM* or *XRD* characterization of the Fe–SiO_2 aerogel was reported, extensive *SQUID* and Mössbauer spectroscopy indicated that the collective magnetic behavior of the interacting pure nanoparticles and of the nanocomposite were similar. Due to the large relative pore size with respect to the nanoparticle size, aggregates of interacting nanoparticles were formed within the matrix; therefore, the method was not successful in isolating the individual magnetic nanoparticles.

It is important to point out that the additional problem of metal oxidation is very common in the preparation of nanocomposites containing non-noble metals. Moreover, as a consequence of lower stability of the zero-valent state, phase separation may be incomplete and interaction with the matrix can occur.

This was also the case in Co–SiO_2 aerogels [36, 67] prepared by *sc-HT drying* of alcogels obtained by cogelation of *TMOS* and cobalt (II) nitrate under basic catalysis. The supercritical drying conditions promoted the formation of cobalt oxide Co_3O_4, and reduction under a stream of H_2 at 500°C was then performed to promote the formation of metallic Co. Those materials were active toward the Fischer–Tropsch process, exhibiting CO conversions from 5.3 to 19.8% and 22.3% as the metal loading in the catalyst was increased from 2, 6, and 10% (w/w). The relatively low conversion observed in the 10% (w/w) nanocomposite was correlated to its structural properties by a combination of *TEM*, *XRD*, and *SQUID*. *XRD* points to the presence of several peaks superimposed on the amorphous silica background, which were attributed to *fcc*-Co, CoO, and $Co_2(Si_2O_5)_2(OH)_2$. The authors were not able to ascribe all phases to the *TEM* data, although needle-like features with ~1 nm diameter were ascribed to metallic Co. Very similar nanostructures have also been observed in silica-based aerogels containing cobalt and nickel, and thanks to extensive structural investigation they have been attributed to metal silicate hydroxides [31].

As a way to stabilize metallic magnetic phases, metal alloys have been prepared. In particular, iron–cobalt alloys have attracted much attention due to their technological relevance to magnetic storage. In the bulk they are soft magnetic materials with saturation induction 15% larger than pure iron and with magnetization values dependent on their composition.

FeCo–SiO_2 aerogels were obtained by *sc-HT drying* of alcogels prepared by cogelation of *TEOS* and iron and cobalt nitrates under acid catalysis [68, 69]. Formation of the metal phase is obtained by reduction under H_2 flow at temperatures higher than 600°C. The as-obtained aerogels have large surface areas (700–1,000 $m^2\ g^{-1}$) and they are relatively dense as expected for gels obtained under an all-acidic procedure. In particular, by varying the supercritical drying conditions either *microporous* or composite *micro-* and *mesoporous* nanocomposites were obtained, as pointed out by N_2-physisorption measurements. The alloy loading in the nanocomposites was 10% w/w, and the relative iron:cobalt molar ratio was varied from pure Fe to pure Co.

TEM shows that the nanoparticles with diameter ~10 nm were homogeneously distributed within the matrix; *XRD* data indicated that best results were obtained in the *microporous* aerogel where after reduction *fcc*-Co and *bcc*-Fe were formed respectively in the composites containing only one metal. In the FeCo–SiO_2 nanocomposites a *bcc* phase expected for the FeCo alloy was detected by *XRD*, and the main peak shifted according to the expected variation of the cell parameter for the FeCo alloy.

As the *XRD* patterns of FeCo and of Fe were quite similar, definitive assessment of the formation of the alloy was achieved by taking advantage of the selectivity of X-ray absorption spectroscopy in studying independently the Fe and Co environment during the nanocomposite formation [70, 71]. *EXAFS* results show that both Fe and Co are in a *bcc* environment, corroborating therefore that the FeCo alloy is actually formed; moreover, it points out that there is no interaction between the nanoparticles and the matrix. The *EXAFS* investigation of the aerogels prior to the reduction treatment also showed that iron and cobalt were initially present in two separate nanocrystalline phases, Co_3O_4 and ferryhidrite, suggesting that the alloy was formed upon an interdiffusion process.

In order to obtain highly porous FeCo–SiO_2 aerogels, we have made use of a urea-assisted cogelation procedure previously developed for multicomponent oxide nanocomposites [27]. By this route, very low-density nanocomposites with tunable alloy loading and composition were successfully obtained [72], as shown in Figure 16.12. Extensive structural investigation by *EXAFS*, *HREM*, *EDX*, *EELS* has been used to provide compositional information at the nanometer scale for these nanocomposites [73, 74], showing that the nanoparticles were individual *bcc* nanocrystals, that the alloy composition at the nanoscale was very close to the nominal values (see Figures. 16.13 and 16.14), and that the distribution of Fe and Co from *EELS* analysis profiles of nanoparticles was homogeneous with no evidence of oxides within the limit of detection (i.e., 0.2 nm).

It is important to point out that the role of the silica matrix was not limited to the stabilization and isolation of the alloy nanoparticles. In fact, the collective properties of those functional nanocomposites could be tuned as a function of the concentration of the magnetic phase, as well as the relative distance between the nanoparticles, which in turn

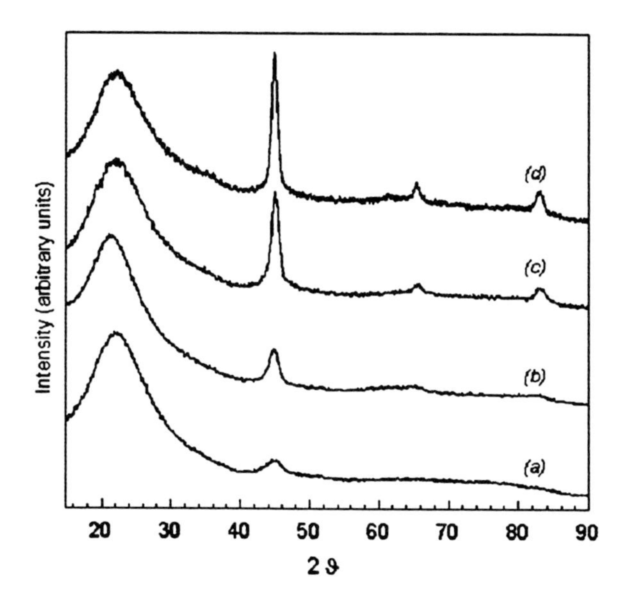

Figure 16.12. XRD patterns of FeCo–SiO_2 nanocomposites with different FeCo loadings: **A.** 3% (w/w), **B.** 5% (w/w), **C.** 8% (w/w), **D.** 10% (w/w) (reproduced from [72] by permission of The Royal Society of Chemistry).

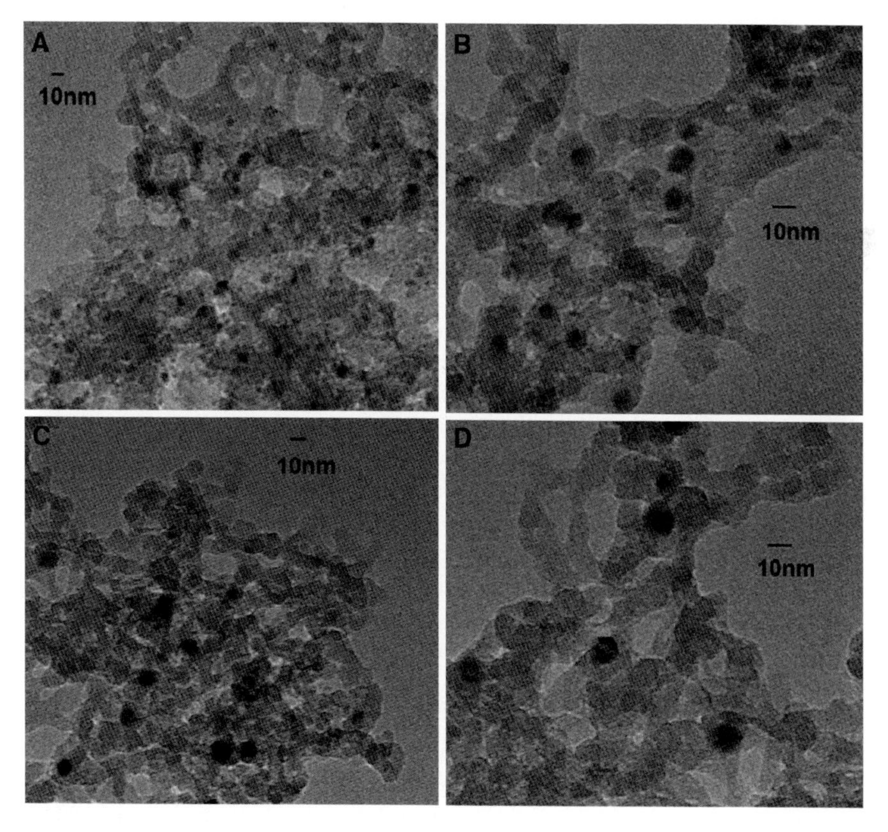

Figure 16.13. BF STEM images of FeCo–SiO_2 nanocomposites with Fe/Co ratio equal to 1 **A**, **B.** and 2 **C**, **D.** (reproduced from [74] by permission of Cambridge University Press).

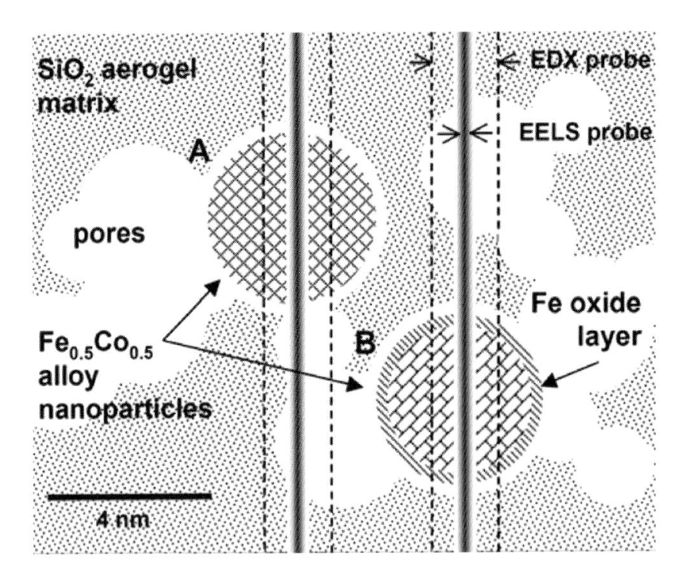

Figure 16.14. Possible scenarios for the microstructure of a 4 nm diameter FeCo alloy nanoparticle: (**A**) nanoparticle consists of a single homogenous alloy phase; (**B**) nanoparticle is covered with a 0.2 nm layer of iron oxide determining a slightly reduced concentration of Fe in the core of the alloy (reproduced from [74] by permission of Cambridge University Press).

determined the kind of interaction that will be present. We have demonstrated this idea by extensive magnetic characterization of relatively dense or highly porous FeCo–SiO_2 nanocomposite aerogels prepared either by the acidic or urea-assisted procedure [72]. The comparison showed that nanocomposites having the same alloy loading, the same alloy composition and similar nanoparticle size but different matrix porosities had very different magnetic behavior. The denser aerogel had a larger blocking temperature, which could be ascribed to the occurrence of stronger interactions related to the closer distance between magnetic centers. Those results are an example of the importance of porosity in the design of magnetic nanocomposites, where interparticle interaction is mediated by the matrix.

Another example on the potential of supported metal alloys for the design of tailored functional materials is provided by Hund et al., who describe a procedure for the preparation of AgAu–SiO_2 nanocomposite aerogels [75]. A silica aquogel prepared from *TEOS* under basic catalysis after repeated solvent exchange was soaked into a solution of $AgNO_3$ and $KAuCl_4$ and then reduced by the addition of formaldehyde under basic conditions. Monolithic aerogels with pore diameters around 7 nm and typical surface areas of 700 $m^2\ g^{-1}$ were obtained by drying under *sc-CO_2* and both *TEM* and spectroscopic investigation confirmed alloying. The authors, however, pointed out the difficulty in obtaining homogeneously loaded gels, and identified kinetic control of the chemical reduction as the key step. Formaldehyde was found to be a suitable reducing agent since its oxidation rate can be adjusted by the addition of alkali, but the size and size distribution control achieved by this room temperature chemical reduction route was poor with respect to other methods. It is noteworthy, however, that by preparing alloys with different molar ratios the optical properties could be tuned, observing the expected shift of the surface plasmon to lower energies as the Au concentration was increased.

Alumina has also been proposed as a suitable matrix for the preparation of functional nanocomposite aerogels. FeCo–Al_2O_3 aerogels were obtained by cogelation of iron and cobalt nitrates and aluminum tri-*sec*-butoxide in ethanol, followed by *sc-HT drying* and reduction under H_2 flow at 700–800°C [76, 77]. The preparation procedure is fast and leads to the formation of highly porous aerogels (density 0.03 $g\,cm^{-3}$) which after reduction exhibit surface areas around 300 $m^2\ g^{-1}$ and *mesopores* with a diameter in the 20–40 nm range. The overall alloy loading in the nanocomposite is 10% (w/w) and the alloy composition is equimolar.

The structural evolution of the nanocomposites upon thermal treatment is very interesting, as formation of the alloy takes place concurrently with crystallization and phase transition of the alumina matrix. Pure Al_2O_3 aerogels obtained by this procedure exhibited a layered pseudo-boehmite structure (in particular, the ethyl derivative of boehmite was formed as a consequence of esterification during supercritical drying), which upon thermal treatment was converted to nanocrystalline γ-Al_2O_3, and finally at 1,000°C, the thermodynamically stable polymorph α-Al_2O_3 started to crystallize.

Aerogels containing both metals exhibited a similar evolution with temperature; however, the boehmite phase formed at low temperatures was more disordered and upon thermal treatment *XRD* showed peaks at the same positions as γ-Al_2O_3, but with different intensity ratios, suggesting that a phase similar to γ-Al_2O_3 was formed with metal ions partially filling the vacancies. By performing the reduction at 700°C or higher, FeCo alloy nanocrystals that dispersed into γ-Al_2O_3 were successfully obtained. A detailed study of the oxidation state and structure of intermediate phases was performed by *EXAFS*, *XANES* [78], and Mössbauer spectroscopy [79]. Results indicated that after reduction treatment most of the metals were in the zero-valent state with a *bcc* environment, whereas some Co(II) and Fe(II, III) were present and located in the tetrahedral vacancies of the γ-Al_2O_3 structure. Those results were not surprising taking into account that γ-Al_2O_3 has a cation-deficient spinel structure, which can easily accommodate metal ions. *SQUID* investigation showed that the FeCo alloy nanoparticles had a superparamagnetic behavior. The magnetic behavior of the nanoparticles could be varied by changing the temperature and the time of the thermal treatment, which affected the average size, the amount, and relative distance of the magnetic nanocrystals.

Chromia has also been used as a support for noble metals in view of catalytic applications, thanks to its rich oxidation chemistry that makes it very suitable for the catalytic reduction of volatile organic pollutants. Pt– and Au–chromia aerogels with a loading of 0.5 and 0.3% (w/w), respectively, were obtained by incipient wetness impregnation of chromia aerogels with H_2PtCl_6 or $HAuCl_4$ and reduction at 250°C [44]. The chromia aerogel, obtained by *sc-HT drying* of wet gels produced by urea-assisted gelation of chromium nitrate, consisted of 1.5–2 nm CrOOH primary particles giving rise to a highly porous network. It is worth noting that those aerogels had a larger surface area (610 $m^2\ g^{-1}$), wider pore diameter distribution (average diameter 24 nm), and larger pore volumes with respect to the corresponding aerogel obtained by *sc-CO_2 drying*. The nanocomposite aerogels had surface areas lower than 500 $m^2\ g^{-1}$ and displayed an increased activity of the chromia support (which was converted to CrO_2 during the catalytic test) toward oxidations by a factor of 1.25–2.7. Pt-based nanocomposites were also prepared by impregnation of chromia aerogels doped by ceria (10%, w/w): the resulting composite aerogels had surface areas ~400 $m^2\ g^{-1}$ and displayed enhanced catalytic performance thanks to the additional effect of Ce, which improved the redox efficiency of the Cr-aerogels by increasing the concentration of oxygen vacancies. The results were interpreted in terms of the textural and surface properties of the different catalysts, rather than in terms of structural/microstructural differences.

As an alternative to inorganic matrices, the use of cellulose has been suggested by Cai et al. as support for silver, gold, and platinum nanoparticles [45]. The nanocomposite aerogels were prepared by *sc-CO_2 drying* of ethanol-exchanged hydrogels obtained by cogelation of the metal precursors and polysaccharides. The glycol groups of the polysaccharides also acted as reducing agents for the metal. In the case of Au and Pt, however, the use of $NaBH_4$ was required to achieve full chemical reduction. By this procedure, low-cost antimicrobial nanocomposites have been obtained with surface areas up to 400 m^2 g^{-1}, which nevertheless had limited processing thermal and chemical stability compared to silica-based aerogels but higher mechanical strength.

16.4. Concluding Remarks

Considerable improvement has been achieved in the design of functional nanocomposites based on aerogel dielectric matrices, in terms of range of accessible compositions, control over nanoparticle size, distribution and loading, and tailored porous texture.

Since every preparation route has specific advantages and drawbacks, it is not possible to define a general strategy for obtaining nanocomposite aerogels. Nevertheless, metal and metal oxide nanoparticles supported on aerogel matrices with given compositional and microstructural features have been effectively prepared.

The flexibility of the sol–gel route in the chemical design of materials opens up possibilities for new applications for aerogels in very diverse fields ranging from thermal and acoustic insulation to catalysis to magnetism.

Acknowledgments

The authors are very grateful to D. Loche, A. Falqui, D. Carta, G. Navarra, and G. Mountjoy for discussion and proof reading.

References

1. Ko I (1998) Aerogels, Kirk-Othmer Encyclopedia of Chemical Technology John Wiley & Sons, 1–3
2. Husing N, Schubert U (1998) Aerogels-Airy Materials: Chemistry, Structure, and Properties. Angew Chem Int Ed 37:22–45
3. Piccaluga G, Corrias A, Ennas G, Musinu A (2000) Sol-Gel Preparation and Characterization of Metal-Silica and Metal Oxide-Silica Nanocomposites. Mater Res Found 13:1–56
4. Baumann TF, Satcher JH Jr, (2003) Homogeneous Incorporation of Metal Nanoparticles into Ordered Macroporous Carbons. Chem Mater 15:3745–3747
5. Baumann TF, Fox GA, Satcher JH Jr Yoshizawa N, Fu R, Dresselhaus MS (2002) Synthesis and Characterization of Copper-Doped Carbon Aerogels. Langmuir 18:7073–7076
6. Casas LI, Roig A, Rodriguez E, Molins E, Tejada J, Sort J (2001) Silica aerogel-iron oxide nanocomposites: structural and magnetic properties. J Non-Cryst Solids 285:37–43
7. Casula MF, Corrias A, Paschina G (2001) Iron oxide-silica aerogel and aerogel nanocomposite materials. J Non-Cryst Solids 293–295:25–31

8. Cannas C, Casula MF, Concas G, Corrias A, Gatteschi D, Falqui A, Musinu A, Sangregorio C, Spano G (2001) Magnetic Properties of γ-Fe_2O_3-SiO_2 Aerogel and Xerogel Nanocomposite Materials. J Mater Chem 11:3180–3187
9. Del Monte F, Morales MP Levy D, Fernandez A, Ocana M, Roig A, Molins E, O'Grady K, Serna CJ (1997) Formation of γ-Fe_2O_3 isolated nanoparticles in a silica matrix. Langmuir 13:3627–3634
10. Casas LI, Roig A, Molins E, Greneche JM, Asenjo J, Tejada J (2002) Iron oxide nanoparticles hosted in silica aerogels. Appl Phys A 74:591–597
11. van Raap MBF, Sanchez FH, Torres CER, Casas L, Roig A, Molins E, (2005) Detailed magnetic dynamic behaviour of nanocomposite iron oxide aerogels. J Phys Condens Matter 17:6519–6531
12. Popovici M, Gich M, Roig A, Casas L, Molins E, Savii C, Becherescu D, Sort J, Surinach S, Munoz JS, Baro MD, Nogues J (2004) Ultraporous single phase iron oxide-silica nanostructured aerogels from ferrous precursors. Langmuir 20:1425–1429
13. Lancok A, Zaveta K, Popovici M, Savii C, Gich M, Roig A, Molins E, Barcova K (2005) Mössbauer studies on ultraporous Fe-Oxide/SiO_2 aerogel. Hyperfine Interact 165:203–208
14. van Raap MBF, Sanchez FH, Leyva AG, Japas ML, Cabanillas E, Troiani H (2007) Synthesis and magnetic properties of iron oxide-silica aerogel nanocomposites. Physica B 398:229–234
15. Fabrizioli P, Burgi T, Burgener M, van Doorslaer S, Baiker A (2002) Synthesis, structural and chemical properties of iron oxide-silica aerogels. J Mater Chem 12:619–630
16. Clapsaddle BJ, Gash AE, Satcher JH, Simpson RL (2003) Silicon oxide in an iron(III) oxide matrix: the sol-gel synthesis and characterization of Fe-Si mixed oxide nanocomposites that contain iron oxide as the major phase. J Non-Cryst Solids 331:190–201
17. Gash AE, Tillotson T.M., Satcher J.H. Jr., Poco J.F., Hrubesh L.W., Simpson R.L. (2001) Use of epoxides in the sol-gel synthesis of porous iron(III) oxide monoliths from Fe(III) salts. Chem Mater 13:999–1007
18. Casula MF, Corrias A, Paschina G (2000) Nickel oxide-silica and nickel-silica aerogel and aerogel nanocomposite materials. J Mater Res 15:2187–2194
19. Mo CM, Li YH, Liu YS, Zhang Y, Zhang LD (1998) Enhancement effect of photoluminescence in assembles of nano-ZnO particles/silica aerogels. J Appl Phys 83:4389–4391
20. Amlouk A, El Mir L, Kraiem S, Alaya S (2006) Elaboration and characterization of TiO_2 nanoparticles incorporated in SiO_2 host matrix. J Phys Chem Solids 67:1464–1468
21. El Mir L, Amlouk A, Barthou C (2006) Visible luminescence of Al_2O_3 nanoparticles embedded in silica glass host matrix. J Phys Chem Solids 67:2395–2399
22. El Mir L, Amlouk A, Barthou C, Alaya S (2008) Luminescence of composites based on oxide aerogels incorporated in silica glass host matrix. Mater Sci Eng C 28:771–776
23. Wei TY, Kuo CY, Hsu YJ, Lu SY, Chang, YC (2008) Tin oxide nanocrystals embedded in silica aerogel: Photoluminescence and photocatalysis. Microporous Mesoporous Mater 112:580–588
24. Kucheyev SO, Biener J, Wang YM, Baumann TF, Wu KJ, van Buuren T, Hamza AV, Satcher JH, Elam JW, Pellin MJ (2005) Atomic layer deposition of ZnO on ultralow-density nanoporous silica aerogel monoliths. Appl Phys Lett 86:083108
25. Yao N, Cao SL, Yeung KL (2009) Mesoporous TiO_2-SiO_2 aerogels with hierarchal pore structures. Microporous Mesoporous Mater 117:570–579
26. Brinker CJ, Lu Y, Sellinger A, Fan H (1999) Evaporation-induced self-assembly: nanostructures made easy. Adv Mater 11:579–585
27. Casula MF, Loche D, Marras S, Paschina G, Corrias A (2007) Role of urea in the preparation of highly porous nanocomposite aerogels. Langmuir 23:3509–3512
28. Casu A, Casula MF, Corrias A, Falqui A, Loche D, Marras S (2007) Magnetic and structural investigation of highly porous $CoFe_2O_4$-SiO_2 nanocomposite aerogels. J Phys Chem C 111:916–922
29. Carta D, Corrias A, Mountjoy G, Navarra G (2007) Structural study of highly porous nanocomposite aerogels. J Non-Cryst Solids 353:1785–1788
30. Carta D, Mountjoy G, Navarra G, Casula MF, Loche D, Marras S, Corrias A (2007) X-ray absorption investigation of the formation of cobalt ferrite nanoparticles in an aerogel silica matrix. J Phys Chem C 111:6308–6317
31. Carta D, Casula MF, Corrias A, Falqui A, Loche D, Mountjoy G, Wang P (2009) Structural and Magnetic Characterization of Co and Ni Silicate Hydroxides in Bulk and in Nanostructures within Silica Aerogels. Chem Mater 21:945–953
32. Loche D, Casula MF, Falqui A, Marras S, Corrias A (2010) Preparation of Mn, Ni, Co ferrite nanocomposite aerogels by an urea-assisted sol-gel procedure. J Nanosci Nanotechnol 10:1008–1016. doi:10.1166/jnn.2010.1907

33. Carta D, Loche D, Mountjoy G, Navarra G, Corrias A (2008) $NiFe_2O_4$ nanoparticles dispersed in an aerogel silica matrix: An X-ray absorption study. J Phys Chem C 112:15623–15630
34. Carta D, Casula MF, Mountjoy G, Corrias A (2008) Formation and cation distribution in supported manganese ferrite nanoparticles: an X-ray absorption study. Phys Chem Chem Phys 10:3108–3117
35. Carta D, Casula MF, Falqui A, Loche D, Mountjoy G, Sangregorio C, Corrias A (2009) A Structural and Magnetic Investigation of the Inversion Degree in Ferrite Nanocrystals MFe_2O_4 (M = Mn, Co, Ni). J Phys Chem C 113:8606–8615
36. Dutta P, Dunn BC, Eyring EM, Shah N, Huffman GP, Manivannan A, Seehra S (2005) Characteristics of cobalt nanoneedles in 10% Co/Aerogel fischer-tropsch catalyst. Chem Mater 17:5183–5186
37. Leventis N, Chandrasekaran N, Sadekar AG, Sotiriou-Leventis C, Lu HB (2009) One-Pot Synthesis of Interpenetrating Inorganic/Organic Networks of CuO/Resorcinol-Formaldehyde Aerogels: Nanostructured Energetic Materials. J Am Chem Soc 131:4576–4577
38. Al-Mutaseb SA, Ritter JA (2003) Preparation and Properties of Resorcinol-Formaldehyde Organic and Carbon Gels. Adv Mater 15:101–114
39. Leventis N, Chandrasekaran N, Sotirou-Leventis C, Mumtaz A (2009) Smelting in the age of nano: iron aerogels. J Mater Chem 19:63–65
40. Balkis Ameen K, Rajasekar K, Rajasekharan T (2007) Silver nanoparticles in mesoporous aerogel exhibiting selective catalytic oxidation of benzene in CO_2 free air. Catal Lett 119:289–295
41. Tai Y, Murakami J, Tajiri K, Ohashi F, Date M, Tsubota S (2004) Oxidation of carbon monoxide on Au nanoparticles in titania and titania-coated silica aerogels. Appl Catal A 268:183–187
42. Anderson K, Fernandez SC, Hardacre C, Marr PC (2004) Preparation of nanoparticulate metal catalysts in porous supports using an ionic liquid route; hydrogenation and C-C coupling. Inorg Chem Comm 7:73–76
43. Martinez S, Moreno-Manas M, Vallribera A, Schubert U, Roig A, Molins E (2006) Highly dispersed nickel and palladium nanoparticle silica aerogels: sol-gel processing of tethered metal complexes and application as catalysts in the Mizoroki-Heck reaction. New J Chem 30:1093–1097
44. Rotter H, Landau MV, Carrera M, Goldfarb D, Herskowitz M (2004) High surface area chromia aerogel efficient catalyst and catalyst support for ethylacetate combustion. Appl Catal B 47:111–126
45. Cai J, Kimura S, Wada M, Kuga S (2009) Nanoporous Cellulose as Metal Nanoparticles Support. Biomacromolecules 10:87–94
46. Ameen KB, Rajasekharan T, Rajasekharan MV (2006) Grain size dependence of physico-optical properties of nanometallic silver in silica aerogel matrix. J Non-Cryst Sol 352:737–746
47. Ayers MR, Song XY, Hunt AJ (1996) Preparation of nanocomposite materials containing WS_2, δ-WN,Fe_3O_4, or Fe_9S_{10} in a silica aerogel host. J Mater Sci 31:6251–6257
48. Biener J, Baumann TF, Wang YM, Nelson EJ, Kucheyev SO, Hamza AV, Kemell M, Ritala M, Leskela M (2007) Ruthenium/aerogel nanocomposites via atomic layer deposition. Nanotechnology 18:055303
49. Kuthirummal N, Dean A, Yao C, Risen W (2008) Photo-formation of gold nanoparticles: Photoacoustic studies on solid monoliths of Au(III)-chitosan-silica aerogels. Spectrochim Acta, Part A 70:700–703
50. Morley KS, Marr PC, Webb PB, Berry AR, Allison FJ, Moldovan G, Brown PD, Howdle SM (2002) Clean preparation of nanoparticulate metals in porous supports: a supercritical route. J Mater Chem 12:1898–1905
51. Morley KS, Licence P, Marr PC, Hyde JR, Brown PD, Mokaya R, Xia YD, Howdle SM (2004) Supercritical fluids: A route to palladium-aerogel nanocomposites. J Mater Chem 14:1212–1217
52. Zhang Y, Kang DF, Saquing C, Aindow M, Erkey, C (2005) Supported platinum nanoparticles by supercritical deposition. Ind Eng Chem Res 44:4161–4164
53. Martinez S, Vallribera A, Cotet CL, Popovici M, Martin L, Roig A, Moreno-Manas M, Molins E (2005) Nanosized metallic particles embedded in silica and carbon aerogels as catalysts in the Mizoroki-Heck coupling reaction. New J Chem 29:1342–1345
54. Moerke W, Lamber R, Schubert U, Breitscheidel B (1994) Metal Complexes in Inorganic Matrixes. 11. Composition of Highly Dispersed Bimetallic Ni, Pd Alloy Particles Prepared by Sol-Gel Processing: Electron Microscopy and FMR Study. Chem Mater 6:1659–1666
55. Dai S, Ju YU, Gao HJ, Lin JS, Pennycook SJ, Barnes CE (2000) Preparation of silica aerogel using ionic liquids as solvents. Chem Commun 243–244
56. Smith DD, Sibille L, Cronise RJ, Noever DA (1998) Surface plasmon resonance evaluation of colloidal silver aerogel filters. J Non-Cryst Solids 225:330–334
57. Tai Y, Watanabe M, Murakami J, Tajiri K (2007) Composite formation of thiol-capped Au nanoparticles and mesoporous silica prepared by a sol-gel method. J Mater Sci 42:1285–1292

58. Creighton JA, Blatchford CG, Albrecht MG (1979) Plasma resonance enhancement of Raman scattering by pyridine adsorbed on silver or gold sol particles of size comparable to the excitation wavelength. J Chem Soc Faraday Trans 2:790–798
59. Brust M, Walker M, Bethell D, Sciffrin DJ, Whyman R (1994) Synthesis of Thiol Derivatised Gold Nanoparticles in a Two Phase Liquid/Liquid System. J Chem Soc Chem Commun 801–802
60. Tai Y, Tajiri K (2008) Preparation, thermal stability, and CO oxidation activity of highly loaded Au/titania-coated silica aerogel catalysts. Appl Catal A 342:113–118
61. Anderson M L, Morris C A, Stroud R M, Merzbacher C I, Rolison D R (1999) Colloidal gold aerogels: Preparation, properties, and characterization. Langmuir 15:674–681
62. Morris C A, Anderson M L, Stroud R M, Merzbacher C I, Rolison D R (1999) Silica sol as a nanoglue: Flexible synthesis of composite aerogels. Science 284:622–624
63. Wallace J M, Stroud R M, Pietron J J, Long J W, Rolison D R (2004) The effect of particle size and protein content on nanoparticle-gold-nucleated cytochrome c superstructures encapsulated in silica nanoarchitectures. J Non-Cryst Solids 350:31–38
64. Wallace J M, Rice J K, Pietron J J, Stroud R M, Long J W, Rolison D R (2003) Silica nanoarchitectures incorporating self-organized protein superstructures with gas-phase bioactivity (their is also a mistake for the pages for this reference. NanoLett 3:1463–1467
65. Leventis N, Elder I A, Long G J, Rolison, D R (2002) Using nanoscopic hosts, magnetic guests, and field alignment to create anisotropic composite gels and aerogels. NanoLett 2:63–67
66. Racka K, Gich M, Slawska-Waniewska A, Roig A, Molins E (2005) Magnetic properties of Fe nanoparticle systems. J Magn Magn Mater 290:127–130
67. Dunn BC, Cole P, Covington D, Webster MC, Pugmire RJ, Ernst RD, Eyring EM, Shah N, Huffman GP (2005) Silica aerogel supported catalysts for Fischer-Tropsch synthesis. Appl Catal A 278:233–238
68. Casula MF, Corrias A, Paschina G (2003) Iron-cobalt-silica aerogel nanocomposite materials. J Sol-Gel Sci Technol 26:667–670
69. Casula MF, Corrias A, Paschina G (2002) FeCo-SiO_2 nanocomposite aerogels by high temperature supercritical drying. J Mater Chem 12:1505–1510
70. Corrias A, Casula MF, Ennas G, Marras S, Navarra G, Mountjoy G (2003) X-ray absorption spectroscopy study of FeCo-SiO_2 nanocomposites prepared by the sol-gel method. J Phys Chem B 107:3030–3039
71. Casula MF, Corrias A, Navarra G (2003) An EXAFS study on iron-cobalt-silica nanocomposite materials prepared by the sol-gel method. J Sol-Gel Sci Technol 26:453–456
72. Casu A, Casula MF, Corrias A, Falqui A, Loche D, Marras S, Sangregorio C (2008) The influence of composition and porosity on the magnetic properties of FeCo-SiO_2 nanocomposite aerogels. Phys Chem Chem Phys 10:1043–1052
73. Carta D, Mountjoy G, Gass M, Navarra G, Casula MF, Corrias A (2007) Structural characterization study of FeCo alloy nanoparticles in a highly porous aerogel silica matrix. J Chem Phys 127: 204705
74. Falqui A, Corrias A, Gass M, Mountjoy G (2009) A Transmission Electron Microscopy Study of Fe-Co Alloy Nanoparticles in Silica Aerogel Matrix Using HREM, EDX, and EELS. Micros Microan 15:114–124
75. Hund JF, Bertino MF, Zhang G, Sotiriou-Leventis C, Leventis N (2004) Synthesis of homogeneous alloy metal nanoparticles in silica aerogels. J Non-Cryst Solids 350:9–13
76. Corrias A, Casula MF, Falqui A, Paschina G (2004) Preparation and characterization of FeCo-Al_2O_3 and Al_2O_3 aerogels. J Sol-Gel Sci Technol 31:83–86
77. Corrias A, Casula MF, Falqui A, Paschina G (2004) Evolution of the structure and magnetic properties of FeCo nanoparticles in an alumina aerogel matrix. Chem Mater 16:3130–3138
78. Corrias A, Navarra G, Casula MF, Marras S, Mountjoy, G (2005) An X-ray absorption spectroscopy investigation of the formation of FeCo alloy nanoparticles in Al_2O_3 xerogel and aerogel matrixes. J Phys Chem B 109:13964–13970
79. Casula MF, Concas G, Congiu F, Corrias A, Falqui A, Spano G (2005) Near equiatomic FeCo nanocrystalline alloy embedded in an alumina aerogel matrix: Microstructural features and related magnetic properties. J Phys Chem B 109:23888–23895

Part VI

Materials and Processing: Exotic Aerogels

17

Chalcogenide Aerogels

Stephanie L. Brock and Hongtao Yu

Abstract A new class of aerogels based exclusively on *metal chalcogenide* frameworks has recently been developed, opening up a range of exciting properties and applications not encompassed by their oxide brethren. The optical semiconducting properties are tunable over a wide range from the *UV* through to the *IR* depending on the chemical composition, and gels prepared from nanoparticle assembly exhibit the characteristic *quantum confinement* effects of their nanoparticle building blocks. The soft *Lewis* basic characteristics of the framework and the presence of an interconnected pore-network result in unique sorption properties that may be suitable for environmental *remediation* or gas-separation. This chapter presents a detailed description of the advances in chalcogenide aerogels since they were initially reported in 2004, focusing on the different methods of synthesis developed and the consequent physicochemical properties.

17.1. Introduction

Traditionally, aerogels have been compositionally limited to oxide and carbon/organic frameworks, or composites of the two, limiting the range of properties that can be accessed in these unique architectures (Chap. 1). Motivated by the physicochemical properties of metal chalcogenides (i.e., S^{2-}, Se^{2-}, or Te^{2-}), which are distinct from oxides and carbon/organics and include direct band-gap semiconductivity that spans the solar spectrum, accessible redox states that facilitate catalysis, and soft *Lewis* basicity, a recent effort has been mounted to create aerogels based on chalcogenide frameworks. Although *chalcogenide aerogels* were not reported until 2004, the compositional scope and properties discovered in the past few years suggest that this is a fertile area for study. Coupling of the metal chalcogenide properties to the high-surface area interconnected porosity that is the aerogel framework is promising for a variety of applications, including photo-activated processes (solar cells, photocatalysis, sensing, etc.), the catalytic removal of sulfur impurities from fossil fuels (hydrodesulfurization), environmental *remediation* of heavy metals, and gas separation, among others. To date, there are three different approaches to the formation of

---

S. L. Brock and H. Yu • Department of Chemistry, Wayne State University, 5101 Cass Avenue, Detroit, MI 48202, USA
e-mail: sbrock@chem.wayne.edu; hongtao.yu@asu.edu

M.A. Aegerter et al. (eds.), *Aerogels Handbook*, Advances in Sol-Gel Derived Materials and Technologies, DOI 10.1007/978-1-4419-7589-8_17,

aerogels (1) *thiolysis reactions* of molecular metal precursors; (2) *condensation reactions* between small negatively charged metal chalcogenide clusters and linking cations; and (3) condensation reactions of discrete metal chalcogenide nanoparticles. In this chapter, we describe the state of the science in metal chalcogenide aerogels, as it currently stands, as a function of these three approaches, and then discuss the future prospects of these unique materials. The reader is also referred to several excellent reviews in this area [1–3].

17.2. Thiolysis Routes to Chalcogenide Aerogels: GeS_2

As a counterpoint synthetic pathway to the traditional *hydrolysis/condensation* route to silica and metal oxide gels, *thiolysis* can be applied to generate a number of metal sulfide gels. In this approach, H_2S gas is employed instead of H_2O, leading to sulfur-linked gels (Figure 17.1). Similar to metal oxide gels, the final product morphology of metal sulfides can be either a precipitate or a gel structure, depending on the relative reaction kinetics of *thiolysis* and condensation. Careful choice of the *precursor* and tuning of the reaction kinetics can yield a number of metal sulfide gels, including TiS_2 [4, 5], NbS_2 [6], YS_x [7], LaS_2 [8], WS_x [9], ZnS [10], and GeS_x [11–13]. However, work to date has been largely motivated by formation of dense monoliths or thin films; thus little attention has been paid to strategies for generating gels and retaining their pore structures during drying. To test if the thiolyis route is amenable to aerogel formation, Brock and co-workers targeted GeS_x compositions [14]. This system was chosen because it is known to yield monolithic xerogels with high surface areas (ca 500 m^2/g), suggesting the presence of a porous network.

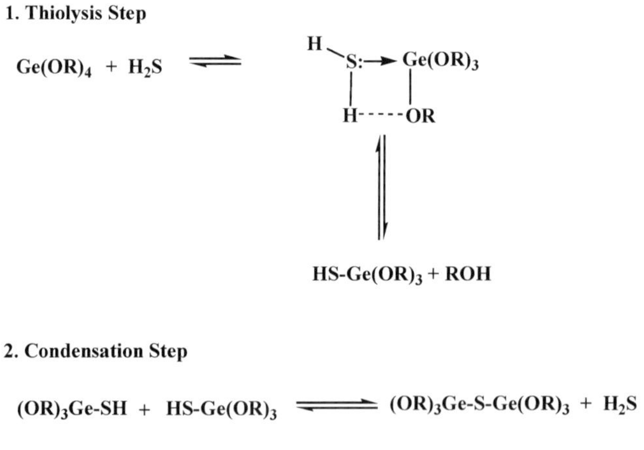

1. Thiolysis Step

$$Ge(OR)_4 + H_2S \rightleftharpoons H\!-\!\ddot{S}\!\rightarrow Ge(OR)_3 \cdots H \cdots OR \rightleftharpoons HS\text{-}Ge(OR)_3 + ROH$$

2. Condensation Step

$$(OR)_3Ge\text{-}SH + HS\text{-}Ge(OR)_3 \rightleftharpoons (OR)_3Ge\text{-}S\text{-}Ge(OR)_3 + H_2S$$

Figure 17.1. Illustration of the thiolysis and condensation steps that give rise to Ge–S–Ge bond formation, and subsequent formation of GeS_x gels. Reproduced with permission from J. Non Cryst. Solids, 352: 232–240 (2006). Copyright 2006 Elsevier.

Reaction of $Ge(OEt)_4$ with H_2S under strictly inert conditions led to gels over a period of hours to days, and cold (CO_2) supercritical drying afforded powdered white aerogels of composition $GeS_{2.4}$ consistent with a sulfur-rich GeS_2 glass. Data from powder X-ray diffraction confirm the amorphous (i.e., glassy) nature of the aerogel, and transmission electron microscopy (*TEM*) and surface area analysis show a porous interconnected colloidal network structure (similar to base-catalyzed silica aerogels) with a Brunauer–Emmett–Teller (*BET*) surface area (up to 755 m^2/g) that is augmented compared to the originally made GeS_x xerogels. Although it is clear that there are extensive opportunities to create new aerogels using this approach, the method is limited by difficulties in handling and syntheses, with a key issue being the need for strict elimination of oxygen and water from the reaction. Indeed, when wet GeS_x gels were exposed to adventitious air during gelation and aging, the resultant aerogels were sulfur-poor and contained crystalline GeO_2. The stability problems are limited to the wet gel stage; once dried, GeS_x gels can be handled in air without problems.

17.3. Cluster-Linking Routes to Chalcogenide Aerogels

In 2007, Kanatzidis and co-workers reported a new route to metal chalcogenide gels (*chalcogels*) by linking anionic *Zintl* clusters, such as $[Ge_4S_{10}]^{4-}$, with transition metal cations, such as Pt^{2+} (see Figure 17.2) in a *metathesis* (partner-switching) *reaction* (17.1); *supercritical drying* with CO_2 yielded the corresponding aerogels [15]. Previously, the combination of chalcogenido *Zintl* clusters with metal ions was used to synthesize mesostructured chalcogenides, with *surfactant* structures serving as *templates* [16–18]. However, attempts to remove the *surfactants* by washing or heating resulted in the collapse of the internal pore structure. In contrast, the new "sol–gel" method yields disordered structures (aerogels) but with accessible pore volume. Importantly, the synthetic approach has considerable flexibility, owing to the diverse building blocks that can be used. The different geometry of the chalcogenido clusters and preferred coordination environment of the metal impact the physicochemical characteristics of this new class of chalcogenide aerogels, such as morphology and structure, surface area and porosity, and optical, catalytic, and gas-sorption properties. To date, Kanatzidis and co-workers have considered two main systems: main group chalcogenido clusters (see Figure 17.2) *cross-linked* with Pt^{2+} in aqueous media and MS_4^{2-} (M = Mo, W) *cross-linked* with Ni^{2+} or Co^{2+} (17.2) in formamide.

$$(Me_4N)_4[Ge_4S_{10}] + 2K_2PtCl_4 \rightarrow Pt_2[Ge_4S_{10}] + 4KCl + 4Me_4NCl \qquad (17.1)$$

$$(NH_4)_2MoS_4 + Ni(NO_3)_2 \cdot 6H_2O \rightarrow NiMoS_4 + 2NH_4NO_3 + 6H_2O \qquad (17.2)$$

17.3.1. Aerogels from Main Group Chalcogenide Clusters and Pt^{2+}

The chalcogenide framework of the aerogels prepared from main group *Zintl* clusters and Pt^{2+} has a colloidal structure and a disordered pore system, similar to that seen in base-catalyzed silica, as exemplified by $Pt_2[Ge_4S_{10}]$ (Figure 17.3) [15]. Like the GeS_x aerogels discussed in Sect. 17.2, the solid framework is amorphous. However, atomic pair distribution function (*PDF*) analysis of the Bragg and diffuse scattering of $Pt_2[Ge_4S_{10}]$ does show a degree of local order in the gel, and correlation lengths of 3.5 Å are consistent with the presence of intact $Ge_4S_{10}^{4-}$ chalcogenido clusters linked to square-planar Pt^{2+} ions (Figure 17.2) [15]. The physicochemical properties of the chalcogenide aerogels vary considerably depending on the nature of the *Zintl* ion cluster. In contrast to conventional metal oxide aerogels, the bandgaps of these chalcogenide aerogels fall mainly in the visible

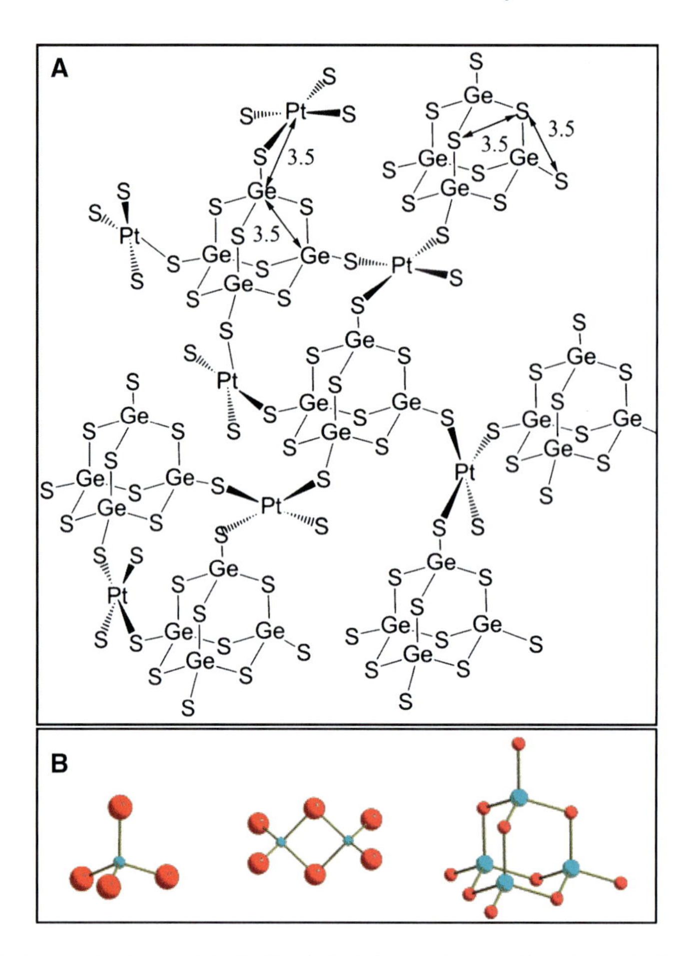

Figure 17.2. **A.** Proposed structure of the $Pt_2[Ge_4S_{10}]$ chalcogenide aerogel based on pair distribution function (PDF) analysis of X-ray data. Relevant inter-atomic distances are indicated. **B.** Different types of Zintl clusters (*teal balls* = Ge, Sn; *red balls* = S, Se) employed in chalcogel formation with Pt^{2+}. Reproduced with permission from Science, 317: 490–493 (2007). Copyright 2007, American Association for the Advancement of Science.

to infrared spectrum, rather than the UV. In the present case, this varies from 0.2 eV (Pt_2SnSe_4) to 2.0 eV ($Pt_2[Ge_4S_{10}]$) [15]. The corresponding surface areas are large, particularly when the large formula mass of the chalcogenide aerogels is considered, and range from ca 100 to 325 m^2/g [15]. Indeed, it is the large surface area combined with the soft *Lewis* basic character of the framework that results in one of the intriguing characteristics of these systems: their ability to selectively sorb heavy metal ions, as demonstrated with Hg^{2+}, even in the presence of a comparable concentration of lighter ions, Zn^{2+} [15]. This is shown schematically in Figure 17.4 [1]. The selectivity can be attributed to the chalcophilicity of the heavy ions, which prefer to bind to softer *Lewis* bases (sulfur, selenium) over hard *Lewis* bases (oxygen). As Zn^{2+} is less chalcophilic, it is not as strongly sorbed.

Classic sorption materials are oxide-based (e.g., zeolites), but these are not effective in removal of heavy, polarizable ions. In order to achieve similar distribution coefficients, K_d^{Hg} (ppm adsorbed Hg^{2+} per gram adsorbent over ppm Hg^{2+} remaining per milliliter solution), to the chalcogenide aerogels (K_d^{Hg} ~10^7 mL/g), analogous mesoporous silica phases must be

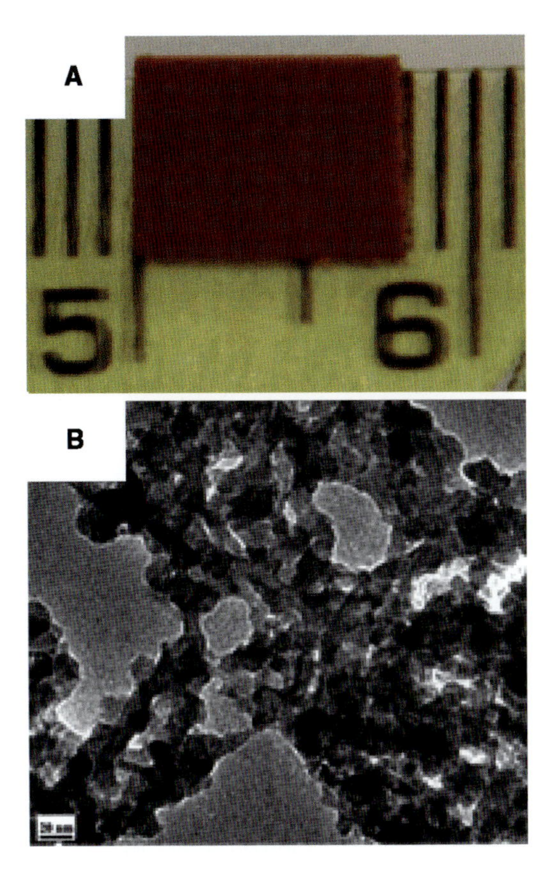

Figure 17.3. Two views of a $Pt_2[Ge_4S_{10}]$ chalcogenide aerogel **A.** macroscopic view of a monolith; the *pinkish brown color* is associated with a band gap of 2.0 eV; **B.** microscopic view of the framework acquired by electron microscopy showing the colloidal framework and the presence of meso- and macropores. Reproduced with permission from Science, 317: 490–493 (2007). Copyright 2007, American Association for the Advancement of Science.

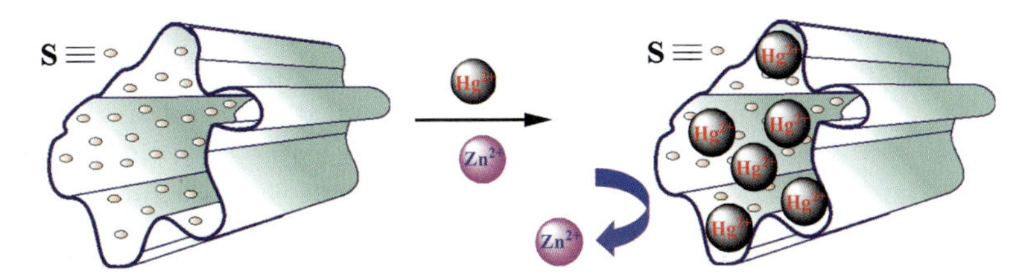

Figure 17.4. Scheme showing how heavy metals (Hg^{2+}) are selectively adsorbed over lighter metals (Zn^{2+}) due to the chalcogenide (S) sites making up the pores of the chalcogenide aerogel; 99.9% Hg^{2+}, at the expense of only 40% Zn^{2+}, can be effectively removed from starting solution concentrations of ca 90 ppm each. Reproduced with permission from J. Mater. Chem., 18: 3628–3632 (2008). Copyright 2008 Royal Society of Chemistry.

thiolated. Intriguingly, the sorption capacity is not limited to metal ions; chalcogenide aerogels readily sorb *hydrophobic* aromatic organic molecules as has been demonstrated using porphyrins [15]. These data suggest possible applications of chalcogenide aerogels in environmental *remediation*.

17.3.2. Aerogels from MS_4^{2-} (M = Mo, W) Ions and Ni^{2+} (Co^{2+})

Similar to the *Zintl* ion/Pt^{2+} system (Sect. 17.3.1), aerogels prepared from crosslinking of MoS_4^{2-} or WS_4^{2-} with Ni^{2+} and/or Co^{2+} have a colloidal morphology, high surface areas (340–530 m^2/g), an amorphous framework, and are effective sorbents for Hg^{2+} and porphyrins [19]. Moreover, the highly polarizable surface and large surface areas also make these chalcogenide aerogels good media for gas separation [20]. Depending on their polarizability, guest molecules are expected to interact with the aerogel surface differently. Indeed, Kanatzidis and co-workers found that CO_2 exhibits a dramatic preference over H_2 in the interaction with the chalcogenide aerogel surface (Figure 17.5). This suggests these materials might be effective for separation of H_2 from CO_2, an important process in hydrogen generation via the water–gas shift reaction (17.3).

$$CO + H_2O \rightarrow H_2 + CO_2 \tag{17.3}$$

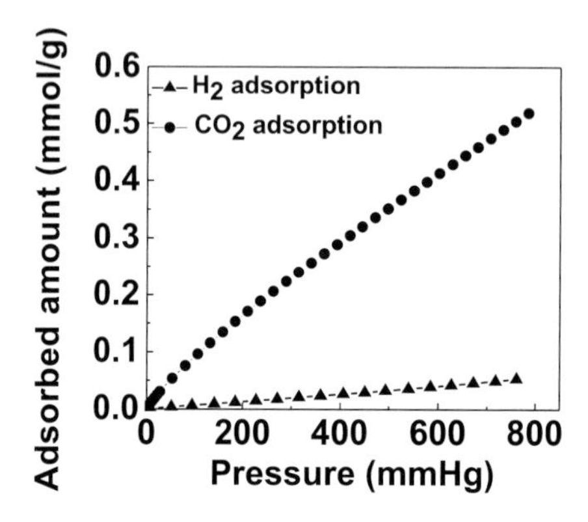

Figure 17.5. Plot showing adsorption isotherms of CO_2 and H_2 at 273 K on a $CoMoS_4$ aerogel with a surface area of 340 m^2/g. The chalcogenide aerogel shows a selectivity of CO_2 over H_2 of nearly 16 times. Reproduced with permission from Nat. Chem., 1: 217–224 (2009). Copyright 2009 Macmillan Publishers Ltd.

One unique attribute of the Mo sulfide materials relative to the main-group cluster analogs is the *redox activity* that accompanies gelation. *PDF*, *XPS*, and *IR* data all suggest that reduction of Mo has occurred from Mo^{6+} in the *precursor* to an oxidation state of 5+/4+ in the Ni^{2+} (or Co^{2+}) linked *chalcogel*, and this is accompanied by formation of polysulfides ($nS^{2-} \rightarrow S_n^{2-} + ne^-$) [1]. The facile redox in these systems is characteristic of Co/Ni/MoS_x composites and one of the reasons they are such effective catalysts for hydrodesulfurization (*HDS*, removal of sulfur compounds from fossil fuels). A comparison of *HDS* activity of an unsupported Co-Mo sulfide aerogel with a traditional Co-promoted sulfided molybdenum catalyst supported on Al_2O_3 showed that the aerogel exhibited over twice the activity of the conventional catalyst and good stability over time [19].

17.4. Nanoparticle Assembly Routes to Chalcogenide Aerogels

An alternate approach to chalcogenide aerogel synthesis is to start with discrete nanoparticles of metal chalcogenides and condense them together to form a gel network. A method for doing just that was reported in 1997 by Gacoin and coworkers [21–25].

Concentrated solutions of CdS nanoparticles, surface modified with 4-fluorophenylthiolate, were found to form gels upon standing for long periods of time. Gacoin and co-workers determined that the process involved air oxidation of the surface *thiolate groups*, forming disulfides (and other oxidized products), and that the partially decomplexed nanoparticles would then form spanning aggregates, and eventually, gels (Figure 17.6). The process could be controllably reproduced by using hydrogen peroxide in lieu of atmospheric oxygen. The gels themselves were found to retain the size-dependent optical properties of the semiconducting nanoparticle building block (i.e., they remained quantum confined), with similar bandgaps to those of the initial particles.

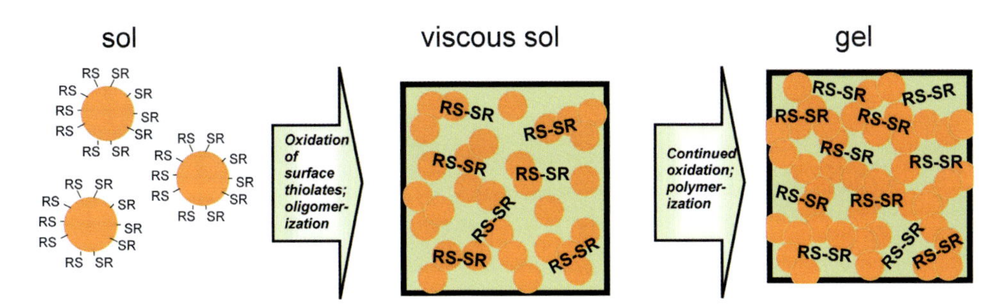

Figure 17.6. Scheme showing the nanoparticle assembly route for gel formation, induced by oxidation of surface thiolate capping groups. Reproduced with permission from Acc. Chem. Res., 40: 801–809 (2007). Copyright 2007, American Chemical Society.

17.4.1. CdS Aerogels

In 2004, Brock and co-workers reported that CdS wet gels prepared á la Gacoin could be transformed into aerogels by CO_2 *supercritical drying* [26], and in 2005, this method was extended to PbS, ZnS, and CdSe [27]. As shown in Figure 17.7A, the effect of *supercritical drying* on the gel structure is quite dramatic; xerogels formed from evaporation of acetone-exchanged wet gels at ambient temperatures and pressures exhibited significant shrinkage due to pore collapse induced by capillary forces, whereas monoliths prepared from *cold supercritical drying* of identically processed wet gels retained most of the volume of the starting gel. This observation correlates nicely with experimentally determined surface areas; the aerogel surface area is five times larger than that of the xerogel (245 vs. 47 m^2/g, respectively). Electron microscopy shows that the aerogel structure is colloidal, with pores clearly evident, similar to a base-catalyzed silica aerogel (Figure 17.7B). Unlike silica, however, the chalcogenide nanoparticles remain crystalline, with very little growth apparent when crystallite size is evaluated by Scherrer analysis of X-ray powder diffraction data. Crystallinity and faceting can also be noted in high-resolution *TEM* images of the aerogel, particularly those subjected to thermal treatment at 100°C (Figure 17.7C). The optical properties of the aerogels are a sensitive function of the size of the nanoscale building block, and can be tuned by heat treatment [26, 27], or by using a different size of particle *precursor*, consistent with quantum confinement effects [28]. The quantum confinement effect for nanoscale semiconductors is

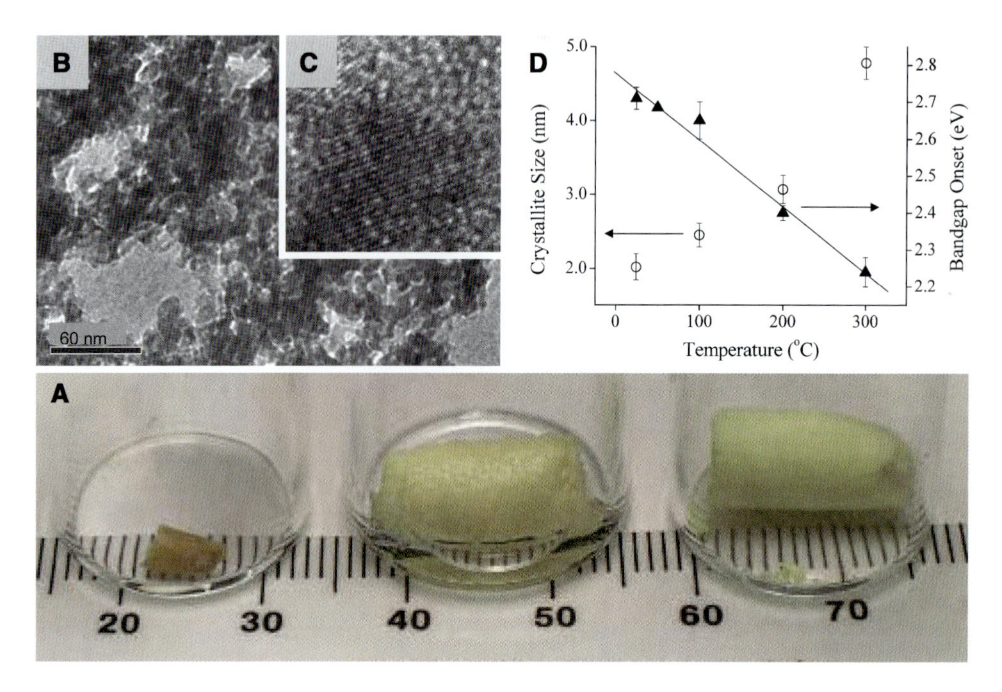

Figure 17.7. A. Effect of drying of CdS wet gels (**center**) on monolith density: benchtop drying (xerogel, **left**) versus supercritical drying (aerogel, **right**); **B.** electron micrograph showing the colloidal morphology of CdS aerogels; **C.** electron micrograph showing hexagonal faceting and lattice fringes in CdS aerogels heated at 100°C; **D.** relationship between bandgap and particle size in temperature-processed CdS aerogels. Reproduced with permission from Science, 307: 397–400 (2005). Copyright 2005 American Association for the Advancement of Science.

characterized by an increased bandgap for materials that are spatially limited in one or more dimensions to a size less than the Bohr radius of the electron–hole pair (exciton). The Bohr excitonic radius is material dependent, but typically on the order of 2–20 nm, hence the interest in semiconducting "nano" materials. It was found that heating aerogels of CdS led to crystallite growth, a process that could be followed by the breadth of the diffraction peaks in the *PXRD*. This growth, in turn, correlates with a systematic decrease in the observed bandgap (Figure 17.7D), as well as a decrease in surface area from ca 240 to 140 m^2/g.

Modest heating (100°C) of CdS aerogel enables one of the other characteristic features of semiconductor nanoparticles to be recovered: band-edge luminescence. Photoluminescence (*PL*) at the bandgap energy is predicated on an absence of defect states that enable electrons and holes to be recombined nonradiatively. In the case of nanocrystals, these can be due to poor crystallinity within the particle, but are often on the surface of the particle where terminating ligands are present. Thiolates are particularly good hole-trap agents for CdS, and the native nanoparticles exhibit only a modest band-edge *PL* peak, with a large broad peak to the red due to the surface trap states. As-prepared aerogels of CdS have *PL* spectra dominated by the trap-state. However, if heated to 100°C, some intensity can be recovered at the band-edge [28].

17.4.2. Application of the Nanoparticle Assembly Route to PbS, ZnS and CdSe: Effect of Oxidant on CdSe Gelation

The nanoparticle assembly route to *chalcogels* is not limited to CdS. In 2005, Mohanan et al. showed that this could be extended to other sulfides, zinc and lead, and even to selenides, CdSe [27]. This extended the bandgap range accessed by the aerogels throughout the visible (CdS, CdSe) and into the infrared (PbS) and UV (ZnS), as shown in Figure 17.8.

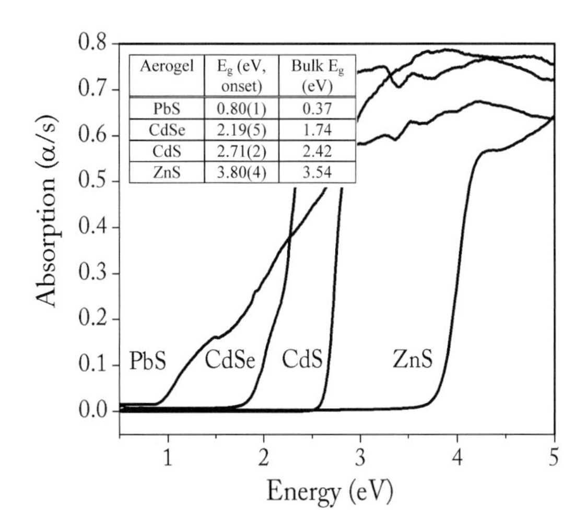

Figure 17.8. Absorbance spectra (converted from solid state reflectance) and band-gap data for aerogels of PbS, CdSe, CdS, and ZnS. Reproduced with permission from Science, 307: 397–400 (2005). Copyright 2005 American Association for the Advancement of Science.

In studying CdSe, it was discovered that H_2O_2 is not an innocent oxidizing agent, as it appears to be for the sulfides. Treatment of red CdSe sols with hydrogen peroxide did result in gel formation, but this was accompanied by a color change to bright yellow. Moreover, after *supercritical drying*, *PXRD* showed the presence of crystalline $CdCO_3$, suggesting CO_2 insertion into Cd–O bonds. It was postulated that this occurred via exchange of surface *selenols* (Cd–SeH) with hydroxyls (Cd–OH) produced as a byproduct of H_2O_2 oxidation. To test the possible role of hydroxide in the color change and subsequent formation of $CdCO_3$, yellow gels were treated with dilute solutions of HCl, which resulted in a color change back to the original red-orange of the sol. Upon *supercritical drying*, the *PXRD* showed only peaks from CdSe. We postulated that the interference of hydroxide was greater for selenides than sulfides because of the relative energy differences of the Cd–O, Cd–S, and Cd–Se bonds. Cd–O is the thermodynamically most stable, and Cd–Se the least. It would appear that at equilibrium there is far more selenol exchange than thiol exchange for the hydroxyl groups.

As an alternate to acid washing, we sought oxidants that could be active without forming hydroxide. Tetranitromethane is a powerful nonoxygen transferring oxidant, producing only nitrites as by-products. When used, no color change is observed during the sol-to-gel transition, nor is there evidence of carbonate formation after *supercritical drying*. Additionally, it was found that sols left under ambient lighting would also form gels, although the rate of gel formation was much slower. This was attributed to a photooxidation effect in the presence of oxygen. Thus, gelation can be achieved with a wide range of oxidative stimuli.

17.4.3. Influence of Density and Dimensionality on Quantum Confinement Effects

Because gels are a unique class of inorganic polymer with a low density, asymmetric, microscopically porous network of noninteger *fractal dimensionality*, the interaction between quantum dots, and hence the extent of *quantum confinement* in the gel network, is expected to be governed by the network density and dimensionality. This is evident when the optical properties of corresponding xerogels and aerogels are compared [29]. In all cases, xerogels exhibited smaller bandgaps than did their aerogel counterparts (Figure 17.9A). This is attributed to an increase in the dimensionality of interaction in the more dense xerogel, leading to a decrease in the extent of *quantum confinement* in the system.

In order to systematically test the relationship between density, dimensionality, and the extent of *quantum confinement* exhibited by the gel network, a series of experiments were conducted within the CdSe system [30]. In one case, different amounts of the oxidizing agent were added into concentration-identical CdSe nanoparticle sols to directly vary the degree of interconnectivity among CdSe quantum dots, thereby forming CdSe monolithic gels with different densities. Alternatively, different drying techniques were employed to generate different kinds of gels: aerogel, *ambigel*, and *xerogel*. Like xerogels, ambigels are dried under ambient conditions, but they employ a poorly wetting solvent, reducing capillary forces, and thus have densities intermediate between the corresponding aerogels and xerogels [31].

Experimental results indeed suggest that the band gap values, obtained from the onsets of the reflectance spectra (converted to absorbance), of the resultant gels can be tuned by varying the density of gel framework. As shown in Figure 17.9B, a gradual and almost linear band gap decrease for CdSe aerogels was observed when increasing volumes of 3% tetranitromethane (2.5–20 μL) were added into CdSe nanoparticle sols to yield monoliths of increasing density. The different drying techniques lead to a nonlinear band gap decrease from CdSe aerogels, to ambigels, and eventually to xerogels (Figure 17.9C). This observation suggests that there might be a critical density where the band gap becomes very sensitive to aggregate size, i.e., dimensionality effects no longer dominate. This is attributed to an increase in dipole–dipole interactions between particles in the dense, aggregated structures.

Thus, it appears that the bandgap change has two contributors, a density-driven change in the dimensionality (extent of *quantum confinement*) and dipole–dipole forces in secondary aggregates. This conclusion is borne out by small angle X-ray scattering (*SAXS*) studies (Figure 17.9D). Ambi and xerogels, where the aggregation is extensive, exhibit no mass-fractal characteristics at all, suggesting little contribution of dimensionality effects to the bandgap evolution. Aerogels, on the other hand, are fractal objects with mass fractal dimensionalities of ca 2.3, [30]. However, it appears that the dimensionality difference between different density aerogels, if any, is too small to detect with the resolution of the *SAXS* instrument.

17.4.4. Optimizing Photoemission Characteristics

Much of the interest in semiconducting nanoparticles such as CdSe is based on the sharp, high-intensity *PL* that can be achieved. Because the energy of the *PL* can be tuned by adjusting the crystallite size (hence the bandgap), there is significant flexibility. However, as indicated previously, the presence of defects within the particle, or redox active sites on the surface, gives rise to a broad emission due to trap states. To address this issue, an improved synthesis route enabling highly crystalline CdSe nanoparticles to be prepared prior to

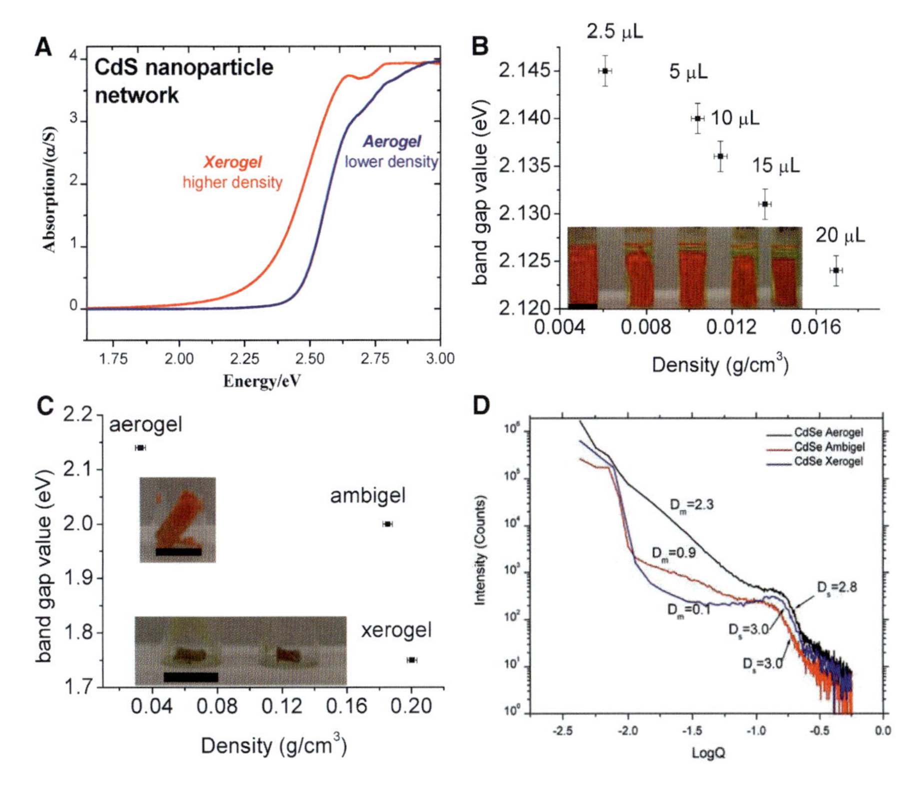

Figure 17.9. A. Optical absorbance data (converted from reflectance) for a CdS aerogel and xerogel. The larger bandgap in the aerogel is attributed to a higher extent of quantum confinement due to the low-dimensional framework; **B.** plot of bandgap as a function of density for CdSe aerogels prepared with different volumes of 3% tetranitromethane (the *inset* shows the wet gels from which the aerogels were derived, with the amount of oxidant increasing from **left** to **right**; *scale bar* = 1 cm); **C.** plot of bandgap versus monolith density for a CdSe aerogel, ambigel, and xerogel (pictured in *inset*) prepared using identical amounts of oxidant; **D.** small-angle X-ray scattering intensity as a function of the scattering vector (LogQ) for the CdSe aerogel, ambigel, and xerogel; fractal characteristics are indicated (D_m = mass fractal density; D_s = surface fractal density). Part **A.** reproduced with permission from Chem. Mater., 17: 6644–6650 (2005). Copyright 2005, American Chemical Society; Parts **B**, **C**, and **D** reproduced with permission from ACS Nano, 3: 2000–2006 (2009). Copyright 2009, American Chemical Society.

gelation, was investigated. This was achieved using a high-temperature arrested precipitation reaction [32], instead of the room temperature inverse micelle synthesis. However, although some *PL* was retained in the aerogels, the emission remained weak [33].

Because surface passivation of semiconducting nanocrystals can minimize surface defects, thereby enhancing *PL* [34], Brock and co-workers explored organic and inorganic surface modification techniques, as shown in Figure 17.10A. Organic surface modification sought to remove residual thiolates from the wet gel by replacing them with pyridine, which is a less effective hole trap for CdSe. Exchange of the wet CdSe gel with pyridine several times prior to acetone exchange and *supercritical drying* yielded aerogels with sulfur contents about half those of unexchanged gels. As shown in Figure 17.10B, although a significant trap state remains, the intensity of the band-edge peak is significantly augmented [33].

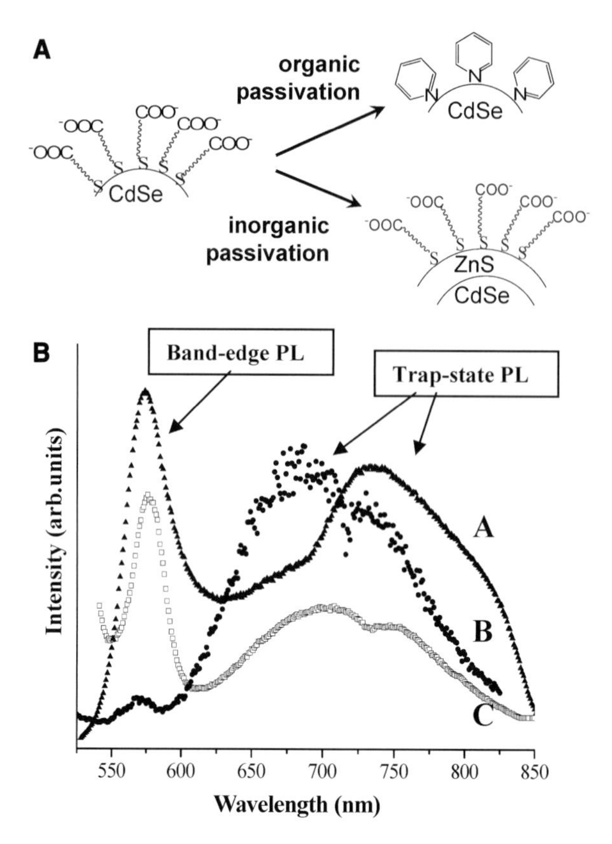

Figure 17.10. A. Scheme showing two approaches to minimizing surface trap-state emission and promoting band-edge emission by surface passivation with pyridine (organic) or ZnS (inorganic); **B.** Photoluminescence spectra of A. CdSe nanoparticles, **B.** conventionally prepared CdSe aerogels, and (**C**) CdSe aerogels prepared from pyridine-exchanged wet gels. Part (**A**) reproduced with permission from Acc. Chem. Res., 40: 801–809 (2007). Copyright 2007, American Chemical Society; Part **B.** reproduced with permission from J. Am. Chem. Soc., 128: 7964–7971 (2006). Copyright 2006 American Chemical Society.

Second, Brock and co-workers reported the effect of inorganic passivation on the emissive properties of CdSe by using CdSe/ZnS (core/shell) nanoparticles as the gelation *precursors*. The inorganic shell ensures that surface atoms on CdSe remain in an inorganic framework, and the large bandgap of ZnS ensures that both the electron and hole produced in the excitation are localized in the core. Thus, ZnS facilitates radiative recombination and acts as a window, letting light in and out. As shown in Figure 17.11A, the resultant aerogels are highly emissive, even when excited with a handheld UV lamp [35]. Furthermore, in contrast to "naked" CdSe nanoparticle aerogel networks, the density of the aerogel network prepared from core–shell nanoparticles has little influence on the optical and electronic properties of the resultant framework. Both *core–shell aerogels* and xerogels exhibit optical absorption and emission spectra that are essentially identical to the primary core–shell building blocks (Figure 17.11B). This is because the CdSe nanoparticles are effectively separated from each other by the intervening ZnS shell, ensuring that they remain essentially zero-dimensional, regardless of the framework density.

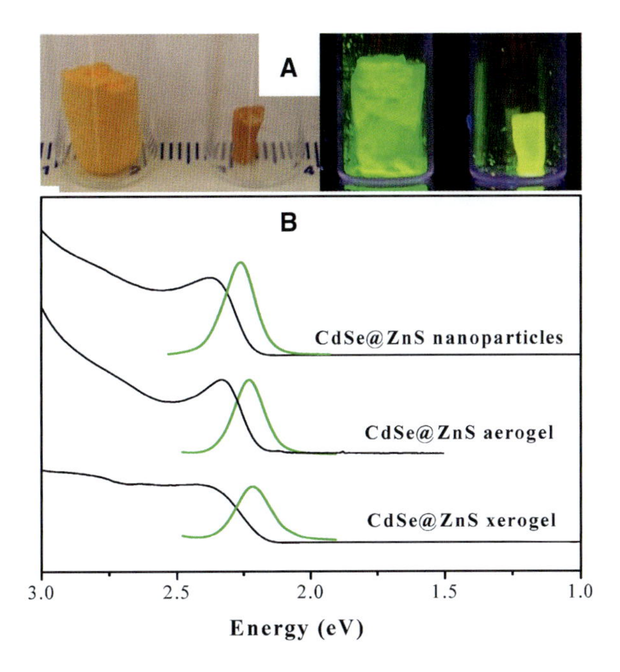

Figure 17.11. **A.** Photos of CdSe aerogels (large monolith) and xerogels (small monolith) under ambient (**left**) and UV (**right**) illumination; **B.** Optical absorption data (*black*, converted from diffuse reflectance) and photoluminescence (*green*) for CdSe/ZnS core/shell nanoparticles and corresponding aerogels and xerogels. Reproduced with permission from J. Am. Chem. Soc., 129: 1840–1841 (2007). Copyright 2007 American Chemical Society.

17.4.5. Controlling Morphology in CdSe Aerogels

Prior to 2008, chalcogenide aerogels were characterized by a colloidal morphology, a result of the spherical nanoparticle building blocks from which they are assembled. This is in direct contrast to silica aerogels (Chap. 2), where the rates of particle formation and particle assembly can be tuned by the use of an acid or base catalyst, yielding polymeric or colloidal gels, respectively. The morphology is known to have a significant effect on the surface area and strength of the gel network. Accordingly, Yu et al. investigated the use of differently shaped CdSe nanoparticle *precursors* for accessing different gel morphologies [36, 37]. To synthesize the differently shaped building blocks, such as dots, rods, branched, and hyperbranched CdSe nanoparticles, diverse synthetic techniques used in classic colloidal synthesis, such as monomer concentration control, growth regime control, coordination ligand ratio control, and growth time control [38, 39], were exploited. The single CdSe particles produced are shown as insets of Figure 17.12. Thiolate exchange, oxidative gelation and *supercritical drying* led to the corresponding aerogels. From the *TEM* images in Figure 17.12, a clear transition in gel morphology from *colloidal-type* to *polymeric-type* is evident when the primary CdSe building block changes from an isotropic dot to anisotropic rod, and eventually to the three-dimensional branched nanoparticle. The morphology of the CdSe aerogel assembled from hyperbranched nanoparticles appears to be intermediate between the colloidal and the polymeric types, as shown in Figure 17.12D.

To investigate the influence of the morphology on the surface area and porosity of resultant aerogels, porosimetry was conducted via nitrogen physisorption isotherms at 77 K. Experimental data based on *BET* surface areas and *BJH* pore distributions (based on

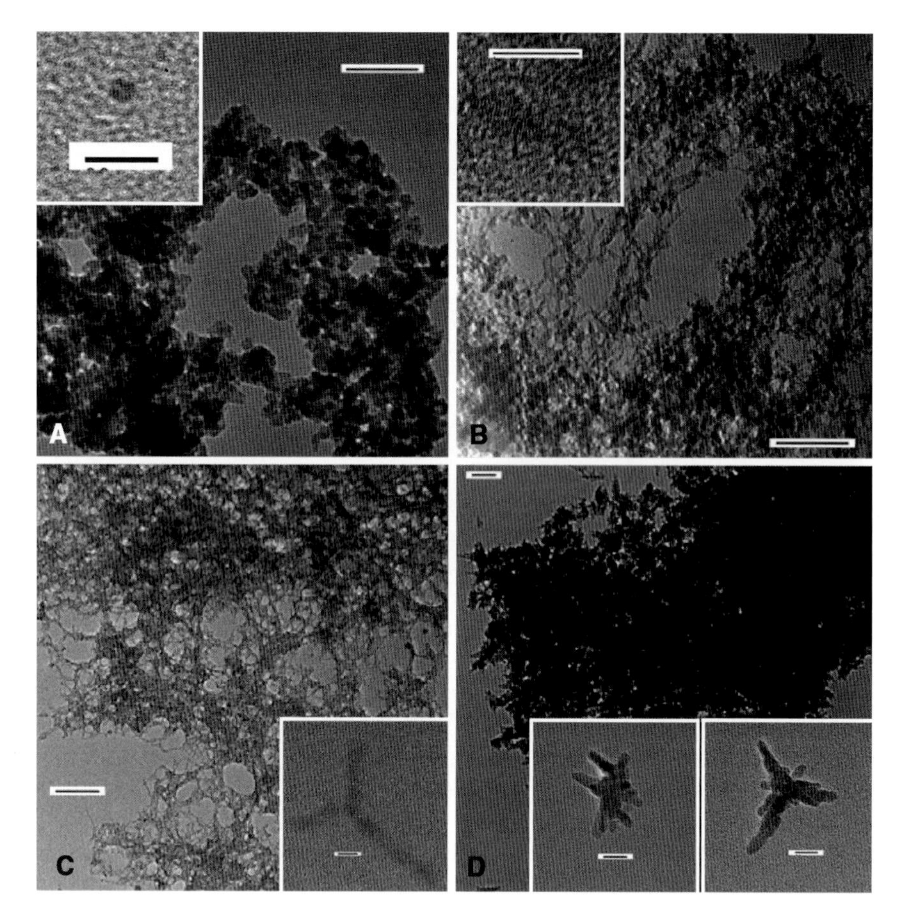

Figure 17.12. TEM images of CdSe aerogels prepared from **A.** dots, **B.** rods, **C.** branched nanoparticles, and **D.** hyperbranched nanoparticles. The *insets* show the single building blocks from which the gels were assembled. The *scale bars* on the main images are 100 nm and on the *insets* they are 20 nm. Reproduced with permission from ACS Nano, 2: 1562–1570 (2008). Copyright 2008, American Chemical Society.

cylindrical pores) showed that the surface areas of the rod or branched CdSe aerogel (average value: 237 m^2/g) are twice as high as the value of the dot aerogel (average value: 108 m^2/g). The low surface area value of the hyperbranched CdSe aerogel (average value: 80 m^2/g) might be due to the bigger size of the hyperbranch-shaped building block and the lower number of surface *thiolate* ligands obtained for this shape compared with the others. Data from X-ray diffraction, band gap, and emission measurements show that the resultant aerogels exhibit the identical crystalline phase (hexagonal CdSe) as their differently shaped building blocks and almost fully maintain the degree of *quantum confinement* observed in the nanoscale semiconducting building blocks [37].

Another interesting phenomenon observed is that the *polymeric aerogel* appears to be mechanically stronger than the colloidal aerogel. From viscosity measurements of *PDMS aerogel* composites, the viscosity enhancement is nearly 100% for the rod aerogel relative to pure *PDMS*, double that obtained from a dot aerogel. Moreover, the difference in enhancement of rod aerogels vs. dot aerogels does not appear to be due to the building blocks themselves, because data from discrete dots and rods suggest nearly identical viscosity enhancement for the two *PDMS* nanoparticle composites [36].

17.4.6. Expanding the Methodology: Ion Exchange

A key issue with the synthesis of new chalcogenide aerogels based on nanoparticle assembly methods is the need to first establish the syntheses of discrete nanoparticles of different phases, and develop appropriate *thiolate* capping protocols. In 2009, Yao et al. developed a work-around for this restriction [40], exploiting ion-exchange methods initially developed for chalcogenide nanoparticles. The methodology is based on the fact that ion-exchange processes that might be kinetically limited in bulk samples are quite facile when diffusion lengths need only to be a few nanometers. Thus, the treatment of CdSe nanoparticles with silver solutions results in rapid exchange, forming Ag_2Se nanoparticles. The same appears to be true for CdSe gels. Treatment of wet gels with solutions of Ag^+, Pb^{2+}, or Cu^{2+} leads to gels of Ag_2Se, PbSe, and CuSe, respectively. The process, illustrated schematically for Ag^+ in Figure 17.13A, can be followed by the rapid color change from orange to black as the exchange occurs. Intriguingly, the rapid exchange can be spatially limited by the concentration of exchanging ions. Thus, use of an insufficient quantity of Ag^+ ions leads to monoliths in which the outer portion has exchanged and been transformed, while the inner

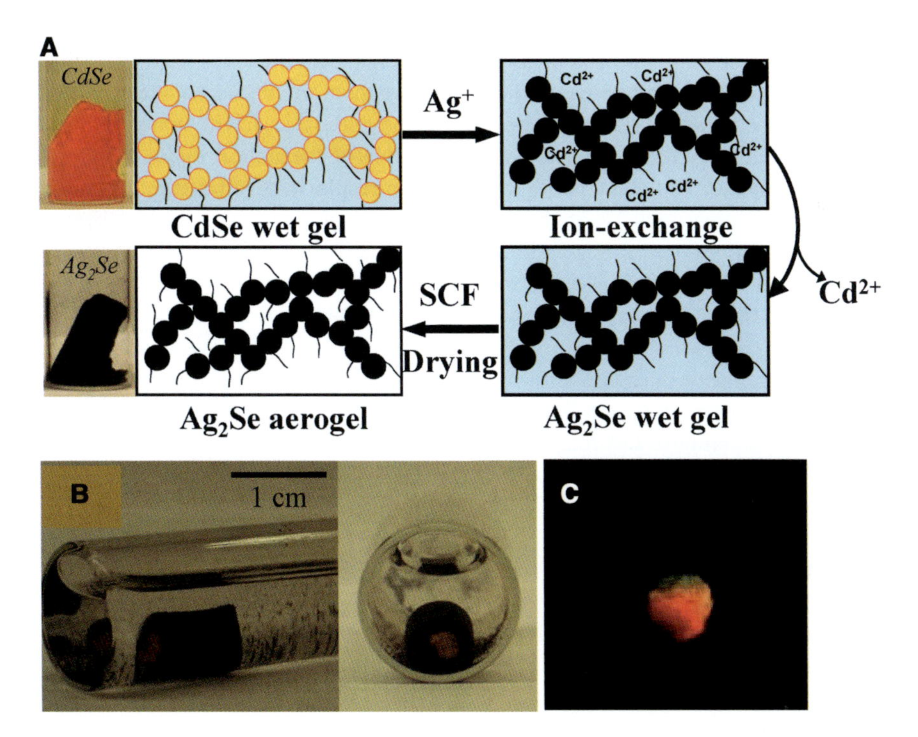

Figure 17.13. **A.** Scheme of the ion exchange process for formation of Ag_2Se aerogels from CdSe wet gels; **B.** photograph of a CdSe wet-gel exchanged with an insufficient quantity of Ag^+ (cut in half); **C.** photograph of a CdTe aerogel under UV illumination. The gradation of color is due to a range of particle sizes within the aerogel. Parts **A** and **B.** reproduced with permission from J. Am. Chem. Soc., 131: 2800–2801 (2009). Copyright 2009, American Chemical Society; Part **C.** reproduced with permission from Adv. Mater., 20: 4257–4262 (2008). Copyright 2008 Wiley-VCH.

portion is still CdSe, as shown in Figure 17.13B. The Ag_2Se gels are particularly interesting because they adopt the disordered α-phase, which exhibits fast ion mobility of the Ag^+ lattice, but which is not stable at room temperature in bulk form. The 3-D nanostructured network may thus prove advantageous for Ag-ion battery applications.

An intriguing aspect of the whole process is the way in which the monolith appears to be physically unaffected by the exchange. There is no change in the shape and no evidence of dissolution during the exchange. The electron micrographs on aerogels prepared from supercritical CO_2 extraction show that the colloidal network is not significantly changed, and the surface area, if anything, is larger when compared on a per mole basis (taking into account the larger mass of Ag_2Se relative to CdSe). The fact that this methodology can be extended to other ions (and divalent ions at that), such as Pb^{2+} and Cu^{2+}, suggests that the methodology is potentially general, enabling access to a wide range of new phases.

17.4.7. Expanding the Methodology: Tellurides

Original reports of nanoparticle gelation were limited to sulfides and then extended in 2005 to selenides. In 2008, Eychmüller and co-workers showed that the method is also applicable to tellurides with the report of CdTe gels and aerogels. Three routes to gel formation were explored. In one case, aqueous solutions of CdTe nanocrystals stabilized with thioglycolic or mercaptopropionic acid were allowed to age in the dark for long periods of time, several years in one case. Gelation was accelerated when some of the *thiolate* groups were removed, as occurs in size-selective precipitation. Alternatively, chemical oxidation using hydrogen peroxide can be used to facilitate removal of surface *thiolates*, although it should only be used in small quantities because of the relative ease with which tellurides are oxidized, and because of luminescence quenching. Finally, the authors explored a controlled photooxidation process, in which bandgap excitation results in hole trapping by the surface *thiolate* and subsequent etching of the surface. In both the chemical and photooxidation processes, the authors found that gelation could be facilitated by slow centrifugation at the viscous sol stage, leading to densification of the colloid. Syneresis is promoted and robust gels formed, and these are transformed into aerogels by supercritical CO_2 drying. These last two methods have the advantage of yielding aerogels in short time periods (days instead of years) with a high degree of reproducibility. An impressive outcome of the aerogels obtained here is the high degree of native fluorescence, as shown in Figure 17.13C. This study is notable because it demonstrates that aerogel formation, traditionally limited to oxides, can be extended to all the nonmetal Group 16 elements.

17.5. Conclusions

The formation of aerogels based entirely on metal chalcogenide (sulfide, selenide, or telluride) frameworks has been achieved using three approaches: thiolysis, cluster-linking, and nanoparticle assembly. These materials couple a semiconducting framework of soft *Lewis* basicity with a highly porous structure, and have already been demonstrated to be powerful agents for selective sorption and catalysis. The fact that many of the semiconductors absorb in the solar spectrum further suggests these architectures could be powerful photocatalysts or components in photovoltaic devices. With so many potential applications, and so much synthesis space to explore, the creation and exploitation of chalcogenide aerogels promises to be an area of continued focus for some time to come.

References

1. Bag S, Arachchige IU, Kanatzidis MG (2008) Aerogels from metal chalcogenides and their emerging unique properties. J Mater Chem 18: 3628–3632
2. Arachchige IU, Brock SL (2007) Sol-gel methods for the assembly of metal chalcogenide quantum dots. Acc Chem Res 40: 801–809
3. Brock SL, Arachchige IU, Kalebaila KK (2006) Metal chalcogenide gels, xerogels and aerogels. Comm Inorg Chem 27: 103–126
4. Sriram MA, Kumta PN (1998) The thio-sol-gel synthesis of titanium disulfide and niobium disulfide. J Mater Chem 8: 2453–2463
5. Carmalt CJ, Dinnage CW, Parkin IP (2000) Thio sol-gel synthesis of titanium disulfide from titanium thiolates. J Mater Chem 10: 2823–2826
6. Carmalt CJ, Dinnage CW, Parkin IP, et al. (2002) Synthesis of a homoleptic niobium(v) *thiolate* complex and the preparation of niobium sulfide via thio "sol–gel" and vapor phase thin-film experiments. Inorg Chem 41: 3668–3672
7. Purdy AP, Berry AD, George CF (1997) Synthesis, structure, and thiolysis reactions of pyridine soluble alkaline earth and yttrium *thiolates*. Inorg Chem 36: 3370–3375
8. Dunleavy M, Allen GC, Paul M (1992) Characterization of lanthanum sulphides. Adv Mater 4: 424–427
9. Stanić V, Pierre AC, Etsell TH, et al. (1997) Preparation of tungsten sulfides by sol-gel processing. J Non-Cryst Solids 220: 58–62
10. Stanić V, Etsell TH, Pierre AC, et al. (1997) Sol-gel processing of ZnS. Mater Lett 31: 35–38
11. Stanić V, Pierre AC, Etsell TH, et al. (2000) Influence of reaction parameters on the microstructure of the germanium disulfide gel. J Am Ceram Soc 83: 1790–1796
12. Stanić V, Etsell TH, Pierre AC, et al. (1997) Metal sulfide preparation from a sol-gel product and sulfur. J Mater Chem 7: 105–107
13. Stanić V, Pierre AC, Etsell TH, et al. (2001) Chemical kinetics study of the sol–gel processing of GeS_2. J Phys Chem A 105: 6136–6143
14. Kalebaila KK, Georgiev DG, Brock SL (2006) Synthesis and characterization of germanium sulfide aerogels. J Non-Cryst Solids 352: 232–240
15. Bag S, Trikalitis PN, Chupas PJ, et al. (2007) Porous semiconducting gels and aerogels from chalcogenide clusters. Science 317: 490–493
16. Trikalitis PN, Rangan KK, Bakas T, et al. (2001) Varied pore organization in mesostructured semiconductors based on the $[SnSe_4]^{4-}$ anion. Nature 410: 671–675
17. Maclachlan MJ, Coombs N, Ozin GA (1999) Non-aqueous supramolecular assembly of mesostructured metal germanium sulfides from $(Ge_4S_{10})^{4-}$ clusters. Nature 397: 681–684
18. Korlann SD, Riley AE, Kirsch BL, et al. (2005) Chemical tuning of the electronic properties in a periodic *surfactant-templated* nanostructured semiconductor. J Am Chem Soc 127: 12516–12527
19. Bag S, Gaudette AF, Bussell ME, et al. (2009) Spongy *chalcogels* of non-platinum metals acts as effective hydrodesulfurization catalysts. Nat Chem 1: 217–224
20. Armatas GS, Kanatzidis MG (2009) Mesoporous germanium-rich chalcogenido frameworks with highly polarizable surfaces and relevance to gas separation. Nat Mater 8: 217–222
21. Gacoin T, Malier L, Boilot J-P (1997) New transparent chalcogenide materials using a sol-gel process. Chem Mater 9: 1502–1504
22. Gacoin T, Malier L, Boilot J-P (1997) Sol-gel transition in cds colloids. J Mater Chem 7: 859–860
23. Gacoin T, Lahlil K, Larregaray P, et al. (2001) Transformation of CdS colloids: Sols, gels, and precipitates. J Phys Chem B 105: 10228–10235
24. Capoen B, Gacoin T, Nedelec JM, et al. (2001) Spectroscopic investigations of CdS nanoparticles in sol-gel derived polymeric thin films and bulk silica matrices. J Mater Sci 36: 2565–2570
25. Malier L, Boilot J-P, Gacoin T (1998) Sulfide gels and films: Products of non-oxide gelation. J Sol-Gel Sci Tech 13: 61–64
26. Mohanan JL, Brock SL (2004) A new addition to the aerogel community: Unsupported CdS aerogels with tunable optical properties. J Non-Cryst Solids 350: 1–8
27. Mohanan JL, Arachchige IU, Brock SL (2005) Porous semiconductor chalcogenide aerogels. Science 307: 397–400
28. Mohanan JL, Brock SL (2006) CdS aerogels: Effect of concentration and primary particle size on surface area and opto-electronic properties. J Sol-Gel Sci Tech 40: 341–350

29. Arachchige IU, Mohanan JL, Brock SL (2005) Sol-gel processing of semiconducting metal chalcogenide xerogels: Influence of dimensionality on quantum confinement effects in a nanoparticle network. Chem Mater 17: 6644–6650
30. Yu H, Liu Y, Brock SL (2009) Tuning the optical band gap of quantum dot assemblies by varying network density. ACS Nano 3: 2000–2006
31. Rolison DR, Dunn B (2001) Electrically conductive oxide aerogels: New materials in electrochemistry. J Mater Chem 11: 963–980
32. Peng ZAP, Peng X (2001) Formation of high-quality CdTe, CdSe and CdS nanocrystals using CdO as precursor. J Am Chem Soc 123: 183–184
33. Arachchige IU, Brock SL (2006) Sol-gel assembly of CdSe nanoparticles to form porous aerogel networks. J Am Chem Soc 128: 7964–7971
34. Trindale TO, O'Brien P, Pickett NL (2001) Nanocrystalline semiconductors: Synthesis, properties and perspectives. Chem Mater 13: 3843–3858
35. Arachchige IU, Brock SL (2007) Highly luminescent quantum-dot monoliths. J Am Chem Soc 129: 1840–1841
36. Yu H, Bellair R, Kannan RM, et al. (2008) Engineering strength, porosity, and emission intensity of nanostructured CdSe networks by altering the building-block shape. J Am Chem Soc 130: 5054–5055
37. Yu H, Brock SL (2008) Effects of nanoparticle shape on the morphology and properties of porous CdSe assemblies (aerogels). ACS Nano 2: 1563–1570
38. Peng ZAP, Peng X (2001) Mechanisms of the shape evolution of CdSe nanocrystals. J Am Chem Soc 123: 1389–1395
39. Kanaras AG, Soennichsen C, Liu H, et al. (2005) Controlled synthesis of hyperbranched inorganic nanocrystals with rich three-dimensional structures. Nano Lett 5: 2164–2167
40. Yao Q, Arachchige IU, Brock SL (2009) Expanding the repertoire of chalcogenide nanocrystal networks: Ag_2Se gels and aerogels by cation exchange reactions. J Am Chem Soc 131: 2800–2801

18

Biopolymer-Containing Aerogels: Chitosan-Silica Hybrid Aerogels

Chunhua Jennifer Yao, Xipeng Liu, and William M. Risen

Abstract Hybrid aerogels comprising bioderived polymers and metal oxides have been explored in the case of chitosan and silica. The physical properties of homogeneous aerogels have been explored extensively and structural models have been formulated on the basis of the data. One attractive feature of these materials is that the chemically functional groups on chitosan can be employed to develop transparent transition- and lanthanide-containing monolithic aerogels. The functionality and chemical properties also enable formation of aerogels with a range of magnetic properties, with gold and other nanoparticles, and with attached polymeric structures. The structure and chemistry make it possible to explore diffusion control of reactions and physico-chemical control of nanoparticle growth. Some biocompatibility data are described.

18.1. Introduction

One class of organic–inorganic hybrid aerogels is a combination of a bioderived organic polymer and silica. Of particular interest in this class are aerogels that comprise chitosan or other polysaccharides which have functional groups that enable them to interact both with the silica-based components of the aerogel and with compounds that are added subsequently. The latter can be metal ions or any of a range of species that can react with or coordinate to pendant $-NH_2$ or carbohydrate –OH groups, in the case of *chitosan*. In the case of the polysaccharides pectic acid and alginic acid, the pendant groups are carboxyl, –COOH, and –OH, while in gellan gum and carrageenans, neutralized carboxylic groups, –COOM (typically, $M = Na^+, K^+, Ca^{2+}$), are present.

The idea of attempting to synthesize *chitosan*-silica aerogels was pursued beginning in the late 1990s by groups at Brown University and at the E. O. Lawrence Berkeley National

C. J. Yao • Research and Development Department, Stepan Company, 22 W Frontage Road, Northfield, IL 60093, USA
e-mail: jyao@stepan.com
X. Liu • Northboro Research and Development Center, Saint Gobain, 9 Goddard Road, Northboro, MA 01532, USA
e-mail: xipeng.liu@saint-gobain.com
W. M. Risen (✉) • Department of Chemistry, Brown University, Providence, RI 02912, USA
e-mail: William_Risen_Jr@brown.edu

M.A. Aegerter et al. (eds.), *Aerogels Handbook*, Advances in Sol-Gel Derived Materials and Technologies, DOI 10.1007/978-1-4419-7589-8_18,

Laboratory. Interestingly, they approached the issue from rather different perspectives. The Brown group under W. M. Risen [1] took the approach of incorporating a functional polymer homogeneously in a silica-based network and exploring its chemistry. The Berkeley group under A. J. Hunt [2] reported coming to the problem as a way to make a high surface-area material with chitosan, which could not be made in aerogel form by itself. They examined hydrophobic examples and their biocompatibility.

The general approach taken was to incorporate chitosan and silica-forming compounds, such as tetraethyl orthosilicate (*TEOS*), to form hybrid aerogels under acidic conditions. These conditions lead to open silica structures. *Chitosan*, shown in Figure 18.1, has a random coil configuration under these conditions [3], so it can be present throughout the open silica structure. This is shown schematically in Figure 18.2.

In this figure, the broad line represents chitosan and the square dots are intended to show that there are unshared electron pairs on its amine groups. These amine groups and their physical availability are keys to the types of chemical reactions the chitosan-silica aerogels undergo. Together with the types of –OH groups on silica and chitosan, quite a range of chemistries are possible. Some of them are described in this chapter.

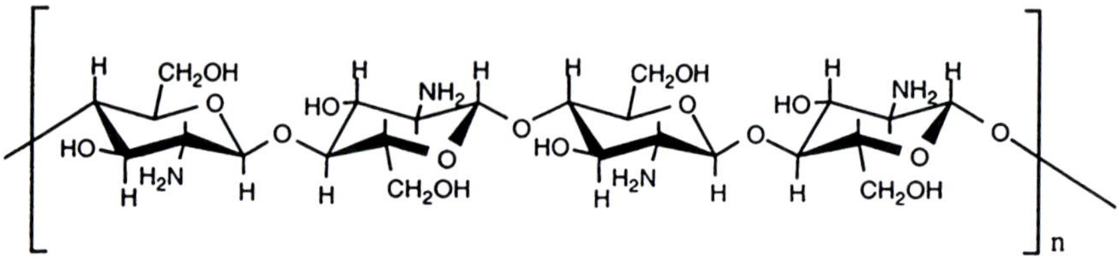

Figure 18.1. The structure of chitosan as fully deacylated chitin.

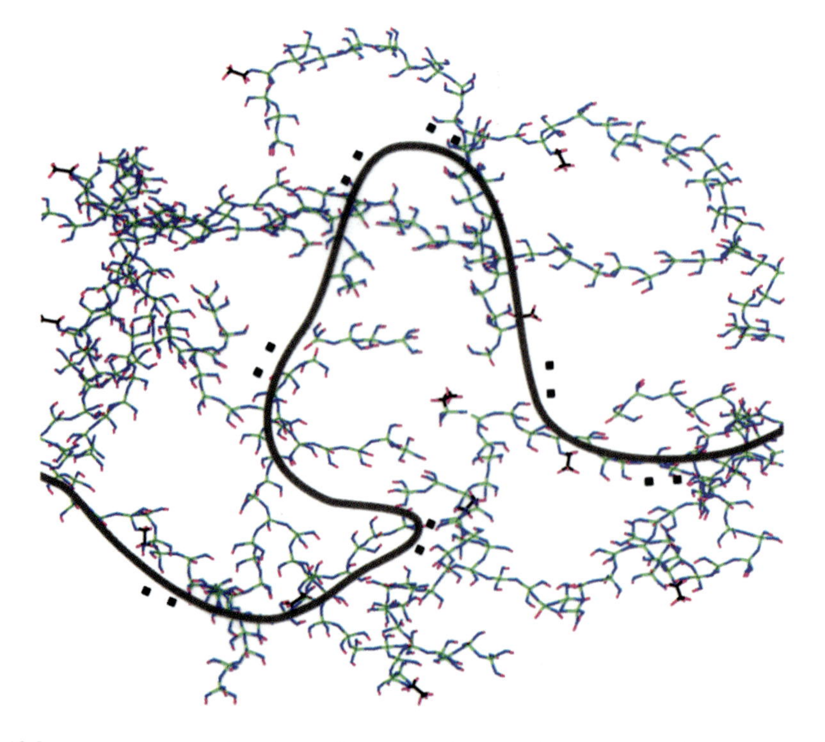

Figure 18.2. Conceptual model of a chitosan-silica aerogel. The *solid line* represents chitosan.

18.2. Syntheses

Four synthetic approaches have been described in the literature. That used by Ayers and Hunt [2] involved combining chitosan (75–80% deacylated chitin) in a 3:1 HCl/HF solution of water and ethanol with *TEOS*. They found that the mixture became a gel in 1–4 h, but they allowed it to stand for 48 h before further treatment. It was washed continuously in flowing water for 3 days, then the water was replaced by absolute ethyl alcohol using 5 days of washes, and then the gels were dried with supercritical CO_2. In some cases, hexamethyl disilazane (*HMDS*) was added to the gels being soaked in ethanol before CO_2 drying. Without the *HMDS* treatment, the aerogels were opaque, white, cracked materials, but with it they were un-cracked, hydrophobic, white, and opaque monoliths.

The three approaches taken by Risen's group are outlined in Figure 18.3. The chitosans used in most of the samples described here have measured molecular weights of 7.3×10^5 Da (Med.) or 2.08×10^6 Da (High), and are 80–82% deacylated forms of chitin. Other experiments have been carried out with 100% deacylated chitin with molecular weight 1.06×10^6 Da [4]. The acid used to dissolve chitosan is acetic acid, and the gels are neutralized with ammonia and then washed in absolute ethyl alcohol before drying in supercritical CO_2. This process allows the acetic acid to be removed as ammonium acetate and the chitosan amines to become unprotonated to a significant, pH-dependent degree.

The degree of deprotonation, that is, the extent to which unprotonated $-NH_2$ groups are present, is an interesting aspect of chitosan in the aerogels. Despite the fact that these are primary amines, the pK_a of chitosan in solution has been determined by the careful studies of Domard [5] to be 6.5. This anomaly, which is ascribed to polymer electrostatic and structural effects, makes it possible for the amine groups to coordinate metallic ions in

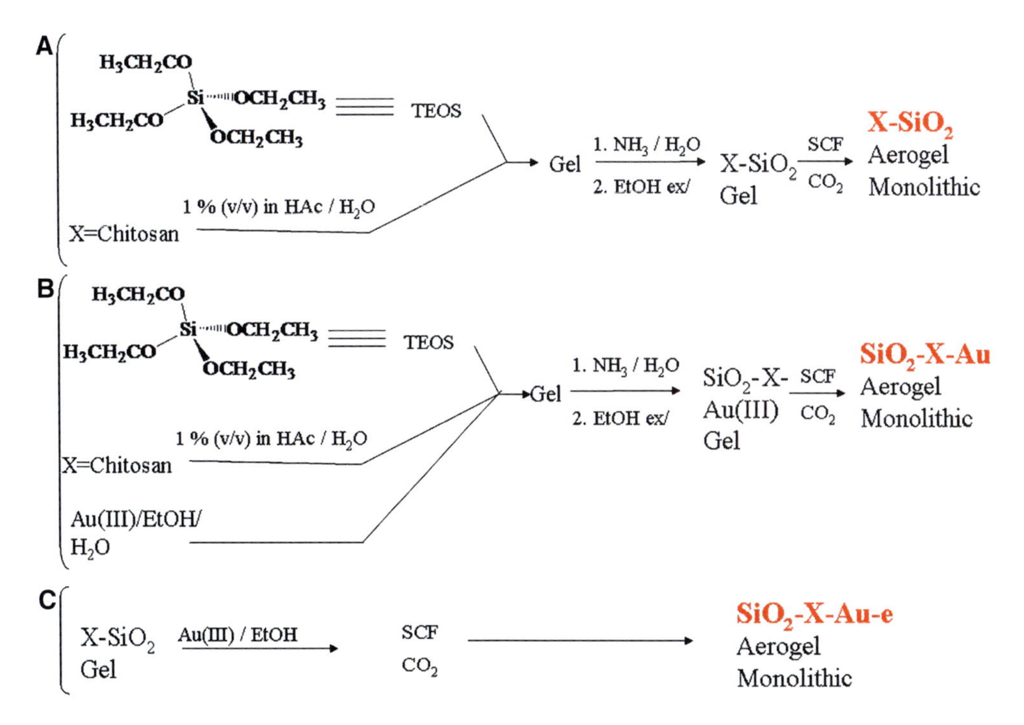

Figure 18.3. Reactions leading to chitosan-silica aerogels and their metal-containing forms.

the pH range of about 4–6 [1, 3, 6]. It also makes it possible to prepare chitosan-silica hybrid aerogels with reactive $-NH_2$ groups.

As shown in equations B and C of Figure 18.3, there are two ways to introduce metal ions. In one, B, the metal ions are present in the gelling solution. In the other, C, the ions in solution are added to the already formed gel. As will be seen below in the context of gold-ion-containing aerogels, hereafter SiO_2-X-Au(III), method B gives a more homogeneous distribution of Au(III) ions in the final aerogel.

One of the initial objectives of the research on SiO_2-X-M aerogels was to prepare materials that have good optical properties, including low refractive indices and fascinating optical characteristics related to those of the metal ions. For example, Er(III) is an important near infrared emitter for optical fiber communications, and Tb(III) is a valuable fluorescent center in the visible range. Another is to prepare materials with valuable catalytic properties, such as analogs containing Pd, Pt, Rh and Ru, as well as having high surface areas and high friability. Ultimately, additional discoveries were made, as described below.

In Figure 18.4, a photograph of a set of monolithic SiO_2-X-M aerogels made in this manner is shown. Each piece was placed on a piece of paper on which the metallic element's

Figure 18.4. Monolithic metal-containing transparent chitosan-silica aerogels.

symbol was written and the photograph was taken from above with two passes of the light through the aerogel monoliths.

The actual visible spectra of these and other analogs have been reported by Hu [7] and Ji [8]. In the case of the SiO_2-X-Co(II) samples, the green–blue one was prepared using the chloride and has tetrahedral Co(II), and the pink one has octahedral Co(II) since it was prepared using the nitrate.

These SiO_2-X and SiO_2-X-M aerogel samples not only are clear (not opaque) in the visible range, they have quite interesting infrared spectra. As shown in Figure 18.5, the spectrum taken by transmission through a *ca.* 0.5 mm monolith of SiO_2-X aerogel reveals features due to silica, chitosan (including the amine region near 1,600 cm^{-1}), and oriented H_2O molecules. Since the amine region is observed, coordination of transition metal ions to chitosan can be confirmed readily [1]. Just as helpful is the fact that the IR spectrum in the ca. 1,400–3,000 cm^{-1} region has low enough absorptivity that one can follow the course of reactions with reactants and reaction products that have characteristic absorptions in this region.

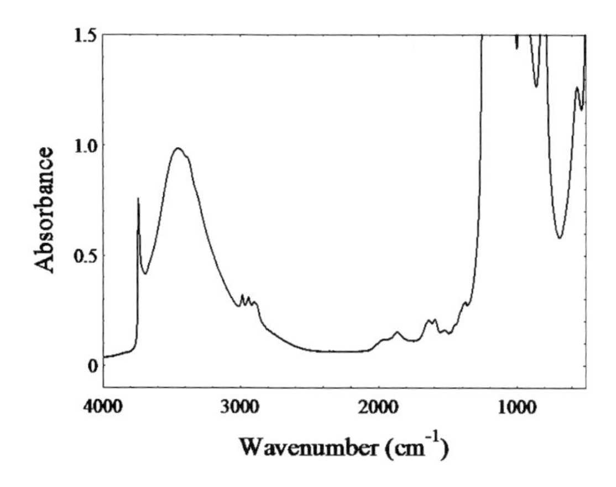

Figure 18.5. The infrared spectrum by transmission of a *ca.* 0.5 mm thick chitosan-silica aerogel [7].

18.3. Properties

Through series of studies, the physical structure of SiO_2-X aerogels has been elucidated. Hunt and Ayers measured the surface areas of their gels to be in the 450–750 m^2/g range, while Ji [8] found his SiO_2-10%X aerogel to have surface area of 555 m^2/g. Additional *BET* measurements yielded values of 570–975 m^2/g for the SiO_2-10%X and Pd(II), Rh(III) and Ru(III) SiO_2-X-M aerogels. This range includes most measurements, which were gathered around 850–900 m^2/g. Some outlying materials had values that were higher and lower. Systematic studies of the detailed dependence of the surface area on preparative conditions have not been completed. The pore size distributions all ranged from 1.5 to 12 nm with peaks in the *ca.* 7 nm range and average size of *ca.* 4–5 nm [1, 8].

Hu and coworkers further investigated the structure by small angle neutron scattering (*SANS*) and found that they have fractal dimensionalities, D_f, of 2.47 for SiO_2-15%X to 2.69 for SiO_2-5%X aerogels [9]. Recall that for three-dimensional systems with no density variation, $D_f = 3$. For a mass fractal cluster, D_f is <3, and the magnitude of D_f is related to the openness of the structure. The lower the value of D_f, the more open the structure is. The original analysis reported by Hu et al. [9] also yielded values for the exponential cutoff length, ς, which essentially shows the distance through a structure over which there is information coherence. The values of, ς, for these materials vary from 3.8 to 5.9 nm, while those for the SiO_2-X-M aerogels vary over a slightly wider range.

Ji [8] and Hu [7] also studied the ^{13}C NMR spectra and found that the role of chitosan in the aerogels is somewhat modified from the pure material by its interaction with the silica framework. They also analyzed the ^{129}Xe NMR spectra of Xe in SiO_2-X aerogels as a function of temperature and pressure, and they showed that the Xe behavior in pores is consistent with the pore size measurements.

These physical data, together with measured densities in the 0.23–0.32 g/cm^3 range, averaging 0.27 g/cm^3 for a representative set of clear monoliths, and the stoichiometric data, were used to construct a model for a SiO_2-X aerogel. This model, shown in Figure 18.6, is consistent with all of the data, but it is not unique since the material is not crystalline.

Further characterizations of the metal-containing SiO_2-X-M aerogels have been reported through studies of their infrared [1, 10–15], near infrared [8], visible [7, 8, 16],

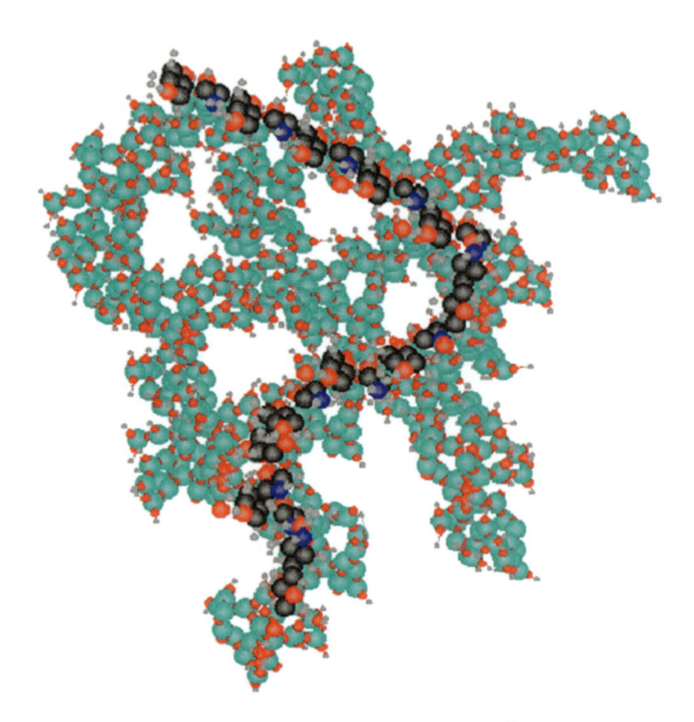

Figure 18.6. Structural model of chitosan-silica aerogel; density 0.27 g/cm^3, 10% chitosan.

fluorescence, electron spin resonance (*ESR*) [8], magnetic (*SQUID* magnetometer, Superconducting QUantum Interference Device) [8, 15, 17], refractive index [8], photoacoustic [18] and other properties. An interesting example is the variation of the magnetic susceptibility of SiO_2-10%X-Ru(III) aerogels with the Ru concentration.

As shown in Figure 18.7, the value of μ_{eff} in units of BM (Bohr magnetons), evaluated from both *SQUID* and *ESR* data, is plotted versus the weight percent of Ru ions present in the aerogels. Clearly, μ_{eff} per Ru decreases greatly as the Ru concentration increases. Ji [8] has ascribed the decrease to the formation of Ru–O–Ru antiferromagnetic species, whose probability of formation increases with the concentration. The value of μ_{eff} extrapolates at high Ru concentration to a value near 0.25 rather than 0, however, so something else must be taken into account. As it turns out, Ru(III) salts, such as the $RuCl_3$ used in this study, actually contain about 93.2% Ru(III), 3.4% Ru(II) and 3.4% Ru(IV), so if all of the Ru(III) are bound in antiferromagnetic complexes bridging one Ru(III) chitosan unit to another, the value of μ_{eff} is expected to be 0.25 from the residual Ru(II) and Ru(IV). Taking these factors into account, the value of μ_{eff} also extrapolates to the expected value for Ru(III) and residuals as the concentration of Ru tends to zero.

The biological and chemical properties of SiO_2-X aerogels are of considerable interest. Several groups have been interested in using materials containing both silica and chitosan to take-up drugs and release them slowly in a controlled manner. Both of these constituents, silica and chitosan, are sold in various pharmaceutical and nutritional formulations and have been widely studied [19–21], so the hope has been that aerogel combinations of them would be suitable for that purpose.

Ayers and Hunt [2] found that the cytotoxicity of their SiO_2-X aerogels was measured to be 1 on a 0–4 scale, where 4 means severe cell damage and 0 is no cell damage.

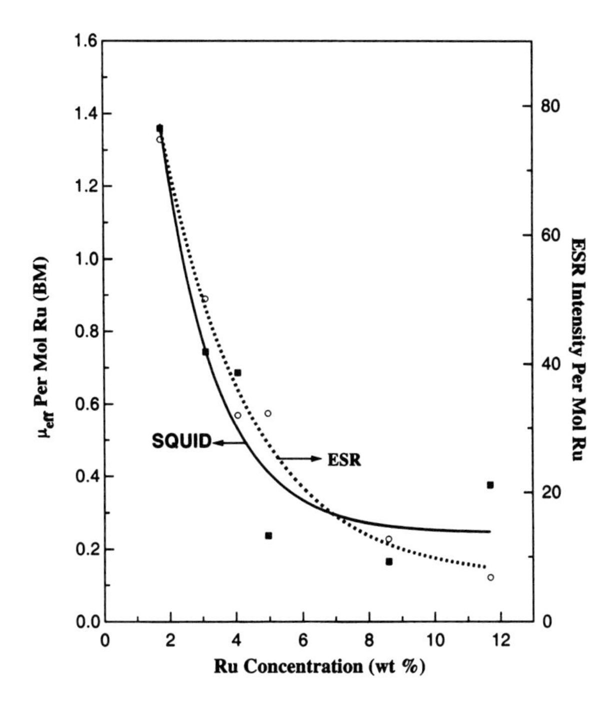

Figure 18.7. Magnetic susceptibility versus Ru concentration in a series of SiO_2-X-Ru aerogels [8].

A hemolysis screening gave a result of 19.7% hemolysis, which is high, but the authors ascribed the result to residual fluoride ion, because HF was used in their synthesis, or to a physical absorptive effect toward certain blood components. It is clear that further study on clean SiO_2-X aerogels is required to characterize their biocompatibility more reliably.

This is important in light of the work by J. Liu et al. [22] on the use of SiO_2 aerogel/chitosan composite materials as potential drug carriers. Although these materials were made differently, by adding chitosan to SiO_2 aerogels and heating them to 450°C, the composites exhibited better drug loading and slower release than silica, at least for sulfate gentamicin. It also relates to the ability of SiO_2-X aerogels to absorb dyes and other charged molecules from aqueous solutions and dispersions [16, 23–25].

18.4. Chemical Properties and Novel Aerogel Materials

The chemical properties of SiO_2-X aerogels have been studied extensively. These studies fall into several categories. One involves chemistry of metallic species, while another is directed toward the interaction of aerogel particles with the external environment.

Three types of reaction chemistry involving metal ions in SiO_2-X-M aerogels will be discussed. The first involves the formation of iron-containing species and their resultant magnetic properties. The second concerns the chemical reactions of metal ions, such as Rh(III) and Ru(III), with external reactants, and the third has to do with photochemistry and nanochemistry of Au particles.

18.4.1. Iron-Containing Chitosan-Silica Aerogels

The reaction scheme in Figure 18.8 shows how either of two types of these aerogels can react first with $Fe(CO)_5(g)$ and then react through various treatments to form SiO_2-X-iron-containing aerogel products [15, 17]. When the SiO_2-X is treated with iron carbonyl on the vacuum line and then heated at 190°C, it forms SiO_2-X-Fe(0), an aerogel-containing ferromagnetic particles. If it is treated according to the steps depicted in the scheme of Figure 18.8, SiO_2-X-iron-containing species of the indicated types and characterization are obtained. X-ray diffraction (*XRD*) and X-ray photoelectron spectroscopy (*XPS*) were used to characterize the resulting composite materials. Of particular interest are the SiO_2-X-Fe_3O_4 aerogels (labeled Fe_3O_4–SiO_2 in Figure 18.8), which were prepared by heating "X-SiO_2-FeOx" (Figure 18.8) under nitrogen. Their *XRD* data show the formation of this iron oxide, and their magnetic properties, shown in Figure 18.9, especially the very small coercivity and remanence magnetization, demonstrate that these are superparamagnetic aerogels. Their superparamagnetism is clear for all materials after thermal treatment above 250°C, as described in detail by Yao and coworkers, using these data and data from field cooling (FC) and zero field cooling (ZFC) measurements at 200 °C. This indicates a blocking temperature of about 75 K and a ZFC/FC separation temperature of 110 K [15, 17].

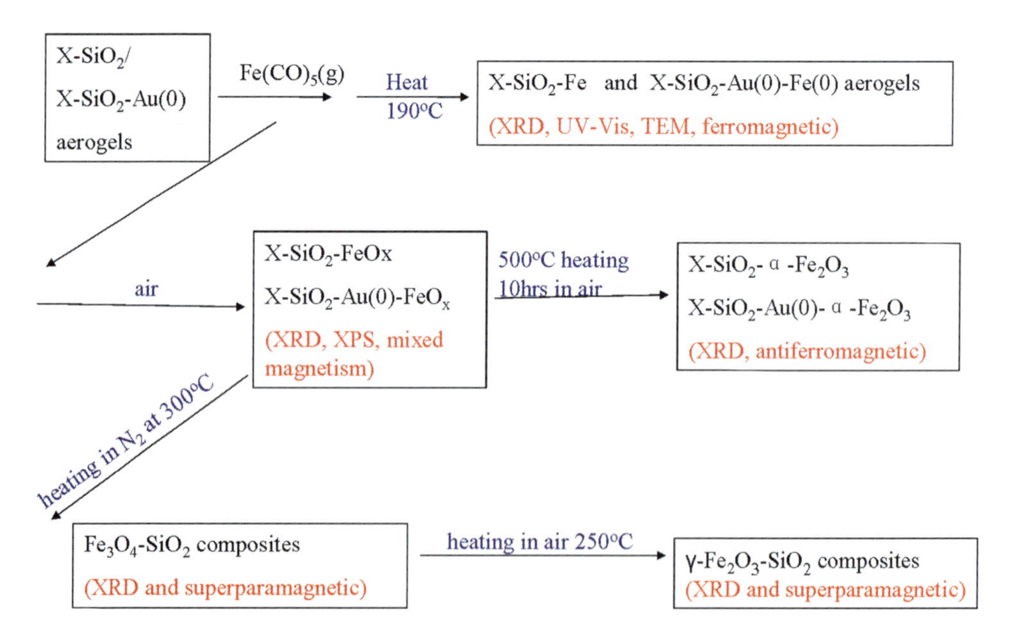

Figure 18.8. Reaction scheme for preparing iron-containing magnetic SiO_2-X-M aerogels [17].

18.4.2. Transition Metal-Containing Aerogel Chemistry

The second chemical reaction approach for all of the SiO_2-X-M aerogels is particularly intriguing. If such aerogels are placed in a vacuum to remove most of the water of hydration around the metal ions, these ions can be thought of as essentially hanging in space. In some ways, then, they are on high area surfaces and are prepared to function catalytically. In other ways they are ions without the surrounding medium they would have

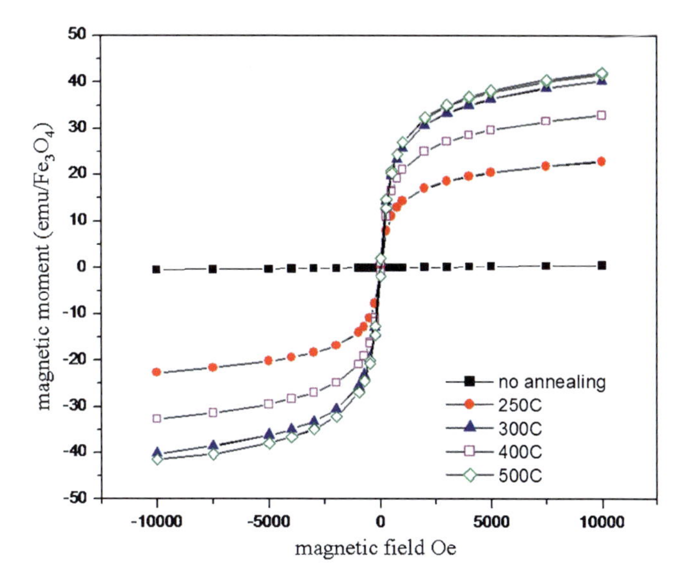

Figure 18.9. Magnetic properties of SiO_2-X-Fe_3O_4 aerogels prepared under N_2 at the temperatures given in the key with the measurements made by *SQUID* magnetometer at 25°C [17].

in solution, and this makes possible certain reactions that would require quite different conditions in the bulk.

One example of this is shown in Figure 18.10, which is the infrared spectrum of SiO_2-X-Ru(III) aerogel in a closed chamber, evacuated and then exposed to CO(g). The

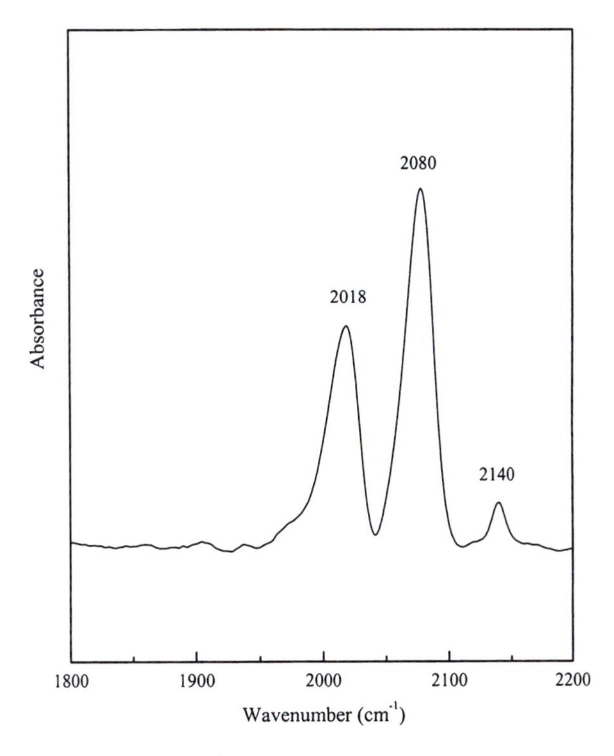

Figure 18.10. IR spectrum in the 2,000 cm^{-1} region of SiO_2-X-Ru(III) aerogel exposed to CO(g) [8].

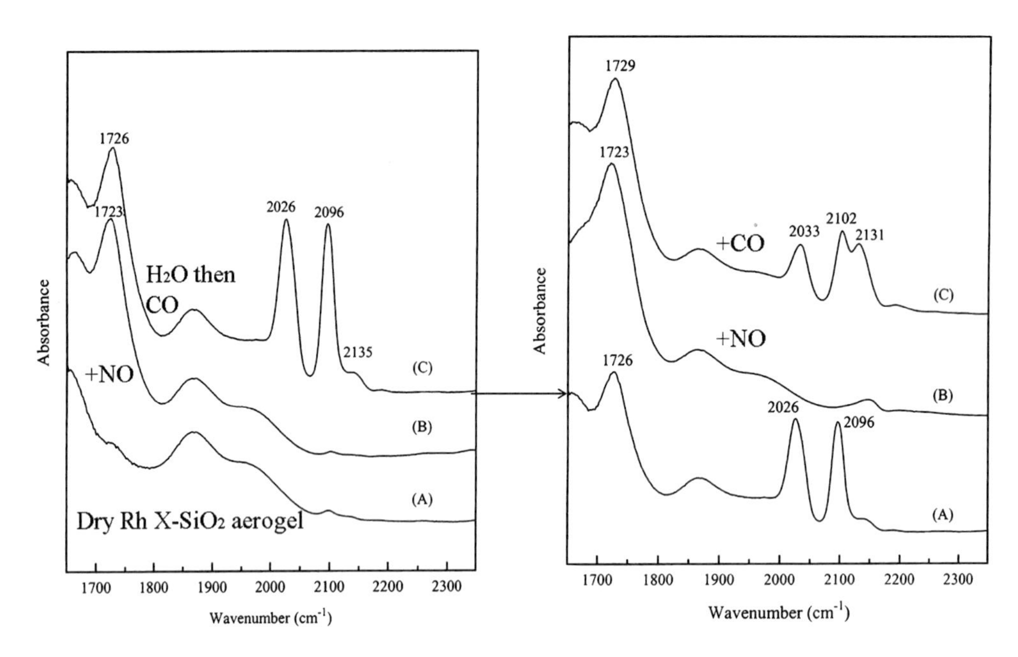

Figure 18.11. Infrared spectra of a rhodium-form chitosan-silica aerogel through a succession of treatments from A in **left panel** through C in the **right panel**, as indicated [8].

spectrum was taken through the aerogel monolith in transmission mode. The bands have been analyzed by Ji and coworkers and assigned to υ (C–O) vibrations of a rhodium carbonyl formed in the aerogel [1, 8].

The observation of the formation of this metal carbonyl shows that chemical reactions in the aerogel monoliths can be followed. A good example is a series of reactions with CO(g) and NO(g) in a SiO_2-X-Rh(III) aerogel. As shown in Figure 18.11, such a series was carried out as follows. In Figure 18.11 (left) the spectrum of a SiO_2-X-2.5 wt% Ru(III) aerogel that had been dehydrated (A) is shown, and in (B) the spectrum of the same material after exposure to NO(g) is shown. The band appearing at 1,723 cm^{-1} is assigned to the υ(N–O) of a ruthenium nitrosyl formed in the aerogel. Subsequent exposure to H_2O(g) and then CO(g) leads to the spectrum (C), which shows additional bands at 2,026, 2,096 and 2,135 cm^{-1} due to the formation of ruthenium carbonyl complexes with reduced Ru sites. The reduction of Ru was achieved by the water–gas shift reaction, in which water at the metal ion sites and CO combined to yield effectively hydrogen and carbon dioxide and the hydrogen caused reduction of the ruthenium. Further findings by Ji Figure 18.11 (right) showed that the addition of NO(g) at this point caused the removal of the υ(C–O) bands as the NO caused oxidation. Moreover, addition of CO(g) to that product leads to the appearance of carbonyl vibrational bands with different relative intensities, showing that the 2,135 cm^{-1} band belongs to a different species than do the other two carbonyl bands.

18.4.3. Chemistry of Gold-Containing Chitosan-Silica Aerogels

The reactions of gold-ion-containing aerogels, SiO_2-X-Au(III), involves a range of chemistries. As presented above (Figure 18.3), they can be made as either SiO_2-X-Au from the gel or as SiO_2-X-Au-e using an external process to introduce the Au(III) into an X-SiO_2 gel.

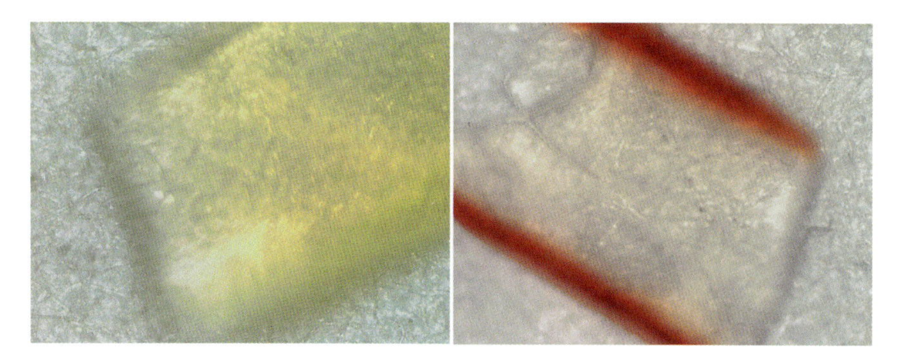

Figure 18.12. SiO_2-X-Au(III)-e aerogel before (**left**) and after ultraviolet photolysis [17].

Both aerogels appear yellow due to the presence of Au(III), as shown in Figure 18.12 (left). The texture seen is due to the substrate on which it was photographed.

When that SiO_2-X-Au-e aerogel was exposed to a weak ultraviolet light source (25 W Hg lamp at *ca.* 10 cm distance) [17, 26], the material turned red on the surfaces and along the edges (Figure 18.12, right). When this experiment was repeated with SiO_2-X-Au(III) aerogels, which have the Au(III) ions dispersed homogeneously, the photolysis products have the red appearance as shown in Figure 18.13. The fraction next to each photograph indicates the ratio of Au(III) to chitosan amine group, Au(III)/NH_2, in the sample. The visible range transmission spectra of these samples are also seen in the figure. The photographs, visible spectra, *XRD* and *TEM* data (not shown) reveal that Au(0) nanoparticles have been formed in this process.

The sizes of the Au(0) nanoparticles grown in these aerogels, measured by *TEM*, depend on the Au(III)/NH_2 ratio, as shown in the top line (solid) of Figure 18.14. All of the observations are consistent. The particles of the *TEM*-measured sizes have nanoparticle plasmon resonances in the region of the absorption maxima shown in the spectra in Figure 18.13. The Au(III)/NH_2 ratio is a measure of the spatial concentration of Au(III)

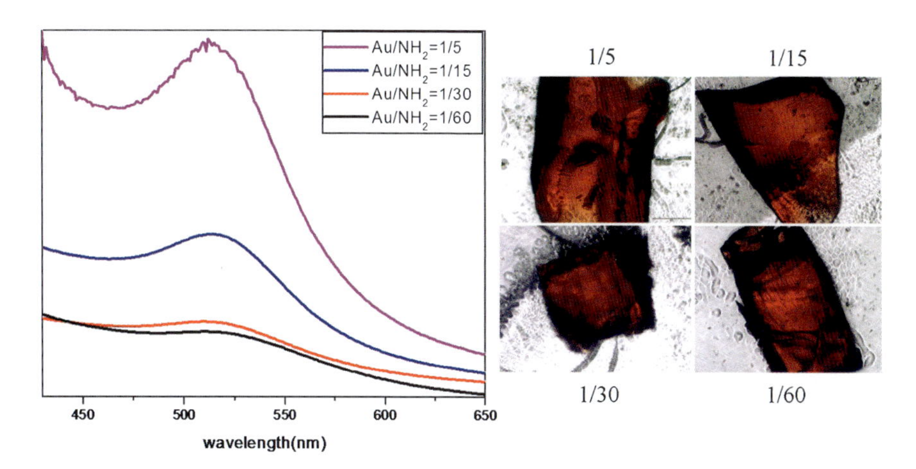

Figure 18.13. The visible spectra and photographs of chitosan-silica aerogels [X-SiO_2-Au(0)] after UV photolysis [17].

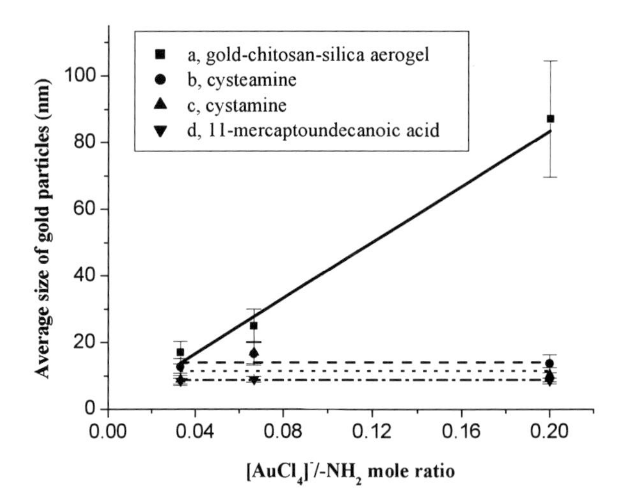

Figure 18.14. Average sizes of gold nanoparticles formed by UV photolysis of gold-chitosan-silica [13] aerogels containing 10% chitosan and the indicated ratio of Au(III) (as $AuCl_4^-$) to amine. The *squares* denote those aerogels, the others denote them with the indicated molecule present during the photolysis.

ions. The mechanism of growth of Au(0) is limited by the available gold species in the neighborhood of the growth center (surrounding a nucleus), so larger particles should result from aerogels with higher Au(III)/NH_2 values if all else is constant. This includes the type and concentration of chitosan, since a small amount of it is oxidized to provide the electrons for the photo-reduction of Au(III) to Au(0).

The condition, "if all else is constant", suggests the question: what if other entities are present? This is especially important in the field of nanochemistry/nanotechnology, where metal-containing nanoparticles usually are covered with small molecules to prevent them from agglomerating instead of staying separate. Typically, a surfactant is applied to cover the nanoparticles as they are being formed.

In the case of SiO_2-X-Au(0) aerogels, however, no surface-coating molecules are used to keep the Au(0) particles apart; the aerogel structure and the growth mechanism play that role. The experimental question then becomes; what happens if molecules that interact with Au(0) are present as they are grown photolytically in the aerogels?

X.Liu addressed this question using biologically relevant molecules and model compounds that can interact with Au(0) surfaces; cystamine, cysteamine, biotinylated thiol-6 and biotinylated disulfide-5, as well as the thiols 11-mercapto-1-undecanol and 11-mercapto undecanoic acid [26]. In each case, the molecule was added to a dry SiO_2-X-Au(III) aerogel before photolysis.

The results of these experiments are shown in Figures 18.14 and 18.15. In the former, it is clear that the Au(0) particles are smaller when grown in the presence of these three molecules than when grown in their absence and that the size does not vary with Au(III)/NH_2 value in the precursor. In Figure 18.15, the size comparison is extended to all of the Au(0)-surface-coating molecules. The comparison shown there is between the average size (diameter) of Au (0) particles grown in SiO_2-X-Au(III) alone and those grown analogously in the same aerogel containing one of the molecules labeled.

It is clear that these molecules limit the growth of Au(0) particles. The interpretation presented by X. Liu and coworkers [26, 27] is that the Au(0) particles grow to the size at which the attaching molecules have effectively covered the surface. Once that is

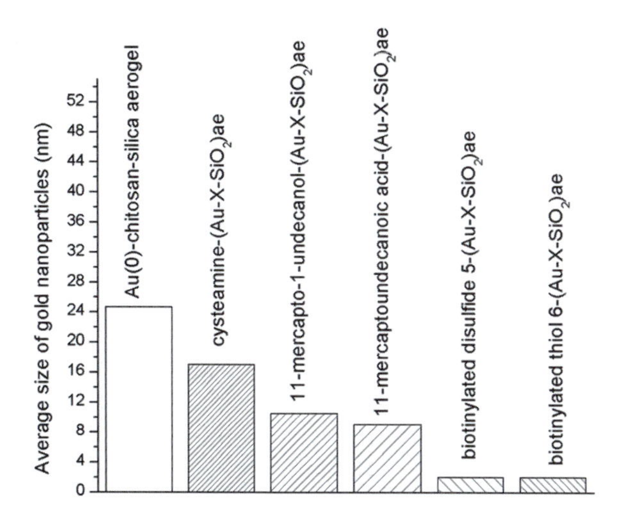

Figure 18.15. Average sizes of gold nanoparticles formed by photolysis of gold-containing chitosan-silica aerogels with 10% chitosan and Au(III)/NH_2 of 1/15 [13].

accomplished, they stop growing. For that reason, they grow to a certain size for all values of Au(III)/NH_2 for a given molecule. In addition, note that the larger molecules, with large surface cone-angles, lead to small coated Au(0) particles, while the smaller ones, such as cysteamine, do not limit growth as much as the larger ones. This relationship between particle size and coating molecule could lead to a useful analytical approach.

Since the Au(0)-containing aerogels can interact with such molecules, another potential opportunity can be considered. Could they be manipulated magnetically? If so, aerogel particles might be used in in vivo applications and moved or removed magnetically. Fortunately, Yao and coworkers [15, 17] found that the Au(0) particles are essentially unaffected by the iron chemistry required to make them ferromagnetic, as shown in Figure 18.16, wherein the powders in vials are attracted to a small permanent magnet.

Figure 18.16. Vials containing SiO_2-X-Fe(0) (**left**) and SiO_2-X-Au(0)-Fe(0) (**right**) powders responding to the presence of a small magnet [17].

18.4.4. Diffusion Control of Chemical Reactions in Nanodomains

Both of the previous examples show that chemistry within the small pores or nanodomains of these aerogels can be quite fascinating, in part because of the chemistry the chitosan enables and in part because of the external but controlled access to the domains. Thus, the particle-growth limitations set by the concentrations (Au(III)/NH_2) in the aerogel and by the addition of external molecules showed that local chemistry is important. That, in turn, suggests another way in which aerogels can have an interesting role in chemistry; namely, by controlling access and local concentrations through aerogel structure.

To test the idea that the aerogels can exert diffusion-control through their structure in such a way as to control chemistry, X. Liu [26, 27] devised an intriguing set of experiments based on the fact that the product of the reaction between Au(III) and L-cysteine in solution depends on the stoichiometric ratio between the reactants and their concentrations. When they are dissolved in solution in the mole ratio of 1:3::[Au(III)]:[L-cysteine], they react to form a white precipitate, which is an Au(I) compound. However, when they are dissolved at ratios in which the Au(III) is in significant excess the precipitate is red because Au(0) particles are formed [28–30]. In a SiO_2-X-Au(III) aerogel, the Au(III) concentration can be locally high when the access to the nanodomains is strongly diffusion controlled. They can be locally low if the concentration of L-cysteine is high. By controlling the volumes of L-cysteine solution, it is possible, of course, to hold the amount constant while varying the concentration.

Through this approach, X. Liu [26, 27] was able to run a series of experiments in which a global ratio of 1:3 was maintained in each reaction and the Au(III) concentration was always the same, but the aerogel structure controlled the diffusion such that the local [Au(III)]:[L-cysteine] ratio was varied from 1:3 to 25:1. The result was that as the local concentration ratio in the nanodomains changed, the product changed from the white Au(I) product to the red Au(0) product, even though the stoichiometric ratio between reactants and the concentration of Au(III) were held constant globally.

18.4.5. Attachment of Chitosan-Silica Aerogels to Polymers and Other Entities

The amine groups in the chitosan-silica aerogels are available for chemical reactions when they are in the –NH_2 form. That is important in coordinating metal ions, and it can be valuable in a range of other reactions. Among them are the formation of urea- and amide-linkages that provide a facile way to make covalent linkages to a substantial portion of the aerogel structure. They are formed through reactions of isocyanate groups or anhydride groups with –NH_2 and are favored because they are much faster than those with either the –OH groups on the silica or those on chitosan.

X. Liu, M. Wang and W. M. Risen, Jr. [12, 31] examined the interactions between SiO_2-X based aerogels and various isocyanates, anhydrides (succinic anhydride), and a sulfonyl chloride (dansyl chloride). They found that the reactions with the amine groups predominated. In fact, as long as the amounts of these reactants were less than the amounts of amine groups, the only reactions observed were with them. Neither water nor the –OH groups in the aerogels competed successfully with the –NH_2 groups under mild conditions.

In the case of reactions of *HMDI* (bis(4-isocyanatocyclohexyl)methane) with SiO_2-X aerogels, they reported that one *HMDI* isocyanate reacts with an amine on chitosan but the other one does not. The assumption was that this was due to steric (distance) requirements. That made the aerogel –NCO composite able to attach to other entities in a controlled manner.

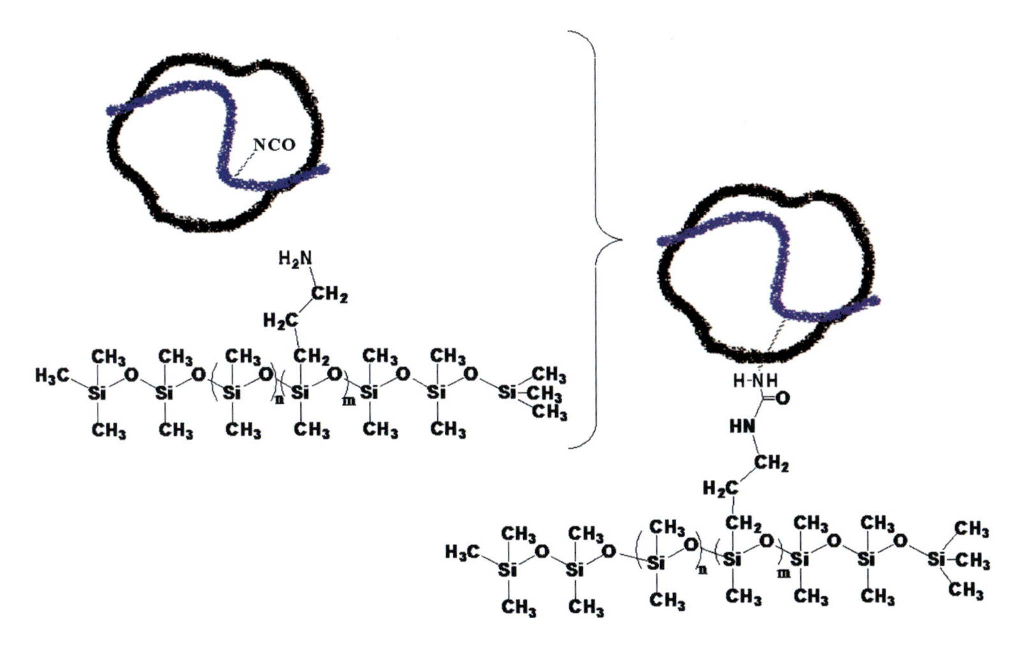

Figure 18.17. Reaction scheme for attachment of an amine substituted silicone polymer to an isocyanate-substituted chitosan-silica aerogel [26].

In Figure 18.17, the aerogel with –NCO attached to the chitosan polymer therein is represented in the diagram in the upper left. It is shown reacting with an amine substituted silicone polymer to form a urea-linkage-attached silicone-aerogel composite site.

Liu [26] also found that isocyanatoethyl methacrylate (*IEMA*) reacts through the –NCO group to chitosan $–NH_2$ groups to attach methacrylate moieties to the SiO_2-X aerogel, as shown in Figure 18.18. The attachment of these groups also was achieved by the combination of hydroxyethyl methacrylate (*HEMA*) to the –NCO functionalized SiO_2-X aerogels.

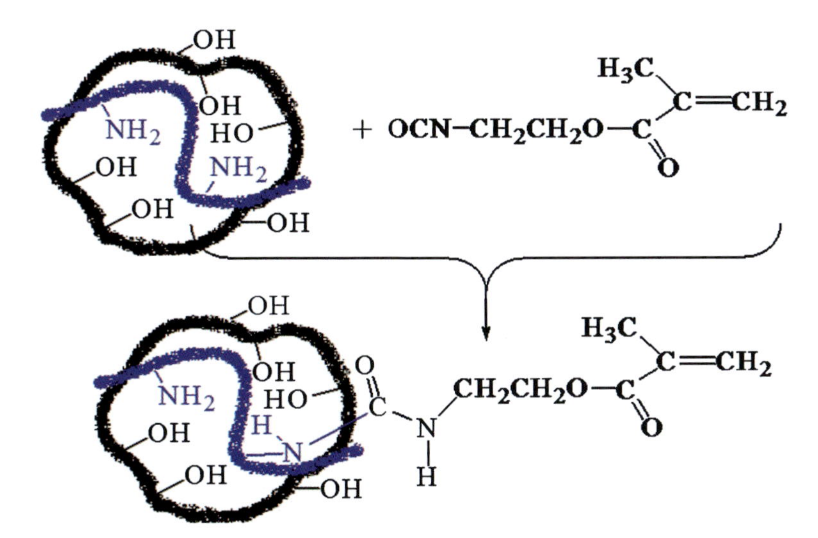

Figure 18.18. Reaction scheme used to attach a methacrylate group onto a chitosan-silica aerogel [26].

18.5. Conclusion

The incorporation of chitosan in an open structure silica-based aerogel produces a class of materials that exhibit a wide range of physical and chemical properties that can be employed in new applications and that help to demonstrate the role of nanostructure in these low density solids.

One example of the use of these properties is the attachment of aerogel particles to natural and synthetic fibers by employing the –NCO functional X-SiO_2-based composites [31]. This can lead to incorporating the gas-absorbant abilities into fibers, clothing and masks for use in hostile chemical environments.

Another example is illustrated in the work of Ji [8], who photopolymerized acrylates in X-SiO_2 aerogel monoliths as a demonstration of the potential for storing information holographically.

The use of X-SiO_2-based aerogels for coatings in ink-jet print media has been explored and found to be successful [23–25]. Their small molecule (dye) absorbant characteristics and their hydrophilic nature make them very promising as a water-fast archival coating component.

Another application being explored is the use of these X-SiO_2-based aerogels, with and without added ions or oxidants, as novel grinding powders in CMP (Chemical Mechanical Planarization) in advanced electronics applications [32]. They provide some or all of the required grinding and add the ability to absorb and remove substantially all of the ions of Cu, Ta and W produced in the CMP process. This can lead to a much cleaner and less feature-perturbing way of planarizing the layers in the chip production.

The catalytic applications of the metal-containing X-SiO_2-based aerogels has been demonstrated in an important case by Ji [8], who showed that the aerogel in the Pd-form can be reduced and used to catalyze the selective hydrogenation of acetylene in the presence of ethylene about as well as – and perhaps superior to – the known commercial catalysts. Since the methods for the production of ethylene-based polymers employ catalysts that can be poisoned by acetylene, this is a key reaction. Moreover, its selectivity is essential, so that the ethylene itself is not hydrogenated.

Further reactions and uses of the chitosan-functionalized silica aerogels are being explored. Hopefully, they will continue to provide fascinating results.

References

1. Hu X, Ji S, Littrell K and Risen WM Jr (2000) Transparent monolithic metal ion containing nanophase aerogels. Mater. Res. Soc. Symp. Proc. 581: 353–362
2. Ayers MR, Hunt AJ (2001) Synthesis and properties of chitosan-silica hybrid aerogels. J. Non-Cryst. Solids 285: 123–127
3. Brack HP, Tirmizi SA, Risen WM Jr (1997) A spectroscopic and viscometric study of the metal ion-induced gelation of the biopolymer chitosan. Polymer 38: 2351–2362
4. Tirmizi SA (1992) Study of Reactions of Platinum Complexes with Chitosan. Ph. D. Thesis, Brown University
5. Domard A, (1987) pH and CD measurements on a fully deacylated chitosan: applications to copper(I)-polymer interactions. Int. J. Biol. Macromol. 8: 98
6. Park JW, Park MD, Park KK (1984) Mechanism of metal binding to chitosan in solution. Cooperative inter- and intramolecular chelations Bull. Korean Chem. Soc. 5(3): 108–112
7. Hu X (2002) Preparation and Characterization of Silica-Polymer Transition Metal Hybrid Aerogels. Ph. D. Thesis, Brown University

8. Ji S (2001) Synthesis and Studies of Transparent, Monolithic Polymer-Silica Hybrid Aerogel Materials. Ph. D. Thesis, Brown University
9. Hu X, Littrel K, Ji S, Pickles DG, Risen WM Jr (2001) Characterization of Silica-polymer aerogel composites by small-angle neutron scattering and transmission electron microscopy, with, J. Non-Cryst. Solids, 288: 184–190
10. Hu X, Ji S, Risen WM Jr (2002) Biopolymer-silica hybrid aerogels containing transition metal species; Structure, Properties and Reactions. Mater. Res. Soc. Symp. Series, Vol. 702: 65–76
11. Wang M, Liu X, Risen WM Jr (2002) A new hybrid aerogel approach to modification of bioderived polymers for materials applications. Mater. Res. Soc. Symp. Proc. 702: 77–85
12. Liu X, Wang M, Risen WM Jr (2003) Polymer-attached functional inorganic-organic hybrid nano-composite aerogels. Mater. Res. Soc. Symp. Proc., 740, I12.24.1–6
13. Liu X, Zhu Y, Yao C, Risen WM Jr (2003) Photo-formed Metal Nanoparticle Arrays in Monolithic Silica-Biopolymer Aerogels. Mater. Res. Symp. Proc. 788: 13–24
14. Yao C, Risen WM Jr (2005) Formation and reaction of metal-containing nanoparticles in organic/inorganic hybrid aerogels. Mater. Res. Soc. Symp. Proc. 847: 515–520
15. Yao C, Liu X, Risen WM Jr (2006) Novel ferromagnetic aerogel composite materials with nanoparticle formation and chemistry in response to light. Mater. Res. Soc. Symp. Proc. (2006), 888: 131–136
16. Risen WM Jr, Zhang R, Hu X, Ji S (2001) Aerogel Materials and Detectors, Liquid and Gas Absorbing Objects, and Optical Devices Comprising Same. U. S. Patent 6,303,046 B1
17. Yao C (2006) Nanoparticles in Aerogels and Cellulose Composites. Ph. D. Thesis, Brown University
18. Kuthirummal N, Dean A, Yao C, Risen WM Jr (2008) Photo-formation of gold nanoparticles: Photoacoustic studies on solid monoliths of Au(III)-chitosan-silica aerogels. Spectrochim. Acta A Molecular and Biomolecular Spectroscopy A70: 700
19. Muzzarelli RAA, Muzzarelli C (2007) Chitosan as a dietary supplement and a food technology agent. Functional Food. Carbohydrates 215–247
20. Muzzarelli RAA (2000) Recent results in the oral administration of chitosan. Adv. in Chitin Sci. 4(EUCHI'99): 212–216
21. Muzzarelli RAA (1999) Clinical and biochemical evaluation of chitosan for hypercholesterolemia and overweight control. 87(Chitin and Chitinases) 293–304. (CODEN: EXSEE7 ISSN:1023–294X
22. Liu J, Zeng M, Shi F, Tang N Wei L (2007) Gougneng Cailiao 38(9), 1527–30 (the title is in Chinese)
23. Risen WM Jr, Zhang R, Hu X, Ji S (2003) Printing Medium Comprising Aerogel Materials U.S. Patent 6,602,336
24. Risen WM Jr, Zhang R, Hu X, Ji (2006) Printing Media Comprising aerogel materials. U.S. Patent 7,037,366
25. Risen WM Jr, Zhang R, Hu X, Ji (2006) Printing Media Comprising aerogel materials. U.S. Patent 7,147,701
26. Liu X (2005) Chemistry in an Inorganic-Organic Hybrid Aerogel: Chitosan-Silica Aerogel. Ph. D. Thesis, Brown University
27. Liu X, Yao C, Risen WM Jr (2006) Chemically selective reactions in confined spaces in hybrid aerogels. Mat. Res. Soc. Symp. Proc.: Vol. 899E (MRS Electronic) http://www.mrs.org/s_mrs/bin.asp?CID=6230&DID=170516&DOC=FILE.PDF. Accessed 7 July 2009
28. Shaw CF III (1999) Gold Based therapeutic agents. Chem. Rev. 99: 2589–2600
29. Shaw, CF III, Cancro MP, Witkiewicz PL, Eldridge JE (1980) Gold(III) oxidation of disulfides in aqueous solution. Inorg. Chem. 19(10): 3198–3201
30. Shaw CF III, Schmitz, GP (1982) (L-Cysteinato)gold. Inorg. Syntheses 21, 31–33
31. Risen WM Jr, Liu X (2004) Chitosan biopolymer-silica hybrid aerogels. In: Wallenberger FT and Weston N (ed) Natural fibers, plastics and composites. Kluwer, Boston
32. Caldwell M, Yao C, Liu X, Risen W M Jr, to be published

19

Anisotropic Aerogels by Photolithography

Massimo Bertino

Abstract A general method is presented that allows fabrication of sol–gel materials with anisotropic physical properties. The gelation solvent is exchanged with a suitable solution of precursors and suitable chemical reactions are triggered in irradiated regions of the monoliths. The physical properties of the exposed regions can be varied almost at leisure by changing the precursors. For example, metal and sulfide nanoparticles can be formed inside the pores of the matrices and these nanoparticles change optical absorption, emission and index of refraction of the exposed regions. Polymers can be attached to the walls of the matrix pores, and this allows modulation of mechanical strength, hydrophobicity and optical properties. The character of the patterns can be adapted to the specific applications by varying the precursor solution. Single-photon reactions are used to generate patterns that start on the surface of the monoliths and extend within the bulk of the monoliths. Precursors that react when exposed to ionizing radiation are employed to create high aspect ratio patterns, and precursors that are dissociated by multiphoton processes are used to produce three-dimensional architectures. Physical properties and possible applications of the monoliths are discussed.

19.1. Introduction

Sol–gel techniques are being employed to fabricate components not only for mainstream applications such as photonics, thermal insulation, electronics and microfluidics, but also for more exotic applications such as space dust and radiation collectors [1]. Methods have been developed to tailor the physical properties of sol–gel materials to the requirements of a specific application. For example, porosity and pore size distribution can be controlled by forming micelles in a sol [2–4]; gels can be made hydrophobic by derivatizing the otherwise hydrophilic pore walls with hydrophobic moieties [5]; superhydrophilicity can be obtained by ultraviolet irradiation [6, 7]; mechanical strength can be increased by cross-linking the oxide nanoparticles that make up the gel [1, 8, 9], and optical properties can be controlled by adding chromophores and nanoparticles to control index of refraction, absorption and luminescence [10–12].

M. Bertino • Department of Physics, Virginia Commonwealth University, Richmond, VA 23284, USA
e-mail: mfbertino@vcu.edu

M.A. Aegerter et al. (eds.), *Aerogels Handbook*, Advances in Sol-Gel Derived Materials and Technologies, DOI 10.1007/978-1-4419-7589-8_19,

Till date, most of the work has focused on the fabrication of porous matrices whose properties are modified uniformly throughout the volume. Sol–gel materials with spatially modulated properties have been comparatively seldom reported and the research has been mostly limited to thin films. Modulation is most frequently obtained by etching; however, holographic techniques have also been reported [13]. One of the first reports of modulation within bulk monoliths came from Dunn's group [10] where three-dimensional Ag patterns were produced by multiphoton reduction of Ag^+ in water. Subsequently, a more general lithographic method was developed by our group soon thereafter allowing to produce a wide variety of sol–gel matrices with anisotropic physical properties [11, 12, 14–18]. That method is reviewed in this chapter and it is demonstrated how anisotropic monoliths containing two- and three-dimensional patterns, including high aspect ratio (~400) patterns, can be produced with ease. The patterns can be made of metal or semiconductor nanoparticles, or polymers that cross-link the oxide backbone. Examples are presented showing that optical absorption, emission, index of refraction, porosity, hydrophobicity and mechanical strength can be tuned in the patterned regions by simply varying the type and concentration of the precursors. Possible applications are discussed.

19.2. General Principle

In our method, a porous monolith is prepared following conventional sol–gel procedures reported in the literature. Most results presented here focus on silica gels prepared by base-catalyzed hydrolyzation of silicon alkoxides. However, other synthesis procedures such as acid-catalyzed sol-gel synthesis, and matrices other than silica can also be employed. After gelation, the solvent is exchanged several times with methanol to remove organic residues, which can interfere with the photochemical reactions or react with the precursors. For example, it has been noted that Au nanoparticles can form spontaneously when the solvent is exchanged with a gold chloride solution after gelation without methanol exchange [19]. Precursors can be chosen to fulfill the needs of several applications. If composites containing metal nanoparticles are desired, the metal ions of water-soluble metal precursors such as $AgNO_3$ can be reduced by thermally dissociating a suitable reducing agent such as formaldehyde or by ionizing the solvent to generate strongly reducing solvated electrons. Metal chalcogenides are produced by photodissociating sulfur-containing molecules such as 2-mercaptoethanol, thioacetamide or selenourea. Polymer photocross-linking requires derivatization of the pore walls prior to polymerization. Such derivatization is achieved in most cases by adding an alkoxide carrying a suitable group to the gelation solution, e.g., vinyltrimethoxysilane. The gelation solvent is then exchanged with a solution of a monomer and a photoinitiator. In the exposed regions polymerization engages the moiety dangling from the pore walls giving rise to cross-linking of the silica nanoparticles that make up the backbone of the porous materials (for polymer cross-linked aerogels refer to Chap. 13).

Our method can also produce patterns with a variety of different light sources, which is relevant for rapid prototyping. For example, an infrared source such as the first harmonic of a Nd:Yag laser (1,064 nm) can be used to locally heat a monolith and induce either dissociation of chelates or thermal polymerization (Sect. 19.3.1). Visible light can be used for photopolymerization. Ultraviolet light can be used to dissociate organics whose fragments react with metal ions or trigger polymerization (Sect. 19.3.2). Ionizing radiation can be used to create highly reactive radicals by dissociating water or organic molecules and to cross-link polymers (Sect. 19.3.3).

Further, our method allows fabrication of patterns with different characters. Surface patterns are produced using short-wavelength ultraviolet light that is strongly absorbed by both the matrix and the organics, e.g., the fourth harmonic of Nd:YAG lasers at 266 nm. Patterns that originate on the surface of monoliths and also protrude into their bulk can be generated using light with longer wavelengths, such as near-ultraviolet and visible light, or using ionizing, deeply penetrating radiations such as collimated X-rays (Sect. 19.3.3). Three-dimensional patterns are produced by multiphoton dissociation of suitable solvents and precursors (Sect. 19.3.4).

19.3. Synthesis of Nanoparticles Within the Matrix Pores

This section demonstrates anisotropic sol–gel materials by synthesizing nanoparticles inside the pores of the matrix. Emphasis is placed on semiconductor and metal nanoparticles showing two- and three-dimensional patterns that can be produced with the use of a wide array of radiation sources.

19.3.1. Infrared Lithography

For infrared lithography the wet-gel filling solvent is exchanged with a solution of precursors that react spontaneously albeit slowly at room temperature, such as $Cd(NO_3)_2$ and thiourea. The reaction proceeds in two steps, hydrolysis of thiourea in basic environment (19.1), followed by reaction of sulfur and Cd^{2+} ions (19.2). The overall reaction is summarized by (19.3).

$$H_4N_2CS + OH^- \rightarrow HS^- + H_2NCN + H_2O \tag{19.1}$$

$$HS^- + Cd^{2+} + OH^- \rightarrow CdS + H_2O \tag{19.2}$$

$$Cd^{2+} + H_4N_2CS + 2OH^- \rightarrow CdS + H_2NCN + 2H_2O \tag{19.3}$$

The idea behind infrared lithography is that the rate of reaction 1 and therefore 3, is low at room temperature but it is greatly increased when the temperature is raised [20, 21]. Refrigeration and chelation are required to prevent nanoparticle formation during the solvent exchange step. In our experiments, Cd^{2+} were chelated by adding NH_4OH to the parent solutions; precursor concentrations of the order of 0.5 mol l^{-1} were employed, which generated nanoparticles in a few seconds at temperatures above 40°C. Such a moderate temperature increase is easily attained by focusing an infrared or even visible laser beam onto a sample. For example, a continuous-wave Nd:YAG laser emitting at 1.064 nm and a power of about 1 W heated the exposed regions to 50–60°C, as measured by a thermocouple inserted in the exposed region of the gel. For systems other than chalcogenides, other concentrations and chelating agents can be employed, provided that the complexes dissociate at temperatures below the boiling point of the solvent of the parent solution. Patterns that formed on the gel surface and extended for a few hundred microns inside the bulk were obtained by translating the sample in front of the focused beam as shown in Figure 19.1A. Patterns with a more three-dimensional character were obtained by focusing a high-power beam inside a gel with a short focal length lens or with a microscope objective, as shown in Figure 19.1B. Samples could be dried in supercritical CO_2 to yield aerogels without loss of the patterns, or alterations of their shape and spatial resolution. The procedure used for CdS was also extended to the fabrication of noble metal nanoparticle patterns. Metal ions were first chelated with a suitable agent (e.g., amines for Ag^+); a concentrated (1 mol l^{-1}) reducing agent like formaldehyde was then added, which reacted rapidly with the metal ions when the sample was heated.

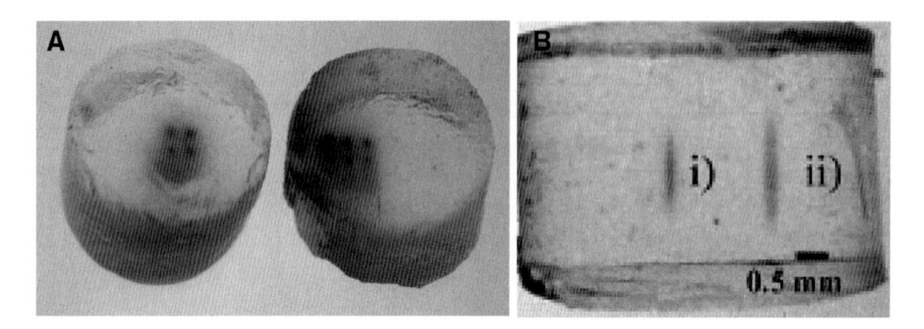

Figure 19.1. Examples of monoliths patterned with CdS quantum dots. The scale bar represents 0.5 mm for both the figures. From [11].

The patterned regions were analyzed with a series of standard analytical techniques such as transmission electron microscopy (TEM), X-ray diffraction (XRD), absorption, emission, Raman and X-ray photoelectron (XPS) spectroscopies, and Brunauer–Emmett–Teller (BET) surface area and porosity measurements. All these techniques confirmed that CdS or chalcogenides and metals (depending on the precursors) were formed in the exposed regions.

Absorption spectra have shown a broad excitonic band at energies characteristic of II–VI chalcogenides. The band narrowed and shifted toward shorter wavelengths upon addition of surfactants to the parent solution, as shown in Figure 19.2. The shift and

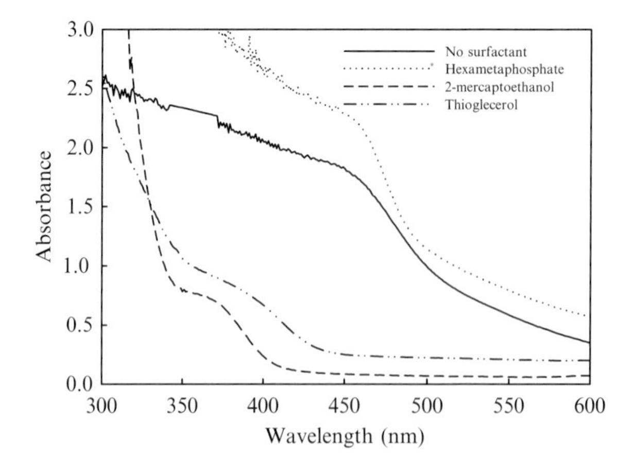

Figure 19.2. Absorption spectra of hydrogels patterned with CdS nanoparticles. From [15].

narrowing of the excitonic bands are a clear indication that surfactants help control particle size. Not surprisingly, smaller particles were obtained when strongly coordinating surfactants such as thiols were employed.

19.3.2. Ultraviolet Lithography

In infrared lithography the precursors are dissociated thermally by focused infrared light (IR). Heat diffusion, however, makes use of masks impractical. Patterns can only be produced by translating the sample in front of the incident beam. Ultraviolet light does not present those issues of infrared light and is preferable for applications that require high spatial resolution or masking. For quantum dot lithography, two classes of precursors can be employed.

In one class of precursors, metal thiolates are formed by adding a thiol (RSH) like 2-mercaptoethanol to a Cd^{2+} solution. Previous work [22] has shown that Cd^{2+} (but also other metal) ions in the presence of thiols form polynuclear species complexed with RS^{-} with the formula $[Cd(RS_5Cd_3)_n]^{(n+2)+}$. Composition and solubility of the complexes depend not only on metal and thiol concentration but also on pH, as reported previously by Hayes et al. [23] UV–Vis spectra of Cd thiolate complexes exhibit an absorption maximum around 250 nm [23, 24] as shown in Figure 19.3. Upon ultraviolet (254 nm) irradiation, the absorption peak at 250 nm decreased, while a second absorption band in the 270–290 nm appeared.

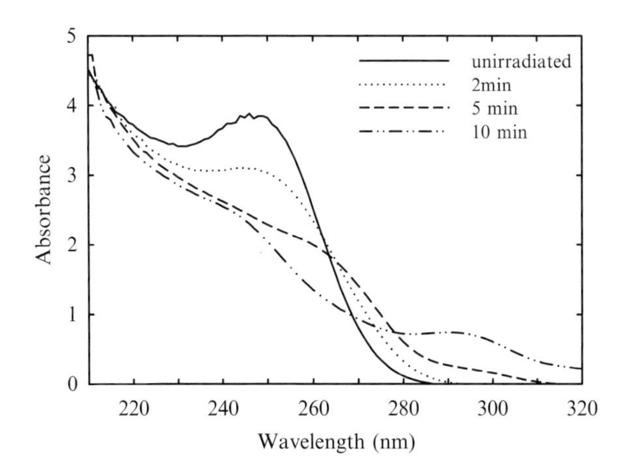

Figure 19.3. Optical absorption spectra of an aqueous solution with $[CdSO_4] = 1.25 \times 10^{-4}$ M, and [2-mercaptoethanol] $= 1.25 \times 10^{-3}$ M at the indicated exposure times. NH_4OH was added to keep the pH of the solution around 11. The solutions were illuminated with a high-pressure, unfiltered 100 W Hg lamp. From [14].

For sufficiently long irradiation times (15–30 min), a yellow precipitate appeared at the bottom of the vial. XRD showed that the precipitate was cubic CdS. Formation of CdS precipitates for long irradiation times suggested that the absorption band in the 270–290 nm region was due to very small CdS nanoparticles. Formation of CdS in the illuminated solutions is likely related to the photodissociation of the Cd–thiolate complexes. It is known that the outer ligands of Cd thiolates [25, 26] and thiols bound to the surface of CdS particles [27] are detached by exposure to ultraviolet light.

The photodetached fragments are typically thiol radicals, $RS^{\bullet}$. These radicals usually react to form disulfides $(RS)_2$, but they can also form anions RS^-, and sulfide ions S^{2-} [28–30]. Those latter anions react promptly with Cd^{2+} and explain the formation of CdS nanoparticles. Thus, aerogels were patterned with quantum dots starting from the same thiolate precursors. However, two disadvantages quickly became evident during experimentation. One was the limited solubility of these precursors in the sol; the second disadvantage was the comparatively poor efficiency of the photodissociation process, which required lengthy exposures with high power sources. More efficient precursors were sought and found in thiourea, thioacetamide and selenourea [12]. The overall reaction scheme leading to chalcogenide nanoparticles is shown in (19.4) and (19.5) or thioacetamide [31].

$$CH_3CSNH_2 + h\nu \rightarrow CH_3CN + H_2S \tag{19.4}$$

$$Cd^{2+} + H_2S \rightarrow CdS + 2H^+ \tag{19.5}$$

For selenourea patterning, oxidation of Se^{2-} [32] was prevented by adding reducing agents to the precursor solution. Thiourea, thioacetamide and selenourea have several advantages over metal thiolate precursors. First, these precursors are more easily photodissociated than thiolates. In samples containing the same precursor concentration exposure times are reduced by a factor 5–10 when selenourea is used instead of Cd thiolates. In addition, thioacetamide, thiourea and selenourea allow to operate at concentrations as high as 0.5 M, which increases nanoparticle concentration and therefore index contrast in the patterned regions.

19.3.3. X-Ray Lithography

For X-ray lithography, the key reagents are the solvated electrons that are generated when ionizing radiation interacts with water and other solvents [33]:

$$H_2O + h\nu \rightarrow e^-_{(aq)} H^{\cdot}, OH^{\cdot}, H_3O^+, \ldots \tag{19.6}$$

Solvated electrons have a reduction potential of −2.7 eV, which is sufficient to reduce most metals. Metal nanoparticles are formed by direct reduction of ions:

$$Me^{n+} + ne^- \rightarrow Me^0 \tag{19.7}$$

$$Me^0 + Me^0 + \rightarrow Me_n \tag{19.8}$$

Using collimated X-rays and a suitable masking system, patterns can be created within sol–gel monoliths. Figure 19.4 shows an example of a grid pattern obtained by radiolytic reduction of highly concentrated (3 mol l^{-1}) $AgNO_3$.

The features in Figure 19.4 had a size of the order of tens of microns, which is the typical feature size attainable with the X-rays that were employed in the experiments. Those X-rays had an energy between 5 and 10 keV, and for these energies masks can only be fabricated with feature sizes of the order of tens of microns. The features in the gels were measured with an optical microscope and they were found to replicate the patterning mask to better than 1 μm. Most importantly, the patterns penetrated deeply inside the

Figure 19.4. Ag patterning of silica aerogels with collimated X-rays. The cell walls have a thickness of 50 μm and the samples are 3-mm thick. From [34]. The images were obtained with a digital camera equipped with a macro objective and they were taken from the same sample.

gel monoliths. Patterns up to 20 mm deep were obtained, which correspond to an aspect ratio of 400 for 50 μm wide features. The deep penetration of the patterns is explained by the comparatively high energy of the incident X-rays (a few keV) and by the low density of the matrix. Sol–gel materials such as the silica gels prepared by our group are up to 90% porous, so that the penetration depth of the X-rays is close to the penetration depth in pure solvents. Patterns with the same geometrical dimensions and aspect ratio can also be fabricated out of metal chalcogenide nanoparticles [12]. In this case, the gelation solvent is exchanged with a solution of group VI metal ions and 2-mercaptoethanol or thioglycerol. These thiols are very efficient scavengers of the solvated electrons produced by water ionization and liberate sulfur species in the process [35, 36]. The overall reaction leading to metal chalcogenides is reported in (19.9–19.11). Reduction of the metal ions by the solvated electrons, (19.12), is prevented by working in a large excess of thiol, typically 10–100 times the metal ion concentration.

$$H_2O + hv \rightarrow e^-_{(aq)}\, H^{\cdot}, OH^{\cdot}, H_3O^+, \cdots \tag{19.9}$$

$$HOCH_2CH_2SH + e^-_{(aq)} \rightarrow SH^- + {}^{\cdot}CH_2CH_2OH \tag{19.10}$$

$$SH^- + Pb^{2+} \rightarrow PbS + H^+ \tag{19.11}$$

$$Pb^{2+} + 2e^-_{(aq)} \rightarrow Pb^0 \tag{19.12}$$

Patterns were obtained out of metal chalcogenides which were comparable in size and penetration depth to those obtained for metals. Nanoparticles grew in size with irradiation time. Shallow patterns consisted of small, highly luminescent nanoparticles. Deep patterns, where long irradiation times were required, contained large particles with a comparatively low luminescence quantum yield.

X-ray lithography is therefore an ideal candidate to prepare high aspect ratio patterns in aerogels. The technique is also quite efficient. Typically, an exposure to 85 mA min was sufficient to generate optically dense Ag or CdS patterns. This exposure is very low when compared to conventional X-ray lithography processes such as the LIthografie, Galvanoformung, Abformung (*LIGA*) process. These processes are usually based on radiation-induced cross-linking of polymers like poly(methyl metacrylate) (PMMA) which require lengthy exposures. On the same set-up, PMMA structures were obtained after exposures of the order of 40,000 mA min.

19.3.4. Three-Dimensional Patterning

As it became apparent in the previous section, an issue relevant for applications is related to the possibility of producing three-dimensional patterns, i.e., patterns formed in the bulk of monoliths but not on their surface. Three-dimensional chalcogenide patterns can be produced by multiphoton ionization as well as X-rays. The technique described here is different from those shown in Figures 19.1 and 19.4. Alternatively, 3D patterns can be obtained via a single photon process by focusing the incident light inside a monolith with a long-working distance microscope objective. For three-dimensional patterning, the solvent of a silica hydrogel is exchanged with a solution of metal and chalcogenide precursors, e.g., $Pb(NO_3)_2$ and 2-mercaptoethanol, and then the samples are exposed to the focused light of a pulsed laser. Patterns are formed in the illuminated regions and their depth within the monolith is varied by moving the sample in front of the focused beam. Typical results are shown in Figure 19.5. Damage of the gel matrix was observed when the laser was operated at powers higher than 0.1 TW cm^{-2}. Figure 19.5D shows a typical patterned region obtained at a laser power slightly above the damage threshold. The beam was incident from the top. Damage is noticeable in the focal region in the form of a broad, cloud-like feature. An elongated dark feature extending in the direction of the beam axis is also clearly

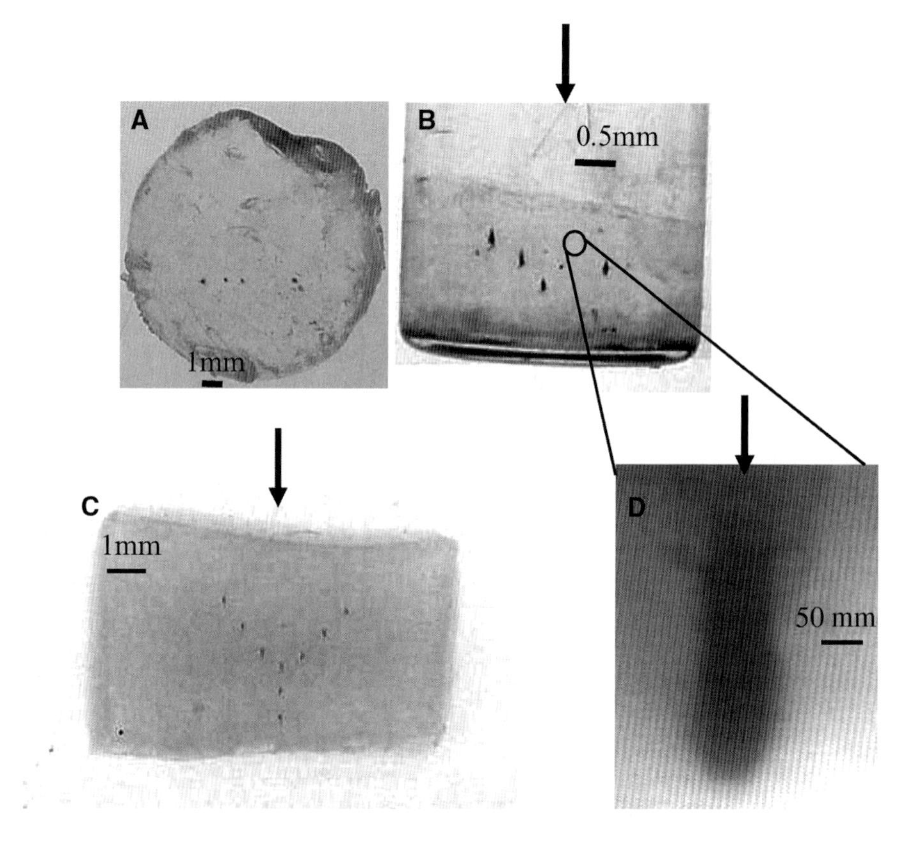

Figure 19.5. **A. Top view** and **B. side view** of PbS patterns. The **side view** of a pattern obtained by mounting a sample on a translation stage is shown in **C. D.** Optical microscope image of a patterned region. *Arrows* indicate the incident beam direction. The regularly spaced vertical lines in some of the figures are an artifact of the digitization process. From [37].

noticeable. The elongated shape of the pattern is likely due to beam self-focusing induced by an increased refraction index in the exposed region [38–40]. The diameter and length of the patterns could be varied by varying the magnification of the microscope objective or by increasing the exposure time.

The mechanism of nanoparticle formation in the exposed regions was investigated with a series of control experiments. Thermal and single-photon processes were ruled out by exposing samples to intense continuous-wave lasers. Patterns did not form upon long exposures (up to 8 h) to intense continuous-wave visible and infrared lasers with a power of the order of 1.5 W focused by a microscope objective. Further, to ensure that the sulfur precursor was not directly dissociated, 2-mercaptoethanol was replaced with other precursors, which are often used to generate metal sulfides such as thioacetamide and thiourea. None of these precursors generated nanoparticles, suggesting that multiphoton dissociation of sulfur-containing organic molecules does not play a relevant role. Therefore, dissociation of the sulfur precursors likely followed an indirect path and was probably triggered by the radicals liberated by water ionization. The power in the focal spot was between 0.01 and 0.1 TW cm^{-2}, which is sufficient to ionize water [35], at which point the reaction likely proceeds following the radiolytic steps of (19.9–19.12).

19.4. Anisotropy by Polymer Photocross-linking

The nanoparticle patterning techniques described above are very powerful but they have some limitations. Nanoparticles affect the optical properties, but not, for example, the mechanical properties or hydrophilicity. In addition, nanoparticles can clog the matrix pores and limit diffusion of fluids and analytes through the monoliths [4].

Our lithographic method can be used to fabricate spatially anisotropic monoliths by selective photocross-linking. In this variation of our method, an olefin such as vinyl or methacrylate is first attached to the pore walls; the gelation solvent is then exchanged with a solution of a monomer and a photoinitiator. In exposed regions the initiator can react with the surface-bound olefins. Polymeric tethers that cross-link the oxide nanoparticles are generated by the monomer reacting with the surface-bound radicals, as shown in Figure 19.1. In addition, the flexibility of polymer chemistry allows to insert a wide range of functionalities into the monoliths.

Photocross-linked aerogel monoliths prepared by uniform illumination of wet gels have physical properties that are comparable to those of cross-linked materials prepared by thermally activated cross-linking, i.e., low density and high mechanical strength [8]. Materials have been prepared by photocross-linking with a modulus of up to 1 GPa and surface areas of the order of 100 m^2 g^{-1} [18]. Use of hydrophobic cross-linkers such as styrene yields hydrophobic materials, consistent with previous reports [35, 36]. AFM (Figure 19.6) and SEM show that the smallest aggregations, well evident in native (non-derivatized) samples, are not identifiable in polymer cross-linked samples, confirming the coating of the pore walls by a polymer layer.

Anisotropic aerogels are produced by scanning a sample in front of a collimated beam as shown in Figure 19.7. Very recently, a promising development has been obtained by photopolymerization with visible light [41]. In this case, methyl methacrylate is polymerized with a green (532 nm), continuous-wave laser using an Eosin-tertiary amine initiator [42, 43]. Different from UV light, visible light is not strongly absorbed or

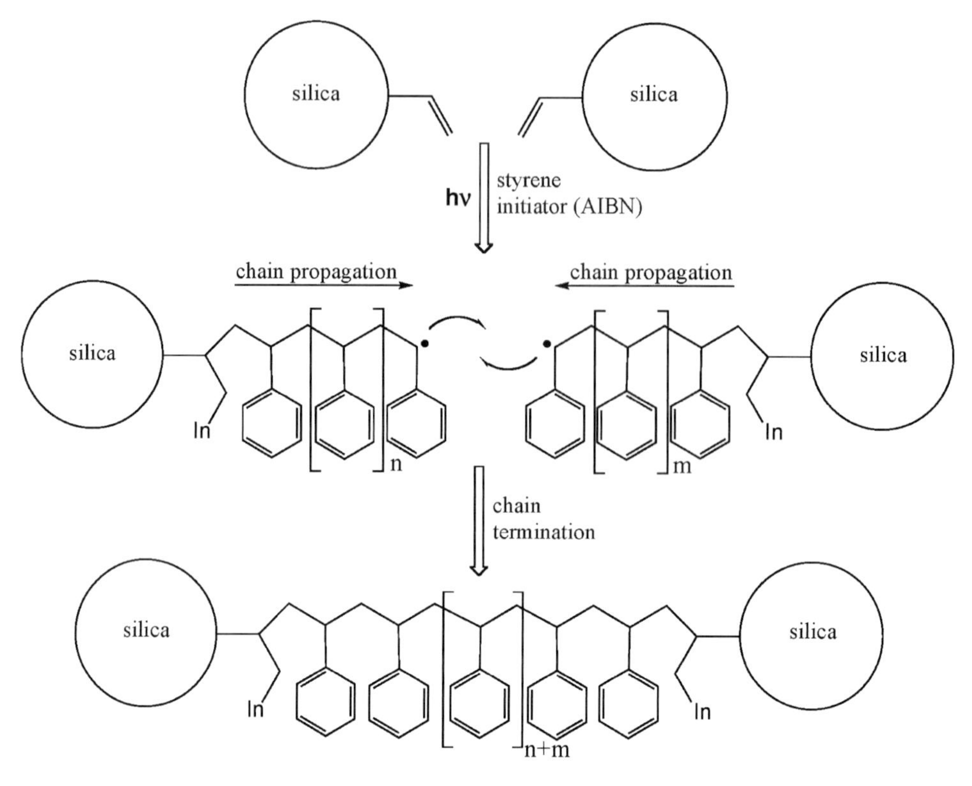

Scheme 19.1. Photocross-linking mechanism of skeletal vinyl-modified silica nanoparticles with polystyrene.

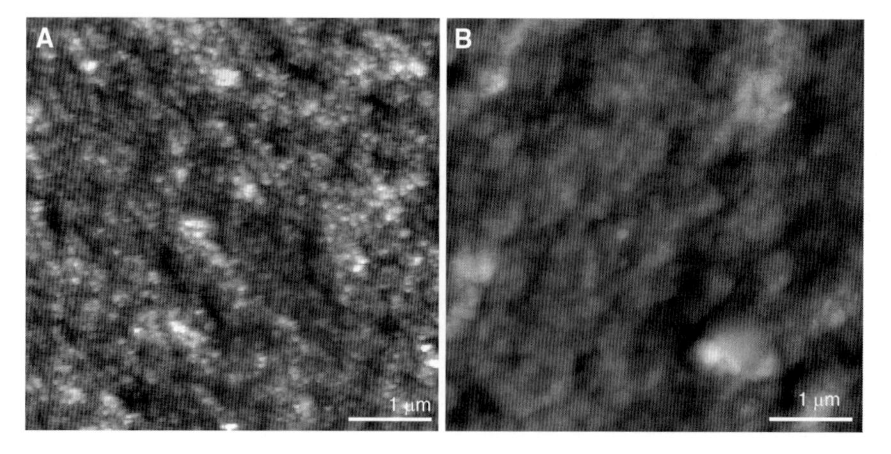

Figure 19.6. AFM images (5 μm × 5 μm) of vinyl-derivatized silica xerogel **A.** before and **B.** after photocross-linking. **A.** shows the typical morphology of a vinyl-derivatized sample before photopolymerization. Granular aggregates with a size of tens of nanometers could be resolved, which likely corresponded to the secondary aggregations. After polymerization, the size of these aggregations increased, **B.** The scale bar represents 1 μm. From [18].

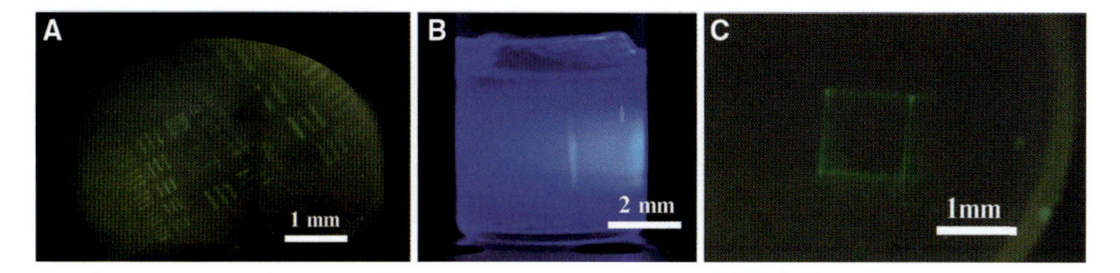

Figure 19.7 Digital camera images of monoliths photocross-linked with 9-vinylanthracene. **A.** Surface pattern produced by masking 266 nm light with a standard USAF 1951 resolution calibration mask. **B.** Pattern produced by focusing with a 100 mm focal length lens, the third harmonic (355 nm) of a pulsed Nd:Yag laser inside a monolith. **C.** Pattern produced by focusing 355 nm light inside a gel monolith with a microscope objective. Luminescence was excited in all cases by the 325 nm light of a He–Cd laser. The *blue hue* in **B.** is due to laser glare and to the luminescence of the solvent (acetonitrile) impregnating the wet gel. The solvent of samples **A.** and **C.** had been exchanged with nonluminescent methanol, and a high-pass filter was employed to cut the laser glare. From [18].

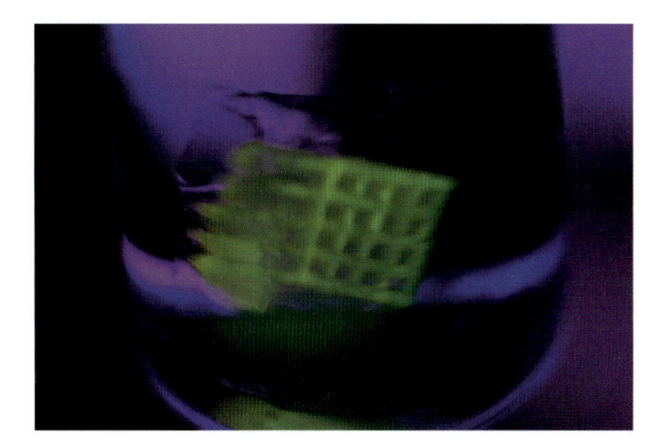

Figure 19.8 Grid structure obtained by photopolymerization. A grid with a spacing of 1 mm and wall thickness of 250 μm was fabricated inside a wet gel. The gel had a thickness of 6 mm and the grid penetrated throughout the gel. A luminescent dye (eosin Y) was incorporated in the polymerized regions for visualization. From [41].

scattered by the organics or by the matrix, and therefore it can penetrate deeply inside the gels. Figure 19.8 shows an example of a grid pattern obtained by masking the laser beam with a 250 μm pinhole.

19.5. Physical Properties

19.5.1. Absorption and Emission

Figure 19.2 shows that the optical absorption of composites can be tuned over a very wide range of wavelengths by adding a suitable surfactant or by varying its concentration. Because of quantum confinement, variation of the mean nanoparticle size shifts the excitonic recombination band and the onset of absorption. Luminescence is dominated by extrinsic recombination and is pretty weak in as-grown samples. Typical quantum yields were well below 1%, which is typical for quantum dots grown in aqueous environment.

The luminescence properties of the composites, however, can be improved by photocorrosion. In photocorrosion, quantum dots are illuminated with light above their band gap, typically near ultraviolet light. Oxidation processes are triggered by illumination (19.13) [44–47]. Since surface defects are photooxidized preferentially, photoactivation is a convenient technique to remove such defects and improve the photoluminescence quantum yield.

$$CdSe + O_2 + hv \rightarrow Cd^{2+} + SeO_2 \quad (19.13)$$

Figure 19.9 shows an example of photoactivation of a CdSe-patterned sample. The emission of as-grown samples increased by more than 300 times after photoactivation. The emission maximum was around 580 nm before photoactivation and around 560 nm after photoactivation. The blue shift of the emission indicates that the mean size of the nanoparticles is reduced by photoactivation. This is not surprising, since (19.13) is a photocorrosion reaction. Quantum yields were measured which were in excess of 30%. This high value of the quantum yield is in substantial agreement with other reports of photoactivation of quantum dots. For example, citrate-capped CdSe suspensions can be photoactivated to reach a quantum yield as high as 59% [47]. The quantum yield of our composites is comparable or higher than the quantum yield of commercially available polymer/quantum dot composites, which is of the order of 10% [48]. Patterned regions were clearly visible to the naked eye when illuminated with a commercial handheld light, as shown in Figure 19.10.

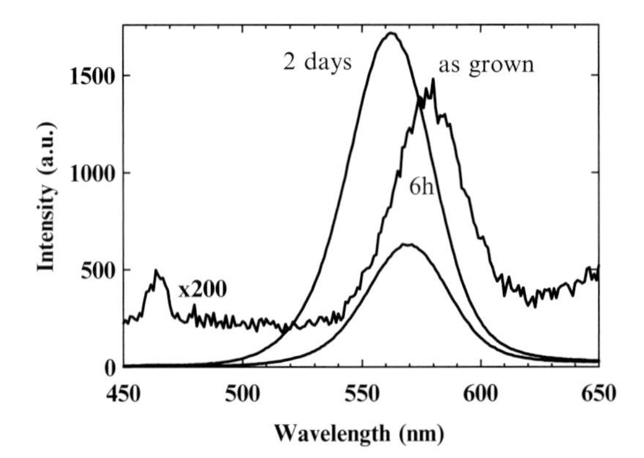

Figure 19.9. Photoactivation of a CdSe/silica composite. The parent solution contained 1.6×10^{-3} mol l^{-1} cadmium perchlorate, 3.5×10^{-3} mol l^{-1} sodium citrate and 4×10^{-4} mol l^{-1} selenourea. Samples were photoactivated for the indicated times with a black light with a power of 15 W. Samples were excited with 400 nm light. From [12].

19.5.2. Index of Refraction

The highly luminescent patterns that can be obtained by our lithographic technique are clearly promising for optical applications. The question is whether the density of the chromophores is sufficient to give rise to guiding effects, thus enabling lithography of optical circuits. We will now present estimates showing that nanoparticle patterns prepared out of parent solutions with a relatively modest concentration of precursors yield patterns that have an increased index of refraction which is sufficient to guide light.

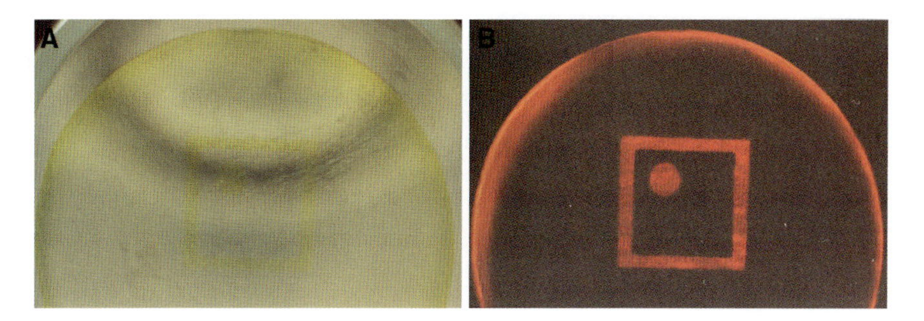

Figure 19.10. **A.** CdSe patterns obtained with UV lithography. **B.** shows the photoluminescence of sample **A.**, obtained by illuminating the patterned sample with the 457.9 nm line of an Ar^{+} laser. From [12].

The dielectric constant $\varepsilon_{\mathrm{eff}}$ of composites can be calculated in first approximation as a weighted average of the constants of the matrix and the nanoparticles:

$$\varepsilon_{\mathrm{eff}} = (1 - V_{\mathrm{f}})\varepsilon_{\mathrm{matrix}} + V_{\mathrm{f}}\varepsilon_{\mathrm{nano}} \tag{19.14}$$

or using more refined approximations such as the Lichteneker approximation:

$$\ln(\varepsilon_{\mathrm{eff}}) = V_{\mathrm{f}}\ln(\varepsilon_{\mathrm{nano}}) + (1 - V_{\mathrm{f}})\ln(\varepsilon_{\mathrm{matrix}}) \tag{19.15}$$

where V_{f} is the nanoparticle volume fraction, $\varepsilon_{\mathrm{nano}}$ the dielectric constant and $\varepsilon_{\mathrm{matrix}}$ is the dielectric constant of the host matrix [49]. For a parent solution containing a concentration c of precursor with a molecular weight MW, yielding nanoparticles with density ρ, the volume fraction of nanoparticles (assuming 100% conversion) is:

$$V_{\mathrm{f}} = \frac{MW \times c}{\rho} \tag{19.16}$$

using values for PbS and silica ($\varepsilon_{\mathrm{PbS}}$= 17.1, $\varepsilon_{\mathrm{matrix}}$ ~1.69) and a precursor concentration $c = 10$ mmol l^{-1}, the index contrast estimated from (19.15) and (19.16) is $\Delta n = 6.6 \times 10^{-4}$, which is sufficient for light guiding [50].

19.5.3. Mechanical Properties

Anisotropic composites with a grid-like reinforcing cross-linked structure combine several attractive properties in a unique way. For example, honeycomb structures can be used to produce ultralight materials with density and thermal conductivity comparable to those of aerogels and mechanical properties comparable to those of aramide fiber composites. We now focus on the physical characteristics of honeycombs, which can be estimated based on volume and surface ratios. The area occupied by the cross-linked aerogel is proportional to $[(2t + a)^2 - a^2]$, where a is the size of the cell filled with native aerogel and t is the thickness of the cross-linked walls. The area occupied by the material between the strong regions is proportional to a^2. Assuming $a = nt$, where n is the cell-to-wall ratio, the density $\rho_{\mathrm{honeycomb}}$ of the composites is:

$$\rho_{\text{honeycomb}} = \frac{\rho_1[(n+2)^2 - n^2] + \rho_2 n^2}{(n+2)^2}, \tag{19.17}$$

where ρ_1 is the density of the cross-linked materials and ρ_2 the density of the filler material. A similar expression can be derived for the compressive strength. The thermal conductivity can be calculated by treating the walls and the filler as parallel resistors. Assuming the cross-linked sections to have a conductivity k_1 and an area A_1, the native silica aerogel sections to have a conductivity k_2 and a cross-sectional area A_2, the conductivity $k_{\text{honeycomb}}$ of the composite is given by:

$$k_{\text{honeycomb}} = \frac{k_1 A_1 + k_2 A_2}{A_1 + A_2},$$

which also yields an expression similar to (19.17). Physical constants of native and cross-linked aerogels have been measured. For cross-linked materials with a density around 0.55 g cm^{-3}, the thermal conductivity is 42 W m^{-1} K^{-1}, the yield strength is 70 MPa and the ultimate compressive strength is about 700 MPa; for native silica aerogels the density is around 0.1 g cm^{-3}, the conductivity is 12 W m^{-1} K^{-1} and the compressive strength is negligible [8]. Density, thermal conductivity and compressive strength were calculated using (19.17) for different values of the cell-to-wall ratio n and they are reported in Table 19.1. The results show that honeycombs have density and thermal conductivity that are close to that of polyurethane foams (~0.1 g cm^{-3}, ~20 W m^{-1} K^{-1}) with a mechanical stability which is orders of magnitude better than the foam's (<1 MPa). The results also show that cross-linked honeycombs have a compressive strength comparable to that of the commercially available Nomex honeycomb.

Table 19.1 Estimated physical properties of honeycomb composites for different values of the cell-to-wall ratio n

Cell-to-wall ratio n	Density (g cm^{-3})	Thermal conductivity (W m^{-1} K^{-1})	Compressive strength (MPa)
5	0.32	26	366
10	0.23	21	228
40	0.16	15	67.5
Nomex (n = 40)	0.01	100	120

Values for $n = 40$ were calculated to compare with Nomex. The Nomex values are for a wall thickness of 250 μm and $n = 40$ [51].

19.6. Conclusions

The techniques presented here allow to produce ultralight materials with anisotropic properties. Optical absorption and emission, hydrophobicity mechanical strength and index of refraction can be tuned within the same monolith. Conceivable applications of our technique include fabrication of photonic devices, membranes, radiation collimators, fuel cell electrodes as well as hierarchically structured materials that combine mechanical strength with the acoustic and thermal insulation properties of conventional aerogels.

References

1. For a recent review on the synthesis and applications of aerogels, see: Pierre A C, Pajonk G M, (2002) Chemistry of Aerogels and Their Applications. Chem. Rev. 21: 4243. For space applications, see Jones S M (2006) Aerogel: Space exploration applications. J Sol-Gel Sci Technol 40: 351
2. Zhao D, Feng J, Huo Q, Melosh N, Fredrickson G H, Chmelka B, Stucky G D, (1998) Triblock copolymer syntheses of mesoporous silica with periodic 50 to 300 angstrom pores. Science 279: 548
3. Amatani T, Nakanishi K, Hirao K, Kodaira T, (2005) Monolithic Periodic Mesoporous Silica with Well-Defined Macropores. Chem Mater 17: 114
4. Heckman B, Martin L, Bertino M F, Leventis N, Tokuhiro A T, (2008) Sol-gel materials for high capacity, rapid removal of metal contaminants. Separ Sci 43: 1474
5. Brandhuber D, Peterlik H, Husing N, (2005) Simultaneous drying and chemical modification of hierarchically organized silica monoliths with organofunctional silanes. J Mater Chem 15: 3896
6. Langlet M, Permpoon S, Riassetto D, Berthome G, Pernot E, Joud J C, (2006) Photocatalytic activity and photo-induced superhydrophilicity of sol–gel derived TiO_2 films. J Photoch Photobio.A 181: 203
7. Bertino M F, Smarsly B, Stocco A, and Stark A, (2009) Densification of nanoparticles with visible light, Adv Funct Mater 19: 1–6
8. Leventis N, (2007) Three-dimensional core-shell superstructures: Mechanically strong aerogels. Acc Chem Res 40: 874–884
9. Leventis N, Sotiriou-Leventis C, Zhang G, Rawashdeh A-M, (2002) Nanoenginering Strong Silica Aerogels. Nano Lett 2: 957–960
10. Wu P-W, Cheng W, Martini I B, Dunn B, Schwartz B J, Yablonovitch E, (2000) Two-Photon Photographic Production of Three-Dimensional Metallic Structures within Dielectric Matrix Adv Mater 12: 1438
11. Bertino M F, Gadipalli R R, Story J G, Williams C G, Zhang Z, Sotiriou-Leventis C, Tokuhiro A T, Guha S, Leventis N, (2004) Laser Writing of Semiconductor Nanoparticles and Quantum Dots. Appl Phys Lett 85: 6007
12. Bertino M F, Gadipalli R R, Martin L A, Rich L E, Yamilov A, Heckman B R, Leventis N, Guha S, Katsoudas J, Divan R, Mancini D C, (2007) Quantum dots by ultraviolet and X-ray lithography. Nanotechnology 18: 315603
13. Muir A C, Mailis S, Eason R W, (2007) Ultraviolet laser-induced submicron spatially resolved superhydrophilicity on single crystal lithium niobate surfaces. J Appl Phys 101: 104916
14. Gadipalli R R, Martin L A, Heckman B, Story J G, Bertino M F, Leventis N, Fraundorf P, Guha S, (2006) Patterning porous matrices and planar substrates with quantum dots. J Sol-Gel Sci Technol 39: 299
15. Gadipalli R R, Martin L A, Heckman B, Story J G, Bertino M F, Leventis N, Fraundorf P, Guha S, (2006) Infra Red Quantum Dot Photolithography. J. Sol-Gel Sci Technol 40: 101
16. Bertino M F, Hund J F, Sosa J, Zhang G, Sotiriou-Leventis C, Leventis, N, Tokuhiro A T, Terry J, (2004) Room Temperature Synthesis of Noble Metal Clusters in the Mesopores of Mechanically Strong Silica-Polymer Aerogel Composites. J Sol-Gel Sci Tech 30: 43–48
17. Hund J F, Bertino M F, Zhang G, Sotiriou-Leventis C, Leventis N, Tokuhiro A T, Farmer J, (2003) Formation and Entrapment of Noble Metal Clusters in Silica Aerogel Monoliths by γ-Radiolysis. J Phys Chem B107: 465
18. C Wingfield, A Baski, M F Bertino, N Leventis, D P Mohite, and H Lu, (2009) Fabrication of sol-gel materials with anisotropic physical properties. Chem Mater .21: 2108–2114
19. Hund J F, Bertino M F, Zhang G, Sotiriou-Leventis C, Leventis N, (2004) High resolution patterning of silica aerogels. J Non-Cryst Solids 350: 9
20. Kitaev G A, Uritskaya A A, and Mokrushin S G (1965) Kinetics of Cadmium Sulfide Precipitation from Aqueous Thiourea Solutions. Zhurnal Fizicheskoi Khimii 38: 2065
21. Kitaev G A, Uritskaya A A, and Mokrushin S G, (1965) Conditions for the chemical deposition of thin films of cadmium sulphide on a solid surface. Russ J Phys Ch 39: 1101–1102
22. De Brabander H F and Van Poucke L C, (1974) Polymeric complexes between Cadmium (II) and 2-mercaptoethanol and 3-mercapto 1, 2 propanediol. J Coord Chem 3: 301; Said FF and Tuck, DG (1982). Inorg Chim Acta 59: 1
23. Hayes D, Mitit O I, Nenadovit M T, Swayambunathan V, and Meisel D, (1989) Radiolytic Production and Properties of Ultrasmall CdS Particles. J Phys Chem 93: 4603
24. Mostafavi M, Liu Y P, Pernot P, and Belloni J, (2000) Dose rate effect on size of CdS clusters induced by irradiation. Radiat Phys Chem 59: 49
25. Turk T, Resch U, Fox M A, and Vogler A, (1992) Spectroscopic Studies of Zinc Benzenethiolate Complexes: Electron Transfer to Methyl Viologen. Inorg Chem. 31: 1854

26. Turk T, Resch U, Fox M A, and Vogler A, (1992) Cadmium Benzenethiolate Clusters of Various Size: Molecular Models for Metal Chalcogenide Semiconductors. J Phys Chem 96: 3818
27. Fischer Ch-H and Henglein A, (1989) Photochemistry of Colloidal Semiconductors. 31. Preparation and Photolysis of CdS Sols in Organic Solvents. J Phys Chem 93: 5578
28. Knight A R, The Chemistry of the Thiol Group, Part 1; Patai, S, Ed. John Wiley & Sons Ltd.: London, 1974; Chapter 10
29. Mirkovic T, Hines M A, Nair P S, and Scholes G D, (2005) Single-Source Precursor Route for the Synthesis of EuS Nanocrystals. Chem Mater 17: 3451
30. Hasegawa Y, Afzaal M, O'Brien P, Wada Y, and Yanagida S, (2005) A novel method for synthesizing EuS nanocrystals from a single-source precursor under white LED irradiation. Chem Commun 242–243
31. Zhao W-B, Zhu J-J, and Chen H-Y, (2003) Photochemical preparation of rectangular PbSe and CdSe nanoparticles. J Cryst Growth 252:587
32. Lin Y-W, Hsieh M-M, Liu C-P, and Chang H-T, (2005) Photoassisted Synthesis of CdSe and Core-Shell CdSe/CdS Quantum Dots. Langmuir 21 728
33. Henglein A, (1993) Chemical and Optical Properties of Small Metal Particles in Aqueous Solution Israel J Chem. 33: 77
34. Bertino M F, Katsoudas J, Leventis N, in preparation
35. Ilhan U F, Fabrizio E F, McCorkle L, Scheiman D, Dass A, Palzer A, Meador M A B, and Leventis N, (2006) Hydrophobic Monolithic Aerogels by Nanocasting Polystyrene on Amine-Modified Silica. J Mater Chem 16: 3046–3054
36. Mulik S, Sotiriou-Leventis C, Churu G, Lu H, Leventis N, (2008) Crosslinking 3D Assemblies of Nanoparticles into Mechanically Strong Aerogels by Surface-Initiated Free Radical Polymerization. Chem Mater 20:5035–5046
37. Bertino M F, Gadipalli R R, Leventis N, in preparation
38. Glezer E N, Mazur E, (1997) Ultrafast-laser driven micro-explosions in transparent materials. Appl Phys Lett 71: 882
39. Cheng G, Wang Y, White J D, Liu Q, Zhao W, Chen G, (2003) Demonstration of high-density three-dimensional storage in fused silica by femtosecond laser pulses. J Appl Phys 94: 1304
40. Kanehira S, Si J, Qiu J, Fujita K, Hirao K, (2005) Periodic Nanovoid Structures via Femtosecond Laser Irradiation. Nano Letters 5: 1591
41. Wingfield C, Bertino M F, Leventis N, in preparation
42. Fouassier J P, Allonas X, Burget D, (2003) Photopolymerization of thiol–allyl ether and thiol–acrylate coatings with visible light photosensitive systems. Progress in Organic Coatings 47:16–36
43. Cramer N B, Bowman C N, (2001) Toward an Enhanced Understanding and Implementation of Photopolymerization Reactions. J Polym Sci A 39: 3311–3319
44. Matsumoto H, Sakata T, Mori H, and Yoneyama H, (1996) Preparation of Monodisperse CdS Nanocrystals by Size Selective Photocorrosion. J Phys Chem 100: 13781
45. Torimoto T, Kontani H, Shibutani Y, Kuwabata S, Sakata T, Mori H, and Yoneyama H, (2001) Characterization of Ultrasmall CdS Nanoparticles Prepared by the Size-Selective Photoetching Technique. J Phys Chem B 105: 6838
46. Wang Y, Tang Z, Correa-Duarte M A, Pastoriza-Santos I, Giersig M, Kotov N A, and Liz-Marzan L M, (2004) Mechanism of Strong Luminescence Photoactivation of Citrate-Stabilized Water-Soluble Nanoparticles with CdSe Cores. J Phys Chem B 108: 15461
47. Rogach A L, Kornowski A, Gao M, Eychmüller A, and Weller H (1999) Thiol-Capping of CdTe Nanocrystals: An Alternative to Organometallic Synthetic Routes. J Phys Chem B 103: 3065
48. Sheng W, Kim S, Lee J, Kim S-W, Jensen K, and Bawendi M G, (2006). Langmuir 22: 3782
49. For a recent review, see George S, Sebastian M T, (2009) Three-phase polymer-ceramic-metal composite for embedded capacitor applications. Composites Science and Technology 69:1298–1302
50. Chan J W, Huser T R, Risbud S H, Hayden J S, Krol D M, (2003) Waveguide fabrication in phosphate glasses using femtosecond laser pulses. Appl Phys Lett 82: 2371
51. http://www2.dupont.com/Nomex/en_US/uses_apps/index.html, last accessed February 15, 2009

20

Aerogels Synthesis by Sonocatalysis: *Sonogels*

Luis Esquivias, M. Piñero, V. Morales-Flórez, and Nicolas de la Rosa-Fox

Abstract High power ultrasound applied to liquids produces cavities that attain very high temperatures and pressures ("hot spots"). When an alkoxide/water mixture is *sonicated*, the cavities act as nanoreactors, where the hydrolysis reaction starts. The products (alcohol, water, and silanol) help continue the dissolution of that immiscible mixture. The reactions depend on catalyst content, temperature bath, and alkyl group length. When the resultant *sonosol* gels, it produces a *sonogel*; it is denser, with finer and more homogeneous porosity than that of a *classic* counterpart. Thus, acoustic *cavitation* makes it possible to obtain nanostructured materials. *Sono-aerogels* have a high surface-to-volume ratio and are built by small particles (~1 nm radius) and a highly cross-linked network with low surface coverage of –OH radicals. The processing as well as their short-range order at an atomic scale and at a micrometric scale is presented in this chapter. Finally, these materials find application, among others as biomaterials for tissue engineering and for CO_2 sequestration.

20.1. The *Sonogel* Approach

20.1.1. An Insight into *Cavitation*

A number of reactions, especially those in heterogeneous systems, such as alkoxide/water mixtures, have been shown to give enhanced reaction rates and yields under *sonication*. The process is driven by acoustic *cavitation* [1–4].

This phenomenon is based on the creation of cavities within the liquid caused by the propagation of a pressure wave through the liquid [5]. During the rarefaction period, the microbubbles increase in size according to the gas compression law. Inside the bubble, the gas pressure tends to expand it. On the other hand, the surface tension at the

L. Esquivias • Departamento de Física de la Materia Condensada, Facultad de Física, Instituto de Ciencias de Materiales de Sevilla (CSIC), Universidad de Sevilla, 41012 Seville, Spain
e-mail: luisesquivias@us.es
M. Piñero • Departamento de Física Aplicada, CASEM, Universidad de Cádiz, 11510 Cádiz, Spain
e-mail: manolo.piniero@uca.es
V. Morales-Flórez • Departamento de Física de la Materia Condensada, Facultad de Física, Instituto de Ciencia de Materiales de Sevilla ICMSE, CSIC-US, 41012 Seville, Spain
e-mail: victor.morales@icmse.csic.es
N. de la Rosa-Fox • Departamento de Física de la Materia Condensada, Universidad de Cádiz, 11510 Cádiz, Spain
e-mail: nicolas.rosafox@uca.es

M.A. Aegerter et al. (eds.), *Aerogels Handbook*, Advances in Sol-Gel Derived Materials and Technologies, DOI 10.1007/978-1-4419-7589-8_20,

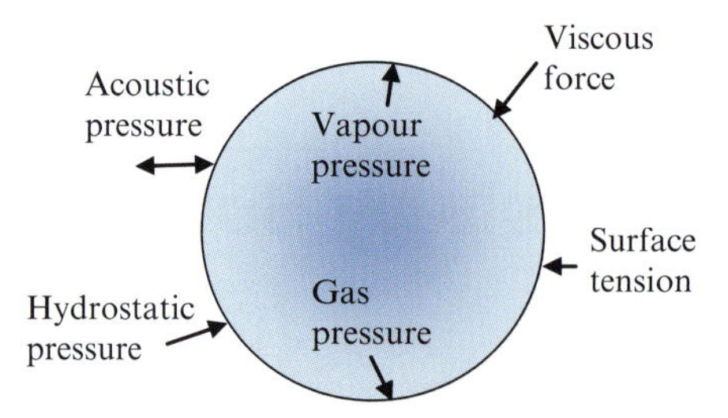

Figure 20.1. Scheme of the forces and pressures acting on a cavitation bubble surface.

bubble–liquid interface and the external pressure drive the bubble to compress (Figure 20.1). Since the interface surface area is slightly higher during the expansion hemicycle, the diffusion toward the inner surface is higher than that during the compression phase; accordingly, the cavities continue to grow over a few cycles. Nevertheless, often the stable microbubbles grow beyond the so-called stable *cavitation* threshold, acquiring a transitory character. Thus, with the cavity radius having attained a critical size, the surface tension, which is inversely proportional to the particle radius, will be insufficient to balance the resulting sound wave pressure and the internal pressure. It is during this period that the known effects of *cavitation* take place. The bubble collapses, producing a shockwave that propagates throughout the liquid. The implosive collapse produces intense, nearly adiabatic, local heating, high pressures, and very short lifetimes; these transient, short-lived localized "hot spots" [6, 7] deliver enough energy to activate chemical reactions. At the same time, a very rapid flow of solvent molecules around the *cavitation* bubble and the intense shock-waves which emanate upon collapse cause liquid motion in the vicinity of the bubble.

The severe conditions generated upon the bubble collapse lead to the production of excited states, to bond breakage, and to the formation of free radicals [8]. The center of the hot spot is where the primary chemistry, which consists of atom and radical recombinations, takes place [9]. The species generated inside the bubble may diffuse outward and react with reagents in the liquid. At this point, the interfacial region has very large temperature gradients, pressure, surface tension, and rapid motion of molecules, leading to efficient mixing [2, 5].

In short, *cavitation* serves as a means of concentrating the diffuse energy of sound into unique chemical nanoreactors to produce unusual materials from dissolved (and generally volatile) precursors.

20.1.2. *Sonogels*

As with standard sol-gel methods, the most popular approach to forming the initial gel by *sonochemistry* is the "alkoxide method." Tarasevich [10] in 1984 described the approach of applying ultrasound to sol-gel processing, which avoids the use of additional solvent (alcohol) in the *TEOS*–water mixture [11] (Chap. 1). In the presence of any acid catalyst, the alkoxides undergo hydrolysis inside the *cavitation* bubbles producing alcohol as a by-product that serves as solvent. Then, the polycondensation reactions, which lead to the formation of the metal oxide, take place. Therefore, the product of the reaction will be the same as in the absence of ultrasound (*classic* or *conventional* method). However, the textural and structural properties of gels synthesized by *sonication* of metal alkoxides $M(OR)_n$ (M is a metal and R an alkyl group), and water mixtures differ from those prepared in a standard way in alcoholic media.

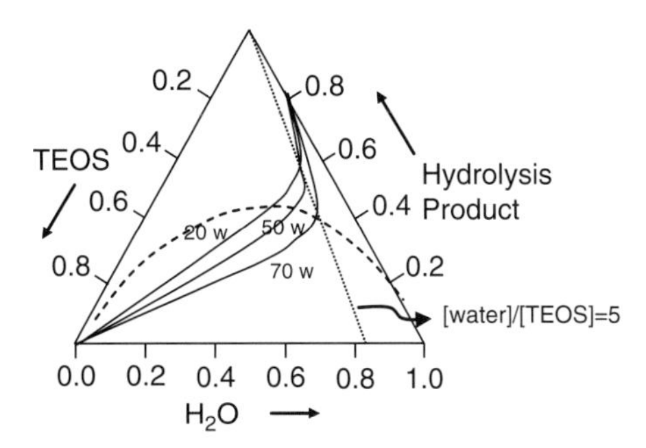

Figure 20.2. Equilibrium diagram of TEOS/H_2O/EtOH system showing the immiscibility area below the dotted line. Sonogels for n = water/TEOS molar ratio = $R_w = 5$ is indicated. The reaction pathway of the heterogeneous TEOS hydrolysis as a function of the ultrasound power as represented in a ternary diagram of the molar fraction of TEOS, water, and product of the reaction (ethanol + $Si(OH)_4$). Adapted with permission from [12].

Figure 20.2 shows the immiscibility area in the phase diagram of the *TEOS*/H_2O/EtOH system. The *sonogel* method permits to obtain homogeneous mixtures below the dotted line.

The Zarzycki and Esquivias groups conducted early extensive work to establish the practical consequences of this approach on the kinetic and textural characteristics of the *sonogels* [13, 14]. These studies encompass numerous systems on pure SiO_2 [15–24], SiO_2–P_2O_5 [16, 25, 26], SiO_2–TiO_2 [27–33], ZrO_2 [34, 35], SiO_2–Al_2O_3–MgO [36, 37] (cordierite), and SiO_2/SiO_2 composites [38, 39].

20.1.3. Processing *Sonogels*

Before proceeding with the operation, the energy provided by the electrostrictive device of the ultrasonic apparatus is calibrated by the temperature increase of a fixed volume (v) of distilled water while applying ultrasound by the relationship

$$W = dQ/dt = mc_w dT/dt, \tag{20.1}$$

where m_w and c_w are the mass and specific heat of water, respectively.

Then, the energy density delivered by the ultrasounds per unit volume will be:

$$U_s = W \cdot /v \approx K \cdot t\,(\mathrm{J \cdot cm^{-3}}) \tag{20.2}$$

t being the *sonication* time in minutes. K values in W cm^{-3} must be calculated for each experimental device, keeping in mind the nominal output power, the tip diameter, size of the beaker, and eventually the temperature of the thermostatic bath for the beaker.

Typical $K\cdot v$ values range from 50 to 100 W [18, 40–43]. Many factors can affect these K values, such as room temperature, catalyst, molar ratios of water/*TEOS*, solvent/*TEOS*, beaker diameter and volume, horn tip immersion in the liquid solution and its diameter.

The basic procedure for the preparation of a silica gel (M = Si) in the laboratory follows four steps.

1. The corresponding mixture of the silicon alkoxide (typically, *TEOS*) and acidic water is placed in a double-volume beaker and a two-phase system is formed with the alkoxide above the water. The water must be acidic because the hydrolysis of

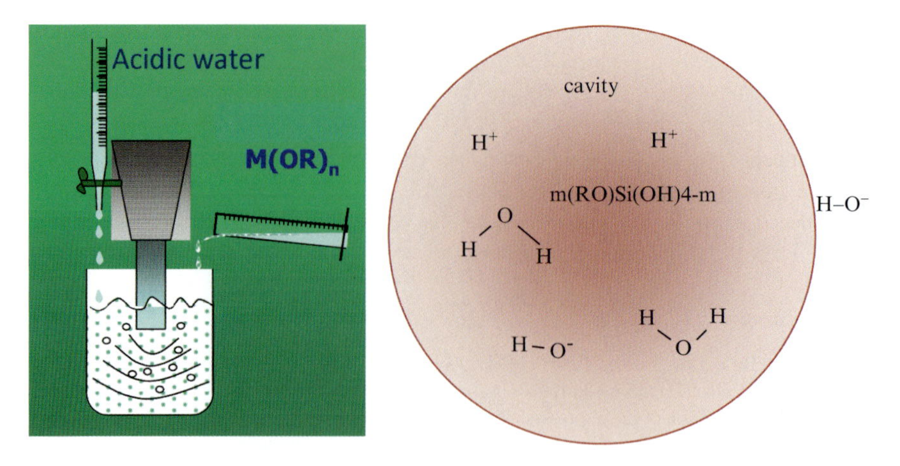

Figure 20.3. Procedure for high power ultrasound treatment of an alkoxide–water mixture. A cavity as a nanoreactor for an alkoxide + water mixture.

TEOS is strongly pH dependent [44] and the silica solubility increases suddenly in alkaline solution [45].

Slow reaction and flocculated products have been observed with neutral or basic water.

At this point, the tip of the ultrasonic device is immersed some millimeters into the liquid. This biphasic system is subjected to ultrasound waves, as shown in Figure 20.3 which emulsifies the mixture. The compression–rarefaction cycles give rise to the formation of *cavitation* bubbles that contain vapors of both reagents.

2. At this moment, the liquid blows up abruptly and its temperature increases drastically, indicating the onset of the hydrolysis reaction [42].

Once the liquid mixture blows up and a homogeneous solution is observed, the energy dose is taken as the threshold of the reaction, U_s(min).

3. Then the polycondensation reactions, which lead to the formation of the metal oxide, take place and the system finally gels.

There is a *sonication* energy dose at which the liquid gels in the container, U_s(max).

In this case, a transparent soft solid gel with a tip hole on top can be observed. Between U_s(min) and U_s(max), one can decide the energy dose to be applied.

4. Having exposed the sol to a chosen ultrasonic energy dose, the liquid sol is kept in a hermetic container and allowed to gel at the corresponding temperature. Since the sol is a low-viscosity liquid, it can be cast in a mold with a selected shape.

20.1.4. Physicochemical Aspects of the Hydrolysis

The acid hydrolysis of *TEOS* may be formulated as: $SiOR + H_2O \xrightarrow{k_o} SiOH + ROH$ where R represents the ethyl group and k_o is a second-order rate constant.

When the container is installed in air at room temperature, the ultrasound makes the temperature attain a value as high as 80°C and then stabilize at a lower temperature (~70°C). The measured temperature should be attributed to the boiling point of the ethanol release, accounting for the exothermic hydrolysis reaction.

The reaction beaker could be surrounded by a thermostatic bath to ensure local temperature control [42, 46, 47]. *TEOS*:water:HCl mixtures of constant volume (35 ml) with a molar ratio of 1:5:0.0089 and a pH ~1.5 and a bath temperature held at 26.5°C present their exothermic peak after 3–4 min of *sonication* (20 kHz). This time depends on the dissipated ultrasound power, the maximum occurring for $K \sim 0.9\ \mathrm{W\,cm^{-3}}$. The coupling between the ultrasound and the heterogeneous mixtures does not seem too good for higher ultrasound power [46]. The oxalic acid catalyzed *TEOS* hydrolysis peak shows a less pronounced form when compared to the HCl catalyzed one [47].

The reaction rate is practically zero during an initial induction period due to the miscibility gap of the *TEOS*/water system. From the heterogeneous reaction pathway drawn in a ternary diagram (Figure 20.2), the alcohol produced at the early hydrolysis due to the *sonication* action enhances further hydrolysis through a parallel autocatalytic path that increases with the ultrasound power [46].

The parallel autocatalytic path that helps homogenize the initial system and favors the subsequent condensation reaction is

$$\mathrm{SiOR} + \mathrm{H_2O} + \mathrm{ROH} \xrightarrow{k'} \mathrm{SiOH} + 2\mathrm{ROH} \tag{20.3}$$

where k' is a pseudo-second-order rate constant alcohol concentration ([ROH]) dependent. Table 20.1 shows the parameters k_o and k'/[ROH] obtained for several temperatures.

Table 20.1. The temperature dependence of the hydrolysis rate constants

T_R (°C)	k_o ($\mathrm{M^{-1}min^{-1}}$)	k'/[ROH] ($\mathrm{M^{-2}\,min^{-1}}$)
10.5	5.0×10^{-4}	5.8×10^{-3}
19.5	6.4×10^{-4}	10.6×10^{-3}
24.0	5.5×10^{-4}	12.1×10^{-3}
32.0	5.5×10^{-4}	19.4×10^{-3}
39.0	5.7×10^{-4}	28.0×10^{-3}
45.5	7.4×10^{-4}	45.7×10^{-3}
55.5	14.3×10^{-4}	61.5×10^{-3}
63.4	12.1×10^{-4}	86.3×10^{-3}

T_R is the thermostatic bath temperature. Reprinted with permission from [40].

The overall hydrolysis process is thermally activated, with apparent activation energy of 36.4 kJ/mol.

The *TEOS*–water system exhibits a pH-dependent onset time for the hydrolysis thermal peak. Donatti and Vollet [46] have found $k = k_o/[\mathrm{H^+}] = (4.6 \pm 0.4)\,\mathrm{M^{-1}\,min^{-1}\,[H^+]^{-1}}$ at 39°C in the range of log $[\mathrm{HCl}]^{-1}$ between 0.8 and 2.0. The dependence of the hydrolysis rate constant on $[\mathrm{H^+}]$ was found to be in agreement with the proton catalyzed reaction.

No activation period for the hydrolysis is apparent in the pure *TMOS*–water system, so the reaction picks up practically at its maximum rate when the ultrasound is turned on [46].

The mixed *TMOS* and *TEOS* system exhibits two independent hydrolysis peaks, corresponding to each of the respective alkoxides. The start time for the latter *TEOS* hydrolysis is diminished as a result of the improved homogenization resulting from the earlier *TMOS* hydrolysis [45].

The threshold of the reaction, U_s(min) (Table 20.2), is related to the hydrolysis reaction. At the other limit, U_s(max) that is related to the polycondensation rate accounts for a greater reactivity of the *TMOS* as a precursor.

Table 20.2. Threshold and limiting of the ultrasonic energy dose for $K = 0.6$ W cm^{-3}

U_s (J cm^{-3})	*TEOS*/water	*TMOS*/water
U_s(min)	80	70
U_s(max)	1,360	350

The actual values for these limits are strongly dependent on the *cavitation* intensity reached and can be modified by varying the reacting system and its conditions [48, 49]. In consequence, alkoxide type, pH, water concentration, additives, static ambient temperature, container diameter, degrees of tip immersion, and other parameters should be controlled to obtain reproducible results.

20.1.5. Experimental Alternatives

An alternative experimental setup involves the use of an ultrasonic bath [50]. In these cases, a much lower ultrasonic intensity is supplied to the reactants and, consequently, the resulting samples do not offer the characteristic *sonogel* features.

Recently, Ocotlán-Flores et al. [51, 52] presented an alternative ultrasonic-activated synthetic method for the preparation of SiO_2 with *TEOS* and neutral water in an inert atmosphere. Several days after the end of the *sonication*, the liquid condensed at room temperature to form SiO_2 gel. A hydrolysis-like reaction (which they call *sonolysis*) is produced time separated from the polycondensation reaction, facilitating independent control of each one.

Another approach that was proposed as a solventless route for gel processing is by vigorous magnetic stirring of the initial mixture [53]. In this case, it was shown that a high speed rotating blender operating at 20,000 rpm induces fast hydrolysis of *TEOS*–water mixtures. Foaming occurs during agitation once the threshold energy (~750 J cm^{-3} in 16 s) is surpassed [39].

20.1.6. *Sonogel* Gelation

The gelation time is taken as the time interval between the end of the ultrasonic treatment and the transition from a viscous fluid to an elastic solid. The first outstanding effect of ultrasound on the gelation process is the drastic decrease of this time in comparison with the *classic* techniques. This fast transition can be estimated visually when the solution surface inside a tilted container is no longer horizontal.

Both the temperature and the increase in the ultrasonic energy dose activate the gelation process (Figure 20.4). Two regimes are present, separated by a crossover at $U_s = 600\ \mathrm{J\,cm^{-3}}$. This point indicates the rapid increase in the polycondensation rate. Assuming an Arrhenius behavior for t_G versus $1/T$ (inset of Figure 20.4), the corresponding linear fit gives an activation energy of 50–70 kJ $\mathrm{mol^{-1}}$ for *TEOS*.

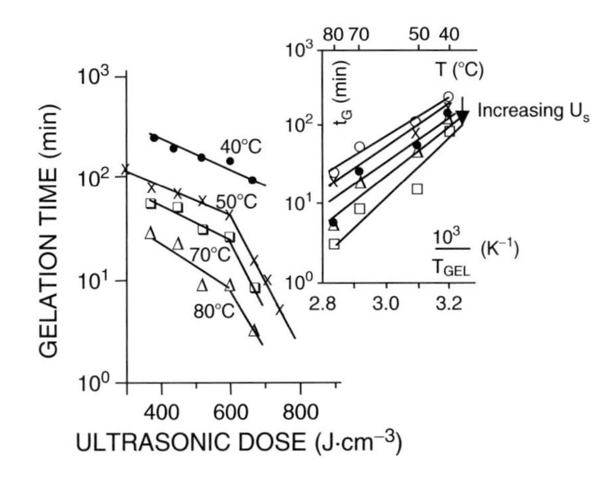

Figure 20.4. Evolution of the gelation time for water/TEOS = 4 pure sonogels as a function of the ultrasonic energy dose. *Inset*: Evolution of the gelation time for water/TEOS = 4 pure sonogels as a function of $1/T$ for different ultrasonic energy dose. Adapted with permission from [17].

When the applied energy density is near U_s(max), the gelation takes place in a few seconds, whereas in other cases this process takes several minutes or hours, depending on the ultrasonic energy dose.

The gelation time t_G presents a minimum value (~1.4 W/cm^{-3}) for an ultrasound power of 50 W [12] (Figure 20.5).

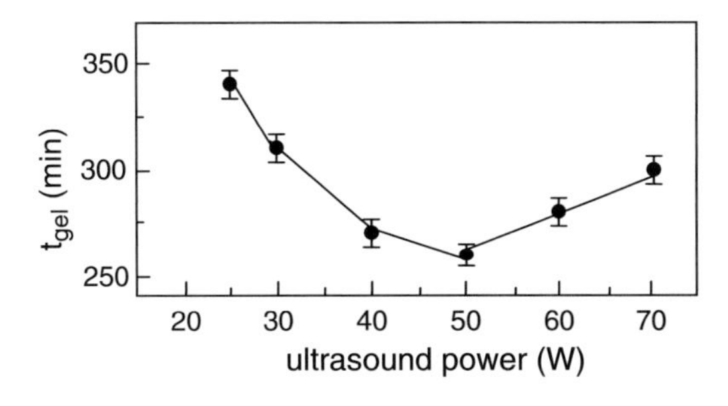

Figure 20.5. The gelation time as a function of the ultrasound power for a TEOS:water:HCl mixtures of constant volume (35 ml), molar ratio 1:5:0.0089, respectively, and pH ~1.5. Adapted from [12].

The gelation time reduction by pH adjustment is shown in Figure 20.6 (left) for *sonosols* submitted to a constant dose ($U_s = 682$ J/cm^3). R_b = [NH_4OH]/[*TEOS*] is the amount of basic solution (6.62×10^{-6} M) that must be added for gelation during the

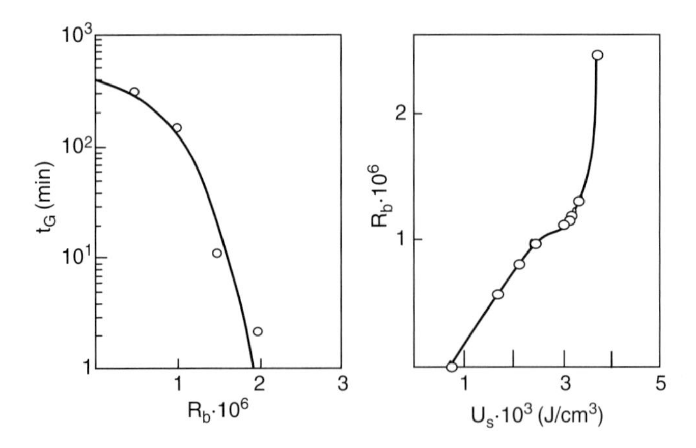

Figure 20.6. Left: Gelation time reduction by pH adjustment by adding diluted base for sonosols submitted to an energy dose $U_s = 682$ J/cm^3 at 50 W. The pH increase is effective for the sonosol since it is poorly polycondensed. **Right**: Calculated equivalence between pH adjustments for supplied dose of energy and U_s. *Lines* are visual guides only. Adapted with permission from [39].

experiment (gelation in situ). Adding a basic solution increases the condensation rate of the system left on its own, and has the same effect as to increase the ultrasound energy (Figure 20.6, right).

Kun-Hong et al. [54] studied the gelation time as a function of the pH value. However, they did not give any detail of the ultrasound transducer device. They prepared *sonosols* with $R_w = 10$ with acidic water (HCl). As a first step, the pH was set to 2.7 and then increased to 4.7 with NH_4OH. Instantaneous gelation occurred at a pH = 5.5 or higher.

Nuclear magnetic resonance (NMR) spectroscopy of ^{29}Si results shows that via *sonochemistry* the number of states of bridging (SiO_4) groups [55] during gelation is very high. Figure 20.7 shows the concentration of species Q^n as a function of time, where n is the number of Si–O–Si bonds per silicon atom. After 20 min of *sonication*, the *sonosol* consisted of 42% of network-forming species ($Q^3 + Q^4$) whereas in the corresponding *classic* sol there

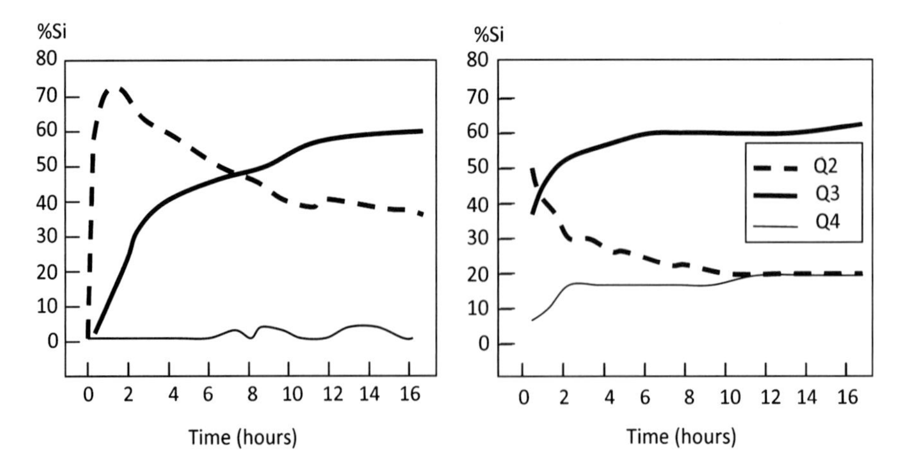

Figure 20.7. Evolution with time of the relative concentration of silica species deduced from ^{29}Si NMR spectra. **Left**, classic sol; **right**, sonosol. Adapted with permission from [56].

is 10% [56]. In terms of reaction mechanism, this confirms that *sonocatalysis* promotes hydrolysis [39]. The concentration of silanol groups is higher and consequently the rate of bridging oxygen formation is also higher.

20.1.7. *Sono-Ormosils*

It is possible to apply the *sonogel* approach in the field of *OIHM* based on ormosils (Chap. 13). This approach to these materials incorporates an organic phase, very commonly *PDMS*, in the inorganic precursor (*TEOS* or *TMOS*) sol in combination with *sonication* of the precursors, producing a *sono-ormosil* (also called *hard ormosil*) after gelation [41, 57–59]. The processing and properties of *sono-ormosils* and derived materials have been recently reviewed [60, 61].

20.2. Structure

The understanding of sol-gel structures has been approached from several points of view, from angstroms to microns, from atomistic descriptions explaining how the network is formed as well as the chemical composition of the nanoparticles, to the micrometric particle aggregation and pore description considering the particles as bulk matter.

20.2.1. From Sol to Gel

The aggregation state for different times of *sonosols* from *TEOS* networks before the gel point ($t < t_G$) was monitored by *SAXS* [26, 27].

The basic theory behind this technique is as follows:

The scattered intensities (I) are given as a function of the scattering vector modulus q. Measuring the scattered intensity at momentum transfer q is equivalent to analyzing the real-space density distribution with a resolution $2\pi/q$.

In a ln $I(q)$ versus ln q plot, three regions can be theoretically distinguished (Figure 20.8):

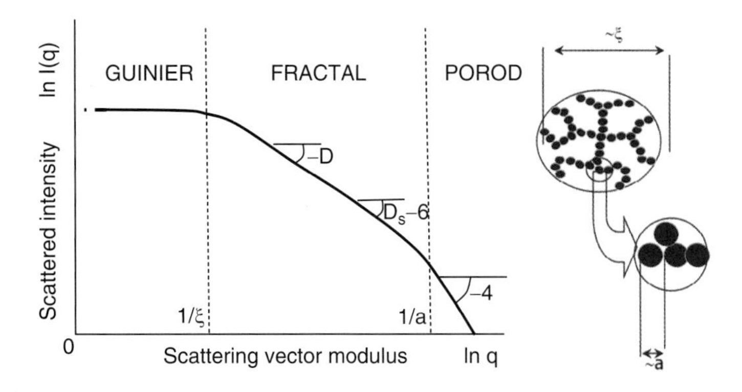

Figure 20.8. Various regions for an SAXS intensity curve and a description of the structural meaning of the crossover positions.

(a) Guinier region. For $q \ll 1/\ \xi$ where ξ is the correlation length, ln $I(q)$ is almost constant; this corresponds to the so-called Guinier law $I(q) = I(0)\exp\left(-R_g^2 q^2/3\right)$ [62], which provides information about the overall size of scatterers through R_g (the gyration radius of the aggregates). The Guinier's plot also provides, by extrapolation, the intensity at $q = 0$, $I(0)$, and a geometrical parameter related to the average volume of the scattering centers is calculated:

$$V_C = 2\pi^2 I(0)/Q_0 \tag{20.4}$$

where $Q_0 = \int_0^\infty I(q)q^2 dq$ is known as Porod's invariant, independent of the scatterer shape.

(b) Fractal region. Thus, $q = q_1 \approx 1/\xi$ is the crossover between Guinier and fractal scattering. For $1/\xi < q < 1/a$. $I(q)$ exhibits a potential dependence [63]. The mass-*fractal dimension*, D, is obtained from the slope: $d \ln I(q)/d \ln q = -D$. Zarzycki [64] demonstrated that

In scattering experiments, even for an ideal object, the fractal range must be at least of the order of $\xi/a \sim 10$ for fractal behavior to be directly detectable.

(c) Porod's region. For $1/a << Q$, $I(q)$ tends progressively to a q^{-4} dependence when the particles are smooth with a well-defined interface. In the case of a surface fractality being present, an intermediate behavior can be observed between the fractal and Porod regions. The slope of the $\log I - \log q$ plot would be $-2D + D_S$, where D_S is the *surface fractal dimension.*

Porod's limiting law [65] states, when the interface solid/pore is sharp, that

$$\lim_{q\to\infty}\left(q^4 I(q)\right) \to \frac{1}{\phi_p \phi_s}\frac{S}{V}\frac{Q_0}{\pi}, \tag{20.5}$$

where ϕ_p and ϕ_s correspond to the volume fractions of solid and pore phases. In the case of a dense system, this relation gives an estimate of the area of the interface per unit volume of the system S/V.

The position of the crossover between fractal and Porod's region defines the size $q = q_2 \approx 1/a$ of the primary particles which comprise the fractal aggregates.

Ramírez-del-Solar et al. have proposed a model of statistical balls or polymeric growing clusters [26, 27] (Figure 20.9), R_g. For $t = t_G$, $R_g \sim 1.5$ nm and $a \sim 0.25$ nm.

The addition of Ti to a silica sol (Chap. 7) magnifies the structural effects induced by the ultrasound, that is, the formation of small statistical balls [26].

Donatti, Vollet, and their collaborators carried out a systematic study [66–69] of wet *sonogels*[1] prepared in excess of water, by previous pH adjusting with different amounts of diluted NH_4OH.

[1]The samples were prepared from TEOS/water mixtures and 0.1 N HC1 as a catalyst. The resulting pH of the mixture was about 1.5–2.0. The hydrolysis was promoted during 10 min under a constant power (60 W) of ultrasonic radiation. The pH of the resulting sol was adjusted to 4.5 by addition of $NH_4(OH)$. The final water/TEOS molar ratio was equal to 14.4.

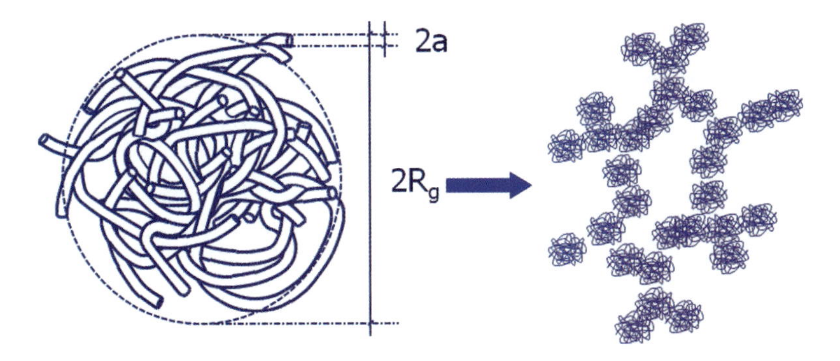

Figure 20.9. Schematic illustration of the proposed model that grows during the transition from sol to gel. The agglomeration of these elementary units (**left**) gives rise to the particulate and microporous structure of sonogel (**right**). Adapted with permission from [26].

20.2.2. Dense Inorganic *Sono-Aerogels*

The *sonogels* are autoclave dried in supercritical conditions of ethanol (Chap. 2). In this section, the effect of drying on the structure is discussed in terms of *WAXS*, N_2 physisorption, and *SAXS*.

Short-Range Order

WAXS studies on the structure of *sono-aerogels* established that the atomic *sonogel* structure is very cross-linked. Figure 20.10 represents the *sonogel* and silica glass reduced *RDF*,

$$rG(r) = 4\ \pi r^2 \rho(r) - 4\ \pi r^2 \rho_0 \tag{20.6}$$

where ρ_0 (also known as skeletal density) represents the mean density and $\rho(r)$ is the local atomic density at a distance r from an arbitrary atom taken as a reference. $rG(r)$ represents the deviation of the *RDF* from a uniform distribution [71]. The *RDF*'s nth peak position indicates the most probable distance of the nth neighbors. The average number of atoms per length unit is given by the area beneath the nth peak.

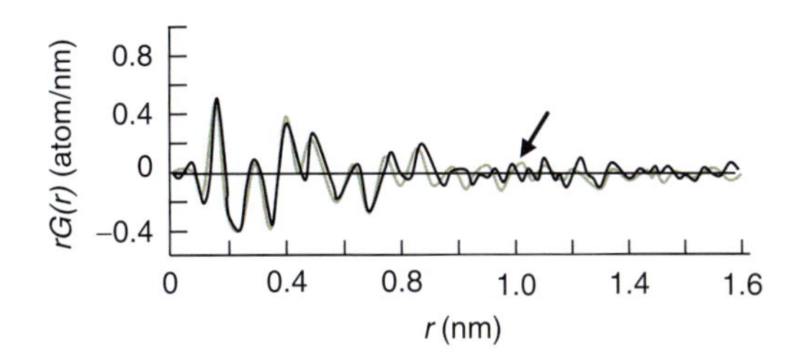

Figure 20.10. Reduced RDF of a silica sonogel (*black line*). The *arrow* indicates where the curve oscillations start to differ from those of the silica glass RDF (*clear line*), included for comparison. Adapted with permission from [70].

Figure 20.10 shows that beyond $r \sim 1$ nm the correspondence between *sonogel* and silica bulk glass *RDF* maxima deviates sharply [70]. This is evidence for *sonogels* being formed by monosized elementary particles of ~1 nm radius.

The skeletal densities of the solid phases were evaluated by the maximum entropy method [72] and found to be $\rho_0 = 2.09 \pm 0.02$ g/cm^3 and 2.19 ± 0.02 g/cm^3 for silica glass. These differences come from the presence of nonbridging oxygens on the pore–matrix interface in the aerogels causing lengthening of the average Si–O bonds [73] (Table 20.3).

Table 20.3. Structural parameters deduced from silica glass, silica gels, and silica *sonogel*

	Area (atom)	First peak position	Second peak position	Si–O–Si average bond angle ϕ (degree)
Silica glass	2.99	1.61	3.10	148.6
Sono-aerogel ($R_w = 4$)	2.97	1.64	3.10	141.9

Adapted with permission from [22].

The atomic arrangements around Ti atoms in titania-doped (5 mol%) silica *sono-aerogel* was deduced from Ti K-edge *XAS* [74]. The calculated *PRDF* is shown in Figure 20.11. The *sonogel* presents two well-defined peaks, corresponding to Ti–O and Ti–cation distances (Table 20.4). The network is formed up to the second neighboring level.

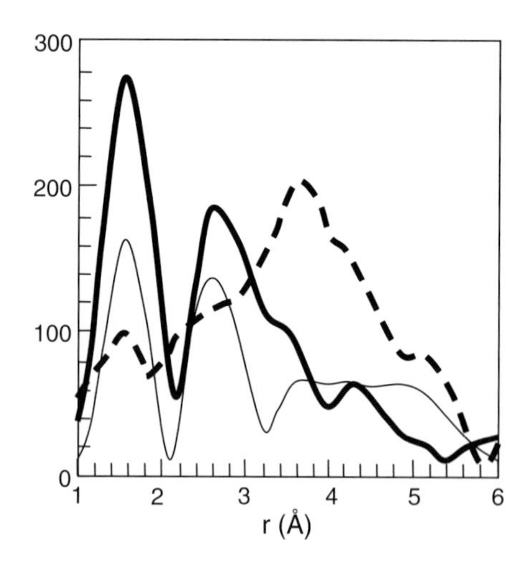

Figure 20.11. PRDF Ti for silica aerogels doped with titania at 5 mol% sonogel (*continuous thin line*), classic gel (*dashed line*) and anatase (*continuous bold line*). Adapted with permission from [32].

Table 20.4. *EXAFS* result of silica gels doped with titania 5 mol%

Sample	Bonding	Coordination number	Bond length (Å)
Sonogel	Ti–O	2	1.91
	Ti–O–Ti	6	2.95
	Ti–O–Si	6	2.90
Classic	Ti–O	1	1.87
Anatase	Ti–O	6	1.97

Adapted with permission from [32].

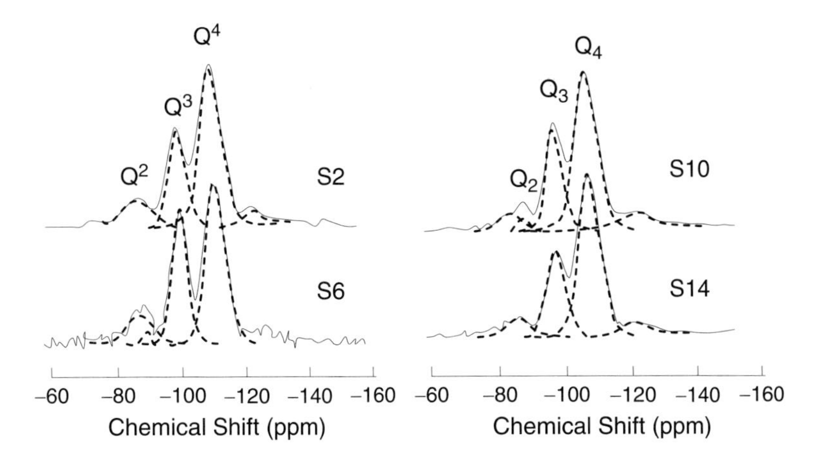

Figure 20.12. MAS-NMR spectra of sono-aerogels prepared with different amounts of hydrolysis water. Reprinted with permission from [25].

The *sonogel* route produces network aerogel structure close to that one of a bulk glass.

The ^{29}Si *MAS-NMR* results (Figure 20.12) indicate that the *sonogel* structure (Sn, $n = R_w$) is highly cross-linked. The spectra are interpreted by means of the Q^n components (see Sect. 20.1.6). The chemical shift (δ) gives a measure of the deviation of the chemical environment of each Si atom.

The absence of the Q^1 component indicates a polymeric formation. However, the existence of noncomplete Si–O–Si bridges in the gels indicates the existence of some –OH surface coverage given by the number of OH per Si atom (n_{OH}/n_{Si}). It decreases when the *sonogel* was prepared with a water concentration much greater or much less than the stoichiometric one.

Accordingly, the amount of hydrolysis water affects only slightly the siloxane condensation and degree of cross-linking.

A number of –OH radicals could be buried in the *sonogel* structure [22].

These hydroxyls are difficult to eliminate, causing problems when a complete densification to form a glass is intended [17].

The effect of ultrasonic mixing on the synthesis of phosphosilicate aerogels also has been studied [25] as a way to improve the homogeneity of the gels prepared in alcoholic solution. The phosphate distribution is not homogeneous in the gels [75–77] although heterogeneities vanish upon heat treatment at temperatures higher than 950°C. The P–O–Si bonds only appear upon further heat treatment at high temperature. A large phosphorus loss (around 50%) was observed upon drying. However, *sono-aerogels* retained larger amounts than their homologous equivalents prepared without *sonication*. Heat treatment causes an increase in the number of Si–O–P in the glass. Above 500°C, a number of characteristic peaks of the Si-5-$(PO_4)_6$ crystalline phase appear.

Textural Characteristics

Table 20.5 shows some textural parameters of *sonogel* samples calculated from N_2 physisorption data evaluated by the *BET* method [78]. The samples, designated SLθ and SHθ, refer, respectively, to those gels prepared at the temperature θ°C with either a low ($L \sim 350$ J cm^{-3}) or high ($H = 700$ J cm^{-3}) dose of ultrasonic energy. In each case, *TEOS* was hydrolyzed with $R_w = 4$ mol H_2O/mol *TEOS* in acid catalysis (pH ~1.5). Figure 20.13 shows the density of the resulting aerogels measured by mercury volumetry for the series corresponding to increasing radiation dose U_s and gelled at $\theta = 40$°C and 80°C, respectively.

Table 20.5. Textural characteristics of the aerogels

	ρ_a (g cm^{-3})	S_{BET} (m^2 g^{-1})	S_V[a] (m^2/cm^3)	V_p[b] (cm^3 g^{-1})	$S_{BET}/V_p \times 10^{-2}$ (Å^{-1})
SL50	0.65	461	300	1.06	4.35
SL80	0.65	370	241	1.06	3.45
SH50	0.82	407	334	0.74	5.56
SH80	0.83	387	321	0.73	5.26
C4-1.0	0.38	517	197	2.15	2.38

[a]$S_V = S_{BET} \cdot \rho_a$; [b]$V_p = 1/\rho_a - 1/\rho_s$. C4-1.0 is a *conventional* aerogel prepared with $R_w = 4$ and equivolume of ethanol/TEOS.

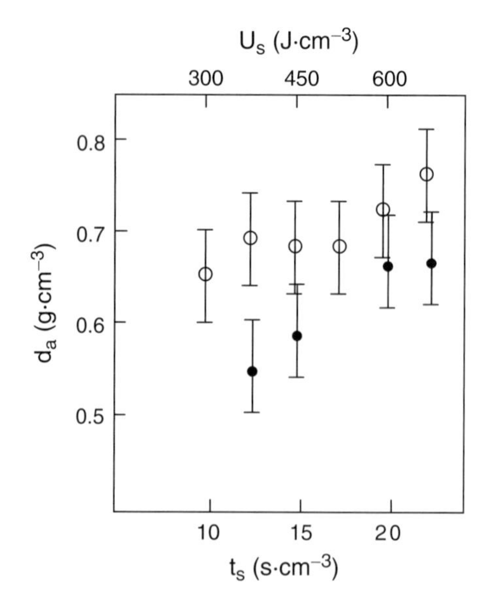

Figure 20.13. Density of aerogel obtained from TEOS with [H2O]/[TEOS] = 4 gelled at (*open circle*) 40°C and (*closed circle*) 80°C for various sonication doses U_s, or sonication time t_s. Reprinted with permission from [17].

Textural parameters are very sensitive to the supplied sonic power and gelation temperature.

- The density increases with U_s.
- A high ultrasonic dose combined with a high gelling temperature leads to a finer porosity.
- The special characteristic of *sonogels* after supercritical drying is that they present a homogeneous and uniform particulate structure.

In short, *sonogels* have a narrow pore size distribution, high bulk density, and a surface-to-volume ratio two or three times higher than gels prepared in alcohol solutions.

The larger the ultrasonic dose, the finer the porosity and the higher the homogeneity.

The particulate structure of the S4 *sonogel* consisting of a network of particle ~5–6 nm size can be seen in Figure 20.14. This makes it reasonable to think of it (see Sect. 20.2.2.1) in terms of a hierarchical structure of agglomerates of elementary particles ~2 nm (Figure 20.15) (Chaps. 21 and 24).

Figure 20.14. TEM micrograph of a sonogel counterpart applying 300 J/cm^3 of ultrasound energy to the TEOS + water mixture. The homogeneous distribution of uniform particles can be seen.

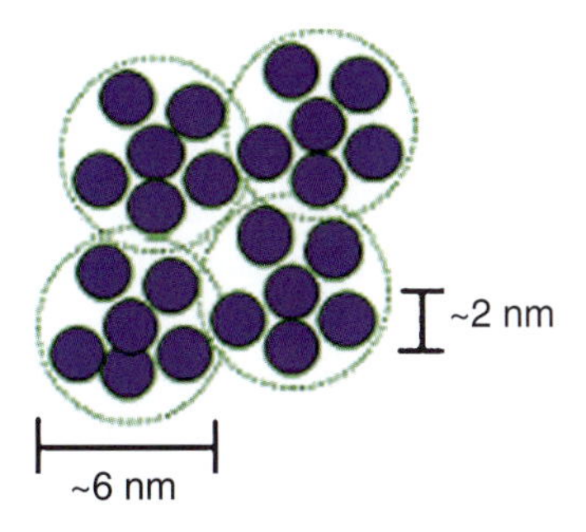

Figure 20.15. Gel structure hypothesis based on a homogeneous distribution of aggregates ~3 nm radius of particles ~1 nm.

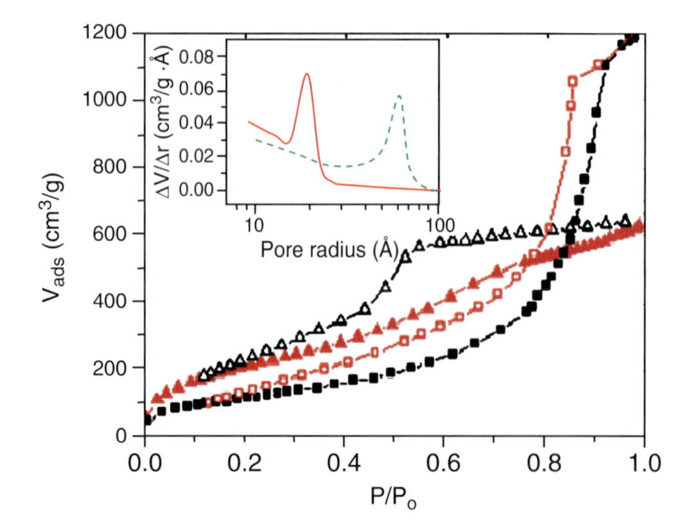

Figure 20.16. N_2 physisorption isotherm for sono-aerogel (*triangles*) and classic gel (*squares*). *Solid symbols* correspond to the adsorption branch and *open symbols* to the desorption one. The *inset* shows the pore size distribution of the same samples using the BJH model. The *continuous line* is for the sono- and *dotted line* for the classic gel.

The shape of the nitrogen adsorption isotherm corresponds to type IV, according to IUPAC classification (Figure 20.16). The brusque increase of the adsorbed volume at low relative pressure (<0.05) is characteristic of microporosity. The *PSD*, calculated from the *BJH* algorithm shows a narrow peak (inset in Figure 20.16) with its maximum at ~3 nm.

The *sono-aerogel* has a high surface-to-volume ratio.

The hysteresis loop can be identified with an H2 type behavior [79], characteristic of pore connectivity effects. In this case the desorption branch occurs at low relative pressure, which is related to the uniform channel-like structure.

Vollet et al. studied light *sono-aerogels* ($\rho_a \sim 0.3$–0.4 g cm^{-3}) prepared from different proportions of *TEOS* and *TMOS* [79], showing the fundamental role of the $R_{TMOS} = [TMOS]/[(TMOS + TEOS)]$ on the structure of the aerogels: the porosity and the pore mean size increases as R_{TMOS} increases.

SAXS Analysis

Sonogels present a wide plateau in the low q region of the log $I(q)$ versus log q plots (Figure 20.17) and, hence, a well-defined mean gyration radius, which is characteristic for a very homogeneous system (Chap. 21). R_g values are found between 2.5 and 2.9 nm (Table 20.6), which correspond to spheres of 3.2–3.7 nm radius.

The position $q \approx 1/\xi$ (Figure 20.8) shifts toward greater q-values with the ultrasonic dose as well as with the gelation temperature. The structural meaning of R_g in dense system is the correlation length above which the system can be considered homogeneous. Consequently, again:

Sonogels are very homogeneous and have a continuous network on a hundred angstroms length scale.

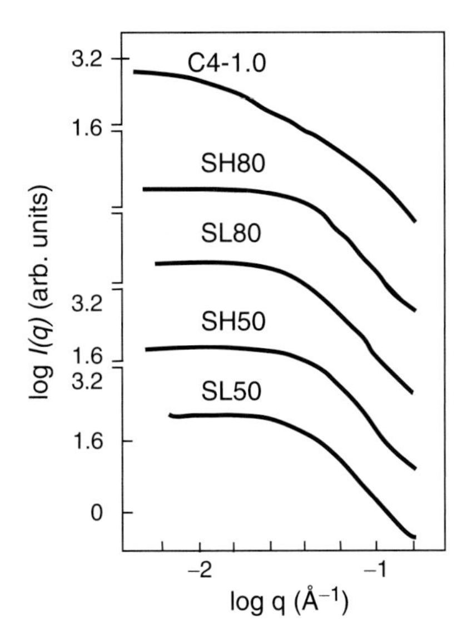

Figure 20.17. Log $I(q)$ versus log q for the series of sono- and a classic aerogel (for the labeling of the samples, see Table 20.5). Adapted with permission from [19].

Table 20.6. *SAXS* parameters for aerogels from experimental intensities

	R_g (nm)	R' (nm)
SL50	2.8	2.4
SL80	2.9	2.5
SH50	2.5	2.2
SH80	2.5	2.2
C4-1.0	8.2	4.2

For the labeling of the samples, see Sect. 20.2.2.2.

The R_g values of the *sonogels* are in agreement with a spherical geometry for the correlation volume V_c (20.4). These results are summarized in Table 20.6 where R' represents the mean correlation radius if we assume a spherical geometry for the average volume of the scattering centers.

The surface-to-volume ratios calculated from (20.5) give bigger values than physisorption experiments using the *BET* model. Consequently:

Dense *sono-aerogels* have a closed porosity and very fine micropores (<1 nm).

Figure 20.18 represents schematically the standard scattering data obtained from a *SAXS* experiment with a SiO_2 *sonogel* prepared with the stoichiometric amount of water ($R_w = 4$). It indicates two aggregation levels. Absolute values of the slopes ($|\alpha_1|$) between q_1 and q_2 (at the 10 nm size range) greater than $|-4|$ are found, which are characteristic of "fuzzy" pore–solid boundaries due to strong electronic fluctuations. In the outer q region, starting at q_2, the slopes (α_2), with $|\alpha_2| < |-4|$, indicate diffused yet smoother boundaries.

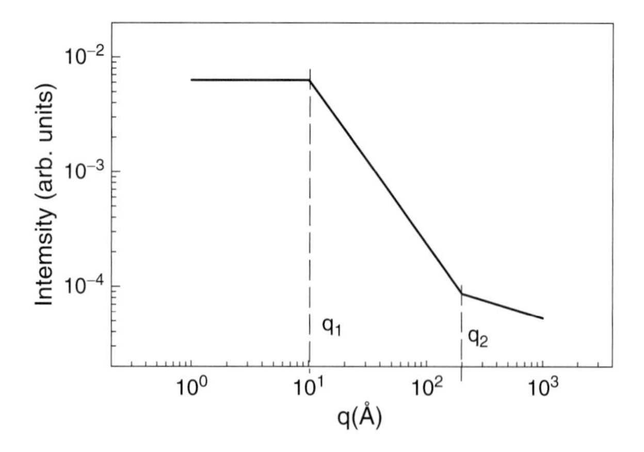

Figure 20.18. Schematic representation of a standard log *I*–log *q* curve of a SAXS experiment for a silica sonogel.

α_1 increases upon heating as a consequence of the close pores that concentrate silanol groups inside them. This fact makes the final consolidation into a glass difficult.

Another effect of the power law behavior of these dense systems is that they deviate positively from Porod's law ($I \propto q^{-4}$), which may be considered as a general rule for this kind of solid. This is induced by density fluctuations within the phases due to their internal microstructure of them.

20.2.3. Light *Sono-Aerogels*

The group from the Universidade Estadual Paulista (Brasil) prepared *sono-aerogels* from their *sonosols* obtained in an excess of water (see footnote 1). Prior to drying, the solvent was exchanged with liquid CO_2 and subsequently extracted under supercritical conditions. These fresh *sono-aerogels* have densities (0.3–0.4 g/cm^3), which are comparable to the gels prepared in alcoholic solution from a *TEOS*/water mixture.

The fundamental role of the increase of *R*w on the structure of resulting aerogels is to increase the pore mean size and the pore volume, without, however, substantially modifying the average size of the silica particles forming the solid network of the aerogels [80].

Their isotherms are classified as type IV with most of the pores in the mesopore region. Table 20.7 shows the structural parameters of these *sono-aerogels* as a function of temperature as determined from nitrogen adsorption data. The sample densities ρ were evaluated from the total pore volume per mass unit from V_p measured from the N_2 physisorption experiments. The mean pore size is evaluated as $l_p = 4V_p/S_{BET}$. Table 20.7 shows that the

Table 20.7. Structural parameters of *sono-aerogels* as a function of temperature as determined from nitrogen adsorption data

T (°C)	ρ (g/cm^3)	V_p (cm^3/g)	S_{BET} (m^2/g)	l_p (nm)
200	0.47 ± 0.02	169 ± 5	848 ± 15	7.9 ± 0.3
400	0.49 ± 0.02	160 ± 5	828 ± 15	7.7 ± 0.3
600	0.55 ± 0.02	137 ± 4	703 ± 13	7.8 ± 0.3
800	0.67 ± 0.02	105 ± 3	538 ± 10	7.8 ± 0.3

Adapted from [69] with permission.

pore volume and the specific surface are reduced with temperature, while the mean pore size remains practically constant. The average pore size (~10 nm) calculated from the *PSD* is almost one order of magnitude larger than that of the dense *sono-aerogels*, the average pore radius of which, from physisorption data, is ~2 nm.

This group also studied light *sono-aerogels*,[2] which were prepared with different amounts of isopropyl alcohol (*IPA*) added [81]. The fundamental effect of the increase of the *IPA* addition on the structure of the resulting aerogels is to diminish the porosity, the pore mean size, and the specific surface area (from 845 to 774 m^2/g) and increase the density (from 0.46 to 0.66 g/cm^3). The supercritical process and degassing the samples at 200°C apparently increase the mass *fractal dimension* (*D* ~2.7), although the fractal range is too short and not very well defined.

20.3. From Gel to Glass

Donatti and Vollet's group reported a complete structural evolution of their light *sono-aerogels* as a function of the heat treatment up to 1,100°C (Figure 20.19) [68]. These aerogels exhibit pronounced structural differences with those formerly described. They present a slope $|\alpha_1|<|-4|$ between q_1 and q_2 that, sometimes with lenient criteria, is described as a surface or mass fractal, having "neat" pore–solid boundaries due to weak electronic fluctuations.

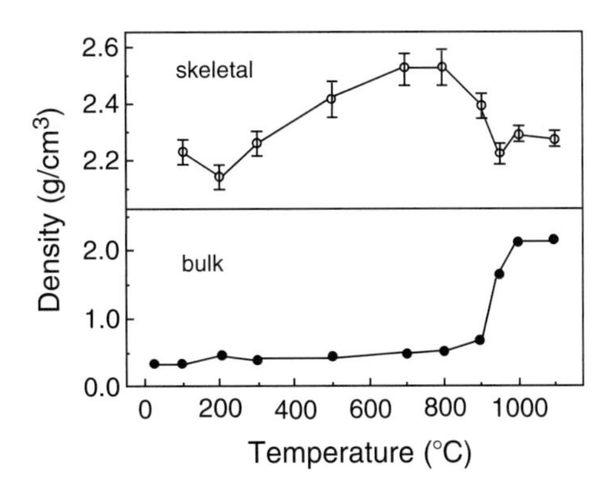

Figure 20.19. Skeletal and bulk density for light sono-aerogels held for 10 h at temperatures up to 1,100°C. Reprinted with permission from [68].

The *SAXS* structural parameters *D*, ξ, *a*, and the evaluated ratio ξ/a, are shown in Table 20.8 extracted from the curves *I*(*q*) for *sono-aerogel* samples heat treated at 500°C and 900°C under atmospheric conditions for 10 h (Figure 20.20).

The aerogel treated at 500°C practically abides by the condition of one order of magnitude for the ratio ξ/a, corresponding to a *D* value obtained to standard preparation found for acid-catalyzed *classic* silica aerogels [82]. Densification studies have shown that dense *sono-aerogels* can be readily converted into dense glasses, provided sufficient care is

[2]From wet silica gels with ~1.4 × 10^{-3} mol SiO_2/cm^3 and ~92 vol% liquid phase obtained from sonohydrolysis of TEOS.

Table 20.8. Mass fractal structural parameters for aerogels as determined by *SAXS*

Temperature (°C)	D	ξ (nm)	a (nm)	ξ/a
500	2.40 ± 0.02	5.26 ± 0.04	0.75 ± 0.05	7.1 ± 0.6
900	2.88 ± 0.06	2.84 ± 0.08	1.1 ± 0.1	2.6 ± 0.3

Reprinted from [67] with permission.

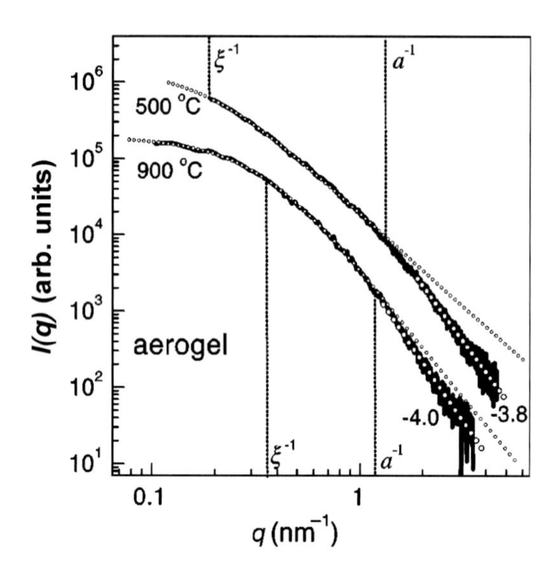

Figure 20.20. SAXS intensity as a function of the modulus of scattering vector q for a light sono-aerogel. The *small circle lines* are the fitting of the mass fractal approach to the experimental data at low- and medium-q region. Reprinted from [67] with permission.

taken to eliminate the organic residues and traces of water (Chap. 29). In the case of a high R_w ratio, bloating occurs, probably due to a closed pore structure [17].

Figure 20.21 show the results obtained [19, 83] for the series SHθ and SLθ. All the aerogels were submitted to oxidation in O_2 atmosphere and, then, dehydrated in a CCl_4 rich atmosphere by consecutive treatments at 500°C for 5 and 8 h, respectively. The densification process consisted of successive treatments for 5 h each at 650°C, 770°C, 870°C, and 950°C, followed by 15-min annealing at 1,040°C and 1,140°C. The SLθ samples (see Sect. 20.2.2.2) have a lower relative density than the SHθ ones up to 800°C, and, above this temperature, the sintering process markedly starts and the gels densify in a very similar way. The specific surface areas present the same trend and their reduction above 700°C provides the source of energy for viscous sintering, which is the predominant mechanism in the densification of *sonogels*.

In terms of *SAXS* data, q_2 (Figure 20.18) decreases on heating, that is to say, the correlation length increases, corresponding to the growth of the micropores due to the formation of larger aggregates. The spherical shape of the particles (or pores) stands along the sintering process [84]. In short, two effects take place during the sintering of *sonogels*.

1. The spherical aggregates of ~3 nm radius particles easily sinter above 700°C due to the viscous flow of these small clusters.
2. The sintering is more complex at the length scale of 10 nm because of the closed porosity that contains residual OH groups.

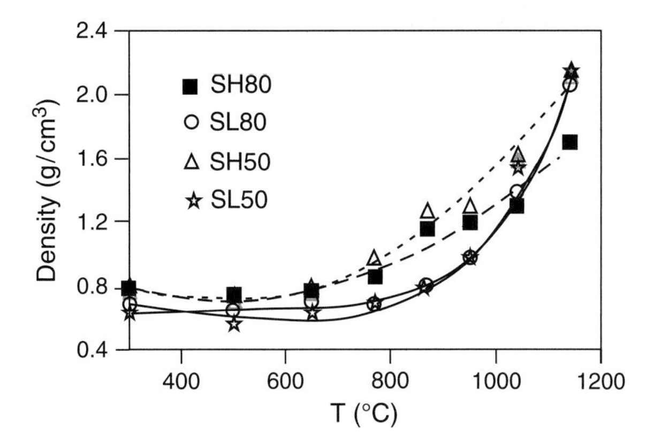

Figure 20.21. Behavior of the apparent density of silica sonogels during sintering. *Lines* are second-order linear regression curves included as a visual guide only.

Densification of a mixed titania–silica aerogel[3] occurs as well via viscous flow mechanism [29]. The reduction of average sizes l_p of pores (derived from *SAXS* data) on sintering (Figure 20.22, left) points to that the thermal evolution of the *sono-aerogel* is dominated by the coalescence of solid particles. So, coalescence of small *sonogel* particles leads to compact clusters. Figure 20.22 (right) is a model of the sintering kinetics.

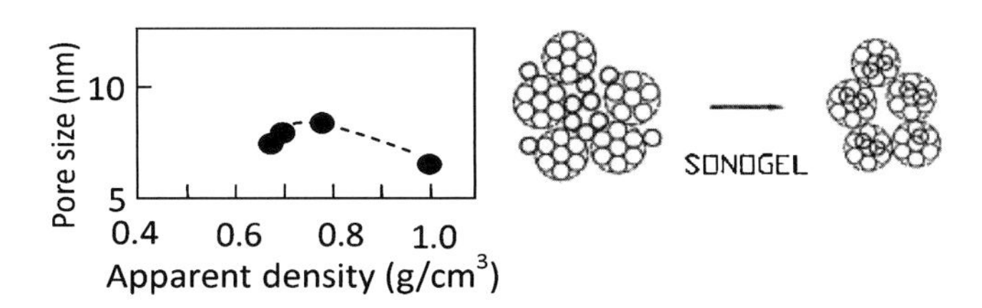

Figure 20.22. Left: Change of the average size of pores (calculated from SAXS data) of mixed titania–silica aerogels upon first sintering stages. **Right**: simplified model of the evolution of sono-aerogels upon first densification steps. Adapted with permission from [30].

Kun-Hong et al. [54] have also transformed aerogels into glasses. They followed a sintering schedule with a long stop (4 h) at 150°C in an inert atmosphere, to let the physisorbed water evaporate. They slowly raised the temperature (50°C/h) up to 600°C to remove by oxidation the unreacted organics. Finally, they raised the temperature (~150°C/h) up to 1,100°C with a short stop (30 min) at 900°C to remove the condensation by-products. The total duration of sintering was 16.5 h. The optimum conditions for sintering, however, depend on the microstructure of the corresponding aerogels. Complete drying is a prerequisite for successful sintering.

[3]Silica gel doped with a 5% molar titania was elaborated from TEOS and TBOT (tetrabutylorthotitanate). The chemical reactions were carried out under acidic conditions with a pH[HC1] = 1.5.

20.4. Mechanical Properties

The elastic modulus of dense *sono-aerogel* is several orders of magnitude higher than for *classic* ones (Chap. 22). Table 20.9 summarizes results obtained for the bulk modulus and apparent density.

Table 20.9. Bulk modulus and apparent density of *sono-aerogels* elaborated by hydrolysis ($R_w = 4$, pH [HCl] = 1.5) and polycondensation of *TMOS*

T(°C)		600	900	1,100
STMH	K_o (MPa)	2.12×10^3	2.93×10^3	2.79×10^3
	ρ_a (g cm^{-3})	0.79	0.62	1.95
STMM	K_o (MPa)	1.23×10^3	1.56×10^3	3.14×10^3
	ρ_a (g cm^{-3})	0.66	0.75	2.21

A device delivering to the system 0.6 W cm^{-3} of ultrasound power was employed for 500 s and 250 s, respectively, for the STMH and STMM samples. They were heat-treated at 600°C, 900°C, and 1,100°C.

It is worth noting the higher reticulation provided by the sono-approach, as revealed by the relatively higher bulk modulus, ~1–3×10^3 MPa, compared with their *conventional* counterparts ~8.0 MPa [84].

20.5. Applications of *Sono-Aerogels*

20.5.1. Biomaterials

Hybrid CaO–SiO_2–*PDMS* bioactive *sono-aerogels* with low density (0.4–0.8 g cm^{-3}) and high specific surface area have been investigated for use as biomaterials [85, 86]. In vitro bioactivity has been proved in *SBF* [87] (Chap. 30). *SEM* images and the corresponding EDS spectra for S20Ca20 (aerogel containing 20% *PDMS* and 20% CaO), before and after 28 days soaked in *SBF*, are shown in Figure 20.23. Before immersion, a smooth surface composed of silicon, calcium, and oxygen can be observed (Figure 20.23A). However, after the in vitro test, a significant change occurred because the aerogel surface appeared covered by a new material composed of globular agglomerates of flake-like particles (Figure 20.23B), more clearly visible in the micrograph at higher magnification (Figure 20.23C). In addition, by tilting the sample the layer thickness close to 5 μm is observed (Figure 20.23D). The corresponding EDS spectra show that the layer is mainly composed of Ca and P in a Ca/P molar ratio of 1.37.

Standard stress–strain curves for both aerogels under uniaxial compression show a first linear region up to 4% strain making it possible to estimate an elastic modulus of 72.4 MPa for S20Ca20, which is close to the low range for cancellous human bone (50–500 MPa). The strain to rupture occurred at 17% and the corresponding stress was 9 MPa.

20.5.2. Nanocomposites for CO_2 Sequestration

Sono-aerogel/calcium silicate composites (Figure 20.24) have proven good candidates for fast CO_2 sequestration (Chap. 12) [88–91]. The composite SAWC40, which contains

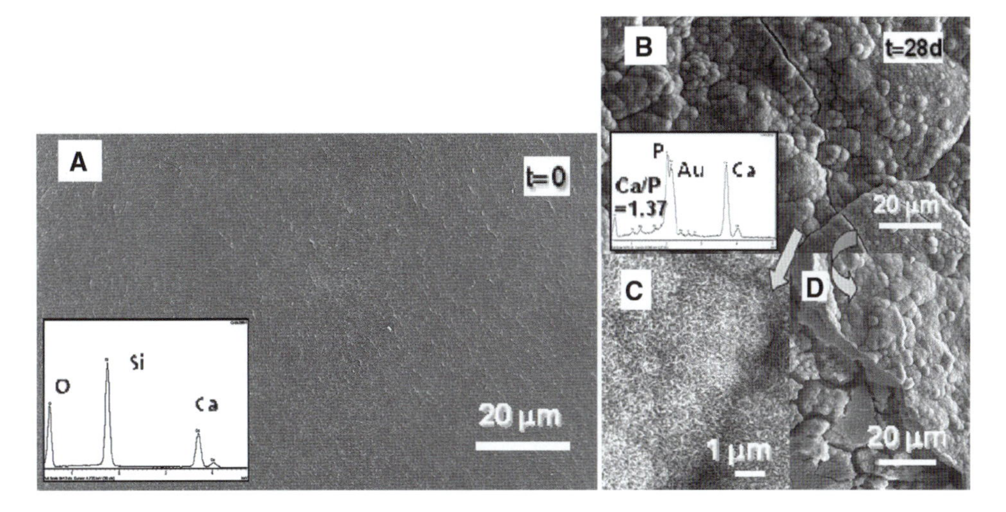

Figure 20.23. SEM micrographs and EDS spectra of S20Ca20 **A.** before and **B.** after 28 days soaking in SBF: **C.** and **D.** images are, respectively, the EDS micrographs at higher magnification and tilting the sample, to see the layer thickness. Reprinted with permission from [86].

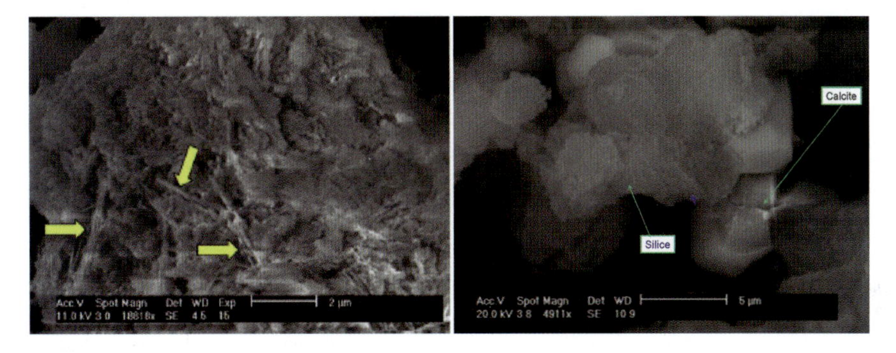

Figure 20.24. SEM image of SAWC40 (which contains CaO 40 wt%) **Left**: aerogel–wollastonite composite. *Arrows* point to embedded "needles" of wollastonite. **Right**: by-products of the same composite after exposure to 30 min of CO_2 flow. The individual monocrystals of carbonates together with unreacted silica gel can be observed. Reprinted with permission from [88].

CaO 40 wt%, gives a better performance than pure *wollastonite* powder due to its porosity and high specific surface area. In this way, without any special experimental conditions, a very short time of carbonation is required, ~30 min, which represents a yield of 0.43 kg of CO_2/(kg of $CaSiO_3$ h).

20.6. Conclusion

High power ultrasound applied to an alkoxide/water mixture makes it possible to obtain nanostructured materials. The process is driven by the acoustic *cavitation* phenomenon. The cavities act as nanoreactors, where the hydrolysis reaction starts. The products (alcohol, water, and silanol) help continue the dissolution of that immiscible mixture.

The process and the derived aerogels present the following features:

1. Slow reaction and flocculated products have been observed with neutral or basic water.
2. There is a *sonication* energy dose at which the alkoxide/water mixture becomes homogeneous. This is taken as the threshold of the reaction.
3. There is a *sonication* energy dose at which the liquid gels in the container.
4. The amount of hydrolysis water affects only slightly the siloxane condensation and degree of cross-linking.
5. The *sonogel* route produces network aerogel structure close to that one of a bulk glass.
6. The hydroxyls are difficult to eliminate, causing problems when a complete densification to form a glass is intended.
7. *Sonogels* have a narrow pore size distribution, high bulk density, and a surface-to-volume ratio two or three times higher than the gels prepared in alcohol solutions.
8. The larger the ultrasonic dose, the finer the porosity and the higher the homogeneity.
9. *Sonogels* are very homogeneous and have a continuous network on a hundred angstroms length scale.
10. Dense *sono-aerogels* have a closed porosity and very fine micropores (<1 nm).
11. The spherical aggregates of ~3 nm radius particles easily sinter above 700°C due to the viscous flow of these small clusters.
12. The sintering is more complex at the length scale of 10 nm because of the closed porosity that contains residual OH groups.

References

1. Suslick KS see some of his Introductory Reviews quoted at http://www.scs.uiuc.edu/suslick/ Accessed 31 July 2009
2. Suslick KS, Price JP (1999) Applications of ultrasound to materials chemistry. Annu Rev Mater Sci 29:295–326
3. Lorimer JP and Mason TJ (1987) Sonochemistry. Part 1—The physical aspects. Chem Soc Rev 16: 239–274
4. Luche JL, (1992) in Current Trends in Sonochemistry (GJ Price, Ed.), p. 34. RSC, Cambridge
5. Suslick KS, Didenko Y, Fang MM, Hyeon T, Kolbeck KJ, McNamara III WB, Mdleleni MM and Wong M, (1999) Acoustic *cavitation* and its chemical consequences. Phil Trans R Soc Lond A, 357:335–353
6. Suslick KS (1990), Sonochemistry, Science, 247:1439–1445
7. Mason TJ, (1991) Practical Sonochemistry. Ellis Horwood, Chichester
8. Leighton TG, (1994) The acoustic bubble. London Academic
9. Price GJ (1996) Ultrasonically enhanced polymer synthesis. Ultrasonics Sonochem 3: S229 S238
10. Tarasevich M (1984), Ultrasonic hydrolysis of a metal alkoxide without alcohol. Cer Bull, 63:500 (Abstract only)
11. Boudjouk P, (1986) Electrochemical and sonochemical routes to organosilane precursors. In Hench LL and Ulrich DR (eds) Science of Ceramics Chemical Processing, Wiley, New York
12. Cortes ADS, Donatti DA, Ibáñez Ruiz A and Vollet DR (2007) A kinetic study of the effect of ultrasound power on the sonohydrolysis of tetraethylorthosilicate. Ultrasonics Sonochem 14 (6): 711–716
13. Zarzycki J (1994) Sonogels. Heterogeneous Chemistry Reviews 1: 243
14. Blanco E, Esquivias L, Litrán R, Piñero M, Ramírez-del-Solar M and de la Rosa-Fox N (1999) Sonogels and derived materials, Appl Organomet Chem 13, 399
15. Esquivias L and Zarzycki J (1986) Study of silica gels obtained by ultrasonic treatment of a silicon alkoxide. In: Baró MD and Clavaguera N (eds) Current topics on non crystalline solids, World Scientific, Singapore

16. Esquivias L and Zarzycki J (1988) Sonogels: An alternative method in sol-gel processing, In: Mackenzie JD and Ulrich DR, (eds) Ultrastructure processing of ceramics, glasses and composites, Wiley, New York
17. Zarzycki J (1992) Ultrastructural models. In: Hench LL and Wcst JK (cds) Chemical processing of advanced materials. Wiley, New York, pp. 77–92
18. de la Rosa-Fox N, Esquivias L and Zarzycki J (1987) Glasses from sonogels. Diffusion and Defect Forum 53–54: 363–373
19. de la Rosa-Fox N, Esquivias L and Zarzycki J (1989) Textural characteristic of aerogels obtained from sonogels. Rev Phys Appl 24 (C4): 233–243
20. de la Rosa-Fox N, Esquivias L (1987) Craievich A F and Zarzycki J, Structural study of silica sonogels. J. Non-Cryst. Solids 192: 211–215
21. de la Rosa-Fox N, Esquivias L and Zarzycki J (1991) Silica sonogels with drying control chemical additives. J Mater Sci Lett 10: 1237–1242
22. Zarzycki J (1992) Sonogels – development and perspectives. In: Ulhmann DR and Ulrich DR (eds) Ultrastructure processing of advanced materials, Wiley, New York
23. Barrera-Solano C, de la Rosa-Fox N and Esquivias L (1992) Ultrastructural aspects of silica sonogels. J Non-Cryst Solids 147–148: 194–200
24. Atik M (1990) La *cavitation* et ses effets dans la synthèse des matériaux composites (SiO_2/SiO_2) et (SiO_2-SiO_2) B_2O_3). Étude détaillée du processus du frittage et effets des inclusions rigides, Ph. D. Thesis, Université de Montpellier II, France
25. Fernández-Lorenzo C (1993) Estudio mediante espectroscopía Raman y resonancia magnética nuclear de la obtención y densificación de los geles mixtos, Ph. D. Thesis, Universidad de Cádiz, Spain
26. Fernández-Lorenzo C, Esquivias L, Barboux P, Maquet J, Taulelle F (1994) Sol-gel synthesis of SiO_2-P_2O_5 glasses. J Non-Cryst Solids, 176: 189–199
27. Ramírez-del-Solar M, Esquivias L, Craievich A Fand Zarzycki J (1992) Ultrastructural evolution during gelation of TiO_2-SiO_2 sols. J Non-Cryst Solids 147–148: 206–212
28. Ramírez-del-Solar M (1991) Efectos de los ultrasonidos sobre la transformación sol-gel. Estudio por dispersión de rayos x a bajos ángulos de las modificaciones inducidas en geles del sistema TiO_2 -SiO_2, Ph. D. Thesis, Universidad de Cádiz, Spain
29. Ramírez-del-Solar M, de la Rosa-Fox N, Esquivias L and Zarzycki J (1990) Effect of the method of preparation on the texture of TiO_2-SiO_2 Gels. J Non-Cryst Solids 121: 84–89
30. Ramírez-del-Solar M and Esquivias L (1994) Ultrastructural evolution during sintering of mixed sonogels. J Sol-Gel Sci Technol 3–1: 41–46
31. Ruiz-Rube JM, Ramírez-del-Solar M, de la Rosa-Fox N and Esquivias L (1990) Short-range order of 0.10TiO_2-0.90 SiO_2 aerogels. In: Colmenero J and Alegría A (eds) Basic features of the glassy state, World Scientific, Singapore
32. Calvino JJ, Cauqui MA, Cifredo G, Esquivias L, Pérez J A, Ramírez-del-Solar M and Rodríguez-Izquierdo JM (1993) Ultrasound as a tool for the preparation of gels: effect on the textural properties of TiO_2-SiO_2 aerogels. J Mat Sci 28: 2191–2195
33. Esquivias L and Ramírez-del-Solar (1997) Short-range order of titania doped silica sono-aerogel. J Non-Cryst Solids 220: 45–51
34. Chaumont D, Craievich AF and Zarzycki J (1992) Effect of ultrasound on the formation of Zr_2O sols and wet gels. J Non-Cryst Solids 147–148: 41–46
35. Chaumont D (1992) Étude structurales de la formation des sols et des gels de zircone en présence d'ultrasons, Ph. D. Thesis. Université de Montpellier II, France
36. Piñero M, Atik M and Zarzycki J (1992) Cordierite-Zr_2O and cordierite-Al_2O_3 composites obtained by sonocatalytic method. J Non-Cryst Solids 147&148: 523–531
37. Piñero M (1993) Elaboración de compuestos cerámica-cerámica con matriz cordierita por el método sol-gel, estudio de sus propiedades mecánicas, Ph.D. Thesis, Universidad de Cádiz, Spain
38. Esquivias L, de la Rosa-Fox N, Bejarano M and Mosquera MJ (2004) Structure of hybrid colloid-polymer xerogels. Langmuir 20: 3416–3423
39. Esquivias L, Morales-Flórez V, Mosquera MJ and de la Rosa-Fox N (2008) Changes in the structure of composite colloide-polymer xerogels after cold isostatic pressing. J Sol-Gel Sci Technol 47: 194–202
40. Ramírez-del-Solar M, de la Rosa-Fox N, Esquivias L and Zarzycki J (1990) Kinetic study of gelation of solventless alkoxide–water mixtures. J Non-Cryst Solids 121: 40–44
41. Donatti DA and Vollet DR (1995) A calorimetric study of the ultrasound-stimulated hydrolysis of solventless TEOS-water mixtures, J Sol-Gel Sci Technol 4: 99–105

42. Vollet DR, Donatti DA and Campanha JR (1996) A kinetic model for the ultrasounds catalyzed hydrolysis of solventless TEOS-water mixtures and the role of the initial addition of ethanol. J Sol-Gel Sci Technol 6:57–63
43. Morita K, Hu Y and Mackenzie JD (1994) The effects of ultrasonic irradiation on the preparation and properties of ormosils. J Sol-Gel Sci Technol 3: 109–116
44. Marino I-G, Lottici PP, Bersani D, Raschellà R, Lorenzi A, Montenero A (2005) Micro-Raman monitoring of solvent-free TEOS hydrolysis. J Non-Cryst Solids 351: 495–498
45. Iler R, The Chemistry of Silica, Wiley, New York, 1979
46. Donatti, DA, Vollet, DR (1996) Study of the hydrolysis of TEOS-TMOS mixtures under ultrasound stimulation. J Non-Cryst Solids 204 (3) 301–304
47. Donatti DA, Ibañez Ruiz A, Vicelli MR and Vollet DR (2007) Structural properties of silica gels prepared from oxalic acid catalyzed tetraethoxysilane sonohydrolysis. Phys Stat Sol (a) 204 (4), 1069–1076
48. Couppis EC and Cklinzing GE (1974) Effect of cavitation on reacting systems. AIChE Journal 20(3): 485–491
49. Aerstin FGP, Timmerhaus KD and Fogler HS (1967) Effect of the resonance parameter on a chemical reaction subjected to ultrasonic waves. AIChE J 13 (3): 453–456
50. Fuqua PD, Mansour K, Alvarez D, Marder SR, Perry JW and Dunn BS (1992) Synthesis and nonlinear optical properties of sol-gel materials containing phthalocyanines. In: SPIE Vol 1758, Sol-Gel Optics II
51. Ocotlán-Flores J and Saniger JM (2006) Catalyst-free SiO_2 sonogels. J Sol-Gel Sci Technol 39: 235–240
52. Morales-Saavedra OG, Rivera E, Flores-Flores JO, Castañeda R, Bañuelos JG and Saniger JM (2007) Preparation and optical characterization of catalyst free SiO_2 sonogel hybrid materials. J Sol-Gel Sci Techn 41:277–289
53. Avnir D and Kaufman VR (1987) Alcohol is an unnecessary additive in the silicon alkoxide sol-gel process. J Non-Cryst Solids 92(1): 180–182
54. Kun-Hong L, Yung-Pyo Kand Jae-Gong L (1994) Synthesis of silica glass using solventless sol-gel process. J Sol-Gel Sci Technol 2: 907–912
55. Jonas J, Irwin AD and Holmgen JS (1988) Solid state ^{29}Si and ^{11}B NMR studies of sol-gel derived borosilicates. J Non-Cryst Solids 101: 249–254
56. Pérez-Moreno A, Jiménez-Solís C, Esquivias L and de la Rosa-Fox N (1998) Estudio mediante RMN de la hidrólisis y policondensación de TEOS bajo la acción de ultrasonidos. Bol Soc Esp Ceram V. 37 (1):13–17
57. Mackenzie JD (1994) Structures and properties of Ormosils. J Sol-Gel Sci Technol 2:81–86
58. Morita K, Hu Y and Mackenzie JD (1992) The effect of ultrasonic radiation on gelation and properties of Ormosils. In: MRS (ed) Mat Res Soc Symp Proc 271: 693–698
59. Iwamoto T and Mackenzie JD (1995) Hard Ormosils prepared with ultrasonic radiation. J Sol-Gel Sci Technol 4:141–150
60. de la Rosa-Fox N, Esquivias L and Piñero M (2003) Organic-inorganic hybrid materials from sonogels. In: Nalwa HS (ed) Handbook of organic-inorganic hybrid materials and nanocomposites 2:241–270, American Scientific Publishers, Stevenson Ranch, California, USA
61. de la Rosa-Fox N, Morales-Flórez V, Piñero M Esquivias L (2009) Nanostructured sonogels in progress in sol-gel production. Key Engineering Materials 391: 45–78
62. Glatter O and Kratky O, Small angle x-ray scattering, Academic Press, London, 1982
63. Schaefer DW (1988) Fractal models and the structure of materials. Mater Res Soc Bull 8: 22–27
64. Zarzycki J, (1990) Structural Aspects of Sol-Gel Synthesis J Non-Cryst Solids 121: 110–118
65. Porod G, (1982) Small angle x-ray scattering, edited by O Glatter and O Kratky, Academic, London
66. Donatti D, Ibáñez Ruiz A and Vollet D (2001) From sol to aerogel: a study of the nanostructural characterisitics of TEOS derived sonogels. J Non-Cryst Solids 292 (1–3): 44–49
67. Donatti, DA, Vollet, DR and Ibañez Ruiz A (2004) Comparative study using small-angle x-ray scattering and nitrogen adsorption in the characterization of silica xerogels and aerogels Phys Rev B 69: 064202–1-064202–6
68. Vollet DR, Donatti DA, Ibañez Ruiz A and de Castro WC (2003) Structural evolution of aerogels prepared from TEOS sono-hydrolysis upon heat-treatment up to 1100°C. J Non-Cryst Solids 332:73–79
69. Vollet DR, Torres RR, Donatti DA and Ibañez-Ruiz A (2005) Structural characteristics of gels prepared from sonohydrolysis and conventional hydrolysis of TEOS: an emphasis on the mass fractal as determined from the pore size distribution. Phys Sta Sol (a) 202 (14): 2700–2708
70. Esquivias L, Barrera-Solano MC, de la Rosa-Fox N, Cumbrera FL and Zarzycki J (1992) Determination of the skeletal density of silica gels from wide-angle x-ray diffraction. In: Ulhmann D (ed) Ultrastructure processing of advanced materials, Wiley, New York
71. Warren B E (1969) X-ray diffraction, Adisson-Wesley, New York
72. Wei W (1986) A new approach to high resolution RDF analysis. J Non-Cryst Solids 81: 239–250

73. Rosenthal AB and Garofalini S (1988) Molecular dynamics study of amorphous titanium silicate. J Non-Cryst Solids 107: 65–72
74. Teo BK, (1986) EXAFS: Basic principles and data analysis, inorganic chemistry concepts, Vol. 9, Springer-Verlag, Berlin
75. Szu SP, Klein LC and Greenblatt M (1992) Effect of precursors on the structure of phosphosilicate gels: ^{29}Si and ^{31}P MAS-NMR Study. J Non-Cryst Solids 143: 21–30
76. Prabakar S, Rao KJ and Rao CNR (1991) A MAS NMR investigation of aluminosilicate, silicophosphateand aluminosilicophosphate gels and the evolution of crystalline structures on heating the gels. J Mater Res. 6: 592–601
77. Fernandez-Lorenzo C, Martin J, Esquivias LM and Blanco E (1992) Raman spectroscopy of phosphorous-doped silica gels. In: Hench LL and West JK (eds) Chemical processing of advanced materials, Wiley, New York
78. Brunauer S, Emmett PH and Teller E (1938) Adsorption of gases in multimolecular layers, J Am. Chem Soc 60: 309–319
79. Sing KSW, Everett DH, Haul RAW, Moscou L, Pierotti RA, Rouquerol J and Siemieniewska T (1985) Reporting physisorption data for gas/solid systems with special reference to the determination of surface area and porosity. Pure Appl Chem 57: 603–619
80. Donatti DA, Vollet DR, Ibañez Ruiz A, Mesquita A and Silva TFP (2005) Mass fractal characteristics of silica sonogels as determined by small-angle x-ray scattering and nitrogen adsorption. Phys Rev B 71: 014203–1 014203–7
81. Donatti DA, Ibañez Ruiz A, Kumakawa MM and Vollet DR (2006) Structural characteristics of silica sonogels prepared with additions of isopropyl alcohol. J Phys Chem B 110 (43) 21582–21587
82. Vacher R, Woignier T, Pelous J and Courtens E (1988) Structure and self-similarity of silica aerogels. Phys Rev B 37, 6500–6503
83. de la Rosa-Fox N, Esquivias L, Zarzycki J and Craievich A (1990) Sintering of sonogels. Riv. della Staz Sper Vetro 5: 67–70
84. Perin L, Faivre A, Calas-Étienne S, Woignier T, (2004) Nanostructural damage associated with isostatic compression of silica aerogels, J Non-Cryst Solids 333: 68
85. Esquivias L, Morales-Flórez V, Piñero M, de la Rosa-Fox N, Ramírez J, González-Calbet J, Salinas A, Vallet-Regí M (2005) Bioactive organic-inorganic hybrid aerogels. MRS Proceedings Vol 847: EE12.1- EE12.6 Material Research Society Press
86. Salinas AJ, Vallet-Regí M, Toledo-Fernández JA, Mendoza-Serna R, Piñero L, Esquivias L, Ramírez-Castellanos J and González-Calbet (2009) Nanostructure and bioactivity of hybrid aerogels. Chem Mater 21 (1): 41–47
87. Kokubo T, Kushitani H, Sakka S, Kitsugi T and Yamamuro T (1990) Solutions able to reproduce in vivo surface-structure changes in bioactive glassceramic A-W3. J Biomed Mater Res 24: 721–726
88. Santos A, Ajbary M, Toledo-Fernández JA, Morales-Flórez V, Kherbeche A and Esquivias L, (2008) Reactivity of CO_2 traps in aerogel-*wollastonite* composite. J Sol-Gel Sci Technol 48: 224—230
89. Santos A, Toledo-Fernández JA, Mendoza-Serna R, Gago-Duport L, de la Rosa-Fox N, Piñero M, Esquivias L (2007) Chemically active silica aerogel-*wollastonite* composites for CO_2 fixation by carbonation reactions. Ind Eng Chem Research 46: 103–107
90. Santos A, Ajbary M, Kherbeche A, Piñero M, de la Rosa-Fox N, Esquivias L (2008) Fast CO_2 sequestration by aerogel composites. J Sol-Gel Sci Technol 45: 291–297
91. Santos A, Ajbary M, Piñero M, Esquivias L (2008) Material compuesto de aerogel de sílice y polvo de larnita y su uso en el almacenamiento y fijación de CO_2 Patent # P200802914, Spain
92. Price GJ, Hearn MP, Wallace Eand Patel AM (1996) Ultrasonically assisted synthesis and degradation of poly (dimethyl siloxane). Polymer 37:2303
93. Zarzycki J (1988) Critical stress intensity factors of wet gels. J Non-Cryst Solids 100: 359–363
94. Vollet DR, Nunes LM, Donatti DA Ibáñez Ruiz A and Maceti H (2008) Structural characteristics of silica sonogels prepared with different proportions of TEOS and TMOS. J Non-Cryst Solids 354:1467–1474

Part VII

Properties

21

Structural Characterization of Aerogels

Gudrun Reichenauer

Abstract Determining reliable structural parameters for an aerogel by applying suitable characterization techniques is a key factor in terms of understanding the different synthesis steps and their impact on the resulting aerogel. Combining structural parameters with the physical properties of the material allows optimization for specific applications. It is only the profound knowledge of the structure–properties relationships that provides access to the full potential of this type of material. This chapter presents different characterization methods commonly used and discusses their potential and limitations. Furthermore, more recent developments and new approaches are introduced.

21.1. Introduction

Aerogels are defined by two different phases, i.e., the solid backbone and the pore phase. Both are characterized by a set of basic parameters such as the respective phase fraction, the characteristic extension of each phase, and its connectivity. In addition, the physical and chemical properties of the large interface separating the two phases represent important characteristics of the aerogel.

A first visual and manual inspection of an aerogel usually already provides quite a bit of information:

Is the material transparent or translucent, is it brittle or ductile, how easily can the sample be deformed, how does it sound when you drop a piece of the aerogel under investigation on your table? These simple tests give a hint about the extension of the void and solid phase present, provide information on whether the aerogel backbone is purely inorganic or whether it contains an organic component, or give a first clue about its elastic modulus and thus the connectivity of its backbone.

More detailed information can be derived by applying structural characterization methods. However, one has to keep in mind that this extraordinary class of materials with its unusual properties might not always fulfill the assumptions for a given experimental technique that originally has been developed for materials with either much larger pores or small pores combined with low porosities.

G. Reichenauer • Bavarian Center for Applied Energy Research, Am Hubland, 97074 Würzburg, Germany
e-mail: reichenauer@zae.uni-wuerzburg.de

M.A. Aegerter et al. (eds.), *Aerogels Handbook*, Advances in Sol-Gel Derived Materials and Technologies, DOI 10.1007/978-1-4419-7589-8_21,

21.2. Structural Parameters and Related Experimental Techniques

The key structural parameters of a porous material are the total fraction of the pore and the solid phase, the typical extension of the backbone and the pore phase, the connectivity of the two phases, the characteristics of the interface between the phases, and the molecular structure of the backbone phase (see Figure 21.1). Related to these properties are different characterization techniques (Table 21.1).

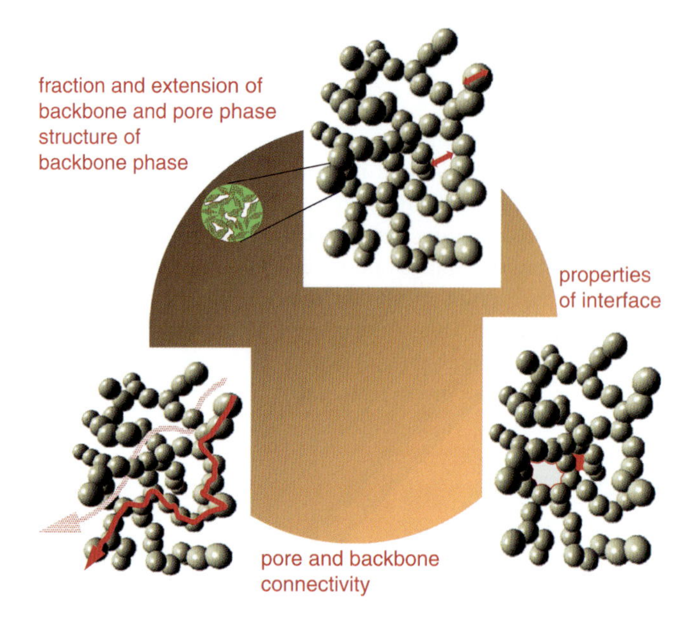

Figure 21.1. Overview over the key structural parameters of a porous solid.

In the following sections, the most common methods are described in detail with special emphasis on their limitations and merits as well as the basic equations applied for the evaluation of the respective experimental data.

21.3. Microscopy

Microscopy provides a visual impression of the aerogel under investigation on length scales down to a few Angstroms. Light microscopy (*LS*) with a resolution limit of about 500 nm, however, misses the features of most aerogels with typical particle sizes well below 1 μm. Atomic force microscopy (*AFM*) can in principle be applied to aerogels, but in practice turns out to be critical; this is due to the high irregularity and depth of the surface topology at a fracture surface of an aerogel.

The two major microscopic methods used to investigate aerogels are therefore scanning electron microscopy (*SEM*) and transmission electron microscopy (*TEM*); while *SEM* with a maximum resolution in the range of about a nanometer is applied to visualize the three-dimensionally interconnected aerogel backbone, the higher resolution of *TEM* is exploited to analyze the substructure of the particles forming the backbone or additional phases incorporated in the aerogel skeleton, such as metal particles.

Table 21.1. Overview over structural properties and related experimental techniques

Solid and void fraction	Pore size	Backbone entity size	Pore connectivity	Backbone connectivity	Interfacial characteristics	Backbone composition and structure
Porosity Mass density	Average pore size Pore size distribution	Particle size Surface area			Surface roughness Surface groups	
Macroscopic volume and mass	Gas sorption	*SEM*	Static or dynamic fluid permeation	Sound velocity measurement	Wetting	*EDX*
Immersion in liquid	Mercury porosimetry	*TEM*	Beam bending	Measurement of the elastic modulus	IR spectroscopy	IR spectrocopy
Three-dimensional scan	Thermoporometry	*AFM*		Beam bending	*EDX*	*XRD*
	Light scattering	Light scattering		Inelastic neutron scattering	Adsorption	Helium pycnometry
	Small-angle scattering	Small-angle scattering		Brillouin scattering	*NMR* spectroscopy	Raman spectroscopy
	NMR relaxation					*NMR*-spectroscopy
	Positron annihilation					Thermal analysis coupled with mass spectroscopy
	Shrinkage upon (de)sorption			Thermal (Chap. 23) or electrical conductivity (Chap. 11)		
	Gaseous thermal conductivity (Chap. 23)					

The methods in the gray marked boxes (except for wetting) are described in the following in details. Dark gray marked are critical techniques requiring special care upon measurement or data treatment when used with aerogels.

Electron microscopy uses a beam of primary electrons that is focused onto the surface of the sample (Figure 21.2); a voltage is applied to accelerate the electrons to energies up to 30 keV (*SEM*) or 100–300 keV (*TEM*). While in the *SEM* mode electrons and X-rays emitted into the hemisphere of the primary beam are analyzed (reflection setup), *TEM* probes the transmission of the sample with respect to high energy primary electrons. All high resolution electron microscopes are operated under a chamber vacuum better than 10^{-4} mbar to reduce parasitic scattering.

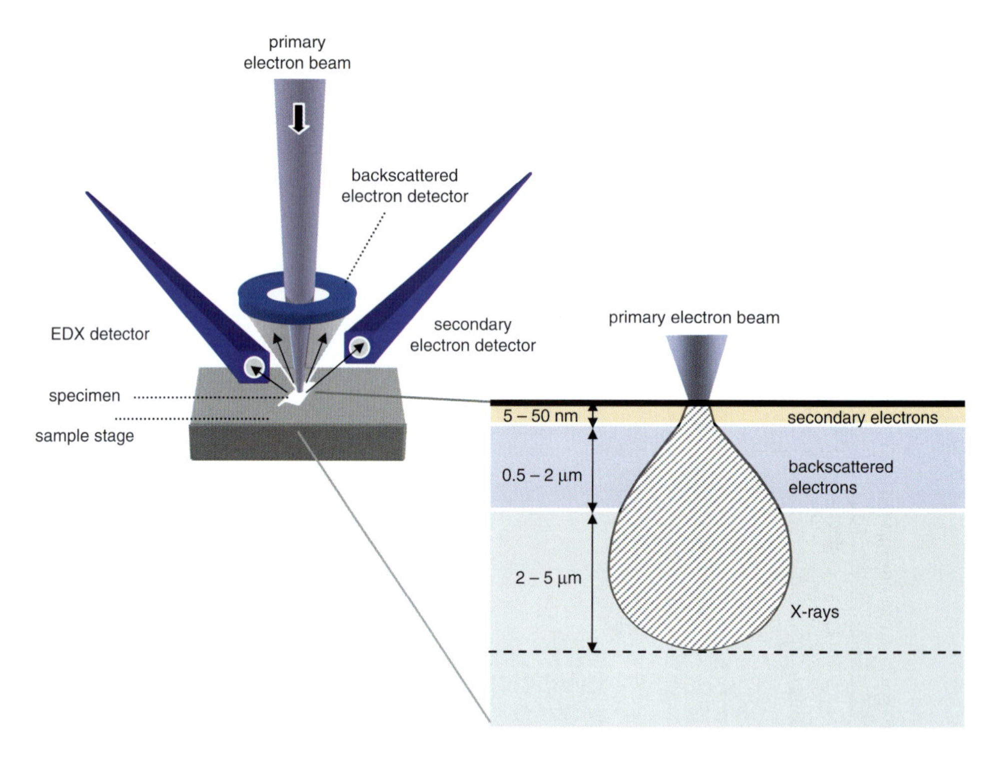

Figure 21.2. Left: Typical *SEM* setup with secondary electron (SE), ring-shaped backscattering (ESB = energy selective backscattering), and *EDX* (energy dispersive X-ray) detector; **right:** depth from which *SE*, *BSE*, and X-rays are typically emitted when exposing the sample to a primary electron beam with an energy of up to 30 keV (after [1], p. 111).

Depending on their energy the primary electrons in the *SEM* mode probe different depths of the sample (Figure 21.2). Figure 21.3 shows an example of the very same spot of an organic aerogel probed with different acceleration voltages at otherwise identical conditions (Chap. 36); the images reveal the switching from a surface to a bulk sensitive imaging mode of *SEM*.

The interaction of the primary electrons with the specimen results in the emission of secondary electrons (*SE*), i.e., inelastically scattered electrons that escape from the sample and reach the respective detector if emitted per definition from a depth less than about 50 nm equivalent of solid phase (Figure 21.2). The resolution of the *SE* is determined by the quality of the focus of the primary beam and the area of the sample surface from which they are emitted.

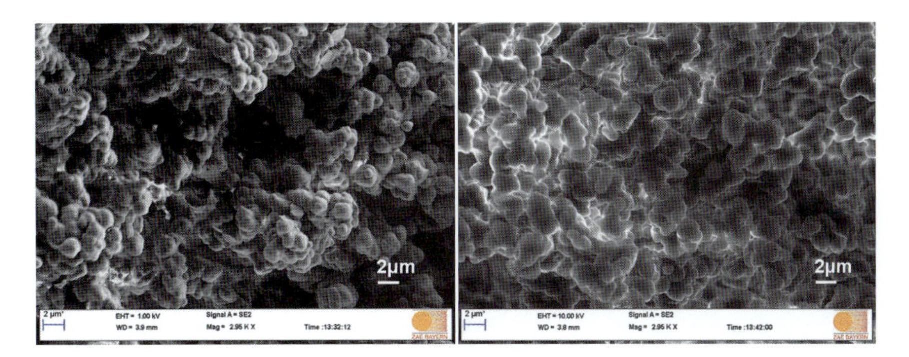

Figure 21.3. *SEM* image of the very same spot of an organic aerogel taken at low (1 kV) and high acceleration voltage (10 kV) at otherwise identical conditions.

Backscattered electrons (*BSE*) are elastically scattered electrons resulting from deeper layers of the sample. Their energy is higher than the one of the *SE* (max. 50 eV); thus a decelerating voltage can be applied to discriminate between *SE* and *BSE*. *BSE* are detected by a ring-shaped four quadrants detector with a center hole to allow the primary beam to pass through (Figure 21.2). As the probability for the electrons to be scattered strongly depends on the atomic number, the *BSE* signal can be used to image the local distribution of different elements within the specimen such as in hybrid aerogels with two solid phases of different chemical composition (Figure 21.4).

Figure 21.4 *SEM* image (**left**) and corresponding image of the same spot of the sample taken with an *ESB* (energy selective backscattering) detector (**right**); the *ESB* signal provides a mapping of the different components present. In this case, a silica infiltrated (*light areas*) carbon aerogel (*dark areas*) is shown.

Inelastic interaction of the outer shell electrons within the specimen with the primary and *SE* results in the emission of X-rays whose energy is a function of the sample density and chemical elements present. The penetration depth of the electrons is hereby element and electron energy specific. An optimum spatial resolution of about 200 nm is reached at an electron energy of about 5 keV depending on the density of the sample. With increasing penetration depth a pear shaped regime of X-ray excitation results (Figure 21.2); the maximum width of this volume is about 1 µm. Using a detector that is able to resolve the energy of the X-rays emitted (energy dispersive X-ray (*EDX*) detector, Figure 21.2), the elementary composition of the sample can be determined with a detection limit of about 1 wt%, depending on the sample under investigation; however, analysis is usually only possible for elements with atomic numbers > 4. Figure 21.5 shows an example for the analysis of a carbon aerogel infiltrated with a carbonate phase.

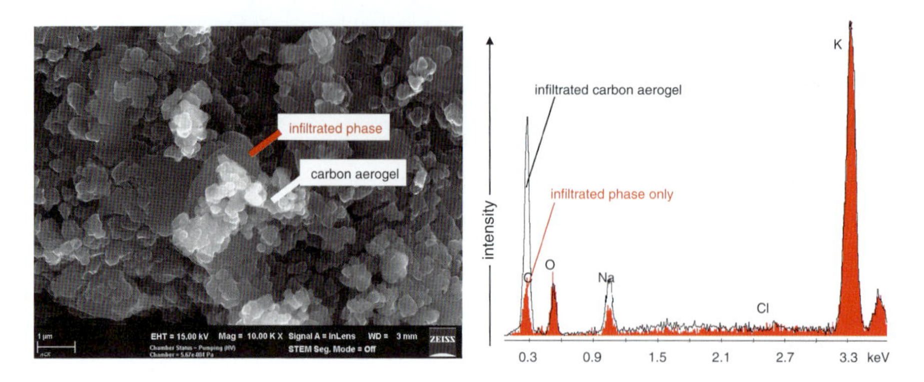

Figure 21.5. **Left:** *SEM* image of a carbon aerogel infiltrated with $NaNO_3/KNO_3$. **Right:** *EDX* spectrum of the composite and the infiltrated phase only.

Upon electron microscopy the sample is exposed to a continuous stream of electrons. If the sample is electrically not well conductive, charges are accumulated on the sample resulting in the formation of an electric field that interferes with the imaging system of the microscope. To prevent artifacts, nonconductive specimens have to be coated for *SEM* imaging with an electrically conductive layer that provides a spreading of the deposited charges out of the area of focus. This is usually performed via sputter coating of the specimen with layers of a few nanometers only; thereby, metals such as gold and palladium or carbon are used. Alternatively, some electron microscopes (such as the ZEISS Ultra Plus) provide a so-called charge compensation feature consisting of small amounts of inert gas injected next to the focus area (Figure 21.6). However, due to additional scattering by the gas, the resolution in this mode is diminished.

SEM imaging provides a two-dimensional projection of SE intensity emitted from different depth of the specimen. The high porosity, typical for aerogels, becomes usually not very obvious in *SEM* pictures unless very thin layers of the samples are prepared.

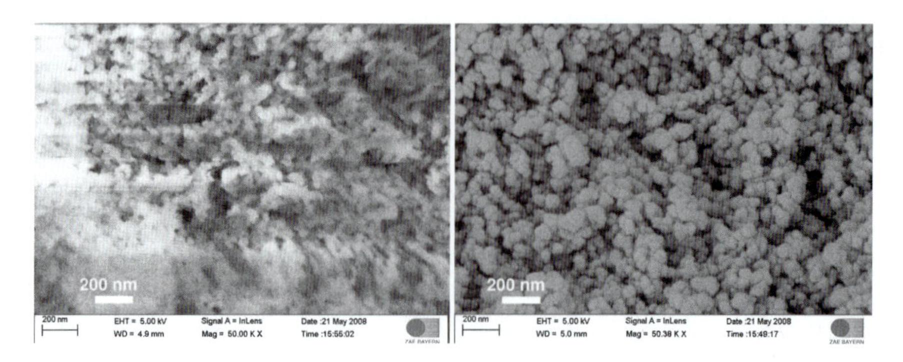

Figure 21.6. *SEM* image of an organic aerogel taken without (**left**) and with (**right**) charge compensation option at otherwise identical conditions.

An alternative for visualizing the sparse aerogel networks and their channel-like pores is to take two *SEM* images of the same spot of the specimen, however with the sample tilted by 3–10° between the two images. Using commercially available software (e.g., Alicona MeX or SIS) the two images can be merged to a stereo image for a set of red–green glasses (Figure 21.7).

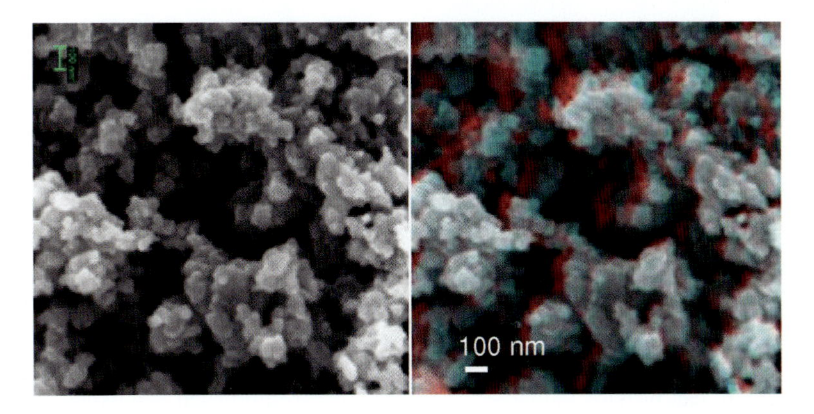

Figure 21.7. Standard *SEM* image (**left**) and corresponding *SEM* stereo image (**right**) of a carbon aerogel; the picture consists of two separately taken pictures, colored in *red* and *cyan*, respectively, of the same spot of the sample, however, with the sample tilted by 9° between the two pictures. Using a pair of *red–cyan* glasses the depth of the pores and the openness of the aerogel backbone becomes evident.

The energy deposited upon *EM* in the area of the specimen investigated can be significant, in particular at high currents. Aerogels with their sparse network represent often good thermal insulators. As a consequence, the energy locally deposited during *EM* cannot be spread fast enough and thus may result in a modification of the sample at the focus of the primary beam. Bourret [2] even reports on the observation of local sintering of silica aerogels during *TEM* imaging. Nevertheless, *TEM* provides extraordinary resolution (see e.g., Figure 21.8) and is particularly helpful to visualize microscopic phases of different crystalline orientation or metal nanoparticles attached to the aerogel backbone (Figure 21.9) [4–18]. A practical guide to *TEM* of aerogels has been compiled by Stroud et al. [19].

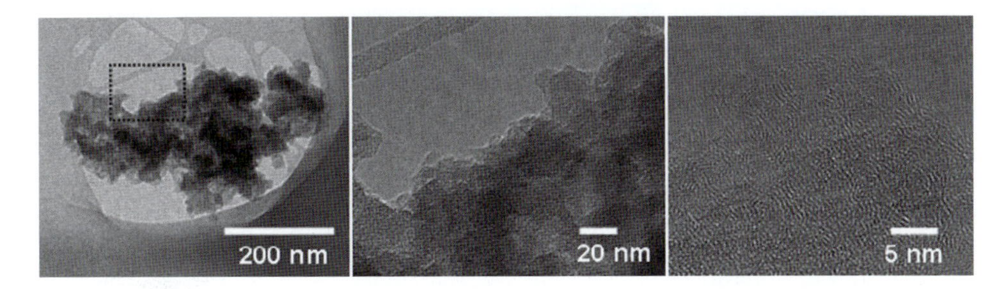

Figure 21.8. *TEM* images of a phenol–formaldehyde based carbon aerogel dispersed on an irregularly shaped copper grid at different magnifications (Courtesy: Evonik Degussa GmbH). The **image in the middle** represents a zoom into the area indicated by the dashed square in the low resolution image (**left**). At the highest resolution the structure of turbostratic carbon within the backbone particles become visible.

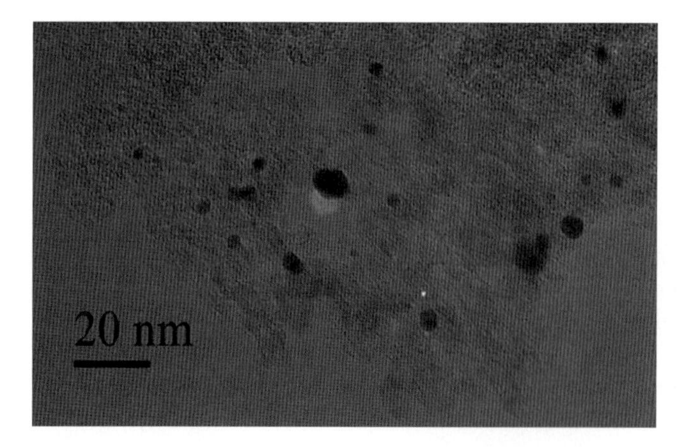

Figure 21.9. Pt-doped carbon aerogel derived via copolymerization of a metal complex with the gel precursors; image reprinted from [3], p. 2339, with permission from Elsevier.

Due to experimental difficulties *AFM* is very seldom applied to investigate aerogels. However, it still can provide valuable information if carefully used. An excellent example is the image taken by Marliere et al. [20] for a silica aerogel (Figure 21.10). The aerogels samples were prepared for *AFM analysis* by "brittle fracturation" [20]. The image reveals a cluster superstructure similar to a stack of snowballs within the aerogel under investigation. Hereby the extension of the clusters as well as the cluster–cluster distance is in the range of about 100 nm. This modulation of the local density also results in a well-pronounced forward scattering in the ultra-small-angle scattering regime as well as a faint scattering in the optical range. The cluster-like structure is suspected to be also

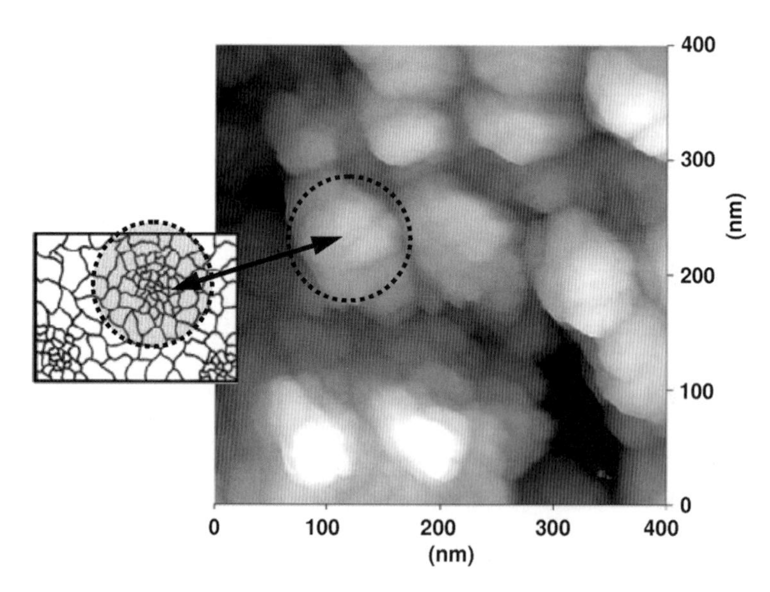

Figure 21.10. *AFM* image of a silica aerogel showing a superstructure formed by the arrangement of porous silica clusters; reprinted from [20], p. 150, with permission from Elsevier. The characteristic distance between the clusters is about 100 nm. The *inset* shows a schematic model of the superstructure as proposed by Beck [21] as a result of the analysis of optical scattering curves. The *dashed circles* indicate corresponding regimes.

responsible for the effect of switching of aerogels from transparent to highly opaque with partial filling of their mesopores [22].

Comment. Classical microscopy methods provide a good visual impression of an aerogel, however, it is impossible to extract directly any information on the pore size or connectivity; usually *SEM* images can be used to deduce an estimate for the average particle size and get a first idea about the distribution of the different phases in case of aerogels whose backbone comprises more than a single solid phase (e.g., carbon backbone with metal particles incorporated). Unless very inhomogeneously distributed phases are present, *EDX* allows for a quantification of the average chemical composition of the aerogel under investigation.

Care has to be taken to not modify the specimen or mask structural features by insufficient spreading of charges or energy deposited.

21.4. Scattering Techniques

Scattering techniques are quantitative, noninvasive, nondestructive tools for structural analysis [23]. They can be applied for the characterization of the aerogel structure as well as for the in situ study of structural changes of aerogels and their precursors, e.g., upon the sol-gel transition, during aging or drying, or during sintering. Inelastic scattering provides information on the eigenmodes of the aerogel backbone and thus the connectivity of the system.

21.4.1. Elastic Scattering

Elastic scattering probes the relative structural arrangement of scatterers (e.g., electrons in case of X-ray scattering) within the specimen under investigation. When an incident monochromatic wave front hits the specimen, a spherical wave is initiated by every constituent of its solid phase; these spherical waves interfere and result in a scattering pattern. The scattered intensity is measured as a function of the angular difference 2θ between the direction of the incident and the scattered beam. To record the scattering pattern either movable counters or two-dimensional detectors (Figure 21.11) are applied.

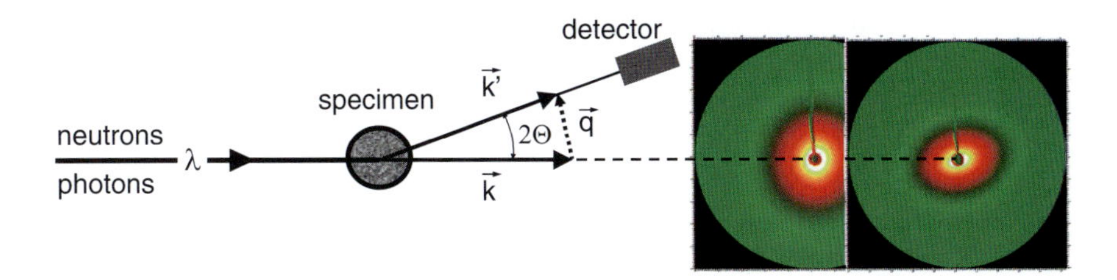

Figure 21.11. General setup of a scattering experiment with the incident beam of photons or neutrons, the specimen under investigation, and the detector for measuring the angular distribution of the scattered intensity. **Right:** Two-dimensional scattering pattern of an isotropic (**left**) and an anisotropic (**right**) specimen, respectively; the image also shows the features of the beam stop hanging from the top blocking the directly transmitted beam and thus protecting the detector.

Since the wavelength (energy) of the incident beam can be varied (e.g., when using a synchrotron as the photon source) the scattering vector rather than the angle Θ is often used as the variable:

$$q = \frac{4\pi}{\lambda} \sin(\theta) \times e \,. \tag{21.1}$$

Here λ is the wavelength of the incident beam, θ is half of the scattering angle, and $\boldsymbol{e}$ is the unit vector characterizing the direction of the vector $\boldsymbol{q}$. The latter is defined by the difference of the wave vectors of the incident and the scattered beam, $\boldsymbol{k}$ ($\boldsymbol{k} = 2\pi/\lambda$) and $\boldsymbol{k}'$, respectively, since we are dealing with elastic scattering $|\boldsymbol{k}|=|\boldsymbol{k}'|$.

To change the resolution in q, or to extend the q-range, the wavelength of the incident beam or the sample-detector distance can be adjusted.[1] The collimation of the incident beam can have either point or slit geometry. In the case of slit geometry, the measured scattered intensity represents the convolution of the scattering pattern of the sample for point collimation with the slit profile [23].

For aerogels, the scattered intensity usually possesses radial symmetry indicating an isotropic specimen; in this case, the angle θ is the only variable in a given experimental setup. If the aerogel sample is anisotropic in nature (e.g., due to anisotropic drying conditions) it shows as anisotropy of its scattering pattern and directional information has to be included to describe the orientation of the anisotropy (see Figure 21.11).

Generally, one distinguishes between small-angle scattering (*SAS*) and wide-angle scattering; in the case of X-rays scattering in a reflection setup the latter is also known as X-ray diffraction (*XRD*). *SAS* covers scattering angles 2θ in the range from 0.001° to about 10°; the wavelengths used are typically in the range of Angstroms to a few nanometers. The corresponding q-range reaches from about 5 to 10^{-4} nm^{-1} (low q limit in ultra-low-angle scattering, *USAXS* and *USANS*). As the q value can be related to a length scale L probed by $q \approx \pi/L$, *SAS* is sensitive to structures on the scale of 10 μm to a few Angstroms.

Light scattering with wavelengths on the order of several 100 nm in combination with scattering angles of a few degrees to 180° yields maximum q values that are overlapping with the one of *SAS* at very low angles (*USAXS* and *USANS*).

In the *SAS* regime, the angular characteristic of the individual scatterer can be disregarded and an angle independent amplitude A_i can be attributed to each wave emitted. The scattered intensity is given by the square of the sum of overall waves emitted from the sample volume irradiated (21.2); as a consequence, the phase shift information between the individual waves is lost and only the autocorrelation function of the system under investigation (in case of a two-phase media, e.g., solid and voids) can be extracted from the experimental data. As the phase information is missing, the autocorrelation function cannot be unequivocally related to a three-dimensional backbone structure.

In practice only part of the scattering is coherent; particularly neutron scattering often contains a strong incoherent background that will be disregarded in the following. The coherent differential scattering cross-section per irradiated volume V_I of the specimen can be written in terms of the sum over the individually emitted waves with amplitude A_i; in a continuous representation, the sum can be replaced by an integral:

[1] Note that the intensity per detector area decreases with the sample-detector distance squared.

$$\frac{d\Sigma}{d\Omega_{coh}}(q) = \left[\frac{1}{V_I}\left|\sum_{i=1}^{N} A_i \times \exp(-i\,qr_i)\right|\right]^2 = \int \rho^2_{SLD}(r)\exp(-i\,qr)d^3r. \quad (21.2)$$

Here $d\Sigma/d\Omega_{coh}$ is the coherent differential cross-section per sample volume[2] irradiated, V_I, $\rho^2_{SLD}(\boldsymbol{r})$ represents the autocorrelation function of the so-called scattering length density, and $\boldsymbol{r}$ is the difference between pairs of position vectors $\boldsymbol{r}_i$.[3] For isotropic systems, the vector in (21.2) can be simplified:

$$\frac{d\Sigma}{d\Omega_{coh}}(q) = \overline{\rho^2_{SLD}(0)} \times V_I \int 4\pi \times r^2\gamma_C(r)\frac{\sin(qr)}{qr}dr, \quad (21.3)$$

with

$$\overline{\rho^2_{SLD}(0)} = \frac{1}{V_I}\int(\rho_{SLD}(r) - \bar{\rho}_{SLD})^2 d^3r \quad (21.4)$$

representing the mean square fluctuation (*MSF*) of the scattering length density. Here $\gamma_C(r)$ denotes the normalized autocorrelation function of the system of scatterers. Figure 21.12 visualizes the meaning of the autocorrelation function using an aerogel-like structure as an example.

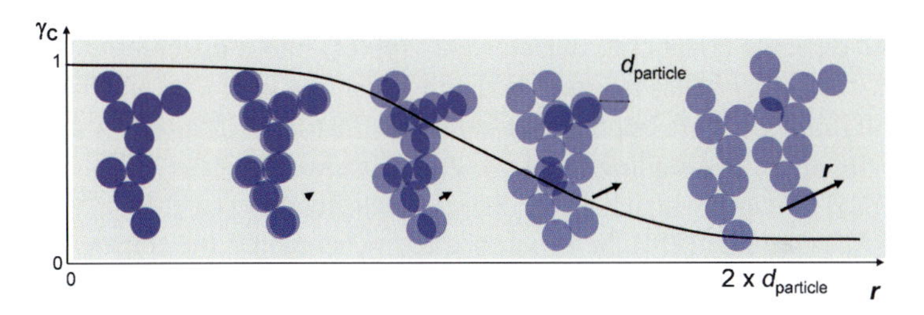

Figure 21.12. Visualization of the autocorrelation function $\gamma_C(\boldsymbol{r})$ (*black line*). The function is derived by shifting the structure under investigation (here: aerogel backbone) relative to itself by a vector $\boldsymbol{r}$. The resulting overlapping (*darker*) area then corresponds to the value of the correlation function γ_C as a function of the vector $\boldsymbol{r}$ (indicated by the *black arrows*).

Since no scattering will be observed unless fluctuations in the scattering length density are present, the difference between the local scattering lengths $\rho_{SLD}(r)$ and the scattering density averaged over the volume under investigation is often used [see (21.4)]. Depending on the probe used for analysis, different *MSF* of the scattering length density apply (Table 21.2). High energy photons (X-rays) interact with the electrons in the system under investigation, long wavelengths photons (e.g., visible light) are probing the fluctuations of the dielectric constant ε [24], and scattering of neutrons is characterized by their interaction with the nuclei. In case of X-ray, the variation of the scattering length is increasing strictly

[2] The volume specific differential cross-section translates into the mass specific differential cross-section by multiplying with one over the bulk density of the sample: $\frac{d\Sigma}{d\Omega}\frac{1}{\rho} = \frac{d\sigma}{d\Omega}$.

[3] Equations (21.2) and (21.3) reveal that $d\Sigma/d\Omega_{coh}(\boldsymbol{q})$, i.e. the scattering pattern, is the Fourier transform of the autocorrelation function.

Table 21.2. Mean square fluctuation (*MSF*) of the scattering length density for different probes when only one phase with a nonzero scattering length density is present

	Neutrons	X-rays	Light
MSF (m^{-4})	$(\rho_s \times \Sigma(b_i \times N_A/M_i))^2 = \rho_s^2 \times C_N^2$	$(\rho_s \times r_e \times N_A \times Z/M)^2 = \rho_s^2 \times C_X^2$	$\frac{\pi}{\lambda^4}\langle \varepsilon^2 \rangle Z$

Here ρ_S is the density of the solid phase, b_i is the coherent neutron scattering length for the element i, Σ is the sum over the different elements in the system, N_A is the Avogardo number, M is the molar mass, r_e is the classical electron radius ($=2.8179 \times 10^{-15}$ m), Z is the atomic number, λ is the wavelength of the light, and $\langle \varepsilon \rangle^2$ is the MSF of the dielectric constant in the system. C_X is a constant with a value of 8.504×10^{11} m/kg, while C_N strongly varies with the nuclei.

monotonously with the atomic number. This is in contrast to neutrons where the scattering length can take both positive and negative values and changes without obvious correlation with the atomic number of the element. This fact can be used for contrast variation.

The differential cross-section is related to the experimentally determined intensity $I(q)$ by

$$\frac{d\Sigma}{d\Omega_{coh}}(q) = I(q) \frac{c}{I_0 \times T_S \times d_S \times A_S \times \Delta\Omega}, \tag{21.5}$$

where I_0 is the normalized intensity of the incident beam, T_S, A_S, and d_S are the transmission, the irradiated area, and the thickness of the sample, respectively, $\Delta\Omega$ is the solid angle element that contains the sample–detector distance, and c is the calibration constant. To convert the detector counts per detector area (pixel) into a differential cross-section in absolute units, the intensity has to be calibrated by a sample with a known differential cross-section. Usually materials are applied as a reference that have a fairly constant differential cross-section in a large q-regime such as glassy carbon in case of small-angle X-ray scattering (*SAXS*) or water with a high incoherent background in case of small-angle neutron scattering (*SANS*) (see e.g., [25]). In a given scattering geometry, the constant c that converts the experimental scattering intensity of the sample investigated into the differential cross-section can be determined by the ratio of the scattering intensity measured for the reference material, I_R, and its theoretical differential cross-section $(d\Sigma/d\Omega)_{theo,R}$:

$$\frac{1}{C} = \frac{I_R(q)/(I_0 \times T_{S,R} \times A_{S,R} \times d_{S,R} \times \Delta\Omega)}{(d\Sigma/d\Omega)_{theo,R}(q)}. \tag{21.6}$$

The scattering pattern of most aerogels shows either the fingerprint of a *mass fractal* (silica and other metal oxide aerogels) or a statistical arrangement of particles (in case of organic and corresponding carbon aerogels).

Figure 21.13 shows a typical scattering curve for a silica aerogel with a fractal, i.e., self-similar backbone structure. The contributions due to the structural features present on different length scales are illustrated by sketches above the plot. In the second graph on the right, the functions describing the scattering of the respective structural entities are plotted separately. The overall scattering curve is given by the product of the respective factors, i.e., the particle form factor $P'(q)$, the structure form factor $S(q)$, and the packing factor $\Phi(q)$ [27]:

$$I \propto \Phi(q) \times S(q) \times P'(q) \tag{21.7}$$

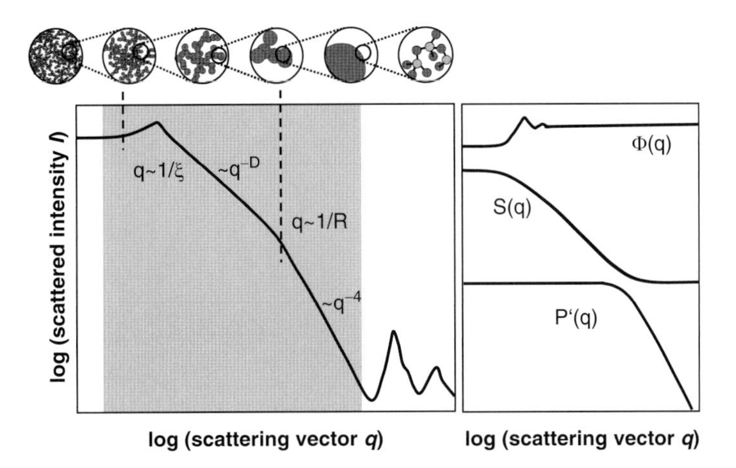

Figure 21.13. Left: Typical scattering pattern of an aerogel with a fractal backbone in the small-angle scattering regime. The *shaded* range corresponds to the q range for which the different contributions are separated in the **right plot**. The *cartoon* above the q^{-D} regime shows the fractally branched clusters. **Right:** Different contributions to the total scattering curve in the small-angle scattering range: $\Phi(q)$ packing factor, $S(q)$ structure factor, $P'(q)$ particle form factor (after [26]).

with

$$\Phi(q) = \frac{1}{1 + 8p'\left(3\frac{\sin(q\xi^*) - (q\xi^*)\times\cos(q\xi^*)}{(q\xi^*)^3}\right)^2}, \quad (21.8a)$$

$$S(q) = 1 + \frac{D \times \Gamma(D-1)}{(qR)^D}\left(1 + \frac{1}{(q\xi)^2}\right)^{(1-D)/2} \sin[(D-1)\arctan(q\xi)], \quad (21.8b)$$

$$P'(q) = \frac{1}{\left(1 + \frac{\sqrt{2}}{3}(qR)^2\right)^2}. \quad (21.8c)$$

Here p' is the packing factor, ξ^* is the radius of a sphere describing the arrangement of the clusters relative to each other, D is the mass fractal exponent, $\Gamma(x)$ is the gamma function, ξ is the cluster radius, and R is the radius of the spherical particles forming the clusters.

The fractality is a consequence of the growth mechanisms involved upon formation of the aerogel backbone during the sol-gel process; the mass fractal exponent D therefore allows identifying the growth mechanism. For silica aerogel typically values between 1.7 and 2.4 are found [25, 28] (Chap. 2).

The scattering curves of organic and carbon aerogels usually show a different characteristic shape (Figure 21.14) [29, 30] (exemption: organic gel prepared in acetone rather than water as solvent, see [31]). This is due to the different mechanism of backbone formation that has been suspected to be dominated by phase separation [32, 33], although this is hard to prove. The scattering curve can be evaluated at low scattering vectors with a Guinier equation

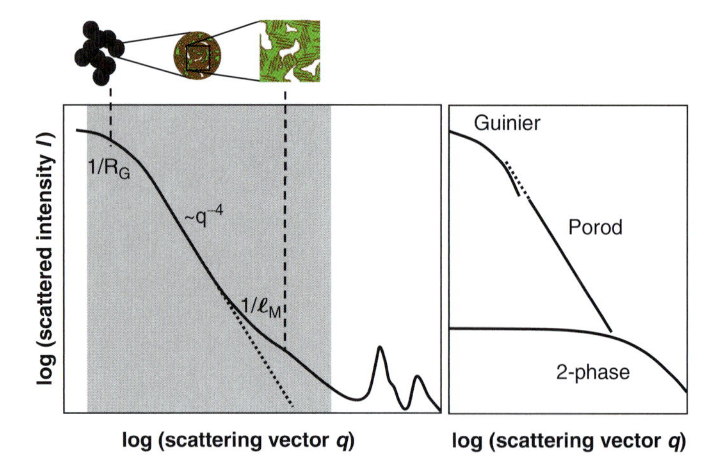

Figure 21.14. Left: Typical scattering pattern of a mesoporous carbon aerogel in the small-angle scattering regime and the range of wide-angle scattering (large q values). The *dashed extrapolation* of the q^{-4} decay of the intensity is representative for the respective organic precursor of the carbon aerogel. The shoulder at $q = 1/\ell_M$ indicates the microporosity (see footnote 4) present in the amorphous carbon backbone. The *shaded range* corresponds to the q range for which the different contributions are separated in the **right plot**. **Right:** Different models applied to describe the scattering curve in the small-angle scattering range.

$$I(q) \propto \exp(-q^2 R_G^2/3). \tag{21.9}$$

This relationship is an approximation for the decay of the scattering curve at low q values. It is valid for all scatterers independent of their geometry. The fit parameter, R_G, is the radius of gyration. Depending on the geometry of the scatterer, R_G can, for example, be converted into the radius of a sphere $R = (5/3)^{0.5} \times R_G$.

In the intermediate q-range, a decay of the intensity with q^{-4} (Porod-law) is observed indicating a smooth backbone surface. Sometimes slightly lower exponents between 3.5 and 4 are detected which is due to the presence of surface roughness. In case the exponent exceeds 4, this is a hint for a gradient in backbone density decreasing from the surface of the skeleton to its core. Such behavior may be, for example, observed for high temperature-treated carbon aerogels.

At q values of about 1 nm^{-1}, a characteristic shoulder is found for carbon aerogels that is not present for their organic precursor. This shoulder represents scattering from the microporous[4] structure within the backbone particles. Towards the wide angle range the shoulder is overlapping with a constant background B.

The scattering in the Porod- and the micropore range can be described by a superposition of the two characteristic contributions (see e.g., [31]):

$$\left(\frac{d\Sigma}{d\Omega}\right)_{coh} = K \times q^{-4} + \frac{b}{\left(1 + q^2 {\ell_M}^2\right)^2} + B. \tag{21.10}$$

[4] Micropores: pores <2 nm, mesopores: pores between 2 and 50 nm, macropores: pores >50 nm (IUPAC definition), see [34].

Here K is the Porod constant, b is a prefactor, ℓ_M the correlation length, and B is the constant background. While for the Porod behavior [first term in (21.10)], no assumption is necessary except for smooth surfaces, the second term is based on the two-phase model. The latter describes the scattering for two statistically distributed phases present with constant scattering length within each phase. For $q \gg 1/\ell_M$, the cross-section approaches a $K' \times q^{-4}$ decay of the intensity (with $K' = \frac{b}{\ell_M^4}$ 21.10). The meaning of K' is equivalent to K, however characterizing the structure on the micropore scale rather than the backbone particles. K and K', respectively, can easily be determined when plotting $(\mathrm{d}\Sigma/\mathrm{d}\Omega)_{\mathrm{coh}}\,(q) \times q^4$ rather than $(\mathrm{d}\Sigma/\mathrm{d}\Omega)_{\mathrm{coh}}\,(q)$ (Figure 21.15). In this representation, K' is masked by the overlapping background.

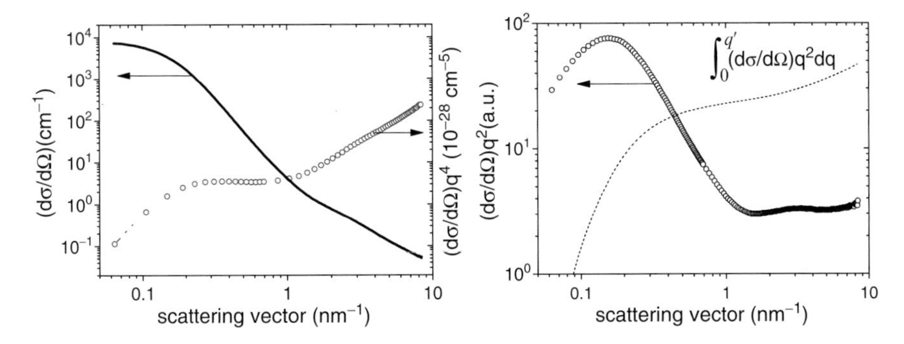

Figure 21.15. **Left:** Experimentally derived differential cross-section and cross-section times q^4 for a carbon aerogel. The *dashed line* indicates the position of K' that is masked in the original dataset by the background B [see (21.10)]. **Right:** Kernel of the integral equation (21.12a, b) and the integral itself as a function of q (*dashed line*).

Independent of the model used to describe the scattering pattern and the underlying structure of the sample, the so-called invariant Q, defined as

$$Q = \int \left(\frac{\mathrm{d}\Sigma}{\mathrm{d}\Omega}\right)_{\mathrm{coh}} q^2 \mathrm{d}q, \tag{21.11}$$

is a characteristic parameter of the system investigated. Due to the general properties of the correlation function it turns out that Q is containing the integral information about the volume fraction of the two phases (e.g., pores and solid) present:

$$Q = 2\pi^2\phi \times (1-\phi) \times (\Delta\rho_{\mathrm{SLD}})^2, \tag{21.12a}$$

which simplifies to

$$Q = 2\pi^2\phi \times (1-\phi) \times \left(\rho_{\mathrm{S}} \times C_{\mathrm{N,X}}\right)^2 \tag{21.12b}$$

when the pores of the porous sample investigated are empty. Here ϕ is the porosity of the sample, $\Delta\rho_{\mathrm{SLD}}$ is the difference of the scattering length densities of the phases (see also Table 21.2), ρ_{s} is the mass density of the solid phase, and C_{N}, C_{X} are the factors to be applied in case of neutron and X-ray scattering, respectively (Table 21.2). Hence (21.12b) allows quantifying the total porosity. Figure 21.15 visualizes the change of the integral as a function of the integration limit q'. Note that the numbers are given on a double-log scale. To extract the total porosity it is necessary that the experimental q range covers all features of the scattering curve that are significantly contributing to the integral value. Therefore, it is required that the crossover of the scattering intensity to a plateau at low q values is included in the experimental q-range. If the invariant can be calculated from the experimental data it can be used to normalized them; this is particularly valuable when the scattering data are not given on an absolute scale (e.g., for nonmonolithic pieces of aerogel).

The Porod constant K can be used to determine the specific surface area S/m either by applying the invariant Q or directly from the experimentally derived scattering cross-section; when no micropores are present the following relations hold:

$$\frac{S}{m}=\frac{S}{V_0}\times\frac{1}{(1-\phi)\times\rho_S}=\frac{K}{Q}\times\pi\times\frac{\phi}{\rho_S}\quad\text{with } K=\lim_{q\to\infty}\left(I(q)\times q^4\right) \tag{21.13a}$$

or

$$K=\lim_{q\to\infty}\left(\mathrm{d}\Sigma/\mathrm{d}\Omega\right)_{\mathrm{coh}}\times q^4=2\pi(\Delta\rho_{\mathrm{SLD}})^2\frac{S}{V_0}=2\pi(\rho_S\times C_{\mathrm{N,X}})^2\frac{S}{V_0}. \tag{21.13b}$$

Here, V_0 is the total volume of the sample, ρ_S is the density of the solid phase, ϕ is the porosity of the sample, $\Delta\rho_{\mathrm{SLD}}$ is the difference in coherent scattering length density between the solid and the pore phase, and C_N, C_X are the factors to be applied in case of neutron and X-ray scattering, respectively (Table 21.2). When two Porod regimes are present [as it is the case for microporous samples (see Figure 21.15)] both can be evaluated according to (21.13a, b). Hereby care has to be taken to correctly match the Porod constants with the respective densities of the solid phase and the surface area that can be extracted.[5]

The correlation length ℓ_M in (21.10) is the geometrical average over the chord length, i.e., the mean extensions of the solid ℓ_S and the void phase ℓ_P:

$$\frac{1}{\ell_M}=\frac{1}{\ell_S}+\frac{1}{\ell_P}\quad\text{and } \ell_S=\ell_M/\phi,\ \ell_P=\ell_M/(1-\phi). \tag{21.14}$$

ℓ_P can be correlated to the width of the micropores in the carbon aerogel backbone. In addition, ℓ_M is related to the ratio of the invariant and the Porod constant by

$$\ell_M=\frac{4}{\pi}\times\frac{Q}{K}. \tag{21.15}$$

When evaluating Q for the micropore term (21.10) only, the density of the backbone particles and thus the micropore volume can be estimated by solving for ϕ in (21.12a, b)[6]. The integral for the respective term in (21.10) yields

$$Q=\frac{\pi}{4}\times\frac{b}{\ell_M^3}. \tag{21.16}$$

[5] In case of carbon aerogels the pairs are: K, density of the interconnected backbone particles ⇨ particle surface, K', ρ_{carbon} ⇨ surface of particles and micropores.

[6] It has to be emphasized that due to the large uncertainties of the background contribution at high q values, the accuracy is on the order of 10–20% only.

With the mean chord or correlation length ℓ_M a chord length distribution can be calculated from the scattering curve. This approach was used by Cohaut et al. to analyze the impact of the type and concentration of catalyst as well as the carbonization temperature on the pore structure of carbon aerogels [35].

Another option for the analysis of scattering data are reconstruction concepts; in these approaches different models are applied and parameters within these models are varied until the simulation of the scattering curves for the modeled structures corresponds to the experimentally derived scattering data [36, 37]. This approach provides three-dimensional visualization of the aerogel structure (Figure 21.16) that can be used for modeling of other properties. However, since the scattering data do not contain unequivocal information the result of the reconstruction depends on the assumptions and the strategy used.

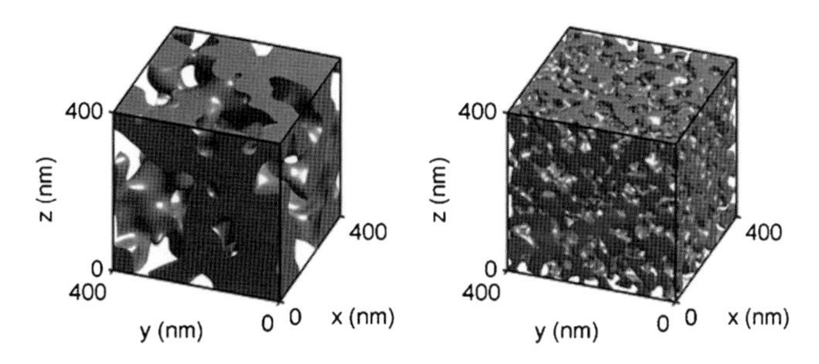

Figure 21.16. Reconstruction of a carbon aerogel macrostructure from *SAXS* data for two different samples with the same total porosity of 85% but different particle sizes [about 150 nm (**left**) and 60 nm (**right**)] (Courtesy: Cedric Gommes, University of Liege).

Approaches for the evaluation of scattering data. Three cases have to be distinguished when evaluating an experimental set of scattering data of an aerogel (Figure 21.17):

(a) The scattering data are not normalized and do not contain any clear structure or crossover in their scattering pattern within the experimentally q-range covered
(b) The scattering data are not normalized, however contain clear crossovers in their scattering pattern within the experimentally covered q-range; in particular the scattering intensity varies only weakly with the q value in the low q-range of the spectrum
(c) The scattering data are normalized, i.e., they are given in terms of a differential scattering cross-section on an absolute scale

In case (a), only the slope of the scattering curve in the double log plot can be determined. If the slope is steeper than -3 the part of the scattering curve shown is likely to be related to the range where the scattering is mainly sensitive to the interface between the two phases, i.e., the solid and void phase present. In Figure 21.17, the slope of -3.4 in the double-log plot indicates a rough surface area. The fact that no crossover to a regime with a different slope is visible indicates that structural entities are larger than about $2/q_{min}$ (here $2/0.8\ nm^{-1}=2.5$ nm) are present.

In case (b), the slope in different regimes and the q values of the crossover between these regimes can be determined. As the scattering curves level off at low q values, also the invariant Q can be determined and used to calibrate the scattering intensity.

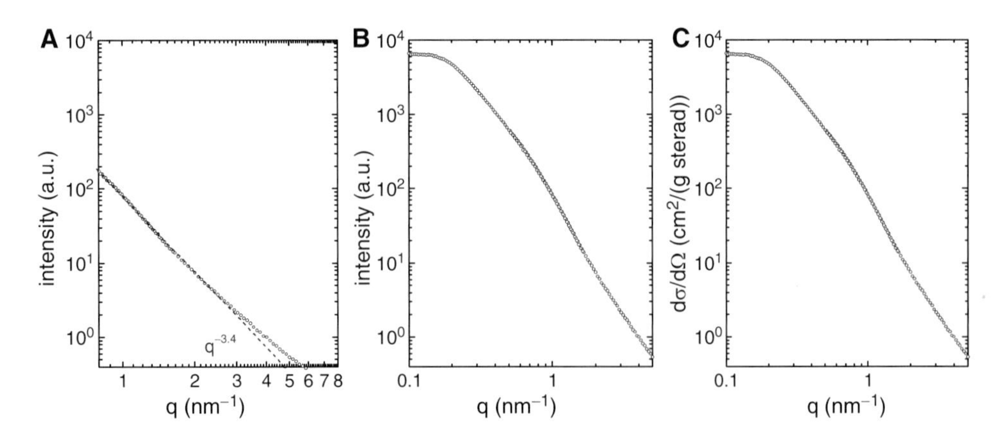

Figure 21.17. Illustration of the three different cases of experimental information available from a scattering experiment (example: *SAXS* curve for a silica aerogel). **A.** Limited experimental q-range with no distinct features, no absolute calibration; the *dashed line* is indicating a slope of -3.4 in the low q-range of the experimental data. **B.** Experimental range covering crossover to almost constant scattering intensities at low q values, **C.** Scattering information available as differential cross-section given on an absolute scale.

In case (c), all information is present in the scattering curve and the individual parts can be analyzed on an absolute scale even without determining the invariant.

Scattering provides even more options than already discussed. For example, Ehrburger et al. [38] as well as Fairen-Jimenez and coworkers [39] extended the q-range from the classical *SAXS* regime to 10^{-2} nm^{-1} by applying ultra small angle X-ray scattering (*USAXS*) to study carbon aerogels; at the same time they included scattering in the wide angle regime, thus providing the database to analyze the structural feature in the range from a few Å to about 100 nm. In particular, Fairen-Jimenez et al. [39] show in detail the evaluation of *SAXS* curves for carbon aerogels.

Emmerling et al. [24] as well as Reim et al. [40] showed that it is possible to combine *SAS* and light scattering curves thus extending the experimental q-range to q values as low as 10^{-4} nm^{-1} and including information on structures in the micron range.

Often an additional increase in scattering intensities at q values $<10^{-1}$ nm^{-1} is observed for aerogels [24, 38–40]. This effect has been sometimes interpreted in terms of macropores; however, when comparing different methods it can be shown that density fluctuations due to clusters (Figure 21.10) rather than large pores are responsible for this forward scattering.

Berthon-Fabry et al. detected significant anisotropies in the *SAXS* pattern of carbon aerogels derived with acetone as a solvent and harsh synthesis conditions; the anisotropy was found for q values $<4 \times 10^{-1}$ nm^{-1}, while the scattering pattern revealed an isotropic structure at larger scattering vectors [41].

SAS has also been applied to investigate in situ the structure formation during the sol-gel process (e.g., [42–44]; see Figure 21.18). This provides direct access to the change of the growth mechanisms with changes in the composition of the initial solution or the process conditions.

In situ investigation of adsorption in a silica aerogel with *SANS* showed the change of microscopic structure with increasing mesopore filling [45]; in this experiment, the scattering length density of the adsorbate (pentane) was adjusted to be zero by mixing deuterated and protonated pentane in the respective ratio. The scattering curve therefore directly reflects

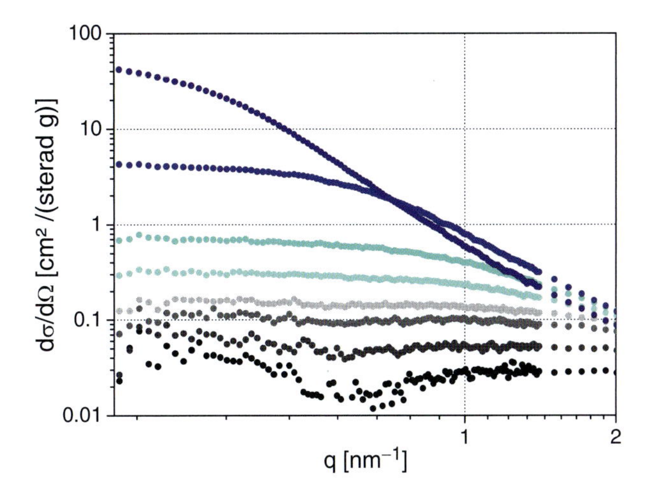

Figure 21.18. Gelation kinetics of an organic gel at 85°C monitored by *SAXS*. The curves were taken at progressing reaction times starting from the initial solution (black to dark blue curve). In the initial phase, the scattering is characterized by an increase in the number of particles smaller than 1 nm while the second phase is dominated by the continuous growth of the particles (*two upper curves*). Courtesy: Christian Scherdel, ZAE Bayern.

the scattering of the silica backbone only and no additional treatment of the scattering data is necessary to account for the fact that three phases, i.e., silica, voids, and pentane are present. The increasing downward bend of the scattering curve (Figure 21.19) at low q values with increasing pore filling indicates the deformation of the mesopores due to capillary forces (see also Chap. 21.6 and (21.36); [45]).

While contrast matching with neutrons can easily be performed over a wide range when protonated components are present, contrast matching in case of X-ray scattering is

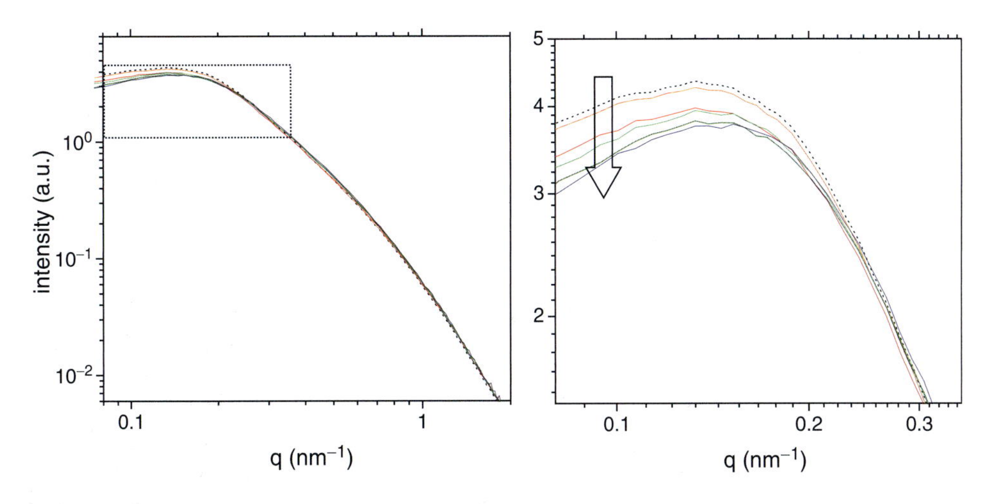

Figure 21.19. In situ *SANS* of a silica aerogel during filling of mesopores with pentane (**left graph**: overview over scattering curve, **right:** zoom into area marked by the *dashed box* in the **left graph**). The data reflect the scattering from the silica backbone only. The increasing downward bend of the scattering curve at low q values with increasing pore filling (in the direction of the arrow) indicates the deformation of the mesopores due to capillary forces (see also Chap. 6 [45]).

essentially restricted to application of anomalous small-angle X-ray scattering (ASAXS). Hereby the scattering pattern of a tertiary system, i.e., an aerogel containing a dispersed metal component (e.g., metal doped carbon aerogel, see Figure 21.9), is recorded at different energies of the incident X-ray beam near the absorption edge of one of the components (e.g., the metal) in the system (Figure 21.20). In that range the atomic form factor f', i.e., the element specific scattering amplitude of the respective component, varies according to the Kramers–Kronig relation. The energies are usually chosen at equidistant positions of f'. When the respective component (e.g., Pt in Figure 21.20) is distributed without correlation to the other phases present, the normalized difference of the scattering curves taken at different energies (here E_1, E_2) should be equal and reflect the scattering of the respective phase only (here: Pt):

$$I_{\mathrm{PT}} = \frac{(\mathrm{d}\Sigma/\mathrm{d}\Omega)_{\mathrm{coh}}(E_1) - (\mathrm{d}\Sigma/\mathrm{d}\Omega)_{\mathrm{coh}}(E_2)}{f'^{\,2}_{\mathrm{Pt}}(E_1) - f'^{\,2}_{\mathrm{Pt}}(E_2)}. \tag{21.17}$$

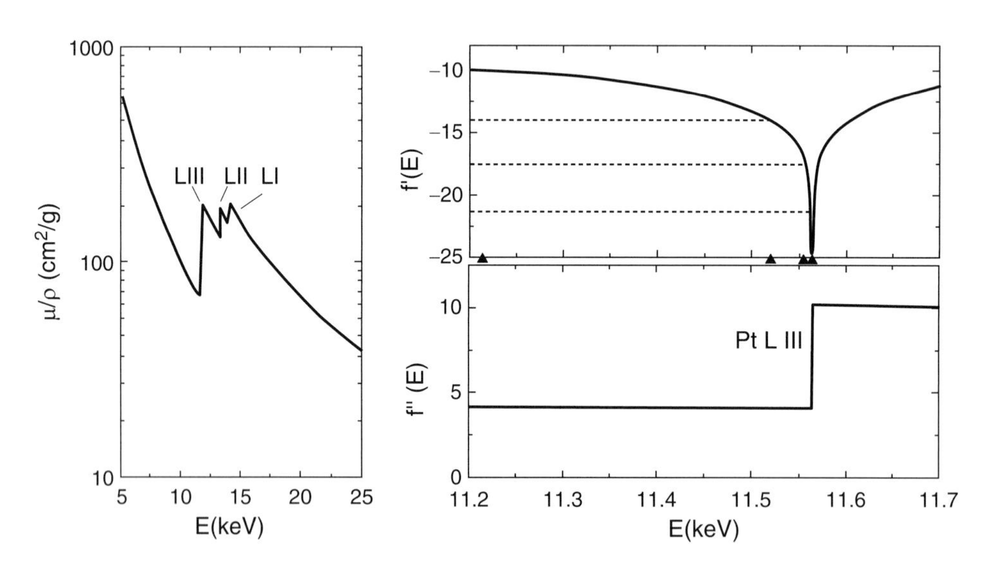

Figure 21.20. Anomalous small-angle X-ray scattering (ASAXS). **Left:** Specific X-ray absorption coefficient as a function of X-ray energy at the Pt adsorption edge; **right:** atomic scattering form factor f' and atomic absorption form factor f'' as a function of energy at the Pt LIII edge. The horizontal lines in the f' plot indicate a set of equidistant f' values corresponding to different energy values E below the absorption edge characterized by the steep increase in f''.

21.4.2. Inelastic Scattering

Inelastic scattering probes motion within the sample under investigation. This can be, for example, diffusional processes during gelation or the vibrational eigenmodes of the aerogel backbone. As in the case of elastic scattering, the inelastically scattered intensity is defined by the Fourier transform of the autocorrelation function of the sample. However, the Fourier transform is now to be performed both in space ($\boldsymbol{x} \Rightarrow \boldsymbol{q}$) and time ($t \Rightarrow \omega$); hereby the frequency ω characterizes the energy transferred from the sample to the probe (i.e., a photon or a neutron) or vice versa. The probability for an energy transfer to take place depends on the eigenmode spectrum, i.e., the vibrational density of states $Z(\omega)$ of the sample, the temperature as well as on the transition rules that apply. The vibrational density of states is providing information on the connectivity of the solid phase in an aerogel. Acoustical modes can be investigated by Brillouin light scattering, i.e., inelastic interaction with acoustic phonons, while Phonon/Raman scattering probes the optical phonons in the system. Calas et al. investigated silica aerogels at different stages of sintering and isostatic compression by Brillouin scattering to monitor the changes in stiffness with the treatment applied [46]. Caponi et al. studied the effect of thermal treatment on silica aerogels and xerogels by Brillouin scattering and found that in the low frequency range the materials react very sensitively to the water content [47].

While Brillouin scattering and Raman scattering are very limited in the frequency range that can be covered, inelastic neutron scattering can probe several orders in magnitude in terms of frequencies thus allowing to determine the complete eigenmode spectrum for an aerogel, ranging from the long-wavelength phonons (frequencies in the upper MHz range) with wavelength far larger than the pores in the aerogel up to molecular vibration within the aerogel backbone with frequencies of THz. Hereby the experimental data derived by using several different inelastic neutron scattering instruments are combined to cover the large frequency range. This tedious work has been performed by different groups to determine the complete density of states of silica aerogels [48–52]. The result is schematically shown in Figure 21.21; the different relevant regimes are identified as the Debye regime at the lowest frequencies, the fracton range, where the vibrational spectrum of the fractally distributed mass in the silica clusters is reflected, the particle regime that is characterized by the eigenmodes of the particles forming the aerogel backbone and finally the molecular vibrations. It has to be pointed out that the modes in the *fracton* range are expected to be fairly localized.

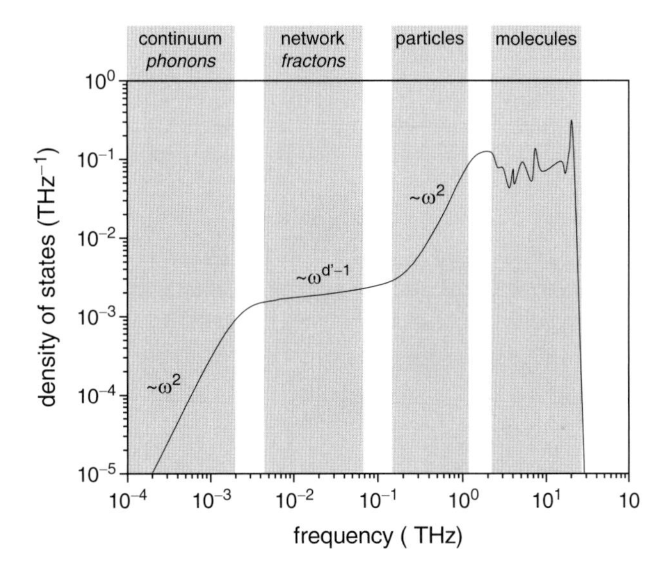

Figure 21.21. Schematic representation of the vibrational density of states $Z(\omega)$ of a silica aerogel with its characteristic regimes indicated on the **top** [53]. The density of states of the aerogel above about a frequency of 0.2 THz is similar to that of amorphous silica. The connectivity of the silica aerogel backbone is reflected by the density of states below 0.2 THz.

Comment. Scattering methods are extremely useful, nondestructive techniques that can be applied in different stages of (aero)gel processing and treatment. It enables in situ analysis on time scales of minutes to hours. In addition, inelastic scattering can be exploited to investigate the kinetics in the system. While inelastic scattering is usually still tedious and requires large-scale facilities (except for inelastic light scattering), elastic scattering analysis is well established and can also be performed with commercially available lab-scale instruments.

The only downside of scattering techniques is the fact that the scattering signal does not provide unequivocal information on the structure. Therefore, data evaluation is not straightforward unless one of the well-known characteristic patterns of aerogels is detected.

21.5. Helium Pycnometry

Helium pycnometry probes the volume of a sample that is inaccessible to helium atoms; hereby a volumetric method is usually applied that is also implemented in different commercially available instruments.

Upon measurement, the sample is placed in a chamber of known volume and purged with helium until both the chamber and the sample can be expected to contain helium only; alternatively, the chamber can be evacuated. In a second chamber (*reference* chamber), also with a well-defined volume, helium is introduced at a pressure higher than the one present in the *sample* chamber (see Figure 21.22). After the *reference* chamber is thermally equilibrated, the gas pressure in the chamber is measured; subsequently, the gas pressure in the *sample* chamber is recorded and eventually a valve is opened to admit helium from the *reference* to the *sample* chamber for gas pressure equilibration. Again the gas pressure is determined after the system is thermally equilibrated and no drift in gas pressure is obvious. Often this measurement is repeated several times in a row to exclude artifacts (e.g., gas desorbing from the sample, insufficient thermal equilibration).

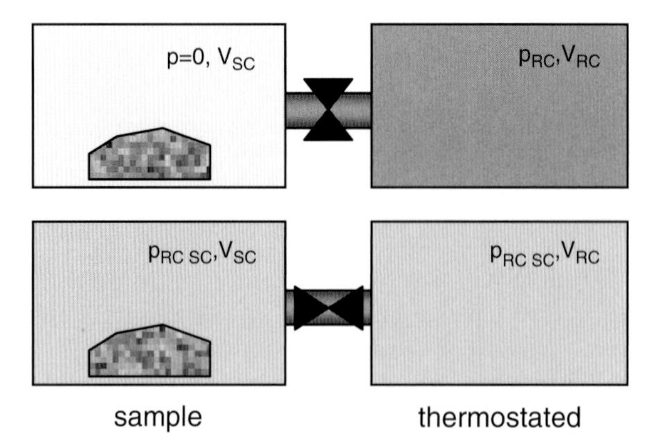

Figure 21.22. Schematic setup for helium pycnometry with an initial pressure in the *sample* chamber of zero (only the difference in gas pressure between the two chambers is relevant; therefore any actual present offset can be subtracted from the pressures in the system). After a difference in helium pressure (p_{RC}) is established between the *reference* and the *sample* chamber (volume V_{RC} and V_{SC}, respectively), the valve between the chambers is opened. At equilibrium the pressure in the two chambers is equal. The setup is kept under thermostated conditions.

Using the ideal gas equation, the volumes of the two chambers involved, and the equilibrium gas pressures prior and after the valve has been opened, the specific volume V_{Sk} of the sample that is not accessible to helium can be calculated:

$$V_{Sk} = \frac{1}{m}\left[\frac{V_{SC} + V_{RC}}{1 - p_{SC}/p_{RC,SC}} - \frac{V_{RC} \times p_{RC} + V_{SC} \times p_{SC}}{p_{RC,SC} - p_{SC}}\right], \qquad (21.18)$$

with m the mass of the sample investigated. In case of $p_{SC} = 0$ (see Figure 21.22) (21.21) simplifies yielding

$$V_{Sk} = \frac{1}{m}\left[V_{SC} + V_{RC} \times (1 - p_{RC}/p_{RC,SC})\right]. \qquad (21.19)$$

For untreated aerogels that possess by definition a well-accessible pore network, the volume determined corresponds to the specific volume of the aerogel backbone, i.e., the inverse of the aerogel backbone density; this is an important figure for the calculation of the actual porosity of the aerogel.

In case of carbon aerogels, where additional microporosity is created upon the carbonization of the organic precursor, the skeletal volume is representing the volume of the amorphous carbon phase when the porosity is well accessible for the helium.

Comment. In many cases the density of the solid phase can well be approximated by using the respective literature values for the given chemical composition. For nonmacroporous samples with well-accessible porosity, the total pore volume and thus the backbone density can also be determined from the macroscopic density of the (degassed!) sample and the respective nitrogen sorption isotherm (see Chap. 21.6).

When actually applying helium pycnometry, care has to be taken that the sample is well degassed and the setup used is at constant temperature. If not, artifacts resulting in backbone densities that are up to 50% off can be the consequence.

Another critical point are pores with narrow entrances that result in only slow pressure equilibration; examples are carbon aerogels that have been treated at temperatures above 1,000°C or samples that have been partially sintered. In these cases, artifacts can arise due to insufficient equilibration of the gas pressure.

21.6. Gas Sorption Porosimetry

Gas sorption analysis probes the interaction between a gas or vapor (adsorptive) and the adsorbent; this interaction can occur upon adsorption in micropores, mono- and multilayer adsorption at the inner surface of the specimen or via liquid/solid interaction when the adsorptive is condensing in the pores of the adsorbent. The probe most commonly used to characterize aerogels is N_2 sorption at 77.3 K. When varying the gas pressure from vacuum to 0.1 MPa (1 bar) rel. gas pressures from almost 0 to 1 can be covered; hereby typically information on specific surface areas down to 0.01 m^2/g and pore widths in the range from 0.3 to 100 nm can be derived.

Recording an isotherm implies that an evacuated and well-degassed sample is exposed to a change in pressure of the probing gas; due to the interaction with the adsorptive the sample takes up a characteristic specific amount of adsorbate. The uptake (in equilibrium) as a function of the rel. gas pressure (p/p_0 = partial pressure over saturation pressure) at fixed temperature is called the adsorption branch of the sorption isotherm; similarly the desorption branch can be measured by reducing the partial gas pressure in subsequent steps. The adsorbate uptake or release is mostly quantified by a manometric (also called volumetric) or a gravimetric method. In volumetric instruments, the uptake is calculated from the gas pressure drop/increase induced by an adsorption/desorption process, respectively, while in the gravimetric method the mass change upon sorption is measured by a balance. In addition to the conventional stepwise dosing approach, the sorption isotherm can also be determined from the change in concentration of the probing gas between in- and outlet of a gas flow passing continuously by the sample. To adjust different partial pressures of the probing gas an inert carrier gas (e.g., He) is mixed with different fractions of the probing gas. The concentration of the probing gas before and after passing the sample can be determined by a gas chromatograph, a mass spectrometer, or any other sensor that is sensitive to different concentrations of the probing gas.

The isotherm is based on the assumption that the system under investigation consisting of the adsorbent, the gas phase (adsorptive), and the adsorbed phase (adsorbate) is in equilibrium at each point taken. It has been shown, in particular for aerogels with their huge specific pore volume, that samples are often insufficiently equilibrated and as a consequence the shape and height of the isotherm in the range of micropores and mesopores are highly erroneous [54–56]. Figure 21.23 shows an example for isotherms measured for exactly the same sample, however with different times allowed for equilibration. Therefore, it is highly recommended to either check the adsorption kinetics in each point if this feature is provided by the instrument manufacturer or repeat the run with two samples from the same batch but significantly different times allowed for equilibration.

Figure 21.24 shows a typical isotherm for an aerogel with a density of about 200 kg/m^3. The arrows indicate the adsorption and desorption branches of the isotherm in the range of capillary condensation. According to the IUPAC classification scheme

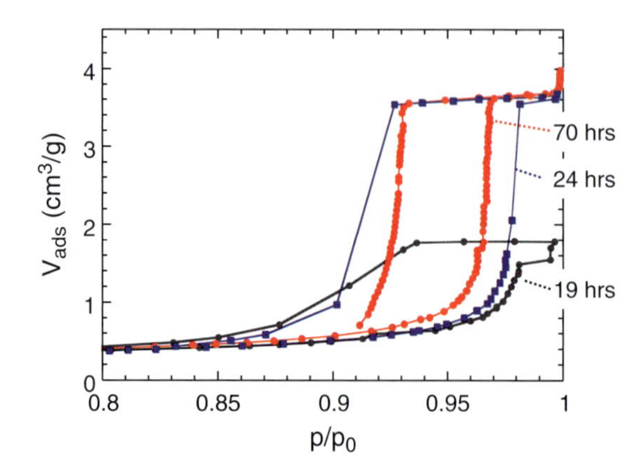

Figure 21.23. Upper rel pressure range for three isotherms taken for the same (slightly sintered) silica aerogel with a density of about 264 kg/m^3. The *figures* indicate the different total run times of the three isotherms.

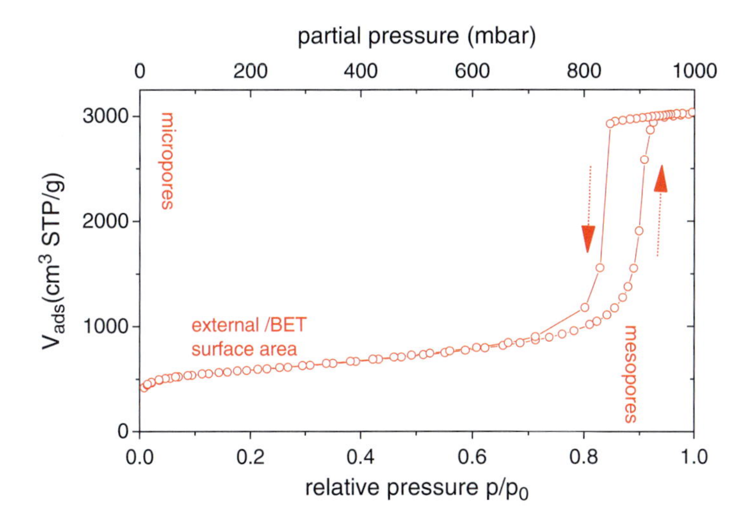

Figure 21.24. Typical type IV isotherm (nitrogen at 77 K) of an aerogel. The different relative pressure regimes correspond to different stages upon ad-/desorption; the *arrows* indicate the adsorption and desorption branch of the hysteresis.

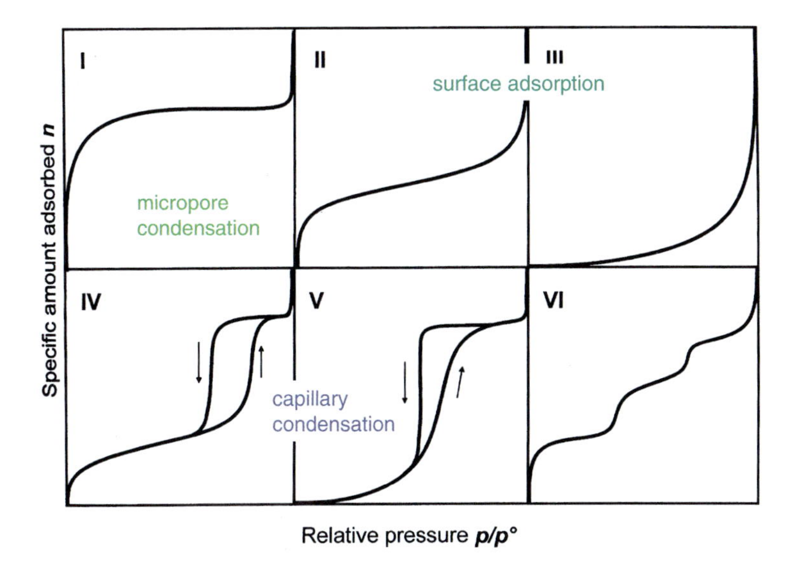

Figure 21.25. Classification of standard isotherms (IUPAC: [34]). Type I: Material that is microporous only, Type II and III, statistical multilayer adsorption on a wetting (II) and a nonwetting surface (III). Type IV (wetting) and V (non-wetting): Mesoporous materials showing a characteristic hysteresis that indicates capillary condensation. VI: Stepwise adsorption in a highly ordered material (e.g., graphite).

(Figure 21.25) the isotherm shows the characteristic type IV shape. The almost parallel adsorption and desorption branches (subtype H_1) indicate a narrow pore size distribution.

Most of the basic information on the sample under investigation can already be deduced qualitatively from the isotherm itself:

A steep increase at low relative pressures indicates the presence of micropores while large slopes in the intermediate range (before a hysteresis takes over) signal high (external) specific surface areas. When a well-pronounced hysteresis is present the sample is mesoporous and the position of the hysteresis in terms of rel. pressure indicates the size of the mesopores (the larger p/p_0 the larger the mesopores). The total accessible porosity in the micro- and mesopore range can be estimated from the plateau value of the hysteresis loop at high relative pressures. If no hysteresis is present the aerogel is macroporous only.

Common evaluation tools that provide quantitative information are the Density Functional Theory (*DFT*) for micropores and mesopores, the *Dubinin–Radushkevich* equation for the extraction of characteristic parameters on micropores, the *t-plot* and the α_S*-plot* for the separation of surface area located in micro- and nonmicropores, the *BET* method to calculate the so-called *BET* surface area and the *BJH* relationship that provides access to the mesopore size distribution.

Total pore volume. The total accessible pore volume in micro- and mesopores can be derived by using the Gurvich rule. The assumption hereby is that the maximum specific amount of gas adsorbed, defined by the value of the plateau in the hysteresis at a rel. pressure of about 0.99, can be converted into a specific pore volume by assuming the adsorbate to be present in a liquid-like state in all pores. As a consequence the specific pore volume $V_{\text{pore, micro+meso}}$ is derived from the amount of gas adsorbed V_{ads} given in cm^3 (STP)/g [7] by

$$V_{\text{pore,micro+meso}}\left[\text{cm}^3/\text{g}\right] = V_{\text{ads}}\left[\text{cm}^3(\text{STP})/\text{g}\right] \times c. \tag{21.20}$$

Here c is a gas-dependent constant that converts gas phase into liquid phase volumes. For nitrogen and CO_2, c is equal to 0.001547 and 0.001816 $\text{cm}^3/(\text{cm}^3$ (STP)), respectively. In case of nonmacroporous samples with well-accessible pores, this specific volume can be used to calculate the macroscopic (bulk) density ρ of the aerogel:

$$\frac{1}{\rho} = V_{\text{total}} = V_{\text{pore, micro+meso}} + \frac{1}{\rho_S}, \tag{21.21}$$

with V_{total} the total mass specific volume and ρ_S the density of the solid phase of the aerogel (e.g., carbon, ρ_S=2,200 kg/m^3 [57], in case of a microporous carbon aerogel). If (21.21) results in a value that is far off the density determined from total mass and macroscopic dimensions of the aerogel one or several of the following reasons may be effective:

- Density value determined macroscopically is higher than the value derived by (21.21): the sample still contains adsorbed gases and vapors; effect can be avoided when sample is properly degassed
- Density value determined macroscopically is lower than value derived by (21.21): $V_{\text{pore,micro+meso}}$ is not representing the total pore volume present as (1) part of the pores is not well accessible, (2) the sample was insufficiently equilibrated [56], (3) macropores are present that are not detectable by the method, or (4) a significant amount of the adsorbate has a lower density than the one of the free liquid

[7] STP stands for standard temperature and pressure; cm^3 STP gives the equivalent volume of a gas at standard temperature (273.15 K) and pressure (101,325.02 Pa).

Micropores and specific surface area. The quantification of micropores with width up to 2 nm is still a challenge as the adsorption isotherm in the relevant range of relative pressure strongly depends on the size and geometry of the probing molecule, the temperature at which the isotherm is recorded, and the size and the geometry of the micropores, as well as the width of their entrance. Micropores in aerogels are usually found in carbon aerogels since microporosity is an intrinsic property of the amorphous carbon backbone; in addition, microporosity is sometimes also detected in inorganic aerogels. For the evaluation of adsorption isotherms in terms of the micropore size and volume, two methods are presented here in more detail, the Dubinin–Raduskevich (*DR*) method and the Density Functional Theory (*DFT*).

Based on the Polanyi potential theory, different approaches to describe the adsorption behavior of a purely microporous material (isotherm type I, Figure 21.25) have been undertaken by Dubinin and Stöckli in collaboration with different other scientists. The simplest relationship that can be considered the base for all other variants is the Dubinin–Radushkevich equation [58]:

$$\frac{V_{\mathrm{ads}}(p/p_0)}{V_{\mathrm{mic},0}} = \exp[-(A/\beta \times E_0)^2]. \tag{21.22}$$

Here V_{ads} is the specific adsorptive volume adsorbed at p/p_0, $V_{\mathrm{mic},0}$ is the so-called limiting micropore volume, β is the gas specific affinity coefficient[8] that empirically includes the properties of the adsorbate, while the characteristic energy E_0 is describing the adsorbent properties only. A is defined as

$$A = -R_{\mathrm{G}} T \ln(p/p_0), \tag{21.23}$$

with R_{G} the gas constant and T the temperature. Further developments of the *DR* equation introduce additional parameters, such as a free exponent n instead of 2 in (21.22), or a discrete or continuous superposition of different contribution equal to (21.22) to account for a pore size distribution.

$V_{\mathrm{mic},0}$ and E_0 are the two free parameters of the *DR* equation (21.22); they provide an estimate for the total amount of micropores present and the width of the micropores. The latter can be deduced from an empirical relationship between E_0 and the pore width d[9] [59]:

$$d[\mathrm{nm}] = \frac{10.8\,\mathrm{kJ/mol}}{E_0 - 11.4\,\mathrm{kJ/mol}}. \tag{21.24}$$

To determine the parameters $V_{\mathrm{mic},0}$ and E_0 the *DR* plot is used; it is essentially a representation of (21.22) in a linearized form (see Figure 21.26). It has to be pointed out that $V_{\mathrm{mic},0}$ is a fit parameter that should not to be confused with the total micropore volume. Several publications indicate that $V_{\mathrm{mic},0}$ is essentially equivalent to the number of molecules adsorbed in a monolayer on the total surface area present (*BET* surface area) [60–62].

An alternative to extract information on micropore sizes is the *DFT* method [63, 64]. Hereby a theoretically derived kernel is used that relates the adsorbate density to the pore

[8] $\beta = 0.33$ for CO_2 at 273 K and 0.35 for N_2 at 77 K.
[9] The relationship was established for pore width d derived from SAXS data.

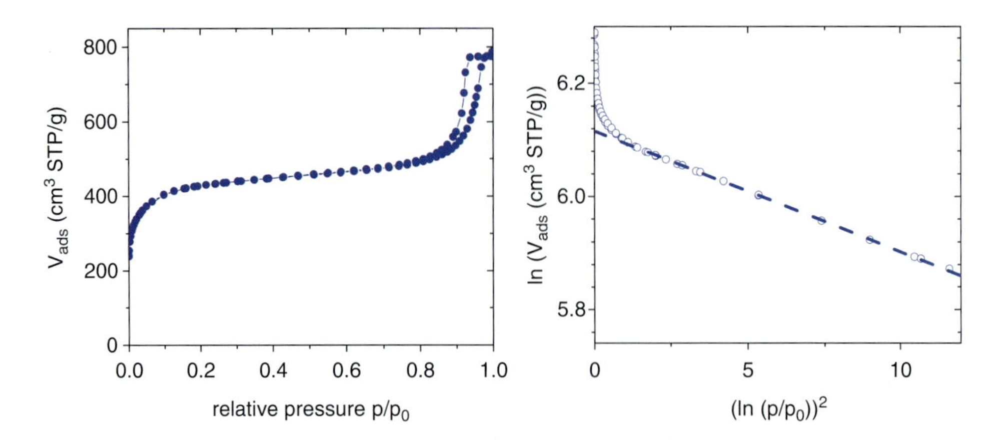

Figure 21.26. Left: Nitrogen sorption isotherm taken for an as-prepared carbon aerogel at 77.3 K. **Right:** DR representation of the same isotherm. The *dashed line* corresponds to a linear fit to the data in the range $(\ln(p/p_0))^2 > 5$ (i.e., $p/p_0 < 0.1$); the fit yields $V_{mic,0} = 0.3\ cm^3/g$ and a characteristic energy E_0 of 21 kJ/mol.

size. Kernels are available for different pore geometries, gases, and temperatures. These kernels are nowadays implemented in the data evaluation software of most commercially available instruments. *DFT* and newer developments such as the nonlocal *DFT (NLDFT)* [65, 66] are suitable for the analysis of micropores as well as mesopores. However, care has to be taken to not overinterpret the data that often are superimposed by artifacts due to the limited number and the accuracy of the experimental data.

To separate contributions due to micropore filling on one hand and the formation of mono- and multilayers on the other hand which are superimposed at relative pressures below 0.2– 0.3, the t-plot or the α_s-plot approach can be applied. Both methods use empirical reference isotherms to be compared with the isotherms taken for the sample under investigation. In the t-plot method, the statistical layer thickness t of a nonmicroporous material is related to the relative pressure p/p_0. One of the most frequently used relationship for the layer thickness is the empirical Harkins–Jura equation [67] derived for metal oxides[10]:

$$t[\mathrm{nm}] = 0.1 \times \sqrt{\frac{13.99}{0.034 - \log(p/p_0)}}. \tag{21.25}$$

For amorphous carbon, Magee [68] proposes the following relationship:

$$t[\mathrm{nm}] = 0.1\left[0.88\left(\frac{p}{p_0}\right)^2 + 6.45\left(\frac{p}{p_0}\right) + 2.98\right]. \tag{21.26}$$

When plotting the isotherm taken for the sample under investigation vs. t rather than p/p_0, a linear relationship is observed that may be combined with an offset (see Figure 21.27). For a nonmicroporous sample a line through the origin is expected, with the slope being proportional to the monolayer volume V_{ml}:

[10] Other equations are provided by Frenkel, Halsey, and Hill (FHH) as well as Broekhoff and de Boer.

$$V_{\text{ads}}(t) = V_{\text{ml}} \times t/t_{\text{ml}}, \tag{21.27}$$

where t_{ml} is the thickness of the monolayer (N_2: 0.354 nm). Therefore (21.27) can be applied to calculate the interfacial area of the meso- and macropores, the so-called specific external surface area S_{ext}, by using the area S_{gas} that a single probing molecule occupies[11]:

$$S_{\text{ext}} = \frac{V_{\text{ml}} \times S_{\text{gas}} \times N_{\text{A}}}{V_{\text{mol,STP}}}. \tag{21.28}$$

Here $V_{\text{mol,STP}}$ is the volume that a mol of gas takes at STP and N_{A} is the Avogadro constant. On the other hand, an offset in the *t*-plot is indicating additional sorption sites filled at low relative pressures, i.e., micropores, and the height of the offset is interpreted as the micropore volume V_{mic}. At large layer thicknesses above 0.6 nm (corresponding to $p/p_0 > 0.4$) an upward bend of the curve indicates the beginning of mesopore filling (in Figure 21.27 the upward bend does not start until $t > 0.9$ nm).

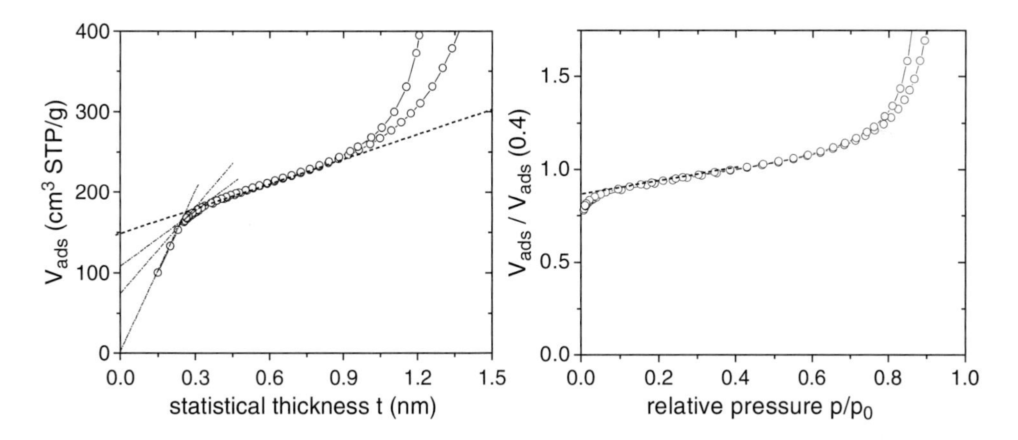

Figure 21.27. *t*-plot (**left**) and α_s-plot (**right**) of the same isotherm of a carbon aerogel. The *t*-plot results in an intercept (*dashed line*) at zero layer thickness of an adsorbed volume of 150 cm^3(STP)/g and a slope corresponding to a monolayer volume of 35.9 cm^3 (STP)/g (equal to a specific surface area of about 155 m^2/g). The *dashed–dotted lines* indicate the series of tangent used for the MP method.

The so-called MP method [69] uses the concept of the *t*-plot not only in the range where the curve saturates at intermediate thicknesses in a linear behavior but also applies a series of tangents to the curve in the *t*-plot starting from a line through the origin to extract $V_{\text{ml}}(t)$.

The most popular quantity used to characterize the surface area of a porous material is the so-called *BET*[12] surface area. The original derivation of the *BET* relation assumes a statistical multilayer coverage of a nonmicroporous surface. Later on it was shown that the evaluation method can also be applied to microporous materials under certain boundary conditions.

The volume adsorbed as a function of relative pressure in the *BET* model is expressed in terms of two parameters, the specific monolayer capacity $V_{\text{ml,BET}}$ and the parameter *C*.

[11] For example, S_{N2}=0.162 nm^2, S_{CO2}=0.170 nm^2.

[12] BET stands for Brunauer, Emmett, and Teller who developed the method.

$$\frac{V_{\text{ads}}(p/p_0)}{V_{\text{ml,BET}}} = \frac{p/p_0}{(1-p/p_0)} \bigg/ \left(\frac{1}{C} + \frac{C-1}{C} p/p_0\right) \tag{21.29}$$

where C provides information about the interaction of the adsorbent surface and the adsorbate:

$$C \approx \exp(\Delta E_{\text{ads}}/R_{\text{G}} \times T), \tag{21.30}$$

with ΔE_{ads} being the net molar energy of adsorption, i.e., the difference in adsorption energy in the first and the higher layers. R_{G} is the gas constant and T is the temperature. For $C = 1$ the interaction of the adsorbate with the surface of the adsorbent is the same as upon condensation of the free adsorptive. $C \ll 1$ results in a type III isotherm while $C \gg 1$ yields a type II shape (see Figure 21.25). For porous silica, Iler [70] provides an empirical relationship between the C parameter and the number density of OH groups per nm^2. C values well above 100 are relatively insensitive to the curvature of the isotherm; very large C values are also a characteristic of microporous samples.

Equation (21.29) in its linearized representation, i.e.,

$$\frac{p/p_0}{V_{\text{ads}} \times (1-p/p_0)} = \frac{1}{V_{\text{ml,BET}} \times C} + \frac{C-1}{V_{\text{ml,BET}} \times C} \times (p/p_0) \tag{21.31}$$

is called the BET equation. In the corresponding plot, the intercept a' with the ordinate provides $1/(V_{\text{ml,BET}} \times \text{C})$ while the slope b' is equal to $(C-1)/(V_{\text{ml,BET}} \times C)$.

Hence

$$V_{\text{ml,BET}} = 1/(a'+b');\ C = \frac{b'}{a'} + 1 \text{ and } S_{\text{BET}} = \frac{V_{\text{ml,BET}} \times S_{\text{Gas}} \times N_{\text{A}}}{V_{\text{mol,STP}}} \tag{21.32}$$

with S_{gas} the area taken by a single adsorbate molecule (see footnote 11), N_{A} the Avogadro constant, and $V_{\text{mol,STP}}$ the molar volume at STP (=22,414 cm^3 (STP)/mol).

Since the result of the fit of the *BET* equation to the isotherm strongly depends on the range used for the fit, it is important to follow recommended rules for its selection. For aerogels that do not contain a significant amount of micropores the range in rel. pressure to be used for the evaluation according to (21.31) is about 0.05–0.25. For highly microporous aerogels, this range will result in negative values of C; for such samples the evaluation range is usually shifted to lower relative pressures until the C value become positive. A good strategy for the selection of the range to use for the fit is to plot $V_{\text{ads}} \times (1 - p/p_0)$ and determine the rel. pressure $(p/p_0)_{\text{max}}$ at which this product reaches its maximum (Figure 21.28). For the fit only relative pressures below $(p/p_0)_{\text{max}}$ should be applied.

Meso- and Macropores. The simplest approach to get an estimate for the average pore size d_{av} in the aerogel under investigation is to determine the specific surface area by gas sorption and the specific total pore volume V_{pore} from the macroscopic and the skeleton density of the aerogel [see (21.21)] and to apply the relationship

$$d_{\text{av}} = \frac{4 \times V_{\text{pore}}}{S_{\text{BET}}}. \tag{21.33}$$

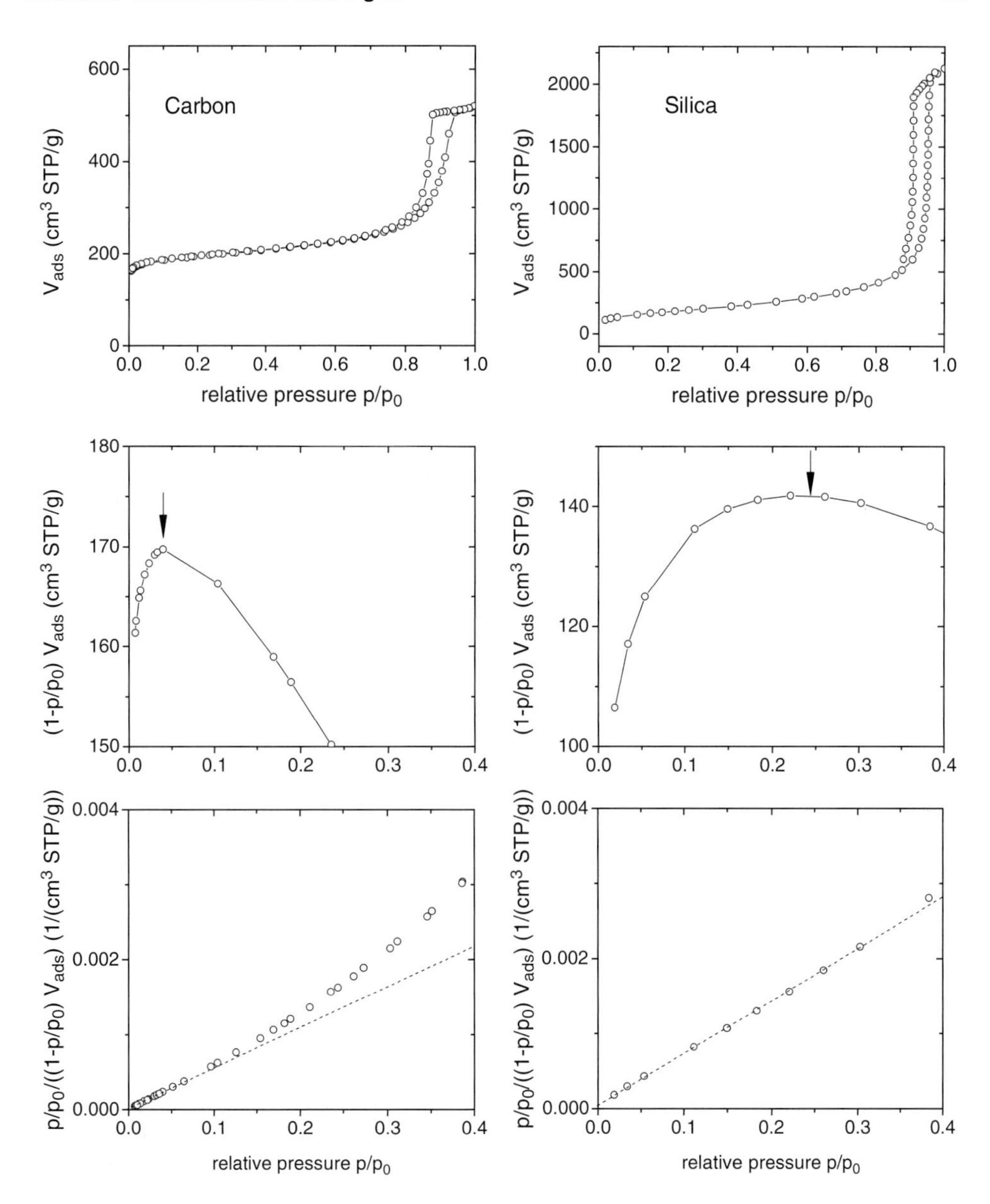

Figure 21.28. Comparison of a microporous carbon aerogel and a nonmicroporous silica aerogel in terms of the BET evaluation of their isotherm (**top row**). **Second row:** Determination of the correct evaluation range to be used for the BET equation; the *arrow* indicates the maximum relative pressure to be used. **Bottom row:** BET-plot with *dashed fit curve*.

To determine the average meso- or macropore size in case of microporous sample, the specific total pore volume V_{pore} has to be replaced by ($V_{pore} - V_{mic}$) and the external surface area has to be used instead of the BET surface area.

The most widely used approach to extract a pore size distribution from a gas sorption isotherm is the BJH model [71], based on the capillary condensation in mesopores.

According to the Kelvin equation the radius r_M of the meniscus formed by a completely wetting liquid upon condensation in a pore is related to the relative pressure by[13]

$$r_M = \frac{2\gamma \times V_{L,\,mol}}{R_G \times T \times \ln(p/p_0)}. \tag{21.34}$$

Here γ and $V_{L,mol}$ are the surface tension and the molar volume of the condensed liquid.[14] As prior to mesopore filling a surface layer of adsorbate t_L is already present, the pore size d is given in case of a cylindrical or spherical pore by

$$d = 2 \cdot (|r_M| + t_L). \tag{21.35}$$

It has to be pointed out that in contrast to a spherical pore, the radius of the meniscus in a cylindrical pore is upon adsorption twice as large as the one effective upon desorption.[15] As a consequence a hysteresis is expected for geometrical reasons. However, other sources for a hysteresis having a more basic thermodynamic background have to be kept in mind.

Combining the volume adsorbed in a relative pressure interval corresponding to a given average value of p/p_0 with the corresponding pore size (21.35) provides a pore volume distribution. For aerogels with low porosity the parallel adsorption and desorption branches vanish and an adsorption branch with a continuous increase is observed, while the desorption branch usually exhibits a sharp drop of the volume adsorbed. Different effects can be responsible for this behavior, one of them being pore blocking or network percolation effect that may result in a sudden emptying of all pores when the so-called breakthrough radius r_{BT} is reached; the latter is defined by the size of the pore that provides access to all – even larger – pores throughout the interconnected pore network.

The evaluation of the adsorption isotherms in terms of the mesopore size distribution assumes that the sample is unchanged during analysis. However, capillary condensation is accompanied by a capillary pressure that is due to the surface tension trying to minimize the interfacial energy. The respective stress σ exerted onto the skeleton of the aerogel can be calculated by [72]

$$\sigma = \ln(p/p_0)\frac{R_G \times T}{V_{L,mol}} \times \frac{V_{ads}}{V_{pore}} \times \Phi. \tag{21.36}$$

Here V_{ads}/V_{pore} is the relative (meso)pore filling and Φ is the (meso)porosity of the sample under investigation. Equation (21.36) can be related to the pores size d using (21.34)

$$\sigma = \frac{4\gamma}{d} \times \frac{V_{ads}}{V_{pore}} \times \Phi. \tag{21.37}$$

Using the surface tension for liquid nitrogen (see footnote 14), the maximum stress arising in a pore with a diameter of 10 nm is 3.5 MPa. This value has to be compared to the

[13] When the adsorbate is only partially wetting a factor of cos (θ_C), with θ_C the contact angle, has to be included on the right side of the equation.

[14] For liquid nitrogen at 77 K, the values for the surface tension and the molar volume are 0.00885 J/m^2 and 34 cm^3/mol.

[15] For adsorption in cylindrical pores therefore $d = 2 \cdot (|r_M|/2 + t_L)$ holds.

typical moduli of compression of aerogels ranging from below 1 MPa to several 10 MPa depending on the aerogel density. It was earlier suspected that aerogels might be deformed during nitrogen sorption analysis and Scherer et al. proved this to be actually true by taking pictures [73] during a sorption run. Later on Reichenauer and Scherer experimentally monitored the deformation and found partly irreversible compressional effects with a volume change of up to 50% for silica aerogels with a density of about 155 kg/m^3 [54, 55]. The macroscopic change of the initial sample volume V_0 as a result of the capillary pressure induced deformation was found to follow the theoretically expected relationship

$$\frac{V}{V_0} = \exp(\sigma/K_0), \text{ with } K_0 = \frac{E_0}{3(1-2v)} \tag{21.38}$$

the modulus of compression of the undeformed sample; K_0 can be calculated from the Young's modulus E_0 and the Poisson's ratio v (about 0.15 for aerogels).

The experimental data also showed that the compression is about twice as large in the initial part of desorption branch compared to the effect close to the completion of the adsorption branch; the reason is the hysteresis and thus the shift in p/p_0 that is directly affecting the stress exerted. In addition, the effective capillary pressure varies with the degree of pore filling thus gradually rising with the volume adsorbed even in case of a steep adsorption branch. Reichenauer and Scherer tried to correct for the compressional effects and calculated a "true isotherm" from the experimental data [72]. Figure 21.29 shows the comparison between the measured and the correct isotherm determined for a silica aerogel. The as-measured hysteresis shows a shape that cannot be related to any isotherm in the standard classification scheme (Figure 21.25); the experimentally determined hysteresis exhibits a continuous slope rather than a flat plateau at the upper end of the sorption branches. When corrected for the simultaneously measured macroscopic deformation, the shape of the isotherm takes the typical type IV characteristics. Detailed analysis showed that the correction

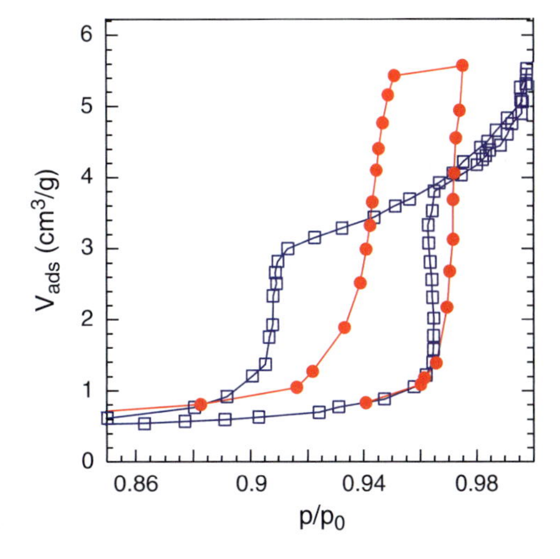

Figure 21.29. Hysteresis of a nitrogen isotherm for a silica aerogel with a density of 155 kg/m^3 (*open squares*) and the corresponding hysteresis corrected for deformational effects (*full circles*) [72].

routine yields pore sizes that are in excellent agreement with other noninvasive techniques such as *SAXS*. It could also be proven that the highly deformed plateau of the isotherm can be used to calculate reliably the elastic modulus of the aerogel as a function of compression (see Figure 21.30). The main reason why this is actually possible is the fact that, in the deformed plateau range, the sample is still completely liquid filled and the change in volume adsorbed is exclusively due to the compression of the liquid filled sample. Although Figure 21.29 shows an extreme case, this concept can also be used for less drastic deformations reflected in an almost linear slope of the isotherm in the plateau range (see e.g., Figure 21.28).

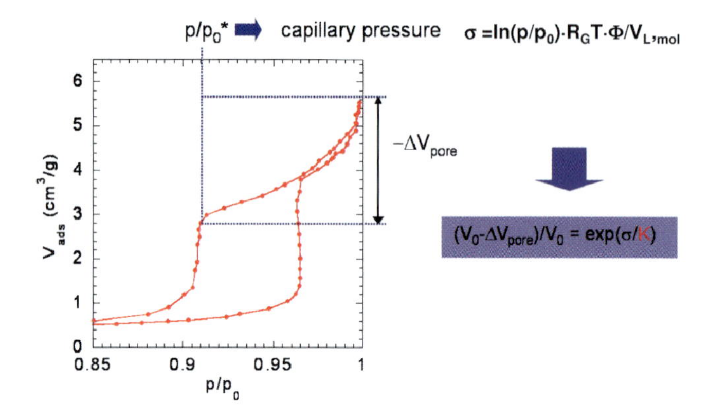

Figure 21.30. Scheme for the calculation of the modulus of compression from the deformed plateau regime of an isotherm. The calculation can be performed at each point of the "plateau" down to a relative pressure of p/p_0*, thus yielding the modulus at different stages of compression; K_0 can be calculated from values near relative pressures of 1. Φ is the porosity of the sample in the respective state (e.g., partially compressed) and V_0 the total specific volume ($=1/\rho$, if the sample is mesoporous only).

It has to be pointed out that the very same process that is observed during desorption upon nitrogen sorption analysis is active when using any other probe, for example when subcritically drying a sample [74, 75]. To use the same evaluation concept for drying shrinkage the fluid related quantities such as the surface tension and the relative pressure and the temperature have to be replaced in the above relationships.

Comment. Gas sorption analysis is a well-established tool for the characterization of open porous solids. For aerogels the method provides reliable information on the surface area. However, care has to be taken in case of microporous aerogels; here a well-equilibrated isotherm in combination with the right choice of the evaluation range will still yield reliable values for the microporosity and the specific surface area. For detailed analysis of microporosity measurements with CO_2 at 273 K are recommended.

In the range of mesopore filling, artifacts can result in highly erroneous numbers. The main prerequisite here is also a well-equilibrated isotherm; in case of aerogels this is often a challenging task as the large specific pore volume can take on the order of 100 h to be filled. When well equilibrated, mesoporous aerogels with low moduli of compression show isotherms that are significantly affected by compressional effects. In this case a simple correction of the average mesopore size can be performed; in addition, the deformation of the isotherm can be exploited to determine the mechanical properties of the aerogel under investigation. If screening of highly porous aerogels is performed it might be more favorable to estimate the average (macro- or) mesopore size from the macroscopic density and the external surface area rather than to invest half a week to measure the true isotherm of one sample only.

21.7. Hg Porosimetry

Mercury (Hg) porosimetry is a method that is able to probe about six orders of magnitude in accessible pore size ranging from about 400 μm down to a few Angstroms [76]. Hereby isostatic pressure is applied to force nonwetting liquid mercury into the pores to be quantified. The external pressure p_{ext} required to access a cylindrical pore with radius R_{pore} is given by the Washburn equation:

$$p_{\mathrm{ext}} = -\frac{2\gamma \times \cos(\Theta)}{R_{\mathrm{pore}}}. \tag{21.39}$$

Here γ is the surface tension of the mercury and Θ is the contact angle between the mercury/gas interface and the pore wall. Typical contact angles range from about 140° for metal oxides to 155° for carbon ([1], p. 314); the surface tension of mercury at 25°C is 0.485 N/m. Using these values the external pressure required for Hg intrusion can be calculated as a function of pore size. Figure 21.31 reveals that the excess pressure p_{ext} has to be adjusted between 10^{-3} and 400 MPa to cover the given range in pore size; to provide high accuracy the pressure range is typically split in two sections, a low and a high pressure range, covering about three decades each. Commercial instruments operate with one sample holder that can be transferred from the low to the high pressure unit of the instrument.

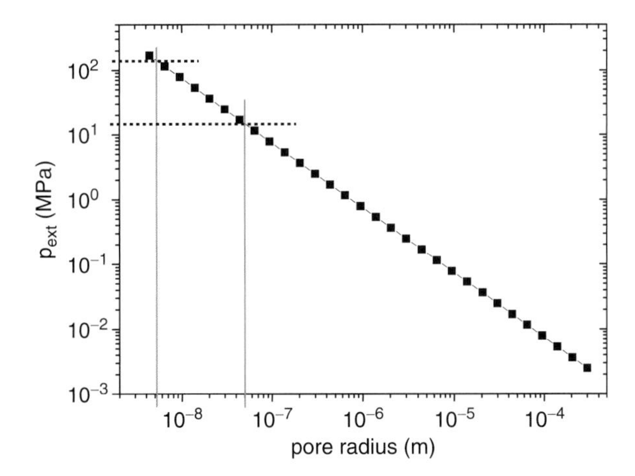

Figure 21.31. Pressure required according to (21.39) to force mercury into cylindrical pores with different radii. *Dashed lines* indicate the excess external pressures required to access 10 and 100 nm wide cylindrical pores.

Prior to the measurement, the sample is introduced into a sample cell that allows for degassing under vacuum. Usually the same port also provides the connections to the low pressure system of the mercury porosimeter, where air pressure is applied onto the pool of mercury to force it into the sample; in contrast, the high pressure port uses a hydraulic system to provide pressure up to 400 or even 600 MPa. The amount of mercury vanishing from the mercury reservoir, $\Delta V_{mercury}$, is recorded together with the net applied pressure p_{ext} and converted into a pore size distribution.

Care has to be taken to choose the rate of pressurization or the time allowed for equilibration (when pulse wise pressurizing) such that the sample is in quasi equilibrium at each pressure; otherwise artifacts will be introduced due to kinetic effects, shifting the calculated pore size to smaller values compared to the actual pore sizes.

Figure 21.31 shows that for 10 and 100 nm wide cylindrical pores, the corresponding pressures p_{ext} required are 150 and 15 MPa, respectively. These pressures are usually not critical when used with stiff porous materials such as ceramics. However, the moduli of compression for aerogels with porosities well above 80% are typically in the range of 0.1 to 50 MPa. As a consequence compression rather than intrusion of the sample can occur and the volume change vs. pressure curve will be misinterpreted as a broad contribution of macropores [77]. The compression effect has been first shown by Brown [78] and has been studied in case of aerogels in detail by Pirard and coworkers (e.g., [79]). It has been shown that aerogels are usually compressed at low to medium external pressures resulting in a densification and thus in an increase of their modulus. The modulus of aerogels increases more than linearly with the density while the pore size decreases inversely with this quantity; as a consequence the behavior switches with increasing pressure from pure compression to intrusion. This effect is shown in Figure 21.32 for an organic aerogel with a porosity of about 80% and a pore size of about 700 nm: Here about a third of the pore volume is compressed before the remaining pores in the stiffened aerogel are finally filled with mercury. The two different regimes show distinctively different characteristics. In this case the pore size distribution above 0.6 μm is a pure artifact due to compression; the pore size distribution below 0.6 μm is representing the properties of the compressed aerogel (by 30 vol%). As a consequence, the average pore size as well as the total volume detected in this range is about 30% lower than the values for the initial (uncompressed) sample. In a second organic aerogel with similar porosity, however, with about a factor of 10 larger macropores, the compression is negligible and the intrusion curve shows the steep intrusion characteristics.

Depending on the average pore size, the modulus of compression, and the type of aerogel under investigation, the compression of the sample can be in part or totally irreversible. In particular, silica-based aerogels without organic surface modification are irreversibly deformed.

When compressional effects occur, one can either try to correct for their contribution or to exploit them to determine the mechanical characteristics of the aerogel under investigation. Pirard compared two intrusion/extrusion curves taken for two different pieces of monolithic precipitated silica taken from the same batch [80]; hereby one sample was wrapped into a mercury impermeable membrane. The experimental curves showed that the first flat slope of the intrusion curve can exclusively be related to compression of the sample while the subsequent steep increase of the intrusion volume with increasing pressure applied is actually dominated by intrusion of the pores. Upon extrusion, the volume detected for the unwrapped sample was not fully recovered; it could be shown that the missing volume upon extrusion was equivalent to the volume compressed in the first stage of pressurization (intrusion).

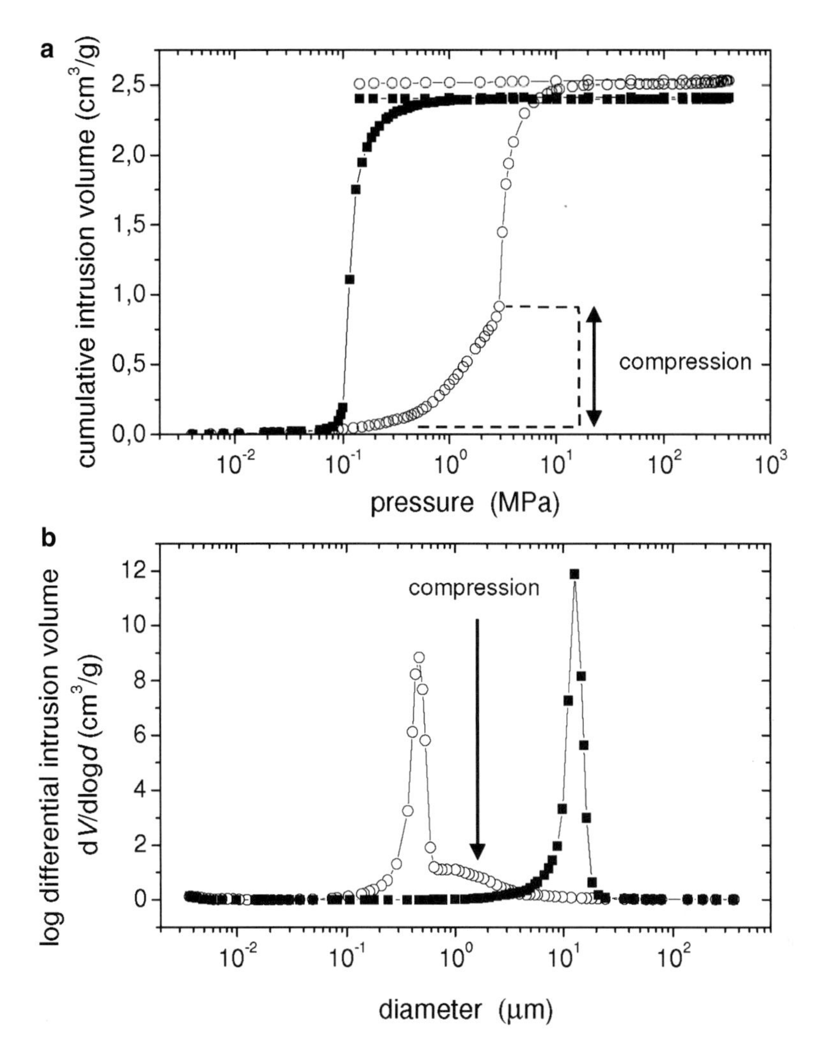

Figure 21.32. **Upper graph:** Mercury intrusion curves for two organic aerogels with similar porosity (80%), however with pores sizes differing by about a factor of 10. **Lower graph:** Corresponding pore size distributions (data: courtesy Bundesanstalt für Materialforschung und – prüfung, BAM, Berlin, Germany). The modulus of compression determined by ultrasonic measurements for the sample corresponding to the smaller pore size is about 12 MPa. This aerogel is compressed by about 30 vol% prior to intrusion.

Pirard identified for aerogels a characteristic relationship between the size of the pores being compressed and the net external pressure applied p_{ext} in the phase of pure compression:

$$2 \times R_{\text{pore}} = k_{\text{S}} / p_{\text{ext}}^{0.25}. \tag{21.40}$$

The structure-dependent constant k_{S} can be calculated from the crossover of the mercury porosimetry data from compression to intrusion using the argument that at this transition point the Washburn equation (21.39) as well as (21.40) have to hold [79].

$$k_{\text{S}} = -\frac{4\gamma \times \cos(\Theta)}{p_{\text{ext}}^{0.75}}. \tag{21.41}$$

The combination of these two equations provides a tool for the calculation of a pore size distribution in the purely compressional regime (Figure 21.33).

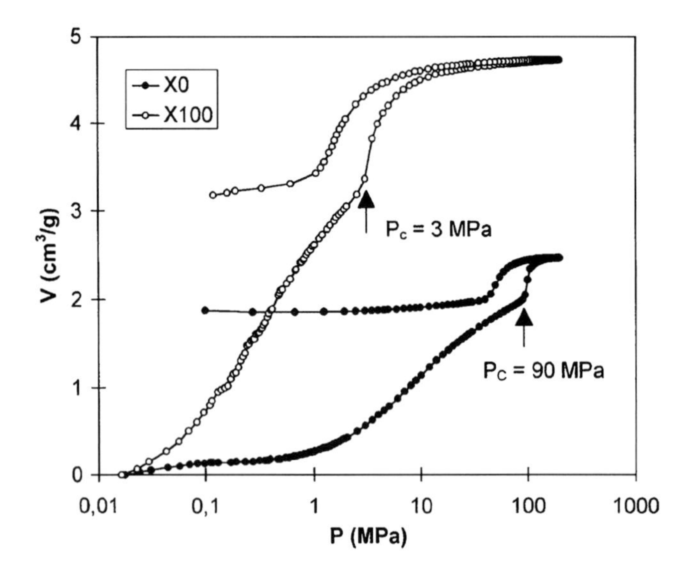

Figure 21.33. Mercury intrusion/extrusion curves for two different TEOS-based silica xerogels. The *arrows* indicate the pressure at which the behavior crosses over from sample compression to mercury intrusion of the sample. The extrusion curves reveal an irreversible deformation of the samples upon intrusion. Plot reproduced from [79], p. 337.

Comment. Hg porosimetry can theoretically be applied to investigate pore size distributions in the range from microns to nanometers. However, in case of compliant materials such as aerogels, significant deformation effects are to be expected unless the relationship between the stiffness of the sample and the pore size is favorable. It has been shown that the data can be corrected for the compressional effects and that even in case of pure compression a pore size distribution can be calculated [81, 82].

Furthermore a mercury porosimeter can be applied purposely to study the mechanical properties under controlled densification, when the specimen is wrapped in a mercury impermeable membrane [83].

Mercury porosimetry has the disadvantage that the sample is often irreversibly deformed and part of the mercury infiltrated is trapped. As a consequence the sample is lost for further investigations and has even to be disposed as toxic waste.

21.8. Thermoporometry

Thermoporometry probes the freezing characteristics of a pore liquid; the data can be used to determine a pore size distribution and the shape of the pores. The method is expected to be sensitive to pores in the range between about 2 and 30 nm. In contrast to classical porosimetry (involving mercury or gas adsorption), the technique can be applied to gels rather than requiring dry samples.

For thermoporometry a pure (noncontaminated) pore liquid that completely wets the inner surface of the sample under investigation is assumed. At equilibrium the chemical potentials of the still liquid and the solidified state within the pores are equal at given temperature. This equivalence provides a relationship between the pore size and the freezing point depression (Gibbs–Thompson effect) that can be used to calculate a pore size from the freezing point temperature shift:

$$\Delta T = T'_{\mathrm{m}} - T_{\mathrm{m}} = -\frac{\gamma_{\mathrm{SL}} \times k_{\mathrm{SL}}}{(S_{\mathrm{L}} - S_{\mathrm{S}})/V_{\mathrm{S}}}. \tag{21.42}$$

Here T_m and T'_m are the melting temperature of the free and the confined liquid, k_{SL} is the curvature of the solid/liquid interface in the pores, and $(S_L - S_S)/V_S$ is the entropy change per unit volume during melting.[16] k_{SL} can be expressed in terms of the principal radii of curvature (r_1,r_2) in the system:

$$k_{\mathrm{SL}} = \frac{1}{r_1} + \frac{1}{r_2}. \tag{21.43}$$

For a spherical pore r_1 is equal to r_2 thus resulting in $k_{SL} = 2/r_1 = 2/R$. As always a liquid-like layer of thickness t_L is expected to be present on the solid backbone of the sample, R is related to the true pore size by $R_{pore}=R + t_L$. Note that by definition the k_{SL} is positive when the center of curvature is inside the solid phase.

When simultaneously the energy release or uptake upon freezing or melting, respectively, is evaluated the pore size distribution for the material under investigation can be extracted. In addition, the method can distinguish between cylindrical and spherical type pores as the latter result in a hysteresis free characteristic when comparing freezing and melting, while for cylindrical pores a hysteresis is expected.

To characterize a sample with thermoporometry a differential scanning calorimeter (DSC) is applied; the signal is typically recorded from temperatures slightly above to well below the freezing point of the free pore liquid. Free excess liquid or liquid confined in macropores results in a signal around the melting point of the free liquid and can be used to define the reference temperature (Figure 21.34).

Phalippou et al. [85] as well as Vollet et al. [84] applied thermoporometry to study the fractal properties of silica alcogels. Comparison of thermoporometry results with *SAXS* data [84] taken for the same gels provided similar trends for different gels investigated with respect to the upper cut-off of the fractal range. Both methods yield fractal dimensions in the range between 2.3 and 2.6, however with different tendencies. Barbieri et al. [31] studied organic aerogels prepared in acetone as a solvent and compared the result from thermoporometry to those from *SAXS*; the authors conclude that thermoporometry is providing "reliable results."

[16] Values for water: $\gamma_{SL} = 40 \times 10^{-3}$ N/m, $\gamma_S/V_S = (S_L - S_S)/V_S = 334 \times 10^3$ J/m^3, $T_m = 0°C$. A 10 nm spherical pore radius causes a shift in melting temperature by about 22°C.

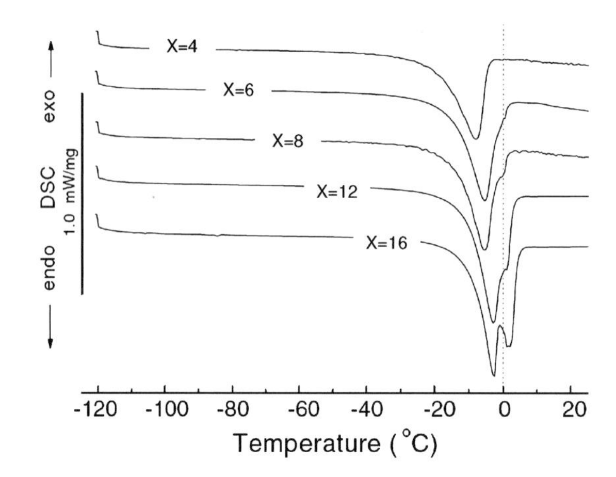

Figure 21.34. DSC signal for a series of acid catalyzed silica gels prepared with different molar water to TEOS ratios *X*; plot adopted from [84], p. 3. Prior to the experiment the pore liquid had been exchanged for pure water. Temperature ramp: 2 K/min.

Comment. Thermoporometry is a method to investigate pore size distributions (limited to small mesopores) in the gel or the rewetted state. As a characterization method for sparse gels, thermoporometry is controversially discussed (see in particular [86], p. 17). It seems to be still unclear how severely the gels are damaged upon characterization and how strongly micro cracks may affect the results.

Independent of damage during freezing and thawing the exchange of the pore liquid by a pure phase can introduce artifacts. For example, exchanging methanol for water introduces cracks in large samples if not performed stepwise.

21.9. Other Characterization Methods

There are many other methods that can be applied to quantify structural characteristics of aerogels. The fact that they are not as popular as the methods described earlier in detail is often merely due to a lack of instrumentation available at the labs or not yet well-established data evaluation routines.

Nevertheless, there are a couple of very useful techniques that may be even better suited for characterization than some of the standard techniques. Their advantage hereby lies in the fact that they are nondestructive techniques that leave the sample unchanged upon analysis.

Fluid permeation. Fluid permeation can be investigated by a classical stationary or a dynamic technique. It provides an effective pore diameter that is the result of the size distribution and the connectivity of the pores; the latter can be expressed in term of the tortuosity τ which is the ratio of the path that the fluid actually takes and the width over which the pressure gradient is applied (usually the sample thickness). If viscous flow rather than molecular diffusion is the dominant mechanism the weight of the pore size distribution with respect to its impact on the fluid permeation transport is shifted to the large pores.

A permeation experiment is relatively easy to implement as it requires only a pressure gradient Δp across the sample and a device to determine the fluid flow rate J (Figure 21.35). The permeability coefficient k_P is then given by

$$J = -k_P \frac{\Delta p}{L} \frac{p_{av}}{\eta} \tag{21.44}$$

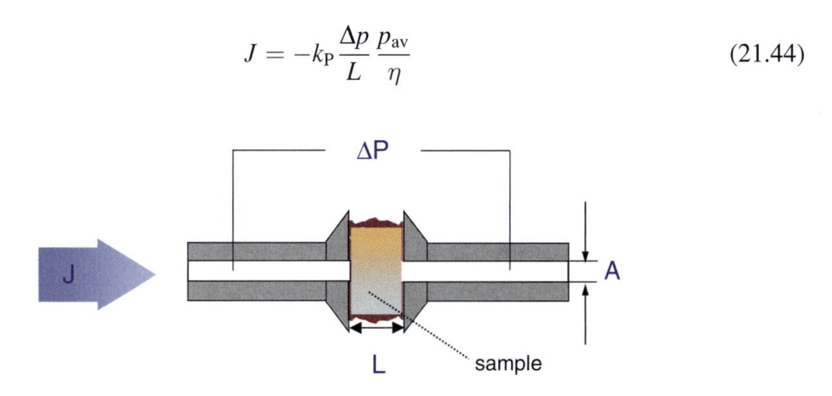

Figure 21.35. Scheme of a setup for a stationary measurement of fluid permeation.

with η the fluid viscosity and p_{av} the average pressure in case of gas permeation. However, it has to be emphasized that (21.44) makes only sense if the permeation is dominated by viscous flow since otherwise the viscosity is not defined.

When measuring the fluid flow *through* the aerogel dead-end pores are ignored while they become effective when flow *into* the sample is analyzed. In the classical setup usually the flow of a fluid through the sample is investigated. This, however, requires a defined flow through rather than around the sample; therefore the perimeter of the specimen has to be sealed. For aerogels wax, silicone or certain epoxy glue can be applied depending on the type of aerogel. Before sealing it should be tested whether the glue is physically or chemically destructing the sample. It also has to be kept in mind that small cracks in the direction of the fluid flow will act as shortcuts and thus result in artifacts. It is therefore recommended to prepare the sample in the shape and size needed for the experiment rather than using additional steps to machine the aerogel.

In case of a nonadsorbing gas as a fluid, the flow through an aerogel with a porosity Φ, a thickness L, and a cross-sectional area A can be described by the following relationship [57]:

$$J = -\left[\frac{\Phi}{\tau}\right]\left(\frac{p_{av}}{32\eta}d^2 + \varepsilon(p_{av}) \times \frac{2}{3}\sqrt{\frac{2R_G T}{\pi \times M}}d\right)\frac{\Delta p}{L}A = P \times A\frac{\Delta p}{L} \tag{21.45a}$$

with

$$\varepsilon(p_{av}) = \frac{1 + 2\sqrt{M/R_G T}(0.5 \times d \times p_{av}/\eta)}{1 + 2.47\sqrt{M/R_G T}(0.5 \times d \times p_{av}/\eta)}. \tag{21.45b}$$

Hereby the pores are modeled by a set of parallel cylindrical pores with diameter d and a tortuosity τ. ε is the so-called Adzumi factor taking values between 0.81 and 1. M is the molar mass of the gas molecules and T and R_G are the temperature and the gas constant. P is the specific permeability, i.e., the permeability normalized to the geometrical dimensions of the sample. Equation (21.45a, b) represents in a good approximation a linear relationship between the flow rate and the average pressure at given boundary conditions with the slope being governed by viscous flow and the offset controlled by molecular diffusion (see Figure 21.36). The ratio of slope and offset provides the effective pore diameter without the need to know the prefactor Φ/τ. However, for pores in the nanometer range, (21.45a, b) yields an almost pressure independent relationship; in that case assumptions have to be made for the tortuosity. For highly porous aerogels this factor was shown to be on the order of 1 to 3 [57]. A good correlation between pore diameters extracted from gas permeation data by modeling the aerogel by parallel cylinders (21.45a, b) and *SAXS* as well as sorption was found for low density mesoporous silica aerogels [57]. Other models, such as the Carman–Kozeny approach for packed beds of spherical particles, do not seem to provide a good description of the gas flow through aerogels. Results on gas permeation through aerogels can be found in references [57, 87–90].

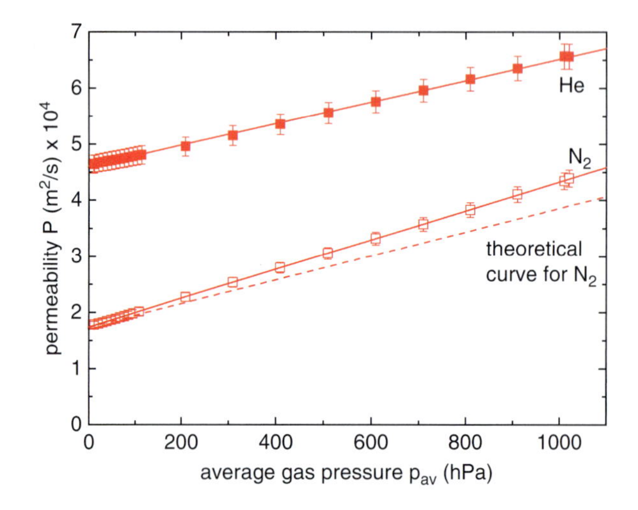

Figure 21.36. Permeabilities $P = (J \times L/(A/\Delta p))$ measured as a function of the average gas pressure with He and N_2 for an organic aerogel with a pore size of about 600 nm (as determined from surface area and pore volume). *Dashed line*: theoretical curve calculated with parameters Φ/τ and d determined from the helium permeation data. The evaluation of the data yields good agreement with results from other methods when a distribution of pore width with a relative standard deviation of 0.5 is assumed (Courtesy: Martin Schormayer, ZAE Bayern).

Experimental setups for investigation of gas transport without the need for sealing were applied by Reichenauer and coworkers [91, 92]; hereby the change in gas pressure is recorded as a function of time after a sudden pressure change in a chamber containing the sample. The latter has to have a well-defined geometry to allow for the extraction of the permeability from the characteristic pressure decay. For this type of measurement, fast pressure transducers or large pieces of samples are required. A similar concept was used by Gross [93] in a high pressure chamber (autoclave).

Studying the transport of liquid in the same way is no real option in case of gel precursors of aerogels as thin layers are required to achieve a high enough flow to be detected. Alternative methods are described in the following section.

Beam bending, Dynamic Mechanical Analysis (DMA). Beam bending [94] or testing of gels with DMA provides an excellent tool for the nondestructive quantification of the effective pore size and the mechanical properties of a gel. The main idea hereby is to initiate a sudden deformation of the gel body and to then monitor the evolution of the deformation (at given load) or the force (at given deformation) with time. As the incompressible liquid in the gel pores is free to move, it will flow to other positions within or around the gel to reestablish a load free state. The time required depends on the size and geometry of the sample as well as the liquid permeability which is a function of the fluid properties and the pore size and connectivity. For a rod-like sample geometry, the time dependence of the load is given by [94]:

$$F_{\mathrm{ext}}(t) = \frac{48 \times E_0 \times (\pi \times a^2)/4 \times \Delta L}{\ell^3}\left[1 + \frac{1-2v}{2(1+v)} f(\theta_{\mathrm{C}})\right] \tag{21.46a}$$

with

$$f(\theta_{\mathrm{C}}) = 1 - 3.00977\theta_{\mathrm{C}}^{0.5} + 3.2698\theta_{\mathrm{C}} - 1.2153\theta_{\mathrm{C}}^{1.5} \quad \text{for } \theta_{\mathrm{C}}<1 \tag{21.46b}$$

$$\theta_{\mathrm{C}} = t \times 4\frac{k'}{\eta_{\mathrm{L}} \times a^2} \times \frac{G_{\mathrm{P}}}{(1-2v)}. \tag{21.46c}$$

Here E_0 is the Young's modulus, a the radius of the rod-like sample, ΔL the impounded deformation, ℓ the length of the rod, and v the Poisson's ratio. k_{P} is the permeability coefficient of the gel [related to the specific permeation by $P = (k_{\mathrm{P}} \times p_{\mathrm{av}}/\eta)$], η is the viscosity of the pore liquid, and G_{p} is the shear modulus. In analogy to the viscous term in (21.45a, b), k_{P} is related to the pore diameter d by

$$k_{\mathrm{P}} = \frac{d^2}{32} \times \frac{\Phi}{\tau}. \tag{21.47}$$

While the eventually pore liquid is reorganizing as a result of the external load (or deformation), the gel skeleton has to take over the external load thus deforming until an equilibrium state is reached. The deformation at given load is then a measure for the stiffness of the gel. As the term in the brackets of (21.46a) tends to 1 for large times the prefactor provides the Young's modulus of the gel in equilibrium if no additional viscoelastic deformation is contributing [95].

A similar concept was applied by Gross et al. in the gas phase [93]. He applied sudden changes in gas pressure in a chamber containing an aerogel and evaluated the response of the sample in terms of length change as well as the change in gas pressure. The analysis was performed in an autoclave. The experiments provided the elastic modulus of the aerogel as well as the pressure dependent gas permeation rate that was further evaluated to determine an effective pore size.

It has to be pointed out that the load for such an experiment does not have to be provided by an external device. In Sect. 21.6 (gas porosimetry), it was shown that upon desorption of adsorbate from the pores, significant pressures can arise resulting in a deformation of the gel. Therefore instead of an external force one can also use the capillary pressure in the early part of drying to set up a similar experiment; however, the capillary pressure (which is a result of a change in relative gas pressure around the sample) needs to be changed rapidly compared to the time required to reestablished equilibrium. The time evolution can be monitored dilatometrically or by recording the relative gas pressure in the vicinity of the sample (Figure 21.37).

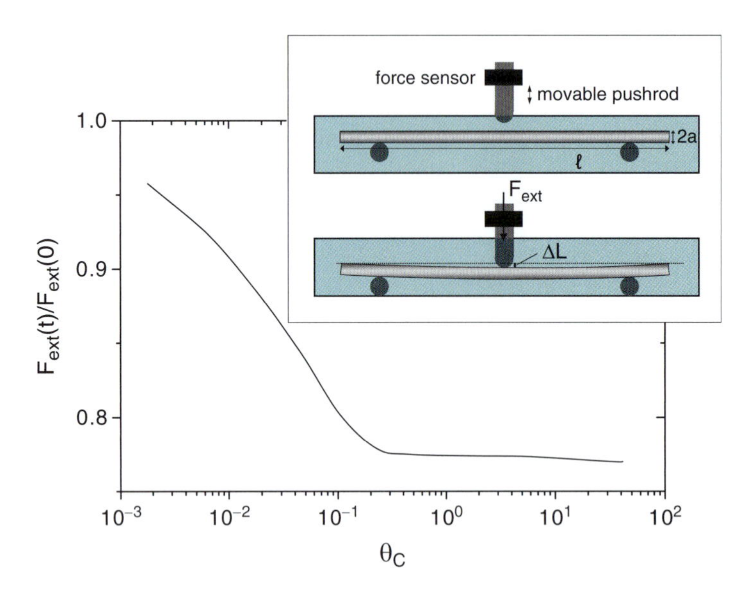

Figure 21.37. Experimental curve (normalized load vs. time) taken by beam bending for a silica gel. *Inset*: typical setup used for a three-point beam bending measurement (see also [94]).

***NMR* imaging and analysis of relaxation times.** While *NMR* spectroscopy is a well-established tool in chemical analysis for the investigation of the time evolution of process steps or the molecular arrangement of chemical elements within a sample, *NMR* imaging is another powerful technique that is rarely applied, as suitable instrumentation is rare and expensive. *NMR* imaging allows in situ investigation of dynamic processes such as fluid exchange steps upon sub- or supercritical drying [96]; in addition, imaging can be applied to locally resolve self-diffusion coefficients within one sample (e.g., for detection of structural gradients) or to perform a screening of the self-diffusion coefficients of the molecules in the pore liquid for many samples at the same time. The self-diffusion coefficient is hereby determined by marking the position of the molecules by applying a magnetic field gradient over a short period of time and probing their new location after a defined time interval with a "reading" gradient.

The self-diffusion coefficient of a molecule in the liquid phase is dominated by the tortuosity τ; here τ characterizes the detour that the molecule has to take due to the obstacle that the gel skeleton imposes, compared to the distance that it would cover by free Brownian motion in the liquid. In particular, it is sensitive to pore accessibility. The self-diffusion coefficient is therefore a characteristic of the pore connectivity which itself is for gels a function of porosity and backbone structure. For small mesopores, the small pore volume to surface ratio is expected to result in an additional reduction of the self-diffusion coefficient due to the interaction of the liquid molecules with the pore surface.

Similarly, the relaxation time T_1 of a spin oriented by a short external magnetic pulse is sensitive to the spin–lattice interaction and thus to the pore size.

Behr et al. used *NMR* imaging to investigate tetramethoxysilane (*TMOS*) based gels prepared under different pH. The analysis of the T_1 values and the self-diffusion coefficients revealed a significant decay with the target density of the gels; in addition, a clear shift in the values between gels with the same target density, however different pH of the initial solution can be identified (see Figure 21.38). *SAXS* measurements performed at the same gels yield for the size of the backbone particles 0.6 nm (pH 4), 2.6 nm (pH 2.6), and 3.8 nm (pH 11.2);

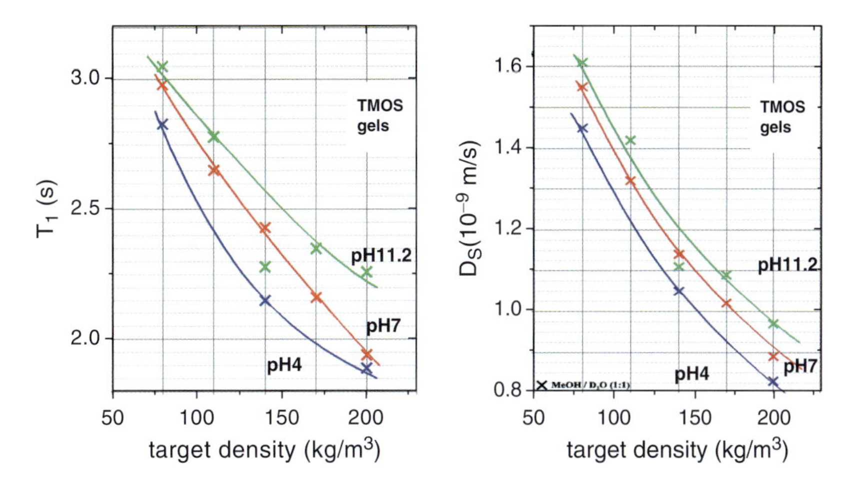

Figure 21.38. Data derived from *NMR* analysis of TMOS-based silica gels prepared at different pH of the starting solution as a function of target density (**left:** T_1 relaxation time), **right:** self-diffusion coefficient of methanol in the pores of the gel (after: [97], courtesy: W. Behr).

at given porosity (i.e., target density) the pores are expected to scale with the particle size. T_1 is sensitive to the pore size while the diffusion coefficient includes pore size and tortuosity; the similar trend and the roughly similar change in the two quantities as a function of target density (factor of 1.5 in the target density range investigated) indicate that the tortuosity is not a strongly changing quantity in this range and that both variables are mainly dominated by the pore size. In particular the gel with the smallest particle and hence pore size (pH 4) also shows the smallest relaxation time and self-diffusion coefficient as expected.

Behr also performed in situ ^{1}H *NMR* imaging during solvent exchange (methanol for liquid CO_2) prior to supercritical drying; therefore he developed an *NMR*-high pressure cell that allows studying cylindrical gel samples with a diameter of about 15 mm at pressures up to 10 MPa [98]. The evaluation of the imaging pictures after purging the autoclave and the methanol soaked sample (pH 7, target density 80 kg/m^3) with liquid CO_2 (Figure 21.39) revealed that the effective transport diffusion coefficient of 2.4×10^{-9} m^2/s is about a factor of 1.5 larger than the self-diffusion coefficient; this is due to the limited length of the specimen and thus the three- rather than the two-dimensional diffusion profile. The self-diffusion coefficient of free methanol is about 2.0×10^{-9} m^2/s under the same conditions and thus less than a factor of 2 larger than the value of methanol confined in the gel pores.

These results indicate that for an estimate of diffusion times in the pore liquid of gels the values for the free liquids provide a good zero order approximation at high gel porosities; the values can be adjusted for higher gel densities by a scaling factor (Figure 21.38). These conclusions were applied to optimize the time allowed for solvent exchange processes.

Comment. The above described methods provide good access to structural information as well as transport properties along the pores and mechanical characteristics of the (aero)gel backbone; although the relationship of the absolute figures determined with data derived by other methods is still under investigation, the approaches can be used to extract trends upon screening of samples. Beam bending or DMA is a tool to analyze mechanical properties and

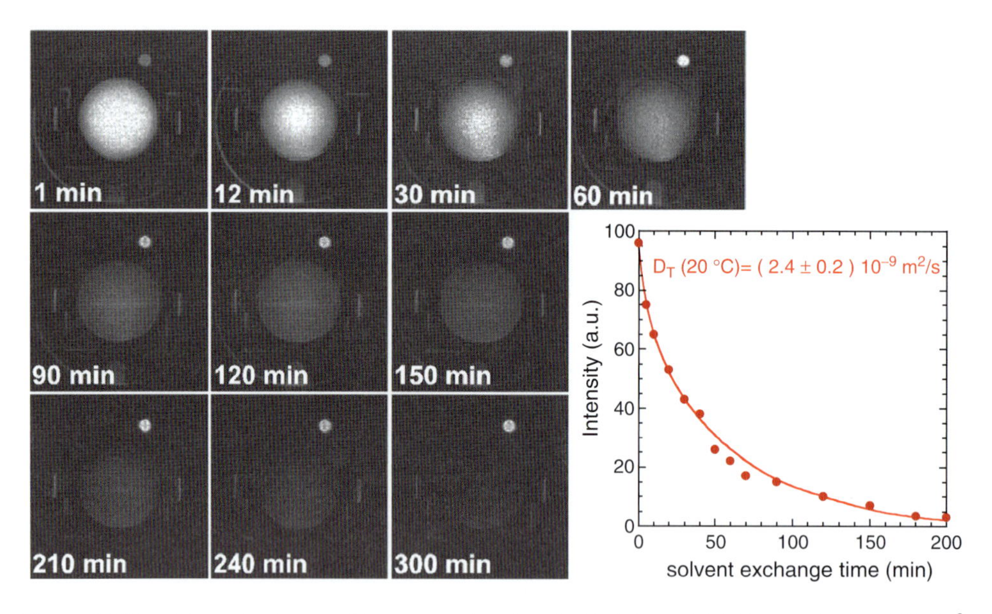

Figure 21.39. *NMR* imaging pictures of the cross-section of a cylindrical silica gel (pH 7, target density =80 kg/m^3) taken in situ at different times after purging the autoclave with liquid CO_2; the pictures are showing the progress of the solvent exchange of methanol for liquid CO_2. The methanol concentration is represented by the *white pixels*. **Right inset**: Integrated methanol concentration calculated from the series of **images on the left-hand side** as a function of time; the *line* is a fit to determine the transport diffusion coefficient D_T of the methanol in the pores of the aerogel [97] (Courtesy: W. Behr).

effective pore width of gels prior to drying while gas permeation is a nondestructive method for aerogels allowing to cover similar pore sizes as mercury porosimetry (i.e., few nanometer to several hundred microns) depending on the resolution of the setup applied.

NMR is a great tool to investigate different properties of gels including transport kinetics. However, imaging instruments are still expensive. Here cooperation with medical centers equipped with such instruments could be one approach to also use the technique for aerogel precursors.

21.10. Conclusions

The past 25 years of structural characterization of aerogels have more than once proven that established characterization methods have to be questioned when applied to new types of materials. In particular, mercury intrusion and nitrogen adsorption have been identified as being only limitedly suitable for the analysis of aerogels. Hereby the significant compression of the sample during analysis of low density aerogels is the most severe effect. In addition, equilibration is another issue that has to be carefully addressed when porosimetry is applied to aerogels.

The most powerful nondestructive quantitative methods are certainly scattering techniques that can be used for studies in the gel as well as in the aerogel state. However, lab instruments are still expensive and experts are needed to perform profound data analysis for new material systems.

In between, additional methods have been developed or adjusted for the structural characterization of aerogels. Particularly interesting is here the beam bending that allows determining mechanical and structural characteristics already in the gel state.

As structural investigation is one of the keys in fundamental research as well as in material R&D further developments of characterization techniques for nanomaterials such as aerogels are needed; hereby in particular well accessible, fast methods that require only small quantities of samples would be a big step forward to support screening when exploring new parameters in (aero) gel development. Furthermore methods that are tailored to structurally characterize aerogel-based composites will be needed as aerogels are increasingly becoming components in novel functional composite materials.

References

1. Schüth F, Sing K S W, Weitkamp J (eds.) (2002) Handbook of Porous Solids (Band 1). Weinheim
2. Bourret A (1988) Low-Density Silica Aerogels Observed by High-Resolution Electron-Microscopy. Europhysics Letters 6: 731–737
3. Job N, Pirard R, Vertruyen B et al (2007) Synthesis of transition metal-doped carbon xerogels by cogelation. Journal of Non-Crystalline Solids 353: 2333–2345
4. Job J, Heinrichs B, Lambert S et al (2006) Carbon Xerogels as Catalyst Supports: Study of Mass Transfer. American Institute of Chemical Engineers: Fluid Mechanics and Transport Phenomena 52: 2663–2676
5. Job N, Heinrichs B, Ferauche F et al (2005) Hydrodechlorination of 1,2-dichloroethane on Pd-Ag catalysts supported on tailored texture carbon xerogels. Catalysis Today 102: 234–241
6. Job N, Pereira M F R, Lambert S et al (2006) Highly dispersed platinum catalysts prepared by impregnation of texture-tailored carbon xerogels. Journal of Catalysis 240: 160–171
7. Padilla-Serrano M N, Maldonado-Hodar F, Moreno-Castilla C (2005) Influence of Pt particle size on catalytic combustion of xylenes on carbon aerogel-supported Pt catalysts. Applied Catalysis B-Environmental 61: 253–258
8. Hara H S, Smirnova A (2005) Method of Preparing Membrane Electrode Assemblies with Aerogel Supported Catalyst. WO/2005/086914
9. Sanchez-Polo M, Rivera-Utrilla J, von Gunten U (2006) Metal-doped carbon aerogels as catalysts during ozonation processes in aqueous solutions. Water Res 40: 3375–3384
10. Guilminot E (2007) Use of cellulose-based carbon aerogels as catalyst support for PEM fuel cell electrodes: Electrochemical characterization. Journal of Power Sources 166: 104–111
11. Soler R, Cacchi S, Fabrizi Get al (2007) Sonogashira Cross-Coupling Using Carbon Aerogel Doped with Palladium Nanoparticles; A Recoverable and Reusable Catalyst. Synthesis 19: 3068–3072
12. Du H, Li B, Kang Fet al (2007) Carbon aerogel supported Pt–Ru catalysts for using as the anode of direct methanol fuel cells. Carbon 45: 429–435
13. Moreno-Castilla C, Maldonado-Hodar F J, Perez-Cadenas A F (2003) Physicochemical surface properties of Fe, Co, Ni, and Cu-doped monolithic organic aerogels. Langmuir 19: 5650–5655
14. Moreno-Castilla C, Maldonado-Hodar F J (2005) Carbon aerogels for catalysis applications: An overview. Carbon 43: 455–465
15. Maldonado-Hodar F J, Moreno-Castilla C, Perez-Cadenas A F (2004) Catalytic combustion of toluene on platinum-containing monolithic carbon aerogels. Applied Catalysis B-Environmental 54: 217–224
16. Maldonado-Hodar F J, Perez-Cadenas A F, Moreno-Castilla C (2003) Morphology of heat-treated tungsten doped monolithic carbon aerogels. Carbon 41: 1291–1299
17. Maldonado-Hodar F J, Ferro-Garcia M A, Rivera-Utrilla J et al (1999) Synthesis and textural characteristics of organic aerogels, transition-metal-containing organic aerogels and their carbonized derivatives. Carbon 37: 1199–1205
18. Maldonado-Hodar F J, Moreno-Castilla C, Perez-Cadenas A F (2004) Surface morphology, metal dispersion, and pore texture of transition metal-doped monolithic carbon aerogels and steam-activated derivatives. Microporous and Mesoporous Materials 69: 119–125
19. Stroud R M, Long J W, Pietron J J et al (2004) A practical guide to transmission electron microscopy of aerogels. Journal of Non-Crystalline Solids 350: 277–284

20. Marliere C, Despetis F, Etienne P et al (2001) Very large-scale structures in sintered silica aerogels as evidenced by atomic force microscopy and ultra-small angle X-ray scattering experiments. Journal of Non-Crystalline Solids 285: 148–153
21. Beck A (1988) Optische Streuuntersuchungen an porösen SiO_2-Systemen. Diploma Thesis, Physikalisches Institut, EP II Würzburg
22. Reichenauer G, Fricke J, Manara J et al (2004) Switching silica aerogels from transparent to opaque. Journal of Non-Crystalline Solids 350: 364–371
23. Glatter O, Kratky O (1982) Small Angle X-ray scattering, London
24. Emmerling A, Petricevic R, Beck A et al (1995) Relationship between Optical Transparency and Nanostructural Features of Silica Aerogels. Journal of Non-Crystalline Solids 185: 240–248
25. Posselt D, Pedersen J S, Mortensen K (1992) A Sans Investigation on Absolute Scale of a Homologous Series of Base-Catalyzed Silica Aerogels. Journal of Non-Crystalline Solids 145: 128–132
26. Emmerling A, Fricke J (1992) Small-Angle Scattering and the Structure of Aerogels. Journal of Non-Crystalline Solids 145: 113–120
27. Kjems J K, Freltoft T, Richter D et al (1986) Neutron-Scattering from Fractals. Physica B & C 136: 285-290
28. Aristov Y I, Lisitsa N, Zaikovski V I et al (1996) Fractal structure in base-catalyzed silica aerogels examined by TEM, SAXS and porosimetry. Reaction Kinetics and Catalysis Letters 58: 367–375
29. Pahl R, Bonse U, Pekala R W et al (1991) Saxs Investigations on Organic Aerogels. Journal of Applied Crystallography 24: 771–776
30. Bock V, Emmerling A, Fricke J (1998) Influence of monomer and catalyst concentration on RF and carbon aerogel structure. Journal of Non-Crystalline Solids 225: 69–73
31. Barbieri O, Ehrburger-Dolle F, Rieker T P et al (2001) Small-angle X-ray scattering of a new series of organic aerogels. Journal of Non-Crystalline Solids 285: 109–115
32. Schaefer D W, Pekala R, Beaucage G (1995) Origin of Porosity in Resorcinol Formaldehyde Aerogels. Journal of Non-Crystalline Solids 186: 159–167
33. Gommes C J, Roberts A P (2008) Structure development of resorcinol-formaldehyde gels: Microphase separation or colloid aggregation. Physical Review E 77: 041409-041421
34. Sing K S W, Everett D H, Haul R A W et al (1985) Reporting physisorption data for gas/solid systems with special reference to the determination of surface area and porosity. Pure Appl. Chem 57: 603–619
35. Cohaut N, Thery A, Guet J M et al (2007) The porous network in carbon aerogels investigated by small angle neutron scattering. Carbon 45: 1185–1192
36. Kainourgiakis M E, Steriotis T A, Kikkinides E S et al (2005) Combination of small angle neutron scattering data and mesoscopic simulation techniques as a tool for the structural characterization and prediction of properties of bi-phasic media. Chemical Physics 317: 298–311
37. Eschricht N, Hoinkis E, Madler F et al (2005) Knowledge-based reconstruction of random porous media. Journal of Colloid and Interface Science 291: 201–213
38. Ehrburger-Dolle F, Fairen-Jimenez D, Berthon-Fabry S et al (2005) Nanoporous carbon materials: Comparison between information obtained by SAXS and WAXS and by gas adsorption. Carbon 43: 3009–3012
39. Fairen-Jimenez D, Carrasco-Marin F, Djurado D et al (2006) Surface area and microporosity of carbon aerogels from gas adsorption and small- and wide-angle x-ray scattering measurements. Journal of Physical Chemistry B 110: 8681–8688
40. Reim M, Reichenauer G, Körner W et al (2004) Silica-aerogel granulate – Structural, optical and thermal properties. Journal of Non-Crystalline Solids 350: 358–363
41. Berthon-Fabry S, Langohr D, Achard P et al (2004) Anisotropic high-surface-area carbon aerogels. Journal of Non-Crystalline Solids 350: 136–144
42. Gommes C, Blacher S, Goderis B et al (2004) In situ SAXS analysis of silica gel formation with an additive. Journal of Physical Chemistry B 108: 8983–991
43. Tamon H, Ishizaka E (1998) SAXS study on gelation process in preparation of resorcinol-formaldehyde aerogel. Journal of Colloid and Interface Science 206: 577–582
44. Gommes C J, Job N, Pirard J Pet al (2008) Critical opalescence points to thermodynamic instability: relevance to small-angle X-ray scattering of resorcinol-formaldehyde gel formation at low pH. Journal of Applied Crystallography 41: 663–668
45. Reichenauer G, Wiener M, Brandt A et al (2009) In-situ monitoring of the deformation of nanopores due to capillary forces upon vapor sorption. HMI Annual Report
46. Calas S, Levelut C, Woignier T et al (1998) Brillouin scattering study of sintered and compressed aerogels. Journal of Non-Crystalline Solids 225: 244–247

47. Caponi S, Fontana A, Mattarelli M et al (2004) Influence of thermal treatment in high and low frequency dynamics of silica porous systems. Journal of Non-Crystalline Solids 345-46: 61–65
48. Buchenau U, Monkenbusch M, Reichenauer G et al (1992) Inelastic Neutron-Scattering from Virgin and Densified Aerogels. Journal of Non-Crystalline Solids 145: 121–127
49. Conrad H, Buchenau U, Schatzler R et al (1990) Crossover in the Vibrational Density of States of Silica Aerogels Studied by High-Resolution Neutron Spectroscopy. Physical Review B 41: 2573–2576
50. Conrad H, Fricke J, Reichenauer G (1989) High-Resolution Neutron Spectroscopy of the Crossover in the Vibrational Density of States of Silica Aerogels. Journal De Physique 50: C4157-C4162
51. Vacher R, Courtens E, Coddens G et al (1990) Crossovers in the density of states of fractal silica aerogels. Physical Review Letters 65: 1008–1011
52. Courtens E, Lartigue C, Mezei F et al (1990) Measurement of the phonon-fracton crossover in the density of states of silica aerogels Condensed Matter – Zeitschrift für Physik B 79: 1–2
53. Reichenauer G (2008) Aerogels. In: Kirk-Othmer Encyclopedia of Chemical Technology
54. Reichenauer G, Scherer G W (2000) Nitrogen adsorption in compliant materials. Journal of Non-Crystalline Solids 277: 162–172
55. Reichenauer G, Scherer G W (2001) Nitrogen sorption in aerogels. Journal of Non-Crystalline Solids 285: 167–174
56. Reichenauer G, Scherer G W (2001) Effects upon nitrogen sorption analysis in aerogels. Journal of Colloid and Interface Science 236: 385–386
57. Reichenauer G, Stumpf C, Fricke J (1995) Characterization of SiO_2, RF and Carbon Aerogels by Dynamic Gas-Expansion. Journal of Non-Crystalline Solids 186: 334–341
58. Dubinin M M, Zaverina E D, Radushkevich L V (1947) Sorbtsiya I Struktura Aktivnykh Uglei .1. Issledovanie Adsorbtsii Organicheskikh Parov. Zhurnal Fizicheskoi Khimii 21: 1351–1362
59. Stoeckli H F, Rebstein P, Ballerini L (1990) On the Assessment of Microporosity in Active Carbons, a Comparison of Theoretical and Experimental-Data. Carbon 28: 907–909
60. Wood G O (1996) Estimating the micropore volume of activated carbonaceous adsorbents for organic chemical vapors. Proceedings of the European Carbon Conference "Carbon 1996", Newcastle (UK) 606–607
61. Noville F, Gommes C, Doneux C et al (2002) Is it possible to obtain a coherent image of the texture of a porous material? Characterization of Porous Solids VI 144: 419–426
62. Scherdel C, Reichenauer G, Wiener M (2010) Relationship between pore volumes and surface areas derived from the evaluation of N_2-sorption data by DR-, BET- and t-plot, Journal of Micro- and Mesoporous Materials 132: 572–575
63. Chmiel G, Lajtar L, Sokolowski S et al (1994) Adsorption in Energetically Heterogeneous Slit-Like Pores – Comparison of Density-Functional Theory and Computer-Simulations. Journal of the Chemical Society-Faraday Transactions 90: 1153–1156
64. Olivier J P, Conklin W B, Vonszombathely M (1994) Determination of Pore-Size Distribution from Density-Functional Theory – a Comparison of Nitrogen and Argon Results. Characterization of Porous Solids III 87: 81–89
65. Jagiello J, Olivier J P (2009) A Simple Two-Dimensional NLDFT Model of Gas Adsorption in Finite Carbon Pores. Application to Pore Structure Analysis. Journal of Physical Chemistry C 113: 19382–19385
66. Ravikovitch P I, Neimark A V (2001) Characterization of micro- and mesoporosity in SBA-15 materials from adsorption data by the NLDFT method. Journal of Physical Chemistry B 105: 6817–6823
67. Harkins W D, Jura G (1944) Surfaces of Solids. XII. An Absolute Method for the Determination of the Area of a Finely Divided Crystalline Solid. J. Am. Chem. Soc. 66: 1362–1366
68. Magee R W (1994) Evaluation of the external surface area of carbon black by nitrogen adsorption. 68: 590–600
69. Mikhail R S, Brunauer S, Bodor E E (1968) Characterization of micro- and mesoporosity in SBA-15 materials from adsorption data by the NLDFT method. Journal of Colloid and Interface Science 26: 45–53
70. Iler R K (1979) The chemistry of silica: Solubility, Polymerization, Colloid and Surface Properties, and Biochemistry
71. Barrett E P, Joyner L G, Halenda P P (1951) The Determination of Pore Volume and Area Distributions in Porous Substances. 1. Computations from Nitrogen Isotherms. J. Am. Chem. Soc. 73: 373–380
72. Reichenauer G, Scherer G W (2001) Extracting the pore size distribution of compliant materials from nitrogen adsorption. Colloids and Surfaces a-Physicochemical and Engineering Aspects 187: 41–50
73. Scherer G W, Smith D M, Stein D (1995) Deformation of Aerogels during Characterization. Journal of Non-Crystalline Solids 186: 309–315
74. Gommes C J, Noville F, Pirard J P (2007) Characterization of gels via solvent desorption measurements. Adsorption-Journal of the International Adsorption Society 13: 533–540

75. Reichenauer G, Pfrang T, Hofmann M (2007) Drying of meso- and macroporous gels – Length change and drying dynamics. Colloids and Surfaces A: Physicochem. Eng. Aspects 300: 211–215
76. Giesche H (2006) Mercury porosimetry: A general (practical) overview. Particle & Particle Systems Characterization 23: 9–19
77. Broecker F J, Heckmann W, Fischer F et al (1985) Structural Analysis of Granular Silica Aerogels. In: Fricke J (ed) International Symposium on Aerogels 6, Springer, Heidelberg
78. Brown S M, Lard E W (1974) Comparison of Nitrogen and Mercury Pore-Size Distributions of Silicas of Varying Pore Volume. Powder Technology 9: 187–190
79. Pirard R, Heinrichs B, Van Cantfort O et al (1998) Mercury porosimetry applied to low density xerogels; relation between structure and mechanical properties. J. Sol-Gel Sci. Tech. 13: 335–339
80. Pirard R, Pirard J P (2000) Mercury porosimetry applied to precipitated silica. Stud. Surf. Sci. Catal. 128: 603–611
81. Pirard R, Alie C, Pirard J P (2002) Characterization of porous texture of hyperporous materials by mercury porosimetry using densification equation. Powder Technology 128: 242–247
82. Job N, Pirard R, Pirard J P et al (2006) Non intrusive mercury porosimetry: Pyrolysis of resorcinol-formaldehyde xerogels. Particle & Particle Systems Characterization 23: 72–81
83. Scherer G W, Smith D M, Qiu X et al (1995) Compression of aerogels. Journal of Non-Crystalline Solids 186: 316–320
84. Vollet D R, Scalari J P, Donatti D A et al (2008) A thermoporometry and small-angle x-ray scattering study of wet silica sonogels as the pore volume fraction is varied. Journal of Physics-Condensed Matter 20: 1–7
85. Phalippou J, Ayral A, Woignier T et al (1991) Fractal Geometry of Silica Alcogels from Thermoporometry Experiments. Europhysics Letters 14: 249–254
86. Scherer G W (1993) Freezing Gels. Journal of Non-Crystalline Solids 155: 1–25
87. Beurroies I, Bourret D, Sempere R et al (1995) Gas-Permeability of Partially Densified Aerogels. Journal of Non-Crystalline Solids 186: 328–333
88. Hasmy A, Beurroies I, Bourret D et al (1995) Gas-Transport in Porous-Media – Simulations and Experiments on Partially Densified Aerogels. Europhysics Letters 29: 567–572
89. Calas S, Sempere R (1998) Textural properties of densified aerogels. Journal of Non-Crystalline Solids 225: 215–219
90. Kong F M, Lemay J D, Hulsey S S et al (1993) Gas-Permeability of Carbon Aerogels. Journal of Materials Research 8: 3100–3105
91. Reichenauer G, Fella H J, Fricke J (2002) Monitoring fast pressure changes in gas transport and sorption analysis. Characterization of Porous Solids VI 144: 443–449
92. Reichenauer G, Fricke J (1996) Gas Transport in Sol-Gel Derived Porous Carbon Aerogels. Fall Meeting of the Material Research Society, Boston (USA) 345–350
93. Gross J, Scherer G W (2003) Dynamic pressurization: novel method for measuring fluid permeability. Journal of Non-Crystalline Solids 325: 34–47
94. Scherer G W (1996) Bending of gel beams: Effect of deflection rate and Hertzian indentation. Journal of Non-Crystalline Solids 201: 1–25
95. Gross J, Scherer G W, Alviso C T et al (1997) Elastic properties of crosslinked Resorcinol-Formaldehyde gels and aerogels. Journal of Non-Crystalline Solids 211: 132–142
96. Behr W, Haase A, Reichenauer G et al (1998) Self and transport diffusion of fluids in SiO_2 alcogels studied by NMR pulsed gradient spin echo and NMR imaging. Journal of Non-Crystalline Solids 225: 91–95
97. Behr W (1999) NMR-Bildgebung an Silica Alkogelen bei Drücken bis zu 10 MPa. PhD Thesis, Physikalisches Institut der Universität Würzburg
98. Behr W, Haase A, Reichenauer G et al (1999) High-pressure autoclave for multipurpose nuclear magnetic resonance measurements up to 10 MPa. Review of Scientific Instruments 70: 2448–2453

22

Mechanical Characterization of Aerogels

Hongbing Lu, Huiyang Luo, and Nicholas Leventis

Abstract Aerogels are multifunctional porous nanostructured materials (e.g., thermally/acoustically insulating) derived from their vast internal empty space and their high specific surface area. Under certain conditions, aerogels may also have exceptional specific mechanical properties as well. The mechanical characteristics of aerogels are discussed in this chapter. First, we summarize work conducted on the mechanical characterization of traditional aerogels, and second, we describe the mechanical behavior of polymer crosslinked aerogels. In polymer crosslinked aerogels, a few nanometer thick conformal polymer coating is applied on secondary particles without clogging the pores, thus preserving the multifunctionality of the native framework while improving mechanical strength. The mechanical properties were characterized under both quasi-static loading conditions (dynamic mechanical analysis, compression, and flexural bending testing) as well as under high strain rate loading conditions using a split Hopkinson pressure bar. The effects of strain rate, mass density, loading–unloading, moisture concentration, and low temperature on the mechanical properties were evaluated. Digital image correlation was used to measure the surface strains through analysis of images acquired by ultrahigh-speed photography for calculation of properties including dynamic Poisson's ratio. Among remarkable results described herewith, crosslinked vanadia aerogels remain ductile even at $-180°C$, a property derived from interlocking and sintering-like fusion of skeletal nanoworms during compression.

22.1. Introduction

Kistler invented silica aerogel in 1931 [1, 2] and started the era of aerogels through introduction of supercritical drying of wet gels (Chap. 1). When wet gels are dried under ambient temperature and pressure, the resulting high-density materials are referred to as xerogels. During supercritical fluid (*SCF*) drying of hydrogels, gelation solvents are first replaced by a liquid such as CO_2, which is taken supercritical (i.e., above its critical point) and is vented off as a gas without ever inducing on the skeletal network the surface tension

H. Lu and H. Luo • Department of Mechanical Engineering, The University of Texas at Dallas, Richardson, TX 75080, USA
e-mail: hongbing.lu@utdallas.edu; huiyang.luo@utdallas.edu
N. Leventis • Department of Chemistry, Missouri University of Science and Technology, Rolla, MO 65409, USA
e-mail: leventis@mst.edu

M.A. Aegerter et al. (eds.), *Aerogels Handbook*, Advances in Sol-Gel Derived Materials and Technologies, DOI 10.1007/978-1-4419-7589-8_22,

forces associated with solid–liquid–gas interfaces (Chap. 3). The resulting aerogels are characterized by low density and high porosity and have been filled mostly with air. There are three major types of aerogels: inorganic, organic, and carbon aerogels [2–8]. Inorganic aerogels are formed from wet gels synthesized through a sol–gel process by hydrolysis and polycondensation of metal and semimetal alkoxides (Chap. 2). Inorganic aerogels include those based on silica, vanadia, and alumina (Chap. 8). In general, organic aerogels are synthesized from wet gels obtained by the polycondensation of resorcinol with formaldehyde (RF aerogels) in aqueous solutions (Chap. 11). Carbon aerogels are prepared by pyrolyzing organic aerogels in an inert atmosphere (Chap. 36). Incorporation of additives into aerogels results in composite aerogels with enhanced properties. Aerogels have been recognized for their tremendous potential in wide range of applications such as in thermal insulation, catalysis, acoustic insulation, dielectrics, ultrafiltration membranes, controlled drug delivery, oil spill cleanup kits, Cerenkov radiation detectors, and collectors of hypervelocity particles in space [2–14].

Although traditional aerogels (referred to as native aerogels in order to be distinguished from polymer crosslinked (*X-aerogels* in this chapter) can be used successfully in some controlled environments such as in space or in nuclear reactors such as Cerenkov radiation detectors [15, 16], the commercial aerogel applications are limited primarily due to poor mechanical properties such as fragility and hydrophilicity, which impose severe limitations on transportation, handling, machining, and durability [17–19]. Primarily due to the fragility and low mechanical properties, native aerogels have not received much attention for structural applications. Mechanical characterization of native aerogels has been conducted using ultrasonic pulse echo methods or quasi-static loading conditions under tension, compression, bending, torsion, or micro/nanoindentation. Polymer crosslinked aerogels, on the other hand, are mechanically strong, and they have potential for load-bearing applications (Chap. 13). They have been characterized under quasi-static and dynamic loading conditions at various strain rates. Much of the chapter is on the description of mechanical characterization results on this class of aerogels.

22.2. Mechanical Characterization Methods

To qualify an aerogel for load-bearing applications, extensive mechanical characterizations are needed to determine its mechanical behavior under actual service conditions. Such mechanical characterization experiments must be designed to produce loading conditions identical or at least similar to the loading conditions under service conditions. Service conditions for an aerogel can be diverse. As a result, several loading conditions need to be produced in order to determine the mechanical response of an aerogel. These typically involve tension, compression, torsion, bending, and multiaxial stress states under quasi-static, dynamic, and fatigue loading conditions. The following methods are those typically used for those mechanical characterization tests for aerogels.

22.2.1. DSC, DMA, and Nanoindentation

Differential scanning calorimetry (*DSC*) allows the measurement of the glass transition temperature. This is needed especially when aerogels contain viscoelastic components such as organic polymers. To determine the stiffness as a function of temperature, a dynamic mechanical analysis (*DMA*) apparatus applies oscillating loads at a fixed frequency such as 1 Hz or at varying frequencies typically between 0.1 and 20 Hz from low-temperature level such as $-100^{\circ}C$ to a high-temperature one such as $300^{\circ}C$ in a three-point bending test configuration. *DMA* results include the storage and loss moduli as a function of temperature. Peaks in the loss tangent as a function of temperature give the glass transition temperatures. Potential inhomogeneities are determined with a nanoindenter that measures the mechanical properties at many different locations (e.g., 100 points on a sample). The challenge in conducting nanoindentation is on the preparation of the specimen surface and the selection of correct indenter tips. For measuring the effective or average Young's modulus of an aerogel, one needs to use a spherical indenter with a tip radius significantly larger than the pore size of the aerogel [20]. Results from such tests determine the suitability of aerogels for aerospace, defense, and energy applications. *Mercury porosimetry* can also provide the bulk modulus of aerogels [21, 22].

22.2.2. Tension, Compression, and Loading–Unloading Tests

Dog-bone-shaped specimens (e.g., following *ASTM* D638 Standard Test Method for Tensile Properties of Plastics) can be used in tension, while cylindrical specimens (*ASTM* D695 Standard Test Method for Compressive Properties of Rigid Plastics) can be used for compression. In the preparation of specimen, surface has to be made smooth and free of defects. For cylindrical specimens used in compression, the end surfaces have to be parallel to one another. Compression needs to be conducted using compression platens with highly parallel surfaces. A slight misalignment can induce large error in data, especially for brittle aerogels. Typically, a self-aligned compression fixture is used. To determine the strength and the energy absorption capability indicated by the area enclosed by the stress–strain curve, the complete stress–strain relationship up to the point of final failure needs to be determined under extremely low strain rates. Constant deflection–compression set tests (Test-D in ASTM D3574 Standard Test Methods for Flexible Cellular Materials-Slab, Bonded, and Molded Urethane Foams) can be conducted to determine the sample dimensional stability, from the thickness changes of the cylindrical specimens after they are subjected to a set of compression strains at an elevated temperature for a period of time. Selected tests can also be conducted under loading/unloading/reloading conditions at several strain levels (a few percent to nearly the failure strain increments such as 10%) to determine the dissipated energy density in a loading cycle. Those tests can be conducted over a range of temperatures, representing the range of service temperatures.

22.2.3. Creep, Relaxation, and Recovery Tests

Compressive creep tests allow measurement of strain as a function of time when a constant stress is applied. These can be conducted at several stress levels for aerogel of various densities. Loads are removed at the end of the creep test, and strains as a function of time are monitored to determine the recovery behavior. Compressive relaxation tests can be conducted at different strain levels. The relaxation functions determined at the same strain level at different temperatures can be shifted horizontally to determine whether a master curve can be formed for use to determine the long-term behavior. Recovery behavior after relaxation can also be characterized by monitoring the stress as a function of time after removing partially the step strain. For aerogels that contain polymers such as *X-aerogels*

(Chaps. 13 and 15), the effect of physical aging on the mechanical behavior needs to be identified for applications with service temperature much lower than the glass transition temperature [23]. It is well known [24] that the effects of physical aging are significant for polymers (or their composites) used well below their glass transition temperatures, and therefore, they must be characterized. Creep and relaxation tests should be conducted with specimens having experienced different physical aging times. Environmental conditions (e.g., water, moisture, and ultraviolet light) also play critical roles on the properties due to the absorption of environmental media through nanoporous structure. The effects of moisture concentration should be determined in relaxation at one or two strain levels. The effects of UV light on the degradation of materials can be examined with specimens exposed to high intensity UV light for variable time periods. Effects of coupling of environmental effects can be evaluated under cyclic environmental conditions such as under periodic moisture exposure, UV exposure, and temperature histories to simulate service environmental conditions.

22.2.4. Testing at Moderate to High Strain Rates

Material properties such as yield stress, strength, and fracture toughness at high strain rates can be significantly different from those at lower strain rates and thus must be determined for use of aerogels effectively in dynamic loading situations (e.g., armor). For this purpose, a servo-hydraulic system, a drop tower tester, or a split Hopkinson pressure bar (*SHPB*) can be used to determine the stress–strain relationship under high strain rates [25, 26]. To reach a strain rate of ~200 to ~3,000 s^{-1} and a compressive strain of ~80%, an *SHPB* with a relatively long incident bar (~10 m) can be used to apply a slowly rising stress wave over a long duration time. An ultrahigh-speed camera with a frame rate in the neighborhood of 100,000 frames per second is needed to observe deformation and failure behavior when experiments are conducted at strain rates on the order of 10^3 s^{-1}.

22.2.5. Ultrasonic Echo Tests

In these tests, an ultrasonic pulse generated by an ultrasonic transducer is sent to an aerogel plate specimen and the time of flight of the pulse within the aerogel is measured. The ultrasonic transducer serves as both transmitter and receiver: it generates an ultrasonic pulse and also receives its echo. The ultrasonic pulse travels through the specimen and reflects back from the opposite face, it keeps bouncing back and forth in the specimen, and its amplitude is attenuated with time. A coupling material, such as glycerin, is used to give a good contact between the transducer and the specimen, allowing the wave to pass through. The speed of sound in the aerogel is given in terms of twice the specimen thickness divided by the observed round trip transit time of the pulse. In the case where the thickness of the aerogel specimen is less than the lateral dimensions, the longitudinal wave speed C_L and the shear wave speed C_s are given [27, 28] as

$$C_L = \sqrt{\frac{E(1-\nu)}{\rho_b(1+\nu)(1-2\nu)}} \quad and \quad C_s = \sqrt{\frac{E}{2\rho_b(1+\nu)}} = \sqrt{\frac{G}{\rho_b}} \tag{22.1}$$

where E is *Young's modulus*, G is the shear modulus, and ρ_b is the bulk density of the aerogel. Solving for E and ν yields

$$E = 2\rho_b C_s^2(1+\nu) \quad and \quad \nu = \frac{1-2(C_s/C_L)^2}{2-2(C_s/C_L)^2}. \tag{22.2}$$

Thus, from the known density ρ_b and the measured wave speeds C_L and C_s, the elastic constants E and ν can be determined. It is noted that the values are given at a high frequency associated with the ultrasound pulse used. The use of these properties at quasi-static loading conditions requires postexamination in order to evaluate whether the aerogel has frequency- or rate-dependent mechanical behavior.

22.2.6. Fracture and Fatigue Tests

Surface and interior defects such as microvoids and cracks often exist in aerogels. Those defects can significantly reduce the load-carrying capability. To evaluate the load-carrying capability of aerogel components with defects such as cracks, specimens with intentionally placed initial cracks can be prepared following *ASTM* Standards (*ASTM* E1820 Standard Test Method for Measurement of Fracture Toughness and *ASTM* E647 Standard Test Method for Measurement of Fatigue Crack Growth Rates). Fractographs are analyzed to determine the failure mechanism and provide guidance for a revised material design that will increase toughness. For aerogels subjected to periodic loading conditions, fatigue tests can be conducted under cyclic tensile, bending or shear stresses with several minimum to maximum stress ratios. Two kinds of specimens can be used: specimens with smooth surfaces can be tested under fatigue loading conditions to determine the fatigue strength as a function of number of cycles; notched specimens (specimens with initial cracks) can be tested again under cyclic loading in order to determine the crack growth rate as a function of number of cycles. The results allow to predict the crack growth as a function of time and to determine the service life using defect tolerant analysis.

22.3. Mechanical Characterization of Native Aerogels

The native aerogels are usually brittle, and therefore owing to the complexity involved during machining and gripping, it is difficult to prepare specimens and conduct testing under tension, torsion, or under multiaxial stresses. Consequently, flexural and compression tests are the only tests often conducted. Flexural tests conducted properly under three-point bending induce failure under tensile bending stress. The flexural strength of a native aerogel can be used to estimate the tensile strength of the material. Aerogels with porosities greater than 95% have been tested under the three-point bending mode and were found to have a maximum flexural strength close to 0.02 MPa at a mass density of 0.1 g cm^{-3}, giving a specific flexural strength of 0.2 N m g^{-1} [29].

Figure 22.1 shows the compressive stress–strain curves of a silica and a vanadia aerogel. Under compression, native aerogels experience up to three stages of deformation, namely, elastic deformation, compaction, and densification regimes. The collapse strengths (inset in Figure 22.1) are 2.5 MPa/0.2 MPa, and the specific energy absorption values are 32.9 J g^{-1}/6.44 J g^{-1} for the silica aerogel/vanadia aerogel, respectively. The mechanical

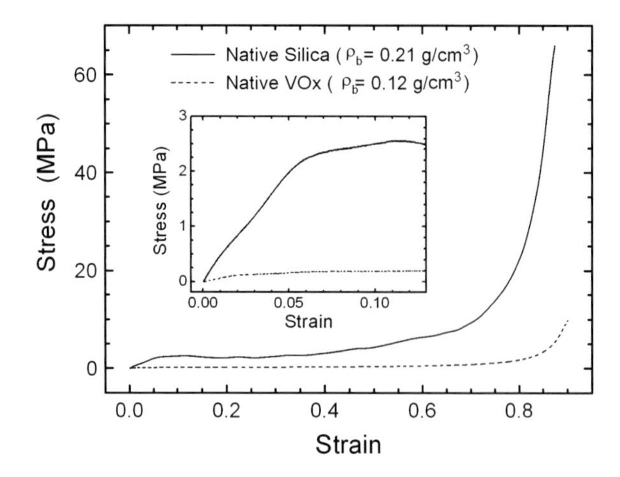

Figure 22.1. Mechanical compression testing of native silica and native vanadia aerogels.

properties (elastic moduli and the flexural strength) of silica alcogels and aerogels have been investigated as a function of the concentration of the silicon precursor, the catalyst, and the time allowed for aging of the wet gels [30]. The mechanical strength of silica over a wide range of porosities (0–95%, v/v) increases by four orders of magnitude [31], and the value depends on the structure and density. To investigate the effects of the water content on the Young's modulus and failure strain, the compressive mechanical behavior of aerogels has been also characterized under different levels of humidity [19]. Through systematical adjustment of the processing variables that affect density and nanoscale morphology, aerogels crosslinked using a small amount of silica were found to be flexible and give up to 40-fold increase in strength over the native structure [32]. Transparent organic–inorganic hybrid aerogels and xerogels show a reversible shrinkage recovery behavior under uniaxial compression [33]. The mechanical behavior of native silica modified with polydimethylsiloxane (*PDMS*) shows gradual transition from brittle to rubbery behavior as a function of the polymeric component. *SANS* (small-angle neutron scattering) measurements were carried out under "in situ" uniaxial compression, indicating rearrangement of the polymer during compression [34]. The bulk modulus shows a power law relationship with initial mass density, with an exponent of 3.6 in the linear elastic regime and 3.2 in the plastic regime [20, 21].

As it was stated above, in general, native aerogels are brittle and break at small tensile strains. Measurement of mechanical properties in the elastic regime imposes challenges. While compression is often used to measure the Young's modulus at small deformations, there are stringent requirements on the specimen preparation and the alignment of the testing apparatus. A slight misalignment induces bending superimposed on compression, leading to significant error, and also potentially localized tensile failure on one side of the specimen that oftentimes propagates rapidly to induce rupture of the entire specimen. To avoid those problems, the Young's modulus of aerogels can be determined effectively using ultrasound as described in Sect. 22.2.5. When an ultrasonic wave generated on one side of a specimen is sent through the material, it will be reflected by the opposite surface back. The time of flight of the wave can be used in connection with the specimen geometry (thickness) to determine the longitudinal wave speed in (22.1), which is related mathematically to the mechanical properties in (22.2).

Thus, the longitudinal wave speeds of silica aerogel with bulk densities of 0.07 and 0.18 g/cm^3 were measured as 120 and 230 m/s, giving *Young's modulus* values of 1 and 10 MPa, respectively [35]. Ultrasonic measurements and small-angle X-ray scattering (*SAXS*) reveal characteristic changes in the elasto-mechanical properties and structural features [34]. The speed and attenuation of ultrasonic waves of native silica aerogel have been further determined as a function of frequency, density, and gas pressure. Data show that the sound speed is independent of frequency, and it decreases with increasing static compression [35]. The elastic properties of silica, carbon, and organic aerogels have been examined by the ultrasonic pulse echo method, showing the scaling behavior of the elastic moduli versus density of the aerogels with the density scaling exponent in the range 2.0–4.0 [36, 37]. In alcogels, longitudinal wave speed remains around the velocity of sound in alcohol itself, whereas, in aerogels, it is significantly lower than the speed in air [38]. To calculate elastic constants (*Young's modulus* and Poisson's ratio), longitudinal and transverse acoustic wave velocities have been measured for silica xerogels as a function of relative humidity [39].

Recently, micro/nanoindentation techniques have also been used for measurement of the mechanical properties when only a small amount of material is available, including aerogels [40–44]. Nanoindentation requires a locally smooth surface, which could impose a challenge for aerogels due to the porous nature of the material. Nevertheless, using the smooth molded surface in a native aerogel, it is possible to obtain reproducible load–-displacement data and determine mechanical properties such as the Young's modulus and hardness using techniques established by Oliver and Pharr [45], without the need for direct imaging of the indent impression. In those cases, Berkovich, spherical, circular-flat punch, or other indenter tip geometries can all be used conveniently. Fracture surfaces of an aerogel, however, are often not locally smooth. In those cases, a relatively large spherical indenter tip with a radius of at least one to two orders of magnitude larger than the average pore size is necessary in order to obtain consistent load–displacement curves representing the local effective mechanical properties of the sample. Thus, the mechanical properties, including *Young's modulus* and hardness of native silica aerogels, have been determined successfully [40]. Nanoindentation has also been used for studying the effects of silicon alkoxide concentration, the preparation protocol, and the powder-activated carbon content on the mechanical properties of silica aerogel [41]. Conducting nanoindentation on silica aerogels using several indenter tips (spherical, pyramidal, and circular-flat tips) suggests that the deformation of the silica aerogel is controlled by elastic bending and fracture of nanoligaments with no signs of plasticity [42]. Nanoindentation on hybrid organic/inorganic silica aerogels has indicated that there is a transition in failure from brittle to ductile. Nanoindentation results have been compared with data obtained using conventional uniaxial compression tests, and a good agreement has been reached. Further, since nanoindentation works with very small amounts of material, it is also suitable for aerogel films with which conventional techniques such as compression are no longer applicable. Under a constant load, nanoindentation with Berkovich tip was conducted for creep testing of hybrid aerogels, confirming that creep is more pronounced with the increase of organic contents and that the creep behavior can be modeled using the Burger's method [43]. Microindentation has also been conducted on alumina aerogel, which shows excellent mechanical properties derived from the interconnected nanoleaflets having a quasi-two-dimensional structure [44].

For nanoindentation with increased spatial resolution for measurement of the nanolocal mechanical properties of aerogels, atomic force microscopy (*AFM*) has been used. Interpretation of *AFM* results on the local properties requires careful calibration of stiffness of each *AFM* tip and appropriate contact mechanics analysis of the tip/aerogel interaction. Once proper calibration has been conducted, *AFM* allows profiling surface topography and determination of load–displacement relationships. A contact model for nanoindentation has been established based on the Hertz model. Analysis of the nanoindentation force–displacement relationships using that model provides the elastic properties of the material. Thus, the Young's modulus of silica aerogels with densities in the 0.15–0.2 g/cm^3 range was determined from 2.7 to 8.6 MPa [46]. Vincent, Babu, and Seala used a scanning force microscope to determine the force–displacement relationship on a silica aerogel optical coating. The surface Young's modulus of the coating was measured between 2.4 and 13.4 GPa, depending on the nature and concentration of the catalyst used in synthesis [47]. In contrast to the brittle behavior under impact in silica aerogels, organic aerogels show ductile behavior and significantly higher impact strength. For example, cellulose aerogels give higher impact strength than silica aerogels under three-point bending [48]. Simulations to link the aerogel nanoscale structures with mechanical properties such as the Young's modulus are sparse. Ma and Roberts et al. modeled the elastic behavior of aerogels using beam elements in the finite element method [49]. Simulations yield a power law relationship between modulus and bulk density with exponents between 3 and 4, consistent with values determined experimentally [50].

Efforts have been made to increase the strength of native aerogels [32, 51–65]. By introducing silica fibers in silica aerogels, Parmenter and Milstein measured the mechanical properties under tension, compression, and shear and identified an increase of elastic modulus by 85% and strength by 26% when 10% (w/w) fiber was introduced [64]. Incorporating 5% (w/w) carbon nanofibers in native silica aerogel improves the compressive modulus by three times and the tensile strength by five times [59]. However, despite these efforts, native aerogels remain fragile materials. Recently, polymer crosslinked aerogels (*X-aerogels*) have been found to significantly improve the mechanical properties under quasi-static [51–62] and high impact loading conditions [63, 65] (Chaps. 13 and 15).

22.4. Mechanical Characterization of *X-Aerogels*

As with typical granular ceramics, the fragility issue of native silica aerogels is traced to the weak links of the inter-nanoparticle necks in the "pearl-necklace" network structure [29, 48, 66]. *SEM* images (Figure 22.2A) show native (noncrosslinked) amine-modified silica aerogel with density 0.19 g/cm^3 [56]. Aerogels could be made mechanically strong and tough through compounding with polymers, referred to as polymer crosslinked aerogels (*X-aerogel*) [51–58] (Chap. 13). Different polymers coat conformably the surface of the secondary particles without clogging the pores [51–59]. In Figure 22.2B, *SEM* images are shown for a Desmodur N3200 isocyanate crosslinked amine-modified silica aerogel with density of 0.48 g/cm^3. After nanoencapsulation, the polymer coating crosslinks (i.e., chemically connects) the nanoparticles and reinforces the structure without clogging the pores. With an increase of bulk density of three times (still low-density materials), *X-aerogels* gain strength (by short beam three-point bending) by 300 times [51].

Hence, crosslinking of nonsilica aerogels and the inorganic framework will affect more on material properties. Owing to their open mesoporous structure, aerogels have been considered as hosts for a variety of desirable functional guests [67]. Furthermore, oxide networks with morphologies different from silica could dissipate load forces differently, resulting in yet stronger materials after crosslinking. One approach is to alternate the whole

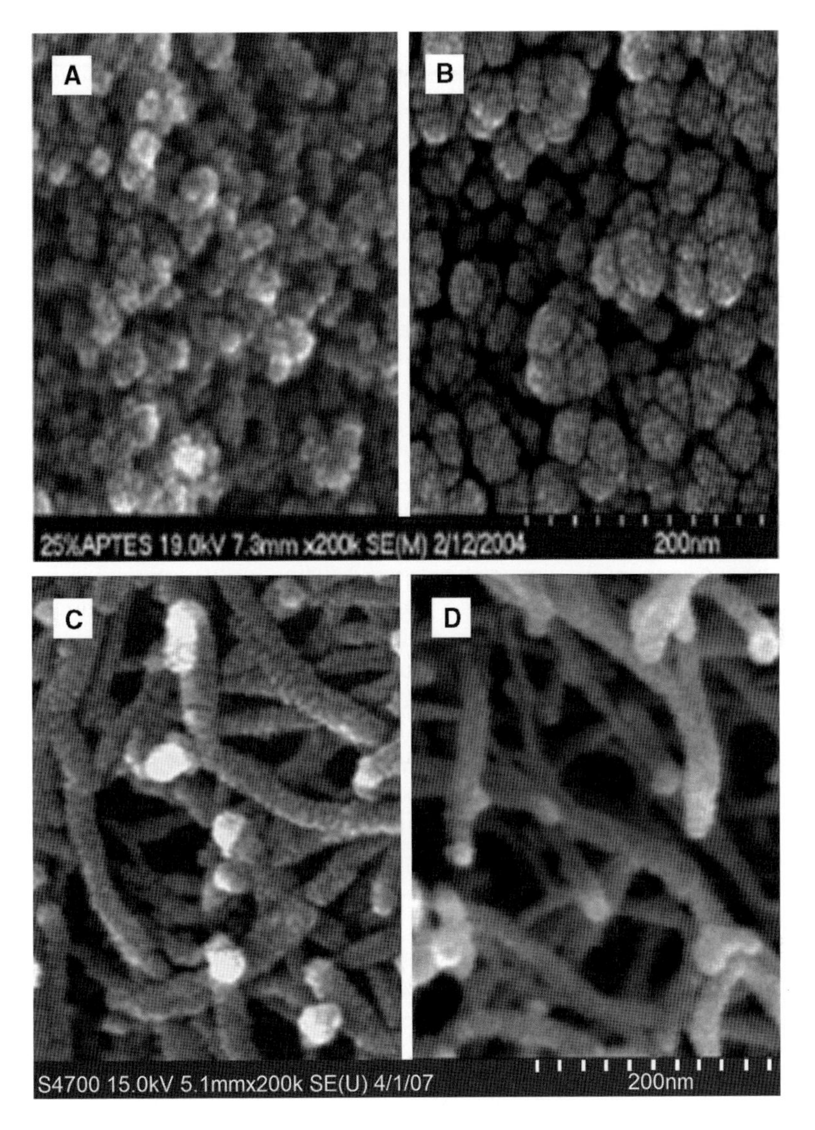

Figure 22.2. Scanning electron micrographs. **A.** A native (noncrosslinked) amine-modified silica aerogel ($\rho_b = 0.19$ g/cm^3); **B.** a Desmodur N3200 isocyanate crosslinked amine-modified silica aerogel ($\rho_b = 0.48$ g/cm^3); **C.** a native vanadia (VO_x) aerogel (bulk density $\rho_b = 0.08$ g/cm^3); **D.** a crosslinked vanadia (*X-VO$_x$*) aerogel ($\rho_b = 0.43$ g/cm^3).

network morphology of mesoporous silica. The bicontinuous meso/microporous template silica nanoencapsulating with a thin layer of polymer can improve the mechanical properties further [60, 61, 68]. However, this crosslinked templated mesoporous silica with robust monoliths and higher strength is not strictly in the definition of traditional aerogel range due to the higher density at 0.6–1.0 g/cm^3, with microscale porosity and very lower *BET* surface area. In other ways, the sol–gel derived vanadia consists of fibers [69, 70] or more accurately of entangled worm-like nanoobjects yielding a 3D structure with morphology akin to bird nests [71]. Mechanically, vanadia aerogels appear to be mechanically weak [69–71], as shown in Figure 22.1. The technique of using nanoencapsulation to provide reinforcement has been applied to more than 30 oxide nanoparticles [53]. Microscopically, most aerogels look very similar to one another, consisting of a random distribution of nanoparticles, similar to silica in Figure 22.2A and having close mechanical properties. Vanadia, however, is an exception. These aerogels consist of entangled worm-like structures [31]. Crosslinking vanadia aerogels with isocyanate increases the density, and as in the case of silica (see Figure 22.2), the polymer coating covers the skeletal framework conformably. Again, the mesoporous space is not compromised. Figure 22.2C shows *SEM* image of a native vanadia aerogel at a mass density 0.08 g/cm^3, and Figure 22.2D shows the *SEM* image of a polyurea crosslinked vanadia with a mass density 0.43 g/cm^3. The interlocking nano-worm structures in crosslinked vanadia aerogels give highly ductile behavior. The mechanical characterizations of crosslinked vanadia are also described in this chapter.

22.4.1. Dynamic Mechanical Analysis

In *DMA*, a sinusoidal load or displacement is applied and the resulting response is measured. Data are analyzed to determine the complex modulus or complex compliance. *DMA* can be conducted in tension, compression, bending, or torsion modes. For polymer crosslinked aerogels, we conducted *DMA* under three-point bending configuration using a *DMA* Model RSA II apparatus. The recorded force and mid-span displacement were converted into the stress and strain history using (22.3) for an Euler–Bernoulli beam, or a slender beam, and were used to calculate the complex modulus. For a viscoelastic material, the strain always lags behind the stress by a phase angle between 0 and $\pi/2$. The flexural fixture had two roller supports with 48 mm span. The temperature was swept at 4°C increments between −130 and 210°C and the flexural frequency was fixed at 1 Hz. Specimens were allowed to shrink freely along the length and the width directions on the fixture. Nominal stress and nominal strain were calculated based on dimensions in the undeformed state and were used in the analysis. Specimens were maintained for 60 s at each temperature prior to data collection.

Figure 22.3A, B depicts the storage modulus, loss modulus, and loss tangent as a function of temperature for polymer crosslinked silica aerogel (*X-SiO$_x$*) with density 0.67 g/cm^3 and polymer crosslinked vanadia aerogel (*X-VO$_x$*) with density 0.65 g/cm^3. At room temperature (23°C), the Young's modulus is 611/673 MPa for *X-SiO$_x$* and *X-VO$_x$*, respectively. As the temperature becomes lower the storage modulus increases. At −119°C, the

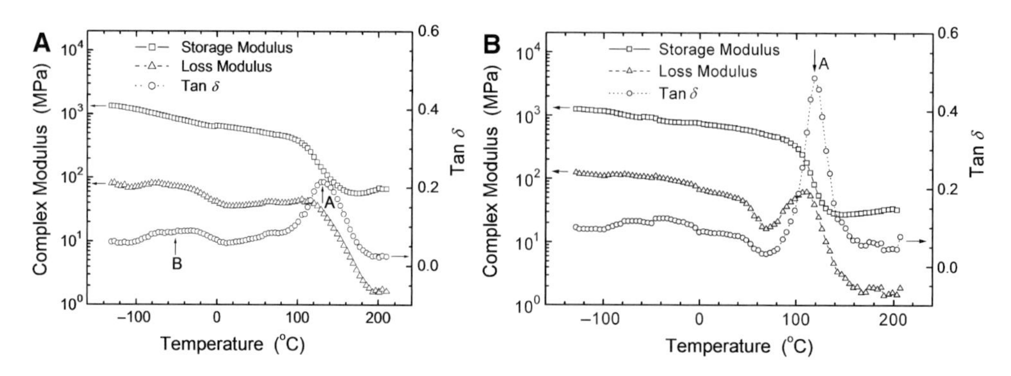

Figure 22.3. Dynamic mechanical analysis in a three-point bending mode at 1 Hz. **A.** *X-SiO$_x$* (ρ_b = 0.67 g/cm^3); **B.** *X-VO$_x$* (ρ_b = 0.65 g/cm^3).

storage modulus values are 1.35/1.24 GPa; at 100°C, the values become 328/315 MPa, for *X-SiO_x* and *X-VO_x*, respectively. Around 120°C, the storage modulus decreases at a steep rate, thereby indicating a primary (or α-) glass transition. After the rapid decrease, the storage modulus shows a slight increase again. At 207°C, the storage modulus values are reduced to 67/24 MPa for *X-SiO_x* and *X-VO_x*, respectively, i.e., less than 2% of the values at −119°C. Figure 22.3 shows the loss modulus and the loss tangent (tanδ), i.e., the tangent of the out-of-phase angle δ, calculated in terms of the ratio of the loss to storage modulus. Figure 22.3 shows that there is a major peak in loss tangent, tanδ, at 130°C for *X-SiO_x* and 118.4°C for *X-VO_x*, marked with "A". This peak corresponds to a high-to-low viscosity transition, indicating that the material undergoes softening around that temperature. This peak is consistent with the rapid reduction in the storage modulus, thus indicating that the major glass transition temperature (α-transition) at 130°C for *X-SiO_x* and 118.4°C for *X-VO_x*.

22.4.2. Flexural Modulus and Strength

Three-point bending on an *X-aerogel* beam specimen allows measurements of flexural modulus and strength. The *ASTM* standard D790 (Standard Test Methods for Flexural Properties of Unreinforced and Reinforced Plastics and Electrical Insulating Materials) specifies a length/width ratio of 16. The stress σ and strain ε are calculated based on the Euler–Bernoulli beam theory as

$$\sigma = \frac{3PL}{(2bh^2)} \quad \text{and} \quad \varepsilon = \frac{6h\Delta}{L^2} \tag{22.3}$$

where P is the applied load, Δ is the displacement at the center of the beam, L is the fulcrum span, b is the width of specimen, and h is height of the specimen. The constant ratio of stress and strain at small deformation is defined as the flexural modulus.

Such high ratio aerogel beam specimens are often difficult to prepare, primarily due to difficulty in handling slender wet gels before crosslinking and preparation of large aerogel plates (if slender beam is to be cut from a plate). To circumvent this problem, we used nonstandard rectangular specimens with a smaller length/width ratio, typically at 4–6. Beams with such ratios are characterized as Timoshenko beam or short beams in which the shear-induced deflection is no longer negligible and must be considered to obtain valid Young's modulus data. The use of the Castigliano's theorem [56] allows consideration of the contribution of the shear-induced deflection in addition to bending-induced deflection. The flexural modulus is then given based on the Timoshenko beam theory as

$$E = \frac{PL^3}{48\Delta I}\left[1 + \frac{12(1+v)}{5}\left(\frac{h}{L}\right)^2\right] \tag{22.4}$$

where I is the moment of inertia of the cross section. This equation is valid for both short and slender beams. In a slender beam, the second term in the brackets is negligibly small.

Three-point bending tests on *X-aerogel* beam specimens with nominal dimensions of $8.6 \times 8.6 \times 48$ mm^3 were conducted on an Instron material testing system at room temperature (21°C) and at −196°C. Figure 22.4 shows the stress–strain curves obtained through three-point bending tests of *X-SiO_x* and *X-VO_x* at room temperature and cryogenic

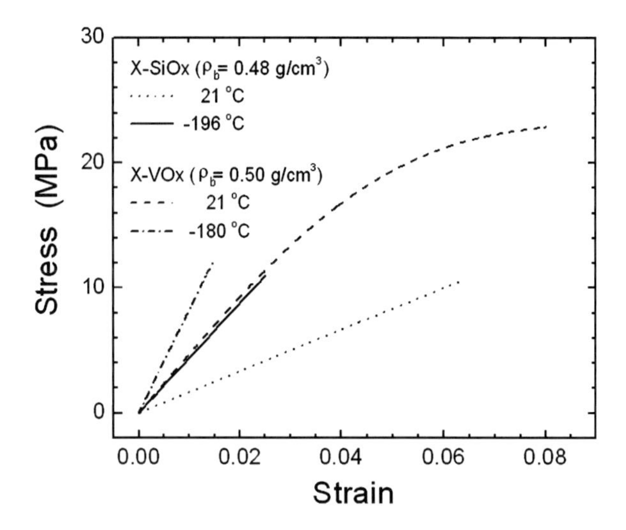

Figure 22.4. Three-point bending results on *X-SiO$_x$* and *X-VO$_x$* aerogels.

temperature using liquid nitrogen. The flexural strength obtained through three-point bending provides an estimate of the tensile strength for a material in the case where tensile tests cannot be conducted readily. The flexural modulus of *X-VO$_x$* with mass density of 0.50 g/cm^3 is calculated as 918/529 MPa at cryogenic/room temperatures, which is slightly higher than *X-SiO$_x$* at a density of 0.48 g/cm^3, giving 748/439 MPa at the cryogenic and room temperature, respectively. The stress–strain curves in Figure 22.4 are linear until 2% strain at room temperature and then become nonlinear from 2 to 8% strain until final rupture with flexural strength of 22.9 MPa. The flexural strength at cryogenic temperatures is 12.07 MPa at 1.5% strain, which is 14% higher than what has been observed with *X-SiO$_x$* with the similar density (0.5 g/cm^3 for *X-VO$_x$* and 0.48 g/cm^3 for *X-SiO$_x$*) [56, 63].

22.4.3. Compression at Low Strain Rates

Currently, standard testing methods for aerogels in compression do not exist. Polymer crosslinked aerogels are mechanically strong (Chap. 13); consequently, we refer to the *ASTM* D695-02a standard (Standard Test Method for Compressive Properties of Rigid Plastics). The standard specifies use of cylindrical specimens with a height to diameter ratio of 2 or 1. Compression testing was conducted on an MTS-810 servo-hydraulic material test machine (Eden Prairie, Minnesota, USA). The recorded force was divided by the initial cross-sectional area to determine the nominal compressive stress. The recorded displacement was corrected by considering the compliance of the test system obtained by compressing two platens directly; the corrected displacement was then divided by the initial length of the specimen to determine the nominal compressive strain. The compressive stress at ultimate failure, the strain at failure, and the *Young's modulus* were extracted from the stress–strain curves. The influence of strain rate on compression of specimens was quantified by comparing the stress–strain curves.

Compression tests were conducted at room temperature (23°C) under 35% relative humidity. The effect of polymer crosslinking was investigated by comparing results from native and *X-aerogels*. The strain recovery behavior on *X-aerogel* specimens was characterized under multiple loading/unloading during the process of compression. The effect of water content on compression of *X-aerogel* was also determined through comparison of results obtained using both dry specimens and water-soaked specimens. The *Young's modulus*, the 0.2% offset yield stress, the ultimate compressive strength, and the ultimate compressive failure strain were determined from the stress–strain curves.

Effect of Polymer Crosslinking

Figure 22.5 shows typical compressive stress–strain curves under quasi-static state for aerogels under unconfined conditions. The porosity and *BET* surface area of native silica, $X\text{-}SiO_x$, native VO_x, and $X\text{-}VO_x$ aerogels tested were 89%/969, 65%/261, 97%/216, and 67%/99.7 m^2/g, respectively. Under extremely slow loading rate, native aerogels can deform by large compressive strain, allowing the deformation to fully include three stages of deformations: elastic, compaction, and densification. Native aerogels, however, are extremely brittle, developing surface cracks that lead to fragmentation as strain rate increases. After crosslinking with isocyanate, the crosslinked aerogels tested, $X\text{-}SiO_x$ and $X\text{-}VO_x$, are ~3 times denser than native specimens. All compressive stress–strain curves show three stages of deformation: a linearly elastic region under small compressive strain (<4%); a yielding regime with slight hardening until 60% compressive strain; and a densification and plastic hardening range until 84% ultimate compressive strain. The inset of Figure 22.5 shows the zoomed-in initial stress–strain curves, indicating that both native aerogel and crosslinked aerogel specimens have similar Young's modulus and yield strength at low strain rates. The addition of small amounts of a polymeric nanocoating did not contribute much to the increase in stiffness and yield strength at small deformations, due to the large difference in modulus between polymer and silica. The addition of polymer, however, increases the stiffness and strength under large deformations as the polymer toughens and enforces the weak links in the aerogel framework (the interparticle necks), preventing disintegration of the material at nanoscales as strain increases, thus improving the mechanical properties significantly under large

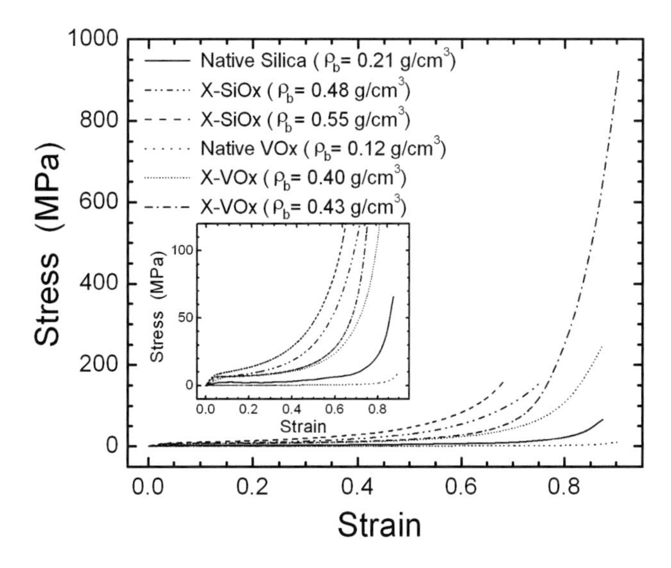

Figure 22.5. Compression testing results of native aerogel and *X-aerogels*.

deformations: crosslinked aerogels have much higher ultimate compressive strength than that of native aerogel. As it is evident in Figure 22.5 inset, native aerogels have lower Young's modulus, yield strength, and ultimate compressive strength than those of *X-aerogels* with higher densities. By comparison, data confirm that native VO_x aerogels are weak materials (Figure 22.1); their specific strength (defined as strength divided by the mass density) and specific energy absorption (calculated from the integrated area under the stress–strain curve divided by the mass density) are less than 10% of the corresponding values of $X\text{-}SiO_x$. We next compare the properties of $X\text{-}VO_x$ and $X\text{-}SiO_x$. At 68% compressive strain, the surface of all $X\text{-}SiO_x$ specimens developed cracks, and chips broke from the specimens leaving behind tire-rim-like cores (Figure 22.6A). In contrast, $X\text{-}VO_x$ specimens do not break into fragments separating from the specimens, and they can still carry high loads even at 91% compressive strain, remaining as single pieces despite the formation of several radial cracks and a half circular surface crack at the center (Figure 22.6B). Clearly, $X\text{-}VO_x$ is a tougher and stronger material than $X\text{-}SiO_x$. For direct comparison of the two materials, the $X\text{-}SiO_x$ results were scaled to the mass density of $X\text{-}VO_x$ (i.e., at $\rho_b = 0.44$ g/cm^3) by using the scaling law identified previously [56, 66], namely, mechanical properties such as the Young's modulus, the yield strength, and the stress–strain curve of $X\text{-}SiO_x$ are proportional to $(\rho_b)^{3.10}$. Although the Young's modulus values of the two materials are comparable, the 2% yield strength of $X\text{-}VO_x$ is twice that of $X\text{-}SiO_x$, the specific ultimate strength of $X\text{-}VO_x$ is 9 times higher, and the specific energy absorption at 91% strain (192 J/g) is 5 times higher than in $X\text{-}SiO_x$ (32 J/g).

A specimen with density 0.40 g/cm^3 expands at ~10% laterally in the elastic regime, but does not expand further until about 60% compressive strain, as the material is mostly

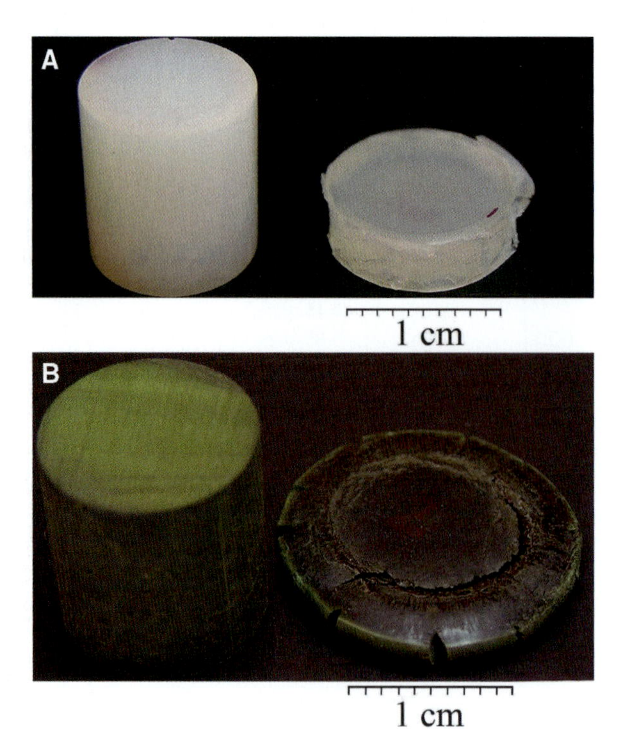

Figure 22.6. $X\text{-}SiO_x$ and $X\text{-}VO_x$ specimens before and after quasi-static compression at room temperature. **A.** $X\text{-}SiO_x$ specimens, **left**, before, and **right**, after testing; **B.** Corresponding $X\text{-}VO_x$ specimens.

absorbed within its own porosity. Figure 22.6B shows typical *X-VO*$_x$ specimens before and after compression testing. At failure, despite the splitting of the sample and the development of numerous cracks emanating in the radial direction, the outer layer of all crosslinked specimens did not shatter into fragments, but instead the sample remained in a single piece. The significant increase in ductility over the native vanadia aerogel can be tracked to the fact that the polymer coating reinforces the weak connection between the fibrous nanostructures.

Effect of Compression Conditions

Figure 22.7 shows the stress–strain curves with loading/unloading/reloading cycles in the nonlinear regime for *X-aerogels*. All unloadings exhibit nonlinear recovery. Prior to each unloading, there was a loading event with increase in stress and strain. At the beginning of each unloading, even though the stress starts decreasing, the strain would keep increasing due to a memory effect, resulting in a negative slope (stiffness) in the initial unloading stage of each cycle. This memory effect is due to the polymer coating, which exhibits viscoelastic behavior. Reloading does not follow the unloading curves, resulting in hysteresis loops seen in each unloading/reloading cycle. The unloading curve becomes steeper at higher compressive strain levels (e.g., between 68 and 76%), indicating a reduced elastic behavior for specimens having experienced higher level of densification by compaction. Figure 22.7A, B also shows a single stress–strain curve of *X-SiO*$_x$ and *X-VO*$_x$ respectively without unloading at the same compression conditions. By comparison, the curve with multiple loading/unloading is lower than the single monotonic loading curve. Again, this is most likely due to viscoelastic effects induced by the polymer nanocoating, which results in a memory effect so that as long as unloading is applied the subsequent curve will be affected and will be adjusted lower than a monotonic loading curve.

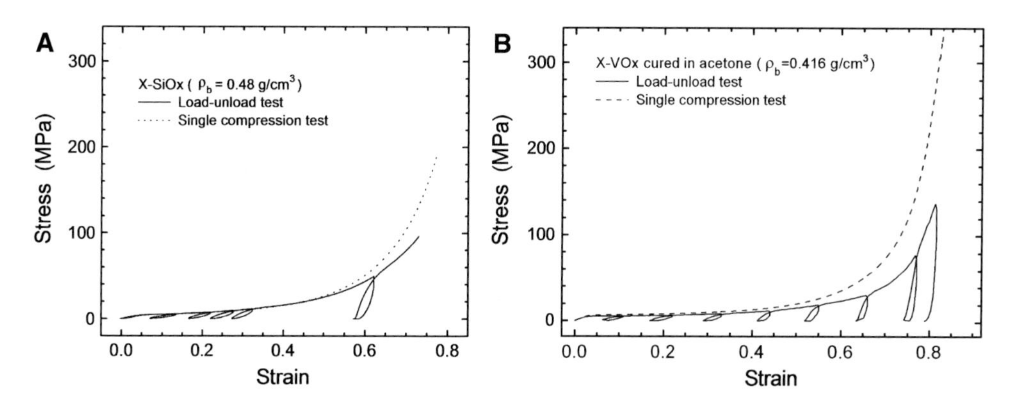

Figure 22.7. Load–unload testing of *X-aerogels* at quasi-static compression. **A.** *X-SiO*$_x$ with density 0.48 g/cm^3; **B.** *X-VO*$_x$ with density 0.416 g/cm^3.

The effect of low strain rate on the mechanical behavior was first evaluated under quasi-static loading conditions. Figure 22.8A shows the stress–strain curves of *X-SiO*$_x$ at strain rates 3.5×10^{-3}, 3.5×10^{-2}, and 0.35 s^{-1} for *X-SiO*$_x$. Figure 22.8B shows the stress–strain curves of *X-VO*$_x$ at strain rates 4×10^{-4}, 4×10^{-3}, and 4×10^{-2} s^{-1} for *X-VO*$_x$, and the inset shows the zoomed-in initial curves at the low strain range. Except the slight mass density effects for *X-VO*$_x$, as the low strain rate changes by two orders of magnitude,

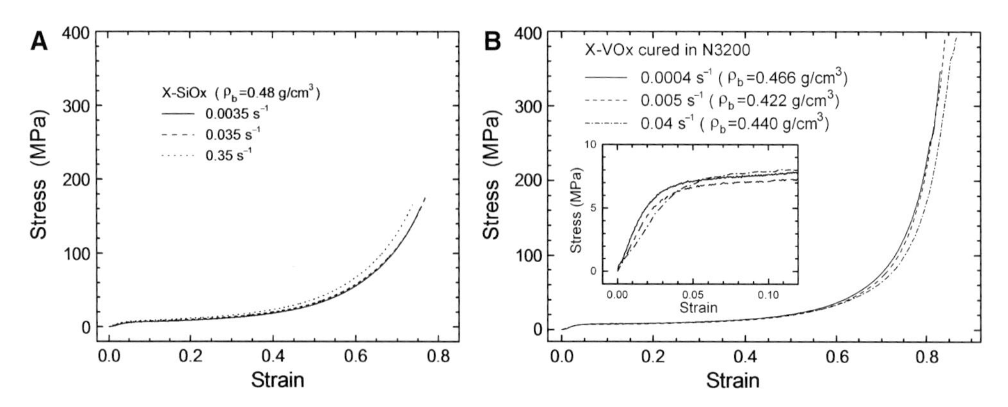

Figure 22.8. Effect of low strain rate on the behavior under compression of *X-aerogels*. **A.** *X-SiO*$_x$; **B.** *X-VO*$_x$. *Inset*: Expanded curves in low strain range.

the overall stress–strain curves do not exhibit significant changes, indicating that at these low strain rates, the mechanical behavior for *X-VO*$_x$ does not have a strong dependence on the low strain rate.

Effect of Water Content and Low Temperatures

The effect of water absorption was investigated by comparing compression testing results using both dry and water-soaked specimens. After immersing a *X-VO*$_x$ sample in water for 2 days (the minimum time for water to saturate), the sample reached fully saturated state and gained weight by 140% with an increase in volume of less than 0.5%, indicating that water enters mostly the pores without causing the worm-like nanofibers to expand or the structure to collapse. Figure 22.9 compares the stress–strain curves for both dry and water-soaked *X-VO*$_x$ specimens at a strain rate of $5 \times 10^{-3}\ s^{-1}$. The inset in Figure 22.9 shows the

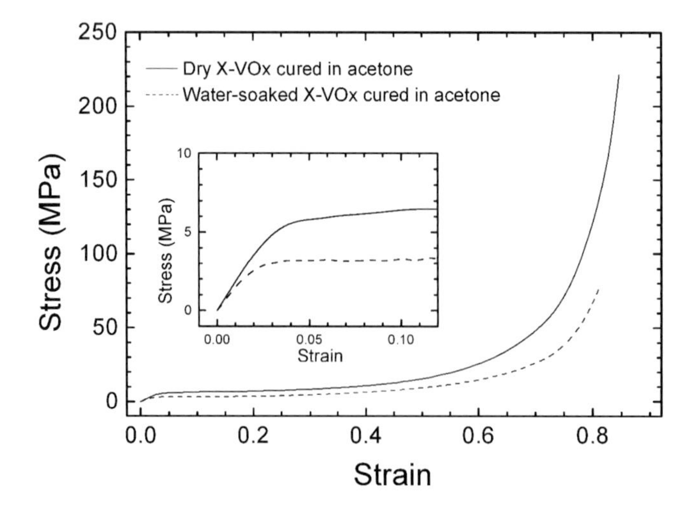

Figure 22.9. The quasi-static compression curves for both dry and water *X-VO*$_x$ cured in acetone (strain rate $0.005\ s^{-1}$). *Inset*: Expanded curves in low strain range.

obvious difference at the initial stress–strain curve at low strain range, indicating that the water-soaked sample is weaker than a dry sample. But remarkably, the water-soaked specimens were still stable, did not dissolve in water, and were still ductile, albeit not as much as the dry specimens. The water-soaked sample failed at an ultimate compressive strength of 78.4 MPa at an ultimate compressive strain of 83%, similar to the dry specimens.

Figure 22.10 shows compression results of *X-aerogels* at low temperatures at low strain rates. As the temperature decreases, the material becomes stiffer, and the stress–strain curve is shifted upwards. This indicates that *X-aerogel* can still absorb similar amounts of energy at lower temperatures (−5 and −50°C). Figure 22.10A shows the stress–strain curves for *X-SiO$_x$* specimens in compression testing at various low temperatures. The material stiffens significantly (the elastic modulus increases to 450 MPa) and suffers premature compressive failure at cryogenic temperatures (e.g., −196°C). While the temperature is decreased to the cryogenic temperature range (−180°C), the *X-VO$_x$* sample still remains ductile until 30% compressive strain and ultimately breaks in buckling failure (Figure 22.10B). The *X-VO$_x$* specimens have higher densities (0.57–0.60 g/cm^3), and under compression at cryogenic temperature (−180°C), they are weaker than *X-VO$_x$* specimens of lower density (0.44 g/cm^3) [62]. This indicates that the length/diameter ratio should be reduced in *SHPB* testing to prevent from buckling in testing at high strain rates.

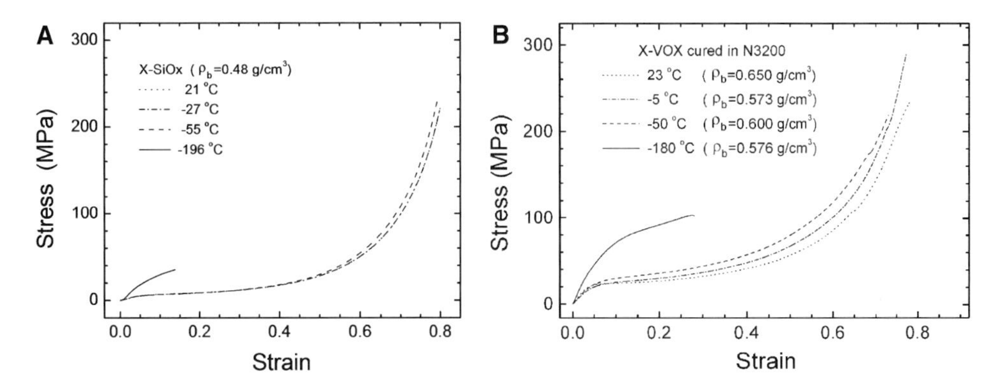

Figure 22.10. Compression results for *X-aerogels* at low temperatures. **A.** *X-SiO$_x$* at strain rate 0.001 s^{-1}; **B.** *X-VO$_x$* at strain rate 0.37 s^{-1}.

Morphology and Porosity Evolution

The morphological changes of *X-aerogel* under static compression were evaluated by *SEM* (Figures 22.11 and 22.12) and porosity/surface area analysis. For this, *X-SiO$_x$* specimens were loaded up to a predetermined strain and subsequently the load was removed and the specimens were analyzed. Curiously, under *SEM*, the material shows some difference from its original state (Figure 22.11A) up to at least 45% strain (only part of which is recovered by removing the load; see Figure 22.11B, C). Surface area analysis, however, shows unambiguously that as the degree of plastic deformation increases, the surface area decreases and the average pore diameter decreases, consistent with loss of mesoporosity as particles are squeezed closer to one another permanently. Ultimately, at failure (77% strain), *SEM* shows (Figure 22.11D) clear signs of collapse, i.e., a nearly total loss of porosity, also supported by surface area analysis. For *X-VO$_x$*, specimens with density 0.64 g/cm^3 were

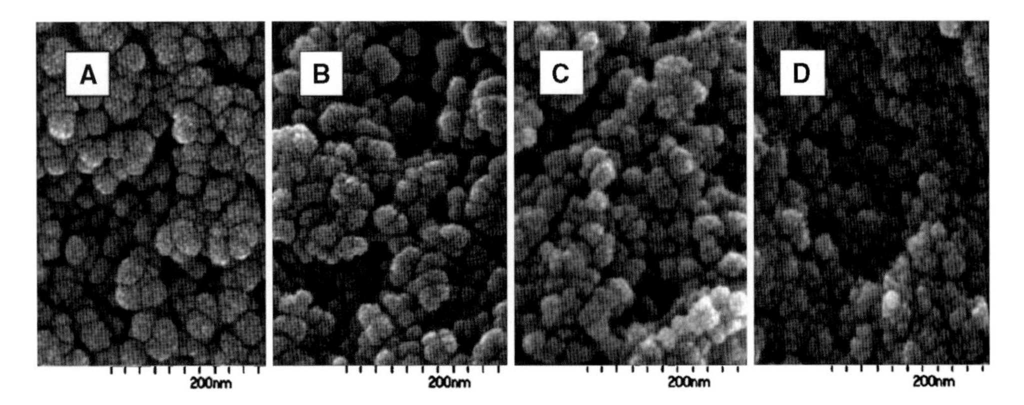

Figure 22.11. *SEM* images showing the morphological changes after compression in N3200 crosslinked APTES-modified silica aerogel cylinder with a diameter 0.5 in and a length of 1 in ($\rho_b \sim 0.48$ g/cm^3). **A.** Initial stage; **B.** at 30% compressive strain; **C.** at 45% strain; **D.** at 77% strain.

loaded at a strain rate of 0.37 s^{-1} up to 0, 25, 50, and 75% strain, and subsequently unloaded and the compressed specimens were analyzed. For intact samples without any deformation, the porous microstructure (Figure 22.12A) of *X-VO$_x$* with high density 0.64 g/cm^3 shows thicker polymer coating on the vanadia nanoworms than on the *X-VO$_x$* samples with lower density. Surface area analysis shows that the surface area and the average pore diameter decrease, as the plastic deformation increases, consistent with a nanoporosity loss as particles are crushed closer with nearly no internal voids remaining between the skeletal worm-like objects under 50% strain (Figure 22.12C). Also, the particle diameter increases gradually from 86.5 nm (Figure 22.12A), to 135.8 nm (Figure 22.12B), to 253.9 nm (Figure 22.12C). This behavior is different from that demonstrated by crosslinked silica aerogels [56], where particles come closer during compression, but their size remains unchanged. Ultimately, at near failure (75% strain), the particle diameter dramatically increases to 5,877.4 nm (Figure 22.12D). Some nanoworms are fused after compression, while the specimen after compression did not break into smaller pieces; rather it remained as one piece (Figure 22.6B). The worm-like objects appear fused together as if the coated polymer melted and solidified again, an interesting phenomenon not observed in *X-SiO$_x$* [56].

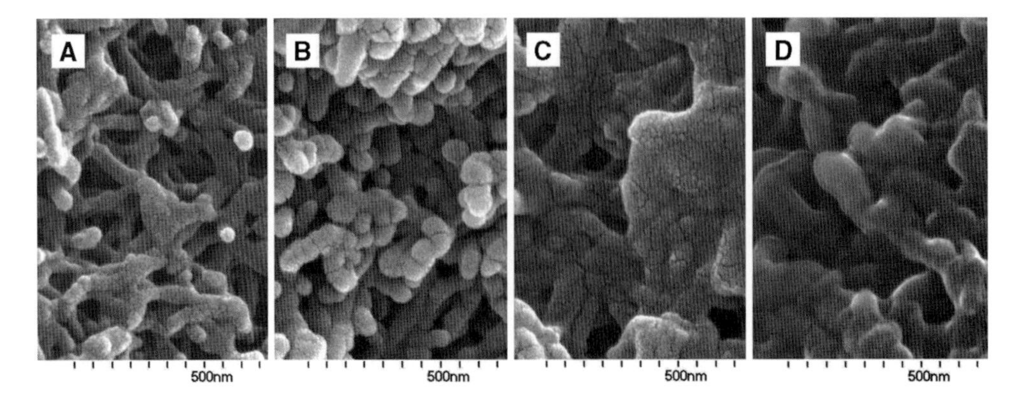

Figure 22.12. *SEM* images of the morphological changes *X-VO$_x$* cured in N3200 after compression ($\rho_b \sim 0.64$ g/cm^3, strain rate 0.365 s^{-1}). **A.** At initial stage (0% strain); **B.** at 25% compressive strain; **C.** at 50% strain; **D.** at 75% strain.

Overall, for quasi-static compression, *X-aerogels* have an average yield strength of 4.3/5.2 MPa, an ultimate compressive strength at 186/370 MPa (at low strain rate range), and an ultimate compressive failure strain as high as 77%/84%, for *X-SiO$_x$* with density 0.48 g/cm^3 and VO$_x$ with density of 0.42 g/cm^3, respectively. Under three-point bending at cryogenic temperatures, the flexural modulus and strength were determined as 918/12.1 MPa for *X-VO$_x$* with density 0.5 g/cm^3, respectively, which are higher values than those of the corresponding values 360/11 MPa obtained for *X-SiO$_x$*. For comparison with other materials, the specific energy absorption determined from the area enclosed by the stress–strain curve divided by the mass density is 192 J/g, indicating that *X-VO$_x$* is especially suitable for force protection such as for energy absorption in automobile collision and armor applications.

22.4.4. Dynamic Compression

X-aerogels have a high energy absorption capability as indicated by the data above, thus opening the way to potential applications under high impact, such as in automobile collision and ballistic protection, where energy absorption at high impact speed is important. Consequently, further evaluation at high strain rates using an *SHPB* would be necessary and the results are described in this section. We used a modified long split Hopkinson pressure bar to generate strain rate within 50–5,000 s^{-1} over a loading duration time of 1.5 ms. The schematic diagram of the *SHPB* is shown in Figure 22.13. The *SHPB* consists of a steel striker bar, incident and transmission bars, and a strain data acquisition system. A disk-shaped *X-aerogel* sample was sandwiched between the incident and transmission bars. In order to increase the weak transmitted signal, a hollow steel bar was employed and a hardened steel end cap was press fit into the tube to support the specimens. Under pulse shaping, the end cap effect can be ignored. The details in experimental setup are well documented in [64, 66]. The pulse shaper was used to help reach a dynamic stress equilibrium state as well as a constant strain rate [72], necessary for a valid *SHPB* experiment.

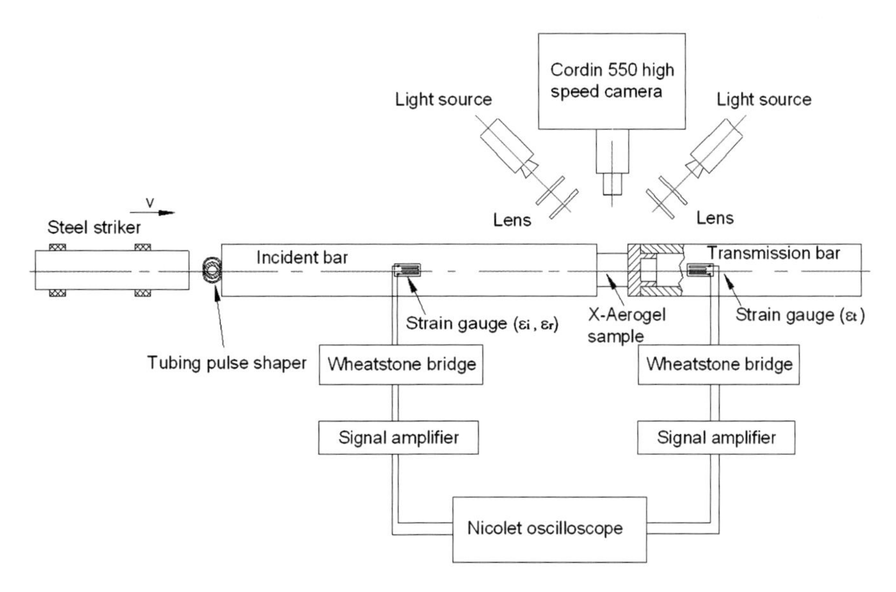

Figure 22.13. Schematic diagram of the *SHPB* setup for testing *X-aerogels*.

The working principle of *SHPB* has been well documented in the literature [26, 73]. Under a valid *SHPB* experiment, the stress applied in a specimen is calculated from the transmission bar signal through

$$\sigma_s(t) = \frac{A_t}{A_s} E_0 \varepsilon_t(t) \tag{22.5}$$

where σ_s is the axial stress applied on the specimen, A_t and A_s are the cross-sectional areas of transmission bar and specimen, respectively, E_0 is the *Young's modulus* of bars, and ε_t is the axial strain on the transmission bar.

When a hollow transmission bar is used, the strain rate on the specimen is determined [74, 75] from the incident bar signal using

$$\dot{\varepsilon}_s(t) = \frac{c_0}{L_s}\left[\left(1 - \frac{A_i}{A_t}\right)\varepsilon_i(t) - \left(1 + \frac{A_i}{A_t}\right)\varepsilon_r(t)\right] \tag{22.6}$$

where $\dot{\varepsilon}_s$ is strain rate of specimen; c_0 is bar wave speed, L_s is the length of a specimen, A_i is the cross-sectional area of the incident bar, and ε_i and ε_r are the incident strain and reflected strain in the incident bar, respectively. Integration of strain rate with respect to time gives the strain history.

In the case of high strain rates, *SHPB* experiments were conducted to determine the effects of strain rate, water content, moisture absorption, and low temperature. In each *SHPB* experiment, five or more specimens were tested to ensure repeatability under the same test conditions.

The incident and reflected waves obtained with *X-SiO$_x$* samples are shown in Figure 22.14A as a function of time. The positive portion indicates compressive waves and the negative portion denotes tensile waves. The transmitted wave is shown as the output signal. In an *SHPB* test, a dynamic equilibrium state, indicated by equal stresses applied on both ends of the specimen, must first be established. Such an experiment is then considered valid and the acquired experimental data are processed to deduce the dynamic stress–strain relationships. To examine the dynamic equilibrium condition, following either the first-wave (transmitted wave) or the second-wave (difference between incident wave and reflected wave) method [26, 66, 73], the front stress (at the end of the specimen in contact

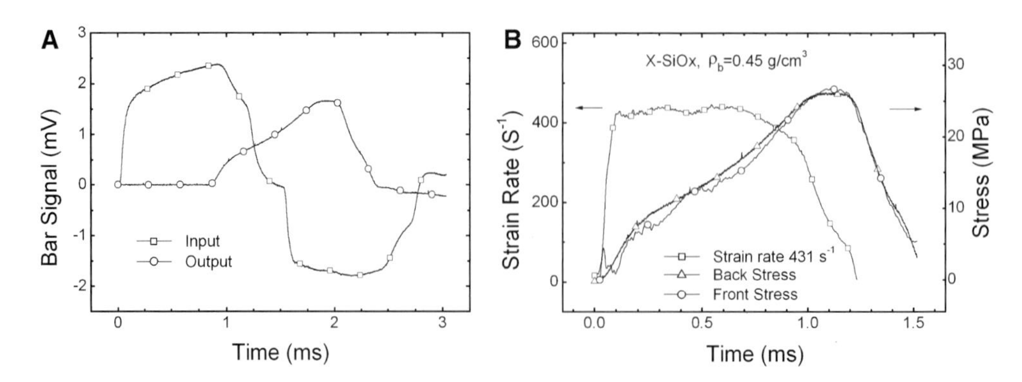

Figure 22.14. Typical *SHPB* results. **A.** Oscilloscope recordings; **B.** dynamic equilibrium state check and strain rate history.

with the incident bar) and the back stress (at the end of the specimen in contact with the transmission bar) are compared as a function of time. The results, together with the strain rate history, are given in Figure 22.14B. The front stress is very close to the back stress, indicating that the dynamic equilibrium condition was nearly established and the specimen was nearly uniformly deformed except during the initial 100 μs rising-up stage. With the use of pulse shaping, the strain rate is nearly constant during most of the loading time, indicating that the specimen deformed somewhat under a constant strain rate. Then the incident, transmitted and reflected signals were processed further to determine the stress–strain relation at high strain rates.

Effect of High Strain Rates

Compressive stress–strain curves for *X-aerogel* under unconfined conditions at several strain rates within the 50–2,500 s^{-1} range are shown in Figure 22.15. In general, at those high strain rates *X-aerogel* specimens are linearly elastic under small strains (<3%). Then they yield at ~3% compressive strain and exhibit uniform compaction until ~50% strain. Subsequently, specimens are densified and fail at ultimate compressive strains of 60–70%. *X-aerogel* specimens become stiffer as strain rate increases. Thus, the mechanical behavior of *X-aerogel* depends highly on the strain rate at high strain rates.

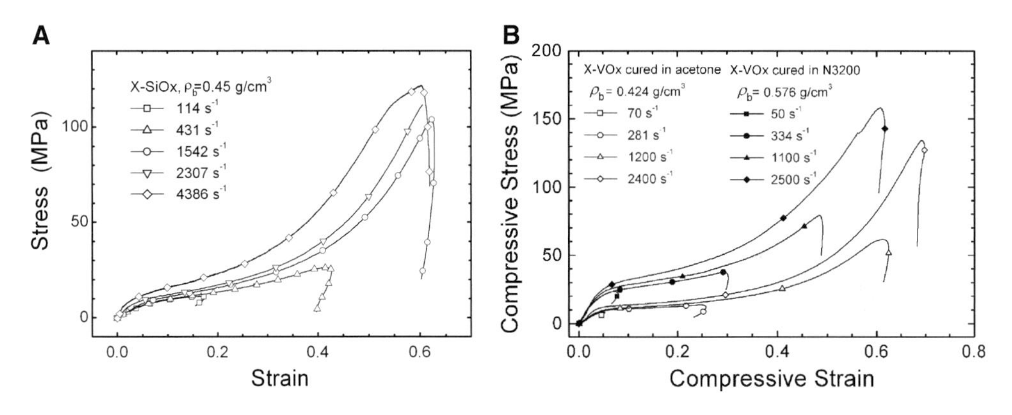

Figure 22.15. Dynamic stress–strain curves of *X-aerogels* with various high strain rates **A.** *X-SiO*$_x$ with density 0.45 g/cm^3; **B.** *X-VO*$_x$ with two densities: 0.42 and 0.58 g/cm^3.

Figure 22.16 shows the *SEM* images before and after *SHPB* testing of *X-aerogel* samples. For *X-SiO*$_x$, the original *X-SiO*$_x$ nanostructure exhibits the usual nanoporous pearl-like framework (Figure 22.16A). After compression by the *SHPB* impact under 43% compressive strain at a strain rate of 431 s^{-1}, most of nanopores disappeared and silica particles with the polymer coating were compacted together (Figure 22.16B). Surface area analysis of two specimens showed that the surface area of intact *X-SiO*$_x$ was 196 m^2/g versus 65 m^2/g of an impacted sample. Clearly, the deformed *X-SiO*$_x$ sample under dynamic loading conditions retained less than one half of the surface area retained under quasi-static loading conditions. For *X-VO*$_x$, the original nanostructure exhibits the usual nanoporous framework (Figure 22.16C). After compression on the *SHPB* impact at 60% plastic strain under a strain rate of 1,200 s^{-1}, most of nanopores disappeared and vanadia nanoworms together with their polymer coating were fused together to form aggregates of larger nanoworms

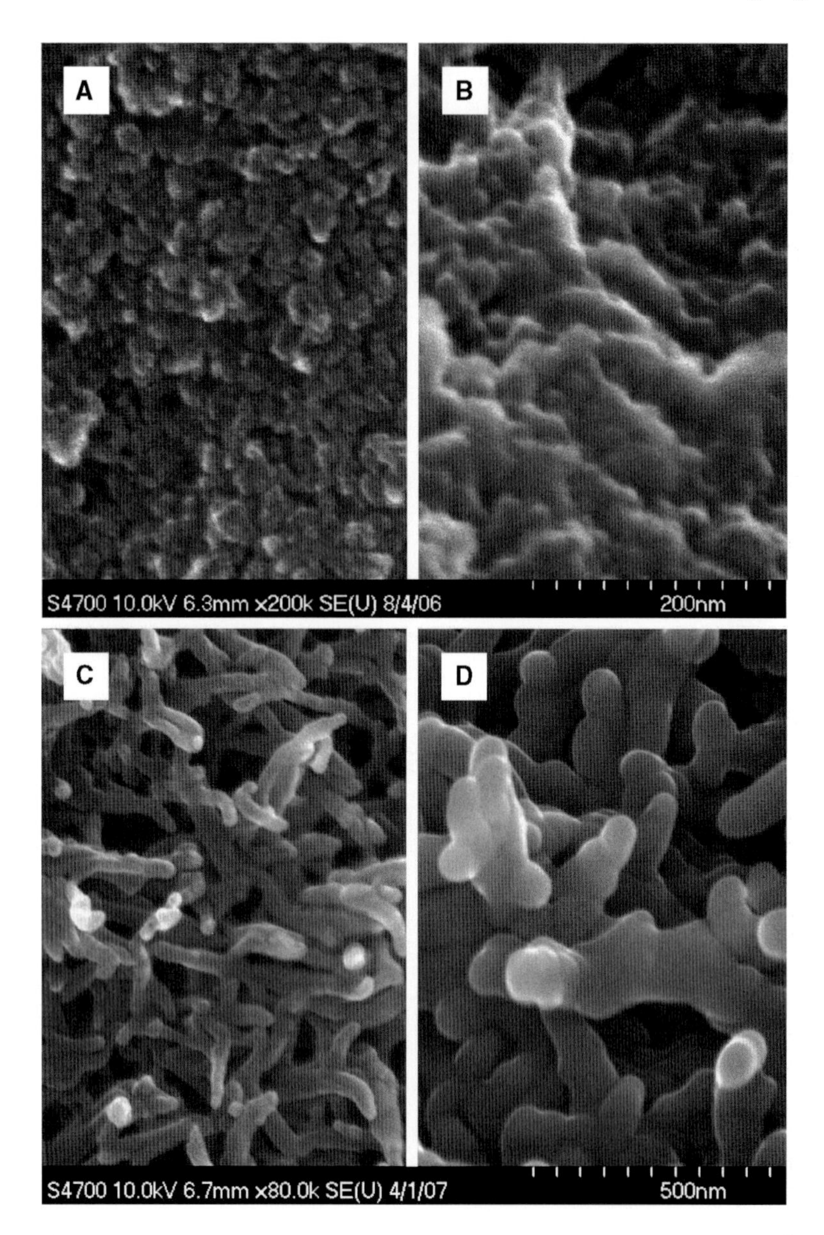

Figure 22.16. Scanning electron micrographs of *X-aerogels*. **A.** *X-SiOx* before impact (ρ_b = 0.45 g/cm^3); **B.** *X-SiO$_x$* after *SHPB* impact (ρ_b = 0.71 g/cm^3, strain rate 431 s^{-1}, maximum compressive strain 43% limited by the loading duration of 1.5 ms limited by the lengths of the *SHPB*.; **C.** *X-VO$_x$* before impact (ρ_b = 0.42 g/cm^3); **D.** *X-VO$_x$* after *SHPB* impact (ρ_b = 0.75 g/cm^3, strain rate 1,200 s^{-1}, and maximum strain 62% limited by *SHPB* bar lengths).

(Figure 22.16D), which are ~600 nm long and ~100–150 nm thick. It is perhaps the fusing of nanoworms that allows *X-VO$_x$* to be much more ductile than *X-SiO$_x$* and not to break into fragments even when radial splitting occurs. After compression at 60% strain, the *BET* surface area was reduced from 54 to 17 m^2/g, indicating a massive structural change. Indeed, the bulk density increased from 0.44 to 0.75 g/cm^3, while the skeletal density remained about unchanged (1.298 g/cm^3 before to 1.293 g/cm^3 after compression). The porosity was reduced from 68% before compression to 42% after compression. Clearly, compression causes collapse of the mesostructure, while the worm-like skeletal objects are nearly incompressible and therefore already compact with no internal voids. SEM after compression (Figure 22.16D) confirms that porosity has practically disappeared while skeletal nanoworms seem fused together even though the bird-nest nanomorphology is still clearly visible.

Both the Young's modulus data and the 0.2% offset compressive yield strengths are plotted in Figure 22.17, as a function of the strain rate. The yield strength increases with the strain rate. The relationship between yield strength and strain rate is fitted into the model given by Chen and Zhou [76],

$$\frac{\sigma_{0.2}}{\sigma_0} = 1 + \frac{2}{\pi} C_1 \tanh\left[\ln\left(\frac{\dot{\varepsilon}}{\dot{\varepsilon}_0}\right)^{m_1}\right] \tag{22.7}$$

where $\sigma_{0.2}$ is the 0.2% offset compressive yield strength, σ_0 is a reference compressive strength, $\dot{\varepsilon}_0$ is the reference strain rate, and C_1 and m_1 are constants. As seen in Figure 22.17,

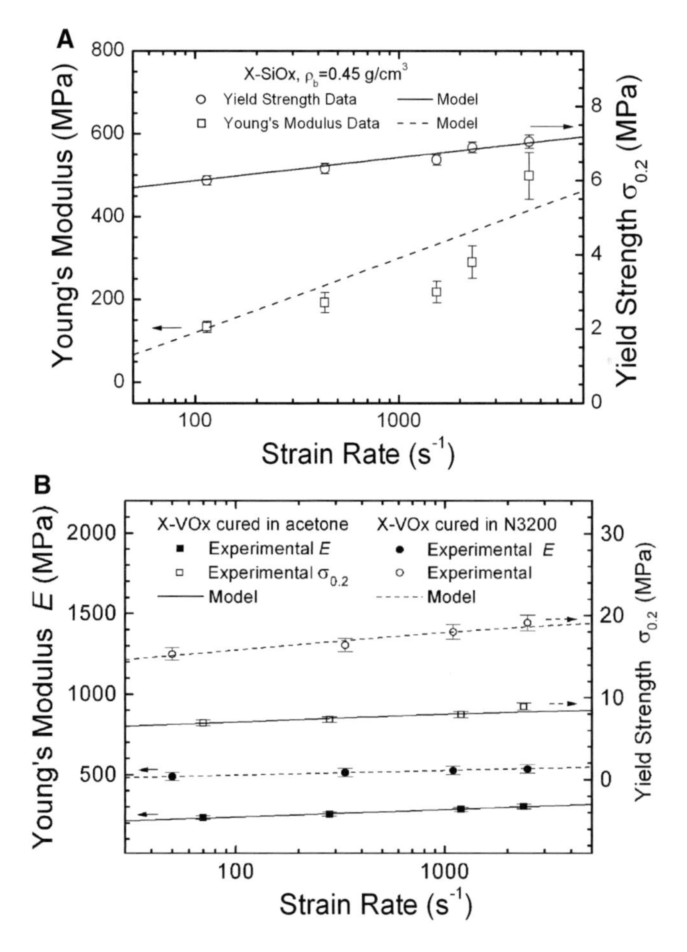

Figure 22.17. *Young's modulus E* and yield strength $\sigma_{0.2}$ as a function of strain rates for *X-aerogels* specimens. **A.** *X-SiO$_x$* with density 0.45 g/cm^3; **B.** *X-VO$_x$* with two densities 0.42 and 0.58 g/cm^3.

the model can describe the yield data reasonably well. The *Young's modulus* also increases with the strain rate. Following the relationship established for other materials [77–79], the Young's modulus is given as a function of the strain rate by

$$\frac{E}{E_0} = C_2\left(1 + \ln\left(\frac{\dot{\varepsilon}}{\dot{\varepsilon}_0}\right)^{m_2}\right) \tag{22.8}$$

where E_0 is a constant, taken as 1 GPa; C_2 and m_2 are constants. From the experimental data, the best-fit constants C_1, C_2, m_1, and m_2 are shown in Table 22.1. This model can also describe the experimental data reasonably well, as seen in Figure 22.17.

Table 22.1. Model parameters used for *Young's modulus* and yield strength

Materials	ρ_b (g/cm^3)	n_1	n_2	m_l	m_2
X-VOx [64]	0.39–0.66	[a]1.87	[a]2.30	0.45	0.083
		[b]1.83	[b]2.03	0.45	0.024
X-SiOx [66]	0.45–0.73	3.10	3.10	0.10	0.40
Silica aerogel [18]	0.08–0.4	3.7 ± 0.2	2.6 ± 0.2	–	–
Silica aerogel [35–37]	0.005–2.7	2.59–3.53	–	–	–
Carbon aerogel [35]	0.06–3.5	2.69 ± 0.05	–	–	–
RF aerogel [35]	0.15–1.5	2.95 ± 0.10	–	–	–
MF aerogel [35]	0.04–1.1	3.13 ± 0.07	–	–	–
Silica aerogel [20, 21, 35]	0.1–2.5	3.6	2.6	–	–

Note: "–" indicates that data were not available.
[a]The exponent at low strain rates from testing on an MTS material test system.
[b]The exponent at high strain rates from testing on *SHPB*.

High-Speed Photography and Digital Image Correlation

A Cordin 550-62 high-speed digital camera (62 color frames at up to 4 million frames per second; 10-bit resolution CCD with 1,000 × 1,000 pixels) was used to acquire images of a specimen surface at high-speed deformation during *SHPB* tests, in order to observe the failure behavior at high strain rates. For this, speckles were firmly imprinted on the outer geometric surfaces of *X-aerogel* specimens, using black silicone rubber dots. A ruler with a grid of 1 × 1 mm was fixed on the base of the *SHPB*, also in order to be recorded by the camera as spatial reference. Two Cordin 605 high intensity Xenon light sources with two sets of lenses were used to illuminate the specimen surface. Surface deformations were determined using the digital image correlation technique. Among the total of 62 color frames, four typical frames of native silica aerogel at a rate of 50,834 frames per second, *X-SiO$_x$* at a rate of 41,317 frames per second, and *X-VO$_x$* at a rate of 71,829 frames per second are shown in Figures. 22.18, 22.19, and 22.20, respectively.

For native aerogel, four images acquired by the ultrahigh-speed digital camera are shown in Figure 22.18 for comparison with results on *X-aerogels*. At time 441.5 μs counted after the incident wave had reached the native aerogel sample, the sample had undergone elastic deformation and remained intact (Figure 22.18B). After that it started to form an axial split crack, which then propagated rapidly through the thickness of the sample. The sample

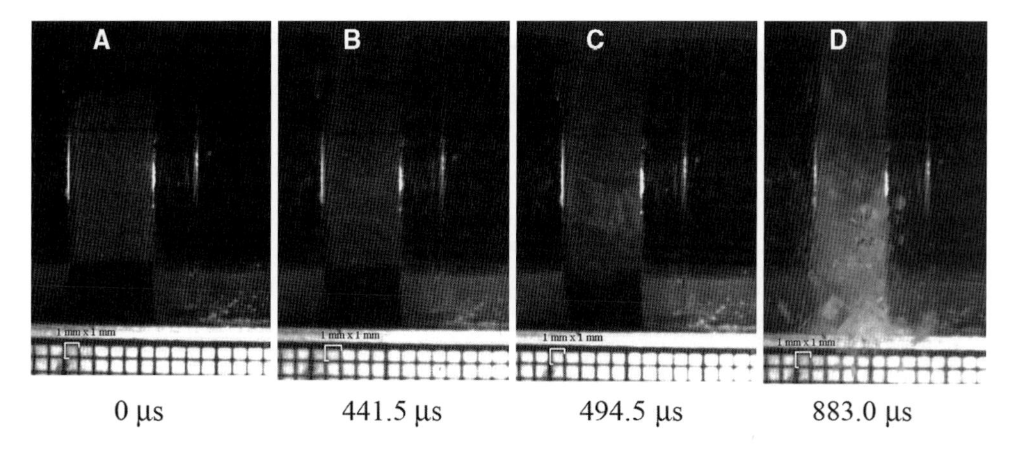

Figure 22.18. Typical images of high-speed photography for native silica aerogel with high density.

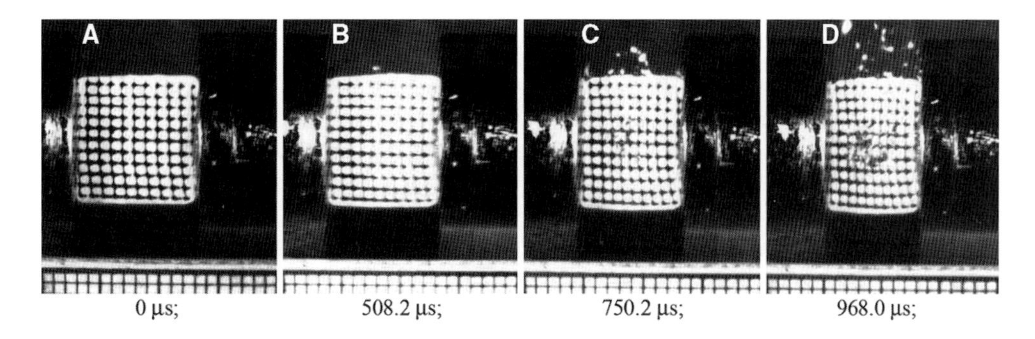

Figure 22.19. Typical images of high-speed photography of *X-*SiO_x (0.45 g/cm^3) on *SHPB* testing.

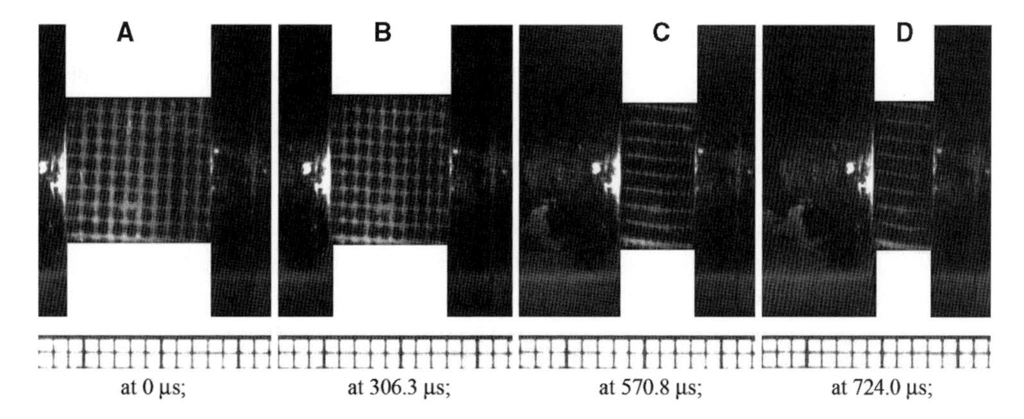

Figure 22.20. Typical images of high-speed photography of *X-*VO_x (0.42 g/cm^3).

then fractured into a large fragment and split at 494.5 μs (Figure 22.18C). Under further impact, the remaining part gradually broke into countless small fragments flying out of the area of impact (Figure 22.18D). After the completion of the impact experiment, there was no large fragment remaining. Overall, the native silica aerogel of density 0.47 g/cm^3 experienced typical brittle failure.

For $X\text{-}SiO_x$, four typical images are shown in Figure 22.19. Figure 22.19A shows the image of a specimen prior to loading at time 0 μs. At time 508.2 μs, a tiny piece broke away from a side of the specimen (Figure 22.19B). With increasing deformation, the specimen periphery gradually split out to form a number of small particles, and the speckles at the center of imaging surface continued to distort and delaminate until unloading (Figure 22.19C, D). To some extent, the dynamic failure is somewhat similar to the failure behavior under quasi-static compression, in which the outer layer $X\text{-}SiO_x$ was shattered into many small pieces, and in the final stage, a sample broke into many pieces with only a core left in the center [56].

For $X\text{-}VO_x$, Figure 22.20A shows the image of the specimen prior to loading at time 0 μs. After 724.0 μs, as shown in Figure 22.20D, the specimen has been compressed by about 55% strain, but it did not show any formation of surface cracks, indicating an extremely ductile behavior under high-speed impact.

Digital Image Correlation (*DIC*), a noncontact full-field deformation technique widely used to measure surface deformations [79–83], was employed to determine the surface strain distribution on specimens. In *DIC*, two images, the reference image and the deformed image, are compared to determine the surface deformations through tracking distinct gray scale patterns. In this work, the *DIC* code developed by Lu and Cary [83], capable of measuring both first-order and second-order displacement gradients, was used to determine the surface deformations.

For $X\text{-}SiO_x$, Figure 22.21A, B shows the displacement contour in the horizontal and vertical (or transverse) directions as determined from *DIC* on deformed image shown in Figure 22.19B, respectively. Contours of both axial displacements are not straight lines, indicating inhomogeneous deformations. Figure 22.21C, D shows the strain contours in the axial (horizontal) and transverse directions, respectively. The nonlinear Poisson's ratio is determined to be 0.16 within regime of nonlinear deformation, very close to the value 0.18 determined under quasi-static compression [56]. However, it is noted that at some locations, strains are far removed from the average values, but they are localized in small areas so that the global stress–strain relation might not be affected much by these localized strains.

Development of surface deformations was determined using Figure 22.20A as the reference image. Under very high compressive strains, as in the case of Figure 22.20D, speckles are too close to be discerned so that deformations can no longer be determined from *DIC*. Consequently, deformed images after Figure 22.20D do not render any useful data points for the *DIC* method. The Poisson's ratio was determined from these results, and its average value is 0.15 at 34% compressive strain. Under such a high compressive strain, most solids have a much higher Poisson's ratio, close to 0.5. The low Poisson's ratio in the $X\text{-}VO_x$ indicates that the material is in general absorbed by its own porosity so that lateral deformation is minimal.

Figure 22.22A, B shows the contours of displacements in the horizontal and transverse directions, respectively, as determined from *DIC* on image "D" of Figure 22.20. Figure 22.22C, D shows the contour of strain field in the horizontal and transverse direction, respectively. These results show that surface deformations on specimens are relatively uniform in the central region, indicating that compaction occurs relatively uniformly throughout the specimen.

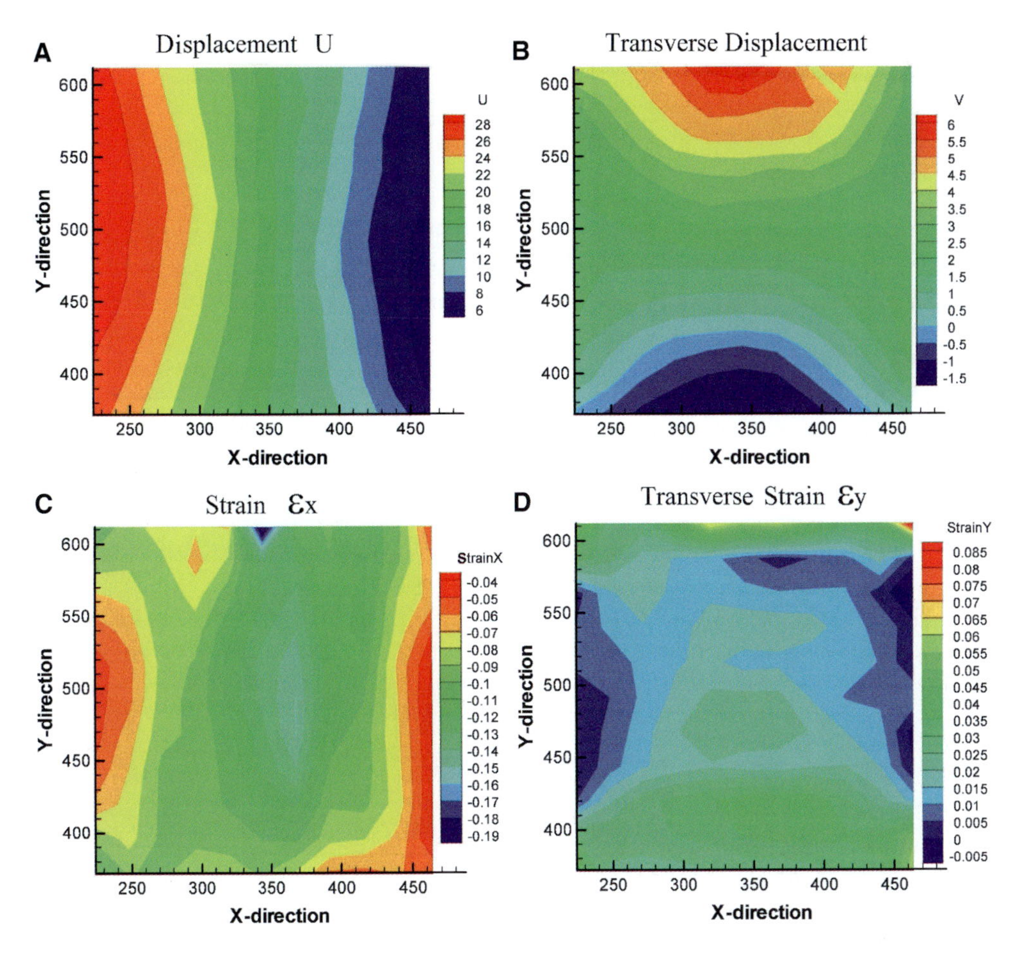

Figure 22.21. Distribution of deformation field of $X\text{-}SiO_x$ from *DIC*. **A.** Axial displacement field; **B.** axial strain field; **C.** transverse displacement field; **D.** transverse strain field. All units for lengths and displacements are in pixels. Each pixel represents a side length of 40.8 μm on the specimen.

Effect of Mass Density

The mechanical behavior of *X-aerogel* depends on their mass density. The effect of density on the stress–strain relationship was investigated at strain rates in the range 300–400 s^{-1}. $X\text{-}VO_x$ specimens with different densities were compressed on the *SHPB* to determine the effect of density on the stress–strain relationships. *X-aerogel* specimens with higher densities were prepared by applying a thicker conformal polymer coating on the skeletal framework.

Since the increase in $X\text{-}SiO_x$ density was achieved through coating with more polymer on the surface of the silica skeletal framework, if a native aerogel sample has the same density as a crosslinked aerogel sample, it would have lower ultimate compressive strength due to brittleness. Figure 22.23A also shows the stress–strain curve of native silica aerogel with density 0.47 g/cm^3. The native silica aerogel has higher Young's modulus (227 MPa) and yield strength (9.6 MPa) than the corresponding values of $X\text{-}SiO_x$ of approximately the

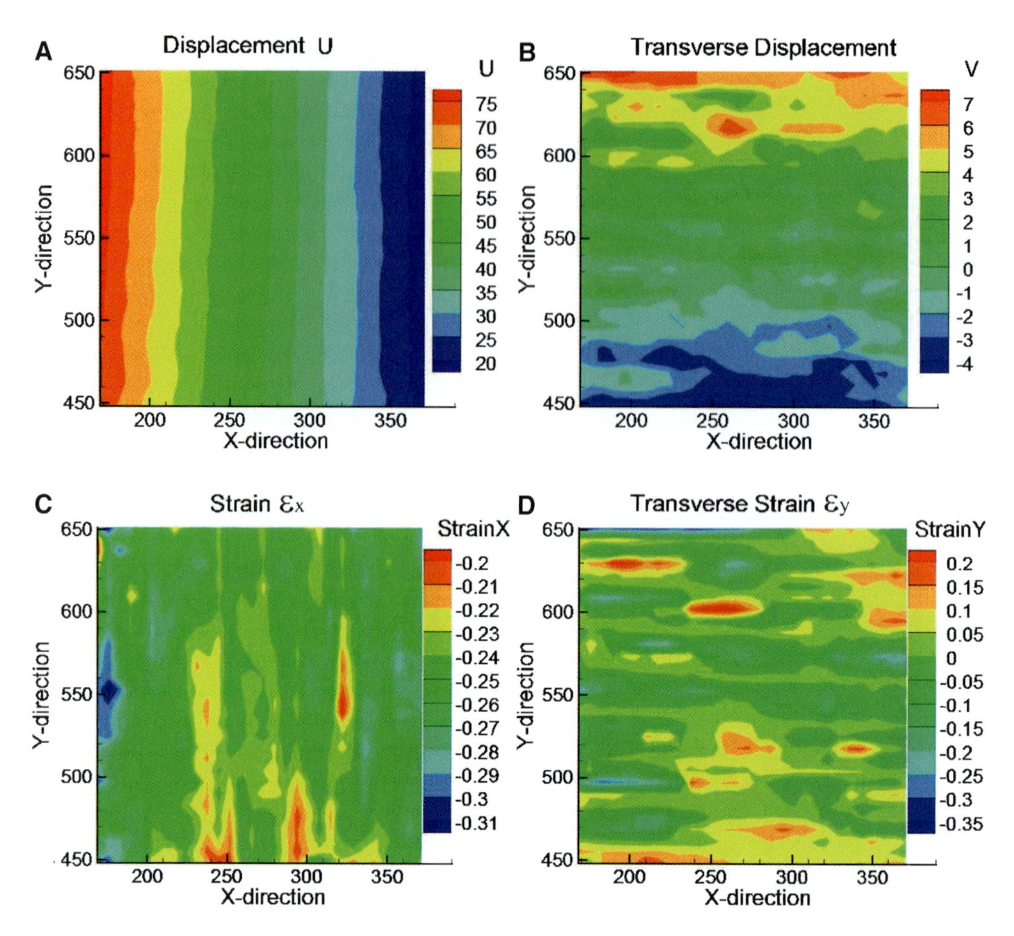

Figure 22.22. Distribution of deformation field of X-VO_x (0.42 g/cm^3) from *DIC*. **A.** Axial displacement field U; **B.** transverse displacement field (1 pixel represents 37.04 μm on specimen); **C.** axial strain field; **D.** transverse strain field. All units are 37.04 μm per pixel.

same density, but the native aerogel has low ultimate compressive strength (14 MPa) and failure compressive strain (18%).

Figure 22.23 shows the stress–strain relations for native VO_x and X-VO_x specimens at four different densities at strain rates in the neighborhood of 300 s^{-1}. It is noted that in dynamic testing, it is difficult to reach exactly the same strain rate in different experiments. The testing at a strain rate near 300 s^{-1} is considered to reveal primarily the effects of density on the stress–strain relationship. The curves are not complete due to the limitation imposed by the 1.5 ms loading duration; therefore, they do not show the complete behavior throughout the failure of the sample. Results indicate that the stress–strain relationship is very sensitive to the mass density at these strain rates. At all densities, the stress–strain curves show similar trends, namely, an elastic range, followed by compaction and by plastic hardening and densification. At higher mass densities, X-VO_x exhibits higher stiffness and higher strength, giving curves above those at lower densities. Figure 22.24 shows the variation of both the Young's modulus and the 0.2% offset compressive yield strength as

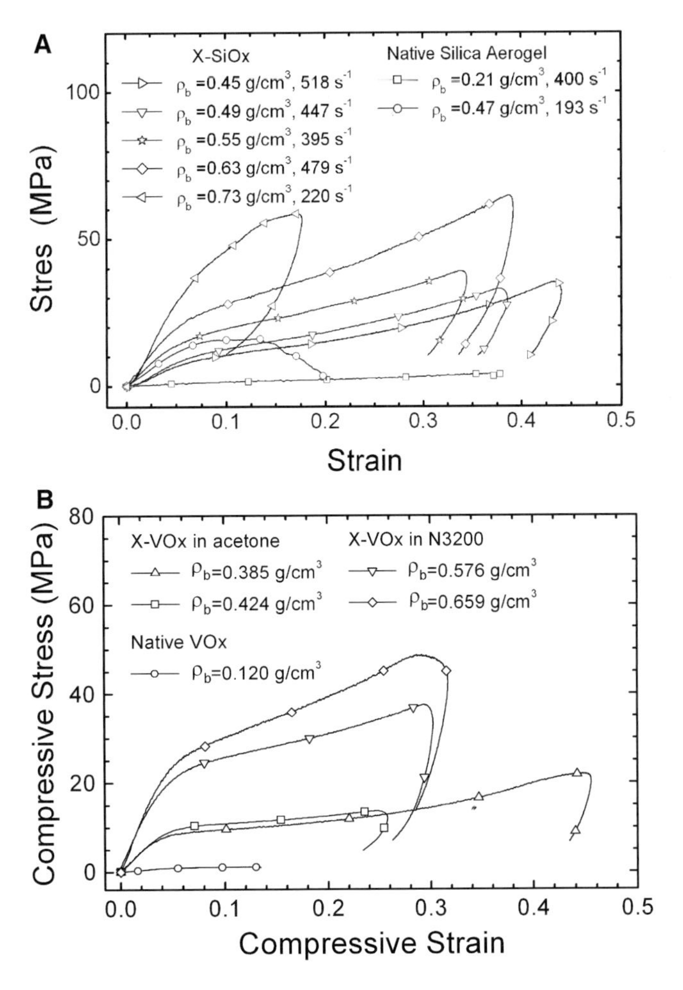

Figure 22.23. Effects of mass density on compressive stress–strain relation of *X-aerogels*. **A.** *X-SiO$_x$* at several strain rates (average is close to 350 s^{-1}); **B.** *X-VO$_x$* at several densities (strain rate ~300 s^{-1}).

a function of mass density. Young's modulus and yield strength (or plastic collapse stress) were fitted into the following power law [35, 37–41, 77, 78] relations:

$$\frac{E_b}{E_s} = C_3\left(\frac{\rho_b}{\rho_s}\right)^{n_1}, \text{ and } \frac{\sigma_b}{\sigma_s} = C_4\left(\frac{\rho_b}{\rho_s}\right)^{n_2} \tag{22.9}$$

where E_b and E_s, σ_b and σ_s, and ρ_b and ρ_s are *Young's modulus*, yield strength, and mass density of the corresponding bulk and skeletal materials, respectively; C_3 and C_4 are constants and depend on the material skeletal density, while n_1 and n_2 are density exponents that do not depend on the material skeletal density. The modulus and the strength data are fitted into (22.9), and the best-fit parameters are listed in Table 22.1 using $E_s = 1$ GPa, $\sigma_s = 10$ MPa, and $\rho_s = 1.0$ g/cm^3. The fitted curves plotted together with the experimental data are shown in Figure 22.24. The best-fit parameters are summarized in Table 22.1. The parameters at low strain rates are close to that at high strain rates. For comparison, the

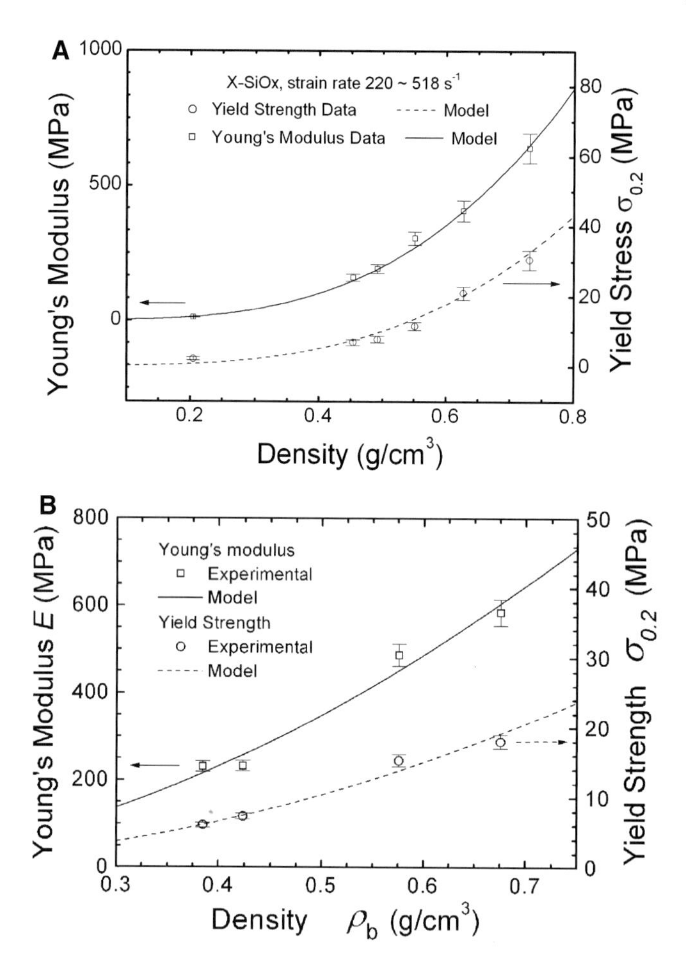

Figure 22.24. Variation of *Young's modulus* and yield strength with mass density. **A.** *X-SiO$_x$*; **B.** *X-VO$_x$*.

exponents for several other aerogels, namely, silica aerogel, carbon aerogel, and organic aerogel, are also listed in Table 22.1. The results show that the mechanical properties of *X-VO$_x$* are less sensitive to the mass (bulk) density than other types of aerogels, such as native, crosslinked silica [56], and carbon aerogels [20, 21, 35, 37, 38]. This different behavior is most likely due to the difference in morphology. *X-VO$_x$* shows a nanoworm nanostructure, while in other types of aerogels, the pearl-necklace type of structure dominates.

Effect of Water and Moisture Concentration

Both water-soaked and moisture-saturated *X-aerogel* specimens were used to measure the stress–strain relationships at a compressive strain rate of 330–430 s^{-1}. Results are compared with those from dry specimens at room temperature. Specimens either fully water saturated or under 97% relative humidity were cut into 2.54 mm thick disks and were tested on the *SHPB*.

We next give results on the effect of water absorption on the stress–strain relationship of *X-SiO$_x$*. Specimens with 12.75 mm in diameter and 2.29 mm thickness and initial mass density 0.46 g/cm^3 at ambient condition were immersed into water. The specimens absorbed moisture and reached a saturated state after about 48 h. The density of the $X\text{-}SiO_x$ after it has reached fully saturated state was 1.11 g/cm^3, representing a weight gain of 152% while the volume increased by only 5%. The water stayed in the $X\text{-}SiO_x$ and did not leak out of the specimens after removing from the water container, indicating that water was generally trapped within the nanopores of the $X\text{-}SiO_x$. The fully saturated specimens were then tested on the *SHPB*. Figure 22.25A shows the stress–strain relationship of two dry and two wet specimens. Results are highly repeatable for the strain rate tested ~400 s^{-1}. There is a clear difference between results from the dry and wet specimens. It is seen that at small strains, the wet specimens have lower yield strength than dry specimens, indicating that the moisture concentration has weakened the interparticle necks of the $X\text{-}SiO_x$. After yielding, the wet $X\text{-}SiO_x$ sample is weaker than the dry sample until 25% strain. At compressive strains of ~25% and higher, the $X\text{-}SiO_x$ specimens have reached the extent of densification that allows the water to carry part of the load so that the wet $X\text{-}SiO_x$ is stronger than the dry one.

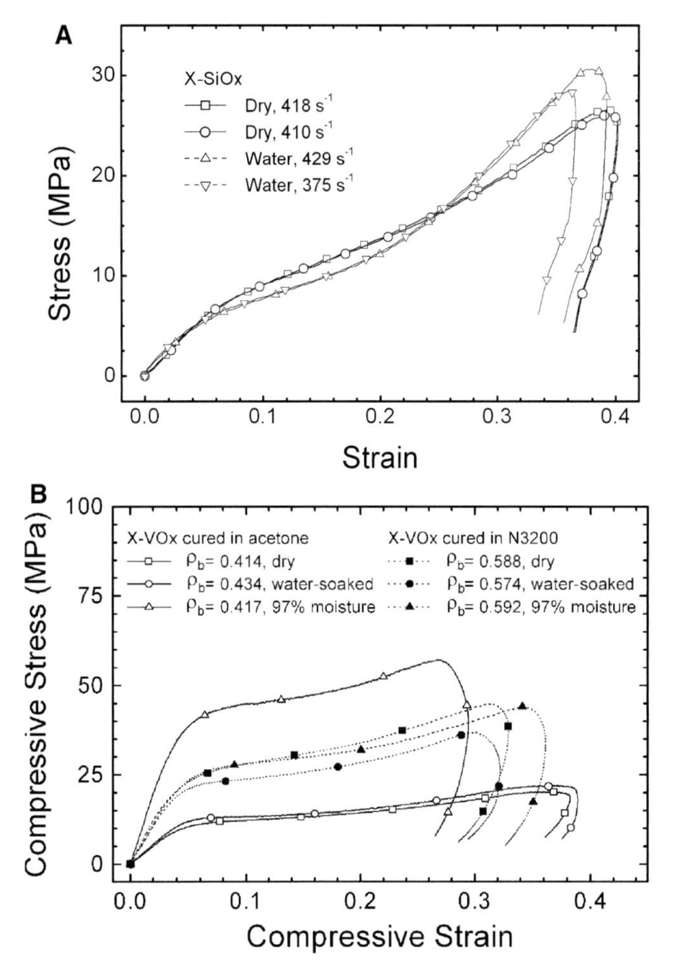

Figure 22.25. The compressive stress–strain relations for dry, water-soaked, and moisturized *X-aerogel* specimens. **A.** $X\text{-}SiO_x$; **B.** $X\text{-}VO_x$.

Figure 22.25B compares the stress–strain curves at strain rates near 300 s^{-1} for the dry, water-soaked, and moisturized $X\text{-}VO_x$ specimens. The shapes of stress–strain curves for three kinds of specimens are similar. The stress–strain curves for the water-soaked specimens are very close to those of dry specimens, indicating that water does not have a very significant effect on the mechanical behavior in compression at high strain rates. By the same token, it may be noted that water absorption has a definite effect on the mechanical behavior at low strain rates, e.g., 5×10^{-3} s^{-1} (see Figure 22.9). At low strain rate, water trapped in the mesopores could escape gradually from the sample so that water-soaked $X\text{-}VO_x$ specimens would be weaker than dry ones, indicating partial collapse of the structure. For $X\text{-}VO_x$ specimens with lower initial density (0.42 g/cm^3), specimens treated under 97% relative humidity have stress–strain relationships different from those of both water-soaked and dry specimens, most probably due to associated volume shrinkage and the 90% increase in density. For $X\text{-}VO_x$ with higher initial density (0.59 g/cm^3), the water absorption weakens only slightly the mechanical properties at high strain rate, while specimens under 97% relative humidity do not have much effect on their mechanical behavior.

Effect of Low Temperatures

A refrigerator was used to cool specimens to −5°C; dry ice (melting point −78°C) was placed in a vacuum flask to cool the specimens to approximately −55°C; liquid N_2 (boiling point: −197°C) in a Dewar was used for reaching cryogenic temperatures (approximately −180°C). All specimens was kept at the set temperature for at least 2 h to reach thermal equilibrium. For *SHPB* experiments, once specimens were ready for testing, they were placed quickly between the precooled incident and transmission bar ends and the testing was completed within 2–3 s.

$X\text{-}VO_x$ disks (2.54 mm thick and 12.70 mm diameter) were tested on the *SHPB* at four temperatures, namely −180, −55, −5°C, and room temperature), for two densities of $X\text{-}VO_x$ (0.41 and 0.55 g/cm^3). The compressive stress–strain relations at strain rate ~1,300 s^{-1} are plotted in Figure 22.26. With the decrease in temperature, mechanical properties such as the

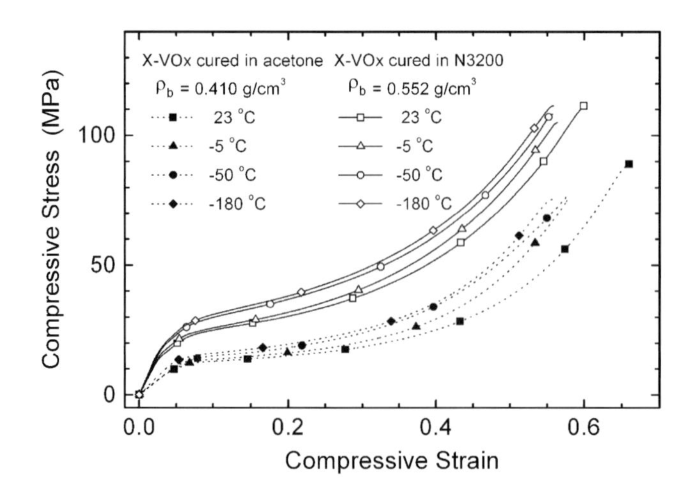

Figure 22.26. Stress–strain curves of $X\text{-}VO_x$ at room and low temperatures (strain rate ~1,300 s^{-1}).

yield strength increases. The mechanical properties of X-VO_x increase with the decrease in temperature. For example, the Young's modulus (or yield strength) at −180°C increases by 44% (22%) from the corresponding data at room temperature for X-VO_x specimens with density 0.55 g/cm^3. Furthermore, for the temperatures investigated, the stress–strain curves are shifted upward as the temperature decreases, indicating that the X-VO_x has superior performance at lower temperatures. At cryogenic temperature, −180°C, the X-VO_x is still ductile, deforming to a compressive strain of more than 55%. The superior ductility of X-VO_x at −180°C is a feature not found with most materials such as BCC metals (e.g., steels) and polymers, which normally become brittle at cryogenic temperatures [84]. The entangled nanoworms in X-VO_x allow the material to stay together under compression. This property makes X-VO_x a material of choice for applications in energy absorption at low temperatures or in liquid hydrogen (LH_2) storage tanks.

At high strain rates (~2,400 s^{-1}), the yield strength (or ultimate compressive strength) is 8.91 (134 MPa) for X-VO_x with densities 0.42. Digital image correlation was used to measure the surface strain through analysis of images acquired by ultrahigh-speed photography. Under compression, the compressive strain in the loading direction is nearly uniform, indicating that the failure of this nanoporous material (or open-cell nanofoam) is different from that of other foams such as polyurethane foams often used for thermal insulation, or as core for fiber-epoxy face sheets, which fails by propagation of instability (compaction) wave. The Poisson's ratio at high strain rates is 0.152. X-VO_x specimens failed at about 60–70% compressive strain through axial splitting, but still remained as single pieces at both room and cryogenic temperatures at high strain rates. At room temperature, the specific energy absorptions of X-VO_x are 87.5 and 69.0 J/g at quasi-static compression and at strain rates ~2,400 s^{-1}. At a strain rate of ~1,300 s^{-1}, the specific energy absorption of X-VO_x is 40.6 J/g at −180°C, which is slightly lower than the value of 46.8 J/g obtained at room temperature. Observation of the morphology indicates that nanoworms in the X-VO_x are fused to form thicker and longer nanoworms, thereby providing enhanced ductility in X-VO_x not found with the native VO_x.

22.5. Conclusion

A list of mechanical characterization tests is summarized for evaluation of the mechanical properties of aerogels. Native aerogels have low mass density; they can reach a specific compressive strength of 314 N m/g, making it possible to sustain an acceleration as high as 300 times of gravitational acceleration g for a silica aerogel plate with a thickness of 10 cm. The strength values for native aerogels, however, are low, preventing them from widespread engineering applications. In recent years, a new class of aerogels, polymer crosslinked aerogels (*X-aerogels*), has been shown to have stiffness and strength comparable to load-bearing engineering polymers while maintaining high porosity, high specific surface areas, and low densities. We described the characterization of polymer crosslinked aerogels under different strain rates and environmental conditions. The glass transition temperatures were determined as 130°C for X-SiO_x and 118.4°C for X-VO_x, respectively. Mechanical behavior of crosslinked aerogel at low strain rate was characterized using three-point bending and compression tests. Using a long split Hopkinson pressure bar, the dynamic behavior of crosslinked aerogel was characterized under high strain rates. The stress–strain relationship at various strain rates was determined to identify the effects of density, strain rates, loading–unloading, moisture absorption, and temperature. In quasi-static compression,

X-aerogels are slightly weakened after it has absorbed water. *X-*VO_x with an entangled nanoworm network has a more ductile behavior at low temperatures than *X-*SiO_x, making them a material with potential multifunctional applications at cryogenic temperatures. The compressive behavior of *X-aerogels* at high strain rates (50–2,500 s^{-1}) was characterized using *SHPB*. The surface full-field deformation of *X-aerogel* specimens was determined using digital image correlation analysis using ultrahigh-speed photography. The axial deformation was relatively uniform and no localized compaction occurred in compression; the Poisson's ratio was determined at 0.15. *X-*SiO_x failed at about 60–70% compressive strain by axial splitting. *X-*VO_x samples remained as one piece at both room and cryogenic temperatures. The highly ductile behavior of *X-*VO_x is traced to the fusion of *X-*VO_x nanoworms during deformation. Under high strain rates, water absorption did not affect the mechanical properties of *X-*VO_x as much as at low strain rates. Moisture absorption has a more significant effect on the compressive behavior of *X-*VO_x specimens with lower densities than with higher ones. All polymer crosslinked aerogels have extremely high specific energy absorption capabilities, reaching values as high as 192 J/g in compression.

References

1. Kistler SS (1931) Coherent expanded aerogels and jellies. Nature 127: 741–741
2. Kistler SS (1932) Coherent expanded-aerogels. J Phys Chem 36: 52–64
3. Kistler SS (1935) The relationship between heat conductivity and structure in silica aerogel. J Phys Chem 39: 79–86
4. Kearby K, Kistler SS, Swann S Jr. (1938) Aerogel catalyst: conversion of alcohols to amines. Ind Eng Chem 30: 1082–1086
5. Kistler SS, Fisher EA, Freeman IR (1943) Sorption and surface area in silica aerogel. J Am Chem Soc 65: 1909–1919
6. Gesser HD, Goswami PC (1989) Aerogels and related porous materials. Chem Rev 89: 765–788
7. Hench LL, West JK (1990) The sol-gel process. Chem Rev 90: 33–72
8. Pierre AC, Pajong GM (2002) Chemistry of aerogels and their applications. Chem Rev 102: 4243–4265
9. Hrubesh LW, Poco JF (1995) Thin aerogel films for optical, thermal, acoustic and electronic applications. J Non-Cryst Solids 188: 46–53
10. Schmidt M, Schwertfeger F (1998) Applications for silica aerogel products. J Non-Cryst Solids 225: 364–368
11. Fricke J, Emmerling A (1998) Aerogels-recent progress in production techniques and novel applications. J Sol-Gel Sci Tech 13: 299–303
12. Akimov YK (2002) Fields of application of aerogels (review). Instrum Exp Tech 46: 287–299
13. Pajonk GM (2003) Some applications of silica aerogels. Colloid Polym Sci 281: 637–651
14. Smirnova I, Suttiruengwong S, Arlt W (2004) Feasibility study of hydrophilic and hydrophobic silica aerogels as drug delivery systems. J Non-Cryst Solids 350: 54–60
15. Jones SM (2006) Aerogel: space exploration applications. J Sol-Gel Sci Techn 40: 351–357
16. Jones SM (2007) A method for producing gradient density aerogel. J Sol-Gel Sci Techn 44: 255–258
17. Fricke J (1988) Aerogels-highly tenuous solids with fascinating properties. J Non-Cryst Solids 100: 169–173
18. Woignier T, Reynes J, Alaoui AH, Beurroies I, Phalippou J (1998) Different kinds of structure in aerogels: relationships with the mechanical properties. J Non-Cryst Solids 241: 45–52
19. Miner MR, Hosticka B, Norris PM (2004) The effects of ambient humidity on the mechanical properties and surface chemistry of hygroscopic silica aerogel. J Non-Cryst Solids 350: 285–289
20. Wingfield C, Baski A, Bertinol MF, Leventis N, Mohite DP, and Lu H (2009) Fabrication of sol-gel materials with anisotropic physical properties by photo-cross-linking. Chem Mater 21: 2108–2114
21. Scherer GW, Smith DM, Qiu X, Anderson LM (1995) Compression of aerogel. J Non-Cryst Solids 186: 316–320
22. Scherer GW (1998) Characterization of aerogels. Adv Colloid Interface Sci 76: 321–339
23. Knauss WG, Emri I, and Lu H (2008) Mechanics of Polymers: Viscoelasticity, in *Handbook of Experimental Solid Mechanics*, pp 49–95, ed. by Sharpe Jr and William N, Springer, USA

24. Struik LCE (1978) *Physical Aging in Amorphous Polymers and Other Materials*. Elsevier, Amsterdam, North-Holland
25. Lu H, Tan G, Chen W (2001) Modeling of constitutive behavior for Epon 828/T-403 at high strain rates. Mech Time-Depend Mater **5**: 119–130
26. Gama BA, Lopatnikov SL, Gillespie JW (2004) Hopkinson bar experimental technique: A critical review. Appl Mech Rev **57**: 223–250
27. Krautkramer K (1969) Ultrasonic Testing of Materi*als*, Springer-Verlag, New York
28. Ensminger D (1988) Ultrasonics: Fundamentals, Technology, Applications, M. Dekker, New York
29. Woignier T, Phalippou J (1988) Mechanical strength of silica aerogels. J Non-Cryst Solids 100: 404–408
30. Woignier T, Phalippou J, Hdach H, Larnac G, Pernot F, Scherer GW (1992) Evolution of mechanical properties during the alcogel-aerogel-glass process. J Non-Cryst Solids 147: 672–680
31. Calas S, Despetis F, Woignier T, Phalippou J (1997) Mechanical Strength Evolution from Aerogels to Silica Glass. J Porous Mater 4: 211–217
32. Capadona LA, Meador MAB, Alunni A, Fabrizio EF, Vassilaras P, Leventis N (2006) Flexible, low-density polymer crosslinked silica aerogels. Polymer 47: 5754–5761
33. Kanamori K, Aizawa M, Nakanishi K, Hanada T (2007) New transparent methylsilsesquioxane aerogels and xerogels with improved mechanical properties. Adv Mater 19: 1589–1593
34. Rosa-Fox NDL, Morales-Florez V, Toledo-Fernandez JA, Pinero M, Esquivias L, Keiderling U (2008) SANS study of hybrid silica aerogels under "in-situ" uniaxial compression. J Sol-Gel Sci Techn 45: 245–250
35. Grob J, Schlief T, Fricke J (1993) Ultrasonic evaluation of elastic properties of silica aerogels. Mater Sci Eng A168: 235–238
36. Fricke J (1990) SiO_2-aerogels: Modification and applications. J Non-Cryst Solids 121: 188–192
37. Grob J, Fricke J (1995) Scaling of elastic properties in highly porous nanostructured aerogels. NanoStruct Mater 6: 905–908
38. Forest L, Gibiat V, Woignier T (1998) Biot's theory of acoustic propagation in porous media applied to aerogels and alcogels. J Non-Cryst Solids 225: 287–292
39. Abramoff B, Klein LC (2005) Elastic Properties of Silica Xerogels. J Am Ceram Soc 73: 3466–3469
40. Moner-Girona M, Roig A, Molins E, Martinez E, Esteve J (1999) Micromechanical properties of silica aerogel. Appl Phys Lett 75:653–655
41. Moner-Girona M, Martinez E, Roig A, Esteve J, Molins (2001) Mechanical properties of silica aerogels measured by microindentation: influence of sol-gel processing parameters and carbon addition. J Non-Cryst Solids 285: 244–250
42. Kucheyev SO, Hamza AV, Satcher Jr JH, Worsley MA (2009) Depth-sensing indentation of low-density brittle nanoporous solid. Acta Mater 57: 3472–3480
43. Rosa-Fox NDL, Morales-Florez V, Toledo-Fernandez JA, Pinero M, Mendoza-Serna R, Esquivias L (2007) Nanoindentation on hybrid organic/inorganic silica aerogel. J Eur Ceram Soc 27: 3311–3316
44. Kucheyev SO, Baumann TF, Cox CA, Wang YM, Bradby JE (2006) Nanoengineering mechanically robust aerogels via control of foam morphology. Appl Phys Lett 89: 041911–3
45. Oliver WC and Pharr GM (1992) An improved technique for determining hardness and elastic modulus using load and displacement sensing indentation experiments. J Mater Res **7**: 1564–1583
46. Stark RW, Drobek T, Weth M, Fricke J, Heckl WM (1998) Determination of elastic properties of single aerogel powder particles with the AFM. Ultramicroscopy 75: 161–169
47. Vincent A, Babu S, Seal S (2007) Surface elastic properties of porous nanosilica coating by scanning force microscopy. Appl Phys Lett 91: 161901–3
48. Tan C, Fung BM, Newman JK, Vu C (2001) Organic aerogels with very high impact strength. Adv Mater 13: 644–646
49. Ma HS, Prevost JH, Jullien R, Scherer GW (2001) Computer simulation of mechanical structure–property relationship of aerogels. J Non-Cryst Solids 285: 216–221
50. Ma HS, Roberts AP, Prévost JH, Jullien R, Scherer GW (2000) Mechanical structure-property relationship of aerogels. J Non-Cryst Solids 277: 127–141
51. Leventis N, Sotiriou-Leventis C, Zhang G, Rawashdeh A-M M, (2002) Nanoengineering strong silica aerogels. Nano Lett 2: 957–960
52. Zhang G, Dass A, Rawashdeh AMM, Thomas J, Counsil JA, Sotiriou-Leventis C, Fabrizio EF, Ilhan F, Vassilaras P, Scheiman DA (2004) Isocyanate-crosslinked silica aerogel monoliths: preparation and characterization. J Non-Cryst Solids 350:152–164
53. Leventis, N (2007) Three dimensional core-shell superstructures: mechanically strong aerogels. Acc Chem Res 40:874–884

54. Bertino MF, Hund JF, Zhang G, Sotiriou-Leventis C, Tokuhiro AT, Leventis N (2004) Room temperature synthesis of noble metal clusters in the mesopores of mechanically strong silica-polymer aerogel composites. J Sol-Gel Sci Techn 30: 43–48
55. Meador MAB, Fabrizio EF, Ilhan F, Dass A, Zhang G, Vassilaras P, Johnston JC, Leventis N (2005) Cross-linking amine-modified silica aerogels with epoxies: mechanically strong lightweight porous materials. Chem Mater 17: 1085–1098
56. Katti A, Shimpi N, Roy S, Lu H, Fabrizio EF, Dass A, Capadona LA, Leventis N (2006) Chemical, physical and mechanical characterization of isocyanate-crosslinked amine-modified silica aerogels. Chem Mater 18: 285–296
57. Meador MAB, Capadona LA, MacCorkle L, Papadopoulos DS, Leventis N (2007) Structure-property relationships in porous 3D nanostructures as a function of preparation conditions: isocyanate cross-linked silica aerogels. Chem Mater 19: 2247–2260
58. Ilhan UF, Fabrizio EF, McCorkle L, Scheiman D, Dass A, Palzer A, Meador MAB, Leventis N (2006) Hydrophobic monolithic aerogels by nanocasting polystyrene on amine-modified silica. J Mater Chem 16: 3046–3054
59. Meador MAB, Vivod SL, McCorkle L, Quade D, Sullivan RM, Ghosn LJ, Clark N, Capadona LA (2008) Reinforcing polymer cross-linked aerogels with carbon nanofibers. J Mater Chem 18: 1843–1852
60. Leventis N, Mulik S, Wang X, Dass A, Sotiriou-Leventis C, Lu H (2007) Stresses at the interface of micro with nano. J Am Chem Soc 129: 10660–10661
61. Leventis N, Mulik, S, Wang X, Dass A, Patil VU, Sotiriou-Leventis C, Lu H, Churu G, Capecelatro A (2008) Polymer nano-encapsulation of templated mesoporous silica monoliths with improved mechanical properties. J Non-Cryst Solids 354; 632–644
62. Leventis N, Sotiriou-Leventis C, Mulik S, Dass A, Schnobrich J, Hobbs A, Fabrizio EF, Luo H, Churu G, Zhang Y, Lu H (2008) Polymer nanoencapsulated mesoporous vanadia with unusual ductility at Cryogenic temperatures. J Mater Chem 18: 2475–2482
63. Luo H, Churu G, Fabrizio EF, Schnobrich J, Hobbs A, Dass A, Mulik S, Zhang Y, Grady BP, Capecelatro A, Sotiriou-Leventis C, Lu H, Leventis N (2008) Synthesis and characterization of the physical, chemical and mechanical properties of isocyanate-crosslinked vanadia aerogels. J Sol-Gel Sci Techn 48: 113–134
64. Parmenter KE, Milstein F (1998) Mechanical properties of silica aerogels. J Non-Cryst Solids 223: 179–189
65. Luo H, Lu H, Leventis N (2006) The compressive behavior of isocyanate-crosslinked silica aerogel at high strain rates. Mech Time-Depend Mater 10: 83–111
66. She JH, Ohji T (2002) Porous mullite ceramics with high strength. J Mater Sci Lett 21: 1833–1834
67. Morris CA, Anderson ML, Stroud RM, Merzbacher CI, Rolison DR (1999) Silica sol as a nanoglue: flexible synthesis of composite aerogels. Science 284: 622–624
68. Amatani T, Nakanishi K, Hirao K, Kodaira T (2005) Monolithic periodic mesoporous silica with well-defined macropores. Chem Mater 17:2114–2119
69. Livage J (1991) Vanadium pentoxide gels. Chem Mater 3: 578–593
70. Sudoh K, Hirashima H (1992) Preparation and physical properties of V_2O_5 aerogel. J Non-Cryst Solids 147: 386–388
71. Sudant G, Baudrin E, Dunn B, Tarascon JM (2004) Synthesis and electrochemical properties of vanadium oxide aerogels prepared by a freeze-drying process. J Electrochem Soc 151: A666–A671
72. Frew DJ, Forrestal MJ, Chen W (2002) Pulse-shaping techniques for testing brittle materials with a split Hopkinson pressure bar. Exp Mech 42: 93–106
73. Gray GT (2000) Classic split-Hopkinson pressure bar technique. Mech Test Eval, ASM Handbook 8: 462–476
74. Chen W, Zhang B, Forrestal MJ (1999) A split Hopkinson bar technique for low-impedance material. Exp Mech 39: 81–85
75. Chen W, Lu F, Cheng M (2002) Tension and compression testing of two polymers under quasi-static and dynamic loading. Polym Test 21: 113–121
76. Chen W, Zhou B (1998) Constitutive behavior of Epon 828/T-403 at various strain rates. Mech Time-Depend Mater 2: 103–111
77. Gibson LJ (2000) Mechanical behavior of metallic foams. Ann Rev Mater Sci 30: 181–227
78. Gibson LJ, Ashby MF (1997) Cellular Solids: Structure and Properties-2nd ed, Cambridge University Press
79. Peters WH, Ranson WF (1982) Digital imaging techniques in experimental stress analysis. Opt Eng 21: 427–432
80. Sutton MA, Wolters WJ, Ranson WF, McNeil SR (1983) Determination of displacements using an improved digital image correlation method. Image Vis Comput 1: 133–139
81. Chu TC, Ranson WF, Sutton MA, Peters WH (1985) Applications of digital-image-correlation techniques to experimental mechanics. Exp Mech 25: 232–244

82. Bruck HA, McNeill SR, Sutton MA, Peters WH (1989) Digital image correlation using Newton-Raphson method of partial differential correction. Exp Mech 29: 261–267
83. Lu H, Cary PD (2000) Deformation measurements by digital image correlation: Implementation of a second-order displacement gradient. Exp Mech 40: 393–400
84. Wigley DA (1971) Mechanical Properties of Materials at Low Temperatures, Plenum Press, New York

23

Thermal Properties of Aerogels

Hans-Peter Ebert

Abstract This chapter provides an insight into the different aspects of heat transfer in aerogels and their thermal properties. In this context, the principal heat transfer mechanisms are discussed and illustrated by exemplary experimental results. Typical thermal conductivity values and radiative properties as well as their dependency on external conditions such as temperature or atmosphere are discussed for different classes of aerogels. The chapter concludes with a brief discussion about the specific heat of aerogels.

23.1. General Aspects of Heat Transfer in Aerogels

The thermal properties of aerogels have been the focus of research ever since R&D was first carried out in the field of these highly porous materials (Chap. 1). The thermal properties generally regarded as most significant are thermal conductivity, followed by specific heat. Kistler [1, 2] sought to develop reliable methods to learn more about structural properties such as pore sizes with sufficient accuracy. The well-established kinetic theory of gases offered him a linear relation between the mean free path of gas molecules and the macroscopic measurand thermal conductivity. Besides the general academic interest in the understanding of heat transfer in aerogel, scientists and engineers recognized early that this diaphanous substance is a perfect thermal insulation material [3]. This is probably the most attractive property of aerogel (Chap. 26), considering that thermal insulation correlates directly to the energy efficiency of buildings, machines and industrial installations, and is now, in the dawn of climate change and limited conventional energy resources, the focus of more research work than ever.

Heat transfer in porous media, like aerogels, is described by the equation of heat transfer. The involved heat transfer mechanisms are schematically illustrated in Figure 23.1. The principal discussion of the equation of heat transfer provides an insight into the nature of the physical material property thermal conductivity. Generally, the equation of heat transfer can be expressed as:

$$\nabla\vec{q} + \Phi = \rho \cdot c \cdot \frac{\partial T}{\partial t} \qquad (23.1)$$

H. P. Ebert • Functional Materials for Energy Technology,
Bavarian Center for Applied Energy Research, Am Hubland, 97074 Würzburg, Germany
e-mail: ebert@zae.uni-wuerzburg.de

M.A. Aegerter et al. (eds.), *Aerogels Handbook*, Advances in Sol-Gel Derived Materials and Technologies, DOI 10.1007/978-1-4419-7589-8_23,

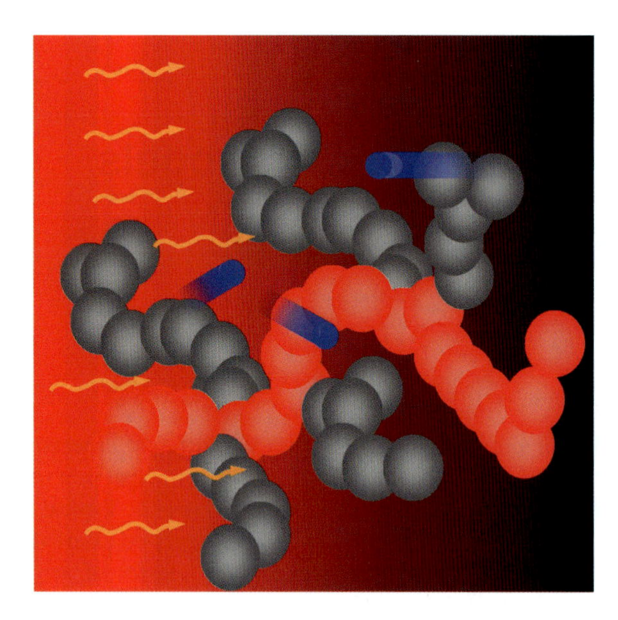

Figure 23.1. Diagram of an aerogel structure and the heat transfer mechanisms involved. Heat is transferred via chains of primary particles forming the solid backbone (indicated by the *red particle chain*), by thermal radiation (*wavy yellow arrows*), and by gas molecules present in the porous structure of the aerogel (*blue dots*).

and

$$\vec{q} = -\lambda \nabla T \tag{23.2}$$

with q: heat flux density, λ: *three dimensional tensor of the thermal conductivity*, density ρ, c: specific heat, Φ: heat source and T: local temperature. Equation (23.1) reflects the law of conservation of energy and balances the heat fluxes across the boundaries of a finite volume. Equation (23.2) is Fourier's law which states that the heat flux density is proportional to the local temperature gradient and defines the thermal conductivity λ. The heat source Φ describes, for example, the influence of phase transitions or sorption processes within an aerogel related to the release or uptake of reaction enthalpies and the gain or loss of thermal radiation. The latter case is of particular importance because it has to be considered explicitly for the treatment of heat transfer in silica aerogels and other infrared optically semitransparent aerogel types. The heat source term Φ, describing thermal radiation, is given by:

$$\Phi = \nabla \vec{q}_{\mathrm{r}} \tag{23.3}$$

with q_{r}: radiative heat flux density, which is in principle a solution of the equation of radiative transfer [4]. For an isotropic aerogel and if the heat transfer depends only on the local temperature gradient (23.1) simplifies to

$$\Delta T = \frac{1}{a} \cdot \frac{\partial T}{\partial t} \tag{23.4}$$

with the thermal diffusivity $a = \lambda/(\rho \bullet c)$. In this case, the experimentally determined thermal conductivity is a material property and does not depend on the experimental conditions of its determination. The product $\rho \bullet c$ is also called the volumetric specific heat and is only relevant for the case of nonstationary heat transfer where temperatures and heat fluxes change with time. For aerogels the heat transfer via the solid structure and the gaseous phase follows Fourier's law (23.2) if it is considered on a macroscopic scale. To calculate the radiative heat transfer, however, one needs to distinguish between the simple case of diffusive radiative heat transfer, where Fourier's law is also valid for radiative heat transfer and a radiative conductivity can be defined, and the case of complex nonlocal radiative heat transfer. Due to the small pore structures within aerogels, natural convection does not have to be considered to describe heat transfer even for extreme conditions [5, 6].

23.2. Effective Thermal Conductivity of Optically Thick Aerogels

In most cases, heat transfer within nonevacuated aerogels is based on three mechanisms: heat conduction via the solid backbone, heat transfer within the gaseous phase present in the open-porous aerogel structure and by radiative heat transfer. In optically thick aerogels, i.e., in the case where the *optical thickness* $\tau >> 1$ [cf. (23.11)], heat transfer within aerogels is governed by diffusion mechanisms: the diffusion of phonons and photons. The magnitude of diffusive heat transfer via the gaseous phase depends on the effective pore size, gas type and gas pressure.

23.2.1. Heat Transfer via the Solid Backbone

The heat transfer via the solid backbone of aerogels depends on the backbone structure and connectivity (Chap. 21), and its chemical composition. For a given temperature gradient within an aerogel, heat is transferred by diffusing phonons via the chains of the aerogel backbone, where the mean free path of the phonons is far below the dimensions of the mostly amorphous, dielectric primary particles. Within the primary particles, the thermal conductivity is a property of the backbone material and is described in terms of the phonon diffusion model by Debye [7], i.e.,:

$$\lambda_0(T) = \frac{1}{3}\rho_0 \cdot c_v(T) \cdot l_{ph}(T) \cdot v_0(T) \quad (23.5)$$

with the backbone thermal conductivity λ_0, the density ρ_0 of the backbone material, the specific heat c_v at constant volume, the frequency-averaged mean free path of phonons l_{ph} and the average velocity of the elastic waves within the backbone material, v_0. Above 100 K the influence of the morphology on the solid thermal conductivity can be taken into account by using a temperature-independent geometrical factor G [8, 9]:

$$\lambda_s(T) = G(\rho) \cdot \lambda_0(T). \quad (23.6)$$

The strongly density-dependent factor G considers the effects of the ineffective dead ends of the solid backbone, the increase in the specific heat due to the enhanced specific surface of an aerogel, the meandering chains along a temperature gradient and the constriction of heat flow at the connecting area between the primary particles within the chains. A proper approach to the description of the density dependence is provided by the *percolation theory* [8, 10]. The solid thermal conductivity follows a scaling law as a function of the density:

$$\lambda_s(T) = \text{const.} \cdot (f - f_c)^\alpha \quad \textit{for } f \geq f_c \tag{23.7}$$

and $\lambda_s(T) = 0$ for $f < f_c$ with f: the solid volume fraction $f = \rho/\rho_0$, f_c: percolation threshold of the solid volume fraction and α: scaling exponent. For aerogels the threshold is close to zero [8]. Therefore, when depicting the solid conductivity of aerogels of one type with different densities versus the solid volume fraction in a log–log plot a linear behavior is expected where the slope represents the scaling exponent (see Figure 23.2). Scaling exponents in the range of 1.2–2 are found in literature [8, 11, 13].

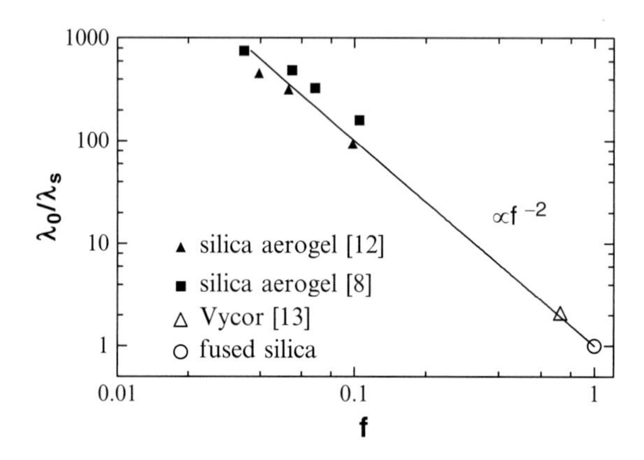

Figure 23.2. Experimentally derived values (*symbols*) of the ratio backbone thermal conductivity/solid thermal conductivity λ_0/λ_s as a function of the solid volume fraction f for different silica aerogels by Lu and Heinemann [8, 11]. Also the ratio is given for porous glass Vycor® [12]. The data indicate that the solid thermal conductivity for the investigated silica aerogel samples follows a scaling law according to (23.7) with the scaling exponent $\alpha = 2$ (*solid line*).

For porous media and under the assumption that the phonon diffusion model is valid, the solid thermal conductivity can be derived by way of an analogy conclusion measuring the sound velocity of the porous media [8, 14]:

$$\lambda_s(T) = \lambda_0 \cdot \frac{\rho}{\rho_0} \cdot \frac{v}{v_0}, \tag{23.8}$$

with v: the sound velocity of the porous media, v_0: sound velocity in the corresponding nonporous (backbone) material, ρ: density of the porous media and ρ_0: density of the backbone material. For the derivation of (23.8) it is assumed that the specific heat capacity of the porous and nonporous material is equal and that the average mean free path of the phonons is small in comparison to the dimensions of the solid structures, e.g., primary particles, of the aerogel.

23.2.2. Heat Transfer via the Gaseous Phase

In general, the characteristic pore size within aerogels is well below the micron range which leads to the fact that the heat transfer via the gaseous phase within the aerogel structure is already reduced compared to heat transfer within the free gas. Therefore, aerogels have the potential to realize total thermal conductivity values far below the thermal conductivity of the free air, i.e., 0.026 $\mathrm{Wm^{-1}\,K^{-1}}$ at ambient temperature. In conjunction with the strongly reduced solid thermal conductivity, these are the key reasons why aerogels have always been considered an ideal thermal insulation material.

The heat transfer in the gaseous phase is characterized by the *Knudsen number* Kn, which is the ratio of the mean free path l_g of the gas molecules and the effective pore dimension D:

$$\mathrm{Kn} = \frac{l_g}{D}. \tag{23.9}$$

In principle, one has to distinguish between three different cases:

Molecular heat transfer, Kn >> 1: The average pore size is significantly smaller than the mean free path of the gas molecules. The gas molecules collide predominantly with the solid backbone of the aerogel and the resulting thermal conductivity contribution is proportional to the number of gas molecules, i.e., the gas pressure.

Diffusive heat transfer, Kn << 1: The mean free path of the gas molecules is much smaller than the average pore size. The gas molecules collide predominantly with each other, which is the classical case of diffusive heat transfer [15]. The resulting thermal conductivity of the gaseous component equals the thermal conductivity of the free gas, which is independent of the gas pressure for ambient and moderate pressures.

Transition regime, Kn $\approx$ 1: The gas molecules collide with both the walls and each other. This effect can be described by introducing a statistical average mean free path which takes both effects into account.

Kaganer provides a closed description of the gas pressure dependence of the effective thermal conductivity of a pore gas, λ_g, which takes into account all three regimes [16]:

$$\lambda_g(p_g, T) = \frac{\Pi \cdot \lambda_{g,0}(T)}{1 + 2 \cdot \beta(T) \cdot \frac{l_g(T)}{D}} = \frac{\Pi \cdot \lambda_{g,0}(T)}{1 + 2 \cdot \beta(T) \cdot \mathrm{Kn}(T)} = \frac{\Pi \cdot \lambda_{g,0}(T)}{1 + \frac{p_{1/2}(T)}{p_g}} \tag{23.10}$$

with Π: meso- and macro porosity of the aerogel, β: coefficient, dependent on type of gas and temperature, $p_{1/2}$: gas pressure at which the thermal conductivity is one-half of $\lambda_{g,0}$ and $\lambda_{g,0}$: thermal conductivity of the nonconvecting free gas. Figure 23.3 shows the ratio $\lambda_g/\lambda_{g,0}$ as a function of the gas pressure for different pore diameters D and $\Pi = 1$. The curves increase in the regime of molecular heat transfer linearly with the gas pressure, followed by the transition regime where the gaseous thermal conductivity increases until the saturation regime is reached. In this case, the thermal conductivity is independent of the gas pressure. The plots also show that with decreasing pore diameters, the half width pressure $p_{1/2}$ is shifted to higher values of the gas pressure. It is obvious that for aerogels with effective pore sizes of less than 0.1 μm the gas pressure-dependent thermal conductivity λ_g is one-half or less of the thermal conductivity of the free gas at ambient pressure. Some examples are depicted in Figure 23.4. At atmospheric pressures, the gas pressure dependence of the effective total thermal conductivity is in the order of 0.5×10^{-5} W(m K Pa)$^{-1}$.

Considering the temperature dependence of (23.10), two reverse effects can be observed: the thermal conductivity of the free gas and the product β•Kn increase with temperature. Considering the thermal conductivity at ambient gas pressure, the increase in

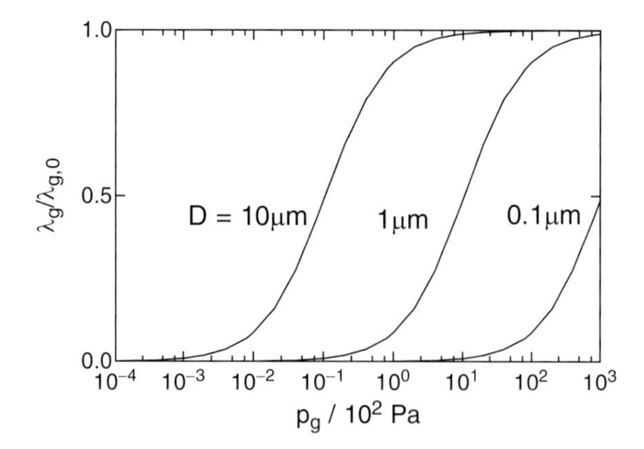

Figure 23.3. Gas pressure dependence of the effective thermal conductivity of the pore gas nitrogen in relation to the thermal conductivity of the free gas according to (23.10) for different pore diameters D and $\Pi = 1$. Typical s-shaped curves can be observed in the linear-log representation.

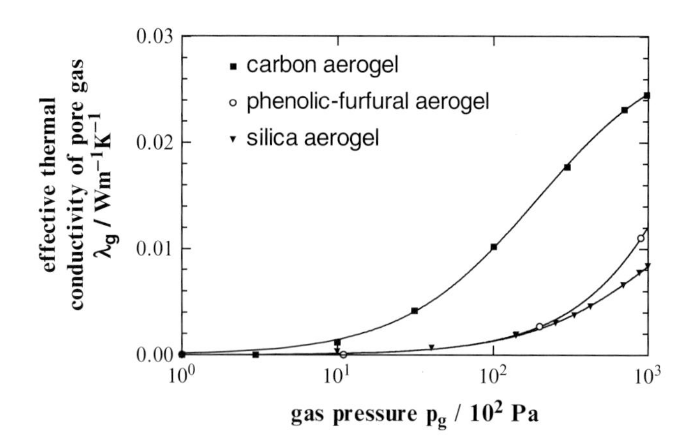

Figure 23.4. Effective thermal conductivity of the pore gas λ_g of a carbon and a silica aerogel in a nitrogen atmosphere [17] and of a phenolic-furfural aerogel in air [13]. The thermal conductivity values λ_g are below the thermal conductivity of still air ($\lambda_{air} = 0.026$ Wm^{-1} K^{-1}) and nitrogen ($\lambda_{N2} = 0.025$ Wm^{-1} K^{-1}).

$\lambda_{g,0}(T)$ can be compensated or even overcompensated by the shift of the half width pressure $p_{1/2}$ toward higher pressures.

The exact description of the contribution of the gaseous phase within aerogels to the thermal conductivity is complicated by the fact that the heat transfer via the solid backbone interacts with that in the gaseous phase. The effect depends on the structure of the aerogel and on the gas pressure and is illustrated for one striking case in Figure 23.5: in the vicinity of the necks between primary particles, diffusing gas molecules generate a thermal shortcut. "Channels" of locally increased heat transfer in dependence of the gas pressure thus occur along the chains of the aerogel backbone. This effect is more pronounced if the thermal conductivity of the backbone material is high in comparison to that of the gas. This coupling effect may therefore drastically enhance the gas pressure-dependent thermal conductivity.

Figure 23.6 shows the experimentally determined thermal conductivity of an organic *resorcinol–formaldehyde aerogel* over a wide argon pressure range [17]. At ambient

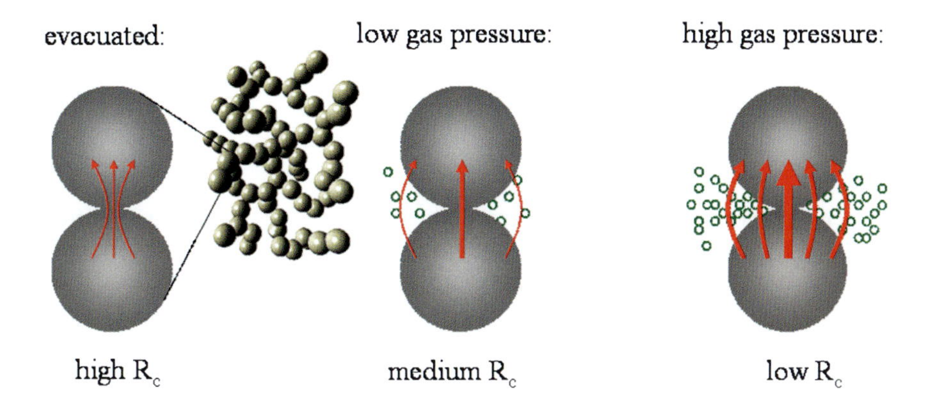

Figure 23.5. Diagram of the coupling effect between the heat transfer within the gaseous and solid phases of an aerogel. The constriction of heat flow lines in the vicinity of the contact area of two primary particles indicates high thermal contact resistance (R_c) values in the case where only the solid component is involved in the heat transfer (i.e., for evacuated aerogels, **left picture**). As the gas pressure increases part of the heat transfer takes place over the gaseous phase which leads to a reduced thermal contact resistance (*picture* in the **middle**). At high gas pressures, a total thermal shortcut can be observed, resulting in low thermal contact resistance values (**right picture**).

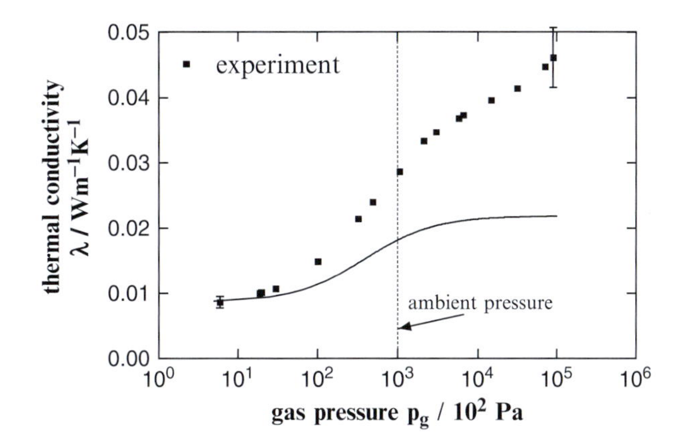

Figure 23.6. Total thermal conductivity of a resorcinol–formaldehyde aerogel (density = 330 kg m^{-3}, porosity = 0.78, average pore size 0.6 μm) as a function of argon gas pressure at 21°C. The *solid line* is the calculated thermal conductivity according to (23.10) without taking a coupling effect into account.

pressure ($p_g = 10^5$ Pa) the total thermal conductivity is 0.029 Wm^{-1} K^{-1} and the sum of the solid thermal conductivity and the radiative conductivity is about 0.0086 Wm^{-1} K^{-1}. Therefore the thermal conductivity contribution of the gas at ambient pressures is about 0.020 Wm^{-1} K^{-1}, which is more than the expected value of approximately 0.011 Wm^{-1} K^{-1} according to Knudsen. The theoretical curve according to the Knudsen model (23.10) is depicted for comparison and shows a nearly constant thermal conductivity for gas pressures above 10^4 Pa. The striking difference between the experiment and the simple theoretical model is due to the coupling effect which influences the gas pressure dependence of the thermal conductivity over the total investigated gas pressure range. A plateau at higher gas pressure was not observed. Furthermore, another steep increase becomes evident at the highest gas pressure values, which can be explained by the intrinsic increase in the thermal conductivity of argon at high pressure.

23.2.3. Radiative Heat Transfer

The nature of radiative heat transfer within an aerogel, i.e., whether it is diffusive or nondiffusive, depends on the optical thickness of the specific type of aerogel, or to be more exact the optical thickness τ of an aerogel layer with the geometrical thickness d [18]:

$$\tau(\Lambda) = \frac{d}{l_{\text{pht}}} = E^*(\Lambda) \cdot d = e^*(\Lambda) \cdot \rho \cdot d \tag{23.11}$$

with $E^*(\Lambda)$: effective extinction coefficient and l_{pht}: mean free path of the photons of the wavelength Λ. The optical thickness is therefore a statistical measure of how often a photon with the mean free path l_{pht} interacts with the material within a given distance d. The wavelength-dependent effective specific extinction $e^*(\Lambda)$ is a material property, i.e., it depends on the chemical composition and structure of the medium, and is a measure of the attenuation of electromagnetic waves by scattering and absorption processes. The unit "square meters per kilogram" represents a mass specific cross-section for electromagnetic radiation. The asterisk indicates that the extinction coefficient and the specific extinction are effective values which take also anisotropic scattering of thermal radiation into account. The wavelength range which has to be taken into consideration to describe radiative heat transfer depends on the temperatures at which the heat transfer takes place and can be derived using the *fractional function of the first kind* $f_{0-\Lambda}(T)$ [19].

For optically thick aerogels, e.g., for most organic, opacified or carbon aerogels, radiative heat transfer is described by the diffusion of photons. The photons interact within short distances in comparison to the macroscopic dimension of the aerogel with its solid backbone. A corresponding solution to the diffusion equation for photons can be derived in analogy to the diffusion of phonons by way of:

$$q_{\text{r}} = -\lambda_{\text{r}}(T) \cdot \nabla T = -\frac{16}{3} \cdot \frac{n^2 \cdot T_{\text{r}}^3}{\rho \cdot e_{\text{R}}^*(T)} \nabla T \tag{23.12}$$

with λ_{r}: radiative conductivity, n: effective index of refraction, ρ: density, $e_{\text{R}}^*(T)$: temperature-dependent effective specific extinction and T_{r}: mean radiative temperature. The effective index of refraction n can be approximately calculated using the Clausius–Mossotti relation [20, 21]. This relation correlates the optical constants of a rarified medium with its density. The temperature-dependent effective specific extinction can be calculated by using the Rosseland weighting function $f_{\text{R}}(z)$:

$$\frac{1}{e_{\text{R}}^*(T)} = \int_0^{\infty} \frac{1}{e^*(\Lambda)} f_{\text{R}}(z) dz \tag{23.13}$$

with

$$f_{\text{R}}(z) = \frac{15}{4\pi^4 z^6} \cdot \frac{e^{\frac{1}{z}}}{\left(e^{\frac{1}{z}} - 1\right)^2} \tag{23.14}$$

and

$$z = \frac{\Lambda \cdot T \cdot k_B}{h \cdot c} \tag{23.15}$$

with h: Planck constant, c: velocity of light and k_B: Boltzmann constant. The spectrum of thermal radiation responsible for radiative heat transfer at a given temperature is defined by the Rosseland weighting function. The Rosseland mean can be calculated assuming two conditions: firstly, that the radiative heat transfer takes place sufficiently far from the boundaries, and secondly, that a given temperature distribution varies only slightly in relation to the optical thickness [22].

To improve the thermal insulation properties of silica aerogels, so-called infrared opacifiers are embedded into the aerogel matrix in order to enhance optical thickness and therefore to reduce radiative heat transfer [23]. The objective is to integrate highly absorbing or scattering particles, e.g., carbon black or titanium dioxide, in a low concentration within the aerogel for enhanced radiation extinction without significantly increasing the solid thermal conductivity of the backbone. Figure 23.7 shows the effective specific extinction of nonopacified and opacified aerogels as a function of wavelength. In Figure 23.7 the normalized Rosseland weighting function for a temperature of 300 K is depicted with a maximum of thermal radiation of 8 μm. It is obvious that for nonopacified silica aerogels the specific extinction is low below wavelengths of 8 μm. Below this wavelength thermal radiation is transmitted within the aerogel without being significantly attenuated by absorption processes and radiative heat transfer is a nonnegligible heat transfer mechanism. In this case, a local radiative conductivity cannot be defined. Opacified aerogels, e.g., carbon black doped silica aerogels and most organic aerogels, e.g., resorcinol–formaldehyde aerogels, show no "transmission windows" in the wavelength region relevant for heat transfer and radiative heat transfer is a local phenomenon which is characterized by the radiative

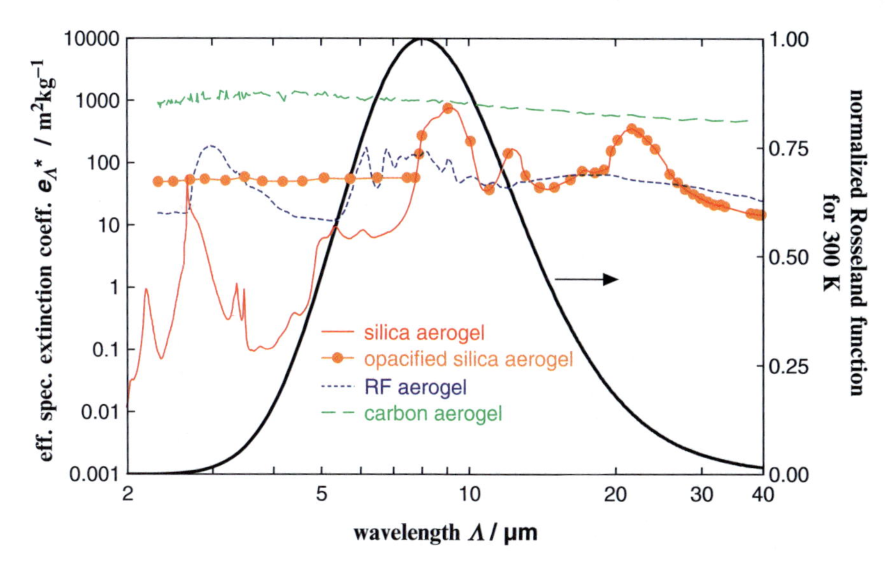

Figure 23.7. Effective specific extinction of a nonopacified and opacified silica aerogel, resorcinol–formaldehyde (RF) aerogel and carbon aerogel as a function of the wavelength. Additionally, the normalized Rosseland weighting function is depicted, which shows the maximum thermal radiation for 300 K at a wavelength of 8 μm [24].

conductivity according to (23.12). The requisite temperature-dependent specific extinction $e^*(T)$ in (23.12) is calculated from the spectral values of the effective specific extinction using the Rosseland weighting function (cf. Figure 23.8). Extremely high effective specific extinction values were observed for carbon aerogels due to the strong absorption properties of carbon. The spectral dependency of the specific extinction of these aerogels is weak and therefore carbon aerogels can be considered as gray media, i.e., radiative properties independent of wavelength. This is also the explanation why the Rosseland-averaged total effective specific extinction is nearly constant as a function of temperature.

It should be mentioned that the attenuation of radiation for nonopacified aerogels is mainly due to absorption processes, because the wavelength of the thermal radiation is much larger than the characteristic lengths of the aerogel structure. In opacified aerogels, besides absorption processes, scattering of thermal radiation can also occur depending on the infrared-optical properties of the opacifier. For example, carbon black is a strong absorber [25], whereas titanium dioxide is a scattering mineral [26].

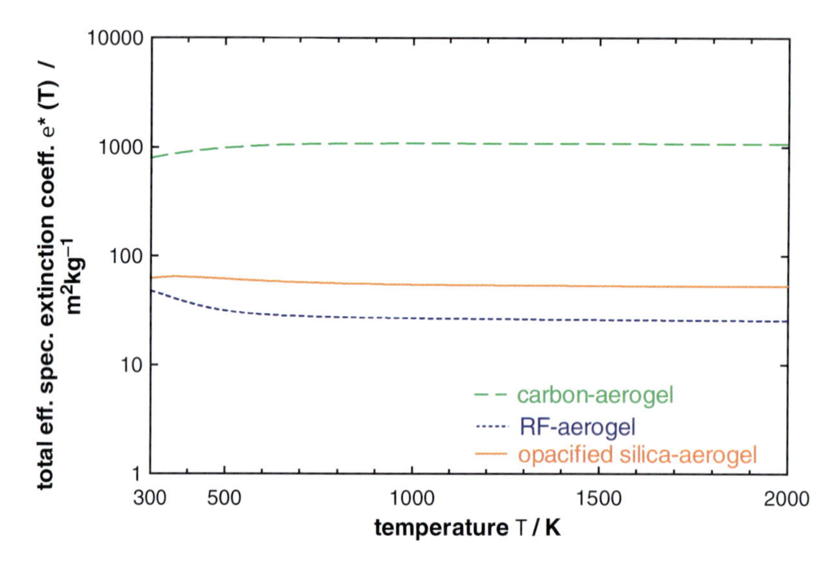

Figure 23.8. Rosseland-averaged total effective specific extinction $e^*(T)$ using the spectral values (Figure 23.7) as a function of temperature of an opacified silica aerogel, resorcinol–formaldehyde (RF) aerogel and carbon aerogel.

23.2.4. Effective Total Thermal Conductivity of Aerogels

In the case where heat transfer in aerogels depends on the local temperature gradient, the *effective total thermal conductivity* λ_{eff} can be expressed as the sum of the solid thermal conductivity of the solid backbone λ_s, the effective thermal conductivity of the gaseous phase λ_g, and finally, the radiative conductivity λ_r:

$$\lambda_{\text{eff}}(T, p_g) = \lambda_s(T) + \lambda_g(T, p_g) + \lambda_r(T). \tag{23.16}$$

Since (23.16) is independent of direction, it is only valid for aerogels with isotropic thermal properties. Many experimentally derived values for the thermal conductivity of aerogels under various conditions are available. Hrubesh et al. discussed the superposition principle of (23.16) and calculated optimal densities to derive minimum total effective thermal conductivity values for different types of nonevacuated organic and inorganic aerogels [27]. The authors concluded that it is important to use an organic or inorganic material system with a small backbone thermal conductivity, whereby organic components have smaller values. Furthermore, reducing the effective pore size is essential to suppress the heat transfer via the gaseous phase, and finally, if opacification is needed, an excellent IR opacifier should be used. It should be mentioned that Hrubesh et al. did not take into account any coupling effect between the heat transfer within the gaseous and solid phase of an aerogel, which could dramatically increase the effective total thermal conductivity even at ambient conditions (cf. Figure 23.6). Therefore, it is also important to synthesize aerogel structures which reduce coupling effects to a minimum [17] to achieve low thermal conductivity values for nonevacuated aerogels.

Opacified Silica Aerogels

The heat transfer properties of silica aerogels (Chap. 2) have always been the focus of scientific and industrial research as they make the material so promising for thermal insulation. Since silica aerogels are transparent for thermal radiation in a wide wavelength range (cf. Figure 23.7) even at ambient temperatures, researchers have explored ways of reducing radiative heat transfer without significantly influencing the other remaining heat transfer mechanisms. In this vein, silica aerogels have been opacified or "doped" by infrared active oxides, e.g., iron oxide or titanium oxide, carbides, e.g., silicon carbide, and carbon blacks to reduce the radiative heat transfer by limiting the mean free path of the photons. Silica aerogels were first successfully opacified within the framework of experiments conducted by White in 1939 [3]. Kuhn et al. investigated the influence of the absorption and scattering properties of various mineral opacifiers (TiO_2, ilmenite, Fe_3O_4, B_4C, SiC) on the extinction spectra of silica aerogel for different amounts of the opacifier [26]. Zeng et al. calculated the efficiency of the opacification of silica aerogels with carbon blacks [28]. Experimental data for carbon black doped silica aerogels were provided by Lu et al. and Lee et al. [25, 29].

Figure 23.9 shows the total effective thermal conductivity of evacuated silica aerogels, opacified with carbon black, as a function of temperature. In one case the corresponding solid thermal conductivity and the radiative conductivity are also shown. The solid thermal conductivity is only slightly dependent on temperature, according to the temperature dependence of the specific heat of silica. The radiative conductivity becomes significant at temperatures above 300 K and is proportional to T^3. This effect is also demonstrated in Figure 23.10 where the thermal conductivity values are plotted versus T^3. At higher temperatures where radiative heat transfer is dominant, a linear increase in this representation can be observed in the effective total thermal conductivity. The specific extinction of the opacified aerogel can be derived according to (23.12) from the slope of the corresponding linear regression curve (cf. Figure 23.10).

Organic Aerogels

Organic aerogels (Part IV) are of special interest due to their thermal insulation properties. The organic solid backbone has a low solid thermal conductivity in comparison to inorganic aerogels [27]. Heat transfer within resorcinol–formaldehyde (RF) aerogels

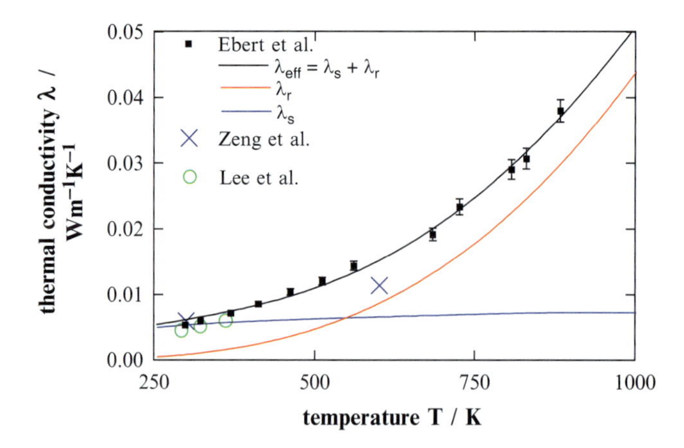

Figure 23.9. Thermal conductivity values of evacuated silica aerogels ($p_g < 0.01$ Pa) as a function of temperature. *Black symbols*: total effective thermal conductivity of a silica aerogel ($\rho = 153$ kg m^{-3}) opacified with 2.8% carbon black and its correspondingly derived solid thermal conductivity (*blue line*), radiative conductivity (*red line*), and the superposition of both (*black line*) [30]. Also depicted are the predicted effective total thermal conductivity values given by Zeng et al. for an optimal loading level of carbon black [28] and experimental data provided by Lee et al. [29].

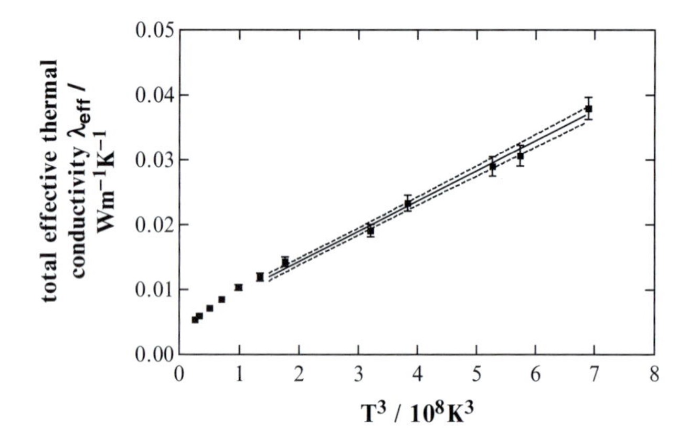

Figure 23.10. Total effective thermal conductivity of an evacuated opacified silica aerogel ($\rho = 153$ kg m^{-3}, opacifier 2.8% carbon black) as function of T^3 (gas pressure $p_g < 0.01$ Pa). A linear increase in this representation can be observed with increasing temperatures due to the diffusive radiative heat transfer according to (23.12). *Solid line*: linear regression curve, *dashed lines*: corresponding 95% coincidence interval.

[31, 32] and *phenolic-furfural (PF) aerogels* [13] have been investigated in detail. Other organic aerogel systems are also described in the literature and have been discussed with respect to their thermal properties, e.g., polydicyclopentadiene (pDCPD) based aerogels and polyisocyanurate aerogels [33], polyurethane based aerogels [33–35] and cellulose based aerogels [36].

Very low thermal conductivity values in the range of 0.011 Wm^{-1} K^{-1} were determined [37] for RF-aerogels. The morphology of RF aerogels can be influenced by the mass ratio *R*/*C* of resorcinol to the catalyst sodium carbonate and by the density of the RF aerogel. Differences in the morphology of RF aerogels are indicated by different colors of the macroscopic material (cf. Figure 23.11). At low R/C ratios, i.e., high catalyst concentration, the primary particles have small diameters with a large thermal contact area, i.e., large necks, between the particles. High R/C values, in contrast, result in large backbone-particle diameters with small necks and a distinct thermal contact resistance between the single particles. Furthermore, the higher the density and the lower the R/C ratio, the smaller the effective pore sizes of the RF aerogels. Lu et al. systematically varied the R/C ratio and densities of RF aerogels to experimentally characterize the influence of these parameters on the heat transfer. Figure 23.12 shows the effective total thermal conductivity of RF aerogels with three different R/C ratios as a function of density. Typical minima can be observed, resulting from the increase in thermal radiation with decreasing densities as λ_r is inversely proportional to the density ρ according to (23.12), and from a solid thermal conductivity λ_s, that increases with density following the scaling law according to (23.7) with $\alpha = 1.2 \ldots 1.5$.

Figure 23.11. Three different types of resorcinol–formaldehyde aerogel tiles. The *different colors* indicate different sizes of primary particles and pore sizes [38].

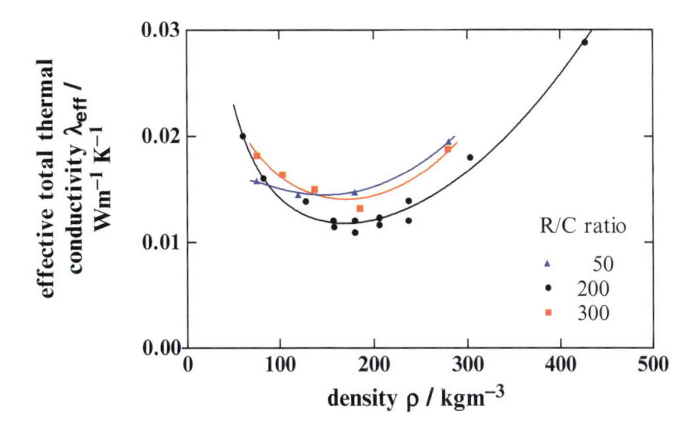

Figure 23.12. Effective total thermal conductivity of resorcinol–formaldehyde aerogels with different R/C ratios as a function of aerogel density at room temperature [24, 37].

Additionally, the gas pressure-dependent thermal conductivity λ_g decreases with increasing density because of the aforementioned influence of the pore size on density and because of the decreasing porosity (cf. Figure 23.13). It is also obvious that the smaller the R/C ratio, i.e., the smaller the thermal contact resistance between the primary particles of the RF aerogels, the smaller the coupling effect and therefore the slighter the effect of the pore gas on the effective total thermal conductivity. The lowest effective total thermal conductivity values were determined for an R/C ratio of 200 (cf. Figure 23.12).

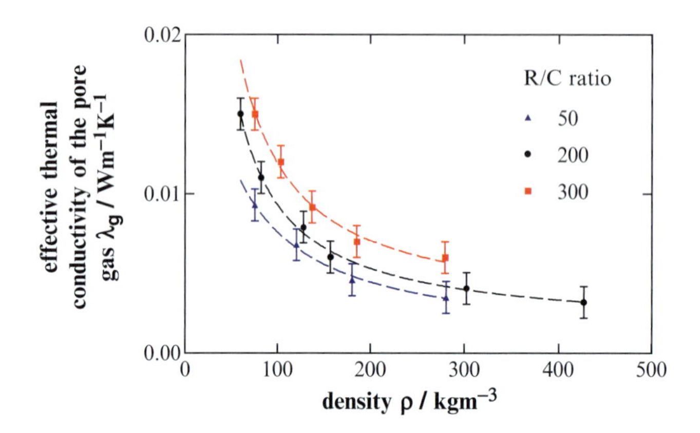

Figure 23.13. Effective thermal conductivity of the pore gas, λ_g, of resorcinol–formaldehyde aerogels with different R/C ratios as a function of aerogel density at room temperature and ambient atmospheric pressure [24, 31].

Carbon Aerogels

Carbon aerogels synthesized, for example, by pyrolyzing resorcinol–formaldehyde aerogels (Chap. 36), are very promising high-temperature insulation materials [39]. Besides the low solid thermal conductivity of the aerogel backbone and the suppressed contribution of the thermal conductivity of the pore gas, the extremely efficient infrared-opacifier carbon is intrinsically distributed homogeneously over the total aerogel volume, leading to a substantial reduction in radiative heat transfer. These advantages encouraged researchers to examine the thermal properties of carbon aerogels at high temperatures in detail [40–42]. Figure 23.14 shows a tile of carbon aerogel optimized for thermal insulation at high temperatures. The maximum operating temperature, i.e., the highest temperature under which the carbon aerogel is stable in inert atmosphere, depends on the maximum temperature used in the *pyrolysis process* and can be up to 2,500°C.

The structure of a carbon aerogel is illustrated in Figure 23.15. A special property of carbon aerogels is that the primary particles contain micropores, i.e., pores with an average pore size of less than 2 nm resulting from voids in between stacks of graphene like layers. The order and size of these layers grow with the maximum pyrolysis temperature and the maximum process time at this temperature [43]. As the ordered graphene layers grow, the phonon transport within the primary particles becomes very efficient and therefore a strong increase in the solid thermal conductivity can be observed with increasing pyrolysis temperature (cf. Figure 23.16).

Figure 23.14. Tile of a carbon aerogel for high-temperature thermal insulation with carbon-fiber paper covering layers on both sides to enhance mechanical stability [38].

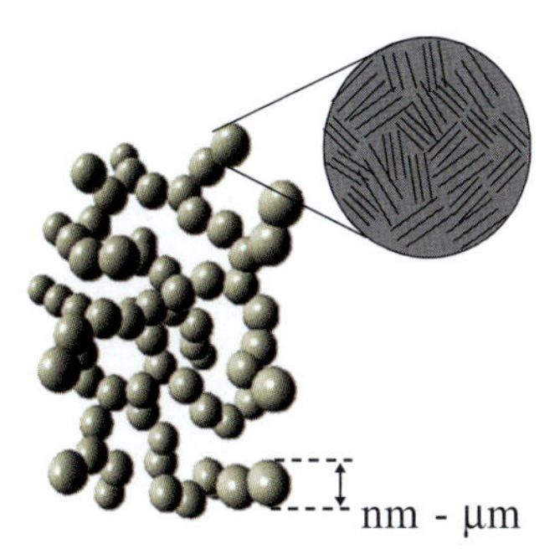

Figure 23.15. Diagram of the structure of a carbon aerogel [38] with its typical network of interconnected primary particles. The primary particles contain micropores and ordered stacks of polycrystalline graphene layers. The size of these microcrystallites depends on the maximum treatment temperature.

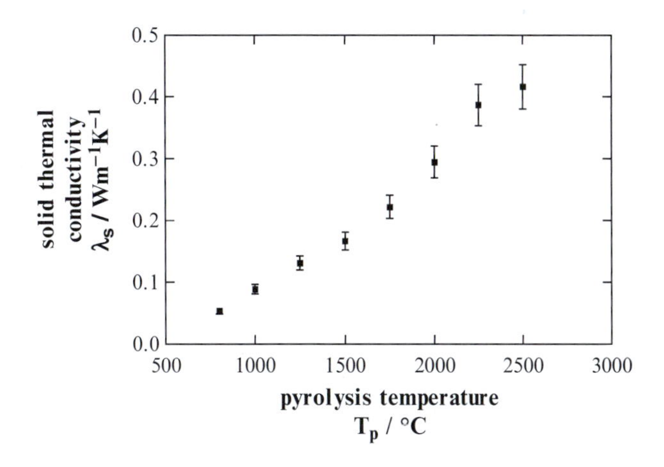

Figure 23.16. Solid thermal conductivity determined for a set of carbon aerogels treated at different temperatures. The values were determined for a temperature of 300°C under vacuum [43].

Another quality of carbon aerogels is that electrons contribute to the heat transfer. However, it was shown that this contribution is negligible in comparison to the other "classical" heat transfer mechanisms [41, 43, 44] involved. The effective total thermal conductivity of a carbon aerogel with a density of $\rho = 225$ kg m^{-3} under vacuum and in an argon atmosphere is depicted in Figure 23.17. Also shown are the contributions of the solid and radiative thermal conductivity, which represent in total the values for the evacuated sample. It is obvious that the radiative thermal conductivity is efficiently reduced because of the high total effective specific extinction coefficient values for carbon aerogels (cf. Figure 23.8). The solid thermal conductivity is the dominant contribution to the effective total thermal conductivity of this carbon aerogel. If the thermal conductivity of the evacuated sample is compared with that of the sample in a 0.1 MPa argon atmosphere, it becomes evident that the difference is nearly constant with temperature and does not increase as expected for a free gas. This indicates that the Knudsen number increases with temperature as the mean free path of gas molecules increases, and this compensates for the increase in the thermal conductivity of the free gas [cf. (23.10)].

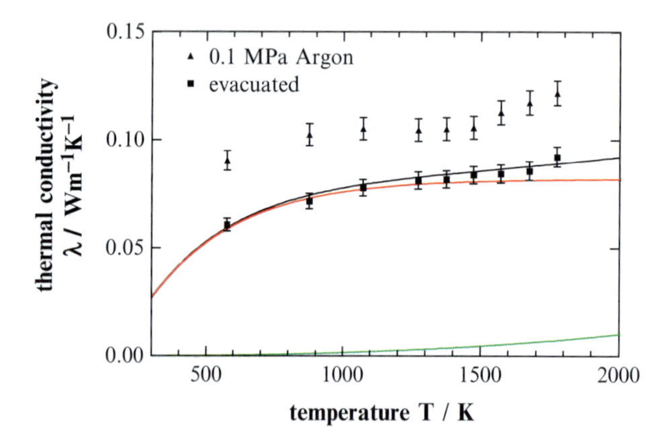

Figure 23.17. Effective total thermal conductivity of a carbon aerogel (density $\rho = 225$ kg m^{-3}, size of primary particles 100–150 nm, average pore size 570 nm, pyrolysis temperature 1,800°C) under different atmospheres and the corresponding solid (*red line*) and radiative (*green line*) thermal conductivity as well as the sum of both (*black line*) as a function of temperature [44].

Fiber-Reinforced Carbon Aerogels

Carbon aerogels are not particularly mechanically stable. Their mechanical stability can be improved by lining the surface with carbon-fiber fleece (cf. Figure 23.14) or by integrating temperature-stable fibers, e.g., carbon fibers. Drach et al. discussed the thermal properties of such a carbon-aerogel/fiber composite [45]. A carbon felt with a density $\rho = 100$ kg m^{-3} was infiltrated with the precursors of a resorcinol–formaldehyde aerogel and pyrolized after drying. The derived carbon-aerogel/carbon-fiber composite had a density of $\rho = 250$ kg m^{-3}. Due to a large amount of fibers with a high bulk thermal conductivity and due to the anisotropic structure of the felt as well as the fact that felt consists of a pinned stack of fiber mats, the effective thermal conductivity depends significantly on the direction. The carbon-aerogel/felt composite has an anisotropy factor

($\lambda_{parallel}/\lambda_{perpendicular}$) of about 2. In one direction, parallel to the carbon-fiber mats, the solid thermal conductivity is about 0.83 $Wm^{-1}\ K^{-1}$, perpendicular to this direction the value is about 0.44 $Wm^{-1}\ K^{-1}$ (Figure 23.18).

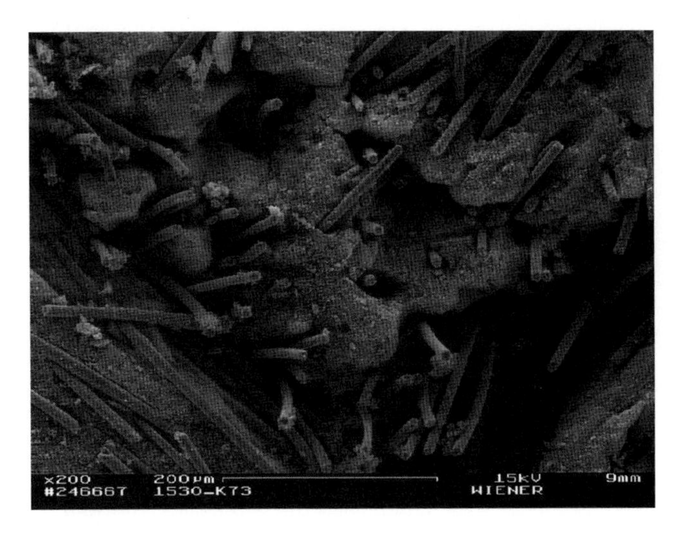

Figure 23.18. *SEM* image of a carbon-fiber felt with carbon-aerogel incorporated [45].

Graded Organic and Carbon Aerogels

Graded resorcinol–formaldehyde and related carbon aerogels were first successfully synthesized in 2009 [46] and offer a new potential to customize the thermal properties to given boundary temperatures. To minimize the thermal conductivity it is advantageous to realize low densities in regions with lower temperatures to reduce the heat transfer via the solid backbone and also to provide high densities in high-temperature regions to efficiently minimize the dominant radiative heat transfer.

Figure 23.19 shows the prepared resorcinol–formaldehyde aerogels with a gradient in structure (particle size, density and average pore size) along the cylindrical axis of depicted samples. The obvious change in the color indicates the structural changes of the aerogel as a function of the position. The variation of the morphology and the density of the samples are compiled in Figure 23.20. The particle size and the average pore size increase over two orders of magnitude from the position 0–15 mm inside the sample. In the same direction, the local density of the graded carbon aerogel decreases from about 360 to 250 kg m^{-3}. Finally, the resulting effective total thermal conductivity values are shown for the pyrolized graded materials as a function of the position along the cylindrical axis in argon atmosphere and under vacuum at 300°C and 600°C. As long as the structures, i.e., particle size and average pore size, are small, the contribution of the pore gas to the total effective thermal conductivity is negligible and the values for argon atmosphere are close to the thermal conductivity values measured under vacuum. The influence of the pore gas on the total heat transfer increases dramatically with the increase of the structure sizes. Since heat transfer via the solid backbone increases more or less with temperature, the thermal conductivity values for 600°C were always higher than the corresponding values for 300°C (Figure 23.21).

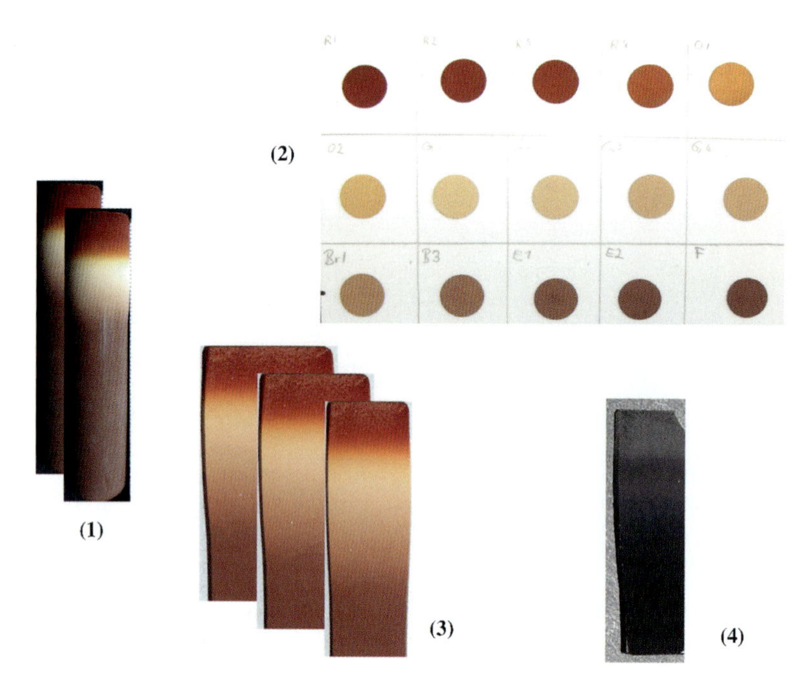

Figure 23.19. Cylinders of functionally graded resorcinol–formaldehyde aerogel (1), disc shaped slides cut from these cylinders perpendicular to the cylinder axis (2) (the change in color is a macroscopic indication of the structural gradient), slides parallel to the cylinder axis (3) and the corresponding pyrolized carbon aerogel (4).

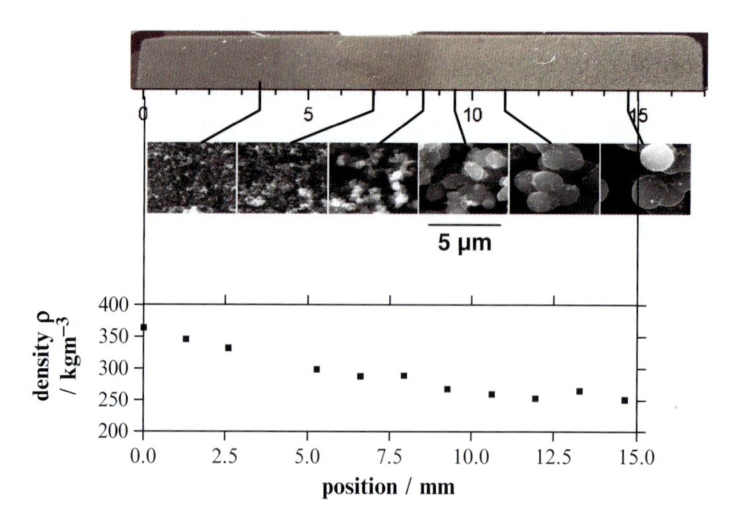

Figure 23.20. Slide of graded carbon aerogel from Figure 23.19 illustrating the change of morphology and density with the position along the cylindrical axis.

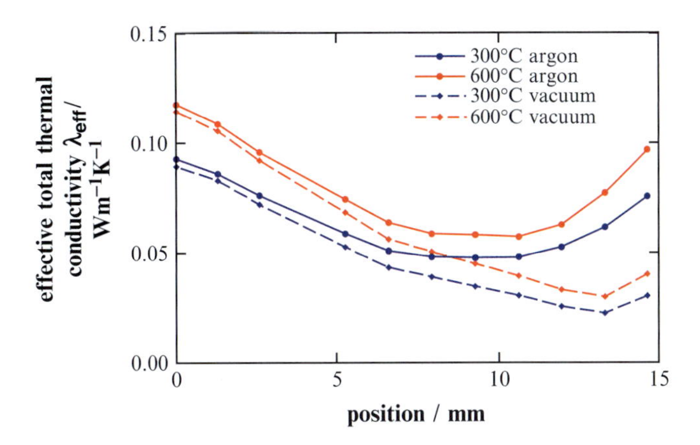

Figure 23.21. Effective total thermal conductivity of the graded carbon aerogel (Figure 23.20) as a function of the position along the cylindrical axis for measurements in argon atmosphere and under vacuum at 300°C and 600°C.

23.3. Heat Transfer Properties of Optically Thin Aerogels

Optically thin aerogels are defined by the nonlocal radiative heat transfer, i.e., the mean free path of the corresponding photons is in the magnitude of the macroscopic dimensions of the aerogel for certain wavelengths in the spectrum of thermal radiation. Radiative heat transfer in nonopacified silica aerogels is of such a nonlocal nature. For wavelengths smaller than 8 μm, the silica aerogel is nearly transparent for thermal radiation (cf. Figure 23.7) and is therefore a semitransparent material. In this case, determining the thermal conductivity value defined by Fourier's law (23.2) is no longer possible. The heat transfer resulting from a given temperature gradient over the aerogel also depends, for example, on the dimension of the aerogel and the infrared-optical properties of the boundaries. Therefore, in this case any experimental method applied to determine the thermal conductivity will influence the results to a certain degree, meaning only so-called "apparent" thermal conductivity values can be determined. These values are not material properties and must be interpreted very carefully with respect to their physical meaning.

In Figure 23.22 a striking example is illustrated: the values of the *thermal conductance* Λ for a silica aerogel tile determined by way of a steady-state guarded hot plate experiment as a function of T^3. The measurements were performed under vacuum (p_g <0.1 Pa) and in nitrogen atmosphere with two different emissivities (0.04 and 0.74) of the boundary plates of the guarded hot plate apparatus. In the case of an optically thick sample, multiplying the thermal conductance Λ with the sample thickness d would only result in the thermal conductivity. Varying the boundary emissivities in the case of an optically thin aerogel, however, influences the experimental results dramatically and it is quite clear that the derived values are not material properties, but system properties, i.e., they are not valid under other experimental conditions or for other specimens. This influence can be observed even at ambient temperature and increases at higher temperatures where the maximum thermal radiation shifts to a smaller wavelength and the silica aerogel becomes more transparent for thermal radiation. At ambient temperatures the Λ-value for the silica aerogel with a density of 50 kg m^{-3} determined in a 0.1 MPa nitrogen atmosphere is about 10% lower for the low emissivity boundaries ($\varepsilon = 0.04$) than the value obtained with boundary emissivities of ε

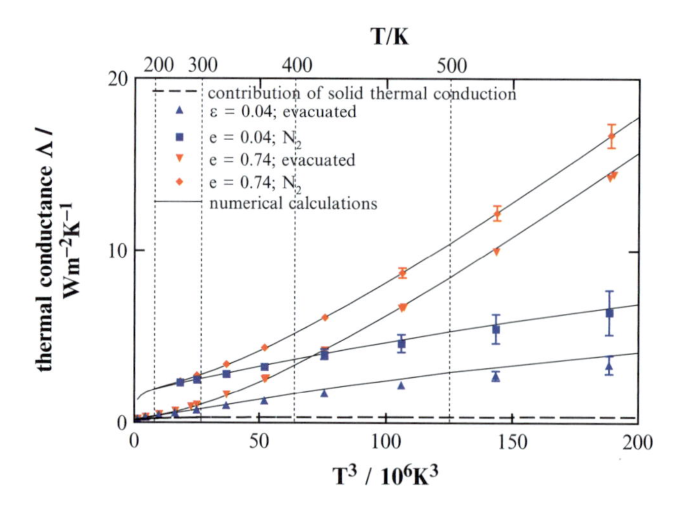

Figure 23.22. Measured and calculated thermal conductance Λ as a function of T^3 for a silica aerogel tile with a density of 50 kg m^{-3} and a thickness d of 6.7 mm [47]. The measurements were performed under different atmospheres and with two different boundary emissivities (0.04: *blue symbols*, 0.74: *red symbols*).

= 0.74 (Figure 23.22). This difference decreases to less than 2% at ambient temperatures for a silica aerogel tile with a density of 220 kg m^{-3} and a thickness of 7.9 mm [47].

The physical–mathematical description of the heat transfer in optically thin aerogels leads to coupled nonlinear integrodifferential equations which cannot be treated analytically. In the heat transfer equation, (23.1), the heat source term Φ considers the net radiative heat flux into the infinitesimal volume element where this equation is defined:

$$\Phi(\vec{x}) = \nabla \vec{q}_{\mathrm{r}}(\vec{x}). \tag{23.17}$$

The radiative heat flux density q_{r} is a solution to the radiative heat transfer equation. The problem of coupled radiation–conduction heat transfer was solved for planar aerogel systems under steady-state conditions by Zeng et al. and Heinemann et al. [28, 47]. In Figure 23.22 the result of the numerical simulations is compared with the experimental results and show excellent agreement between theory and experiment. Analytical approximation solutions can also be applied to estimate this coupling effect. In this vein, Fricke et al. developed an approximation solution for gray, scattering and absorbing media, which takes into account the coupled heat transfer by radiation and conduction in the case of a planar medium with equal emissivities on both boundaries [48]. The influence of the *dynamic hot-wire method* (widely used to ascertain the thermal conductivity of aerogels) when determining the thermal conductivity of optically thin silica aerogels was investigated by Ebert and Fricke [49].

The effective total thermal conductivity of nonopacified silica aerogels at cryogenic temperatures (Figure 23.23) was determined by Bernasconi et al. and Scheuerpflug et al. [50, 51]. Due to the strong nonlinear decrease in the radiative heat transfer toward lower temperatures, the derived thermal conductivity values most closely correspond to real solid thermal conductivity values. Generally, the solid thermal conductivity of silica aerogel at low temperatures can be described with the knowledge of the *density of vibrational states* $g(\omega)$ of the aerogel [52], the frequency- and temperature-dependent average velocity of the

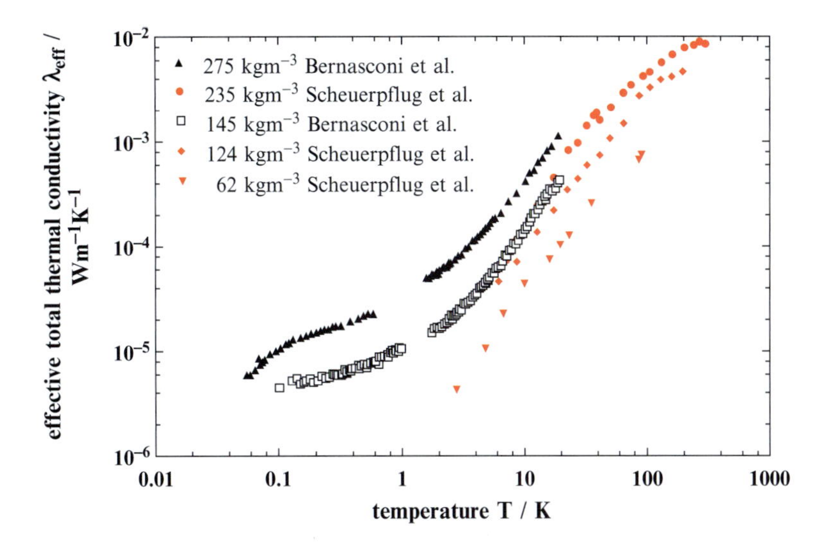

Figure 23.23. Effective total thermal conductivity of silica aerogels with different densities ρ below 300 K [50, 51].

elastic waves, v, and the frequency- and temperature-dependent mean free path of the phonons [50]:

$$\lambda_s(T) = \frac{1}{3}\int_0^{\infty} \rho \cdot g(\omega) \cdot k_B \frac{(\hbar\omega/k_BT)^2 \exp(\hbar\omega/k_BT)}{[\exp(\hbar\omega/k_BT) - 1]^2} \cdot v(\omega, T) \cdot l_{ph}(\omega, T)d\omega \qquad (23.18)$$

with k_B: Boltzmann constant, $\hbar$: reduced Planck constant and ω: angular frequency. Scheuerpflug et al. distinguish four different temperature regions: Below 0.01 K the wavelengths of phonons exceed the largest aerogel pores and the aerogel can be treated as a homogeneous medium. From 0.01 to 1 K the phonon wavelength is in the magnitude of the chains. A temperature region can be recognized where the thermal conductivity is only slightly temperature dependent (λ plateau) because of the saturation of phonon modes. The plateau for aerogels is significantly shifted to lower temperatures in comparison to the λ plateau of amorphous silica, which is in the range of $2 < T/\mathrm{K} < 10$ [50]. For the temperature range $1 < T/\mathrm{K} < 50$, volume and surface modes of the primary particles are excited and the propagation of phonons is about half the phonon speed in vitreous silica [52–54]. Above 50 K the phononic heat transfer takes place on the same length scale as in vitreous silica, i.e., diffusing phonons are responsible for heat transfer within the primary particles. The solid thermal conductivity is proportional to the specific heat.

23.4. Thermal Conductivity of Aerogels Powders, Granulates, and Aerogel Composites

Beds of aerogel granulate or powder are used for technical thermal insulation (Part 15), because they offer specific advantages in comparison to standard insulation materials, e.g., they can be poured into complicated shaped cavities. Opacified aerogel powders and granulates are used as thermal insulation in space applications [55], automotive applications

[56] and in buildings [57]. In the latter case, nonopacified silica aerogel is also used as transparent thermal insulation. Grinding aerogel monoliths into dispersed powders or granulates, or synthesizing aerogel particles via a direct route has a significant influence on the heat transfer. On the one hand, the interparticular voids in the aerogel bed enhance the contribution of the gas to the total heat transfer, because the average pore size of the interparticular voids is always larger than the mesopores of the aerogel. On the other hand, the contact points between the powder or granulate particles act as additional thermal contact resistances, which decrease the solid thermal conductivity. This contact resistance depends on the contact area which is a function of the external load on the dispersed system. A model provided by Kaganer for the effective solid thermal conductivity of a bed of hard spheres can be applied to describe this effect [16]:

$$\lambda_s = \lambda_0 \cdot 3.4 \cdot (1 - \Pi)^{4/3} \sqrt[3]{\frac{1 - \mu^2}{Y}}\, p_{ext}^{1/3}. \tag{23.19}$$

with λ_0: "bulk" thermal conductivity of the monolithic aerogel, Π: porosity, μ: the Poisson's ratio, Y: the elastic modulus and p_{ext}: the external pressure load. Other models besides Kaganer's can also be used here, such as the previously mentioned percolation model [see (23.7)], cell models [58] or models based on a perturbation method, as suggested by Klemens [59].

Figure 23.24 depicts the effective total thermal conductivity values of three silica aerogel powders in the temperature range from 13 to 275 K. The powders were doped with different amounts of carbon black (Printex 60, Degussa, Hanau, Germany) and were loaded with an external pressure of 0.1 MPa. The silica aerogel powder was synthesized by milling silica aerogel granulate from BASF, Ludwigshafen, Germany. The bulk density of the granules was 150 kg m^{-3}. The highest values were determined for the pure silica aerogel powder. The opacification obviously reduces the effective total thermal conductivity. In principle, the higher the carbon black content the lower the measured thermal conductivity values and the slope of the curves. Since the radiative conductivity is proportional to the temperature cubed the influence of thermal radiation only became remarkable above 100 K. Below 100 K the effective total thermal conductivity is only slightly higher than the solid

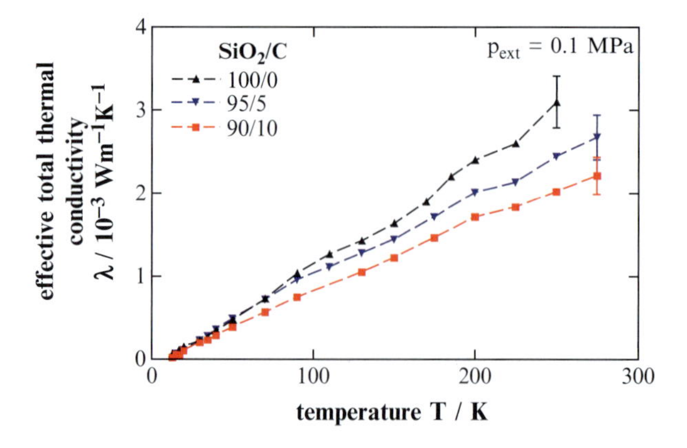

Figure 23.24. Effective total thermal conductivity of evacuated opacified silica aerogel powders (gas pressure $p_g < 0.01$ Pa) with different weight percents of opacifier (0/5/10% carbon black). The specimens were loaded with an external pressure of 0.1 MPa [60, 61].

thermal conductivity, which is within the experimental uncertainties. The lowest total effective thermal conductivity values were measured for the silica aerogel powder with 10% carbon black. The thermal conductivity values increase from 0.02×10^{-3} Wm^{-1} K^{-1} to about 2.2×10^{-3} Wm^{-1} K^{-1}. The values derived for the powder systems at an external load of 0.1 MPa were also compared with total thermal conductivity values determined for corresponding opacified aerogel monoliths. The authors found that the solid conductivity is reduced by a factor of about 4 in comparison to aerogel monoliths [60].

Silica aerogel granulate can also be used as transparent insulation in building applications (Chap. 38), where it is advantageous to transmit sunlight into the interior of buildings, but the façade needs to provide excellent thermal insulation [57]. In this case, the translucent granulate is filled in PMMA double skin sheets. Typical application temperatures are around 300 K. Therefore, the nonopacified, dispersed silica aerogel system has to be treated as a semitransparent medium and, as already mentioned, only system-related heat transfer properties can be determined, i.e., the thermal conductance. From a theoretical point of view, it is of interest to discuss the gas pressure-dependent thermal conductance, because the different influences of macropores (i.e., intergranular voids) and mesopores on the heat transfer can be studied [57, 62]. Figure 23.25 shows the thermal conductance and the corresponding apparent thermal conductivity of a bed of silica aerogel granulate with a thickness of 22.3 mm as a function of the nitrogen gas pressure. The bimodal pore size distribution, consisting of macropores and mesopores, results in the superposition of two typically s-shaped curves expected from theory [cf. (23.10)]. The intergranular voids are responsible for a steplike increase in the determined thermal conductance of 0.5 Wm^{-2} K^{-1}, starting at a gas pressure of about 1 Pa. These intergranular voids are the reason why the excellent thermal insulation values of powders cannot be achieved, even if the granulates are opacified.

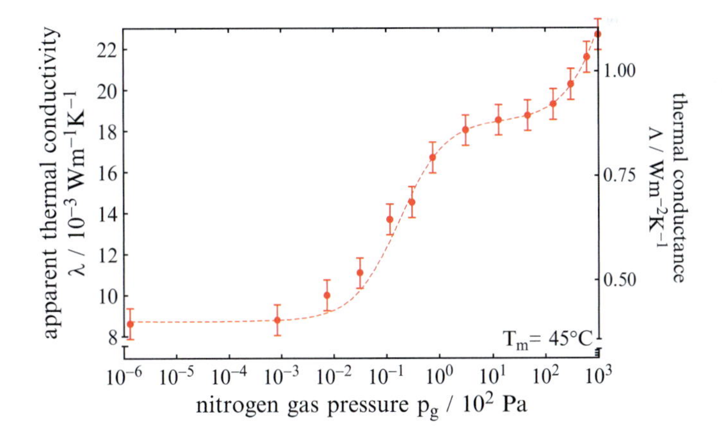

Figure 23.25. Thermal conductance and the corresponding apparent thermal conductivity of a bed of silica aerogel granulate with a thickness of 22.3 mm as a function of gas pressure. The *dashed line* is the result of the theoretical modeling of the heat transfer [57].

Commercially available flexible aerogel blankets – which are a composite material of silica aerogel synthesized within a fiber matrix – show excellent thermal insulation performance from cryogenic to high temperatures. The heat transfer within these material systems can be described in the same manner as for optically thick aerogels. Additionally, degradation in the thermal conductivity can be observed in some cases, if the flexible blankets are bent, which leads to microcracks in the homogeneous aerogel phase. Figure 23.26 depicts the effective thermal conductivity of such a degraded aerogel blanket versus gas pressure. The incident microcracks are responsible for the slight plateau starting at a gas pressure of 1 Pa and increase the total effective thermal conductivity of the investigated specimen by about 0.002 $Wm^{-1} K^{-1}$ [63].

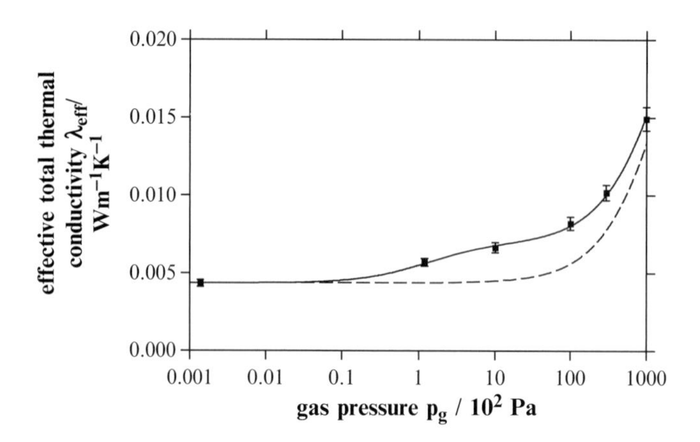

Figure 23.26. Effective total thermal conductivity as a function of gas pressure p_g of an aerogel blanket with microcracks at a temperature of 311 K in a nitrogen atmosphere. The investigated thermal specimen was not loaded with an external pressure [63]. The *dashed line* shows the expected thermal conductivity values for this aerogel blanket without of microcracks.

23.5. Specific Heat of Aerogels

The physical property "specific heat" is the measure of the amount of heat Q required to increase the temperature of a unit mass m of a substance by 1 K. One differentiates between specific heat under constant pressure, c_p, and at constant volume c_v. The following applies for specific heat under constant pressure:

$$c_p = \frac{1}{m}\left(\frac{\partial Q}{\partial T}\right)_p. \tag{23.20}$$

The specific heat not only describes the capacity to store heat, but also influences the dynamics of heat transfer within aerogels according to the equation of heat transfer (23.1) and (23.4). The higher the specific heat, the slower the heat propagation within a material, if the density and the effective thermal conductivity remain constant. In the late eighties and early nineties several authors investigated the specific heat of silica aerogels in detail [9, 51, 52, 54, 64–67]. The focus of their research work was the study of the density of vibrational states $g(\omega)$ correlated to the specific heat and the solid thermal conductivity by phonons:

$$c_p(T) = \int_0^\infty g(\omega) \cdot k_B \frac{(\hbar\omega/k_BT)^2 \exp(\hbar\omega/k_BT)}{[\exp(\hbar\omega/k_BT) - 1]^2} d\omega. \tag{23.21}$$

with k_B: Boltzmann constant, ħ: reduced Planck constant, and ω: angular frequency.

The specific heat of silica aerogels increases monotonously with temperature (Figure 23.27) and, in the range of $T < 2$ K, is on the order of two magnitudes higher than the specific heat of amorphous silica. As can be seen from the log–log plot the temperature dependence of the specific heat follows a power law:

$$c_p(T) \propto T^\delta \tag{23.22}$$

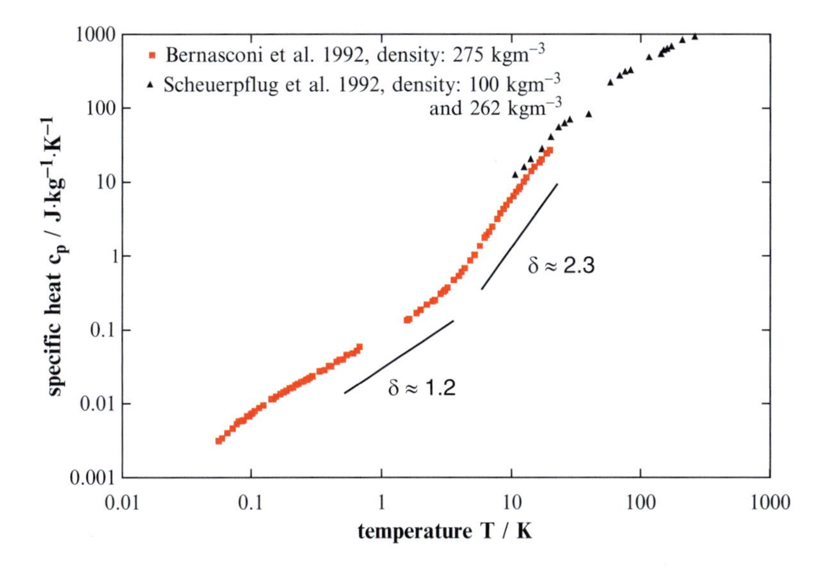

Figure 23.27. Specific heat of silica aerogels at constant pressure as a function of temperature determined by Bernasconi et al. [50] and Scheuerpflug et al. [51]. The density of the investigated aerogel specimens varied between 100 kg m^{-3} and 275 kg m^{-3}.

with different exponents for different temperature regions. At higher temperatures the temperature dependence decreases and the specific heat converges asymptotically to the limit of the *Dulong–Petit law* ($\delta = 0$) [51].

Data for the specific heat of carbon aerogels in the temperature range from 300 to 1,800 K are provided by Wiener et al. [44]. The specific heat values are described by an empirical analytical equation:

$$c_p(T) = -C1 + C2\left(1 - \exp\left(-\frac{T}{C3}\right)\right) \tag{23.23}$$

with $C1 = 1{,}440$ J kg^{-1} K^{-1}, $C2 = 3{,}528$ J kg^{-1} K^{-1} and $C3 = 319$ K. These data are used to calculate effective total thermal conductivity values from the experimental determined values of the thermal diffusivity, a, at high temperatures:

$$\lambda(T) = a(T) \cdot \rho(T) \cdot c_p(T). \tag{23.24}$$

For the determination of the thermal diffusivity the Laserflash method is applied [40].

23.6. Conclusion

In the course of this chapter the principal mechanisms of heat transfer in aerogels were discussed and illustrated by experimental results. The mesoporous structure of aerogels is the main reason for their remarkable thermal properties, e.g., thermal conductivity values at ambient conditions far below the values of still air for low density silica or organic aerogels. For optically thick aerogels, i.e., in the case of diffusive radiative heat transfer, an effective total thermal conductivity can be defined which is a real material property. However, for optically thin, i.e., semitransparent, aerogels, such as nonopacified silica aerogels, radiative heat transfer becomes a nonlocal phenomenon. Thus an experimentally determined apparent thermal conductivity value depends on the experimental set up and is therefore no longer a material property.

In the past the general interest to determine thermal properties of aerogels also promoted the development of more precise measurement techniques for the determination of low thermal conductivity values and of theoretical models for a deeper understanding of heat transfer in semitransparent media. The demand for advanced functional materials with improved thermal properties and the nowadays available knowledge about aerogel synthesis lead to a continuous development of new types of aerogels and aerogel based composite materials where no reference values of thermophysical properties are available in literature. Therefore it will be still a challenging task in the future to determine reliable thermophysical data for aerogels and to understand the physics behind these data in order to allow a target-oriented material development.

References

1. Kistler, S.S., The Relation between Heat Conductivity and Structure in Silica Aerogel. J. Phys. Chem., 1935. **39** (1): p. 79–86.
2. Kistler, S.S., The calculation of the surface area of microporous solids from measurements of heat conductivity. Journal of Physical Chemistry, 1942. **46**(1): p. 19–31.
3. White, J.F., Silica Aerogel: Effect of Variables on Its Thermal Conductivity. Industrial and Engineering Chemistry, 1939. **31**(7): p. 827–831.
4. Özisik, M., Radiative Transfer. 1973, New York: John Wiley & Sons.
5. Bankvall, C.G., Natural Convection in Vertical Permeable Space. Wärme- und Stoffübertragung, 1974. **7**: p. 22–30.
6. Clyne, T.W., Golosnoy, I.O., Tan, J.C. and Markaki, A.E., Porous materials for thermal management under extreme conditions. Philosophical Transactions of the Royal Society a-Mathematical Physical and Engineering Sciences, 2006. **364**(1838): p. 125–146.
7. Debye, P., Vorträge über die Kinetische Theorie der Materie und der Elektrizität. 1914, Berlin: B. G. Teubner.
8. Lu, X., Transport Properties of Porous Media, PhD thesis, University of Würzburg, Germany, 1991.
9. Scheuerpflug, P., Morper, H.J., Neubert, G. and Fricke, J., Low-Temperature Thermal Transport in Silica Aerogels. Journal of Physics D-Applied Physics, 1991. **24**(8): p. 1395–1403.
10. Stauffer, D., Introduction to Percolation Theory. 1985, London: Taylor and Francis.
11. Heinemann, U., Wärmetransport in semitransparenten nichtgrauen Medien am Beispiel von SiO2-Aerogelen, Dissertation, University of Würzburg, 1993.
12. Cahill, D.G., Stephens, R.B., Tait, R.H., Watson, S.K. and Pohl, R.O. Thermal Conductivity and Lattice Vibration in Glasses. in Thermal Conductivity. 1990: Plenum Press New York.
13. Pekala, R.W., Alviso, C.T., Lu, X., Gross, J. and Fricke, J., New Organic Aerogels Based Upon a Phenolic-Furfural Reaction. Journal of Non-Crystalline Solids, 1995. **188**(1-2): p. 34–40.
14. Nilsson, O., Rüschenpöhler, G., Gross, J. and Fricke, J., Correlation between thermal conductivity and elasto-mechanical properties of compressed porous media. High Temperatures - High Pressures, 1989. **21**: p. 267.

15. Reif, F., Fundamentals of Statistical and Thermal Physics. 1965: McGraw-Hill, Inc.
16. Kaganer, M.G., Thermal Insulation in Cryogenic Engineering. 1969, IPST Press: Jerusalem, Israel.
17. Swimm, K., Reichenauer, G., Vidi, S. and Ebert, H.-P., Gas Pressure Dependence of the Heat Transport in Solids with Pores Smaller than 10μm. International Journal of Thermophysics, 2009. **30**(4): p. 1329–1342
18. Siegel, R. and Howell, J.R., Thermal Radiation heat transfer. 1972, Tokyo: McGraw-Hill Kogakushka, Ltd.
19. Cerny, M., Walther, A., Tables of Fractional Functions for the Planck Radiation Law. 1961, Berlin: Springer Verlag.
20. Jackson, J.D., Classical Electrodynamics. third edition ed. 1999, New York: John Wiley & Sons.
21. Poelz, G. Aerogel in High Energy Physics. in First International Symposium on Aerogels (1st ISA). 1985. Würzburg, Germany: Springer-Verlag.
22. Pomraning, G.C., Prinja, A. K., Shokair, I. R., J.Quant. Spectrosc. Radiat. Transfer, 1981. **26**: p. 199–213.
23. Caps, R. and Fricke, J. Radiative Heat Transfer in Silica Aerogel. in First International Symposium on Aerogels (1st ISA). 1985. Würzburg, Germany: Springer-Verlag.
24. Ebert, H.-P., Caps, R., Heinemann, U. and Fricke, J. Aerogels – open-pored nanostructered insulating materials. in International Centre for Heat and Mass Transfer (ICHTM) Symposium on Molecular and Microscale Heat Transfer in Materials Processing another Applications. 1996. Yokohama, Japan.
25. Lu, X., Wang, P., Buttner, D., Heinemann, U., Nilsson, O., Kuhn, J. and Fricke, J., Thermal transport in opacified monolithic silica aerogels. High Temperatures - High Pressures, 1991. **23**(4): p. 431–436.
26. Kuhn, J., Gleissner, T., Arduinischuster, M.C., Korder, S. and Fricke, J., Integration of Mineral Powders into SiO_2 Aerogels. Journal of Non-Crystalline Solids, 1995. **186**: p. 291–295.
27. Hrubesh, L.W. and Pekala, R.W., Thermal-Properties of Organic and Inorganic Aerogels. Journal of Materials Research, 1994. **9**(3): p. 731–738.
28. Zeng, S.Q., Hunt, A. and Greif, R., Theoretical Modeling of Carbon Content to Minimize Heat-Transfer in Silica Aerogel. Journal of Non-Crystalline Solids, 1995. **186**: p. 271–277.
29. Lee, D., Stevens, P.C., Zeng, S.Q. and Hunt, A.J., Thermal Characterization of Carbon-Opacified Silica Aerogels. Journal of Non-Crystalline Solids, 1995. **186**: p. 285–290.
30. Ebert, H.-P., Bock, V., Nilsson, O. and Fricke, J. Errors From Radiative Heat Transfer In The Determination Of The Thermal Conductivity In Semitransparent Media Using The Hot-Wire Method. in 21th Eurotherm Seminar: Heat Transfer in Semi-Transparent Media. 1992. Lyon.
31. Lu, X., Caps, R., Fricke, J., Alviso, C.T. and Pekala, R.W., Correlation between Structure and Thermal-Conductivity of Organic Aerogels. Journal of Non-Crystalline Solids, 1995. **188**(3): p. 226–234.
32. Lu, X., Arduini-Schuster, M.C., Kuhn, J., Nilsson, O., Fricke, J. and Pekala, R.W., Thermal Conductivity of Monolithic Organic Aerogels. Science, 1992. **255**(5047): p. 971–972.
33. Lee, J.K. and Gould, G.L., Polydicyclopentadiene based aerogel: a new insulation material. J Sol-Gel Sci Technol, 2007. **44**: p. 29–40.
34. Lee, J.K., Gould, G.L. and Rhine, W., Polyurea based aerogel for a high performance thermal insulation material. J Sol-Gel Sci Technol, 2009. **49**(2): p. 209–220.
35. Biesmans, G., Randall, D., Francais, E. and Perrut, M., Polyurethane-based Organic Aerogels Thermal Performance. J. Non-cryst. Solids, 1998. **225**: p. 36.
36. Fischer, F., Rigacci, A., Pirard, R., Berthon-Fabry, S. and Achard, P., Cellulose-based aerogels. Polymer, 2006. **47**(22): p. 7636–7645.
37. Lu, X., Wang, P., Arduinischuster, M.C., Kuhn, J., Büttner, D., Nilsson, O., Heinemann, U. and Fricke, J., Thermal Transport in Organic and Opacified Silica Monolithic Aerogels. Journal of Non-Crystalline Solids, 1992. **145**(1-3): p. 207–210.
38. ZAE Bayern. 2009: Würzburg, Germany.
39. Fricke, J. and Petricevic, R., Aerogels - Carbon Aerogels, in *Handbook of Porous Solids*, F. Schüth, K.S.W. Sing, and J. Weitkamp, Editors. 2002, Wiley VCH. p. 2037–2062.
40. Bock, V., Nilsson, O., Blumm, J. and Fricke, J., Thermal Properties of Carbon Aerogels. Journal of Non-Crystalline Solids, 1995. **185**(3): p. 233–239.
41. Lu, X.P., Nilsson, O., Fricke, J. and Pekala, R.W., Thermal and Electrical-Conductivity of Monolithic Carbon Aerogels. Journal of Applied Physics, 1993. **73**(2): p. 581–584.
42. Nilsson, O., Bock, V. and Fricke, J., High Temperature Thermal Properties of Carbon Aerogels, in 22nd Thermal Conductivity Conference. 1993: Tempe, Arizona, USA.
43. Wiener, M., Reichenauer, G., Hemberger, F. and Ebert, H.-P., Thermal conductivity of monolithic synthetic hard carbons as a function of pyrolysis temperature. International Journal of Thermophysics, 2006. **27**(6): p. 1826–1843.

44. Wiener, M., Reichenauer, G., Braxmeier, S., Hemberger, F. and Ebert, H.P., Carbon Aerogel-Based High-Temperature Thermal Insulation. International Journal of Thermophysics, 2009. **30**(4): p. 1372.
45. Drach, V., Wiener, M., Reichenauer, G., Ebert, H.P. and Fricke, J., Determination of the Anisotropic Thermal Conductivity of a Carbon Aerogel–Fiber Composite by a Non-contact Thermographic Technique. International Journal of Thermophysics, 2007. **28**(4): p. 1542–1562.
46. Hemberger, F., Weis, S., Reichenauer, G. and Ebert, H.P., Thermal Transport Properties of Functionally Graded Carbon Aerogels. International Journal of Thermophysics, 2009. **30**(4): p. 1357–1371
47. Heinemann, U., Caps, R. and Fricke, J., Radiation-conduction interaction : An investigation on silica aerogels. International Journal of Heat and Mass Transfer, 1996. **39**(10): p. 2115–2130.
48. Fricke, J., Caps, R., Hümmer, E., Döll, G., Arduini-Schuster, M. and De Ponte, F., Optically Thin Fibrous Insulations, in *Insulation Materials, Testing, and Applications*, D.L. McElroy, Editor. 1990, American Society for Testing and Materials ASTM STP 1030: Philadelphia. p. 575–586.
49. Ebert, H.P. and Fricke, J., Influence of Radiative Transport on Hot-Wire Thermal Conductivity Measurement. High Temperatures - High Pressures, 1998. **30**: p. 655–669.
50. Bernasconi, A., Sleator, T., Posselt, D., Kjems, J.K. and Ott, H.R., Dynamic properties of silica aerogels as deduced from specific-heat and thermal-conductivity measurements. Physical Review B, 1992. **45**(18): p. 10363–10376.
51. Scheuerpflug, P., Hauck, M. and Fricke, J., Thermal Properties of Silica Aerogels between 1.4 and 330 K. J. Non-cryst. Solids, 1992. **145**: p. 196–201.
52. Conrad, H., Buchenau, U., Schatzler, R., Reichenauer, G. and Fricke, J., Crossover in the Vibrational Density of States of Silica Aerogels Studied by High-Resolution Neutron Spectroscopy. Physical Review B, 1990. **41** (4): p. 2573–2576.
53. Tsujimi, Y., Courtens, E., Pelous, J. and Vacher, R., Raman-Scattering Measurements of Acoustic Superlocalization in Silica Aerogels. Physical Review Letters, 1988. **60**(26): p. 2757–2760.
54. Buchenau, U., Monkenbusch, M., Reichenauer, G. and Frick, B., Inelastic Neutron-Scattering from Virgin and Densified Aerogels. Journal of Non-Crystalline Solids, 1992. **145**(1-3): p. 121–127.
55. Fesmire, J.E., Aerogel insulation systems for space launch applications. Cryogenics, 2006. **46**: p. 111–117.
56. Blüher, P., Latentwärmespeicher erhöht den Fahrkomfort und die Fahrsicherheit. Automobiltechnische Zeitschrift ATZ, 1991. **93**(10).
57. Reim, M., Korner, W., Manara, J., Korder, S., Arduini-Schuster, M., Ebert, H.P. and Fricke, J., Silica aerogel granulate material for thermal insulation and daylighting. Solar Energy, 2005. **79**(2): p. 131–139.
58. Reim, M., Reichenauer, G., Korner, W., Manara, J., Arduini-Schuster, M., Korder, S., Beck, A. and Fricke, J., Silica-aerogel granulate - Structural, optical and thermal properties. Journal of Non-Crystalline Solids, 2004. **350**: p. 358–363.
59. Klemens, P.G., in *Thermal Conductivity 21*, C.J. Cremers and H.A. Fine, Editors. 1990, Plenum: New York. p. 383.
60. Rettelbach, T., Saeuberlich, J., Korder, S. and Fricke, J., Thermal conductivity of IR-opacified silica aerogel powders between 10 K and 275 K. Journal of Physics D: Applied Physics, 1995. **28**(3): p. 581–587.
61. Rettelbach, T., Der Wärmetransport in evakuierten Pulvern bei Temperaturen zwischen 10 K und 275 K, Dissertation, University of Würzburg / Germany, 1996
62. Bisson, A., Rigacci, A., D., L. and Achard, P., Effective thermal conductivity of divided silica xerogel beds. Journal of Non-Crystalline Solids, 2004. **350**: p. 379–384.
63. ZAE Bayern. 2003: Würzburg, Germany.
64. Calemczuk, R., De Goer, A.M., Salce, B., Maynard, R. and Zarembowitch, A., Low-Temperature Properties of Silica Aerogels. Europhys. Lett., 1987. **3**: p. 1205–1211.
65. Maynard, R., Calemczuk, R., De Goer, A.M., Salce, B., Bon, J., Bonjour, E. and Bourret, A., Low Energy Excitations in Silica Aerogels. Revue de Physique Appliquêe Colloque C4, 1989. **supplément au n°4**(24): p. 107–112.
66. Sleator, T., Bernasconi, A., Posselt, D., Kjems, J.K. and Ott, H.R., Low-Temperature Specific Heat and Thermal Conductivity of Silica Aerogels. Physical Review Letters, 1991. **66**(8): p. 1070–1073.
67. Nilsson, O., Fransson, A. and Sandberg, O. Thermal Properties of Silica Aerogel. in First International Symposium on Aerogels. 1985. Würzburg: Springer-Verlag.

24

Simulation and Modeling of Aerogels Using Atomistic and Mesoscale Methods

Lev D. Gelb

Abstract Molecular modeling and simulation are now widely used in many areas of materials science. In this chapter, we consider the application of these techniques to developing a better understanding of the structure and properties of aerogels. Both atomistic simulations and "coarse-grained" models are reviewed, and the challenges and possible solutions facing this field are also discussed. We focus on silica aerogels, as the great majority of simulation work in this area has been directed at understanding these materials.

24.1. Introduction

The preparation and characteristics of aerogels are reviewed in detail earlier in this volume, but we summarize the points here of particular relevance to modeling. Gels are highly branched, mechanically rigid, and system-spanning polymer networks that are produced by the aggregation of polyfunctional monomer species in solution. In the case of silica gels, the monomer species are alkoxide precursors $Si(OR)_3$ [most often tetramethoxysilane (*TMOS*) or tetraethoxysilane (*TEOS*)] in a water/alcohol solution. These species undergo hydrolysis to form reactive silanol groups, which then undergo condensation, forming Si–O–Si bridging bonds. Under appropriate reaction conditions one rapidly obtains a solution ("sol") of small, dense silica particles. These aggregate into clusters through additional condensation reactions and eventually form a gel. Aerogels are then prepared by removing the solvent without collapsing the gel structure, most often through supercritical drying [1–5].

Aerogels are known to have mass and surface distributions consistent with fractal behavior over certain length scales. At sufficiently small (atomic) length scales their structure is determined by chemical considerations and is therefore not fractal, and at sufficiently large length scales they are homogeneous [2, 5, 6]. The evolution of this structure, and the corresponding effects of relaxation, chemical kinetics, and atomic coordination have been of particular interest in modeling and simulation studies. Other topics that have been the focus of modeling studies include the prediction and analysis of gelation

L. D. Gelb • Department of Materials Science and Engineering, University of Texas at Dallas,
RL 10, Richardson, TX 75080, USA
e-mail: lev.gelb@utdallas.edu

M.A. Aegerter et al. (eds.), *Aerogels Handbook*, Advances in Sol-Gel Derived Materials and Technologies, DOI 10.1007/978-1-4419-7589-8_24,

kinetics and aerogel mechanical properties, for which a considerable amount of experimental data is available [7–9].

Computational methods can be used to address a number of the outstanding questions concerning aerogel structure, preparation, and properties. In a computational model, the material structure is known exactly and completely, and so structure/property relationships can be determined and understood directly. The most challenging part of such a computational study is obtaining the model structure itself; experimental characterization does not provide sufficient information to completely specify a model structure at the atomic scale, and so other methods must be used. Techniques applied in the case of aerogels include both "mimetic" simulations, in which the experimental preparation of an aerogel is imitated using dynamical simulations, and reconstructions, in which available experimental data is used to generate a statistically representative structure.

Having prepared a model structure, simulation studies have generally focussed on structural characterization and the relationship between aerogel structure and mechanical properties. Global measures including fractal dimensions, surface areas, and pore size distributions can be directly calculated from the model structure and compared with experimental data. Microscopic measurements such as the distributions of bond lengths and bond angles and the number of bridging oxygens bound to each silicon atom are also measured. Finally mechanical properties, including moduli, shrinkage upon drying, and the vibrational density of states can be determined and correlated with the gel structure.

Simulating the formation and properties of aerogels, like all other applications of molecular simulation, involves choosing both a model and a simulation technique. The model specifies the actual "fundamental" objects simulated, be they electrons, atoms, molecules, or sol particles, and their interactions, expressed through some kind of potential energy function. Models discussed below include both atomistic descriptions, in which each atom is treated individually, and coarse-grained descriptions, which treat larger objects. At the atomic scale, there are two classes of potential used. In quantum-mechanical potentials, calculation of the energy of a configuration of atoms is accomplished by determining the electronic wavefunction (or density, in the case of density functional theory [10]) and associated energy [11, 12]. In empirical potentials, also known as force fields, the energy is built up as a sum over different types of interactions: core-repulsions, bond stretches and bends, torsions, coulombic interactions, hydrogen bonds, dispersion forces, etc., each of which is described using a relatively simple function that has been parametrized against either quantum-mechanical results or experimental data [13–15].

Once a model has been specified, different calculations may be performed. Of particular relevance here are dynamical calculations, which generate a trajectory according to specified equations of motion. In the simplest case, Newton's equations are used. The trajectory thus generated conserves total energy and, under equilibrium conditions, samples the microcanonical ensemble. The equations of motion can be modified to enforce constant-temperature and/or constant-pressure conditions. These are collectively referred to as molecular dynamics simulations. Stochastic dynamics, based on the Langevin equation (or derived results), are used in order to avoid the simulation of solvent molecules [13, 15]. In these techniques, friction terms and random impulses are used to model the interaction of solute molecules with the solvent; in dilute solutions, this can reduce the number of objects simulated by several orders of magnitude. In the particular case of sol–gel processing in aqueous media, the use of such an "implicit" solvent is problematic because water itself is both a product of and catalyst for the siloxane condensation reaction. Other, "nonsimulation" operations include energy minimization and transition-state location, and are primarily used in quantum-mechanical studies of chemical reactions [16].

These different computational and simulation techniques can access different length and time-scales. In general, the less empirical the method, the smaller the length and time-scales that can be treated. For instance, fully quantum-mechanical calculations can be very accurate, but in most circumstances, they can be applied only to a few molecules, and when used in dynamical calculations they can access only picosecond time-scales. These limitations are imposed by the high computational cost of calculating the total energy of an assembly of atoms using either wavefunction-based or density-functional techniques. Because this cost rises steeply with increasing system size, such methods are restricted to relatively small systems. Because the time required for such a calculation can be very long, dynamical simulations, which require a long sequence of many such calculations, are likewise restricted to very short trajectories.

Atomistic modeling with empirical potentials is much less expensive and also scales better with increasing system size. Such calculations are therefore capable of accessing larger systems, typically as many as 10^5 atoms, and occasionally much larger. Likewise, they can be extended to longer times; simulations of more than a microsecond have been accomplished using commonly available computers, although most force-field-based molecular dynamics simulations to date are shorter than 10 ns or so. Because sol–gel processing involves the making and breaking of chemical bonds, the empirical potentials used in such studies must be able to describe such processes. The well-known "standard" force fields widely used in biochemical (and other) simulations [15] do not incorporable breakable bonds, and are therefore not suitable. Furthermore, the participation of the solvent (water) in these reactions means that the great majority of the more than 60 water potentials now available in the literature are also not suitable. Many of the empirical modeling studies described below, therefore, also required the development and validation of new potentials.

To access length scales larger than tens of nanometers and time scales longer than nanoseconds, simulators turn to coarse-grained models. In such simulations the "primary" particle simulated is no longer a single atom or small molecule, but a greatly simplified representation of an assembly of many atoms or molecules. These objects interact with each other through effective potentials obtained from either theoretical considerations, matching to higher-resolution atomistic simulations, or fitting against experimental data. Such models can be extended to much larger length and time scales, for several reasons. Reduction in the number of degrees of freedom in the simulation greatly reduces the computational effort required for energy and force calculations. Furthermore, because these particles are heavy (compared with atoms or small molecules) and interact through somewhat softer potentials, the equations of motion can be integrated with much larger time steps.

Although many hybrids of these basic approaches are also known, the physical and chemical processes underlying aerogel preparation and properties are thought to be relatively friendly to the separation of scales and methods just described, and studies to date have exploited this. The aqueous hydrolysis/condensation chemistry underlying silicate precursor oligomerization has been studied using quantum mechanical methods. Formation of larger oligomers and complete gelation of dense aerogels has been studied with atomistic simulations of empirical force-field models. Since such simulations cannot access the time or length scales associated with the gelation of low-concentration precursors, a variety of coarse-grained models have been proposed for these processes. In this chapter we review these studies and discuss both the methods and models used and the conclusions reached.

An alternative approach to modeling complex porous materials is *reconstruction*, in which experimental data are inverted to generate a computational realization of the material structure. There is insufficient data in any experimental characterization short of high-resolution three-dimensional tomography to exactly reconstruct an amorphous material. Indeed, most experiments provide either a one-dimensional data set, such as a structure factor or adsorption isotherm, or a two-dimensional data set, such as an electron microscope image. *Stochastic reconstruction* generates representative structures that fit such limited input data but are otherwise random and isotropic [17–19]. There are in general many possible model structures that will fit the input data equally well, which is known as the "problem of uniqueness." Although we do not review these techniques or their applications in detail, we do note that they have been applied to sol–gel materials in a number of recent studies. Quintanilla et al. applied the method of Gaussian random fields to generate stochastic reconstructions of *TEOS*-derived aerogels [20]. Eschricht et al. used evolutionary optimization to reconstruct the commercially available "GelSil 200" material from a combination of small-angle neutron scattering data (*SANS*) and a pore-size distribution derived from gas adsorption data [21]. Steriotis et al. used *SANS* data to guide a "process-based" atomic-scale reconstruction of a silica xerogel [22].

Finally, an entirely different approach to simulating gelation is the "Dynamic Monte Carlo" (*DMC*) method, in which chemical reactions are modeled by stochastic integration of phenomenological kinetic rate laws [23]. This has been used successfully to understand the onset of gel formation, first-shell substitution effects, and the influence of cyclization in silicon alkoxide systems [24–26]. However, this approach has not so far been extended to include the instantaneous positions and diffusion of each oligomer, which would be necessary in order for the calculation to generate an actual model of an aerogel that could be used in subsequent simulations.

24.2. Atomistic Modeling

Atomistic modeling and simulation have been used both in studies of the hydrolysis/condensation chemistry underlying silica sol–gel processing, and in dynamical simulations of the formation of sol particles and even gels. Studies of reaction mechanisms and energetics have made extensive use of quantum mechanical methods, while the larger length scales and time scales of oligomerization and gelation have been handled using empirical potentials, as reviewed in this section.

24.2.1. Underlying Chemistry

Pereira et al. used density functional theory (*DFT*) and the *COSMO* effective-dielectric solvent model to study the condensation of silicic acid monomers, considering two possible reaction mechanisms [27]. The first is an S_N2-type nucleophilic attack in which the nucleophile (the oxygen in $Si(OH)_4$) attacks from the side opposite to the leaving group (water), and the other is a lateral attack in which the leaving group is adjacent to the nucleophile. This study concluded that the S_N2 path is favored over the lateral path, though noted that a variety of less energetically and statistically favorable mechanisms are also possible that under solution conditions may also be significant. The activation energies found were in reasonable agreement with experimental values. Lasaga and Gibbs used Hartree–Fock theory to study the hydrolysis of H_3SiOH (effectively, the exchange of one hydroxyl group for another), finding a four-membered cyclic transition state [28]. Okumoto et al. used a combination of Hartree–Fock and *DFT* calculations to examine both hydrolysis and condensation of alkoxylsilanes [29]. They found that the most favorable neutral hydrolysis reaction mechanisms contained two or three water molecules in a cyclic geometry, and that the energetic barriers to hydrolysis were very low. The lowest-barrier neutral condensation pathway also contained two water molecules (besides that produced in the reaction). Hydrolysis, but not condensation, was studied under acidic and neutral conditions, with the finding that hydrolysis of protonated and deprotonated silica species was strongly influenced by hydration effects. More recently, Mora-Fonz et al., with a focus on zeolite-producing reaction conditions, used *DFT* to obtain the reaction free energies for a large selection of deprotonation and condensation reactions [30]. In this work both explicit waters (and sodium ions) as well as a continuum solvent model were used, the combination of which was found to give quite good results in all cases. Schaffer and Thompson also used a hybrid explicit/implicit approach to solvation to obtain condensation energies for selected reactions involving species with as many as eight silicon atoms, including cyclizations [31]. In agreement with previous studies, these calculations favored an S_N2 mechanism under neutral conditions and supported a two-step mechanism with a stable pentacoordinated intermediate under acidic conditions.

"First-principles" simulations, in which forces and energies are calculated using density functional theory, have also been used in studies of condensation chemistry. Doltsinis et al. studied (multiple) silica monomers, dimers, and trimers dissolved in water under conditions of high temperature and pressure. Even at 1,200 K, the extremely short time scale accessible to simulations of this type (approximately 10 ps) meant that no chemical reactions were observed; this work instead focussed on structural analysis and diffusion constants. Trinh et al. used constraint dynamics to obtain the free energy profiles along the reaction pathways for a variety of condensation reactions involving species as large as tetramers [32–34]. In this method a particular reaction coordinate, such as the distance between a silicon atom and an oxygen atom from another oligomer, is held constant while the rest of the system evolves as usual; the integral of the necessary constraint force with distance yields a reaction free energy profile. Only anionic mechanisms appropriate to high pH were considered, and the initial deprotonation of one hydroxyl was not itself studied. These calculations confirmed the findings of earlier studies: that water molecules both participate in the reaction itself and that solvation effects substantially alter the free energy profile of reaction. In particular, water was found to participate differently in the reactions forming dimers and trimers, mediating the final transfer of a proton to the leaving hydroxyl in trimerization, but not in dimerization [32, 33]. The presence and type of counterions were also found to have substantial effects on condensation chemistry. Both lithium and ammonium cations were found to substantially stabilize the reactants, increasing the barrier to reaction, and ammonium itself was found to participate in proton transfer in the second stage of the reaction [34]. Although generally supporting the picture developed in calculations using isolated clusters and/or continuum models, this work underscores the complexity of the chemistry that occurs in solution and the necessity of explicit treatment of water in these systems.

Finally, there have also been many quantum-mechanical studies of the structure, solvation, and relative energetics of small silica oligomers [35–39]. In addition to these quantum-mechanical studies, Pereira et al. used nonreactive empirical potentials to simulate small silica clusters and solutions of water, methanol, ethanol, *TMOS*, and *TEOS* [40, 41], finding that the structures and densities of these systems were generally in good agreement with experiment.

24.2.2. Simulations of Oligomerization and Gelation

Despite the relatively long time scales and large length scales involved in gelation of aqueous silicate precursors, several groups have used molecular dynamics simulations and atomistic potentials to study the sol–gel process and sol–gel derived silica materials.

The first molecular dynamics studies of oligomerization and the silica sol–gel process were carried out by Garofalini and Melman [42]. A solution consisting of water, silicic acid monomers, and silicic acid dimers was simulated. The potential used was developed specifically for this problem. This consisted of a modified Born–Mayer–Huggins (*BMH*) model for the Si–Si, Si–O, and O–O interactions in nonwater molecules, and a modified Rahman–Stillinger–Lemberg (*RSL2*) potential [43] for O–O, O–H, and H–H interactions in water molecules. Experimental studies suggest that under neutral conditions deprotonation of silicic acid (H_4SiO_4) monomers is the step leading to the water-producing condensation reaction [44]. However, although deprotonations were observed in the simulations, no condensation reactions occurred; instead, monomers associated to form stable complexes with two pentacoordinated silicon atoms.

In 1990, Feuston and Garofalini revised the model in a subsequent study [45]. The modified *BMH* potential was applied to interactions between all atoms, with H–X interactions supplemented by terms from the *RSL2* potential [43]. A three-body (angular) potential, which steers bond angles toward target values, was also introduced in order to improve the bonding geometry displayed by the model. Target Si–O–Si, O–Si–O, and Si–O–H bond angles were set to the tetrahedral angle, and the H–O–H angle to the experimental value of 104.5°. The ground state structures of the water monomer, dimer, trimer, and a water–silicic acid cluster obtained with this potential were in good agreement with previous ab initio calculations. Direct simulations of polymerization were then performed using molecular dynamics. A system containing 27 silicic acid monomers was simulated at temperatures between 1,500 and 2,500 K, in order to promote reactions on a simulation-accessible time scale. Condensation reactions were observed at 1,500 K and above, with mechanisms consistent with previous expectations: deprotonation of silicic acid followed by formation of a penta-coordinated silicon intermediate, and subsequent loss of either oxygen or hydroxide. Oligomers containing up to six silicon atoms were observed in the longest (20 ps) simulations.

A subsequent study by Garofalini and Martin extended this to larger systems (216 silicic acid monomers) at densities between 1.4 and 1.6 g/cm^3 [46]. A more complete picture of the species that appear during the initial stages of oligomerization was obtained; of particular interest were the populations and rates of appearance of linear and cyclic oligomers. The time evolution of the concentrations of Q_n species in the system was also obtained. A Q_n silicon is one bound to n bridging oxygen atoms; these results were in qualitative agreement with experimental data [47]. Martin and Garofalini have also further studied the details of the condensation reaction in this model [48], observing that when there are multiple bridging oxygens in the intermediate, there is a tendency to break a silanol bond in creating a leaving group, rather than lose a nonbridging oxygen. This is consistent with the experimental finding that acid-catalyzed conditions favor the formation of linear clusters [44]. Yamahara and Okazaki added methoxy groups to the Feuston–Garofalini model and performed simulations under conditions similar to those used by Garofalini and Martin [46]. They were able to distinguish between slow increase in the average cluster size at short times and rapid increase at larger times, the latter due to aggregation of oligomers. In these simulations nearly all monomers were incorporated in a single, system-spanning cluster within a few hundred picoseconds.

The systems described in all these studies were quite small ($N \leq 216$) and of relatively high density (1.3 g/cm^3 and above). The temperature and pressure conditions used are comparable with an autoclave process used to produce aerogels [44, 46, 48]. The initial configuration contained no water. These are not typical laboratory conditions for the preparation of sol–gel derived materials. More commonly, sol–gel chemistry is performed in solution, at near-ambient conditions. The alkoxide concentrations used are relatively low, between 5% and 50% by volume, and the water to silicon ratio, r, is always greater than 1.0. Rao and Gelb used the Feuston and Garofalini model in simulations closer to these conditions [49]. In this work, 729 silicic acid monomers were placed in water at liquid density with r between 0 and 26, with most work performed at $r = 11$. Simulation times were as long as 12 ns. High temperatures (1,500–2,500 K) were again used to promote reaction, and the system volume held fixed. A series of snapshots from one of these simulations is shown in Figure 24.1. These studies confirmed the action of water as a catalyst, and further

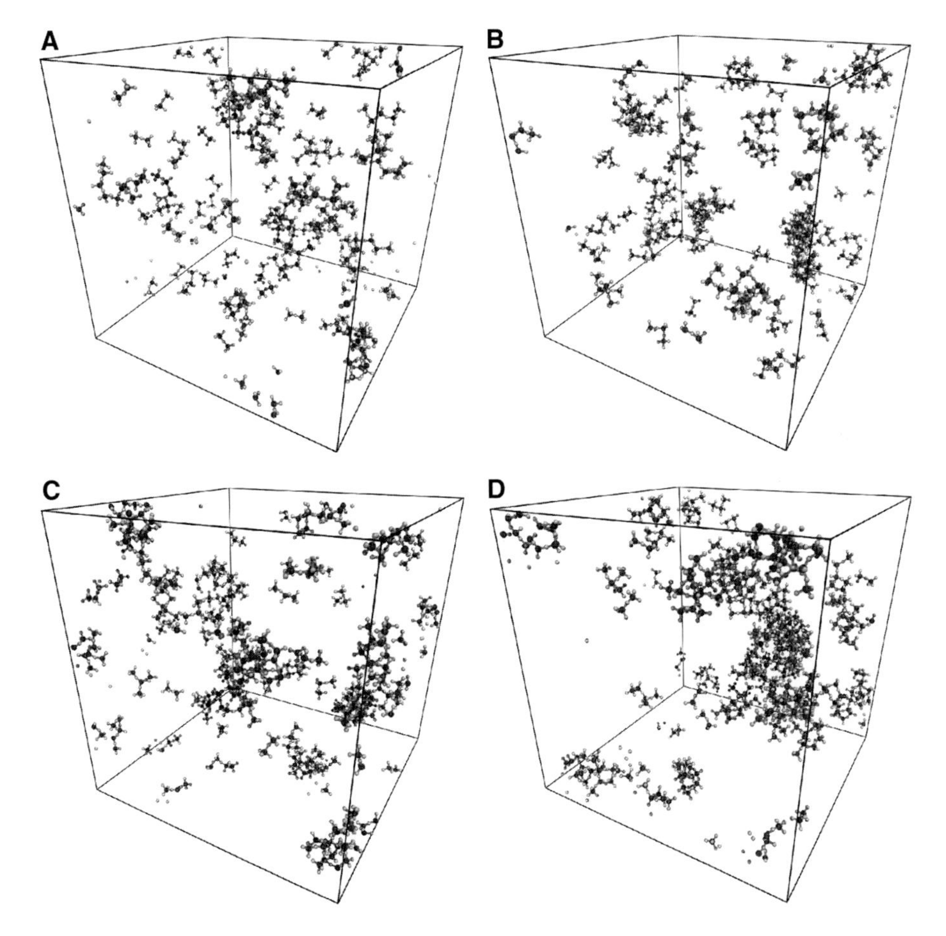

Figure 24.1. Snapshots from molecular dynamics simulations of oligomerization in aqueous silicic acid with $r = 11$. Only silica polymorphs containing at least two silicon atoms are shown; all silicic acid monomers and water molecules (which completely fill the remainder of the simulation cell) are removed for clarity. Only silicon and oxygen atoms are shown; terminal hydrogens have been removed. Snapshots are taken from the simulation at times **A.** 1.0, **B.** 2.0, **C.** 3.0, and **D.** 4.8 ns. The evolution of small oligomers into compact sol particles is clearly visible. Reprinted with permission from [49]. Copyright 2007 American Chemical Society.

illustrated the processes involved in the conversion of monomer precursors to "sol" particles. At short times, monomers quickly dimerized, and further growth occurring via monomer addition. At longer times (after several nanoseconds), condensation between larger oligomers was also observed. Finally, the in-solution dimerization kinetics were analyzed and found to give an activation energy in good agreement with ambient-condition experimental values, despite the high temperatures used.

The model developed by Feuston and Garofalini, while successfully encompassing much of the important chemistry involved in sol–gel processing, could nonetheless be improved in many ways. In particular, the description of water in this model is rather unrealistic; because full formal charges are assigned to each molecule (hydrogen ions all carry a "+1" charge, oxygens "−2" and silicons "+4") the water molecules have unreasonably high dipole moments. Furthermore, both Feuston and Garofalini [45] and Rao and Gelb [49] noted the spontaneous autodissociation of water even at near-ambient conditions, which indicates that this model is quite significantly more acidic than real water.

Bhattacharya and Kieffer introduced a more realistic all-atom reactive "charge-transfer" potential for simulating silica sol–gel processing [50, 51]. In this approach, a certain amount of charge is transferred between bonded atoms; this allows for more realistic electrostatic interactions between molecules, and correctly charged ions upon dissociation. In the first study, up to 512 silicic acid molecules were simulated at temperatures between 300 and 700 K and at various densities, comparable with autoclave-process conditions and earlier simulation studies. As in previous work, linear chains were found to form before cyclic structures. Also, the morphology of the silica network formed at late stages depended strongly on the total density, with independent clusters formed at low densities and spanning structures formed at high densities. The morphologies of the final structures, particularly as characterized by fractal dimension, were also compared with the results of "expansion" simulations (described in detail below), but using the new potential. Significant differences were observed, with the expanded-structure fractal dimensions being closer to 3 (less fractal) than the simulated-gelation structures, which were attributed to the structure obtained in the expansion simulations being unduly influenced by a single "fracture event." Finally, Bhattacharya and Kieffer also studied the effects of removal of water (produced in the condensation reactions) on the gel structure, concluding that the drying process has no significant effect on the fractal dimension.

In the second study [51] water was introduced at the start of the simulations, and the effects of water concentration on gel morphology and growth kinetics were analyzed in detail. Increasing the water concentration retarded condensation kinetics, and appeared to limit the equilibrium conversion. Different characteristic morphologies were observed with different water and precursor concentrations, with isolated compact clusters (as in [49]) at low precursor concentration and high water concentration, percolating structures at high precursor concentration and low water concentration, and distinct, branched structures at high precursor/high water conditions. Note that the water concentration changes through the simulation as condensation occurs. The mechanical strength of the final gel structures produced was evaluated by monitoring changes in fractal dimension and collapse during drying, with the finding that the percolated structures were the most robust.

Much earlier, Kieffer and Angell [52] had used an alternative approach to generating high-porosity, aerogel-like structures, which did not rely on direct simulation of the gelation process. In this work a dense silica glass, itself prepared using a rapid quench procedure, was forcibly expanded in several steps, allowing for short relaxation periods at each step. This produced structures of specified density with complex porous networks and aerogel-like characteristics. Silica was modeled using the Born–Mayer empirical potential, and up to 1,500 atoms were included in the simulations. The quantity of greatest interest in this work was the fractal dimension, which was found to vary strongly with density over a range from about 1.8 (highly fractal) up to 3 (nonfractal), with values at intermediate densities near 2.4, comparable with those of base-catalyzed silica aerogels. Kieffer and Angell also varied the "step size" of the expansion procedure and concluded that, since the resulting fractal dimensions hardly differed, "the fractal dimension of a silica structure is unique and describes the configuration of least resistance to rupturing under isotropic tensile stress" [52].

Very similar simulation protocols were used in subsequent work by Nakano et al. [53, 54]. In these studies, however, simulation cells containing over 40,000 atoms were used, such that much larger length scales could be probed. An example of these large-system results is shown in Figure 24.2. Somewhat more complex silica potentials were also used. Results for fractal dimensions obtained were similar to the previous study. However, varying the temperature during expansion (rather than the expansion step size) lead to considerably different fractal dimensions and morphology, suggesting that the morphology produced by this process was determined by the interplay of diffusion and local relaxation. Additional properties discussed in this work included predicted neutron diffraction patterns, surface areas, and pore size distributions.

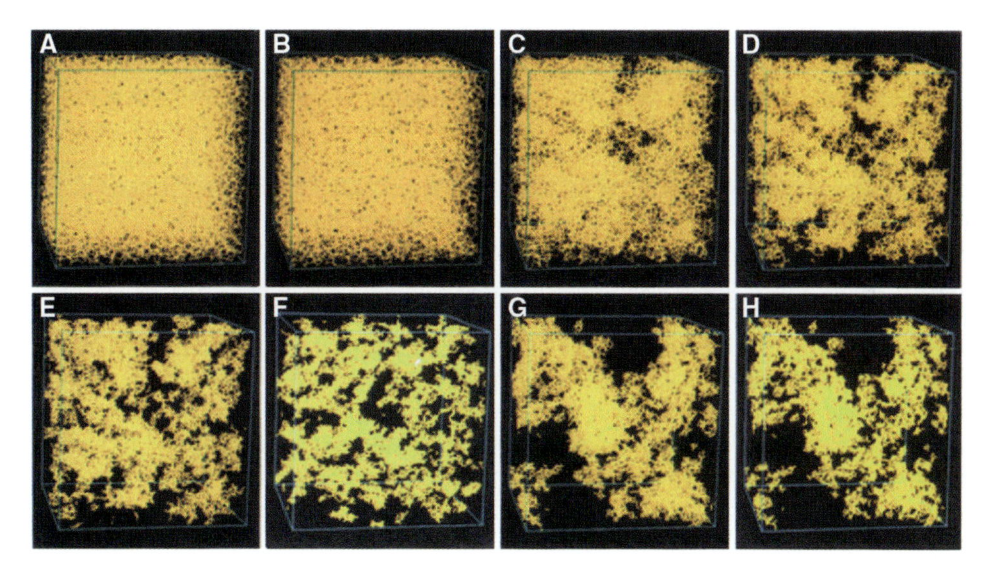

Figure 24.2. Snapshots of molecular dynamics of porous SiO_2 glasses **A.** 2.2, **B.** 1.6, **C.** 0.8, **D.** 0.4, **E.** 0.2, and **F.** 0.1 g/cm^3 prepared at 300 K, and **G.** 0.2 and **H.** 0.1 g/cm^3 prepared at 1,000 K. Yellow lines represent Si–O bonds. Reproduced with permission from [54]. Copyright 1993 American Physical Society. http://prl.aps.org/abstract/PRL/v71/i1/p85_1.

Pohl et al. compared several approaches to creating atomistically detailed aerogel-like structures [55]. These included an expansion procedure similar to those just described, the packing of colloidal silica spheres of varying radius and coordination, and the assembly of very small silica oligomers into an aggregate structure as guided by certain experimental information. The results indicate that aerogel-like surface areas and pore volumes could indeed be obtained using a sphere-based model, provided that the spheres are about 3 nm in diameter and themselves porous. The aggregate-based model appeared to be the most structurally realistic, but was not easily extended to a periodic material usable in subsequent simulations.

We note that the fractal dimensions found in the (low-temperature) expansion simulations of Nakano et al. [54] and Bhattacharya and Kieffer [50] are not in perfect agreement, again suggesting that either or both of the potential and the simulation protocol can also affect the final network morphology.

24.3. Coarse-Grained Simulations

As discussed above, the large length- and time-scales involved in most experimental realizations of gelation and other aspects of aerogel preparation are not directly accessible by molecular simulations. An alternative approach is to use coarse-grained models, which alleviate these scale problems at the expense of the loss of atomistic detail. The construction of coarse-grained models affords considerable freedom in choice of the primary objects simulated, their interaction, and the associated dynamics; so much so, in fact, that much work has focussed on the "simplest possible" models, both to expose the most fundamental physics involved and to avoid laborious and possibly underdetermined parametrization problems.

The aggregation of sol particles into a gel is a subclass of aggregation phenomena in general, and the preparation of silica gels from whence aerogels are derived is itself a subclass of sol–gel phenomena. Likewise, the literature on computer modeling of aggregation in general is rather larger than that focussing on modeling of sol–gel materials. Here we try to focus on the specific case of silica aerogels, but recommend for a more general picture the excellent reviews by Meakin [56, 57] and by Poon and Haw [58].

Coarse-grained computer models of silica aerogels (and other gels) may be roughly divided into two categories, depending on the simulation algorithm used and type of interactions included. In hard-sphere aggregation models, which have been extensively studied, aggregates are formed out of simple particles according to one of several procedures. Nonbonded particles interact as "billiard balls," without soft attractive or repulsive forces, while bonded particles are held rigidly together at the point of contact. In flexible models, equations of motion similar to those used in atomistic simulations are applied to objects with rather more complex interactions, including soft nonbonded interactions and deformable and/or breakable bonds.

24.3.1. Hard-Sphere Aggregation Models

The early literature on aggregation introduced a series of simple models in which particles were moved into contact according to various criteria and then irreversibly and rigidly bound together. The earliest such studies used two-dimensional "on-lattice" models, in which objects moved along lattice coordinates [59, 60], with subsequent work on both on- and off-lattice models in three (and more) dimensions [57]. The models relevant to aerogels include diffusion-limited aggregation (*DLA*), reaction-limited aggregation (*RLA*), diffusion-limited cluster aggregation (*DLCA*), and reaction-limited cluster aggregation (*RLCA*). In *DLA* and *RLA*, clusters grow by addition of monomers. In *DLA* there is no "reaction barrier" to monomer addition, and so the reaction rate is diffusion-limited, while in *RLA* not every collision leads to a reaction. In *DLCA* and *RLCA*, clusters are themselves mobile and may meet and aggregate.

In two important papers, Hasmy et al. showed that *DLCA* models can reproduce many of the structural properties (as measured by *SANS*) of colloidal silica aerogels [61, 62]. In the first, concerning the short-range structure of aerogels, a variety of different aggregation models were used to grow single aggregates [62]. Diffusion-limited, ballistic, and reaction-limited growth, with and without restructuring, were considered. Restructuring was found to produce much more compact (but still quite fractal) aggregates. Overall, diffusion-limited aggregates compared best with experimental results, though the agreement was only qualitative. In the second paper, periodic boundary conditions were used in the simulations to generate model aerogels according to a *DLCA*-type algorithm [61]. In these simulations monomers aggregate into smaller clusters, which themselves aggregate into a system-spanning gel structure which is homogeneous at sufficiently large length scales. The simulation proceeds by choosing individual clusters according to the inverse of their effective diffusion constant and displacing them randomly; this ensures that large clusters diffuse more slowly than smaller ones. An example of a very low-density gel model grown according to this algorithm is shown in Figure 24.3. This work demonstrated directly the crossover between scattering behavior dominated by the properties of individual aggregates and that of the gel, in which intercluster correlations are significant and the scattering behavior at large wavelengths is substantially different.

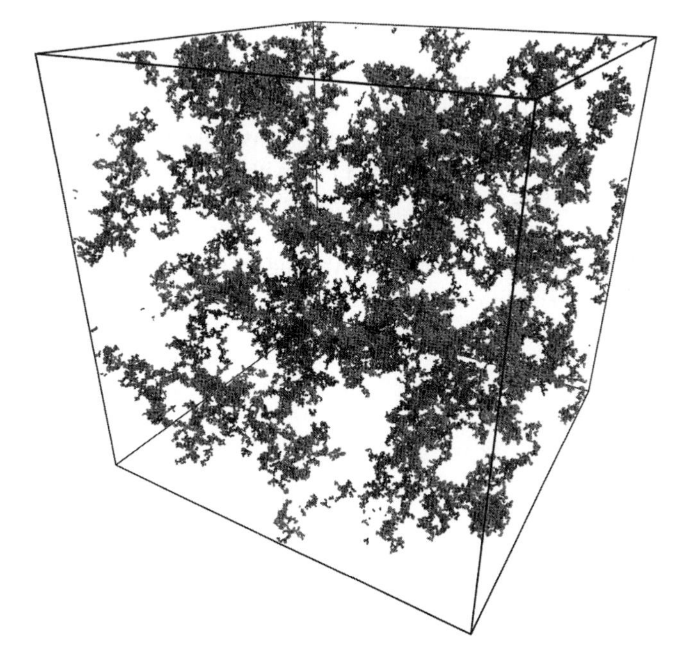

Figure 24.3. *DLCA* aerogel structure at 0.5% gel volume fraction, prepared (by the author) according to the algorithm proposed by Hasmy et al. [61].

In the comparison of *DLCA* results with experimental data on colloidal aerogels, in which the primary particles are relatively large and monodisperse, excellent agreement was obtained in nearly all respects. Basic and neutral aerogels were not fit as well, which was hypothesized due to polydispersity and the neglect of deformations and restructuring effects. Finally, different measures for the average size of the clusters within the gel network were compared in some detail. In subsequent work, this group addressed these concerns with refinements of the model. Incorporation of polydispersity did indeed improve agreement with experimental *SANS* data for base-catalyzed aerogels [63]. In a return to lattice-based modeling, a certain degree of flexibility was incorporated which proved effective in making contact with experiments on neutral aerogels [64].

A more realistic dynamics was incorporated in *DLCA* modeling by Olivi-Tran et al. [65], in studies of zirconia gels. Here the Langevin dynamics technique common in molecular simulations was adapted to the case of rigid aggregation of hard spheres; rather than being individually displaced, all objects moved continuously under the influence of both friction and thermal (Brownian) forces. In such simulations the progress of aggregation and the evolution of the structure factor in real time can be monitored. Interestingly, these authors reported an effective fractal dimension of 1.6, lower than the value of 1.8 commonly reported in *DLCA*. Pierce et al. also studied the incorporation of an "Epstein" drag force (proportional to the cross-sectional area of the cluster) in *DLCA* at the move-probability level, finding that the resulting aggregates had fractal dimension of 1.8, but different power-law growth kinetics [66].

The gel models obtained by *DLCA* and related methods do not, strictly speaking, have any mechanical properties; they are completely rigid (infinite modulus) in the simulations. Ma et al. addressed this as follows [67]. First, they modified *DLCA* with a "dangling bond deflection" algorithm in which "branches" were rotated so as to form many closed loops. The resulting structures were then used as input to a finite-element modeling (*FEM*) procedure, in which each pair of contacting particles was considered as a beam element. In this way the scaling of modulus with density was examined, with the finding that modulus was proportional to density raised to the power 3.6, in reasonable agreement with experiment. In addition, the distribution of strain energy in the gel was examined, with the finding that most of the strain was borne by only a small fraction of bonds, making the structure mechanically inefficient, and explaining the sharp drop-off in modulus with decreasing density. In recent related work, Lu et al. have applied the related Material Point Method to study the mechanical properties of aerogel structures obtained from X-ray tomography, finding good agreement with experimental stress–strain data in the elastic region [68].

24.3.2. Flexible Coarse-Grained Models

Flexible models, which allow for the effects of relaxation and fluctuations on gel structure, have been used in the studies of colloidal gels and flocculation in food colloids. In these systems the primary particle size is often much larger, and the interparticle interactions much softer, than in silica aerogels. Such models are simulated dynamically with methods similar to those used by Olivi-Tran et al. [65] and well-known in atomistic simulations; the difference here is that motions internal to clusters are incorporated naturally, with amplitudes and time-scales determined by the interaction potentials. Bijsterbosch et al. [69] used Brownian dynamics to simulate 1 μm particles described by an unbreakable square-well-type bonding model, in which bonds act between points fixed to the particle surface. The primary quantities tracked were the fractal dimension and particle coordination number distribution. Different conditions and bond-creation parameters led to varying "fineness" of the gel structure. One significant finding of this work was that the fractal dimension is not a "complete" characterization of the gel network and, particularly, is not necessarily sensitive to crosslinking effects which can substantially affect rheological properties.

Whittle and Dickinson used a more elaborate model, incorporating harmonic bonds, short-ranged repulsive forces, and a shifted-center Lennard-Jones potential to study aggregates of soft particles typical of food colloids [70]. These authors also studied the failure of a colloidal gel under strain [71]. Similar studies, using either hard spheres or repulsive-core particles, have also been performed by other groups [72–75].

We extended this approach to the case of silica aerogels by proposing a more complex potential for the interparticle bonds [76], reminiscent of some terms in the Feuston–Garofalini model described earlier [45]. This potential includes angular and torsional terms, and uses a (dissociable) Morse potential to describe bond stretching. The model was parametrized for 2 nm silica sol particles, and Langevin dynamics were used in simulations of gelation, relaxation, and mechanical deformation. Because the effective forward rate constant for bond creation, P, was an adjustable parameter in this work, a range of behavior could be observed. A series of snapshots of gels grown at different P is shown in Figure 24.4. Very substantial fluctuations and relaxation in aggregate structure were observed during these simulations. In all cases, the pressure in the final gel structure was found to be slightly negative; that is, gelation was accompanied by a small amount of strain. This is physically plausible, since large-amplitude extensions of aggregate structure (breathing motions) are most likely to make contact with other aggregates and form new bonds; if this behavior also occurs in experimental systems, it may explain the shrinkage observed upon even supercritical drying in real aerogels. Under sufficiently low-reactivity conditions, the bond formation rates were decoupled from gel relaxation; that is, reactions took place infrequently, between well-relaxed clusters. Further reduction in reactivity P thus increased only the gelation time, without changing the properties

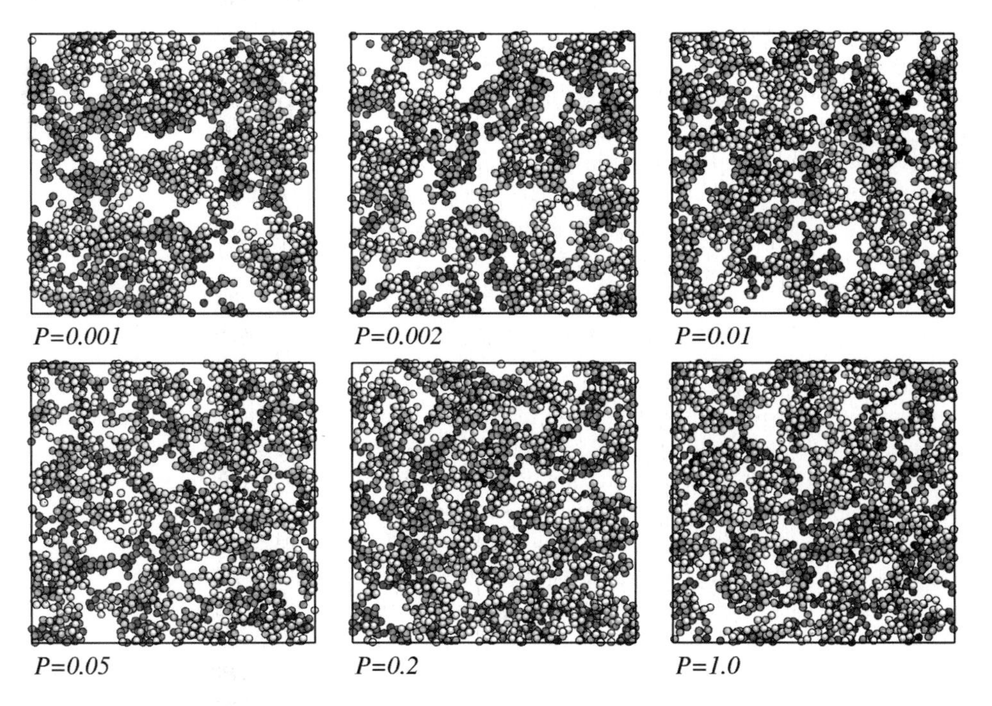

Figure 24.4. Final configurations from gelation simulations performed with different values of P. Configurations are shown for $P = 0.001, 0.002, 0.01, 0.05, 0.2$, and 1.0; $P = 1.0$ corresponds to *DLCA*-like behavior, with reaction always occurring upon suitable contact. Only a 15 nm thick "slab" of each simulation cell is shown, with particles shaded by depth. Reprinted with permission from [76]. Copyright 2007 American Chemical Society.

of the resulting gel. At higher reactivity, the time-scales for aggregation and relaxation were comparable and so cluster growth "locked in" unrelaxed structures, with the result that upon relaxation at zero pressure, aerogels prepared at high P shrunk substantially (up to 6% by volume) while those prepared at low P shrunk by only 2–3%. Likewise, high P gels had lower fractal dimensions (were more diffuse) than those at high P.

The bulk moduli of these gels were found to be in the 1–3 bar range, approximately 5–10× lower than those of real silica aerogels of comparable density, which was felt to be a fault of the parametrization used. Due to the coarse-graining, the time- and length-scales in this study are much larger than those of the atomistic simulations to which they are otherwise closely related; simulation cells measured $(75\ \text{nm})^3$ and simulations ran as long as 2.5 μs. Although structural properties (fractal dimension and mean pore size) appeared well-converged in these systems, reproducibility of bulk modulus between "samples" was found to be a problem, suggesting that even larger cells, or averaging over many replicas, would be required for reliable values.

24.4. Conclusions and Outlook

Recent work at all levels of description, from quantum mechanical to empirical atomistic modeling to coarse-grained modeling, has confirmed the broad picture developed by previous studies but also illustrated the many challenges remaining in developing a complete description and understanding of silica aerogels. The latest first-principles calculations clearly indicate that the roles of water and counterions in the underlying chemistry are more important and more complex than previously thought. Atomistic simulations have illustrated that structure at small scales is dramatically affected by degree of hydration and that the small clusters produced by short-time oligomerization have a complex distribution of size, shape, and structure. Our recent attempt at more quantitative coarse-graining indicates that quite large system sizes are required to obtain reliable results for mechanical properties of even reasonably dense aerogels. In addition, the neglect of structural fluctuations and of substantial intra- and intercluster relaxation in nearly all prior studies is now seen to be a questionable approximation.

Although atomistic empirical (or even quantum-mechanical) simulations of increasingly large scale and length would seem to be a route forward, there is no practical way that these can access the actual time- and length-scales required. Near-complete gelation was observed in many atomistic simulations, but this was only possible due to the tiny simulation cells and high densities used; with more reasonable system sizes, gelation would not have occurred on a "simulable" time-scale. In addition, important parameters, including the pH and pK_as, of many of the proposed atomistic models are unknown, which makes correlation with experiment difficult.

Recent experimental work suggests that the *DLCA*-type structures seen in silica aerogels are not universal, with tubular and other morphologies appearing in materials of other composition [77]. Simulation of such materials will have to begin at the atomistic level in order to understand the growth and relaxation mechanisms that yield nonspherical particle morphologies, and development of coarse-grained models appropriate for these materials will clearly be more challenging than for silica aerogels. Likewise, extension of modeling to resorcinol–formaldehyde-derived carbon aerogels, composite organic/inorganic aerogels, and other chemically complex materials, although possible in principle, would require considerable effort and creativity.

It appears clear that further quantitative improvements in computational aerogel models are most likely to result from a *multiscale* approach, incorporating all three levels of descriptions studied here. Only with such combinations will modelers be able to reach the necessary system sizes and simulation times while still connected to an atomistic description of the underlying chemistry. Empirical models could be further improved by parametrization against quantum-mechanical simulation results. Large-scale atomistic simulations could be used to generate chemically realistic distributions of sol particles, even if gelation cannot be directly observed. Finally, atomistic models of all types could be used as reference systems for developing better coarse-graining techniques. Multiscale techniques are of considerable interest in the wider simulation community at this time, and developments in other areas may be expected to help further the goal of realistic simulation and modeling of aerogels.

References

1. Fricke J (1992) Aerogels and their applications. J Non-Cryst Solids 147&148: 356–362.
2. Fricke J, Emmerling A (1998) Aerogels - recent progress in production techniques and novel applications. J Sol-Gel Sci Tech 13: 299–303.
3. Gesser HD, Goswami PC (1989) Aerogels and related porous materials. Chem Rev 89: 765–788.
4. Nicolaon GA, Teichner SJ (1968) Préparation des aérogels de silice à partir d'orthosilicate de méthyle en milieu alcoolique et leurs propriétés. Bull Soc Chim France 5: 1906.
5. Scherer GW (1998) Characterization of aerogels. Adv Coll Int Sci 76-77: 321–339.
6. Himmel B, Burger H, Gerber T, Olbertz A (1995) Structural characterization of SiO_2 aerogels. J Non-Cryst Solids 185: 56–66.
7. Duffours L, Woignier T, Phalippou J (1995) Plastic behaviour of aerogels under isostatic pressure. J Non-Cryst Solids 186: 321–327.
8. Scherer GW, Smith DM, Qiu X, Anderson JM (1995) Compression of aerogels. J Non-Cryst Solids 186: 316–320.
9. Woignier T, Duffours L, Alaoui A, Faivre A, Calas-Etienne S, Phalippou J (2003) Mechanical behaviour of highly porous glasses. J Non-Cryst Solids 316: 160–166.
10. Parr RG, Yang W (1989) Density-Functional Theory of Atoms and Molecules. Oxford Univ Press, New York.
11. Martin RM (2004) Electronic structure: basic theory and practical methods. Cambridge U Press, Cambridge, UK.
12. Szabo A, Ostlund NS (1996) Modern Quantum Chemistry. Dover.
13. Allen MP, Tildesley DJ (1987) Computer Simulation of Liquids. Clarendon Press, Oxford.
14. Frenkel D, Smit B (1996) Understanding Molecular Simulation. Acad Press, San Diego.
15. Schlick T (2002) Molecular modeling and simulation: an interdisciplinary guide. Springer-Verlag, New York.
16. Cramer CJ (2002) Essentials of Computational Chemistry: Theories and Models. John Wiley & Sons, Chichester, UK.
17. Levitz P (1998) Off-lattice reconstruction of porous media: critical evaluation, geometrical confinement and molecular transport. Adv Coll Int Sci 76–77: 71–106.
18. Torquato S (2001) Random Heterogeneous Materials: Microstructure and Macroscopic Properties. Springer-Verlag, New York.
19. Yeong CLY, Torquato S (1998) Reconstructing random media. Phys Rev E 57: 495–506.
20. Quintanilla J, Reidy RF, Gorman BP, Mueller DW (2003) Gaussian random field models of aerogels. J Appl Phys 93: 4584–4589.
21. Eschricht N, Hoinkis E, Mädler F, Schubert-Bischoff P, Röhl-Kuhn B (2005) Knowledge-based reconstruction of random porous media. J Colloid Int Sci 291: 201–213.
22. Steriotis T, Kikkinides E, Kainourgiakis M, Stubos A, Ramsay JDF (2004) Monitoring adsorption by small angle neutron scattering in tandem with digital reconstruction-simulation techniques. Colloids and Surfaces A: Physicochem Eng Aspects 241: 231–237.
23. Mikeš J, Dušek K (1982) Simulation of polymer network formation by the Monte Carlo method. Macromolecules 15: 93–33.

24. Kasehagen LJ, Rankin SE, McCormick AV, Macosko CW (1997) Modeling of first shell substitution effects and preferred cyclization in sol-gel polymerization. Macromolecules 30: 3921–3939.
25. Rankin SE, Kasehagen LJ, McCormick AV, Macosko CW (2000) Dynamic Monte Carlo simulation of gelation with extensive cyclization. Macromolecules 33: 7639–7648.
26. Šefčík J, Rankin SE (2003) Monte Carlo simulations of size and structure of gel precursors in silica polycondensation. J Phys Chem B 107: 52–60.
27. Pereira JCG, Catlow CRA, Price GD (1998) Silica condensation reaction: an *ab initio* study. Chem Commun 13: 1387–1388.
28. Lasaga AC, Gibbs GV (1990) Ab-initio quantum mechanical calculations of water-rock interactions: adsorption and hydrolysis reactions. Am J Science 290: 263–295.
29. Okumoto S, Fujita N, Yamabe S (1998) Theoretical study of hydrolysis and condensation of silicon alkoxides. J Phys Chem A 102: 3991–3998.
30. Mora-Fonz MJ, Catlow CRA, Lewis DW (2007) Modeling aqueous silica chemistry in alkali media. J Phys Chem C 111: 18,155–18,158.
31. Schaffer CL, Thomson KT (2008) Density functional theory investigation into structure and reactivity of prenucleation silica species. J Phys Chem C 112: 12,653–12,662.
32. Trinh TT, Jansen APJ, van Santen RA, Meijer EJ (2009) The role of water in silicate oligomerization reaction. Phys Chem Chem Phys 11: 5092–5099.
33. Trinh TT, Jansen APJ, van Santen RA, Meijer EJ (2009) Role of water in silica oligomerization. J Phys Chem C 113: 2647–2652.
34. Trinh TT, Jansen APJ, van Santen RA, VandeVondele J, Meijer EJ (2009) Effect of counter ions on the silica oligomerization reaction. ChemPhysChem 10: 1775–1782.
35. Lasaga AC, Gibbs GV (1987) Applications of quantum mechanical potential surfaces to mineral physics calculations. Phys Chem Minerals 14: 107–117.
36. O'Keeffe M, Domenges B, Gibbs GV (1985) Ab initio molecular orbital calculations on phosphates: Comparison with silicates. J Phys Chem 89: 2304–2309.
37. Pereira JCG, Catlow CRA, Price GD (1999) Ab initio studies of silica-based clusters. part I. Energies and conformations of simple clusters. J Phys Chem A 103: 3252–3267.
38. Pereira JCG, Catlow CRA, Price GD (1999) Ab initio studies of silica-based clusters. part II. Structures and energies of complex clusters. J Phys Chem A 103: 3268–3284.
39. Sauer J (1989) Molecular models in ab initio studies of solids and surfaces: From ionic crystals and semiconductors to catalysts. Chem Rev 89: 199–255.
40. Pereira JCG, Catlow CRA, Price GD (2001) Molecular dynamics simulation of liquid H_2O, MeOH, EtOH, $Si(OMe)_4$, and $Si(OEt)_4$, as a function of temperature and pressure. J Phys Chem A 105: 1909–1925.
41. Pereira JCG, Catlow CRA, Price GD (2002) Molecular dynamics simulation of methanolic and ethanolic silica-based sol-gel solutions at ambient temperature and pressure. J Phys Chem A 106: 130–148.
42. Garofalini SH, Melman H (1986) Applications of molecular dynamics simulations to sol-gel processing. In: CJ Brinker, DE Clark, DR Ulrich (eds) Better Ceramics Through Chemistry II 497–505. Materials Research Society, Pittsburgh.
43. Stillinger FH, Rahman A (1978) Revised central force potentials for water. J Chem Phys 68: 666–670.
44. Brinker CJ, Scherer GW (1990) Sol-Gel Science. Acad Press, San Diego.
45. Feuston BP, Garofalini SH (1990) Oligomerization in silica sols. J Phys Chem 94: 5351–5356.
46. Garofalini SH, Martin G (1994) Molecular simulations of the polymerization of silicic acid molecules and network formation. J Phys Chem 98: 1311–1316.
47. Kinrade SD, Swaddle TW (1988) ^{29}Si NMR studies of aqueous silicate solutions. 1. Chemical-shifts and equilibria. Inorg Chem 27: 4253–4259.
48. Martin GE, Garofalini SH (1994) Sol-gel polymerization: analysis of molecular mechanisms and the effect of hydrogen. J Non-Cryst Solids 171: 68–79.
49. Rao NZ, Gelb LD (2004) Molecular dynamics simulations of the polymerization of aqueous silicic acid and analysis of the effects of concentration on silica polymorph distributions, growth mechanisms, and reaction kinetics. J Phys Chem B 108: 12,418–12,428.
50. Bhattacharya S, Kieffer J (2005) Fractal dimensions of silica gels generated using reactive molecular dynamics simulations. J Chem Phys 122: 094715.
51. Bhattacharya S, Kieffer J (2008) Molecular dynamics simulation study of growth regimes during polycondensation of silicic acid: from silica nanoparticles to porous gels. J Phys Chem C 112: 1764–1771.
52. Kieffer J, Angell CA (1988) Generation of fractal structures by negative pressure rupturing of $\{SiO_2\}$ glass. J Non-Cryst Solids 106: 336–342.

53. Nakano A, Bi L, Kalia RK, Vashishta P (1994) Molecular-dynamics study of the structural correlation of porous silica with use of a parallel computer. Phys Rev B 49: 9441–9452.
54. Nakano A, Lingsong B, Kalia RK, Vashishta P (1993) Structural correlations in porous silica: Molecular dynamics simulation on a parallel computer. Phys Rev Lett 71: 85–88.
55. Pohl PI, Faulon JL, Smith DM (1995) Molecular dynamics computer simulations of silica aerogels. J Non-Cryst Solids 186: 349–355.
56. Meakin P (1988) Models for colloidal aggregation. Ann Rev Phys Chem 39: 237–267.
57. Meakin P (1999) A historical introduction to computer models for fractal aggregates. J Sol-Gel Sci Tech 15: 97–117.
58. Poon WCK, Haw MD (1997) Mesoscopic structure formation in colloidal aggregation and gelation. Adv Coll Int Sci 73: 71–126.
59. Meakin P (1983) Formation of fractal clusters and networks by irreversible diffusion-limited aggregation. Phys Rev Lett 51: 1119–1122.
60. Witten TA, Jr, Sander LM (1981) Diffusion-limited aggregation, a kinetic critical phenomenon. Phys Rev Lett 47: 1400–1403.
61. Hasmy A, Anglaret E, Foret M, Pelous J, Jullien R (1994) Small-angle neutron-scattering investigation of long-range correlations in silica aerogels - simulations and experiments. Phys Rev B 50: 6006–6016.
62. Hasmy A, Foret M, Pelous J, Jullien R (1993) Small-angle neutron-scattering investigation of short-range correlations in fractal aerogels - simulations and experiments. Phys Rev B 48: 9345–9353.
63. Hasmy A, Foret M, Anglaret E, Pelous J, Jullien R (1995) Small-angle neutron scattering of aerogels: simulations and experiments. J Non-Cryst Solids 186: 118–130.
64. Jullien R, Hasmy A, Anglaret É (1997) Effect of cluster deformations in the DLCA modeling of the sol-gel process. J Sol-Gel Sci Tech 8: 819–824.
65. Olivi-Tran N, Lenormand P, Lecomte A, Dauger A (2005) Molecular dynamics approach of sol-gel transition: Comparison with experiments. Physica A 354: 10–18.
66. Pierce F, Sorensen CM, Chakrabarti A (2006) Computer simulation of diffusion-limited cluster-cluster aggregation with an epstein drag force. Phys Rev E 74: 021411.
67. Ma HS, Prévost JH, Jullien R, Scherer GW (2001) Computer simulation of mechanical structure-property relationship of aerogels. J Non-Cryst Solids 285: 216–221.
68. Lu H, Fu B, Daphalapurkar N, Hanan J, Soritrou-Leventis C, Leventis N (2008) Simulation of the evolution of the nanostructure of crosslinked silica-aerogels under compression. Polymer Preprints 49: 564–565.
69. Bijsterbosch BH, Bos MTA, Dickinson E, van Opheusden JHJ, Walstra P (1996) Brownian dynamics simulation of particle gel formation: From argon to yoghurt. Faraday Discuss 101: 51–64.
70. Whittle M, Dickinson E (1997) Brownian dynamics simulation of gelation in soft sphere systems with irreversible bond formation. Mol Phys 90: 739–757.
71. Whittle M, Dickinson E (1997) Stress overshoot in a model particle gel. J Chem Phys 107: 10,191–10,200.
72. d'Arjuzon RJM, Frith W, Melrose JR (2003) Brownian dynamics simulations of aging colloidal gels. Phys Rev E 67: 061404.
73. Rzepiela AA, van Opheusden JHJ, van Vliet T (2001) Brownian dynamics simulation of aggregation kinetics of hard spheres with flexible bonds. J Colloid Int Sci 244: 43–50.
74. Rzepiela AA, van Opheusden JHJ, van Vliet T (2002) Large shear deformation of particle gels studied by Brownian Dynamics simulations. Comp Phys Comm 147: 303–306.
75. Rzepiela AA, van Opheusden JHJ, van Vliet T (2004) Large shear deformation of particle gels studied by Brownian Dynamics simulations. J Rheol 48: 863–880.
76. Gelb LD (2007) Simulating silica aerogels with a coarse-grained flexible model and Langevin dynamics. J Phys Chem C 111: 15,792–15,802.
77. Leventis N (2007) Three-dimensional core-shell superstructures: Mechanically strong aerogels. Acc Chem Res 40: 874–884.

Part VIII

Applications: Energy

25

Aerogels and Sol–Gel Composites as Nanostructured Energetic Materials

Alexander E. Gash, Randall L. Simpson, and Joe H. Satcher Jr.

Abstract In the last 10 years there have been a significant number of investigations of the application of aerogels and sol–gel-derived materials and methods to the field of energetic materials (e.g., explosives, propellants, and pyrotechnics) specifically through the synthesis and characterization of nanostructured energetic composites. Aerogels have unique density, composition, porosity, and particle sizes as well as low temperature and benign chemical synthetic methods all of which make them attractive for energetic nanomaterials candidates. The application of these materials and methods to this technology area has resulted in three general types of sol–gel energetic materials (1) sol–gel inorganic oxidizer/metal fuel pyrotechnics (thermite-like composites); (2) sol–gel-derived porous pyrophoric metal powders and films; and (3) sol–gel organic fuel/inorganic oxidizer nanocomposites (propellant and explosive-like composites). This chapter details results from synthesis and characterization research in all three areas. General trends are detailed, analyzed, and discussed. In general, all sol–gel nanostructured energetic material behaviors are highly dependent on several factors including surface area, degree of mixing between phases, the type of mixing (sol–gel or physical mixing), solids loading, and the presence of impurities. Sol–gel methods are attractive to the area of nanostructured energetics because they offer a great deal of many processing options such as monoliths, powders, and films and have broad compositional versatility. These attributes coupled with strong synthetic control of the microstructural properties of the sol–gel matrix enable the preparation of energetic nanocomposites with tunable performance characteristics. Various aspects of the present literature work are reviewed and future challenges for this technological area are presented and discussed.

25.1. Introduction

In order to put the topic of aerogel and sol–gel-derived energetic nanocomposites into the proper perspective, a short discussion of the classes and properties of energetic materials (*EMs*) is appropriate. Energetic materials are separated into three classes (1) explosives; (2) propellants; and (3) pyrotechnics [1]. Materials categorized this way are generally based

---

A. E. Gash, R. L. Simpson, and J. H. Satcher Jr. • Lawrence Livermore National Laboratory,
P.O. Box 808 L-24, Livermore, CA 94551, USA
e-mail: gash2@llnl.gov; simpson5@llnl.gov; satcher1@llnl.gov

M.A. Aegerter et al. (eds.), *Aerogels Handbook*, Advances in Sol-Gel Derived Materials and Technologies, DOI 10.1007/978-1-4419-7589-8_25,

on their speed of reaction, the phase of the resulting reactants, as well as their type of energetic output. Explosives are materials that react at supersonic velocity (detonation) and whose reaction products are primarily gaseous. Propellants react rapidly to give mostly gaseous reaction products, but react subsonically; unlike high explosives. Pyrotechnics tend to react the most slowly of the three *EMs* and generate high-temperature solid reaction products, little, if any, gas, and intense visible light output.

Since the invention of black powder, 1,000 years ago, the technology for making solid *EMs* has remained the physical mixing of solid oxidizers and fuels (e.g., black powder) to make binary energetic materials. Later in history it was recognized that oxidizing and fuel moieties could be incorporated into a single organic molecule (e.g., trinitrotoluene) to yield monomolecular energetic materials. These two types of energetic materials have key thermodynamic and kinetic distinctions.

In the binary systems, desired energy properties can be attained through readily varied ratios of oxidizer and fuels. A complete balance between the oxidizer and fuel may be reached to store energy as densely as 23 kJ/cm^3, nearly twice that of monomolecular materials [2]. However, due to the granular nature of composite energetic materials, reaction kinetics are dictated by the mass and thermal transport rates between the reactant oxidizer and fuel phases. Therefore, composites can have high energy densities; however, their energy release rate is below that which may be attained in a chemical kinetics controlled process.

Conversely, in monomolecular energetic materials, the rate of energy release is primarily controlled by chemical kinetics, and not by mass transport. Thus, monomolecular materials can release their energy more rapidly (i.e., more powerfully) than composite energetic materials. However, a major limitation with these materials is the total energy density achievable. Therefore, it would be highly desirable to combine the excellent thermodynamics of composite energetic materials with the rapid kinetics of the monomolecular energetic materials to enable materials with tailorable energy content and power. One approach to addressing this material challenge is through the application of nanomaterials and nanotechnology.

The field of nanoscience has been one of the most active areas of research in various scientific disciplines for at least two decades and energetic materials have been no exception to this [2]. Through the use of nanomaterials and advanced fabrication techniques nano-energetic composites have been synthesized and fully characterized. For the purpose of this discussion, energetic nanocomposites are defined as mixtures of oxidizer(s) and fuel(s) whose particles sizes have at least one critical dimension less than 100 nm [3]. The size reduction increases the interfacial surface area contact between reactant phases and decreases the limitations of mass and thermal transport present in conventional (i.e., micron-sized) energetic composites. This has been achieved using a variety of methods including vapor condensation methods [4], micellar synthesis, chemical reduction, sono-chemical mixing [5], and mechanical milling methods [6]. The results have been impressive and are beyond the scope of this discussion but are detailed elsewhere [7]. However, as an example, the burning rates for pyrotechnic nanocomposites of Al/MoO_3 were reported to be nearly three orders of magnitude greater than conventional mixtures [8]. Additionally, the particle size dependence of energetic materials to impact sensitivity (an extremely important safety characteristic) implies that energetic materials with smaller particle sizes may be less susceptible to impact ignition and thus have better safety properties [9]. These two important examples provide strong incentive for the application of nanomaterials and technology to energetic nanomaterials. With this in mind, aerogel and other sol–gel-derived materials have been investigated in the last decade as nanostructured energetic materials.

25.2. Attributes of Aerogels and Sol–Gel Processing for Nanostructured Energetic Materials

Aerogels are a special class of open-cell structures that exhibit many interesting properties, such as low mass densities, continuous porosities, and high surface areas [10, 11]. These unique properties are derived from the aerogel microstructure, which typically consists of three-dimensional networks of interconnected nanometer-sized secondary particles. Both inorganic and organic-based aerogel materials are possible via rich and well understood chemistries and both types of aerogels are utilized as sol–gel nanostructured energetics. As this text has numerous fine contributions discussing aerogel syntheses and characterization these topics will not be covered here and the reader is referred to those sections (Part II) and (Part VII). Because of their unusual chemical and textural properties, aerogels have been investigated for a wide variety of applications (Part VIII).

There are several attributes of aerogels and also the sol–gel methodology that are very appealing and compatible for energetic materials. Sol–gel synthesis enables precise control of the composition, density, morphology, and particle size of the target material at the nanometer scale [12]. This is especially important as the microstructural characteristics of an energetic material strongly influence its properties. This is critical for both safety and performance considerations. Sol–gel materials are distinctive in that they typically posses high surface areas, high porosities, and small primary particle size. The properties unique to sol–gel materials lead to their enhanced reactivity. Fine control of the microstructural parameters of energetic composites, enabled by the sol–gel method, allows the chemist the convenience of making *EMs* with tailored properties (e.g., energy output, ignition, safety characteristics).

Additional benefits include the convenience of low-temperature preparation and drying schemes prevent degradation of the energetic molecules, and using general and inexpensive laboratory equipment and starting materials. From a chemical point of view, the method affords easy control over the chemical stoichiometry and homogeneity that conventional methods lack [13]. In addition, one of the integral features of the method is its ability to produce materials with special shapes such as monoliths, fibers, films, coatings, and powders of uniform and very small particle sizes. This permits a great deal of versatility in the processing and final form of the energetic material that is not presently possible, which is critical for the advancement of this field. This type of processing versatility may enable the development of energetic materials for the actuation, ignition, and or propulsion in micro-electromechanical systems [14]. These systems require materials with high energy density, small failure diameters, and must be easily integrated with micro-electronics processing methods and conditions. Since sol–gel chemistry is a "bottom up" chemical approach it allows the extremely intimate mixing of component phases not attainable by most methods. Along with the exquisite mixing the extremely high surface areas of sol–gel materials ensure an energetic nanocomposite with very large interfacial contact areas. All of these favorable attributes have led to active investigating the application of sol–gel chemistry to *EM* research and development over the past decade.

25.3. General Sol–Gel Nanostructured Energetic Materials

In total, the current progress in the application of aerogel and other sol–gel-derived materials to *EMs* is summarized by three general sol–gel energetic nanostructures that are shown in Figure 25.1. The first, shown in Figure 25.1A, is a composite that consists

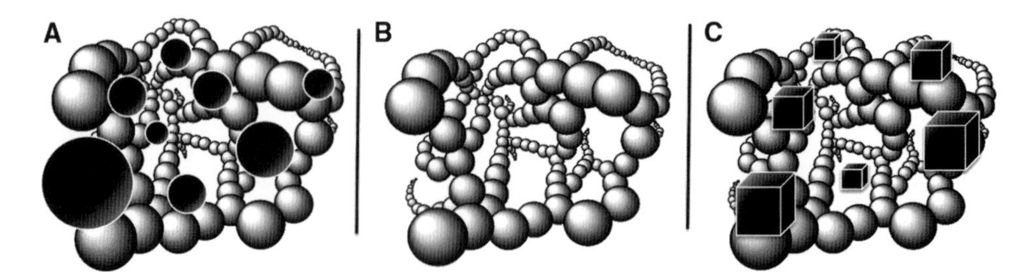

Figure 25.1. Schematic diagram of the three general types of sol–gel nanostructured energetic composite materials described in this chapter **A.** sol–gel oxidizer scaffold with fuel phase integrated into the porous network, this composite generates primarily high-temperature solid phase products on reaction; **B.** sol–gel-derived porous metal scaffold where the metal is combusted in air to generate its energy; and **C.** sol–gel fuel scaffold with inorganic oxidizer phase integrated into the fuel network, this composite generates primarily gaseous products when reacted.

of a sol–gel oxidizer scaffold with fuel particles imbedded in the pores. The second, Figure 25.1B, is a schematic of another energetic sol–gel network in which the network itself is the energetic material (e.g., nanostructured pyrophoric metal). The final idealized material, shown in Figure 25.1C, consists of a sol–gel fuel network with inorganic oxidizer particles imbedded in its nanostructure. The selection of the term "nanostructured" is intentional and is used to convey the distinction between these energetic composite materials, where the phases are intertwined or immobilized, to some extent, in an overall primary scaffold structure (e.g., monolith for aerogels, free-standing films for energetic nano-laminates [15]) derivable from the synthesis process and those *EMs* that are simple physical mixtures of solid phases powders with no scaffold structure.

25.3.1. Inorganic Aerogel Materials as Nanostructured Energetic Composites

Much of the sol–gel energetic work has focused on the development of sol–gel methods to synthesize porous monoliths and powders of nano-sized transition metal oxides (i.e., Fe_2O_3, Cr_2O_3, and NiO). In the past decade the epoxide-initiated sol–gel method has enabled the straightforward nonalkoxide syntheses of many transition metal oxide aerogels [16–20] (Chap. 8). When oxidizers of this type are combined with oxophillic metals such as aluminum, magnesium, or zirconium, ensuing mixtures can undergo the *thermite* reaction, shown below. In the *thermite* reaction the metal oxide (M(1)O(s)) and oxophillic

$$\mathrm{M(1)\,O\,(s) + M(2)\,(s) \rightarrow M\,(1)\,(s) + M(2)\,O\,(s)} + \Delta H \qquad (25.1)$$

metal (M(2)(s)) undergo a solid state reduction/oxidation reaction, which can be rapid and very exothermic, and in many instances self-sustaining [21]. Some thermite reaction temperatures exceed 3,000 K. Such reactions are examples of oxide/metal reactions that provide their own oxygen supply and thus will react in inert environments and are difficult to quench once initiated. Thermitic compositions have found use in a variety of processes and products. They are used as hardware destruction devices, for welding railroad tracks, as torches in underwater cutting, additives to propellants and high explosives, free standing heat sources, airbag ignition materials, and as thermal sources in foundry castings [22]. Historically, *thermites* are prepared by physical mixing of dry or fine component powders, such as ferric oxide and aluminum, by mechanical means. Mixing fine dry metal powders by conventional means can be an extreme fire hazard; sol–gel methods reduce that hazard while achieving ultrafine particle dispersions that are not possible with normal processing methods. In conventional mixing, domains rich in either fuel or oxidizer exist, which limit the mass transport and therefore decrease the efficiency and speed of the reaction. However, due to the nature of their formation sol–gel-derived nanostructured energetic materials are more uniformly mixed, thus reducing the magnitude of this problem.

The issue of homogeneity in an energetic composite appears to influence the characteristics of that material a great deal. Other methods of energetic nanocomposite synthesis involve the physical mixing of fine powders of the component phases in a vessel by stirring or ultrasonicating a mixture in a solvent or by grinding, ball-milling, or tumbling dry powders [4, 5]. Differences in density, wetability, agglomeration, and particle size of the component phases can lead to nonuniform dispersion of the mixed phases that can lead to a less efficient reaction. What is more desirable is a very homogeneous dispersion of one phase in the other that enables a great deal of interfacial contact surface area. This type of mixing is inherent in the sol–gel formulation of nanostructured energetic materials.

Morris et al. first commented on the synthesis of low density nanoscale mesoporous composites readily synthesized using sol–gel methodology [23]. By adding a dispersed solid in an about to gel metal oxide sol the dispersed phase is homogeneously integrated into the structure of the gelling phase and is immobilized when the gelation process is complete. This approach is especially versatile as the mild synthesis conditions of sol–gel enable the mixing of wide variety of dispersed phases up to a very large volume fraction of the composite. Tillotson et al. combined this synthesis approach with the epoxide-initiated sol–gel method and nanometer-sized Al particles to prepare the first sol–gel nanostructured energetic composites [24]. Figure 25.2 shows the photograph of an aerogel nanostructured energetic composite of iron (III) oxide and nanometric Al indicating the size and scale of the monolith. A detailed examination microstructure of this material is shown in the transmission electron micrographs shown in Figure 25.3. The lower magnification image shows the web-like iron (III) oxide sol–gel network wrapped around the spheres of Al (~40 nm in diameter). The higher magnification image shows the interface between the Al and iron (III) aerogel phases. This image is qualitatively like that shown in Figure 25.1A which describes one of the three types of sol–gel energetic material motifs discussed in this chapter. Additional characterization was performed to evaluate the homogeneity of the composite material utilizing elemental

Figure 25.2. Photographs of a monolith aerogel iron (III) oxide/nanometric aluminum energetic composite that is shown with a penny for scale.

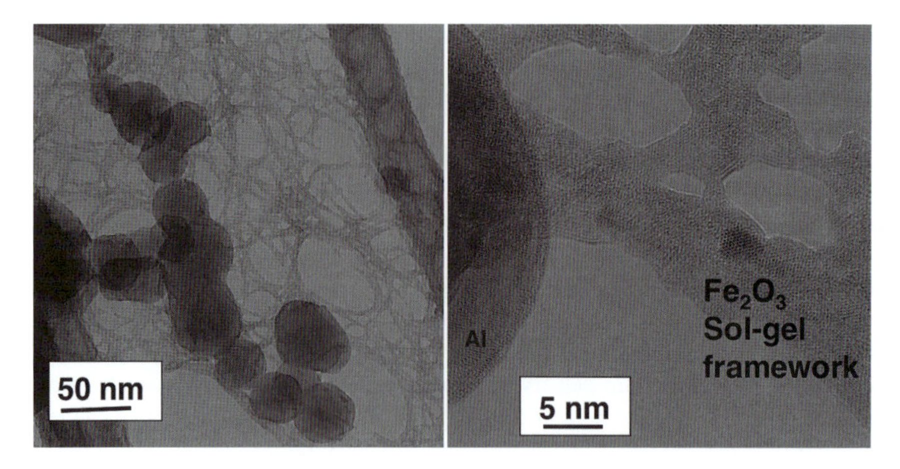

Figure 25.3. Transmission electron micrographs of an aerogel iron (III) oxide/nanometric aluminum energetic composite showing the mixing of the component phases. Note the qualitative correlation between these images and that of Figure 25.1A.

filtered transmission electron microscopy (*EFTEM*) which enables a chemical map of the constituent phase to be superimposed on the micrograph. Figure 25.4 shows the reference *TEM* (left) and transformed *EFTEM* image (right) for both aluminum (red region) and iron (III) oxide (green) respectively. This type of characterization reveals the superb mixing on this length scale with no evidence of agglomeration. An image of this particular composite combusting is shown in Figure 25.5. As mentioned, homogeneous intermixing of component phases in energetic composites significantly affects their reaction behavior by accelerating the rate limiting process of mass diffusion between reactants. This effect was demonstrated by Kim and Zachariah who utilized an aerosolized sol–gel process to assemble composites of iron (III) oxide and nanometric aluminum via electrostatic enhanced assembly [25]. In these composites the fuel/oxidizer assembly was enhanced by opposing electrostatic charges put on

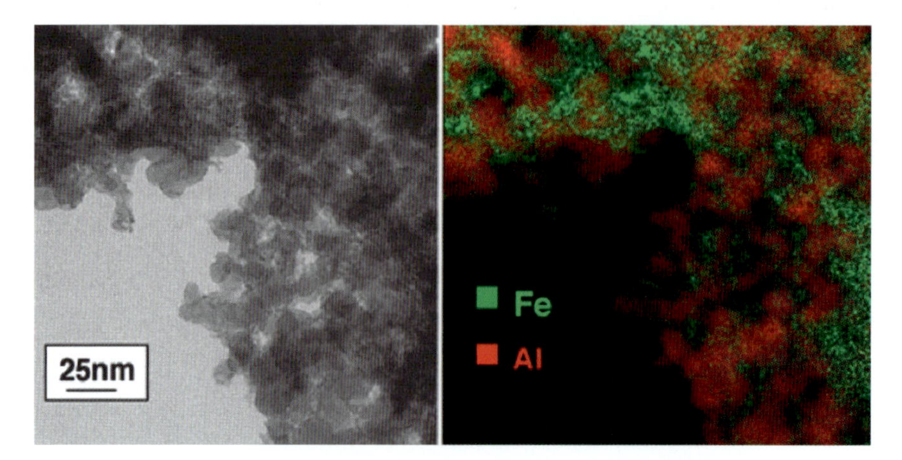

Figure 25.4. TEM and EFTEM images of an iron (III) oxide aerogel/nanometric Al composite. The EFTEM image shows the distribution of iron (III) oxide (*green*) and aluminum (*red*) phases and provides a representation of the homogeneous distribution of the two phases.

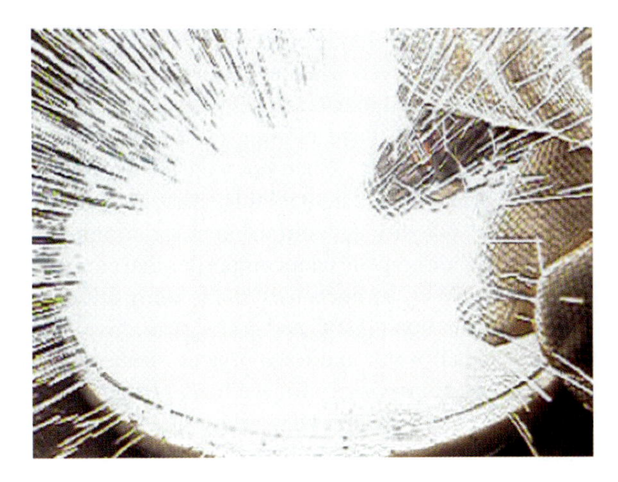

Figure 25.5. Image of the thermal ignition of an aerogel iron (III) oxide/nanometric aluminum energetic composite that shows the production of high-temperature solid combustion products.

the reactants. This approach led to composites with much more rapid energy release rates than composites that relied on Brownian motion mixing. In addition, Plantier et al. observed a 50% increase in combustion velocities for aerogel iron (III) oxide/nanometric Al composites when a small amount of surfactant was used to enhance mixing of the two phases [26]. Presumably, the surfactant improves mixture homogeneity. These two examples demonstrate the importance of a high degree of mixing between phases is a critical requirement to new and potentially enabling nanostructured energetics.

The combustion behavior of a variety of sol–gel-derived energetic materials has also been reported. Selected results are shown in Table 25.1; for all of these data the sol–gel oxidizer was mixed with nanometric Al fuel (~50 nm diameter). In these experiments dried premade sol–gel oxidizer was mixed with nanometric aluminum in hexane solvent while being ultrasonicated. The mixture was then rapidly dried and the loose powder combusted in an open trough or open ended tubes. These materials are referred to as xerogels. Whereas the liquid phase in an aerogel is extracted supercritically it is not in a xerogel. Combustion wave speeds were determined using high speed video capture. In general, there are some initial trends to be discussed. First and foremost, in all of the materials studied, the aerogel form of the material has greatly enhanced combustion wave speeds (on the order of hundreds m/s) relative to mixtures of micron composites (on the order of a few meters per second). Also

Table 25.1. Summary of the burning rate of several aerogel metal oxide/nanometric Al energetic nanocomposites

Oxidizer	Material type	Surface area of oxidizer (m^2/g)	Treatment	Burning rate (m/s)
Fe_2O_3	Aerogel	300	As prepared	12
Fe_2O_3	Aerogel	63	Annealed (410°C)	890
Fe_2O_3	Xerogel	50	Annealed (410°C)	290
Fc_2O_3	Nanopowder	250	NA	640
WO_3	Aerogel	77	Annealed (120°C)	450
WO_3	Aerogel	28	Annealed (400°C)	310
NiO	Aerogel	54	Annealed (415°C)	410

worthy of note, the combustion velocity for an energetic composite containing aerogel Fe_2O_3 is significantly greater than that observed when a commercial nano-sized Fe_2O_3 powder is used as the oxidizer (890 m/s compared to 640 m/s) [26]. In comparing sol–gel materials, the energetic nanocomposites with Fe_2O_3 have burning rates considerably greater than those containing xerogel Fe_2O_3 as an oxidizer. When one considers the lower surface area and porosity of a typical xerogel relative to an aerogel this observation appears to be consistent.

Additional evidence that porosity and surface area are strong drivers of combustion rates is reflected in the data for the energetic nanocomposites that contain tungsten (VI) oxide (WO_3) as the oxidizer. The two WO_3 aerogels used in the study differed by the temperature each was annealed at after preparation (120°C and 400°C respectively) [27]. Annealing was performed to remove any residual water and trace organic species from the supercritically dried WO_3 aerogel. The combustion velocity for the nano-energetic composite using WO_3 aerogel annealed at 120°C had a combustion wave speed of 450 m/s whereas the composite utilizing the WO_3 annealed at 400°C burned at 310 m/s. The difference in these rates can be related to the surface areas of the annealed aerogels which were 77 and 28 m^2/g respectively. As the WO_3 aerogel is annealed to higher temperatures, its porosity and surface area decrease, and particle size increases, all of which limit the interfacial contact area between oxidizer and fuel phases. One will note that all of the aerogel oxidizer materials that gave high burning rates in Table 25.1 were annealed. This was necessary to produce nano-energetic composites that gave high and reproducible burning rate measurements.

The presence of even trace amounts of impurities is known to affect the properties of energetic materials and sol–gel energetic composites are no exception [1]. The need for thermal treatment of the sol–gel oxidizer phase is to remove impurities. This effect is illustrated by comparing the combustion wave speeds for the energetic nanocomposites using the as-prepared iron (III) oxide aerogel and the annealed aerogel which are 12 m/s and 890 m/s respectively. Thermal annealing of the iron (III) aerogel removes adsorbed water, surface hydroxyl groups, and trace organic contaminants that are residual from the epoxide gelling agent and the solvent used in the synthesis. The combustion/decomposition of impurities in the as-prepared aerogel materials act as energy traps for the propagating reaction and act to retard the combustion wave speeds and nonuniformities in the flame propagation. These impurities can be significant in sol–gel materials as elemental analysis confirms the removal of 2–4 wt% organic residue and thermal gravimetric analysis indicates a total mass loss of up to 30 wt%, much of which is assumed to be water.

From the data in Table 25.1 it is clear that the impurity issue is more dominant than purely the surface area of the aerogel component. For example, the combustion wave speed of the composite with the as-prepared iron (III) aerogel is much lower than that of the annealed iron (III) aerogel despite the fact that surface area is approximately fivefold larger (300–63 m^2/g). It has been demonstrated that thermal annealing to temperatures as low as 275°C leads to drastic improvement in combustion velocities. Lower temperature annealing is of course desirable as it results in a decreased degree of sintering and hence higher surface area in the resulting oxidizer. Mehendale et al. prepared nano-energetic composites of mesoporous iron oxide and aluminum and observed comparable combustion rates on samples that were annealed [28].

The impurity problem is the most challenging issue facing inorganic sol–gel nanostructured energetic materials and has yet to be fully addressed. For sol–gel nanostructured energetic materials to yield uniform and high burning rates the materials must be impurity free. This issue precludes an effective evaluation of one of the major purported benefits of sol–gel nanostructured energetics: the extremely homogeneous dispersion of fuel and oxidizer phases. As prepared, sol–gel nanostructured composites have low combustion velocities, attributed to the high degree of impurities. If instead, the aerogel oxidizer phase is first purified by thermal treatment and then physically mixed with the fuel phase, combustion velocities are on the order of hundreds of meters per second. By using this type of mixing, the resulting sol–gel material is susceptible to the same nonhomogeneous mixing that other methods suffer from. It is the opinion of the authors that the degree of dispersion of fuel in the oxidizer phase is critical for energetic nanocomposites to realize their full potential. Unfortunately, thermal treatment of as-prepared sol–gel nanostructured energetic composites leads to the premature oxidation of the aluminum fuel. Presently, the homogeneous mixing benefit is not realized, but the sol–gel composition does enable drastic control of the energy release characteristics of inorganic nanostructured energetic composites.

This aspect of sol–gel nanostructured energetics was well documented by Prentice et al. in a study that examined the influence of mixing an inert diluent SiO_2 into Fe_2O_3/Al nanocomposite by both sol–gel and physical mixing [29]. Silica is often added to thermite reactions to slow down the rate and decrease the temperatures the reaction reaches as SiO_2 does not react as energetically with Al as Fe_2O_3 does. In that study, various compositions (100/0 to 40/60 weight percent) of the mixed oxidizer Fe_2O_3/SiO_2 were prepared by two methods (1) physical mixing of commercial powders and (2) cogellation of silica and iron (III) oxide precursors via the epoxide-initiated method (Chap. 8) [30].

Combustion wave velocities as a function of synthesis technique and oxidizer stoichiometry are summarized in Table 25.2 and illustrate some important aspects of sol–gel energetics. Both of the pure Fe_2O_3/nanometric Al materials burned the most rapidly of all the compositions tested and as the weight percentage of SiO_2 was gradually increased those velocities decreased, as expected. However, the magnitude of that decrease coupled to the respective synthesis technique. The combustion velocities for the sol–gel oxidizer was especially sensitive to composition, decreasing by a factor of nearly 200 (from 40 to 0.23 m/s) for the addition of 20 wt% SiO_2 diluent into the sol–gel nanostructure. Conversely the physical mixture of 20 wt% SiO_2 diluent led to a fourfold decrease in combustion wave velocity. Inspection of the entire data set present in Table 25.2 indicates that this trend is supported at increasing weight percentages of silica. The increased sensitivity of the sol–gel oxidizer composition can be related to the nanostructure of the diluent/oxidizer network that consists of interpenetrating finely mixed matrices [30]. Clearly compositional mixing on that nanometer length scale is sufficient to significantly affect the combustion properties of sol–gel nanostructured energetic materials. A future challenge would be to incorporate a sol–gel component into a mixed system that accelerates a reaction instead of acting as a diluent.

Table 25.2. Burning rate data for the combustion of various stoichiometries for two material mixture types (aerogel and physical mixture) for the reaction of the mixed oxidizer Fe_2O_3/SiO_2 and nanometric Al fuel

Fe_2O_3/SiO_2 (wt/%)	Mixture type	Burning velocity (m/s)
100/0	Aerogel	40
100/0	Physical mix	8.7
80/20	Aerogel	0.23
80/20	Physical mix	2.1
60/40	Aerogel	0.02
60/40	Physical mix	0.22
40/60	Aerogel	0.01
40/60	Physical mix	0.1

Since sol–gel materials and methods are so amenable to the deposition of thin films and coatings, there is an obvious extension of this to sol–gel nanostructured energetic composites. In general, nanostructured energetic composites are good candidates for coatings as their high energy densities and reaction temperatures may enable small critical thickness dimensions needed for self-propagating films, which may have severe thermal losses [14]. Apperson et al. have successfully deposited thermite thin films on substrates by spin coating and demonstrated self-propagation of energetic nanocomposites of CuO/Al with combustion velocities on the order of several hundreds of meters per second [31]. Park and coworkers describe the use of iron (III) oxide sol–gel coatings to deposit a uniform layer of iron metal on a silicon surface through the thermite interfacial reaction [32]. To produce flexible and durable films it is sometimes desirable to incorporate an organic polymer into the coating mixture. This is readily done through modification of the solvent and precursors for sol–gel nanostructured energetic composites as described in an invention by Gash et al. [33]. The process described in this art was used to coat one end of the reactive metal substrate shown in Figure 25.6A. Mechanical ignition of the reactive metal substrate is shown in Figure 25.6B which provides the thermal impulse which ignited the sol–gel nanostructured energetic coating.

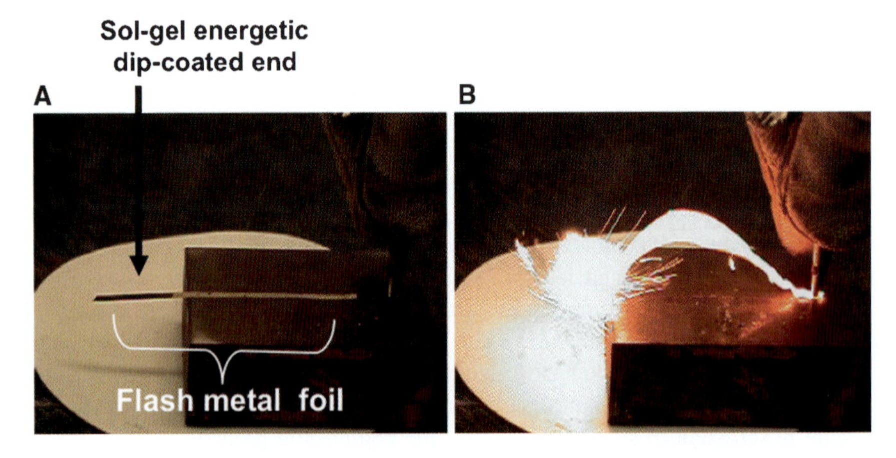

Figure 25.6. A strip of reactive foil that has had one end dip-coated with sol–gel nanostructured iron (III) oxide/ aluminum (**see image on the left**) and an image of the reaction of the foil which provides the thermal stimulus needed to ignite the sol–gel energetic.

25.3.2. Aerogel and Sol–Gel Composites Nanostructured Pyrophoric Materials

The use of pyrophoric materials in military and commercial applications is widespread. Pyrophoric materials are commonly used in pyrotechnic devices as well as being utilized as self-heating materials in commercial products such as ready-to-eat meals and thermal hand warmers. Quite possibly the most significant application of aerogel materials and sol–gel technology to energetic materials has been in the area of nanostructured porous metals. Metals are frequently employed in energetic articles for a few reasons. First, many metals have very high heats of combustion (e.g., aluminum) and thus contribute greatly to the overall temperature of the energetic formulation upon reaction. Additionally some metals can be formulated to be pyrophoric and thus react spontaneously upon exposure to air.

This attribute is particularly useful in applications where the material is utilized in devices whose mechanism of action is dispersal in air. A particularly important example of this is aircraft decoy flares, which include infrared (*IR*) and solid pyrophoric flares [34]. Aircraft pyrophoric decoy flares are solid pyrotechnic devices ejected as a precautionary measure or in response to a missile warning system. The most significant requirement of the device is that it develops a high-intensity, characteristic signature, rapidly. In order to meet this requirement, the energy radiated by the flare is typically provided by a pyrotechnic reaction. The most common composition of a conventional pyrotechnic flare consists of pyrophoric iron. This composition provides the high energy density required for the decoy and also produces solid combustion products for good radiation efficiency. The net reaction is shown below:

$$2\,\mathrm{Fe}\,(\mathrm{s}) + 3/2\,\mathrm{O_2}\,(\mathrm{g}) \rightarrow 2\mathrm{Fe_2O_3}\,(\mathrm{s}) + 412\,\mathrm{kJ/mol} \qquad (25.2)$$

Decoy materials of this composition undergo the above reaction to reach temperatures of 820°C in less than one second and above 750°C for twelve seconds after their exposure to air. Presently this type of material is used in a commercial decoy flare that is composed of pyrophoric iron coated onto steel foil articles [35]. Due to increasingly demanding materials performance, environmental standards, aging, and duty-cycle, there exists a need for continued development of new materials and approaches to achieve pyrophoric materials with tailorable output.

Sine sol–gel synthesis is a "bottom-up" approach, the homogeneous incorporation of various dopants and additives in, for example, mixed composition materials is relatively facile and has been demonstrated numerous times. This attribute is especially important to pyrotechnics as additives to modify the reactivity, spectral, or thermal output are a necessity [36, 37].

The relative ease of modification of attributes such as the density, porosity, morphology, and composition of sol–gel materials make them especially attractive as pyrophoric materials. Since the reaction is driven by the oxidation of metal by gaseous oxygen at the gas/solid interface, it stands to reason that the rate of that reaction is intimately coupled to the surface area, pore size, and pore size distribution of the pyrophoric material. These are all variables that are highly controllable in sol–gel techniques through careful selection of reactants, reaction conditions, and postprocessing methods such as annealing. Therefore, one could envision a class of pyrophoric material with a whole spectrum of rise-times, action-times, spectral outputs, and temperatures based on modifications of sol–gel conditions. In many instances, sol–gel technology and materials may have low environmentally impact through the use of common inorganic salts, benign solvents such as short chain alcohols and water, and common commodity-scale nontoxic commodities such as iron, aluminum, and iron oxide. This is particularly relevant as many energetic materials and processes are highly toxic and generate waste streams with undesirable environmental properties. In that regard, there has been a devoted effort in the last decade to the "greening" of energetic materials and process technologies [38].

There have been several reports of applying sol–gel materials and techniques to the formulation of nanostructured pyrophoric materials. The application of sol–gel materials to this technology area has involved three different approaches, which include (1) use of the aerogel as a framework for the deposition of pyrophoric materials; (2) direct transformation of the aerogel skeleton itself (metal oxide) to a pyrophoric or easily combustible article; and (3) use of an aerogel network as a reactive template for the reduction of a second sol–gel metal oxide phase to a pyrophoric or readily combustible material.

Mezbacher et al. disclosed the invention of thermally emitting iron aerogel composite materials as long-duration emitters of infra-red and visible radiation spontaneously upon exposure to air [39]. This report details two synthetic approaches, the more successful being the metal-organic chemical vapor deposition of iron into the porous framework of resorcinol-formaldehyde (*RF*), carbonized *RF*, and silica aerogels. While this approach led to iron deposition in all three aerogel substrates, only the *RF* and carbonized *RF* materials were pyrophoric, reaching peak temperatures of 600–700°C and having a total combustion duration of up to 15 min. That duration was found to depend critically on aerogel substrate, size, and geometry. Uniform deposition of iron metal into the aerogel was highly dependent on the pore size, and volume as well as the flow conditions and iron precursor.

A second approach disclosed involved the synthesis of mixed iron oxide/silicon oxide aerogels that were subsequently reduced with hydrogen at elevated temperatures. This method was less successful as it did enable the formation of iron metal-doped silica aerogel composites; however, the materials themselves were not pyrophoric. That was attributed to the formation of large Fe crystallites that were coated with an iron oxide or iron silicate layer. More recently, Gash et al. have filed the invention of spontaneously combustible metal-doped activated carbonized RF materials where the doping method involves the impregnation of the activated carbon aerogel network with a liquid solvent containing dissolved dopant [40]. An image depicting the spontaneous combustion of an article made by this route is shown in Figure 25.7. These materials have been shown to reach peak temperatures in tens of seconds upon exposure to ambient atmosphere. In addition to iron, this approach enables the production of pyrophoric articles with other metals, demonstrative examples of such being titanium, nickel, and platinum.

Figure 25.7. Photograph of the pyrophoric response of a nanostructured carbon/iron metal composite immediately after exposure to ambient atmosphere at ambient temperature. This composite has been measured to reach temperatures of ~650°C.

The direct conversion of sol–gel-derived structures to nanometric and pyrophoric iron has been achieved by others. The design of this approach is quite straightforward in that the metal oxide sol–gel structures are directly converted to the native metal by reduction with hydrogen. The intention was to effect a pseudomorphic transformation of the high surface area nano-porous metal oxide aerogel network to a native metal network with the same characteristics. This approach has been utilized by two different research groups and has led to the formation of nanometric and pyrophoric powders and coatings. Both groups have utilized sol–gel iron oxide-derived materials made using the epoxide-initiated gelation method (Chap. 8). In one case, Doorenbos et al. utilized this synthetic method to prepare loose powders and coatings on steel foils or porous substrates of pyrophoric iron. The method utilized the gelation of ferrous (Fe (II)) salts in the presence of organic surfactant followed by reduction in hydrogen at 500°C. These materials ignite spontaneously upon exposure to ambient air and achieve temperatures as high as 900°C with a rise time of one second [41].

In several instances, Gash et al. describe the use of iron (III) aerogel materials as starting materials for the synthesis of nanometric iron metal powders for application as energetic material ingredients [42]. They report complete reduction of iron (III) oxide aerogel to nanometric iron by a 100% hydrogen stream at temperatures as low as 350°C [43]. Consideration of the composition and phase of the initial iron (III) oxide material is extremely important. The presence of trace impurities can affect the properties of materials dramatically. To complex the situation there are several known phases of iron oxides, each of which having distinct properties and chemical characteristics. Sol–gel-derived iron (III) oxide materials made by the epoxide-initiated method have been determined to consist of ferrihydrite phase, $Fe_5HO_8{\bullet}4H_2O$ (formula weight (*FW*) = 480 g/mol). This is a highly hydrated poorly crystalline iron (III) oxide phase that was determined by X-ray diffraction (*XRD*) [44].

Thermogravimetric analysis (*TGA*) is a valuable technique for monitoring the progress of the reduction reaction. This technique permits following the extent of reaction with time and enables one to determine when the reduction is complete. In addition, by examining any weight gain of the sample as it cools, the fidelity of the experimental system to atmospheric impurities (e.g., O_2 or water) can be evaluated. Figure 25.8 shows a representative *TGA* trace for the reduction of an iron (III) oxide aerogel in a 100% hydrogen atmosphere with a flow rate of 50 standard cubic centimeters per minute (*sccm*) at 700°C. From this *TGA* trace it can be seen that the mass loss levels out after ~4 h under these conditions. This is interpreted as the completion of the reduction reaction. Hydrogen levels above ~20% were sufficient to bring about complete reduction of the sol–gel iron (III) oxide material. At hydrogen levels below that reduction was incomplete even at temperatures up to 600°C. From Figure 25.8 it is clear that the sol–gel-derived iron (III) oxide materials had mass loss of ~48% weight

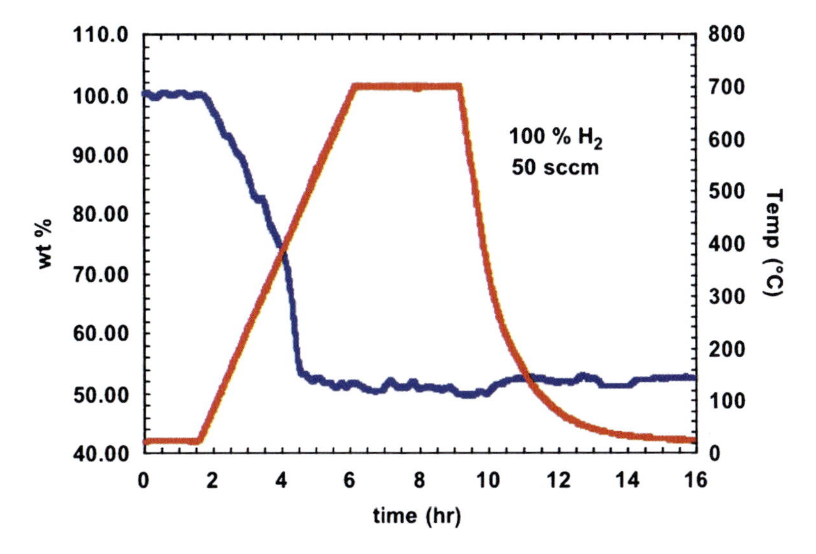

Figure 25.8. TGA trace of mass loss and temperature as a function of time for the reduction of sol–gel iron (III) oxide aerogel to nanostructured iron metal by pure hydrogen at a flow rate of 50 standard cubic centimeters per minute (sccm).

percent upon completion of the reduction. If one considers the iron (III) oxide aerogel material to be ferrihydrite, the reduction of this iron oxide phase to metallic iron would result in a 42% mass loss.

$$Fe_5HO_8 \cdot 4H_2O\,(FW: 480\text{ g/mol}) \rightarrow 5\text{ Fe}\,(FW: 56\text{g/mol}) + H_2O \tag{25.3}$$

As previously stated, elemental analyses indicate a background level of organic contaminant (C and H) of 4–8 wt%. Taking this into account, as well as the reduction mass loss and the dehydration, the expected weight loss of the iron (III) aerogel should range from 46 to 51 wt%. It is therefore reasonable to infer that the mass loss seen in *TGA* is consistent with dehydration and reduction of ferrihydrite aerogel to pure iron metal. In addition to the *TGA* evidence powder X-ray diffraction (*PXRD*) analysis of the product shown in Figure 25.9 indicates the product material is pure iron metal.

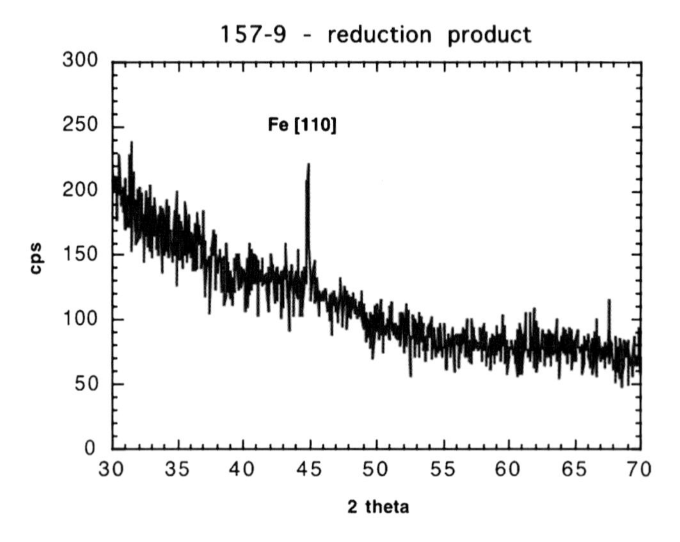

Figure 25.9. XRD of iron (III) oxide aerogel that has been reduced with pure hydrogen and it indicates the conversion of the aerogel to iron metal.

The transformation of high surface area sol–gel iron (III) oxide to iron metal nanoparticles is illustrated via microscopy on aerogel materials before and after reduction. Figure 25.10 contains two *SEM* images. The one in Figure 25.10A is that of the base aerogel materials and the image in Figure 25.10B is that of the reduced material transformed to Fe nanoparticles. The reduced aerogel sample appears to consist of nominally spherical particles with a diameter on the order of approximately 200–500 nm. While these diameters are less than a micron, they are still significantly larger than the primary particle size of the aerogel starting material (~5–20 nm). These images indicate that significant sintering has taken place upon transformation. As a measure of this sintering the surface area before reduction is ~350 m^2/g and after it is 27 m^2/g.

Powders from commercial sources, different phases of iron oxides, as well as sol–gel-derived aerogels and xerogels were evaluated by the *TGA* method. According to the results the iron (III) oxide aerogel reduced to metallic iron the most rapidly under constant conditions (1:1 v/v, H_2/Ar at 450°C). This is possibly related to the extremely high surface

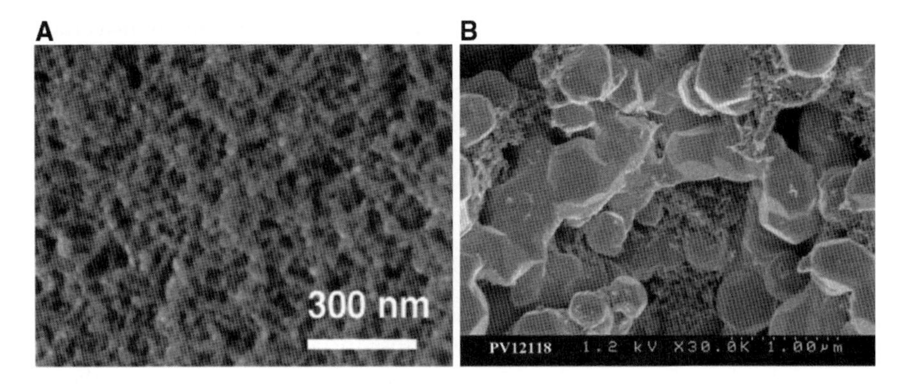

Figure 25.10. Scanning electron micrographs (SEM) of **A.** as-prepared iron (III) oxide aerogel; and **B.** the same aerogel in **A** after reduction with pure hydrogen. Note that the dramatic particle size increases with reduction; however, the resulting iron powder has a surface area of 27 m^2/g.

area of the aerogel material. According to *TGA*, iron produced from the reduction of aerogel iron (III) oxide material oxidizes at ~340°C by in flowing air. This temperature is at least 75°C less than is seen for the oxidation of iron particles made from commercial Fe_2O_3 (Aldrich) (T_{ox}~ 415°C). This is potentially a very interesting result. It is known that ultra-fine grained Al powders prepared by vapor phase condensation oxidize at much lower temperatures than micron sized powders [7]. This type of nanometric Al has shown exceptional enhancement in energy release rates in mixtures with oxidizers and is currently being examined for numerous applications in energetic compositions [45].

When considering the energy release properties of fine metallic iron powders produced by this method, none showed pyrophoric behavior upon exposure to room atmosphere. However, they combusted readily in air with the application of a thermal source (flame or soldering iron). Once ignited, the powders burned smoothly with a blue flame and left behind a red residue of hematite, Fe_2O_3. One possible rationale for the nonpyrophoric behavior is iron particles with a stable passivating surface. Iron metal readily reacts with oxygen or water to passivate its surface and with high surface area powders, the heat generated can be significant enough to ignite the entire iron particle. However, with a suitable oxide coating the iron particles can be very stable. It is possible that the powders reduced by this method get slightly passivated soon after reduction. Nonetheless, the method described enables the use of sol–gel chemistries and aerogel materials for the production of high surface area nanometer-sized metals for energetic applications using environmentally benign chemical precursors and reagents.

Most recently, Leventis et al. have reported the synthesis and characterization of nanostructured metals using a unique approach that begins with the cogelling of metal oxide and *RF* sol–gel networks in a one-pot synthesis (Chap. 14) [46, 47]. This mixed gel with interpenetrating inorganic and organic networks is then dried supercritically and pyrolyzed under flowing argon at 1,000°C. This processing yields a resulting monolithic structure that is entirely the reduced native metal. The organic *RF* network, under pyrolysis conditions, acts as a chemical reductant for the metal oxide sol–gel skeleton. This process is smelting as it was introduced in the Iron Age 3,000 years ago, and is likely governed by the simple reaction shown below for the transformation of copper (II) oxide to copper metal:

$$2\,CuO\,(s) + C(s) \rightarrow 2\,Cu\,(s) + CO_2\,(g) \tag{25.4}$$

While not pyrophoric, the porous copper metal monolith is readily combustible by the application of a flame in air. In addition to copper, this method has been applied to prepare other nanoporous metals such as iron, cobalt, nickel, and tin, as well as carbides of chromium, titanium, and hafnium (Chap. 14). This method is especially powerful as it appears to be applicable to a wide variety of metals and affords articles that are monolithic.

25.3.3. Organic Aerogel Materials as Nanostructured Energetic Composites

So far in this chapter on aerogels as nanostructured energetics, the aerogel framework has been a metal oxide or metal. Further application of aerogels and sol–gel technology has involved the utilization of hydrocarbon or carbonized-based (organic) *scaffolds*. The research described here utilizes an organic gel (as a fuel) *scaffold* with nanoparticles of inorganic oxidizers crystallized in the pores. Another key distinction with these types of *EMs* is that the primary products of the reaction of an organic aerogel fuel network with inorganic oxidizers are gaseous, and not condensed phases. This opens up the possibility of this class of nanostructured energetic being susceptible to detonation. While extensively researched, detonation is not generally considered to occur in *thermite*-type and metal combustion reactions.

It is not within the scope of this chapter to discuss the chemistry and properties of organic aerogels that is reviewed in another section (Chap. 11). While the bulk of energetic nanostructures work utilizing organic aerogels involves resorcinol-formaldehyde (*RF*) materials, it is worthwhile to consider other organics gel networks that are possible by the sol–gel methodology. Examples include pfloroglucinol–formaldehyde, and melamine–formaldehyde gels as well as other emerging organic aerogels (Part IV). In short summary, the result of these types of reaction chemistries under sol–gel conditions is the formation of a porous organic-based (phenolic-type) cross-linked solid that is made up of nanometer-sized clusters linked together [48]. Deposition of an oxidizer into the nanometer-sized pores of this material results in an energetic nanocomposite like that depicted in Figure 25.1C. In such a composite the nanostructured framework of the composite is made up of a hydrocarbon or carbon-based scaffold with the pores containing inorganic oxidizer (e.g., ammonium perchlorate, ammonium nitrate) particles.

For *EM* formulation it is important to have a stoichiometric balance of oxidizer and fuel. This refers to a composite composition where there is enough oxidizer to react completely with all of the fuels to maximize the production of gaseous products. In most cases these products are CO_2, H_2O, and N_2, whose stabilities dictate large heats of reaction and high temperatures on formation. For an energetic material this is referred to as "oxygen balance" and a detailed discussion of this is beyond the scope of this chapter [1]. Such a balanced reaction is shown below for the case of *RF* sol–gel material with the oxidizer ammonium nitrate. For the sake of this equation the average repeating *RF* unit is taken as the fuel. This is based on the assumption that each unit in *RF* consists of a resorcinol core with three cross-linking methylene bridges.

$$C_{7.5}H_6O\,(s) + 18\,NH_4NO_3\,(s) \rightarrow 7.5\,CO_2\,(g) + 39\,H_2O\,(g) + 9N_2\,(g) \qquad (25.5)$$

One will note the high mole percentage of oxidizer needed to balance this reaction which corresponds to a high weight percentage of inter-pore oxidizer salt deposited in the overall composite. The resorcinol-formaldehyde/ammonium nitrate (*RF/AN*) reaction shown above requires ~90% by weight NH_4NO_3 to achieve optimal energy. This requirement represents a challenge to sol–gel chemists as this formulation necessitates incorporation of a large fraction of the composite weight to be contained in the pores and relatively little from the nanostructured organic framework.

One approach to incorporate oxidizing materials into the pores of sol–gel materials entails the use of oxidizers soluble in the initial sol–gel mixture, along with the gel-framework precursors. This is especially effective as the oxidizers of choice have very good solubility in the typical solvents (e.g., water, water–alcohol mix) used in organic sol–gel synthesis. The sol–gel condensation reaction occurs between the organic precursors in a solvent medium that also contains the dissolved oxidizing agent. The fuel nanostructure condenses leaving the solvent, containing dissolved oxidizer, in the pores. As the rigid gel is dried the solvent is evaporated from the pores, which causes the dissolved oxidizer to reach its super-saturation point. Once that point is reached, nucleation and crystal growth (or deposition) occur throughout the pore network. This series of steps leads to composite microstructures such as that shown in Figure 25.11. This figure contains two scanning electron micrographs, of a *RF* xerogel and a *RF* xerogel with oxidizer deposited throughout the network. This is a simple one-step approach that yields composites with a very high degree of mixing between the oxidizer and fuel phases. This approach has been utilized most frequently and enables the synthesis and formulation of high inorganic salt content organic sol–gel composites.

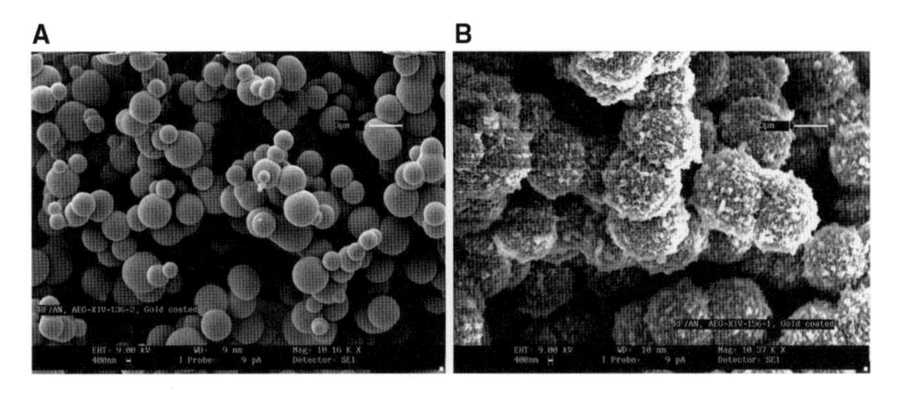

Figure 25.11. Scanning electron micrographs (SEM) of **A.** a resorcinol-formaldehyde (RF) sol–gel scaffold; **B.** a RF sol–gel scaffold with ammonium nitrate. Note that the coating of the RF particles with ammonium nitrate salt and that the images were taken at the same magnification.

This requirement in many organic sol–gel/oxidizer systems has presented a significant challenge to the processing of monolithic aerogel structures. What it has led to is the utilization of freeze-drying of the solvent in wet gels to give of cryogels as the organic framework. Both of these materials are more mechanically robust than aerogels and enable the production of monolithic composite sol–gel materials with high solids loadings. For example, Figure 25.12 is a photograph of a monolith of *RF/AN* material that is 70% *AN* by weight and 30% *RF*. Even higher loadings of *RF/AN* are possible (up to 90% *AN* by mass).

Thermal methods such as differential scanning calorimetry (*DSC*) and *TGA* allow an evaluation of the total energy of this type of composite. Figure 25.13 shows the *DSC* trace and *TGA* scan for a *RF/AN* (1:9 w/w) xerogel composite. The small endothermic peaks at lower temperatures in Figure 25.13A are associated with phase changes of the ammonium

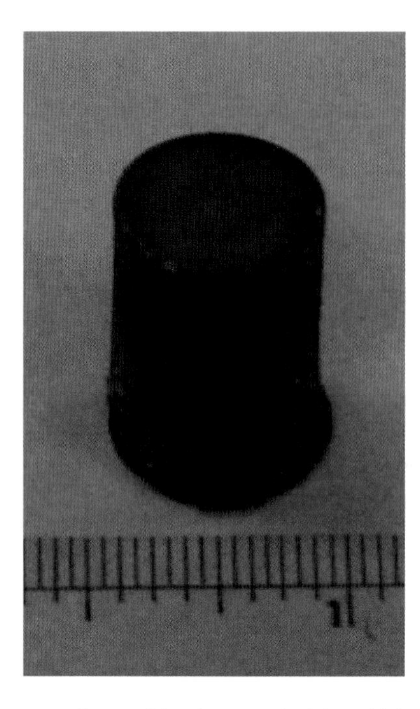

Figure 25.12. Photograph of a xerogel monolith of a resorcinol-formaldehyde/ammonium nitrate (3:7 w/w) nanostructured energetic composite illustrating the capability for sol–gel energetic materials to be cast and dried to well-defined shapes.

nitrate in the composite [53]. However, the large exotherm at 260°C is associated with the decomposition due to interaction of the *AN* and *RF* matrices and the release of a significant amount of energy. Figure 25.13B shows this event in its differential heat response but also indicates the rapid and near quantitative mass loss associated with this thermal event, through its *TGA* trace. This is especially revealing as the 97% weigh loss indicates that essentially all of the composites have been converted to gaseous products at this temperature. The absence of any significant amount of residual fuel is a strong indication that the stoichiometry of the composite is very close to balanced.

There are some important studies of organic sol–gel nanostructured energetic materials to consider. Tappan and Brill utilized sol–gel processing of resorcinol-formaldehyde/hydrazinium diperchlorate nanostructured composites in order to desensitize the energetic material [49]. The investigators were able to achieve solid composite loadings of up to 88% by weight oxidizer by using freeze-drying methods to give cryogels. Similar attempts with ambient air drying were not successful in maintaining monoliths.

The rapid freezing process assisted in forming a homogeneous dispersion of very small oxidizer particles in the sol–gel pores. The investigators characterized the cryogel composites by thermal methods, impact sensitivity, and rapid thermal decomposition and results were compared to physical mixtures of *RF* and oxidizer with the same compositions. In addition, the results showed that the cryogel composites were less sensitive to impact than physical mixtures of the fuel and oxidizer. This is partly attributed to the very fine particle size of the materials which should lead to lower temperature rises according to models of impact response [9]. It is also recognized that the sol–gel matrix affords some isolation of oxidizer particles from one another, which may also affect impact response.

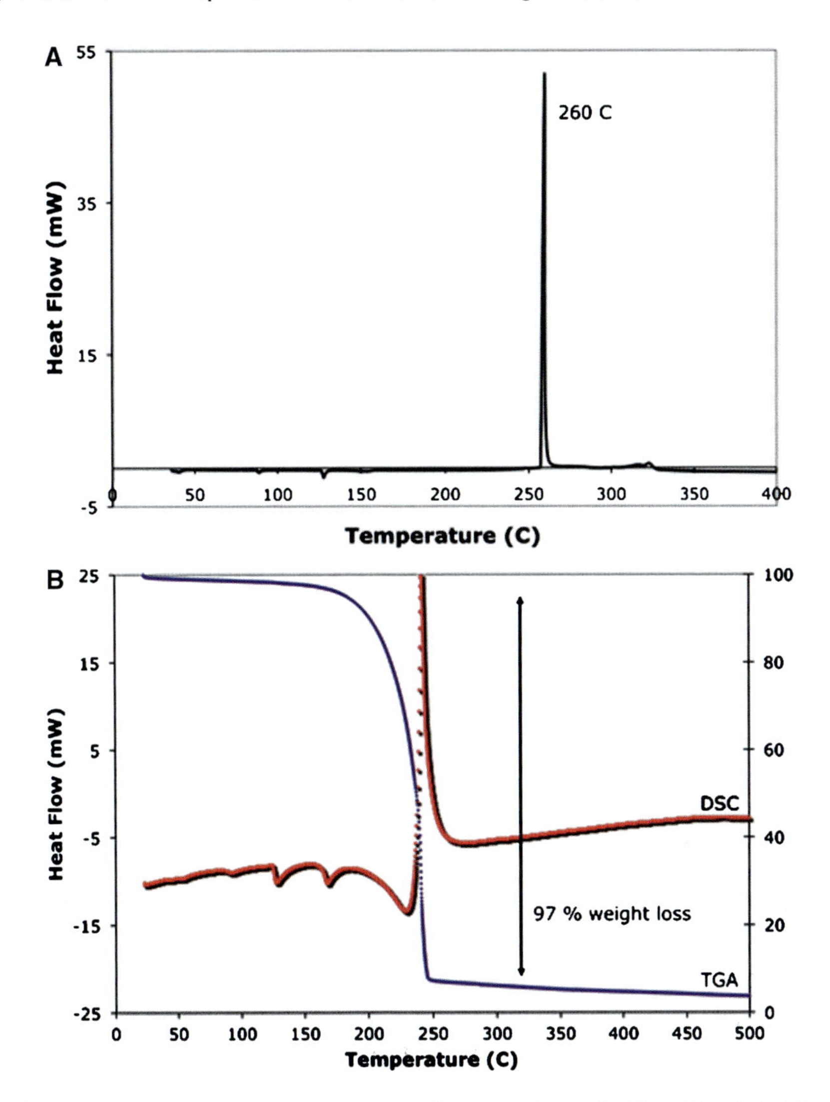

Figure 25.13. Comparison of the DSC and TGA traces for a xerogel resorcinol-formaldehyde (RF)/ammonium nitrate (1:9 w/w) nanostructured energetic composite. Note that the composite loses nearly all of its weight at ~260°C, which is expected for a composite that is oxygen balanced.

Thermal analyses indicated both the decomposition temperature and the time to explosion of cryogel samples heated at 2,000°C/s varied with composition, whereas those values were invariant with physical mixtures of the fuel and oxidizer. In all cases, burning rates of the cryogels were much higher than those of the physical mixtures. In sum, these results indicate a significant degree of tunability, decreased impact sensitivity, and ease of processing afforded by the sol–gel approach, all of which surpass that of physical mixing processing.

The same investigators have successfully applied cryogel processing to prepare composite of a nitrocellulose gel (an energetic gel), polyurethane cross-linked nitrocellulose and the energetic material *CL-20* [50]. Under those processing conditions, all of the components were initially in solution, which enabled the formation of nanoparticles of the energetic gel and the explosive. In several cases, the impact sensitivity of *CL-20* is reduced in the cryogel matrix relative to the bulk explosive.

A more recent study by Cudzilo and Kicinski evaluated the sol–gel synthesis of *RF* and resorcinol–furfural gels with a variety of different oxidizing salts crystallized in the pores [51]. High solid loadings of various perchlorate and nitrate-based oxidizers were achieved via air-drying methods, an improvement over previous approaches that required freeze-drying. Apparently a key issue here is the need for mixed solvent systems with water–methanol and water–*N,N*-dimethylformamide to achieve monolithic energetic composites. As with previous studies the materials displayed violent and rapid deflagration with loud sound and a flash when exposed to an open flame. This type of observation is consistently reported for these types of materials and suggests that the rate of energy release is influenced by reaction kinetics, and it may not rely as heavily on the mass and thermal transport processes that are presumably more prevalent in the physical mixtures of oxidizers and fuels. One material *RF*/ammonium perchlorate (1:4 w/w) was scaled up and a detonability test performed. That material had a measured detonation velocity of 6.9 km/s [52].

A second potential approach to deposition of oxidizing salts in an organic framework represents a future challenge in the area of composite synthesis. This involves the infiltration of a solvent, with dissolved oxidizer in it, into a preformed dried aerogel structure. Subsequent atmospheric or freeze-drying of the solvent leads to deposition of the oxidizer salt in the pores. This approach appears to be more complex and demanding as there is a need for an aerogel structure that can withstand the capillary forces induced on infiltration and drying. Possible candidates here are carbonized *RF* foams that have good mechanical properties (strength). In addition, it is unlikely that one could load a high weight percentage of inorganic oxidizing salt into the pores of such a structure without several infiltration/drying cycles. While this method has not been demonstrated at this point, it does remain a viable route to organic gel/inorganic oxidizer energetic nanostructures.

25.4. Summary

Aerogel and xerogel sol–gel derived materials have been investigated as nanostructured energetic materials over the past decade. The methodology can be used to prepare high surface area nano-sized porous networks of both oxidizers (metal oxides) and fuels (organic aerogels). The straightforward synthetic nature, mild processing conditions, and processing versatility are all attributes of sol–gel science that are compatible with energetic materials synthesis and processing. The textural properties of aerogels enable a great deal of interfacial contact area between phases that reduce the effects of mass and thermal transport and lead to materials with desirable energy release characteristics and are tunable to a degree not possible with current methods. Materials parameters such as surface area, density, porosity, surface impurities, and intimacy of phase mixing are shown to affect the combustion speed and reactivity of sol–gel energetics greatly. Several scientific challenges remain and those are anticipated to be addressed in future studies.

Acknowledgments

Work was performed under the auspices of the U.S. Department of Energy by Lawrence Livermore National Laboratory under Contract DE-AC52-07NA27344.

References

1. Cooper PW (1996) *Explosives Engineering*. New York: Wiley-VCH.
2. Gash AE, Satcher JH, Simpson RL (2002) Direct preparation of nanostructured energetic materials using sol-gel methods. In Miziolek, AW, Karna Sp, Mauro JM, Vaia Ra (Eds.) *Defense applications of nanomateirals*. Washington DC : American Chemical Society.
3. Dagani R (1999) Putting the nano into composites. Chemical & Engineering News 77: 25–31.
4. Wickersham CE, Poole JE (1988) Explosive Crystallization in Zr/Si Multilayers. J. Vac. Sci. Technol. A, 6(3): 1699–1702.
5. Bockman BS, Pantoya ML, Son SF, Asay BW, Mang JT (2005) Combustion velocities and propagation mechanisms of metastable interstitial composites. Journal of Applied Physics 98(6): 064903/1-064903/7.
6. Umbrajkar SM, Schoenitz M, Dreizin EL (2005) Structural refinement in Al-MoO_3 nanocomposites prepared by arrested reactive milling. Thadhani N, Armstrong RW, Gash AE, Wilson WH Materials Research Society Symposium Proceedings Vol. 896; Multifunctional Energetic Materials. Warrendale PA (USA): Material Research Society.
7. Son SF, Yetter YA, Yang V (2007) Introduction: Nanoscale energetic materials. Journal of Power and Propulsion and Power 23(4): 643–644.
8. Aumann CE, Skofronick GL, Martin JA (1993) Oxidation behavior of aluminum nanopowders. J Vac Sci Technol B 13(2) : 1178–1183.
9. Armstrong RW, Coffey CS, DeVost VF, Elban WL (1990) Crystal size dependences for the impact initiation of cyclotrimethylenetrimitramine explosive. J Appl Phys 68: 979–984.
10. Hüsing N, Schubert U (1998) Aerogels-airy materials: Chemistry, structure, and properties. Angew Chem Int Ed 37: 22–47.
11. Pierre AC, Pajonk GM (2002) Chemistry of aerogels and their applications. Chem Rev 102: 4243–4265.
12. Brinker CJ, Scherer GW (1990) Sol-gel science: The physics and chemistry of sol-gel processing. Academic Press, USA.
13. Teipel U (Ed) (2005) *Energetic materials particle processing and characterization*. Weinheim: Wiley-VCH.
14. Rossi, C, Zhang K, Esteve D, Alphonse P, Tailhades P, Vahlas C (2007) Nanoenergetic materials for MEMS: A review. J Microelectromechanical Systems 16(4): 919–931.
15. Barbee TW, Weihs TP (1996) US Patent 5,547,715.
16. Tillotson TM, Sunderland WE, Thomas IM, Hrubesh LW (1994) Synthesis of lanthanide and lanthanide-silicate aerogels. J Sol-Gel Sci Tech 1: 241–249.
17. Gash AE, Satcher JH, Simpson RL (2004) Monolithic nickel(II)-based aerogels using an organic epoxide: The importance of the counterion. J Non-Cryst Solids 350: 145–151.
18. Gash AE, Tillotson TM, Satcher JH, Poco JF, Hrubesh LW, Simpson RL (2001) Use of epoxides in the sol-gel synthesis of porous iron oxide monoliths from Fe(III) salts. Chem Mater 13: 999–1007.
19. Gash AE, Tillotson TM, Satcher JH, Hrubesh LW, Simpson RL (2001) New sol-gel synthetic route to transition and main-group metal oxide aerogels using inorganic salt precursors. J Non-Cryst Solids 285: 22–28.
20. Gash AE, Satcher JH, Simpson RL (2003) Strong akaganeite aerogel monolith using epoxides: Synthesis and characterization. Chem Mater 15: 3268–3275.
21. Goldschmidt H (1908) New thermite reactions Zeitsch Elektro Angew Physik Chem 14: 558–564.
22. Fisher SH, Grubelich MC (1998) Theoretical energy release of thermites, intermetallics, and combustible metals. Proceedings from the 24th International Pyrotechnic Seminar. Monterey CA: IPS USA, Inc.
23. Morris CA, Anderson ML, Stroud RM, Merzbacher CI, Rolison DR (1999) Silica Sol as a nanoglue : Flexible synthesis of composite aerogels. Science 284: 622–624.
24. Tillotson TM, Gash AE, Simpson RL, Hrubesh LW, Satcher JH, Poco JF (2001) Nanostructured energetic materials using sol-gel methodologies. J Non-Cryst Solids 285: 338–345.
25. Kim S, Zachariah MR (2004) Enhancing the rate of energy release from nanoenergetic materials by electrostatically enhanced assembly. Adv. Mater. 16(20): 1821–1825.
26. Plantier KB, Pantoya ML, Gash AE (2005) Combustion wave speeds of nanocomposite Al/Fe_2O_3 : The effects of Fe_2O_3 particle synthesis techniques. Combustion and Flame 140: 299–309.
27. Prentice D, Pantoya ML, Gash AE (2006) Combustion wave speeds of sol-gel-synthesized tungsten trioxide and nano-aluminum: The effect of impurities on flame propagation. Energy & Fuels 20: 2370–2376.
28. Mehendale B , Shende R, Subremanian S, Gangopadhyay S, Redner P, Kapoor D, Nicolich S Nanoenergetic composite of mesoporous iron oxide and nanometric aluminum. Journal of Energetic Materials 24: 341–360.

29. Prentice D, Pantoya ML, Clapsaddle BJ (2005) Effect of nanocomposite synthesis on the combustion performance of ternary thermite. J Phys Chem B 109: 20180–20185.
30. Clapsaddle BJ, Gash AE, Satcher JH, Simpson RL (2003) Silicon oxide in an iron(III) oxide matrix: The sol-gel synthesis and characterization of Fe-Si mixed oxide nanocomposites that contain iron oxide as a major phase. J Non-Cryst Solids 331: 190–201.
31. Apperson S, Bhattacharys S, Ga Y, Subramanian S, Hasan S, Hossain M, Shende R, Kapoor D, Nicolich S, Gangopadhyay K, Gangopadhyay S On-chip initiation and burn rate measurements of thermite energetic reactions. Thadhani N, Armstrong RW, Gash AE, Wilson WH Materials Research Society Symposium Proceedings Vol. 896; Multifunctional Energetic Materials. Warrendale PA (USA): Material Research Society.
32. Park C, Walker J, Tannenbaum R, Stiegman AE, Frydrych F, Machala L (2009) Sol-gel derived iron oxide thin films on silicon: Surface properties and interfacial chemistry Applied Materials and Interfaces 1(9): 1843–1946.
33. Gash AE, Satcher JH, Simpson RL (2004) Inorganic metal oxide/polymer nanocomposite and method thereof. (2004) US Patent 6,712917B1.
34. Koch EC (2006) Pyrotechnic countermeasures: II. Advanced aerial infrared countermeasures. Propellants, Explosives, and Pyrotechnics 31(1): 3–19.
35. Baldi AL (1995) Pyrophoric materials and methods for making the same. US Patent 5,464,699.
36. McLain JH (1980) *Pyrotechnics*. Philadelphia: The Franklin Institute.
37. Conklin JA (1985) *Chemistry of Pyrotechnics : Basic Principles and Theory*. New York: Marcel Dekker, Inc.
38. Gash AE, Satcher JH, Simpson RL (2002) Safe and environmentally acceptable sol-gel derived pyrotechnics (fact sheet-wp-1276). Retrieved from http://serdp.org/research/WP-propellants-pyrotechnics-and-explosives.cfm.
39. Merzbacher C, Limparis K, Bernstein R, Rolison D, Homrighaus ZJ, Berry AD (2001) Long duration infrared-emitting material. US Patent 6,296,678 B1.
40. Gash AE, Satcher JH, Simpson RL, Baumann TF, Worsley MA (2009) Pyrophoric metal/carbon composites and method of making same. US Patent pending.
41. Doorenbos Z, Vats A, Puszynski J, Shende R, Kapoor D, Martin, D. Haines C (2008) Pyrophoric films based on nano-sized iron. Proceedings of the American Institute of Chemical Engineers Conference.
42. Gash AE, Satcher JH, Simpson RL (2006) Preparation of porous pyrophoric iron using sol-gel method. US Patent application 20060042412 Appl. No. 11/165734.
43. Gash AE, Satcher JH, Simpson RL, Metcalf P, Hubble W, Stevenson B (2002) Safe and environmentally acceptable sol-gel derived pyrotechnics (Final Report-1276). Retrieved from http://serdp.org/research/WP-propellants-pyrotechnics-and-explosives.cfm.
44. Duraes L, Costa BFO, Vasques J, Campos J, Portugal A (2005) Phase investigation of as-prepared iron/oxide/hydroxide produced by sol-gel synthesis. Mater Lett 59: 859–863.
45. Armstrong RW, Baschung B, Booth DW, Samirant (2003) Enhanced propellant combustion with nanoparticles. Nano Lett 3(2): 253–255.
46. Leventis N, Chandrasekaran N, Sotiriou-Leventis C, Mumtaz A (2009) Smelting in the age of nano: Iron aerogels. J Mater Chem 19: 63–65.
47. Leventis N, Chandrasekaran N, Sadekar AG, Sotiriou-Leventis C, Lu H (2009) One-pot synthsis of interpenetrating inorganic/organic networks of CuO/resorcinol-formaldehyde aerogels: Nanostructured energetic materials. J Amer Chem Soc 131: 4576–4577.
48. Pekala RW (1989) Organic aerogels from the polycondensation of resorcinol with formaldehyde. J Mater Sci 24: 3221–3227.
49. Tappan BC, Brill TB (2003) Thermal decomposition of energetic materials 85: Cryogels of nanoscale hydrazinium perchlorate in resorcinol-formaldehyde. Propellants Explosives and Pyrotechnics 28(2): 72–76.
50. Li J, Brill TB (2005) Nanostructured energetic composites of *CL-20* and binders by sol-gel methods. Propellants Explosives and Pyrotechnics 31(1): 61–69.
51. Cudzilo S, Kicinski (2009) Preparation and characterization of energetic nanocomposites of organic gel – inorganic oxidizers. Propellants, Explosives, and Pyrotechnics 34: 155–160.
52. Cudzilo S, Trzcinski WA, Kicinski W (2009) Detonation parameters of resorcinol-formaldehyde/ammonium chlorate (VII) nanocomposites. Proceedings of the 40th International Annual Conference of the ICT Karlshrule, Germany.
53. Herrmann MJ, Engle W (1997) Phase transitions and lattice dynamics of ammonium nitrate. Propopellants, Explosives, and Pyrotechnics 22: 143–147.

26

Aerogels for Superinsulation: A Synoptic View

Matthias M. Koebel, Arnaud Rigacci, and Patrick Achard

Abstract The present chapter is focused on describing the intimate link which exists between aerogels and thermal superinsulation. For long, this applied field has been considered as the most promising potential market for these nanostructured materials. Most likely this old vision will become reality in the near future.

Following a short presentation of the global need for superinsulation together with a closer look at the specific situation in the building sector, we propose within this synopsis a brief analysis of (1) the world's insulation markets, (2) superinsulating aerogel materials and their alternatives, (3) commercial aerogel insulation products available today, and (4) our estimation of their most likely applications worldwide in the future. We conclude this chapter with some first considerations on health, toxicity, and environmental aspects.

Based on recent developments in the field, it can be stated that aerogels still offer the greatest potential for nonevacuated superinsulation systems and consequently must be considered as an amazing opportunity for sustainable development. This chapter of the handbook bridges the gap between those dealing with thermal insulation properties of aerogel materials in general (Chap. 21) and the various commercial products described in Part XV.

26.1. Superinsulation: Global Necessity and Building Specificity

26.1.1. Why Superinsulation?

Ever since the first global oil crisis in the 1970s, the scarcity of fossil fuels, which is the number one resource for our chemical industry and energy carrier, has underlined the dependence of modern society on cheap energy and resources [1]. Over short or long that very fact is forcing humanity to rethink global energy strategies and politics. In addition to a limited supply of carbon-based fuels worldwide, the effect of the carbon footprint, i.e., the influence of a rising carbon dioxide (CO_2) concentration in the earth's atmosphere [2] and its effect on the global climate [3, 4], has become indisputably clear: wide media coverage has

M. M. Koebel • Swiss Federal Laboratories for Materials Testing and Research (Empa), Überlandstrasse 129, CH-8600 Dübendorf, Switzerland
e-mail: matthias.koebel@empa.ch
A. Rigacci and P. Achard (✉) • MINES ParisTech, CEP, Centre Energétique et Procédés, BP 207, F-06904 Sophia Antipolis, France
e-mail: arnaud.rigacci@mines-paristech.fr; patrick.achard@mines-paristech.fr

M.A. Aegerter et al. (eds.), *Aerogels Handbook*, Advances in Sol-Gel Derived Materials and Technologies, DOI 10.1007/978-1-4419-7589-8_26,

inseminated public awareness at the turn of the millennium. Pictures displaying the melting polar cap went around the world leaving its spectator in a state of awe: the effects of the humans triumph through stellar advancements in the technological age can no longer be denied. Suddenly it becomes clear that our precious technology and free market economy are threatening the continuity of our human race.

Having come to this realization, international political efforts are asking for immediate solutions to the global warming and climate change problems. A goal often cited in this context is the stabilization of the atmospheric CO_2 concentration below 500 ppm [5]. The necessity for action and a rapidly increasing demand for renewable energy sources (sources with no net emission of CO_2) led to an overestimation of the potential of such technologies, a statement which describes the current situation quite adequately. Even though absolutely necessary for our advancement and an essential investment in our future, the development of alternative energy schemes is an arduous task and will take decades if not centuries to completely replace current technology. Our dependence on oil, gas, and coal (>85% of the world energy demand in 2003) is much deeper than most people feel comfortable to admit. It is therefore completely unrealistic to assume that renewable energies will be able, like certain groups claim, to replace a significant fraction of carbon-based energy carriers in the next 10–20 years. The main reasons against immediate implementation and for a delayed action are technological difficulties and economic barriers [6]. This realization automatically leads to a limited number of reasonable strategies for a global reduction of CO_2 and greenhouse gas emissions. An ideal course of action requires short-, medium-, and long-term strategies to bridge gaps while developing long-term renewable energy supply systems for the blue planet.

Figure 26.1 illustrates a realistic scenario including necessary steps required to stay on track with the ambitious goal to stabilize the atmospheric CO_2 concentration at 500 ppm by the year 2100. Those measures are divided into the main CO_2 emitters, namely, mobility (vehicles, ships, planes including transport of goods), power plants/heavy industry, and buildings. Surprisingly to most people, buildings worldwide account for approximately 40% of the global energy consumption, a larger share than that of the entire mobility sector.

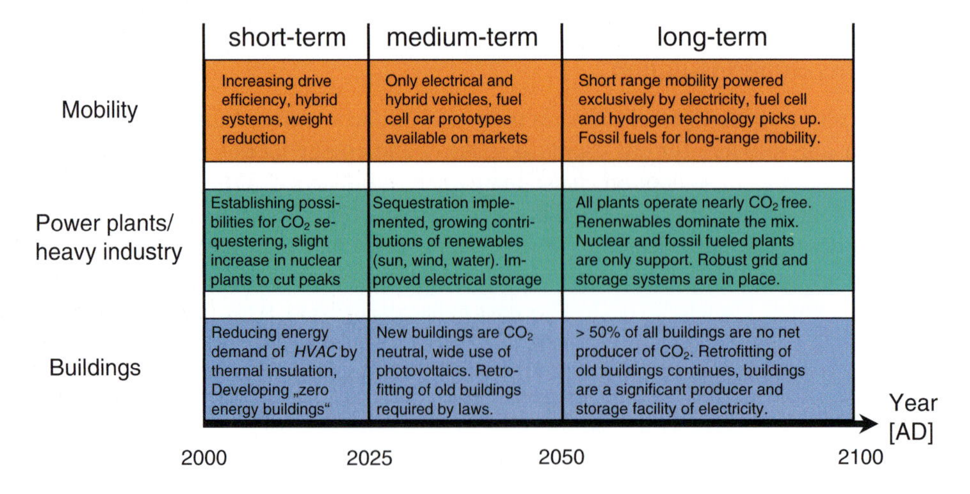

Figure 26.1. Appropriate strategies for halting global climate change sorted by time and source. Immediate action and a proper combination of short-, medium-, and long-term measures is essential for the future habitability of the earth.

With it comes a CO_2 footprint of a similar magnitude. The buildings sector is marked by another key feature: whereas current technologies used in transportation, power plants, and heavy industries are optimized to a large extent, the same is not true for the energy saving potential of buildings [7]. *HVAC* (heating, ventilation, and air conditioning) has the lion's share of a buildings total energy demand; however, in many areas of the world (almost everywhere except in central and northern Europe) the actual thermal insulation standards leave much to be desired. Hence, the vast stock of poorly insulated buildings around the globe adds up to a tremendous short-term potential for reducing CO_2 emissions right away, here and now.

26.1.2. Zoom on Thermal Insulation for Buildings

Working out ways to propel the use of renewable energy sources and to render transportation even more efficient are great challenges and will require several decades to be realized commercially. Technologies need to be developed and upscaled if this vision is to become a reality. However, in order to stay on track with climate change protocols, immediate measures are required to start curtailing the CO_2 output now. One of the most promising ways to achieve this is, of course, a reduction of the energy consumption of buildings' *HVAC* systems. This can be done with minimal effort by putting in place proper thermal insulation. If the climate change is taken seriously, society first off demands that improved insulation solutions are going to be put in place right now. This concerns both new and already existing buildings, even though the savings potential in the case of older structures is considerably larger. The most economical way to achieve a reduction in thermal losses of buildings is to install thicker layers of conventional insulation materials. There is of course an aesthetic disadvantage associated with such a cumbersome facade construction: the insulated object takes up more space and the volume fraction of inhabitable living space decreases. In most cases, cost and insulation performance are the primary parameters for new constructions and retrofitting of buildings. Figure 26.2 shows a simplified view of the correlation between cost, performance, and the market share in the insulation sector. The attribute "space saving" goes hand in hand with cost and insulation performance as criteria for choosing building insulation. The conventional or standard insulation products

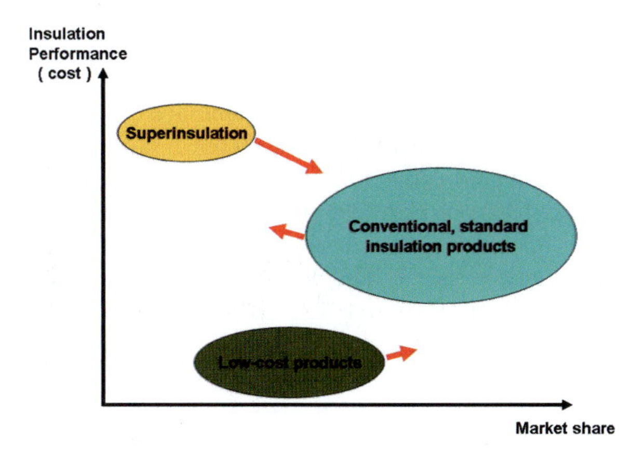

Figure 26.2. Market share displayed by cost/performance: high-performance (superinsulation) and low-cost products have smaller market shares. The *red arrows* symbolize the expected development of each segment in the future.

sector [8] is the most representative of the average building owner's need when shopping for insulation as it offers the best performance per unit cost. Superinsulation offers improved performance but at a significantly higher cost. Low-cost products offer poor performance (and durability) for an extremely low price.

The use of superinsulation is restricted to areas where it can offer a cost advantage due to a space-saving effect, improved service-life (lower servicing/support cost), or advanced product properties (resistance against chemicals, high or low temperature, etc.). In the building sector, space saving is the number one reason for the use of superinsulation. Typical examples include side-on balcony and accessible roof balcony constructions, interior insulation solutions for building retrofit as well as slim façade insulation for the renovation of historical buildings. A number of other niche insulation markets outside of the building sector such as thermal insulation for apparel, aerospace, petrochemical pipelines, and pumping fluid media in industry applications as well as low-temperature processes are worth mentioning and are discussed in more detail later on.

26.2. High-Performance Insulation or Superinsulation: The Basics

26.2.1. Range of Thermal Conductivity Values and the Physics of Heat Transport

Thus far, we have established definitions for markets and applications for *superinsulation* products. But perhaps the key question has not been addressed this far: What is the definition of *superinsulation*? One of the simplest ways to define the two synonyms "*superinsulation*" or "*high-performance insulation*" is via the *thermal conductivity* λ (lambda) measured under standard (ambient) conditions. The *thermal conductivity* is an intrinsic materials property and is defined by the heat flow through a slab of material of area A and thickness d with an effective temperature difference ΔT acting on the two surfaces. The units of λ are [W m^{-1} K^{-1}]. Metals are extremely good thermal conductors with λ values in the tens to hundreds of W m^{-1} K^{-1} range. Glass, sand, and minerals have typical *thermal conductivities* in the single digit W m^{-1} K^{-1} range. Polymers, plastics, and organic materials range in the tenths of W m^{-1} K^{-1} regime. Typical thermal insulators and related products exhibit λ values which are well below 0.1 W m^{-1} K^{-1}. Table 26.1 gives an overview over the performance of common insulation materials based on their respective λ-values: conventional insulation materials highlighted in red have typical λ-values on the order of 0.030–0.040 W m^{-1} K^{-1}. There are several definitions of superinsulation which are commonly used. A misleading and rather meaningless one is based on the total heat transfer coefficient or *U-value* of a well-defined insulating layer (single workpiece or element). It refers simply to a thick layer of high total thermal resistance and it is a poor definition because there is no correlation with the materials insulation performance (per unit thickness). This is certainly not the idea behind superinsulation and hence it makes sense to adhere to a more meaningful *thermal conductivity*-based definition rather than a resistance/ workpiece specific one. Probably, the most common one is defined by a maximum λ-value for superinsulating materials of 0.025 W m^{-1} K^{-1}. High performance *foam* insulation, such as polyurethane or phenolic resins, defines a transitional area between conventional and superinsulation products (see Table 26.1). In our opinion it is today more sensible to set the limit for superinsulation materials at 0.020 W m^{-1} K^{-1}, a definition which will be used in this chapter from here on: this choice of limits restricts the use of the term to advanced

Table 26.1. Overview of insulation materials sorted by their ambient condition thermal conductivities: conventional (red) and superinsulation materials and components (turquoise) can be classified by their respective *thermal conductivity* values. Phenolic and polyurethane *foams* mark a transition area between those two families (pink)

Insulation product	Chemical composition	λ [W m^{-1} K^{-1}]
Mineral wool	Inorganic oxides	0.034–0.045
Glass wool	Silicon dioxide	0.031–0.043
Foam glass	Silicon dioxide	0.038–0.050
Expanded polystyrene (EPS)	Polymer *foam*	0.029–0.055
Extruded polystyrene (XPS)	Polymer *foam*	0.029–0.048
Phenolic resin *foam*	Polymer *foam*	0.021–0.025
Polyurethane *foam*	Polymer *foam*	0.020–0.029
Silica aerogels	SiO_2 based aerogel	
Organic aerogels	Aerogels derived from organic compounds	0.012–0.020
Vacuum insulation panels (VIP)	Silica core sealed and evacuated in laminate foil	0.003–0.011[a]
Vacuum glazing (VG)	Double glazing unit with evacuated space and support pillars	0.0001–0.0005[b] 0.003–0.008[c]

[a]Vacuum insulation panels age with time. As the pressure inside the evacuated element rises because of envelope permeability, so does the *thermal conductivity* of the assembly. Equivalent conductivity values of 30-year aged *VIP*s depend strongly on the product and materials used and are typically on the order of 0.006–0.010 W m^{-1} K^{-1}.

[b]*Vacuum glazing* performance (in terms of equivalent *thermal conductivity*) determined from potential *U-values* on the order of 0.5–0.3 W m^{-2} K^{-1} and a thickness of the evacuated gap ($d = 0.3 \ldots 1.0$ mm) without taking into account the thickness of the glass panes.

[c]λ-Values determined from typical *U-values* (0.5–0.3 W m^{-2} K^{-1}) and the total thickness $d' = 10 \ldots 16$ mm under consideration of the window panes.

insulation materials and solutions which are intrinsically different from conventional *foam* and fiber products and represent only state-of-the-art technology. According to this second definition, the group of superinsulators comprises only three kinds of products: aerogels, *vacuum insulation panels* (*VIP*), and *vacuum glazings* (*VG*). They are discussed in more detail below.

VIP and *VG* offer outstanding thermal resistance because evacuation of the porous core material or the glazing cavity, respectively, results in a drastic reduction of heat transport by gas molecules. Aerogels on the contrary are nonevacuated superinsulators. Their low *thermal conductivity* is correlated with the pore structure of these materials. Before we briefly touch on the effect of the unique structural features of aerogels on heat transport, let us recapitulate the basics. Generally, heat is transferred by conduction, convection, and radiation. In porous materials there are five possible contributions to the total heat transfer, namely:

- Conduction through the solid material
- Conduction through the pore medium (interstitial fluid, e.g., air, water)
- Convective transport by the pore medium
- Radiative transport from solid surfaces through the pore fluid
- Radiative transport from the solid through the solid network or bulk

Superinsulating SiO_2 aerogels are rather light nanomaterials with a porosity >90% and typical *mesopore* diameters between 4 and 20 nm (which can add up to more than 90% of their whole porous volume). They are the most commonly used type of aerogel in the insulation business today. Aerogels owe their extremely low *thermal conductivity* to the combination of low density and small pores. Briefly, the small pores effectively limit conductive and convective gas transport, the low density on the other hand implies that the solid network is delicate providing only limited pathways for conduction though the solid network. Even if there is a complete section dedicated to the thermal properties of aerogels (Chap. 23), let us still look at the individual contributions of the total heat transfer in more detail:

On one hand, convection by the pore fluid, i.e., air is reduced already significantly in porous materials with micrometer pore sizes (such as polymer *foam* insulators) so that only conduction and radiation remain. Because of the small pore sizes, however, the conductive heat transfer by the pore fluid, in this case air, equals no longer that of a free gas but follows the *Knudsen* formula [9] and thus results in an effective gas conduction contribution to the total *thermal conductivity* much lower than for a free gas. This condition is met when the mean-free path of the gas molecules is effectively constrained by the pore structure or size. The problem of simultaneous conduction and radiation between parallel plates is formulated in terms of an integro-differential equation which can only be solved numerically [10]. For semitransparent "gray" materials like aerogels, an analytical solution based on the concept of effective emissivity has been proposed [11]. Radiative transfer through the pore fluid is comparable to a conventional *foam* insulation material and depends strongly on the density (or more precisely the network connectivity which is directly linked to the density). Radiative heat transfer through the solid network is generally neglected but can sometimes represent a few tens of percent of the total *thermal conductivity* in aerogels. What remains is the conductive transfer in the solid, the dominant contribution for mesoporous aerogels of raw densities >200 kg m^{-3} at 300 K. Its magnitude strongly depends on the nanostructure as well as the bulk density. An expression for the thermal conduction λ_{solid} in a solid is known from the kinetic transport theory

$$\lambda_{\text{solid}} = \frac{1}{3}\left(c_e \ell_e v_e + c_{ph} \ell_{ph} v_{ph}\right) \tag{26.1}$$

where c is the specific heat capacity per unit volume in J m^{-3} K^{-1}, ℓ the mean free path, and v is the velocity with the subscripts denoting electronic (e) and phononic (ph) contributions. For electrical insulators such as SiO_2, the electronic term in (26.1) can be neglected so that conduction is mostly due to phonons. The propagation of phonons in a solid skeleton is determined by the network *tortuosity* as a pathway for conduction. For silica aerogels consisting of SiO_2 particles generally in the 3–7 nm size range, Bernasconi et al. [12] found that phonon heat transport can be attributed to particle modes and molecular oscillations. SAXS and SANS measurements have often been used to correlate the overall network structure of aerogels with their thermal transport properties [13–15] (For a more detailed discussion of structural effects, see also Chap. 23).

Most commercially relevant superinsulating SiO_2 aerogels have densities between 80 and 200 kg m^{-3}. Their *thermal conductivity* values are dominated by conduction through the solid silica particle network at high densities and a combination of radiation and gaseous conduction through the air inside the pores at low densities. To produce the lowest conductivity materials, it is necessary to find an optimum between those two. It seems apparent that the areogel density has established itself as a central parameter in the discussion of thermal transport properties of these materials. For a more detailed discussion on this topic, the reader is referred to Chap. 23. Note that for reasons of clarity and simplicity we are basing our discussion solely on room temperature *thermal conductivity* values. They are relevant for ambient applications, which, led by building insulation, are representative of the main share of the world's insulation markets. Because of a growing interest in superinsulation products, we shall also provide a brief description of the other two evacuated superinsulation systems which are complementary to areogels, namely, *VIP*s and *VG* to complete the picture (Chap. 41).

26.2.2. Vacuum Insulation Panels, Vacuum Glazing, and Aerogel Glazing

Vacuum Insulation Panels

The idea of achieving superinsulating properties by evacuating a vessel and thus limiting gaseous heat transport is nothing new. The first Dewar flask, an evacuated, hollow-walled glass bottle, was invented by Sir James Dewar more than 100 years ago. The space in between the glass walls is evacuated to about 10^{-4} Pa (10^{-6} Torr) in order to eliminate gaseous convection and conduction. The inner glass surfaces are coated with an IR reflecting thin metal film which reduces radiative heat transport. One realizes quickly that in the case of an evacuated vessel, a high quality vacuum is necessary to curb gaseous heat transport. However, if a porous material was to be placed inside and evacuated, heat transfer by gas molecules would already be effectively suppressed at much higher pressures (lower quality vacuum levels). However, the conduction through the bulk material needs to be minimized, which is synonymous with the requirement of a high porosity of the evacuated solid. In conventional insulation materials like glass or mineral wool or *foamed* polymers, heat transfer is dominated by the gas contribution within the pores.

A large potential for improved insulation properties can be realized by reducing or even completely eliminating the gaseous conduction contribution. In porous solids, this term is determined by the number density of gas molecules as a transfer medium as well as by the number of "walls" or solid skeleton connection pathways between hot and cold sides. At a high pressure, the mean free path of the gas molecules is much smaller than the size of the pores. This means that the collision (momentum transfer) between the gas particles is limiting the total heat transfer. Under atmospheric conditions, this is true for most conventional porous insulation materials. If one were now to reduce the gas pressure by evacuation, the gaseous conductivity remains more or less constant until the mean free path of the gas molecules attains values which are on the order of the size of the pores of the solid. At this point, communication or heat exchange through gas molecules drops significantly because of collisions with the pore walls. This means that an evacuated solid with small enough pores can become a superinsulating material.

VIPs are based on exactly that principle: a core of a pressed mesoporous powder, typically fumed silica, is wrapped in a multilayer laminate barrier foil and evacuated to submillibar pressures and sealed [16]. The barrier foil consists of a multilayer polymer foil sandwich construction with Al diffusion barrier layers in between them. Such foils are extremely sensitive to mechanical piercing or other forms of damage (*aging*, loss of integrity, delamination). The great sensitivity and *aging* effects which are due to gas permeability (as the pressure inside the panel rises, so does the *thermal conductivity*) are one of the main disadvantages of vacuum insulation. Where suitable, *VIP*s are extremely powerful insulation systems if properly implemented and mounted. Current *VIP* technology uses mostly pyrogenic silica as a core material. It is the only evacuated system available today which can fulfill the requirements for long-term applications in building insulation with a service life period of 30 years [17]. Other mineral-based materials such as glass fibers, porous minerals, or expanded clays are used as alternative inexpensive core materials and/or fillers.

Vacuum and Aerogel Glazing

Substitution of conventional double glazing units with *VG* can bring about a reduction of a factor of 2–5 of heating and cooling energy per area of glazed building surface [18]. In a conventional state-of-the-art double glazing, the two glass panes are spaced by a gap of typically a few cm which is filled with a poor heat conducting gas such as Ar or Kr. A *VG* unit consists of two sheets of glass separated by a much thinner (200–800 μm) gap and held together by an absolutely hermetic edge seal. Evacuation of the space in between eliminates heat transport from the hot to the cold pane through gaseous conduction altogether. Surprisingly, the heat transfer through the evacuated cavity is almost independent of the gap thickness. A cavity pressure on the order of $\sim 10^{-1}$ to 10^{-2} Pa or 10^{-3}–10^{-4} Torr is required for optimal insulation performance; at higher pressures the thermal performance of a VG rapidly deteriorates. The high vacuum environment must be maintained inside the glazing cavity over a prospected service life period of 30 years without the need for additional pumping or servicing. Those stringent requirements make a practical realization of *VG* on a commercial scale a *Gordian task*. From a practical point of view, the lack of suitable edge sealing technologies has inhibited the large-scale commercial realization of superinsulating *VG* systems with the desired performance. Currently, the only commercial *VG* products which are being manufactured in Asia (Japan, China, and Korea) have an insulation performance which is significantly inferior to that of Ar or Kr filled state-of-the-art commercial double glazings. To conclude, VG has great potential to revolutionize the glazing and building insulation markets, however new, advanced edge-sealing and integrated manufacture concepts are necessary. At best, the realization and implementation of distinct technological advances could reach the public markets within the next 10 years. Alternative highly insulating translucent glazing systems are aerogel-filled windows which are already available on the market (Chap. 41). There exist other commercial sources of similar products as well.

26.3. Overview of the World's Insulation Markets

Insulation products are being sold worldwide for a number of applications such as building construction and HVAC systems, industrial processes as well as niche markets, and high performance products or systems. The latter sector holds the main market share of superinsulation products. Generally, an insulating layer reduces heat transfer between two workpieces which are held at different temperatures and thus helps to reduce energy losses. Without proper insulation, these losses need to be compensated by an additional input of energy into the system. Apart from the energy savings and the related reduction in CO_2 emissions, insulation also offers additional comfort: in an insulated building or industrial piping system one encounters a more uniform temperature distribution which results in a more agreeable living experience or improved plant operation, respectively.

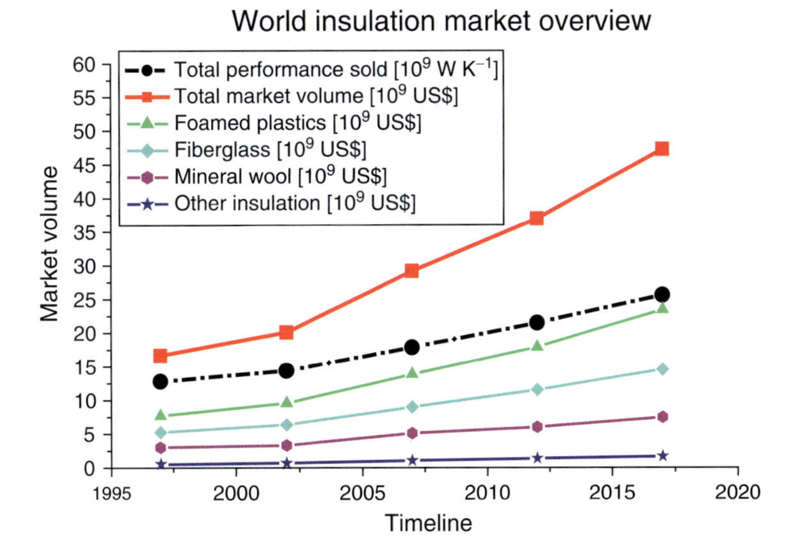

Figure 26.3. Overview over global thermal insulation markets including projected developments and respective market shares of individual insulation product sectors from [19].

The world's insulation markets [19] are still an expanding industry trade with a projected growth rate around 5% until 2012 when measured by sales. Figure 26.3 gives an overview of the global markets from 1997 until today and the projected annual volumes until the year 2017, sorted by product species: the total of sold insulation performance in units of 10^9 W K^{-1} is represented by the dotted black curve with black circles. The thick red line with red squares represents the dollar value of total insulation sold per year in billion US$. This curve again shows a more or less continuous 5% annual growth rate, with a below average performance from 1997 to 2002, a superior performance between 2002 and 2007 and finally a slight deterioration in sales after 2007 which is partly due to the subprime initiated global economic crisis. The slope of the red line is steeper than that of its black, dotted analog, the reason for this being an increasing cost per unit of insulation product performance. The finer lines represent the individual insulation product sectors which in total add up to the total volume (red squares) which has just been discussed. On a global scale, *foamed* plastics (green triangles) hold the majority of the total insulation products market with a share of approximately 48%. Until 2017 they are expected to slightly extend their leading position to approximately 50% market share. Fiber glass (light blue squares) and mineral wool products (purple hexagons) hold second and third place with roughly 31% and 17% shares, respectively. Their particular market relevance is expected to decline slightly over the next decade, covering for the gain of polymer *foam* products. Depending on location, the respective roles can be redistributed when looking at continental market data. In Western Europe, for example, fiberglass products are market leader with an almost 40% share, followed by extruded polystyrene *foam EPS* (22%) and mineral wool (16%). Eastern European markets are dominated by mineral wool products which account for almost 48% of local markets. *EPS* and fiberglass insulation are worth 29% and 13%, respectively. Such differences in local product hierarchies can be explained in part by resource availability and supply arguments.

Going back to the global insulation data, niche products, which are denoted as "other insulation" (dark blue stars) in the graph, account for less than 4% of the global market. Still, this sector has an annual turnover in excess of US$ 1.1 billion (2008). Within this category, aerogels account currently for approximately 5%.

In 2008, aerogel materials and products sales were in excess of US$ 80 million [20], with thermal insulation products making up for approximately 60% of this number or US$ 50 million. The aerogels sector being an emerging market, tremendous annual growth rates from 50 to 75% have been seen in recent years and similar developments are expected for the near future. In 2004 for example, global aerogel sales accounted for only US$ 25 million. Recent market studies project worldwide sales of aerogel products in excess of US$ 500 million for the year 2013. This development is consistent with a number of new production facilities which have recently been coming online. These developments shall be taken as a sign that global markets are ready for high-priced, aerogel-based superinsulation products. It is expected that, starting from today, the aerogel markets will continue to grow much more rapidly for at least a decade than the "conventional" insulation business up to the point when markets will begin to saturate. Saturation will certainly depend on a continuous reduction of product cost which goes hand in hand with increasing production capacities. A current benchmark value for a cubic meter of silica aerogel is on the order of US$ 4,000. With increasing commercialization, this value could drop below the US$ 1,500 mark by the year 2020. To gain more insight into the materials side, let us discuss super-insulating aerogels, their chemical composition and synthesis as well as application-geared improvement strategies.

26.4. Current Status of the Superinsulating Aerogels and Associated Components

26.4.1. Superinsulating Silica Aerogels

Basics

Even though an entire section of this handbook is dedicated to silica-based aerogels (Part II), we feel that it is necessary to briefly summarize the basics about their synthesis and structure.

For a long time, aerogels have been defined as nanostructured materials (Figure 26.4) obtained exclusively by *supercritical drying* of sol-gel derived gels [21]. If the same gels were dried via a subcritical or evaporative route, the resulting materials were often named *xerogels*. With improved *subcritical drying* and synthetic strategies developed over the last two decades, materials with very similar structural features when compared to supercritically dried "classical" aerogels can be obtained. Subcritically dried aerogels (assuming they have a low density and a high porosity to qualify as such) are preferably named ambient-dried aerogels or *ambigels* nowadays. Whatever the drying route, it is the sol-gel synthesis which permits to control the aggregation of the *precursors* and thus to tailor the nanostructuration of the final materials (Figure 26.5). Silica aerogels are obtained when silicon-based compounds are used as *precursors*. Those are, for example, sodium silicate – so-called *waterglass* [22] – or silicon *alkoxides* [23, 24]. Drying of the silica gels is marginally modifying their nanostructuration but has a strong influence on their morphology because of mechanical stresses (shrinkage, warping, cracking, etc.) [21]. After drying,

Figure 26.4. Scanning electron microscopy of silica aerogel presenting a density of 0.18 g cm^{-3} (SEM-FEG microscopy from F. Charlot, INPG-CMTC, Grenoble, France), with courtesy of Masmoudi Y. MINES ParisTech/CEP.

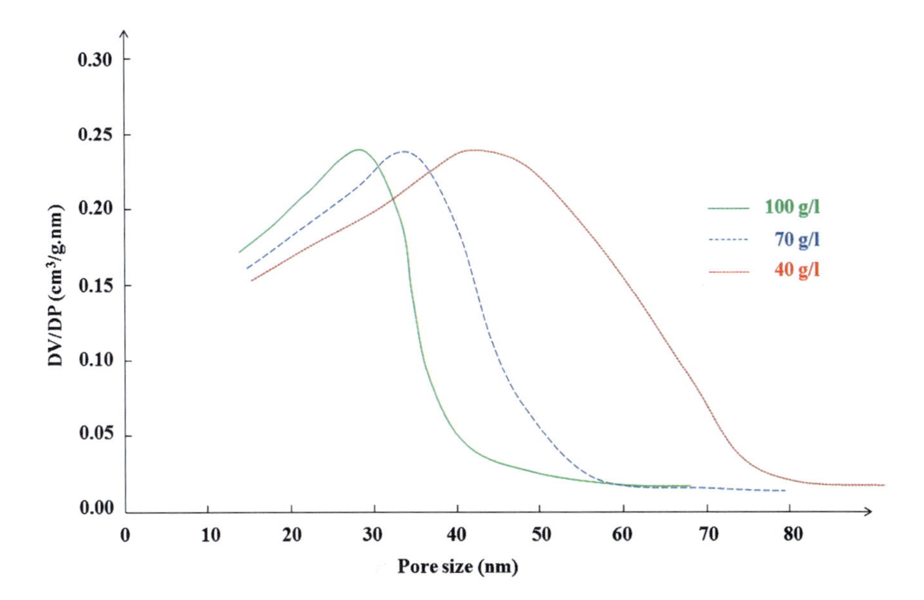

Figure 26.5. Influence of the silica concentration in the sol (Sg varied from 40 to 100 g/l) on the pore size distribution DV/DP (PSD, characterized by nonintrusive mercury porosimetry by R. Pirard at Liège University, Belgium), with courtesy of Bisson A. MINES ParisTech/CEP.

silica materials can conserve their initial monolithic form but given their friable appearance, they tend to break up into small pieces yielding granular or powdered samples. Whatever its macroscopic shape, an untreated silica aerogel is *hydrophilic* because of residual *silanol* groups (≡Si–OH). A typical surface concentration is between 3 and 6 ≡Si–OH groups nm^{-2} [25, 26].

All these properties have a strong influence on the thermal characteristics. For example, as previously described in this handbook by H.-P. Ebert (Chap. 23) and already underlined in the previous paragraphs of this chapter, the effective *thermal conductivity* is strongly dependent on the aerogel's apparent density [27]. Indeed superinsulating samples (e.g., λ <0.020 W m^{-1} K^{-1}) can only be found within a rather short range of densities (Figure 26.6), usually located between 0.10 and 0.20 g cm^{-3}.

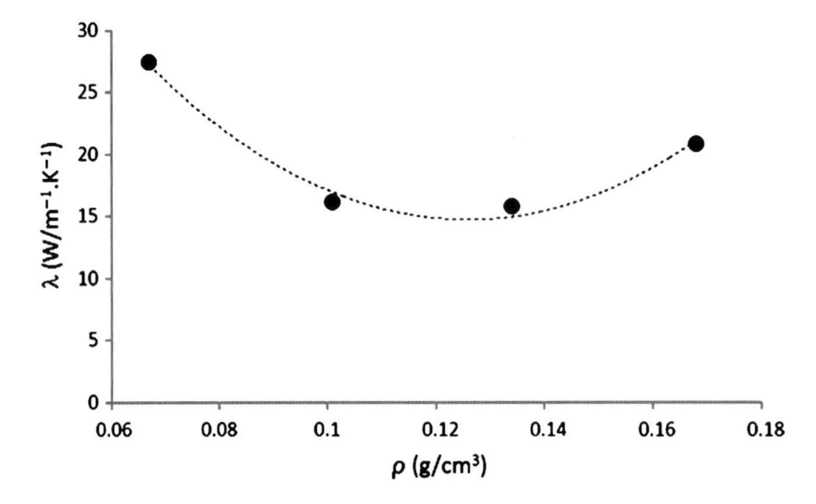

Figure 26.6. Evolution of the effective *thermal conductivity* λ of some monolithic silica aerogels (as measured by the *hot-disk method* under ambient conditions) [28].

Finally, it must be added that the flexibility of the sol-gel process also permits to disperse particulate matter into the sol to improve both thermal and mechanical characteristics of the resulting insulators [29]: for example, powders and nanoparticles (e.g., C, TiO_2, ZrO_2) for infrared opacification or fibers (e.g., ceramics, organics, glass) or inorganic clays such as kaolin or attapulgite for mechanical strengthening.

Large Monoliths

Because of its inherent property to minimize capillary stresses, *supercritical drying* is the royal way to elaborate large monolithic superinsulating silica aerogel plates (Figure 26.7). Combining very low *thermal conductivity* and good optical properties, particularly transparency (Figure 26.8), such silica aerogels are of high interest for potential applications in the transparent thermal insulation field, for example in monolithic transparent glazings [71].

Granular Materials

Even if *supercritical drying* permits to obtain perfectly monolithic and transparent silica-based thermal insulators, the process to obtain such crack-free large plates remains still too far from industrial large-scale commercialization. This was the initial reason why different processes have been studied to develop subcritical routes to access rapid massive commercialization. Among the various studies, aging of the gels in silica *precursor* containing solution has permitted to reach room temperature *thermal conductivities* as low as

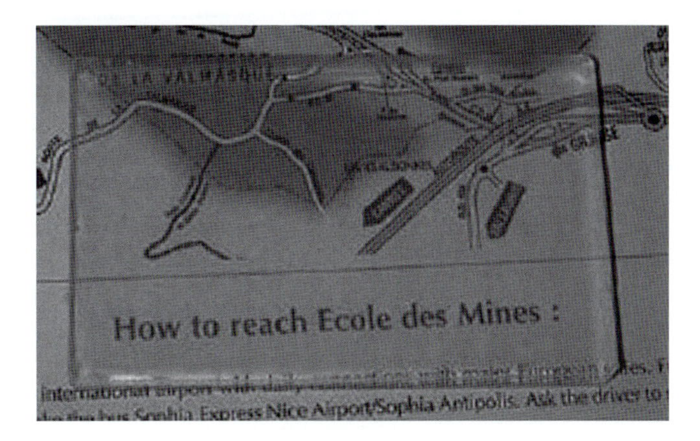

Figure 26.7. Monolithic silica aerogel monolith synthesized by a two-step catalytic process from TEOS [30], presenting a density of 0.18 g cm^{-3} and an effective *thermal conductivity* of 0.015 W m^{-1} K^{-1} (as respectively measured by *mercury pycnometry* and with the *hot-wire method* at room temperature) with courtesy of Rigacci A. and Achard P.

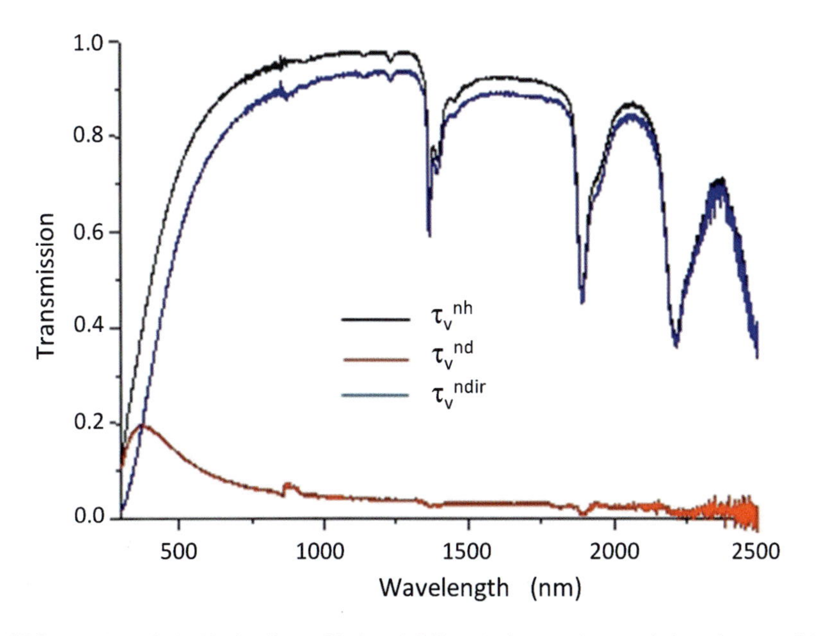

Figure 26.8. Hemispherical (*black*), direct (*blue*) and diffuse (*red*) normal transmission of a monolithic silica aerogel shown in Figure 26.7 presenting a transparency ratio $\%TR = (\tau_v^{nh} - \tau_v^{nd})/\tau_v^{nh}$ above 90% (as calculated from measurements performed on OPTORA platform by B. Chevalier at CSTB, Grenoble, France), with courtesy of Rigacci A. and Achard P.

0.020 ± 0.002 W m^{-1} K^{-1} at the laboratory scale [31]. The most efficient routes, however, which until now have not made it past the pilot scale are based on chemical modification of the pore surface after the gelation step [32] and/or employ silylated *precursors* (such as MTMS) together with TMOS and/or TEOS within a hybrid sol-gel process [33]. It is believed that these chemical methods permit to perform simple evaporative drying at atmospheric pressure without prejudicial densification. Thanks to the induced spring-back effect [34], *thermal conductivities* equivalent to these of large silica aerogel plates can be achieved with ambient dried analogs today. Furthermore, as a consequence of these direct/indirect *silylation* processes, the dried materials exhibit hydrophobic properties. Indeed, because of replacements of *hydrophilic silanols* by nonpolar ≡Si–R groups (with –R being an alkyl group) the adsorption and uptake of water decrease drastically (Figure 26.9). This has significant consequences on thermal insulation applications under real-life conditions. The main difference between supercritical superinsulating aerogels lies in the fact that so far the subcritical route only permits to obtain granular pieces of material (Figure 26.10).

As previously described in detail by Ebert (Chap. 23), packed beds of granular beads present effective *thermal conductivities* slightly larger than their monolithic parents at atmospheric pressure and within the ambient pressure to low-vacuum range because of the trapped air, even if the granules themselves offer the same *thermal conductivity* values as the monoliths (Figure 26.11). Of course, the overall packed bed conductivity can be significantly decreased by simple compression and/or packing which reduces the volume fraction of air [36].

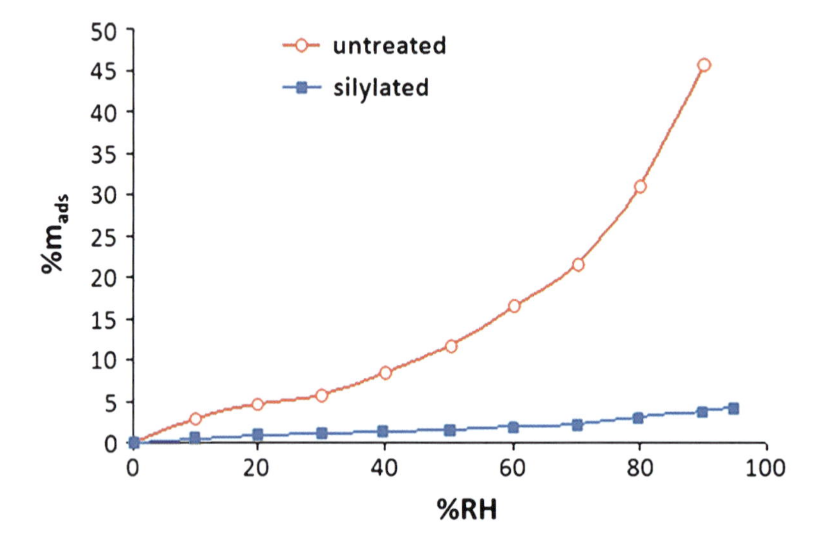

Figure 26.9. Water adsorption isotherm (expressed as mass uptake $\%M_{ads}$ with relative humidity %RH) of untreated (*red curve*) and silylated (*blue curve*) ambient-dried silica aerogels measured under standard conditions [28].

Figure 26.10. Granular silica *ambigels* in a beaker and individual granules (**top** and **bottom** *photographs* respectively) obtained by evaporative drying at moderate temperature of silica gels synthesized via a two-step acid/base catalyzed process [35], with courtesy of Rigacci A. and Achard P.

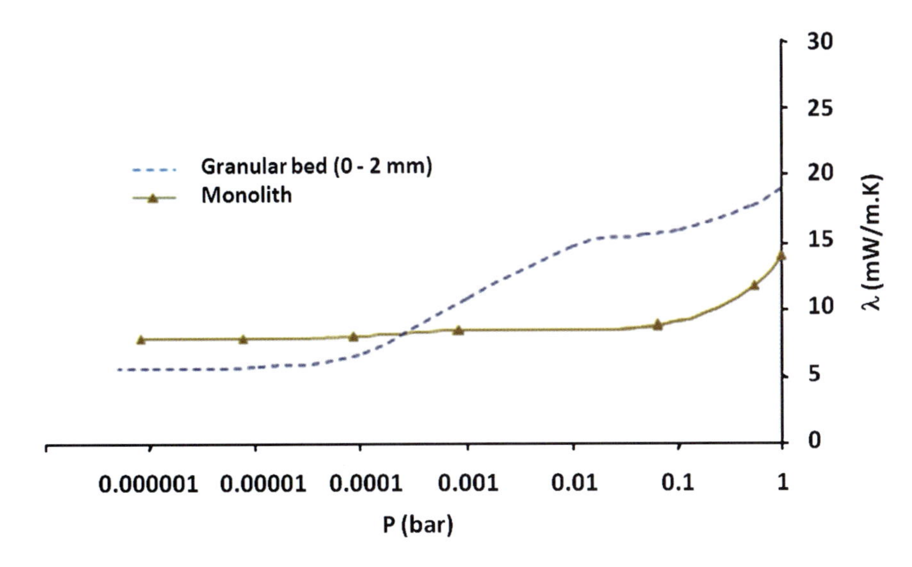

Figure 26.11. Influence of the system pressure (vacuum quality) on the effective *thermal conductivity* of granular and monolithic silica aerogel (as measured by the hot-wire technique under ambient conditions) [28].

Blankets

Within the field of superinsulating sol-gel silica materials, the most recent studies concern fibrous composites. Even if monolithic and granular aerogels can be considered as elastic materials with respect to their mechanical behavior [37], their inherent mechanical limitations prevent them from providing real flexibility (Chap. 22). This is the initial reason why brand new composites, such as blanket-type forms, have been developed more recently [38]. First, the silica sol is cast onto a low-density, unwoven fibrous batting (usually but not exclusively a polymer or mineral fiber [39]). Generally, the drying of the wet composite is then performed under supercritical conditions but some first attempts of evaporative drying have been reported as well [40]. During evaporation, the silylated gel cracks but the fibers permit to maintain the global cohesion (Figure 26.12) while offering macroscopic flexibility (Figure 26.13). Further characterizations have shown that the increase of the *thermal conductivity* under isostatic compression of 1.2 MPa can be negligible [41]. At present the main physical drawback of superinsulating aerogel blankets is still their dust release behavior.

26.4.2. Superinsulating Organic Aerogels

Organic aerogels have not been developed or adapted for insulation purposes but are typically used as carbon aerogel *precursors* [42] (Chap. 36). Because parallel studies have shown that some of them (e.g., Chap. 11) could yield very low *thermal conductivity* values [43] combined with mechanical properties superior to silica aerogels [44] and a significantly smaller tendency of dust release, a number of more recent studies have directly targeted their insulation application. Among these works, in the late 1990s, some remarkable studies were centered on polyisocyanurate aerogels [45] (Chap. 10). The first results were excellent [46] but later studies have led to *thermal conductivities* above 0.020 W m^{-1} K^{-1} [47] partly because of microphase-segregation issues (Figure 26.14).

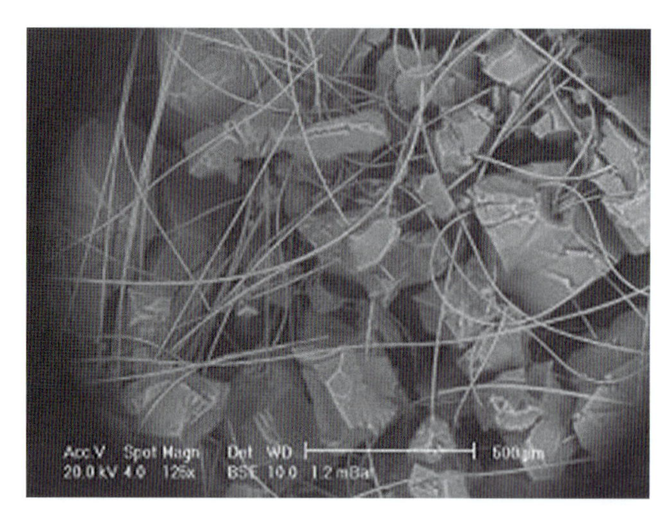

Figure 26.12. Typical SEM image of an ambient-dried silica aerogel-based blanket featuring a density of 0.1 g cm^{-3} and an effective *thermal conductivity* of 0.014 W m^{-1} K^{-1} at ambient conditions (measured in a guarded hot plate setup at CSTB, Grenoble, France), with courtesy of Rigacci A. and Achard P.

Figure 26.13. Photograph of the blanket observed in Figure 26.12, with courtesy of Bonnardel P.A, PCAS, Longjumeau, France and Sallée H. CSTB, Grenoble, France.

Since then, this system has been intensively studied. By introducing, for example, polyamines in the sol together with polyol hardeners, some polyurea aerogels offering conductivities as low as 0.013 W m^{-1} K^{-1} at ambient conditions have recently been obtained [48]. Among the few other works initiated during this period on organic aerogel-like materials for thermal insulation, a number of studies performed on the melamine–formaldehyde system are worth mentioning. Through a microemulsion templating sol-gel route [49], some *thermal conductivities* slightly below 0.025 W m^{-1} K^{-1} have been obtained [50].

Figure 26.14. Scanning electron microscopy of a polyurethane aerogel (SEM-FEG microscopy from S. Jacomet, MINES ParisTech/CEMEF, Sophia Antipolis, France), with courtesy of Rigacci A. and Achard P.

Because all these synthetic pathways use quite costly and also harmful reagents (resorcinol, formaldehyde, isocyanate, etc.) more recent endeavors in this field have been concentrating on the utilization of inexpensive and/or non- or less-toxic *precursors*. Among the most recent studies, the work by J.K. Lee and G.L Gould on polycyclopentadiene-based aerogels should be pointed out [51]. As a petrochemical industry byproduct, the monomer (dicyclopentadiene) is readily available and thus rather inexpensive. A sol-gel synthesis with this *precursor* can be conducted in alcohols according to a quite simple ring opening metathesis polymerization process which is catalyzed by transition metal complexes of Ruthenium. The corresponding aerogels can present effective *thermal conductivities* lower than 0.015 $W\ m^{-1}\ K^{-1}$ at room temperature. They consequently appear extremely promising for thermal insulation applications but one must note here that, whatever the organic sol-gel synthetic method, no subcritically dried superinsulating organic aerogels are currently available even at the laboratory scale.

In parallel, recent progress on nanostructured sol-gel cellulosic materials (Chaps. 9 and 10) reveals a certain potential of this type of green aerogels for thermal insulation product applications.

26.4.3. Composites and Hybrids

For various reasons, it seems at present that superinsulating solutions at atmospheric pressure could merge from composites and hybrids and, more particularly, from coupling inorganic and organic chemistry (Part V). Among others, pure silica systems are mechanically too weak, blankets are dusty, and purely organic derived materials are flammable. Combining the advantages from both ends, inorganic–organic composites and/or hybrids could permit to obtain superior superinsulators, particularly from mechanical and thermo-mechanical points of view.

As an illustration of studies involving organo-mineral composites, the following approaches must be mentioned: (1) preparation of *foam*-based composites through elaboration of superinsulating silica aerogels directly within the pore network of open organic *foams* (polyolefin, polyurethane, etc.) [52], (2) dispersion of finely divided silica aerogels in organic sols (which have previously proven to lead to intrinsically low conductivities [53]), and (3) polymeric binding of superinsulating silica granules (for example, with polytetrafluoroethylene (PTFE) [54] or epoxides [55]). All these types of composites can have extremely low effective *thermal conductivities* under ambient conditions.

Among the various hybridization approaches performed with the primary goal to reinforce the tenuous solid skeleton of silica aerogels by an organic chemical modification without degrading the superinsulating property, we shall begin with the *Ormosil* (short for organically modified silica) methodology. The primary works of Mackenzie [56] have been significantly improved towards superinsulation applications, for example, via the use of specific trialkoxysilyl [57] or polymethacrylate [58] derivatization agents. Conductivities as low as 0.015 $W\ m^{-1}\ K^{-1}$ have been obtained after supercritical CO_2 drying. Besides, some recent sol-gel studies performed with methyl silsesquioxane, assisted with surfactants (CTAB and poly-ethylene glycol) and urea, have permitted to obtain large monoliths with a typical bulk density of 0.10 $g\ cm^{-3}$ through ambient pressure drying [59]. This could also represent a promising way to elaborate large superinsulating plates via a subcritical *Ormosil* process. Parallel to these promising studies, some cross-linking methods have recently been adapted for use in the aerospace/aeronautic thermal insulation field (Chaps. 13 and 15). They consist in reacting the remaining *silanols* present at the surface of the pores of silica aerogels with reactive molecules such as isocyanate and/or amines so that the fragile skeleton's inner surfaces end up coated with an organic layer [60, 61]. For the moment, such a technique has demonstrated its efficiency through drastically superior mechanical properties such as stress at failure and flexural modulus; however, the density of this type of material also increases significantly. In order to improve on this aspect, some hybrid cogelation works have been performed with modified alkoxysilanes comprising either at least one isocyanate group or, in an alternative approach, at least one amine group [62]. The so-promoted urea cross-linking permits to improve significantly the initial mechanics of the silica aerogel while maintaining *thermal conductivity* as low as 0.013 $W\ m^{-}\ K^{-1}$ at room temperature.

It must be stressed here that some fascinating activities have recently been initiated on new types of mineral aerogel-like materials and their associated composites. These studies are dealing with clay-based aerogels so-called *Aeroclay*® [63] (for some of them reinforced with cellulose nanowhiskers [64]) and Xonotlite–silica aerogel composites [65]. This far, these two families have exhibited too high *thermal conductivity* values to be considered superinsulators but calculations suggest that within a certain range of composition and properties, rapid improvements could be envisaged.

26.4.4. Commercial Products

With a large number of synthetic methods, in principle, available to obtain superinsulating aerogels, let us now focus on products that are available commercially today. Those are almost exclusively based on silica. The first commercial aerogels were based on SiO_2, following a recipe similar to Kistler's original procedure and were produced after 1940 by the Monsanto Chemical Corp., Everett, Massachusetts, USA. Production stopped in the 1960s for more or less two decades but started again in the early 1980s

(Thermalux L.P., Airglass A.B.) and during the 1980s (BASF, Hoechst). Different types of aerogels were then produced for various applications (silica aerogels for the agricultural sector [66] and paints [67], carbon aerogels for energy storage and conversion devices, etc.) but a large-scale commercial adaptation by the thermal insulation trade took place only recently. At present, the two main players involved in this field are the North American industrials Cabot Corporation and Aspen Aerogels. They both offer silica-based aerogel products which are available in the form of a granular material or flexible blankets. CABOT has developed a family of silica aerogels under the trade name of *Nanogel*™ (Chap. 38) while Aspen is focussing on flexible aerogel blankets. *Nanogel*™ materials are obtained from both *subcritical and supercritical drying* of silylated gels synthesized from waterglass. ASPEN aerogels blanket products are obtained by CO_2 *supercritical drying* of gels generally made from alkoxysilanes. More recent competitors such as Nanuo (Nano Hi-Tech) in China (Chap. 40) and EM-Power (Korea) are also supplying sol-gel-based superinsulating materials; however, their production volume at present does not compare with the ones of CABOT and ASPEN. In addition, there are a number of specialty aerogel companies with extremely small sale volumes. One example is AIRGLASS AB (Sweden) which can supply large monolithic flat plates of silica aerogel. These monoliths have been designed and developed initially for very specific scientific applications such as elementary particle physics counters at CERN, Geneva, Switzerland; however, in a later stage the properties were optimized for applications in supersinsulating glazings, so-called aerogel glazings. In May 2010 a new player, the German Rockwool, has publically declared their activity in the field of aerogel-based insulation. In that context, the launch of a new mineral wool fiber [72] aerogel composite product family was announced for 2011.

Additional developments are going on worldwide to bring forward improved insulating components and systems based on current silica aerogel technology, both at the materials and component levels. As far as application is concerned, *Nanogel*™ granules can be directly used as superinsulating filling materials (such as, for example, in daylighting panels) but CABOT is also marketing *Nanogel*™-based components (such as *Thermal Wrap*™ and *Compression Pack*™) for specific applications like pipe in pipe and cryogenic insulation systems. For its part, ASPEN is commercializing blanket-based components for building insulation (so-called Spaceloft) as well as materials optimized for use in extremely hot and cold environments (Pyrogel and Cryogel products). More detailed information can be found in Part XV.

26.5. Applications for Aerogel-Based Products

The role of superinsulation as niche products in the insulation business and their particular importance for space-saving applications has been presented previously. A lower thermal conductivity means that the same insulation performance is achieved with a thinner insulation layer or that the same thickness of superinsulation is used to ensure far better insulation performance. This advantage is contrasted by a significantly higher cost per installed performance as shown below. This assessment is based on data from 2009.

In order for a high priced superinsulation product to become competitive with conventional ones, the value of the "saved space" and other beneficial factors must compensate significantly for the added cost of the material. In this context, the key question to be addressed is how can one put a dollar value on space savings? Depending on the type of application, a reduction in space, weight, or volume affects overall assembly and operation cost (Figure 26.15). The aerogel insulation business must find small niche markets to set foot in, which are best evaluated according to their potential for capitalizing on their benefits. Such niche markets can later serve as a basis for expanding activities into other fields of business. Table 26.2 gives an overview of today's typical aerogel insulation applications and their economic relevance.

For most aerogel insulation products, the reduction of the operating cost is higher than the savings accrued during installation. However, in the case of off-shore oil and gas pipeline

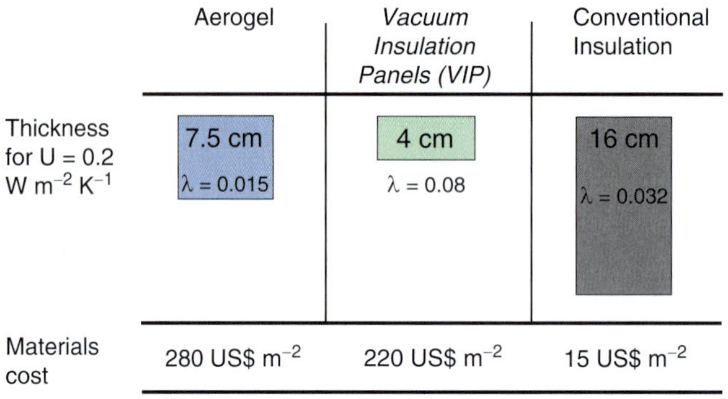

	Aerogel	*Vacuum Insulation Panels (VIP)*	Conventional Insulation
Thickness for U = 0.2 W m^{-2} K^{-1}	7.5 cm λ = 0.015	4 cm λ = 0.08	16 cm λ = 0.032
Materials cost	280 US$ m^{-2}	220 US$ m^{-2}	15 US$ m^{-2}

Figure 26.15. Comparison of space savings and cost of conventional insulation, aerogel, and vacuum insulation systems. Units of the *thermal conductivity* λ are in W m^{-1} K^{-1}.

Table 26.2. Typical aerogel insulation product branches, their space savings benefit, and economic relevance

Application	Installation and assembly cost	Operating cost	Economic potential
Off-shore oil and gas	Smaller pipe diameter, lower weight, more pipes per installation round trip, fewer trips	Superior lifetime, improved degradation resistance	++++
Aeronautics/ aerospace	Simplification of overall design, light construction, size reduction lowers materials/assembly cost	Smaller gross weight results in fuel savings or additional capacity	++(+)
Building insulation	Comparable to conventional insulation, currently more elaborate due to lack of experience	Reduction of heating/cooling energy and/or larger useable building/exterior volume	+++
High temperature insulation	Smaller overall pipe diameter or exterior dimensions, easier installation	Reduced surface area per unit length, lower radiative losses, improved resistance and lifetime	+++
Cryogenic applications	Smaller overall pipe diameter or exterior dimensions, easier installation	Reduced sensitivity to cryo-embrittlement, increased lifetime, energy and/or space savings	++
Appliances/ apparel	Significantly more complex than standard technology	Energy savings/increased thermal comfort for lightweight extreme performance personal wear/gear	++

insulation, the installation part is the key factor in favor of aerogel-based products: With significantly smaller pipe diameters and reduced weight, assembly ships can carry significantly more pipeline with each trip, thus significantly cutting down on the number of trips and the overall installation cost. This is a model example of how an extremely specialized application can open up opportunities for high-performance high-cost products. In the following, we shall briefly discuss the various types of aerogel insulation products which are available commercially today. Of course, it is again worth mentioning that all current commercial products are silica based.

26.5.1. Off-Shore Oil and Gas

This is one of the oldest fields of activity of the Aspen Aerogels Spaceloft thermal insulation blanket. Cabot Corporation is offering their *Nanogel*-filled Compression Pack as a promising alternative solution to the blanket product. The superior thermal performance and improved chemical/pressure resistance of aerogels combined with assembly cost savings make them ideal candidates for this type of application. To this day, this is the undisputed model example for finding a niche market for aerogel insulation products and exploring it successfully. More detailed information can be found in Chap. 38.

26.5.2. Aeronautics and Aerospace Applications

One of the traditional fields of aerogels during their revival in the 1960s following Kistler's discovery in 1925, the materials were extensively studied by the U.S. space administration NASA. Large monolithic aerogel blocks are being used to collect stellar dust and particles in outer space (see also Chap. 32), on the Mars Rover roving vehicle silica aerogels were used for thermal insulation. Given their space savings potential and relatively low weight, aerogel insulation systems could also find applications in civil and military aviation, however only some activity is known in this field so far. The main factors which prevent aerogels from being used widely in civil aviation technology are dust evolution (certification) and cost reasons.

26.5.3. High Temperature

There are a number of industrial processes which require pumping hot fluids from one place to another. Many older plants were designed and built with little or no insulation, especially at times where energy was all abundant and cheap. Nowadays, significant savings can be generated by insulating the piping systems of steam cycles, chemical processes, or oil and gas processing refineries. This opens up a large potential for high performance insulation systems, particularly for older or already existing plants with restricted access and tightly spaced arrangements of individual pipes. Aspen's Pyrogel is a product which was developed especially for high-temperature applications.

26.5.4. Cryogenic Applications

In the field of cryotechnology, which includes, for example, transport and storage of liquefied gases or frozen biomedical specimens, aerogel insulation offers numerous advantages. Besides its superinsulation properties, aerogels tend to embrittle less with decreasing temperature than, for example, polymer *foam* insulation. Aspen Aerogels have developed a special low-temperature aerogel blanket insulation product, the Cryogel Z. Granular products such as Cabot's *Nanogel* offer great versatility for lining nontrivial vessel and piping geometries which are part of more complex cryogenic systems. Cryogenic aerogel insulation is also investigated by the NASA's rocket engine research teams for liquid hydrogen and liquid oxygen propelled drive systems [68].

26.5.5. Apparel and Appliances (Refrigeration Systems, Outdoor Clothing, and Shoes)

Superinsulation properties can find use in a number of everyday products. Particularly, personal apparel such as performance outdoors equipment, for example, clothing, shoes, gloves, footwarmers, tents, sleeping bags, etc., could be outfitted with aerogel thermal insulating layers. The same holds for household appliances such as refrigerators or outdoors cooling boxes. A number of such products have recently reached consumer markets. With increasing market volume and increasing customer awareness, such products have a significant potential to compete in the high-priced products sector. However, it is unlikely that aerogel-based consumer products will conquer the middle- and low-cost sectors in the near future (Chap. 37).

26.5.6. A Closer Look at Aerogels for Building Insulation: Startup and Testing Phase

In the beginning of this chapter, the need for superinsulation was based on the necessity for a global CO_2 reduction. It was shown that insulating the existing building stock can bring about substantial energy savings and that it is amongst the most sensible short-term measures to curb CO_2 emissions immediately. Newly constructed buildings and most renovations will use conventional insulation products. Aerogels have only very hesitantly found their way into the building insulation sector with just a few demonstration objects realized today.

Together with a number of other European countries, Germany, Austria, and Switzerland have taken a leading role in energy and environment related issues. Supported by their government's energy policy, innovation in the field of energy efficient building and construction is considered indispensable. This allows for new, high-priced, high-performance products and solutions make their way into selected demonstration units which are often funded in part by the government. Two examples of aerogel insulated demonstration projects are shown in Figure 26.16. On the right is the gymnastic hall of the high-school "Buchwiesen" in Zürich which is outfitted with a transparent aerogel insulated roof construction. On the left is a prestige "green" single family home building in Meilen, ZH. Demonstrator units like those receive some media coverage where they are portrayed as energy efficient building solutions of the future. Obviously, aerogel superinsulation is not ready for a broad application in the building market yet, mostly because of its price. Nevertheless, there is a great potential for aerogels in the building industry in the not too distant future. With falling aerogel prices, the saved space will be able to largely compensate for the extra cost.

Figure 26.16. Examples of two aerogel insulation demonstration projects in Switzerland: Transparent insulation (**right**, high-school gymnastic hall in Zürich) and façade blanket insulation (**left**, single family home in Meilen, ZH).

The most typical space saving applications in the building industry are as follows:

- Renovation of historical buildings (exterior insulation)
- Interior and translucent insulation
- Flat-roof balcony constructions (aerogels as a less damage sensitive alternative to VIP)
- Elegant, esthetic architecture with slim exterior insulation (analogous to the object in Meilen shown in Figure 26.16)

Given the tremendous volume of insulation materials needed for building applications, this is clearly one of the fields with a huge potential for the worlds growing aerogel insulation markets. We look forward to seeing a rising number of aerogel insulated homes over the next few decades to come.

26.6. Toxicity, Health, and Environmental Considerations

Nanomaterials, in general, are currently being investigated worldwide for their potential health and environmental risks [69]. As nanostructured materials, aerogels are nanomaterials and hence need close observation. Silica aerogels, the most prominent aerogel insulation materials, are made from colloidal silica nanoparticle building blocks. Now the question arises whether there is a significant health risk for end-users and the immediate environment where the product is applied. From a risk assessment point of view, the potential danger of a new nanomaterial is given by the product of intrinsic toxicity of the nanomaterial (nanotoxicity) and risk (probability) of exposure.

The toxicity of nanoparticulate silica has been widely studied in recent years [70]. Generally, amorphous silica nanoparticles are considered relatively safe; however, the surface functionalization must be taken into account when assessing the specific (for example, cell-) toxicity. Crystalline SiO_2 nanoparticles are accompanied by a significantly higher risk for cell *mutagenesis*; however, sol-gel processes produce amorphous materials unless they are sintered for longer periods of time at elevated temperature which can induce crystallization at the nanodomain level. From an exposure point of view, the risk of exposure to significant quantities of nanoparticulate silica when working with aerogels such as by inhalation is also quite low. Given the three-dimensional network structure of aerogels, the resulting dust contains only very few individual particles in the nanometer size regime. By far more likely is the formation of chunks or fragments of the entire network in the micrometer size regime, consisting of many particles still covalently bound to each other. The dust produced in this way is in a way more similar to typical household dust.

Even though nanosafety does not seem an immediate concern for aerogel insulation, prototypes of each next generation product must be tested thoroughly. Manufacturers and sales staff alike must keep in mind the tremendous impact on the entire aerogels industry which can come from a single negative example. Because nanotoxicity is such a new field with a lot of unknowns, the power of bad news and their potential impact as a market inhibitor is tremendous. Therefore, proper action must be taken so that the nanotoxicity problematic be addressed adequately in all instances. Of course, when more developed, organic or composite/hybrid aerogel insulators must be studied in the same way. Service life aspects and renewability aspects of aerogels, in general, are quite appealing: both Aspen aerogels and Cabot *Nanogel*™ product families have been awarded the MBDC Silver cradle to cradle certificate.

26.7. Conclusions

Aerogel-based insulation products have established a place in various niche markets, particularly in the insulation products trade, over the past 10 years. Aerogel superinsulation affords superior insulation performance (lower *thermal conductivity* values) compared to conventional materials which translates into slimmer installed constructs and/or improved insulation performance. Motivated by global CO_2 emission and space saving restrictions, aerogel superinsulation products will continue their advance into new market segments. The development of improved aerogel insulation systems and rapidly advancing process technology will allow manufacturers to promote next generation custom-tailored products. Inorganic (particularly SiO_2 based) materials are likely to play a key role, also for years or decades to come. It is expected that in addition to the pure silica systems, organic aerogels and a variety of composites and/or hybrids (such as organic–inorganic or fiber-reinforced) materials and products will become available. Overall, the aerogels insulation products have a tremendous growth potential in excess of 50% per annum. For comparison, conventional insulation products grow at an annual rate on the order of 5%. In 2008, aerogel insulation products contributed less than 0.3% to the total insulation products market volume.

Today, the materials cost for areogel insulation is roughly twenty times higher (per installed performance) than that of standard insulation products. The main justification for the use of aerogel insulation systems therefore is space saving, reducing operating cost, longevity, and chemical resistance. Off-shore oil, high-temperature and building insulation as well as aeronautics/aerospace applications are the industry trades with the most significant growth potential for aerogel products in terms of estimated sales volume. Of all these, the building has certainly by far the largest growth potential but is also more sensitive to the materials cost than, for example, the aeronautics/aerospace field.

A current reference market price for a cubic meter of silica aerogel is on the order of US$ 4,000 (2008). With increasing commercialization, this value could drop below the US$ 1,500 mark by the year 2020. Lower prices will cause an equilibration or stabilization of the aerogels insulation market. The advance of organic and hybrid systems could invigorate this development process additionally. In the meantime, we shall look forward to seeing new aerogel insulation products and manufacturers appear on the markets.

References

1. Barsky RB, Kilian L (2004) Oil and the Macroeconomy Since the 1970s, J Econ Persp 18(4):115–134
2. Woodwell GM (1978) The carbon dioxide question, Scientific American 238:34–43
3. Cox PM, Betts RA, Jones CD, Spall SA, Totterdell I (2000) Acceleration of global warming due to carbon cycle feedbacks in a coupled climate model, Nature 408:184–187
4. Houghton JT, Jenkins GJ, Ephraums JJ (1990) Climate change – The IPCC scientific assessment, Cambridge University Press, Cambridge England and New York
5. Caldeira K, Jain AK, Hoffert MI (2003) Climate sensitivity uncertainty and the need for energy without CO_2 emission, Science 299(5615):2052–2054
6. Weber L, (1997) Some reflections on barriers to the efficient use of energy, Energy Policy 25(10):833–835
7. Janda KB, Busch JF (1994) Worldwide status of energy standards for buildings, Energy 19(1):27–44
8. Papadopoulos AM, (2005) State of the art in thermal insulation materials and aims for future developments, Energy and Buildings 37(1):77–86
9. Lee OJ, Lee KH, Yim TJ, Kim SY, Yoo KP (2002) Determination of mesopore size of aerogels from thermal conductivity measurements, J Non-Cryst Solids 298:287–292
10. Viskanta R, Gosh RJ (1962) Heat transfer by simultaneous conduction and radiation in an absorbing medium, J Heat Trans 2:63–71
11. Scheuerpflug P, Caps R, Büttner D, Fricke J (1985) Apparent thermal conductivity of evacuated SiO_2 aerogel tiles under variations of radiative boundary conditions, Int J Heat mass Transfer 28:2299–2306
12. Bernasconi A, Sleator T, Posselt D, Kjems JK, Ott HR (1992) Dynamic properties of silica aerogels as deduced from specific-heat and thermal-conductivity measurements, Phys Rev B 45:10363–10376
13. Vacher R, Woignier T, Pelous J (1988) Structure and self-similarity of silica aerogels, Phys Rev B 37:6500–6503
14. Craievich A, Aegerter MA, dos Santos DI, Woignier T, Zarzycki J (1986) A SAXS study of silica aerogels, J Non-Cryst Solids, 86:394–406
15. Hasmy A, Foret M, Anglaret E, Pelous J, Vacher R, Jullien R (1995) Small-angle neutron scattering of aerogels: simulations and experiments, J Non-Cryst Solids 186:118–130
16. Simmler H, Brunner S (2005) Aging and service life of VIP in buildings, Energy and Buildings 37(11): 1122–1131

17. Caps R, Heinemann U, Ehrmanntraut M, Fricke J (2001) Evacuated insulation panels filled with pyrogenic silica powders – Properties and Applications, High Temperatures – High Pressures 33(2):151–156
18. Manz H (2008) On minimizing heat transport in architectural glazing, Renewable Energy 33(1):119–128
19. Freedonia market study #2434 (2009) Freedonia group, Cleveland, OH, USA
20. BCC market study #AVM052B (2009) BCC research Inc, Wellesley, MA, USA
21. Brinker CJ, Scherer GW (1990) Sol-Gel Science: The Physics and Chemistry of Sol-Gel Processing, Academic Press, New York, USA
22. Kistler SS (1932) Coherent expanded aerogels, J Phys Chem 36:52–64
23. Teichner SJ, Nicolaon GA, Vicarini MA, Gardes GEE (1976) Inorganic oxide aerogels, Advances in Colloid and Interfaces Science 5:245–273
24. Mehrotra MC (1992) Precursors for aerogels, J Non-Cryst Solids 145:1–10
25. Iler RK (1979) The Chemistry of Silica, John Wiley & Sons, New York USA
26. Calas S (1997) Surface et porosité dans les aérogels de silice : étude structurale et texturale. PhD thesis Université de Montpellier (France)
27. Hrubesh LW, Pekala RW (1994) Thermal properties of organic and inorganic aerogels, Journal of Materials Research 9:731–738
28. Bisson A, Rigacci A, Lecomte D, Achard P (2004) Effective thermal conductivity of divided silica xerogels beds, J Non-Cryst Solids 350:379–384
29. Deng Z, Wang J, Wu A, Shen J, Zhou B (1998) High strength SiO2 aerogel insulation, J Non-Cryst Solids 225:101–104
30. Pajonk GM, Elaloui E, Achard P, Chevalier B, Chevalier JL, Durant M (1995) Physical properties of silica gels and aerogels prepared with new polymeric precursors, J Non-Cryst Solids 186:1–8
31. Haereid S (1993) Preparation and characterization of transparent monolithic silica xerogels with low density, PhD thesis NTNU (Norway)
32. Schwertfeger F, Frank D, Schmidt M (1998) Hydrophobic waterglass based aerogels without solvent exchange or supercritical drying, J Non-Cryst Solids 225:24–29
33. Rao VA, Bhagat SD, Hirashima H, Pajonk GM (2006) Synthesis of flexible silica aerogels using methyltrimethoxysilane (MTMS) precursor, J Colloid Interface Sci 300:279–285
34. Smith DM, Deshpande R, Brinker CJ (1992) Preparation of low-density aerogels at ambient pressure, Mat Res Soc Symp Proc Vol. 271 567–572
35. Bisson A, Rigacci A, Achard P, De Candido M, Florent P, Pouleyrn G, Bonnardel P (2006) Procédé d'élaboration de xérogels de silice hydrophobes, FR2873677
36. Smith DM, Maskara A, Boes U (1998) Aerogel-based thermal insulation, J Non-Cryst Solids 225:254–259
37. Woignier T, Phalippou J (1989) Scaling law variation of the mechanical properties of silica aerogels, Rev Phys Appl C4:179–184
38. Ryu J (2000) Flexible aerogel superinsulation and its manufacture, US Pat. # 6068882
39. Trifu R, Bhobho N (2007) Flexible coherent insulating structures, US2007173157
40. Chandradass J, Kang S, Bae D-S (2008) Synthesis of silica aerogel blanket by ambient drying method using waterglass based precursor and glass wool modified alumina sol, J Non-Cryst Solids 354:4115–4119
41. Bardy ER, Mollendorf JC, Pendergast DR (2007) Thermal conductivity and compressive strain of aerogel insulation blankets under applied hydrostatic pressure, Journal of Heat Transfer 129:232–235
42. Pekala RW, Kong FM (1992) Resorcinol-formaldehyde aerogels and their carbonised derivatives, Polym Prepr 30:221–223
43. Lu X, Caps R, Fricke J, Alviso CT, Pekala RW (1995) Correlation between structure and thermal conductivity of organic aerogels, J Non-Cryst Solids 188:226–234
44. Pekala RW, Alviso CT, LeMay JD (1990) Organic aerogels: microstructural dependence of mechanical properties in compression, J Non-Cryst Solids 125:67–75
45. Biesmans GL (1999) Polyisocyanate based aerogel, US Pat. # 5990184
46. Biesmans G, Randall D, Francais E, Perrut M (1998) Polyurethane-based organic aerogels' thermal performance, J Non-Cryst Solids 225:36–40
47. Rigacci A, Maréchal JC, Repoux M, Moreno M, Achard P (2004) Elaboration of aerogels and xerogels of polyurethane for thermal insulation, J Non-Cryst Solids 350:372–378
48. Lee JK, Gould GK, Rhine W (2009) Polyurea based aerogel for high performance thermal insulation material, J Sol-Gel Sci Technol 49:209–220
49. Egger CC, du Fresne C, Schmidt D, Yang J, Schädler V (2008) Design of highly porous melamine-based networks through a bicontinuous microemulsion templating strategy, J Sol-Gel Sci Technol 48:86–94

50. du Fresne C, Schmidt DF, Egger C, Schädler V (2007) Supramolecular templating of organic xerogels, XVth International Sol-Gel Conferencc, Montpellier, France (september 2–7):129
51. Lee JK, Gould GL (2007) Polycyclopentadiene based aerogel: a new insulation material, J Sol-Gel Sci Technol 44:29–40
52. Tang Y, Polli A, Bilgrien CJ, Young DR, Rhine WE, Gould GL (2007) Aerogel-foam composites, WO Pat. # 2007146945
53. Lee JK (2007) Organic aerogels reinforced with inorganic fillers, US Pat. # 2007259979
54. Ristic-Lehmann C, Farnworh B, Dutta A, Reis BE (2008) Aerogel/PTFE composite insulating material
55. Mensahi J, Bauer U, Pothmann E, Peterson AA, Wilkins AK, Anton M, Doshi D, Dalzell W (2007) Aerogel based composites WO Pat. # 2007047970
56. Mackenzie JD, Chung YJ, Hu Y (1992) Rubbery ormosils and their applications, J Non-Cryst Solids 147&148:271–279
57. Ou DL, Gould GL (2005) Ormosil aerogels containing silicon bonded linear polymers, WO Pat. # 2005068361
58. Ou DL, Gould GL, Stepanian CJ (2006) Ormosil aerogels containing silicon bonded polymethacrylate, WO Pat. # 2005098553
59. Kanamori K, Aizawa M, Nakanishi K, Hanada T (2008) Elastic organic-inorganic hybrid aerogels and xerogels, J Sol-Gel Sci Technol 48:172–181
60. Capadona LA, Meador MA, Alunni A, Fabrizio EF, Vassilaras P, Leventis N (2006) Flexible, low-density polymer cross-linked silica aerogels, Polymer 47:5754–5761
61. Leventis N, Mulik S, Wang X, Dass A, Patil VU, Sotiriou-Leventis C, Lu H, Churu G, Capecelatro A (2008) Polymer nano-encapsulation of template mesoporous silica monoliths with improved mechanical properties, J Non-Cryst Solids 354:632–644
62. Rhine WE, Ou, DL, Sonn JH (2007) Hybrid organic-inorganic materials and methods of preparing the same, WO Pat. # 2007126410
63. Hostler SR, Abramson AR, Gawryla MD, Bandi SA, Schiraldi DA (2008) Thermal conductivity of a clay-based aerogel, International Journal of Heat and Mass Transfer 52:665–669
64. Gawryla MD, van den Berg O, Weder C, Schiraldi DA (2009) J Mater Chem 19:2118–2124
65. Wei G, Zhang X, Yu F (2009) Effective thermal conductivity of Xonotlite-aerogel composite insulation material, Journal of Thermal Science 18:142–149
66. Frisch G, Zimmermann A, Schwertfeger F (1997) Use of aerogels in agriculture, MX Pat. # 9706411
67. Vukasovich MS (1970) Fluorescent pigment, GB Pat. # 1191483
68. Wang, X-Y.; Harpster, G.; Hunter, J.; Nasa TM-report #214675, 2007
69. Savolainen, K.; Pylkkaenen, L.; Norppa, H.; Falck, G.; Lindberg, H.; Tuomi, T.; Vippola, M.; Alenius, H.; Brouwer, D.; Mark, D.; Bard, D.; Berges, M.; Jankowska, E.; Posniak, M.; Farmer, P.; Singh, R.; Krombach, F.; Toxicology Letters, 2008, 180, S21
70. Nel A, Xia T, Mädler L, Li N (2006) Toxic potential of materials at the nanolevel, Science 311:622–627
71. Jensen KI, Schultz JM, Kristiansen FH (2004) Development of windows based on highly insulating aerogel glazings, J Non-Cryst Solids 350:351–357
72. Schultz JM, Jensen KI, Kristiansen FH (2005) Superinsulating aerogel glazing, Solar Energy Materials and Solar cells 89:275–285

Part IX

Applications: Chemistry and Physics

27

Aerogels as Platforms for Chemical Sensors

Mary K. Carroll and Ann M. Anderson

Abstract Sensing of chemical species in air, in water and in other solvents is important for a wide variety of applications, including but not limited to monitoring chemical species that might have environmental, health, forensic, manufacturing, or security implications. The unusual properties of aerogels – very high surface area, high porosity, low density – render them particularly appealing for sensing applications. In this chapter, we survey the published reports of the application of aerogels to chemical sensing. These include sensors based on silica, silica composite, titania, carbon and clay aerogels, with spectroscopic and conductimetric detection methods.

27.1. Introduction: Why Use Aerogels for Sensor Applications?

A *sensor* is a device that responds to an external stimulus with a measurable response. In this chapter, we restrict our discussion to *sensors* that respond to chemical species, including oxygen and water vapor (humidity). There are many applications of chemical *sensors*, including monitoring gas-phase species for air quality and safety applications, and measuring pH or other ion concentrations in water samples. *Sensors* also find use in healthcare, forensic, security, and consumer product applications.

The ideal chemical *sensor* would respond instantaneously to the relevant *analyte*, with high specificity and sensitivity, be usable over a wide range of *analyte* concentrations and be repeatable. In addition, the ideal *sensor* device would be inexpensive to produce, simple to use, rugged, and both reusable and recyclable.

There is considerable literature on the use of sol-gel materials in chemical sensing applications. Sol-gel materials are attractive for sensing applications due to their porosity and their relatively low density; in addition, some sol-gel materials (silica in particular) have optical properties that render them useful in optical *sensors*. A recent literature search yielded more than 2,700 references for "sol-gel *sensor*." The vast majority of these

M. K. Carroll • Department of Chemistry, Union College, Schenectady, NY 12308, USA
e-mail: carrollm@union.edu
A. M. Anderson • Department of Mechanical Engineering, Union College, Schenectady, NY 12308, USA
e-mail: andersoa@union.edu

M.A. Aegerter et al. (eds.), *Aerogels Handbook*, Advances in Sol-Gel Derived Materials and Technologies, DOI 10.1007/978-1-4419-7589-8_27,

references are to *sensors* based on xerogels, gels dried under ambient conditions that result in pore collapse. Relatively few of these papers focus on aerogel platform *sensors*.

There are many excellent review articles related to sol-gel *sensors*, a few of which are described briefly here. In 1994, Dave et al. [1] described the use of sol-gel encapsulation for biosensor design, whereby *probes* incorporated into the precursor mixture become entrapped within the sol-gel matrix but remain accessible to smaller *analyte* species. The use of organically modified precursors to tailor the properties of sol-gel *sensors* was described by Lev et al. [2] the following year. Dunn and Zink [3] surveyed the literature on the use of molecular probes to report on the environments experienced by probes within the pores of sol-gel materials, and the interactions between the probes and the matrix itself. Keeling-Tucker and Brennan [4] reviewed the use of fluorescent probes in elucidating both local structure and dynamics in sol-gels. More recently, Lukowiak and Strek [5] reviewed the use of sol-gel materials for sensing applications and Walcarius and Collinson [6] presented a review of analytical chemistry uses of silica sol-gels, including sensing.

Aerogels are sol-gel materials from which the solvent byproducts of the polycondensation reactions have been removed from the pores without effecting pore collapse. Aerogel materials have a combination of physical properties that render them particularly attractive for chemical *sensor* applications: 98–99% porosity, unusually high surface area (as high as 1,000 m^2/g) and, in some cases, good optical translucency (85% transmittance). There is a wealth of literature on the preparation and modification of sol-gel materials, including aerogels. In 2002, Pierre and Pajonk [7] published a concise survey of the chemistry of aerogels and their applications (see also Chap. 1).

Aerogels result when wet sol-gels are dried under conditions that minimize (ideally, eliminate) shrinkage of the sol-gel matrix. The drying methods include (a) CO_2 supercritical extraction (*CSCE*), in which one exchanges the solvent in the pores for liquid CO_2, then takes the CO_2 above its supercritical point in a critical point dryer; (b) alcohol supercritical extraction (*ASCE*) in a critical point dryer; (c) rapid supercritical extraction (*RSCE*) methods which occur in confined molds within a pressure vessel or hydraulic hot press; (d) ambient drying following surface modification; and (e) freeze drying. These methods have been described in detail in a previous chapter of this Aerogel Handbook (Chap. 2).

Because aerogel materials have interesting properties and the techniques for preparing them are well established, these materials are finding application in a variety of different *sensor* applications. In this chapter, we provide a detailed review of the use of aerogels in sensing. We note that we have restricted the scope of this review to reports published in the scientific literature, including proceedings of conferences; patents and the considerable number of conference abstracts on this topic are not included. To date, there are very few papers that describe complete *sensor* devices, so we have included work that demonstrates the suitability of aerogel materials for sensing applications. In the following sections, we describe optical *sensors* based on silica aerogel platforms, featuring different approaches to the incorporation of probe species into the aerogels, conductimetric *sensors* based on silica and carbon aerogel platforms, and the promise of titania and clay aerogels for sensing applications.

27.2. Optical Sensors Based on Silica Aerogel Platforms

In much of the work to date on aerogel *sensors*, silica aerogel platforms are employed. There are two likely reasons for this. First, much of the prior literature on xerogel-based *sensors* is based on silica sol-gels, and silica sol-gel chemistry is well understood. Second, monolithic silica sol-gels can be fabricated with excellent transparency in the UV/visible region of the electromagnetic spectrum, which renders them appropriate for use with optical sources.

Unless otherwise noted, the aerogels were interrogated using standard spectroscopic instrumentation. The *sensor* responses are recorded using a number of different types of spectroscopic measurements, including absorption spectroscopy, *luminescence* emission and lifetime measurements, and *Raman scattering* techniques.

27.2.1. Photoluminescent Modification of the Aerogel Itself

One of the first reported uses of silica aerogels in optical sensing was by Ayers and Hunt [8]. Aerogels were prepared via a "standard" method: they employed a base-catalyzed reaction with Silbond-5 to create wet gels, then used CSCE to produce aerogels. *Photoluminescent* silica aerogels were generated by exposing the "standard" aerogels to energized ammonia, a reducing gas. Oxygen *quenches* the *luminescence* of the aerogel. As the collision frequency between gas-phase oxygen molecules and the luminescent carriers generated in the aerogel increases, the emission intensity decreases due to energy transfer to the oxygen molecules.

If an emission signal is due to a single type of luminescent species experiencing one microenvironment, the emission signal can be related to the concentration of quencher ($[Q]$) by the *Stern–Volmer equation* [9]:

$$I_0/I = 1 + K_{\mathrm{SV}}[Q]$$

where I_0 is the *luminescence* intensity in the absence of quencher; I is the *luminescence* intensity at a given concentration of quencher; and K_{SV} is the *Stern–Volmer quenching constant*. If the luminescent species is present in two different environments, each of which has a different accessibility to the quencher, a somewhat more complicated equation is required [10]. Figure 27.1 shows Stern–Volmer plots for simulated data.

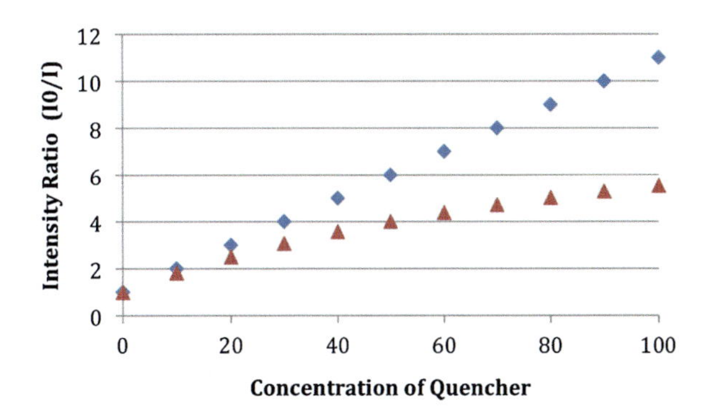

Figure 27.1. Stern–Volmer plots of simulated data. Data shown as *blue diamonds* are for a luminescent species in one microenvironment with $K_{SV} = 0.10$. *Red triangles* are for the same luminescent species in two microenvironments with 90% of the species having $K_{SV} = 0.10$ and the other 10% being nearly inaccessible to the quencher, with $K_{SV} = 0.0001$.

Ayers and Hunt constructed a *sensor* device using a mercury penlamp as the UV light source, a monolithic aerogel disk as the sensing platform, a cut-off filter to select for *luminescence*, and a photodiode detector. The response of the *sensor* to 1–760 Torr O_2

was measured, Stern–Volmer (*S–V*) plots [9] were prepared, and K_{SV} values were calculated to be 0.0155 $Torr^{-1}$ using a linear S–V model [9]. (These data might have been better fit to a two-site model [10], as there appear to be systematic deviations from linearity.)

Ayers and Hunt noted that silica aerogels collapse and dissolve when brought into contact with liquids. They prepared hydrophobic *photoluminescent* aerogels through post-treatment reaction of the *photoluminescent* aerogels with hexamethyldisilazane at 70°C. Post-treatment was required because the hydrophobic surface groups of hydrophobic "standard" aerogels would char if exposed to the microwaves used in the photosensitizing process. The value of K_{SV} for hydrophobic *photoluminescent* aerogels, 2.4×10^{-3} $Torr^{-1}$ indicated that these materials were less sensitive than those that did not undergo hydrophobic modification, but the authors postulated that this disadvantage would be offset by the possibility of using these *sensors* in contact with water.

27.2.2. Covalent Attachment of Probe Species

Another way to produce an aerogel-based *sensor* is to incorporate the sensing moiety directly into the backbone of the aerogel through covalent attachment, as shown schematically in Figure 27.2. If probe moieties are covalently attached to the silica matrix of the aerogel they cannot *leach* from the aerogel. This is a major advantage.

Figure 27.2. Schematic representation of covalent attachment of a sensing species (*dark blue spheres*) to the silica sol (*light blue spheres*). Not to scale.

Leventis et al. [11] synthesized a tetraethoxysilane (*TEOS*) derivative in which the fluorescent moiety 2,7-diazapyrene is tethered to the silicon atom by a short alkyl chain: *N*-(3-trimethoxysilylpropyl)-2,7-diazapyrenium bromide (*DAP*). They took two approaches to incorporating the *DAP* into *TMOS*-based silica gels. "*Bulk-modified*" *DAP*-doped silica aerogels were prepared by mixing a solution containing *DAP* with one containing *TMOS* to prepare the sol, followed by gelation and *CSCE* processing. This resulted in a uniform distribution of sensing moieties throughout the aerogel. "*Postdoped*" aerogels were prepared by soaking aged, ethanol-rinsed tetramethylorthosilicate (*TMOS*)-based sol-gels in a solution containing *DAP* and either acid or base. These "*postdoped*" materials showed luminescence only in the exterior portions, which is promising for applications of thin aerogel films. Addition of *DAP* had no significant impact on the bulk properties of the silica aerogels (surface area, microstructure, bulk density).

The *DAP*-doped aerogel monoliths are sensitive to oxygen, which quenches the fluorescence of the diazapyrene group. Response times within 15 s were achieved for the aerogel monoliths, but the response was actually shorter than the measured value because the researchers were limited by the speed at which the gas streams could be altered in the experiment. This was considerably faster than the response times observed for xerogels prepared from the same precursor recipes and allowed to dry under ambient conditions. It appears that oxygen permeated the aerogel matrix at speeds approaching the speed of open-air diffusion, resulting in a nearly ideal *sensor* response time. Fluorescence lifetime analysis indicates that the *DAP* moieties are in a single environment. *Bulk-modified* and *postdoped* materials showed the same lifetime, indicating that the probe is at the surface of the colloid, not interspersed within it.

This approach is effective, but a significant amount of synthetic expertise and effort is required. For each luminescent probe to be incorporated into an aerogel, one must first tether the probe to a silica precursor chemical, and do so in a way that does not compromise the emission characteristics of the probe.

27.2.3. Electrostatic Attachment of Probe Species

Another approach is to exploit the negative charges on the silica aerogel surface to attract and hold probe species via electrostatic attraction. Cations are attracted to the surface via electrostatic force, as is shown schematically in Figure 27.3.

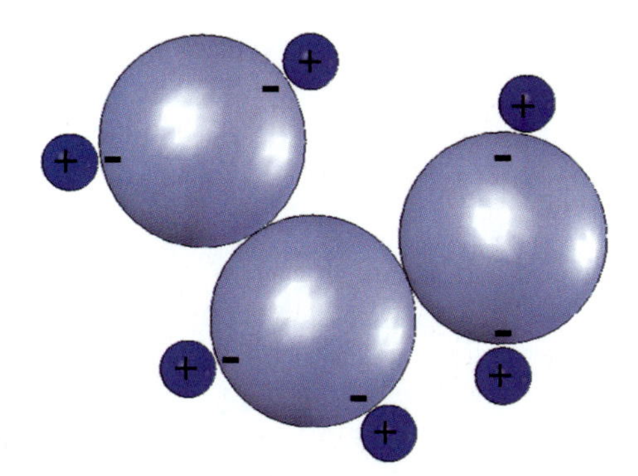

Figure 27.3. Schematic representation of electrostatic attachment of a cationic sensing species (*dark blue spheres*) to the anionic silica sol (*light blue spheres*). Not to scale.

In 2004, Leventis et al. [12] presented a study in which they demonstrated that *probe-doped* silica aerogels could be prepared by *post-doping* cationic ruthenium probes and cationic probe-electron donor *dyads* into silica sol-gel monoliths. Ruthenium complexes have well-described properties and have been studied extensively by others (as noted in [4, 13]).

The probe ruthenium(II) *tris*-(1,10-phenanthroline) and the electron donors 4-benzoyl-*N*-methylpyridinium and *N*-benzyl-*N*′-methyl viologen were employed by Leventis et al. [12]. A base-catalyzed *TMOS* precursor recipe was used to prepare silica sol-gel monoliths. The wet gels were then soaked in methanol baths containing one of the *dyads*, or the ruthenium complex, each of which is a cationic species that was taken up by the sol-gels to such an extent that no color was observed in the remaining bath solution. The doped gels were then processed using a standard *CSCE* approach. No leaching was observed in the solvent exchange baths, because of the strong electrostatic attraction of the dopants for the silica matrix, and the resultant aerogels were visibly colored.

The materials were then fully characterized as oxygen *sensors*. Aerogels doped with 4-benzoyl-*N*-methyl-pyridinium-based *dyads* had a higher dynamic range than either the $[Ru(2,2'\text{-bipyridine})_3]^{2+}$-doped aerogels or the other *dyad*-doped aerogels, and were suitable for both room-temperature and low-temperature (77 K) oxygen sensing applications. Response times were similar to the previous work with covalent attachment [11]. The Stern–Volmer plot data were fitted to a two-site model, indicating that the probes are found in different microenvironments within the aerogels.

The *postdoping* method is very simple and might be readily applicable to other cationic probe species. However, the *dyad* synthesis (described in detail in [12]) is complex.

27.2.4. Entrapment of Probe Species

Physically entrapping probe species within the matrix is another way to incorporate probes into aerogels. Because this approach does not require synthetic modification of either the probe or the precursor used, it is potentially a more universal method of incorporating probes. In this case, the luminescent probe is simply dissolved in one of the reagents used to prepare the sol-gel. A *probe-doped* gel forms. This is an extremely common means of preparing doped xerogels (see [1–6]). When dried under ambient conditions, the solvents within the pores evaporate, but the nonvolatile probes remain entrapped within the xerogel, as shown schematically in Figure 27.4. The solvent extraction methods employed when preparing aerogels can make entrapment a more challenging approach for doped aerogel preparation.

Figure 27.4. Schematic representation of entrapment of a sensing species (*dark blue spheres*) in the silica sol (*light blue spheres*). Not to scale.

Of major concern is the potential for leaching of the probe from the sol-gel matrix during the processing steps required to prepare the aerogel from the wet sol-gel. As was noted by Pierre and Pajonk [7] and Leventis et al. [11], among others, the multiple solvent-exchange steps, including several exchanges with liquid CO_2 that are employed prior to conventional supercritical CO_2 extraction (*CSCE*) methods provide the probe species with ample opportunity to diffuse from the matrix. Thus, the concentration of entrapped probe will be much lower than the concentration in the initial wet sol-gel. Moreover, the probes located in regions of the gel that will be most accessible to *analyte* are the most likely to *leach* from the matrix, and this reduces the sensitivity of the *sensor*.

Nevertheless, Morris et al. [14] demonstrated that molecular species, such as acid–base indicators, can be trapped noncovalently within silica composite aerogels prepared using *CSCE* extraction methods. (This work is described in more detail in Sect. 27.2.6.)

One way to overcome the issue of leaching is to entrap species that are so physically large that it is difficult or impossible for them to *leach* from the matrix. Wallace et al. [15] were able to entrap within silica aerogels protein superstructures of *cytochrome c* (*cyt c*) around gold nanoparticles. They demonstrated the applicability of these materials for gas-phase NO sensing. The size of the *cyt c* superstructures is such that the resulting material can be considered a composite; consequently, this work is described in more detail in Sect. 27.2.6.

Another way to avoid leaching of probe during aerogel formation is to eliminate the solvent-exchange steps from the extraction process. In our own work, Plata et al. [16] demonstrated that it is possible to entrap oxygen-sensitive luminescent probes into silica aerogels prepared via a one-step precursor-to-aerogel rapid supercritical extraction (*RSCE*) method. Because the process contains no solvent-exchange steps, the probe has no opportunity to exchange out of the matrix. We used a standard, base-catalyzed *TMOS* recipe and successfully entrapped three oxygen-sensitive probes: *tris*(2,2′-bipyridyl) ruthenium(II) $[Ru(bpy)_3]^{+2}$, *tris*-4,7-diphenyl-1,10-phenanthroline ruthenium (II) $[Ru(dpp)_3]^{+2}$, and platinum (II) octaethylporphine [*PtOEP*] within *RSCE* aerogels and in xerogels prepared from the same precursor mixtures. These materials had high sensitivity to oxygen. Response times of the doped aerogel monoliths to oxygen were in the 10–20 s range, with the times limited by our ability to switch the gas composition (similar to Leventis et al. [11]). The response times of doped xerogel monoliths were somewhat longer, ca. 50 s, but still sufficiently rapid for gas *sensor* applications. It is important to note that the *RSCE* process occurs at high temperature as well as high pressure because it is necessary to exceed the supercritical point of the alcohol byproduct of the sol-gel formation reactions. This means that only thermally stable probes can be entrapped via this method.

A second disadvantage to the use of physical entrapment methods is the relative lack of control of where the probes are located within the resulting aerogel. Because the probe is present during sol formation and/or gelation, probe moieties can become entrapped within different microenvironments of the matrix. The spectral properties and the response of a molecular probe can depend on the environment experienced by the probe. There is considerable evidence that this occurs. In the Plata et al. [16] work, for example, Stern–Volmer plots are clearly nonlinear at higher %oxygen, and the data fit well to a two-site model [10]. The majority of the probe species are entrapped in one type of site, in which they are readily accessible to oxygen gas. The other probe species show little or no response to oxygen. *Luminescence* lifetime measurements confirm multiple lifetimes for the probes, indicating multiple environments.

Recently, we have compared *probe-doped RSCE* aerogels, *CSCE* aerogels and xerogels, all prepared from the same precursor mixtures [17]. The oxygen-sensitive probes *PtOEP* and platinum (II) mesotetrafluoro phenylporphyrin (*PtTFPP*) were included in the *TMOS* precursor mixture, and then aliquots of the mixture were processed in three different ways.

The resulting xerogels had the highest concentration of entrapped probe, due to the matrix shrinkage and to the fact that no leaching or thermal decomposition occurred. Based on the color of the *RSCE* aerogels, it appeared that some fraction of the probe decomposed during processing; however, the spectral properties of the remaining probes agreed well with those of the xerogels. For the *CSCE* aerogel fabrication process, we noted that the methanol and acetone used for the solvent exchanges became pink in color, indicating that the probe readily leached from the sol-gels. These porphyrin complexes are cationic, but it was clear that electrostatic forces between the complexes and the silica matrix were not sufficiently strong to eliminate leaching as in [12]. The resulting *CSCE* aerogels did not visibly contain the probe; however, *luminescence* spectra showed small quantities of probe had remained in the matrix during the processing steps. A photograph of these sol-gels is shown in Figure 27.5.

Figure 27.5. Photograph of *PtOEP*-doped sol-gel materials: xerogel, *RSCE* aerogel, *CSCE* aerogel (**left to right**). Note that the xerogel and *CSCE* aerogel have the shape of the test tube in which gelation occurred, whereas the cuvette holding the *RSCE* aerogel sample contains pieces taken from a larger cylindrical monolith. The discoloration of the *CSCE* aerogel is consistent with results obtained by Leventis for acetone-washed gels [11].

In 100% nitrogen, the emission signal from the *PtOEP* or *PtTFPP* remaining in the *CSCE* aerogels was significantly smaller than that of the probe in either the *RSCE* aerogels or xerogels, with only about a twofold enhancement of signal relative to air (ca. 21% oxygen). The relative response of the *CSCE* aerogels was low enough that full Stern–Volmer plots could not be prepared, but the lack of response is consistent with the majority of the entrapped probes being relatively inaccessible to the quencher. In contrast, the ratios for *probe-doped* xerogel and *RSCE* aerogel monoliths were in the 20–65 range (intensity in 100%N_2:intensity in air). Moreover, when the solvent exchanges are performed with methanol and acetone solutions that contain the probe, the resulting *CSCE* aerogels yield comparable response to the *RSCE* aerogels but lower response than for the xerogels. In all cases for which Stern–Volmer plots could be prepared, the data fit to a two-site model, with the majority of probe moieties found to be accessible to oxygen.

27.2.5. Silica Aerogels as Sample Holders for Raman Scattering Measurements

Yet another approach is simply to use the silica aerogel as a host into which gas-phase *analyte* molecules permeate and are detected. Konorov and coworkers [18, 19] have demonstrated the ability to sense both gas-phase and condensed-phase species within silica aerogels, using coherent anti-Stokes Raman scattering (*CARS*). Here, the silica aerogel matrix serves as a sample holder for the *analyte* of interest. The optical transparency, porosity and low scattering cross-section of silica aerogels renders them well-suited for this application. The optical properties of the silica gel permit the laser-based interrogation without significant background interference. The *analyte* is simply allowed to equilibrate into the aerogel matrix, and then the optical measurements are performed.

In initial work [18], aerogels were prepared using a *TMOS*-based recipe and *CSCE* extraction. They used reagent-grade toluene in solution to demonstrate that *CARS* measurements on liquids within aerogel pores could be obtained, and nitrogen in ambient air to demonstrate the ability to use *CARS* for gas-phase species within the aerogel matrix.

Subsequently, time-resolved *CARS* was employed [19]. A commercial silica aerogel was used, with ethanol as the model liquid and nitrogen as the model gas. The authors note that, in principle, *CARS* could be used to discriminate between adsorbed *analyte* molecules and free *analyte* molecules within the aerogel matrix; however, for the small *analytes* studied the data fitted to a single relaxation time.

It is noteworthy that in both cases, the relative concentration of the species being detected was high. The applicability of this approach for trace analysis has not yet been demonstrated.

27.2.6. Silica Composite Materials

Composite materials are prepared by combining two different materials. In general, composites are employed in order to exploit the advantages of each type of material, and minimize the disadvantages. For example, as is noted by many authors, silica aerogels are fragile (Chap. 22). Including another component in the material produced might result in a stronger material that has the desirable optical properties, high surface area, and low density of silica aerogels.

The Rolison group has been active in the area of nanoscale engineering of silica composites for sensing applications [14, 15, 20]. In Morris et al. [14], they described a "silica-modified silica composite" prepared by *postgelation modification* of a base-catalyzed silica sol-gel with another silica sol, this one prepared using acid catalysis. These base-catalyzed, acid-modified (*BCAM*) gels are then processed using the *CSCE* method to produce aerogels. The resulting *BCAM* aerogel monoliths have the desirable bulk properties of base-catalyzed silica aerogels, including good transparency, in concert with surface properties more typical of acid-catalyzed aerogel materials. Consequently, it is possible to noncovalently entrap molecular species, including acid–base indicators, at the interface for use in sensing applications.

In Anderson et al. [20], they reported composite aerogels of colloidal metal (either gold or platinum) and silica that had the optical transparency of silica aerogels, combined with the surface and optical properties of the metal colloid. The metal colloids are dispersed throughout the mixture and, therefore, isolated from each other. At the same time, the porosity of the silica matrix renders the metal colloids accessible to species that traverse the matrix. The metal colloid surface can be modified either before or after gelation to tailor the optical properties of the metal.

The Rolison group subsequently applied this method to the preparation of aerogel monoliths doped with the protein *cyt c* [15]. In buffered solution, the protein forms a superstructure containing thousands of individual protein molecules around a single colloidal gold particle. The modified gold particles are then reacted with a base-catalyzed *TMOS* sol to produce a composite material, as in [20]. Although *cyt c* moieties on the outside of the superstructure were damaged during the solvent exchange and extraction process, most of the interior proteins survived the extraction process intact, in a buffer-like environment around the gold particle. These aerogel monoliths retain some of the reactivity of *cyt c*, as was demonstrated by their response to gas-phase NO, which was monitored as a change in absorbance value over time.

We note that the relatively low temperature of the *CSCE* extraction process was important to the retention of protein function in this application; an *RSCE* process would not be expected to yield comparable results because the protein would not likely survive the higher temperatures required.

In a different approach, Boday and Loy [21] have reported preliminary results that demonstrate that inclusion of polyaniline nanofibers into a *TMOS*-based *CSCE* silica aerogel results in higher-strength materials with potential application in sensing gas-phase acids and bases. They found that including about 6 wt% of polyaniline increased the aerogel strength threefold while still yielding a low-density (0.088 g/mL) material. They applied gold electrodes to the surface of a composite aerogel and noted a dramatic decrease in resistance when the aerogel was exposed to HCl vapor.

27.3. Conductimetric Sensors Based on Aerogel Platforms

The optical properties of silica aerogels are not the only ones that can be exploited for *sensor* applications. In this section, we review the literature on aerogel *sensors* based on measurements of conductance.

27.3.1. Silica Aerogel Platforms as Conductimetric Sensors

Silica aerogels are electrical insulators. These can be used to advantage in optical sensing applications in environments that contain significant sources of electrical noise. If enough water adsorbs onto a silica aerogel surface, the movement of charge carriers in the water can result in increased conductivity, and this can be the basis of a different type of silica aerogel *sensor*.

Wang et al. [22] and Wang and Wu [23] prepared mesoporous silica aerogel thin films for use as humidity *sensors*. They either spin-coated [22] or dip-coated [23] sol-gel-derived silica colloids onto gold electrode-patterned alumina substrates, then used *CSCE* extraction to obtain aerogel films. The response of the aerogel films to applied electric fields was evaluated over a range of plausible ambient temperatures and relative humidities. When water adsorbed onto the *sensor* surface, the conductivity of the material increased.

The average pore diameter and surface area of the spin-coated aerogel films were comparable to those achieved for silica aerogel monoliths [22]. Under conditions of relative humidity of 40% or higher, the electrical resistance decreased as humidity increased. *Sensor* performance depended on the uniformity and the porosity of the film.

The dip-coated materials [23] were suited for operation over a wider range of relative humidity (20–90%) and ambient temperature (15–35°C). Unsurprisingly, as the film thickness decreased, surface conductivity increased and the response time increased. However, thicker films recovered more rapidly. The response could be tailored by controlling the film thickness and pore structure of the materials.

27.3.2. Carbon-Based Aerogel Composites as Conductimetric Sensors

Carbon aerogels cannot readily be used for optical *sensors*, because they absorb light strongly across the visible region (they appear black) (Chap. 36). But many organic species adsorb onto carbon surfaces, so measurements based on changes in an electrical response are possible if the material is conductive.

Zhang et al. [24] fabricated conductive polymer composites by filling polystyrene polymer with carbon aerogel. They then tested the resistance of these composites when exposed to a variety of gas-phase organic vapors. The observed response depended on both the carbon aerogel content of the composite and the adsorption of the various *analytes*. In comparison to carbon black/polystyrene composites, the aerogel materials showed better response, which the authors attribute to the higher surface area of the aerogels.

Bryning et al. [25] have reported the preparation of carbon nanotube aerogels, reinforced with polyvinyl alcohol. These materials conduct and have structural integrity, and show promise for sensing applications.

27.4. Other Aerogel Platforms that Show Promise for Sensing Applications

In this section, we review the literature reports that demonstrate the suitability of titania and clay aerogel matrices for sensing applications.

27.4.1. Titania Aerogels as Sensor Platforms

Titania aerogels are generally considerably more opaque than are silica aerogels; however, titania has other attractive properties including high photocatalytic ability. Many of the *sensor* approaches described above for silica aerogels are also applicable to titania aerogels (Chap. 7). To date, work in this area has been preliminary, more proof-of-concept rather than development of *sensor* devices.

Morris et al. [14] demonstrated that the base-catalyzed, acid-modified (*BCAM*) approach they used for silica aerogels, and which showed considerable promise for entrapment of *sensor* moieties, could also be applied to titania aerogel preparation.

Howell and Fox [26] have shown that it is possible to entrap molecular probes within titania aerogels. They used the fluorescence of pyrenyl groups, both tethered and unbound, to interrogate the porous TiO_2 matrix. These probes were selected because pyrene and its derivatives are sensitive to the polarity of their environment and can form excited-state dimers (excimers) due to π stacking; moreover, the fluorescence properties of pyrenyl monomers and excimers are well known [4]. Howell and Fox showed that there are local interactions between the dye species within the titania aerogel. They also used organic modification (methylation and phenylation) of the titania matrix to tailor the materials, and found that π stacking of tethered pyrenyl groups was inhibited by the presence of methyl or phenyl groups. Thus, the placement of probe species is controlled somewhat by the matrix composition. This can be exploited in *sensor* design.

Baia et al. [27] prepared porous composite materials containing colloidal gold particles and TiO_2 aerogel for use as *sensors* in water quality. When Raman-active species adsorb onto rough metal surfaces, an enhancement in scattering is observed; this is known as surface-enhanced Raman scattering (or surface-enhanced Raman spectroscopy, *SERS*). Baia et al. demonstrated the utility of their titania/gold composites as *sensors* by exposing the materials to aqueous test solutions containing thioacetamide, Crystal Violet or Rhodamine 6G, obtaining detection limits for the dyes as low as 10^{-7} M by this method.

27.4.2. Clay Aerogels for Sensing Applications

Schiraldi and coworkers prepared hydrophilic aerogels from silicate clays and then used these to make organic polymer–aerogel composites [28, 29]. They converted montmorillonite and hectorite gel suspensions into aerogels using a freeze–drying process. They have incorporated organic polymers into the clay aerogels in two different ways.

The first of these was polymerization of *N*-isopropylacrylamide within the clay aerogel [28]. This *in situ* method resulted in stable, low-density composites with interpenetrating organic–inorganic phases. The presence of the organic polymer rendered the aerogel structure stable in water, and the clay aerogel gave the organic polymer additional structural integrity. In this work, reversible morphological changes with different amounts of hydration were observed, indicating that the composite might be suitable for humidity *sensor* development.

In the subsequent work [29], polyvinyl alcohol was introduced to the clay suspension prior to freeze–drying to produce a polymer/aerogel composite. A similar approach was used to incorporate nickel particles into a clay aerogel, yielding a metal/aerogel composite with magnetic properties.

Clay aerogel composites show considerable promise as platforms for chemical *sensors* due to the ease with which the polymer/clay and metal/clay aerogel composites were formed, combined with the extensive literature on modification of clays to yield functional materials.

27.5. Summary and Future Directions

Although there has been considerable research published on the preparation and applications of aerogels to date there have been relatively few published papers in which the authors focus on the use of aerogels in sensing applications. From the papers reviewed in this chapter, it is apparent that researchers are taking a variety of approaches to the use of aerogels as platforms for chemical *sensors*.

Long and Rolison [30, 31] have suggested that scientists should consider applying an architectural design approach to the design of aerogels and aerogel-like nanostructures for specific applications, including sensing. They emphasize that there is a wealth of prior work demonstrating that aerogel materials can be tailored for desired properties. These include but are not limited to modification of the precursor chemistry, entrapping particles within a gel, adding chemical modifiers to the gel, making alterations to the drying process, coating the resulting solid matrix, as well as combinations of these.

One example of this approach is the recent work by Arachchige and Brock [32], who applied sol-gel chemistry and surface modification to create architectures based on CdSe quantum dots, resulting in highly luminous monoliths. Sensing is an obvious application for these materials.

Clearly, the development of aerogel-platform chemical *sensors* is a research area whose potential has not yet been fully tapped. Researchers have demonstrated the utility of aerogels for rapid optical and conductimetric sensing of gas-phase *analytes*. Several types of aerogels have been employed in this work and a handful of *analytes* have been investigated. We anticipate seeing increased activity in this research area, with novel aerogel-based materials tailored to sensing a wider range of *analyte* species in both the gas and solution phase. There is extensive literature on xerogel-based *sensors*, aerogel structure, aerogel composite materials, and chemical probes for *sensor* applications that can be brought to bear on the development of new aerogel *sensor* materials. Within the next decade, we expect to see a dramatic increase in the number of published reports of aerogel-platform chemical *sensors*.

Acknowledgments

Our own work with aerogels as platforms for chemical sensors has been funded by grants from the National Science Foundation (NSF MRI CTS-0216153, NSF RUI CHE-0514527, NSF MRI CMMI-0722842, and NSF RUI CHE-0847901) and the American Chemical Society's Petroleum Research Fund (ACS PRF 39796-B10). Any opinions, findings, and conclusions or recommendations expressed in this material are those of the authors and do not necessarily reflect the views of the National Science Foundation.

References

1. Dave B C, Dunn B, Selverstone Valentine J, Zink J I (1994) Sol-gel encapsulation methods for Biosensors. Anal Chem 66: 1120A–1127A.
2. Lev O, Tsionsky M, Rabinovich L, Glezer V, Sampath S, Pankratov I, Gun J (1995) Organically modified sol-gel sensors. Anal Chem 67: 22A–30A.
3. Dunn B, Zink J I (1997) Probes of pore environment and molecule-matrix interactions in sol-gel materials. Chem Mater 9: 2280–2291.
4. Keeling-Tucker T, Brennan J D (2001) Fluorescent probes as reporters on the local structure and dynamics in sol-gel-derived nanocomposite materials. Chem Mater 13: 3331–3350.
5. Lukowiak A, Strek W (2009) Sensing abilities of materials prepared by sol-gel technology. J Sol-Gel Sci Tech 50: 201–215.
6. Walcarius A, Collinson M M (2009) Analytical chemistry with silica sol-gels: traditional routes to new materials for chemical analysis. Ann Rev Anal Chem 2: 121–143.
7. Pierre A C, Pajonk, G M (2002) Chemistry of aerogels and their applications. Chem Rev 102: 4243–4265.
8. Ayers M R, Hunt A J (1998) Molecular oxygen sensors based on photoluminescent silica aerogels. J Non-Cryst Solids 225: 343–347.
9. Lakowicz J R (1999) Principles of Fluorescence Spectroscopy, 2nd Ed, Kluwer Academic/Plenum Publishers.
10. Demas J N, DeGraff B A, Xu W (1995) Modeling of Luminescence Quenching-Based Sensors: Comparison of Multisite and Nonlinear Gas Solubility Models. Anal Chem 67: 1377–1380.

11. Leventis N, Elder I A, Rolison D R, Anderson M L, Merzbacher C I (1999) Durable modification of silica aerogel monoliths with fluorescent 2,7-diazapyrenium moieties. Sensing oxygen near the speed of open-air diffusion. Chem Mater 11: 2837–2845.
12. Leventis N, Rawashdeh A-M M, Elder I A, Yang J, Dass A, Sotiriou-Leventis C (2004) Synthesis and characterization of Ru(II) tris(1,10-phenanthroline)-electron acceptor dyads incorporating the 4-benzoyl-*N*-methylpyridinium cation or *N*-benzyl-*N*′-methyl viologen. Improving the dynamic range, sensitivity, and response time of sol-gel-based optical oxygen sensors. Chem Mater 16: 1493–1506.
13. Innocenzi P, Kozuka H, Yoko T (1997) Fluorescence properties of the $[Ru(bpy)_3]^{2+}$ complex incorporated in sol-gel-derived silica coating films. J Phys Chem B 101:2285–2291.
14. Morris C A, Rolison D R, Swider-Lyons K E, Osburn-Atkinson E J, Merzbacher C I (2001) Modifying nanoscale silica with itself: a method to control surface properties of silica aerogels independently of bulk structure. J Non-Cryst Solids 285: 29–36.
15. Wallace J M, Rice J K, Pietron J J, Stroud R M, Long J W, Rolison D R (2003) Silica nanoarchitectures incorporating self-organized protein superstructures with gas-phase bioactivity. Nano Lett 3: 1463–1467.
16. Plata D L, Briones Y J, Wolfe R L, Carroll M K, Bakrania S D, Mandel S G, Anderson A M (2004) Aerogel-platform optical sensors for oxygen gas. J Non-Cryst Solids 350: 326–335.
17. Carroll M K, Barrow A J, Ferrarone J R, Phillips A F, Baig S, Anderson A M (forthcoming) Silica sol gels as platforms for chemical sensors: Spectroscopic comparison of materials prepared via two supercritical extraction methods and ambient drying. *In preparation.*
18. Konorov S O, Mitrokhin V P, Smirnova I V, Fedotov A B, Sidorov-Biryukov D A, Zheltikov A M (2004) Gas- and condensed-phase sensing by coherent anti-Stokes Raman scattering in a mesoporous silica aerogel host. Chem Phys Lett 394: 1–4.
19. Konorov S O, Turner R F B, Blades M W (2007) Background-free coherent anti-Stokes Raman scattering of gas- and liquid-phase samples in a mesoporous silica aerogel host. Appl Spectrosc 61: 486–489.
20. Anderson M L, Rolison D R, Merzbacher C I (1999) Composite aerogels for sensing applications. Proc SPIE 3790: 38–42.
21. Boday D J, Loy D A (2009) Poly aniline nanofiber/silica aerogel composites with improved strength and sensor applications. Polymer Preprints 50: 282.
22. Wang C-T, Wu C-L, Chen I-C, Huang Y-H (2005) Humidity sensors based on silica nanoparticle aerogel thin films. Sensor Actuator B107(1): 402–410.
23. Wang C-T, Wu C-L (2006) Electrical sensing properties of silica aerogel thin films to humidity. Thin Solid Films 496: 658–664.
24. Zhang B, Dong X, Song W, Wu D, Fu R, Zhao B, Zhang M (2008) Electrical response and adsorption performance of novel composites from polystyrene filled with carbon aerogel in organic vapors. Sensor Actuator B132(1): 60–66.
25. Bryning M B, Milkie D E, Islam M F, Hough L A, Kikkawa J M, Yodh A G (2007) Carbon nanotube aerogels. Adv Mater 19: 661–664.
26. Howell A R, Fox M A (2003) Steady-state fluorescence of dye-sensitized TiO_2 xerogels and aerogels as a probe for local chromophore aggregation. J Phys Chem A 107: 3300–3304.
27. Baia M, Danciu V, Cosoveanu V, Baia L (2008) Porous nanoarchitectures based on TiO_2 aerogels and Au particles as potential SERS sensor for monitoring of water quality. Vib Spectrosc 48: 206–209.
28. Bandi S, Bell M, Schiraldi D A (2005) Temperature-responsive clay aerogel-polymer composites. Macromolecules 38: 9216–9220.
29. Schiraldi D A, Gawryla M D, Johnson J R III, Griebel J (2007) Functional materials based on clay aerogels. Polymer Preprints 48: 988–989.
30. Long J W, Rolison D R (2007) Architectural design, interior decoration, and three-dimensional plumbing en route to multifunctional nanoarchitectures. Acc Chem Res 40: 854–862.
31. Rolison D R, Long J W (2008) Architectural design en route to scaleable 3D multifunctional nanomaterials. Polymer Preprints 49: 502–503.
32. Arachchige I U, Brock S L (2007) Sol-gel methods for the assembly of metal chalcogenide quantum dots. Acc Chem Res 40: 801–809.

28

Transparent Silica Aerogel Blocks for High-Energy Physics Research

Hiroshi Yokogawa

Abstract A transparent silica aerogel with an extremely low refractive index has been often used as a medium of *Cherenkov counter* in high-energy physics research. The durability of silica aerogel for the moisture is essential because some large-scale scientific experiments in this field often take very long time for installments, measurements, and analyses. In this section, the performance of silica aerogel *Cherenkov counters* and examples of specifications of hydrophobic silica aerogel products are introduced.

28.1. Introduction

In high-energy physics research, transparent silica aerogel blocks are often used for *Cherenkov counter*s because of its very low refractive index. As a medium of *Cherenkov counter*, its hydrophobic property (Chaps. 3 and 4) and durability for the moisture are very important. Panasonic Electric Works Co., Ltd. (*P-EW*) had been engaged in the development of a hydrophobic silica aerogel, which has high transparency and low refractive index, and collaborated with High Energy Accelerator Research Organization in Japan (*KEK*) toward *KEK*'s B-factory project. After the collaborated development, *P-EW* has been arranging and supplying silica aerogel blocks for some high-energy physics researchers.

In this section, examples of specifications of *P-EW*'s hydrophobic silica aerogel products are introduced and motivation and performance of a silica aerogel *Cherenkov counter* are also introduced. The astronomical data collected in B-factory experiment in which hydrophobic silica aerogel was installed as a *Cherenkov counter* in BELLE lead to Makoto Kobayashi and Toshihide Maskawa winning the Nobel prize in 2008.

28.2. Hydrophobic Silica Aerogel Blocks

The durability of silica aerogel for the moisture is very important when this material is used in the atmosphere. When a silica aerogel is used as a medium of a *Cherenkov counter*, the durability for the moisture, hydrophobicity, is essential because some large-scale

H. Yokogawa • Advanced Materials Development Department, New Product Technologies Development Department, Panasonic Electric Works Co., Ltd., 1048 Kadoma, Osaka 571-8686, Japan
e-mail: yokogawa@panasonic-denko.co.jp

M.A. Aegerter et al. (eds.), *Aerogels Handbook*, Advances in Sol-Gel Derived Materials and Technologies, DOI 10.1007/978-1-4419-7589-8_28,

scientific experiments in this field often take a very long time for installments, measurements, and analyses. Silica aerogel prepared by conventional sol–gel method is usually damaged by the moisture. This type of silica aerogel gradually absorbs the moisture and shrinks; therefore, a refractive index increases and transparency is decreased. The author and colleagues have succeeded in preparing hydrophobic silica aerogels (*TMSA*: trimethylsilyl-modified aerogel) that show no temporal degradation [1]. Surface of silica nanoparticles that compose the units of the silica aerogel skeleton was modified by trimethylsilyl (*TMS*) groups. The manufacturing process of *TMSA* is registered as the Patent [2]. Figure 28.1 shows a drop of water on the surface of a *TMSA* block. The contact angle of the drop is over 100° because of the excellent hydrophobicity of the surface of the sample.

Figure 28.1. A drop of water on the surface of a hydrophobic silica aerogel block.

28.2.1. Manufacturing Process

An *alcogel* is obtained by the sol–gel method using a commercially available oligomer of tetramethylorthosilicate as an initial material. Because of the use of this oligomer, it became more convenient to use the two-step method of aerogel processing for industrialization [3]. After mixing the oligomer, water, and alkaline catalyst in alcoholic solution, the solution is gelled to become an *alcogel*. The *alcogel* is rinsed by alcohol and immersed in a solution of hexamethyldisilazane (*HMDS*). Figure 28.2 shows the reaction model of *TMS* modification on the surface of the silica nanoparticles. After the *TMS* modification, the *TMS*-modified *alcogel* is immersed again in alcohol to remove the by-products inside the *alcogel*.

The *TMS*-modified *alcogel* is then subjected to supercritical drying. The supercritical drying process is based on the carbon dioxide extraction method [4]. The standard process parameters pressure and temperature versus time are shown in Figure 28.3. The substitution of carbon dioxide for alcohol was performed at a supercritical condition of carbon dioxide. The temperature of substitution is 80°C and the pressure is 16 MPa. At this condition, alcohol is soluble in supercritical state carbon dioxide with every concentration. The whole process of supercritical drying takes ordinarily about 2 days. The maximum size of *TMSA* obtained by the above process is approximately 200×300 mm^2.

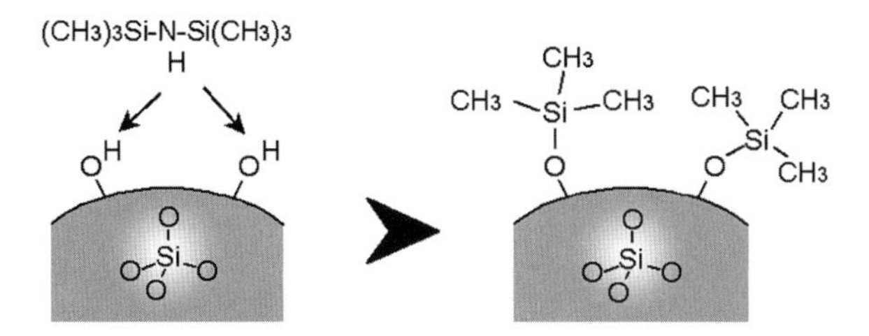

Figure 28.2. TMS (trimethylsilyl) modification by hexamethyldisilazane on the surface of SiO_2 nanoparticles composing silica aerogels.

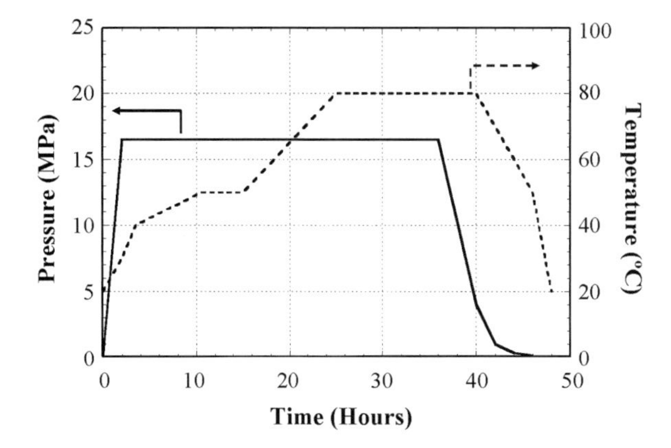

Figure 28.3. Temperature and pressure patterns of CO_2 extraction method of supercritical drying process.

28.2.2. Optical Properties

TMSA, as mentioned above, is obtained as highly transparent monolithic blocks. The photograph of representative block is shown in Figure 28.4. In Figure 28.4, the red and green lines of laser paths appear by scattering occurring within the nanoparticles composing the *TMSA*. The transmittance of 10 mm thick blocks is approximately 90% or more at the center of the visible wavelength range and approximately 70% or more at the wavelength of 400 nm.

The refractive index of *TMSA* can be designed in the range of 1.008–1.06 by controlling their densities. Figure 28.5 shows the relation between density and refractive index of *TMSA*. Relative value ($n - 1$) is almost in proportion to the densities in the range of high porosity as aerogel materials. This result follows the theory of "*Maxwell–Garnet* relation" applied to the nanocomposite composed by organically modified silica and air [5].

28.3. Aerogel *Cherenkov Counter*

When a charged particle passes through a transparent medium with a velocity faster than that of light in the material, Cherenkov light is emitted [6]. Thc Cherenkov light is radiated at the Huygens wavefront as shown in Figure 28.6. Here, the particles passes through a medium with speed v and the ratio between the speed of the particle and the speed of light is defined as $\beta = v/c$, where c is the speed of light. n is the refractive index of

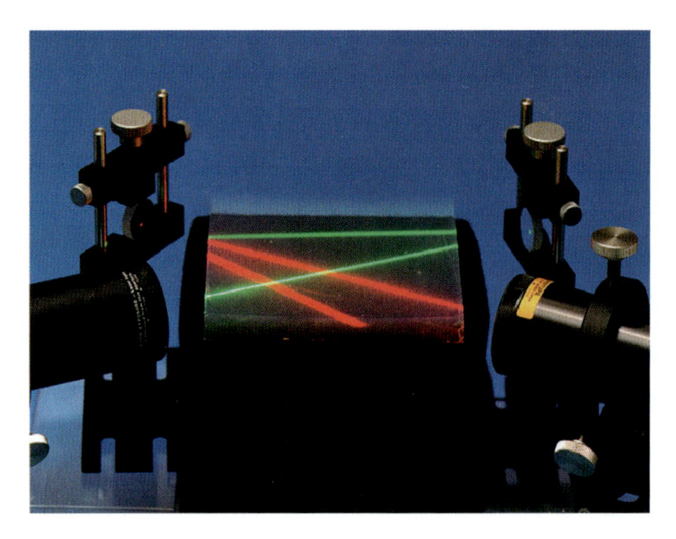

Figure 28.4. Photograph of 110 × 110 × 10 mm^3 hydrophbic silica aerogel block (Panasonic Electric Works Co., Ltd. P-EW).

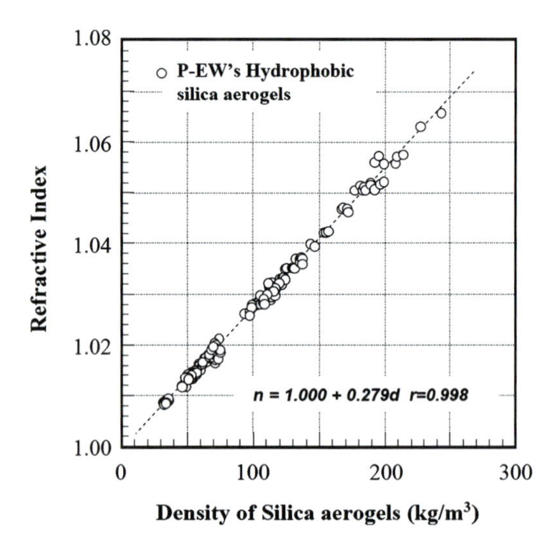

Figure 28.5. Correlation between densities and refractive indices of P-EW's hydrophobic silica aerogels.

the medium; therefore, the emitted light waves has a speed of c/n. The angle between the direction of the emitted light and that of the particle is defined as θ (Cherenkov angle) and expressed as follows:

$$\cos\theta = \frac{(c/n)\Delta t}{v\,\Delta t} = \frac{1}{n\beta}.$$

In high-energy physics research, observations and analysis of the Cherenkov light produced in transparent silica aerogel blocks can identify charged particles. Since each particle has a characteristic mass, any particle can be identified by a simultaneous

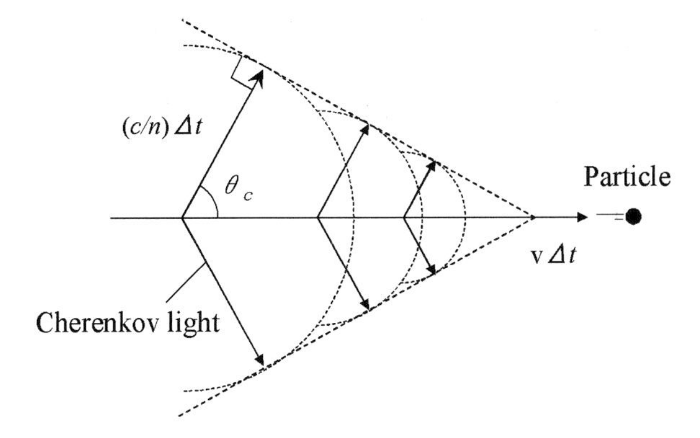

Figure 28.6. Cherenkov light radiation in a transparent medium.

measurement of its velocity and momentum. Hence, information of the velocity obtained by the Cherenkov light is directly translated into mass information [7], and these identifications have led to the success of high-energy physics experiments.

There are two methods that make use of Cherenkov light for particle identification: one is known as the "threshold" type and the other is the "Ring Imaging" type (*RICH*). The former method identifies particle species according to whether Cherenkov light is emitted or not in the medium, while the latter measures the Cherenkov angle by observing the position of light on a detector.

28.3.1. Threshold-Type *Cherenkov Counter*

Figure 28.7 shows a typical model of a threshold-type *Cherenkov counter*. Silica aerogel is filled in a box to which photomultiplier tubes (*PMT*s) are attached. The refractive index of aerogel is set so that the Cherenkov light is emitted for pions but not for kaons [8, 9]. In this way, it is known that kaon/pion/proton separation in the momentum region of 1.0–3.5 GeV/c is possible using aerogel with low refractive index such as 1.01–1.03.

This type of *TMSA Cherenkov counter* module was tested using a 3.5 GeV/c negative pion beam at *KEK* [10]. *TMSA* with refractive index is 1.015 and two 2.5″ *PMT*s were arranged. Typical pulse height distributions for 3.5 GeV/c pions and protons observed are

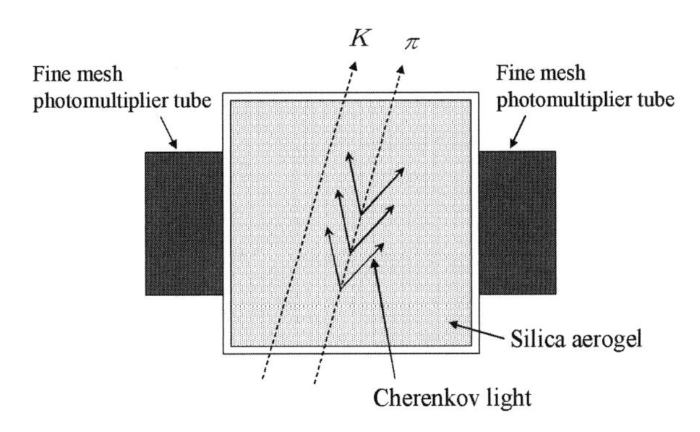

Figure 28.7. Schematic view of a typical threshold-type silica aerogel Cherenkov counter.

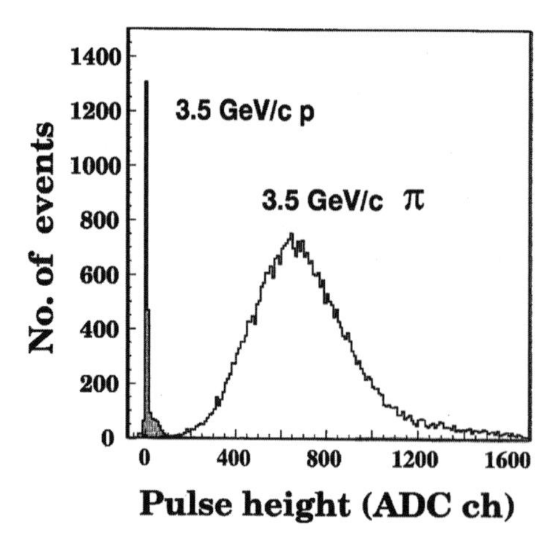

Figure 28.8. Pulse height spectra for 3.5 GeV/c pions (above threshold) and protons (below threshold) obtained by a single module of aerogel Cherenkov counter, in which $n = 1.015$ TMSA blocks were stacked.

shown in Figure 28.8. Pions (above threshold) and protons (below threshold) are clearly separated by more than 3σ (standard deviation), and this separation is maintained even in a high magnetic field (1.5 T).

28.3.2. Ring Imaging *Cherenkov Counter*

Figure 28.9 shows a schematic view of a typical *RICH*. Its principle is as follows: Cherenkov photons emitted from the radiator with Cherenkov angle (θ) hit the photocathode. Then photoelectrons are released from the CsI photocathode surface. The photoelectrons drift to the anode wire and are multiplied around the wire. Then electrical charges are induced on segmented cathode pads. From the radius of a reconstructed ring image, particle information (Cherenkov angle (θ), i.e., particle speed (β)) can be deduced. Figure 28.10

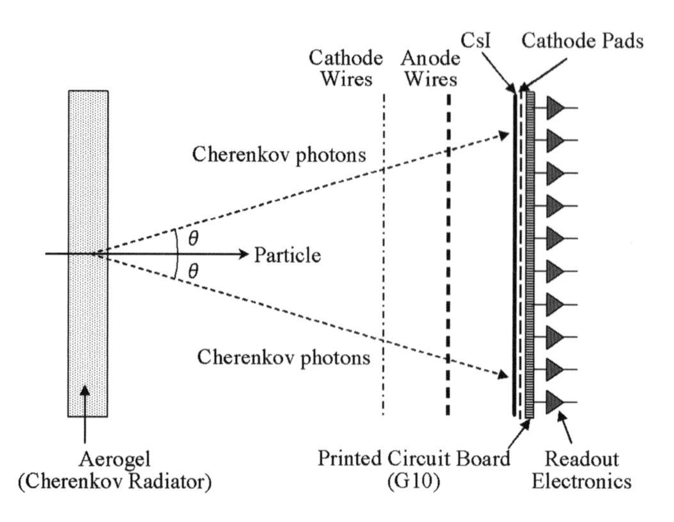

Figure 28.9. Schematic view of a ring image Cherenkov counter using silica aerogel blocks.

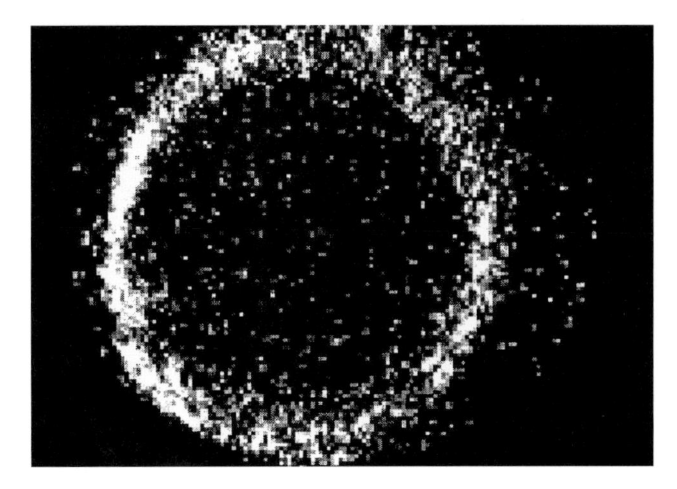

Figure 28.10. Ring image of Cherenkov light radiated by 3.5 GeV/c negative pions in the TMSA of $n = 1.015$ and 50 mm thick observed by an image intensifier on which the Cherenkov light was focused by a concave mirror.

shows an example of a Cherenkov ring image [9]. The ring image from the 50 mm thick *TMSA* radiator whose refractive index is 1.03 was clearly observed. The Cherenkov photons were focused on an image intensifier by a concave mirror.

28.4. *KEK* B-Factory Experiment

28.4.1. Objective

One of the most intriguing puzzles of nature is why the Universe is composed only of matter in contradiction with the cosmological theory, which suggests that an equal amount of particles and antiparticles have been produced in the "big-bang." A simple explanation of this phenomenon requires the violation of matter–antimatter symmetry (so-called CP symmetry). In order to elucidate this interesting physics, several B-factories had been proposed and were constructed around the world, where large numbers ($\sim 10^{7-8}$ per year) of B-meson decays would be examined for the study of CP violation. Figure 28.11 shows decay patterns of a B-meson pair. A B-meson pair would appear by the collision of electron and positron, and B- and anti-B-meson would respectively decay to pions, kaons, and electrons.

In such B-factories, the separation of pions from kaons is vital for the identification of B- or anti-B-meson and the selection of rare decays. It is important to measure the decay time from the appearance of B-mesons to the final state.

28.4.2. Aerogel *Cherenkov Counter* of BELLE Detector

A typical single threshold-type aerogel *Cherenkov counter* (*ACC*) module looks as shown in Figure 28.12. *TMSA* tiles are stacked in a thin aluminum box of approximate dimensions $120 \times 120 \times 120$ mm^3. In order to detect the Cherenkov light effectively, two fine mesh-type *PMT*s are attached directly to the aerogels at both sides of the box. The inner surface of the box (except for the phototube windows) is lined with a diffuse reflector

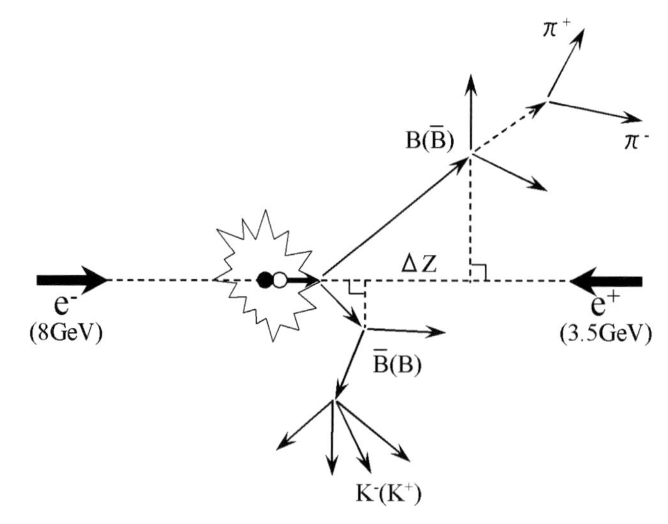

Figure 28.11. Decay pattern of moving B-meson pairs.

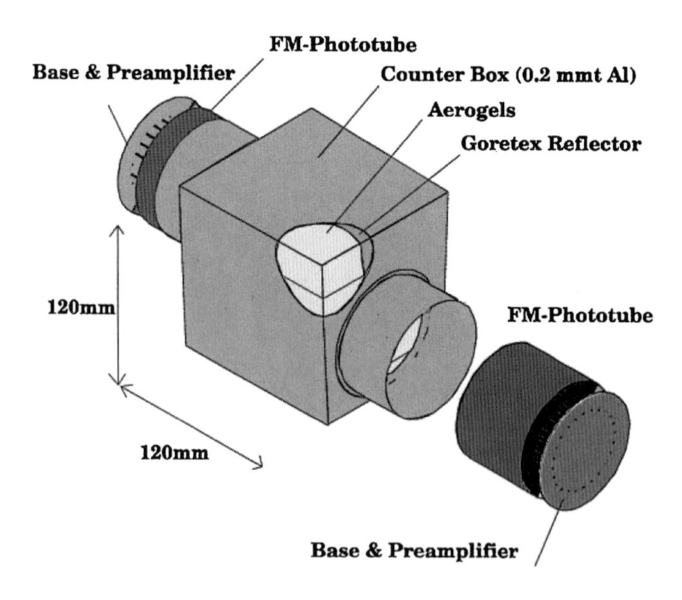

Figure 28.12. Schematic drawing of a single module of TMSA's Cherenkov counter for BELLE.

(Goretex sheet) to obtain a uniform response. Figure 28.13 shows the configuration of the *ACC* in a central part of the BELLE detector at *KEK* B-factory [11, 12]. The *ACC* consists of 960 counter modules for the barrel part and 228 modules for the end-cap part of the detector. In order to obtain good pion/kaon separation for the whole kinematical range, refractive indices of aerogels are selected between 1.01 and 1.03, depending on their polar angle region. The total volume of *TMSA* for the *ACC* was approximately 2.0 m^3.

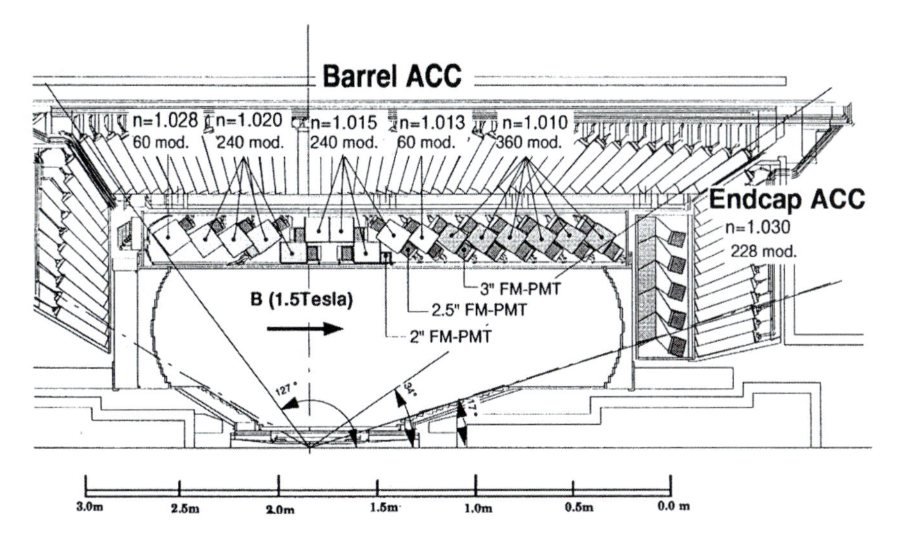

Figure 28.13. Arrangement of the TMSA's Cherenkov counters at the central part of BELLE detector.

28.4.3. Results of B-Factory [13, 14]

B-factory had been installed since 1994 in *KEK* and began to be operated in 1998. B-factory is composed of the *KEK* B accelerator, which is 3 km around, and the BELLE detector, which has 1,400 ton weight and 8 m width and length. Figure 28.14 shows the scheme and picture of BELLE detector. *ACC* is positioned outside of the CDC shown as Figure 28.14A. BELLE is composed of approximately 200,000 channels that sense the positions, times of flight, and energies of particles. Approximately 350 researchers participate in B-factory experiments and have improved the accelerator to improve the luminosity and measured and analyzed astronomical volume of data. Through the observations of the decay times of B-meson pairs, the difference of the decay time distribution between B-mesons and anti-B-mesons has been clarified as shown in Figure 28.15. This result can clearly support the "Violation of CP symmetry" theory and was decisive for the awarding of the 2008 Nobel prize in Physics to Makoto Toshihide Maskawa.

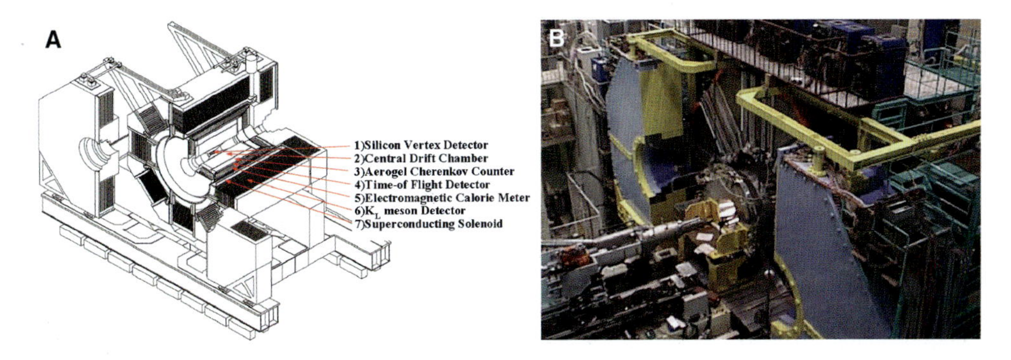

Figure 28.14. Scheme and picture of the BELLE detector. **A.** Internal structure of BELLE, **B.** photograph of the BELLE detector.

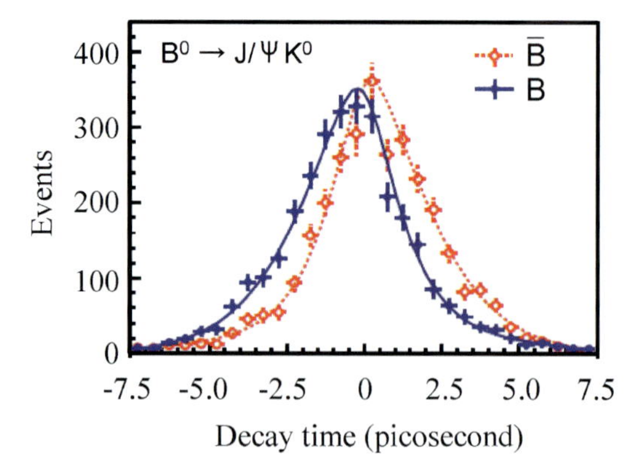

Figure 28.15. "Violation of CP symmetry" measured by BELLE in 2006: difference of the decay time distribution between B-mesons and anti-B-mesons.

28.5. Achievements of Other Experiments

Since *KEK* B's successful mass production of silica aerogels, several experimental groups decided to use silica aerogels for their particle identification devices as below.

Some examples of the threshold-type Cherenkov detectors that separate kaons and pions, similar to *KEK* B, have been discussed in [15–17]. A balloon-borne detector equipped with aerogels (*BESS*) was launched to an altitude of around 10,000 m to observe antiparticles in primary cosmic rays [18, 19]. Antiprotons are separated from other negatively charged particles with a 10^{-3} rejection by using information from the ACCs. For a search of antimatter nuclei in space, silica aerogel *Cherenkov counter*s are also used for particle separation [20–23]. HERMES experiments have been performed using aerogels as the radiator for their *RICH* detectors [24–29]. It was reported that a clean image of the Cherenkov ring could be observed thanks to the sufficient transparency of aerogels.

28.6. Specifications of "Panasonic" Silica Aerogels

P-EW has been supplying *TMSA* blocks to high-energy physics research fields since around 1997. From the experiences of manufacturing and supplying of *TMSA*, three types of standard specifications whose refractive index is 1.015, 1.030, and 1.050 are set up as shown in Table 28.1. SP-30 whose refractive index is 1.03 has the highest transparency (>92% at 550 nm wavelength), and the values for SP-15 and SP-50 (>90% and >88%, respectively, at

Table 28.1. Standard specifications of P-EW's TMSA blocks

Grade		SP-15	SP-30	SP-50
Form	Shape	Monolithic block (square)		
	Size	110 mm × 110 mm	113 mm × 113 mm	
Refractive index (–)		1.015	1.030	1.050
Density (kg/m^3)		60	110	190
Transmittance (%)	(@550 mm)	>90%	>92%	>88%
(10 mm thickness)	(@400 mm)	>65%	>75%	>62%

550 nm wavelength) are a little lower. A transparency of silica aerogel depends on sizes and concentration of silica nanoparticles and their uniformity. The middle density of SP-30 can show the optimum for their factors for the transparency. The standard size of a block is approximately 110 × 110 × 10 mm^3. We can usually manufacture 100 *TMSA* blocks with standard specification in a single batch. If a different size or optical properties are required, it is possible to arrange separately for their specifications such as size, thickness, and refractive index with reliable transparency.

Figure 28.16 shows the distributions of the refractive index of *P-EW*'s *TMSA* blocks, SP-30. These data were obtained using about 1,300 pieces of manufactured blocks. The refractive index is usually measured by the prism method [19]. The refractive index was in the range of 1.029–1.033 and almost all the blocks are within 1.029–1.031. This dispersion is caused by the differences between batches of manufacturing, solutions for alcogels, supercritical drying processes, measuring conditions of the refractive index, etc.

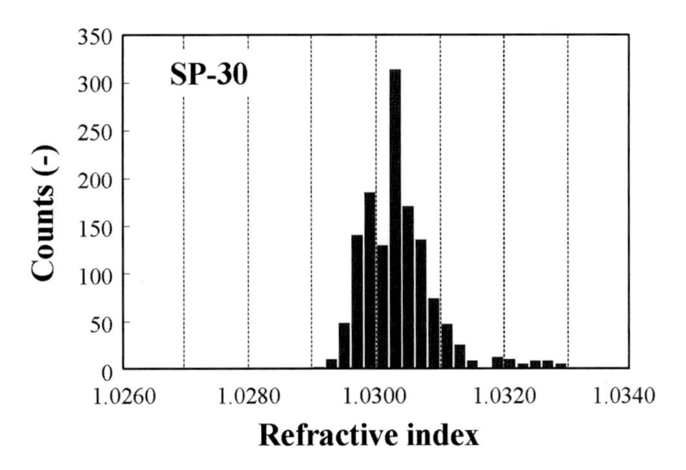

Figure 28.16. Distributions of a refractive index of P-EW's $n = 1.030$ TMSA blocks.

TMSA has an additional feature for the realization of various kinds of *Cherenkov counters*. Because of blocks' excellent hydrophobicity, it is possible to cut them by water-jet processing and secondarily to arrange them to the required shape or size of *Cherenkov counter*. Figure 28.17 shows the samples cut down from standard blocks by water-jet. The properties and hydrophobicity are not depreciating after the cut processing.

28.7. Conclusions

Silica aerogels whose particle surface is modified by *TMS* show excellent hydrophobicity. Since it has superior properties, transparency, extreme low refractive index, and moisture resistance, it has been often used as a *Cherenkov counter*. Although silica aerogel monolithic blocks produced by supercritical drying methods are rather expensive for industrial applications, they have contributed much toward progress in science fields such as high-energy physics research. The progress of science has always contributed to the improvement of research and development in industrial world, so it is much expected that

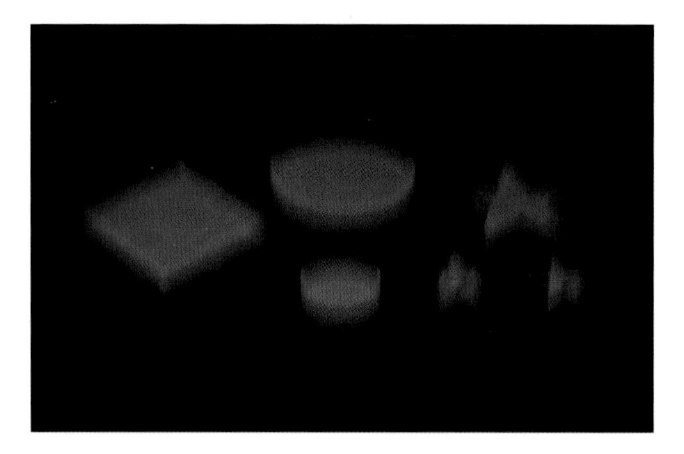

Figure 28.17. TMSA samples cut down from the P-EW's standard TMSA blocks by water-jet processing.

aerogels could keep pioneering the new technologies such as nanocomposites, optics, cosmic development, energy devices, and so on.

Acknowledgments

The works mentioned above were partially supported by a collaborative research program between *KEK* and *P-EW*. I am very grateful to F. Takasaki, T. Sumiyoshi, and I. Adachi. I have also referred to a large number of reports describing B-factory experiments which I cannot put on the references list. I am thankful to many researchers who have been concerned in the B-factory project.

References

1. Yokogawa H, Yokoyama M (1995) Hydrophobic silica aerogels. J Non-Cryst Solids 186: 23–29.
2. Yokogawa H, Takahama K, Yokoyama M, Uegaki Y (1992) Process for hydrophobic aerogel. US Patent No 5830387.
3. Tillotson TM, Hrubesh LW (1992) Transparent ultralow-density silica aerogels prepared by a two-step sol-gel process. J Non-Cryst Solids 145: 44–50.
4. Yokoyama M (1991) Materials prepared by supercritically drying process. Development and applications of porous ceramics II CMC: 295–309.
5. Maxwell Garnett JC (1904) Colours in metal glasses and in metallic films. Philosophical Transactions Royal Society of London A203: 385–420.
6. Cherenkov PA (1937) Visible radiation produced by electrons moving in a medium with velocities exceeding that of light. Phys Rev 52: 378–379.
7. Frank IM, Tamm IE (1937) Coherent visible radiation of fast electrons passing through matter. Compt Rend Acad Sci USSR: 109–14.
8. Carlson P (1986) Aerogel Cherenkov counters: construction principles and applications. Nucl Instrum Methods Phys Res Sect A248: 110–117.
9. Gunter P (1986) Aerogel Cherenkov counters at DESY. Nucl Instr and Meth A248: 118–129
10. Sumiyoshi T, Adachi I, Enomoto R, Iijima T, Suda R, Yokoyama M, Yokogawa H (1998) J Non-Cryst Solids 225: 369–374.
11. Suda R, Watanabe M, Enomoto R, Iijima T, Adachi I, Hattori H, Kuniya T, Ooba T, Sumiyoshi T, Yoshida (1998) Monte-Carlo simulation for aerogel Cherenkov counter. Nucl Instrum Methods Phys Res Sect A406: 213–226.

12. Sumiyoshi T, Adachi I, Enomoto R, Iijima T, Suda R, Leonidopoulos C, Marlow DR, Prebys E, Kawabata R, Kawai H, Ooba T, Nanao M, Suzuki K, Ogawa S, Murakami A, Khan MHR (1999) Silica aerogel Cherenkov counter for the KEK B-factory experiment. Nucl Instrum Methods Phys Res Sect A433: 385–391.
13. Nishida S (2008) SOKENDAI Journal, The Graduate University for Advanced Studies, Special Number: 10-15.
14. KEK: KEKB Quest for CPV, KEKB_brochure, http://kekb.jp/ (2008).
15. Ishino M, Chiba J, En'yo H, Funahashi H, Ichikawa A, Ieiri M, Kanda H, Masaike A, Mihara S, Miyashita T, Murakami T, Nakamura A, Naruki M, Muto R, Ozawa K, Sato HD, Sekimoto M, Tabaru T, Tanaka KH, Yoshimura Y, Yokkaichi S, Yokoyama M, Yokogawa H (2001) Mass production of hydrophobic silica aerogel and readout optics of Cherenkov light. Nucl Instrum Methods Phys Res Sect A457: 581–587.
16. Asaturyan R, Ent R, Fenker H, Gaskell D, Huber GM, Jones M, Mack D, Mkrtchyan H, Metzger B, Novikoff N, Tadevosyan V, Vulcan W, Wood S (2005) The aerogel threshold Cherenkov detector for the High Momentum Spectrometer in Hall C at Jefferson Lab. Nucl Instrum Methods Phys Res Sect A548: 364–374.
17. Siudak R, Budzanowski A, Chatterjee A, Clement H, Dorochkevitsch E, Ernst J, Hawranek P, Hinterberger F, Jahn R, Jarczyk L, Joosten R, Kilian K, Kirillov Di, Kliczewski S, Kolev D, Kravcikova M, Machner H, Magiera A, Martinska G, Massmann F, Munkel J, Piskunov N, Protic D, Ritman J, Possen P von, Roy B, Sitnik I, Slepnev I, Smyrski J, Tsenov R, Ulbrich K, Urban J, Vankova G, Wrgner GJ, Ziegler R (2008) A threshold Cherenkov detector for K^+/π^+ separation using silica aerogel. Nucl Instrum Methods Phys Res Sect A596: 311–316.
18. Shikaze Y, Orito S, Mitsui T, Yoshimura K, Matsumoto H, Matsunaga H, Nozaki M, Sonoda T, Ueda I, Yoshida T (2000) Large-area scintillator hodoscope with 50 ps timing resolution onboard BESS. Nucl Instrum Methods Phys Res Sect A455: 596–606.
19. Yoshimura K, Abe K, Fuke H, Haino S, Hams T, Hasegawa M, Horikoshi A, Kim KC, Kumazawa T, Kusumoto A, Lee MH, Makida Y, Matsuda S, Matsukawa Y, Mitchell JW, Moiseev AA, Nishimura J, Nozaki M, Orito R, Ormes JF, Sakai K, Sasaki M, Seo ES, Shikaze Y, Shinoda R, Streitmatter RE, Suzuki J, Takeuchi K, Thakur N, Tanaka K, Yamagami T, Yamamoto A, Yoshida T (2008). BESS-Polar experiment: Progress and future prospects. Adv Space Res 42: 1664–1669.
20. Gougas AK, Ilie D, Ilie S, Pojidaev V (1999). Behavior of hydrophobic aerogel used as a Cherenkov medium. Nucl Instrum Methods Phys Res Sect A421: 249–255.
21. Baret B, Aguayo P, Aguilar Benitez M, Arruda L, Barao F, Barrau A, Belmont E, Berdugo J, Boudoul G, Borges J, Buenerd M, Casadei D, Casaus J, Delgado C, Diaz C, Derome L, Eraud L, Gallin-Martel L, Giovacchini F, Goncalves P, Lanciotti E, Laurenti G, Malinine A, Mana C, Marin J, Martinez G, Menchaca-Rocha A, Palomares C, Pimenta M, Protasov K, Sanchez E, Seo ES, Sevilla I, Torrento A, Vargas-Trevino M (2004). In-Beam tests of the AMS *RICH* prototype with 20 A Gev/c secondary ions. Nucl Instrum Methods Phys Res Sect A525: 126–131.
22. Martinez-Davalos A, Belmont-Moreno E, Menchaca-Rocha A (2005). Optical and ageing studies of aerogel samples for RICH applications in space. Nucl Instrum Methods Phys Res Sect A553: 177–181.
23. Sallaz-Damaz Y, Barrau A, Bazer-Bachi R, Bourrion O, Bouvier J, Boyer B, Buenerd M, Derome L, Eraud L, Foglio R, Gallin-Martel L, Ganel O, Han JH Kim KC Lee MH, Lutz L, Mangin-Brinet M, Malinine A, Menchaca-Rocha A, Perie JN, Putze A Scordilis JP, Seo ES, Walpole P, Yoo JH, Yoon YS, Zinn SY (2008). CHERCAM: The Cherenkov imager of the CREAM experiment. Nucl Instrum Methods Phys Res Sect A595: 62–66.
24. Carter P (1999). The aerogel radiator of the HERMES RICH. Nucl Instrum Methods Phys Res Sect A433: 392–395.
25. Sakemi Y (2000). Dual radiator RICH of HERMES for π, K, p identification. Nucl Instrum Methods Phys Res Sect A453: 284–288.
26. Alemi M, Bellunato T, Braem A, Calvi M, Chesi E, Duane A, Joram C, Liko D, Matteuzzi C, Negri P, Neufeld N, Paganoni M, Seguinot J, Voillat D, Wotton S, Ypsilantis T (2002). Testbeam results on particle identification with aerogel used as RICH radiator. Nucl Instrum Methods Phys Res Sect A478: 348–352.
27. Miyauchi Y (2003). The HERMES RICH aerogel radiator. Nucl Instrum Methods Phys Res Sect A502: 222–226.
28. Jackson HE (2005). The HERMES dual radiator RICH – Performance and impact. Nucl Instrum Methods Phys Res Sect A553: 205–209.
29. Raffaele De Leo (2008). Long-term operational experience with the HERMES aerogel RICH detector. Nucl Instrum Methods Phys Res Sect A595: 19–22.

29

Sintering of Silica Aerogels for Glass Synthesis: Application to Nuclear Waste Containment

Thierry Woignier, Jerome Reynes, and Jean Phalippou

Abstract To ensure long-term storage, long-life nuclear wastes (actinides) must be incorporated in a matrix exhibiting excellent chemical durability. Among the usual glasses, silica glass, which does not contain alkali and boron, is expected to present high durability. Associated with good mechanical properties, a low thermal expansion, and consequently a good thermal shock resistance, silica glass is a promising candidate. However, the glass preparation requires a high temperature (1,800°C) melting process. Nevertheless, silica glasses can be synthesized from silica aerogels followed by a sintering in a temperature range (1,000°C) close to half of that of the glass melting process. The impregnation of aerogels by actinide solutions followed by aerogels sintering has been proposed. After sintering, the aerogel silica matrix traps the nuclear waste and protects them against water erosion. The resulting material is a nanocomposite material (glass-ceramic) composed of nuclear wastes embedded in the silica matrix. The dissolution rate of such a composite in water is almost 10^2 times lower than that of the common nuclear glasses. This result clearly proves the improvement of the chemical durability of the glass-ceramic made from aerogels and, thanks to the high mechanical properties, these new materials can be suitable matrices for the containment of actinides.

This contribution describes the different steps of the gel–aerogel–glass transformations and the characterization of the durability and mechanical properties of such glass-ceramics trapping the nuclear waste. The interest of silica aerogel as host matrix for nuclear waste containment is emphasized.

29.1. Introduction

The physical and chemical properties of conventional glasses have been studied for a long time and their applications are well known. Glasses are used traditionally in the field of optical materials and containers and also as fibers for mechanical reinforcement and some "exotic" applications such as nuclear waste storage [1–5].

T. Woignier • UMR CNRS 6116-UMR IRD 193- IMEP- IRD-PRAM, Le Lamentin, Martinique, France
and
CNRS-Université Montpellier II, Place E. Bataillon, 34095 Montpellier Cedex 5, France
e-mail: woignier@univ-montp2.fr
J. Reynes and J. Phalippou • CNRS-Université Montpellier II, Place E. Bataillon, 34095 Montpellier Cedex 5, France
e-mail: reynes.jerome@neuf.fr; jean.phalippou698@orange.fr

M.A. Aegerter et al. (eds.), *Aerogels Handbook*, Advances in Sol-Gel Derived Materials and Technologies, DOI 10.1007/978-1-4419-7589-8_29,

In the literature, a number of materials to encapsulate high-level radioactive wastes have been considered [1–11]. These materials include glasses of various compositions, ceramics of a wide variety of formulations; glass-ceramic compounds; cements; and particles in metal matrices. Based on production technology, the two best candidate materials are glasses and crystalline ceramics. Titanates-based minerals phases (Synrocs) [6–8] have been studied for sometime as possible nuclear wastes storage materials. The constituent's phases of Synroc (zirconolite, titanite, pyrochlore) suggest that these materials are highly corrosion resistant. Zirconolite ($CaZrTi_2O_7$) based glass-ceramics, in which the crystalline phase is embedded in an aluminosilicate glassy matrix, is envisaged as waste form candidates. They consist of a highly durable crystalline phase homogeneously dispersed in a glassy matrix exhibiting also a good chemical durability [9–11]; glass-ceramics incorporating preferentially radionuclides in the crystalline phase are particularly interesting (double containment principle).

Glass is a solid in which a wide range of waste can be dissolved and a number of countries have developed successful industrial-scale vitrification technologies to solidify their wastes. Glass can accept a wider range of wastes than ceramic and the processing feasibility has inclined much to the research for developing borosilicate glass materials for immobilizing waste solutes [1, 2]. In this process, the radioactive elements are mixed and melted with a glass fritt (a borosilicate glass containing sodium). This borosilicate nuclear glass has become the reference in nuclear waste conditioning.

One of the major issues related to the expanded use of nuclear power is the fate of plutonium and actinides. Although there are a number of fission product radionuclides of high activity (^{137}Cs and ^{90}Sr) and long half-life (^{99}Tc, 200,000 years; ^{129}I, 1.6×10^7 years) in spent nuclear fuel, actinides account for most of the radiotoxicity of nuclear waste because, after several hundred years, the radiotoxicity is dominated by ^{239}Pu (half-life = 24,100 years) and ^{237}Np (half-life = 2,000,000 years). Thus, a major part of the long-term risk is directly related to the fate of these two actinides in the geosphere.

There are two basic strategies for the disposal of actinides: transmuting the actinides using nuclear reactors or accelerators or trapping them in chemically durable, radiation-resistant materials that are suitable for geological disposal. Therefore, researches are investigating new containment matrices with high chemical durability, because it is important to limit the possible release of radionuclides if the matrix is destroyed by aqueous erosion.

In the case of oxide glasses, the erosion starts in an initially dense glass by the extraction of ions having high mobility in the glass and high water solubility, such as Na, Ca, or B. The erosion process leaves a skeleton with a chemical composition essentially based on the glass formers such as silica and alumina. So, high chemical durability is achieved by high content of silica and low content of oxides modifier. Moreover, silica glass also offers good mechanical strength and thermal shock resistance.

Thus, the use of "pure" silica glass will optimize the properties which characterize a suitable glassy matrix for the fixation of actinides but silica glass preparation requires solving problems associated with high melting temperature (~1,800°C). In the literature [5], an elegant method has been proposed: high silica content porous glasses are prepared from a phase separated borosilicate glass and is used as host matrix for radioactive wastes. The fixation of nuclear waste is achieved in the porous structure, followed by the pore collapse.

In the last decades, a new kind of glass synthesis has been extensively described in the literature. These glasses are no longer prepared by the classical melting and refining processes but by a sol-gel method [12]. The structure of these materials results from the aggregation of colloidal oxides building blocks such as SiO_2, Al_2O_3, ZrO_2, B_2O_3, TiO_2, etc.

The aggregation mechanism does not allow filling of the entire space, leading to a porous and amorphous material. These gels are transformed into aerogels by supercritical drying and a further sintering step collapses the porosity [13, 14].

This contribution deals with actinide-bearing glass-ceramics derived from sintered silica aerogels. First, the different steps of the gel–aerogel–glass transformations are described, and then the interest of silica aerogel as precursors (host matrix) for nuclear waste containment is specially detailed. Finally, the durability and mechanical properties of glass-ceramics trapping the nuclear waste are characterized.

29.2. Glasses Obtained by the Sol-Gel Process

Since 30 years, it is demonstrated that "sol-gel processes" could be a new way to synthesize materials as different as ceramics, glasses, or composites. Brinker and Scherer [12] have shown the possibilities of this technique to allow the synthesis of materials in the form of films, fibers, or bulk products.

The classical procedure for making glasses includes a step at elevated temperature, which ensures that the raw materials have reacted; the liquid structure (amorphous) is preserved by a rapid cooling of the melt. In a sol-gel process, this high temperature step is eliminated. The homogenization is achieved in solution at room temperature and particle cross-linking and preservation of the amorphous structure are accomplished by the gelation step. Further heat treatments only serve to remove organic species, hydroxyls, and porosity [11, 13]. The sol-gel process has several flexibilities and unique features that are of significant importance in preparing pure homogeneous glasses: (1) the high purity of the starting compounds which can be preserved during the process; (2) for multicomponent systems, rapid homogenization of the different chemical species can be achieved at the molecular level by chemical reactions in the solution; (3) in the case of compositions leading to a glass, the gel is amorphous and can be transformed into a glass without melting or refining. The sintering treatment is carried out at temperatures one-third to one-half lower than those required for melting of the glass forming components; (4) the control of the morphology of the product, film, fiber, and bulk materials by the adjustment of the viscosity at room temperature.

The challenge of this process is then to obtain a solid, dense, mineral, and amorphous material from a room temperature liquid solution which is generally organic. So, several transformations of the gelling solution will be necessary [12–14]. The first step is the formation of a gel from the solution and there are two classical ways of obtaining silica-based gels (Chap. 2): the destabilization of silica sols or the hydrolysis and polycondensation reactions of an organometallic compound of silicon [12]. Both methods lead to noncrystalline materials containing substantial amounts of water and organic liquids, which can be eliminated by suitable drying treatments. The drying process can be carried out at ambient temperature but during this drying stage considerable shrinkage occurs converting the wet and soft gel into a several pieces of solid without any control of the size, morphology, and porous properties of the dried samples. To avoid these problems supercritical drying can be the appropriate solution. After this process, the aerogel is a solid, amorphous but extremely porous (75–99% of porosity) material. The last step of the transformation is a densification heat treatment necessary to convert by sintering the aerogels into solid glasses devoid of porosity, i.e., with a *relative density* ρ_r, the ratio between the aerogel bulk density and the silica glass density (2.2 g/cm^3) equals to 1. Figure 29.1 shows the typical evolution of

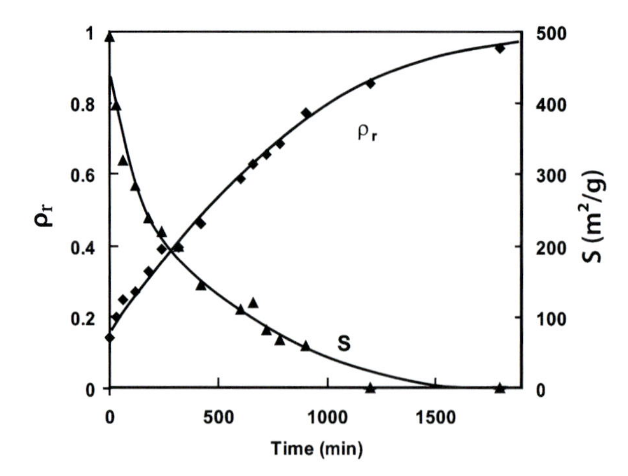

Figure 29.1. Evolution of the aerogel *relative density* ρ_r (*filled diamond*) and specific surface area *S* (*filled triangle*) versus the sintering time at 1,000°C.

the *relative density* and specific surface area during a sintering heat treatment. These curves are obviously strongly dependent on the heat treatment temperature but also on the aerogel hydroxyl content which will affect the aerogel viscosity [14].

Gels which are originally noncrystalline may crystallize during a later heat treatment. The successful formation of a glass is the result of a competition between the phenomena which lead to densification and those which promote crystallization [14, 15].

Figure 29.2 shows the different steps of sintering between the aerogel sample on the right and the sample on the left is silica glass; an aerogel after a complete sintering.

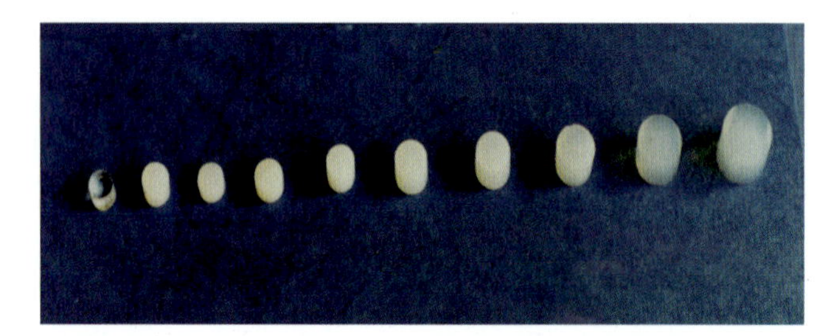

Figure 29.2. Set of partially sintered aerogels: from aerogel (*relative density* is 0.15, for the sample **on the right**) to fully sintered aerogel (transparent silica glass **on the left**, *relative density* is 1).

29.3. Principle of the Containment Process

The actinides generated by the nuclear fuel cycle can be provided in the form of salts in aqueous solutions. We will take advantage of the totally open pore structure of the aerogel to allow the migration of the liquid species (salt in solution) throughout the entire volume of the aerogel. Then, the liquid phase is eliminated, and the porous composite (aerogel + salt) is fully sintered giving rise to the synthesis of a multicomponent material (Figure 29.3). The porous structure of the aerogel is used as a volume host.

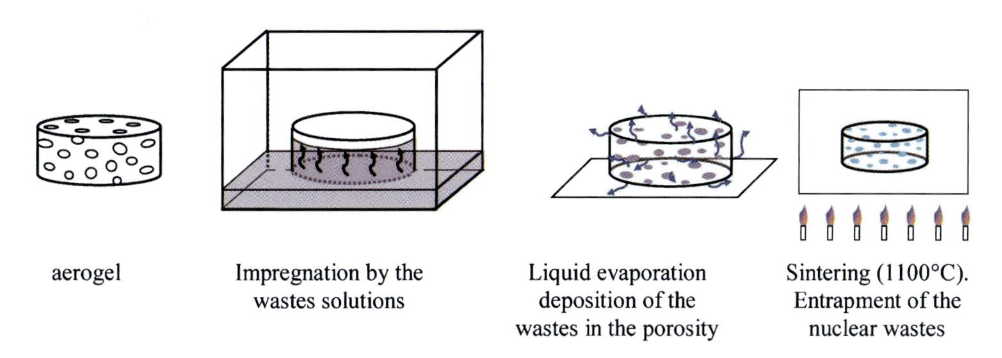

Figure 29.3. The different steps of the containment process.

According to the small size of aerogel pores we expect to prepare a nanocomposite by a very simple process. The size of actinides domains will depend on the aerogel pore size and actinides content in liquid.

However, if we try to fill the aerogel with a liquid such as water, capillary forces induce the fracture of the aerogel [16]. Due to the complexity of the aerogel texture, a detailed calculation of the local stresses during filling is difficult, but they are dependent on the liquid–vapor surface energy, γ, and the inverse of the pore size. Thus to avoid cracking during filling, different strategies can be proposed: (1) the synthesis of aerogels with large pores which will diminish the magnitude of the capillary forces, (2) the improvement of the aerogel mechanical strength, and (3) surface functionalization.

29.4. Synthesis of Silica Aerogel Host Materials

The small-angle scattering techniques (X-ray or neutron scattering) give information on the structure and compactness in the cluster forming the gel network. The aerogel network is often described as an assembly of clusters (~50 nm). The clusters can be fractal ($Df \sim 2$) built by the aggregation of small particles (~1–2 nm) [17–20]. This 3D network is totally open and spans over a range of pore size from the micro to the macro-porosity. Although the aerogel networks are rather different from the structure of silica glasses, Raman, infrared, and NMR spectroscopy used to characterize the molecular structure exhibit signatures identical to those observed in vitreous silica [21]. These kinds of porous materials can be considered as porous glasses although they largely differ in their way to attain the "glassy" state.

In the proposed containment process, the large pore volume is used as a sponge to incorporate chemical species with the objective to get a two phases material. However, the drawback of this porosity is the poor mechanical properties with the consequence that unmodified aerogels crack during filling. Another important parameter is the permeability. A high permeability is generally an advantage because it means that the fluid and thus the chemical species of interest migrate easily into the porous network and lead to a homogeneous distribution of the chemical species. However because of the large microporosity, the classical aerogel permeability is poor, typically a few nm^2 [12, 22]. Consequently, the impregnation time could be long. In summary, mechanical properties and permeability are the most important parameters for the loading process.

29.4.1. Partially Sintered Aerogels

One way to control the mechanical properties is to perform a partial sintering. Depending on the duration of the heat treatment, the microporosity is progressively eliminated and partially sintered samples can be obtained in the range of relative densities between 0.15 (classical aerogel) and 1 (silica glass). The sintering has several effects: it increases the connectivity and mechanical strength and collapses the small pores. In this way, sintering improves the ability of the porous network to resist a liquid filling step. However, the permeability D decreases with the sintering [16, 22]. A compromise should be found in terms of *relative density*, it should be high enough to get a matrix with acceptable mechanical properties but not too high, to have still sufficient permeability. This compromise corresponds to a *relative density* in the range of 0.4–0.5 [16].

29.4.2. Composite Aerogels

It is generally admitted that inclusion of particles or fibers in the material could improve the mechanical properties. So, it is possible to adjust the apparent density, the mechanical property, and the permeability by the addition of silica powder (silica soot such as aerosil) in the monomer solution, just before gelling [23, 24]. The mechanical properties (Chap. 21) rapidly increase with increasing silica powder percentage [24] and a composite aerogel with a *relative density* close to 0.18 is able to resist the capillary stresses during the pore filling by an aqueous solution. The permeability increases with the addition of silica powder because the composite aerogel structure is vastly different from the classical aerogel network. In the case of the classical aerogel, the network consists of one huge molecule in which the pores look as they were formed by the aggregation mechanisms. In the case of the composite aerogel, gelling results from the sticking between the silica soot particles (40 nm diameter) [24]. The natural spaces between them form larger pores and improve the permeability.

29.4.3. Permeability

The permeability of the composite aerogel can be estimated from the *Carman–Kozeny equation* applied to a gel [25], an approximation that relates the permeability to the pore size and which can be written as:

$$D = (1 - \rho_r) r_w{}^2 / 4K_w \qquad (29.1)$$

where ρ_r is the *relative density*, r_w is the hydraulic radius (characteristic pore size), and K_w is the so-called *Kozeny constant*. However, in the case of gels, K_w is a function of ρ_r [25]. Over the range of density used in this study, it can be approximated by:

$$K_w = 2.03 + 2.56\rho_r, \quad 0.08 \leq \rho_r \leq 0.4. \qquad (29.2)$$

Thus, from the *relative density* and the mean pore radius, one can estimate the permeability of the partially sintered aerogel and the composite aerogel. The permeability of the composite aerogel is roughly 60 times higher than for the sintered aerogel (Table 29.1).

Table 29.1. *Relative density* (ρ_r), mean pore size, and calculated permeability of partially sintered aerogel and composite aerogel

Kind of aerogel	ρ_r	r_w (nm)	D (nm^2)
Partially sintered aerogel	0.45	≈10	≈4.5
Composite aerogel	0.18	≈60	≈290

This new kind of composite aerogels is the more appropriate as host matrix because of its improved mechanical properties and permeability.

29.5. Synthesis of the Nuclear Glass-Ceramics

The impregnation of these peculiar aerogels is proposed to trap the nuclear wastes in a silica matrix. After a soaking treatment, the host matrix is dried, then sintered. The expected material is a nanocomposite constituted by nuclear wastes crystals embedded into a silica matrix. For safety reasons the lanthanides, Ce and Nd, are used to simulate the III and IV valent actinides [26–28] [for example, Am(III), Cm(III), Np(IV), Pu(IV)]. Nd and Ce as actinides surrogates have been also chosen because of their dissimilar affinity to silica. Hence, Nd and Ce will lead to different kinds of glass-ceramics, simulating the possible behaviors of the actinides in the presence of silica depending on their silica affinity.

Cerium and neodymium nitrates are dissolved in water and after soaking, the samples are dried and heat treated in such a way to complete the drying and to decompose the nitrates. Further heating at ~1,100°C fully sinters the silica network by viscous flow [14, 29].

The dense solids consist of a silica matrix in which lanthanide oxides are trapped. Weight differences before soaking and after sintering allow measuring the surrogate loading in weight percent. Figure 29.4 shows that the neodymium and cerium oxide loading obviously increases with the concentration of the surrogate nitrate solution; it can be as high as 20%.

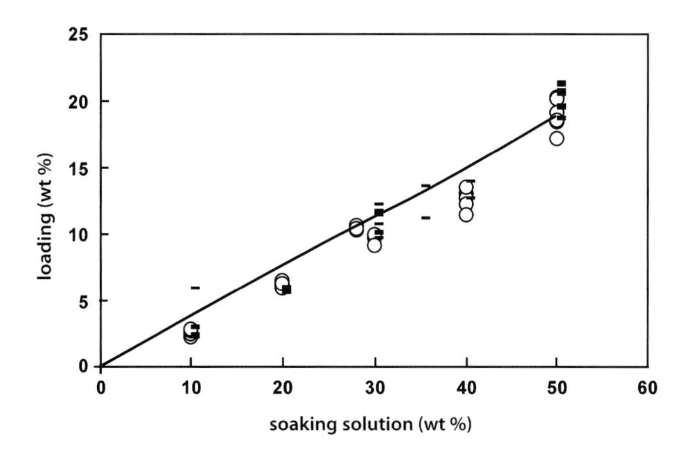

Figure 29.4. Evolution of the loading rate versus the soaking solution concentration for Nd (*open circle*) and Ce (*dash*).

It must be noted that the final sintering procedure is affected by the loading solution. Figure 29.5 shows the dilatometer curves measured on the samples loaded with 13% of

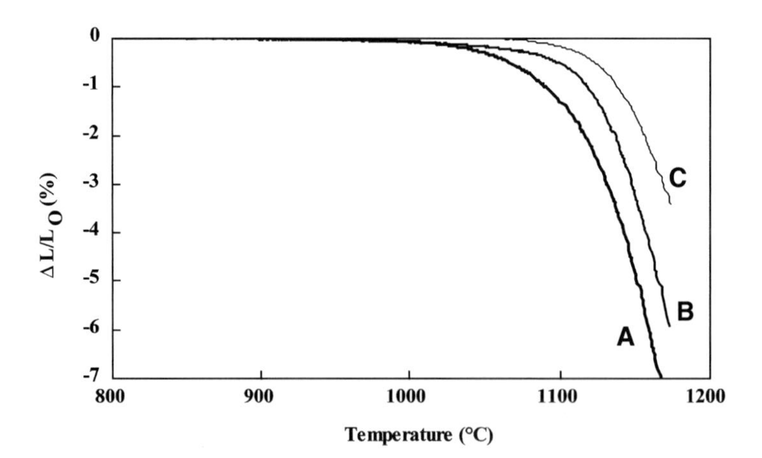

Figure 29.5. Sintering shrinkage ($\Delta L/L_0$ %) of composite aerogel **A.** composite aerogel loaded with 13% of CeO_2 **B.** and composite aerogel loaded with 13% of Nd_2O_3 **C.** versus sintering temperature.

Nd_2O_3 and CeO_2.The dilatometer response which expresses the sintering shrinkage ($\Delta L/L_0$ %) evidences that the loading increases the temperature range in which the shrinkage is maximized by up to 100°C. This maximum shrinkage temperature is close to 1,150°C for the native silica aerogel composite and close to 1,200°C for the silica aerogel containing Nd_2O_3. The loading by the surrogate oxides hinders the sintering, because crystalline phases (CeO_2 and Nd_2O_3) are not playing an active role in the viscous flow mechanism responsible for the sintering in the case of amorphous materials. The effect on the sintering rate is more pronounced in the case of Nd because of the more complicated structure of the glass-ceramic which contains vitreous silica and Nd_2O_3 but can also contain neodymium silicate and disilicate (see Sect. 29.6.1). Consequently, the sintering step must be adapted to the composition of the loaded porous glass and can vary by 50–100°C.

29.6. Characterization of the Glass-Ceramic

29.6.1. Structure

The final structure of the fully sintered materials achieved with this process is that of a composite material with surrogate loading as high as 20 wt%. The X-ray diffraction pattern (not shown) demonstrates that the sintered Ce loaded material is clearly a biphasic compound (SiO_2–CeO_2). On the other hand, the Nd-loaded sample shows three different crystalline phases, neodymium oxide (Nd_2O_3) but also neodymium mono (Nd_2SiO_5) and disilicate ($Nd_2Si_2O_7$) (Figure 29.6)

The evolution of the X-rays patterns shows that the structure of the crystalline phases is essentially Nd_2O_3 for the low loading percentage (2–10%) and tends towards $Nd_2Si_2O_7$ for concentration higher than 14%. For Nd loading below 2%, the sintered aerogel is amorphous leading to the conclusion that Nd is associated to Si to form a neodymium silicate glass. Figure 29.7 shows the measured bulk density of the Ce and Nd glass-ceramics versus the loading percentage. The density calculated from a biphasic description (weight additive law) is also plotted in Figure 29.7:

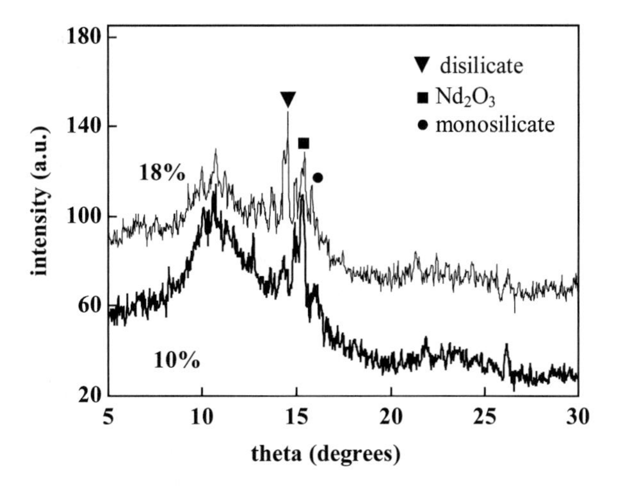

Figure 29.6. XRD of glass-ceramics loaded with neodymium (10 and 18 wt%).

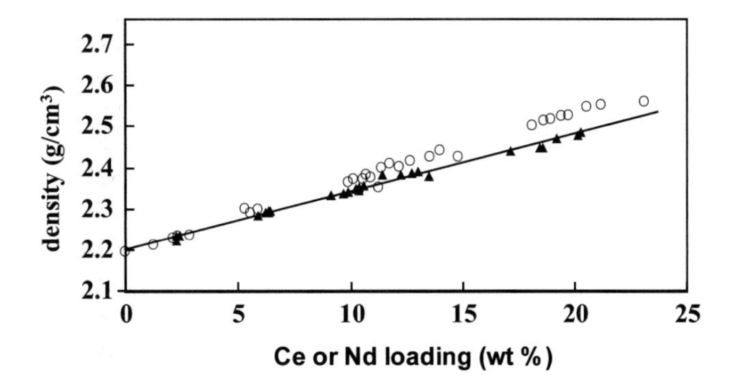

Figure 29.7. Density of the set of glass-ceramics from aerogels loaded with Nd (*open circle*) and Ce (*filled triangle*) versus the loading percent.

$$\rho = 2.2\,(\%\mathrm{SiO_2}) + 7.3\,(\%\mathrm{Nd_2O_3}) \tag{29.3}$$

and

$$\rho = 2.2\,(\%\mathrm{SiO_2}) + 7.1\,(\%\mathrm{CeO_2}). \tag{29.4}$$

Because of the close density of Nd_2O_3 (7.3 g/cm^3) and CeO_2 (7.1 g/cm^3) the two straight lines are not distinguishable.

Thanks to its biphasic composite structure, the cerium loaded glass-ceramics densities closely follow a straight line. Conversely, the bulk density of the Nd loaded materials is systematically higher than the straight line in Figure 29.7 which confirms the multiphase structure. The crystallization of neodymium mono (Nd_2SiO_5) and disilicate ($Nd_2Si_2O_7$) besides neodymium oxide (Nd_2O_3) and silica glass increases the density of the Nd loaded glass-ceramics. These structural differences are the result of the affinity of Ce and Nd ions to Si. In the case of Ce, the formation of a binary glass in a melting process is difficult,

Ce(IV) generally forms crystalline CeO_2 [28, 30]. Moreover, the crystalline phases of cerium silicates like $Ce_2Si_2O_7$ are not stable under 1,400°C and transformed in SiO_2 and CeO_2 [31, 32]. In the case of Nd, glasses with weight loading between 2 and 5% have been obtained [33, 34]. The phase diagram shows also that, besides the oxide (Nd_2O_3), the different Nd silicates (Nd_2SiO_5 and $Nd_2Si_2O_7$) are stable at room temperature [35]. The size of the Ce and Nd domains is in the range 20–100 nm increasing with the loading percentage.

29.6.2. Aqueous Erosion Behavior

In the case of long-lived nuclear wastes, it is important to limit a possible release of actinides if the glass structure is destroyed by an alteration process [36]. We have seen that the Nd glass-ceramics have a structure more complicated than the Ce glass-ceramics. Because of these structures we can expect different resistance to aqueous corrosion. The chemical durability of the glass-ceramics is measured with a conventional Soxhlet device consisting of a boiler containing ultra pure water and a condenser system [16]. A sample monolith was placed in a recipient trough in which condensed steam was flowed continuously. The test was conducted at 100°C, and the sample analyzed after 28 days of leaching. The kinetics of glass corrosion were determined from the analysis of the silica mass loss. The normalized silica loss $NL(SiO_2)$ characterizes the glass network destruction by the glass former dissolution. $NL(SiO_2)$ (g/m^2) was calculated using the following formula:

$$NL(SiO_2) = (C(Si) \times fc)/(\%SiO_2 \times S/V) \quad (29.5)$$

where $C(Si)$ is the Si element concentration in solution (g/m^3), fc is the oxide molar mass ($fc = 2.139$ for silica), $\%SiO_2$ is the mass fraction of SiO_2 in the glass-ceramic, and S/V (m^{-1}) is the ratio of the glass surface (m^2) to solution volume (m^3). It has been explained in the introduction that vitreous silica has a high chemical durability. The mechanisms of alteration of the Nd and Ce loaded glass-ceramics with vitreous silica and the classical nuclear borosilicate glass are now compared. For this study, we have chosen two glass-ceramics loaded with 7 wt% of Nd_2O_3 and of CeO_2. Figure 29.8 shows the normalized SiO_2 loss and the surrogates loss measured on the two kinds of loaded material and on pure SiO_2.

From Figure 29.8, one can calculate the aqueous corrosion rate (V_0) for vitreous silica glass and for the glass-ceramics loaded with the surrogate oxides. The corrosion rate (V_0) for pure silica is equal to 0.015 g/m^2 per day, 100 times lower than the corrosion rate of the usual nuclear waste glasses for which V_0 is equal to 2 g/m^2 per day [37]. V_0 for the glass-ceramics loaded with the Ce and Nd is, respectively, equal to 0.035 g/m^2 per day and to 0.25 g/m^2 per day. This result evidences the improvement of the chemical durability of the glass-ceramics compared to the standard nuclear waste glasses. Thanks to its simple structure, the corrosion rate of the Ce-glass-ceramics is quite close to that of the pure silica, the small difference can be attributed to higher silica dissolution at the interface with the cerium oxide domain, the corrosion could be activated by the presence of thermomechanical stresses [38, 39]. The corrosion rate of the Nd-glass-ceramic is 8 times higher than for pure SiO_2. This result confirms that a part of the glass matrix is made of a neodymium silicate glass which is less durable than SiO_2.

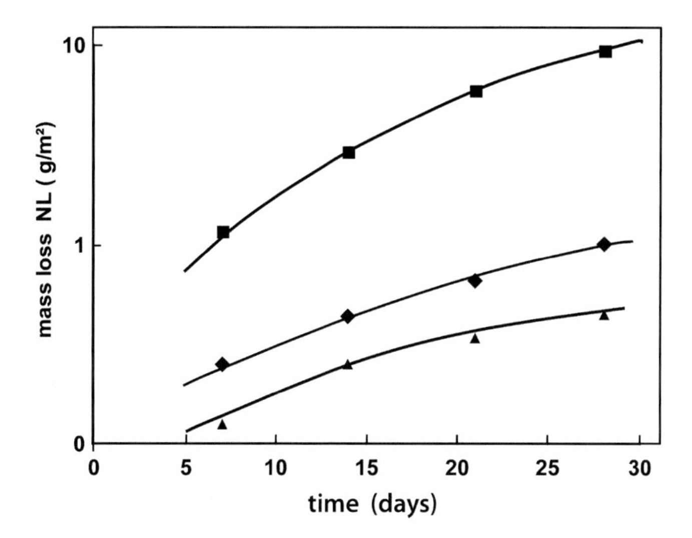

Figure 29.8. Normalized SiO_2 loss measured on the two kinds of loaded material (Nd *filled square*, Ce *filled diamond*) and on vitreous SiO_2 (*filled triangle*) versus time.

The comparison of the dissolution rate of sintered aerogel glass-ceramics with other matrices considered as highly corrosion resistant shows that V_0 of the glass-ceramics loaded with the Ce is of the same order of magnitude as that of the Synroc zirconolite (~0.01 g/m^2 per day) [40, 41] and of one order of magnitude lower than that of the zirconolite-based glass-ceramics (~0.5–1 g/m^2 per day) [11]. Because of the presence of boron (5%) in the glass matrix of host porous borosilicate [5] it is likely that the dissolution rate should be close but higher than the V_0 of the sintered aerogels glass-ceramics.

In order to give a more complete description of the chemical durability, the long-term aqueous corrosion should be characterized by tests under saturation conditions. In this case one measures the Si concentration in the solution as a function of the normalized parameter tS/V, where t is the duration of the test, S being the geometrical surface of the sample, and V the volume of water. In this kind of test, the ratio S/V is high so that the erosion rapidly increases and attains the conditions of saturation in the solution. To perform such test, Ce and Nd glass-ceramics have been crushed and the powder specific surface area was precisely measured by Kr BET technique. The results of Figure 29.9 confirm the close behaviors of the Ce glass-ceramics and silica glass and the differences with the Nd sample.

More interestingly, the results suggest also that for these new matrices a limit of solubility is attained stopping the material alteration, which is not the case for the classical nuclear waste glass for which the process of erosion never stops in water [42, 43]. The dissolution rate of amorphous silica and loaded glass-ceramics is still lower than that of nuclear glass in stagnant, silica-saturated solutions.

Because of their different affinity with silica, the two used surrogates, Nd and Ce, indicate what kind of behavior we can expect when actinides are able to form silicate phases or binary glasses and, on the opposite, when no silicate crystalline or glassy phases can exist. In the latter case, the chemical durability is improved and is close to that of vitreous silica. In an open system, flowing groundwater in which solubility limits are not reached, the leach rate for silica and loaded glass-ceramics used in this comparison is two orders of magnitude lower than the leach rate of borosilicate glasses.

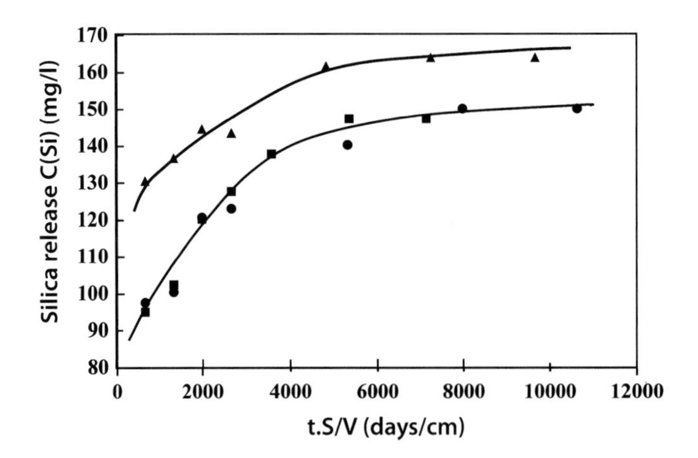

Figure 29.9. Silica released in conditions of saturation for the two kinds of loaded material (Nd *filled triangle*, Ce *filled square*) and on vitreous SiO_2 (*filled circle*) versus tS/V (t is the duration of the test, S the geometrical surface of the sample, and V the volume of water).

29.6.3. Mechanical Properties of the Nuclear Glass-Ceramics

A disadvantage of such a composite structure not discussed so far are the thermomechanical stresses appearing during cooling. Owing to the thermal expansion mismatch, thermal stresses can induce cracks between the crystalline phase and the glass. That is ascribed to the fact that the amorphous matrix material is sintered at a temperature corresponding to a high viscosity and that the Nd ions cannot diffuse easily in the glass structure at such a temperature and during a short cooling time. The mechanical properties of the various glass-ceramics obtained from aerogels are important to validate this process. Indeed, the mechanical behavior of matrices used to contain nuclear wastes should be as high as possible because the fracture of the matrix will lead to an increase of the corrosion rate (increase of the contact surface between the solid and water). Figures 29.10 and 29.11 show the elastic modulus (E) and the rupture strength (σ) of the different glass-ceramics obtained with

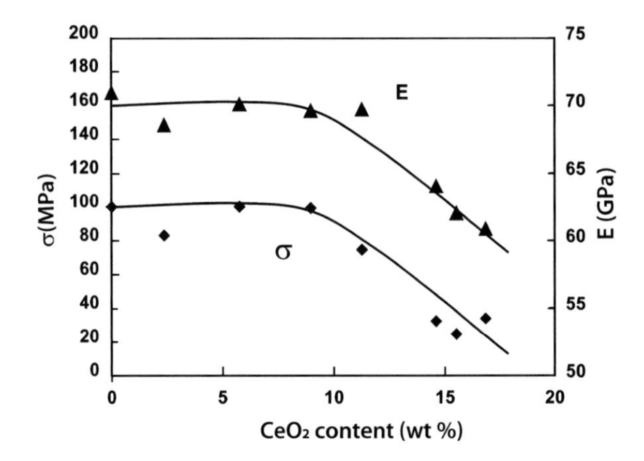

Figure 29.10. Young's modulus E (*filled triangle*) and mechanical strength σ (*filled diamond*) of the Ce-loaded glass-ceramics.

Ce and Nd, respectively. *E* and *σ* have been measured by three-point bending techniques and the reported values are the mean of 4 or 5 data points. Figure 29.10 shows that for a Ce loading higher than 11% the mechanical properties drop. In the case of the Nd-loaded glass-ceramics, the loss of mechanical properties is progressive (Figure 29.11).

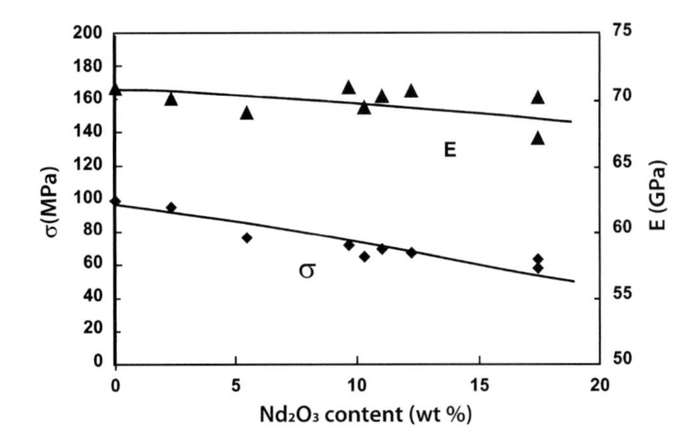

Figure 29.11. Young's modulus *E* (*filled triangle*) and mechanical strength *σ* (*filled diamond*) of the Nd-loaded glass-ceramics.

This weakening of the materials structure is the result of the differences between the thermal expansion coefficients and the elastic modulus of the different compounds present in the glass-ceramics. For example, the thermal expansion coefficient of vitreous silica ($0.5 \times 10^{-6\circ}C^{-1}$ [44]) is 20 times lower than the one of the CeO_2 ($11.5 \times 10^{-6\circ}C^{-1}$) and Nd_2O_3 crystallites ($12 \times 10^{-6\circ}C^{-1}$ [45]). The elastic modulus of the silica glass (73 GPa) [46] is much lower than the elastic modulus of CeO_2 (184 GPa) and Nd_2O_3 (418 GPa). After sintering and during cooling, local stresses can occur at the boundaries of CeO_2 and Nd_2O_3 crystallites, because of their important differences in thermoelastic properties. The net result is the formation of flaws, which weaken the glass-ceramics.

29.7. Conclusion

The disposal of actinides requires effective containment of waste generated by the nuclear fuel cycle. Because actinides (e.g., ^{239}Pu and ^{237}Np) are long lived, they have a major impact on risk assessments of geologic repositories [46, 47]. Current research concerns new specific matrices designed for the immobilization of these long-lived radio nuclides.

For this purpose, glass-ceramic waste forms are good candidates if they consist of actinide crystalline phases homogeneously dispersed in a glassy matrix. Aerogel-based glass-ceramics which contain a long-lived radionuclides phase embedded in a durable silica glassy matrix are expected to be good candidates. These waste matrixes are prepared by controlled sintering of aerogels containing large amounts of surrogates such as Nd_2O_3 and CeO_2. The final structure of the fully sintered material is that of a composite material glass plus the corresponding surrogate oxide (in crystalline form and mixed phases). Thanks to the low sintering temperature process, the liquid state is avoided during the aerogel–glass transformation and the dissolution of the surrogate in the silica glass matrix is difficult. Consequently, the surrogate elements are not included in the glass structure but rather embedded in a glass matrix. High waste loading (20 wt%) can be realized since segregation cannot take place and the solubility limit of the surrogate no longer imposes restrictions.

One aspect of the process validation is the determination of long-term physical and chemical durability which refers to a wide variety of properties: mechanical strength, slow kinetics for corrosion processes, and self-irradiation. In this work, one has shown that owing to its simple structure, the corrosion rate of the Ce-glass-ceramics is low and close to that of the pure silica. More interestingly, the results suggest also that for these new matrices a limit of solubility is attained stopping the material alteration which is not the case for the classical nuclear waste glass.

In addition, the mechanical behavior of such glass-ceramics has also been characterized. It is clear that, because of the differences in the thermal expansion coefficients for high Ce or Nd loaded composites, mechanical stresses exist during the cooling and these stresses could lead to the formation of flaws at the interface between the surrogate oxide and the silica glass. Mechanical characterization, toughness, and Weibull modulus measurements could provide information about the magnitude of the damages.

Concerning the self-irradiation behavior, a previous study [48] has compared the aqueous corrosion in both radiation damaged and annealed nuclear waste glasses doped with curium. The nuclear waste glasses that received a cumulative α-recoil dose of 9×10^{24} α decays/m^3 were leach tested under static conditions. Test results indicate a difference of less than a factor of 2 between annealed and radiation-damaged glass in the release of ^{244}Cm, glass-forming constituents, and simulated waste products. We could conclude that the self-irradiation does not strongly affect the chemical durability.

The process described here combines the flexibility of glass technology, which permits to prepare glasses with a large and continuous domain of composition [49], contrary to crystalline materials, with the flexibility of a sol-gel process, i.e., a great liberty to control and adjust the pore and structural features. The association of these two advantages allows tailoring the physical and chemical properties of the porous solid to attain specific objectives and applications giving rise to superior long-term storage properties.

References

1. Jacquet-Francillon N, Bonniaud R, and Sombret C (1978) Glass a material for the final disposal of fission products. Radiochimica Acta 25: 231–240
2. Ewing RC (2001) Nuclear waste form glasses: the evaluation of very long term behaviour. Mat Tech Adv Perf Mater 16: 30–36
3. Glasser FP (1985) The role of ceramics, cement and glass in the immobilization of radioactive wastes. Br. Ceram. Trans. J. 84: 1–8
4. Donald IW, Metcalfe BL, Taylor RNJ (1997) The immobilization of high level radioactive wastes using ceramics and glasses. J. Mat. Science 32:5851–5887
5. Simmons JH, Macedo P B, Barkatt A, Litovitz TA (1979) Fixation of radioactive waste in high silica glasses. Nature 278: 725–731
6. Oversby VM, Ringwood WE (1982) Leaching studies on Synroc at 95°C and 200°C Rad. Waste Mgmt 2:223–237
7. Lumpkin GR, Ewing RC (1995) Geochemical alteration of pyrochlore group minerals . Amer.Min. 80:732–743

8. McGlin PJ, Hart KP, Loi EH, Vance ER (1995) Ph dependence of the aqueous dissolution rates of perovskite and zirconolite at 90°C.Mater.Res.Soc.Symp.Proc. 353:847–854
9. Hayward P J (1988) The use of glass-ceramics for immobilising high level wastes from nuclearfuel recycling. Glass Technol. 29(4):122–136
10. Loiseau P, Caurent D, Baffier N, Mazerolles L, Fillet C (2001) Development of zirconolite-based glass-ceramics for the conditioning of actinides, , in Scientific Basis for Nuclear Waste Management XXI V, edited by K.P.Hart, G.R.Lumpkin (Mater. Res. Soc. Symp. Proc. 663 : 179–189, Warrendale, PA :
11. Frugier P, Martin C, Ribet I, Advocat T, Gin S (2005) The effect of composition on the leaching of three nuclear waste glasses R7T7,AVM and VRZ. J.of Nuclear Mater. 346;194–207
12. Brinker JF, Scherer GW (1990) Sol-Gel Science, Academic Press , N.Y.
13. Prassas M, Woignier T, Phalippou J (1990) Glasses from aerogels part 1 The synthesis of monolithic silica aerogel. J Mater Sci 24: 3111–3117
14. Woignier T, Phalippou J, and Prassas M (1990) Glasses from aerogels part 2: The aerogel glass transformation. J Mater Sci 25: 3118–3126
15. Zanotto ED (1992) The formation of unusual glasses by sol-gel processing. J Non-Cryst Solids 147&148: 820–823
16. Woignier T, Reynes J, Phalippou J, Dussossoy JL, Jacquet-Francillon N (1998) Sintered silica aerogel: a host matrix for long live nuclear wastes. J Non-Cryst Solids 225: 353–357
17. Lours T, Zarzycki J, Craiewich A, Dos Santos D, Aegerter M, (1987) SAXS and BET studies of aging and densification of silica aerogels. J Non-Cryst Solids 95&96: 1151–1158
18. Sempere R, Bourret D, WoignierT, Phalippou J, Jullien R (1993) Scaling theory and numerical applications of aerogel sintering. Phys Rev Lett 71: 3307–3312
19. Bourret A (1988) Low density silica aerogels observed by high resolution electron microscopy .Europhysics lett 6(8):731–737
20. Woignier T, Phalippou J, Pelous J, Courtens E (1990) Different kinds of fractal structures in Si02 aerogels. J Non-Cryst Solids 121:198–204
21. Woignier T , Fernandez Lorenzo C, Sauvajol JL, Schmidt JF, Phalippou J Sempere R (1995) Raman study of structural defects in SiO_2 aerogels. J Sol-Gel Sci and Techn 5:167–172
22. Reynes J, Woignier T, Phalippou J, 2001. Permeability measurements in composites aerogels: application to nuclear waste storage. J Non-Cryst Solids 285: 323–327
23. Toki M, Miyashita S, Takeuchi T, Kande S, Kochi A (1988) A large-size silica glass produced by a new sol-gel process. J Non-Cryst Solids 100: 479–482
24. Marlière C, Woignier T, Dieudonné P, Primera J, Lamy M, Phalippou J (2001) Two fractal structure in aerogel. J Non-Cryst Solids 285:175–181
25. Scherer GW (1994) Hydraulic Radius and Mesh Size of Gels J Sol-Gel Sci and Techn 1: 285–291
26. Fillet C, Marillet J, Dussossoy JL, Pacaud F, Jacquet-Francillon N, Phalippou J (1997) Titanite and zirconolite glass-ceramics for long-lived nuclear wastes. Environmental Issues and wastes Management Technologies in the Ceramic and nuclear Industries III 87:531-535, 99th Annual Meeting on Ceramics of the American Ceramic Society, Cincinati
27. Advocat T, Fillet C, Marillet J, Leturcq G, Boubals JM , Bonnetier A, (1997) Nd-Doped Zirconolite Ceramic and Glass Ceramic Synthesized by Melting and Controlled Cooling , in Scientific Basis for Nuclear Waste Management XXI, edited by Ian G. McKinley. Charles McCombie (Mater. Res. Soc. Symp. Proc. 506:55–62 Warrendale, PA
28. Lopez C, Deschanel X, Auver CD, Cachia JN, Peuget S, Bart JM, (2005) X-ray absorption studies of borosilicate glasses containing dissolved actinides or surrogates. Phys. Scr 115: 342–345
29. Scherer GW (1977) Sintering of low density glasses: Experimental study. J Amer Ceram Soc 60: 239–243
30. Haire RG, Assefa Z, Stump N (1998) Fundamental science of elements in selected immobilization glasses: Significance for TRU disposal schemes." Mat Res Soc Symp Proc V 506: 153–160
31. Felsche J, Hirsiger W (1969) The polymorphs of the rare earth pyrosilicates RE, Si, O, [RE: La, Ce, Pr, Nd, Sm]. J Less-Common Met 18: 131–37
32. Van Hal H A M, Hintzen H T (1992) Compound formation in the Ce_2O_3-SiO_2 system. J of Alloys and Compounds 179: 77–85
33. Thomas I M, Payne S A, Wilke G D (1992) Optical properties and laser demonstration of Nd-doped sol-gel silica glasses. J Non-Cryst Solids 151:183–190
34. Pope EA, Mackenzie JD (1993) Sol gel processing of Neodymia –silica glass. J Am Ceram Soc. 76(5): 1325–1329
35. Miller RO, Rase DE (1964) Phase equilibrium in the system Nd_20_3-$Si0_2$. J Am Ceram Soc 47(12): 65–654

36. Lutze W, Malow G, Ewing R C, Jercinovic M J, Keil K (1985) Alteration of basalt glasses: implications for modelling the long-term stability of nuclear waste glasses. Nature 314: 252–255
37. Nogues J.L, (1984) PhD thesis « Les mécanismes de corrosion des verres de confinement des produits de fission, Université de Montpellier
38. Bunker R C (1994) Molecular mechanisms for corrosion of silica and silicate glasses. J Non-Cryst Solids 179: 300–308
39. Martin C (2002) Etude du comportement à long terme des vitrocristallin sà base de Zirconolite . PhD thesis, Université de Montpellier 2, France
40. Buck EC, Ebbinghaus B, Backel AJ, Bates JK (1997) characterization of a Pu-Bearing zirconolite –rich synroc . Mat. Res. Soc Sym. Proc. 465:1259–1266
41. Ji H , Rouxel T, Abdelouas A, Grambow B, Jollivet P (2005) Mechanical behaviour of a borosilicate glass under aqueous corrosion. J Am Ceram Soc 88 [11]:3256–3259
42. Berme L O, Björner I K, Bart G, Zwicky H-U, Grambow B, Lutze W, Ewing C, Magrabi C (1990) Chemical corrosion of highly radioactive borosilicate nuclear waste glass under simulated repository conditions. J Mat Res 5:1130–1146
43. Lutze W, Ewing RC , eds (1988) Radioactive Waste Forms for the Future , North–Holland, Amsterdam
44. Bansal NP, Doremus DR (1986) "Handbook of Glass Properties", Academic press
45. Warshaw I, Roy R, (1961) Thermal expansion measurements from nonindexed high temperature X-Rays Powder patterns J Am Ceram Soc,c44(8) 421– 428
46. Verney-Carron A, Gin S, Libourel G (2008) A fractured roman glass block altered for 1800 years in sea water : analogy with nuclear waste glass in deep geological repository. Geochim and Cosmochim Acta 72: 5372–5385
47. Plodinec MJ, Wicks GG, (1994) application of hydration thermodynamics to in situ results. Mat Res Soc Symp Proc, V 303: 145–157
48. Wald JW, Roberts W, Lackey J (2006) Comparison of short term leaching in both radiation damaged and annealed nuclear wastes glasses. J Amer Ceram Soc 67 (4):c69–c70
49. James PF (1988) The gel to glass transition: chemical and microstructural evolution. J Non-Cryst Solids 100: 93–114

Part X

Applications: Biomedical and Pharmaceutical

30

Biomedical Applications of Aerogels

Wei Yin and David A. Rubenstein

Abstract This section highlights a few applications of aerogels in a biological context, as a biomaterial. Some aerogel formulations have been shown to have compatibility with the cardiovascular system and others have been able to induce apatite formation for potential bone growth. Others have provided proof that proteins can be embedded within aerogel samples maintaining their biological functionality, therefore aerogels may be used in a drug delivery system. At this point, more work is needed to determine how aerogels can be applied to the biological systems; however, with the improvements in aerogel processing along with a better understanding of biomaterials the use of aerogels in biological applications will be significant in the future.

30.1. Introduction

Synthetic materials have been used in biological environments for approximately 3,000 years. As new materials are developed and new material processing methods are designed, the biological effect of materials has been continually investigated to determine whether or not the material has any potential to be used in the human body. Over the past 100 years, our understanding of biological processes combined with our ability to precisely fabricate materials has led to the fast growing field of biomaterials. It was estimated that in the year 2000, about 20 million individuals had medical devices implanted, incurring an approximate annual cost of $300 billion (including hospitalization and surgical costs) [1]. The use of novel biomaterials in medical applications has significantly enhanced the longevity of patients. However, the failure of biomaterials can lead to symptoms, diseases, side effects, or sudden patient death, which leaves room for improvement in the current biomaterial properties and applications.

Silica aerogel is a well-known lightweight solid. Even though it was invented in 1931, it only started to draw interest from materials scientists about 15 years ago, due to the improvements in the use of sol-gel processes to expedite the synthesis time (Chap. 1). However, traditional aerogels have never been used as a biomaterial. Along with the significant improvement in the mechanical strength, modern aerogels have demonstrated

W. Yin (✉) and D. A. Rubenstein • School of Mechanical and Aerospace Engineering, Oklahoma State University, Stillwater, OK 74078, USA
e-mail: wei.yin@okstate.edu; david.rubenstein@okstate.edu

M.A. Aegerter et al. (eds.), *Aerogels Handbook*, Advances in Sol-Gel Derived Materials and Technologies, DOI 10.1007/978-1-4419-7589-8_30,

certain properties required for biomedical applications. Thus, investigations on the biocompatibility of aerogels have been recently initiated. This contribution highlights some of the current work being carried out to extend aerogel applications in the biomaterials field (Chap. 31).

Cardiovascular diseases are currently the number one killer in the Western world and there is a significant need for new devices to augment these conditions. Tissue engineering is currently hindered by the design of new porous but strong biocompatible materials that can be used to promote cell growth throughout the material. These new materials can act as a template for new cell growth. New materials are also needed to design targeted-controlled release systems that can be used to treat various diseases, including genetic diseases. In the following sections, the use of various aerogel materials as cardiovascular implantable devices, as tissue engineering scaffolds, and as drug delivery vehicles is described.

Before we get into the details of the above-mentioned applications, we need to keep in mind that for any implantable device, there can be unwanted biological responses, which can be initiated by the device itself, and unwanted material responses, which can be initiated by the biological system. In general, there are four common biological changes that occur in response to foreign material implantation. These include the inflammatory response, coagulation/hemolysis, the allergic response, and the carcinogenic response. Inflammation is a nonspecific response to tissue trauma, infection, local cell death, or intrusion of a foreign material. Inflammation is typically characterized by redness, swelling, pain, and heat. Coagulation and hemolysis are nonspecific responses that occur when there is damage to blood vessels or when blood comes in contact with a foreign material. Coagulation is the activation and aggregation of platelets, and hemolysis is the destruction of red blood cells. The allergic response is a specific response carried out by the immune system when there is a recognized foreign particle within the body. This particle is typically attacked by white blood cells, T-cells, B-cells, and antibodies to be removed. Certain implantable materials could express or degrade into compounds that are recognized by this system. The last response occurs when an implantable material causes damage to a host cell's DNA that induces cancer formation. Cancer is characterized by the uncontrollable cell growth and spreading throughout the body. Material responses can include swelling or leaching, corrosion, wear, deformations, and the formation of new biological compounds (i.e., organometallic compounds, which are a combination of a metal with an organic part). To investigate the potential applications of aerogels in the biomedical field, all the responses discussed earlier need to be considered.

30.2. Aerogels Used for Cardiovascular Implantable Devices

Cardiovascular diseases and complications are the leading cause of death in the United States, which creates a large demand for biomaterials suitable for cardiovascular implantable devices. In general, in order for a biomaterial to be usable in a certain organ system in the human body, the material must be compatible with that system over the device's entire service life period. For a cardiovascular system, this includes compatibility with the patient's blood, especially with respect to immune responses and hemostatic/thrombotic responses and compatibility with blood vessels, if the device is in contact with them.

When in contact with blood, a foreign material of the implantable device may induce a rapid deposition or adsorption of plasma proteins onto the surface, followed by the adhesion and activation of platelets [2]. Plasma protein deposition onto a material may trigger acute immune responses, characterized by the attack of leukocytes [3]. Platelet activation and coagulation can lead to clot formation or thrombosis. A clot that forms on the surface of a blood implantable device can severely affect the performance of the device. Furthermore, inflammation may also be triggered by foreign material–blood contact [4]. An ideal blood implantable polymeric material should not induce plasma protein adhesion/adsorption or trigger an acute immune response or platelet activation. It should also not initiate cell death for those cells that it is in contact with.

As one of the most commonly known blood implantable devices, artificial heart valves are used to replace diseased or damaged heart valves. There are approximately 300,000 artificial heart valves implanted every year, about 60% of which being mechanical heart valves. Tissue valves are the most common alternative to mechanical heart valves, accounting for the remaining prosthetic heart valve market. Polymeric heart valves have been investigated for more than 10 years, though their performance is far from satisfactory. They are usually made of polyurethane and in the shape of a native heart valve (with soft leaflets). Their overall success rate is fairly low due to numerous problems, including material degradation and poor mechanical strength.

The popularity of mechanical heart valves is closely related to their mechanical strength, durability, and good hemodynamic performance. However, valve-induced *cavitation* (vaporous bubble formation due to a transient pressure drop below the liquid vapor pressure) may curb the performance of mechanical heart valves. *Cavitation* occurs during valve closure when the fluid mass is squeezed through the valve opening. The fluid reaches very high velocities and is subjected to a large pressure drop over a small distance. When vaporous bubbles inside the fluid collapse near the blood vessel wall, the wall boundary is impacted with very high local pressures, which can result in elevated shear stress and lead to platelet activation. The transvalvular pressure gradient is critical to *cavitation* and is directly related to the leaflet inertia [5]. Employing a low density material for leaflets may be the potential solution to *cavitation* problems. In summary, materials suitable for mechanical heart valves must be blood compatible, blood vessel wall endothelial cell compatible, strong, durable, and light. With a significantly improved mechanical strength, new aerogels may be considered as candidate materials for artificial heart valve leaflets, where high mechanical strength and low inertia are required to enhance valve durability and minimize the occurrence of cavitation. Furthermore, aerogels are generally inexpensive to formulate, which may benefit many potential prosthetic heart valve recipients.

Recent work started by Yin and Rubenstein at Oklahoma State University has aimed to investigate the compatibility of various formulations of aerogels within the cardiovascular system. The type of aerogel used in their initial studies was a surfactant-templated polyurea-nanoencapsulated macroporous silica aerogel (referred to as cross-linked aerogel: *X-MP4-T045*) [6] (Chap. 13). *X-MP4-T045* was made using tetramethylorthosilicate (*TMOS*) and tri-block copolymer $PEO_{20}PPO_{70}PEO_{20}$ (*Pluronic P123*) following the Leventis approach [7]. It had a mass density of 0.66 g/cm^3 and a porosity of 50.0%. The average pore size is approximately 5 μm. Its effects on plasma anaphylatoxin generation (*C3a* is a very important plasma inflammation marker), plasma protein adsorption, platelet activation, coagulation, and vascular endothelial cell normal functions were investigated.

First, it was determined whether or not aerogel samples would induce platelet activation upon contact. To investigate this, washed platelets were incubated with aerogel samples for up to 6 h. The expression of phosphatidylserine (*PS*) and P-selectin (*CD-62P*) was measured to determine how platelets responded to aerogel contact. Platelets are a critical component of the hemostatic system and play a salient role in clot formation.

Platelets undergo two parallel processes within the body: activation and aggregation. Platelet activation involves the expression of the negative phospholipid, phosphatidylserine, along with the release of multiple procoagulant proteins. Active platelets participate in the coagulation cascade, which results in stable fibrin formation. Platelet aggregation involves the expression of many adhesion molecules and the end product is stable platelet–platelet adhesion or platelet–endothelial cell adhesion. The results demonstrated that contact with aerogel under static conditions did not trigger platelet activation and was reduced as compared to calcium ionophore (*A23187*) activated platelets (Figure 30.1). This indicates that this particular aerogel formulation (*X-MP4-T045*) was compatible with platelets.

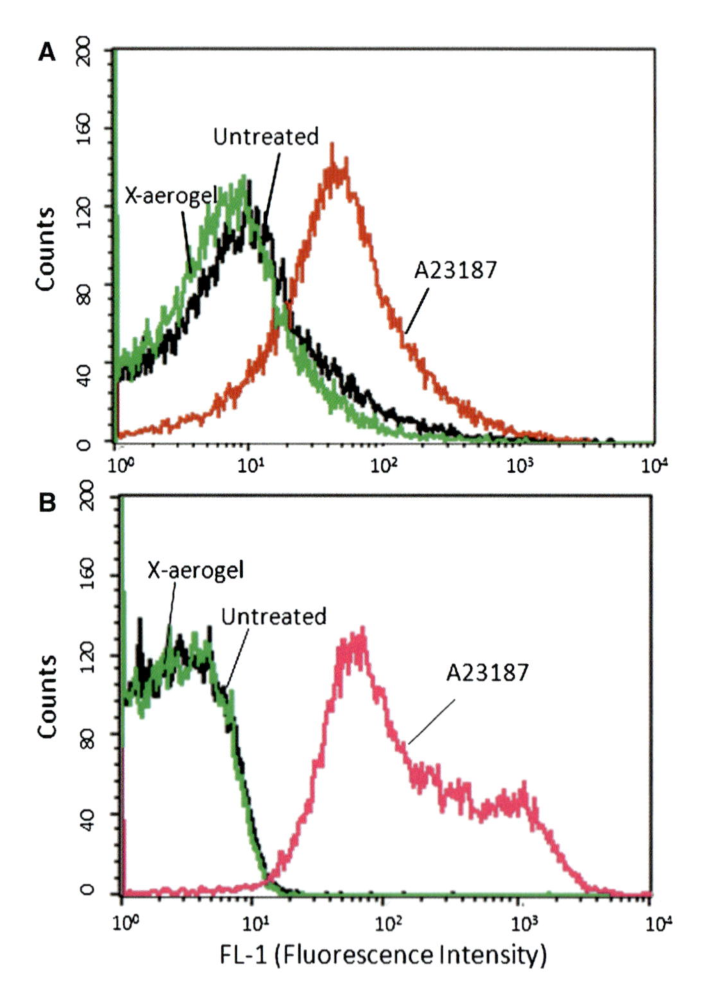

Figure 30.1. There was no significant difference in CD62P **A.** and phosphatidylserin (*PS*) **B.** expression on untreated and X-aerogel treated platelets after 6-h incubation, while A23187 treatment induced a much higher CD62P and *PS* expression on washed platelets.

To answer the question if platelet normal functions are altered by this particular aerogel, platelet activation and aggregation were monitored after aerogel contact using the modified prothrombinase assay and optical platelet aggregation. For aggregation experiments, platelets were incubated with aerogel samples and then antagonized with thrombin receptor agonist peptide 6 (*$TRAP_6$*) or *ADP*. Within 6 h, aerogel incubation did not enhance or impair the normal aggregation process (Figure 30.2A). Using the modified prothrombinase assay [8], aerogel incubation did not enhance or inhibit platelet participation in the coagulation cascade (Figure 30.2B) measured through thrombin generation. Combined, these data suggest that surfactant-templated polyurea-nanoencapsulated macroporous silica aerogel is compatible with platelets [6].

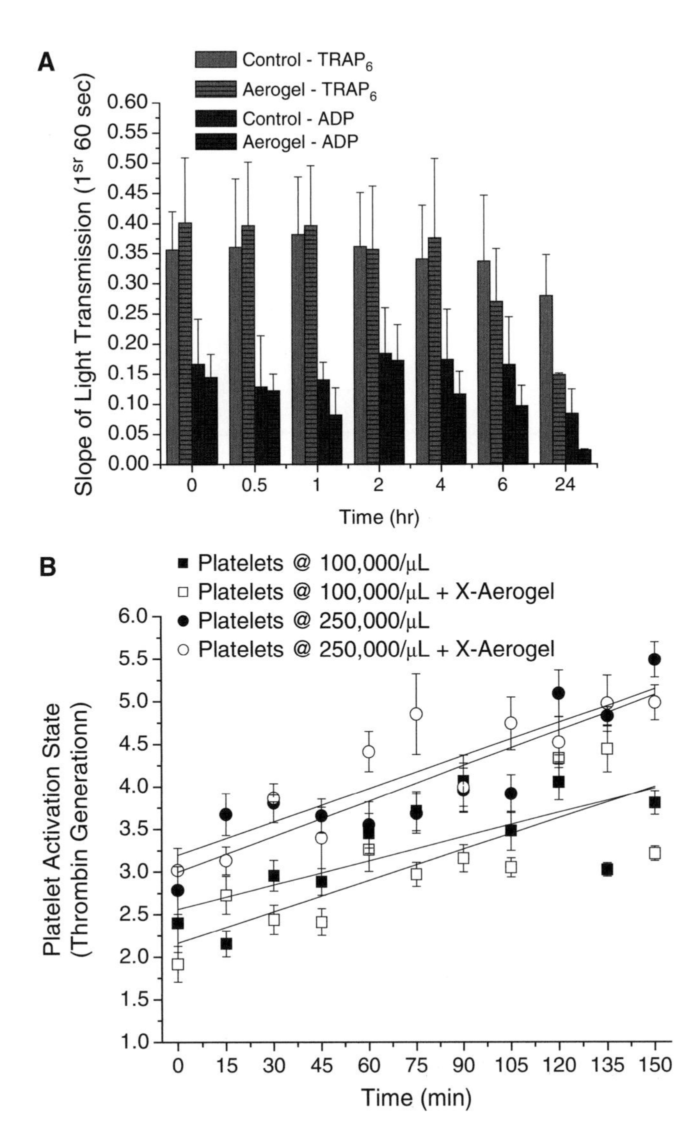

Figure 30.2. **A.** The effect of X-aerogel on platelet aggregation rate towards $TRAP_6$ and ADP (Data presented as mean + standard error of the mean). **B.** The effect of X-aerogel on platelet participation within the coagulation cascade, as measured by thrombin generation (Data presented as mean ± standard error of the mean). Different platelet concentrations were used to determine if enhanced platelet–platelet contact affects the biocompatibility.

The complement system is an important contributor in inflammation, whose activation may be triggered after foreign material contact with blood. Complement activation can lead to the generation of physiologically relevant levels of *anaphylatoxins*, *C3a* and *C5a*, which support the recruitment of leukocytes and can enhance vascular wall permeability. As the major complement proteins involved in inflammation, *C3a* and *C5a* can induce production and release of cytokines and cause cell death. They are usually considered biomarkers for inflammatory conditions. To determine if cross-linked aerogel *X-MP4-T045* would trigger immune responses, the aerogel samples were incubated with normal human plasma at room temperature for up to 24 h. The aerogel induced plasma anaphylatoxin level was measured. The results indicated that aerogels, as a foreign material, induced acute inflammatory responses and led to *C3a* level increase. However, this increase was not statistically significant, compared to that observed with control plasma (plasma that was not treated with aerogels) (Figure 30.3).

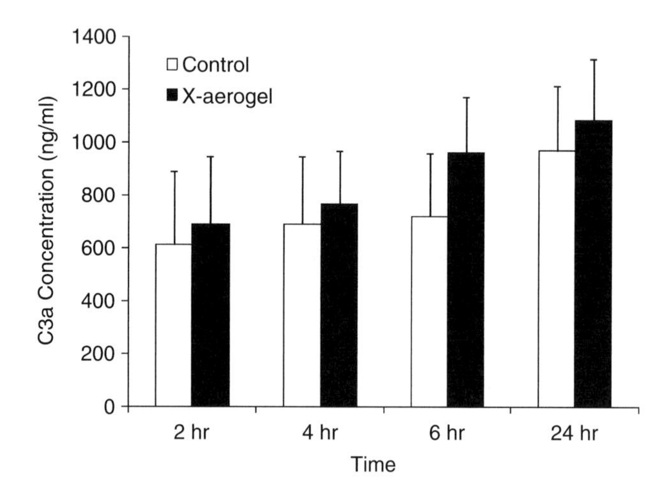

Figure 30.3. Generation (ng/ml) after plasma incubation with X-aerogel for 2, 4, 6, and 24 h. All data were presented as mean + standard error of the mean ($n = 6$). No significant difference was detected between X-aerogel treated and untreated plasma.

To further investigate the compatibility of cross-linked aerogel *X-MP4-T045* with the cardiovascular system, human bone marrow microvascular endothelial cells and human umbilical vein endothelial cells were cultured in the presence of aerogel samples. Endothelial cells are the cell type that lines the interior of all blood vessels and any blood implantable device would come into contact with these cells. It is required for a material used in blood implantable device that after its incubation with endothelial cells, there should be no change in endothelial cell function, growth, viability, or migration. Endothelial cell culture parameters were investigated using a *live/dead cell cytotoxicity assay*, which consists of calcein and ethidium. After 5 days of endothelial cell culture with aerogel, there were no differences in cell culture parameters (Figure 30.4). Combining all the data regarding cardiovascular compatibility, it has been shown that crosslinked aerogel *X-MP4-T045* is compatible with the inflammatory system, platelets, and endothelial cells [6]. Current work is being carried out on whether or not this type of aerogel is compatible with red blood cells.

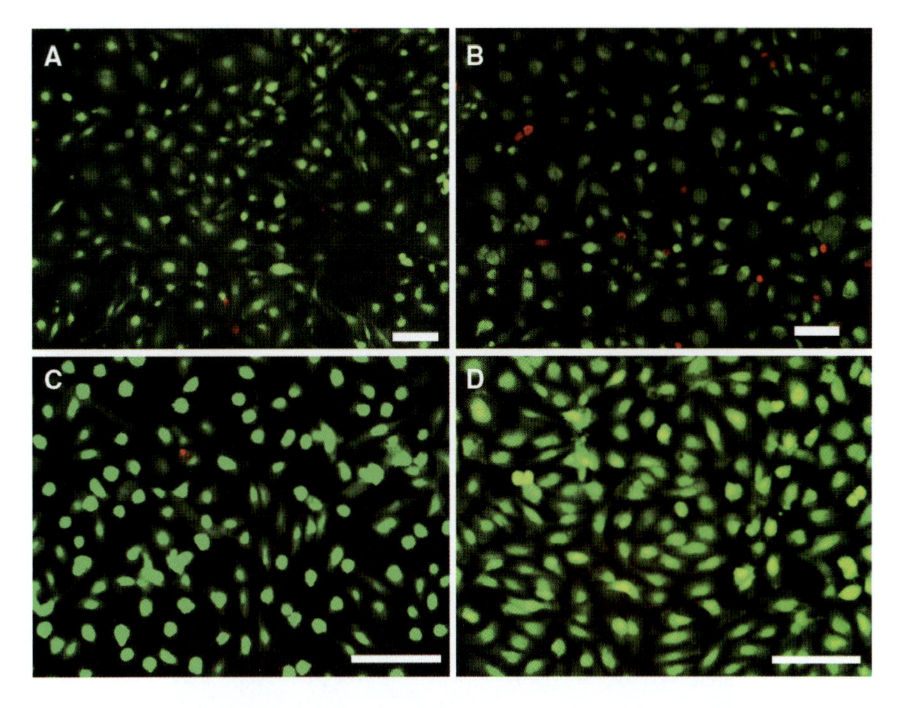

Figure 30.4. Digital images of bone marrow microvascular endothelial cells (Panel **A/B**) or human umbilical vein endothelial cells (Panel **C/D**) cultured without X-aerogel samples (Panel **A/C**) or with X-aerogel samples (Panel **B/D**). Cells were stained with calcein (*green*, live) or ethidium (*red*, dead). All scale bars are 100 μm.

However, with all materials, slight changes in the material composition and/or the processing of the material can drastically affect the compatibility of the material within a biological system of interest. For instance, a chitosan–silica hybrid aerogel is not compatible with red blood cells and induces a significant amount of hemolysis [9] (Chap. 18). This is interesting because silica aerogels are compatible within the cardiovascular system and chitosan, processed in other forms, is a common biomaterial for cardiovascular devices. Therefore, it is important to classify the compatibility of individual aerogel formulations within the biological system. There are a number of research groups currently interested in determining the compatibility of various aerogels (silica, polymeric) within the cardiovascular system with the long-term goal of fabricating cardiovascular implantable devices, with the majority of the current work focused on determining the compatibility of different aerogel samples within the vascular system.

Our main interest is the formation of novel mechanical heart valve leaflets composed of a functionalized cross-linked silica aerogel. As discussed previously, the silica aerogel of interest has a very low bulk density combined with a high mechanical stiffness, which makes it ideal for a heart valve leaflet. These properties would minimize *cavitation* formation and would increase the working life time of the valve. Current work is focused on designing and fabricating leaflets and leaflet housing for future use in cardiovascular implant system. Figure 30.5 depicts a flow chamber with two prototype aerogel monoleaflet heart valves installed. The two valves are located on the opposite sides of the flow chamber and are placed in opposing directions to control the flow direction under dynamic conditions. The flow chamber (made of Teflon) is connected to a peristaltic pump which can generate

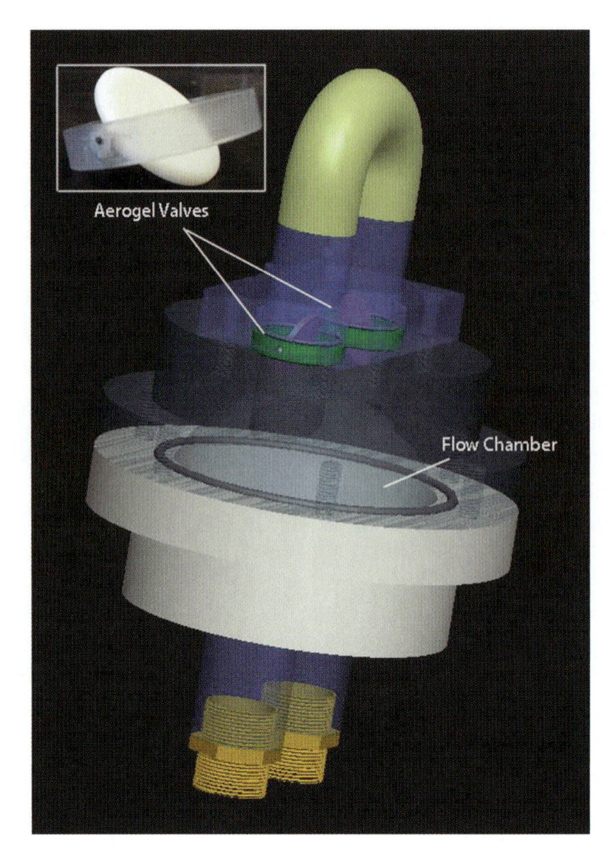

Figure 30.5. Proengineering drawing of a flow chamber with two prototype aerogel monoleaflet heart valves. A picture of the prototype monoleaflet valve is shown in the **left top corner**.

physiological flow conditions, mimicking that of the heart. The interaction between the prototype aerogel heart valve and blood/plasma under dynamic conditions will be tested in this flow chamber. Preliminary studies indicated that neither the prototype aerogel valves nor the flow chamber itself activate or damage blood cells under static conditions.

30.3. Aerogels as Tissue Engineering Substrates

Tissue engineering is concerned with the design and manufacturing of replacement tissues and organs. In order to successfully tissue engineer an organ, it is required that the tissue that is being fabricated be housed within a biocompatible *scaffold*. These scaffolds act to support and promote the tissue growth. There are two major approaches in tissue engineering scaffolds. The first one is to have the scaffold mimic the native extracellular matrix of the tissue that is being formed. With this approach, the forming tissue would recognize native topographies and the tissue would not need to form its own extracellular matrix. The *scaffold* provided would be used during the entire lifetime of the engineered tissue and hence would need to be compatible over the entire lifetime. The second approach is to fabricate a *scaffold* which promotes cell infiltration but is also degradable. Using this approach, cell growth into tissues would be promoted within the scaffold but then the cells would need to degrade this scaffold and fabricate their own extracellular matrix. Both approaches have distinct advantages and disadvantages and it has recently become clear that the choice of approach is dictated by the type of tissue which is being fabricated.

Aerogels have various physical properties which make them ideal candidates as tissue engineering *scaffolds*. The first is a high porosity. The native extracellular matrix is on the order of 80% porous and this is difficult to mimic with many biomaterials. The second property is good mechanical strength (Chap. 22). Most biomaterials that are highly porous are very weak mechanically (i.e., electrospun scaffolds). This typically hinders cell growth and migration throughout the tissue engineered scaffolds because the mechanical stiffness of the native extracellular matrix is on the order of 100 MPa. Most aerogel formulations can easily match this property. The last critical parameter is the chemical composition. With a better understanding of materials science, it has become easy to fabricate aerogel samples from various materials. Many of these materials have acceptable biocompatibilities in their polymer form, which would suggest that in the form of an aerogel the compatibility should be acceptable as well, since biocompatibility is typically determined by chemical composition and degradation products. However, the compatibility testing of these materials must be conducted one by one.

Surfactant-templated polyurea-nanoencapsulated macroporous silica aerogels (*X-MP4-T045*) are compatible with endothelial cells, suggesting they may be used as a scaffold for vascular tissue engineering. However, the formulation of the scaffold would need to be changed to allow for the migration of endothelial cells throughout the *scaffolds*. The chitosan–silica hybrid aerogel, which showed a high value of *hemolysis*, also showed a low value for *cytotoxicity* [9]. While it was not reported which cell type was investigated, it suggests that if cytotoxicity is the primary concern for biomaterial scaffolding, this chitosan–silica aerogel may be considered as a suitable tissue engineering substrate.

Bone replacements are currently one of the most investigated tissue engineered product. These replacements can be used for patients with severe osteoporosis or a compound break of one of the load bearing bones. Bone has a very high mechanical stiffness in compression (Chap. 22) but is also highly porous, which demonstrates two important analogies to aerogels. Bone is also a very dynamic tissue which is constantly undergoing remodeling. Therefore, any scaffold to be used for bone tissue engineering needs to replicate this property. Recently, the group of Toledo-Fernandez has investigated the bioactivity of wollastonite aerogels (composites), which matches bone in mechanical properties and porosity, within simulated body fluid [10]. After 25 days of incubation within such a fluid, a uniform apatite layer is formed on the surface of the aerogel. This layer is the most common additive to bone implantable devices (i.e., cobalt–chromium–molybdenum implants) to promote bone cell growth into the scaffold. Due to the formation of this layer on the aerogel surface, it makes it an ideal candidate for bone tissue engineered scaffolds. The next steps for real bone tissue growth would be to determine whether or not bone cells (osteocytes, osteoblasts, and osteoclasts) are compatible with the apatite-coated aerogel samples. Also, it would need to be determined whether or not the bone cells could then remodel the aerogel and incorporate into this material. If this were the case, it would be possible for the bone cell remodeled aerogel to be implanted into a patient.

It is also important to note that highly porous scaffolds are very desirable in tissue engineering to promote cell migration. Although the tissue compatibility of most aerogels has not been extensively studied, it is worthwhile to mention that highly porous (>95%) aerogel scaffolds can be formed [11]. These scaffolds had a relatively low pore diameter

(<1 μm), but with different process parameters, these scaffolds may be able to be fine tuned to be useful for tissue engineered applications, especially if the pore size could be increased. At a pore diameter of less than 1 μm, cells would not be able to permeate the aerogel sample. However, with increasing pore size, on the order of a 5–10 μm diameter, cells would begin to be able to invade the scaffold and potentially use it as a template for cell growth.

30.4. Aerogels as Drug Delivery Systems

The last major biomedical field that aerogels have been applied to is drug delivery systems. Not much work has been done yet within this field, but what has been investigated shows promising uses of aerogels as drug delivery systems (Chap. 31). Drug delivery is primarily concerned with administering a drug to a targeted region within the body. As an example, aspirin is very effective as a pain reliever but there are many unwanted side effects associated with aspirin treatment. A perfect drug delivery system would locate the region of the body which needs the drug, would deliver the drug to that region, and would not affect any other locations within the body. In this way, side effects will be significantly reduced.

There are a few principles that need to be discussed regarding drug delivery systems before we will show how aerogels can be used as a drug delivery system. When designing a drug delivery system, the first critical parameter to consider is the desired release profile of the drug. In general, there are three types of release profiles: a *bolus release*, a sustained release, and a controlled release. The *bolus release* is the easiest system to design and this would include the complete release of the entire drug payload within a short amount of time (<30 min). Any prescription that states, "take x pills every y hours and not to exceed z pills in 24 h" is an example of a bolus release system. Sustained release systems, generally have one dose that lasts for at least 4 h. An example of this medication is the melatonin release system that aids in sleeping. One dose of this medication lasts for approximately 8 h, so that the patient can sleep throughout the entire night. The last system is a controlled release drug delivery system and this is some combination of a bolus release system and a sustained release system. These systems are also known as smart release systems, because they can recognize that a dose is needed and the delivery system releases an appropriate dose within a short amount of time. An example of this type of system would be the smart insulin delivery systems that sense the levels of insulin within the blood stream and release insulin accordingly. The second critical parameter for drug delivery systems is the type of release mechanism used. For instance, a pill can be designed that completely dissolves within the body (can be *bolus* or sustained) or the pill can make use of an osmotic pump to eject the drug from its reservoir. Differences in these systems would arise due to different polymers that the drug can be embedded in.

There has been some work documenting that aerogel samples can be prepared with a protein payload encapsulated within the aerogel [12]. In this work, a red fluorescent protein was loaded into a silica aerogel. On 5 subsequent days after the aerogel preparation, the stability of the encapsulated protein was examined by quantifying the fluorescence intensity. After 24 h in the aerogel formulation, there was a drop in intensity (to ~85%), which then remained constant for the remaining of the observation time. Although there was no compatibility testing associated with this work and the protein investigated has no immediate biological applications, this study was a proof of principle that a protein can be embedded within aerogel samples and more importantly, the protein was viable after material processing. Using a similar method, there is a high potential to employ aerogels in drug delivery systems.

A second study investigated the immobilization of lipase (a protein that breaks down lipids) onto methyl-modified silica aerogels [13]. This work showed that a protein can be placed on the surface of aerogel samples and that this protein remains partially active after immobilization (there was an approximate 50% reduction in enzyme activity). While this is a significant reduction in activity, it suggests that functional groups can be placed onto aerogel samples and these groups can retain some biological activity (perhaps due to the random nature of this adsorption). In drug delivery devices, this would be critical for targeted delivery systems. In these types of systems, a functional group is added to a delivery system which typically prevents drug release until some condition is met. For instance, such a functional group could be sensitive to pH and would prevent an oral drug from being released within the stomach where the drug could potentially be degraded. More work is needed within this research field, but again, this report acts as a proof of principle for protein immobilization onto aerogels, which can later be released in a controlled manner.

The TAASI Corporation (Delaware, OH, USA) has begun to investigate the development of their *Pristina*™ *aerogels* for targeted controlled release of cancer pharmaceuticals for the potential application of a drug delivery system to treat cancer. In their system, aerogel particles were coated with ligands that would target cancerous cells. Upon ligand–receptor binding, pharmaceutical agents (melitin) loaded within the aerogel particles would be released through a specific pore structure within the aerogel. In two cell culture experiments, these loaded aerogel particles were able to kill 97% of the targeted cells and only 1% of the nontargeted cells, which did not express the specific ligands receptor. The efficacy of this product needs to undergo more testing, such as a coculture system where there are two (or more) cell types that either do or do not express the receptor of interest.

30.5. The Future of Aerogels in Biomedical Applications

This contribution has discussed the few reports that have investigated aerogels in a biological context. Some aerogel formulations have demonstrated good compatibility with the cardiovascular system and others have been able to induce apatite formation for potential bone growth. Others have provided evidence that proteins can be embedded within aerogel samples and maintain their functionality after protein removal. At this point, more work is needed to determine how aerogels can be applied to various biological systems, however, with the improvements in aerogel processing along with our better understanding of biomaterials; it is believed that the use of aerogels in biological applications will be significant in the future. As discussed earlier, there is a direct link for silica aerogels to be used as a cardiovascular implantable device. The potential for aerogels as a tissue engineering scaffold and a drug delivery system was also discussed. There is also a potential to adapt aerogels to be used as filters in extracorporeal devices, due to their high porosity, nonfouling behavior, and ability to be functionalized. Aerogel samples can potentially be used as dental substitutes and can be used as components in many other implantable systems. However, before that step is taken, the compatibility of aerogel samples must be confirmed under application relevant conditions.

30.6. Conclusion

Modern cross-linked aerogels have significantly improved mechanical properties. They are strong and more durable, while they maintain their traditional characteristics of being light and porous. The findings discussed here indicated that by adjusting chemical components and manufacturing procedure, aerogels can be made suitable for many biomedical applications. More is about to be explored and we believe aerogels will have a promising future in the biomaterials market.

References

1. Lysaght MJ and O'Loughlin JA (2000) Demographic scope and economic magnitude of contemporary organ replacement therapies. ASAIO J 46:515–521
2. Andrade JD and Hlady V (1986) Protein Adsorption and Materials Biocompatibility - A Tutorial Review and Suggested Hypotheses. Advances in Polymer Science 79:1–63
3. Wu Y and Meyerhoff ME (2008) Nitric oxide-releasing/generating polymers for the development of implantable chemical sensors with enhanced biocompatibility. Talanta 75:642–650
4. Burd J, Noetzel V and Tamerius J (1993) Rapid testing of biomaterials for complement activation using in vitro complement immunoassays. 19th Annual Meeting of the Society of Biomaterials, Birmingham, Alabama
5. Maines BH and Brennen CE (2005) Lumped parameter model for computing the minimum pressure during mechanical heart valve closure. J Biomech Eng 127:648–655
6. Yin W, Venkitachalam SM, Jarrett E, Staggs S, Leventis N, Lu H and Rubenstein DA (2010) Biocompatibility of surfactant-templated polyurea-nanoencapsulated macroporous silica aerogels with plasma platelets and endothelial cells. J Biomed Mater Res A 92A:1431–1439
7. Leventis N, Mulik S, Wang X, Dass A, Patil VU, Sotiriou-Leventis C, Lu H, Churu G and Capecelatro A (2008) Conformal polymer nano-encapsulation of ordered mesoporous silica monoliths for improved mechanical properties. Journal of Non-Crystalline Solids 354:632–644
8. Jesty J and Bluestein D (1999) Acetylated prothrombin as a substrate in the measurement of the procoagulant activity of platelets: elimination of the feedback activation of platelets by thrombin. Anal Biochem 272:64–70
9. Ayers MR and Hunt AJ (2001) Synthesis and properties of chitosan-silica hybrid aerogels. Journal of Non-Crystalline Solids 285:123–127
10. Toledo-Fernandez JA, Mendoza-Serna R, Morales V, de la Rosa-Fox N, Pinero M, Santos A and Esquivias L (2008) Bioactivity of wollastonite/aerogels composites obtained from a TEOS-MTES matrix. J Mater Sci Mater Med 19:2207–2213
11. Gavillon R and Budtova T (2008) Aerocellulose: new highly porous cellulose prepared from cellulose-NaOH aqueous solutions. Biomacromolecules 9:269–277
12. Li YK, Chou MJ, Wu TY, Jinn TR, Chen-Yang YW (2008) A novel method for preparing a protein-encapsulated bioaerogel: using a red fluorescent protein as model. Acta Biomater 4:725–732
13. Gao S, Wang Y, Wang T, Luo G and Dai Y (2009) Immobilization of lipase on methyl-modified silica aerogels by physical adsorption. Bioresour Technol 100:996–999

31

Pharmaceutical Applications of Aerogels

Irina Smirnova

Abstract At present, an interest has grown in the field of biocompatible aerogels and composite aerogel materials in life science, especially in biomedical and pharmaceutical applications. Due to their large surface area, open pore structure, and biocompatibility, aerogels are very promising candidates for drug delivery systems. The use of both inorganic and organic aerogels as carriers for pharmaceutically active compounds is discussed in this chapter. Both the stability and the release kinetics of the drug can be significantly improved by loading them into aerogels. First attempts to prepare semisolid and solid pharmaceutical formulations have been made. Furthermore, aerogels can be used as a host matrix for bioactive compounds (enzymes and proteins), which improve or enable their performance. Taking into account all research activities in the area of aerogels, a number of promising pharmaceutical applications can be expected in future.

31.1. Introduction

Silica aerogels were used in daily life products since 1960s, when Monsanto's aerogels were introduced as additives for cosmetics and toothpaste. At present, an interest has grown in the field of biocompatible aerogels and composite aerogel materials in life sciences, especially for biomedical and pharmaceutical applications. In principle, biocompatible aerogels can be made from any organic compound, whose properties such as toxicity and biodegradation are suitable for the given application. In the case of aerogel composites, the final material consists of an aerogel matrix and one or more additional phases (of any composition or scale), which influence the properties of the final product. Thus, at least one phase has a physical structure with dimensions in the order of nanometres (the particles and pores of the aerogel). The aerogel composites can be prepared in two different ways: through the addition of a target compound during the sol–gel process (Figure 31.1A) and by post-treatment of the dried aerogels (Figure 31.1B), for example, by adsorption, vapour phase deposition, or reactive gas treatment.

The first method (a) is attractive by its simplicity and flexibility, since virtually every material can be easily added to the sol solution before gelation. This added material can be of

I. Smirnova • Technische Universität Hamburg-Harburg, Institut für Thermische Verfahrenstechnik-V8, Eißendorferstr. 38, 21073 Hamburg, Germany
e-mail: irina.smirnova@tu-harburg.de

M.A. Aegerter et al. (eds.), *Aerogels Handbook*, Advances in Sol-Gel Derived Materials and Technologies, DOI 10.1007/978-1-4419-7589-8_31,

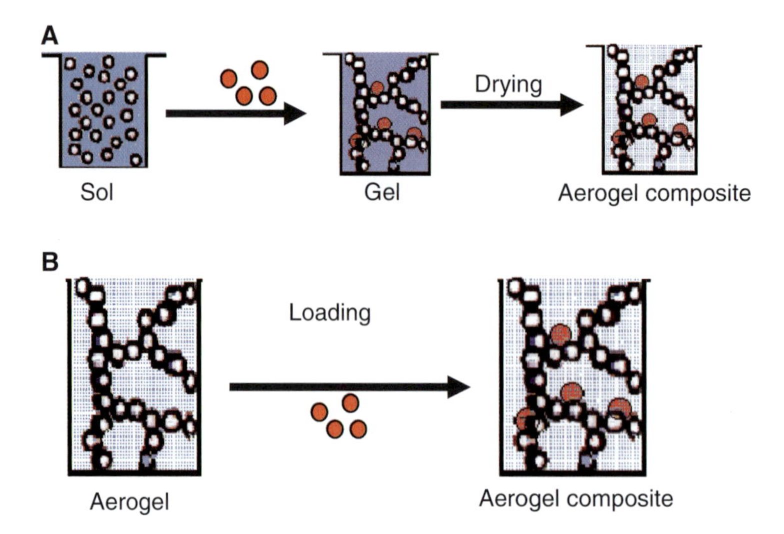

Figure 31.1. Preparation of aerogel composites and their mode of incorporation: **A.** during the sol–gel process and **B.** through adsorption/precipitation in the dry aerogel.

various natures including soluble organic or inorganic compounds, insoluble powders, polymers, biomaterials, and pharmaceuticals. In all cases, the additional components must be able to withstand the subsequent processing steps during the aerogel synthesis. The most critical one is the drying process. The added material may be destroyed by high temperatures or, in case of low-temperature supercritical drying (extraction by supercritical CO_2), may simply be washed out by the CO_2. An option in this case would be the subcritical drying of drug-loaded silica aerogels (at least hydrophobic ones), although no respective studies were published so far. If the added components are bulk, insoluble materials, the settling of the insoluble phase before gelation should be prevented by special treatments. Also, it is important to prove how the added component influences the gelation process, especially the buildup of the network and its structural properties.

The second method (b) implies the penetration of the target compound into the aerogel pores from the liquid or gaseous phase. This method is limited by diffusion and requires an additional step, if the carrier of the target component was a liquid.

Both approaches have been used by several research groups in order to prepare nanocomposites based on silica aerogels that are used for many different applications discussed in this handbook: chitosan-containing aerogels (Chap. 18), polymer-crosslinked aerogels (Chap. 13), polymer-reinforced aerogels (Chap. 15), and aerogels containing metal nanoparticles (Chap. 16). In this chapter, we focus on the applications of biocompatible aerogels and aerogel composites in pharmacy and related areas.

31.2. Silica Aerogels as Host Matrix for Drugs (*Drug Carriers*)

31.2.1. Loading of Aerogels by Adsorption

The high adsorption capability of aerogels can be used to load them with pharmaceutical compounds. However, the use of aerogels as drug delivery systems depends on their toxicity. Although sol–gel derived materials such as silica xerogels have been widely studied as a carrier matrix for various drugs, a complete toxicity study and risk analysis of silica aerogels are currently not yet available. Nevertheless, the chemical composition of silica aerogels is close to that of fumed silica, which is produced by combustion/hydrolysis of silicon tetrachloride. Amorphous silicon oxide has been widely used in the pharmaceutical industry and cosmetics for many years. The corresponding product is called "Aerosil" and has been an exclusive product of the German company "Degussa" since 1940. Aerosil has passed all clinical tests and has been found to be non-harmful to the human body [1]. It has been shown that orally administered Aerosil passes through the gastrointestinal tract without being resorbed in detectable quantities. It is expected that silica aerogels having an amorphous structure as Aerosil have similar clinical characteristics. Aerosil has an average surface area around 200 m^2/g [1]. Silica aerogels have much larger internal surface areas (500–1,000 m^2/g), thus exhibiting superior adsorptive properties compared to Aerosil. Several principal ideas for the use of aerogels in pharmaceutical applications already exist. It has been reported [2] that aerogel powder can be used as a free flow agent. The flow characteristics of powders are of particular importance in the pharmaceutical industry for pellet production. It has been shown that the addition of 0.5% aerogel powder to lactose powder improves the flow properties of the resulting powder more than the addition of the same amount of the conventional products used for this purpose (Aerosil and Sipernat) [2]. In combination with the ability of aerogels to adsorb and encapsulate chemical substances, this fact could allow the aerogels to be used simultaneously as free flow agent and drug delivery system (DDS). One of the most important characteristics of a DDS is the release rate of the loaded substances. "Release" describes the dissolution of the corresponding substance with particular attention to its time dependence. Depending on the drug type, different drug release rates (immediate, prolonged, and controlled release) are desirable [3]. The specific surface area is one of the most important parameters controlling both the dissolution rate of the drug and its absorption in the body. A large specific surface area allows a fast dissolution and, thus, an effective absorption within the body [3]. The specific surface area of the drug may be increased either by particle micronization or by increasing the surface area through adsorption of a drug onto a carrier or substrate. Since aerogels have an extremely large surface area, it is expected that the drug dispersed or adsorbed in the aerogel can get improved dissolution characteristics. Schwertfeger et al. [4] loaded both hydrophobic and hydrophilic silica aerogels with pharmaceuticals by means of adsorption from corresponding liquid solutions. Ambient dried aerogels having a relatively high density ($\rho > 0.1$ g/cm^3) were used for this purpose. The authors equilibrated the aerogels with the solution of the target drug and then filtered the mixture to get the loaded aerogel. The resulting powder was dried and could be used as a drug delivery system. Moreover, by choosing a suitable hydrophilic or hydrophobic aerogel, the substance with which the aerogel is loaded was shown to be released in an accelerated or a delayed form [4]. Still such kind of loading is contingent upon the stability of aerogels in the solvent used for loading. In the case that this cannot be guaranteed, the second approach, adsorption from supercritical fluids, is quite promising. Our group has extensively studied adsorption of different drugs and other organic compounds onto silica aerogels from supercritical solutions [5–8]. *Adsorption isotherms* were measured for several drugs; a typical isotherm for miconazole (antimycotic drug) is shown in Figure 31.2 [5]. Isotherms of most of the studied drugs have a Langmuir form, showing an increase of the loading with drug concentration in the bulk phase up to the saturation, where drug can be no more adsorbed.

The maximum loadings that could be achieved for a number of model compounds (drugs) by adsorption from their solutions in supercritical CO_2 are shown in Table 31.1.

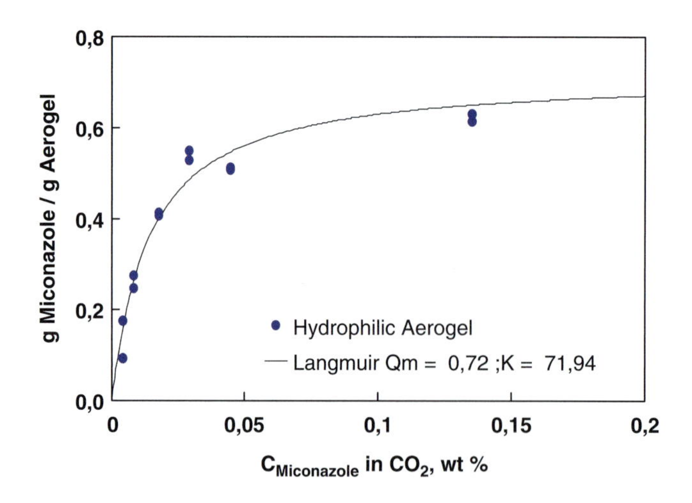

Figure 31.2. Adsorption of miconazole on hydrophilic silica aerogel with a density 0.03 g/cm^3 at 180 bar, 40°C. Q_m and K are the parameters of Langmuir isotherm [5].

Table 31.1. Maximum (saturation) loading of hydrophilic silica aerogels with some exemplary drugs at 40°C, 180 bar. Density of the aerogels used is 0.03 g/cm^3

	Molar mass (g/mol)	Loading (wt%)	Loading (mmol/g)
Miconazole	416.1	60	1.44
Ibuprofen	206.3	70	4.36
Flurbiprofen	244.3	24	0.98
Dithranol	226.0	10	0.2
Ketoprofen	254.3	30	1.2

The chemical nature of the respective drugs does not change during the loading procedure. As evidenced by X-ray diffraction analysis, no crystallites of the drug arc present in drug–aerogel formulations, and therefore, no long-range order is established upon adsorption of the drugs on silica aerogels, reaching the conclusion that intact drug molecules adsorb as a mono- or multilayer on the surface of the silica aerogel.

Consequently, no changes in the appearance of the aerogel samples are observed after adsorption, as can be seen in Figure 31.3 for adsorption of menthol in both hydrophilic and hydrophobic silica aerogels.

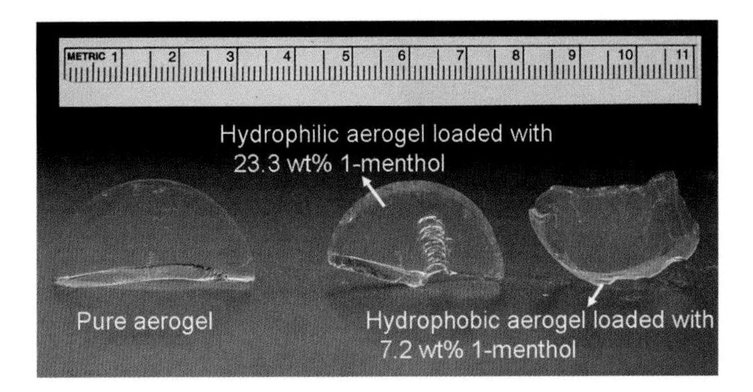

Figure 31.3. Physical appearance of pure aerogel, hydrophilic aerogel loaded with 23.3 wt% menthol, and hydrophobic aerogel loaded with 7.2 wt% menthol [9].

The drug loading depends, on the one hand, on the solubility of drugs in supercritical CO_2 and, on the other hand, on the affinity of the specific drug to a given aerogel surface. Therefore, the structural properties of aerogels such as density, pore size, and surface area and the functionality of the surface influence the adsorption process and thus the maximum loading as well. For a given drug, the loading increases with increasing surface area and with the volume of the mesopores of the aerogel [10, 11]. Usually, higher density aerogels, having larger surface areas, adsorb more drugs than samples of lower density [7]. Generally, in case of drug adsorption from the supercritical solutions, the final drug concentration in the aerogel is explicitly determined by the temperature, bulk concentration of the drug in the supercritical phase, and the properties of the aerogel. It is important to note that the contact time does not play any role, once the adsorption equilibrium is reached.

31.2.2. Release of the Drugs from Silica Aerogels

The release kinetics of drugs can be drastically affected by their adsorption onto aerogels. The use of hydrophilic aerogels as carriers results in a very fast release of the drugs [4, 6]. For instance, the release of ketoprofen from the ketoprofen–aerogel formulation (in powdered form) was determined experimentally (paddle method according to USP) and compared to that of crystalline ketoprofen (Figure 31.4) [6].

The release of ketoprofen from the ketoprofen–aerogel formulation, prepared by the adsorption from supercritical CO_2 solution, is much faster than the dissolution of

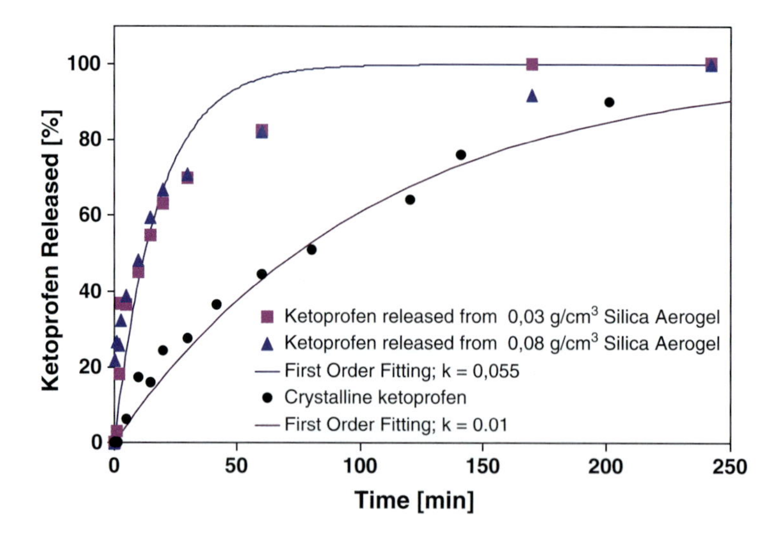

Figure 31.4. *Dissolution profiles* of crystalline ketoprofen and aerogel–ketoprofen formulation in 0.1 M HCl solution at 37°C. Particle size of crystalline ketoprofen is 0.7–15 μm. Aerogel was ground up to the average size of ca. 10–30 μm and loaded with ketoprofen by adsorption from saturated CO_2 solution at 180 bar, 40°C for 24 h [6].

crystalline ketoprofen. Eighty percent of ketoprofen is released from the aerogel after 50 min, whereas in the case of the crystalline drug dissolution of 80% takes 200 min. The release rate was obtained by fitting the first-order release kinetic model to the experimental data (Figure 31.4). Although in reality more complex behaviour is expected, this fitting gives a first idea about the release rate: the release rate of ketoprofen from hydrophilic aerogel is five times higher than that of the crystalline drug. In case of another drug, griseofulvin, the release from silica aerogels was compared to that of the dissolution of griseofulvin micro- and nanoparticles [11]. As can be seen from Figure 31.5, the release rate increases with decreasing particle size and is by far the highest in case of the powdered aerogel formulation.

Several effects play a role in this *release enhancement*. First, the specific surface area of the drug is significantly enlarged due to the adsorption on silica aerogel. Second, the hydrophilic silica aerogel rapidly collapses in water due to the capillary forces, which are exerted by the surface tension when liquid water enters a nanoscale pore of the aerogel. As a result, the solid silica backbone is fractured completely and the aerogel loses its integrity as a solid. So, the drug molecules adsorbed on the aerogel network are immediately surrounded by water molecules and thus dissolve faster. Finally, as discussed above the drug adsorbed on the aerogel does not have a crystalline structure, so no energy is needed to destroy the crystal lattice of the drug, as in the case of dissolution of its crystalline form. Only the energy of desorption of drug molecules from the silica surface (enthalpy of desorption) is to be considered in this case.

The effect of the aerogel density on the release rate can be seen in Figure 31.4, where the release rates of ketoprofen from two aerogels with different densities (0.03 and 0.08 g/ cm^3) are compared. Obviously, the release rate of ketoprofen does not depend on the aerogel density. This confirms the suggestion that the fast release of ketoprofen from hydrophilic aerogels is the result of the collapse of the aerogel network upon contact with aqueous media. Thus, the hydrophilic nature of such aerogel formulations is key for the enhancement of drug dissolution effects.

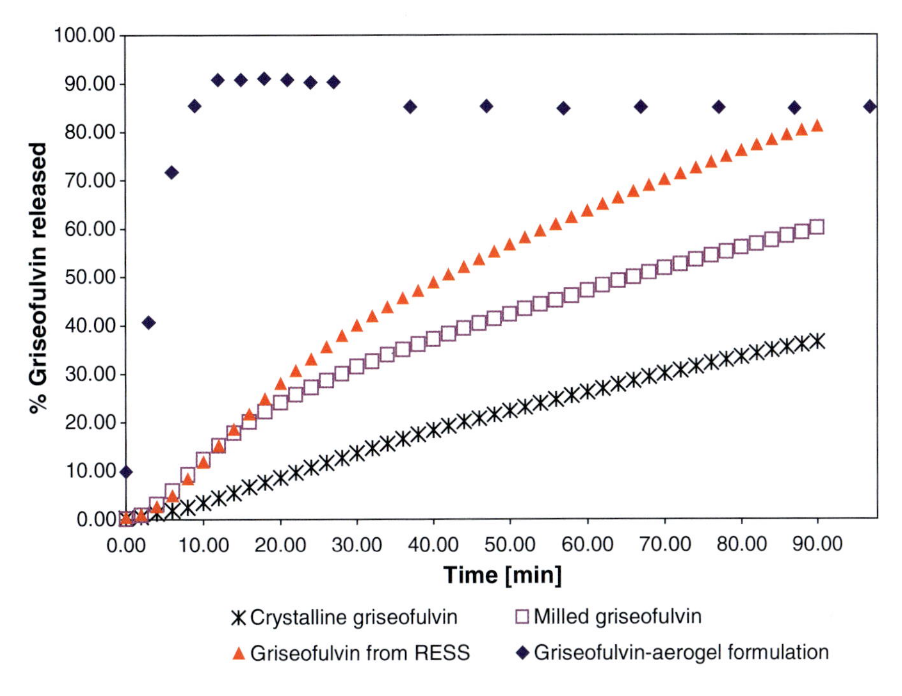

Figure 31.5. *Dissolution profiles* of griseofulvin in phosphate buffer (artificial gut fluid, pH 7.4). Particle size (d_{50}): crystalline – 177 μm; milled – 5.3 μm; RESS – 0.43 μm; aerogel powder – 0.7–10 μm.

31.3. Modified Silica Aerogels: Influence of Functional Groups on the Drug Adsorption and Release Kinetics

31.3.1. Adsorption

Besides the structural properties of the aerogels, both the adsorption and the release rate of the drugs from the aerogel formulations depend on the functional groups present on the aerogel's surface. Ideally, both properties can be "tailored" for each specific drug compound by the specific functionalization of the aerogels matrix. It is well known that originally hydrophilic aerogels can be transferred into hydrophobic ones by esterification of the free hydroxyl groups [5]. Since the OH groups can be regarded as active adsorption cites, one could generally expect the decrease of the adsorption capacity of the aerogel after the esterification. This was proven experimentally for a number of drugs and is exemplarily demonstrated in Figure 31.6 for ketoprofen. Also the adsorption of pure solvent – in this case CO_2 – is influenced by the esterification of the aerogels in the same manner [12]. It should also be noted that a considerable amount of CO_2 is adsorbed on the aerogel even in comparison to standard adsorbents, such as zeolites and activated carbon (Figure 31.7). Thus, adsorption of the drugs from CO_2 solution is always a competitive process with regard to the solvent CO_2.

Still, the decrease of the adsorbed amount of drugs after esterification of the aerogel can be very unfavourable for the production of pharmaceutical formulations, since the aerogel with lower adsorption capacity requires higher amounts of the carrier for a given therapeutic dose. In this case, it would be favourable to modify the silica aerogels with functional groups exhibiting strong interactions with a given drug. For instance, in case of drugs with a carboxylic group,

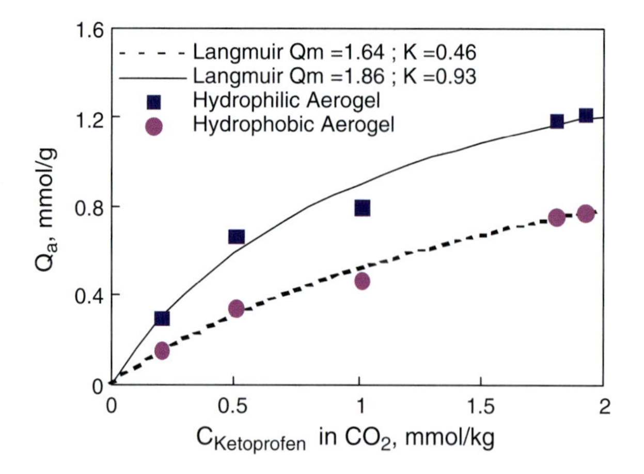

Figure 31.6. Adsorption of ketoprofen on hydrophobic and hydrophilic aerogels. Adsorption conditions: saturated supercritical CO_2 solution, 40°C, 180 bar. Q_m and K are the parameters of the Langmuir isotherm.

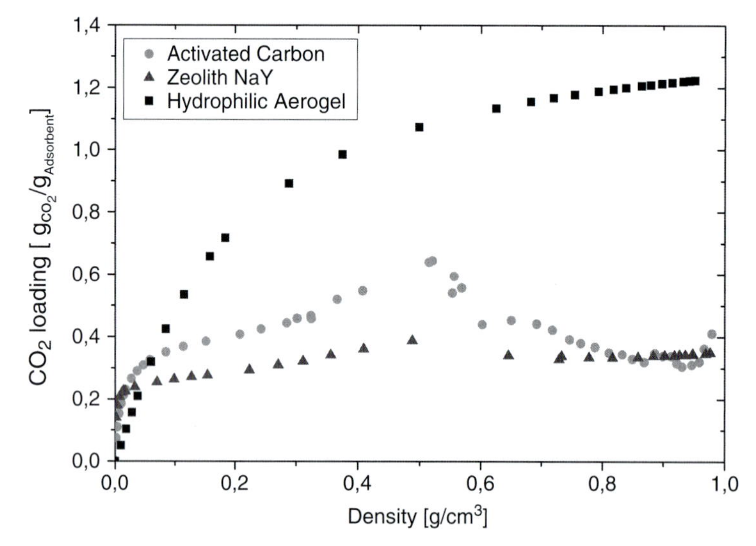

Figure 31.7. *Adsorption isotherms* of CO_2 on silica aerogels in comparison to those of zeolite and activated carbon.

modification of silica aerogels with amino groups allows us to increase the adsorption capacity, as shown in Figure 31.8 [13]. Thus, the adsorption capacity can be easily influenced by the functionalization of the aerogels. The nature of the modification for each specific drug can be determined analogous to chromatographic separation applications, where the retention time of the drug on the given stationary phase is given by the solute–stationary phase interactions. Since most stationary phases in gas and liquid chromatography are based on modified silica gels, this huge experience can be directly used for the chemical modification of silica aerogels.

31.3.2. Release Kinetics

An important question arising from the aerogel modification is: how are the release kinetics of the drug influenced by the functional groups? The release kinetics depend on

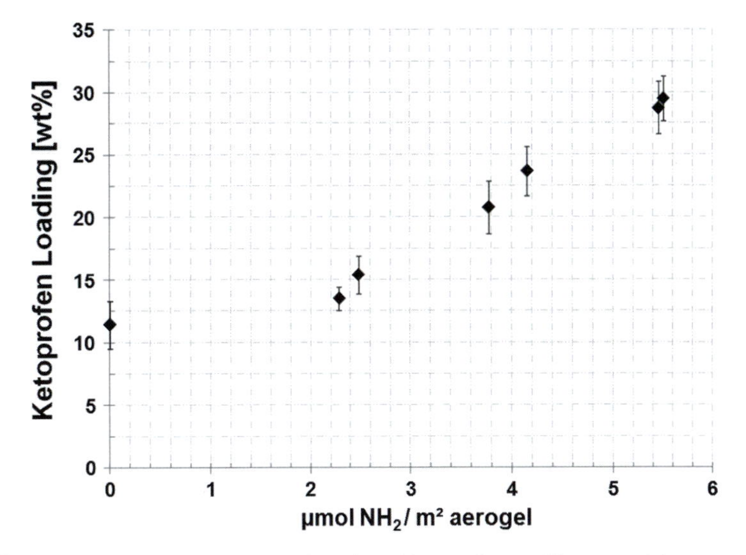

Figure 31.8. Effect of amino groups on the adsorption of ketoprofen on silica aerogel from supercritical CO_2 solution at 180 bar, 40°C.

both the structural properties and the functionality of the aerogels. Depending on the modification process, structural properties might change, so that in this case the comparison of the release kinetics with that of the original aerogels is rather difficult. However, in the case of functionalization from the gas phase [7], it is possible to preserve most structural properties of the aerogels such as density, pore volume, and size, so that a direct comparison between original and modified aerogels is possible. In Figure 31.9, the

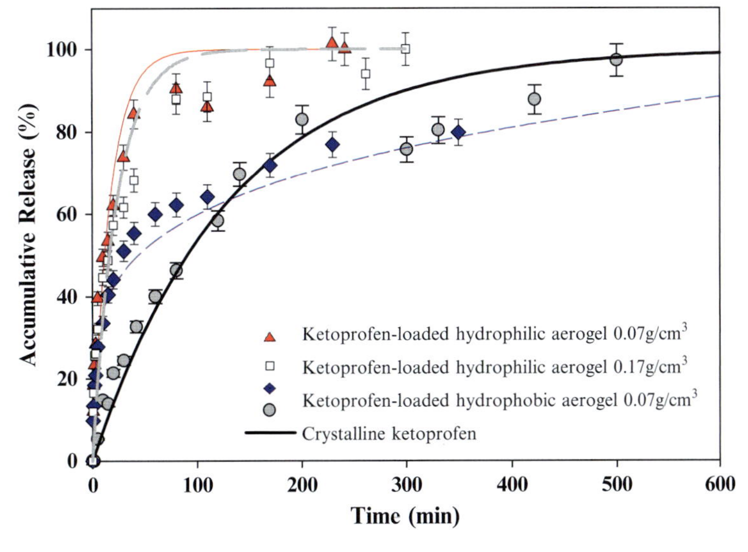

Figure 31.9. Release of ketoprofen from hydrophilic and hydrophobic aerogels of different densities. Dissolution medium: 0.1 N HCl, 37°C. Paddle method.

release of ketoprofen from hydrophilic and hydrophobic (esterified) silica aerogels of two different densities is compared. Hydrophilic silica aerogels are rapidly wetted with water and lose their structural integrity as discussed above, so that the drug molecules are surrounded with water, which translates into a fast dissolution of the drugs. If the same drug is adsorbed on hydrophobic aerogels, its dissolution becomes slower. At the beginning of the dissolution process, the drug dissolves from the surface of the aerogel and then later diffuses through its pores. Because the structure of hydrophobic aerogels is much more stable in water than that of hydrophilic samples, the release process from hydrophobic aerogels is governed to a greater extent by the slower permeation of water into the SiO_2 network, which leads to a slower release. Thus, both the loading of silica aerogels with drugs and their release rate can be influenced by the surface properties of silica aerogels.

In contrast to esterification, implementing NH_2 groups into silica aerogels does not change the hydrophobicity (and thus the wettability) of the aerogels significantly, so no significant changes in the release properties are observed (Figure 31.10).

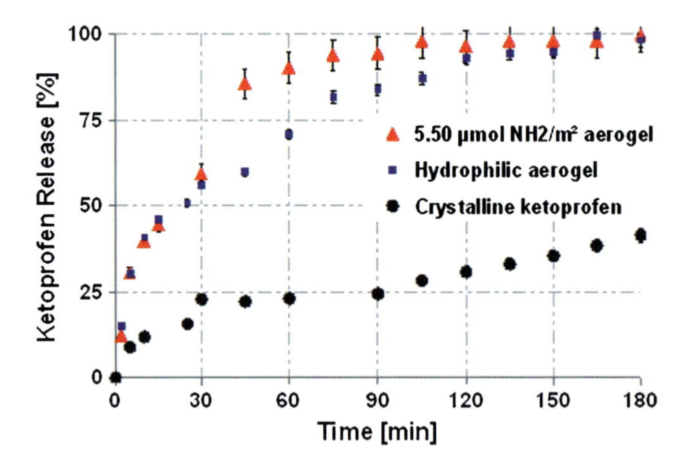

Figure 31.10. Release of ketoprofen from the amino-modified aerogels in comparison to the dissolution of its pure crystalline form. Dissolution medium: 0.1 N HCl, 37°C, paddle method.

These findings allow the release rate to be adjusted to some extent according to the desired application. Hydrophilic aerogels used as a carrier material accelerate the dissolution of poorly soluble drugs and thus improve their bioavailability. Hydrophobic aerogels slow down the dissolution rate and thus can be used for "long-term" drug release systems.

31.4. Pharmaceutical Formulations with Silica Aerogels

Since it was demonstrated that many drugs can be successfully adsorbed on silica aerogels and their adsorption and release properties can even be tailored, the next step is to evaluate the possibility of producing pharmaceutical formulations containing silica aerogels. The kind of the formulations depends primarily on the applications of the specific drug. The drugs tested with aerogels so far can be applied for both dermal and oral delivery. Thus, suitable solid and semisolid formulations have to be developed.

31.4.1. Semisolid Formulations

First attempts to produce semisolid formulations with drug-loaded aerogels were made using the model drug dithranol. Dithranol (1,8-dihydroxy-9(10H)-anthracenone, CAS-No. 1143-38-0) is a drug used in the treatment of psoriasis vulgaris; however, staining and skin irritation restrict their application. Another disadvantage of dithranol is its instability: in the presence of oxidative agents, polymerization to different coloured polymers takes place. In addition, it is well known that only a few percent of dithranol is released from a white soft paraffin suspension, which is one of the common formulations used in dermatology, and that most of this fraction persists in the stratum corneum (SC) without reaching deeper epidermal regions, which is a prerequisite for a therapeutic effect. Therefore, dithranol has an effect on mitochondrial function and structure and, hence, it is essential to obtain therapeutic drug concentrations in viable skin layers [14]. Adsorption onto silica aerogels helps overcome some of these handicaps.

Several semisolid formulations containing dithranol-loaded aerogels were prepared and evaluated concerning the release and penetration of dithranol into different membranes (since dermis cells are targeted). Based on the release studies of the drug from the aerogel into the formulation and the penetration experiments with artificial membranes, the most suitable semisolid formulation for the dithranol–aerogel system was found to be a hydrophilic ointment. Penetration of dithranol into two different artificial membranes both hydrophobic and hydrophilic as well as in the human stratum corneum was studied in comparison to a suspension of the crystalline drug in white soft paraffin, whereas FTIR-ATR spectroscopy was used to determine the corresponding drug concentrations [14]. Higher flux and a shorter lag time could be achieved if dithranol was adsorbed on silica aerogels (Figure 31.11). Furthermore, the amount of dithranol in the steady state was higher if the aerogel formulation was utilized. One reason is again the state of the drug within the formulation. In the case of crystalline dithranol, the drug is suspended in the vehicle, whereas dithranol adsorbed on the aerogel exhibits a non-crystalline structure as proved by X-ray diffraction experiments and electron microscopy. Dermal absorption of drugs

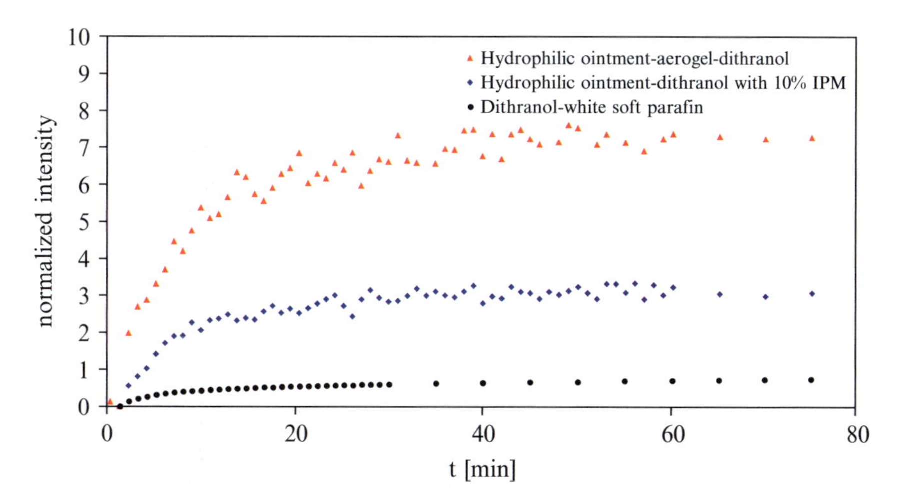

Figure 31.11. Penetration of dithranol from different formulations in dodecanol-collodium membrane determined by FTIR-ATR spectroscopy ($\nu = 1{,}603\ \text{cm}^{-1}$).

depends on their solubility within the vehicle as well as in the skin. Substances can only penetrate into a membrane – artificial or natural – in a dissolved state. Thus, at least a partial dissolution of the drug in the vehicle is essential. As discussed above, drugs in a non-crystalline state present in the aerogel are better soluble than crystalline ones; hence, the initial situation for the absorption of dithranol is advantageous if drug-loaded aerogels are used. The enhanced penetration properties of novel formulations were reproducible for months after production [14].

These effects were also confirmed for another drug, ibuprofen, by comparison of the drug release from different semisolid formulations into dodecanol collodion membrane (Figure 31.12) [15].

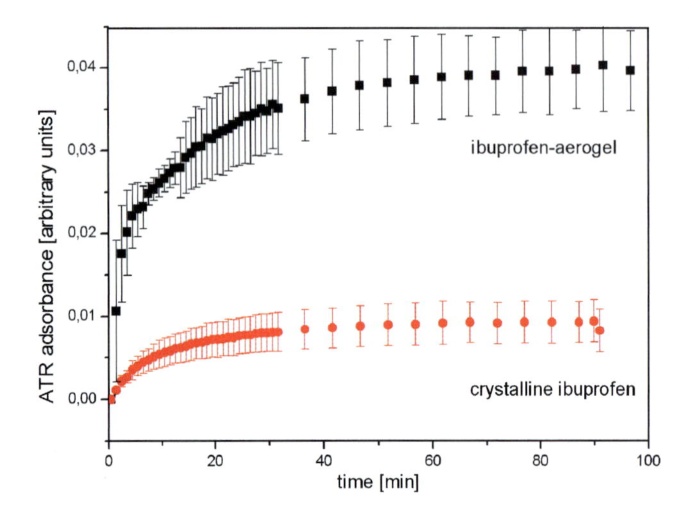

Figure 31.12. Penetration of ibuprofen from poloxamer gel into dodecanol collodion membrane [15]. Integration of the ibuprofen band from 1,504 to 1,525 cm^{-1} and correlating to time.

31.4.2. Solid Formulations

First attempt to prepare tablets out of silica aerogels was done. Hydrophilic silica aerogels loaded with ibuprofen were tabletted together with microcrystalline cellulose and anhydrous lactose as additives. The comparison of the release profiles of the corresponding tablets shows that the *release enhancement* by using hydrophilic aerogels is preserved in the form of tablets as well (Figure 31.13). These first positive results show the high potential of using aerogels in different areas of pharmaceutical technology.

Thermal Release of Active Compounds from Silica Aerogels

So far the release of the adsorbed compounds from silica aerogels triggered by the addition of water to the systems was discussed, since it is the most common case in pharmaceutical applications. However, other release triggers are also possible, one of them being temperature. This kind of release is well applicable to extremely volatile organic compounds (flavours and perfumes). In general, organic compounds exhibit a much lower

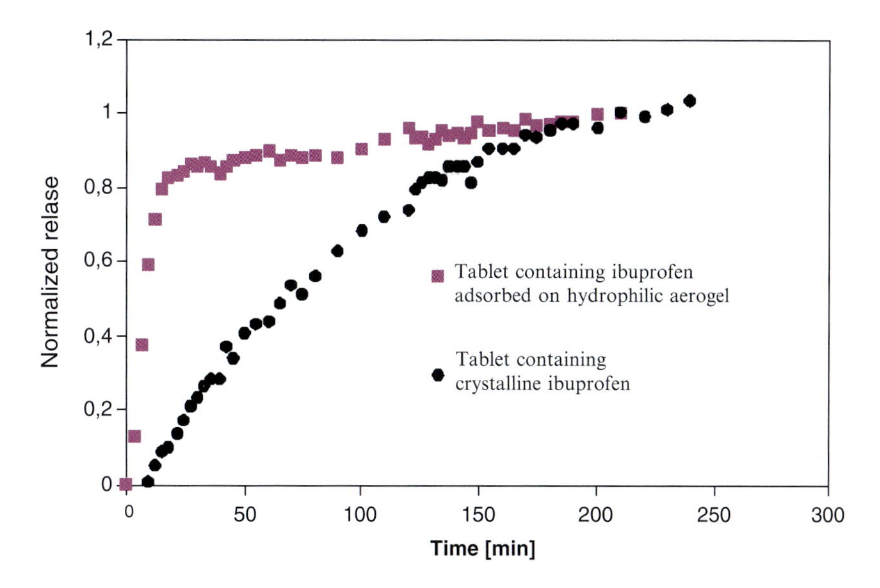

Figure 31.13. Release of ibuprofen from different tablets. Each tablet contains: 50% of ibuprofen-loaded aerogel or 50% of crystalline ibuprofen correspondingly. Additives: MCC/Lactose 75/25. Final ibuprofen concentration in both tablets: 16%.

vapour (sublimation) pressure when adsorbed onto a porous matrix. This effect can be used to reduce the loss of the volatile active agent during storage. Because of their high porosity, aerogels can also be utilized for this purpose. Menthol was used as a model compound and adsorbed onto silica aerogels and the thermal stability of the composite demonstrated [16]. The corresponding mass loss of menthol from various formulations upon heating (sublimation) is shown in Figure 31.14. Pure menthol rapidly sublimates as the temperature is ramped

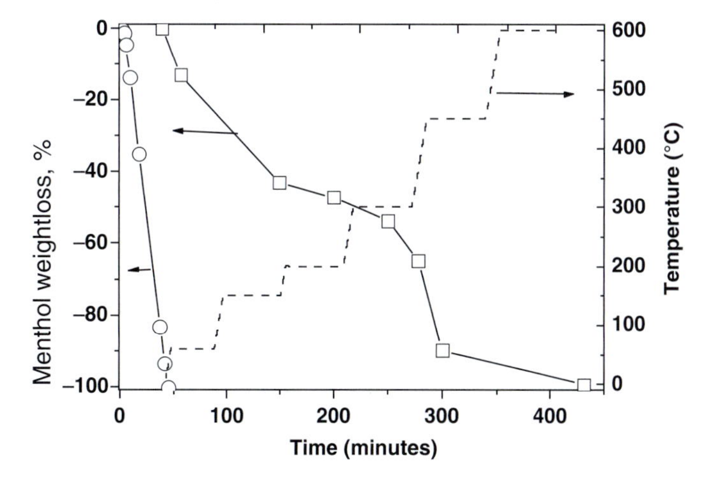

Figure 31.14. Release of menthol observed during temperature ramping TGA experiment (*dashed line*): 23.3 wt% menthol-loaded hydrophilic aerogel (*open squares*) and pure menthol (*open circles*); menthol loss is calculated as 100 × [(wt(t) − wt(t0))/wt(t0)], where wt(t0) = weight of menthol in aerogel at time = 0, wt(t) = weight of menthol in aerogel at time = t. The TGA data for the loss of water from a virgin hydrophilic aerogel are subtracted from the TGA menthol loss data; so the data shown here account only for the menthol weight loss.

to 60°C, whereas the release of menthol from the hydrophobic aerogel is delayed. The rate of menthol release is accelerated as the temperature is ramped from 60 to 150°C. Virtually all of the menthol is released from the hydrophobic aerogel at 150°C, while close to 50% of the menthol remains loaded in the hydrophilic aerogel at the same conditions. This observation suggests that a desired temperature-release profile of an adsorbate can be designed by partial functionalization of the internal surfaces of a hydrophilic aerogel. Hydrophilic aerogels possess two types of –OH groups: surface-active (or free) and bulk (or bound). The chemical structure of the adsorbate has a significant effect on the release characteristics, especially when the aerogel contains a large amount of free surface hydroxyl groups available for interaction with the adsorbate. The *thermal release* of the adsorbate reveals a two-step release process: weakly bound adsorbate that interacts with the bound –OH groups is released first at temperatures above the adsorbate normal melting temperature and strongly bound adsorbate that interacts with free –OH groups is released at very high temperatures, sometimes in excess of ~400°C. Generally, volatile compounds can be stabilized in silica aerogels even at elevated temperatures in excess of several hundred degrees in some cases. Hence, aerogel carriers could find application in the pharmaceutical and food industry for stabilization, retention, and/or release of flavour compounds during thermal treatment. Further, the thermally triggered release of an adsorbate-loaded aerogel might be useful as temperature sensor [16].

31.5. Crystallization/Precipitation of Drugs in Aerogels

In the previous sections, adsorption from supercritical CO_2 was discussed as a preferable method to load the aerogels with drugs. However, the loading capacity of aerogels is sometimes not large enough (compare Table 31.1), which is a serious disadvantage for their application as a drug delivery system since the needed amount of the carrier is too large. To overcome this problem, crystallization or precipitation of drugs in aerogels as an alternative to adsorption can be used. Besides the higher loading, particles formed inside the pores of aerogels are less sensitive to oxidation and less reactive compared to the pure micro – or nanoparticles of the same compound because of transport limitations in the pores. Furthermore, the main problem of the handling of small particles, their agglomeration, can be avoided since the particles are separated from each other by the walls of the pores. Precipitation of solutes inside the pores of silica aerogels should be controlled not only by process parameters, such as pressure, temperature, and bulk concentration of the target compound, but also, at a more fundamental level, by the physico-chemical properties of the carrier aerogels [12]. Thus, a fundamental understanding of the effect of the adsorptive properties of the carrier (silica aerogels) on the crystallization or precipitation of solute in its pores is needed. Recently, a term "adsorptive crystallization" was introduced to describe this process [9, 12], which is discussed as a tool to adjust the loading and physical state of solutes/drugs, for the future potential applications in pharmaceutical, food, storage, and other related industries. Especially, for the pharmaceutical industry, the bioavailability of the drug is an important factor, which is influenced by the state of the drug, i.e., amorphous forms of drugs have higher bioavailability compared to their corresponding crystalline form. The instability of the amorphized pharmaceuticals is a major factor precluding their more widespread use in solid dosing system. Through precipitation inside the pores of aerogels, amorphous forms of pharmaceuticals can be stabilized up to certain concentration [12] due to strong interactions with the aerogel matrix.

The precipitation of various organic substances ranging from polar to nonpolar (1-menthol, 2-methoxy pyrazine, naphthalene, benzoic acid, octacosane, dodecane, and ibuprofen) inside the pores of silica aerogels having different physical and surface properties was investigated [9, 12, 16]. Supercritical CO_2 was again used as a solvent since high supersaturation rates are needed to initiate the precipitation, which can be achieved by a rapid depressurization. The physical appearance of various aerogel samples after a typical precipitation experiment is shown in Figure 31.15.

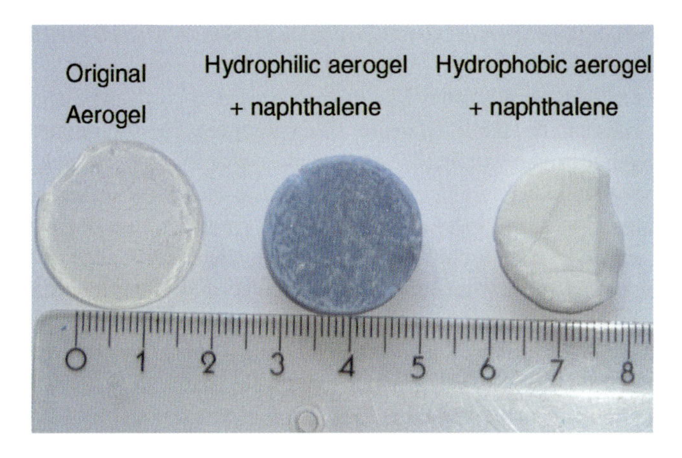

Figure 31.15. Naphthalene precipitated in hydrophilic and hydrophobic aerogels.

Whereas the transparency of the aerogel after adsorption process is nearly unchanged, it is completely lost after the precipitation of the particles in the pores of the aerogel because of the larger size of the crystallites.

Still the adsorptive properties of aerogels play a major role for crystallization process: strong interactions between the aerogel surface and solutes lead to amorphous precipitate, whereas weak interactions results in crystal formation inside the aerogel pores [12]. In the case of strong adsorption (combination of polar solute–polar aerogel surface), the absolute loading is very high and amorphous precipitates are preferably formed. This behaviour was observed for menthol and benzoic acid: up to 40 wt% amorphous menthol and 31 wt% of amorphous benzoic acid could be stabilized in the hydrophilic (polar) aerogels. When the same substances were adsorbed on less polar aerogels (esterified ones), benzoic acid tends to crystallize at lower concentrations (above 15 wt% loading solute tends to be crystalline). The adsorbed solute acted as a kind of nucleus or active surface during crystallization, i.e., higher amounts of solute are precipitated inside the pores of a polar aerogel.

In the case of loading of moderately polar solutes in aerogels (combination of moderately polar solute–polar aerogel surface), the absolute loading is lower compared to polar solutes, and the solute starts to crystallize even at low loadings due to smaller interactions with the aerogel surface. This behaviour was observed for naphthalene: up to 20 wt% amorphous form of naphthalene can be stabilized in hydrophilic aerogels. When the same was adsorbed in a less polar aerogel, naphthalene tends to crystallize at lower concentrations starting from approximately 12 wt% loading. As for the loading with nonpolar solutes (combination of nonpolar solute–polar aerogel surface), the absolute loading of the solute, which can be achieved, is very low compared to polar and moderately polar solutes, and a crystalline precipitate is obtained. This behaviour was observed, for example, for octacosane: above 1.5 wt% loading octacosane is incorporated into its crystalline form. When the same substance was loaded into a less polar aerogel, no significant changes in the loading were observed, due to similar types of interactions (hydrophobic substance–hydrophobic surface) [12]. Thus, adsorptive crystallization appears to be a useful tool to produce stable pharmaceutical formulations containing amorphous drug forms.

31.6. Silica Aerogels as Carriers for Enzymes and Proteins

Many modern drugs are protein-based. In this case, the question arises: if protein or protein-like materials can be stabilized inside silica aerogels. The stabilization of enzymes in silica aerogels was intensively studied in the literature over the last 10 years. First, aimed for enzymatic catalysis, this idea can be easily extended to the carrier system for enzyme and protein-based drugs. Sol–gel support materials for enzymes were known earlier [17], so the main idea of the encapsulation technique was adopted to aerogels. One of the well-studied enzymatic catalysts, lipase, was incorporated into silica and aluminosilicate gels during a sol–gel process. It was proven that neither the gelation nor the following supercritical drying damages the lipase, so that aerogels loaded with an intact lipase could be prepared [18–22]. The biocatalytical activity of the final product was studied using the esterification reaction of 1-octanol with lauric acid. In all cases, the lipase immobilized in aerogels showed very high catalytic activity, in most cases much higher than that of free lipase. This fact was also proved for the other esterification reactions in supercritical fluids (CO_2 and propane) [23]. It can be explained by the fact that lipase, being dispersed in solid aerogel matrix, has a lower tendency to aggregate than free lipase, so that the amount of active adsorption sites available for the reactants is much higher than in the free state lipase. Also, the comparison of the lipase immobilized in aerogels and xerogels reveals that its catalytic activity is higher if an aerogel is used [18]. A good catalytic activity of lipase is closely related to its ability to change the conformation. It can be easily provided in case of aerogels because the gel shrinkage during drying is mostly avoided by the use of supercritical conditions, so that the lipase remains flexible inside the pores. In case of xerogels it is not so easy, since the large shrinkage during conventional drying results in a denser structure, where the enzyme is more or less fixed. Furthermore, the capillary stress might partially crush the enzyme, leading to loss of activity as well [18]. In addition, the influence of the aerogel structure and hydrophobic–hydrophilic balance on the enzyme performance in different solvents was studied in detail [19–22]. By controlling the aerogel hydrophobicity, the transport of the solvents of different polarity inside the aerogel matrix can be controlled. For instance, polar substances or products, in particular water, may have some difficulties to penetrate the pores of very hydrophobic aerogels or to be removed from them. In this case, the transport phenomena influence the reaction rate significantly [20]. This work reveals the possibility to produce aerogel-supported enzymatic catalysts, which are tailored for certain reaction conditions. The encapsulation of enzymes during a sol–gel process also seems to be more beneficial compared to the simple adsorption of enzymes on the aerogels after supercritical drying (post-treatment). Basso et al. [24] report that the catalytical activity of the enzymes PGA, thermolysin, and chymotrypsin – adsorbed on aerogels by soaking in the enzyme suspension – is rather low. It might be due to the fact that the enzymes are not really integrated into the aerogel structure, so that aggregation cannot be avoided. Moreover, some adsorption sites of the enzyme can be blocked by the interaction with the aerogel surface.

Recently, it was shown that proteins (cytochrome *c*) can also be stabilized in an aerogel matrix using the so-called nanogluing approach [25, 26], which implies the formation of a stable protein superstructure around colloidal gold nanoparticles. Proteins together with nanoparticulate silver or gold colloids are added to the sol solution close to the gelation point. After the replacement of the solvent, the gel is supercritically dried to get a protein-noble metal nanoparticle–aerogel composite. The authors suggest that the specific adsorption of the protein at colloidal gold nanoparticles initiates formation of a stabilizing superstructure followed by a rapid assembling of protein molecules in a large superstructure, which is captured in the silica network during gelation [25]. The presence of the colloidal gold allows to preserve proteins for the most part from being damaged by solvent exchange and supercritical drying steps, so that over 80% of the protein remains active in the final aerogel composite [25]. The amount of intact protein depends on the protein to gold nanoparticle ratio. The resulting composites show better stability at a greater range of temperatures than previously reported biocomposite xerogels, which require the storage at low temperatures (4°C) to maintain protein viability, and exhibit an extremely rapid sensing of gas-phase analytes. This example shows the principal possibility to design different complex biofunctionalities within an artificial matrix [26].

First efforts were done to prove the applicability of an aerogel matrix as a membrane support. Planar phospholipid bilayers were deposited on silica aerogel surfaces and their lateral mobility and homogeneity were studied [27]. It is expected that the permeable nature of the porous materials, such as aerogels, provide new advantages for supported lipid bilayer systems, such as for example, better accommodation of proteins.

31.7. Organic Aerogels as Drug Delivery Systems

For applications in pharmaceutical industry, biocompatibility and biodegradability are essential in many cases. Unfortunately, silica aerogels are biocompatible, but they are not biodegradable. This demand can be met in matrices made of organic material. The loading of organic aerogels with pharmaceutical substances was principally claimed by several authors in patent applications. Berg et al. [28] describe the loading of organic (resorcinol-formaldehyde) aerogels with testosterone adipate and 5-fluorouracil. However, adsorption of drugs from liquid solutions was used, which leads to the partial collapse of the aerogel structure, especially in the case of hydrophilic aerogels. Lee and Gould [29] have loaded organic aerogels with drugs such as methadone and naltrexone by co-gelling method before supercritical drying. The resulting aerogel powder was proposed to be used as part of an aerosol for inhalation since the low density of the product allows the particles to be carried by the aerosol stream. 5-Fluorouracil could also be incorporated into resorcinol-formaldehyde aerogels by this method [28].

Other promising matrices for drug delivery are polysaccharides. Because of their low toxicity, polysaccharides such as starch and alginate are quite common additives in food and drug formulations. The combination of the outstanding structural properties of aerogels with the physiological compatibility and acceptability of polysaccharides results in high potential drug delivery systems. For alginate [30] and starch [31, 32], the feasibility of producing highly porous structures by supercritical drying with CO_2 has been reported. The specific surfaces for high amylose corn starch reach up to 180 m^2/g [33] for alginate depending on the composition up to 500 m^2/g [30, 34]. By comparing the properties of polysaccharide aerogels with those of silica aerogels, it should be mentioned that the bulk densities of the starch matrices are much higher [31]. Alginate aerogels, however, exhibit bulk densities in

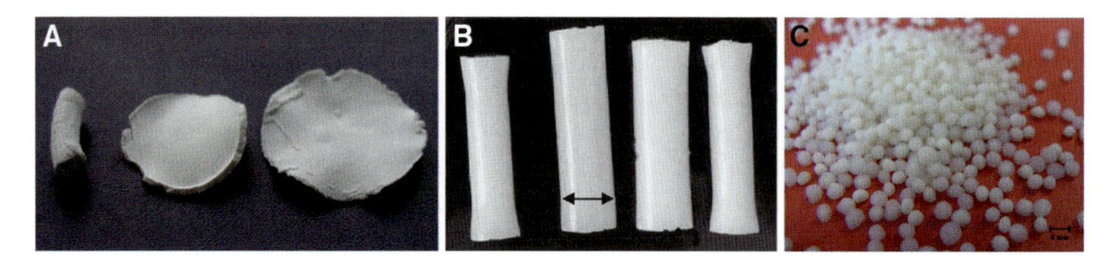

Figure 31.16. Pictures of some polysaccharides aerogels **A.** starch aerogel, **B.** monolithic alginate aerogels, **C.** alginate aerogels in form of spherical particles.

the range of that for silica aerogels (0.003–0.27 g/cm^3). Consequently, the porosity of alginate aerogels is comparable to that of silica aerogels. Both types of aerogels can principally be produced either as (micro)spheres [30] or as bulk materials [35], as shown in Figure 31.16. The surface area of the starch aerogels depends on many factors. These include the type of starch as well as the conditions used during gelation and solvent exchange. Especially, the temperature during the gelation step exhibits a significant influence: higher temperature causes a greater degree of destruction of the granular structure and an increased concentration of free amylose in solution. The resulting aerogel structure can be considered as an amylose matrix with intermediate granules, whose ratio is dependent on the temperature reached during gelation. With increasing gelation temperature, the granular structure is more and more converted to a network matrix, which causes an increase in the surface area. Also, an increase in the degree of retrogradation is attended by an enlargement of the surface area. In case of alginate gels, the alginate source (ratio of guluronic acid to mannuronic acid) as well as alginate and salt concentrations play a major role in defining the structural properties [35]. An increase in density with increasing $CaCO_3$ concentration is ascribed to a greater amount of linkages between the alginate – chains due to Ca^{2+} ions. Increasing the alginate concentration leads to a decrease of the specific surface area [34].

Loading of polysaccharide aerogels with drugs and their release were recently studied in our group [35]. Analogous to the case of silica aerogels, two different ways of loading were used: (a) adding of the drug during dehydration, prior to CO_2 extraction (the final solution is replaced by drug saturated ethanol solution, from which the loading takes place) and (b) adsorption of the drug dissolved in supercritical CO_2 after completing CO_2 extraction. Two different types of starch (potato starch and modified starch, Eurylon7) and two different drugs, ibuprofen and paracetamol, were used. The resulting loadings are shown in Table 31.2.

Table 31.2. Maximal loading of aerogels with drugs [35]

Kind of aerogel	Ibuprofen [% (w/w)]	Paracetamol [% (w/w)]	Specific drug loading (g/m^2)
Potato starch	10 ± 0.3	10 ± 0.3	$10–20 \times 10^{-4}$
Eurylon7 starch	22 ± 0.7	25 ± 0.8	$20–40 \times 10^{-4}$
Alginate	21 ± 0.6 26 ± 0.8	–	$7–11 \times 10^{-4}$

The values reached for ibuprofen and paracetamol loading [% (w/w)] for a specific matrix are nearly equal. The reached drug fractions for alginate and Eurylon7 aerogels are in the same range, while it is less for potato starch aerogels. Regarding specific loading (adsorbed drug per specific aerogel surface area), Eurylon7 starch shows about double the value reached for potato starch and alginate aerogels. The increase of specific drug loadings observed for Eurylon7 aerogels compared to other systems might be due to the pore structure: the average pore radius of Eurylon7 aerogels is 1.9 nm. For potato starch aerogels and alginate aerogels, the pores are significantly larger, with an average radius of 7.2 and 11 nm, respectively. Smaller pore sizes enhance capillary forces; this might be one of the reasons for capillary condensation and thus higher drug loading. The ibuprofen loading of alginate aerogels is comparable to starch aerogel Eurylon7. Since the structure of the alginate aerogels is dominated by mesopores, capillary condensation can be excluded. The resulting loading can be attributed to the high surface area. Regarding specific loading, the organic structures are on par with silica matrices.

31.7.1. Drug Release

The release kinetics of the different drug–matrix combinations are quite distinctive as shown in Figure 31.17. The release of paracetamol from both types of starch aerogel is completed after 2 h, similar to pure paracetamol. The release of ibuprofen in contrast is dependent on the origin of the starch: a fast and prolonged release for Eurylon7 or potato starch aerogels, respectively, can be achieved. The dissolution of ibuprofen from alginate aerogels is very fast, almost reaching the release rate of silica aerogels. Since paracetamol is readily soluble in aqueous media at the given pH, the dissolution process requires a short time for all aerogels. It was observed that some part of paracetamol is present in the form of small crystals loosely connected with the matrix and that these crystals dissolve very fast, independent of the matrix properties. On the contrary, the release of ibuprofen is dependent on the kind of the matrix, since ibuprofen is adsorbed at a molecular level. Further, the solubility of ibuprofen at a given pH is much smaller than that of paracetamol, so that the differences in the release kinetics are more evident on this time scale. The structure of Eurylon7 aerogels collapses relatively fast, because of poor mechanical properties, providing very fast release of ibuprofen. The same effect is observed for hydrophilic silica

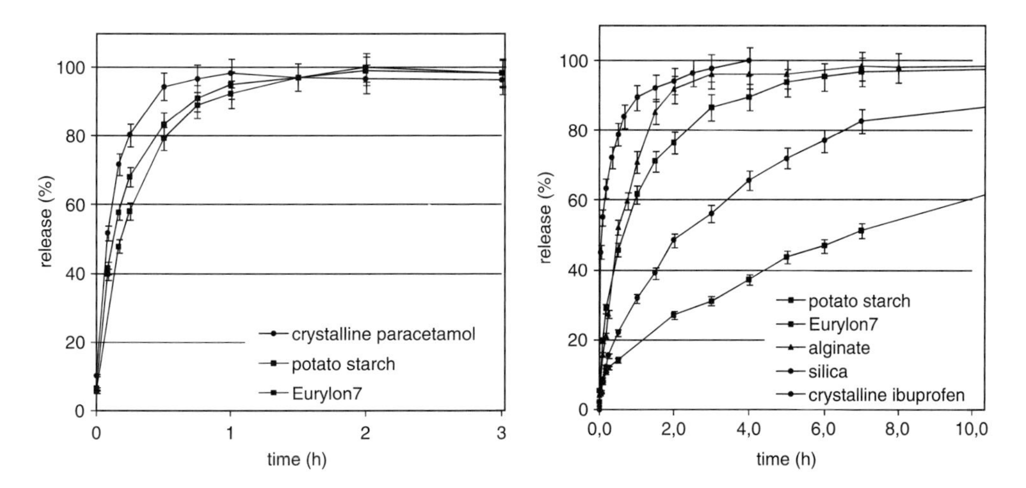

Figure 31.17. Release of paracetamol and ibuprofen from polysaccharide aerogels in comparison to the dissolution of the crystalline drug form. Release conditions: paddle method, 37°C, 0.2 M phosphate buffer at pH 7.2.

aerogels. On the contrary, potato starch aerogels are more stable, resulting in a slower drug release. Also, the release of ibuprofen from alginate matrices is quite fast, which can be ascribed to the high affinity of alginate to water. Furthermore, larger pores allow more efficient transport of the drug to the surrounding.

So far we can conclude that the release rate from polysaccharide aerogels mainly depends on the stability of the aerogel matrix in water and on the crystallinity of the drug inside the matrix, which is similar to the case of silica aerogels. Since the matrix properties can be influenced by the polysaccharide source and by the gelation procedure, it is possible to tailor the release properties depending on the desired application. Both prolonged and immediate release can be achieved for the same drug by varying the properties of the aerogel matrix.

31.8. Aerogels Based on Biopolymers as *Drug Carriers*

Biopolymers are widely used in pharmacy as drug delivery systems. The preparation of biopolymers in the form of aerogel could result in a product having high porosity and high surface area, which is favourable for many applications. One of the most abundant organic polymers is chitin, a polyglucosamide, which contains exoskeletons of insects, cell walls, etc. Chitin itself is not applicable for sol–gel processes, but its derivative, chitosan, being obtained by alkaline deacetylation of chitin can be used for this purpose [36, 37]. Chitosan is commercially available and its biocompatibility is proved. Although the preparation of pure chitosan aerogel was not successful due to the high shrinkage of the gel, composite aerogel material (silica–chitosan) could be prepared (Chap. 18). The biocompatibility of these aerogels was proven by cytotoxicity and hemolysis tests, in which case cytotoxicity tests were passed with very little cell damage [37]. The chelating effect of chitosan was used for incorporating transition metal ions [36, 38] into chitosan–silica aerogels, since these metal ions can coordinate with the amine group of chitosan. This makes chitosan-based aerogels quite interesting for a number of pharmaceutical and biomedical applications [39]. To give proof of principle, first studies of chitosan–silica aerogels as drug delivery systems were carried out: it was shown that up to 17 wt% gentamicin could be encapsulated in this hybrid material [38, 40].

Another promising type of bioderived aerogels for pharmaceutical applications is an aerogel based on cellulose and its derivatives. Details of cellulose aerogel preparation are given in Chap. 9.

Recently, aerogels based on peptides were synthesized [41] as well. Self-assembling of the fibril network was observed and chiral, nanostructured low-density aerogels out of the resulting gel. This product can be used for sensing, tissue engineering, and drug release.

Poly(lactic acid) (PLA) aerogels have also shown promising applications in these fields [42, 43], due to their porosity, 3D network regularity, and mechanical properties. Since PLA particles are widely used as carriers for drugs applied by inhalation, it is in principle possible to use PLA aerogels for the same purpose. PLA aerogels with specific surface area up to 140 m^2/g could be produced in our group (unpublished results). This high value suggests high adsorption capacity, which is currently proved in further experiments. Since PLA aerogels were produced in a bulk monolithic form so far, the future trend in this area would be the production of PLA aerogels in form of microspheres with narrow particle size distributions.

A principal use of other kinds of organic aerogels as a biodegradable matrix for drug delivery is discussed by Berg et al. [28].

31.9. Conclusion

In this chapter, a number of pharmaceutically relevant systems containing aerogels were discussed; a short overview of the systems described above is given in Table 31.3.

These results show a huge potential of aerogel-based materials in the field of pharmaceutical science and further life science applications (biomedical applications are discussed in Chap. 30). It is not only the unique properties of the existing aerogels materials, but especially the flexibility of the sol–gel chemistry, which plays here an important role and must be emphasized. Adsorption and release of active agents (drugs and flavours) can be controlled by structural properties (surface area and pore volume) and functionalization of the corresponding aerogels. Depending on the properties of the matrix and the precursor (organic/inorganic), the desirable properties can be "tailored" for any specific applications. Further on, composite aerogels provide additional "degrees of freedom" to meet the targeted formulation properties. In accordance with the suggestion of Kistler, aerogels can be produced from virtually any organic and inorganic gel, and almost every year we observe that new types of aerogels and aerogel composites appear. Taking into account the large ensemble of research activities in the area of aerogels, a number of promising applications in life sciences can be expected in the future.

Table 31.3. Overview about aerogel-based systems and their applications

Type of aerogels	Benefit of using aerogels	Applications
Silica (hydrophilic)	Very fast drug release (up to five times faster as standard dosage form)	Immediately acting drug dosage forms
Silica (hydrophobic)	Prolonged drug release	Delayed drug forms
Functionalized silica	Both fast and prolonged drug release; high adsorption capacity tailored for specific drugs	Immediately acting and delayed drug delivery formulations
Polysaccharide (starch, alginate), PLA	Pharmaceutically accepted matrices: both fast and prolonged drug release controlled by the matrix properties	Immediately acting and delayed drug delivery formulations
Silica (hydrophobic and hydrophilic)	*Thermal release* of highly volatile compounds and gases (CO_2)	Stabilization, retention, and release of flavour compounds in food industry during thermal treatment. Temperature sensor
Silica/biopolymer composites (chitosan, etc.)	High biocompatibility	Antimicrobial applications
Silica/ silica–metal composites	Encapsulation of enzymes and proteins: very high enzyme activity is observed	Biocatalysis, biosensors

References

1. Technical Bulletin Aerosil & Silanes (2001) Company publication, Degussa, Düsseldorf
2. German patent application DE 19653758 A1 (1996)
3. Stricker H, (1998) Physikalische Pharmazie, Wissenschaftliche Verlagsgesellschaft GmbH, Stuttgart, 3rd ed.
4. Schwertfeger F, Zimmermann A, Krempel H (2001) Use of inorganic aerogels in pharmacy. US Patent 6, 280,744
5. Smirnova I, Mamic J, Arlt W (2003) Adsorption of drugs on silica aerogels, Langmuir, 19,20: 8521–8525
6. Smirnova I, Suttiruengwong S, Seiler M, Arlt W (2004) Improvement of the solubility of poor soluble drugs by adsorption on silica aerogels , Pharm. Dev. Techn. 9 (4): 443–452
7. Smirnova I, Suttiruengwong, S, Arlt, W (2004) Feasibility study of hydrophilic and hydrophobic silica aerogels as drug delivery systems. J Non-Cryst Solids 350: 54–60
8. Smirnova I, Tuerk M, Wischumerski R, Wahl M A (2005) Comparison of different methods for enhancing the dissolution rate of poorly soluble drugs: Case of griseofulvin. Eng. Life Sci 5(3): 277–280
9. Gorle S, Smirnova I, Dragan M, Dragan S, Arlt W (2008) Crystallization of naphthalene in silica aerogels from supercritical CO_2. J Supercrit Fluids 44: 78
10. Suttiruengwong S (2005) Silica Aerogels and Hyperbranched Polymers as Drug Delivery Systems. Dissertation, Universitaet Erlangen-Nuernberg,
11. Smirnova I, Suttiruengwong S, Arlt W (2005) Aerogels: tailor-made carriers for immediate and prolonged drug release. KONA 23: 86–97
12. Gorle S (2009) Adsorptive Crystallization of Organic Substances in Silica Aerogels from Supercritical solutions, Dissertation, Universitaet Erlangen-Nuernberg
13. Alnaief M, Smirnova I (2010) Effect of surface functionalization of silica aerogel on their adsorptive and release properties. J Non-Cryst Solids 356 (33–34): 1644–1649
14. Günther U, Smirnova I, Neubert RHH (2008) Hydrophilic silica aerogels as dermal drug delivery systems - dithranol as a model drug. Eur J Pharm Biopharm 69 (3): 935–942
15. Günther U, Smirnova I, Arlt W, Neubert RHH (2008) Applications of silica aerogels as drug delivery systems. Proceedings of 6th World Meeting on Pharmaceutics, Biopharmaceutics and Pharmaceutical Technology, Barcelona
16. Gorle S, Smirnova I, McHugh M (2009) Adsorption and thermal release of highly volatile compounds in silica aerogels. J Supercritical Fluids 48: 85–92
17. Reetz M T, Zonta A, Simpelkamp (1995) Effiziente heterogene Biokatalysatoren durch den Einschluß von Lipasen in hydrophoben Sol-Gel-Materialien. J Angew Chem 107: 373–376
18. Buisson P, Hernandez C, Pierre M, Pierre AC (2001) Encapsulation of Lipases in Aerogels. J Non-Cryst Solids 285: 295–302
19. Maury S, Buisson P, Pierre AC (2001) Porous Texture Modification of Silica Aerogels in Liquid Media and Its Effect on the Activity of a Lipase. Langmuir 17: 6443–6446
20. El Rassy H , Perrard A, Pierre AC (2003) Behavior Of Silica Aerogel Networks as Highly Porous Solid Solvent Media for Lipases in a Model Transesterification Reaction. ChemBioChem 4: 203–210
21. El Rassy H, Maury S, Buisson P, Pierre AC (2004) Hydrophobic silica aerogel–lipase biocatalysts: Possible interactions between the enzyme and the gel. J Non-Cryst Solids 350: 23–30
22. Maury S, Buisson P, Perrad A, Pierre AC (2005) Compared esterification kinetics of the lipase from Burkholderia cepacia either free or encapsulated in a silica aerogel. J Mol Catal B: Enzymatic 32 (5–6): 193–203
23. Novak Z, Habulin M, Krmelj V, Knez Z (2003) Silica aerogels as supports for lipase catalyzed esterifications at sub- and supercritical conditions. J Supercrit Fluids 27: 169–178
24. Basso A, De Martin L, Ebert C, Gardossi L, Tomat A, Casarci M, Rosi OL (2000) A Novel Support for Enzyme Adsorption: Properties and Applications of Aerogels in Low Water Media. Tetrahedron Lett 41: 8627–8630
25. Wallace Jean Marie, Rice Jane K, Pietron Jeremy J, Stroud Rhonda M, Long Jeffrey W, Rolison Debra R (2003) Silica Nanoarchitectures Incorporating Self-Organized Protein Superstructures with Gas-Phase Bioactivity. Nano Lett 3(10): 1463–1467
26. Wallace Jean Marie, Dening Brett M, Eden Kristin B, Stroud Rhonda M, Long Jeffrey W, Rolison Debra R (2004) Silver-Colloid-Nucleated Cytochrome c Superstructures Encapsulated in Silica Nanoarchitectures. Langmuir 20 (21): 9276–9281
27. Weng KC, Stalgren JR, Risbud SH, Frank CW (2004) Planar bilayer lipid membranes supported on mesoporous aerogels, xerogels, and Vycor glass: an epifluorescence microscopy study. J Non-Cryst Solids 350: 46–53

28. Berg A, Droege MW, Fellmann JD, Klaveness J, Rongved P (1995) Medical use of organic erogels and biodegradable organic aerogels. WO 95/01165
29. Lee K, Gould G (2001) Aerogel powder therapeutic agents. WO 02/051389 A2
30. Valentin R, Molvinger K, Quignard F, Di Renzo F (2005) Acidity of alginate aerogels studied by FTIR spectroscopy of probe molecules. Macromolecular Symp 230: 71–77
31. Glenn GM, Irving DW (1995) Starch-based Microcellular Foams. Cereal Chem 72: 155–161
32. Glenn GM, Stern DJ (1999) US Patent, 5,958,589
33. Doi S, Clark JH, Macquarrie DJ, Milkowski K (2002) New materials based on renewable resources: chemically modified expanded corn starches as catalysts for liquid phase organic reactions. Chem Com 22: 2632–3
34. Trens P, Valentin R, Quignard F (2007) Cation enhanced hydrophilic character of textured alginate gel beads Colloids Surf A 296: 230–237
35. Mehling T, Smirnova I, Guenther U, Neubert RHH (2009) Polysaccharide-based aerogels as drug carriers. J Non-Cryst Solids 355 (50–51): 2472–2479
36. Hu X, Littrel K, Ji S, Pickles DG, Risen WM (2001) Characterization of silica-polymer aerogel composites by small-angle neutron scattering and transmission electron microscopy. J Non-Cryst Solids 288 (1–3): 184–190
37. Ayers MR, Hunt AJ (2001) Synthesis and properties of chitosan-silica hybrid aerogels. J Non-Cryst Solids 285 (1–3): 123–127
38. Liu X, Yao C, Risen W M Jr (2006) Chemically selective reactions in confined spaces in hybrid aerogels. Mat Res Soc Symp Proc 899 E (Dynamics in Small Confining Systems VIII), Paper: 0899-N06-06
39. Singh J, Dutta PK, Dutta J, Hunt A J, Macquarrie D J, Clark J H (2009) Preparation and properties of highly soluble chitosan -L-glutamic acid aerogel derivative. Carbohydrate Polymers 76:2, 188–195
40. Zeng M, Liu J, Shi F, Wang J, Guo S (2006) Preparation of SiO2 aerogel/chitosan composite material for biomedical applications. Dalian Qinggongye Xueyuan Xuebao 25:2, 121–123
41. Scanlon S, Aggeli A, Boden N, Koopmans R, Brydson R, Rayner C (2007) Peptide aerogels comprising self-assembling nanofibrils. Micro Nanolett 2 (2): 24–29
42. Reverchon E , Cardea S., Rapuano C (2008) A new supercritical fluid-based process to produce scaffolds for tissue replacement. J Supercrit Fluids, 45 (3): 365–373
43. Schugens Ch, Maquet V, Grandfils Ch, Jerome R, Teyssie Ph (1996) Polylactide macroporous biodegradable implants for cell transplantation. II. Preparation of polylactide foams by liquid-liquid phase separation. J Biomed Mat Res, 30 (4): 449–461

Part XI

Applications: Space and Airborne

32

Applications of Aerogels in Space Exploration

Steven M. Jones and Jeffrey Sakamoto

Abstract NASA has used aerogel in several space exploration missions over the last two decades. Aerogel has been used as a *hypervelocity particle capture medium* (Stardust) and as thermal insulation for the Mars Pathfinder, Mars Exploration Rovers, and Mars Science Lander. Future applications of aerogel are also discussed and include the proposed use of aerogel as a sample collection medium to return upper atmosphere particles from Mars to earth and as thermal insulation in thermal-to-electric generators for future space missions and terrestrial waste-heat recovery technology.

32.1. Introduction

Aerogels are micro- and mesoporous networks composed of randomly interlocking nanoscale filaments. This results in a highly porous, open cell structure that gives aerogels many unique physical properties. Despite the fact that aerogels are up to more than 99% by volume open pore space, they are solid materials. Among the more potentially useful properties of silica aerogels are optical transparency, low density, high porosity, high surface area, low thermal conductivity, low refractive index, and low dielectric constant. However, the aspect of aerogels that makes them truly unique is that they exhibit many uniquely valued physical properties concurrently. And since many of these properties are correlated to the density and the density is easily controlled, then so too are these physical properties.

Aerogels are usually produced by the hydrolysis and condensation of a metal alkoxide, but can also be produced from metal salts. These reactions result in a continuously connected, solid network that is filled with solvent. To remove the solvent while leaving the fragile wet gel network in its initial expanded state, the gel is pressurized above the vapor pressure of the solvent, heated to a temperature above the supercritical temperature of the solvent, and then depressurized to withdraw the supercritical solvent from the interstitial space of the network. This process is known as high-temperature supercritical

S. M. Jones • Jet Propulsion Laboratory, California Institute of Technology, 4800 Oak Grove Drive, MS 125-109, Pasadena, CA 91109, USA
e-mail: steven.m.jones@jpl.nasa.gov
J. Sakamoto • Chemical Engineering and Materials Science Department, Michigan State University, 2527 Engineering Building, East Lansing, MI 48824, USA
e-mail: jsakamot@egr.msu.edu

M.A. Aegerter et al. (eds.), *Aerogels Handbook*, Advances in Sol-Gel Derived Materials and Technologies, DOI 10.1007/978-1-4419-7589-8_32,

solvent extraction. Aerogels can also be dried by exchanging the initial liquid solvent with supercritical carbon dioxide and then removing the carbon dioxide, which does not require high temperatures. Many reviews about the production and properties of aerogels have been published in past few decades [1–4].

The proposed applications of aerogel are numerous and vary a great deal; yet most remain unrealized. Commercial applications such as thermal window insulation, acoustic insulation, optical coatings, capacitor electrodes, low dielectric constant layers in integrated circuits, piezoelectric transducers, and catalytic supports have all been proposed, but little in the way of actual use has resulted [1, 5–11]. However, silica aerogel monoliths have been used extensively in Cerenkov radiation detectors in high-energy physics experiments [12–16].

NASA has used aerogel in three recent space exploration missions. Silica aerogels were used in the Mars Pathfinder rover (thermal insulation), the Stardust Missions (particle capture medium), and in the 2003 Mars Exploration Rovers (thermal insulation). It will also be used in the Heat Exchangers in the Multi-Mission Radioisotope Thermoelectric Generator (MMRTG) on the upcoming Mars Science Laboratory (MSL) Mission. These missions utilize the unique thermal and mechanical properties of aerogel and have sparked continued interest in the further use of aerogel for future space science applications.

32.2. Hypervelocity Particle Capture

32.2.1. Initial on Orbit Studies

After conducting many ground-based experiments on the feasibility of using aerogels as a *hypervelocity particle capture media*, aerogels were first flown on the Space Transport System (STS) in 1992 to test their survival and particle capture properties in space [17]. Aerogel panels were placed on the lids of Get-Away-Special (GAS) Canisters in the payload bay of the STS-42 [18]. The aerogels proved to be sufficiently durable to survive the launch and reentry environments and captured three particles during the flight. Subsequent to that initial flight, aerogels were flown successfully on the lids of many other GAS Canisters.

In the mid-1990s, aerogels were also flown on Spacelab 2 and the Mir Space Station. The Orbital Debris Collection Experiment consisted of 0.63 m^2 of single density, 20 mg/cm^3 silica aerogel mounted on Mir, and stayed on orbit for 18 months. After being retrieved from the space station, it was returned to the Johnson Space Center for analysis. A wide range of impacts were observed in the aerogel from both man-made, for example, paint flakes and solder, and naturally occurring, for example, cosmic dust and objects. Many of the objects captured were removed from the aerogel and analyzed [19], thus paving the way for a mission based on using aerogel for hypervelocity particle capture and return.

32.2.2. The Stardust Mission

The Stardust Mission was designed around the fact that low-density aerogel had been demonstrated to be an excellent *hypervelocity particle capture medium* [17]. The mission plan was to transport a grid of aerogel cells into space, rendezvous with a comet, capture material from the comet in the aerogel, and return the collected samples to the earth [20]. The primary science requirement of the mission was that 1,000 cometary particles, 15 μm or larger in diameter, be captured and returned to Earth.

The Stardust spacecraft was launched successfully in February of 1999 from Kennedy Space Center on a Delta launch vehicle. During the cruise phase prior to the comet encounter, there were two extended periods when the collection grids were deployed in an attempt to capture interstellar particles. The orientation of the spacecraft was such that any interstellar particles captured would be collected in the interstellar grid of aerogel, which was specially designed for this purpose. The aerogel collection grids were deployed again in January of 2004 for the encounter with the comet 81P Wild 2. During the comet encounter, the aerogel grid was deployed such that particles from the coma of the comet impacted with the cometary aerogel and embedded themselves in the porous network of the aerogel (Figure 32.1).

Figure 32.1. The aerogel is seen deployed extending above the spacecraft (**left of center**) for particle capture in this artist's image of the Stardust spacecraft.

Having captured the cometary particles, the aerogel collection grid was retracted into the Sample Return Capsule (SRC) and the SRC was closed and latched. After cruising through space for 2 years, the spacecraft returned to Earth and released the SRC, which parachuted successfully to the Utah Testing and Training Range.

The primary instrument of the mission, the sample collection grids, was made up of two distinct arrays of aerogel blocks: one for the collection of particles from the comet during the comet encounter and one for the collection of interstellar particles during the cruise phase. The two grids were mounted back to back so that one could collect particles while the other was facing away from the particle stream during each particular collection period. The comet particle collection aerogels were all gradient density blocks, while the interstellar aerogels were a mixture of both gradient density and single density pieces (Figure 32.2). The cells for comet particle capture were approximately 4 by 2 by 3 cm, while those for the interstellar particles are 4 by 2 by 1 cm.

Aerogel is an excellent *hypervelocity particle capture medium* due to the fact that it is a highly porous material with a microstructure made up of nanoscale filaments (Figure 32.3). The particles impacting the aerogel are microns in diameter, while the filaments of the aerogel are tens of nanometers in length and width. Therefore, the filaments yield to the force exerted by the particles. As the individual filaments are crushed, the kinetic energy of the particle is continuously transferred to the aerogel network as it is converted to thermal and mechanical energy over the course of the capture track. Thus, the particle is slowed and finally stopped, largely intact (Figure 32.4). When small dense particles are captured in low-density aerogel, they produce a characteristic long, gradually tapering

Figure 32.2. The cometary (*facing forward*) and interstellar aerogel grids were assembled back to back for separate cometary and interstellar particle capture periods. The aerogel cells shown are gradient density typically 10–50 mg/cc, as explained in the text.

Figure 32.3. The nanometer-scale filaments composing a silica aerogel network can be seen in this scanning electron microscope image. The aerogel density is 13 mg/cc.

penetration track. These are commonly known as a "carrot" or "stylus" tracks, with a particle at the terminus. Impact features known as "pits," which are relatively shallow and wide, have also been observed in many aerogels impact tested in light gas guns, in samples returned from low earth orbit capture experiments, and in the Stardust aerogels. Pits often have several short stylus tracks leading from them, with terminal particles. The additional observation of blunt-nosed, cylindrical impact features with no terminal particles suggests that there is an entire spectrum of the morphology of such features [21].

Gradient density silica aerogel was developed and produced for the Stardust particle capture grids, since it was considered to have superior capture properties. When high-velocity particles begin to penetrate the *gradient density aerogel*, they first encounter

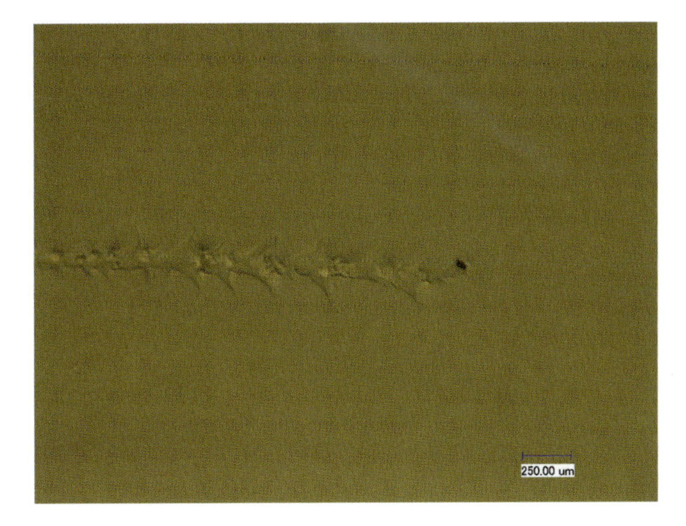

Figure 32.4. Entry track produced by the capture of 50 μm hypervelocity fine particle in silica aerogel. The aerogel shown was gradient density 15–50 mg/cc.

the low-density material that begins the deceleration process. As the particle moves further into the aerogel, gradually slowing as it progresses, it continuously encounters higher density material, which increases the resistance on the particle. The cometary aerogel had a gradient profile that changes from approximately10 mg/cm^3 at the impact surface to 50 mg/cm^3 in the lower half of the cell. The specific density profile varied from cell to cell, with some composed of more low-density aerogel, while others had much more high-density aerogel.

To produce the aerogel, a sol (a suspension of submicron-sized particles in a liquid) was made from tetraethyl ortho silicate (TEOS) (Chap. 2) [22]. This sol was produced by refluxing TEOS in ethanol with water and a trace amount of nitric acid. Once the reflux was complete, most of the ethanol was removed by distillation and the resulting sol was diluted with acetonitrile. To form the aerogel precursor mixture, the sol was mixed with acetonitrile, water, and ammonium hydroxide. The density gradient was achieved by a method that gradually diluted the gel precursor and then pumped the mixture into the molds in which the aerogels were cast [23]. To accomplish this, a precursor that would result in a high-density aerogel was pumped initially to the base of the molds from stock high-density precursor. A precursor mixture that would result in a low-density aerogel was then slowly pumped to the stock high-density precursor, where they were stirred together. This resulted in a gradual dilution of the high-density precursor. As the dilution took place, the mixture was continuously pumped to the mold. During the dilution process, the concentration of the hydrolyzed and partially condensed silica decreased and thus the final density of the aerogel from the bottom of the profile to the top decreased. Once the pumping was complete, the precursor was allowed to gel and it was then dried by high-temperature supercritical solvent extraction in a semiautomated autoclave.

By varying the rate at which the precursors were combined and the rate at which the mixture was pumped to the mold, different gradients were produced. The density of the final aerogel was determined by measuring the refractive index of the aerogel. A laser with a line generator was passed through the corner of a piece of aerogel, which acts as a prism, thus deflecting the laser beam. Since the density of the material and the refractive index are correlated, the density was then calculated from the value of the deflection.

While this method of pumping together aerogel precursors for different density aerogels was developed for producing *gradient density aerogel*, it should be noted that the method could be used to produce materials with gradient thermal, optical, acoustic, and dielectric properties, since these properties are density dependent. It could also be used to produce gradient oxide or dopant compositions. Materials with novel compositional gradients such as these could also someday be used in future space exploration missions.

The Stardust Mission control requirements of the flight aerogel were that it be capable of efficiently capturing hypervelocity particles, that it be the proper shape and size, that it be free of significant flaws, for example, cracks and inclusions, that it be of reasonably high purity, and that it be capable of surviving launch and landing environments. Efficient *hypervelocity particle capture* was verified primarily by testing with light gas gun hypervelocity impact testing. Many tests were conducted in which fine particles (10–100 s of microns in diameter of a variety of compositions) were impact tested in aerogels and then imaged and extracted. Analyses of the capture tracks led to an initial understanding of the hypervelocity capture process in aerogel. Suitable quartz molds and processing methods were developed to produce aerogel blocks that were of the correct shape and size, were free of significant flaws, and were of acceptable purity. The shape and size of the aerogel was crucial since the aerogel was held in the flight grid by compression of the individual aerogel cells. The ribs between the spaces for the aerogel cells in the flight grid had a draft angle of 1.455°, so that the size of the space at the impact *surface* was slightly smaller than that at the back of each grid. Aerogel cells were produced that approximately matched the size of the back of the space and therefore were compressed as they were inserted into the spaces. The launch and landing survivability of the aerogel was tested by conducting random shake, landing pyro shock tests, as well as by launches and landings of aerogel grids that were flown on the Space Shuttle. During the Stardust launch, the aerogel also had to survive the transition from atmospheric pressure to the vacuum of space. Since aerogel is an open cell network, interstitial gases are able to flow out of the network. Testing was conducted to demonstrate that pressure changes experienced during launch and reentry would not damage the aerogel in any way.

The Stardust spacecraft was successfully launched in February of 1999 and encountered the comet Wild 2 in January of 2004. The encounter with Wild 2 went according to the mission design, collecting both particles from the comet and close-up images of the comet. In 2006, the SRC containing the aerogel grids parachuted to Earth successfully. The samples returned by Stardust are the first material samples from a specified extraterrestrial body, other than the Moon. The analyses done thus far indicate that comets contain organic molecules not observed in other extraterrestrial materials, for example, meteorites, and that the composition of comets are more complex than previously believed [24, 25]. Comparisons with existing data gathered from meteorites, micrometeorites, and interplanetary dust particles are ongoing and are anticipated to improve models of the origin of the solar system and of silicate materials in our solar system. The results gathered from the samples collected by Stardust have already indicated that there was significantly more radial mixing of the solar nebula than previously known [26]. This conclusion has been arrived at by the fact that high-temperature minerals have been identified in the samples. These minerals must have been formed relatively close to the sun and then been transported to the outer regions of the solar nebula and incorporated into comets.

The analysis of the aerogel grids that were deployed to collect interstellar dust during the cruise phase of the mission is currently under way. Interstellar dust is a key component of the interstellar medium, since it is ubiquitous and interacts extensively with solar radiation. Identification and analyses of contemporary interstellar dust samples would provide valuable information regarding galactic chemical evolution and *stellar nucleosynthesis*. The search for captured interstellar grains is being done by first scanning the aerogel cells with an automated microscope, which takes digital images of minute portions of the cells. The images are being analyzed by thousands of volunteers over the Internet [27]. Tracks identified by the volunteers have been extracted from the aerogel and in many cases analyzed for elemental composition. While no interstellar particles have been definitively identified, the process for locating them, extracting them, and analyzing them has been demonstrated.

The aerogel cells used on the Stardust collection grids withstood the rigorous testing that they were subjected to and the launch and landing environments. None of the cells exhibited any damage from these environments or that of space. This demonstrates quite clearly that aerogel, despite its relatively fragile nature, can survive highly stressful conditions if it is properly contained. In this case, the containment structure was a simple aluminum frame.

32.2.3. The SCIM Mission (Proposed)

The proposed SCIM (Sample Collection of the Investigation of Mars) Mission design is based on the sample capture and return concept developed for the Stardust Mission. It would fly a spacecraft through the upper Martian atmosphere, collect dust particles that are suspended in the atmosphere in aerogels, and return the sample to Earth (Figure 32.5) [28, 29]. This would be an impressive feat, in that it would be the first sample collection and return mission from another planet. And since the mission design would not require descent to and ascent from the planetary surface, it is a low-cost, low-risk mission. The proposed sample collection using aerogel and the design of the SRC are based largely on those designed, tested, and produced for the Stardust Mission [30]. Like the Stardust Mission, returning extraterrestrial samples to Earth for study would make available the entire

Figure 32.5. Artist's image of the proposed SCIM spacecraft exiting the Martian atmosphere after the sample collection period.

spectrum of handling and analytical techniques available, rather than the extremely limited techniques that can be transported to the sample's location. One advantage of this scheme for sample collection over landing on the planetary surface would be that since the dust in the Martian atmosphere represents a global sample of the surface of Mars, this mission would return a global sample rather than a sample primarily specific to one particular location. SCIM would also collect a sample of the Martian atmosphere, which would also be returned to Earth for analysis.

The science goals of the mission would be to (1) determine the extent of aqueous processing of *Martian crustal materials*, (2) quantitatively establish the chemical, isotopic, and mineralogical composition of the Martian dust, (3) quantify the size and compositional distribution of dust in the Martian atmosphere, (4) provide ground truth for past and future remote sensing observations of Martian surface materials, (5) determine whether or not Martian SNC (shergottite, nakhlite, and chassigny) meteorites originated on Mars, (6) find evidence relevant to Mars' original volatile chemical inventory and its subsequent evolution, and (7) investigate current atmospheric chemical composition. Accomplishing these goals would make possible basic advances in our understanding of the geological and the climatic history of Mars and, ultimately, its habitability by humans.

Since the amount of dust suspended in the Martian atmosphere varies a great deal by season, by altitude, and by latitude, the SCIM Mission would be planned so that it would arrive during the period of greatest atmospheric dust content and at the best altitude and latitude for maximum dust collection. The presence of a significant dust storm on Mars cannot be accurately predicted; however, a dust storm would only increase the atmospheric dust content and thus the amount of sample collected and returned.

Unlike the Stardust Mission, SCIM would use two trains of individual aerogel collection modules (Figure 32.6). Each train of collector modules, located on opposing sides of the spacecraft, would be launched in place with ejectable covers protecting the aerogel. Prior to the aeropass through the Martian atmosphere, the covers would be ejected. After the aeropass has been completed, the aerogel would be pulled back into a sterilization oven, where it would be heated at 250°C for several hours. This would be done to satisfy Planetary Protection Protocols, ensuring that no live organisms that might possibly be

Figure 32.6. Proposed SCIM collection modules containing silica aerogel with a density of 20 mg/cc for hypervelocity particle capture.

present in or on the dust would survive the return to Earth. The aerogel cells would then be stowed in the SRC for Earth return. Based on the successful return scenario of the Stardust samples, the SRC would be released as the spacecraft approached the earth and would parachute to the Earth for retrieval. The aerogel collection modules would be taken to a specially designed curatorial facility for preliminary analyses. The techniques developed for the analysis of the Stardust samples and the lessons learned in employing those techniques would be brought to bear on the returned Martian samples.

Since the SCIM spacecraft would be passing through the Martian atmosphere at a velocity of several kilometers per second during the collection phase, the aerogel would be required to survive significant heating during the collection period. Computer models were used to simulate the atmospheric gas flow field around the spacecraft and over the aerogel collectors. The data obtained from these models were used to calculate the surface heating of the aerogel collectors. Test assemblies were built of simulated collection modules containing aerogel and tested in the arc jet facility at the Ames Research Center Aerodynamic Heating Facility (AHF). The test assemblies were subjected to heating rates of between 1.5 and 10 W/cm^2 for 120 s. Silica aerogel samples that were tested at heating rates of less than 6.5 W/cm^2 exhibited no significant surface melting. To ensure that the heating of the aerogel would be less than 6.5 W/cm^2, the design of the spacecraft was further refined such that the aerogel collection modules would be housed in two bands on opposite sides of the aeroshell. Particles greater in diameter than 2 μm would travel in a straight path into the aerogel, while the atmosphere would flow past the sloped aeroshell and the aerogel collector modules. Extensive modeling of the dust content of the Martian atmosphere has indicated that even under poor collection conditions, the proposed SCIM Mission would collect more than the Science Requirement of 1,000 particles.

The SCIM Mission was first proposed as a 2003 Mars Scout Mission, proceeded through a Phase A study period, but was not selected as the final flight project. It was proposed a second time as a 2006 Mars Scout Mission, but was again not selected as the final flight mission. Since sample return from Mars is considered a high priority goal for the Mars Exploration Program, a mission such as SCIM should continue to remain as a viable candidate for Mars sample return. Since Stardust proved to be highly successful, its technical successes could be applied to other sample capture and return missions. In the 10 years since the delivery of the aerogel for the Stardust mission, the production and testing of nonsilica aerogel have advanced considerably. By using newer, more highly developed types of aerogels for the collectors in potential future missions, such as a SCIM-type mission, the science return from these missions could be expanded.

32.2.4. Nonsilica Aerogels

The study of the comet particles returned by the Stardust Mission could have been made easier and more comprehensive if aerogels other than silicon dioxide had been used. This is due to the fact that since silicon was the major element in the capture medium and most of the Wild 2 particles fragmented and mixed at the microscopic scale with the aerogel, it was not possible to determine elemental ratios using silicon as the standard, for example, iron to silicon ratio and nickel to silicon ratio. Silicon is typically used as the standard for the determination of elemental abundances in geochemistry and cosmochemistry, since it is found throughout our planet and the solar system. Since silicon could not be used and iron was used as the standard element for the Stardust geochemical analyses, comparison of the Wild 2 analyses and existing reference data of other samples was complicated. Using both silica and nonsilica aerogels as capture materials would mitigate this problem. Unfortunately, at the time of the Stardust silica aerogel flight production, nonsilica aerogels were not sufficiently developed to be used for this purpose.

Since the time of the delivery of the flight aerogel for the Stardust Mission, studies have been done to determine the suitability of nonsilica aerogel as efficient hypervelocity particle capture media. Scanning electron microscope (SEM) images of the networks of various nonsilica aerogels indicate that they are similar in microstructure to silica aerogels (Figure 32.7). However, the fact that the networks of the various aerogels are similar to that of silica aerogel does not verify that they are efficient hypervelocity capture media. This could only be verified by actually conducting hypervelocity impact tests.

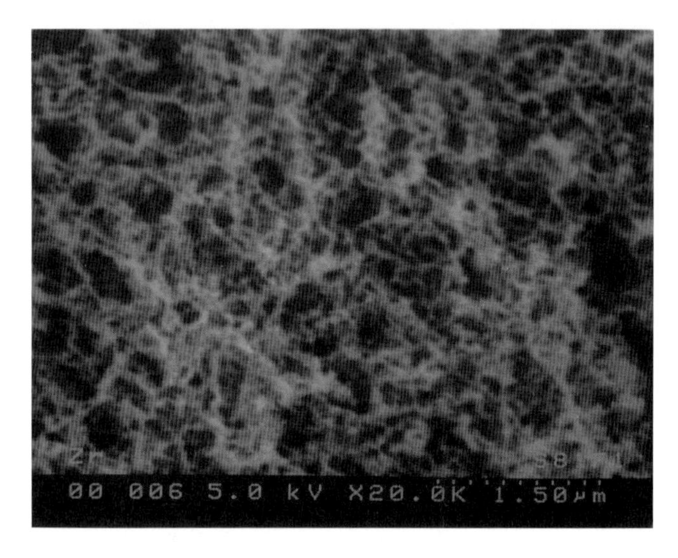

Figure 32.7. Zirconia aerogel network as seen in a scanning electron microscope. Note similarity to silica network shown in Figure 32.3. The density of this aerogel is 45 mg/cc.

Initial hypervelocity impact tests on nonsilica aerogel were done using 20, 50, and 100 μm glass microspheres as the projectiles. The microspheres were launched with a light gas gun and captured in different types of nonsilica aerogels. These tests indicated that resorcinol–formaldehyde (RF) (an organic polymeric aerogel) (Chap. 11), alumina, and zirconia (Chap. 6) aerogels were the most promising candidate materials based on their structural robustness and their ability to efficiently capture glass microspheres at high velocities [31].

Further testing was then conducted using meteoritic fragments as the projectiles. Meteoritic fragments were considered to be reasonably good analogs of the type of extraterrestrial materials that might be captured during future missions using nonsilica aerogels. To be able to compare the results of these tests to the analyses done on the returned Stardust comet samples, the same methods of in situ analysis and particle extraction were used. *Synchrotron X-ray microprobe* (SXRM) was used to locate and conduct in situ compositional analyses of the captured particles. The captured meteoritic particles were located by identifying their high iron content, whereupon chemical analyses of their compositions were done. The RF and alumina aerogels proved to be the materials best suited for this analytical technique. This is due to the fact that the X-ray background scattering of these materials is low due to their low atomic numbers, that are, 6 and 13, respectively, while zirconia has an atomic number of 40. The smallest particles that could be analyzed in the zirconia aerogel were 22 μm in diameter, whereas particles as small as 7 μm in diameter could be analyzed in the RF and alumina aerogels. The latter results are comparable to those obtained for silica aerogels.

Cutting small, triangular wedges, known as keystones, of aerogel around the track(s) and terminal particle(s) is the accepted technique for extracting particles captured in the Stardust aerogel. This is done by micromachining the aerogel cells and removing the wedge with a microscopic fork [32]. A keystone was cut and removed from a sample of RF aerogel that had been impact tested (Figure 32.8). The composition of RF aerogel proved to be quite suitable for this technique. Alumina aerogel tended to crumble into small pieces when the micromachining required to form a keystone was attempted. Having demonstrated that RF aerogel can capture hypervelocity particles efficiently, that the captured particles can be analyzed in situ with SXRM, and that keystones can be cut and removed effectively, this material should be considered for future hypervelocity sample capture and return missions. While the carbon content would preclude any analyses of organic composition of captured particles, it could be used to complement silica aerogel so that silicon could be used as the standard on elemental ratios analyses.

Figure 32.8. After hypervelocity capture of meteoritic particle, this keystone was cut and removed from the bulk sample of RF aerogel. Approximate dimensions are 600 by 250 by 75 μm.

32.2.5. Calorimetric Aerogel

Nonsilica aerogels have been developed for the capture, identification, and analysis of interstellar grains, which are extremely important to current astronomical studies. However, due to the small size of these grains, typically less than 1 μm in diameter, identifying them once they have been captured in aerogel is extremely difficult. Work involving the production of nonsilica and mixed oxide aerogels has been done on the development of novel aerogel materials that could greatly facilitate the identification process. Alumina/silica aerogels doped with transition metals, for example, titanium, terbium, and gadolinium, have been produced and tested as capture media for interstellar grains [33, 34]. When a hypervelocity particle is captured in a porous material, such as low-density aerogel, much of the particle's kinetic energy is converted to heat. When this happens in these doped aerogels, the heat generated causes the aerogel immediately adjacent to the penetration track to undergo a permanent phase change. By correctly choosing the matrix and dopant materials, the resultant phase-changed materials are highly fluorescent. Thus, the capture of a hypervelocity particle results in a distinct fluorescent signature that is easily identified with a low power fluorescent microscope (Figure 32.9). Since the strength of the fluorescent signal is proportional to the kinetic energy of the particle, the aerogel becomes a calorimeter. A variety of compositions of these types of aerogels have been produced and impact tested, and proposals for their use in a flight detector are being developed.

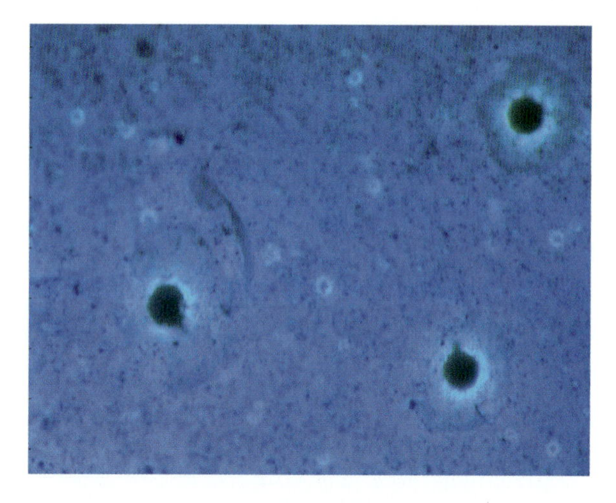

Figure 32.9. Fluorescent image of impact features of hypervelocity particles into calorimetric aerogel. Particle velocity was 4.7 km/s.

New specialty aerogels such as these could greatly enhance the use of aerogel in astronomical research. For example, a grid of such aerogel could be flown in low Earth orbit on the International Space Station to capture extraterrestrial particles and orbital debris. Since interstellar grains have velocities much greater than *anthropogenic* orbital debris, the fluorescent signatures they produce during their capture could be used to identify the impact locations of particles of potentially interstellar origin. Further analysis would be required to definitively state that they are interstellar grains. Since these materials are calorimetric, they could be used in a variety of future sample capture experiments to determine the velocities of different types of interplanetary and interstellar particles.

32.3. Thermal Insulation

32.3.1. 2003 Mars Exploration Rovers

In May and June of 2003, the Mars Exploration Rovers, Spirit and Opportunity, were launched from Kennedy Space Center. They successfully landed and were deployed in January of 2004 (Figure 32.10). Their exploration missions began in February of 2004 and for more than 5 years they have been exploring a variety of geological features on the surface of Mars.

Silica aerogel was used as the thermal insulation material (Chap. 23) in the Warm Electronics Boxes (*WEB*) of each of the rovers to meet the rigorous thermal requirements

Figure 32.10. Artist image of a 2003 Mars Exploration Rover on the surface of Mars.

of surviving the Martian environment for an extended period of time [35]. The thermal insulation was used in addition to the heat generated by Radioisotope Heater Units (*RHU*s) to keep the rover electronics at a relatively steady temperature during the approximately 100°C variations in temperature between Martian days to nights. Aerogel was used since it is very low weight and is an extremely efficient thermal insulator.

Aerogel was first used as the thermal insulation material in an exploration rover in the 1996 Mars Pathfinder rover, Sojourner. Since the aerogel used in Sojourner was found to be successful, aerogel was also selected to be used in the 2003 rovers. However, the aerogel used in Pathfinder was plain silica aerogel, which is transparent to radiative thermal transport. To minimize radiative thermal transport, the aerogel for the 2003 rovers was rendered opaque by adding 0.4% (w/w) graphite. Due to its high volume nano- and mesoporosity, silica aerogels have extremely low conductive and convective thermal transport properties. However, since it is largely transparent to infrared radiation, it does not prevent radiative thermal transport. By making the aerogel opaque, radiative thermal transport is minimized because thermal energy is absorbed at the surface and then either irradiated back out of the aerogel or slowly transported into the aerogel.

Since the precise dimensions of the pieces needed were not known when the aerogel production began, relatively large panels, for example, 38 by 28 by 2.5 cm, were produced and then were cut and shaped to produce the final pieces (Figure 32.11). A two-step process was used to produce the aerogel, similar to the one described in Sect. 32.2.2 [22]. The graphite was added by mixing colloidal graphite into the liquid aerogel precursor. The graphite incorporated into the gel was found to significantly change the drying characteristics of the wet gels, so changes in the processing of the sol were necessary. After extensive testing, it was discovered that to produce panels without significant cracking and shrinking taking place during the supercritical drying process, the sol had to be refluxed for extended periods of time.

The 2003 MER landings were accomplished by the deployment of balloons around the spacecraft and bouncing the spacecraft along the Martian surface until it came to a stop. Since the aerogel appears to have functioned successfully after the landings, it is assumed

Figure 32.11. Silica aerogel (density of 20 mg/cc) opacified with graphite was cut and attached to the walls. The aerogel is seen as the *bluish-gray materials* under the gold Kapton.

that it survived these impacts intact. So, once again it has been demonstrated that if low-density aerogel is well contained, it can withstand violent conditions and remain intact and fully functional.

The 2003 Mars Exploration Rovers are still traversing the surface of Mars, 5 years after being deployed. In part, the longevity of the rovers is attributed to the successful performance of the aerogel. Since the aerogel has acted as such an efficient thermal insulator, with the heat from the *RHU*s, the power required from the batteries to keep the *WEB* warm has been kept to a minimum.

32.3.2. Mars Science Laboratory

The MSL will be using graphite-doped silica aerogel as a component of its thermal control systems. However, the aerogel will not be used as the thermal insulation material in the MSL rover chassis, as it was in the chassis of the 2003 Mars Exploration Rovers. Since MSL is much larger than the 2003 rovers, its thermal and power architectures are quite different. The 2003 rovers used solar panels to generate electrical power for on-board power requirements. MSL will generate electrical power with a Multi-Mission Radio Isotope Thermoelectric Generator (MMRTG) specially designed to work on a planetary surface. The MMRTG generates electrical power by the conversion of heat that comes from a General Purpose Heat Source (GPHS) to electrical power (see Sect. 32.3.3). On MSL, the aerogel will be used as thermal insulation in the heat exchangers that extract heat from the MMRTG. To maintain the thermal balance requirement of the system, waste heat must be rejected. This is done by heat exchangers, which absorb the waste heat on a hot side and reject the heat on a cool side. The aerogel-based thermal insulation is being used to separate the hot and cold sides of the heat exchangers. Graphite-doped silica aerogel will be used to fill the open spaces of a resin-based honeycomb core (Figure 32.12). The honeycomb core provides structural support for the panels, while the aerogel acts as the thermal barrier between the hot and cold sides.

Figure 32.12. Heat exchanger panel for thermal stability of MSL MMRTG. MSL is scheduled to launch from the Kennedy Space Center in 2011 and land on Mars in 2012.

32.3.3. Thermoelectrics

Over the last four decades, NASA has used Radioisotope Thermoelectric Generators (RTGs) to provide electrical power on-board deep space, lunar, Mars, and Earth orbit spacecraft. Because RTG technology does not involve moving parts or require exposure to sunlight, it is reliable and has successfully provided continuous electrical power for the Apollo, Voyager, Galileo, and Cassini missions to name a few. However, at present the supply of plutonium (the radioisotope, Pu) is limited; thus NASA and the Department of Energy (DOE) are interested in improving the efficiency of RTGs to reduce the required mass of Pu. Additionally, DOE (Energy Efficiency and Renewable Energy or EERE) is interested in reducing energy consumption and the use of fossil fuels, which has created the impetus to develop new energy-efficient technologies [36]. Thermoelectric technology is an example of a technology that can improve energy efficiency of industrial processes and internal combustion engines used in transportation [37, 38].

In most cases, thermoelectric technology requires heat collectors that have significantly larger cross sections compared to the electrodes to achieve heat fluxes in the range of 1–10 s of W/cm^2. Thus, thermal insulation with extremely low thermal conductivity must be used between the heat collector and the cold side. The ratio of heat channeled through the thermoelectric elements to the heat lost to parasitic heat losses is referred to as thermal efficiency and is >90% for NASA's RTG technology for use in space vacuum. *Heritage RTG* technology employs Multilayer Insulation (MLI) since it is the most effective insulation in vacuum. However, planets such as Mars, Venus, and, of course, Earth have significant atmospheric pressures such that MLI cannot be used. To this end, aerogel-based thermal insulations have been developed that can be used as effective insulation for thermoelectric technology for use in advanced radioisotope power systems and terrestrial generator technology. In addition to improving thermal efficiency, there are other benefits for employing aerogel-based thermal insulation in thermoelectric technology. For example, aerogel is synthesized through sol–gel chemistry; thus it can be cast into place to achieve molecular scale intimacy with the thermoelectric elements. This level of intimacy is key in mitigating parasitic heat losses that reduce system efficiency. Furthermore, the intrinsic microporosity of aerogel not only reduces gas transport, but also slows the sublimation rate of volatile molecular species. In this section, a brief review of the aerogel development for thermoelectric power generation is given. The work presented here focuses on aerogel integration, thermal characterization, and aerogel sublimation suppression.

Advanced RTG technology as well as novel thermoelectric generators (*TEG*) for terrestrial applications will require new and improved heat transfer technology that can operate in space vacuum (as in the past) as well as on other planets with atmospheres including Earth. DOE's EERE Program and NASA are considering new thermoelectric device technology such as the schematic depicting a Cascaded Multicouple Thermoelectric Module (CTMM) pursued under NASA's Project Prometheus earlier this decade (Figure 32.13). As heat exchanger technology improves, the resulting heat fluxes are likely to increase; thus, the demand for better thermal insulation is necessary. For example, the heat collector shown in Figure 32.13 has a larger footprint than the stack of thermoelectric elements; thus, effective insulation occupying the open space underneath the heat collector is required. Additionally, the space between thermoelectric electrodes (approximately 0.1–1 mm) must also be filled to prevent parasitic heat losses and suppress and volatile molecular species subliming from the hot side of the electrodes and condensing on the cold side, which causes short circuiting. Lastly, the insulation must be stable up to the maximum operating temperature of the RTG or *TEG*, which can be as high as 1,000°C. Simply stated, new thermoelectric device technology requires advanced thermal insulation that (1) can be cast into place to fill micron to centimeter scale voids, (2) has ultralow thermal conductivity, (3) has high thermal stability, and (4) can prevent or slow the sublimation of volatile molecular species. The work conducted over the last decade has identified silica-based aerogel as a material that can meet these four criteria, each of which is discussed in detail below.

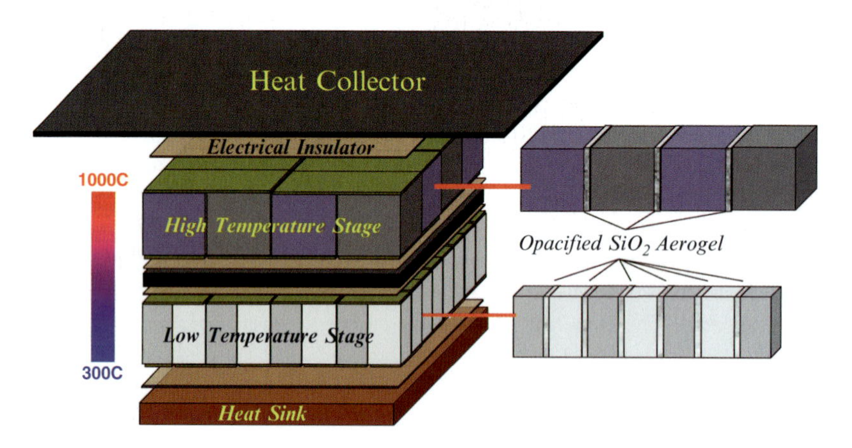

Figure 32.13. A proposed thermoelectric module developed under Project Prometheus.

Thermoelectric electrodes typically range from microns to centimeters in scale. At this length scale, bulk high-temperature insulation cannot be used, because it cannot be cast into place. Aerogel is synthesized through sol–gel chemistry; hence it can be cast into place and the thermoelectric device along with the aerogel can be simultaneously dried in a supercritical solvent extraction system. The cast into place aspect enables intimate contact to be made between the thermoelectric electrodes and the aerogel insulation. The integration of aerogel insulation is intended to be one of the last steps in fabrication; thus, all free volume is filled with ultralow thermal conductivity insulation to improve thermal efficiency.

An example of recent work involving the use of thermoelectric technology to capture waste heat from the diesel trucks is shown in Figure 32.14. The prototypical *TEG* was tested on a test bed, whereby simulated exhaust gas (heated nitrogen gas) passed through the open volume at the center of the box-shaped *TEG*. Two hundred and fifty thermoelectric electrodes were encapsulated with opacified aerogel and evenly distributed on the four panels that comprise the box. The aerogel was cast into place such that only the heat collectors were exposed and the thermoelectric electrodes were completely encapsulated. In this particular application, the inherently low thermal conductivity of the silica aerogel insulation enables the placement of thermoelectric heat collector/electrode assemblies directly in the exhaust stream. Because the hot side temperatures of terrestrial and space *TEG*s ranges from 400 to 1,000°C, pure silica aerogel cannot be used since it is more or less transparent to radiation. It is known that at temperatures significantly greater than 25°C radiative heat transfer dominates; thus, silica aerogel must be opacified. Work conducted by Fricke et al. [39] demonstrated the efficacy of adding TiO_2 powder to silica-based aerogel to significantly reduce radiative heat transfer up to 500°C in air.

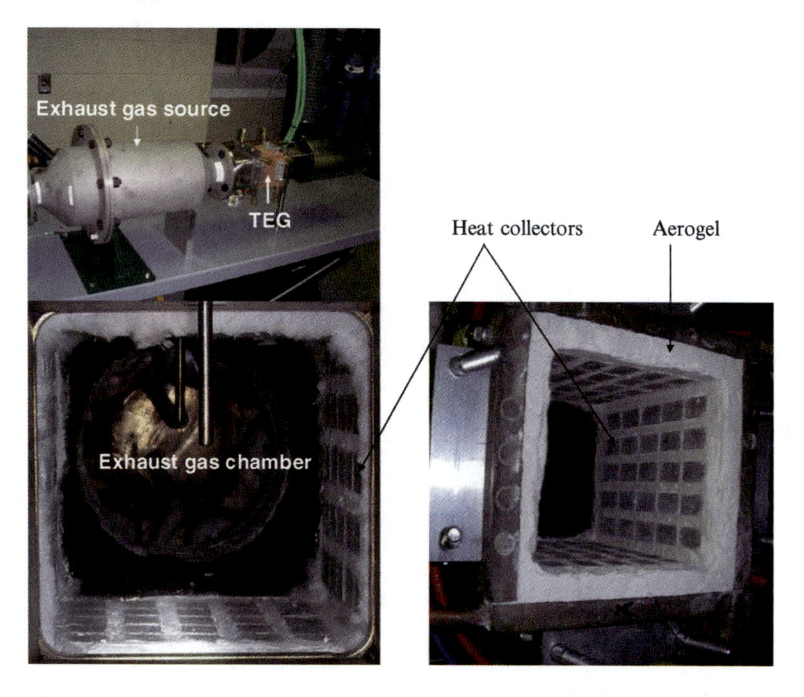

Figure 32.14. Aerogel-encapsulated *TEG* for integration into waste-heat recovery applications.

Recent efforts have focused on characterizing the effect of TiO_2 concentration on the thermal conductivity of silica-based aerogel up to 800°C in vacuum. There are limited published thermal conductivity data in this range as it can be difficult to manufacture aerogel coupons big and mechanically robust enough to withstand the rigors of *ASTM* standard techniques used to measure thermal conductivity. In this work, the knowledge gained from our efforts to improve the thermal and mechanical integrity of aerogels, some of which is described below, was used to enable the fabrication and testing of several precision fabricated 4 cm diameter coupons using an *ASTM* calorimetric technique. The goal of this work was to determine the optimum concentration of TiO_2 that would give the lowest thermal conductivity at elevated temperatures primarily in high vacuum.

Several silica aerogel coupons were fabricated and consisted of quartz fiber felt, various concentrations of 1 μm TiO_2 powder, quartz powder, and fumed silica powder. For the most part, all the constituents were kept the same except for the concentration of TiO_2 powder. Basically, the powders were combined in acetonitrile, agitated and combined with the two-step sol–gel process described in Sect. 32.2.2. The sol was then cast into molds prefilled with the quartz fiber felt, allowed to gel, aged for 24 h, supercritically dried, and the thermal conductivity was measured.

Since it is known that radiative heat transfer is the dominant heat transfer mechanism in porous materials at high temperatures, the intent was to study the effect of TiO_2 concentration (the opacifier) on thermal conductivity in silica-based aerogel. The range of TiO_2 concentrations varied from 0 to 200 mg/cm^{3}aerogel in 50 mg/cm^3 increments. The thermal conductivity values for these coupons tested in high vacuum (10^{-5} torr) are shown in Figure 32.15. There are two important points to be made from these data. First, at 297 K, radiative heat transfer is not the dominant heat transfer mechanism; thus, all compositions have approximately the same thermal conductivity. However, as the temperature increase above 573 K, radiative heat transfer dominates as is apparent from the sample with no TiO_2 opacification. As TiO_2 is added, for example, 50 mg/cm^3, the radiative heat transfer (and therefore the overall thermal conductivity) is significantly reduced. At the 100 mg/cm^3 concentration of $TiO_{2,}$ the radiative heat transfer is further reduced and is believed to be the optimal concentration as the thermal conductivity is the lowest for all the coupons tested. The net effect of adding more than 100 mg/cm^3 TiO_2 results in the further reduction of radiative heat transfer; however, above this concentration the TiO_2 particles form a percolative network to the point that solid-state conduction increases. Thus, the thermal conductivity increases above 100 mg/cm^3 TiO_2. Second, we believe that the thermal conductivity values for the 100 mg/cm^3 TiO_2 sample are the lowest ever reported for any material other than MLI in high vacuum. These data bolster

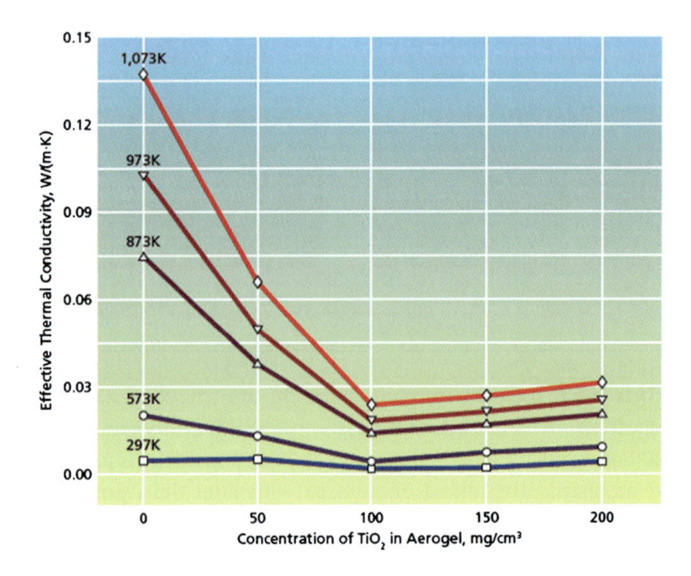

Figure 32.15. Thermal conductivity of silica aerogel opacified with TiO_2 powder in varying concentrations. The aerogel density was 50 mg/cc.

the notion that silica aerogel has the lowest thermal conductivity of any material in its class from low to high temperatures. Furthermore, although the data are not reported in this section, we measured the thermal conductivity of 100 mg/cm^3 TiO_2 aerogel in 1 atm Argon and found a relatively small increase of 16% in thermal conductivity at 1,073 K, which demonstrates the efficacy of blocking gas conduction with aerogel.

Since terrestrial and space *TEG*s are to be operated at elevated temperatures for extended periods of time, it is important to characterize the thermal stability of the above-mentioned opacified aerogel formulation. Thus, a study was conducted to characterize the thermal stability of various aerogel formulations. Aerogel coupons were prepared using the same methodology and dimensions as described in Sect. 32.2.2. Basically, coupons were heated in air and in vacuum from 100 to 1,000°C for 4 h at a time after which the largest dimension was measured to determine the degree of isotropic shrinkage. The standard pure silica aerogel shrank 5 and 45% at 700 and 1,000°C in air, respectively. By adding the quartz fiber felt, *opacifying* agent, and other powders, as described above, the shrinkage at 1,000°C was negligible and was less than 3% in air. It is believed that the fiber reinforcement and perhaps the addition of the powder constituents reinforce the silica aerogel network mitigating the effects of sintering at high temperatures especially in air compared to vacuum. As part of another effort to develop aerogel-based thermal insulation for *Stirling engine* technology (Sect. 32.3.4), tests are currently under way to characterize the thermal stability as a function of months at elevated temperatures.

Although the rationale for integrating aerogel into thermoelectric technology for thermal insulation is clear, aerogel can serve another purpose of equal significance: suppressing sublimation in thermoelectric electrodes. One of the primary modes of degradation in thermoelectric power generation is related to sublimation of volatile molecular species (Figure 32.16A). As sublimation occurs, the cross section of the thermoelectric electrodes is reduced, resulting in an increase in electrical resistance and thus a decrease in electrical power output. *Heritage RTG* technology employing SiGe electrodes used thin ceramic films to suppress sublimation of Ge. The efficacy of the ceramic coatings was sufficient to slow the sublimation rate to a level commensurate with the power output degradation rate to match the life of the mission (typically in the 10–30+ year range). Recently, work was conducted to develop thin ceramic coatings to suppress sublimation of advanced thermoelectric electrodes containing antimony (Sb, which readily sublimes

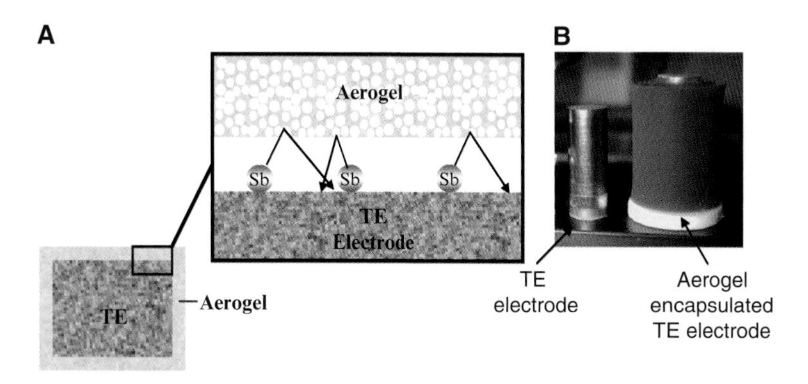

Figure 32.16. Aerogel-based sublimation coatings suppress sublimation of volatile metals by impeding vapor transport **A.** Aerogel coatings are cast around thermoelectric electrodes to achieve intimate contact between the electrode and the aerogel coating **B.**

at higher temperatures); however, the coatings were not chemically and mechanically robust enough to remain adhered to the surface. As an alternative approach, related work has identified silica-based aerogel as an effective Sb sublimation barrier for use with advanced thermoelectric electrodes containing Sb. First, silica is chemically stable against Sb vapor. Second, aerogel has an intrinsically low elastic modulus; thus, it can expand and contract along with the thermoelectric electrode to maintain intimate contact through multiple thermal cycles. Finally, the micro- and mesoporous nature of aerogel creates a highly tortuous path for Sb vapor such that sublimation rates can be significantly slowed [40].

The standard two-step sol–gel process [22] was used to encapsulate thermoelectric electrode coupons (Figure 32.16B) in opacified aerogel. The coupons were evacuated in quartz ampoules to simulate space vacuum and heated in a gradient to allow Sb vapor to sublime from the coupon and condense on a colder spot within the ampoule. The goal of this work was to control the sublimation rate to reduce the power degradation associated with sublimation over a 10-year mission. The aerogel-coated thermoelectric electrodes sublimed 1,000 times slower compared to uncoated coupons over tests conducted over thousands of hours [40]. The fact that aerogel is an effective sublimation barrier is expected since aerogel is known to significantly reduce thermal transport associated with gas convection. The inherently tortuous pore network impedes the transport of metal vapors, thus establishing a microequilibrium environment at the interface between the thermoelectric electrode and the aerogel coating. Ongoing work is being conducted to increase the tortuosity of the aerogel coatings by adding ceramic powders to further reduce the sublimation rates.

32.3.4. Advanced Stirling Radioisotope Generators

Advanced high-temperature thermal insulation aerogel composites have been developed and are being proposed for use in Advanced Stirling Radioisotope Generators (ASRGs). Unlike RTGs, which generate electrical power from a temperature gradient by nonmechanical means, ASRGs generate electrical power from temperature gradients by the motion of pistons. Temperature gradients are used to drive the motion of the pistons that in turn drive a generator, which produces electrical power. The sources of the thermal power are General Purpose Heat Source (GPHS) modules, which contain four iridium-clad plutonium 238 fuel pellets. ASRGs produce roughly four times more power than RTGs from the same heat source and are therefore being considered for use in the designs of flagship space missions to the outer planets.

The aerogel composites would be used in thermal insulation of the GPHS [41]. By surrounding the GPHS with a very low thermal conductivity material, the primary path for heat to escape the source would be through the *Stirling engine* pistons. Aerogel composites have been found to have thermal conductivities roughly the same as the best commercially available high-temperature thermal insulation materials, for example, 2.4×10^{-3} W/mK at 23°C in a vacuum. However, they have lower bulk densities, which is important when choosing materials for space missions, where mass is of critical importance.

Since the temperature of the heat source would be many hundreds of degrees Celsius, for example, 700–800°C, pure silica aerogel cannot be used as the thermal insulation material. To make silica aerogel a good high-temperature thermal insulation, it must be made opaque to radiative thermal transport. In this case, this was accomplished by adding oxide powders to the aerogel matrix. Also, at these temperatures silica aerogel shrinks and compacts significantly and is rendered a poor insulation material. By making an aerogel composite of silica aerogel oxide powders and a low-density glass fiber felt, the new materials becomes an excellent high-temperature thermal barrier and does not shrink significantly at these elevated temperatures. The glass felt provides overall mechanical stability to the composite, while the titania and silica powders prevent radiative thermal transport through the material. The powders also provide some mechanical stability.

The aerogel composites are produced by combining a typical silica aerogel precursor, that is, silica sol, water, catalyst, and organic solvent, with a dispersion of titania and silica powders in the same organic solvent. This mixture is then poured into the appropriate molds over the glass fiber felt. The mixture forms a wet gel due to the hydrolysis and condensation of the silica particles in the sol and in so doing traps the oxide powder particles in the silica matrix formed. The wet gel is then dried by taking the solvent to the supercritical state and depressurizing the system at a temperature above the supercritical temperature. Since the initial mixture is a liquid, the aerogel composites can be formed into any shape required by the overall system design. By producing many precisely shaped panels, an insulation package can be assembled to house a radioisotope heat source (Figure 32.17).

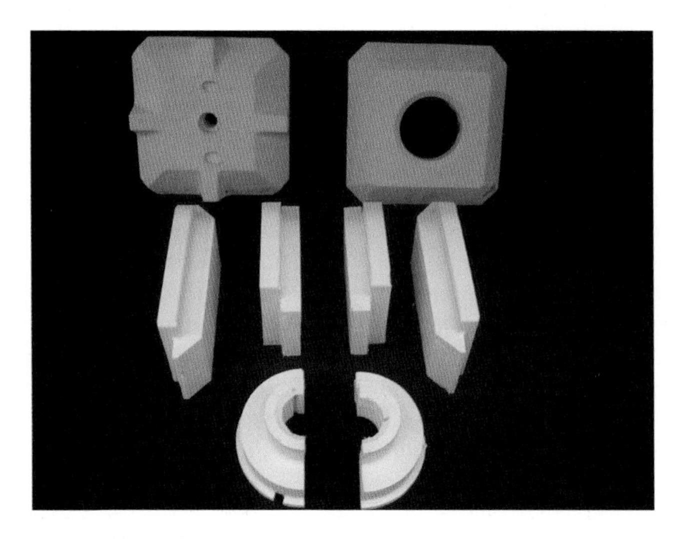

Figure 32.17. Aerogel composite panels for thermal insulation of a Stirling radioisotope generator heat source.

Aerogel composites have the added working advantage that they shrink significantly at temperatures of over 1,200°C. If one, or both, of a pair of Stirling pistons should stop functioning, the heat from the GPHS would not be carried away from the source through the piston(s). Under this scenario, the temperature of the radioisotope source could reach well over 1,200°C. At these temperatures, the cladding on the radioisotope pellets becomes unstable. To prevent the heat source from reaching these temperatures, the thermal insulation must fail and allow the excess heat to escape. By shrinking at these elevated temperatures, aerogel composites do exactly that. Due to the anisotropic nature in which the glass fiber felt is incorporated into the composites, the shrinkages are anisotropic. The felt is cut into thin panels and the final aerogel composite is formed layer by layer, until the bulk material is formed. When the samples are heated to 1,300°C for 30 min and then 1,200°C for 2 h, the shrinkage has been observed to be roughly 40% in the directions parallel to the glass fiber felt and roughly 50% in the directions perpendicular to the plane of the fiber felt. These tests demonstrated that the insulation panels will shrink at these temperatures and allow heat to escape, thus reducing the temperature to acceptable levels.

32.4. Cryogenic Fluid Containment

The STEP (Satellite Test of the Equivalence Principle) Mission proposes to place a spacecraft in earth orbit that would be equipped to test the ratio of inertial mass and gravitational mass [42, 43]. Any difference found in the values of the inertial and gravitational mass would violate Einstein's Theory of General Relativity. By measuring the constant acceleration of specially designed test masses while on orbit, the equivalence of inertial and gravitational mass can be determined using differential accelerometers. Since the spacecraft would be held on a constant orbit around the earth and since these conditions would provide for a thoroughly isolated environment, the fundamental constant of gravity, G, could also be determined with very high precision. To maintain the stability of the test system and the detectors, they must be cooled to a few degrees Kelvin. To maintain these extremely low temperatures, the test masses and detectors would be located in a central column of a Dewar filled with liquid helium.

However, the liquid helium required to establish and maintain this cryogenic environment poses a new technical challenge in that any bulk flow of the liquid would introduce inertial changes to the spacecraft that could obscure experimental results. To prevent this from occurring, the liquid helium must be contained in a porous material that would confine the liquid helium in a microscale porous network. The network must have an open cell configuration so that the helium is allowed to flow into and subsequently out of the network. Silica aerogel is composed of a microscopic porous network that would allow the liquid helium to fill the network initially and then gradually bleed out from it as the mission proceeded and yet prevent any bulk flow of the liquid helium. The helium that diffuses from the aerogel would be bled from the Dewar and directed through spacecraft thrusters, which would be used to maintain the constant orbit required by the experiment.

The basic shape of the aerogel required for the mission would be an annular cylinder, since the design calls for a cylindrical 200 l Dewar, with a central column running down the middle of it. The central column would contain the measurement components and thus be cooled by the liquid helium. The initial test pieces of aerogel were cast in the form of annular cylinders (Figure 32.18). The procedure used to form the wet gels was a two-step

Figure 32.18. Twenty-five centimeter diameter annular cylinder of silica aerogel for testing.

process that was described previously (see Sect. 32.2.2). Initially, 100 mg/cm^3 silica aerogel was to be used, although to reduce the total mass of the system it was later decided to use 50 mg/cm^3. Since an annular cylinder of aerogel large enough to fill the Dewar could not be produced, it was necessary to produce precisely shaped pieces that could be assembled into the structure needed. The individual blocks were rounded trapezoids of approximately 2.5 l in volume (Figure 32.19).

Figure 32.19. Rounded trapezoidal block of silica aerogel for assembly into larger ring. Block is 12 cm in height.

Six of these units would be assembled into an annular ring (Figure 32.20). These six piece rings would then be stacked one on top of another to form an annular cylinder. Secondary rings, with larger radii of curvature, would then surround the first sets of rings to form an assembly that would completely fill the Dewar. It was proposed that the individual rings of aerogel blocks would be held together by encircling them with lightweight, flexible bands, thus slightly compressing the blocks together. This work demonstrated that, despite the fact that very large, for example, tens of liters, pieces of aerogel cannot be produced, smaller precisely cast pieces can be produced and assembled into much larger structures.

Figure 32.20. Prototype assembly of six rounded trapezoids. Outer diameter is approximately 50 cm.

The addition of silica aerogel to the Dewar posed various potential risks to the overall mission success. Ground testing was conducted to mitigate these risks. The primary concerns regarding the introduction of the aerogel to the system were that the relatively large pieces of aerogel would not fill with liquid helium and that the aerogel could not survive the launch vibration environment while filled with liquid helium. To test whether large samples of silica aerogel would fill with liquid helium, an annular test cylinder of aerogel that was 50 mm tall and 150 mm outer diameter, with a 38 mm inner annular bore diameter, was placed in a test Dewar and cooled with liquid helium. Visual observations indicated that the aerogel filled with liquid helium and a thermal probe showed a temperature of 2.6 K. The liquid helium was then removed from the aerogel. To test whether silica aerogel would survive a launch vibration environment while filed with liquid helium, a similar piece of aerogel was manually inserted into an aluminum sleeve and attached to the bottom of an experimental shake Dewar. After cooling the Dewar and filling the aerogel with liquid helium, the entire assembly was subjected to the launch vibration levels of a Delta II launch vehicle. The aerogel was then removed from the assembly and observed visually. No signs of damage or degradation were observed after the testing [44].

The STEP Mission proposal was not selected for flight when the final NASA 2003 Small Experiments Mission selection was made. However, work on the technology of the mission continues and with the success of Gravity Probe B, which employs similar technology, it is hoped that STEP will someday be using aerogel to conduct fundamental physics experiments in space.

32.5. Conclusion

The use of aerogel in space exploration began with its use in the Mars Pathfinder rover, Sojourner, in 1996. Since that time, several other missions have either used aerogel or have been planned around an aerogel-based primary instrument. As the types of aerogel available expand beyond silica, for example, carbon, nonsilicate oxides, and nonoxides, and the expertise in the production of these aerogels, as well as specially doped aerogels, becomes more extensive, the use of aerogel in space missions could expand. Many unique properties of aerogel have yet to be utilized in the quest to explore space. The fact that aerogels exhibit several unique physical properties concurrently has not even been explored yet, when considering the potential space science applications of aerogels.

The aerogel development, testing, and production done thus far dispel one commonly held idea, which is the extreme fragility of aerogels; particularly low-density aerogels are too fragile for all but a few very specialized applications. If aerogel can stand up to the rigors of space flight testing and actual launches and landings, then it can be deemed a viable material for almost any challenging environment.

Acknowledgments

The research described in this publication was carried out at the Jet Propulsion Laboratory, California Institute of Technology, under a contract with the National Aeronautics and Space Administration.

References

1. Pierre A C, Pajonk G M (2002) Chemistry of aerogels and their applications. Chem Rev, 102: 4243–4265
2. Livage J, Sanchez C (1992) Sol-gel chemistry. J Non-Cryst Solids, 145: 11–19
3. Hench L L, West J K (1990) The Sol-Gel Process. Chem. Rev., 90(1): 33–72
4. Brinker C J, Scherer G W (1990) Sol-Gel Science: The Physics and Chemistry of Sol-Gel Processing. Academic Press, San Diego
5. Akimov Y K (2003) Fields of applications of aerogels. Instr Exper Techniques 46(3): 287–299
6. Herrmann G, Iden R, Mielke M, Teich F, Ziegler B (1995) On the way to commercial production of silica aerogel. J Non-Cryst Solids 186: 380–387
7. Hrubesh L W (1998) Aerogel applications. J Non-Cryst Solids 225: 335–342
8. Schmidt M, Schwertfeger F (1998) Applications for silica aerogel products. J Non-Cryst Solids 225: 364–368
9. Ulrich D R (1990) Prospects for sol-gel processes. J Non-Cryst Solids 121: 465–479
10. Fricke J, Tillotson T (1997) Aerogels: production, characteriza- tion, and applications. Thin Solid Films 297: 212–223
11. MacKenzie J D (1988) Applications of the sol-gel process. J Non-Cryst Solids 100: 162–168
12. Adachi I, Sumiyoshi T, Hayashi K, Iida N, Enomoto R, Tsukada K, Suda R, Matsumoto S, Natori K, Yokoyama M, Yokogawa H (1995) Study of a threshold Cherenkov counter based on silica aerogels with low refractive index. Nucl Instr Meth Phys Res A, 335: 390–398
13. Asner D, Butler F, Dominick J, Fadeyev V, Masek G, Nemati B, Skubic P, Strynowski R (1996) Experimental study of aerogel Cherenkov detectors for particle identification. Nucl Instr Meth Phys Re. A 374: 286–292
14. Sumiyoshi T, Adachi I, Enomotoi R, Iijima T, Suda R, Yoko yama M, Yokogawa H (1998) Silica aerogels in high energy physics. J Non-Cryst Solids, 225: 369–374
15. Ishino M, Chiba J, En'yo H, Funahashi H, Ichikawa A, Ieiri M, Kanda H, Masaike A, Mihara S, Miyashita T, Murakami T, Na kamura A, Naruki M, Muto M, Ozawa K, Sato H D, Sekimoto M, Tabaru T, Tanaka K H, Yoshimura Y, Yokkaichi S, Yoko yama M Yokgawa H (2001) Mass production of hydrophobic silica aerogel and readout optics of Cherenkov light. Nucl Instr Meth Phys Res A, 457: 581–587
16. DeLeo R, Lagamba L, Manzari V, Nappi E, Scognetti T, Alemi M, Becker H, Forty R, Adachi I, Suda R, Sumiyoshi T, Leone A, Perrino R, Matteuzzi C, Seguinot J, Ypsilantis T, Cisbani E, Frullani S, Garibaldi F, Iodice M, Uriuoli GM (1997) Electronic detection of focused Cherenkov rings from aerogel. Nucl Instr Meth Phys Res A 401: 187–205
17. Tsou P (1995) Silica aerogel captures cosmic dust intact. J Non- Cryst Solids 186: 415–427
18. Tsou P, Brownlee D E, Sandford S A, Horz F, Zolensky M E (2003) Wild 2 and interstellar sample collection and earth return. J Geophys Res 108(E10): 1–21
19. Horz F, Zolensky M E, Bernhard R P, See T H, Warren J L (2000) Impact features and projectile residues in aerogel ex-posed on Mir. Icarus 147(2): 559–579
20. Brownlee D E, Tsou P, Atkins K L, Yen C-W, Vellinga J M, Price S, Clark B C (1996) Stardust: finessing expensive cometary sample returns. Acta Astronautic, 39 (1- 4): 51- 60
21. Horz F, et al (2006) Impact features on Stardust: implications for comet 81P/Wild 2 Dust. Science 314: 1716–1719
22. Tillotson T M, Hrubesh L W (1992) Transparent ultra low-density silica aerogels prepared by a two-step sol-gel process. J Non-Cryst Solids 145: 44–50
23. Jones S M (2007) A method for producing gradient density aerogel. J Sol-Gel Sci Technol 44: 255–258
24. Sandford S A, et al (2006) Organics captured from comet 81P/Wild 2 by the Stardust spacecraft. Science 314: 1720–1724
25. Zolensky M E, et al (2006) Mineralogy and petrology of comet 81P/Wild 2 nucleus samples. Science, 314: 1735–1739
26. Brownlee D E, et al (2006) Comet 81P/Wild 2 under a micro-scope. Science 314: 1711–1716
27. Westphal A J, Butterworth A L, Snead C J, Craig N, Anderson D, Jones S M, Brownlee D E, Farnsworth R, Zolensky M E (2005) Stardust at Home: A massively distributed public search for interstellar dust in the Stardust interstellar dust collector. Lunar Planetary Sci. Con. XXXVI, Lunar Planetary Institue, Houston, TX Abstract 1908
28. Leshin L A, Yen A, Bomba J, Clarke B, Epp C, Fourney L, Gamber T, Grave C, Hupp J, Jones S, Jurewicz A J G, Oakman K, Rea J, Richardson M, Romeo K, Sharp T, Sutter B, Thiemens M, Thornton J, Vicker D, Willcockson W, Zolensky M (2002) Sample collection for investigation of Mars (SCIM): An early Mars

sample return mission through the Mars Scout Program. Lunar Planetary Sci. Conf. XXXIII, Lunar Planetary Institute, Houston, TX Abstract 1721

29. Leshin L A, Clark B C, Forney L, Jones S M, Jurewicz A J G, Greeley R, McSween H Y, Richardson M, Sharp T, Thiemens M, Wadhwa M, Wiens R C, Yen A, Zolensky M (2003) Scienti-fic benefit of a Mars dust sample capture and Earth return with SCIM Lunar Planetary Sci. Conf.,XXXIV, Lunar Plane-tary Institute, Houston, TX Abstract 1288
30. Jurewicz A J G, Forney L, Bomba J, Vicker D, Jones S, Yen A, Clark B, Gamber T, Leshin L A, Richardson R, Sharpe T, Thie- mens M, Thornton J M, Zolensky M (2002) Investigating the use of aerogel collectors for the SCIM martian dust sample return. Lunar Planetary Sci. Conf. XXXIII, Lunar Planetary In- stitute, Houston, TX., 2002 Abstract 1703
31. Jones S M, Non-silica aerogel as a hypervelocity capture material (Accepted by Meteor Planet Sci)
32. Westphal A J, et al (2004) Aerogel keystones: Extraction of complete hypervelocity impact events from aerogel collectors. Meteor Planet Sci 39(8): 1375–1386
33. Dominguez G, Westphal A, Phillips M, Jones S (2003) A fluorescent aerogel for capture and identification of inter-planetary and interstellar dust. Astrophysical Journal 592: 631–635
34. Dominquez G, Westphal A J, Jones S M, Phillips M L F (2004) Fluorescent impact cavities in a titanium-doped Al2O3-SiO2 aerogel: implications for the velocity resolution of calorimetric aerogels. J Non-Crystalline Solids 350: 385–390
35. Novak K S, Phillips C J, Burir G C, Sunada E T, Pauken M T (2003) Development of a thermal control architecture for the Mars Exploration Rovers. Space Technology Applications Inter- national Forum 2003 Feb. 2–6, 2003
36. Tritt M, M A Subramanian (2006) Thermoelectric Materials, Phenomena, and Applications. MRS Bulletin, 36: 188–229
37. Rowe D M, ed. (1995) CRC Handbook on Thermoelectrics. CRC Press, Boca Raton, FL
38. Stabler F (2002) Automotive Applications of High Efficiency Thermolectrics. DARPA/ONR Program Review and DOE High Efficiency Thermoelectric Workshop San Diego, CA, March 24–27
39. Kuhn J, Gleissner T, Arduini-Schuster M C, Korder S, Fricke J (1995) Integration of mineral powders into SiO_2 aerogels. J Non-Cryst Solids, 186: 291–295
40. Sakamoto J, Caillat T, Fleurial J P, Jones, S, Paik J, Dong W (2006) Improving thermoelectric device performance and du-rability through the integration of advanced aerogel-based ce- ramics. Ceramic Transactions 196: 275–290
41. Paik J-A, Jones S M, Sakamoto J Composite Aerogels for high temperature thermal insulation (In preparation)
42. Worden P, Torii R, Mester J C, Everitt C W F (2000) The STEP Payload and Experiment. Adv Space Res. 25(6): 1205–1208
43. Mester J, Torii R, Worden P, Lockerbie N, Vitale S, Everitt C W F (2001) The STEP mission: principles and baseline design. Class Quantum Grav. 18(13): 2475–2486
44. Wang S, Torii R, Vitale S (2001) Silica aerogel vibration test- ing Class. Quantum Grav. 18(13):2551–2559

33

Airborne Ultrasonic Transducer

Hidetomo Nagahara and Masahiko Hashimoto

Abstract A high-sensitivity airborne ultrasonic transducer was produced by applying an aerogel to its acoustic matching layer, because an aerogel's low density and low acoustic velocity result in extremely low acoustic impedance. We estimated the acoustic properties of this transducer and also fabricated an ultrasonic transducer having two acoustic matching layers made of silica aerogel and porous ceramic. The sensitivity of this novel ultrasonic transducer was found to be about 20 times higher than that of a conventional airborne ultrasonic transducer.

33.1. Transducers for Ultrasonic Sensing

Bats have highly advanced *sonar systems* (biosonar) that they use in order to fly and hunt in dark environments [1, 2]. The biosonar systems of bats greatly outperform artificial echo-sensing systems in terms of efficiency and flexibility. Bats can produce high-energy ultrasonic pulses in a very efficient manner and hear the echoes returned from objects in the surrounding environment with super sensitivity. For example, the acoustic pressure level generated by bats' transmitters is approximately 80–120 dB with nearly flat response over the frequency range of 20–200 kHz, which cannot be achieved by commercially available *ultrasonic transducers*. On the other hand, artificial ultrasonic sensing systems are used in various tasks such as range finding and positioning and in flowmeters. Ultrasonic transducers are key devices for determining the performance of such sensing systems. Few commercially available ultrasonic transducers have sufficient sensitivity or *bandwidth* to advance *sonar systems* to the level of the *biosonar systems* of bats.

Today, many of the *ultrasonic transducers* available commercially are piezoelectric in nature. Piezoelectric ceramics such as lead zirconate titanate (*PZT*) have been generally used as piezoelectric elements for such transducers. Polyvinylidene difluoride (*PVDF*) piezoelectric polymer films are also available. Ultrasonic transducers of high sensitivity are roughly classified into two types. The first has unimorph or bimorph vibration and a radial cone construction. A radial cone is attached to the point of maximum displacement of the vibrator. Because the radial cone vibrates as a rigid body with high amplitude, it can transmit and receive acoustic waves with high sensitivity. The second type of highly

---

H. Nagahara (✉)**and M. Hashimoto** • Advanced Technology Research Laboratory, Panasonic Corporation, 3-4 Hikaridai, Seika-cho, Soraku-gun, Kyoto 619-0237, Japan
e-mail: nagahara.hidetomo@jp.panasonic.com; hashimoto.masahiko@jp.panasonic.com

M.A. Aegerter et al. (eds.), *Aerogels Handbook*, Advances in Sol-Gel Derived Materials and Technologies, DOI 10.1007/978-1-4419-7589-8_33,

sensitive ultrasonic transducers has a bulk layer called the acoustic matching layer on the front of the piezoelectric element. This type of transducer, using the principle of *acoustic impedance* matching, can transmit and receive acoustic waves with high efficiency.

To ensure good vibration characteristics in the ultrasonic range, the radial cone needs to be lightweight and have high acoustic velocity. In contrast, the acoustic matching layer is placed between the air (as an acoustic propagation medium) and the piezoelectric element (as a vibrator), so that there is a five orders of magnitude difference between their *acoustic impedances*, and because an intermediate acoustic impedance is also required. In order to apply an aerogel to the ultrasonic transducer, we measured and analyzed its acoustic characteristics and researched the potential of the acoustic matching layer. Figure 33.1 shows a schematic view of the typical *ultrasonic transducer* with acoustic matching layer. The following section explains the conditions necessary for the acoustic matching layer to achieve high sensitivity.

Acoustic impedance is the most important parameter for configuring the acoustic matching layer. Using acoustic velocity c and density ρ of the propagation material, acoustic impedance Z can be represented by the following (33.1).

$$Z = \rho c \tag{33.1}$$

When the acoustic waves move into the boundary of the propagation media, acoustic impedance determines basic characteristics such as reflection or transmission. Referring to Figure 33.1, if Z_p is the acoustic impedance of the piezoelectric element, Z_a is the acoustic impedance of air, Z_m is the acoustic impedance of the acoustic matching layer, and d is the thickness of the acoustic matching layer, acoustic impedance matching conditions can be represented by (33.2), with thickness being represented by (33.3) [3].

$$Z_m = \sqrt{Z_a Z_p} \tag{33.2}$$

$$d = \frac{\lambda}{4} \tag{33.3}$$

Here, λ is the ultrasonic wavelength in an acoustic matching layer. If (33.2) and (33.3) are satisfied, the transmission rate of the plane wave will be maximum. If the acoustic matching layers are of a multilayered structure, it is necessary for (33.2) and (33.3) to be satisfied at any continuous three layers.

Assuming the transmission and reception of acoustic waves in an air atmosphere, the acoustic impedance of air is about 400 kg/(m^2 s) (density 1.2 kg/m^3, acoustic velocity 343 m/s at 1 atm, 20°C), and the acoustic impedance of piezoelectric ceramic as a typical

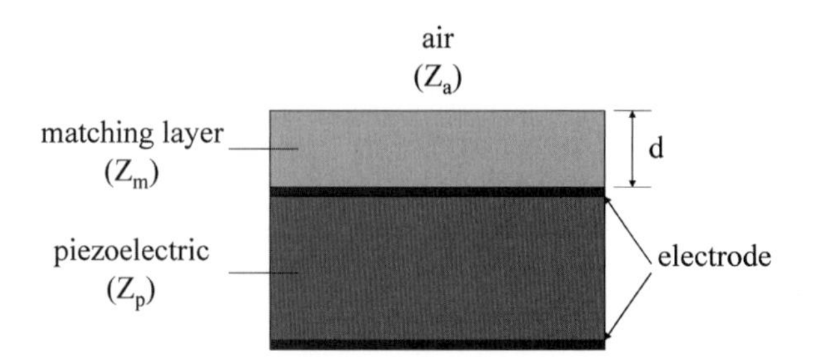

Figure 33.1. Schematic view of the typical matching layer-type ultrasonic transducer.

piezoelectric body is 30×10^6 kg/(m^2 s) (density 7,800 kg/m^3, acoustic velocity 3,800 m/s). There is a five orders of magnitude difference between the acoustic impedances.

If there is a significant difference between the acoustic impedance of the media, the propagation of acoustic waves from piezoelectric body to air is extremely difficult. Moreover, it becomes increasingly difficult to achieve acoustic wave propagation from the air to the piezoelectric body, and as a result the ultrasonic transducer cannot achieve high sensitivity. As shown in Figure 33.1, from the introduction of the acoustic matching layer, with acoustic impedance matching between the piezoelectric body and air in order to achieve efficient transmission and reception of acoustic waves, it can be found that the acoustic impedance of the acoustic matching layer must be around 100×10^3 kg/(m^2 s) according to (33.2).

In fact, solid materials with such low acoustic impedance are extremely rare. For the acoustic matching layer of an airborne ultrasonic transducer, non-woven fabrics made of polymer or polymer materials packed in small hollow glass spheres are generally used.

Acoustic impedance of these materials is around $1{,}000–2{,}000 \times 10^3$ kg/(m^2 s), more than ten times the value of ideal acoustic impedance. That is, the current performance of ultrasonic transducers with an acoustic matching layer is restricted by the limitations of the acoustic characteristics of its acoustic matching layer. Silica aerogel has an extremely low density as a solid. We focused on this fact and researched the potential regarding the acoustic matching layer of ultrasonic transducers, as we describe in more detail in the next section.

33.2. Acoustic Properties of Aerogels

To understand the acoustic properties of silica aerogels (Chap. 2), silica aerogels of differing densities were prepared and their acoustic velocities were measured [4, 5]. The differing densities were achieved by changing the blending ratio of the raw materials, with other conditions such as the process of creating aerogels being the same. The silica aerogels were prepared by drying wet silica gels synthesized from tetramethoxysilane (*TMOS*) by the sol–gel process. The detailed procedure is given below.

1. Gelation: The raw material solution for gelation was prepared by mixing TMOS, ethanol, and ammonia water at predetermined ratios. By maintaining the temperature of the solution at 40°C for 24 h, a hydrophilic silica wet gel with water and ethanol was obtained.
2. Silation: Next, in order to introduce a hydrophobic methyl group to the silanol group of the gel (Chap. 3), it was soaked in a mixed solution of isopropyl alcohol (*IPA*), dimethoxydimethylsilane (*DMDMS*), and ammonia water. To remove water from the gel, solvent exchange was carried out in IPA at 40°C for 24 h.
3. Supercritical drying: The *supercritical drying* technique was used to avoid shrinkage of the gel due to the strong capillary pressure derived from its very small pores. The wet gel was held in the autoclave of a supercritical drier at 20°C and 12 MPa by introducing liquid carbon dioxide (CO_2) into the autoclave, *IPA* in the wet gel was replaced with the liquid CO_2. Then, by raising the temperature in the autoclave to 50°C at 12 MPa, the supercritical state of CO_2 was realized. Next, the pressure was slowly lowered to atmospheric pressure at 50°C, and the state of the CO_2 was directly changed from supercritical to gas. Through the drying technique, CO_2 was removed from the wet gel without capillary pressure, and a nonshrinkage dry gel was obtained.

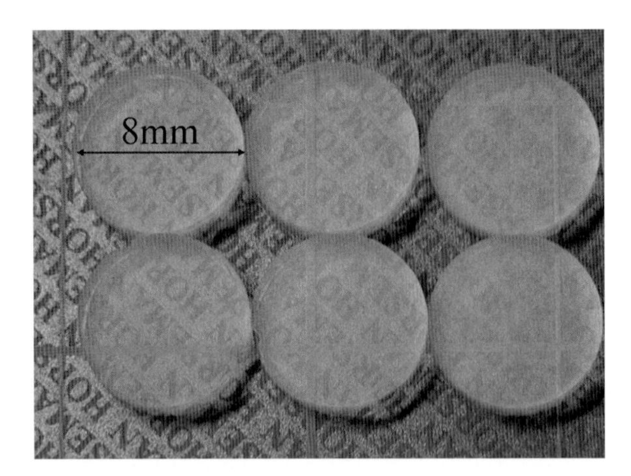

Figure 33.2. Appearance of aerogels.

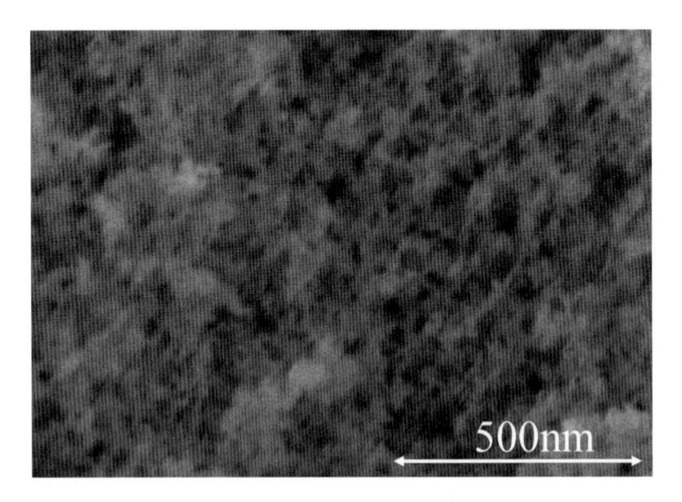

Figure 33.3. SEM image of aerogel.

4. Characteristics: Figures 33.2 and 33.3 show the appearance and the scanning electron microscope (SEM) image of an aerogel synthesized by the above process, respectively. Density was calculated from the volume and weight. The acoustic velocity was calculated from thicknesses and ultrasonic propagation time (about 500 kHz).

Figure 33.4 shows the method used for measuring the acoustic velocity.

Figure 33.5 shows the measurement results of the density and acoustic velocity of the aerogels.

From Figure 33.5, the acoustic velocity and density of silica aerogel was determined to have an approximately linear relationship in the range of this experiment. The densities of the silica aerogels were $0.06–0.27 \times 10^3$ kg/m^3, and the corresponding acoustic velocities were 80–220 m/s. *Acoustic impedance* values calculated from these measurements were $4.8–59.4 \times 10^3$ kg/(m^2 s).

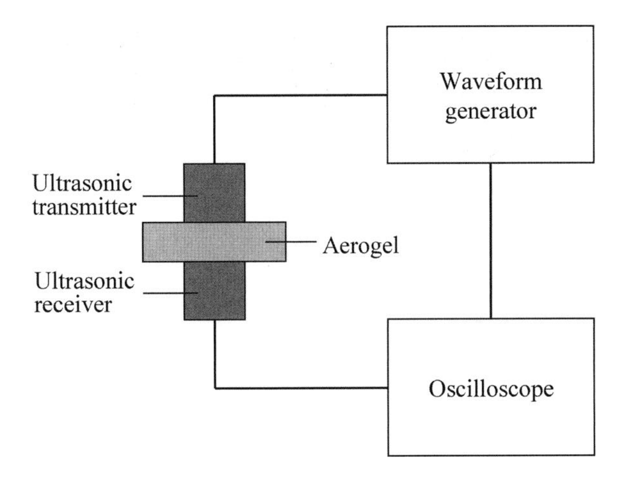

Figure 33.4. Measurement method for obtaining acoustic velocities.

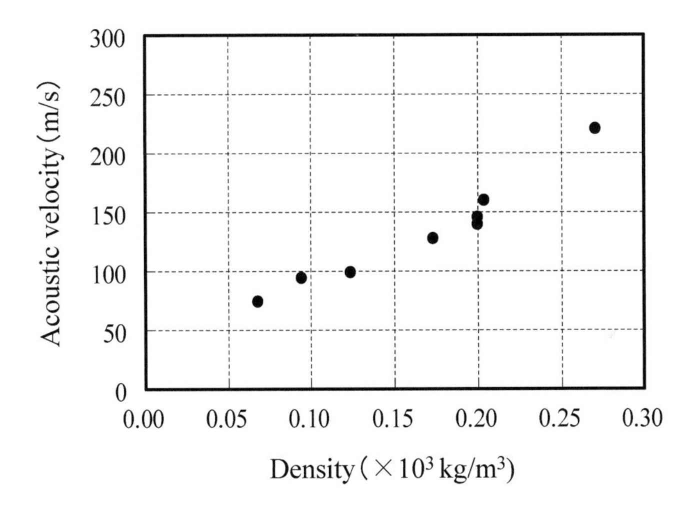

Figure 33.5. Measurement results of the density and acoustic velocity of the aerogels.

The most notable feature of silica aerogel was its extremely low acoustic velocity. Even though the silica aerogel is made of an inorganic material with an acoustic velocity of about 5,000 m/s in the case of solid [6], its acoustic velocity is extremely low as a bulk material.

From these basic experiments, it was found that silica aerogels had very low acoustic impedances compared to those of the current matching layers [7–9], and are therefore very promising as materials for the acoustic matching layers of high-sensitivity airborne ultrasonic transducers.

33.3. Design of Ultrasonic Transducer

The acoustic properties of ultrasonic transducers with an aerogel acoustic matching layer were studied using a one-dimensional Krimholtz–Leedom–Matthaei (*KLM*) equivalent circuit [10, 11] to simulate using computer. Figure 33.6 shows the structure of the

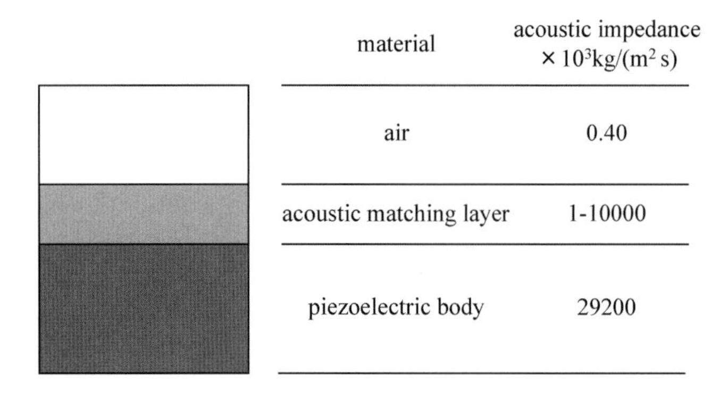

material	acoustic impedance × 10³kg/(m² s)
air	0.40
acoustic matching layer	1-10000
piezoelectric body	29200

Figure 33.6. Simulation model and material properties.

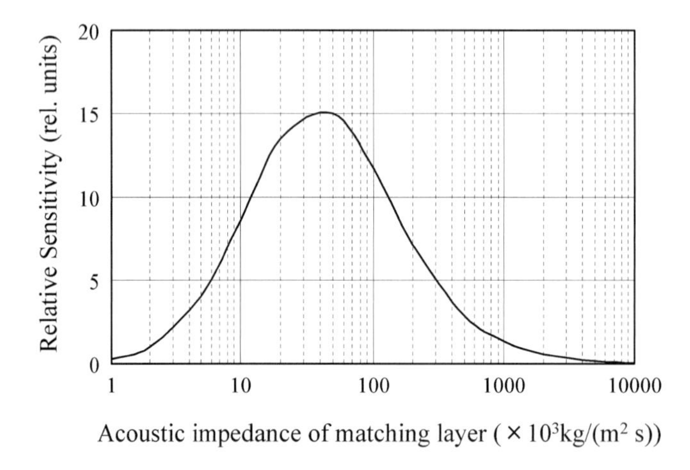

Figure 33.7. Simulation result of aerogel ultrasonic transducer.

simulation model and the material properties, and Figure 33.7 shows the results of the computer simulation.

In Figure 33.7, the horizontal axis represents the acoustic impedance of the acoustic matching layer, and the vertical axis represents the transmitted and received sensitivities. The vertical axis (sensitivity) is normalized using the sensitivity of an ultrasonic transducer that has a conventional acoustic matching layer (acoustic impedance is $1{,}300 \times 10^3$ kg/m^2 s) consisting of epoxy resin and a hollow glass balloon. From Figure 33.7, when the acoustic impedance of the aerogel is about 40–50 $\times$ 10^3 kg/(m^2 s), the sensitivity of an aerogel ultrasonic transducer may be 15 times as high as a conventional ultrasonic transducer.

However, silica aerogel can attain acoustic impedance lower than the sensitivity maximum point in Figure 33.7. With single matching layer structure, the feature of low acoustic impedance of the silica aerogel is not fully utilized. For high sensitivity and broad bandwidth, a multilayered acoustic matching layer is generally used in ultrasonographs [12]. It is well known that sensitivity and bandwidth are improved by the multilayered acoustic matching layer. However, a multilayered acoustic matching layer type has not been previously used in airborne ultrasonic transducers, because no material with sufficiently low acoustic impedance for airborne ultrasonic transducers has been available. Figure 33.8 shows the simulation model of the two-acoustic matching layer and its material properties.

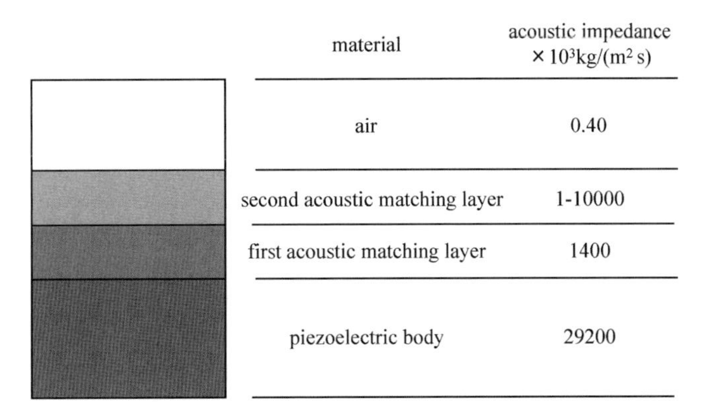

Figure 33.8. Simulation model of two-acoustic matching layer and material properties.

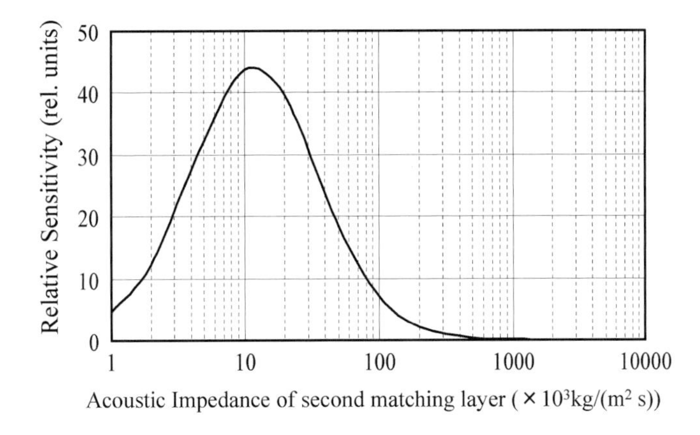

Figure 33.9. Simulation result of two-acoustic matching layer ultrasonic transducer.

The acoustic matching layer, next to the piezoelectric element, is called the first matching layer, and the other layer is called the second matching layer. Figure 33.9 shows the results of the computer simulations of the two-acoustic matching layer type of ultrasonic transducer.

From Figure 33.9, when the acoustic impedance of the aerogel is about 10–20 $\times$ 10^3 kg/ (m^2 s), the sensitivity of an aerogel ultrasonic transducer may be 45 times as high as a conventional ultrasonic transducer.

33.4. Fabrication of Aerogel Acoustic Matching Layer

The results of the simulations in the previous chapter suggest that the performance of the ultrasonic transducer would be significantly improved by applying an aerogel to its acoustic matching layer. However, to achieve such high performance, it is necessary to bond the aerogel acoustic matching layer solidly with another material and ensure the good propagation of acoustic waves between these materials, in addition to optimizing the acoustic impedance and thickness of the layer.

In particular, an aerogel is easily damaged because of its extremely low density and must be made hydrophobic in order for it to withstand moisture and other environmental changes. It is therefore almost impossible for the aerogel to be bonded using the adhesive process conventionally used for ultrasonic transducers. We thus devised a new form of acoustic matching layer that was able to resolve these problems. Figure 33.10 shows the structure of the acoustic matching layer consisting of an aerogel and a porous ceramic (hereafter denoted as *AC-ML*). The *AC-ML* is abbreviation of Aerogel Composite acoustic Matching Layer.

AC-ML has two acoustic matching layers. In the cross section of the structure in Figure 33.10, the lower matching layer consists of a porous ceramic filled with an aerogel, and the upper matching layer consists of aerogel only. *AC-ML* has the following characteristics:

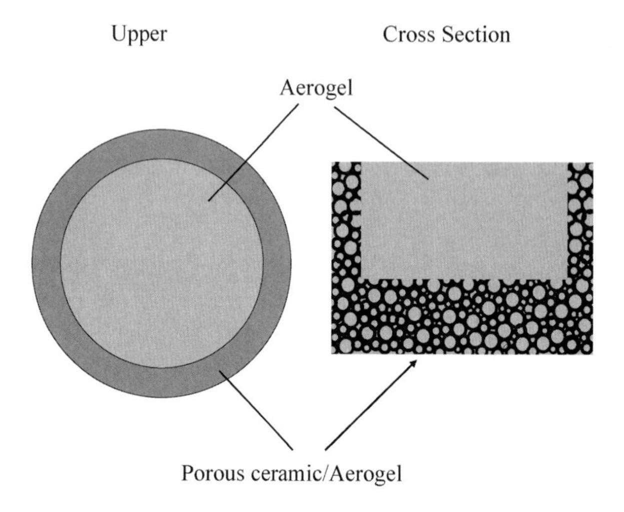

Figure 33.10. Schematic view of aerogel composite acoustic matching layer consisting of aerogel and a porous ceramic.

1. The main part of *AC-ML* is a high-strength porous ceramic filled with an aerogel.
2. The aerogels in the two matching layers form a continuous structure, and it is therefore difficult to delaminate the two layers.
3. Various devices can be realized by bonding electronic components with the porous ceramic surface.
4. The porous ceramic portion, which protects the aerogel of the second matching layer and regulates its thickness, makes it possible to fabricate stable ultrasonic transducers.
5. As *AC-ML* is multilayered, the time-domain response of the ultrasonic transducer can be improved.

With the above features, *AC-ML* is suitable for the matching layer of the airborne ultrasonic transducer. AC-ML with the above advantages was fabricated as follows:

1. Preparation of porous ceramic: The porous ceramic was made from a mixed powder of silica, glass, and acrylate resin, which was pressed in a mold and sintered. A porous ceramic was made from silica and glass, which has about 30–50 μm pores and pore frames. An acrylic resin was used to make pores, and was then removed in the sintering process. A porous ceramic was machined to the

shape of a circular disk 11 mm in diameter and 1.0 mm in thickness. Furthermore, the concave part of 9 mm diameter and 0.1 mm depth used as the second matching layer (aerogel layer) was machined into the center.

2. Creation of aerogel layer: *AC-ML* consisting of a porous ceramic and an aerogel was made by the same process used for aerogel only. A mixed solution of *TMOS*/ethanol/ammonia water was prepared, and the porous ceramic already machined was soaked in the solution. Defoaming was carried out by decompression to completely fill the porous ceramic with the solution. *AC-ML* was dried by supercritical drying after the gelation of the solution, and an aerogel was made in the concave part and inside the porous ceramic.

Figure 33.11 shows the appearance of the *AC-ML* made by the above process. As the aerogel is transparent and the porous ceramic is white, it is not easy to observe the aerogel in Figure 33.11. In this study, we used the above process to synthesize an aerogel whose density and acoustic velocity were about 0.20×10^3 kg/m^3 and 150 m/s, respectively. Although it was slightly bigger than ideal acoustic impedance [10–20 $\times$ 10^3 kg/(m^2 s)], we adopted this condition because it is easier to produce.

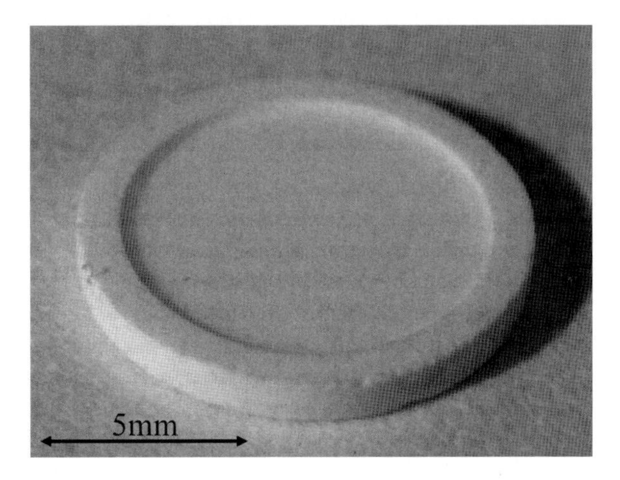

Figure 33.11. Appearance of aerogel composite acoustic matching layer.

33.5. Aerogel Ultrasonic Transducer

An ultrasonic transducer was fabricated with AC-ML and is called the aerogel ultrasonic transducer. Figure 33.12 shows the structure of the ultrasonic transducer. In order to allow easy handling of the ultrasonic transducer, a metal casing was set between the piezoelectric element and *AC-ML*, as shown in Figure 33.12. *AC-ML* and the piezoelectric element were tightly bonded on the metal casing using an epoxy resin. The thickness of the metal casing was set to about one-fiftieth the wavelength in the metal casing and did not significantly obstruct the transmission and reception of ultrasonic waves. Figure 33.13 shows the top-view of the aerogel ultrasonic transducer. The aerogel layer is transparent, and is hence not visible in Figure 33.13; however, a dent on the porous ceramic can be observed.

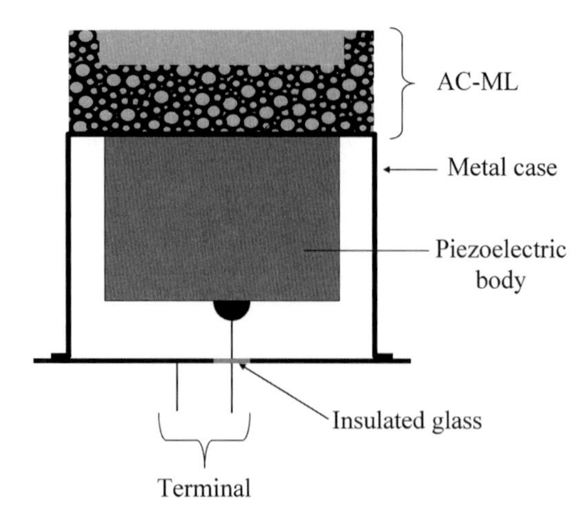

Figure 33.12. Structure of aerogel ultrasonic transducer.

Figure 33.13. Appearance of aerogel ultrasonic transducer.

Table 33.1 shows the material properties of the main structures of the aerogel ultrasonic transducer.

Figure 33.14 shows the calculated results of the time-domain response of the aerogel ultrasonic transducer using the parameters shown in Table 33.1.

Red line and black line are the calculated results of aerogel ultrasonic transducer and ultrasonic transducer with conventional matching layer respectively. The properties of conventional acoustic matching layer are 0.54×10^3 kg/m^3 density, 2,400 m/s acoustic velocity, and $1{,}296 \times 10^3$ kg/m^2 s acoustic impedance.

Calculation was performed under the following conditions: the driving signal was a rectangular wave with three wavelengths at the frequency of 500 kHz, and the input impedance of the receiver was 50 Ω.

The thickness of each layer of *AC-ML* and conventional acoustic matching layer were optimized so that it could efficiently transmit and receive ultrasonic waves at a frequency of 500 kHz.

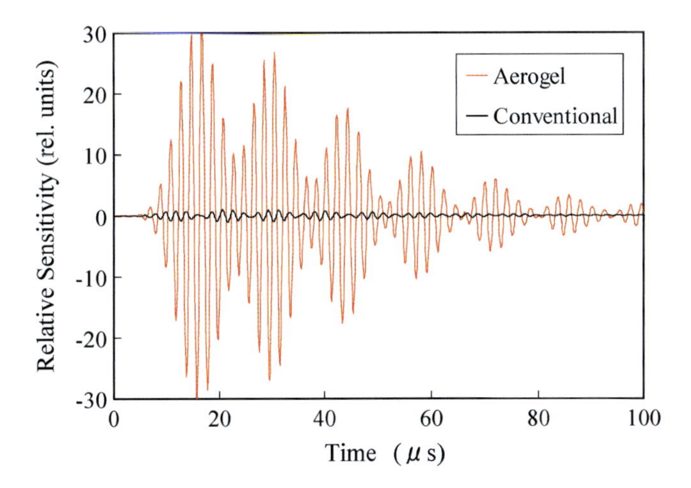

Figure 33.14. Calculated results of time-domain response of aerogel and conventional transducer.

Table 33.1. Material properties of aerogel ultrasonic transducer

	Piezoelectric	Metal case	Porous ceramic/aerogel	Aerogel
Density [10^3 kg/m^3]	7.7	7.9	0.76	0.2
Acoustic velocity [m/s]	3,800	5,500	1,800	150
Acoustic impedance [10^3 kg/(m^2 s)]	29,200	43,500	1,370	30

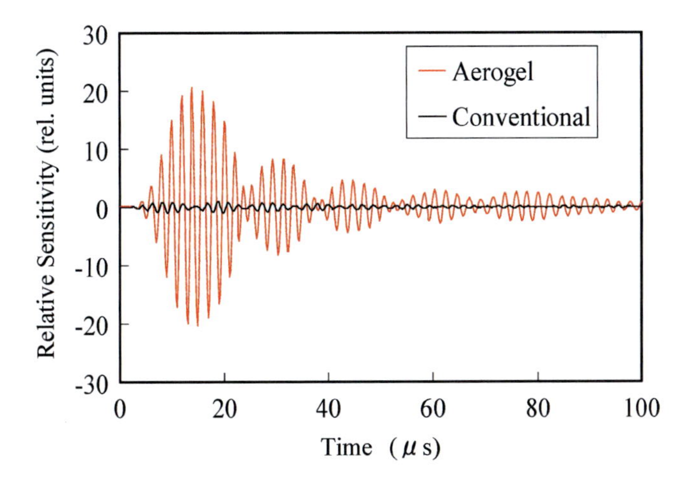

Figure 33.15. Experimental results of time-domain response.

From Figure 33.14, it was predicted that the sensitivity of an aerogel ultrasonic transducer may be 30 times higher than the conventional ultrasonic transducer.

Figures 33.15 and 33.16 show the experimental results of the time-domain and frequency-domain responses. The red line and the black line show the aerogel ultrasonic transducer and conventional ultrasonic transducer. An acoustic matching layer of

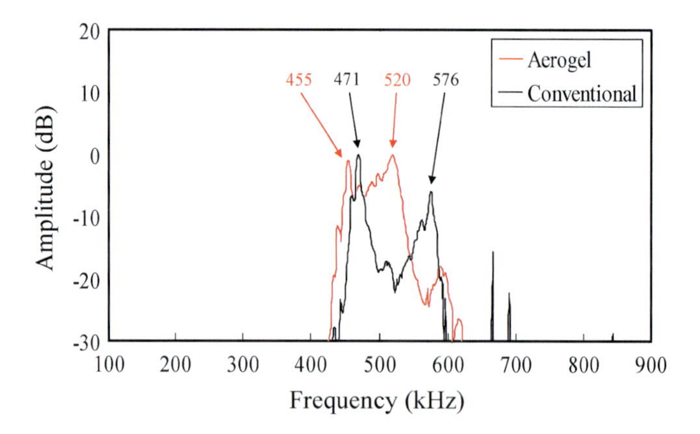

Figure 33.16. Experimental results of frequency-domain response.

conventional ultrasonic transducer was consisted of a hollow glass sphere and epoxy resin. The results of Figures 33.15 and 33.16 were measured by two ultrasonic transducers of the same design, which were placed opposite each other, using one of them as a transmitter and the other as a receiver.

The vertical axis in Figure 33.15 is the relative value based on the amplitude of the time-domain response of the conventional ultrasonic transducer. The vertical axis in Figure 33.16 is in dB, based on each maximum response level of the aerogel ultrasonic transducer and conventional ultrasonic transducer.

As shown in Figure 33.15, the sensitivity of the aerogel ultrasonic transducer is about 20 times as high as that of the conventional airborne ultrasonic transducer. Thus, high sensitivity at the same level as the calculation result was verified.

Moreover, the time-domain responses of the transducers were evaluated. The duration ($t_{-20\mathrm{dB}}$) of the aerogel ultrasonic transducer was about 86 ms in Figure 33.15.

Figure 33.17 shows the time-domain response of the conventional ultrasonic transducer on vertical and horizontal scales different from those in Figure 33.15. The duration ($t_{-20\mathrm{dB}}$) of the conventional ultrasonic transducer was about 140 ms in Figure 33.17; therefore, it was

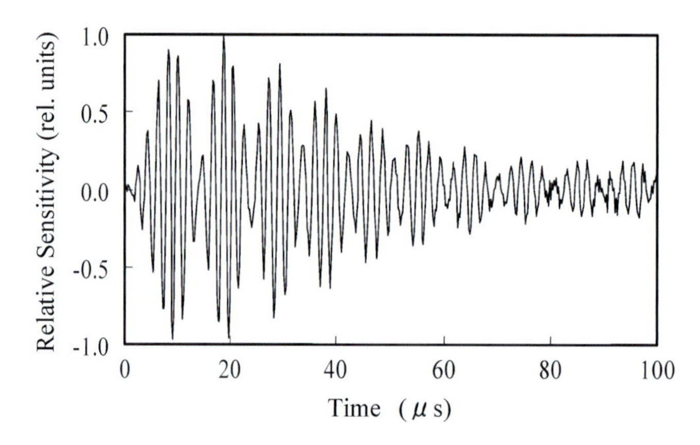

Figure 33.17. Experimental results of time-domain response of conventional ultrasonic transducer.

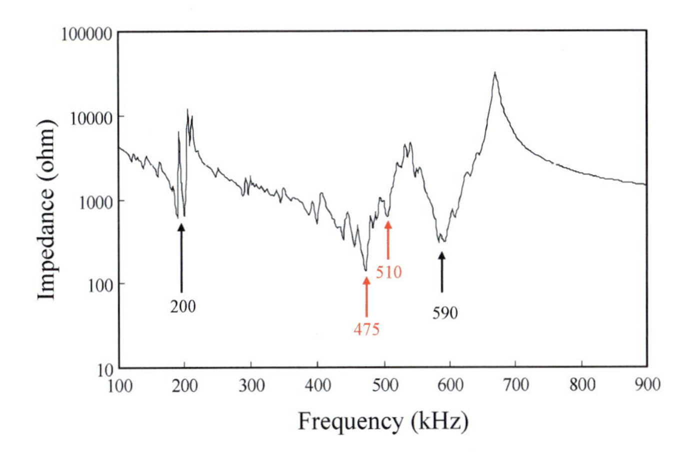

Figure 33.18. Experimental results of electric impedance of aerogel ultrasonic transducer.

determined that the time-domain response had improved from the experimental result. Although the duration ($t_{-20\mathrm{dB}}$) of the experimental result of the aerogel ultrasonic transducer was at the same level as that of the calculated result, the duration of the experimental result of the conventional ultrasonic transducer was markedly longer than that of the calculated result. The low sensitivity of the conventional ultrasonic transducer was considered to be due to noise.

A fundamental vibration of about 500 kHz (about 2 ms) and a beat vibration of about 60 kHz (about 16 ms) can be observed for the aerogel ultrasonic transducer in Figure 33.15. It was considered that the beat vibration was based on the difference (about 65 kHz) between the two peak frequencies at about 455 kHz and 520 kHz shown by the red arrows in Figure 33.16.

Similarly, we can observe a fundamental vibration of about 500 kHz (about 2 ms) and a beat vibration of about 100 kHz (about 10 ms) for the conventional ultrasonic transducer in Figure 33.17. It was considered that this beat vibration was based on the difference (about 105 kHz) between the two peak frequencies at about 471 kHz and 576 kHz shown by the black arrows in Figure 33.16. Figure 33.18 shows the measurement results of the electric impedance of the aerogel ultrasonic transducer. The results of acoustic measurements in Figures 33.15 and 33.16 and the electrical measurement results in Figure 33.18 were compared.

We can observe two resonance frequencies at about 475 kHz and 510 kHz based on the two acoustic matching layers, as shown by the red arrows in Figure 33.18. These two peaks are resonances that generated the beat vibration in Figure 33.15. As previously mentioned, the resonance frequencies are 455 kHz and 520 kHz in Figure 33.16. The resonance frequencies between the result in Figure 33.16 and that in Figure 33.18 differed slightly, because it was considered that the two aerogel ultrasonic transducers being measured had slightly different properties. It was also verified from the electrical measurement that *AC-ML* effectively worked as two acoustic matching layers. The resonance of about 590 kHz shown by the black arrow in Figure 33.18 is based on piezoelectric vibration only. In addition, the resonance of 200 kHz shown by the black arrow in Figure 33.18 is the lateral piezoelectric vibration; it negligibly influenced the ultrasonic transmission and reception.

33.6. Conclusion

For applying an aerogel to the acoustic matching layer of an airborne ultrasonic transducer, an aerogel composite matching layer (AC-ML) was devised in order to solve the problems of the low strength and low adhesiveness of an aerogel. An ultrasonic transducer with AC-ML was constructed, and it was verified that its sensitivity was about 20 times as high as that of the conventional ultrasonic transducer and that its time-domain response was improved.

References

1. Altringham JD (1996) Bats Biology and Behaviour, Oxford: Oxford University Press p.79–113
2. Thomas JA, Moss CF, Vater M (2004) Echolocation in Bats and Dolphins, Chicago: University of Chicago Press p. 3–27
3. Toda M (2002) Narrowband Impedance Matching Layer for High Efficiency Thickness Mode Ultrasonic Transducer. IEEE Transactions on Ultrasonics, Ferroelectrics, and Frequency Control 49:299–306
4. Nagahara H, Hashida T, Masa- Suzuki M, Hashimoto M (2005) Development of High-Sensitivity Ultrasonic Transducer in Air with Nanofoam Material. Japanese Journal of Applied Physics 44:4485–4489
5. Nagahara H, Suginouchi T, Hashimoto M (2006) Acoustic Properties of Nanofoam and its Applied Air-Borne Ultrasonic Transducers. IEEE Ultrasonic Symposium p. 1541–1544
6. Gross J, Fricke J, Hrubesh LW (1992) Sound Propagation in SiO_2 Aerogels. J Acoust Soc Am 91:2004–2006
7. Fricke J (1988) Aerogel – Highly Tenuous Solids with Fascinating Properties. Journal of Non-Crystalline Solids 100:169–173
8. Fricke J (1990) SiO_2-Aerogels – Modifications and Applications. Journal of Non-Crystalline Solids 121:188–192
9. Gibat V, Lefeuvre O, Woignier T, Pelous J, Phalippou J (1995) Acoustic Properties and Potential Applications of Silica Aerogels. Journal of Non-Crystalline Solids 186:244–255
10. Krimholtz R, Leedom DA and Matthai G (1970) New equivalent Circuit for Elementary Piezoelectric Transducers. Electronics Letters 6:398
11. Leedom DA, Krimholtz R, Matthai GL (1971) Equivalent Circuit for Transducers Having Arbitrary Even- or Odd-Symmetry Piezoelectric Excitation. IEEE Transactions on Sonics and ultrasonics, SU-18:128–141
12. Lee W, Idriss SF, Wolf PD, Smith SW (2004) A Miniaturized Catheter 2-D Array for Real-Time 3-D Intracardiac Echocardiography. IEEE Transaction on Ultrasonics, Ferroelectrics, and Frequency Control 51:1334–1346

Part XII

Applications: Metal Industry

34

Aerogels for Foundry Applications

Lorenz Ratke and Barbara Milow

Abstract The casting of metals and alloys is very often performed into molds made of sands bonded by polymers. Resins based on, for instance, phenol–formaldehyde build bonding bridges between the sand grains, establishing a macroporous tight and strong sand form having a shape mirroring the workpiece to be cast. Any cavity in a casting is mapped by so-called cores, which are also made of polymeric-bonded sands. Organic aerogels can replace conventional polymers and offer a variety of advantages due to their nanostructure and composition, especially for cores. The development of these organic aerogels for light-metal and nonferrous heavy metal casting is described, their properties elaborated and compared with conventional ones. Transforming especially resorcinol–formaldehyde aerogels into carbon aerogels allows bonding sand grains by amorphous, nanostructured carbon with special advantages. New developments in the last few years are described, revealing that inorganic and organic aerogels in a granular form can replace a part of any sand used in foundries, leading to improved cast parts. In contrast to polymeric aerogels, silica-based ones have been used for more than a decade in solidification engineering to study fundamental aspects of metal solidification and casting. The final section describes various applications of inorganic aerogels with respect to solidification science.

34.1. General Aspects of Mold Preparation for Castings

Castings of liquid metals and alloys have to be done in forms which map a mirror image of the workpiece to be cast, for instance, a motor block or cylinder head or housing for a notebook. There are several types of molds: permanent and lost ones. The permanent ones are typically made from heat resistant steel and are built as a kind of 3D puzzle: the pieces are stacked together, the casting operation performed, and the *mold* disassembled into its pieces and the cast part taken out. *Permanent molds* are used if a huge number of cast pieces are needed, since the costs to machine the 3D puzzle must be profitable. Lost form casting processes use a form, which has to be destroyed after casting and solidification of the alloys. Only then the cast piece can be extracted. Their molds are typically made from sand of

L. Ratke (✉) **and B. Milow** • Institute of Materials Physics in Space DLR, German Aerospace Center, 51147 Cologne, Germany
e-mail: lorenz.ratke@dlr.de; barbara.milow@dlr.de

M.A. Aegerter et al. (eds.), *Aerogels Handbook*, Advances in Sol-Gel Derived Materials and Technologies, DOI 10.1007/978-1-4419-7589-8_34,

various kinds (like quartz, alumina, zirconia, chromium ores, or olivine) which are glued together by special binders. The simplest *binder* is oil. Others are colloidal binders, based on sodium water glass or plastic clays, like *bentonites*, which build firm and tight molds on firing at higher temperatures. In addition to the molds used to keep the liquid metal in shape, any cavity inside a casting must be mapped by some piece of solid materials, which can be extracted easily after casting. This can be quite challenging for complex shaped castings such as cylinder heads, gear boxes, and others, having a lot of complex undercuts. To map the cavities, so-called *cores* are fabricated, most from sand, but salts and ceramics might also be used, especially in precision casting. Modern foundry shops typically use quartz sand, due its low costs, for both the mold and the cores and bind them by means of polymeric binders. Here phenolic-urethane, phenolic-isocyanate binders are used in huge quantities. In the last decade, more environment friendly binders were developed which are based mainly on sodium water glass.

The engineering science and technology of mold materials and manufacturing technologies are well developed and described in detail in the literature [1–3]. There is a continuous evolution of binder recipes for various reasons, such as health and safety issues, economic and commercial ones, modifications to improved alloys and casting processes, governmental regulations with respect to environment, emission of volatile decomposition products, and so on. For nearly all alloy classes, there are special variants of the general recipe of sand bonded by a polymer or silicate ester or something else.

One can make a general subdivision into three classes of alloys: Light-metal castings (aluminum, magnesium), nonferrous heavy metals (copper, brass, bronze), and ferrous metals. In light metal castings, typical temperatures seen by a mold or core is at a maximum around 750°C, for copper and its alloys it is around 900–1200°C, and for cast iron it is usually not larger than 1,350°C, whereas for steels it might be as high as 1,600°C. The different temperature ranges lead to different thermal loads on molds and cores and the organic or inorganic binder used. Upon thermal loading, various reactions occur such as decomposition, baking, carbonization, evolution of volatiles, vaporization of moisture, mechanical degradation, and metal penetration to name only a few.

If one is aiming to replace a conventional binder system with an aerogel-based system, one has to understand the advantages and disadvantages of conventional binders with respect to their application or the alloy class they are used for. This is not the place to describe in detail the mold and core engineering, but a brief review of a few typical technologies, problems, and challenges of casting into sand molds and cores will be given. For a deeper understanding, the reader is referred to the literature [1–3].

34.2. Functional Requirements for Molds and Cores

A foundry molding mixture of sand with a binder passes through many production steps before the cast part can be extracted and the sand with some attached binder be reclaimed. There are many properties a molding and core material must possess, partly contradicting each other. First of all the mixture must have a good *flowability*. The sand–binder mixture should not lump, but behave as if there would be no binder at all. It must easily trickle into the forms. Molds and cores can be made by hand or by machines. Especially cores are typically made by a so-called core shooting machine: the mixture is first fluidized at high pressure and then pushed into a metal form yielding a high pressure compaction. Such operations also require a *compatibility* and a certain *green strength*, since in many operations the compacted mixture has to be taken out of the mold or form. It is handled in the foundry shop either by men or machines. A sufficient green strength is often achieved by chemical hardening, for instance phenolic resins are hardened by gaseous amines or silica esters by carbon dioxide. Alternatively the resins are cured and hardened at higher temperatures up to 250°C. The hardened and dry mold and core material must have an even higher *dry strength*.

As already mentioned, the mold or core must withstand the cast heat of the liquid metal. Different alloys strain the sand–binder system used in a different way. This ability is called *refractoriness*. Being in contact with a hot liquid metal, the mold material must withstand fluid friction during the casting operation, meaning the binder must firmly bind the sand grains and these shall not be entrained in the liquid metal stream passing by at the mold or core surface. Next the sand–binder systems must have certain *deformability*. It should not crumble during the solidification step, which induces severe shrinkage stresses due to the thermal shrinkage and the density difference between solid and liquid states. The mold and core material can heat up considerably leading to the evaporation of water or humidity absorbed by the porous material. The evaporation of one weight percent of moisture leads to a tremendous steam volume, about 30 times the volume of the original solid workpiece. In addition, the heat load induces thermal decomposition of organic binders, which also gives rise to a gas evolution. Molds and cores should therefore have high *gas permeability* otherwise the gas would enter the casting and lead to blowholes and pores especially below the skin. The liquid metals might react with the sand. It can penetrate into the pores or react chemically with the sand forming mixed silicates or more generally mixed oxides and ores (this is especially important for cast iron and steels). Thus, a mold and core must be resistant against *metal penetration*. It is obvious that a cast piece later should have a fine surface finish. This means the mold and core material must have a sufficient *fineness*. This cannot always be achieved by simply using fine sand because, for instance, in nonferrous heavy metal casting, this would lead to metal penetration. After casting it must be easy to destroy the mold and extract the core. The material therefore should have a good *collapsibility* during knocking out operations.

In foundry practice, these requirements are tried to be fulfilled with suitable combinations of sand and binder. The variables at hand of the foundry engineer are sand type (quartz, aluminum oxide, olivine, chromium, zirconium dioxide, artificial sands, like hollow sphere made from mullite), sand grain size and distribution, binder type, binder amount, hardener (amines, carbon dioxide), and curing temperature.

Typical problems occurring with conventional sand–binder systems will not be reviewed here but discussed within the context of the aerogel binders and aerogel-based nanoadditives.

Polymeric aerogels were developed as binders for foundry sands in the last decade to solve certain problems, which are typical for classical organic binders. In the last few years, organic and inorganic aerogels were developed in granular form as additive to foundry sands to improve certain properties such as metal penetration, fineness, and collapsibility.

34.3. Resorcinol–Formaldehyde Aerogels as Binders

Polymeric aerogel based on resorcinol and formaldehyde was first tried by Brück and Ratke [4], to see if they can be used to bind foundry sands. They used the fact that Resorcinol–Formaldehyde (RF) aerogel can be gelled and dried under ambient conditions, if the ratio of resorcinol (R) to catalyst (C) (here sodium carbonate) is larger than 1,000 [5] (Chap. 11). This aspect is important because supercritical drying, often applied to produce super light aerogel, would be too expensive for foundry shops. Brück and Ratke studied the mixing of various sand types, having different grain size with different amounts of RF-aerogel sol, ranging from a few weight percent with respect to the sand weight up to complete filling of the interstices left by the compacted sand (around 40–50 volume percent and 20–25 wt%). They found that the sand could be excellently bonded with RF aerogel. Gelation and drying were performed in a drying cabinet at 40°C within a day. They observed that gelation and drying of the RF aerogel within the interstices of sand was faster compared to a pure, monolithic RF-aerogel solution of the same total volume. The typical *bending strength* was between 1 and 4 MPa, varying with the sand grain size and the binder amount. Importantly, they observed that there is no shrinkage of the sand–RF-sol mixture upon drying. One of their important findings published later [6] is that castings made into RF-aerogel-bonded sands lead to a superfine surface finish although a coarse sand was used. This is shown in Figure 34.1.

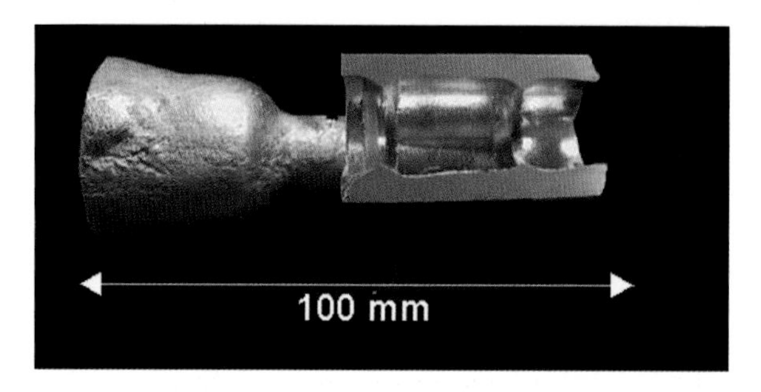

Figure 34.1. Example of an aluminum alloy casting (AlSi7Mg0.6) cast into a sand mold with a core made from a RF-aerogel-bonded sand. The core had two circumferential notches. The core was removed by thermal decomposition at 350°C within 30 min. The inner surface has a shiny and smooth appearance, which is typical for RF-aerogel-bonded sands at higher binder amounts.

Secondly, the RF aerogel was easy to oxidize at low temperatures of around 350°C within half an hour which is below the typical thermal treatment temperatures of Al cast alloys used to increase their mechanical strength. Thus, the collapsibility is almost perfect, especially for Al castings, where the core removal can be a serious problem. The bonding bridges between the sand grains were analyzed carefully by scanning electron microscopy (SEM) and it was found that these bridges exhibit a 3D nanostructure common to that of aerogels and that the particle sizes in these bridges are smaller than in a compact RF aerogel prepared under the same conditions. Figure 34.2 shows a *bonding bridge* between sand grains with a high resolution image inserted exhibiting the nanostructure inside the bridge. Ratke and Brück also coined the acronym *AeroSand* for their RF-aerogel-bonded sands.

34.4. Mechanical Properties of AeroSand

These basic first results were later extended in much detail leading to a quantitative description of the mechanical properties and their dependence on binder amount and sand grain size [4, 6, 7]. The mechanical properties were measured in great detail and depth by

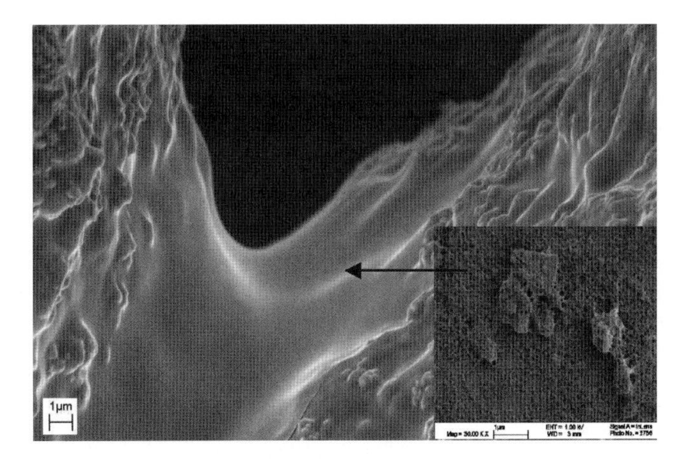

Figure 34.2. Bonding bridge between two sand grains. The RF-aerogel binder wets both sand grains and has built between them originally a liquid bridge, which dried after gelation at 40°C. The bridge shows the typical appearance of an aerogel at very large magnifications: a sponge-like open porous, nanoscaled structure (*see inset*).

Reuß and Ratke [8] and compared with a model setup by Ratke and Brück [7]. They used a standard RF-aerogel solution containing resorcinol and deionized water in a molar ratio of (0.044:1), resorcinol and formaldehyde (37%) in a ratio of 0.72:1, and a ratio of R:C (=Na_2CO_3) of 1512:1. Such a solution was mixed in different weight amounts from 5 to 20 wt% with various sands such as Cerabeads™ (CB), Alodur™, quartz sand, and zirconia. After gelation and drying, the amount of binder reduces itself (loss of water) to approximately 1/4 of the original liquid content. The wet mixture of sand and aerogel solution was pressed into a metal mold to prepare so-called bending bars (100 × 20 × 20 mm). These were dried at 40°C for 24 h and then were annealed under vacuum at temperatures of 250, 500, 750, and 1,000°C for 90 min.

Figure 34.3 shows the bending strength of an RF-AeroSand (CB1450, 17.5 wt% RF solution) in relation to the heat treatment temperature. Surprisingly, with increasing

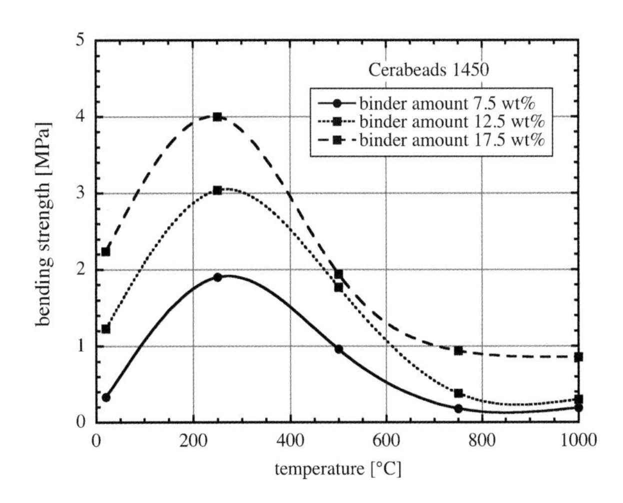

Figure 34.3. Bending strength of a RF-AeroSand (CB1450, 17.5 wt% RF) after heat treatment at different temperatures at constant duration and heating rate after Reuß [8].

annealing temperature, the strength exhibits a maximum at 250°C and with further increasing of the annealing temperature it decreases considerably. An explanation could be that during heat treatment reactions still occur inside the porous network of the aerogel body, in other words the residual fluids strengthen the bonds between the nanoparticles in the binder. Figure 34.4 shows the bending strength of RF-AeroSand made from CB1450 with different binder amounts after different heat treatments. It clearly reveals that the maximum at 250°C is independent of the binder amount.

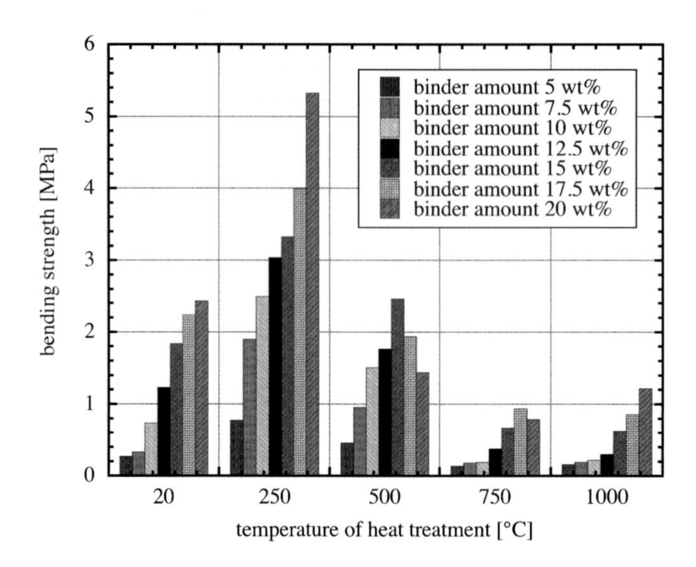

Figure 34.4. Bending strength of RF-AeroSand (CB1450, binder amounts from 5 to 20 wt%) for different heat treatment temperatures after Reuß [8]. Note that in all cases the maximum strength is achieved when the gelation and drying at 40°C is followed by a vacuum drying at 250°C.

AeroSand made from other sands (Alodur, quartz, zirconia) shows the similar behavior. RF-AeroSand made by quartz sand (Sepasil) breaks down, however, already during the heat treatment if the temperature is higher than 500°C. This is caused by the phase transition of quartz from low quartz to high quartz at 573°C, which leads to a destabilization of the sand–binder system as a whole. Independent of the annealing temperature the bending strength of an AeroSand decreases with increasing particle size. The classical *Griffith criterion* for *brittle fracture* of solids can be modified and used to describe the mechanical behavior of AeroSand [7]. Microscopic investigation shows that the fracture path is always along the interface between the sand and the aerogel and thus the interface is the weakest point and not the aerogel itself, which could have been another possibility. Ratke and Brück [7] developed, on the basis of the classical Griffith criterion for fracture of brittle solid, a new criterion for fracture strength, which can describe the fracture strength of the RF-aerogel-bonded sands. Their modified Griffith criterion means with Φ_A the volume fraction of the aerogel inside the AeroSand.

$$\sigma_F^{\mathrm{AeroSand}} = \sigma_0\sqrt{\frac{\Phi_A}{d}} = \frac{1}{\left(\frac{1}{1-\varepsilon_A}\right)^{1/3} - 1}\sqrt{\frac{2E\gamma_{SA}}{\pi\varepsilon_S}}\sqrt{\frac{\Phi_A}{d}}. \tag{34.1}$$

In (34.1), the factor σ_0 is a constant of proportionality detailed with the second equality sign. It depends on *Young's modulus E* of the AeroSand, the interfacial energy between sand and aerogel γ_{SA}, the porosity of the aerogel ε_A, and the volume fraction of sand ε_S [7]. Equation (34.1) predicts that the bending strength should be linearly proportional to the inverse square root of the average particle diameter *d*. Figure 34.5 shows the bending strength of AeroSand made from three different Cerabead sands as a function of the inverse square root of the average grain size *d* at all annealing temperatures. Independent of the heat treatment temperature the modified Griffith criterion is valid.

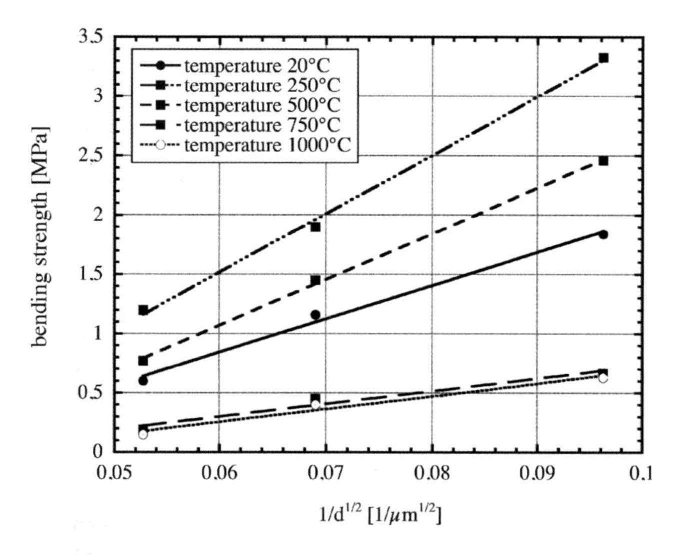

Figure 34.5. Bending strength of RF-AeroSand made from Cerabead sands as a function of the inverse square root of the average grain size *d* at different annealing temperatures. The *dotted lines* show linear fits according to (34.1).

This behavior was observed in all cases of sands bonded by polymeric RF aerogels. The absolute strength values in Figure 34.4 show that very small amounts of RF binder (measured always with respect to the liquid RF solution) are not suitable, since the strength is below one MPa. Values above are required in foundry practice. 7.5 wt% RF solution, which gives around 2 wt% solid aerogel are just sufficient, especially after a vacuum drying at 250°C.

34.5. Drying of RF Aerogel–Sand Mixtures

For practical applications, not only the mechanical strength is important but also processing times, drying behavior, and, as mentioned in the introduction, properties of the AeroSand such as thermal conductivity, gas permeability, thermal decomposition, resistance to metal penetration, gas evolution, and baking behavior. All these parameters were investigated.

Sands bonded by RF aerogels have the disadvantage that their green strength is negligible, since simply a wet fluid bonds the sand grains. Therefore, both the gelation and the drying process should be rapid. Low green strength can only partly be compensated by the addition of small amounts of bentonites [9] or the addition of hydrophilic aerogel or xerogel grains to the sand [10]. Especially important is the drying process. Reuß and Ratke intensively studied the drying process using a simple device allowing mainly for 1D drying [11]. They used various sands as mentioned earlier, varied the amount of binder from 5 to 20 wt% and the drying temperature from 20°C to 70°C. The mixture of sand and RF solution was put into a cylindrical tube and closed on one side. Since the AeroSand does not shrink, drying could only proceed by evaporation of the *volatiles*, especially water at the front side. The length of the AeroSand mixture in the tube was also varied. As a measure of drying they simply measured the variation of weight as a function of time with a digital balance.

The drying time turns out to be linearly dependent on the sample length and the binder amount. All measurements performed point out that the microstructures (sand grain size, RF-aerogel amount) are of minor importance. A slight variation is always within the measurement uncertainties. With the curves shown in Figure 34.6, the drying time can be defined as the time at which the weight of the sample does not change by more than 1% anymore. It turns out that this drying time depends in an exponential way on the temperature, and can be described by an *Arrhenius law* (Figure 34.7). Thus, they conclude that simple evaporation of water being there from the initial aerogel solution and the *polycondensation* reaction inside the RF aerogels determine the drying kinetics. The sand packing seems to have no effect at all. The aerogel filling even at 5 wt% solution added to a sand around 20 vol% builds a continuous network and its pores drain the liquid to the surface such that evaporation of the pore fluid at the surface determines the kinetics.

The drying time can be described in such a model with a simple expression

$$t_d = \rho_F L_0 \sqrt{\frac{2\pi RT}{\mu}} \exp\left(\frac{Q}{RT}\right), \tag{34.2}$$

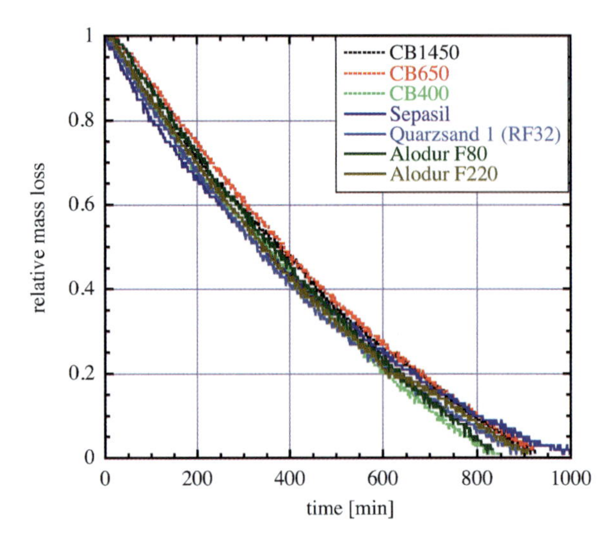

Figure 34.6. Relative weight loss of aerogel-bonded sands (Cerabead sands: CB1450, CB650, CB400; quartz sand 1 and Sepasil and Alodur (alumina): F80, F220) of different average grain sizes, measured during drying at 40°C.

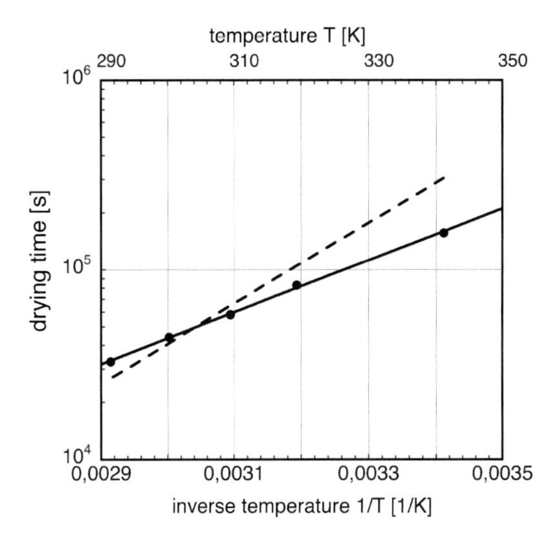

Figure 34.7. Plot of the drying time as a function of inverse drying temperature. The linear relation is expected from simple evaporation kinetics. The *dashed line* shows the evaporation kinetics with the activation energy of water. The nanonetwork inside the aerogelic binder reduces the activation energy after Reuß and Ratke [11].

with ρ_F the density of the RF-aerogel sol, L_0 the sample length, μ the molecular weight, and Q the heat of evaporation. R and T have their usual meaning. Such a relation is exactly observed with activation energy of 26.2 kJ/mol, which is around 2/3 of that of water (40.7 kJ/mol). The difference between both is attributed to capillary stresses in the aerogel network changing the heat of evaporation. Understanding the kinetics of drying helps to improve the processing of RF-bonded sands in foundry shops.

34.6. Thermal Decomposition

The thermal decomposition was measured by Ratke and Brück [6] and later extended by Milow and Ratke [12] using *TG-FTIR* measurement to analyze the thermal decomposition products (Chap. 23). The weight loss is measured as a function of temperature using a NETZSCH TG 209 Iris equipment. The exposed gases are transferred through a thermostatic connection and analyzed by Infrared spectroscopy (BRUKER FTIR SENSOR). The analysis is started at room temperature. Then the sample is heated up with a heating rate of 10 [K/min] up to 1,000°C under a gas flow rate of 40 [mL/min]. A sample of AeroSand was processed under these conditions. The result is shown in Figure 34.8. At the end a mass of 98.9% is kept. Two significant losses of mass are detected. The first one is detected around 93°C and can be attributed to CO_2 and water and the second one around 639°C where in addition CO and methane are detected by FTIR.

The corresponding FTIR spectra at 86°C and 639°C are shown in Figure 34.9. A comparison with library spectra shows that at 93°C, CO_2 and water are the main components. At 639°C beside CO_2 and water also methane and CO are found.

Under nonoxidative conditions, the thermal decomposition of the RF-aerogel binder used leads to BTX-free decomposition products such as CO_2, CO, and methane as expected.

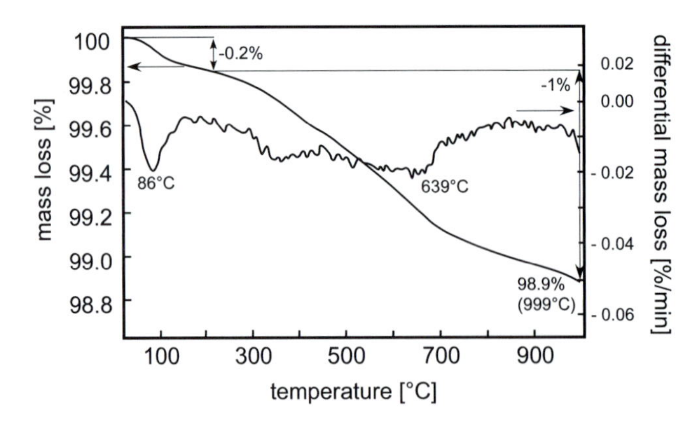

Figure 34.8. TG/DTG curve of an AeroSand core sample. The *black line* shows the mass loss as a function of temperature and the green line the corresponding DTG curve.

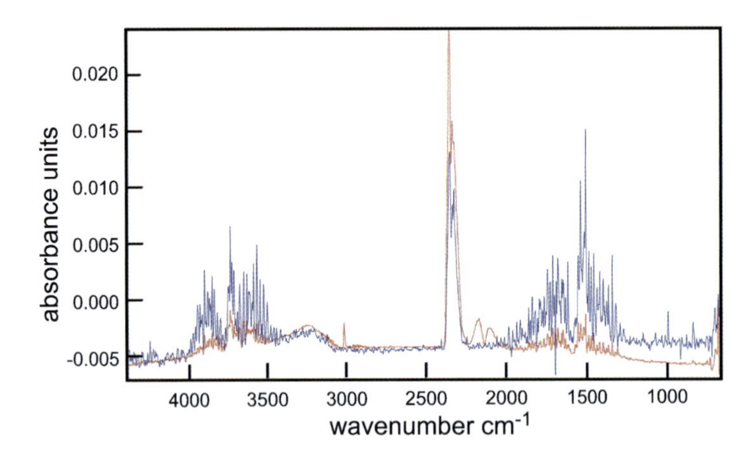

Figure 34.9. FTIR spectra of thermal decomposition product at 93°C (*blue*) and at 639°C (*red*) of an AeroSand core sample. At 93°C the band from 4,000 to 3,500 cm^{-1} and 2,000 to 1,350 cm^{-1} stems from water. The high and sharp peak at 2,300 cm^{-1} is from CO_2. At 639°C the *red curve* also shows two new peaks. At 3,000 cm^{-1} the methane peak appears and to the right of the CO_2 peak a double from CO can be seen.

For casting processes it is necessary to reduce toxic emissions and to develop products in a more environmental acceptable direction. Up to now the development of AeroSand improves on this aspect.

34.7. Gas Permeability

As mentioned molds and cores must have a suitable permeability for gases, since upon casting of a hot liquid metal into a mold, moisture evaporates, the polymeric binder decomposes, and volatiles are formed. These must not harm the casting ingot and therefore should leave the mold via its porous network. Gas permeability is thus essential to avoid casting defects like blowholes and internal porosity, especially under the skin. Molds and cores bonded by RF aerogels have to be analyzed with respect to their permeability because it can be expected that, depending on the amount of aerogelic binder, the permeability decreases. On the other hand, the high specific surface area of aerogels might help to absorb volatiles evolving during the casting operation.

Reuß and Ratke [13] measured the permeability of AeroSand. The gas permeability of a porous body like an AeroSand can be measured using a setup shown in Figure 34.10. The sample is mounted in a stainless steel ring and fixed therein with a special silicon gel such that no gas can pass the porous body aside but passes only through its pores. This steel ring is connected to a flange that itself is connected to a valve.

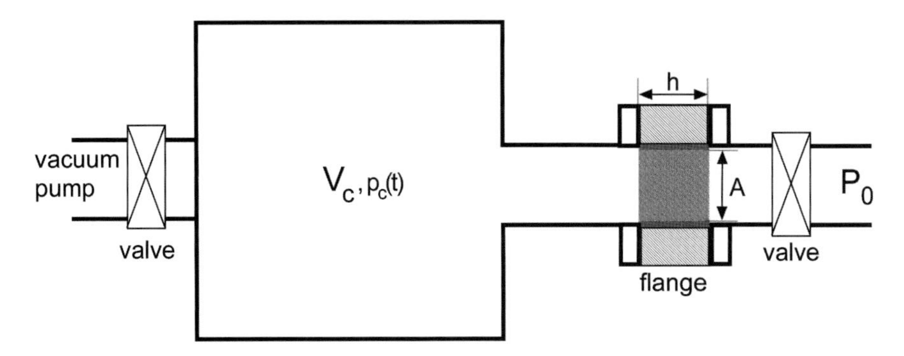

Figure 34.10. Schematic setup for the measurement of the gas permeability of RF-aerogel-bonded sands (AeroSand).

A cylindrical chamber of volume V_c serves as a test volume. For the measurement, the test volume is evacuated to a preset pressure $p_c(0)$ and the valve in front of the AeroSand piece is opened. Then the ambient pressure p_0, being measured, exists on one side of the sample and on the other side initially the lower pressure p_c. The pressure difference drives a gas flow through the porous body. A vacuum gauge measures the pressure increase $p_c(t)$ in the chamber with time. This pressure increase can be calculated to follow a law such as

$$p_c(t) = p_0 \tan h\left(\frac{AKp_0}{2\eta h V_c} t + \operatorname{arctan} h\left(\frac{p_c(0)}{p_0}\right)\right), \tag{34.3}$$

where A is the cross section of the test piece (see Figure 34.10), h its thickness, and η the dynamic viscosity of the gas. K is the permeability as defined by *Darcy's law*

$$\upsilon_D = -K \frac{1}{\eta} \frac{dp}{dx}. \tag{34.4}$$

υ_D is the gas flow velocity through the porous body and dp/dx the applied pressure difference. K has the dimension of a squared length (see also Chap. 21).

The theoretically predicted behavior was also confirmed by the experimentally measured pressure vs. time relation. It turned out that K is not a constant, but varies with pressure difference $p_0 - p_c(0)$. This is in contrast to the expectation that K is determined fully by the geometry of the porous body. Test measurements performed on sintered porous glass samples indicated that the permeability is indeed constant and thus that the pressure difference dependence is a specialty of polymer-bonded sands, and aerogel-bonded sands are in this sense just a subclass. The polymer mixture does not bond all sand grains perfectly and thus the larger the pressure difference the more of the loosely bonded grains can move and fill interstices and block the gas flow. From the pressure difference it is possible to calculate of so-called *rattlers*, i.e., free or weakly bonded sand grains in aerogel-bonded sands could be calculated. One example is given in Figure 34.11.

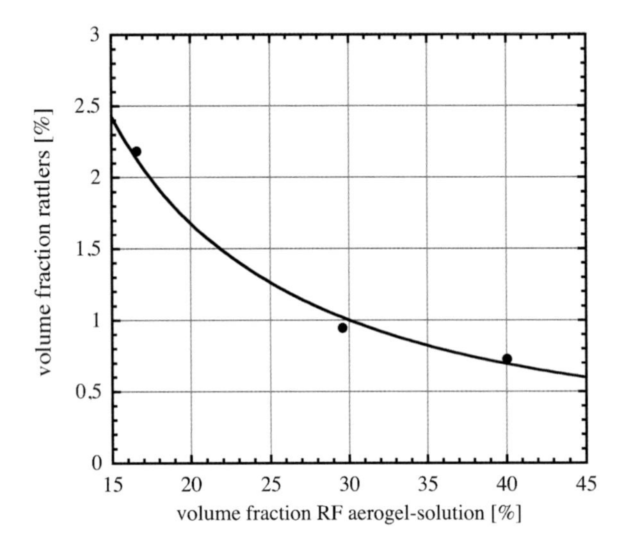

Figure 34.11. Amount of rattlers in aerogel bonded aluminum oxide sand (Alodur 220) as a function of initial RF-solution content. Rattlers are loosely bonded sand grains.

Permeabilities of sands bonded with conventional binders like sodium water glass are typically of the order of 10–80 μm^2, depending on the sand grain size as expected. Aerogel-bonded sands have absolute values being a factor of 2–5 smaller. The amount of rattlers is, however, a factor of 4–5 smaller than with conventional binders. So aerogels bind more sand grains. The reduced permeability compared to conventional binders is in part a result of a different binder amount. To properly bond sand grains, a higher amount of aerogel binders is needed compared to conventional binders. The specific surface area of aerogel-bonded sands is, however, much larger and thus these mold and core materials can better absorb cast gases.

34.8. Carbon Aerogels as Binder Materials

If it would be possible to develop a mold and core material bonded with carbon, the problems of degassing during mold filling and solidification could be eliminated. But conventional crystalline carbon cannot bind sand, since its basal planes easily slip over each other. Glassy or amorphous carbon, however, would be able to establish a stiff matrix for sand grains to be embedded into. But, how to prepare such a material? Based on the known polymeric RF-AeroSands composition described earlier, Voss and Ratke [14] developed so-called *Carbon–AeroSand*. The polymeric RF aerogels are pyrolyzed and are then able to bind conventional foundry sands firmly (see Figure 34.12). The potential application as a core material then leaves open the question of the core removal because carbon usually oxidizes at temperatures above 800°C. This temperature would be too high for thermal core disintegration within cast parts made from light metals. However, Voss made the astonishing observation that Carbon–AeroSand composites oxidize already at temperatures above

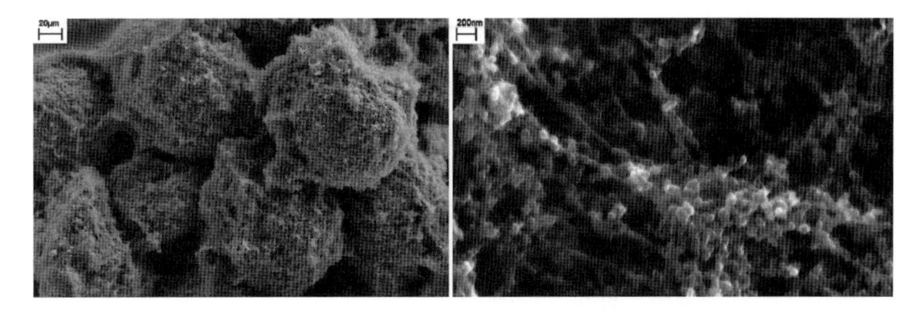

Figure 34.12. SEM picture of an Alodur sand covered completely with carbon aerogel. The bonding bridges between the sand grains are visible. Similar to RF aerogels, they consist of nanostructured carbon aerogel particles as can be seen in the right figure, which is a magnification. The scale bars in the left figure indicates a length of 20 μm, in the right of 200 nm.

450°C being a direct consequence of the nanostructure. Thus, Carbon–AeroSand composites have a high potential as a new core and mold material for foundry applications (Chap. 36).

RF aerogels are the basis of *carbon aerogels*. They are made of the following precursors: formaldehyde, resorcinol, water, and a catalyst, as it is described earlier and in the literature [6, 8].

Carbon–AeroSands are made by mixing the solvents of the RF aerogel in an accurately defined ratio with conventional foundry sands such as aluminum oxide, mullite, or silicon carbide. This mixture is filled into a container and compressed under vibration. The gelation takes place in a sealed, air tight container at 40–50°C for one day. After complete gelation, the cover of the container is removed and the final drying takes place at ambient pressure conditions at 40–50°C. The dried AeroSand mold or core can now be removed from the container. The final step namely the pyrolysis takes place in a vacuum oven at 1,050°C for 2 h under argon.

The characterization of the Carbon–AeroSands was done by Voss [14] for all properties related to foundry applications. All characterizations have been done with respect to the effect of different grain sizes, shapes, and the amount of binder. Good mechanical properties are essential for sand cores due to the casting pressure. The compressive and bending strength of the Carbon–AeroSand was investigated using compression- and three-point-bending tests. The compression and bending strength is typically in the range of 1.5–2.50 MPa. The elastic modulus are around 300–550 MPa. All results in terms of the grain sizes and binder amounts are shown in Figure 34.13 for Alodur sand.

The strengths of Carbon–AeroSands show the following correlation: the strength decreases with reduction of the binder amount and with increasing grain sizes. The strength of sands with round grains is lower than that with splintery.

For several sand types, grain sizes, and binder amounts, the thermal conductivity was measured by Voss with a stationary heat flow arrangement. The thermal conductivity of Carbon–AeroSands was independent of the sand type and carbon aerogel content typically around 0.8 ± 0.2 W/(mK). Within these margins the *thermal conductivity* of Carbon–AeroSands becomes higher with decreasing binder amount. The grain size has no influence on the thermal conductivity. The investigations showed that the Carbon–AeroSands have a good potential as mold and core material and can be influenced by the choice of the sand type. The thermal conductivities are in the range of conventional core materials.

The oxidation of the Carbon–AeroSand composites was investigated at different temperatures ($T_1 = 480$°C, $T_2 = 650$°C) and binder amounts. The loss of weight was

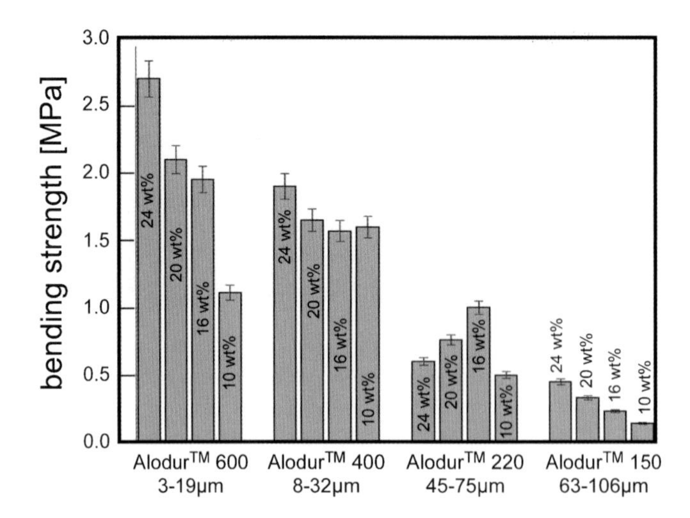

Figure 34.13. Bending strength of Alodur® Carbon–AeroSand composites for different grain sizes and binder amount with respect to the initial RF solution (wt%).

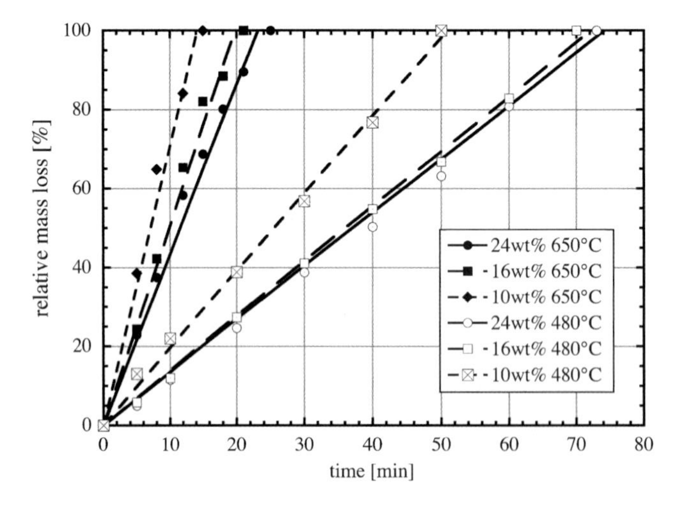

Figure 34.14. Oxidation level of Alodur® based Carbon–AeroSands with a wall thickness of 2 cm. Measured in dependence of the binder amount and temperature.

measured as a function of time. The results are shown in Figure 34.14. The degree of oxidation scales linearly with time and decreases with higher temperatures. Carbon–AeroSands oxidize completely independently of the amount of aerogel and the type of sand. From the measurements of the oxidation at two temperatures, the activation energy was estimated. A value of 112 kJ/mol shows that the activation energy for oxidation of the nanostructured carbon aerogels is lower than those of conventional crystalline graphite (164 kJ/mol), which is attributed to the nanostructure of the aerogels. This explains why appreciable oxidation occurs at the relatively low temperature of 480°C.

The complete oxidation is an outstanding property of the new material and was proved by X-ray scattering. Sand from Carbon–AeroSand is reusable without any further preparation of the material.

34.9. Aerogels as Nanoadditives for Foundry Sands

The techniques to produce cores with bonded sands are well established. Conventional binder systems, organic as well as inorganic ones, are widely used in sand casting foundry applications. There are still many open issues to reduce casting defects associated with the sand–binder system as discussed previously. It was also shown that RF aerogels as binders can overcome some of the problems encountered with conventional polymeric binders.

However, aerogels can be used in a completely different way: not as a gelling and drying sol, which binds sand grains, but as a granular material added to existing sand–binder systems. This improves the sand–binder systems being used in industry and opens a novel field of research and application of nanostructured aerogels. It turns out that adding granular aerogels to sand–binder systems is beneficial to improve the properties of sand molds and cores as well as to improve the properties of the final casting.

Aerogel-based additives are not wetted by alloys melts. Their high internal specific surface area absorbs gases which evolve during casting. They can compensate the 2% volume change due to the quartz inversion effect by their high elastic deformation ability. In principle it is possible to use organic aerogels as well as inorganic aerogels.

In Table 34.1, some tested nanoadditives and their properties are listed.

The aerogels or xerogels used are first pulverized with a cross hammer mill to fit the particle size distribution of the sand grains. A substitution of sand between 2 and 10 vol% by nanoadditives is suitable and sufficient. Sand and nanoadditives are mixed with a given type of binder and are then ready for molding.

Table 34.1. List of aerogel additives

Chemical material	Silica aerogel	m-Silica aerogel	Silica xerogel	m-Silica aerogel	Organic RF aerogel
Wetting behavior	$<30°$	$>160°$	$<30°$	$>160°$	$<30°$
Density [g cm^{-3}]	0.06	0.06	0.4	0.28	0.26

m-Silica stands for chemically modified silica aerogel; RF means resorcinol–formaldehyde; $<30°$ marks onset of the hydrophilic wetting situation and $>160°$ characterizes the wetting angle of super hydrophobicity.

First laboratory tests were simply carried out with the aforementioned bending bars having a dimension of 20 mm $\times$ 20 mm $\times$ 100 mm being prepared manually or using a core shooting machine [15–17]. These test bars are built of various combinations of sand, binder, and nanoadditive. They are used to measure bending strength, abrasion resistance, gas permeability, thermal decomposition with thermogravimetry combined with FTIR, and the fracture path and bonding bridges by SEM. Casting tests are carried out to analyze the quality of the surface finishing, the amount of baking due to binder oxidation and pyrolysis, metal penetration, and ease of core removal.

Figure 34.15 shows some results of bending tests. For these experiments, a mixture of Minelco sand (MIN ASF230) and natural sand with a ratio of 4.8:1 was used. Between 2 and 10 vol% of nanoadditives were additionally added and everything is bonded with 9.6 wt% of RF-aerogel solution as binder (being around 2.5 wt% binder after drying) [9].

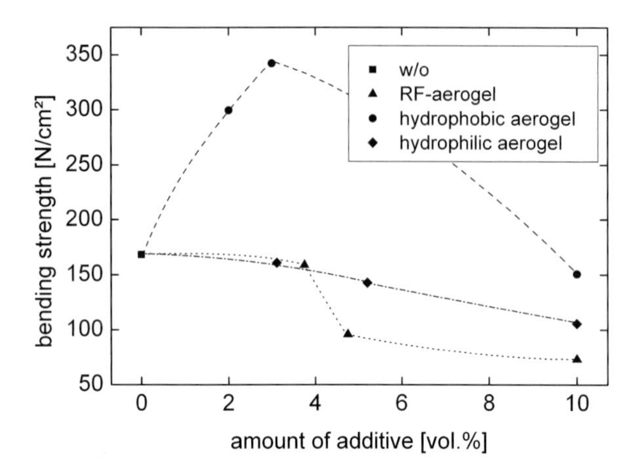

Figure 34.15. Bending strength of MIN ASF230 sand bonded with RF-aerogel solution using different nanoadditives referencing to the additive free material (w/o).

The roughness depth of the casting surface was measured using a MAHR300 profilometer, the stylus instrument enabling 2D tracing of a surface. A traced profile is shown in Figures 34.16 and 34.17. Within the binder–sand system of quartz sand and urea–formaldehyde resin binder, an inorganic aerogelic additive is used and compared with the additive free material. For the cores with nanoadditives, the roughness is twice as less as for the common casting surface.

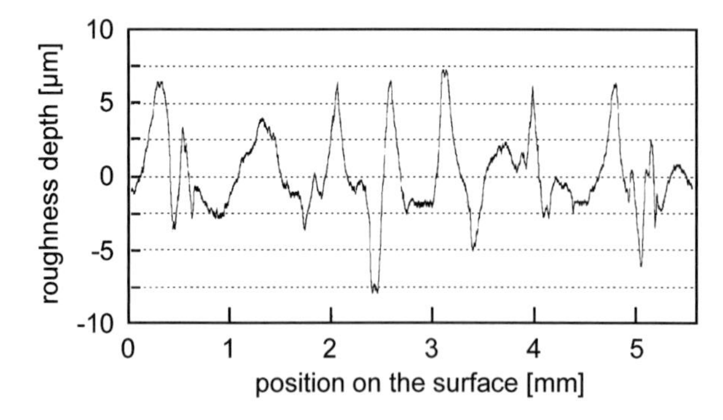

Figure 34.16. Roughness depth profile of the surface of an aluminum casting without addition of an aerogelic nanoadditive.

First experiments were also carried out in an industrial environment using a standard production line and equipment. In Figure 34.18, the production of three cores assembled to a package to cast a water tap is shown.

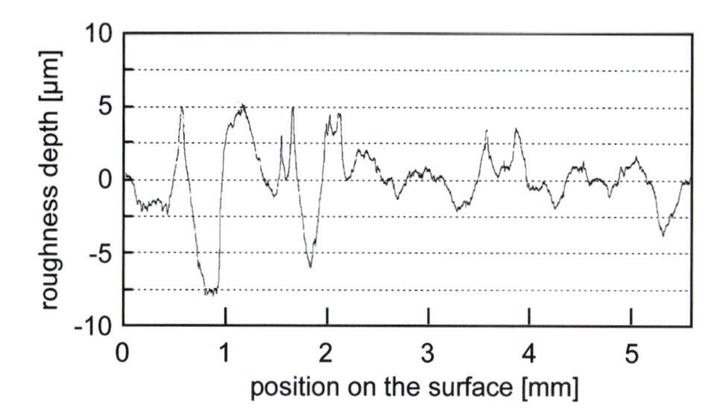

Figure 34.17. Roughness depth profile of the surface of an aluminum casting with the addition of an aerogelic nanoadditive.

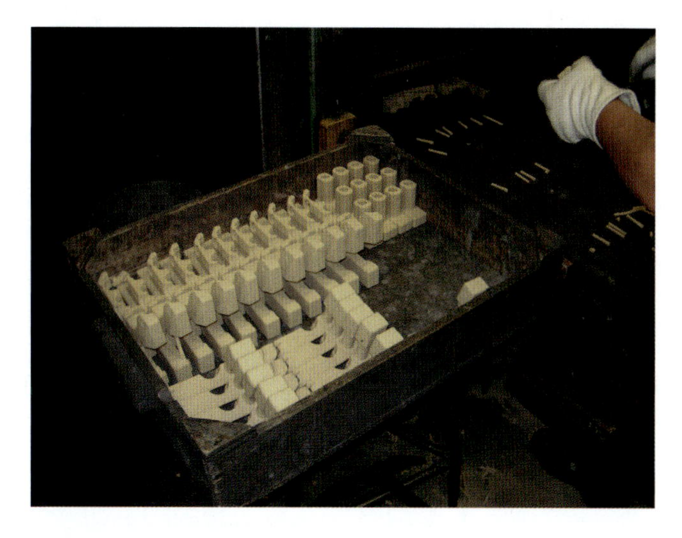

Figure 34.18. Core production in a foundry using m-silica aerogel as a nanoadditive (with permission of the company).

Currently, systematic studies are carried out to optimize the core and casting quality as a function of the used sand–binder system in combination with aerogelic nanoadditives.

34.10. Aerogels in Solidification and Casting Research

Since 1993 silica aerogels have been used as molds for casting operations [18], as it was observed that they withstand molten alloy temperature up to 1,000°C easily without macroscopic sintering effects, in contrast to what is known from the literature about silica aerogels [19, 20] (Chaps. 2 and 29). They also do not react with molten Al alloys, even after hours of direct contact with them, although bulk quartz glass does. Their optical transparency in the wavelength range from about 400 to 2,200 nm allows to directly observe the temperature distribution of a melt cast into a cylindrical cavity inside an aerogel with an infrared camera and to follow its time evolution. In addition it was shown by numerical analysis [21] that the small thermal conductivity of silica aerogels leads to almost flat isotherms in cylindrical castings. Silica aerogels have therefore been used since then as a new type of *crucible* and mold material. This knowledge was used to study form filling using aerogels as molds [22–24] or replacing a part of, for instance, a permanent steel mold by an aerogel [25–28]. Silica aerogel tiles were also used to investigate *hot tearing* susceptibility of Al alloys used for continuous casting [29] and to study solidification and phase decomposition in immiscible alloys [30]. The unique properties of silica aerogels also allowed to develop and construct a new type of directional solidification furnace. In the last decade many studies on solidification of Al-based cast alloys have been performed in these facilities [31–34].

34.10.1. Form Filling

Any kind of mold, be it made from sand, ceramic shells, heat resistant steels, gypsum, or clays, will be filled with a hot melt, which then cools down and solidifies. The filling operation, classically done manually with a more or less "big spoon," is of essential importance for the quality of the cast workpiece [2, 3]. During filling air might be entrapped, the flow might be turbulent, the surface might be wiggled, have protuberances, react with the mold wall, etc. The way a mold is filled with liquid metal determines to a certain extent the casting defects. Mold filling is today modeled by solving the Navier–Stokes equations coupled with a heat balance equation in arbitrary complex geometries. Although everything seems now at hand, at least in computer models and can be visualized, it turns out that the real mold filling as observed by X-ray tomography exhibits filling problems not modeled with conventional mold filling software. Therefore, the experimental investigation of mold filling is of great importance to improve both the software codes and the mold filling in foundry shops. Silica aerogels were used for such purposes in the last years [22–28]. Tscheuschner [24] used aerogel wedge molds, which were prepared from silica-aerogel sheets made by Airglass^{TM} using a CNC milling machine. The wedge type is depicted in Figures 34.19 and 34.20.

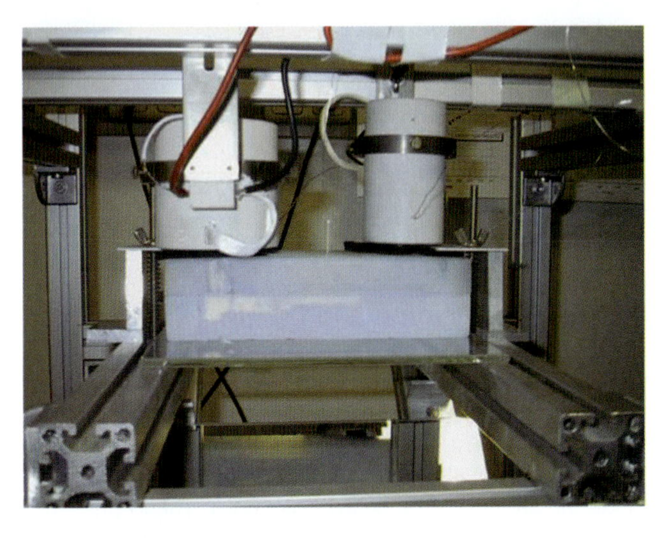

Figure 34.19. The picture shows in the center the Aerogel mold, into which the molten metal enters the wedge-type cavity from the left side (see Figure 34.20, below). A mirror below the shell allows to follow the mold filling and temperature distribution with an infrared camera.

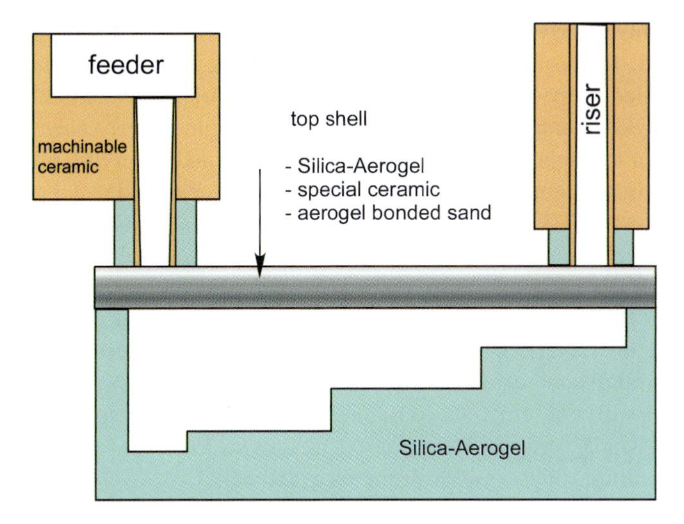

Figure 34.20. Wedge casting of A357 with a silica-aerogel mold and different top shells. The liquid metal is cast into the mold through the feeder. The riser prohibits shrinkage holes in the wedge.

The total length was 150 mm and the width is 62 mm. Three steps have a length of 46 mm and one only of 12 mm. The depth of the right step is 5 mm, the following 10 mm, etc. At the top, different cooling conditions were realized using either silica aerogel, a typical investment casting ceramic shell, or aerogel-bonded sand as described earlier. The ceramic shell could be heated by a fan to temperatures of about 300°C. The mold filling started on the left-hand side in the gate system and ended in the feeding system on the right-hand side. The feeder and gate system consisted of a special ceramic with a low thermal conductivity (≈ 1 W/(Km)) and were heated slightly over the casting temperature. The left area with its right ridge served to reduce the flow and thus to contribute to a nearly uniform filling of the mold cross section. The alloy used was a standard Al cast alloy (AlSi7Mg0.6). The filling and solidification was monitored by an infrared CCD camera mounted below the wedge. Figure 34.21 shows the result. In the right panel, three subsequent images of the real filling process are shown. The melt flows into the cavity with a parabolic interface shape. Thus, it behaves as expected from a Poiseuille flow. The modeling shown in the left panel shows quite a different behavior, especially the interface shape does not match: It exhibits too many disturbances and instabilities.

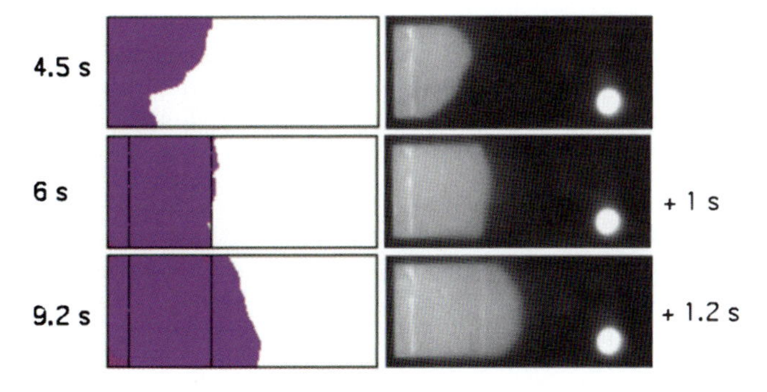

Figure 34.21. Sequence of wedge filling. Left-numerical calculation, right optical observation by the NIR-CCD camera.

These results clearly show why it is important to measure filling processes with a transparent mold. In the case shown below it might be that the interface tension was incorrectly modeled or the Young–Laplace equation not properly taken into account.

Viets and Haferkamp [25–28] investigated the origin of *hot cracks* in die casting, replacing a part of the steel mold with a silica aerogel tile and observing filling and solidification with an infrared camera. The interdependence between the development of temperature gradients at the solid–liquid interface during solidification and formation of local defects demands for thermal investigation. In foundry practice, thermocouples are used to control the die's overall cooling rate, but fluctuations in product quality are frequent.

The use of IR-transparent silica aerogel (Chaps. 2 and 28) as mold material offers a new perspective for a better understanding of solidification process, since *thermography* may start right from the moment of introducing the melt. Viets and co-workers developed a prototype casting with a hybrid mold consisting of the regular steel block with engraving and a front plate of silica aerogel. The infrared images in Figure 34.22 represent the data recorded during filling of the mold. A thermography CCD device with sensitivity up to 1.3 μm was used. The filling sequence shows that first the bottom of the cavity is flushed with liquid metal and thus heated and leading to vertical temperature gradients. After 1 s the branching node is reached and the three arms are filled. Heat still is concentrated in the right side of the sprue and the filling pressure is not really allowing to fill the three arms at the left side equally well even after 1.5 s. The simultaneous action of filling pressure and temperature gradients in the mold allowed Viets and Haferkamp for the first time to improve existing numerical modelling of form filling based on experimental data. According to these measurements a model for the casting process could be derived, subdividing the image scenes into the three relevant stages of solidification: mold filling, hull formation, and shrinkage.

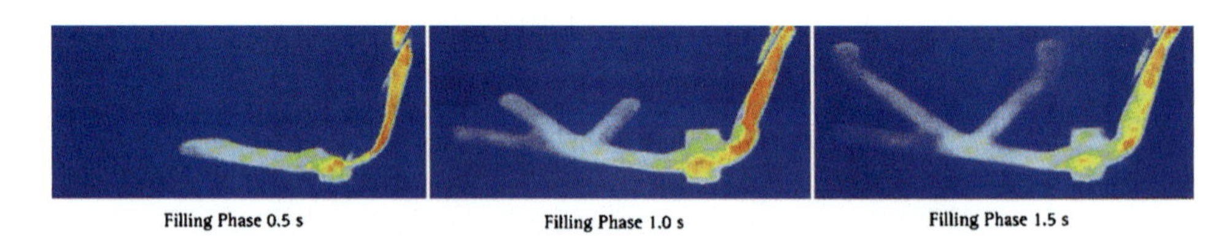

Figure 34.22. Filling of a permanent mold in which a part is replaced by a silica aerogel tile. The figure shows infrared pictures of the filling and the temperature distribution [27].

Herfurth and Engler [29] were concerned with the origin of hot tearing. A few aluminum-based alloys are sensitive to an effect called hot tearing, which means during continuous casting of thick billets cracks run through the sample just in the transition zone between fully liquid and fully solid. They built a special mold. A part of it was then being replaced by a silica aerogel tile to observe isotherms and crack development in situ with an infrared camera.

A schematic of the facility is shown in Figure 34.23. A rectangular piece of an aluminum alloy (they used AlCu5) is fixed in a special device that allows to extract heat at the bottom, the top, and the sides and also to apply a tensile load as indicated.

Thus, the molten zone acts as a stress riser and a crack might go through the sample and ideally along its centerline. Hot tearing is usually attributed to refilling problems in the semisolid region (called a mush). Observation through silica aerogels as indicated in the figure allowed to analyze simultaneously the isotherms and the real crack path and correlate both.

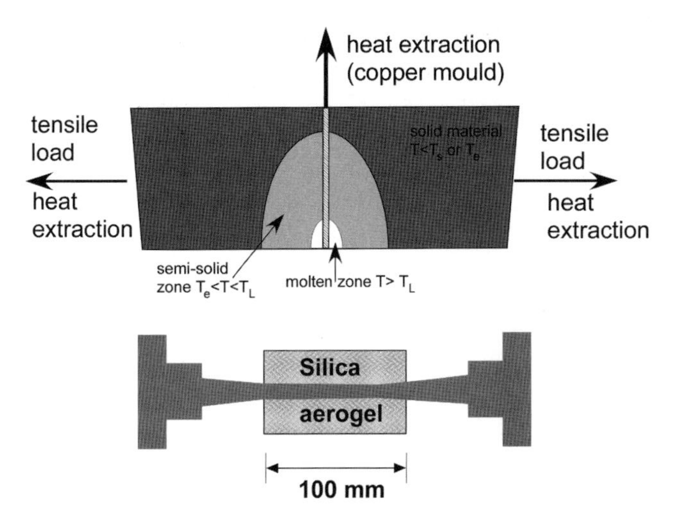

Figure 34.23. Schematic setup of the hot tearing tests performed by Herfurth and Engler [29]. A thick plate of an Al alloy is subjected to several heat sinks inducing a temperature distribution as indicated. After a thermal equilibrium has been established with a small liquid zone, acting as a stress concentrator, a tensile load is applied with a given crosshead speed and fracture goes from the bottom to the top. The sample is covered in the interesting area with a silica aerogel and the test is observed by an infrared camera. Note: T_L means the liquidus temperature, T_S the solidus temperature, T_e the eutectic point.

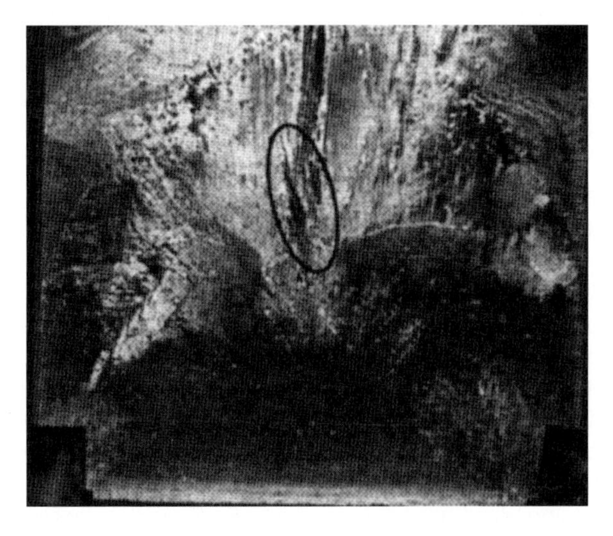

Figure 34.24. Crack in the mushy or semisolid zone of a stressed AlCu alloy. The elliptically marked area denotes a region in which filling by the melt is impeded by a solid bridge. Observation was made through a silica aerogel.

A result is shown in Figure 34.24. An AlCu5 sample is stressed with the tensile load at a crosshead speed of 2.5 mm/s. After 1 s a crack has developed and a solid bridge in the mush impedes a refilling by melt. Note the liquid zone is at the bottom and the crack runs from the bottom to the top.

34.10.2. Aerogels for Directional Solidification

Directional solidification of metals and alloys is often performed with *Bridgman type furnaces* shown schematically in Figure 34.23. Such a furnace consists of a hot and a cold zone separated by an adiabatic zone. The temperature profile in a sample has a typical S-shaped profile (see Figure 34.25, right). The sample, being in a cartridge, is withdrawn with a fixed speed from the hot into the cold zone.

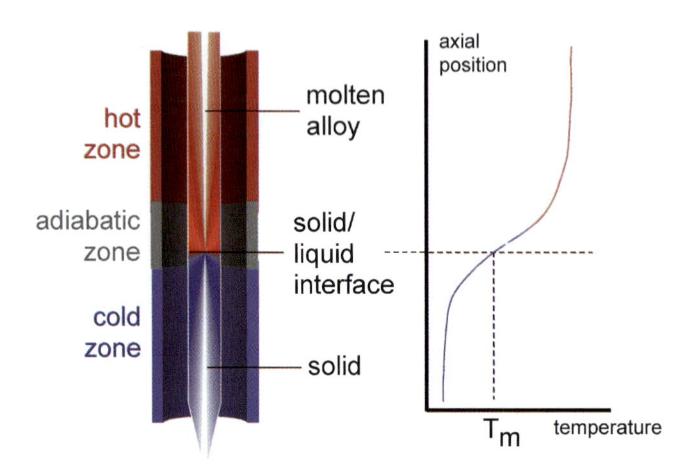

Figure 34.25. Sketch of a Bridgman furnace which consists of a hot and a cold zone interrupted by an adiabatic zone. The temperature profile in a sample is shown schematically in the right profile. The sample, being in a cartridge, is withdrawn as a fixed speed from the hot into the cold zone. An alloy melt should nominally solidify with exactly the pulling speed. The temperature gradient at the melting front is determined by the slope of the temperature profile at the melting temperature T_m.

An alloy melt should nominally solidify with exactly the pulling speed. The temperature gradient at the melting front is determined in principle by the temperature profile and the slope of it at the melting temperature T_m. The information about the temperature field during solidification is usually determined by thermocouples or thermistors, which are placed either in the sample or in the cartridge.

The information that can be obtained by this method is only pointwise and because of the varying conditions during the solidification process often not satisfying. The velocity of the solid–liquid interface may then be determined as a function of position. The temperature field in the sample is calculated from the thermocouple readings.

A new solidification facility was developed by Ratke and co-workers in the last decade utilizing silica aerogels and their properties [21, 31], called ARTEMIS – *AeR*ogel furnace *TE*chnology for *MI*crogravity *S*olidification. Due to their structure the thermal conductivity is extremely low (0.02 W/Km) and can with evacuation of the aerogel be lowered to 0.005 W/Km. Due to their transparency optical temperature measurements can be performed during the experiments, which allow to read whole temperature–position profiles with a good accuracy and time resolution.

Figure 34.26 shows a schematic of the ARTEMIS facility used for directional solidification. The central part of the facility consists of two heating elements, a piece of aerogel as

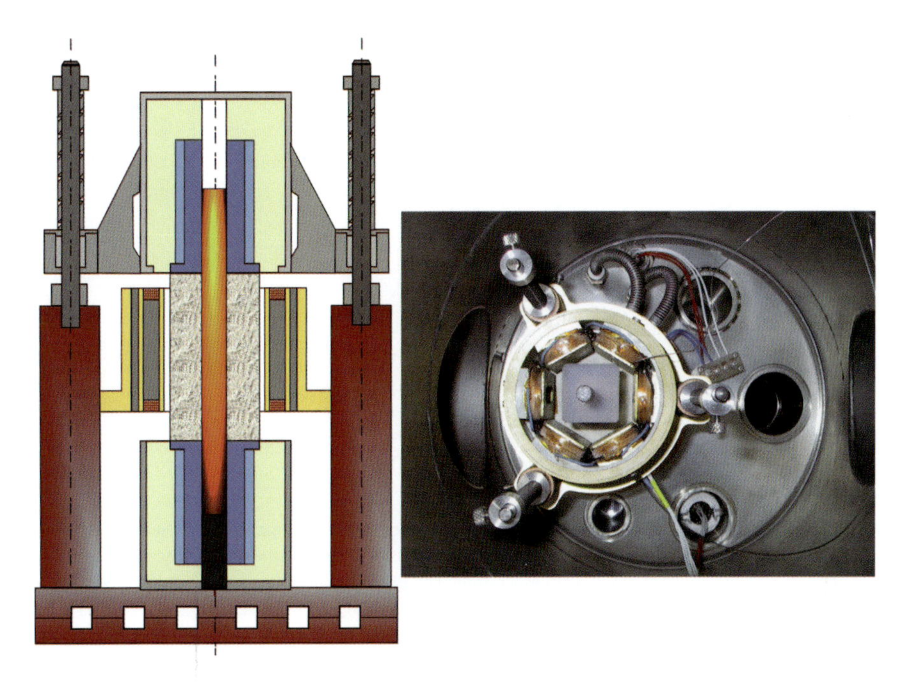

Figure 34.26. The **left picture** shows a schematic of the ARTEMIS furnace facility. A cylindrical sample sticks at the bottom and the top in two independent heaters and in between it is fully surrounded by a silica aerogel having a cylindrical hole. The **right photo** gives a picture of such a furnace assembly mounted in a vacuum chamber. The sample and the surrounding aerogel can be seen. There are six coils, which form an asynchronous motor and thus induce a flow field in the molten alloy.

a thermally insulating region and a cooling element below the bottom heater. Both heaters exchange heat primarily via the sample. To fully control the heat flow from top to bottom, a molybdenum rod is inserted in the bottom furnace and connected with a water cooled copper plate.

Due to the excellent insulation properties of silica aerogels, the isotherms are almost flat since the axial heat flux is typically more than 100 times greater as the radial one (Chap. 23). It can be shown that controlling the cooling rates of both furnaces independently leads to a solidification front moving with constant temperature gradient at the solidification front and constant solidification velocity [21]. Since the aerogel is transparent, the brightness of the alloy along the surface of the sample can be recorded with a line-CCD-camera during the experiment. The position of the solid/liquid interface and the temperature gradient ahead of it can be extracted from intensity versus time profiles directly [21]. Figure 34.27 shows an example of an intensity recording measurements and Figure 34.28 shows the position of the hump in the curves of Figure 34.27 as a function of time, clearly revealing that the solidification front passed through the sample with constant speed.

Aerogels thus allow a perfect control of the solidification process and an analysis of the intensity curves allows to extract further information, like the solid fraction or the eutectic fraction [31]. The photo of such a facility shown also in Figure 34.26, shows an additional feature easily build in: close to the sample three pairs of Helmholtz coils are inserted, allowing to induce rotating magnetic fields and thus to stimulate fluid flow in the samples for mixing [32]. With conventional Bridgman furnaces this is only possible with

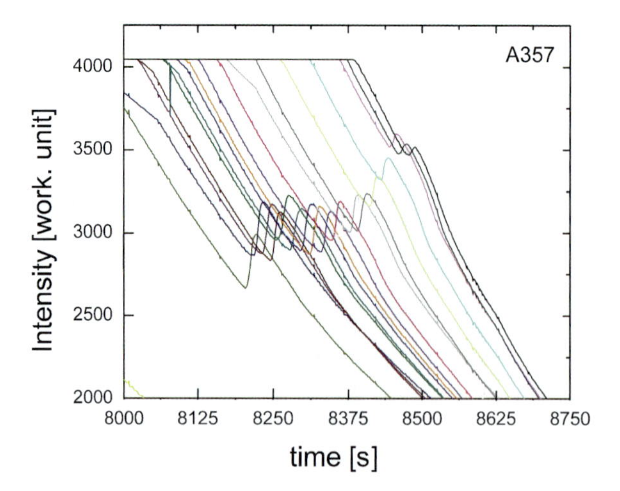

Figure 34.27. The light emitted by the sample surface is recorded by an NIR line camera. Each pixel of the camera corresponds to a point of the sample surface. During solidification the temperature of a given point decreases from high values to lower ones on reaching room temperature. The graphs of several pixels clearly show a pronounced hump which shifts with time through the sample from the bottom to the top. These humps correspond to a change from liquid to solid, since both have different emissivities, especially obvious in silicon containing alloys, since Si is semiconducting in the solid and metallic in the liquid state.

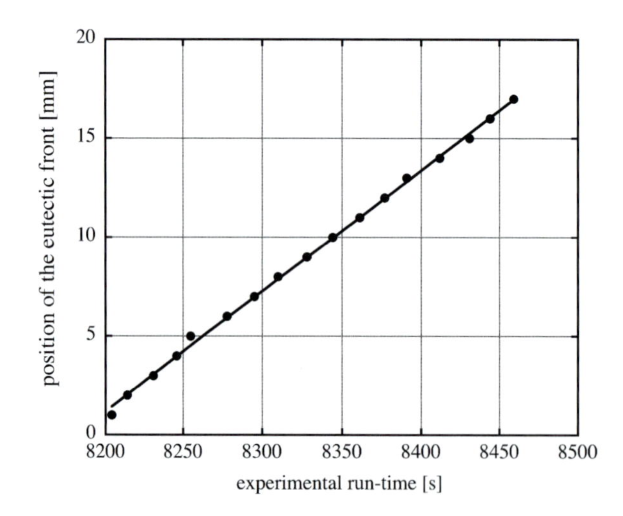

Figure 34.28. Position of the eutectic front as a function of experiment time. The dots are taken from intensity curves as shown in Figure 34.27, where just the hump in the curve starts. A linear fit yields the solidification velocity. This is not only constant over the processing length, but it was also shown that the temperature gradient at the solid/liquid interface is constant [21, 31].

greater technical effort [35]. The new ARTEMIS technology has been successfully used to study solidification phenomena in situ [31–34].

At the moment the technology is developed further to include different sources for electromagnetic stirring like traveling magnetic fields or to built facilities for use in strong magnetic fields.

34.11. Conclusions

Polymeric aerogels based on RF were developed in the last years for application in foundry practice as a new nanostructured binder for foundry sands. Their properties allow, for instance, to replace conventional polymeric binders, since they solve certain problems of core removal, penetration, surface reactions, blow holes, and other casting defects especially in light metal casting. RF aerogels as polymeric binders are at the verge of being used by industry.

A promising development is the use of aerogel granulate materials replacing a part of the foundry sand. Here polymeric aerogels and hydrophilic and hydrophobic silica aerogels can be used. First industrial test cases with aerogelic nanoadditives showed that they are able to solve problems such as metal penetration, rough casting surfaces, mold adhesion in core shooting, and others. In this area, the essential task is to develop cheaper routines for the mass production of aerogel granulates.

Silica aerogel tiles have been used since more than a decade for fundamental studies in solidification engineering. Such applications are a niche and will probably never be a market driver, but the problematic availability of bigger silica monoliths on the market prevents a broader utilization even in the area of scientific investigations such as those presented here on hot tearing sensitivity of aluminum alloys or hot spots in permanent mold casting. Most often scientific investigations on solidification processing utilize the nearly perfect insulation of aerogels and their transparency. It would be highly desirable if monolithic aerogels in sufficient quantity would be available with a much higher sintering temperature than silica aerogels.

References

1. Tilch W, Flemming E (1993) Formstoffe und FormverfahrenDeutscher Verlag für Grundstoffindustrie, Leipzig/Stuttgart
2. Campbell J (2000) Castings. Butterworth-Heinemann, Oxford,Paperback Edition
3. Beeley PR (2001) Foundry Technology. Butterworth-Heinemann,Oxford, 2nd Edition
4. Brück S, Ratke L (2002) RF – Aerogels: A new binding material for foundry application. J Sol-Gel Sci Tech 26:663–666
5. Bock V, Emmerling A, Fricke J (1998) Influence of monomer and catalyst concentration on RF and Carbon aerogel structure. J Non-Cryst Solids 225:69–73
6. Brück S, Ratke L (2004) AeroSande – ein neuer Formstoff für Gießereianwendungen. Giessereiforschung 56:55–65
7. Brück S, Ratke L (2006) Mechanical properties of aerogel composites for casting purposes. J Mat Sci 41:1019–1024
8. Reuß M, Ratke L (2009) Characterization of Carbon-AeroSands. Int J Foundry Res 61:2–11
9. Milow B, Ratke L (2009) German patent application 102009024013
10. Milow B, Ratke L, Nolte E (2006) German patent application 1020060560936
11. Reuß M, Ratke L (2010) Drying of aerogel bonded sands. J Mat Sci 45:3974–3980
12. Milow B, Ratke L (2010) BTX free decomposition of polymeric aerogel binders. Int J Foundry Res, submitted
13. Reuß M, Ratke L, (2010) On the fraction of sand grains bounded in moulding materials – a new measurement technique. Int J Foundry Res, 62:24–29
14. Voss D, Ratke L (2005) Ein neuer, entgasungsarmer und verbrennbarer anorganischer Formstoff für die Gießereiindustrie – Kohlenstoff-Aerogel-Sandverbunde. Giessereiforschung 57: 18–25
15. Milow B, Ratke L (2008) German patent application 102008056856
16. Milow B, Ratke L (2008) German patent application 102008056842
17. Brück S, Ratke L (2005) German patent application 102006003198

18. Alkemper J, Ratke L, Diefenbach S (1993) Chill Casting into Aerogels. Scripta Metall et Mater 29:1495–1500
19. Emmerling A, Lenhard W, Fricke J, Van de Vorst GAL Densification behaviour of silica aerogels upon isothermal sintering (1997) J Sol-Gel Sci Tech 8:837–842
20. Brinker J, Scherer GW, Sol-Gel Science. (1990) Academic Press, San Diego
21. Alkemper J, Sous S, Stöcker C, Ratke L (1998) Directional solidification in an aerogel furnace with high resolution optical temperature measurement. J Crystal Growth 191:252–260
22. Tscheuschner D, Ratke L (1999) Feinguss in Aerogelen. In: Ludwig A (ed) DGM Symposium Erstarrung metallischer Schmelzen in Forschung und Gießereipraxis. Wiley-VCH, Weinheim, p. 257–264
23. Tscheuschner D, Ratke L (2000) Investment Casting in Silica Aerogels. Materials Science Forum 329-330:479–486
24. Tscheuschner D, Ratke L (2001) Wedge casting of AlSiMg alloys in aerogels. In: Stefanescu DM, Ruxanda R, Tierean M, Serban C (eds), Proc Int Conf The Science of Casting and Solidification, Editura Lux Libris, Bukarest pp. 245–251
25. Viets R, Breuer M, Haferkamp H, Krüssel T, Niemeyer M (1999) Solidification process and infrared image characteristics of permanent mold castings. Thermosense XXI, Int. Conf. on Thermal Sensing and Imaging Diagnostic Applications, Orlando
26. Schaper M, Haferkamp H, Niemeyer M, Pelz C, Viets R (1999) Thermal investigation of compound cast steel tools. Thermosense XXI, Int. Conf. on Thermal Sensing and Imaging Diagnostic Applications, Orlando
27. Haferkamp H, Bach FW, Niemeyer M, Viets R (1999) Merkmale von Wärmebildern zur Prozeßüberwachung beim Taktgießen. Aluminium 75:945–953
28. Haferkamp H, Bach FW, Niemeyer M, Viets R, Weber J, Breuer M, Krüssel T (1999) Tracing Thermal Process of Permanent Mould Casting. ISIE '99, Proceedings of the IEEE, Bled-Slovenia, Vol 3, pp. 1442–1447
29. Herfurth KT (1999) Beitrag zur Entwicklung einer Versuchs-technik für die Beobachtung von Warmrissen und anderen mit der Erstarrung verbundenen Erscheinungen. Shaker Verlag, Aachen
30. Ratke L, Korekt G (2000) Solidification of Al-Pb base alloys in low gravity. Z Metallkde 91:919–927
31. Steinbach S, Ratke L (2004) In-Situ optical determination of fraction solid. Scripta Materialia 50:1135–1138
32. Steinbach S, Ratke L (2005) The effect of rotating magnetic fields on the microstructure of directionally solidified Al-Si-Mg Alloys. Mat Sci Eng A413-414:200–204
33. Steinbach S, Ratke L (2007) The Influence of Fluid Flow on the Microstructure of Directionally Solidified AlSi-Base Alloys. Metall. Mater Trans 38A:1388–1394
34. Steinbach S, Ratke L (2007) Experimental study on interaction of fluid flow and solidification in Al–Si–Cu alloys. Int J Cast Metals Research 20:140–144
35. Zimmermann G, Weiss A, Mbaya Z (2005) Effect of forced melt flow on microstructure evolution in AlSi7Mg0.6 alloy during directional solidification. Mat Sci Eng A 413-414:236–242

Part XIII

Applications: Art

35

AER( )SCULPTURE: A Free-Dimensional Space Art

Ioannis Michaloudis

Abstract Functional materials usually find uses in technology. This contribution is, therefore, quite unusual for a scientific and technological handbook as it presents a series of silica aerogel-based *aer( )sculptures* realized by the artist since 2002.

35.1. An Artist View of Aerogels

Between mountains and clouds meeting each other, nearby a lake changing colors every day, this is the place visual artist Ioannis MICHALOU(di)S has chosen to have his first atelier/lab. This cloud-hunter follows Centaurs' and Nymphs' footprints and lies in wait of air streams, grapping pieces of sky, shaping them, molding them, creating "images of forms," and baptizing them as *aerosculptures*.[1]

99.9% air and 0.1% silica is the composition of every *aer( )sculpture*. In space technology, this same composition is named silica aerogel. This material is the lightest solid on planet Earth and is also used by NASA (Chap. 32) as an excellent heat insulator for spacecrafts and stardust collection.

The author, I. MICHALOU(di)S, is the first visual artist worldwide bringing this ethereal material in Art, choosing to hunt with it skies and dreams.

Despite the fact that the space technology required for the creation of the *aer( ) sculptures* costs inevitably a lot in time and funds, the results are always amazing: weightless sculptures having the ability to hover in space, opening new paths toward the Space Art era, where light and immaterial dialogues replace the heavy and voluminous.

Each *aer( )sculpture* is – at the same time – a "ready made" but also a masterpiece, because the inner world of every *aer( )sculpture* is unique. Different *biomimicry* equals to a different cosmos seen in the sculpture: airy clouds, fragments of gold, and orbits of planets creating "spaces in between."

[1] A legend says that these "non-handmade" *aer( )sculptures* made out of pieces of sky had caused the ozone's hole phenomenon!...

I. Michaloudis • Visual Artist, 13, Gortynias Street, 136 75 Athens, Greece
e-mail: info@michalous.com; michalou@alum.mit.edu

M.A. Aegerter et al. (eds.), *Aerogels Handbook*, Advances in Sol-Gel Derived Materials and Technologies, DOI 10.1007/978-1-4419-7589-8_35,

Light and shadow is one more dialogue opened whenever a light beam transpierces each blue *aer( )sculpture* projecting their transparent gold hue shadow in orbit.

If humans are (organic) carbon-based representations, then every *aer( )sculpture* is their immaterial shadow on (inorganic) silica. We know that silica natural glass is a basic component for the fabrication of data storage devices. If we accept that one day one grain of sand (silica) will be the Bank of all human memory, then we can surely see that every *aer( ) sculpture* travels as a Memory Ark.[2]

Past, Present, and Future are melted together into an unknown infinity where Space and Time become Logos. . .an elastic Logos.

Into an endless beginning. . .

35.2. About the Artistic Development and Realization

The author was invited in September 2001 by the artist Otto Piene, founder of "Sky Art," and Stephen A. Benton, physicist, inventor of the "rainbow hologram" and Ex-Director of the Massachusetts Institute of Technology's Center for Advanced Visual Studies, to conduct postdoctorate artistic research with the title $(Nephele)^3$ (*Nephele* to the third power). In Greek *nephele* means "cloud," and the objective of this project was to realize what seems unrealistic and crazy: a cubic cloud, a cloud delimited in space by six square invisible planes. For this research, the author had received a Fulbright Grant and a Research Scholarship from the Hellenic Government. During that collaboration with scientists and engineers at MIT, the author had the chance to see a small piece of silica aerogel. ". . . I was so surprised by the appearance of something that one is not quite sure is there! To believe your eyes, you need your hand, not only to touch, but also to handle, to move around, to press the material... thus you discover that this "frozen smoke" is so lightweight and fragile. Immediately, I thought of creating immaterial, ethereal sculptures with it. I knew nothing about aerogels and the difficulties of its expensive fabrication, but I was certain that I had something important to do with it. I was looking for a cubic *Nephele* and I found the sky!. . ." [1].

In September 2005, the author installed a 20 L high-temperature supercritical drying reactor (Figure 35.1) in Greece. The setup was made by Dr. Michael Droege (Ocellus Technologies, CA, USA) who trained the author. Into this vessel (24 cm diameter, 42 cm height) are created all the *aer( )sculptures* since then. The following photographs are just a few examples of small to rather large sculptures starting from the first one made in 2002 (Figure 35.2) to the most recent ones (Figures 35.3–35.28).

[2] Another legend wants the *aer( )sculptures* to be transferred by spacecrafts to other planets in order to be used for NASA's Sky Cloning Project (cf. http://www.skyforsale.com).

Figure 35.1. The artist is opening the 20 L high-temperature supercritical drying reactor after a 48 h run. The vessel was installed in Greece, September 2005, and all the *aer( )sculptures* are created since then in it (photograph and copyright: Massimo Pizzocaro 2/2008).

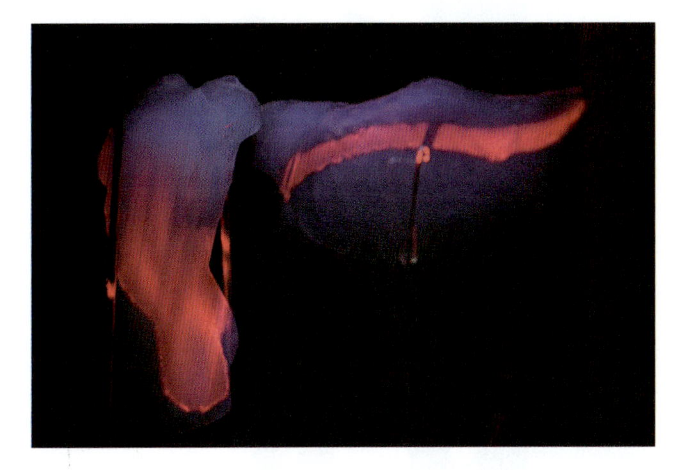

Figure 35.2. "*-Icare, I care...*," silica aerogel, laser liner beam 25 × 25 × 12 cm Livermore, CA, USA (photograph and copyright: MICHALOUS, 10/2002). This is the first ever silica aerogel sculpture, realized by Dr. Michael Droege (Ocellus Inc.).

Figure 35.3. *Leto*, silica aerogel, 25 × 15 × 4 cm. In every *aer( )sculpture*, the orange golden hue is due to Mie scattering phenomenon and the blue one is due to the Raylight scattering phenomenon (photograph and copyright: MICHALOUS, 10/2006).

Figure 35.4. *Dancing couple*, marble, silica aerogel, LED light. 25 × 20 × 4 cm with variable size of the projected shadow. This is the projected shadow of the couple: one made of the *aer( )sculpture*, inspired by the Early Cycladic Art, and the other of the original Cycladic figurine (photograph and copyright: MICHALOUS, 10/2006).

Figure 35.5. *The cup-bearer*. 15 × 10 × 7cm. The silica aerogel sculpture weights 20 g, but the marble sculpture weights 350 g (photograph and copyright: MICHALOUS, 10/2006).

Figure 35.6. *Modigliani* (*detail*). An hydrophobic treatment permits to the *aer( )sculpture* to float on the water (photograph and copyright: MICHALOUS, 10/2006).

Figure 35.7. *Modigliani* (*detail*). A glossy *black* and *white* background permits to the silica aerogel sculpture to reveal – throughout its transparency – the blue and the orange colors due to scattering phenomenon (photograph and copyright: MICHALOUS, 10/2006).

Figure 35.8. *Modigliani* (*detail*). The natural sunlight offers a dramatic sunset's color to the transparent sculpture, presented with ten more *aer( )sculptures* in October 2006 at the Museum of Cycladic Art, Athens, Greece [2] (photograph and copyright: MICHALOUS, 10/2006).

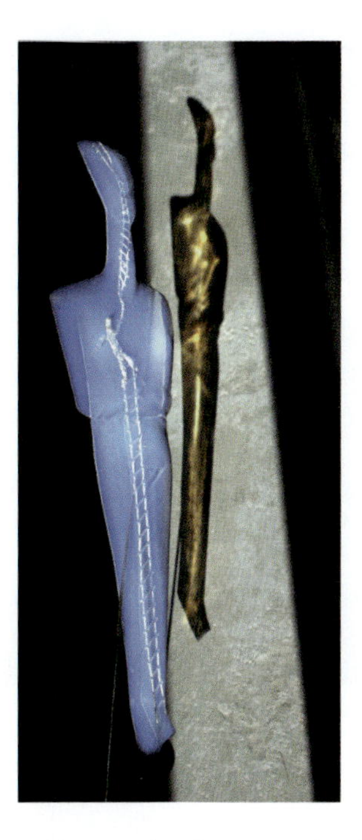

Figure 35.9. *Modigliani* 42 × 14 × 2.5 cm. The scattering phenomenon but also an "inner world" are both obvious on this *aer( )sculpture*, seen as a "sculpture into sculpture". A man is finishing his climbing on a stairway and is ready to catch a new one, a stairway to heaven. This sculpture was presented at the Museum of Cycladic Art under an electromagnetic field and thus it was hovering in the air (photograph and copyright: MICHALOUS, 10/2006).

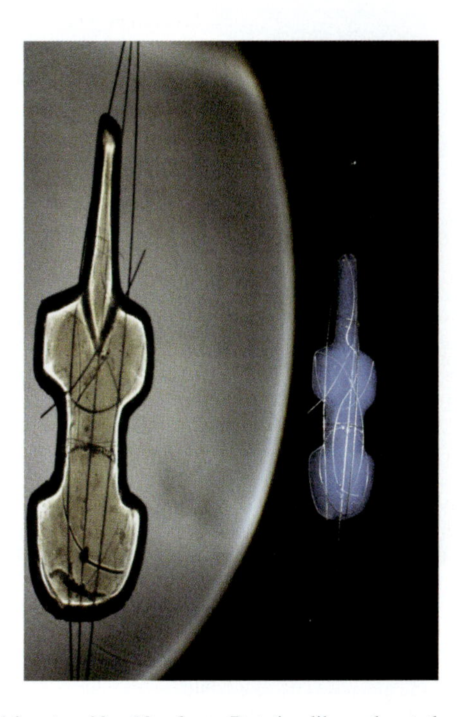

Figure 35.10. *Violin-shaped figurine*. 20 × 12 × 2 cm. Rotating like a planet, the sculpture changes its shadow every second. Permanent collection of "A.S. Onassis Public Foundation" (photograph and copyright: MICHALOUS, 10/2006).

Figure 35.11. *Ephebos and stairway*, 15 × 15 × 2.5 cm. This aerogel plate was first realized at Jet Propulsion Laboratory by Dr. Steven Jones (photograph and copyright: MICHALOUS, 12/2005).

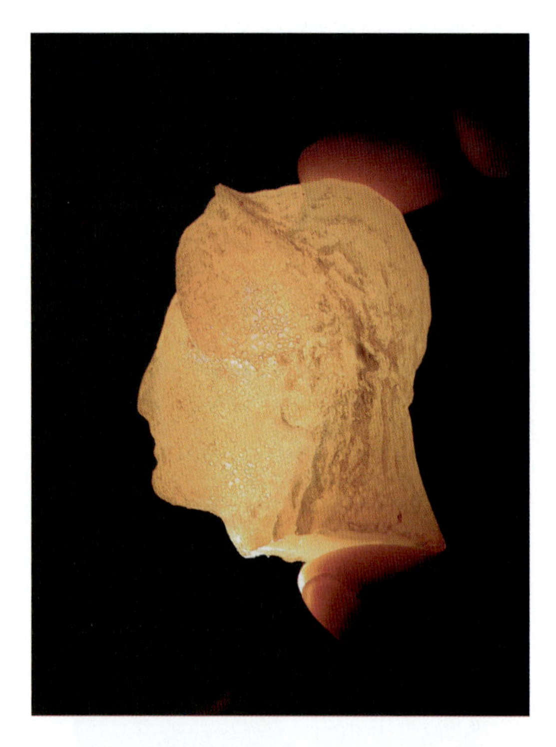

Figure 35.12. *Kore*, 5 × 3 × 3 cm, Realized by Dr. Nicholas Leventis: "It was back illuminated with focused white light from two halogen lamps using a Dolan-Jenner Industries Inc. High Intensity Illuminator Series 180. The photograph was taken with a Nikon CoolPix 5000 digital camera" (photograph and copyright: Dr. Nicholas Leventis, 2008). One of the author's projects in progress is the re-establishment of missing parts on classical sculptures, using silica aerogel. Transparent statues' members out of silica aerogel could offer to the viewer a celestial aspect on the 'wounded' classical statues. And that because these members out of silica aerogel will be almost immaterial and absent. Thus, the new conserved statues' missing parts will not impose themselves as they do now the statues' missing parts made out of plaster and other opaque materials cf. the missing parts of *Venus of Milos*, Louvre Museum, Paris, France.

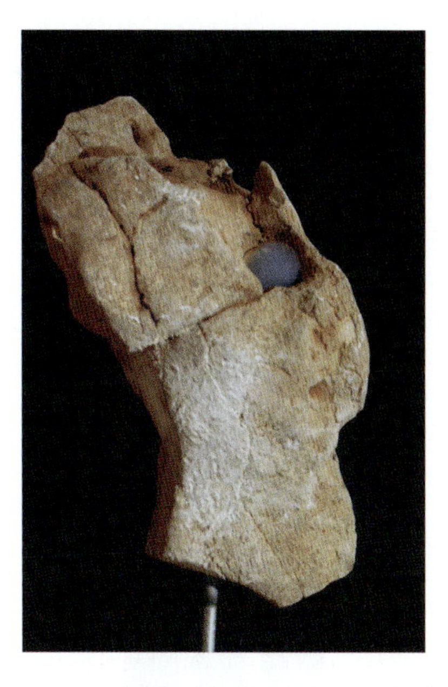

Figure 35.13. *Blue I (eye)*, silica aerogel, rock, 35 × 15 × 16 cm. An aerogel sphere of a diameter of 2 cm was just gently pressed into the natural cavity of the head-form rock, in order to present a dialogue between the heavy and the ethereal sculptural medium (photograph and copyright: MICHALOUS, 1/2008).

Figure 35.14. *Atlas holding the sky*, aerogel, brass, LED light, 18 × 12 × 12 cm (photograph and copyright: MICHALOUS, 11/2003).

Figure 35.15. *(M)other-Earth*, silica aerogel, golden leaf, stainless steel, 50 × 10 × 3 cm (photograph and copyright: MICHALOUS, 1/2008).

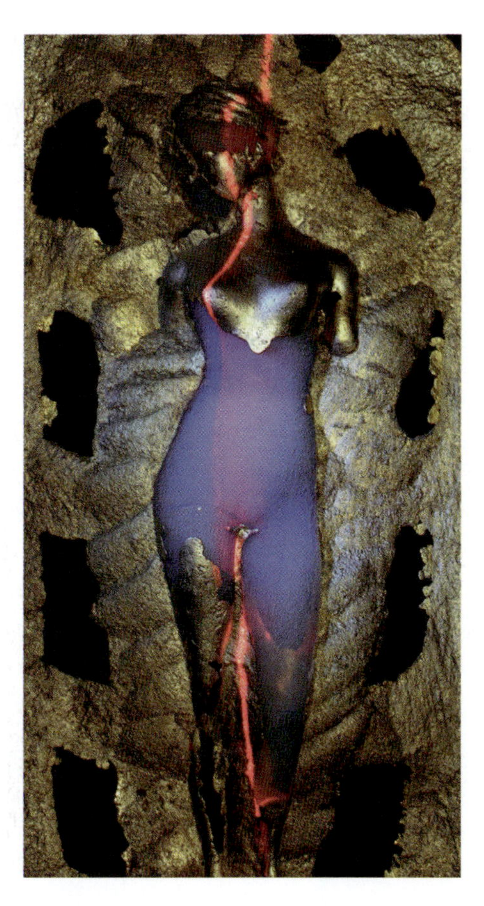

Figure 35.16. *Veria girl*, silica aerogel, brass, laser liner beam. A laser liner creates the red "scanning" light we see. Thus, the bi-dimensional line is transformed in a three-dimensional surface, thanks to the quality of silica aerogel to react as frozen smoke. The *aer( )sculpture* was placed directly into the foundry's sand where the melted brass was poured all around it (photograph and copyright: MICHALOUS, 2006).

Figure 35.17. *M(ed)use*, silica aerogel, LED light, 12 × 12 × 9 cm. Collection of Valentis' family (photograph and copyright: MICHALOUS, 1/2008).

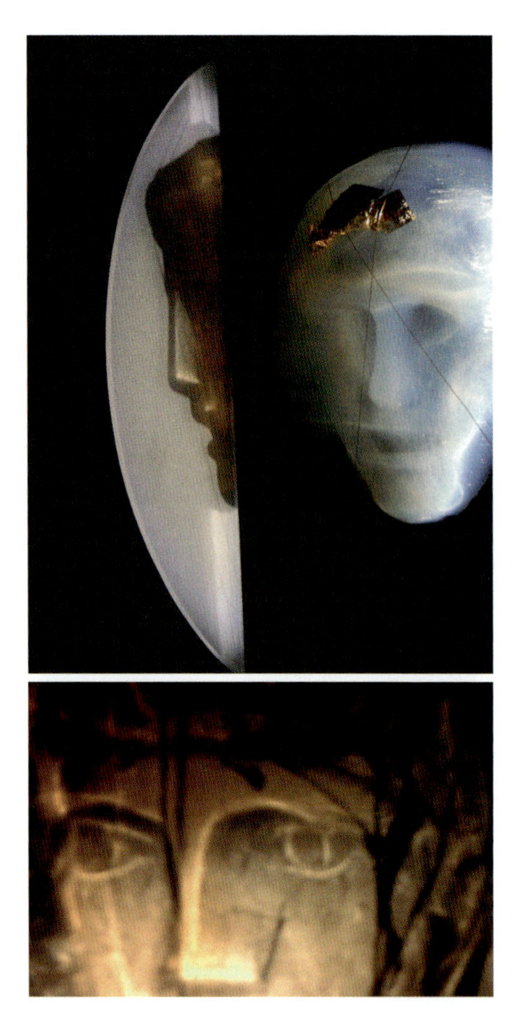

Figure 35.18. *Pho(e)bos*, silica aerogel, LED light, 15 × 15 × 10 cm (photographs and copyright: MICHALOUS, 1/2008).

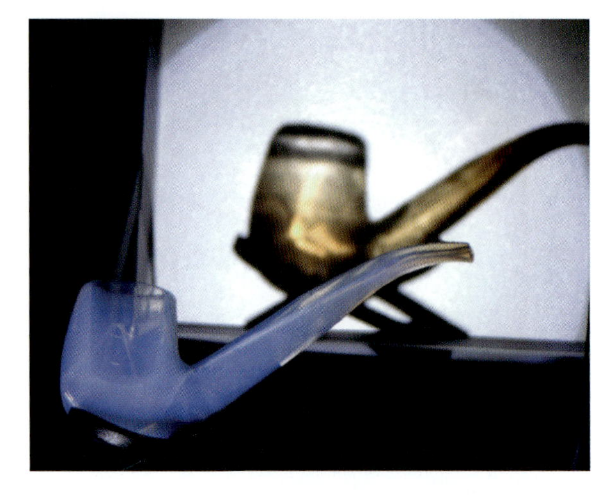

Figure 35.19. *This is not (the smoke of) a pipe*, silica aerogel, LED light, 15 × 5 × 5 cm. Inspired by a masterpiece of the surrealist painter René Magritte "*This is not a pipe*," which represents the painting of a pipe (photograph and copyright: MICHALOUS, 1/2008).

Figure 35.20. *(e.g)gnossis*. silica aerogel plates, aluminum, glass, letters, iron. 220 × 40 × 40 cm. This outdoor installation was made in 2004 with the kind collaboration of Mr. Jorgen Schultz (Technical University of Denmark) and Mr. Leif Cullberg (Airglass, Sweden). There are six plates 25 × 25 × 1.5 cm and in between are placed two Greek letters and the personal symbol of parenthesis ( ), as a letter O. All three double aerogel plates and the letters are forming the Greek word ωον *oon* "egg" that – thanks to the transparency of silica aerogel – can be read also from the end to the beginning as νοω, *noo* "I comprehend" (photographs and copyright: MICHALOUS, 2004).

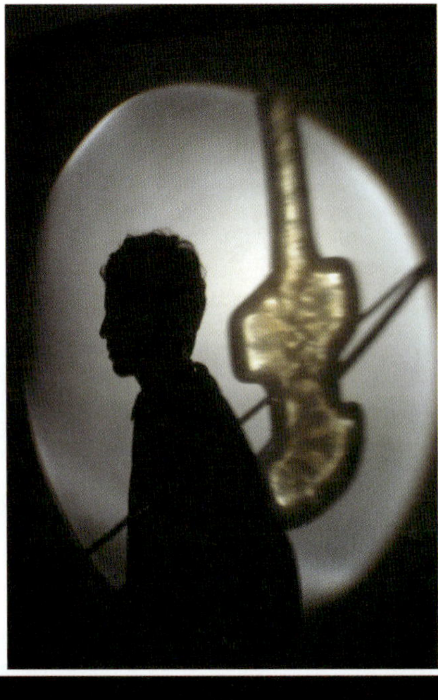

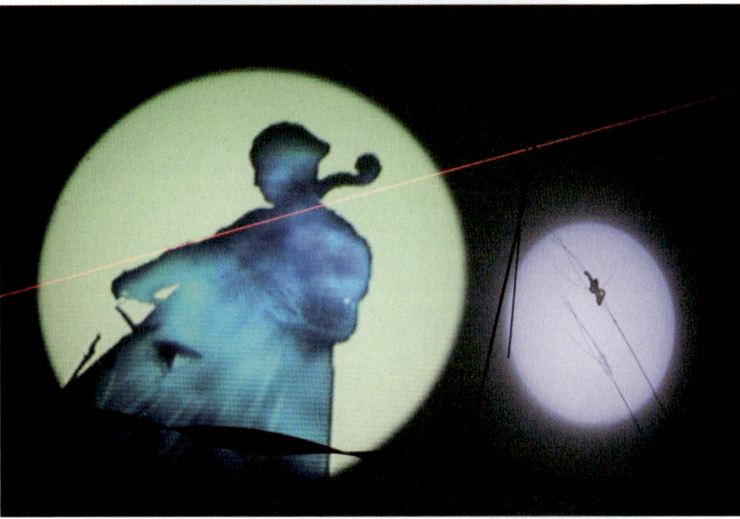

Figure 35.21. *Prelude for cello and three violins,* video installation with three violin-shaped *aer( )sculptures*, their golden hue shadows, and also a 7 min video of the shadow of a female cellist. The indoor work was presented during the XXIV Biennale of Alexandria, Egypt, and had received the 2007 Golden Price (photographs and copyright: MICHALOUS, 2007).

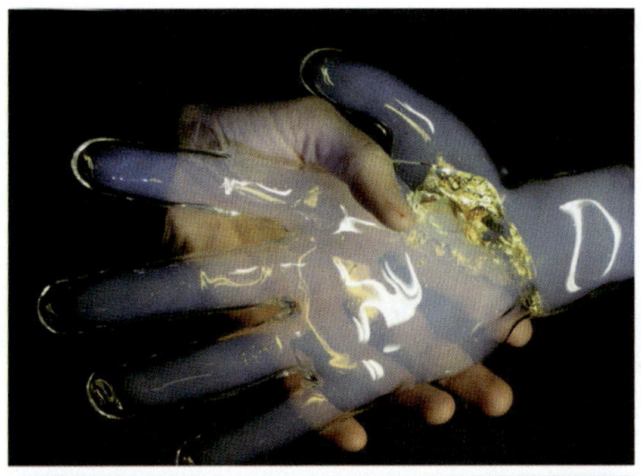

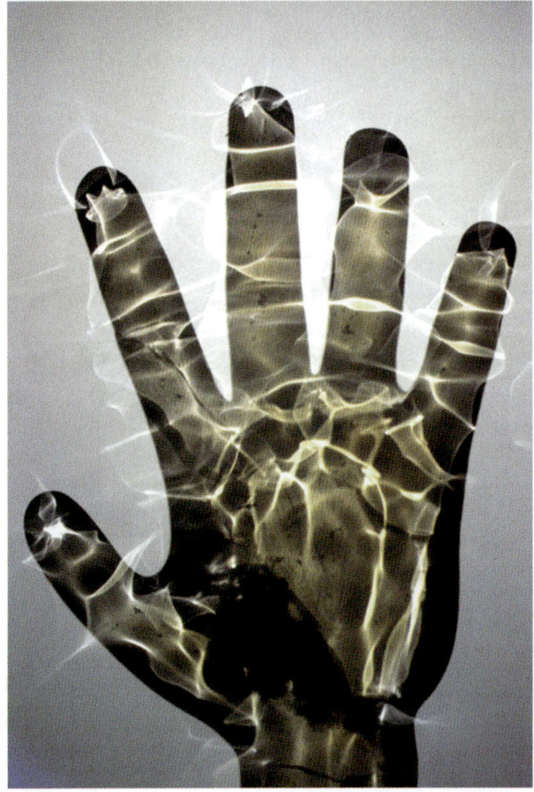

Figure 35.22. *Hand*, handmade blown borosilicate glass, silica aerogel, golden leaf, LED light, 25 × 16 × 8 cm. In 2005, invited as an Artist in Residence by Dr. Himanshu Jain at Lehigh University, PA, USA, the author had visited the "Banana Factory" and their Glass Workhop where he asked them to try to blow a real size hand on borosilicate. Then the author had poured the alcogel. Here, we see the final result (photographs and copyright: MICHALOUS, 2007).

Figure 35.23. *A piece of sky in between your fingers*, In Mars 2004, invited as an Artist in Residence by Dr. A. Venkateswara Rao at Shivaji Univerity, Maharastra, India, during some experiments we made with Dr. Sharad Bhagat, we had arrived to create the first cumulus cloud into a sample of silica aerogel. It was Sunday the 21st of March 2004 (photograph and copyright: MICHALOUS, 2004).

Figure 35.24. *Bottled Sky*, silica aerogel, glass, rubber, 10 × 3 × 3 cm. In September 2008, the author published the Web site http://www.skyforsale.com where he put his last achievement up for sale: vials of 20 and 30 ml of silica aerogel with a unique and different feather cloud into each vial (photographs and copyright: MICHALOUS, 2008).

Figure 35.25. *Bottled Nymphe*, silica aerogel, glass, rubber, $1 \times 3 \times 3$ cm (photograph and copyright: MICHALOUS, 2008).

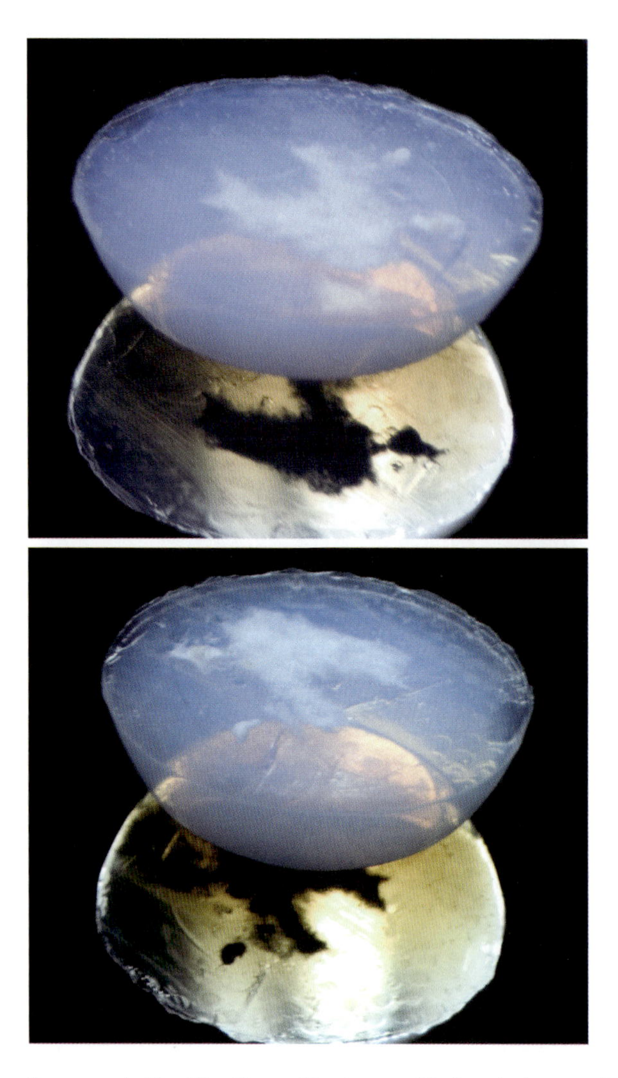

Figure 35.26. *Sky*, silica aerogel, 20 × 20 × 10 cm. Thanks to a black and glossy surface, we can see the magnificent sky-blue and sunset-orange natural colors of the artwork. One of the author's projects in progress is to realize silica aerogel sculptures where the clouds could appear, move, and disappear into the sculpture... Hopefully with the help of the scientific community specialized on aerogels (photographs and copyright: MICHALOUS, 2009).

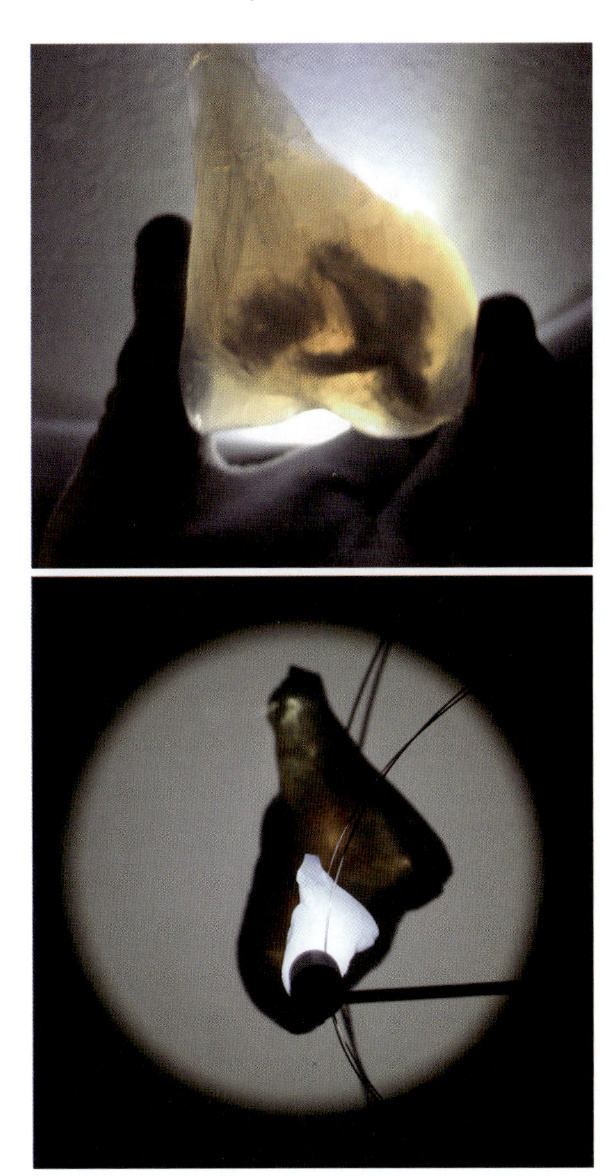

Figure 35.27. *Karditsa's Heart*, Permanent museum installation, silica aerogel, stainless steel rods, LED lights, 100 × 70 × 100 cm. The idea of a heart made of aerogel was of the artist Ms. Yi-Zhou. In August 2008, I tried to make it for her but without success (because of the big volume of the alcogel). In April 2009, Ms. Fenia Lekka, the director of the Museum of the town of Karditsa, asked me to make an artwork for the Museum. As the Greek word *Karditsa* means "little heart" (it is the author's native town in the middle of Greece) and to this museum it belongs the heart of the ever memorable Greek Prime Minister Nicholaos Plastiras, the idea of a silica aerogel heart came again. With a very slow drying the author arrives to create the silica aerogel heart with two clouds into it. On the final display, the heart is placed onto orbits of five planets. Two LEDs are lighting the total: the first one is lighting the silica aerogel heart and thus creates a golden hue shadow and a second heart. The second LED is lighting a stainless steel capsule with the donated heart (photographs and copyright: MICHALOUS, 2009).

Figure 35.28. *(L)imited sky*, four liters of silica aerogel, birdcage, LED white light, 40 × 40 × 25 cm (photograph and copyright MICHALOUS 2008).

Acknowledgments

The author wishes to thank Dr. Michael Droege, Dr. Larry Hrubesh, the Agricultural Cooperative of Kato Ktimeni, the A.S. Onassis Public Benefit Foundation, Mrs. and Mr. Papadimitriou, the Museum of Cycladic Art in Athens (Greece), Mrs. Katerina Koskina, Mrs. Ayako Ono, Ms. Yi Zhou, Dr. Nicholas Leventis, Dr. Michel Aegerter, Dr. Himanshu Jain, Dr. Steven Jones, Dr. Sharad Bhagat, Mrs. Mari and Teti Georgantopoulou, Mrs. Mariliza Papadouri, Mr. Marios Vazaios, Mr. Vassilis Philippatos, Mr. Iannis Roussos, Mr. Alex Liakopoulos, Mrs. Sylvie Blanchet, the Hellenic Union of Physicists, the ERT World Channel, Mr. Maccimo Pizzocaro, Mrs. Elissavet Laloudaki, Mrs. Iris Kritikou, the Museum of Town of Karditsa, Mrs. Fenia Lekka, and all the anonymous collectors of *Bottled Skies*.

References

1. MICHALOU(di)S I (2004) Aer( )sculpture: the enigmatic beauty of aerogel's non-entity in a pilot art & science project Journal of Non-Crystalline Solids 350:61–66
2. http://www.cycladic.gr/frontoffice/portal.asp?cpage=resource&cresrc=99&cnode=77

Part XIV

Applications: Other

36

Preparation and Application of Carbon Aerogels

Jun Shen and Dayong Y. Guan

Abstract Carbon aerogels are nanostructure porous carbon materials with high specific surface areas and relative low electric resistance. In recent years, they have attracted much attention because of their extraordinary properties and potential uses in a wide variety of technological applications. Many studies have been carried out concerning carbon aerogel synthesis and characterization. This contribution summarizes recent developments in the synthesis, properties, and applications of carbon aerogels.

36.1. Introduction

Aerogels (Chap. 1) are highly porous sponge-like materials derived via a sol–gel process, in which the dimensions of both the solid skeleton and the pores are in the nanometer range. Aerogels were first made by Samuel Kistler in the early 1930s [1]. He dried waterglass-derived silica gels, employing solvent exchange and using supercritical conditions to remove the pore liquid [2–4]. In the 1950s, Monsanto adapted Kistler's process and produced silica aerogels [5] (Chap. 2). Milled aerogels were tested as thickening agents in paints and as thermal insulators. A lower cost fumed silica process using silica tetrachloride brought Monsanto's aerogel project to a halt.

Renewed interest in aerogels occurred in the 1960s. Teichner and his coworkers [6, 7] invented a much faster method: They hydrolyzed and gelled a tetramethoxysilane/methanol solution by a molar excess of water in the presence of an acid or a base catalyst. That method was employed in the 1970s to make several cubic meters of monolithic highly transparent silica aerogel tiles for Cerenkov detectors (Chap. 28) [8–10]. The Cerenkov detector application stimulated aerogel research and development. In the 1980s, enough material became available to investigate properties such as optical transmission [11], low thermal conductivity [12] (Chap. 23), and the fascinating mechanical properties [13] (Chap. 22) of

J. Shen (✉) • Department of Physics, Tongji University, 1239 Siping Road,
20092 Shanghai, P.R. China
e-mail: shenjun67@tongji.edu.cn
D. Y. Guan • Shanghai Key Laboratory of Special Artificial Microstructure Materials and Technology,
Department of Physics, Tongji University, 1239 Siping Road, 20092 Shanghai, P.R. China
e-mail: dayong415@163.com

M.A. Aegerter et al. (eds.), *Aerogels Handbook*, Advances in Sol-Gel Derived
Materials and Technologies, DOI 10.1007/978-1-4419-7589-8_36,

SiO_2 aerogels as a function of density and other parameters. Aerogels had therefore become sufficiently important to discuss their characteristics and properties in symposia [9, 14–18].

In general, aerogels are produced when the solvent contained within the voids of a gelatinous structure is exchanged with an alternative solvent, such as liquid CO_2, that can be removed supercritically in the absence of a vapor–liquid interface and thus without any interfacial tension. Ideally, this supercritical drying process leaves the gel structure unchanged with no shrinkage of the internal voids or pores [19, 20]. In contrast, xerogels are produced when the solvent is removed by conventional methods such as evaporation under normal (e.g., ambient) conditions. This typical drying process causes the internal gel structure to collapse because of the tremendous interfacial tension forces by the presence of the vapor–liquid–solid interface, especially in the very small voids or pores [20]. It is worth noting that some authors have been referring to some materials as aerogels even though they are actually xerogels because they have been dried conventionally, not supercritically [21–24]. Using special methods, however, certain xerogels may maintain quite high porosities. Thus, the point here will not be terminology, but rather the physical and chemical properties that are expected to be in general different between carbon aerogels and xerogels.

Organic aerogels (Part IV) were first synthesized in the late 1980s by Pekala and his colleagues at LLNL (Lawrence Livermore National Laboratory) via the polycondensation of resorcinol (*R*) with formaldehyde (*F*) (Chap. 11) through a gelation mechanism similar to the sol–gel processing of silica and subsequently drying supercritically with CO_2 [25–29]. Since then, numerous articles have been published on the properties and potential uses not only of these novel "organic" aerogels [19, 30–37], but also of their pyrolyzed (carbonized) versions referred to as "carbon" aerogels [38–42], and even of the organic and carbon xerogel derivatives [38, 43] of those highly porous materials. Other organic aerogels such as phenolic–furfural (*PF*) and melamine–formaldehyde (*MF*) aerogels have also been synthesized [28].

Carbon aerogels are a special class of aerogels that are derived by the pyrolysis of organic precursors. Carbon aerogels [19, 21–94] have been prepared from the *R–F*, *M–F*, phenolic–furfural (*PF*), polyacrylonitrile, and polyurethane systems [44, 45]. Table 36.1 shows the types and characteristics of such carbon aerogels prepared using those different organic sources. Among these materials, the *R–F* system is the most frequently studied and the reaction parameters are the most well understood [46, 47].

Table 36.1. Types and characteristics of carbon aerogels prepared with different organic sources

Name	Reagent	Density (cm^3/g)	Specific surface area (m^2/g)	Pore size (nm)	References
RF	Resorcinol, formaldehyde	0.03–0.60	400–1,000	~50	[28, 29, 48]
MF	Melamine, formaldehyde	0.10–0.80	875–1,025	~50	[29]
PF	Phenol, furfural	0.10–0.25	350–600	~10	[28]
CF	Cresol, formaldehyde	0.15–0.4	~800	30	[49]
PUR	Polymeric isocyanate	0.12–0.5	300–600	~20	[39, 50]
PM	Phenol, melamine	0.53–0.71	600–800	10	[51]
PVC	Polyvinyl chloride	0.08–0.52	300–700	2–20	[52]

36.2. Synthesis of Carbon Aerogels

36.2.1. Synthesis of RF Aerogels

RF wet gels are generally prepared via a sol–gel process by the aqueous polycondensation of resorcinol with formaldehyde, using sodium carbonate as a base catalyst [28, 29]. The reaction mechanism is shown in Figure 36.1.

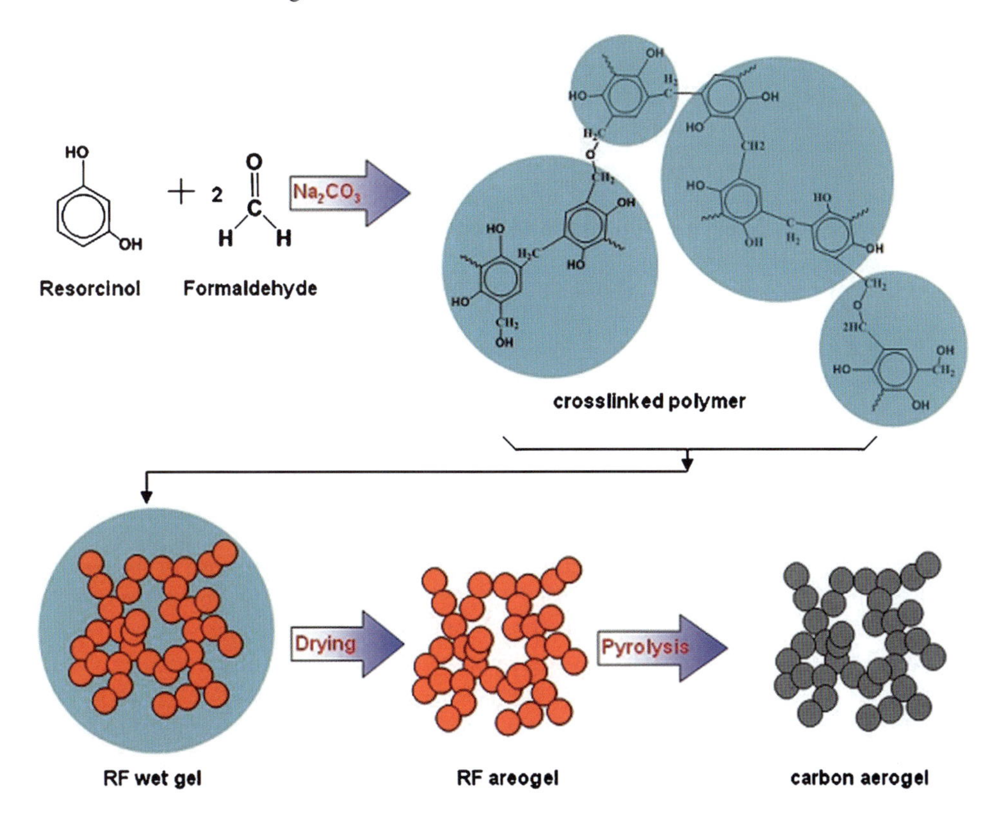

Figure 36.1. Reaction mechanism for the synthesis of resorcinol–formaldehyde (*RF*) wet gels, aerogels, and carbon aerogels.

Resorcinol (*R*) is dissolved in a 37% aqueous formaldehyde solution in a molar ratio of 1:2. An aqueous solution of Na_2CO_3 is added as catalyst [29, 53]. Water is added to the initial mixture to control the density of the final product. The solution is mixed, poured into glass vials, and sealed. Stable *RF* wet gels are formed by heating for 1 day at 25°C, 1 day at 50°C, and 3 days at 90°C [27]. Aerogels are obtained by suitable drying processes, as described in detail below.

The key parameters for controlling the structural properties of *RF* aerogels are the catalyst concentration, defined as the molar ratio of resorcinol to catalyst (*R*/*C* ratio) and the monomer concentration of reactants (*R* + *F*) in solution (*RF* wt%) [53, 54].

The *R*/*C* ratio primarily determines the particle size. Decreasing the catalyst concentration causes building up of increasingly larger particles (Figure 36.2); the density of *RF* aerogels is mainly controlled by the monomer concentration of reactants (*R* + *F*). For example, a large *RF*

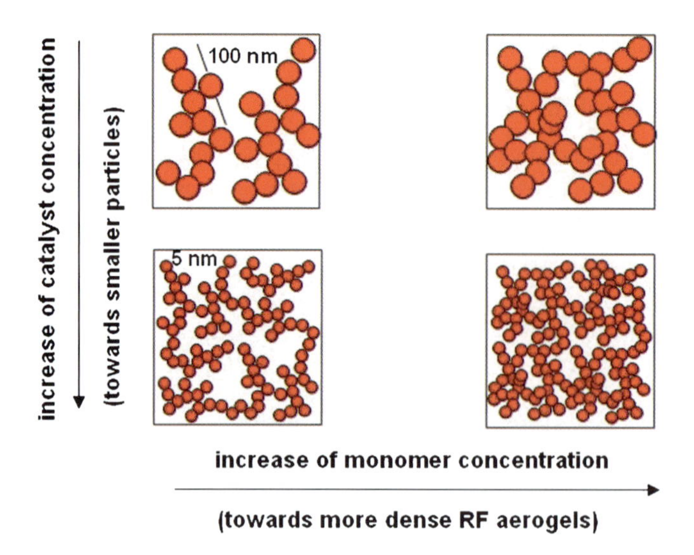

Figure 36.2. Variation of the mesostructure in *RF* and carbon aerogels.

mass concentration can result in high-density aerogels. With those two parameters, the structure and the density of *RF* aerogels can be controlled over a wide range [55, 56].

In this polymerization process, resorcinol behaves as a trifunctional monomer capable of taking up formaldehyde at its 2-, 4-, and 6-ring positions [57]. This monomer along with other di- and trihydroxy benzene derivatives is particularly reactive because of the electron donating and the ortho- and para-directing effects of the attached hydroxyl groups. The substituted resorcinol rings condense with each other to form nanometer-sized clusters in solution. The size of those clusters is regulated by the catalyst concentration in the *RF* mixture. Eventually, clusters crosslink through their surface groups (e.g., $–CH_2OH$) to form a gel. *RF* gels and aerogels are normally dark red in color, possibly as a result of oxidation products formed during polymerization [25, 26].

Resorcinol reacts with formaldehyde under alkaline conditions to form mixtures of addition and condensation products. The major reactions include (1) the formation of hydroxymethyl ($–CH_2OH$) derivatives of resorcinol, (2) the condensation of the hydroxymethyl derivatives to form methylene ($–CH_2–$) and methylene ether ($–CH_2OCH_2–$) bridged compounds, and (3) the disproportionation of methylene ether bridges to form methylene bridges plus formaldehyde as a by-product [29]. These reactions have been studied extensively by NMR [58]. The crosslinked polymer grows into discrete particles, which eventually form a three-dimensional network or a gel (Chap. 13).

36.2.2. Preparation of Carbon Aerogels

From organic precursors, carbon aerogels are obtained by pyrolysis in a tube furnace under flowing argon (0.3–0.5 l/min) at about 1,050°C. A typical corresponding pyrolysis program is shown in Figure 36.3. The shrinkage and mass loss during the pyrolysis process are shown in Table 36.2. The shrinkage of the samples due to pyrolysis is in the range of 20–30%, and the mass loss is about 50% [52].

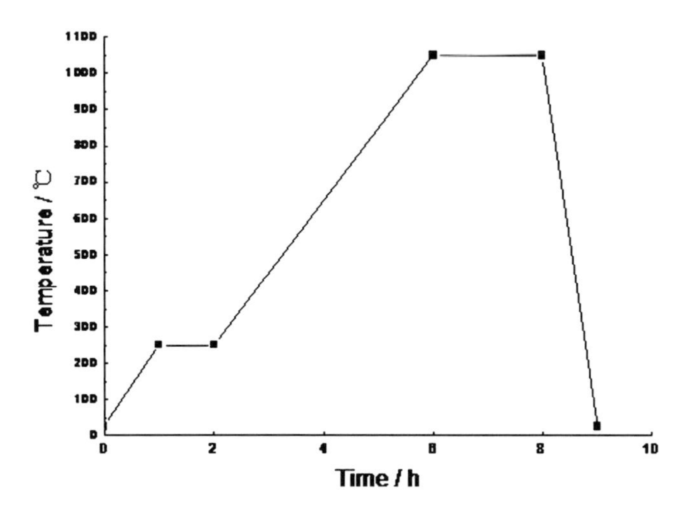

Figure 36.3. Typical pyrolysis program for manufacturing of carbon aerogels.

Table 36.2. Shrinkage and mass loss of carbon aerogels during pyrolysis under Ar

R/*C* ratio	*RF* mass concentration (RF wt%)	Linear shrinkage (%)	Mass loss (%)	References
500	50%	28.2	50.1	[59]
1,000	50%	25.5	50.0	[59]
1,500	50%	25.6	51.3	[59]
1,500	40%	23.2	52.2	[59]
1,500	30%	21.1	50.7	[59]

36.3. Characterization of Carbon Aerogels

The structural properties of carbon aerogels have been investigated by scanning electron microscopy (*SEM*), nitrogen adsorption, and X-ray diffraction methods. Specific surface areas are determined by the standard *BET* (Brunauer–Emmett–Teller) analysis. X-ray diffraction (*XRD*) of the carbon aerogel is performed in a diffraction with CuKα radiation [54].

36.3.1. Scanning Electron Microscopy

SEM pictures of carbon aerogels with different *R*/*C* ratios and *RF* mass concentration are shown in Figure 36.4. It is clear that the particle size increases with increasing *R*/*C* ratio, while with the same *R*/*C* ratio, the particle size of carbon aerogels decreases with increasing RF mass concentration. The carbon aerogels with *R*/*C* ratio of 1,000 and *RF* mass concentration 40% have a compact network with a particle size of about several tens of nanometers while that of the carbon aerogels with R/C ratio of 1,000 and RF mass concentration 30%

Figure 36.4. *SEM* of carbon aerogels with different *R/C* ratios and *RF* mass concentrations. From *top* to *bottom*: **A.** *R/C* = 1,000, *RF* = 30% (w/w), **B.** *R/C* = 1,000, *RF* = 40% (w/w), **C.** *R/C* = 1,500, *RF* = 30% (w/w), **D.** *R/C* = 1,500, *RF* = 40% (w/w).

(w/w) is as large as hundreds of nanometers. The particle size and pore diameter of the carbon aerogels with *R/C* ratio of 1,500 can reach several micrometers or even more [59].

36.3.2. Nitrogen Sorption Measurements

Adsorption and desorption isotherms for carbon aerogels with *R/C* ratio of 1,500 and different *RF* mass concentrations are shown in Figure 36.5. The isotherms are type IV, indicating multilayer adsorption on the surfaces of the carbon aerogels. At the front part of the isotherms, the adsorbed volume increases slowly, as described by the *BET* equation. All isotherms exhibit hysteresis loops, which are associated with the filling and emptying of the mesopores by capillary condensation. As shown in Figure 36.5, the size and shape of the hysteresis loops vary with the *RF* mass concentration, which is indicative of the different mesopore sizes.

The *BET* surface area and pore structures of the RF and carbon aerogels with different R/C ratios and RF mass concentrations are shown in Table 36.3. The surface area of carbon aerogels increases dramatically after pyrolysis, which is due to the creation of micropores as a result of evaporative loss of organic moieties. It is also shown that the micropore area of carbon aerogels decreases with the increase in density [60].

Figure 36.6A, B shows the pore size distributions calculated using the *BJH* method from the desorption branches for carbon aerogels. The pore size distribution of carbon aerogels with *RF* mass concentration 50% and different *R/C* ratios are shown in

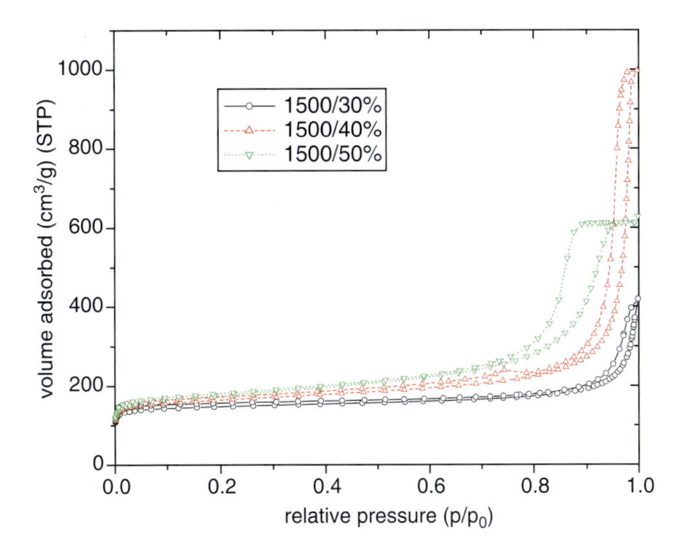

Figure 36.5. Adsorption and desorption isotherms for different carbon aerogels. Inset gives the *R*/*C* ratios and the *RF* mass concentrations (*R* + *F*, % w/w).

Table 36.3. The surface area and pore diameter of various carbon aerogels

Sample series	*R*/*C* ratio	*RF* mass concentration (%)	Density (kg/m^3)	*BET* surface area (m^2/g)	Micropore area (m^2/g)	Average pore diameter (nm)
RF aerogel	500	30	315	188.6	72.6	14.2
RF aerogel	1,500	40	515	159.2	68.4	11.1
Carbon aerogel	500	50	873	585.9	251.8	6.6
Carbon aerogel	1,000	50	812	651.0	381.1	8.4
Carbon aerogel	1,500	50	722	618.7	379.5	11.0
Carbon aerogel	1,500	40	597	581.1	439.3	26.6
Carbon aerogel	1,500	30	324	521.9	463.8	21.5

Figure 36.6A; it indicates that the average pore diameter is 6.6, 8.4, and 11.0 nm with the peak pore size of 8, 11, and 15 nm when the R/C ratio is 500, 1,000, and 1,500, respectively. The pore size increases with the increase of the R/C ratio, which is the same with the *SEM* pictures. Figure 36.6B shows the pore size distribution of samples with the *R*/*C* ratio of 1,500 and different *R*/*F* mass concentrations; it indicates that the peak pore diameter decreases from 60 to 40 nm and 15 nm with the average pore diameter of 21.5, 26.6, and 11.0 nm, respectively, when the *R*/*F* mass concentration increases from 30 to 40 and 50%.

36.3.3. X-Ray Diffraction

XRD spectra of carbon aerogels obtained after processing at different temperatures are shown in the Figure 36.7. Strong C(0 0 2) and weak C(1 0 1) diffraction peaks are located at 2θ of 23.5° and 43.8°, respectively. The width of the peaks implies that carbon aerogels belong to disordered carbon materials.

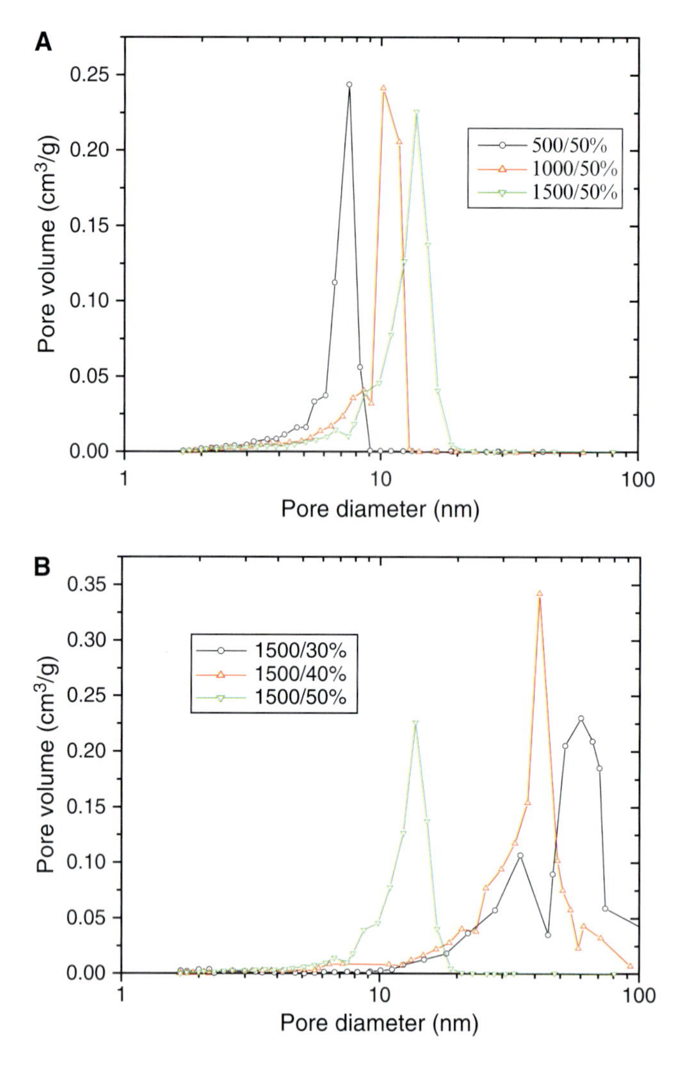

Figure 36.6. The pore size distributions calculated using the BJH method from the desorption branches for carbon aerogels. **Top**: Pore size distribution of carbon aerogels with *RF* mass concentration 50% (w/w). **Bottom**: Pore size distribution of carbon aerogels with *R*/*C* ratio of 1,500.

36.4. Effect of Process Control on the Carbon Aerogel Structure

The two process parameters with direct impact on the carbon aerogels microstructure are drying and carbonization.

36.4.1. The Drying Process

(a) Drying conditions

Once the final crosslinked gel is formed, it becomes necessary to remove the gelation solvent (e.g., water). The different methods used to remove the gelation solvent have dramatic effects on the properties of the RF organic aerogels, as outlined in Table 36.4.

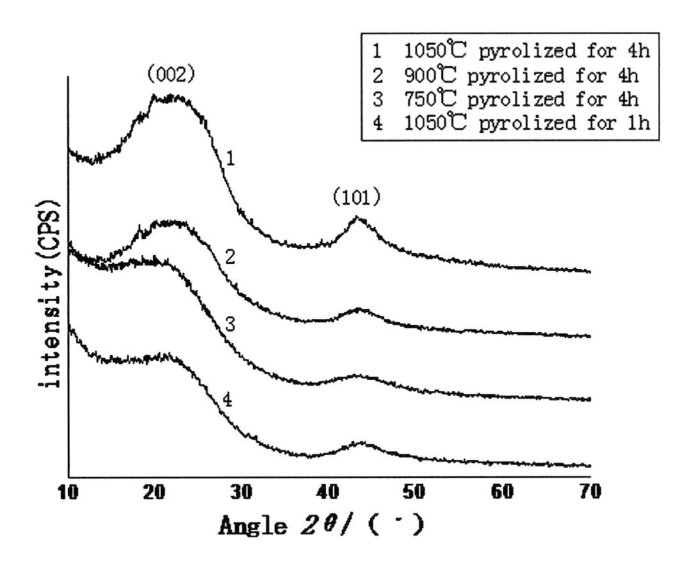

Figure 36.7. *XRD* of the carbon aerogels.

Table 36.4. Solvent-exchange and drying: effects of various factors on the results properties [61]

Factor	Effect	References
Solvent-exchange	Necessary for supercritical drying with CO_2 or freeze-drying	[29] [60–62]
	Facilitates replacement with drying media	
	Reduction of surface tensions upon subcritical evaporation	
Subcritical drying	Production of dried dense polymers called "xerogels"	[23] [63–65]
	Causes significant shrinkage of especially wide pores	
	Effects can be insignificant if gels are synthesized with high mechanical strength	
	Increases lithium-ion charge and discharge capacities	
Supercritical drying with CO_2	Production of dried light polymers called "aerogels"	[66–68]
	Insignificant shrinkage of pore structure	
	High surface areas, pore volumes, and, sometimes, electrochemical capacitances	
	Requires high pressures, long times for exchanging solvent with CO_2	
Supercritical drying with acetone	Such as supercritical drying with CO_2, but with lower pressures	[69–71]
	Eliminates necessity for exchanging solvent with CO_2 shortens processing time significantly	
	Requires high temperatures to shift acetone to supercritical state	
	May cause partial thermal decomposition of dried gels	
Freeze-drying	Production of dried light polymers called "cryogels" based on sublimation of frozen solvents	[69, 72, 73]
	Cryogels mostly mesoporous	
	Density of solvents must be invariant with freezing	

(b) Ambient pressure drying

In recent years, efforts have been made to avoid the supercritical drying step and thus to improve the chances for large-scale application [71]. By increasing the R/C ratio from traditionally $R/C = 50$–200 to $R/C = 500$–2,000 or even higher, and by reducing the surface tension of liquids in the pores, RF aerogels can be dried at ambient pressure with little or negligible shrinkage [23, 74]. Shrinkage of the gel during drying is derived by the capillary pressure, e.g., $P_c = -2\gamma/(r_p - t)$ [75], where γ is the surface tension of the pore liquid, r_p is the pore radius, and t is the thickness of a surface adsorbed layer. Normally, the water in the pores is replaced with acetone so that the surface tension decreases from 72.1 to 24.0 dyne/cm. The small surface tension of the pore liquid and the large pore radius decreases the capillary pressure during the drying process. The increase of particle size and the stronger necks between the particles improve the mechanical properties of the skeleton. The gel can be dried at ambient conditions successfully under those conditions [76].

Drying can be performed at 50°C or room temperature with an open container. Research has shown that the *RF* aerogels with *R/C* ratio more than 500 can be successfully dried at ambient pressure in a large range of densities. All linear shrinkage is smaller than 10%, and in some cases it is even smaller than 2%, which can be considered negligible.

36.4.2. *Pyrolysis* (Carbonization) Technology

During pyrolysis (carbonization), the organic aerogel structure is transformed into a relatively pure carbon structure. The carbonization process is most commonly carried out in a tube furnace under a constant flow of inert gas (~200 $cm^3 min^{-1}$), such as N_2, Ar, or He. The temperature is first held constant at the room level for about an hour (to replace all air in the furnace with inert gas) [77] and then is raised to a level between 600 and 2,100°C [67]. The desired pyrolysis temperature is generally approached gradually using temperature programming. Interestingly, the electrical conductivity of the samples increases significantly during this transformation, as indicated by the common appearance of a broadband, semimetal-like infrared absorbance [78].

Variations in the pyrolysis conditions cause significant changes in the properties of the RF-derived carbon gels. A summary of these changes is shown in Table 36.5. Higher pyrolysis temperatures tend to reduce the surface areas of RF carbon aerogels; they also reduce the electrochemical double-layer capacitance of RF carbon aerogels [68, 79]. It is also found that when the pyrolysis temperature is above 600°C, the surface area of the carbon aerogels decreases slightly with increasing pyrolysis temperature, whereas the surface area increases with increasing pyrolysis temperature when the latter is under 600°C. However, *RF* carbon aerogels may not be electrically conductive unless carbonized above 750°C [29]. The pyrolysis temperature also has significant effects on the charge–discharge behavior of lithium ions from *RF*-derived carbon aerogels; however, the effect also depends on the gel pH and the concentration of the reactants.

The specific volumes of *RF*-derived carbon aerogels are always higher than those of *RF* organic aerogels. In agreement with the last trend, the pore volumes (reciprocal density) of *RF*-derived carbon aerogels decrease slightly with an increase in the pyrolysis temperature. Also, the corresponding skeletal densities also increase and level off at just below the skeletal density of graphite after 1,050°C [43]. This indicates that these particular *RF* carbon aerogels have very few closed pores within the skeletal structure. The electrochemical double-layer capacitance also exhibits a maximum between 800 and 900°C [79]. Meanwhile, the weight loss of *RF* carbon aerogels due to pyrolysis is practically limited

Table 36.5. Pyrolysis and activation: effects of various factors on the resulting properties [61]

Factor	Effect
Increasing pyrolysis temperature	Reduces residual oxygen content
	Reduces surface area of carbon aerogels and xerogels
	Reduces pore volumes of carbon aerogels and xerogels
	Increases macropore size distribution
	Increases micropore size distribution when very low R/C ratios are used
	Increases electrochemical capacitance up to ~850°C, thereafter reduces it, and either increases or decreases lithium-ion charge and discharge capacities, depending on gel pH and reactants concentrations
Increasing thermal activation time	Increases pore sizes, volumes, and surface areas
	Increases electrochemical capacitance up to ~3 h and thereafter reduces it

to the period before reaching ~750°C, after which it is almost constant at ~50% (w/w), which indicates the initiation of structural changes within the carbon gel without further weight loss. Overall, pyrolysis produces almost identical particle size distribution (*PSD*) for *RF* carbon aerogels [80], except at very low *R*/*C* ratios. In the latter case, pyrolysis causes the pore size distribution to spread further toward the micropore region.

The temperature required for complete graphitization of *RF* aerogels may exceed 2,000°C; nevertheless, aerogels pyrolyzed at ~1,050°C contain district graphitic structures. The endothermic pyrolytic reactions are most active during the programmed temperature rise. The first endothermic reaction, which occurs during the first period of heating from room temperature to ~250°C, is attributed to the release of absorbed water (as indicated by the disappearance of the OH group infrared bands); this corresponds to a mass loss of ~3%. The other endothermic reactions, which occur during the second period of heating, are attributed to the release of organic compounds. However, very slight endothermic and exothermic reactions may also occur after these periods during the constant temperature carbonization step.

The pyrolysis step reduces the number of macropores (possibly due to shrinkage) and increase in the surface area of *RF* carbon aerogels, especially at low pyrolysis temperatures (in accordance with the previously discussed trends). This effect is a result of the burnout of the organic groups, which leads to the creation of new pores or voids in the gel. Overall, shrinkage and mass reduction of 20 and 50%, respectively, can be expected due to pyrolysis of subcritically dried *RF* carbon aerogels. Nevertheless, unlike the pore sizes and pore volumes, which both decrease upon pyrolysis, the surface area of the carbon aerogels may increase after pyrolysis especially with dilute sols, i.e., when the catalyst to water ratio (*C*/*W*) (or alternatively the resorcinol to water ratio (*R*/*W*)) is low [81, 82] and shrinkage and mass reduction due to pyrolysis decrease with increasing *R*/*C* ratio [83]. Moreover, in addition to the effect on the gelation rate and also pore structure of carbon xerogels, addition of organic fibers to the sol may also minimize shrinkage during pyrolysis [24].

36.5. Applications

36.5.1. Electrical Applications

(a) Double-layer capacitors

Supercapacitors are electrochemical double-layer capacitors (*EDLC*s) with moderate energy storage capability but high power. Their field of application extends from backup power storage devices to peak power sources for electrical vehicles [23, 84]. The state of development and applied materials for these devices have been reviewed [23, 33, 53, 59, 67–69]. Among various high surface area carbons, carbon aerogels may be promising electrode materials [37, 84].

Because of their monolithic structure, high specific surface area (400–1,100 m^2/g), and high electrical conductivity, carbon aerogels can be used as electrodes for electrochemical double-layer capacitors. The basic principle of *EDLC*s is that energy can be stored via separation of charges across a polarized electrode/electrolyte interface. The stability of charges in the electrode/electrolyte interface will influence the characteristics of *EDLC*s, such as self-discharge. Another requirement is that electrodes should be good conducting materials and that they should not participate in faradaic reactions with the electrolyte at the applied voltage.

(b) Lithium-ion batteries

If high-energy densities are required, lithium-ion batteries are generally used. In the field of rechargeable batteries in recent years, the main field of interest shifted from the NiCd batteries to the Ni–metal hydride system and lately to Li-ion accumulators. Even though these accumulators have significantly higher charge capacities than NiCd batteries, for the application in electric vehicles the energy densities are still not high enough. Usually, carbon aerogels are used as the negative electrodes [85, 86]. The advantage of carbon is the higher negative redox potential compared to metal oxides and polymers. Moreover, most of the carbon materials are stable even over hundreds of charging/discharging cycles, thus improving the performance of the cells. The possibility of charge storage into carbon materials strongly depends on the microstructure and surface properties [59, 87]. The carbon structure should prevent the so-called co-intercalation of solvated lithium ions, because they are one reason for large leakage currents. In practice, mostly graphitic carbons with well-developed graphitic structures are used. These materials show mass specific charge capacities of 350 mAh/g and the best reversibility of all carbons, because they prevent effectively parasitic co-intercalation. For disordered carbons, even higher charge capacities are expected theoretically (up to 500 mAh/g). This is because in contrast to graphitic carbons where lithium intercalates between two graphene layers, here the process is merely an adsorption on both sides of graphene layers (house of cards model) [88]. However, up to now, the poor reversibility and high leakage currents have prevented the use of such materials in battery applications.

Carbon aerogels belong to the latter type of carbon materials. The degree of disorder can be adjusted with the pyrolysis temperature. Using coarse structured carbons, it has been shown that those materials are able to intercalate lithium and that via modification of the pyrolysis conditions the intercalation capacity can be improved and the irreversible intercalation reduced [29, 37, 56].

36.5.2. Hydrogen Storage and Adsorption

(a) Hydrogen Storage

New types of hydrogen storage technologies and hydrogen storage materials are major topics of current interest [89–97]. According to statistics, about 50% in all costs of hydrogen research of the U.S. Department of Energy is for hydrogen storage research. Research around the globe is trying to find economic, safe, practical, and convenient methods for hydrogen storage. In order to achieve the U.S. Department of Energy's programmatic objective of hydrogen storage for development of fuel cell electric vehicles (in an acceptable application temperature (60–120°C) and pressure conditions, the mass fraction of hydrogen storage system should reach 6% (w/w) by 2010 and up to 9% (w/w) by 2015 [98]); physical adsorption is considered the most promising hydrogen storage technology. The physisorption-based mechanism of adsorption of hydrogen storage of porous solid materials, because of the advantages in their low work pressure, storage containers' light weight, enough space in the shape choice, etc., has become the hotspot in hydrogen storage materials' research and development. At present, the porous adsorbents for hydrogen storage are basically divided into four categories [99]: carbon-based porous materials, noncarbon nanotube materials, minerals, and metal-organic porous materials. The basic requirements of hydrogen storage materials for porous solid adsorption are large surface area and high porosity [100]. Nanoporous carbon aerogel is a new type of hydrogen storage material.

Carbon aerogel prepared by the sol–gel process presents some interesting characteristics such as high specific surface area, low-density, nanopore structure, controllable structure, and can be easily doped. However, as a hydrogen storage material, carbon aerogel's research is still in its initial stages. Currently, preliminary findings at LLNL and others such as Houria Kabbour are as follows. At 77 K and under 3 MPa pressure, carbon aerogels with a specific surface area of 3,200 m^2/g reached a maximum mass fraction of hydrogen storage of 5.3% (w/w) [101]. At the same time, the above groups also studied the Ni- and Co-doped carbon aerogels for hydrogen adsorption and found out that doped carbon aerogels' specific surface areas and hydrogen storage capacity increased. However, due to a decline in the whole surface area, the ultimate hydrogen storage capacity was 2.3% (w/w) (Ni-doped) and 2.1% (w/w) (Co-doped). Other hydrogen adsorption results show that the saturated hydrogen storage mass fraction of the carbon aerogel sample (3,212 m^2/g) by the optimal activation process is 3 wt% at 92 K, 3.5 MPa. The hydrogen storage capacity of this sample was 0.84% (w/w) at 303 K, 3.2 MPa.

Therefore, further research should be pursued in studying the effect of doping on carbon aerogels' hydrogen adsorption properties and adopting more effective methods to improve the hydrogen storage capacity of carbon aerogels.

(b) Adsorption

Spherical carbon aerogels (*SCAs*) with controlled particle size and mesopore size were synthesized by an emulsified sol–gel polymerization of phenol, melamine, and formaldehyde [102–112]. The adsorption rate and capacity for biomolecules with different molecular dimensions, including L-phenylalanine (*Phe*), vitamin B_{12} (*VB*), α-chymotrypsin (*Chy*), and bovine serum albumin (*BSA*), onto *SCAs* was investigated. The mesopore size can be easily tuned in the range from 5 to 10 nm by simply adjusting catalyst concentration in the initial solution and the spherical particle size can be controlled in 50–500 μm by changing stirring speed. The as-prepared *SCAs* have high specific surface area (>600 m^2/g) and large pore volume (>1 cm^3/g). The hardness of *SCAs* is ca. ten times as large as that of commercial spherical activated carbon particles. The adsorption rate of *VB* is strongly depended on the mesopore size and particle size and shows an increasing trend with the increase of mesopore size and the decrease of particle size. For small molecule Phe, the specific surface area is key factor to determine the adsorption capacity, but the adsorption capacity of large molecules (*VB*, *Chy*, and *BSA*) is dependent on the pore size of *SCAs*, which should be suitably larger than the molecule size of biomolecules [113].

36.5.3. Catalyst Supports

Metal-doped carbon aerogels are emerging materials in the field of heterogeneous catalysis and their use will likely increase in the near future due to the versatility in their pore texture and form [114–121]. An important feature of these materials is that metal-doped monolithic organic aerogels can be readily prepared, for instance, by adding a soluble metal salt to the initial mixture (sol). After gelation, the metal salt is trapped within the gel structure and the metal ions can be chelated by the functional groups of the polymer matrix. The added metal salts can also catalyze to a varying degree the polymerization or gelation process, which can in turn affect the morphology and pore texture of the organic aerogel. After carbonization of this aerogel, the metal-containing phase is distributed throughout the porosity of the carbon phase. This method yields a metal-doped monolithic carbon aerogel, which can be used as a catalyst in different reactions or can find other applications such as electrodes in fuel cells and electrical double-layer capacitors. Furthermore, the metal catalyst can be deposited on the monolithic carbon aerogel by other methods, such as impregnation or adsorption [122].

36.5.4. Materials for Thermal Insulation

Because of their nanostructure and porosity, all types of aerogels show a high thermal resistance. Under ambient pressure the pores, 1–100 nm wide, are capable of reducing the gaseous heat conduction through the material. In comparison with the gaseous thermal conductivity of free nonconvecting air ($\lambda_g = 0.026\ \mathrm{W\ m^{-1}\ K^{-1}}$) and the thermal conductivity within the pores (lies between 0.005 and 0.010 $\mathrm{W\ m^{-1}\ K^{-1}}$), the thermal conductivity of carbon aerogels lies between 0.03 and 0.05 $\mathrm{W\ m^{-1}\ K^{-1}}$ [9, 14, 15, 27] (Chap. 23). To be used as high-temperature insulators, carbon aerogels have to be evacuated, since they would burn in an oxidizing atmosphere. In an evacuated sample, the gaseous thermal conduction is suppressed completely and the carbon skeleton leads to a very efficient reduction of the radiative heat transfer. Due to the high porosity, the solid thermal conductivity is also very small [47].

36.5.5. Other Applications

Aerogels' acoustic impedance can change in a wide range with the change of density, so it is an ideal acoustic impedance coupling material. It can lower the transmission efficiency of sound waves and lower the signal-to-noise ratio in device applications. If one uses suitable density-gradient aerogels, the acoustic coupling will be greatly enhanced. Of course, its longitudinal acoustic propagation rate is extremely low. So it is also an ideal acoustics delay material and high-temperature deadening [123].

Carbon aerogels are compatible with living organisms. They can be used in the manufacture of man-made biological tissue, artificial organs and organ components, medical diagnostic agents and carrier of parenteral drug delivery systems, or sustained-release pharmaceutical systems (Chaps. 30 and 31). They can also be used for the production of nontoxic and effective pesticides, cosmetic deodorant, and so on.

In recent years, another new electrochemical application of carbon aerogels as electrode material for the capacitive deionization in water phase was reported. J.C. Farmer and other researchers in the United States did an electric desalting experiments using carbon aerogels and found improved performance with such materials relative to those obtained with other carbons [79, 84, 124].

36.6. Conclusion

Carbon aerogels show great promise for applications in variety of technological areas where special structure and physical properties are required. Substantial progress has been made in the development, processing, characterization, and application of carbon aerogels over the recent years. Special attention has been paid to the use of inexpensive precursors and improving the drying technology to make the production commercial. However, the actual application of carbon aerogels is virtually untapped. In the near future, with researchers from all over the world continuing their work on carbon aerogels, the scientific and technological innovations they bring forward could have a significant effect on this area and realize the important applications of carbon aerogels.

References

1. Kistler SS (1931) Coherent expanded aerogels and jellies. Nature 127: 741.
2. Kistler SS (1932) Coherent expanded aerogels. J Phys Chem 36:52–64.
3. Kistler SS (1941) Aerogels. Patent US 2249767 assigned to Monsanto Chemical Co.
4. Kistler SS (1952) Water Repellent Aerogels. Patent US 258970.
5. Smith DM, Stein D, Anderson JM, Ackermann W (1995) Preparation of low-density xerogels at ambient pressure. J Non-Cryst Solids 186: 104.
6. Teichner SJ (1986) Aerogels of Inorganic Oxides, Springer Proc Phys 6:22–30.
7. Brinker CJ, Ward KJ, Keefer KD, Holupka E, Bray PJ, Pearson RK (1986) Synthesis and structure of borate based aerogel. In Aerogels. Springer Proc Phys 6:57–67.
8. Henning S, Svensson L (1981) Production of Silica Aerogel. Phys Scr 23: 697.
9. Fricke J (ed) (1986) Proceedings of the First International Symposium on Aerogels (ISA 1), Springer Proc Phys 6:167–173. Springer-Verlag, Berlin.
10. Fricke J (ed) (1986) Proceedings of the First International Symposium on Aerogels (ISA 1), Springer Proc Phys 6: 38. Springer-Verlag, Berlin.
11. Fricke J (ed) (1986) Proceedings of the First International Symposium on Aerogels (ISA 1), Springer Proc Phys 6: 127. Springer-Verlag, Berlin.
12. Fricke J (ed) (1986) Proceedings of the First International Symposium on Aerogels (ISA 1), Springer Proc Phys 6: 94. Springer-Verlag, Berlin.
13. Fricke J (ed) (1986) Proceedings of the First International Symposium on Aerogels (ISA 1), Springer Proc Phys 6: 167. Springer-Verlag, Berlin.
14. Vacher R, Phalippou J, Pelous J, Woignier T (eds) (1989) Proceedings of the Second International Symposium on Aerogels (ISA2), Rev Phys Appl Colloq 24-C4.
15. Fricke J (ed) (1992) Proceedings of the Third International Symposium on Aerogels (ISA 3), J Non-Cryst Solids 145:141–145.
16. Pekala RW, Hrubesh LW (eds) (1995) Proceedings of the Fourth International Symposium on Aerogels (ISA 4), J Non-Cryst Solids 186:159–167.
17. Phalippou J, Vacher R (eds) (1998) Proceedings of the Fifth International Symposium on Aerogels (ISA 5), J Non-Cryst Solids 225:220–225.
18. Ashley CS, Brinker CJ, Smith DM (eds) (2001) Proceedings of the Sixth International Symposium on Aerogels (ISA6), Albuquerque, NM, USA (2000) J Non-Cryst Solids 285:295.
19. Fricke J, Emmerling A (1992) Aerogels - preparation, properties, applications. Struct Bonding (Berlin) 77:37–87.
20. Brinker CJ, Scherer GW (1990) Sol-Gel Science. The Physics and Chemistry of Sol-Gel Processing, Academic Press, New-York, USA: (a) p461; (b) p523; (c) p97.
21. Farmer JC, Fix DV, Mack GV, Pekala RW, Poco JF (1996) Capacitive deionization of NH4ClO4 solutions with carbon aerogel electrodes. J Appl Electrochem 26: 1007–1018.
22. Farmer JC, Bahowick SM, Harrar JE, Fix DV, Martinelli RE, Vu AK, Carroll KL (1997) Electrosorption of chromium ions on carbon aerogel electrodes as a means of remediating ground water. Energy Fuels 11:337–347.

23. Saliger R, Fischer U, Herta C, Fricke J (1998) High surface area carbon aerogels for supercapacitors. J Non-Cryst Solids 225: 81–85.
24. Glora M, Wiener M, Petricˇevic´ R, Pro¨bstle H, Fricke J (2001) Integration of Carbon Aerogels in PEM Fuel Cells. J Non-Cryst Solids 285: 283–287.
25. Pekala RW, Kong FM (1989) A synthetic route to organic aerogels – mechanism, structure, and properties. J Phys Colleg Suppl 50:c4–33.
26. Pekala RW, Alviso CT, Lemay JD (1990) Organic aerogels: microstructural dependence of mechanical properties in compression. J Non-Cryst Solids 125: 67–75.
27. Pekala RW, Alviso CT (1992) Carbon aerogels and xerogels. Mat Res Soc Symp Proc 270: 3–23.
28. Pekala RW, Alviso CT, Lu X, Gross J, Fricke J (1995) New organic aerogels based upon a phenolic-furfural reaction. J Non-Cryst Solids 188: 34–40.
29. Pekala RW (1989) Organic aerogels from the polycondensation of resorcinol with formaldehyde. J Mater Sci 24: 3221–3227.
30. Ruben GC, Pekala RW, Tillotson TM, Hrubesh LW (1992) Imaging Aerogels at the Molecular Level. J Mater Sci 27:4341.
31. Hench LL, West FK (eds) (1992) Chemical Processing of Advanced Materials; John Wiley & Sons: New York.
32. Attia YA (ed) (1994) Pekala RW, Mayer ST, Kaschmitter JL, Kong FM (1994) Carbon aerogels: an update on structure, properties, and applications. Plenum, New-York, Sol-gel Process Appls 369–377.
33. Miller JM, Dunn B, Tran TD, Pekala RW (1997) Deposition of ruthenium nanoparticles on carbon aerogels for high energy density supercapacitor electrodes. J Electrochem Soc 144: 309–311.
34. Tamon H, Ishizaka H, Mikami M, Okazaki M (1997) Porous structure of organic and carbon aerogels synthesized by sol-gel polycondensation of resorcinol with formaldehyde. Carbon 35: 791–796.
35. Attia SM, Wang J, Wu GM, Shen J, Ma JH (2002) Review on Sol-gel Derived Coatings:Process, Techniques and Optical Applications. J Mar Sci Technol 18: 211–218.
36. Reichenauer G, Emmerling A, Fricke J, Pekala RW (1998) Microporosity in carbon aerogels. J Non-Cryst Solids 225: 210–214.
37. Wang J, Zhang SQ, Guo YZ, Shen J, Attia SM, Zhou B, Zheng GZ, Gui YS (2001) Morphological Effects on the Electrical and Electrochemical Properties of Carbon Aerogels, J Electrochem Soc 148:75–77.
38. Lin C, Ritter JA (1997) Effect of synthesis pH on the structure of carbon xerogels. Carbon 35:1271–1278.
39. Biesmans G, Mertens A, Duffours L, Woignier T, Phalippou J (1998) Polyurethane based organic aerogels and their transformation into carbon aerogels. J Non-Cryst Solids 225(1):64–68.
40. Shen J, Wang J, Guo YZ, Attia SM, Zhou B (2000) Resorcinol-formaldehyde-derived carbon aerogel films. Proc SPIE 4086: 811–814.
41. Shen J, Hou JQ, Guo YZ, Xue H, Wu GM, Zhou B (2005) Microstructure Control of RF and Carbon Aerogels Prepared by Sol-Gel Process. J Sol-Gel Sci Tech 36:131–136.
42. Shen J, Wang J, Zhai JW, Guo YZ, Wu GM, Zhou B, Ni XY (2004) Carbon Aerogel Films Synthesized at Ambient Conditions. J Sol-Gel Sci Tech 31:209–213.
43. Lin C, Ritter JA (2000) Carbonization and activation of solgel derived carbon xerogels. Carbon 38: 849–861.
44. Biesmans G, Mertens A, Duffours L, Woignier T, Phallipou J (1998) Polyurethane based organic aerogels and their transformation into carbon aerogels. J Non-Cryst Solids 225: 64–68.
45. Pekala RW (1990) Synthetic control of molecular structure in organic aerogels. Mater Res Soc Proc 171: 285–291.
46. Zhang SQ, Wang J, Shen J, Deng ZS, Lai ZQ, Zhou B, Attia SM, Chen LY (1999) The investigation of the adsorption character of carbon aerogels. Nanostruct Mater 11:375-381.
47. Bock V, Nilsson O, Blumm J, Fricke J (1995) Thermal properties of carbon aerogels. J Non-Cryst Solids 185: 233–239.
48. Alviso CT, Pekala RW, Gross J, Lu X, Caps R, Fricke J (1996) Resorcinol–formaldehyde and carbon aerogel microspheres. Micropor Macropor Mater 431: 521–525.
49. Li W, Reichenauer G, Fricke J (2002) Carbon aerogels derived from cresol–resorcinol–formaldehyde for supercapacitors. Carbon 40: 2955–2959.
50. Biesmans G, Randall D, Francis E, Perrut M (1998) Polyurethane-based organic aerogels' thermal performance. J Non-Cryst Solids 225:36–40.
51. Zhang R, Lu Y, Zhan L, Liang X, Wu G, Ling L (2002) Monolithic carbon aerogels from sol–gel polymerization of phenolic resoles and methylolated melamine. Carbon 41:1660–1663.
52. Yamashita J, Ojima T, Shioya M, Hatori H, Yamada Y (2003) Organic and carbon aerogels derived from poly (vinyl chloride). Carbon 41:285–294.
53. Wang J, Golra M, Petricevic R (2001) Carbon Cloth Reinforced Carbon Aerogel Films Derived from Resorcinol Formaldehyde. J Porous Mater 8: 159–165.

54. Bock V, Emmerling A, Saliger R, Fricke J (1997) Structural Investigation of Resorcinol Formaldehyde and Carbon Aerogels Using SAXS and BET. J Porous Mater 4: 287–294.
55. Lu X, Caps R, Fricke J, Alviso CT, Pekala RW (1995) Correlation between structure and thermal-conductivity of organic aerogels. J Non-Cryst Solids 188: 226–234.
56. Wang J, Zhang SQ, Shen J, Guo YZ, Attia SM, Zhou B, Lai ZQ, Zheng GZ, Gui YS (2001) Electrical transport properties of carbon aerogels. J Porous Mater 8:167–170.
57. Pekala RW, Hrubesh LW (eds) (1995) Proceedings of the Fourth International Symposium on Aerogels (ISA 4), J Non-Cryst Solids 186:159–162.
58. Mark HF, Othmer DF, Overberger CG, Serborg GT (eds) (1981) Encyclopedia of Chemial Technology, 13:39, Wiley, New York.
59. Shen J, Han WN, Mi YJ, Ou YL, Wu GM, Zhou B, Zhang ZH, Ni XY, Niu XX, Wang GQ, Wang PQ,.Wang QF (2008) Nanostructure Control of Carbon Aerogels and the Application in Lithium Ion Cells. 2nd IEEE International Nanoelectronics Conference (INEC) (2008):74–77.
60. Pekala RW (1989) Low density, resorcinol-formaldehyde aerogels. US Patent 4873218 assigned to The United States Department of Energy.
61. Al-Muhtaseb SA, Ritter JA (2003) Preparation and Properties of Resorcinol-Formaldehyde Organic and Carbon Gels. Adv Mater 15:101–114.
62. Despetis F, Barral K, Kocon L, Phalippou J (2000) Effect of aging on mechanical properties of resorcinol-formaldehyde gels. J Sol-Gel Sci Tech 19: 829-831.
63. LeMay JD, Tillotson TM, Hrubesh LW, Pekala RW (1990) Microstructural dependence of aerogel mechanical properties. Mat Res Soc Sym Proc 180:321–324.
64. Tamon H, Ishizaka H, Yamamoto T, Suzuki T (2000) Influence of freeze-drying conditions on the mesoporosity of organic gels as carbon precursors. Carbon 38: 1099–1105.
65. Petricevic R, Glora M, Fricke J (2001) Planar fibre reinforced carbon aerogels for application in PEM fuel cells. Carbon 39: 857–867.
66. Zanto EJ, Al-Muhtaseb SA, Ritter JA (2002) Sol-gel-derived carbon aerogels and xerogels: Design of experiments approach to materials synthesis. Ind Eng Chem Res 41: 3151–3162.
67. Kaschmitter JL, Mayer ST, Pekala RW (1998) Process for producing carbon foams for energy storage devices. US Patent 5789338 assigned to Regents of the University of California.
68. Mayer ST, Pekala RW, Kaschmitter JL (1993) The aerocapacitor: an electrochemical double-layer energy-storage device. J Electrochem Soc 140: 446–451.
69. Liang CH, Sha GY, Guo SC (2000) Resorcinol-formaldehyde aerogels prepared by supercritical acetone drying. J Non-Cryst Solids 271: 167–170.
70. Berthon S, Barbieri O, Ehrburger-Dolle F, Geissler E, Achard P, Bley F, Hecht AM, Livet F, Pajonk GM, Pinto N, Rigaci A, Rochas C (2001) DLS and SAXS investigations of organic gels and aerogels. J Non-Cryst Solids 285: 154–161.
71. Guo YZ, Shen J, Wang J (2001) Carbon aerogels dried at ambient conditions. New Carbon 16: 55-57.
72. Tamon H, Ishizaka H, Yamamoto T, Suzuki T (1999) Preparation of mesoporous carbon by freeze drying. Carbon 37: 2049–2055.
73. Kocklenberg R, Mathieu B, Blacher S, Pirard R, Pirard JP, Sobry R, Van den Bossche G (1998) Texture control of freeze-dried resorcinol-formaldehyde gels. J Non-Cryst Solids 225: 8–13.
74. Wang J, Saliger R, Fricke J (2001) Carbon cloth reinforced carbon aerogel films derived from resorcinol formaldehyde. J Porous Mat 8:159–165.
75. Brinker CJ, Scherer GW (eds) (1990) Sol-Gel Science: The Physics and Chemistry of Sol-Gel Processing, Academic Press Inc, USA.
76. Fricke J, Emmerling A (1998) Aerogels - Recent progress in production techniques and novel applications. J Sol-Gel Sci Technol. 13:299–303.
77. Yamamoto Y, Nishimura T, Suzuki T, Tamon H (2001) Control of mesoporosity of carbon gels prepared by sol-gel polycondensation and freeze drying. J Non-Cryst Solids 288: 46–55.
78. Kuhn J, Brandt R, Mehling H, Petricevic R, Fricke J (1998) In situ infrared observation of the pyrolysis process of carbon aerogels. J Non-Cryst Solids 225: 58–63.
79. Pekala RW, Farmer JC, Alviso CT, Tran TD, Mayer ST, Miller JM, Dunn B (1998) Carbon aerogels for electrochemical applications. J Non-Cryst Solids 225: 74–80.
80. Tamon H, Ishizaka H, Okazaki M (1997) Porous structure of organic and carbon aerogels synthesized by sol-gel polycondensation of resorcinol with formaldehyde. Carbon 35:791–796.
81. Yamamoto Y, Nishimura T, Suzuki T, Tamon H (2000) Control of mesoporosity of carbon gels prepared by sol-gel polycondensation and freeze drying. J Non-Cryst Solids 288: 46–55.

82. Yamamoto T, Yoshida T, Suzuki T, Mukai SR and Tamon H (2002) Dynamic and static light scattering study on the sol–gel transition of resorcinol–formaldehyde aqueous solution. J Colloid Interface Sci 245: 391–396.
83. Saliger R, Bock V, Petricevic R, Tilloston T, Geis S, Fricke J (1997) Carbon aerogels from dilute catalysis of resorcinol with formaldehyde. J Non-Cryst Solids 221: 144–150.
84. Frackowiak E. Beguin F (2001) Carbon materials for the electrochemical storage of energy in capacitors. Carbon 39: 937–950.
85. Fauteux D, Koksbang R (1993) Rechargeable lithium battery anodes. Alternatives to metallic lithium. J Appl Chem 23: 1–10.
86. Winter M, Besenhard JO, Spahr ME, Novak P (1998) Insertion electrode materials for rechargeable lithium batteries. Adv Mater 10: 725–763.
87. Dahn JR, Zheng T, Liu YH, Xue JS (1995) Mechanisms for lithium insertion in carbonaceous materials. Science 270: 590–593.
88. Zheng T, Xing W, Dahn JR (1996) Carbons prepared from coals for anodes of lithium-ion cells. Carbon 34: 1501–1507.
89. Liu W, Zhao YH , Nguyen J , Jiang Y Li, Q , Lavernia EJ (2009) Electric field induced reversible switch in hydrogen storage based on single-layer and bilayer graphenes. Carbon 47: 3452–3460.
90. Liu C , Fan YY , Liu M , Cong H T , Cheng H M , Dresselhaus M S (1999) Hydrogen storage in single-walled carbon nanotubes at room temperature. Science 286: 1127–1129.
91. Morris RE, Wheatley PS (2008) Gas storage in nanoporous materials. Angew Chem, Int Ed 47: 4966–81.
92. Meng S, Kaxiras E, Zhang ZY (2007) Metal-diboride nanotubes as high-capacity hydrogen storage media. Nano Lett 7: 663–697.
93. Siegel DJ, Wolverton C, Ozolin V (2007) Thermodynamic guidelines for the prediction of hydrogen storage reactions and their application to destabilized hydride mixtures. Phys Rev B 76: 134102-134107.
94. Dillon AC , Heben MJ (2001) Hydrogen storage using carbon adsorbents: past, present and future. Appl Phys A: Mater Sci Process 72:133–142.
95. Oostinga JB, Heersche HB, Liu XL, Morpurgo AF, Vandersypen LMK (2008) Gate-induced insulating state in bilayer grapheme devices. Nature Mater 7: 151–157.
96. Geim AK, Novoselov KS (2007) The rise of grapheme. Nature Mater 6: 183–191.
97. Lherbier A, Blase X, Niquet YM, Triozon F, Roche S (2008) Charge transport in chemically doped 2D graphene. Phys Rev Lett 101: 036808–036811.
98. Li Z (2000) Theory and large-scale application prospect of carbon-based adsorptive hydrogen storage materials. Mater Rev 14: 3–5.
99. Shang FL, Yang HT, Han HT (2006) Progress in porous hydrogen storage materials. Chem Eng J 20: 58–61.
100. Mark Thomas K (2007) Hydrogen adsorption and storage on porous materials. Catal Today 120: 389–398.
101. Kabbour H, Baumann TF, Satcher Jr JH, Saulnier A, Ahn CC (2006) Toward new candidates for hydrogen storage: High-surface-area carbon aerogels. Chem Mater 18: 6085–6087.
102. Horikawa T, Hayashi J, Muroyama K (2004) Size control and characterization of spherical carbon aerogel particles from resorcinol-formaldehyde resin. Carbon 42: 169–175.
103. Kima J, Gratea JW, Wang P (2006) Nanostructures for enzyme stabilization. Chem Eng Sci 61: 1017-1026.
104. Hartmann M (2005) Ordered Mesoporous Materials for Bioadsorption and Biocatalysis. Chem Mater 17: 4577–4593.
105. Lee J, Kim J, Hyeon T (2006) Recent progress in the synthesis of porous carbon materials. Adv Mater 18: 2073–2094.
106. Hu ZH , Srinivasan MP, Ni Y (2000) Preparation of mesoporous high-surface-area activated carbon. Adv Mater 12: 62.
107. Yang T, Lua AC (2003) Characteristics of activated carbon prepared from pistachio-nut shell by zinc chloride activation under nitrogen and vacuum conditions. J Colloid Interface Sci 267: 408-417.
108. Pekala RW, Kong F M (1989)Resorcinol-Formaldehyde Aerogels and their Carbonized Derivatives. Polym Prepr 30: 221–223.
109. Fu R, Zheng B, Liu J, Dresselhaus MS, Dresselhaus G, Satcher JH (2003)The Fabrication and Characterization of Carbon Aerogels by Gelation and Supercritical Drying in Isopropanol. Adv Funct Mater 13: 558–562.
110. Ryoo R, Joo SH, Jun S (1999) Synthesis of highly ordered carbon molecular sieves via template-mediated structural transformation. J Phys Chem B 103: 7743–7746.
111. Tanaka S, Nishiyama N, Egashira Y, Ueyama K (2005) Synthesis of ordered mesoporous carbons with channel structure from an organic-organic nanocomposite. Chem Commun 16: 2125–2127.
112. Vinu A, Streb C, Hartmann M (2003) Adsorption of cytochrome c on new mesoporous carbon molecular sieves. J Phys Chem B 107: 8297–8299.

113. Long DH, Zhang R, Qiao WM, Zhang L, Liang XY, Ling LC (2009) Biomolecular adsorption behavior on spherical carbon aerogels with various mesopore sizes. J Colloid Interface Sci 331: 40–46.
114. Guilminot E, Fischer F, Chatenet M, Rigacci A, Berthon-Fabry S, Achard P, Chainet E (2007) Use of cellulose-based carbon aerogels as catalyst support for PEM fuel cell electrodes: Electrochemical characterization. J Power Sources 166: 104–111.
115. Du H, Li B, Kang F, Fu R, Zeng Y (2007) Carbon aerogel supported Pt–Ru catalysts for using as the anode of direct methanol fuel cells. Carbon 45: 429–435.
116. Smirnova A, Dong X, Hara H, Vasiliev A, Sammes N (2005) Novel carbon aerogel-supported catalysts for PEM fuel cell application. Int. J Hydrol Eng 30: 149–158.
117. Kalpana D , Omkumar KS , Kumar SS , Renganathan NG (2006) A novel high power symmetric ZnO/carbon aerogel composite electrode for electrochemical supercapacitor. Electrochem Acta 52: 1309–1315.
118. Zhang SQ, Huang CG, Zhou ZY, Li Z (2002) Investigation of the microwave absorbing properties of carbon aerogels. Mater Sci Eng B 90: 38–41.
119. Wu DC, Fu RW (2006) Synthesis of organic and carbon aerogels from phenol–furfural by two-step polymerization. Microporous Mesoporous Mater 96: 115–121.
120. Czakkel O, Marthi K, Geissler E, Laszlo K (2005) Influence of drying on the morphology of resorcinol–formaldehyde-based carbon gels. Microporous Mesoporous Mater 86: 124–133.
121. Job N , Thery A , Pirard R , Marien J , Kocon L , Rouzaud J N , Beguin F , Pirard JP (2005) Carbon aerogels, cryogels and xerogels: Influence of the drying method on the textural properties of porous carbon materials. Carbon 43: 2481–2494.
122. Moreno-Castilla C, Maldonado-Ho´dar F J (2005) Carbon aerogels for catalysis applications: An overview. Carbon 43 :455–465.
123. Caps R, Doell G, Fricke J, Heinemann E, Hetfleisch J (1989) Thermal transport in monolithic silica aeroge.lIn Vacher R, Phalippou J, Pelous J, Woignier T (eds) Proceedings of the Second International Symposium on Aerogels (ISA2), Rev Phys Appl Colloq 24-C4:113–118.
124. Farmer JC, Fix DV, Mack GV (1996) Capacitive deionization of NaCl and $NaNO_3$ solutions with carbon aerogel electrodes. J Electrochem Soc 143: 159–169.

Part XV

Commercial Products

37

Insights and Analysis of Manufacturing and Marketing Consumer Products with Aerogel Materials

Bruce McCormick

Abstract Aerogel material offers several potential benefits for use in consumer products. The superior insulating value of these materials compares very favorably with insulating materials currently in use in the apparel and insulated container industries. There are several key challenges to overcome, before aerogel insulating materials are widely used in consumer products. Consumer awareness of aerogel materials is very low; there are technical challenges in replacing conventional insulating materials with aerogel, and the cost versus performance of aerogel must be properly marketed to consumers.

37.1. Introduction

Aerogel-based materials offer fantastic properties for a wide variety of consumer product applications. One of the principal consumer benefits of these materials is the unparalleled thermal performance, which can be achieved with aerogel insulation. Aerogel materials can deliver thermal performance several times better than conventional materials. However, aerogel-based products also present several important manufacturing and marketing challenges for the manufacturers and retailers of consumer goods, one of which is higher cost. As shown below, aerogel is already being successfully sold in the consumer goods market at premium prices. The successful growth of aerogel in this market depends on addressing four key elements:

1. Creating awareness of aerogel with product designers
2. Overcoming the technical challenges of integrating aerogel into current products
3. Employing a marketing strategy to convey the value of aerogel in product to consumer
4. Creating a product with a reasonable price/performance ratio

37.2. Insulating Solutions

37.2.1. Current Insulating Materials

Consumer products that range from clothing, footwear, cooking and outdoor equipment to food and beverage containers currently feature conventional insulating

B. McCormick • Savsu Technologies, 1 High Country Rd, Santa Fe, NM 87508, USA
e-mail: b.mccormick@savsu.com

M.A. Aegerter et al. (eds.), *Aerogels Handbook*, Advances in Sol-Gel Derived Materials and Technologies, DOI 10.1007/978-1-4419-7589-8_37,

materials. Common insulation materials, which are used in the clothing and footwear markets, include natural fibers such as wools and synthetic fibers such as polyesters and polyolefins [which for example are found in very popular brands such as Thinsulate (3M St Paul, MN, USA) and Primaloft (Primaloft Albany, NY, USA)]. The kitchen apparel market uses materials such as silicon and leather. The outdoor equipment market utilizes a broad range of insulating materials, including synthetic fibers, molded foams, and gels, and even encapsulated air pockets. Food and beverage containers often are designed to provide vacuum barriers in order to achieve the thermal insulation desired. The well-known Thermos brand is an example of a vacuum container. Many containers use foams, such as polystyrene, polyurethane, and polyisocyanurate, to provide the desired insulation. Aerogel materials have recently been tested to provide significantly better performance than these foams.

The multibillion dollar a year outerwear apparel industry currently incorporates a broad use of insulating materials into their products. Some examples that currently use conventional insulation include gloves, hats, jackets, pants, socks, boots, shoes, and undergarments. Outdoor equipment also utilizes conventional insulation in products such as sleeping bags, tents, sleeping pads, waders, wet suits, and seat pads. The list of products utilizing conventional insulation is very long, indeed. Whenever humans encounter outdoor conditions, chances are that they will need some form of thermal protection from the elements.

37.2.2. The Synthetic Revolution

Conventional insulation, which dominates the consumer market, includes popular brands of insulation such as 3M, Thinsulate and Hollofil, PolarTec, PolarFleece, and PrimaLoft. Prior to the development of these synthetic insulations, insulation needs were met by use of natural materials such as wools, downs, skins, and furs. Even today, the use of these natural materials is still desirable. During the1970s, synthetic apparel insulation became popular material used in the outdoor apparel markets. In the 1970s, Thinsulate was launched and first incorporated into skiwear, and hence, a revolution was begun. At that time, the manufacturing world, as well as consumers, saw that warm clothing could be made using materials that in many respects outperformed the natural insulating materials such as wools, feathers, and skins. This revolution allowed for garments to be designed that were light, comfortable, and, as far as commercialization is concerned, very importantly – fashionable.

Insulation that is used in the consumer apparel market must exhibit several important performance characteristics in addition to being cost effective. Generally, insulation used in garments must

- Be lightweight
- Provide effective insulation within practical thicknesses
- Be compatible with the final products production processes
- Allow easy connection between inner and outer linings
- Exhibit hydrophobic properties
- Be soft to the touch
- Be compatible with other fabrics as they must move, stretch, recover, and remain durable under typical service conditions and for extended periods of time

One interesting example of this challenge is seen in footwear products. Insulated footwear has two main areas for insulation need. The shoe side of the upper area needs to be insulated with a material, which must move easily and recover with each step, and the underfoot area of the shoe must withstand the constant compressive force exerted on the material with each step. Conventional insulation is very popular within the upper area of footwear, but the underfoot area of footwear is generally left unprotected from thermal loss. This is due to the fact that the conventional materials do not perform well when they are compressed. These materials require an unacceptable amount of thickness in order to be effective.

Obviously, different applications will require different insulation materials. Specialized apparel-specific insulation applications also exist within the consumer markets. Firefighters, North Slope oilfield workers, and military personnel, for example, all have very distinct and often critical insulation needs.

Since the development of synthetic materials, there has been a huge growth in the outerwear industry as a result of the technical innovation that synthetic insulation has offered, as well as the ease of application of these materials into fashionable products that consumers could be persuaded to purchase. However, since this time, there have been only incremental performance improvements in synthetic insulation. As far as insulation is concerned, consumer apparel insulation has not made any significant performance improvements in almost 40 years. Are these needs being met with the best insulating technology available? Can we honestly say that in this age of rapid change that the best insulating products are available to consumers? Is it time for the next revolution? Answers to these questions are all positive and aerogel-based materials are perfectly suited to capture some portion of this large market.

37.3. Market Opportunities for Aerogel Products

The successful introduction of aerogel materials into the consumer markets depends on the value that the aerogel materials bring to the product as compared to the insulating materials currently being used. This is basically a cost versus benefit analysis. Is there a market for improved thermal performance in products that need thermal insulation? The answer for this is quite simply, "yes," but the more important question is, "what is the price" that consumers will pay for this improved insulation? Knowing the answer to this question will allow consumer product manufacturers and marketers to produce and sell products successfully.

When analyzing the consumer markets, it is very important to recognize that technical improvement alone will not translate into sales success. Consumer behavior is largely impacted on brand awareness and "perceived performance" versus actual performance. By recognizing that many consumer products are somewhat, if not entirely, discretionary in terms of need, an analysis of what stimulates the consumer to buy these products in the first place must be conducted. The answer often lies in marketing and advertising.

37.3.1. Innovation Diffusion of Aerogel Products

If marketing and advertising are major factors in creating consumer demand, then how do new technologies such as aerogel insulation receive acceptance in the market place? This answer is tied to the competition manufacturers who want to invest in new technologies

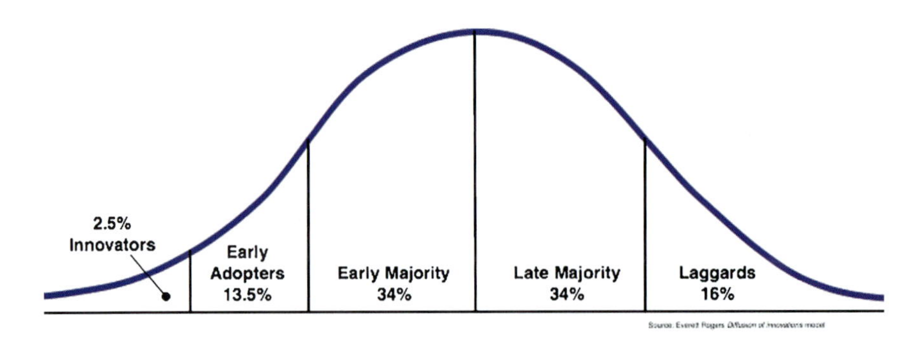

Figure 37.1. Typical consumer behavior pattern for new product acceptance.

in order to possibly gain a competitive advantage over their current competitors. The challenge, at this point, is the cost to promote this new technology to the public, who in many cases have some form of emotional attachment to the technology currently being used. The reason for the emotional attachment is because the consumer has spent money on products with the current technology! Perhaps also the consumer has grown to like a product and identifies him/herself with the product. Even if there is something better out there at a lower price, he/she still prefers the old one. This logic certainly seems somewhat circular, but there is a place for innovation within the consumer market, and it starts with, what in marketing terms is called the "innovator." The innovator is the consumer who wants "the latest and greatest" of all things. They do not get stuck with emotional attachments to the current products they own. They want the newest and the best available. Prove to them that your product is the best, and they will buy it (Figure 37.1). (Source: Rogers, Everett M. Diffusion of Innovations, 1995) [1].

Whether aerogel products follow this pattern of consumer acceptance is very dependent on several factors. Cost, convenience, and performance are perhaps the key factors in the exact shape that the above curve will take. Clearly, the first stage of any product's sales success will require the acceptance of the "Innovators." The next group of consumers, called "Early Adopters," is critical to broad acceptance of a consumer product. To get this group of consumers to purchase a product, the challenges of pricing and convenience become critical. How much of a premium, if any, this group will pay for a product will depend on how much value (performance) the "Early Adopters" perceive the product offers compared to alternative products. Another challenge relates to the convenience for the consumer to obtain the product at the time they need it. This challenge is typically dealt with by getting products on the shelves of appropriate stores. In order for this to happen, of course, the product maker must convince the store buyer that the product will sell. Convincing the store buyer that a new product, which has little sales history, will sell is difficult and is typically dealt with by the manufacturer offering the retailer an advertising program to promote the product and draw customers to the store.

The ability of aerogel product manufacturers to support an advertising campaign to gain retail space is largely dependent on the size of this manufacturer. A national marketing campaign for a product launch costs a minimum of $5 million and could easily be $15 million. Here, we get back to the issue of the cart before the horse. Will a large manufacturer want to invest these sums into a product or line of products that have not been proven to be successful in the retail environment? It is entirely possible and even likely that a collection of small companies will need to do the dirty work of creating consumer awareness. When a few of them have been successful in the retail market launching innovative aerogel products, the big companies are likely to develop and promote their own aerogel lines that rapidly broaden consumer awareness of aerogel.

37.3.2. The Internet and Aerogel

In today's fast-paced retailing world, the traditional cycle of product growth and success can be significantly disturbed or influenced due to the enormous growth of e-commerce on the Internet. In today's retail markets, products are promoted on the Internet as soon as they appear publicly. It is becoming more and more common for goods makers to offer to sell their products online directly to consumers. Online product information and promotion of goods is relatively inexpensive and the reach is worldwide. Another advantage, which the Internet offers to and for complex new products, is the ability to provide in-depth information about these new products. Common formats to provide information now include video, as well as the standard text-based web pages.

Aerogel-based products are very well positioned to take advantage of the Internet. Products made with aerogel can be demonstrated to show the dramatic performance benefits of the product. Video can be used to document various test data, which will go a long way in helping launch a high performance product. There are demonstration videos currently on the web using dramatic scenes of aerogel insulating against high-temperature flames. This is the type of information that can be well leveraged to promote fast growth of new products.

Another benefit of Internet marketing of aerogel products relates to the fantastic ability to offer shopping convenience to customers around the world. Toasty Feet insoles for example are sold in thousands of retail stores around the world, but this only begins to touch the potential market. With e-commerce, the same Toasty Feet insoles are available to customers 24 h a day all around the globe. Long after the product has been removed from the shelves of retailers when the spring merchandise is brought into stores, Toasty Feet insoles are still available to consumers online. It is amazing that insulated insoles are sold all year long on the Internet. The tremendous reach of e-commerce is a powerful resource, particularly for new and emerging products, and it lends itself very well to aerogel-based products.

37.3.3. Production Costs and Obstacles

Manufacturers who choose to lead the market in terms of producing products with advanced performance, must also consider the cost of changing production methods. Two of the most popular aerogel materials that are currently available for this type of production are materials made by Aspen Aerogel and by Cabot Corporation (Chap. 38). These materials come in blanket form or loose beads and are available in configurations, which are common in the insulation business (rolls of materials in standard widths or in bulk weight). Aerogel materials delivered in this way are easily integrated by manufacturers into their manufacturing processes.

One adverse aspect of aerogel insulation worth mentioning is the dust-release behavior of silica aerogels. This dust, which is released especially during cutting and handling, is actually the base aerogel material. In the case of the blanket material manufactured by Aspen Aerogel, the aerogel is infused into a blanket substrate. The blanket material used by Cabot incorporates "beads"or granules of Aerogel, which are sandwiched inside a polyester blanket; these beads also create dust as they are freed from the blanket, which happens upon cutting the material.

The aerogel dust that comes from both of these materials can cause significant problems in a typical manufacturing process. These problems include contamination of surrounding equipment and even possible damage to the fine working parts. Manufacturers of consumer goods need to create streamlined production processes that deliver products of reliable quality without constant interruption by adverse affects by, for example, dust and contamination. Obviously, there are solutions to these challenges, but it is important to recognize that these solutions represent yet another challenge (and a cost barrier) to the use of aerogel-based insulating materials in the consumer markets.

There are a number of factors that can influence, and possibly limit, the use of superior technology into the consumer markets. These factors include the following:

1. Awareness of new technology by product developers and potential customers
2. Interest in producing higher performance products
3. Reasonable ease of implementation into existing manufacturing processes
4. Acceptable cost versus benefit
5. Ability to convey value of new technology to the consumer
6. Cost to convey value to consumer
7. Benefits of being a leader in technology implementation
8. Ability of new technology to solve real problems

37.4. "Low Hanging Fruit" and Aerogel Products

As discussed previously, the consumer products market that uses insulation is of considerable size. Aerogel producers and product makers must focus on products and markets that can most readily benefit from the high performance of an aerogel. These markets most likely include products, which can best benefit from maximum thermal performance within a minimal amount of space.

Since standard insulating materials are very inexpensive compared to aerogel-based materials, there is little incentive for products to use more expensive aerogel insulation unless the amount of insulation is very small or there is a substantial need for improved insulation performance within a very limited thickness.

One example of a space-limited application is in footwear (Figure 37.2). Current models of insulated footwear require loft to provide insulating protection. However, loft cannot be practically maintained in the construction of a shoe or boot. The insulating materials used in footwear, such as Thinsulate, do not have the insulating performance of aerogel blanket materials, such as the material produced by Aspen Aerogel. In fact, tests have shown that, per unit thickness, the Aspen Aerogel material is 425% better than Thinsulate. Simply using more Thinsulate in footwear cannot make up this difference in performance. In a practical sense, there are limitations to how thick a shoe or boot can be constructed before it becomes unbearable or at least undesirable from an aesthetic point of view (Figure 37.2C).

Figure 37.2C illustrates the cutaway view of a very heavily insulated boot, and it can be seen that the upper of the boot contains a very large amount of Thinsulate insulation. This application of insulation measures approximately 1″ in thickness, and the result of this is an extremely large outer boot. This pursuit of extreme thermal insulation may result in a product that may not gain significant market acceptance due to the fact that the user will perceive that the boots are too large on their feet. This boot likely represents the limits that traditional insulation technology offers in terms of practical insulation for boot uppers. If a same amount of insulating capability can be achieved by one fourth the amount of thickness

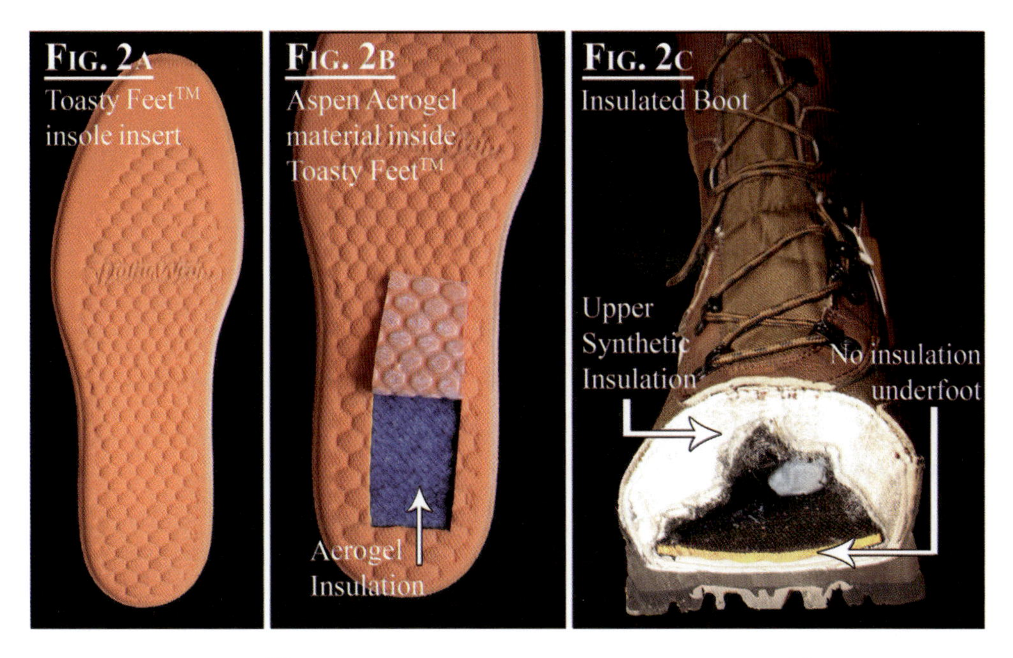

Figure 37.2. **A.** Toasty feet™ insole insert **B.** Aspen Aerogel materials inside Toasty Feet™ **C.** Cutaway of boot constructed with 2,000 g of Thinsulate.

of insulating material, the question is, would this product have value in the consumer market? Aerogel insulating materials can achieve the same thermal performance of this boot while not looking any different than a standard boot.

In addition to the upper insulation, it is important to note the essential lack of insulation under the foot in this boot's (Thinsulate model) construction. This is due to the fact that conventional insulating materials require loft, and when these materials are placed under foot, the loft is compressed from this insulating material and its effectiveness diminishes dramatically. As a result, there is very little practical insulation used "under foot" in shoe or boot design. The use of aerogel insulation in footwear represents several possibilities for successful penetration into the consumer markets:

1. The amount of insulation per shoe is small.
2. The thermal performance can be improved within a limited thickness.
3. The material (Aspen Aerogel Spaceloft 2250) can withstand the pressures experienced underfoot without losing insulating value.

Other consumer products that can achieve these types of advantages with aerogel insulation are likely to represent the "Low Hanging Fruit" within the consumer insulation market.

37.4.1. Wet and Under Pressure Test Results

Table 37.1 represents test data that compares the thermal performance of Aspen Aerogel spaceloft material with typical materials used in the footwear industry. This data demonstrates performance opportunities, which product designers and producers should clearly be able to capitalize on to produce high performance products in the footwear and outerwear markets.

Table 37.1. Test data comparing the thermal performance of Aspen Aerogel Spaceloft material with typical materials used in the footwear industry

Samples	Thermal resistance $-R$ (m^2 K/W) (Δ%) comment	Clo (Δ%) comment	Thermal conductivity $-K$ (W/m K) (Δ%) comment	Dry weight (g)	Gain in weight (g)	% Gain in weight
1. Thinsulate	0.040	0.259	0.018	21.13	5.77	27.3
2. Aerogel	0.189 (372.5) superior	1.217 (369.9) superior	0.012 (−33.1) superior	42.68	0.77	1.8
3. Polyurethane	0.031 (−22.5) inferior	0.201 (−22.4) inferior	0.020 (11.1) inferior	34.2	17.93	52.4

37.4.2. Aerogel Products in Wal-Mart?

PolarWrap introduced a line of insulated insoles into the consumer market in 2005. The insoles, called Toasty Feet, were originally introduced through a small number of consumer catalogs. The success with sales through this channel led to the pursuit of broader retail outlets, and the insoles were offered to Wal-Mart to test the general consumer market. Wal-Mart agreed to test the product in 500 stores. The Wal-Mart test run was successful and this led to a wider distribution of Toasty Feet insoles throughout geographically appropriate U.S. Wal-Mart stores. The Wal-Mart customer profile is that of a price conscious shopper, not generally considered as an "Innovator." Why then would this line of distribution prove to be so successful? The answer may be due to the nature of the problem (cold feet), which the product solves, more than the answer as to how the problem is solved (with aerogel). If this analysis is correct, then other products may also be able to leap into the mass markets if they indeed provide unique solutions to common problems.

Cost will constantly be an obstacle to the success of aerogel use in consumer goods, particularly where the product is simply exchanging aerogel for traditional and widely accepted insulation materials. In the case of Toasty Feet, this product generally represents a new category of product – an insulated insole that provides *warmth*. There are no popular insoles in the category of warmth. Since this category has resonated with the consumer, the acceptance in the market has accelerated without the obstacle of challenging traditional products, which would, most likely, be less expensive.

The Elite Nanogelite water bottle (Figure 37.3B), which is marketed to the bicycle accessories market, is not unlike any ordinary water bottle. The bottle's Nanogel insulation has thermal properties that provide up to 4 h of insulation. This bottle is squeezable like a typical water bottle and yet provides advanced insulation not found in similar bottles. The material used in these bottles is Nanogel, produced by Cabot Corporation (Chap. 38).

Toasty Feet insoles (Figure 37.3A) are removable shoe insoles that are easily placed into ordinary shoes and boots. The insoles are made with Aspen Aerogel's Spaceloft insulation material. These insoles have been successfully used on summits of Mt. Everest and endurance runs through Death Valley. The use of aerogel insulation underfoot provides unparalleled thermal protection in both cold and hot conditions. Due to the limited amount of available space, which a shoe insole can practically use, the implementation of aerogel insulation in this area demonstrates a very tangible benefit to the user. The Toasty Feet insoles have been tested under pressures equal to those experienced under foot, and when compared to traditional materials, such as urethanes and polyesters, these insoles offer a 500% performance improvement.

Figure 37.3. Examples of aerogel consumer products currently in the market: **A.** Toasty Feet insoles, **B.** Elite Nanogelite bottle, **C.** Dunlap Aerogel 200 tennis racket.

The Dunlap Aerogel 200 tennis racket (Figure 37.3C) takes advantage of another unique property of aerogel materials. The racket fills the tubing of the racket frame with Cabot's Nanogel material, thus providing uniform stiffness to the frame when under load while striking a tennis ball.

37.4.3. Consumer Awareness of Aerogel

The success of any consumer product that uses aerogel materials is significantly related to the awareness of the benefits or performance of these materials. Since aerogel materials offer a wide range of functional benefits, this message is not so simple. One product may be utilizing energy absorption, another could be promoting thermal insulation, and a third could be touting acoustical dampening. As a result, the consumer may become confused as to how this material performs. Since aerogels represent a family of products, perhaps the simple use of the name does not represent enough clarity to the consumer. Unfortunately, the use of a wide variety of product names such as Aspen Aerogel's Spaceloft or Cabot's Nanogel will not allow for easy leverage of product awareness. As a result, each manufacturer of consumer goods using aerogel bears the responsibility for making the consumer aware of the virtues of their own product. Unfortunately, this will result in "recreating the wheel" at a significant expense. There are additional perils in this, not the least of which is the potential for misinformation on aerogel being promoted to the public, thereby damaging the public's opinion of aerogel.

The image in (Figure 37.4) is visually attractive, but does not generally convey a message that can easily be converted into marketing value. This is but one challenge that consumer product manufacturers have in marketing the virtues of aerogel materials.

37.4.4. Fashion Versus Performance

Since many opportunities lie ahead with aerogel being used in consumer apparel, the attitude and orientation of product developers and designers must be considered. Currently, there are no foreseeable applications of aerogel materials being used to directly enhance the

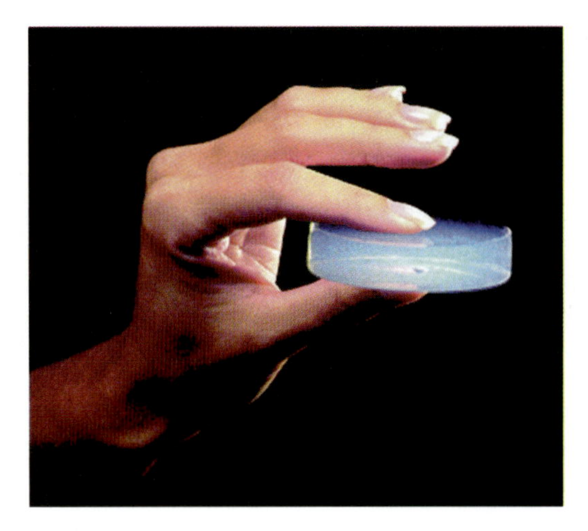

Figure 37.4. Example of Aerogel Information. Source: Aspen Aerogel.

fashionability of products. There is an indirect fashion implication in that the material can potentially offer comparable thermal protection in a thinner garment/product. This may have value to the fashion conscience, but the primary virtue of the material is in the performance it offers.

The consumer outerwear market is heavily influenced by fashion (i.e., the latest colors, finishes, styles, etc.). Even market segments that are allegedly performance oriented, such as skiwear, remain slaves to fashion. There is little reason to believe that this trend will change. As a result, the adoption of aerogel materials into these markets depends on several important factors.

1. The use of the material must not diminish the fashionability of the end product.
2. The message of the enhanced thermal performance of the end product must be communicated clearly and dramatically to the consumer.
3. There must be a strong and valid justification for any cost differential between products made with traditional insulation versus that product made with aerogel insulation.
4. Innovator consumers must support the aerogel products.

37.4.5. The Cost Factor

Consumer goods are generally produced and sold in competitive markets. For this reason, a simple cost factor of four times of the material cost can be used to project the impact a material cost will have on the retail price of a product. For example, if a jacket manufacturer currently uses a traditional material to insulate a jacket, and that jacket uses 10 sq. ft. of insulation, which costs $0.20 per sq. ft., then that insulation cost for the jacket is $2.00. Multiply this by four to obtain the insulation impact on the retail cost of the jacket – $8.00. Manufacturing the same jacket using aerogel may cost $1.50 per sq. ft. with a total insulation cost of $15.00; multiplying by four provides an increased retail value of $60.00. Compare the retail cost of $60.00 versus $8.00 and it indicates the challenges that aerogel insulation has in the consumer market.

In this case, the challenge is to convince the consumer to pay a premium of $52.00 to have the same jacket made with aerogel versus traditional insulation. In today's extremely cost conscience retail environment, this represents a significant challenge. The extra $52.00 for that jacket may be a real bargain in terms of performance, but the challenge is convincing the consuming public of this fact.

Products that require small amounts of insulation material have an advantage in the early adoption of aerogel materials. As an example, Toasty Feet insoles have an approximate yield of 0.6 sq. ft. of aerogel. The cost difference between 0.6 sq. ft. of aerogel and a traditional material this size is approximately $.72. Using the 4× metrics, this results in an additional cost of $2.88. That the consumer can be convinced that the insoles, which are 400% more effective, are worth $2.88 more than if they were made with traditional materials is evident in the success of the product. This product illustrates the great potential in small products, at least while the overall consumer awareness of aerogel increases.

37.5. Summary of Commercialization of Aerogel in Consumer Markets

Aerogel materials can be employed in consumer products to take advantage of the materials' many high performance features. The consumer markets currently contain numerous products that need and utilize a variety of insulating materials. Due to the advanced thermal insulating performance of aerogel, it seems likely that this material will be integrated into many consumer products that now use various "low performance" insulation materials. One of the vast opportunities in the consumer market is within the apparel and footwear industry. These are multibillion dollar a year markets.

One of the challenges to entering these markets includes creating an awareness of the materials' advanced performance to product developers. The product developers must then solve any technical problems related to the use of aerogel in their production process. Next, the product company must develop a marketing strategy to convey to the consumer the value of this new aerogel material within the finished product. The final challenge to the product developer is to produce an aerogel-based product that can successfully compete in terms of price. At this point in time, aerogel materials are considerably more expensive than materials that they may replace. The performance of the aerogel material is much higher than traditional insulating materials, and the question remains how much the consumer is willing to pay for this advanced performance.

Currently, there are a small number of aerogel products that are sold to consumers. Toasty Feet insoles have shown initial success of aerogel-based products in the mass consumer retail markets. The way is being paved for other aerogel products to enter the consumer market, and with each successful new product entrant, the barriers to the consumer market lessen. As consumer awareness of aerogel's potential performance grows, the cost challenges, which currently exist, will be lowered. For some time, this combination of factors will prove a valuable marketing advantage, which will be capitalized by savvy brands.

Reference

1. Rogers, Everett M. (1995). *Diffusion of Innovations*. 5th ed. New York: Free Press

38

Aerogel by Cabot Corporation: Versatile Properties for Many Applications

Hilary Thorne-Banda and Tom Miller

Abstract Cabot Corporation manufactures a hydrophobic aerogel in granular form sold worldwide. This contribution outlines the history of the company, the basic Cabot process, and several but not exhaustive applications in the field of architectural daylighting, building insulation, oil and gas pipelines, LNG and cryogenic applications, outdoor gear and apparel, coatings and specialty chemicals, and personal care. Four different products are described and their properties are discussed.

38.1. Introduction

Sometimes called "frozen smoke," aerogel is the lightest and best insulating solid in the world. Cabot's aerogel is a *hydrophobic* aerogel produced in a granular form. Each granule consists largely of air (>90%) contained in a nanostructure with pore sizes less than the mean free path of air molecules, which severely inhibits heat transfer through the material. Cabot aerogel granules can be contained in various ways to facilitate incorporation into a wide range of systems including pipe-in-pipe systems, LNG and other cryogenic gas transportation and storage systems, insulative coatings, daylighting panels, sporting equipment, clothing, and others. Cabot's production process is environmentally friendly and Cabot produces aerogel in its state-of-the-art manufacturing facility located near Frankfurt, Germany, where it began commercial production in 2003.

38.2. Cabot Aerogel

Cabot Aerogel is a business of Cabot Corporation, a global performance materials company headquartered in Boston, MA (USA). Cabot's major products in addition to aerogel include carbon black, fumed silica, inkjet colorants, capacitor materials, and cesium formate drilling fluids. The company owns over 50 patent families related to aerogel manufacturing and aerogel in foams, blankets, composites, and coatings.

H. Thorne-Banda and T. Miller • Cabot Corporation, 157 Concord Road, Billerica, MA 01821, USA
e-mail: hilary_banda@cabot-corp.com; tom_miller@cabot-corp.com

M.A. Aegerter et al. (eds.), *Aerogels Handbook*, Advances in Sol-Gel Derived Materials and Technologies, DOI 10.1007/978-1-4419-7589-8_38,

Cabot is as committed to protecting the environment as it is to innovation. These ideals come together in the processing and use of aerogel as Cabot continually strives to reduce its ecological impact and improve efficiency in the manufacturing process.

More information can be found at www.cabotaerogel.com.

38.3. History

Several companies tried to commercialize aerogel from 1950 to 1990, but ultimately failed due to the high manufacturing costs, limited batch production, and the dangers of using the supercritical drying technique with flammable solvents.

38.3.1. Timeline: Cabot Pioneers Atmospheric Aerogel Production

- 1992: Researchers at the University of New Mexico discovered how to control gel shrinkage via silylation of organogels [1, 2].
- 1993: Cabot, Dow Corning, Hoechst, and Nanopore each began research programs aimed at controlling gel shrinkage through surface modification technology. It was during this time that an alternative drying technique was developed that enabled the production of aerogel at a reasonable cost through a continuous, sustainable, and safe manufacturing process.
- 1996–1998: Cabot, Dow, and Hoechst all filed a series of patents on direct silylation routes to control gel shrinkage. Cabot then purchased and/or licensed the patents from Dow, Nanopore, and Hoechst [3].

Using its four decades of expertise in modifying the surface of fumed silica, Cabot revolutionized the manufacture of aerogel and was the first and only company to develop a commercialized process that allows production of the material under ambient drying conditions. This process allows Cabot to control the material's porosity, pore size, and distribution and bypasses the high-cost traditional method of supercritical drying.

Cabot's process (Figure 38.1) includes a water-based silica solution, which is destablized to create a gel. The water in the formulation is then replaced with an organic solvent. Finally, the solvent is removed via a drying process, leaving behind a treated, porous aerogel, which is >90% air.

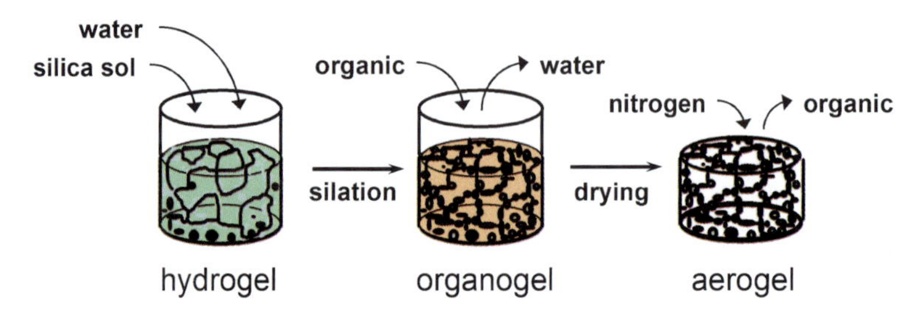

Figure 38.1. Schematic of Cabot's aerogel manufacturing process.

38.4. Applications

Because of its unique properties, Cabot's aerogel is extremely versatile and can be easily adapted into various products and forms for a wide range of applications, including

- Architectural daylighting
- Building insulation
- Oil and gas pipelines
- Industrial and cryogenic applications
- Outdoor gear and apparel
- Specialty chemicals and coatings
- Personal care

38.4.1. Architectural Daylighting

Architects and building owners face challenges in balancing aesthetics and daylighting design needs with increasingly stringent building and energy code requirements. Aerogel-insulated daylighting systems are the perfect solutions for combating these issues by maintaining and enhancing energy efficiency while enabling a wide range of commercial and residential building design choices. Unlike traditional insulation materials, Cabot's aerogel granules can be incorporated into structural polycarbonate or composite systems (Part V), u-channel and insulated glass, skylights, smoke vents, and tension fabric roofing. The inclusion of aerogel in daylighting systems (Chap. 41) virtually eliminates the historical trade-off of insulation versus natural light by providing 3–6 times the thermal performance of traditional, poorly insulated fenestration products while maintaining, even improving, light transmission. Benefits reach beyond thermal and light transmission; aerogel systems also increase sound dampening effects and provide complete hydrophobicity to resist mold and mildew. An example of use is shown in Figure 38.2.

Figure 38.2. Cabot's aerogel in daylighting, Souchais Sports Complex, Carquefou, France.

38.4.2. Building Insulation

Cabot's aerogel is also the solution for building owners and architects when it comes to the building insulation itself (Chap. 26). Using a traditional insulation material often means having to accept increasingly thick layers of insulation in walls, floors, and roofs in order to comply with energy standards. This consumes valuable floor space in new construction. In renovation projects, if it is even possible to increase insulation thickness at all, significant aesthetic and functional compromises are often required to retrofit more insulation on the inside or outside of the building envelope. With a thermal efficiency that is 2–4 times greater than traditional materials such as polystyrene, mineral wool, and cellulose (Figure 38.3), aerogel delivers more insulation performance with less thickness, while resisting the settling, aging, and moisture damage that plagues other materials. Along with the high-efficiency daylighting systems mentioned above, the use of aerogel in walls, floors, and roofs can help a building meet even the most difficult requirements of the Passivhaus Institut/Passive House Institute US and LEED Platinum standards, as well as building and energy codes such as Part L in the UK, Energie Einsparverordnung (EnEV 2009) in Germany, and RT2005 and RT2012 in France.

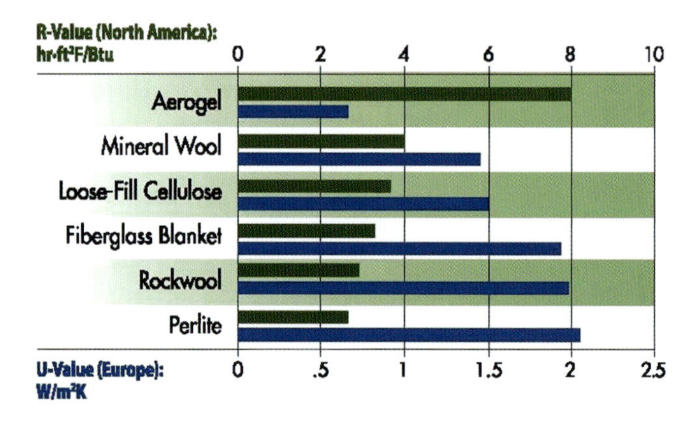

Figure 38.3. Building insulation: aerogel thermal advantage versus traditional building insulation materials.

38.4.3. Oil and Gas Pipelines

Oil and gas operators also face historical compromises between insulation performance, easy installation, and precise, consistent final forms. Rarely are these benefits available in one product for the oil and gas industry. Traditional materials cannot match the thermal performance of aerogel and can be difficult to install. Products for oil and gas pipelines, including the Cabot aerogel Compression Pack (Figure 38.4) and Expansion Pack, are easy to fabricate super-low U-value systems ready for use in frontier regions such as deep/ultra deep waters, oil sands, and arctic environments. These durable products enable longer tiebacks, smaller jacket pipes, longer cooldown, faster fabrication, and broader design options, all resulting in improved project economics.

38.4.4. Industrial and Cryogenic Applications

Industrial and cryogenic applications create challenges comparable to those in the oil and gas industry, and thus, are easily met by Cabot's aerogel products. Thermal performance, durability, and ease of application are top concerns when insulating piping, transport

Figure 38.4. Cabot's aerogel Compression Pack on a pipe-in-pipe system.

and stationary tanks, ships, rail cars, fittings, and trailers aerogel is ideal for these and other load-bearing applications due to its performance consistency, and even improvement, under compression. Aerogel insulation provides much higher thermal efficiency than other materials in use today at the same thickness and/or a significantly reduced insulation profile and allows easy application and stability in the harshest conditions over a wide range of temperatures (Figure 38.5).

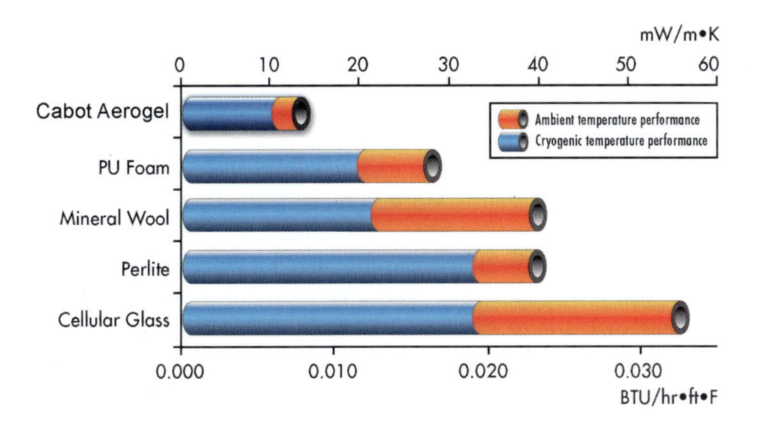

Figure 38.5. Cabot's aerogel performance in industrial applications versus traditional industrial insulation materials.

38.4.5. Outdoor Gear and Apparel

Similar to other applications, getting the best thermal performance with the least thickness is of great significance to outdoor product manufacturers, who can now engineer completely hydrophobic footwear, outerwear, sleeping pads, gloves, hats, and accessories with half the thickness and bulk of common insulating materials. The Cabot aerogel Thermal Wrap, an ultrathin nonwoven fabric that retains insulating properties when wet, is twice as warm as leading insulators and up to 12 times warmer when compressed (Figure 38.6). Because the nanosized pores in aerogel block heat transfer far more efficiently than traditional fibrous insulators, Thermal Wrap insulation does not have to be bulky to be efficient. This allows outdoor product designers to engineer sleeker, ultrathin products, which can increase speed, dexterity, and agility for outdoor athletes (Chap. 37).

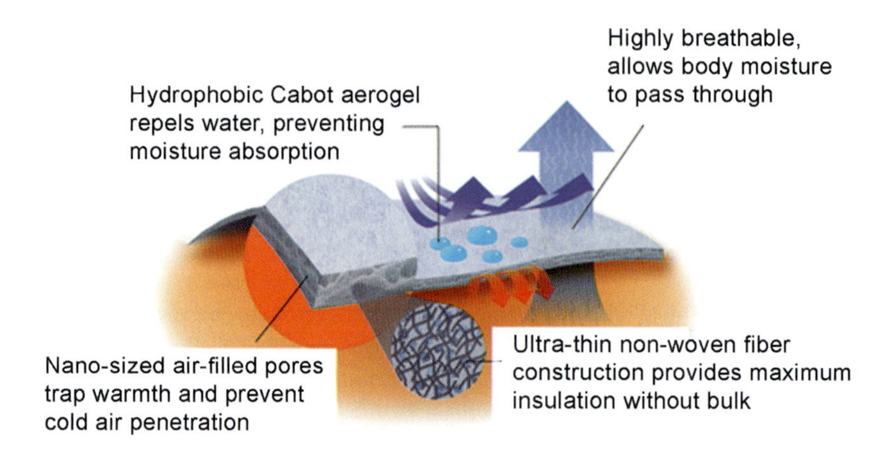

Figure 38.6. Diagram of Thermal Wrap in apparel applications.

38.4.6. Specialty Chemicals and Coatings

For coatings and specialty chemical manufacturers, aerogel is a key ingredient in a wide variety of formulations. Because only a small amount of aerogel is required to deliver superior performance benefits, formulators can enjoy the freedom to include additional additives for other properties. Even at these low loading levels, Cabot's aerogel provides improved clarity, hydrophobicity, matte effects, increased operating temperature flexibility, and extension of the shelf life of the formulation. At higher loading levels in water-based formulations, aerogel enables high-performance thermally insulative coatings. Figure 38.7 shows the performance of aerogel as a rheology modifier.

38.4.7. Personal Care

Personal care formulations require comparable attention to loading levels to achieve desired effects. Cabot's aerogel offers exceptional oil and sebum absorption, soft focus effects, rheology modification, and controlled release. Compared to other additives, it is also a superior thickener and matting agent, providing low gloss at low loading in a variety of personal care applications from lotions and foundations to sunblock, lipstick, and more.

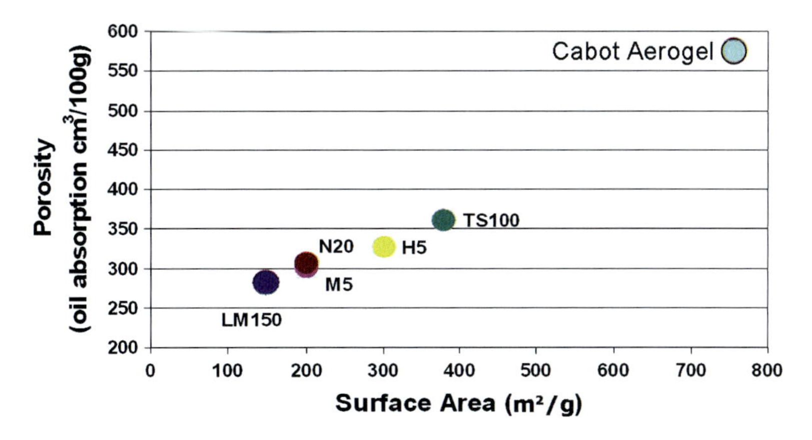

Figure 38.7. Cabot aerogel performance as a rheology modifier.

38.5. Products

Cabot Aerogel offers four unique aerogel insulation products:

- Infill – also known as loose-fill or granules – can be used to insulate annular spaces of almost any size or dimension. The granules flow like water and can be densely packed for long life and high-performance use – with no settling, shifting, or residual air gaps. Cabot aerogel can be handled with the same installation techniques as traditional loose-fill materials for maximum flexibility and efficiency (Figure 38.8).
- Thermal Wrap – an easy to handle aerogel fabric blanket delivered in rolls that provide multiple options for rapid and automated installation in various applications. The wrap is compressible and conformable, without sacrificing tensile strength and resilience (Figure 38.9).
- Compression Pack – for pipeline projects – consists of custom-sized packs of compressed aerogel granules, preattached to a protective outer layer of high-density polyethylene. Packs are designed for ultrahigh performance and flexibility, while variable thicknesses allow precise U-value targeting, delivering optimal thermal performance across the insulation layer (Figure 38.10).
- Expansion Pack – also for pipeline projects – installs easily and quickly without air gaps, creating a mechanical bond that dynamically adapts to stresses with no significant degradation of system performance over time. Packs feature superior thermal performance and robust mechanical strength (Figure 38.11).

38.5.1. Properties

With a focus on fine particle technology, Cabot is able to control optical properties, granule size, distribution, and porosity during the manufacturing of aerogel – the end result is a highly adaptive material (Table 38.1).

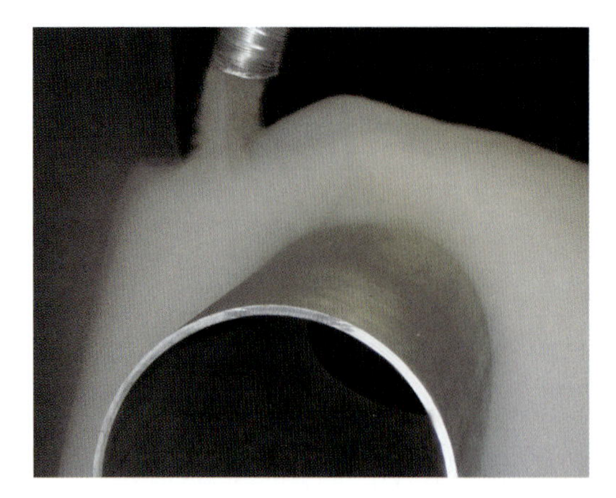

Figure 38.8. Cabot aerogel infill.

Figure 38.9. Cabot aerogel Thermal Wrap.

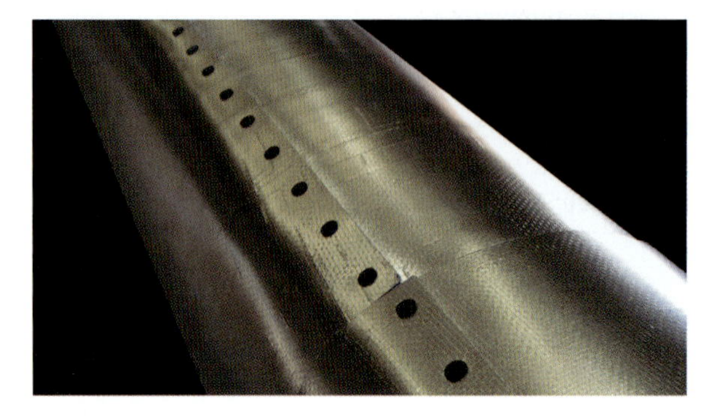

Figure 38.10. Cabot aerogel Compression Pack.

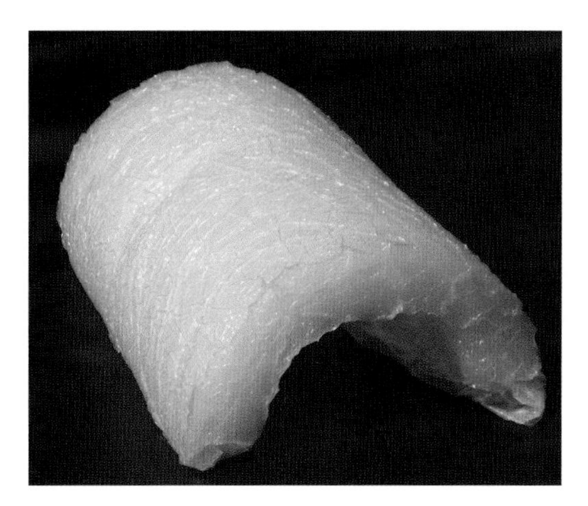

Figure 38.11. Cabot aerogel Expansion Pack.

Table 38.1. Properties

Extremely low thermal conductivity	9–12 mW/mK
High porosity	>90% (v/v) air
Nanosized pores	20–40 nm in diameter
High surface area	~750 m^2/g
Very low tap density	30–100 kg/m^3
High oil absorption capacity (DBP)	540/100 g
Specific heat capacity	0.7–1.15 kJ/Kg
Variety of granule sizes	5–4 mm
Surface chemistry	Completely hydrophobic
Opacity	Translucent, IR opacified, and opaque

[a]Characteristics vary depending on application, temperature, and form.

Properties include

- *High porosity*: Highly porous structure provides high reinforcement and matting efficiency at low concentrations.
- *Flowability*: Excellent flow properties.
- *Hydrophobic surface*: Granules repel water but can be easily wetted by oleophilic substances such as oils.
- *Open porosity*: Pores are open to each other within the structure.
- *Particle size*: Available in a wide range of particle sizes.
- *Surface treatment*: Surface groups are chemically anchored to the silica skeleton and thus are stable over a wide range of temperatures and pH.

38.5.2. Green Material

Cabot aerogel is recyclable, reusable, and safe for human and ecological systems, contains no Volatile Organic Compounds (VOCs), and is created through a closed loop process with little to no impact on the environment. It is manufactured from one of the world's most plentiful raw materials, silica. In 2008, Cabot aerogel was awarded the updated Silver Cradle to Cradle^{SM} certification from McDonough Braungart Design Chemistry.

38.6. Conclusion

Although significant advancements have been made with Cabot aerogel products in the architectural daylighting, building insulation, oil and gas pipelines, LNG and cryogenic, outdoor gear and apparel, coatings and specialty chemical, and personal care markets, the highly adaptive nature of aerogel means limitless possibilities for future forms and additional industries.

References

1. Prakash SS; Brinker CJ and Hurd AJ Silica aerogel films at ambient pressure, Journal of Non-Crystalline Solids; Oct. 1995; 190, 264–75
2. Smith DM, Deshpande R and Brinker CJ, in: Better Ceramics Through Chemistry V, ed. M. J. Hampden-Smith, W. G. Klemperer and C. J. Brinker, Mater Res Soc,Symp Proc., Vol. 271 (Materials Research Society, Pittsburgh, PA, 1992) p. 567
3. See for example, US patent Nos. 7297718, 6378229, 6475561 and 6156386

39

American Aerogel Corporation: Organic Aerogel Commercialization

Robert Mendenhall

Abstract American Aerogel produces an organic monolithic aerogel panel called *Aerocore*. Its properties have been optimized for the mass production of vacuum insulation panels (*VIP*s). This contribution summarizes the history of the company and describes the Company's materials.

39.1. Introduction

American Aerogel currently produces *Aerocore*, an organic monolithic panel, which is manufactured in sizes up to 21 × 26 in. × (0.25–1.25) in. in thickness. The properties of *Aerocore* have been optimized for the mass production of vacuum insulation panels (*VIP*s). A *VIP* is a porous solid inside an impermeable gas barrier material, wherein a low gas pressure is maintained to decrease gaseous thermal conductivity. It was important to balance the cost versus performance of *Aerocore* in order to compete with commodity materials such as polystyrene and polyurethane. Although the focus of the Company is now the commercialization of *VIP*s, *Aerocore* technology may be exploited for other applications similar to mainstream aerogel materials (thermal insulation, acoustic attenuation, charge storage, chemical catalysis/filtration, mechanical energy absorption, etc.).

39.2. History

In the 1990s, research teams at Lawrence Livermore National Labs (*LLNL*) created a route for the rapid production of inorganic aerogels (Rapid Supercritical Extraction) [1] as well as the first organic aerogel materials [2]. Combining these two ideas, Greg Andrews and Robert Mendenhall founded the Company in 1997. Shortly thereafter, they began collaborating with Donald Albert and Joseph Bruno (Wesleyan University) to create organic aerogels using a rapid supercritical method [3].

It became apparent that the rapid techniques and materials from the above developments were still not cost effective for mass production. New chemistries and techniques were then

R. Mendenhall • American Aerogel Corporation, 460 Buffalo Road, Rochester, NY 14611, USA
e-mail: rmendenhall@americanaerogel.com

M.A. Aegerter et al. (eds.), *Aerogels Handbook*, Advances in Sol-Gel Derived Materials and Technologies, DOI 10.1007/978-1-4419-7589-8_39,

Table 39.1. Selected *Aerocore* physical properties

Thermal conductivity	
Radiative	1 mW/mK
Solid	1 mW/mK
Gaseous	From <1 to 30 mW/mK
Specific heat	~1 J/g/°C
Young's modulus	>70 Mpa
Pyrolyzed electrical resistivity	<0.11 (ohm-cm)
Density	0.10–1 g cm^{-3} (~0.18 for VIPs)

pursued, unrestricted by the boundaries of traditional processes. The primary driving philosophy was the commercial needs of customers. The resultant *Aerocore* material can be manufactured with properties spanning those of traditional aerogels as well as other unique properties. For example, *Aerocore* can be made into large sizes and various complex shapes. One large monolith has been made that is about 4 ft^2 by 1.5 in. thick and another was about 1 ft^2 by 6 in. thick. Other samples have been made "in situ", that is, within the walls of a plastic mold and dried within these walls, wherein the syneresis has been minimal. Another property is its high mechanical strength, which has allowed for its commercial use as a core material in VIPs at densities of well below 0.2 g cm^{-3}. *Aerocore* is also naturally black in color, combined with high carbon content, allowing for a very low (ca. 1 mW/mK) radiative thermal conductivity (Table 39.1).

In 2003, after several years of research on the material properties and processes to produce *Aerocore*, the Company began small-scale manufacturing in Rochester, New York. Select customers were using *Aerocore VIP*s for the temperature-sensitive shipping of biopharmaceuticals. The Company continued to refine and expand its production capabilities, and a pilot plant was completed in 2009 with anticipated volumes of over 20,000 board feet per month.

American Aerogel is currently based in Rochester, New York, with more than 30,000 sq.ft. of manufacturing space. The Company maintains and continues to pursue an extensive set of patents. Additional market segments are also being investigated in addition to *VIP*s.

39.3. *Aerocore* Description: Small Pore Area Material

Aerocore is an organic open cellular foam based upon the sol–gel polymerization and crosslinking of a hydroxylated aromatic with an electrophilic linking agent carried out within a hydrogen bonding carboxylic acid solvent. It can also be carbonized after pyrolysis, wherein it maintains its net shape. The resultant material can be made into monoliths of several feet, or greater, in all dimensions, without cracking and only minimal shrinkage.

There is disagreement within the aerogel community about the definitions of aerogel materials. Some definitions include porosity restrictions such as mesoporosity requirements (2–20 nm), others require supercritical drying techniques, and some include density restrictions. Even measuring these properties can lead to uncertainties and controversy.

Acknowledging these difficulties, researchers at *LLNL* proposed a nomenclature for open cellular foam structures called Low-Density Microcellular Materials (*LDMM*s) [4] and described them as combining the properties of low density and small pore structures. These materials break the correlation between density and pore size, being able to change one property without a corresponding change in the other. Consequently, a subset of *LDMM*s is defined as aerogels. Readers are encouraged to study the above reference by *LLNL* for more detailed information about LDMMs. For traditional foam materials such as polystyrene and polyurethane, density and porosity are closely but inversely linked. Without regard to the definition differences, processing commonalities still exist, such as the need for sol–gel processing and then drying to form an open, predominantly nanoporous cellular structure.

Since there is no acknowledged institution such as *IEEE* or *NIST* to resolve these differences and motivated by the need to protect the intellectual property of the Company, we decided to create our own set of definitions outlined in our patents. We have defined our material as a "small pore area material" (*SPAM*) based in part upon the measurement of the average areas of pores within aerogel materials via microscopy using the image analysis program ImageJ (freely available from *NIST*). A SPAM is a foam with densities of less than 1 g cm^{-3} and an average pore area of less than 500 μm^2. The average pore area is the average of the pore areas of at least the 20 largest pores identified by visual examination of images generated by the *SEM* images and measured using the ImageJ program. We then continued to define an *LDMM* as a type of *SPAM* with average pore areas of less than 0.8 μm^2 and an Aerogel as a type of *LDMM* and thus as a subcategory of *SPAM* [5].

We arrived at this definition after discovering that other techniques, such as calculating pore size distributions from the *BJH* method via nitrogen isotherms, do not account for a large portion of the total pore volumes within these types of materials. The amorphous structures seen in *SEM* images indicate a porosity that is difficult to describe mathematically, whereas the field of microscopy has given us tools that are readily available, which help us to quantify these structures.

Aerocore has pore sizes, correlated with *SAXS* data, which can range from mesoporous to microporous and densities from roughly 0.050 to 1 g cm^{-3}. The morphology is typically of the "pearl necklace" variety with larger beads and small necks. *SEM* images show it to be a fractal structure. It can be made in a monolith, powder, and thin film.

The following *SEM* images illustrate some of the variety in *Aerocore* morphology. Figure 39.1 shows two different length scale images for *Aerocore* that is optimized for use in VIPs. Figures 39.2 and 39.3 demonstrate the range of pore size distributions for *Aerocore*.

While not an exhaustive list, measurements have also been made for the properties listed in Table 39.1 above.

Figure 39.4 also shows a comparison of the values for thermal insulation of *Aerocore* versus commonly used insulation materials proving the superior property of this material.

Good *VIP* insulation has *low density* (to reduce solid thermal conductivity), *small pore size* (to reduce gas thermal conductivity), *high opacity* (to reduce radiation thermal conductivity), and must be very *rigid* (to keep its shape while being evacuated). Because *Aerocore* is an organic *LDMM*, it has all four qualities – making it the only material that combines all the necessary components for the perfect *VIP* insulation.

Figure 39.5 presents a typical *Aerocore* cutaway showing the core material within a gas barrier film, and Figure 39.6 shows a typical *VIP* box featuring *Aerocore* optimized for the needs of safely and inexpensively shipping temperature-sensitive biomedical and pharmaceutical products. The use of such systems can cut the shipping costs by as much as 50%.

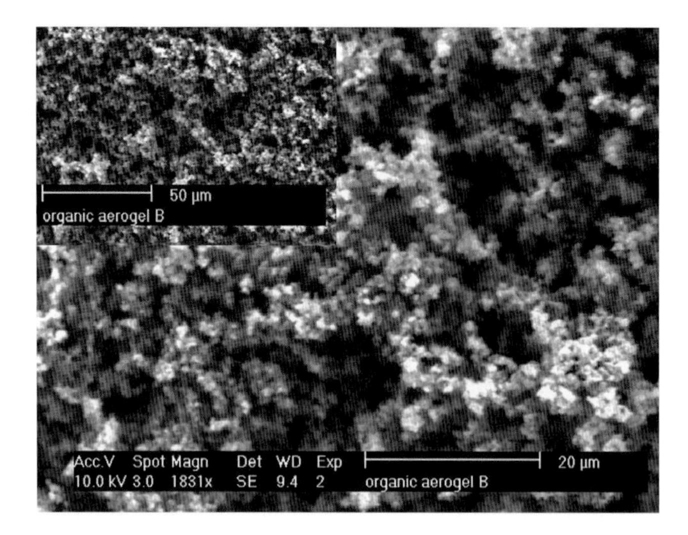

Figure 39.1. *Aerocore* optimized for vacuum insulation panel use.

Figure 39.2. *Aerocore* with decreased pore size structure.

39.4. Observations on Commercialization

There are a small number of people involved in aerogel research, compared to other mature fields of study. Considering this, with the vastly different disciplines and motivations of researchers, it is not surprising to see why aerogels are not exploited commercially on a large scale. A perusal of the literature shows aerogels as a class of materials having a long history of abandoned commercial ventures with little success, while aerogel-related advances in academic and government laboratories have been generally unexploited.

Figure 39.3. *Aerocore* with increased pore size structure.

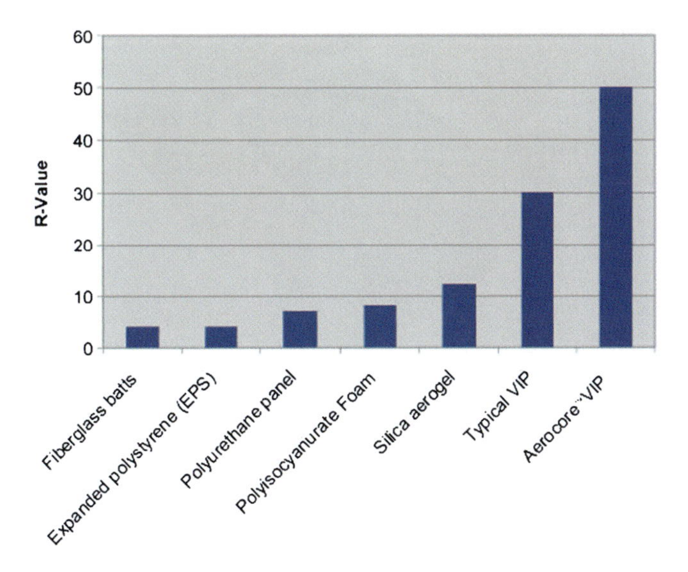

Figure 39.4. Comparison of thermal insulation values *Aerocore* versus common insulation materials (R-value in units of ft.$^{2\circ}$ F h Btu^{-1}).

As opposed to other material industries such as plastics and metals, there are no large-scale institutions devoted to aerogels. Degrees in aerogel engineering do not exist. Historically, researchers ranging from biologists to chemists and physicists tend to discover aerogel materials only as it relates to their individual tasks or projects. A biologist may need a material to study bacteria, a chemist may need a catalyst support, and a physicist may need a material for a laser target. Whatever the reason, these researchers come to aerogels from their own disciplines and not generally with the goal of advancing or exploiting aerogels themselves.

Figure 39.5. *Aerocore* VIP cut-away showing core material within gas barrier film.

Figure 39.6. *Aerocore* VIP Box.

This has led to a fragmented base of knowledge within the broader aerogel community without a large effort or critical mass of people to move aerogel materials toward commercialization.

39.5. Conclusion

By focusing on customer-driven development, American Aerogel has successfully commercialized a form of aerogel materials. *Aerocore*, a black organic monolithic open cellular rigid foam, can be produced with a wide range of properties and sizes. It can be used as a thermal insulator, acoustic attenuator, chemical support media, electrical conductor, and mechanical shock absorber.

References

1. JF Poco, PR Coronado, RW Pekala, LW Hrubesh, A Rapid Supercritical Extraction Process for the Production of Silica Aerogels MRS Apr 8–12 1996 Spring Meeting
2. RW Pekala, CT Alviso, X Lu, J Gross, J Fricke, New organic aerogels based upon a phenolic-furfural reaction J Non-Cryst Solids 188 (1995) 34–40
3. US Pat Nos. 6,090,861 and 6,077,876
4. JD LeMay, RW Hopper, LW Hrubesh, RW Pekala Low-Density Microcellular Materials UCRL-JC-104935 Preprint for MRS Bulletin Dec 1990
5. US Pat No. 7,005,181

40

Aerogels Super-thermal Insulation Materials by Nano Hi-tech

Chengli Jin

Abstract Nano High-tech Co. Ltd is the first, largest and most robust research and production base of silica aerogel nanoporous super-thermal insulation materials in China. The company has developed several series of products, such as super-thermal insulation flexible felts, thermal insulation panels, thermal insulation cylinders and special-shaped parts, translucent thermal insulation panels, silica aerogel powders, particles and monoliths. These products can find wide-spread application in aerospace, aeronautics, metallurgy, chemical plant installations, oil pipes, refrigerators, cars, ships, buildings, etc. Besides silica aerogels, Nano High-tech also developes organic, carbon, alumina and titania aerogels.

40.1. About Nano High-Tech

Nano High-tech Co. Ltd, which is fully invested by Zhejiang Zhonglian Construction Group is located in the historic and cultural city of Shaoxing Zhejiang, P. R. China (Figure 40.1). The Zhejiang Zhonglian Construction Group set up in 1957 is one of the National Construction Special Grade Enterprises, and among the top 500 of Chinese Construction Enterprises and Chinese Enormous Private Enterprises.

Nano High-tech Co. Ltd was set up in April 2004 with 50 million RMB registered capital and an area of 200,000 m^2, which is the earliest, largest and strongest research and production facility of silica aerogel nanoporous super-thermal insulation materials (Chaps. 2 and 26). The company has passed ISO9001:2000 International Quality System and Zhejiang High-tech Product certification (Figure 40.2).

Nano High-tech is becoming the new important economic growth point for the Zhejiang Zhonglian Construction Group, it will be the flagship of domestic aerogel production in China and will occupy an important position in R&D, production and sales of aerogels worldwide.

C. Jin • R&D Center, Nano Hi-tech Co.Ltd, Shanxi Road 488, Shaoxing, Zhejiang, China
e-mail: jin_cl@163.com

M.A. Aegerter et al. (eds.), *Aerogels Handbook*, Advances in Sol-Gel Derived Materials and Technologies, DOI 10.1007/978-1-4419-7589-8_40,

Figure 40.1. Office building of Nano High-tech.

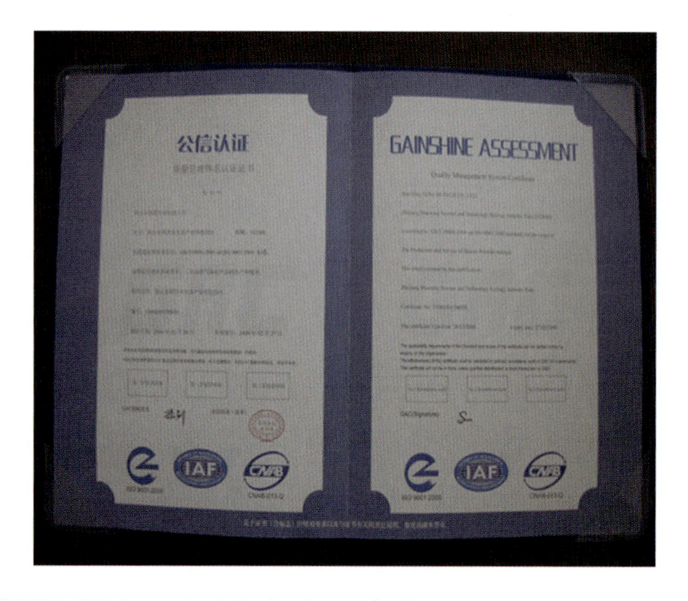

Figure 40.2. ISO9001: 2000 International Quality System Certificate.

40.1.1. Chronology of Nano High-Tech

2004: Cooperation with Tsinghua University and becomes the first domestic enterprise working on the industrialization of silica aerogels.

2006: Set up of the first domestic production line of silica aerogel powders. It passed the authentication of Zhejiang Science and Technology Bureau. A committee made up of academicians evaluated it and announced that the technology of Nano High-tech is at international advanced level and possesses independent intellectual property.

2006: Set up of the first Chinese enterprise research center of silica aerogels.

2007: Cooperation with Tongji University and set up of Tongji • Nano High-tech Combined Institute.

2007: Command the key technology of producing silica aerogel composite materials, the property of which is at international advanced level.

2008: The Tongji • Nano High-tech Combined Institute is upgraded to be a provincial research center. Meanwhile, Nano High-tech exploits a series of silica aerogel products.

2009: Preparation to construct a production line to realize 150,000 m^2 silica aerogel products per year.

40.2. Main Products

The company has developed several series of products, such as super-thermal insulation flexible felts, thermal insulation panels, thermal insulation cylinders and special-shaped parts, translucent thermal insulation panels, silica aerogel powders and particles. These products can be widely applied in aerospace and aeronautics, metallurgy, chemical plant installations, oil pipes, refrigerators, cars, ships, buildings, etc.

40.2.1. Flexible Thermal Insulation Felt (*FM*)

This product is flexible, waterproof, fireproof, 3–5 times the thermal preservation property of traditional material as well as excellent purification function, and can be widely applied in chemical plant installations, air-condition pipes, oil pipes, aerospace and aeronautics, roofs and so on.

Main characteristic properties:

1. Low thermal conductivity, good insulation property (Chap. 23)
2. Flexible, un-infiltrated, against cracking, waterproof (Chap. 4) (Figures 40.3 and 40.4)
3. Environment friendly, no toxicity, noncorrosive
4. Good properties of sound insulation and shock absorption (Chap. 22)
5. Save material dosage, increase application volume
6. More convenient for constructing

Key material properties are summarized in Table 40.1.

40.2.2. Thermal Insulation Panel (*IP*)

This product has higher compressive strength, high utilization rate of space (1/3–1/5 of the thickness of traditional thermal insulation material), excellent sound insulation characteristics in addition to waterproof and fireproof, and can be widely used in machinery, sea vessels (ships), high-speed trains, refrigeration equipment, etc. (see Figures 40.5 and 40.6).

Main characteristic properties:

1. Low thermal conductivity, high strength
2. Wide service temperature and application field
3. Un-infiltrated, against cracking, waterproof

Figure 40.3. Excellent hydrophobic performance.

Figure 40.4. Flexible thermal insulation felt.

Table 40.1. Main characteristics of *FM*

Type / Property	FM200	FM380	FM650
Service temperature (°C)	80–200	180–380	80–650
Thickness (mm)	3, 6, 10	3, 6, 10	3, 6, 10
Thermal conductivity ($Wm^{-1}K^{-1}$)	0.016	0.015	0.020
Density (kg m^{-3})	≤180	≤180	≤350
Hydrophobicity	Hydrophobic	Hydrophobic	Hydrophobic

Figure 40.5. Thermal insulation panel.

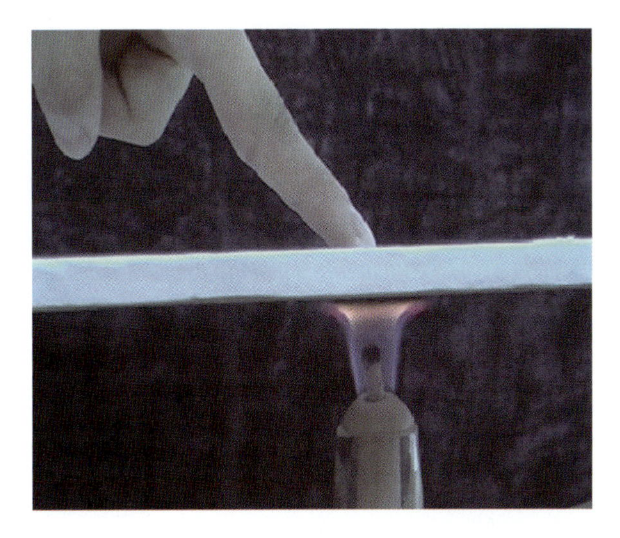

Figure 40.6. Excellent fire resistance.

4. Environmentally friendly, nontoxic, corrosion-resistant
5. Good properties of compressive resistant, shock-resistant and sound insulation (Chap. 22)

Key material properties are summarized in Table 40.2.

40.2.3. Cylinders and Special-Shaped Parts for Thermal Insulation (*CS*)

This product is mainly used in the thermal insulation of irregular pipes and equipments and has a high utilization rate of space, is waterproof, fireproof and shockproof and can be widely used in fuel batteries, engines, metallurgy, power systems, etc. (Figures 40.7 and 40.8).

Table 40.2. Main physical characteristics of *IP*

Type / Property	IP200	IP380	IP650
Service temperature (°C)	80–200	180–380	80–650
Thickness (mm)	3, 6, 10, 15, 20	3, 6, 10, 15, 20	3, 6, 10, 15, 20
Thermal conductivity ($Wm^{-1}K^{-1}$)	0.016	0.015	0.020
Density ($kg\ m^{-3}$)	$\leq$180	$\leq$180	$\leq$350
Hydrophobicity	Hydrophobic	Hydrophobic	Hydrophobic

Figure 40.7. Thermal insulation cylinders.

Figure 40.8. Thermal insulation special-shaped parts.

Main characteristic properties:

1. Low thermal conductivity, high strength, good hydrophobicity
2. Wide service temperature and application field
3. Environment friendly, no toxicity, noncorrosive
4. Good properties of compressive-resistant, shock-resistant and sound insulation

Key material properties are summarized in Table 40.3.

Table 40.3. Main physical characteristic of *CS*

Type / Property	CS200	CS380	CS650
Service temperature (°C)	80–200	180–380	80–650
Thickness (mm)	3, 6, 10, or tailored	3, 6, 10, or tailored	3, 6, 10, or tailored
Thermal conductivity ($Wm^{-1}K^{-1}$)	0.016	0.015	0.020
Density (kg m^{-3})	$\leq$180	$\leq$180	$\leq$350
Hydrophobicity	Hydrophobic	Hydrophobic	Hydrophobic

40.2.4. Daylighting Panels (*TP*)

This product has daylighting and thermal insulation functions, and gets the maximum use of sunlight while keeping the room at suitable temperature. It can be widely used in large theaters, exhibition centers, airport terminals, special experiment centers, advanced restaurants, villas, solar energy collectors, etc. (Figures 40.9 and 40.10).

Figure 40.9. Daylighting panels.

Figure 40.10. Daylighting panels installed in windows.

Main characteristic properties:

1. High transmittance, can effectively use solar energy, saving electricity for illumination
2. Low thermal conductivity, good insulation ability and hydrophobicity
3. Environment friendly, nontoxic, corrosion-resistant
4. Good compressive strength and shock resistance
5. Safe, light, convenient for constructing, beauty

Key material properties are summarized in Table 40.4.

Table 40.4. Main physical characteristics of *TP*

Type / Property	TP40	TP55	TP70
Transmittance (%)	≥40	≥55	≥70
Thickness (mm)	30	20	10
Thermal conductivity ($Wm^{-1}K^{-1}$)	0.025	0.025	0.025
Service temperature (°C)	60–350	60–350	60–350
Hydrophobicity	Hydrophobic	Hydrophobic	Hydrophobic

40.2.5. Aerogel Powders, Particles (*AP*) and Monoliths

This series of products possess low density, high *specific surface area*, high *porosity* ratio, light blue translucent particles or monoliths. They have low thermal conduction constant, good insulation property, strong adsorptive ability, are environment friendly, nontoxic, fireproof and noncorrosive, with no toxicity to human, and can be widely applied as functional structural interlayer, filling layer and composite layer for thermal insulation and preservation, air purification and water treatment (Figures 40.11 and 40.12).

Figure 40.11. Aerogel particles.

Figure 40.12. Aerogel disk.

Key material properties are summarized in Table 40.5. The microstructure of the product is shown in Figure 40.13.

Table 40.5. Main physical characteristics of *AP*

Type Property	QK	SK
Particle size (mm)	0.5–5	0.5–5
Specific surface area ($m^2\ g^{-1}$)	600	600
Density ($kg\ m^{-3}$)	40–80	40–80
Porosity ratio (%)	90–98	90–98
Pore diameter (nm)	25–45	25–45
Core capacity ($cm^3\ g^{-1}$)	3.0–6.0	3.0–6.0
Hydrophobicity	Hydrophilic	Hydrophobic

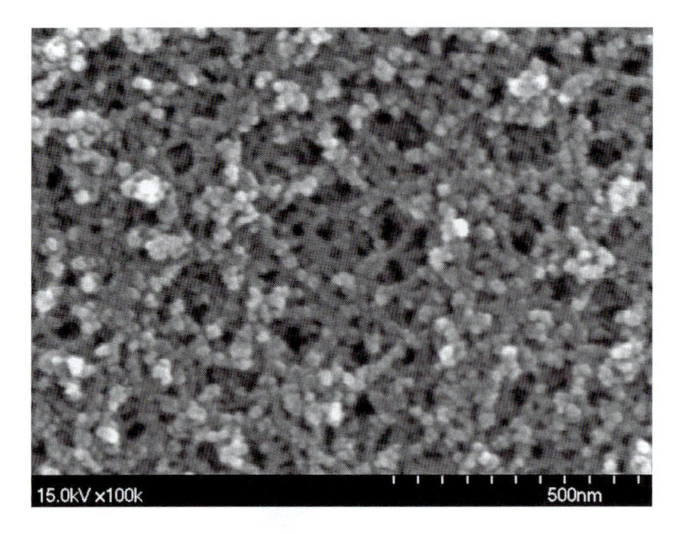

Figure 40.13. SEM image of the product.

40.3. Fields of Application and Customers

With good thermal insulation property, the products can be widely applied as thermal insulation materials in transportation, cryostat, industrial pipe, instrument and equipment, construction, etc.

Transportation: Thermal insulation of large oil tankers and cars.
Cryostat: Thermal preservation of special ultralow temperature refrigerators and refrigeration equipment. For example, it can be used as the insulation layers of the liquid natural gas (*LNG*) storage tanks (Figure 40.14).

Figure 40.14. Liquid natural gas (*LNG*) storage tanks.

Industrial pipe: Thermal insulation of oil and steam pipes (Figure 40.15).
Instrument and equipment: Thermal insulation of precision instruments and metallurgical furnaces (Chap. 34) (Figure 40.16).
Construction: Thermal insulation of roofs, outer walls, floors and windows (Figure 40.17).

So far, the oil pipes, the refrigerators and the construction are the most important markets. Its applications in industrial pipe and transportation will increase rapidly.

Figure 40.15. Oil pipes.

Figure 40.16. Metallurgical furnaces.

Figure 40.17. Building roofs.

40.4. R&D and Future Applications

Nano High-tech spends much time and money on the research and application of silica aerogels. It possesses a building area of 1,600 m^2, a creative team made up of several PhDs and MSs who are led by academicians and professors, and advanced equipments such as imported high precision conductivity tester. The company is cooperating with universities such as Tsinghua University and Tongji University (Figure 40.18) and keeping tight contact

Figure 40.18. The certificate of Tongji • Nano High-tech Combined Institute.

with home and abroad experts. Moreover, it has applied for several provincial scientific projects including a National 863 project and it owned over ten patents [1–10].

Besides silica aerogels, Nano High-tech also produces organic aerogels, carbon, alumina and titania aerogels.

Nano High-tech will expand the application of aerogels from thermal insulation to water treatment and air purification in the future.

40.5. Conclusion

As the producing base of silica aerogel nanoporous super-thermal insulation materials in China, Nano High-tech Co. Ltd mainly produces such products as super-thermal insulation flexible felts, thermal insulation panels, thermal insulation cylinders and special-shaped parts, translucent thermal insulation panels, silica aerogel powders, particles and monoliths. These products have good thermal insulation property, are environmentally friendly, non-toxic and can be widely applied in aerospace and aeronautics, metallurgy, chemical plant installations, oil pipes, refrigerators, cars, ships, buildings, etc.

References

1. Jin CL, Zhang J (2009) The mold to produce vacuum insulation panels internal fill aerogels. China Patent 200820084525.8.
2. Jin CL, Zhang RY, Zhang HB, Yu SJ (2008) Preparation of a oxide nanowires to enhance transparent monolithic aerogels. China Patent 200810061204.0.
3. Jin CL, Chen QP, Zhang RY, Li CS (2008) Preparation of silica aerogels with diatomaceous earth as raw material. China Patent 200810061203.6.
4. Jin CL, Zhang J (2008) A high-strength vacuum insulation panel. China Patent 200810061206.X.
5. Jin CL, Zhang HB, Zhang RY, Yu SJ (2008) Preparation of silica aerogels with rice husk as raw material by ambient pressure drying. China Patent 200810061201.7.
6. Jin CL, Yang XH (2008) A high-strength, flame-retardant vacuum insulation panel and its preparation method. China Patent 200810061205.5.
7. Jin CL, Shen J, Zhang HB, Ni XY, Zhang RY, Wang B, Yu SJ (2008) Preparation of silica aerogels with rice husk ash as a raw material by ambient pressure drying. China Patent 200810061202.1.
8. Jin CL, Zhang HB, Zhang RY, Yu SG (2008) Silica aerogels composites prepared by ambient pressure drying. China Patent 200810062792.X.
9. Jin CL, Shen LQ, Dong MJ (2008) Nano-modified additives high-strength wall insulation material and its preparation method. China Patent 200810062791.5.
10. Zhang RY, Jin CL, Zhang HB, Yu SJ, Xiao Y (2009) A low-cost method of Rapid Preparation of aerogels by hydrothermal synthesis. China Patent 200810181688.2.

41

OKAGEL: High Insulating Day Lighting Systems

Frank Schneider

Abstract Today Insulating Glass used in architectural applications has become a dominant part of the building skin. With the increasing number of glass used in buildings, its technical properties become more and more important. For that reason, the insulating glass plays a major role for the buildings energy consumption. Beside the thermal insulation, the solar heat gain and the light transmission are significant parameters. These parameters have to be designed carefully to achieve an optimal performance. Often even contradictory functions, i.e. solar protection and use of natural daylight have to be realized at the same time. A new alternative to standard insulating glass and opaque walls is the use of translucent insulating materials. Therewith, an improved thermal insulation is possible and the solar radiation can be controlled efficiently.

41.1. Introduction

The energy consumption of new and existing buildings still causes a major part of the worldwide CO_2 emissions. Due to the increasing requirements according to changing energy standards the demands on Insulating Glass become progressively complex and manifold. Therefore, the physical properties of the glass have to be improved continuously. In case of the insulating capacity, this is usually done by low-emissive coatings and special gas fillings. However, today the technical limits in the field of the thermal insulation are very limited. Figure 41.1 shows these developments qualitatively.

An Insulating Glass Unit (*IGU*) consists of a minimum of two glass panes, separated by a spacer as illustrated in Figure 41.2. The spacer is filled with desiccant to absorb rests of moisture, captured inside of the cavity during production. The cavity can be filled by air or by a special gas like Argon or Krypton to improve the thermal insulation capacity. A primary seal of permanent plastic Butyl has the function of a vapour barrier. The secondary seal of Silicone or Polysulfide has a structural purpose to bond the glass–glass sandwich together statically.

F. Schneider • OKALUX GmbH, Am Jöspershecklein 1, 97828 Marktheidenfeld, Germany
e-mail: fschneider@okalux.de

M.A. Aegerter et al. (eds.), *Aerogels Handbook*, Advances in Sol-Gel Derived Materials and Technologies, DOI 10.1007/978-1-4419-7589-8_41,

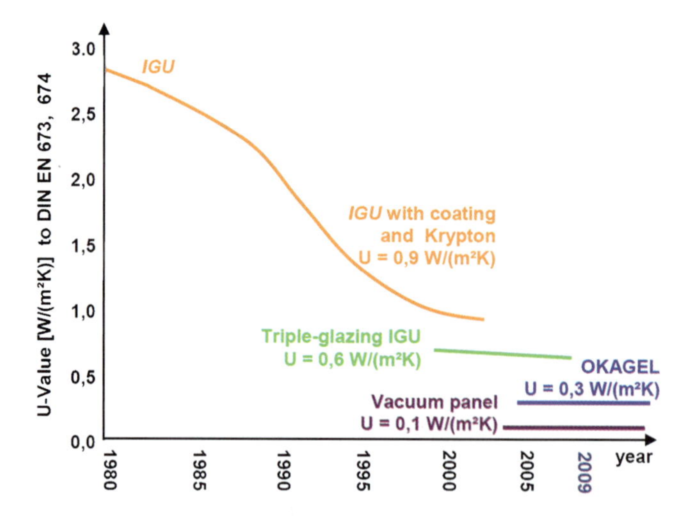

Figure 41.1. Development of thermal transmittance, *U-value*, for *IGU*s over the recent years.

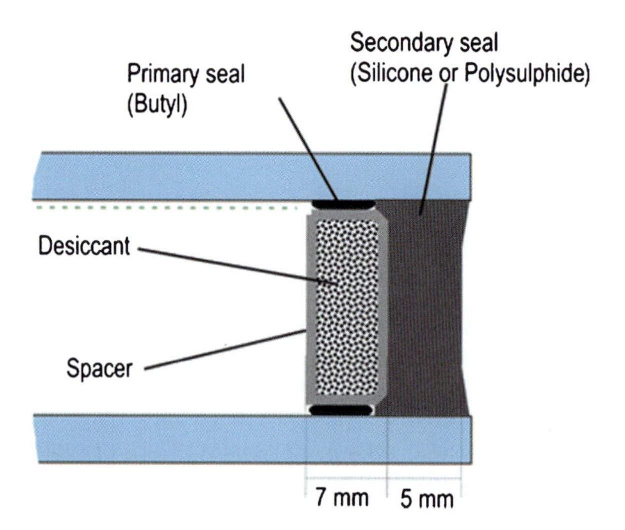

Figure 41.2. Buildup of a typical insulating glass unit.

41.2. Insulating Capacity

The insulating capacity achieved by the *IGU* is described by a physical parameter called thermal transmittance, respectively *U-value*. As illustrated in Figure 41.3, it can be defined by the amount of heat Q, which is transmitted through an area A within the time period t along a temperature gradient ΔT (11.1).

$$U \cong \frac{Q}{A \cdot \Delta T \cdot t}, \quad [U] = \frac{W}{m^2 \cdot K} \tag{41.1}$$

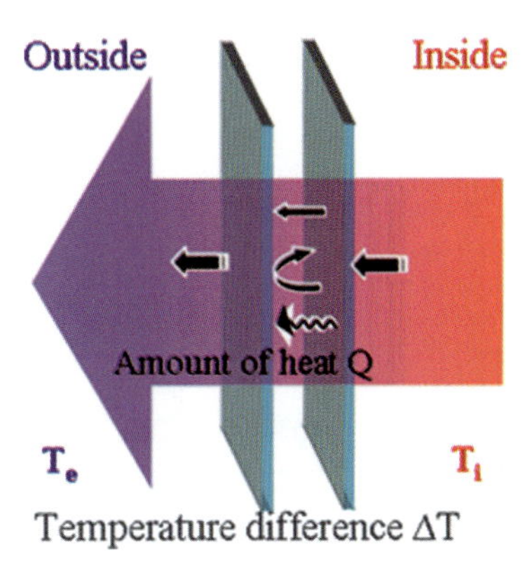

Figure 41.3. Heat transfer from warm inside to cold outside through a double-pane *IGU*.

The insulating capacity is influenced by the internal heat transfer h_i, the heat transfer in the cavity h_t, which can be divided into ~17% heat conduction, ~17% convection and ~66% radiation and the external heat transfer h_e so that the *U-value* can be calculated according to the standard DIN EN 673 [1] (11.2).

$$\frac{1}{U} = \frac{1}{h_i} + \frac{1}{h_t} + \frac{1}{h_e} \tag{41.2}$$

It becomes obvious that the radiation in the cavity plays a dominant role which determines the total heat transfer. Therefore, the development of special coatings reducing the emissivity *e* of the glass surface from $e_{glass} = 0.89$ to $e_{coating} = 0.04–0.02$ was a main step improving the *U-value*. Such coatings reduce the long wave IR-emissivity (heat radiation) effectively.

Before 1960, single glazings with *U-values* of ~6.0 W/(m²K) were standard. By introducing double glazing, the *U-value* could be cut in half. Today a *U-value* of ~1.0 W/(m²K) can be achieved with such low-e(missivity) coatings and gas fillings. Triple glazing units reach a *U-value* of ~0.6 W/(m²K), depending on the glass thickness, the cavity and the inclination installed. Generally, the dependence of the *U-value* on the inclination angle is often not taken into consideration in daily praxis. The *U-value* is defined and measured in vertical position according to the appropriate standard DIN EN 674 [2]. Furthermore the DIN EN 1279 [3] refers to the *U-value* of DIN EN 674 and does not consider the real inclination angle. Therefore, the actual insulation capacity of standard insulating glass can be worse than that expected when used, for example, as a roof glazing. The technical reason for this behaviour is explained in Figure 41.4. Inside a vertical glazing, the thermal convection forms a long, slow loop as warm air rises along the interior side. The convection is not significantly high. In the case of a horizontal glazing, the warm air meets the colder outer side more quickly when rising. The result is the formation of many small rapid circuits and an accelerated airflow. Thus the thermal energy is transferred faster through the cavity and the insulating capacity degrades while the *U-value* rises. Therefore, the performance of a gas-filled standard *IGU* drops significantly, when used in a roof application. The real *U-value* can be more than 50% higher than the *U-value* measured or calculated according to the appropriate code. If this is not considered carefully, the energy consumption of the whole building may rise unexpectedly due to the increased heat loss.

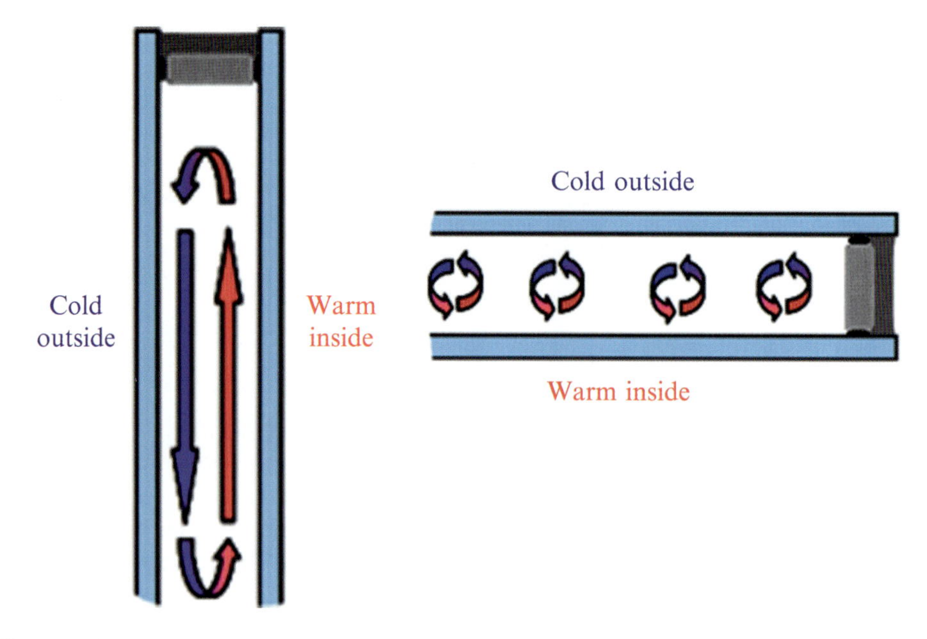

Figure 41.4. Convection inside an *IGU* in the case of a vertical façade application (**left**) and convection inside an *IGU* in the case of a horizontal roof application (**right**).

41.3. Translucent Insulation Materials

Light diffusing insulating material, usually called "Transparent" or better "Translucent" Insulation, is often used to achieve a good *U-value* on one hand and a certain solar energy gain on the other. The working principle of such materials is based on the polar bear coat. The material collects and conducts the light, respectively the solar energy. At the same time a good thermal insulation is achieved by stopping the convection. Such materials can also be used inside an *IGU* to expand the features of the façade. When embedded inside the cavity, inserts like capillary honeycomb slabs improve the thermal performance of the building. By using a 16-mm-thick light diffusing capillary insert (Figure 41.5), the *U-value* could be

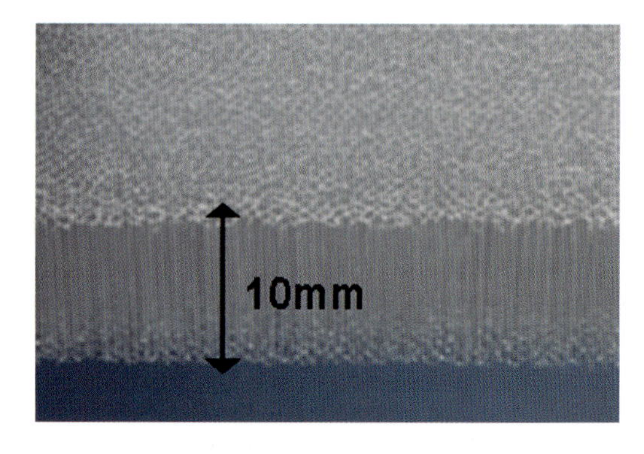

Figure 41.5. Translucent Capillary Honeycomb slabs.

reduced from ~3.0 W/(m^2K) for an air filled cavity to ~2.0 W/(m^2K) without any additional coating or gas filling. A 40 mm Capillary Honeycomb slab has an *U-value* of ~1.3 W/(m^2K). Adequate triple glazing units with an additional gas-filled cavity and a low-e coating reach a *U-value* of even ~0.8 W/(m^2K). This benefit of an *IGU* with a Translucent Insulation insert even increases when used as a roof application. Despite a conventional gas-filled unit, the *U-value* of an element with an adequate insert stays constant, if it is inclined to the vertical (Figure 41.6).

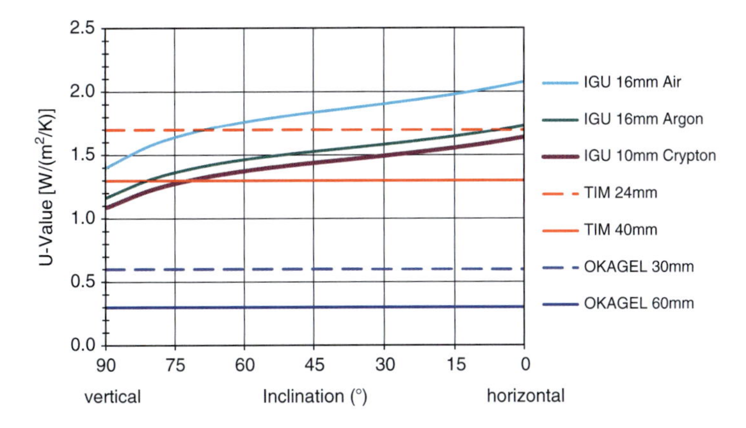

Figure 41.6. Dependance of *U-Value* on the inclination.

Aside the low *U-value* of such materials, their light diffusing characteristic brings the light very deep into the room while the colour rendering index remains by ~100%. For that reason the translucent insert fulfils multifunctional demands and offers additional advantages like sound insulation, the use of glare-free natural daylight and a unique façade design. The natural daylight is used efficiently and without any disturbing glare. As a result, comfortable lighting and heating conditions can be achieved, and direct solar radiation avoided. For that reason the user's comfort increased and the energy consumption for heating, cooling and artificial lighting is reduced. Consequently such materials are often used wherever good light conditions, low and a variable Light and Energy Transmission is necessary.

41.4. Silica Aerogels

In case of the thermal insulation, the technical limits are nearly reached today and a further improvement will affect aspects like the units' transparency (*VIP* – Vacuum Insulation Panels) or their maximum size, not mentioning the costs. A functional transparent vacuum insulating unit with *U-values* of ~0.1 W/(m^2K) is not available by now due to the technical problems of keeping the system permanently gas-tight. Therefore, the use of translucent Silica Aerogels, as a highly efficient insulating material inside the *IGU's* cavity, represents the best possible thermal insulation at the moment.

The granular Silica Aerogel particles have a diameter of 0.7–3.5 mm. The material consists of SiO_2, the basic material of sand or glass (Chap. 2). However, it has an air content of about 97% and weighs only 75 g per litre. This makes the Aerogel the lightest and best insulating solid in the world at the moment. Due to the low solids content inside the granulate, the heat conductivity is minimal. The three dimensional random lattice of the Aerogel structure has a mean pore size of approximately 20 Nm. Gas molecules are enclosed in these pores, reducing their possibility to move and to transfer heat by hitting each other. Hence, not only the convection but also the thermal conduction in the gas phase is efficiently hindered (Chap. 23). As a result the insulating capacity is improved and the speed of sound drastically reduced.

The material is very promising for architects and planers as a building material. Silica Aerogel is translucent and has the optical appearance of ice crystals or white snow (Figure 41.7). It is also 100% hydrophobic and therefore permanently resistant against water and mould (Chap. 3). It offers outstanding sound and thermal insulation uniquely combining them with a high light transmittance and light diffusion. Due to its unique combination of properties, the material offers new design solutions for someone who wants to achieve a maximum of natural day lighting as well as an optimal energy efficiency in terms of thermal insulation.

Figure 41.7. Visual appearance of the Cabot Aerogel granulate (see Chap. 38).

As a high insulating façade element, the translucent Silica Aerogels provide new possibilities. As the Aerogel mainly consists of air, the heat flow within the material itself is minimal. Furthermore, the convection and the conduction are minimized due to its nanostructure. Thus, a cavity with Aerogel-filling behaves like an evacuated volume and reaches *U-values* of $U = 0.3\ \mathrm{W/(m^2K)}$ and less. Again the *U-value* does not rise when used as a roof glazing (compare see Figure 41.6). Like other translucent insulating materials, the light is scattered and brought deep into the building.

41.5. Multifunctional, High Insulating Façade Elements

The manufacturing and the handling of glass elements filled with high porous granulates require new production technologies in the field of the *IGU* production. Therefore, the granulate is filled in a controlled continuous and uninterrupted process, in order to achieve a homogeneous appearance of the translucent façade. Additionally, a hydrostatic stress state is applied to the material by a defined pre-stressing. Due to this compression unintended settlements during the service life can be avoided and constant physical properties of the façade can be achieved for a long time [4].

The translucent facade element *OKAGEL* is a new class of a multifunctional, high insulating façade element (Figure 41.8). This system is able to fulfil various demands at the same time. The light and solar energy transmittance of an *OKAGEL* façade can be designed individually according to the requirement. Glass build-ups with a 60 mm Aerogel layer result in a thermal transmittance of $U \approx 0.3$ W/(m²K). This offers a completely new range of applications by substituting massive walls by lightweight and translucent façade elements and fulfils low-energy requirements at the same time. The total solar energy transmittance (*TSET*) of *OKAGEL*-elements is relatively low, because the secondary energy transmission, caused by infrared radiation, is very low too. The transparent particles inside the *IGU* stay cool themselves and do not warm up when exposed to direct sunlight. The *OKAGEL*-façade has also the effect of diffusing the light and gives an excellent illumination of the room with natural daylight.

Figure 41.8. Sample of *OKAGEL* façade element filled with translucent aerogel.

41.6. Applications

The *OKAGEL* façade element offers new design solutions for those who want to achieve a maximum of daylight as well as an optimum of energy efficiency. Typical applications are museums, production and sport facilities, supermarkets and office buildings.

One existing project, realized with *OKAGEL* is the new complex of the British Antarctic Survey Research Station Halley VI, located on the Brunt Ice Shelf, a floating area of ice that is moving westward by approximately 700 m per year. Since 1956, the first Halley bases were used to conduct research into meteorology, glaciology, seismology, radio astronomy and geospace science. The new station will allow long-running research on global change to continue at the site where the ozone hole was discovered. The new modular station is to be made up of eight individual modules, which are connected together by short, flexible corridors. The modules are kept above the snow surface using hydraulic legs mounted on skis. As well as keeping the buildings above the rising snow level the new design will allow the station to be periodically relocated across distances of many kilometers. The central, red coloured module illustrated in Figure 41.9 accommodates the majority of the stations social areas and therefore consists of double height with a large east-facing window made of *OKAGEL*. The translucent façade has an *U-value* of $U < 0.3$ W/(m²K). Because of the extreme temperatures of −60°C and less, a standard insulation glass would not perform properly. The primary seal of Butyl would be too hard and too brittle to keep the cavity permanently proof. Hence the gas would get lost quickly and humidity may penetrate into the cavity.

Figure 41.9. Left: Animation of the new British Antarctic Survey Research Station Halley VI. The red coloured main module contains 72 m² *OKAGEL* as façade elements with an *U-value* of 0.3 W/(m²K). **Right:** The entire research station at night.

Another project is the day-lighting system of the "*Tengelmann Klimamarkt*" Supermarket in Mühlheim an der Ruhr, Germany, shown in Figure 41.10. This store is the first low-energy Supermarket in Germany. The roof glazing also exists in *OKAGEL* with a *U-value* of $U < 0.3$ W/(m²K) to enable an excellent thermal insulation of the roof. At the same time, the heat introduction could be minimized to save an air-conditioning system. Due to the natural anti-glare lightning, the energy costs for electrical lightning could be reduced too and a comfortable environment could be provided. The use of natural daylight has an additional advantage for shops or supermarkets, since the natural daylight has a positive effect in the sales behaviour of customers. An American study in 1999 has investigated the relationship between daylighting and human performance [5]. This analysis is focused on the correlation between retail and skylights as an illumination source. The study has come to the conclusion that roof skylights,

Figure 41.10. Left: The first low energy Supermarket "Tengelmann Klimamarkt" in Mühlheim an der Ruhr, Germany, with an 82 m^2 *OKAGEL* day-lighting system in the roof. **Right:** Vision from the inside showing the function of the daylighting system.

that enable natural day-lighting, have a significant positive effect on the sales rate. In this case, the natural daylight has led to higher sales of up to 40%.

41.7. Conclusion

High efficient thermal insulating systems will become more and more important in the future. Stronger energy saving regulations and also the changing environmental consciousness of the modern society will accelerate this development, even though the technical limits are nearly reached today. Thus, other considerations will determine the future developments in the field of insulating glass used for buildings envelope. Façade elements filled with translucent aerogels are one possibility to further improve the functionality of the facade, combining excellent thermal insulation with an efficient use of natural daylight and glare protection. Therefore, the traditional way of glazing, as single windows in solid walls, has to change and large-area translucent coverings have to be introduced to allow an optimal performance. Such day lighting have not only a positive influence on the buildings energy consumption, they have also an influence on the users comfort and their subjective well-being. As a result such systems can stimulate the sales behaviour of customers. Moreover, natural day lighting has a positive effect on the general human performance [6].

References

1. DIN EN 673 "Glas im Bauwesen – Bestimmung des Wärmedurchgangskoeffizienten (U-Wert) – Berechnungsverfahren", Beuth, Berlin, 2003.
2. DIN EN 674 "Glas im Bauwesen – Bestimmung des Wärmedurchgangskoeffizienten (U-Wert) – Verfahren mit dem Plattengerät", Beuth, Berlin, 2003.
3. DIN EN 1279 "Glas im Bauwesen – Mehrscheiben-Isolierglas", Beuth, Berlin, 2004.
4. Schneider F, Gertner G, "Use of transparent aerogels for efficient thermal insulation", Intelligent Glass Solutions, Issue 1, S 39-42, Intelligent Publications Limited (IPL), Beijing, London, Munich, New Delhi, 2007

5. HESCHONG MAHONE GROUP "Sky lighting and Retail Sales – An Investigation into the relationship between Day lighting and Human Performance", Pacific Gas and Electric Company on behalf of the California Board of Energy Efficient Third Party Program, 1999.
6. HESCHONG MAHONE GROUP "Day lighting in Schools – An Investigation into the Relationship Between Day lighting and Human Performance", Pacific Gas and Electric Company on behalf of the California Board of Energy Efficient Third Party Program, 1999.

Part XVI

Conclusion

42

Concluding Remarks and Outlook

Michel A. Aegerter, Nicholas Leventis, and Matthias M. Koebel

Aerogels were invented by S. Kistler some 80 years ago, but surprisingly, commercialization lags behind by an unusually long time lapse. This should not have been the case. Browsing through this volume it can be argued that the structure–property relationships of aerogels are well-understood. A large array of aerogels have been synthesized and characterized; "exotic" aerogels based on oxides of aluminum, tin, copper, ferromagnetic metals (iron and nickel), all refractories, all rare earths and selected actinides, organic–inorganic aerogel composites, carbide aerogels, selected pure metallic aerogels and several organic aerogels have been described and can be prepared with passable effort. Why then do not we see aerogels all over, around us, from homes to appliances to transportation?

Historically and for economic reasons, most fundamental and applied research has focused on silica aerogels. But, the starting materials (mostly alkoxides) are expensive, the process (supercritical fluid drying) is also expensive and potentially hazardous and the end-product, the silica aerogel, is environmentally unstable (hydrophilic) and mechanically weak (fragile). Tremendous progress has been made along understanding and controlling those issues. The old technology of using water-glass as the starting material for silica aerogels has been revisited; currently, alkoxides are readily replaced by sodium silicate. Hydrophobic aerogels have simultaneously circumvented both the environmental instability problems and the need for supercritical fluid drying via the so-called spring-back effect. Also, mechanically super-strong polymer cross-linked aerogels have been developed. Does not this mean that all of the main barriers to commercialization have been lifted? In principle, the answer to this question is yes. After all, large-scale production of aerogels in conventional shapes such as monolithic panels, as well as in more innovative application-specific forms, such as aerogel pellets and blankets has become possible. Many products are currently evolving, finding their ways into well-selected niche-markets. Traditional kinds of applications use aerogels as thermal and acoustic insulation.

M. A. Aegerter (✉) • Ch. des Placettes 6, 1041 Bottens, Switzerland
e-mail: michel.aegerter@bluewin.ch
N. Leventis • Department of Chemistry, Missouri University of Science and Technology, Rolla, MO 65409, USA
e-mail: leventiss@mst.edu
M. M. Koebel • Laboratory for Building Technologies, Empa, Swiss Federal Laboratories for Materials Science and Technology, 8600 Dübendorf, Switzerland
e-mail: matthias.koebel@empa.ch

M.A. Aegerter et al. (eds.), *Aerogels Handbook*, Advances in Sol-Gel Derived Materials and Technologies, DOI 10.1007/978-1-4419-7589-8_42,

New applications, unthinkable for aerogels up to recently, including ballistic protection and drug delivery, are being investigated intensely under this new perspective.

However, despite those many recent improvements, market penetration is still low and large-scale commercialization remains elusive. So, what is keeping the industry from putting the spurs to the horse? Is this because of hidden material properties issues? Unlikely. Economics? Probably. Is there anything that can be done? Positively, yes. For this, we should stop antagonizing Nature and instead work with it. There is no doubt that aerogels, as a class of materials, possess properties that cannot be found elsewhere. Silica aerogels, however, have run their trajectory. Kistler himself observed in 1931 that his organic aerogels were more robust materials than their inorganic counterparts. Nevertheless, further development remained dormant until the late 1980s when R. Pekala started publishing his work on resorcinol–formaldehyde resin-based aerogels. Those materials have been investigated extensively for their conversion to high surface area, electrically conducting carbon aerogels and, more recently, use as thermal insulators in the packaging of sensitive biological specimens. For a number of years following resorcinol–formaldehyde aerogels, the term organic and resorcinol–formaldehyde aerogels were practically synonymous. Slowly though, other systems started showing up (polyacrylonitrile, polyurethane, and polyurea aerogels in the 1990s) and this trend has taken an accelerated pace in the last 5 years (2005–2010) with publications/patents appearing on monolithic aerogels based on most major classes of polymers including even conducting polymers (PEDOT, 2010). At this point, it is easily understood that organic aerogels can take advantage of the tremendous progress, which has taken place in Polymer Science over the past five decades and the variety of chemical compositions/properties that can be realized therewith. Organic polymer aerogels may alleviate all at once, most or even all, the economic limitations in terms of synthesis, environmental issues and mechanical properties of traditional silica aerogels: hydrohilicity/hydrophobicity can be tuned by carefully choosing the chemical identity of the polymer; polymer cross-linked aerogels imply that purely polymeric systems with the same morphology and interparticle connectivity as those of the former materials should have similar strength, and above all they may be processable under ambient conditions. So far studies of polymer aerogels are not systematic in nature though; they appear more like as hit-or-miss situations. Issues such as what polymers gel rather than precipitate, and which polymer gels can be dried to aerogels have not been addressed yet in full. The morphology of polymer aerogels is also an open issue; questions such as what conditions yield fibers versus nanoparticles, or what are the properties of chemically identical but morphologically different materials need to be addressed. The current situation reminds of trial and error runs, which are often necessary to initiate work in a new field or with a new class of materials. In that sense, aerogels are just about to be rediscovered for the third time.

In addition, let us also be reminded of the tremendous potential of organic–inorganic composite aerogels. Many examples of organically modified silica compounds (Ormosils) are known. For other metal oxide and mixed metal oxide systems, a similar diversity of composites can be expected. Given the nearly endless possibilities of organic surface modification and cross-linking reactions, the systematic reinforcement of metal oxide aerogels in terms of mechanical properties and with the goal to introduce specific chemical or physical functionalities and/or properties represents a universe of parameter space and great potential for the discovery of plenty of new and fascinating materials.

The definite, undisputed applications of aerogels remain in Cherenkov radiation detectors, thermal insulation and in space exploration. But, as evidenced by this Aerogels Handbook, it is expected to be a short matter of time before the concerted efforts of industry, academia and government lead to the wide use of those remarkable materials in construction, transportation, health care and the National Defense.

The editors

Glossary, Acronyms and Abbreviations

The following words listed in bold are written in italic in the contributions

A

A H_2O/Na_2SiO_3 molar ratio
A23187 Calcium ionophore
Absorption In chemistry, the process by which one substance, such as a solid or liquid, takes up another substance, such as a liquid or gas, through minute pores or spaces between its molecules. In physics, the process in which the incident radiated energy is retained without reflection or transmission on passing through a medium
ACC Aerogel Cherenkov Counter
Acidity Acid strength
AC-ML Aerogel composite acoustic matching layer
Acoustic impedance Ratio of particle velocity and the sound pressure
Acoustic matching layer Antireflective layer of acoustic wave
Acoustic velocity Velocity v of an acoustic wave as it travels across a solid, a liquid, a gas or a mixed material medium
ADP Adenosine di-phosphate, a product which can be used to induce platelet aggregation
Adsorption Surface phenomenon by which molecules of a gas or a liquid are fixed on the surface of a solid, with various types of bonding
Adsorption isotherms Adsorption is usually described through isotherms, i.e., the amount of adsorbate on the adsorbent as a function of its pressure (if gas) or concentration (if liquid) at constant temperature
Aerobic oxidation Oxidation by air under ambient conditions
Aerocellulose Aerogels made from cellulose
Aeroclay Aeroclays are clay-based aerogels generally elaborated by freeze-drying of clay-based hydrogels (for example, gels prepared with sodium-exchanged montmorillonite clay). It is accepted that the freeze-drying step results in a rearrangement of clay sheets to create the lightweight, oriented and porous structure of the final clay-based aerogel
Aerocore An organic monolithic panel manufactured by American Aerogel Corporation
Aerogels Defined by IUPAC as "Gels comprised of a microporous solid in which the dispersed phase is a gas" [see Pure Appl. Chem. 79(10):1801–1809, 2007]. However, the definition retained in this handbook is the one by Kistler, as "gels in which the liquid has been replaced by air, with very moderate shrinkage of the solid network"
Aerogel functionalization Implementation of functional groups into aerogels
AeroSand Sand bonded by aerogels

M.A. Aegerter et al. (eds.), *Aerogels Handbook*, Advances in Sol-Gel Derived Materials and Technologies, DOI 10.1007/978-1-4419-7589-8,

Aerosculpture Word invented by the author I. Michaloudis, from *air* and *sculpture*. A pair of parenthesis – as a personal symbol of an open "space in between" – replaces the letter o and gives the visual word *aer( )sculpture*
Aerosol Suspension of colloidal solid particles or liquid droplets in a gas, such as air
AFM Atomic force microscopy
Ageing (or **Aging**) Experiment where a material is maintained for a certain time in constant environment conditions and its characteristics are slowly changing with time. Specifically speaking with gels, ageing is a curing period where gels are immerged in a specific liquid media where ageing mechanisms (generally, syneresis and/or Ostwald ripening) are promoted or frozen
AIBN Azobisisobutyronitrile
Alcogel A wet gel containing alcohol formed by sol-gel method
ALD Atomic layer deposition
Alkoxide Sol-gel precursor of chemical formula $M(OR)_n$, where M designates a cation, R an alkyl group and OR an alkoxide group. Alkoxides are often available, in a more or less polymerized form, in solution in their parent alcohol (for example ethanol for TEOS or methanol for TMOS)
Allophane Allophane is an amorphous clay from natural soil; it is a natural alumino-silicate originating from the transformation of volcanic ashes and glasses with weathering. Allophane exists for young volcanic soil and is progressively transformed into crystalline clays, halloysite, then into kaolinite
Ambigels Gels dried by evaporation of their liquid with moderate shrinkage. Ambigels are considered as "aerogels" and sometimes termed "ambient pressure drying aerogels," when their shrinkage is moderate and their specific pore volume high
Ambient pressure drying Drying of gels by evaporation of their liquid with moderate shrinkage in room temperature and pressure conditions
Amorphous material An "amorphous material" is a solid in which there is no long-range order of the positions of the atoms
AN Ammonium nitrate
Analyte The chemical species being analyzed
Anaphylatoxins A toxic substance formed in the blood that is associated with bacterial polysaccharides
Anthropogenic A process that results from human activity
AP Aerogel powder and/or particles
APD **Ambient pressure drying**. See also **Ambigels**
Apparent thermal conductivity Thermal conductivity value assigned to a material that exhibits thermal transmission by at least one nondiffusive mode of heat transfer resulting in property variation with specimen thickness, or surface emittance
APTES 3-Aminopropyltriethoxysilane
Arrhenius law Expression describing the exponential dependence of a reaction on an activation energy
ASCE Alcohol supercritical extraction
ASTM American Society for Testing and Materials
Attenuation length Λ (optical) The distance traveled across in an aerogel, where the probability that a photon has not been absorbed has dropped by $1/e$ (where $e = 2.71828$)
AZC Aerogel sulfated zirconia doped with cerium
AZS Aerogel sulfated zirconia
AZSCr Aerogel sulfated zirconia doped with chromium

B

BCAM Base-catalyzed, acid-modified
bcc Body centered cubic structure
Bending strength Strength of a material tested in a bending test most often using rectangular bars supported by two racks separated by a fixed distance. A bar is subjected in the middle by a force and bends thereby. The fracture strength is calculated from the maximum force and the geometry of the test bar
BESS Balloon-borne experiment with a superconducting spectrometer
BET Evaluation method named after Brunauer, Emmett, and Teller to obtain the surface area from adsorption isotherms due to a theory for multilayer adsorption
BF Bright field
Binder A material bonding together e.g., sand grains, very often a polymer or colloidal sodium silicate
Biomimicry From bios, meaning life, and mimesis, meaning to imitate: It is an ancient concept recently returning to scientific thought that examines nature, its models, systems, processes, and elements – and emulates or takes inspiration from them to solve human problems sustainably. Scientific and engineering literature often uses the term *biomimetics* for the process of understanding and applying biological principles to human designs
Biosensor see "**Sensors (biochemical)**"
BJH Barret, Joyner, and Halenda who developed a methodology to determine the pore volume distribution from nitrogen adsorption isotherms using the Kelvin equation, relating the capillary pressure to pore size via the interfacial tension of a fluid condensing in a capillary
BMH Born–Mayer–Huggins model for silicon dioxide, an empirical potential consisting of pairwise-additive repulsive terms and Coulombic interactions
BO Bridging oxygen
Bolus release Administration of a drug over a relatively quick amount of time (less than 5 min) in comparison to a sustained delivery which can last over hours
Bonding bridge Bridge between two sand grains made up by a polymer or an aerogel
Brönsted acidity Chemical compound presenting Brönsted acid sites that are able to lose a proton
Bridgman furnace A furnace with at least three zones being at different temperatures. A hot zone and a cold zone are separated by an adiabatic zone. Directional solidification or crystal growth is achieved by either pulling the sample through the furnace from hot to cold or by moving the furnace
Brittle fracture Fracture of a material being completely elastically stressed (no irreversible plastic deformation)
BSA Bovine serum albumin
BSE Backscattered electrons
BTAC Benzyltrimethylammoniumchloride
BTMSH 1,6-Bis(trimethoxysilyl)hexane
BTMSPA Bis(trimethoxysilyl-propyl) amine
Bulk-modified Sol–gel material prepared by adding probe to the precursor mixture (also known as predoped)

C

C3a A plasma inflammation marker (plasma anaphylatoxin generation) which is part of the complement pathway. C3a is the released degradation product of C3 (complement component 3)

C5a A plasma inflammation marker (plasma anaphylatoxin generation) which is part of the complement pathway. C5a is the released degradation product of C5 (complement component 5)
Capillary Tube of small internal diameter, typically of the order of a few μm or less, in which liquids can raise spontaneously, when one end of the tube is dipped in these liquids
Capillary stresses Mechanical stresses due to surface interaction of a liquid with a capillary and the presence of a gas–liquid meniscus in a capillary pore, responsible for the liquid raise in such **capillary**
Carbon aerogel Aerogel obtained by pyrolysis of an organic (polymeric) aerogel under protective atmosphere
Carbon AeroSand Sand bonded by carbon aerogels
Carman–Kozeny equation This equation is empirical and relates the permeability to the pore size and pore volume:$D = (1 - \rho_r)r_w^2/4K_w$where ρ_r is the relative density, r_w is the hydraulic radius and K_w is the so-called the Kozeny constant close to 5
CARS Coherent anti-Stokes Raman scattering
Cavitation Creation of cavities within a liquid or solid caused by the propagation of a pressure wave through this liquid or solid. It may also mean the formation of vaporous bubbles in a liquid due to a transient pressure drop below its vapor pressure
Cellulose acetate Salt of acetic acid with cellulose
Cellulose carbamate Carbamates are esters of the carbamic acid NH_2COOH. A salt made from cellulose and urethanes (an amine bonded to any organic radical R)
Cerenkov counter see Cherenkov counter
Chalcogel A gel composed of a metal chalcogenide framework
Chalcogenide aerogel Aerogel composed of metal chalcogenide frameworks
Chemical gel According to the definition given by Flory PJ [(1974) Gels and gelling processes. Disc. Faraday 57:7–18], these gels are polymeric gels that present, on the contrary of physical gels, a cohesive structure governed by the presence of covalent bounds
Cherenkov (or Cerenkov) counter Photo counters which permit to identify electrically charged particles in high-energy physics experiments, from their radiative properties when traveling across a transparent material with a refractive index close to 1 (e.g., an aerogel)
Chitosan A polysaccharide derived from chitin
Chy α-Chymotrypsin
Classic or conventional gel Gel obtained by hydrolysis and polycondesention of alkoxides in the presence of a common precursor and solvent (alkoxide and water)
Chlordecone Chlordecone is a very though pesticide (C_{10} Cl_{10} 0) which was used in Africa, Central America and West Indies mainly for the protection of banana plantation. Chlordecone induces cumulative and delayed toxicity
CD-62P P-selectin, a cell adhesion molecule found on active platelet cell membranes
CL-20 Energetic materials lexicon for 2,4,6,8,10,12-hexanitro-2,4,6,8,10,12-hexaazaisowurtzitane, a powerful chemical explosive
cN/tex The unit Centi Newton per tex
Cogelation Gelation in the same **sol** or polymer solution of, at least, two interconnected solid networks due to the presence in this sol of the corresponding precursors and catalysts
Coke Carbonaceous solid deposed on the surface of the solid catalyst
COLD supercritical drying The use of supercritical CO_2, with a critical temperature of 31°C, for aerogel formation. This is in contrast to **hot supercritical** drying employing alcohols, where high temperatures are required to achieve a supercritical state

Collapsibility Ability of a core or mould material to be disintegrated (mechanically, thermally, by vibration or oxidation)
Colloidal-type aerogel Gel obtained by gelation of **colloidal particles**. Aerogels adopting a morphology characterized by spherical nanoparticle components
Compactibility The ability of a sand-binder mixture to become dense homogeneously
Compression Pack™ An insulation system based on Cabot **Nanogel®** in a pre-packaged form commonly used to insulate tubing and pipe systems such as oil pipelines
Compressive strength The maximum compressive stress (Force per unit cross section area) a material can withstand without failure
Condensation In oxide aerogels, a chemical transformation which consists in linking two precursor molecules, one of which at least has been hydrolyzed, to build a M–O–M or a M–(OH)–M bridge by elimination of a H_2O or an alcohol ROH molecule. In organic aerogels, strong –(C–C)– covalent bonds are established between two organic sol–gel precursor molecules, by elimination of another smaller organic molecule. In chalcogenide aerogels, metal–chalcogenide bridges form by release of H_2S when preceded by thiolysis; alternatively, metal chalcogenide particles are linked together under oxidative conditions where the nature of the bonding is yet to be determined
Conductimetric Related to measurement of solution conductivity
Confined mold A closed, fixed volume sample well used in **RSCE** processing
Contact angle Angle between a droplet of liquid and a flat surface
Coprecursor technique or method Method used to make hydrophobic aerogels through incorporation of an additional silane reagent to sol–gel recipe
Core Positive image of a cavity in cast pieces of metal
Core–Shell aerogel Aerogel comprising of core–shell nanoparticles i.e., nanoparticles that have two separate chemical components, one that makes up the inside (core), and one that deposits on the outside (shell)
COSMO "Conductor-like screening model," in which a molecule is embedded in a continuum dielectric meant to represent the surrounding solvent
Cross-linker A molecule being able to link cellulose polymer chains chemically together, replacing the hydrogen bonds by stronger covalent bonds. For chalcogenide aerogels formed from Zintl ions, the cross-linker is a Pt^{2+} ion that forms a covalent linkage between chalcogenide clusters
Crucible Hollow body in which a metal is molten
Cryogel Gel dried by rapid freezing and sublimation of its solidified solvent. A cryogel is considered as an "aerogel" when its shrinkage is moderate and its specific pore volume high
CS Thermal insulation cylinders and special-shaped parts
CSCE CO_2 supercritical extraction
CSH Combined solvent exchange and hydrophobization
C/T Molar ratio between the cellulosic repeat units and **TDI**
CTAC Cetyltrimethylammoniumchloride
CVD Chemical vapor deposition
cyt c Cytochrome c
Cytotoxicity A quantification of the products that may have a toxic effect on cells

D

DABCO Diazobicyclo[2,2,2]octane
DAP *N*-(3-trimethoxysilylpropyl)-2,7-diazapyrenium bromide

Darcy law Relation between the gas flow velocity through a porous body and the pressure gradient across it. It has the form of an Ohm's law and the proportionality constant is called the permeability
DCCA Drying control chemical additives. Additives, in the sol-gel process before gelation, which permit to obtain uncracked dry monoliths with a relative high specific pore volume
ΔT Temperature gradient
Deformability Ability of a dry sand-binder mixture (core or mould) to deform elastically and plastically
Density of vibrational states Number of states available in the frequency interval $[\omega, \omega + d\omega]$
DF Dark field
DFT Density functional theory, a type of electronic structure calculation
DIC Digital image correlation
Differential Scanning Calorimetry (DSC) Technique in which the difference in the amount of heat required to increase the temperature of a sample being analyzed and a reference sample, are measured as a function of temperature
Diffusion Limited Cluster Aggregation (DLCA) Process whereby clusters of particles undergoing a random walk due to Brownian motion, aggregate together. This theory, proposed by Witten and Sander in 1981
Dissolution profile Time dependence of the drug liberation/dissolution from the carrier material
Dissolution/reprecipitation Dissolution of solid material in areas where the solid surface has a small convex radius of curvature, followed by re-precipitation on local areas with a greater convex radius of curvature or on concave local areas
DLA Diffusion-limited-aggregation, an aggregation model in which monomers add to clusters upon contact
DLCA Diffusion-limited cluster aggregation, similar to DLA, but incorporating aggregation of clusters themselves
DMA Dynamic mechanical analysis
DMAc *N,N*-dimethylacetamide
DMC Dynamic Monte Carlo, a stochastic integration of the phenomenological kinetic equations for a chemical system
DMDMS Dimethoxydimethylsilane
DMF Dimethylformamide, common solvent for chemical reactions
DMSO Dimethyl sulfoxide
DoE Design-of-experiments
DP_n Degree of polymerization; number of monomeric units in a polymer chain. Is usually calculated from the total molecular weight of the polymer divided by that of the monomer
DRIFT Diffuse reflectance infrared fourrier transform spectroscopy
Drug carriers Materials loaded with pharmaceutical compounds
Dry strength Strength of the fully hardened sand-binder mixture
DTA Differential thermal analysis
Dtex Unit to measure the density of fibers, defined as the mass in grams per 1,000 m. Dtex = decitex = mass per 10,000 m
Dulong–Petit law A chemical law proposed in 1819 by French physicists and chemists Pierre Louis Dulong and Alexis Thérèse Petit stating the classical expression for the specific

heat capacity of a crystal due to its lattice vibrations; the specific heat equals 3 R/M, corresponding to the high temperature limit of the specific heat. R is the gas constant and M is the molar mass
Dyad Two individual units that work together as a pair
Dynamic hot-wire method It allows to determine the thermal conductivity by embedding a wire into the material to be investigated. The wire deals as heating element and temperature sensor at the same time and is heated with a constant heat power. From the time-dependent temperature increase of the wire the thermal conductivity is derived

E

E **Young's modulus** (E). It is, within the limits of elasticity, the ratio of the linear stress (force to which a solid rod is submitted, divided by its cross section area) to the linear strain (relative elongation or contraction of the rod)
$e_{coating}$ Emissivity of a coated glass surface
e_{glass} Emissivity of a glass surface
EACAC Ethylacetoacetate
EDL Entry, descent, and landing
EDLC Electrochemical double-layer-capacitor
EDX Energy dispersive X-ray spectroscopy, a common spectroscopy technique available on many electron microscopes
EELS Electron energy loss spectroscopy, a common spectroscopy technique available on many electron microscopes
Effective total thermal conductivity Thermal conductivity of a material that exhibits thermal transmission by several modes of diffusive heat transfer
EFTEM Elemental transmission electron microscopy
Elastic modulus see **Young's modulus**
Elastomer Thermoplastic or thermoset polymer that can stretch by a large extent and then return to its original shape without permanent deformation. Elastomers are only slightly cross-linked
Electret Dielectric material which presents a quasi permanent electric dipole
EM Energetic material or electron microscopy
EPS Expanded Polystyrene, a commonly used, inexpensive polymer based insulation. Made from polystyrene beads which are loaded with a solvent such as pentane and then allowed to expand at elevated temperature to produce **foam** insulation
Esterification A reaction of an alcohol ROH (R: organic group) with an acid HA (A: anionic group) to produce an ester ROA and water H_2O
ESR Electron spin resonance
Et_3N Triethylamine
Ethanol Ethyl alcohol (C_2H_5OH)
ETA Emanation thermal analysis
etac Ethyl acetate
EVA Extravehicular activity
EW ratio Ratio between the amine equivalent weight of hydrogen and the isocyanate
EXAFS Extended X-ray absorption fine structure, i.e., the fine structure present in the X-ray absorption spectra beyond the absorption edge

F

F Formaldehyde
F-aerogel Aerogel obtained after a freeze drying process
fcc Face centered cubic structure
FEM Finite-Element Modeling, a computational technique for determining the mechanical properties of objects
Fibril A thin fiber of cellulose in which the polymer chains are arranged like filaments in a rope. Macrofibrils have a diameter of around of 60–400 nm, microfibrils of around 20–30 nm, elementary fibrils 3–10 nm
Fineness Surface roughness of a sand mould or core
Fischer–Tropsch synthesis reaction Is a catalyzed chemical reaction in which synthesis gas, a mixture of carbon monoxide and hydrogen, is converted into liquid hydrocarbons of various forms
Flowability There are two types of flowability. One is the ability of a sand binder mixture to be fluidized by pressurized air and thus can be shot into a form to yield a core; the second one means that the sand-binder mixture trickles easily into a form
FM Flexible thermal insulation felt
Foam A solid or liquid matrix containing macropores or gas bubbles
Formate species Methanoate species or ion $CHOO^-$
Fourier transformed infra red spectroscopy (FTIR) Measurement technique whereby infra-red absorption spectra of materials are analyzed by mathematical Fourier transform methods
4-methylpenta2-ol $C_6H_{14}O$
Fractal dimension In gel networks or aggregates of nanoparticles in which the mass M inside any sphere of radius R, about a centre chosen at random in the gel network or aggregate, increases statistically with R as $M \approx R^f$, the number f is usually a noninteger and is termed the fractal dimension. Aerogels are examples of fractal solids
Fractional function of the first kind $f_{0-\Lambda}(T)$ Fraction of the total black-body radiation intensity having wavelengths between 0 and Λ
Fracton A collective quantized vibration on a substrate with a fractal structure
Free-dimensional space art The author I. Michaloudis is playing with the "three-dimensional" space which in the case of the indefinitive transparency of his aer( ) sculptures becomes a non-Euclidean space. Silica aerogel itself can be considered as a personification of what the French mathematician Henri Poincaré named a "representative space," a space you cannot measure; you just live in with all your senses; for more information see Ioannis MICHALOU(di)S, "Aer( )sculpture: the enigmatic beauty of aerogel's nonentity in a pilot art and science project," Journal of Non-Crystalline Solids (2004) 350:61–66
Freeze drying Drying of a wet sample along the sublimation curve of the solvent. The process avoids the direct liquid–gas transition observed in ordinary drying
FTIR Fourier transform infrared spectroscopy
Functionalization Chemical action consisting in functionalizing (adding functional chemical groups) a molecule or the surface of a solid, by chemical synthesis or grafting methods
FW Formula weight

G

Gas permeability The gas flow velocity through a porous body is proportional to the pressure gradient applied across it. The constant of proportionality between gas flow velocity and pressure gradient is the permeability and thus a measure of the resistance a porous body exerts against gas flow through it
Gel point This is the state of condensation of a precursor or of suspended nanoparticles, in a liquid, where **gelation** is first observed
Gelation Transformation of a liquid polymer solution or colloidal suspension (a sol) to a solid impregnated with liquid, by continuous formation of an open and porous three-dimensional polymer, uniformly throughout the entire solvent, without any formation of a precipitate
GHSV Gas hourly space velocity; it is used for gaseous feed stream with the volumetric flow rate expressed in term of volume per hours, thus GHSV has units per hour
Glassy state A dense monolithic **amorphous** solid state in which the atoms or molecules are not arranged in any regular order, as in a crystal, and which crystallizes only after an extremely long time
g/Nm^3 Unit meaning grams per normal cubic meter; Nm^3 is the gas volume, in cubic meter, at 0°C and 760 mm Hg
Goniometer Instrument used to measure contact angles
Gordian Task A task of epic proportions, of great difficulty, named after the Gordian knot which in Ancient times posed an unsolvable challenge
Gradient density aerogel Aerogel with a continuously changing density profile
Grafting Chemical grafting is a chemical technique to permanently bond some chemical moieties or molecules, onto a solid surface or another macromolecule
Graphene Monolayer of carbon atoms in a hexagonal arrangement
Green strength Strength of a sand-binder mixture directly after for instance core shooting or manual compaction allowing to handle the sand mould or core without any waiting time for hardening to occur
Griffith criterion The Griffith criterion states that the fracture strength of a brittle solid is solely determined by its elastic properties, its surface tension and the crack or flaw size being in the material

H

h_e External heat transfer
h_i Internal heat transfer
h_t Heat transfer in a cavity
HDI Hexamethylene di-isocyanate
Heck coupling Catalytic formation of a substituted alkene starting from an unsaturated halide and an alkene (also called Mizoroki–Heck reaction)
Helium pycnometry Characterization method permitting to determine the skeletal density and specific volume of a porous sample by measuring the pressure change of helium in a calibrated volume
HEMA Hydroxyethyl methacrylate
Hemicellulose A blend of several polysaccharides, like cellulose, xylose, galactose, mannose, and other glucoses
Hemolysis The breakdown of red blood cells with the release of hemo-globin

Heritage RTG (Radioisotope Thermoelectric Generator) Thermoelectric generators used over the past several decades employing radioisotope heat sources
HPLC High pressure liquid chromatography
HMDI Bis(4-isocyanatocyclohexyl)methane
HMDS HexaMethylDiSilazane $(CH_3)_3Si{-}(NH){-}Si(CH_3)_3$
Hot crack Crack appearing during solidification in a casting
Hot disk method (or technique) A well-known transient plane method used to characterizes both thermal conductivity and thermal diffusivity of materials. It is based on the use of a hot-disk probe. In some publications, in honour of the inventor of the technique, it is also referred to as "the Gustafsson probe" method
HOT supercritical drying High temperature supercritical drying process, at temperatures >260°C, in an alcohol
Hot tearing Fracturing of a material at high temperature
HREM High resolution electron microscopy
HVAC Heating, Ventilation and Air Conditioning. It is the main source of energy consumption in a building
Hybrid aerogel Aerogel with a solid network constituted of mixed organic and inorganic (often silica) moieties
Hydraulic hotpress Machine used to seal and heat confined mold in **RSCE** processing
Hydrolysis Chemical transformation of a sol-gel precursor, which consists in replacing a ligand OR (in an alkoxide) or OH_2 (in a hydrated metal salt), by an OH ligand
Hydrophilic aerogel Aerogel for which the contact angle of a water droplet with the porous external surface is <90° (i.e., water attracting)
Hydrophilicity Ability of a material to be wetted by water
Hydrophobic aerogel Aerogel for which the contact angle of a water droplet on a porous external surface is >90° (i.e., water repelling)
Hydrophobicity Literally, "water-hating." Ability of a material which cannot be wetted by water; they are usually terminated with nonpolar organics
Hydroxyl –OH terminal group
High-performance insulation A common synonym for superinsulation. Materials and/or systems with superior thermal insulation performance when compared with conventional ones
Hygroscopy The ability of a material to attract and absorb water from the environment (moisture, humidity)
Hyper velocity particle capture The capture of high velocity (>1 km/s), micrometer scale particles in a porous material

I

I Luminescence intensity at a given concentration of quencher
I_0 Luminescence intensity in the absence of quencher
ICF An "Inertial confinement Fusion" experiment in which the fluid submitted to nuclear fusion experiments (deuterium and tritium) is confined inside the pores of carbon or silica aerogels
IEEE Institute of Electrical and Electronics Engineers
IEMA Isocyanatoethyl Methacrylate
IGU Insulating glass unit

Impact strength Strength measured during an impact test as e.g., using a pendulum hammer falling onto a prenotched sample from a certain height. After passing the sample and fracturing it, the hammer has lost kinetic energy that is easy to measure. This loss is the energy absorbed by the material upon fracture and called the impact strength. The same approach can be taken for films by measuring the potential energy of a ball falling onto a sample need to crack it (height variation at constant weight)
IP A thermal insulation panel
IPA Isopropyl alcohol
Ionic liquid Any salt that has a relatively low melting point. Ionic liquids can be used as solvents when water or traditional organic solvents cannot be used
IR Infra red
IRVE Inflatable reentry vehicle experiment

K

K **Bulk modulus** (K) of a substance that measures the substance's resistance to uniform compression. It is defined as the pressure increase dP, needed to cause a given relative decrease in dV/V (with dV <0). It is defined as $K = \frac{-\mathrm{d}P}{\frac{\mathrm{d}V}{V}}$
KEK High Energy Accelerator Research Organization (in Japan)
Kelvin probe Probe used to measure the contact potential difference between tow materials by Kelvin method
K_d^{Hg} Ratio of the ppm adsorbed Hg^{2+} per gram adsorbent over the ppm Hg^{2+} remaining per mL solution
KLM Krimholtz–Leedom–Matthaei equivalent circuit
Knudsen number (Kn) Dimensionless number defined as the ratio of the mean free path length of gas atoms/molecules to a representative physical length scale, e.g., pore size; the number is named after Danish physicist Martin Knudsen (1871–1949)
K_{SV} See Stern–Volmer quenching constant

L

LDMM Low density microcellular materials, a type of **SPAM**
Leach Removal of a soluble species from a sample due to washing with solvent
LED Light-emitting diode, an electronic light source
Lewis acid A chemical compound, A, presenting a site that can accept a pair of electrons from a **Lewis base**, B, that acts as an electron-pair donor, forming an adduct
Lewis base A molecule or ion with a lone pair of electrons capable of acting as a donor to an electron-deficient molecule or ion, such as a metal cation. Lewis bases can be classified as hard or soft, depending on the extent of polarizability. Hard bases have minimal polarizability and form more thermodynamically stable complexes with compact metal ions, whereas soft bases are more polarizable and form more stable complexes with large metal ions. Examples are the F^- anion and the O atom in water H_2O
LIGA German acronym for LIthographie, Galvanoformung, Abformung (Lithography, Electroplating and Molding); it is a technique used to produce high aspect ratio components. In LIGA, selected regions of a polymer slab are cross-linked by exposure to collimated X-rays. Unexposed regions are removed with suitable solvents, and metal is electrodeposited

in the voids. After electrodeposition the remaining polymer is removed and the metal mold is used as a mold for polymer and ceramic components. Because of the large penetration of X-rays within the polymer, aspect ratios can be achieved with the Liga method that are otherwise not accessible to any other technique
Lignin Chemical compound derived from wood and plant cell walls. It is a complex biopolymer made of a network of furan and benzol rings with many OH and MeO groups attached
Lignocellulose Mixture of cellulose, hemicellulose and lignin
Live/dead cell cytotoxicity assay A measurement technique to quantify cell viability though the amount of live and dead cells within the culture. Only live cells uptake calcein which is hydrolyzed by intracellular esterases to fluoresce green. Only cells with a compromised cell membrane (dead or dying cells) can uptake ethidium which binds to DNA to fluoresce red
LLNL Lawrence Livermore National Laboratories
Lost form mould Mould made from a material being destroyed after casting (materials are typically bonded sand, clays and gypsum)
LNG Liquid natural gas
LS Light scattering
Luminescence Emission of light, often following absorption of a photon

M

Macropores According to IUPAC convention, macropores are pores with a characteristic size (e.g., diameter when meaningful) above 50 nm
Macroporous Material with pores than >50 nm
Martian crustal materials Ubiquitous, fine grained dust found on the surface of Mars believed to consist of silicates, iron oxide, carbonates and clays
MAS-NMR Magic angle spinning-nuclear magnetic resonance
Maxwell–Garnet relation A theory by which a refractive index of nano-composite is determined by the ratio of compositions
MCF Mesoporous cellular foams
Mercury porosimetry (or intrusion) Measurement of the specific porous volume and of the pore size distribution function by applying a continuous increasing pressure on liquid mercury such that an immersed or submerged porous solid is penetrated by mercury. If the porous body can withstand the pressure without fracture the Washburn equation, relating capillary pressure to capillary diameter allows converting the pressure penetration curves into a size distribution curve. If a sample is contracted without mercury intrusion, a specific mechanical model based on the buckling theory must be used
Mesopores According to IUPAC convention, mesopores are pores with a characteristic size (e.g., diameter when meaningful) comprised between 2 and 50 nm
Mesoporous Material with pores with size in the 2–50 nm range
Metal Chalcogenide A material composed of metal cations and chalcogenide (sulfide, selenide, or telluride) anions
Metal penetration Liquid metals can intrude into the open porous surface of a sand-binder mixture. This penetration leads to casting defects. Metal penetration also means a reaction occurring between the melt and an oxidic sand that can lead to the formation of silicates
Metathesis reaction Partner switching
Methoxylation Reaction with methanol to convert Si–OH groups to Si–OCH_3 groups

MF Melamine–formaldehyde aerogels
Micropores According to IUPAC convention, micropores are pores with a characteristic size (e.g., diameter when meaningful) below 2 nm
Microporous Material with pores size in the 0.1–2 nm range
MLI Multilayer insulation
MMA Methylmethacrylate
Mould A body having a more or less complex geometrically shaped cavity in which a liquid metal is cast and solidifies
MOR The Modulus of Rupture; it is the force per unit cross-section area, necessary to break a given material bar supported flatwise at two points 1 ft apart. It is expressed by 18 times the load which is required to break a bar, when the loaded is applied in the middle between the points of support
MOx Metal oxide aerogels
MRI NMR imaging
MS Mass spectrometry
MSF Mean square fluctuation
MTMS Methyltrimethoxysilane $(CH_3)Si(OCH_3)_3$
MTES Methyltriethoxysilane
Multiscale approach A simulation strategy in which different models and methods are used at different length and time scales
Mutagenesis A process which causes irreversible changes of and/or damage to the genetic material (DNA) of a cell or an entire organism

N

Nano-additives Aerogels are a nano-structured granular additive to foundry sand improving mould and core properties
Nanogel® A translucent silica aerogel manufacturered by Cabot Corporation
Nanostructuration Control of a solid structure and texture on a scale from 1 to 100 nm
NBO Nonbridging oxygen
***n*-Butane isomerization** Reaction allowing the rearrangement of the *n*-butane molecule to isobutane
Nematic liquid crystal One of the most common liquid crystal phases, where the molecules of a liquid have no positional order, but they have long-range orientational order inside domains. They all point in the same direction within each domain
Network connectivity The degree to which a polymer or a nano-particle has connections with neighbor polymers or nanoparticles, over a gel network
NIJ National Institute of Justice
Nitrogen adsorption Measurement of the adsorption isotherm of a porous material at liquid nitrogen temperature. The monolayer coverage by nitrogen measures the specific surface area
NMMO *N*-methylmorpholine *N*-oxide
NMP *N*-methylpyrolidone
Non-intrusive mercury porosimetry Characterization technique to study the porous texture of a material by collapsing this material under an isostatic mercury pressure, at pressures low enough to not induce mercury intrusion in the pores
NOVA laser High-power laser built at the Lawrence Livermore National Laboratory (USA) in 1984 which conducted advanced inertial confinement fusion (ICF) experiments

Nucleation center Local area where a particle of a new material phase is formed
Nucleophilic atom Atom holding a partial positive electronic charge δ^+

O

OIHM Organic–inorganic hybrid materials
OKAGEL High insulating day lighting aerogel developed by OKALUX GmbH (Germany)
Oleophilic Oil attracting
Oligomer A polymer that consists of two, three, or four monomers
Opacify To scatter visible or infrared radiation
Optical thickness A statistical measure of how often a photon with the mean free path l_{pht} interacts with the material within a given distance d
Optical transmission Transmission of an optical signal by transfer of light
Optical transparency Optical property of a piece of matter to let a picture or design be visible, eventually by projection on a screen, when light shines through it
Organosilane Derivative of an inorganic silane that contains one or more organic moieties bonded to a central Si atom
Ormosil ORganically MOdified SILicates: hybrid silica aerogels obtained by gelation of a solid network with functionalized silica precursors on which some specific organic moieties have been grafted
Ostwald ripening Ageing evolution of particles in a liquid by **dissolution re-precipitation**, where the smaller particle dissolve and re-precipitate on the larger ones

P

P Porosity
PAN Poly(acrylonitrile)
PDF Atomic pair distribution function obtained from the analysis of the Bragg and diffuse scattering
PDMS Polydimethylsiloxane, a macromolecular silica precursor belonging to the Ormosil precursors family and called polydimethyl siloxane
PDMS aerogel A composite between **PDMS** and a **metal chalcogenide aerogel**
PEG Polyethyleneglycol
Percolation theory Describes the formation of clusters defined by a random variation in the degree of connectivity of the cluster elements
Permanent mould Mould made mainly from heat resistant steel that can be re-used many times
Permeability (*D*) The rate of flow of a liquid or a gas through a porous material
PEM Proton exchange membrane
P-EW Panasonic Electric Works Co., Ltd
PF See Phenolic-furfural aerogels
Phe l-Phenylalanine
Phenolic-furfural (PF) aerogels Aerogel obtained by the base or acid catalyzed reaction of phenol and formaldehyde following the sol-gel route
Photocatalytic Capable of catalyzing a reaction initiated by light absorption
Photoluminescence Emission of light following excitation due to absorption of a photon
PL Photoluminescence
Pluronic P123 Poly(ethyleneoxide)-*block*-poly(propyleneoxide)-*block*-poly(ethyleneoxide

PMCNA Poly(methylcyanocrylate)
P-MDI Polymethylene polyphenylene polyisocyanate
PMMA Poly(methylmethacrylate)
PMT A photo-multiplier tube
Polarity Refer to the distribution of electrons in a chemical bond
Polycondensation Step growth polymerization in which small molecules on each reaction between monomers are released (e.g., water)
Polymeric Aerogel Aerogel made from polymers like resorcinol–formaldehyde, melamine–formaldehyde, cellulose or polyurethanes
Polymerization The bonding of two or more simple identical molecules termed monomers to form a polymer. Polymerization maybe by **Condensation**
Polymer-type aerogel Aerogel adopting a fibrous, polymeric morphology
Porometry Characterization technique to determine the minimum pore diameter (hence its pore size distribution) in a material, by flowing a fluid (e.g., mercury) through its capillary porous network.
Porosimetry Technique to determine the total pore volume of a material. Porosimetry and **Porometry** are often used as synonymous terms
Porosity (%) Ratio of the whole volume to the whole volume of a porous material
Postdoped Sol–gel material to which a **probe** is added after the matrix has formed
Postgelation modification Chemical modification of a wet sol–gel matrix prior to the solvent extraction process to prepare an aerogel
PRDF Pseudo Radial Distribution Function obtained from the EXAFS spectrum and related to the radial distribution of atoms surrounding a particular atom with a shift caused by the scattering phase
Precursor (sol–gel) General term for a molecule or complex ion that is used to generate a gel. It can be a chemical compound which contains a cation M and/or an organic group X from which a ceramic aerogel (mostly oxide aerogel) or a hybrid aerogel will be made
Prehydrolyzed silica (or silicon) precursor A precursor submitted to a partial hydrolysis before packaging, storage and further use to synthesize a gel
Pristina aerogel™ Trade mark of an aerogel developed by TAASI Corporation
Probe Chemical species that responds to other species and/or its environment
Probe-doped Sol–gel material into which a chemical probe has been incorporated
Protective Lipid layer on insects Layer of cuticular (surface) lipids. Lipids gather organic substances such as a fat, oil or wax that dissolves in alcohol but not in water
Proton scavenger A reactant, such as an epoxide, which slowly capture a hydrogen atom from a structural H_2O ligand bound to a hydrated metallic salt
PS Polystyrene or for Phosphatidylserine, a cell membrane phospholipid with a overall negative charge
PSD Pore size distribution or particle size distribution
PtOEP Platinum (II) octaethylporphine
PtTFPP Platinum (II) mesotetrafluoro-phenylporphyrin
PUA Polyurea
Pulping Pulping is a process to convert materials containing cellulose (wood, cotton, bast, etc.) into to a pulp of fibers. Pulping can be done mechanically (grinding), chemically (dissolution) and by extracting the noncellulose parts from the natural resources, like lignin, pectin and others
PVC Poly(vinylchloride)
P2VP Poly(2-vinylpyridine)

PVDF Polyvinylidene difluoride polymer
PXRD Powder X-ray diffraction
Pyrolysis Conversion by thermal decomposition of an organic material into pure carbon under an inert gas atmosphere
Pyroshock Deliberate actuation of spacecraft componentry achieved by an explosive device
PZT Lead zirconate titanate ceramics

Q

Q Amount of heat or quencher
Q_n A silicon atom (in silica) which is bound to n oxygen atoms also bound to other silicon atoms
Quantum confinement A characteristic effect observed in semiconductors when the size of the particle is reduced beneath the Bohr radius of the electron-hole pair (exciton) generated by optical absorption. The effect manifests as an increase in optical bandgap with a decrease in particle size
Quench To reduce the emission intensity or lifetime of a chemical species due to energy transfer to another chemical species called the quencher

R

r Ratio of water to silicon-containing monomers in an aqueous sol–gel precursor solution
R Resorcinol
Raman Inelastic scattering of incident radiation; the Raman photon differs in energy from the incident photon by an amount equal to a vibrational transition of the molecule
Rapid Supercritical Extraction Method See **RSCE**
Rattler Loosely bonded particle in a sand bed
Rayleigh scattering Elastic scattering of light or other electromagnetic radiation by particles much smaller than the wavelength of the light, which may be individual atoms or molecules
R/C-ratio Molar ratio of resorcinol to formaldehyde
R/F Abreviation for resorcinol–formaldehyde solution (sol)
R/W-ratio Molar ratio of resorcinol to water
RDF Radial distribution function. Average number of atoms per length unit situated at a distance r from an arbitrary atom taken as a reference
RE Rare earth
Reconstruction Generation of a computer model of a material (or other object, such as an image) from analysis of experimental data
Redox activity Activity pertaining to the gain or loss of electrons
Refractive index Ratio of the velocity of light in a vacuum to that in a medium
Refractoriness Ability of a dry sand core or mould to withstand the hot liquid metal temperatures
Regeneration Physically or chemically modified cellulose can be regenerated in suitable solvents (for instance xanthates in sulfuric acid) leading to pure cellulose
Relative density (ρ_r) The ratio between the aerogel bulk density and the bulk glass density (e.g., for silica 2.2 g/cm^3)

Release enhancement Increase of the rate of the drug liberation/dissolution from the carrier
Relative dielectric constant Ratio of the dielectric permittivity of a medium to the permittivity of free space
Remediation A process whereby impurities are removed, as when water is purified
Resorcinol–formaldehyde aerogel Aerogel obtained by the base or acid catalyzed reaction of resorcinol and formaldehyde following the sol-gel route; also called RF aerogels
RF Resorcinol–formaldehyde aerogels
RF aerogel Aerogel made from resorcinol and formaldehyde
RF/AN Resorcinol–formaldehyde/ammonium nitrate
RF-MOx Resorcinol–formaldehyde/metal oxide interpenetrating networks
RHU Radioisotope heat source. A heater that radiates heat that results from radioisotope decay
RICH counter A ring imaging Cerenkov counter. When a high energy physics particles travels inside an aerogel, a radiation is emitted in a direction located on a cone which is observable as a ring
RLA Reaction-limited aggregation, an aggregation model similar to DLA, except with a finite probability of forming a bond upon contact between objects
RLCA Reaction-limited cluster aggregation, an aggregation model similar to DLCA, except with a finite probability of forming a bond upon contact between objects
RSCE Rapid supercritical extraction method where the sol-gel precursors themselves brought to a supercritical state inside a pressurized mould
RSL2 The (revised) Rahman–Stillinger–Lemberg model for water, consisting of hydrogen and oxygen atoms interaction only through pair potentials
$[Ru(bpy)_3]^{+2}$ Tris(2,2′-bipyridyl) ruthenium(II)
$[Ru(dpp)_3]^{+2}$ Tris(4,7-diphenyl-1,10-phenanthroline ruthenium (II)

S

S-aerogel An aerogel obtained after supercritical CO_2 (l) extraction
Salt-Hydrate melt A concentrated solution of salts in water at higher temperature leading to salt-hydrates like $Ca(SCN)_2 \cdot 4H_2O$, $LiClO_4 \cdot 3H_2O$, $ZnCl_2 \cdot 4H_2O$ being mostly solid at room temperature
SANS Small angle neutron scattering
SAS Small angle scattering
SAXS Small angle X-ray scattering
SBF Simulated body fluid. An acellular aqueous solution with inorganic ion composition almost equal to human plasma
Scaffold A three dimensional structure that can be used to direct cell growth or serve as a location for cell growth
SCAs Spherical carbon aerogels
Scattering General physical process where some forms of radiations, such as light, sound or moving particles, are forced to deviate from a straight trajectory by one or more localized nonuniformities in the medium through which they pass
Sc-CO_2 Supercritical CO_2
Sc-CO_2 drying Supercritical drying of the gel performed by either substituting the solvent present in the gel with liquid CO_2 which is then taken to the supercritical state and removed,

or by directly using supercritical CO_2 to remove the solvent. T_C of CO_2 is about 31°C, much lower than the T_C of common sol–gel solvents
SCD Supercritical drying
SCF Supercritical fluid
Sc-HT drying Supercritical drying of the gel performed by removing the solvent present in the gel after taking it to the supercritical state, i.e., at temperature higher than the critical temperature T_C and pressure higher than the critical pressure P_C
SEM Scanning electronic microscopy or microscope
SEM-FEG Field emission gun scanning electron microscopy
Semitransparent material Material partially transparent for thermal radiation in certain wavelength regions; the results of its measurement concerning the radiative heat transfer are influenced by the surface emittance of the boundaries
Sensor (chemical or biochemical) Device that measure a physical characteristics linked to the concentration of a chemical or biochemical and converts it into a signal which can be read by an observer or by an instrument
SERS Surface-enhanced raman scattering *or* Surface-enhanced raman spectroscopy
Shock impedance The total reaction of a medium to the transmission of shock wave through it, expressed as the ratio of shock pressure to particle velocity at a given point in the medium
SHPB Split Hopkinson Pressure Bar
^{29}Si-NMR Nuclear magnetic resonance of ^{29}Si atoms
Silanol Hydroxyl terminal group bonded to a Si atom: ≡Si–OH
Silica Aerogel Aerogel obtained by hydrolyzation and condensation of a silicic alkoxide
Siloxane bridge Bond ≡Si–O–Si≡ between two Si atoms by the intermediate of one oxygen atom
Silylation Method used to make hydrophobic aerogels via post gelation derivatization with an additional silane compounds
SiOx Silica aerogels
SOFCs Solid oxide fuel cells
Solidification First order phase transformation, in which a liquid (melt) crystallizes
Sol–gel transition Transition of a colloidal solution into a solid network extending through the whole sample space. The network is made up of the colloidal particles attached to each other mechanically and might be bonded by various potentials. At the point of contact so-called necks establish. The sol–gel transition might be considered as exactly that point when a percolating net of particles (a cluster) spans through the wet fluid. See also gelation
Solvent A liquid substance capable of dissolving other substances
Solvent exchange Replacement of one liquid in a solid–liquid mixture by another; a solvent inside a wet gel is replaced by e.g., ethanol or carbon dioxide
Sonar system Sound navigation and ranging system
Sono-aerogel Aerogel obtained from sonogel
Sonochemistry Study of the effect of sonic waves and wave properties on chemical systems
Sonication Method that applies ultrasound energy in a homogenenous or heterogeneous system
Sonogel Gel obtained when the mixing of the precursor, solvent, and water is enhanced by ultrasonic vibrations (sonication)

Sono-ormosil Ormosil prepared by a sonogel approach, also called "hard Ormosil." It incorporates an organic phase in the inorganic precursor sol in combination with sonication of the precursors
Sonosol Sol resulting from applying ultrasound to an alkoxide–water mixture
SPAM Small pore area material, a foam with density of less than 1 g/cm^3 and an average pore area of less than 500 μm^2 produced by American Aerogel Corporation
Specific surface area Surface area per unit volume of a material or per unit mass. Both are convertible via the density of the material
SQUID Superconducting quantum interference device, a very sensitive magnetometer
sscm Standard cubic centimeters per minute
Stellar nucleosynthesis Nuclear reactions within a star that create elements with an atomic number >1
STEM Scanning transmission electron microscopy
Stern–Volmer equation Plot of the intensity ratio in the absence and presence of quencher (I_0/I) as a function of quencher concentration; used as a calibration curve for quenching-based sensors
Stern–Volmer quenching constant (K_{SV}) Constant for a given quenching process; it is the product of the bimolecular rate constant for the reaction of quencher and the luminescent species and the luminescence lifetime in the absence of quencher
Stirling engine A technology that involves the adiabatic expansion and contraction of a gas to move a piston as a result of a thermal gradient. The mechanical is typically converted to electrical power through an alternator
Stochastic reconstruction A type of **reconstruction** in which limited experimental data is used to generate a statistically representative model of a material
Strength See bonding-, impact-, tensile strength
Subcritical conditions Temperature T and pressure P conditions below the critical point temperature T_c and pressure P_c of a fluid
Subcritical crack growth In a piece of solid material, a crack will begin to spontaneously grow when the strain energy (plastic + elastic deformation) which is released during crack growth is $\geq$ the gain in surface energy. Subcritical crack growth relates to crack growth below this critical limit, in particular by corrosion mechanisms
Subcritical drying A drying process performed at a temperature T and a pressure P below the critical point temperature T_c and the pressure P_c of the fluid contained in the porosity of the wet gel
Supercritical drying Drying of a wet gel by application of temperature–pressure conditions beyond the critical point of the solvent (T_c, P_c in the P–T phase diagram) avoiding thus liquid–gas interfaces in the pore space of the gel body. The interfacial tension in capillaries of different pore radii would establish stress gradients in the wet gel leading to rupture of the fragile network. The capillary stresses might reach values up to a few MPa
Supercritical fluid Fluid brought at a temperature and a pressure beyond the critical temperature T_c and pressure P_c
Superfluid A special phase of matter such as helium at temperatures below 2.18 K in which, when cooled to temperatures near absolute zero, the molecules exhibit special quantum effects. In the superfluid state, flow proceeds without any friction and the fluid viscosity becomes zero
Superhydrophobic Surface on which a water droplet appears to sit without any significant surface wetting and exhibits a contact angle in excess of 150°

Superhydrophobic aerogel Aerogel on which the contact angle of a water droplet with the porous external surface is $>150°$
Superinsulation Insulation systems based on the use of superinsulating products and/or components and/or materials. A superinsulating materials is commonly defined by a thermal conductivity lower than the one of air (e.g., 0.025 W/m K in room conditions) and more recently lower than 0.020 W/m K
Superparamagnetic behavior Magnetic behavior of superparamagnetic materials
Superparamagnetic materials Nanosized ferro-, antiferro-, or ferrimagnetic materials below the magnetic ordering temperature, with a single magnetic moment directed up or down on a preferential (anisotropy) axis. When the thermal energy is sufficient to overcome the magnetic anisotropy energy barrier, the transition from blocked to superparamagnetic regime is observed
Superparamagnetic relaxation Transition from blocked to superpara magnetic regime
Surfactant Substance changing the wetting behavior of a surface, generally lowering the surface tension. Molecules with long alkyl chains capable of forming supramolecular assemblies (e.g., micelles). The supramolecular assemblies can act as **templates** for the formation of porous inorganic materials
S–V Stern–Volmer plot
SXRFM See Synchroton X-ray microprobe
Sylilation Operation which consists in grafting ≡Si–X hydrophobic functionalities, with the help of silica precursor carrying such functionalities (e.g., trichloromethyl silane). Generally, R is an alkyl (linear or branched) or a perfluoroalky chain (for example, $CH_2C_4F_{11}$) and X, an halogenide (–F, –Cl or –I) or an amine (–NH2) or an alkoxy (–OCH_3, –OC_2H_5 etc.) group
Synchrotron X-ray microprobe A technique involving synchrotron radiation to detect elements in low (parts per million) concentrations (SXRFM)
Syneresis Exudation of part of the liquid component of a gel contained in its porosity, which, when immersed in a liquid, shrinks homogenously while keeping its shape (e.g., cylindrical or conical . . .) under the continuation of the condensation reactions after the gel point

T

Tank baffles Straight flat plates that run inside and along the straight side of vertically oriented cylindrical tanks or vessels
***t*-amylalcohol** 2-Methyl-2-butanol $C_5H_{12}O$
***t*-butanol** *tert*-Butyl alcohol $(CH_3)_3C$–OH
T/C Inverse of **C/T** ratio
TDI Tolylene-2,4-diisocyanate
TEA Triethylamine
TEG Thermoelectric generator. A general term for describing an electrical generator that generates heat from a heat source
TEM Transmission electronic microscopy
Template Molecule or surfactant that dictate the morphology of the resultant material, often by moderating growth. For instance a DNA molecule can serves as a pattern for the synthesis of another macromolecule such as RNA. Also any porous solid such as a porous polymer that serves as a pattern to the synthesis of another or porous solid of a different nature (e.g., silica)

Tensile strength The maximum tensile stress (Force per unit cross section area) a material can withstand without failure
TEOS Tetraethyl orthosilicate $Si(OC_2H_5)_4$, also termed silicon tetra ethoxysilane: a common SiO_2 sol–gel precursor
Tensile strength Strength measured in a tensile test. A long thin sheet of a material is subjected to a force stretching the sample. In a diagram of force needed to pull the sample against elongation typically a maximum is shown, which marks the point of tensile strength. Several norms (DIN, ISO, and ASTM) are available defining details of such a test for any type of materials
TEP Triethylphosphate
Texture Physical characteristics of a solid with respect to the size, shape, pore surface, and arrangement of its solid parts
TG Thermogravimetry, a measurement of a sample's weight during heating
t_g Gelation time
TGA Thermogravimetric analysis
Thermal conductivity The thermal conductivity k is a characteristic property of a matter (solid, liquid, or gas) indicating its ability to conduct heat. It is defined as $k = \frac{\mathrm{d}Q}{\mathrm{d}t}\frac{1}{A}\frac{1}{\mathrm{d}T/\mathrm{d}x}$ where, $\mathrm{d}Q$ is the heat quantity (in Joule) passing during the time $\mathrm{d}t$, across an area A of a slice of matter of thickness $\mathrm{d}x$, when the temperature difference between both faces of the matter slice is $\mathrm{d}T$. It is the proportionality constant between the heat flux and the temperature gradient across a sample. The unit in the International System is $\mathrm{Wm}^{-1}\,\mathrm{K}^{-1}$
Thermal release Liberation of the compound from an aerogel matrix induced by heat
Thermal Wrap™ Nanogel thermal wrap is an insulating blanket based on Cabot's Nanogel™ material
Thermites Pyrotechnic composition comprising a metal nanopowder (e.g., Al) dispersed in an oxide, such as an oxide aerogel
Thermography Imaging of a hot surface by an infrared sensitive camera
Thermoporometry (Thermoporosimetry) A calorimetric method for characterizing the pore texture from the melting or freezing point depression of a liquid confined in a pore, by reason of the added contribution of surface curvature to the phase-transition free energy
THF Tetrahydrofuran
Three-dimensional tensor of the thermal conductivity Tensor which, when multiplied by the temperature gradient vector according to the rules of matrix multiplication, gives the heat flux density vector, i.e., heat flux density and temperature gradient must not have the same orientation, which is the case in media with an anisotropic thermal conductivity
Thiolate An organic molecule terminated with an S^- functionality. An example is 4-fluorophenylthiolate, $FC_5H_4S^-$
Thyolysis reaction Analogous to hydrolysis, thiolysis involves reaction of H_2S, in lieu of H_2O, to form metal thiols, which subsequently condense to form metal sulfides
TMB Trimethylbenzene
TMCS Trimethylchlorosilane
TMOS Tetramethyl orthosilicate $Si(OCH_3)_4$, also named silicon tetra methoxysilane: a common SiO_2 sol–gel precursor
TMS Trimethylsilyl group
TMSA Trimethylsilyl modified silica aerogel
TMSPM 3-(Trimethoxysilyl)propyl methacrylate

TNB Ti(O–$nC_4H_9)_4$
Tortuosity A property of a packed column of particles that indicates the degree of unevenness of the path followed by the solute molecule as it passes down the column. The most mathematic method to estimate tortuosity τ is the arc-chord ratio i.e., the ratio of the length of the curve L to the distance between the ends of it C; $\tau = L/C$
TP Daylightning panels
TPR Temperature programmed reduction
$TRAP_6$ Thrombin receptor agonist peptide 6, which contains residues 42–47 of the thrombin receptor
TSET Total solar energy transmittance
Transesterification Reaction of exchange of the organic group R″ of an ester R″O–(CO)–R with the organic group R′ of an alcohol R′OH to produce another alcohol R″OH and another ester R′O–(CO)–R
Two-step process Acid catalyzed hydrolysis followed by base catalyzed condensation of a silica precursor

U

U *U*-value, thermal transmittance
Ultrasonic transducer Mechano-electrical transduction device
USANS Ultra small angle neutron scattering
USAXS Ultra small angle X-ray scattering
UV Ultra violet

V

VB Vitamin B12
VG Vacuum glazing
VIP Vacuum insulation panel
Viscose process A chemical process to produce viscose filaments of cellophane from cellulose pulp (e.g., treatment of cellulose with NaOH and CS_2) yielding a highly viscous cellulose xanthate solution; extrusion with a spinneret or splitting into a sulfuric acid bath followed by washing in water generates fibers or cellophane
Viscosity The viscosity of a fluid is a measure of its resistance to flow. It describes the internal friction of a moving fluid
Volatile Substance (e.g., water, alcohols) becoming easily gaseous on heating (e.g., by a hot metal)
Volatility Measure of the ease of a substance to vaporize
VOx Vanadia aerogels
VTMS Vinyltrimethoxysilane

W

Water glass Sodium silicate is the common name for a compound sodium metasilicate, Na_2SiO_3, also known as water glass or liquid glass. It is available in aqueous solution and in solid form and is used in cements, passive fire protection, refractories, textile and lumber processing, and automobiles
WAXS Wide angle X-ray scattering

WEB Warm electronics box, a thermally insulated box used to house the electronics on the Mars Exploration Rovers
Wet spinning Spinning of a viscous fluid into a liquid bath (e.g., **xanthate** solution into sulfuric acid, salt-hydrate cellulose melt into ethanol)
Wollastonite Calcium silicate, $CaSiO_3$
Work function The minimum energy (usually measured in electron volts) needed to remove an electron from a solid to a point immediately outside the solid surface

X

XANES X-ray absorption near edge structure, i.e., the fine structure present in the X-ray absorption spectra around the absorption edge
Xanthate Cellulose xanthate is a salt of cellulose being a result of a reaction with a solution of NaOH and CS_2 in which the cellulose is bonded to –OC(=S)S–
X-aerogels Polymer crosslinked aerogels
XAS X-ray absorption spectroscopy
Xerogel Defined by IUPAC as an "Open network formed by the removal of all swelling agents from a gel." However, the definition retained in this chapter is the one first introduced by Freundlich to designate "shrinking" (or swelling) gels, as a consequence of drying
Xonotlite–silica aerogel composites Xonotlite–silica aerogel composites are materials formed of a xonotlite-type calcium silicate and a silica aerogel phase. Xonotlite-type calcium silicates are porous materials synthesized by hydrothermal processing of quartz powder and limestone
XPS X-ray photoelectron spectroscopy
X-MP4-T045 Cross-linked aerogel made from a surfactant-templated polyurea-nanoencapsulated macroporous silica
X-RF-MOx Polymer crosslinked RF-MOx networks
X-SiOx Crosslinked silica aerogels
X-VOx Crosslinked vanadia aerogels
XRD X-ray diffraction
Xylan Generic term for complex polysaccharides
XZC Xerogel Sulfated Zirconia doped with Cerium

Y

Young modulus Proportionality constant between stress and strain in linear elasticity theory
YSZ Yttria-stabilized Zirconia

Z

Zeolite A natural or synthetic hydrated crystalline aluminosilicate with a framework structure enclosing interconnected cavities occupied by large metal cations (positively charged ions) – generally sodium, potassium, magnesium, calcium, and barium – and water molecules. The ease of movement of ions and water within the framework allows reversible dehydration and cation exchange
ZFC-FC Magnetization curves collected under different conditions. The zero field cooled (ZFC) magnetization curve is collected by heating the sample in the presence of a weak magnetic external field after cooling down it without external field. The FC magnetization

curve is recorded by cooling down the sample in the presence of a weak magnetic external field

Zintl clusters/anions Clusters or anions of intermetallics wherein the bonding is electron precise and arranged so that each component achieves an octet of electrons. An example is the $SnSe_4^{4-}$ Zintl ion. Sn donates 4 valence electrons and each Se donates 6 for a total of 28 electrons. This is increased by 4 electrons from the charge on the anion, yielding 32 electrons. A tetrahedral arrangement of 4 Se atoms around a central Sn atom enables all 5 atoms to achieve an octet with the 32 electrons ($8 \times 4 = 32$). These ions can be linked together by Pt^{4+} cations to form chalcogels

$Zr(OCH_2CH_2CH_2CH_3)_4$ Chemical formula of Zirconium *n*-butoxide

$ZrO(NO_3)_2 \cdot 5H_2O$ Chemical formula of Zirconyl nitrate

$Zr(OR)_4$ Chemical formula for a Zirconium alkoxide

Subject Index

The subject index has been worked out by the editors and is partly based from the author's suggestions; it should assist the readers to find references to a particular topic. Effort has been made to realize an index as comprehensive and useful as possible.

Entries may have several levels of reading. Cross-references are also given to terms of related interest and are found after the entry for the first-level term to which they apply. The numbers in parenthesis correspond to a chapter; when there is a substantial discussion in a chapter, the number appears in bold.

M.A. Aegerter et al. (eds.), *Aerogels Handbook*, Advances in Sol-Gel Derived Materials and Technologies, DOI 10.1007/978-1-4419-7589-8,

Printed in the United States of America